T0181722

Calculus and Linear Algebra in Recipes

Christian Karpfinger

Calculus and Linear Algebra in Recipes

Terms, phrases and numerous examples in short learning units

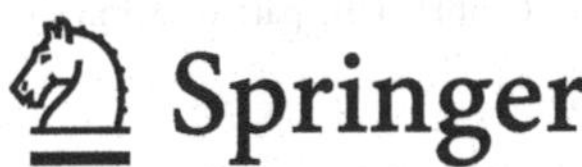

Christian Karpfinger
Technische Universität München
Zentrum Mathematik
München, Germany

ISBN 978-3-662-65457-6 ISBN 978-3-662-65458-3 (eBook)
https://doi.org/10.1007/978-3-662-65458-3

The translation was done with the help of artificial intelligence (machine translation by the service DeepL.com). A subsequent human revision was done primarily in terms of content.

This Springer imprint is published by the registered company Springer-Verlag GmbH, DE, part of Springer Nature.
The registered company address is: Heidelberger Platz 3, 14197 Berlin, Germany

Foreword to the Third Edition

In the present third edition, we have revised and extended the recipe book: all known errors have been corrected, new tasks have been added, and further topics have been added, namely residual element estimation in Taylor expansion, the numerical solution of boundary value problems, and the solution of first-order partial differential equations by means of the characteristic method. Thus, we have presented further topics on numerical mathematics or differential equations, which are important for the engineer, in a proven and understandable manner.

On the website of the book you will find, besides the previous add-ons, video animations and apps that illustrate some of the mathematical contents of the recipe book. The website for this book can be found via

http://www.springer-spektrum.de/

Munich, Germany
April 2017

Christian Karpfinger

Preface to the Second Edition

The main new features in this second edition are the chapter on the solution of partial differential equations using integral transformations and a section on the numerical solution of the wave equation. This is intended to introduce further important methods for finding solutions of the partial differential equations which are so fundamental in the natural sciences and technology.

Many small improvements or additions to explanations or examples have found their way into the second edition, and of course all known erroneous passages in the text have been corrected. We have also increased the number of exercises, especially in the later chapters, in order to promote the practice of the formulas and the understanding of the later and more difficult topics.

On the website for this book, as a special extra, you will find video recordings of many lectures that follow the present text. In addition, we have supplemented the script "Introduction to MATLAB", which can be found on the website for this book, with the part "MATLAB—a great calculator". Mr. Benjamin Rüth contributed significantly to the creation of this script, and I would like to thank him very much. Also the MATLAB codes, which are used in the recipe book as well as in the accompanying workbook, and also the solutions to most of the exercises can be found on the mentioned website. The link to the website can be found via

http://www.springer-spektrum.de/

Further comments from the readership are always welcome.

Munich, Germany
April 2014

Christian Karpfinger

Preface to the First Edition

Joining the many books on higher mathematics is another, the present book *Calculus and Linear Algebra in Recipes*. In writing the book, the author had the following aspects in mind:

- Many typical problems in higher mathematics can be solved in recipes. The book provides a collection of the most important recipes.
- A clear presentation of the topics that can be covered in four semesters of Calculus and Linear Algebra.
- An inexpensive book for undergraduates that covers all the major content.
- Numerous examples that reinforce the theories and practice using the recipes.
- A division of the material into many roughly equally short teaching or learning units (each chapter can be covered in about a 90-minute lecture).
- Omitting content that is usually only actually understood by about 10 percent of the student body and that is of lesser importance for practice.
- Numerical mathematics and also the use of MATLAB are an integral part of the content.

It is customary, but perhaps not entirely correct, to teach higher mathematics as proof-completely as possible. The disadvantages are obvious: desperate students who then quickly realize that the exams can largely be passed without the proofs. It might make more sense to use the time gained by omitting proofs to cover the topics that are so important for practice, such as numerics and MATLAB. We cover a few topics in numerical mathematics, punctuating them with numerous examples, and always showing how to use MATLAB as a great calculator in the engineering mathematics problems covered. Occasionally, especially in the assignments, we also solve programming problems using MATLAB. Thereby hardly any previous knowledge for MATLAB are necessary. We put on the Internet page to this book under

http://www.springer-spektrum.de/

a short introduction course to MATLAB on a few pages.

The inputs to MATLAB we always formulate with an excellent `font type`. And instead of a comma we do like MATLAB we also put a dot, so we write 1.25 for 5/4. Occasionally we calculate with MATLAB symbolically, which is possible thanks to the SYMBOLIC MATH TOOLBOX this is also possible, note that you also have this toolbox installed.

We summarize the special features of this book once again:

- We do not attempt to erect the abstract edifice of mathematics, created over millennia, in a few 100 pages as comprehensively and proof-completely as possible. We address the topics of mathematics that are important to engineers, make concepts and rules plausible if only that is feasible, and use examples and many problems to learn how to solve problems.
- We divide the topics of higher mathematics into nearly 100 chapters of roughly equal length and formulate numerous problem-solving strategies in a recipe-like fashion. Each chapter covers about the material of a 90-minute lecture. This provides an overview and the opportunity to plan, both for students and lecturers.
- We use the computer, in particular MATLAB as a powerful calculator to deal with realistic examples instead of the usual academic examples.

At the end of the chapters some exercises are given, which are recommended to work on. These exercises can be used to check the understanding of the presented recipes and methods. On the Internet page to the book under

http://www.springer-spektrum.de/

we have provided detailed solution suggestions for a lot of the exercises. The exercises and solutions are also printed in the accompanying workbook.

The creation of this comprehensive book was not possible without the help of many colleagues and collaborators. For proof-reading, for numerous hints, suggestions, proposals for improvement, tasks, examples, sketches and MATLAB programs, I would like to thank Dr. L. Barnerßoi, Prof. Dr. D. Castrigiano, S. Dorn, F. Ellensohn, Dr. H.-J. Flad, P. Gerhard, S. Held, Dr. F. Himstedt, Dr. F. Hofmaier, Prof. Dr. O. Junge, Dr. S. Kaniber, B. Kleinherne, Y. Kochnova, A. Köhler, Dr. M. Kohls, Dr. P. Koltai, A. Kreisel, Prof. Dr. C. Lasser, Dr. D. Meidner, N. Michels, S. Otten, M. Perner, P. Piprek, Dr. M. Prähofer, F. Reimers, Dr. K.-.D. Reinsch, Prof. Dr. P. Rentrop, B. Rüth, M. Ritter, Th. Simon, A. Schreiber, Dr. Th. Stolte, Prof. Dr. B. Vexler, Dr. H. Vogel, J. Wenzel and E. Zindan.

Special thanks go to Dr. Ch. Ludwig, who not only always had an open ear for my questions, whether during the day or at night, he also always had a solution ready. Finally, my thanks also go to Th. Epp, who created most of the pictures, and to B. Alton and Dr. A. Rüdinger of Springer Spektrum, who accompanied the creation of the book with numerous pieces of advice.

Munich, Germany
August 2013

Christian Karpfinger

Contents

1 Speech, Symbols and Sets 1
- 1.1 Speech Patterns and Symbols in Mathematics 1
- 1.2 Summation and Product Symbol 4
- 1.3 Powers and Roots 5
- 1.4 Symbols of Set Theory 6
- 1.5 Exercises 8

2 The Natural Numbers, Integers and Rational Numbers 11
- 2.1 The Natural Numbers 11
- 2.2 The Integers 15
- 2.3 The Rational Numbers 15
- 2.4 Exercises 17

3 The Real Numbers 19
- 3.1 Basics 19
- 3.2 Real Intervals 20
- 3.3 The Absolute Value of a Real Number 21
- 3.4 n-th Roots 22
- 3.5 Solving Equations and Inequalities 23
- 3.6 Maximum, Minimum, Supremum and Infimum 25
- 3.7 Exercises 26

4 Machine Numbers 29
- 4.1 b-adic Representation of Real Numbers 29
- 4.2 Floating Point Numbers 31
- 4.3 Exercises 35

5 Polynomials 37
- 5.1 Polynomials: Multiplication and Division 37
- 5.2 Factorization of Polynomials 41

5.3 Evaluating Polynomials 43
5.4 Partial Fraction Decomposition 44
5.5 Exercises 48

6 Trigonometric Functions 49
6.1 Sine and Cosine 49
6.2 Tangent and Cotangent 53
6.3 The Inverse Functions of the Trigonometric Functions 55
6.4 Exercises 57

7 Complex Numbers: Cartesian Coordinates 59
7.1 Construction of $\mathbb{C}$ 59
7.2 The Imaginary Unit and Other Terms 61
7.3 The Fundamental Theorem of Algebra 63
7.4 Exercises 65

8 Complex Numbers: Polar Coordinates 67
8.1 The Polar Representation 67
8.2 Applications of the Polar Representation 69
8.3 Exercises 74

9 Linear Systems of Equations 75
9.1 The Gaussian Elimination Method 75
9.2 The Rank of a Matrix 80
9.3 Homogeneous Linear Systems of Equations 82
9.4 Exercises 84

10 Calculating with Matrices 87
10.1 Definition of Matrices and Some Special Matrices 87
10.2 Arithmetic Operations 89
10.3 Inverting Matrices 94
10.4 Calculation Rules 96
10.5 Exercises 98

11 *LR*-Decomposition of a Matrix 101
11.1 Motivation 101
11.2 The $L\,R$-Decomposition: Simplified Variant 102
11.3 The $L\,R$-Decomposition: General Variant 105
11.4 The $L\,R$-Decomposition-with Column Pivot Search 108
11.5 Exercises 109

12 The Determinant ... 111
12.1 Definition of the Determinant ... 111
12.2 Calculation of the Determinant ... 113
12.3 Applications of the Determinant ... 117
12.4 Exercises ... 120

13 Vector Spaces ... 123
13.1 Definition and Important Examples ... 123
13.2 Subspaces ... 126
13.3 Exercises ... 128

14 Generating Systems and Linear (In)Dependence ... 131
14.1 Linear Combinations ... 131
14.2 The Span of X ... 134
14.3 Linear (In)Dependence ... 135
14.4 Exercises ... 138

15 Bases of Vector Spaces ... 141
15.1 Bases ... 141
15.2 Applications to Matrices and Systems of Linear Equations ... 146
15.3 Exercises ... 150

16 Orthogonality I ... 153
16.1 Scalar Products ... 153
16.2 Length, Distance, Angle and Orthogonality ... 156
16.3 Orthonormal Bases ... 157
16.4 Orthogonal Decomposition and Linear Combination with Respect to an ONB ... 159
16.5 Orthogonal Matrices ... 161
16.6 Exercises ... 164

17 Orthogonality II ... 165
17.1 The Orthonormalization Method of Gram and Schmidt ... 165
17.2 The Vector Product and the (Scalar) Triple Product ... 168
17.3 The Orthogonal Projection ... 172
17.4 Exercises ... 174

18 The Linear Equalization Problem ... 177
18.1 The Linear Equalization Problem and Its Solution ... 177
18.2 The Orthogonal Projection ... 178
18.3 Solution of an Over-Determined Linear System of Equations ... 179

18.4 The Method of Least Squares ... 181
18.5 Exercises ... 185

19 The *QR*-Decomposition of a Matrix ... 189
19.1 Full and Reduced QR-Decomposition ... 189
19.2 Construction of the QR-Decomposition ... 190
19.3 Applications of the QR-Decomposition ... 196
19.4 Exercises ... 198

20 Sequences ... 201
20.1 Terms ... 201
20.2 Convergence and Divergence of Sequences ... 204
20.3 Exercises ... 208

21 Calculation of Limits of Sequences ... 209
21.1 Determining Limits of Explicit Sequences ... 209
21.2 Determining Limits of Recursive Sequences ... 212
21.3 Exercises ... 215

22 Series ... 217
22.1 Definition and Examples ... 217
22.2 Convergence Criteria ... 220
22.3 Exercises ... 225

23 Mappings ... 227
23.1 Terms and Examples ... 227
23.2 Composition, Injective, Surjective, Bijective ... 229
23.3 The Inverse Mapping ... 234
23.4 Bounded and Monotone Functions ... 237
23.5 Exercises ... 238

24 Power Series ... 241
24.1 The Domain of Convergence of Real Power Series ... 241
24.2 The Domain of Convergence of Complex Power Series ... 246
24.3 The Exponential and the Logarithmic Function ... 248
24.4 The Hyperbolic Functions ... 250
24.5 Exercises ... 252

25 Limits and Continuity ... 255
25.1 Limits of Functions ... 255
25.2 Asymptotes of Functions ... 259
25.3 Continuity ... 261

25.4 Important Theorems about Continuous Functions 262
25.5 The Bisection Method 263
25.6 Exercises 265

26 Differentiation 269
26.1 The Derivative and the Derivative Function 269
26.2 Derivation Rules 272
26.3 Numerical Differentiation 277
26.4 Exercises 278

27 Applications of Differential Calculus I 281
27.1 Monotonicity 281
27.2 Local and Global Extrema 282
27.3 Determination of Extrema and Extremal Points 285
27.4 Convexity 289
27.5 The Rule of L'Hospital 291
27.6 Exercises 293

28 Applications of Differential Calculus II 295
28.1 The Newton Method 295
28.2 Taylor Expansion 299
28.3 Remainder Estimates 302
28.4 Determination of Taylor Series 306
28.5 Exercises 309

29 Polynomial and Spline Interpolation 311
29.1 Polynomial Interpolation 311
29.2 Construction of Cubic Splines 315
29.3 Exercises 319

30 Integration I 321
30.1 The Definite Integral 321
30.2 The Indefinite Integral 326
30.3 Exercises 333

31 Integration II 335
31.1 Integration of Rational Functions 335
31.2 Rational Functions in Sine and Cosine 338
31.3 Numerical Integration 341
31.4 Volumes and Surfaces of Solids of Revolution 343
31.5 Exercises 345

32 Improper Integrals 347
32.1 Calculation of Improper Integrals 347
32.2 The Comparison Test for Improper Integrals 350
32.3 Exercises 352

33 Separable and Linear Differential Equations of First Order 355
33.1 First Differential Equations 355
33.2 Separable Differential Equations 357
33.3 The Linear Differential Equation of First Order 361
33.4 Exercises 364

34 Linear Differential Equations with Constant Coefficients 367
34.1 Homogeneous Linear Differential Equations with Constant Coefficients 367
34.2 Inhomogeneous Linear Differential Equations with Constant Coefficients 372
34.3 Exercises 379

35 Some Special Types of Differential Equations 381
35.1 The Homogeneous Differential Equation 381
35.2 The Euler Differential Equation 383
35.3 Bernoulli's Differential Equation 385
35.4 The Riccati Differential Equation 387
35.5 The Power Series Approach 389
35.6 Exercises 392

36 Numerics of Ordinary Differential Equations I 395
36.1 First Procedure 395
36.2 Runge-Kutta Method 399
36.3 Multistep Methods 402
36.4 Exercises 405

37 Linear Mappings and Transformation Matrices 407
37.1 Definitions and Examples 407
37.2 Image, Kernel and the Dimensional Formula 410
37.3 Coordinate Vectors 412
37.4 Transformation Matrices 413
37.5 Exercises 416

38 Base Transformation 419
38.1 The Tansformation Matrix of the Composition of Linear Mappings 419
38.2 Base Transformation 420

38.3 The Two Methods for Determining Transformation Matrices 422
38.4 Exercises 425

39 Diagonalization: Eigenvalues and Eigenvectors 429
39.1 Eigenvalues and Eigenvectors of Matrices 429
39.2 Diagonalizing Matrices 431
39.3 The Characteristic Polynomial of a Matrix 433
39.4 Diagonalization of Real Symmetric Matrices 439
39.5 Exercises 441

40 Numerical Calculation of Eigenvalues and Eigenvectors 445
40.1 Gerschgorin Circles 445
40.2 Vector Iteration 447
40.3 The Jacobian Method 449
40.4 The QR-Method 453
40.5 Exercises 456

41 Quadrics 457
41.1 Terms and First Examples 457
41.2 Transformation to Normal Form 461
41.3 Exercises 466

42 Schur Decomposition and Singular Value Decomposition 469
42.1 The Schur Decomposition 469
42.2 Calculation of the Schur Decomposition 470
42.3 Singular Value Decomposition 474
42.4 Determination of the Singular Value Decomposition 475
42.5 Exercises 479

43 The Jordan Normal Form I 481
43.1 Existence of the Jordan Normal Form 481
43.2 Generalized Eigenspaces 484
43.3 Exercises 489

44 The Jordan Normal Form II 491
44.1 Construction of a Jordan Base 491
44.2 Number and Size of Jordan Boxes 499
44.3 Exercises 500

45 Definiteness and Matrix Norms 503
45.1 Definiteness of Matrices 503
45.2 Matrix Norms 507
45.3 Exercises 513

46 Functions of Several Variables 515
46.1 The Functions and Their Representations 515
46.2 Some Topological Terms 519
46.3 Consequences, Limits, Continuity 522
46.4 Exercises 525

47 Partial Differentiation: Gradient, Hessian Matrix, Jacobian Matrix 527
47.1 The Gradient 527
47.2 The Hessian Matrix 533
47.3 The Jacobian Matrix 535
47.4 Exercises 538

48 Applications of Partial Derivatives 541
48.1 The (Multidimensional) Newton Method 541
48.2 Taylor Development 544
48.3 Exercises 551

49 Extreme Value Determination 553
49.1 Local and Global Extrema 553
49.2 Determination of Extrema and Extremal Points 556
49.3 Exercises 563

50 Extreme Value Determination Under Constraints 565
50.1 Extrema Under Constraints 565
50.2 The Substitution Method 567
50.3 The Method of Lagrange Multipliers 569
50.4 Extrema Under Multiple Constraints 574
50.5 Exercise 576

51 Total Differentiation, Differential Operators 579
51.1 Total Differentiability 579
51.2 The Total Differential 581
51.3 Differential Operators 583
51.4 Exercises 587

52 Implicit Functions 591
52.1 Implicit Functions: The Simple Case 591
52.2 Implicit Functions: The General Case 596
52.3 Exercises 599

53 Coordinate Transformations 603
53.1 Transformations and Transformation Matrices 603
53.2 Polar, Cylindrical and Spherical Coordinates 604
53.3 The Differential Operators in Cartesian Cylindrical and Spherical Coordinates 606
53.4 Conversion of Vector Fields and Scalar Fields 610
53.5 Exercises 613

54 Curves I 615
54.1 Terms 615
54.2 Length of a Curve 622
54.3 Exercises 625

55 Curves II 627
55.1 Reparameterization of a Curve 627
55.2 Frenet–Serret Frame, Curvature and Torsion 630
55.3 The Leibniz Sector Formula 633
55.4 Exercises 634

56 Line Integrals 637
56.1 Scalar and Vector Line Integrals 637
56.2 Applications of the Line Integrals 642
56.3 Exercises 644

57 Gradient Fields 645
57.1 Definitions 645
57.2 Existence of a Primitive Function 647
57.3 Determination of a Primitive Function 650
57.4 Exercises 652

58 Multiple Integrals 655
58.1 Integration Over Rectangles or Cuboids 655
58.2 Normal Domains 658
58.3 Integration Over Normal Domains 659
58.4 Exercises 663

59 Substitution for Multiple Variables 665
59.1 Integration via Polar, Cylindrical, Spherical and Other Coordinates 665
59.2 Application: Mass and Center of Gravity 670
59.3 Exercises 671

60 Surfaces and Surface Integrals 675
60.1 Regular Surfaces 675
60.2 Surface Integrals 678
60.3 Overview of the Integrals 682
60.4 Exercises 682

61 Integral Theorems I 685
61.1 Green's Theorem 685
61.2 The Plane Theorem of Gauss 688
61.3 Exercises 691

62 Integral Theorems II 693
62.1 The Divergence Theorem of Gauss 693
62.2 Stokes' Theorem 697
62.3 Exercises 701

63 General Information on Differential Equations 705
63.1 The Directional Field 705
63.2 Existence and Uniqueness of Solutions 706
63.3 Transformation to 1st Order Systems 708
63.4 Exercises 711

64 The Exact Differential Equation 713
64.1 Definition of Exact ODEs 713
64.2 The Solution Procedure 714
64.3 Exercises 718

65 Linear Differential Equation Systems I 721
65.1 The Exponential Function for Matrices 721
65.2 The Exponential Function as a Solution of Linear ODE Systems 725
65.3 The Solution for a Diagonalizable A 726
65.4 Exercises 730

66 Linear Differential Equation Systems II 731
66.1 The Exponential Function as a Solution of Linear ODE Systems 731
66.2 The Solution for a Non-Diagonalizable A 735
66.3 Exercises 736

67 Linear Differential Equation Systems III 739
67.1 Solving ODE Systems 739
67.2 Stability 743
67.3 Exercises 749

68 Boundary Value Problems ... 751
68.1 Types of Boundary Value Problems ... 751
68.2 First Solution Methods ... 752
68.3 Linear Boundary Value Problems ... 753
68.4 The Method with Green's Function ... 756
68.5 Exercises ... 759

69 Basic Concepts of Numerics ... 761
69.1 Condition ... 761
69.2 The Big O Notation ... 764
69.3 Stability ... 765
69.4 Exercises ... 767

70 Fixed Point Iteration ... 769
70.1 The Fixed Point Equation ... 769
70.2 The Convergence of Iteration Methods ... 771
70.3 Implementation ... 775
70.4 Rate of Convergence ... 775
70.5 Exercises ... 776

71 Iterative Methods for Systems of Linear Equations ... 779
71.1 Solving Systems of Equations by Fixed Point Iteration ... 779
71.2 The Jacobian Method ... 780
71.3 The Gauss-Seidel Method ... 782
71.4 Relaxation ... 784
71.5 Exercises ... 786

72 Optimization ... 787
72.1 The Optimum ... 787
72.2 The Gradient Method ... 788
72.3 Newton's Method ... 789
72.4 Exercises ... 791

73 Numerics of Ordinary Differential Equations II ... 793
73.1 Solution Methods for ODE Systems ... 793
73.2 Consistency and Convergence of One-Step Methods ... 795
73.3 Stiff Differential Equations ... 799
73.4 Boundary Value Problems ... 802
73.5 Exercises ... 810

74 Fourier Series: Calculation of Fourier Coefficients 813
74.1 Periodic Functions 813
74.2 The Admissible Functions 816
74.3 Expanding in Fourier Series—Real Version 817
74.4 Application: Calculation of Series Values 821
74.5 Expanding in Fourier Series: Complex Version 822
74.6 Exercises 826

75 Fourier Series: Background, Theorems and Application 829
75.1 The Orthonormal System $1/\sqrt{2}$, $\cos(kx)$, $\sin(kx)$ 829
75.2 Theorems and Rules 831
75.3 Application to Linear Differential Equations 835
75.4 Exercises 838

76 Fourier Transform I 841
76.1 The Fourier Transform 841
76.2 The Inverse Fourier Transform 848
76.3 Exercise 850

77 Fourier Transform II 851
77.1 The Rules and Theorems for the Fourier Transform 851
77.2 Application to Linear Differential Equations 855
77.3 Exercises 859

78 Discrete Fourier Transform 863
78.1 Approximate Determination of the Fourier Coefficients 863
78.2 The Inverse Discrete Fourier Transform 867
78.3 Trigonometric Interpolation 867
78.4 Exercise 872

79 The Laplace Transformation 875
79.1 The Laplacian Transformation 875
79.2 The Rules and Theorems for the Laplace Transformation 879
79.3 Applications 883
79.4 Exercises 890

80 Holomorphic Functions 893
80.1 Complex Functions 893
80.2 Complex Differentiability and Holomorphy 900
80.3 Exercises 903

81 Complex Integration 905
81.1 Complex Curves 905
81.2 Complex Line Integrals 907
81.3 The Cauchy Integral Theorem and the Cauchy Integral Formula 911
81.4 Exercises 917

82 Laurent Series 919
82.1 Singularities 919
82.2 Laurent Series 921
82.3 Laurent Series Development 924
82.4 Exercises 927

83 The Residual Calculus 929
83.1 The Residue Theorem 929
83.2 Calculation of Real Integrals 934
83.3 Exercises 938

84 Conformal Mappings 941
84.1 Generalities of Conformal Mappings 941
84.2 Möbius Transformations 943
84.3 Exercises 949

85 Harmonic Functions and the Dirichlet Boundary Value Problem 951
85.1 Harmonic Functions 951
85.2 The Dirichlet Boundary Value Problem 954
85.3 Exercises 961

86 Partial Differential Equations of First Order 963
86.1 Linear PDEs of First Order with Constant Coefficients 964
86.2 Linear PDEs of First Order 967
86.3 The First Order Quasi Linear PDE 970
86.4 The Characteristics Method 971
86.5 Exercises 975

87 Partial Differential Equations of Second Order: General 977
87.1 First Terms 977
87.2 The Type Classification 979
87.3 Solution Methods 981
87.4 Exercises 983

88 The Laplace or Poisson Equation 985
88.1 Boundary Value Problems for the Poisson Equation 985
88.2 Solutions of the Laplace Equation 986
88.3 The Dirichlet Boundary Value Problem for a Circle 988
88.4 Numerical Solution 989
88.5 Exercises 993

89 The Heat Conduction Equation 995
89.1 Initial Boundary Value Problems for the Heat Conduction Equation 995
89.2 Solutions of the Equation 996
89.3 Zero Boundary Condition: Solution with Fourier Series 998
89.4 Numerical Solution 1001
89.5 Exercises 1004

90 The Wave Equation 1005
90.1 Initial Boundary Value Problems for the Wave Equation 1005
90.2 Solutions of the Equation 1006
90.3 The Vibrating String: Solution with Fourier Series 1008
90.4 Numerical Solution 1010
90.5 Exercises 1014

91 Solving PDEs with Fourier and Laplace Transforms 1015
91.1 An Introductory Example 1015
91.2 The General Procedure 1017
91.3 Exercises 1021

Index 1025

1 Speech, Symbols and Sets

In this first chapter we get an overview of the ways of speaking and symbols of mathematics and consider sets in the naive and for our purposes completely sufficient sense as summaries of well-differentiated elements together with the mostly familiar set operations from school time.

The list of terms we will be confronted with in this first chapter is an agreement for us (i.e. readers and writers): we stick to these notations and use these ways of speaking and symbols until the last page of this book and even further on until eternity in order to always be sure that we are talking about one and the same thing: about mathematics, its rules, its applications, ...

1.1 Speech Patterns and Symbols in Mathematics

In mathematics, statements are formulated and examined for their truth content. In simplified terms, a *statement* is a declarative sentence to which one of the two truth values FALSE or TRUE can be unambiguously assigned. As examples serve

$$\textit{it is raining} \quad \text{or} \quad \sqrt{2} > 1.12\,.$$

1.1.1 Junctors

With junctors link *simple* statements to a *complex* statement. We consider the five (most important) junctors

NOT AND OR Implication Equivalence.

C. Karpfinger, *Calculus and Linear Algebra in Recipes*,
https://doi.org/10.1007/978-3-662-65458-3_1

Junctors

- If A is a statement, then $\neg A$ is the **negation** of A.
- If A and B are statements, then $A \wedge B$ consider; one calls $\wedge$ the **AND-junctor**. We have
  - $A \wedge B$ is true if both statements are true.
  - $A \wedge B$ is false if one of the two statements is false.
- If A and B are statements, then $A \vee B$ consider; one calls $\vee$ the **OR-junctor**. We have
  - $A \vee B$ is true if one of the statements is true.
  - $A \vee B$ is false if both statements are false.
- **If** A, **then** B, short $A \Rightarrow B$. One calls $\Rightarrow$ **implication**.
- A **if and only if** B, in short: $A \Rightarrow B$ and $B \Rightarrow A$, even shorter : $A \Leftrightarrow B$. One calls $\Leftrightarrow$ **equivalence**.

Remark OR is not exclusive—both statements may be true. Excluding is EITHER OR.

Example 1.1

- $\neg$(It is raining today) means: it is not raining today.
- $\neg(x \geq 5)$ means: $x<5$.
- $\neg$(All cars are green) means: There are cars that are not green (negation of an *all statement* is an *existential statement*).
- $\neg$(Besides the earth there are other inhabited planets) means: All planets, except the earth, are uninhabited (negation of an *existence statement* is an *all statement*).
- $\neg$(For all $x,\ y \in M$ we have $f(x+y) = f(x) + f(y)$) means: There are $x,\ y \in M$ with $f(x+y) \neq f(x) + f(y)$.
- If $A:\ x \leq 5$ and $B:\ x \in \mathbb{N}$, then $A \wedge B$ means $x \in \{1, 2, 3, 4, 5\}$.
- If $A:\ x \in \mathbb{R} \wedge x \geq 2 \wedge x \leq 4$ and $B:\ x \in \{2, 3, 7\}$, then $A \vee B$ means $x \in [2, 4] \cup \{7\}$.
- **When** it rains, **then** the road is wet; in short:

$$\text{It's raining} \Rightarrow \text{The road is wet.}$$

- **If** m is an even natural number, **then** $m \cdot n$ for every $n \in \mathbb{N}$ is an even natural number; in short:

$$m \text{ even} \Rightarrow m \cdot n \text{ even } (n \in \mathbb{N}).$$

  Because: m even $\Rightarrow m = 2 \cdot m',\ m' \in \mathbb{N} \Rightarrow m \cdot n = 2 \cdot m' \cdot n,\ m' \cdot n \in \mathbb{N} \Rightarrow m \cdot n$ even.

- The road is wet, **if and only if**
  - it rains,
  - the road is cleaned,
  - snow melts,
  - a bucket of water is spilled,
  - ...
- For $x \in \mathbb{R}$ we have:

$$x \leq 5 \wedge x \in \mathbb{N} \Leftrightarrow x \in \{1, 2, 3, 4, 5\}.$$

- For $x \in \mathbb{R}$ applies:

$$x \in \mathbb{Q} \Leftrightarrow \text{ there is a } n \in \mathbb{N} \text{ with } n \cdot x \in \mathbb{Z}.$$

Since: $x \in \mathbb{Q} \Leftrightarrow x = \frac{p}{q}$ with $p \in \mathbb{Z}$ and $q \in \mathbb{N} \Leftrightarrow q \cdot x = p$ with $p \in \mathbb{Z}$ and $q \in \mathbb{N}$. Choose $n = q$.

- If $m, n \in \mathbb{N}$, then we have:

$$m \cdot n \text{ is even } \Leftrightarrow m \text{ is even } \vee n \text{ is even}.$$

$\Leftarrow$: m even or n even $\Rightarrow m \cdot n$ even (see above).
$\Rightarrow$: $m \cdot n$ even. Suppose neither m nor n are even. Then

$$m = 2 \cdot m' + 1 \text{ and } n = 2 \cdot n' + 1$$

with $m', n' \in \mathbb{N}$. It follows $m \cdot n = 4 \cdot m' \cdot n' + 2 \cdot (m' + n') + 1 = 2 \cdot k + 1$ with a $k \in \mathbb{N}$. This is a contradiction to $m \cdot n$ is even. ■

1.1.2 Quantifiers

Quantifiers capture variables quantitatively. We consider four quantifiers:

Quantifiers

- $\forall$ *to each* or *for all*.
- $\exists$ *there is*,

(continued)

- $\exists_1$ there is *exactly one*,
- $\nexists$ there is *no*.

Example 1.2

- For *every real number x there is a natural number n which is greater than x can be written briefly as*

$$\forall x \in \mathbb{R}\, \exists n \in \mathbb{N} : n \geq x\,.$$

  Note the order, the statement

$$\exists n \in \mathbb{N} : n \geq x\, \forall x \in \mathbb{R}$$

  is obviously false.
- If $A = \{1, 2, 3\}$ and $B = \{1, 4, 9\}$, then

$$\forall b \in B\, \exists_1 a \in A : a^2 = b\,.$$

■

1.2 Summation and Product Symbol

The summation symbol $\sum$ and the product symbol $\prod$ are useful abbreviations, you put

$$a_1 + a_2 + \cdots + a_n = \sum_{i=1}^{n} a_i \quad \text{and} \quad a_1 \cdot a_2 \cdots a_n = \prod_{i=1}^{n} a_i\,.$$

Examples

- $\sum\limits_{i=1}^{100} 2^i = 2 + 2^2 + 2^3 + \cdots + 2^{100}$.
- $\prod\limits_{i=1}^{100} \frac{1}{i^2} = 1 \cdot \frac{1}{4} \cdot \frac{1}{9} \cdots \frac{1}{10000}$.
- $\sum\limits_{i=1}^{10} \prod\limits_{j=1}^{5} i \cdot j = 1 \cdot 2 \cdot 3 \cdot 4 \cdot 5 + 2 \cdot 4 \cdot 6 \cdot 8 \cdot 10 + \cdots + 10 \cdot 20 \cdot 30 \cdot 40 \cdot 50$.
- $\sum\limits_{i=0}^{n} a_i = a_0 + \sum\limits_{l=1}^{n-1} a_l + a_n$.

Occasionally you also need the *empty sum* or the empty *product*, meaning that the upper limit is smaller than the lower limit. We define the empty sum as 0 and the empty product as 1, e.g.

$$\sum_{k=1}^{0} a_k = 0 \text{ and } \prod_{k=2}^{-1} b_k = 1 .$$

1.3 Powers and Roots

We form powers and roots from real numbers. In doing so, we assume (for now) that the following *sets of numbers are* known:

$$\mathbb{N} \subseteq \mathbb{N}_0 \subseteq \mathbb{Z} \subseteq \mathbb{Q} \subseteq \mathbb{R} .$$

$\mathbb{N}$ denotes the natural numbers, $\mathbb{N}_0$ the natural numbers including 0, $\mathbb{Z}$ the integers, $\mathbb{Q}$ the rational numbers, and $\mathbb{R}$ the real numbers.

Further, we know the following notations:

- $\forall a \in \mathbb{R}\ \forall n \in \mathbb{N} : a^n = a \cdots a$ *(n times)*—the *n-th power of a*.
- $\forall a \in \mathbb{R} \setminus \{0\}\ \forall n \in \mathbb{N} : a^{-n} = (a^{-1})^n$.
- $\forall a \in \mathbb{R} : a^0 = 1$; in particular $0^0 = 1$.
- $\forall a \in \mathbb{R}_{>0},\ \forall n \in \mathbb{N} : a^{\frac{1}{n}} = \sqrt[n]{a}$—is the *n-th root of a*, i.e. the (uniquely determined) positive solution of the equation $x^n = a$.
- $\forall a \in \mathbb{R}_{>0},\ \forall n \in \mathbb{N} : a^{-\frac{1}{n}} = (a^{-1})^{\frac{1}{n}}$.
- $\forall a \in \mathbb{R}_{>0},\ \forall \frac{m}{n} \in \mathbb{Q} : a^{\frac{m}{n}} = (a^{\frac{1}{n}})^m$.

With these agreements we have the following rules:

Power Rules
$\forall a, b \in \mathbb{R}_{>0}, \forall r, s \in \mathbb{Q}$ we have:

- $a^r a^s = a^{r+s}$,
- $a^r b^r = (a\,b)^r$,
- $(a^r)^s = a^{r\,s}$.

Example 1.3 $\forall a, b > 0$ we have:

$$\frac{\sqrt[10]{a^6}\,\sqrt[5]{b^{-2}}}{\sqrt[5]{a^{-2}}\,\sqrt[15]{b^9}} = a^{\frac{6}{10}} b^{-\frac{2}{5}} a^{\frac{2}{5}} b^{-\frac{9}{15}} = a^{\frac{10}{10}} b^{-\frac{15}{15}} = \frac{a}{b}\,.$$

■

1.4 Symbols of Set Theory

Under a **set** we understand a summary of certain well-differentiated objects, which we call elements of this set:

$$\underset{\text{set}}{\underset{\uparrow}{A}} = \{\underset{\text{elements}}{\underset{\nwarrow\uparrow\nearrow}{a,\ b,\ c}}, \ldots\}.$$

Basically, there are two different ways of writing down sets:

- One can describe sets by explicitly stating the elements:

$$A = \{a,\ b,\ c\} \qquad \text{or} \qquad \mathbb{N} = \{1,\ 2,\ 3,\ \ldots\}.$$

- One can specify properties that characterize the elements:

$$A = \{n \in \mathbb{N} \mid 1 \leq n \leq 5\} \qquad \text{or} \qquad B = \{n \in \mathbb{N} \mid 2^n + 1 \text{ is prime}\}.$$

The vertical bar introduces the conditions that the elements must satisfy, and is read as *with the property*.

Example 1.4 The elements of a set can be explicitly stated, such as

$$A = \{1, \sqrt{2}, 13, \text{Angela Merkel}\}\,,$$

or be explained by properties

$$A = \{n \in \mathbb{N} \mid n \text{ is odd }\} = 2\,\mathbb{N} - 1\,.$$

■

We list some self-explanatory or already known notations:

Terms and Notations for Sets

- $a \in A$: a is an **element** of A,
- $a \notin A$: a **is not an element of** A,
- $A \subseteq B$: A is a **subset** of B: $a \in A \Rightarrow a \in B$,
- $A \nsubseteq B$: A is **not a subset of** B: $\exists a \in A : a \notin B$,
- $A \subsetneq B$: A is a **proper subset** of B: $a \in A \Rightarrow a \in B \wedge \exists b \in B : b \notin A$,
- A=B: A **is equal** B: $A \subseteq B \wedge B \subseteq A$,
- $\emptyset$: the **empty set**, a set with no elements: $\emptyset = \{n \in \mathbb{N} \mid n < -1\}$,
- $A \cap B = \{x \mid x \in A \wedge x \in B\}$—the **average** from A and B,
- $A \cup B = \{x \mid x \in A \vee x \in B\}$—the **union** from A and B,
- $A \setminus B = \{x \mid x \in A \wedge x \notin B\}$—the **difference** A without B,
- $C_B(A) = B \setminus A$ if $A \subseteq B$—the **complement** of A in B,
- $A \times B = \{(a, b) \mid a \in A \wedge b \in B\}$—the **cartesian product** of A and B,
- $A^n = A \times \cdots \times A = \{(a_1, \ldots, a_n) \mid a_i \in A \, \forall i\}$ with $n \in \mathbb{N}$—the **set of n-tuples** $(a_1, \ldots, a_n)$ via A.
- $|A|$—the **cardinality** of A, i.e. the number of elements of A, if A is finite and ∞ else.

Compare Fig. 1.1 to some of the set operations listed.

Example 1.5 We consider the sets

$$A = \{1, 2, 5, 7\}, \; B = \{n \in \mathbb{N} \mid n \text{ is odd}\}, \; C = \{B, 2, \sqrt{2}\}, \; D = \{1, 5, 7\}.$$

With the operations addressed above, we have:

- $D \subseteq A$,$D \subsetneq A$,$D \subseteq B$,$D \subsetneq B$,$C \nsubseteq B$,$B \nsubseteq C$,$B \in C$,
- $A \cap B = D$,$C \cap D = \emptyset$,$C \cap B = \emptyset$,
- $B \setminus C = B$,$B \setminus A = B \setminus D = C_B(D) = \{n \in \mathbb{N} \mid n\}$ is odd and $n \geq 9 \cup \{3\}$,

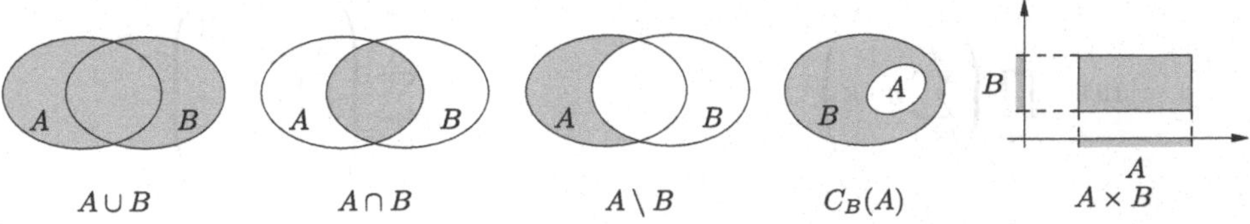

Fig. 1.1 Diagrams of set operations

- $C \times D = \{(B, 1), (B, 5), (B, 7), (2, 1), (2, 5), (2, 7), (\sqrt{2}, 1), (\sqrt{2}, 5), (\sqrt{2}, 7)\}$,
- $|C \times D| = 9 = |C| \cdot |D|$. ■

By the way, two sets A and B with $A \cap B = \emptyset$ are called **disjoint**. Obviously, the following rules apply:

Rules for Set Operations
If A and B are finite sets, then

- $|A \times B| = |A| \cdot |B|$,
- $|A \cup B| = |A| + |B| - |A \cap B|$,
- $|A \setminus B| = |A| - |A \cap B|$,
- $|A \cap B| = 0 \Leftrightarrow A, B$ disjoint.

For arbitrary sets A, B and C we have:

- $\emptyset \subseteq B$,
- $A \setminus (B \cup C) = (A \setminus B) \cap (A \setminus C)$,
- $(A \cap B) \cap C = A \cap (B \cap C)$ and $(A \cup B) \cup C = A \cup (B \cup C)$,
- $A \cap (B \cap C) = (A \cap B) \cap (A \cap C)$ and $A \cup (B \cap C) = (A \cup B) \cap (A \cup C)$.

Justifying these last four statements is not difficult (see Exercise 1.5).

1.5 Exercises

1.1 In each of the following statements, reverse the order of the quantifiers $\forall$ and $\exists$ and check the sense and correctness of the resulting statements:

(a) $\forall n \notin \mathbb{P} \cup \{1\} \, \exists k \notin \{1, n\} : k \mid n$ ($\mathbb{P}$ denotes the set of all prime numbers),
(b) $\exists N \in \mathbb{N} \, \forall n \geq N : \frac{1}{n} \leq 0.001$.

1.2 Write out the following expressions:

(a) $$\prod_{n=1}^{j} \left(\sum_{k=1}^{j} n \cdot k \right),$$ (b) $$\sum_{n=1}^{5} \left(\prod_{i=1}^{n} \frac{\sum_{k=1}^{3} k^3}{5^i} \right).$$

1.3 Write the following expressions in the form $\sum_{n=1}^{k} a_n$ or $\prod_{n=1}^{k} a_n$:

(a) $\frac{3}{1} + \frac{5}{4} + \frac{9}{9} + \frac{17}{16} + \ldots + \frac{1073741825}{900}$,
(b) $\frac{2}{3} + \frac{4}{9} + \frac{6}{27} + \frac{8}{81} + \ldots + \frac{18}{19683}$,
(c) $\frac{6}{1} \cdot \frac{9}{2} \cdot \frac{12}{3} \cdot \ldots \cdot \frac{300}{99}$,
(d) $1 + \frac{1 \cdot 2}{1 \cdot 3} + \frac{1 \cdot 2 \cdot 3}{1 \cdot 3 \cdot 5} + \cdots + \frac{1 \cdot 2 \cdot 3 \cdot 13}{1 \cdot 3 \cdot 5 \cdot 25}$.

1.4 Given the following subsets of the set of real numbers $\mathbb{R}$:

$$A = \{x \in \mathbb{R} \,|\, 5 > x > -2\}, \quad B = \{x \in \mathbb{R} \,|\, 1 > x\}, \quad C = \{x \in \mathbb{R} \,|\, -1 < x \le 1\}.$$

Determine the following sets and sketch them:

(a) $A \cap C$, (b) $B \setminus A$, (c) $(\mathbb{R} \setminus C) \cup B$.

1.5 Given the following subsets of the real numbers:

$$A = \{x \in \mathbb{R} \,|\, -2 < x < 5\}, \quad B = \{x \in \mathbb{R} \,|\, 1 \ge x\},$$
$$C = \{x \in \mathbb{R} \,|\, x^2 \le 4\}, \quad D = \{x \in \mathbb{R} \,|\, x^2 > 1\}.$$

Determine each of the following sets and sketch them on the number line:

(a) $A \cap B$,
(b) $A \cup D$,
(c) $B \setminus C$,
(d) $D \setminus (A \cup B)$,
(e) $C \cap (A \cup B)$,
(f) $(\mathbb{R} \setminus (A \cup B)) \cup (C \cap D)$.

1.6 Let be $A = \{a, b, c, d\}$ and $B = \{M \,|\, M \subseteq A\}$. Decide and justify which of the following statements are true and which are false:

(a) $a \in B$,
(b) $\{b\} \in B$,
(c) $\{a\} \in A$,
(d) $a \in B$,
(e) $A \subseteq B$,
(f) $\{a\} \subseteq A$,
(g) $\emptyset \in B$,
(h) $\emptyset \subseteq B$,
(i) $\{\emptyset\} \subseteq B$.

1.7 Show, that for any sets A, B and C we have:

(a) $\emptyset \subseteq B$,
(b) $A \setminus (B \cup C) = (A \setminus B) \cap (A \setminus C)$,
(c) $(A \cap B) \cap C = A \cap (B \cap C)$, and $(A \cup B) \cup C = A \cup (B \cup C)$,
(d) $A \cap (B \cap C) = (A \cap B) \cap (A \cap C)$ and $A \cup (B \cap C) = (A \cup B) \cap (A \cup C)$.

The Natural Numbers, Integers and Rational Numbers

2

The number sets $\mathbb{N}$, $\mathbb{Z}$, $\mathbb{Q}$ and $\mathbb{R}$ of the natural, integer, rational, and real numbers are familiar from school days. In this chapter we briefly consider a few aspects concerning the natural numbers, integers and rational numbers, as far as we need them in engineering mathematics. The largest space here is taken up by induction, which usually causes problems for beginners. Often it helps to just stubbornly do the recipe, understanding comes over time.

The real numbers take more space, we will take care of them in the next chapter.

2.1 The Natural Numbers

$\mathbb{N} = \{1, 2, 3, \ldots\}$ is the set of **natural numbers**. If we also want to include the zero, we write $\mathbb{N}_0 = \mathbb{N} \cup \{0\} = \{0, 1, 2, \ldots\}$.

Mathematicians *explain* the natural numbers starting from the empty set, we, on the other hand, take the natural numbers along with the familiar arrangement, addition and multiplication of these numbers as *given* and no longer want to question this.

In later chapters, we will repeatedly face the problem of finding a statement for all natural numbers $n \in \mathbb{N}_0$ or for all natural numbers n from a $n_0 \in \mathbb{N}$ respectively.

Example 2.1

- For all natural numbers $n \geq 1$ we have $\sum_{k=1}^{n} k = \frac{n(n+1)}{2}$.
- For all natural numbers $n \geq 0$ and $q \in \mathbb{R} \setminus \{1\}$ we have $\sum_{k=0}^{n} q^k = \frac{1-q^{n+1}}{1-q}$.
- For all natural number $n \geq 1$, the number $a_n = 5^n - 1$ is a multiple of 4.

Induction is a method that can often be used to justify such statements.

C. Karpfinger, *Calculus and Linear Algebra in Recipes*,
https://doi.org/10.1007/978-3-662-65458-3_2

Recipe: Induction

Given for a $n \in \mathbb{N}_0$ the statement $A(n)$. To justify the statement $A(n)$ for all $n \geq n_0 \in \mathbb{N}_0$ proceed as follows:

1. **Base case:** Show the statement $A(n_0)$.
2. **Induction hypothesis:** Assume that we have the statement $A(n)$ for a $n \in \mathbb{N}_0$ with $n \geq n_0$.
3. **Induction step:** Show the statement $A(n+1)$.

We abbreviate the base case with BC, the induction hypothesis with IH and the induction step with IS.

Before we explain how induction justifies $A(n)$ *for all* $n \geq n_0$ we consider examples (cf. Example 2.1):

Example 2.2

- We show with induction:

$$\sum_{k=1}^{n} k = \frac{n(n+1)}{2} \quad \text{for all } n \in \mathbb{N}.$$

(1) BC: The formula is correct for $n_0 = 1$, since $\sum_{k=1}^{1} k = 1$ and $\frac{1(1+1)}{2} = 1$.
(2) IH: We assume that the formula $\sum_{k=1}^{n} k = \frac{n(n+1)}{2}$ is correct for a $n \in \mathbb{N}_0$ with $n \geq 1$.
(3) IS: Since the formula for $n \in \mathbb{N}_0$ is correct, we obtain for $n+1$:

$$\sum_{k=1}^{n+1} k = n+1+\sum_{k=1}^{n} k = n+1+\frac{n(n+1)}{2} = \frac{(n+1)(n+2)}{2}.$$

- We show with induction:

$$\sum_{k=0}^{n} q^k = \frac{1-q^{n+1}}{1-q} \quad \text{for all } n \in \mathbb{N}_0 \text{ and } q \in \mathbb{R} \setminus \{1\}.$$

(1) BC: The statement is true for $n_0 = 0$, since $\sum_{k=0}^{0} q^k = q^0 = 1$ and $\frac{1-q^1}{1-q} = 1$.
(2) IH: We assume that the formula $\sum_{k=0}^{n} q^k = \frac{1-q^{n+1}}{1-q}$ is correct for a $n \in \mathbb{N}_0$ with $n \geq 0$.

(3) IS: Since the formula for $n \in \mathbb{N}_0$ is correct, we obtain for $n+1$:

$$\sum_{k=0}^{n+1} q^k = q^{n+1} + \sum_{k=0}^{n} q^k = q^{n+1} + \frac{1-q^{n+1}}{1-q}$$
$$= \frac{q^{n+1} - q^{n+2} + 1 - q^{n+1}}{1-q} = \frac{1-q^{n+2}}{1-q}.$$

- We show with induction:

$$\text{For all } n \in \mathbb{N}: \; a_n = 5^n - 1 = 4 \cdot k \quad \text{for a } k \in \mathbb{N}.$$

(1) BC: The statement is true for $n_0 = 1$, since $a_1 = 5^1 - 1 = 4 = 4 \cdot 1$, choose $k = 1$.
(2) IH: We assume that $a_n = 5^n - 1 = 4 \cdot k$ for a $n \in \mathbb{N}$ and a $k \in \mathbb{N}$.
(3) IS: Since the formula is correct for $n \in \mathbb{N}$, we obtain for $n + 1$:

$$a_{n+1} = 5^{n+1} - 1 = 5 \cdot 5^n - 1 = (4+1) \cdot 5^n - 1 = 4 \cdot 5^n + 5^n - 1$$
$$= 4 \cdot 5^n + 4 \cdot k = 4 \cdot (5^n + k).$$

With $k' = 5^n + k$ you get $a_n = 4 \cdot k'$. ■

Why does it work? At the beginning of the induction (BC) it is shown that the statement $A(n_0)$ is valid for the *first* n_0. Since the statement $A(n_0)$ is valid, after the induction step also $A(n_0 + 1)$. And since now we have $A(n_0 + 1)$, again after the induction step we also have $A(n_0 + 2)$ etc.

Play this out in the above examples and realize that interposing the *induction hypothesis* is an extremely clever move to justify the statement for infinitely many natural numbers with only one induction step.

We justify another formula with induction. To do this, we need two terms:

- The **factorial** of $n \in \mathbb{N}$ is defined as the product of the numbers from 1 to n:

$$n! = n \cdot (n-1) \cdot \ldots \cdot 2 \cdot 1 \qquad \text{and} \qquad 0! = 1.$$

For example:

$$1! = 1,\; 2! = 2 \cdot 1 = 2,\; 3! = 3 \cdot 2 \cdot 1 = 6,\; 4! = 4 \cdot 3 \cdot 2 \cdot 1 = 24,\; 5! = 5 \cdot 4 \cdot 3 \cdot 2 \cdot 1 = 120.$$

- For $n, k \in \mathbb{N}_0$ with $k \leq n$ the number

$$\binom{n}{k} = \frac{n!}{k!(n-k)!}$$

is called **binomial coefficient** *n choose k*. It indicates how many different k-element subsets a set with n elements has. We have:

$$\binom{3}{0} = \frac{3!}{0! \cdot 3!} = 1\,, \quad \binom{3}{1} = \frac{3!}{1! \cdot 2!} = 3\,, \quad \binom{3}{2} = 3\,, \quad \binom{3}{3} = 1\,.$$

Example 2.3 We show with induction:

$$\text{For all } a,\ b \in \mathbb{R} \text{ and all } n \in \mathbb{N}_0:\ (a+b)^n = \sum_{k=0}^{n} \binom{n}{k} a^k b^{n-k}\,.$$

(1) BC: The formula is correct for $n_0 = 0$ since $(a+b)^0 = 1$ and $\sum_{k=0}^{0} \binom{0}{k} a^k b^{0-k} = 1$.
(2) IH: We assume that $(a+b)^n = \sum_{k=0}^{n} \binom{n}{k} a^k b^{n-k}$ is true for a $n \in \mathbb{N}$.
(3) IS: Since we have the formula for $n \in \mathbb{N}$, we obtain for $n+1$ with Exercise 2.2(c):

$$\begin{aligned}
(a+b)^{n+1} &= (a+b) \cdot (a+b)^n \overset{\text{IH}}{=} (a+b) \cdot \sum_{k=0}^{n} \binom{n}{k} a^k b^{n-k} \\
&= a \cdot \sum_{k=0}^{n} \binom{n}{k} a^k b^{n-k} + b \cdot \sum_{k=0}^{n} \binom{n}{k} a^k b^{n-k} \\
&= \sum_{k=0}^{n} \binom{n}{k} a^{k+1} b^{n-k} + \sum_{k=0}^{n} \binom{n}{k} a^k b^{n-k+1} \\
&= \sum_{k=1}^{n+1} \binom{n}{k-1} a^k b^{n-k+1} + \sum_{k=0}^{n} \binom{n}{k} a^k b^{n-k+1} \\
&= \binom{n}{n} a^{n+1} b^0 + \binom{n}{0} a^0 b^{n+1} + \sum_{k=1}^{n} \left(\binom{n}{k-1} + \binom{n}{k} \right) a^k b^{n-k+1} \\
&= \binom{n+1}{n+1} a^{n+1} b^0 + \binom{n+1}{0} a^0 b^{n+1} + \sum_{k=1}^{n} \binom{n+1}{k} a^k b^{n+1-k} \\
&= \sum_{k=0}^{n+1} \binom{n+1}{k} a^k b^{n+1-k}.
\end{aligned}$$

■

We have learned about and reasoned about three important formulas in this section, and we summarize them:

Important Formulas
For all natural numbers $n \in \mathbb{N}_0$ and $q \in \mathbb{R}$ and $a,\ b \in \mathbb{R}$ we have:

- $\sum\limits_{k=1}^{n} k = \frac{n(n+1)}{2}$ **(Gaussian summation formula).**
- $\sum\limits_{k=0}^{n} q^k = \begin{cases} \frac{1-q^{n+1}}{1-q}, & \text{if } q \neq 1 \\ n+1, & \text{if } q = 1 \end{cases}$ **(geometric summation formula).**
- $(a+b)^n = \sum\limits_{k=0}^{n} \binom{n}{k} a^k b^{n-k}$ **(Binomial formula).**

2.2 The Integers

The set $\mathbb{Z}$ the **integers**

$$\mathbb{Z} = \{\ldots, -3, -2, -1, 0, 1, 2, 3, \ldots\} = \{0, \pm 1, \pm 2, \pm 3, \ldots\}$$

is known with its ordering $\cdots < -2 < -1 < 0 < 1 < 2 \cdots$ and addition and multiplication of integers known from school days, as well as the following rules:

- $-a = (-1)\,a$ for all $a \in \mathbb{Z}$,
- $(-a)(-b) = a\,b$ for all $a,\ b \in \mathbb{Z}$,
- $a < b \Rightarrow -a > -b$,
- $ab > 0 \Rightarrow (a, b > 0 \vee a, b < 0)$ for $a,\ b \in \mathbb{Z}$,
- $a + x = b$ has the solution $x = b + (-a) = b - a$ for all $a,\ b \in \mathbb{Z}$.

2.3 The Rational Numbers

Also the set $\mathbb{Q}$ of the **rational numbers**

$$\mathbb{Q} = \left\{ \frac{m}{n} \mid m \in \mathbb{Z},\ n \in \mathbb{N} \right\}$$

is with their ordering

$$\frac{m}{n} < \frac{m'}{n'} \Leftrightarrow mn' < m'n$$

and the addition and multiplication of rational numbers

$$\frac{m}{n} + \frac{m'}{n'} = \frac{mn' + m'n}{nn'} \text{ and } \frac{m}{n} \cdot \frac{m'}{n'} = \frac{mm'}{nn'}$$

familiar from school days, as are the following rules:

- $\frac{m}{n} = \frac{m'}{n'} \Leftrightarrow m \cdot n' = m' \cdot n$,
- $\frac{m}{n} = \frac{mk}{nk}$ for all $k \neq 0$, in particular therefore $\frac{m}{-n} = \frac{-m}{n}$,
- $ax = b$ has the solution $x = \frac{b}{a}$ if $a \neq 0$.

Every rational number $\frac{m}{n} \in \mathbb{Q}$ can be represented as **decimal** representation. This representation is either **finite**,

$$\frac{m}{n} = a.a_1a_2 \ldots a_k\,,$$

or **periodic**,

$$\frac{m}{n} = a.a_1a_2 \ldots a_k\overline{b_1b_2 \ldots b_\ell}\,,$$

with $a \in \mathbb{Z}$ and $a_i,\ b_i \in \{0, \ldots, 9\}$.

Example 2.4 By successive division we get:

$$\frac{9}{8} = 1.125, \qquad \frac{12}{33} = 0.\overline{36} = 0.36363636\ldots \qquad \blacksquare$$

The decimal representation of $\frac{m}{n}$ can always be calculated by dividing m by n. But how do you find the fraction representation $\frac{m}{n}$ from the decimal representation?

We show this by examples, the general procedure is then immediately clear:

Example 2.5 For finite decimal numbers, expanding and reducing helps:

$$1.125 = 1.125 \cdot \frac{1000}{1000} = \frac{1125}{1000} = \frac{225}{200} = \frac{45}{40} = \frac{9}{8}\,.$$

For periodic decimal representations, we make do with a trick: We set $a = 0.\overline{36}$. We then have:

$$100 \cdot a - a = 36 \Leftrightarrow (100-1) \cdot a = 36 \Leftrightarrow 99 \cdot a = 36 \Leftrightarrow a = \frac{36}{99} = \frac{12}{33}.$$

Or, if we have a fractional representation $\frac{m}{n}$ of $a = 0.25\overline{54}$:

$$\underbrace{10000 \cdot a - 100 \cdot a}_{=9900 \cdot a} = 2554.\overline{54} - 25.\overline{54} = 2529 \Leftrightarrow a = \frac{2529}{9900} = \frac{281}{1100}. \qquad \blacksquare$$

2.4 Exercises

2.1 Prove the following statements by induction:

(a) For all $n \in \mathbb{N}$ we have: n elements can be arranged in $1 \cdot 2 \cdots n = n!$ different ways.
(b) The sum over the first n odd numbers yields for all $n \in \mathbb{N}$ the value n^2.
(c) Bernoulli's inequality $(1+x)^n \geq 1 + nx$ is valid for all real numbers $x \geq -1$ and all $n \in \mathbb{N}$.
(d) For every $n \in \mathbb{N}$ the number $4^{2n+1} + 3^{n+2}$ is divisible by 13.
(e) For all $n \in \mathbb{N}$: $\sum_{i=1}^{n}(i^2 - 1) = \frac{1}{6}(2n^3 + 3n^2 - 5n)$.
(f) For all $n \in \mathbb{N}$: $\sum_{k=1}^{n} k \cdot k! = (n+1)! - 1$.
(g) For all $n \in \mathbb{N}$: $\sum_{i=0}^{n} 2^i = 2^{n+1} - 1$.
(h) For all $n \in \mathbb{N}_{>4}$: $2^n > n^2$.
(i) The Fibonacci numbers $F_0, F_1, F_2, \ldots$ are recursively defined by $F_0 = 0$, $F_1 = 1$ and $F_n = F_{n-1} + F_{n-2}$ for $n \geq 2$. For all $n \in \mathbb{N}$ we have: $\sum_{i=1}^{n}(F_i)^2 = F_n \cdot F_{n+1}$.

2.2 Show that the following rules of calculation apply to the binomial coefficients, where are $k, n \in \mathbb{N}_0$ with $k \leq n$:

(a) $\binom{n}{k} = \binom{n}{n-k}$, (b) $\binom{n}{n} = 1 = \binom{n}{0}$, (c) $\binom{n+1}{k} = \binom{n}{k} + \binom{n}{k-1}$.

2.3 Represent the following decimal numbers x in the form $x = \frac{p}{q}$ with $p \in \mathbb{Z}$ and $q \in \mathbb{N}$:

(a) $x = 10.1\overline{24}$, (b) $x = 0.0\overline{9}$, (c) $x = 0,\overline{142857}$.

2.4 In a new housing development, a total of 4380 housing units were completed within a period of about 12 years. One apartment per day was ready for occupancy. From the day the first apartment was occupied until the day after the last unit was handed over, the residents consumed a total of $1.8709 \cdot 10^8$ kWh of electricity was consumed by the residents. Determine the average consumption per day and per apartment.

2.5 A mortgage loan for €100,000 bears interest at 7% per annum and is repaid at a constant rate A (annuity) at the end of each year.

How large must A be if the loan is to be repaid in full by the 20th repayment instalment?

3 The Real Numbers

The set of rational numbers is full of *holes* in a way that can be described in more detail, but which is of no further interest to us. These *holes* are plugged by the *irrational* numbers. The totality of the rational and irrational numbers forms the set of *real numbers* and thus the familiar *number line*.

The real numbers form the foundation of (real) analysis and thus also of engineering mathematics. The handling of the real numbers must be practiced and should not cause any difficulties. Here, we mainly consider the resolution of equations and inequalities with and without absolute values. Such calculations are necessary again and again until the end of the study and beyond.

3.1 Basics

We denote by $\mathbb{R}$ the set of **real numbers**, that is, the set of **all** decimal numbers. An illustrative representation of the real numbers is formed by the **number line** (cf. Fig. 3.1).

Every real number is a point on the number line, and every point on the number line is a real number. The number line also expresses the familiar ordering of the real numbers: a is *greater* than b, $a > b$, if a lies to the right of b on the number line.

Since the rational numbers have a finite or periodic decimal representation, or since the rational numbers naturally also lie on the number line, the following applies $\mathbb{Q} \subseteq \mathbb{R}$. In fact, even $\mathbb{Q} \subsetneq \mathbb{R}$, in the following list we summarize these and other interesting facts:

- $\mathbb{Q}$ is the set of rational numbers,
- we have $\mathbb{Q} \subsetneq \mathbb{R}$,
- $\mathbb{R} \setminus \mathbb{Q}$ is the set of **irrational** numbers,
- $\sqrt{2}$, the circular number π and the Eulerian number e are irrational numbers,

C. Karpfinger, *Calculus and Linear Algebra in Recipes*,
https://doi.org/10.1007/978-3-662-65458-3_3

Fig. 3.1 The number line $\mathbb{R}$

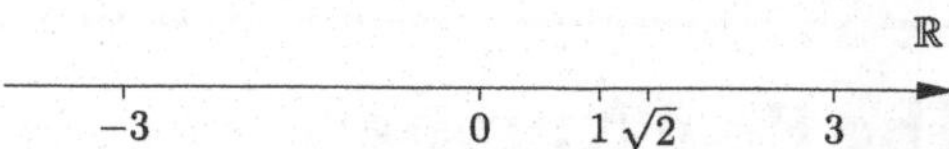

- $|\mathbb{R}| = |\mathbb{R} \setminus \mathbb{Q}| > |\mathbb{Q}|$. (We have not captured this distinction of different infinities with our agreement of Sect. 1.2.)

The fact that $\sqrt{2}$ is irrational, can be easily justified (Exercise 3.7).

For the sake of completeness, we also give the following calculation rules for the well-known addition and multiplication of real numbers, which also apply to all real numbers and thus in particular to all rational, integer and natural numbers:

Associative, Commutative and Distributive Laws

- For all $a,\ b,\ c \in \mathbb{R}$ we have the **associative laws**

$$a + (b + c) = (a + b) + c \text{ and } a \cdot (b \cdot c) = (a \cdot b) \cdot c\,.$$

- For all $a,\ b \in \mathbb{R}$ we have the **commutative laws**

$$a + b = b + a \text{ and } a \cdot b = b \cdot a\,.$$

- For all $a,\ b,\ c \in \mathbb{R}$ we have the **distributive law**

$$a \cdot (b + c) = a \cdot b + a \cdot c\,.$$

From now on we write more simply $a\,b$ instead of $a \cdot b$.

3.2 Real Intervals

Real calculus will essentially take place on *intervals*. We distinguish the following types of *intervals*:

Intervals

For $a, b \in \mathbb{R}$, $a < b$, one explains the **bounded intervals**:

(continued)

- $[a, b] = \{x \in \mathbb{R} \mid a \leq x \leq b\}$—**closed interval**.
- $(a, b) = \{x \in \mathbb{R} \mid a < x < b\}$—**open interval**.
- $(a, b] = \{x \in \mathbb{R} \mid a < x \leq b\}$ or $[b, a) = \{x \in \mathbb{R} \mid a \leq x < b\}$—**half-open intervals**.

Analogously, for $a \in \mathbb{R}$ the **unbounded intervals** are defined:

- $\mathbb{R}_{\geq a} = [a, \infty) = [a, \infty[= \{x \in \mathbb{R} \mid a \leq x\}$,
- $\mathbb{R}_{>a} = (a, \infty) =]a, \infty[= \{x \in \mathbb{R} \mid a < x\}$,
- $\mathbb{R}_{\leq a} = (-\infty, a] =]-\infty, a] = \{x \in \mathbb{R} \mid x \leq a\}$,
- $\mathbb{R}_{<a} = (-\infty, a) =]-\infty, a[= \{x \in \mathbb{R} \mid x < a\}$.

It is $(-\infty, \infty) = \mathbb{R}$. Where ∞ and $-\infty$ are symbols and not numbers.

3.3 The Absolute Value of a Real Number

For every $a \in \mathbb{R}$ one calls

$$|a| = \begin{cases} a\,, & \text{if } a \geq 0 \\ -a\,, & \text{if } a < 0 \end{cases}$$

the **absolute value** of a, e.g. we have:

$$|0| = 0\,, \quad |3| = 3\,, \quad |-4| = 4\,, \quad |\sqrt{\pi}| = \sqrt{\pi}\,.$$

The following rules apply to the absolute value of a real number:

Rules for the Absolute Value

- $\forall a \in \mathbb{R}: \ |a| \geq 0$.
- $\forall a \in \mathbb{R}: \ |a| = \sqrt{a^2}$.
- $\forall a, b \in \mathbb{R}: \ |a - b| =$ distance of the numbers a and b on the number line.
- $\forall a, b \in \mathbb{R}: \ |a - b| = |b - a|$.
- $\forall a, b \in \mathbb{R}: \ |a\,b| = |a|\,|b|$.
- $\forall a, b \in \mathbb{R}: \ |a + b| \leq |a| + |b|$—**triangle inequality**.
- $\forall a, b \in \mathbb{R}: \ |a - b| \geq |a| - |b|$.

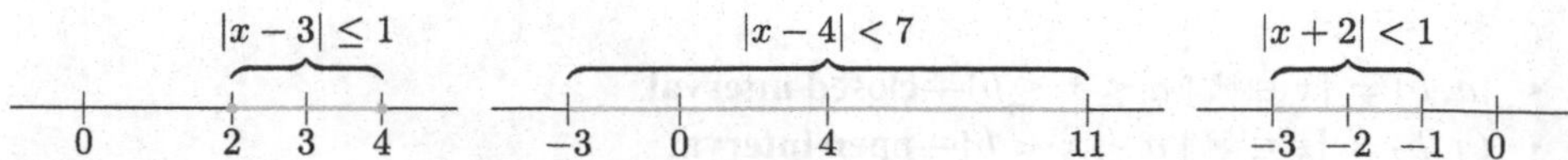

Fig. 3.2 The intervals as inequalities

It is usual to describe bounded open or closed intervals with absolute values, namely:

$$[a, b] = \left\{ x \in \mathbb{R} \mid |x - \frac{(a+b)}{2}| \leq \frac{(b-a)}{2} \right\}$$

and

$$(a, b) = \left\{ x \in \mathbb{R} \mid |x - \frac{(a+b)}{2}| < \frac{(b-a)}{2} \right\}.$$

In the notation one is usually careless and writes short $|x - c| < r$ instead of $\{x \in \mathbb{R} \mid |x - c| < r\}$.

Example 3.1 For the closed interval $[2, 4]$ one can also write $|x - 3| \leq 1$.

For the open interval $(-3, 11)$ one can also write $|x - 4| < 7$.

For the open interval $(-3, -1)$ one can also write $|x + 2| < 1$.

Cf. Fig. 3.2. ■

3.4 *n*-th Roots

For every natural number $n \in \mathbb{N}$ and every real number $a \geq 0$ the equation $x^n = a$ has exactly one real solution $x \geq 0$. We write for this $x = \sqrt[n]{a}$, in the case $n = 2$ we write more simply $\sqrt[2]{a} = \sqrt{a}$ and speak of the ***n*-th root** of a.

Using the root, we can now give the real solutions of the equation $x^n = a$:

Existence and Number of Solutions of $x^n = a, n \in \mathbb{N}$

$a \in \mathbb{R}_{>0},\ n$ even $\Rightarrow x^n = a$ has exactly 2 distinct solutions: $\pm \sqrt[n]{a}$.
$a \in \mathbb{R}_{>0},\ n$ odd $\Rightarrow x^n = a$ has exactly 1 solution: $\sqrt[n]{a}$.
$a \in \mathbb{R}_{<0},\ n$ even $\Rightarrow x^n = a$ has no solution.
$a \in \mathbb{R}_{<0},\ n$ odd $\Rightarrow x^n = a$ has exactly 1 solution: $-\sqrt[n]{-a}$.
$a = 0,\ n \in \mathbb{N}$ any $\Rightarrow x^n = a$ has exactly 1 solution: 0.

Remark This somewhat confused situation becomes much clearer with the complex numbers: *Every complex number not equal to 0 has n different n-th roots.*

3.5 Solving Equations and Inequalities

When solving equations or inequalities one determines the set L of all $x \in \mathbb{R}$ that satisfy the equation or inequality.

We solve some equations and inequalities keeping in mind the following recipe-like rules:

Recipe: Solving Equations and Inequalities with or without Absolute Values

- Can we simply solve the equation or inequality for x, i.e. can we write $x = \ldots, x < \ldots, x \geq \ldots$? In that way you get the solution set right away.
- For absolute values, make case distinctions: $|x - a| = -x + a$ if $x < a$ and $|x - a| = x - a$ if $x \geq a$. Then solve for x if possible.
- The inequality sign turns around when you
  - multiply an inequality by a negative number, or
  - inverts the inequality and both sides of the inequality have the same sign.
- If x occurs in a higher than first power in an inequality, a transformation to an inequality of the form $a_n x^n + \cdots + a_1 x + a_0 \lesseqgtr 0$ and a factorization of the left side,

$$a_n x^n + \cdots + a_1 x + a_0 = a_n (x - b_1)^{\nu_1} \cdots (x - b_r)^{\nu_r},$$

  often helps to decide when this product is positive or negative.

Example 3.2

- The equation $ax^2 + bx + c = 0$ has the two solutions $x_{1/2} = \frac{-b \pm \sqrt{b^2 - 4ac}}{2a}$, if $b^2 - 4ac \geq 0$, so that $L = \{x_1, x_2\}$.
- The inequality $3x - 7 \leq x + 2$ can easily be converted to x :

$$3x - 7 \leq x + 2 \Leftrightarrow 2x \leq 9 \Leftrightarrow x \leq 9/2 \Leftrightarrow x \in (-\infty, 9/2]\,,$$

  so that $L = (-\infty, 9/2]$.

- We can use the inequality $x^2 - 4x + 3 > 0$ by factorizing the left-hand side. We have:

$$x^2 - 4x + 3 > 0 \Leftrightarrow (x-1)(x-3) > 0 \Leftrightarrow (x > 1 \wedge x > 3) \vee (x < 1 \wedge x < 3)$$
$$\Leftrightarrow x > 3 \vee x < 1 \Leftrightarrow x \in (-\infty, 1) \cup (3, \infty)\,,$$

so that $L = (-\infty, 1) \cup (3, \infty)$.
- The inequality $\frac{1}{x} \leq \frac{1}{x+2}$ we solve by expanding both fractions to the common denominator:

$$\frac{1}{x} \leq \frac{1}{x+2} \Leftrightarrow \frac{1}{x} - \frac{1}{x+2} \leq 0 \Leftrightarrow \frac{x+2-x}{x(x+2)} \leq 0 \Leftrightarrow \frac{2}{x(x+2)} \leq 0$$
$$\Leftrightarrow (x < 0 \wedge x > -2) \vee (x > 0 \wedge x < -2)$$
$$\Leftrightarrow x \in (-2, 0) \cup \emptyset = (-2, 0)\,,$$

so that $L = (-2, 0)$.
- In the inequality $|x-1| + \frac{1}{|x-1|} \leq |x+1|$ the absolute values have to be resolved first. This also requires some case distinctions:

1. Case: $x < -1$. It is then $|x-1| = -x+1$ and $|x+1| = -x-1$. Thus:

$$|x-1| + \frac{1}{|x-1|} \leq |x+1| \Leftrightarrow -x+1 - \frac{1}{x-1} \leq -x-1$$
$$\Leftrightarrow 2 \leq \underbrace{\frac{1}{x-1}}_{<0} \Leftrightarrow x \in \emptyset\,.$$

2. Case: $x > 1$. It is then $|x-1| = x-1$ and $|x+1| = x+1$. Thus:

$$|x-1| + \frac{1}{|x-1|} \leq |x+1| \Leftrightarrow x-1 + \frac{1}{x-1} \leq x+1$$
$$\Leftrightarrow \frac{1}{x-1} \leq 2 \Leftrightarrow x-1 \geq \frac{1}{2}$$
$$\Leftrightarrow x \geq \frac{3}{2} \Leftrightarrow x \in [3/2, \infty)\,.$$

3. Case: $x \in [-1, 1)$. $x = 1$ is excluded because division by 0 is not possible. We have: $|x-1| = -x+1$ and $|x+1| = x+1$ and thus:

$$|x-1| + \frac{1}{|x-1|} \leq |x+1| \Leftrightarrow -x+1 - \frac{1}{x-1} \leq x+1$$
$$\Leftrightarrow -\frac{1}{x-1} \leq 2x \Leftrightarrow -1 \geq 2x(x-1)$$

$$\Leftrightarrow 2x^2 - 2x + 1 \leq 0 \Leftrightarrow x^2 + (x-1)^2 \leq 0$$

$$\Leftrightarrow x \in \emptyset\,.$$

In total, the solution set of the inequality is therefore $L = [3/2, \infty)$. ■

3.6 Maximum, Minimum, Supremum and Infimum

The **maximum** or the **minimum** of a set $M \subseteq \mathbb{R}$ is the greatest or smallest element of M, if indeed there is one. We write for this $\max(M)$ and $\min(M)$.

Example 3.3

- For the set $M = [1, 2] \subseteq \mathbb{R}$ we have $\min(M) = 1$ and $\max(M) = 2$.
- For the set $M = [1, \infty[\subseteq \mathbb{R}$ we have $\min(M) = 1$, a maximum does not exist.
- The set $M = (1, 2) \subseteq \mathbb{R}$ has neither a maximum nor a minimum. ■

A subset $M \subseteq \mathbb{R}$ is called **bounded from below**, if there is a $a \in \mathbb{R}$ with $a \leq x$ for all $x \in M$, and **bounded from above**, if there is a $b \in \mathbb{R}$ with $x \leq b$ for all $x \in M$. The set M is called **bounded** if it is from below and above bounded.

Example 3.4

- The set $M = (-\infty, 2)$ is bounded from above (by $a = 2, 3, 4, \ldots$), but not from below.
- The set $M = (-12, 38]$ on the other hand is bounded from below (e.g. by $a = -12$) and also from above (e.g. by $b = 38$). Thus, it is bounded. ■

Maximum and minimum of a nonempty subset of $\mathbb{R}$ need not exist, even if the set is bounded. But there is always a smallest upper bound and a greatest lower bound, respectively; these bounds get their own names:

Supremum and Infimum

- Every from above bounded non-empty set $M \subseteq \mathbb{R}$ has a smallest upper bound

$$\sup(M) \quad - \text{ the } \textbf{supremum} \text{ of } M.$$

- Every from below bounded non-empty set $M \subseteq \mathbb{R}$ has a greatest lower bound

$$\inf(M) \quad - \text{ the } \textbf{infimum} \text{ of } M.$$

Example 3.5

- For the set $M = [1, 12)$ we have:

$$\min(M) = 1 = \inf(M) \quad \text{and} \quad \nexists \max(M), \quad \text{but} \quad \sup(M) = 12\,.$$

- The set $M = [1, 2] \cup (3, \infty)$ on the other hand, has neither maximum nor supremum, since it is not bounded from above. It is $\inf(M) = \min(M) = 1$.
- The set

$$M = \{1/n \mid n \in \mathbb{N}\} = \{1,\ 1/2,\ 1/3,\ \ldots\}$$

has no minimum, but an infimum,

$$\inf(M) = 0, \text{ and } \sup(M) = \max(M) = 1\,. \qquad \blacksquare$$

Clearly, if M has a maximum, then $\sup(M) = \max(M)$; and if M has a minimum, then $\inf(M) = \min(M)$.

3.7 Exercises

3.1 Show that $\sqrt{2} \notin \mathbb{Q}$.
Note: Assume that $\sqrt{2} = \frac{m}{n}$, where $\frac{m}{n}$ is completely truncated.

3.2 In each of the following cases, determine the set of all $x \in \mathbb{R}$ satisfying the inequalities and sketch these sets on the number line:

(a) $\frac{x-1}{x+1} < 1$, (d) $|1 - x| \leq 1 + 2x$, (g) $\frac{x|x|}{2} = 8$,
(b) $x^2 + x + 1 \geq 0$, (e) $15x^2 \leq 7x + 2$, (h) $x|x| = \frac{1}{2}x^3$,
(c) $x^3 - x^2 < 2x - 2$, (f) $|x + 1| + |5x - 2| = 6$, (i) $|x - 4| > x^2$.

3.3 Given rational numbers p, q and irrational numbers r, s. Prove or disprove the following statements:

(a) $x = p + q$ is a rational number.
(b) $y = r + s$ is an irrational number.
(c) $z = p + r$ is an irrational number.

3.4 Which of the following statements are correct? Give reasons for your answer!

(a) For all $x, y \in \mathbb{R}$ we have $|x - y| \leq |x| - |y|$.
(b) For all $x, y \in \mathbb{R}$ we have the equation $|x - y| = ||x| - |y|$.
(c) For all $x, y \in \mathbb{R}$ we have $|x| - |y|| \leq |x - y|$.

3.5 Examine the sets

(a) $M = \{x \in \mathbb{R} \mid x = n/(n+1),\ n \in \mathbb{N}\}$,
(b) $M = \{x \in \mathbb{R} \mid x = 1/(n+1) + (1 + (-1)^n)/2n,\ n \in \mathbb{N}\}$,
(c) $M = \{n^2/2^n \mid n \in \mathbb{N}\}$

for boundedness and, if necessary, determine the infimum, supremum, minimum and maximum.

Machine Numbers

4

Computer algebra systems like MAPLE or MATHEMATICA can calculate symbolically, e.g. with $\sqrt{2}$ as a positive solution of $x^2 - 2 = 0$ circumvent. In the following we refrain from this symbolic calculation and consider *machine numbers*.

Machine numbers are those numbers that are stored in a computer. Due to a finite memory, only a finite number of numbers can be represented on a calculator. This has far-reaching consequences, since every real number that is not a machine number must be rounded to a machine number so that the computer can continue to calculate with it. This results in *rounding errors*, which in some cases strongly falsify the result or make it unusable.

The storage of machine numbers is partially standardized, e.g. by the IEEE 754 standard, based on the *binary representation of* real numbers.

4.1 *b*-adic Representation of Real Numbers

If $b \geq 2$ is a natural number, then any real number x in a **b-adic number representation** specify

$$x = a_k b^k + a_{k-1} b^{k-1} + \cdots + a_1 b + a_0 + a_{-1} b^{-1} + a_{-2} b^{-2} + \cdots$$

with $a_i \in \{0, 1, \ldots, b-1\}$. We abbreviate this with

$$x = (a_k a_{k-1} \ldots a_1 a_0 \,.\, a_{-1} a_{-2} \ldots)_b \,.$$

In the case $b = 10$ we get the familiar **decimal representation** and in case $b = 2$ the otherwise important **binary representation** or **dual representation** of x.

C. Karpfinger, *Calculus and Linear Algebra in Recipes*,
https://doi.org/10.1007/978-3-662-65458-3_4

To obtain the *b-adic* representation of a real number x, proceed as follows:

Recipe: *b-adic* Representation of a Real Number x
We determine the *b-adic* representation $(a_k \dots a_1 a_0 . a_{-1} a_{-2} \dots)_b$ of the real number x and assume without restriction $x > 0$.

(1) Set $x = n + x_0$ with $n \in \mathbb{N}_0$ and $x_0 \in [0, 1)$.
(2) Determine $a_k, \dots, a_1, a_0$ from

$$n = \sum_{i=0}^{k} a_i b^i \text{ with } a_i \in \{0, 1, \dots, b-1\}$$

and get the *b-adic* digit representation of $n = (a_k \dots a_1 a_0)_b$.
(3) In the case $x_0 \neq 0$ determine $a_{-1}, a_{-2}, \dots$ from

$$x_0 = \sum_{i=1}^{\infty} a_{-i} b^{-i} \text{ with } a_{-i} \in \{0, 1, \dots, b-1\}$$

and get the *b-adic* digit representation of $x_0 = (0 . a_{-1} a_{-2} \dots)_g$.
(4) It is then $x = (a_k \dots a_1 a_0 . a_{-1} a_{-2} \dots)_b$ the *b-adic* digit representation of x.

Example 4.1

- For $x = 28$ we obtain the binary representation $x = (11100)_2$ because of

$$28 = 1 \cdot 2^4 + 1 \cdot 2^3 + 1 \cdot 2^2 + 0 \cdot 2^1 + 0 \cdot 2^0 .$$

- For $x = 0.25$, we obtain the binary representation $x = (0.01)_2$ because of

$$0.25 = 0 \cdot 2^{-1} + 1 \cdot 2^{-2} .$$

- For $x = 13.625$ we obtain the binary representation $x = (1101.101)_2$ because of

$$13 = 1 \cdot 2^3 + 1 \cdot 2^2 + 0 \cdot 2^1 + 1 \cdot 2^0$$

and

$$0.625 = 1 \cdot 2^{-1} + 0 \cdot 2^{-2} + 1 \cdot 2^{-3} .$$

■

MATLAB In MATLAB you get for natural numbers a with

- `dec2bin(a)` the binary representation of the decimal number a and with
- `bin2dec(a)` the decimal representation of the binary number a.

4.2 Floating Point Numbers

The set of real numbers is unbounded from above and below and *dense* in itself, i.e., for every two distinct real numbers there is a real number lying between them; in particular, $\mathbb{R}$ is not finite. A calculator, on the other hand, has only a finite number of memory cells. Thus one can also represent only finitely many numbers on a computer, if one disregards the possibility of symbolic computation. The numbers that can be represented on a computer are the *machine numbers*, these are special *floating point numbers*.

Let $b \in \mathbb{N}_{\geq 2}$ arbitrary. For $t \in \mathbb{N}$ we now consider the *t-digit* **floating point numbers** which are 0 and the numbers of the form

$$s\,\frac{m}{b^t}\,b^e = \pm 0\,.\,a_1 a_2 \cdots a_t \cdot b^e \quad \text{with} \quad a_i \in \{0, \ldots, b-1\},$$

here are

- $s \in \{-1, 1\}$ the **sign**,
- $b \in \mathbb{N}$ the **base** (typically $b=2$),
- $t \in \mathbb{N}$ the **accuracy** or number of **significant digits**,
- $e \in \mathbb{Z}$ the **exponent**,
- $m \in \mathbb{N}$ the **mantissa**, $1 \leq m \leq b^t - 1$.

A floating point number $\neq 0$ is **normalized**, if $b^{t-1} \leq m \leq b^t - 1$, so in this case $a_1 \neq 0$. We also take the 0 to be normalized.

We consider only normalized floating point numbers and write for them

$$\mathbb{G}_{b,t} = \{x \in \mathbb{R} \mid x \text{ is } t\text{-digit normalized floating point number to base } b\}.$$

Note that the set of floating point numbers is infinite, since the exponent e can become arbitrarily small or large. We now make restrictions on the exponents and obtain the *machine numbers*:

4.2.1 Machine Numbers

Let $b,\ t,\ e_{\min},\ e_{\max} \in \mathbb{Z}$, $b \geq 2$, $t \geq 1$. The set of **machine numbers** is

$$\mathbb{M}_{b,t,e_{\min},e_{\max}} = \{x \in \mathbb{G}_{b,t} \mid e_{\min} \leq e \leq e_{\max}\} \cup \{\pm\infty,\ \text{NaN}\}.$$

Where NaN is *Not a Number* (which is an undefined or unrepresentable value, such as $\frac{0}{0}$ or $\frac{\infty}{\infty}$).

Remark The leading bit is always equal to 1 for normalized numbers in the case $b = 2$, so it does not need to be stored.

Example 4.2 The positive machine numbers for $t = 1$ and $b = 10$ and $e_{\min} = -2$ and $e_{\max} = 2$ are as follows:

0.0010	0.0100	0.1000	1.0000	10.0000
0.0020	0.0200	0.2000	2.0000	20.0000
0.0030	0.0300	0.3000	3.0000	30.0000
0.0040	0.0400	0.4000	4.0000	40.0000
0.0050	0.0500	0.5000	5.0000	50.0000
0.0060	0.0600	0.6000	6.0000	60.0000
0.0070	0.0700	0.7000	7.0000	70.0000
0.0080	0.0800	0.8000	8.0000	80.0000
0.0090	0.0900	0.9000	9.0000	90.0000

We have these numbers in Fig. 4.1 shown, where the second figure uses a logarithmic scale. ■

The normalized machine numbers are quite simple to count: If t is the precision, b the base and a the number of exponents, then there are

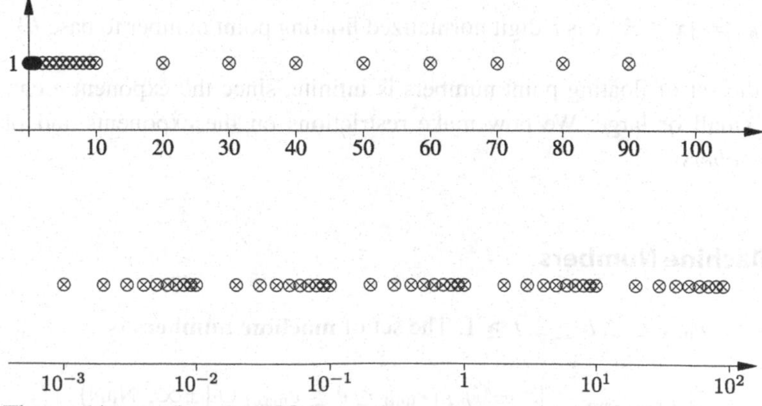

Fig. 4.1 The machine numbers from $\mathbb{M}_{10,1,-2,2}$

$$2 \cdot (b-1) \cdot b^{t-1} \cdot a + 1$$

machine numbers not equal to $\pm\infty$, NaN. The leading 2 comes from plus and minus, the trailing one comes from 0, $b-1$ is the number of ways to occupy the first digit after the decimal point, and b^{t-1} is the number of ways to fill the remaining digits of the mantissa.

In particular, there is a smallest and a largest positive machine number $x_{\min}$ and $x_{\max}$, at MATLAB are these

`xmin=realmin, xmin = 2.2251e-308, xmax=realmax, xmax = 1.79 77e+308.`

We have made a few simplifications in this representation of the number of machine numbers. In fact, MATLAB still distinguishes different zeros and also different NaN.

4.2.2 Machine Epsilon, Rounding and Floating Point Arithmetic

The distance $\varepsilon_{b,t}$ from the machine number 1 to the next larger number in $\mathbb{G}_{b,t}$ is called the **machine epsilon**:

$$\varepsilon_{b,t} = b^{-(t-1)}.$$

Thus, there are 1 and $1 + \varepsilon_{b,t}$ adjacent machine numbers, and there is no other machine number between them.

Example 4.3 For $b = 10$ and $t = 5$ we get

$$1 = +0.10000 \cdot 10^1 \Rightarrow 1 + \varepsilon_{10,5} = +0.10001 \cdot 10^1 .$$

In MATLAB is the machine epsilon eps$= \varepsilon_{2,53} = 2^{-(53-1)} \approx 2 \cdot 10^{-16}$. ■

There are real numbers that are not machine numbers, e.g., any real number in the interval $(1, 1 + \varepsilon_{b,t})$ or $\sqrt{2}$. Thus, to represent such a number as well as possible on the computer, one must *round* by choosing a machine number that *approximates it well*. There are different strategies for rounding, we only explain one obvious method to become aware of the problem that errors are generated when rounding:

By *rounding* a $x \in \mathbb{R}$ is mapped to a *nearby* $\tilde{x} \in \mathbb{G}_{b,t}$ i.e., **rounding** is a mapping

$$\mathrm{fl}_{b,t} : \mathbb{R} \to \mathbb{G}_{b,t} .$$

The calculator supports four types of rounding, all of which are quite complicated. We omit the representation and *round* as we know it from our school days. Roughly speaking, one orders $x \in \mathbb{R}$ the number of machines $\tilde{x}$ to with $|x - \tilde{x}|$ is minimal.

Example 4.4

- For $x = 2.387$ we have $\mathrm{fl}_{10,3}(x) = 2.39 = +0.239 \cdot 10^1$.
- For $x = 0.1234$ and $y = 0.1233$ we have $\mathrm{fl}_{10,3}(x) - \mathrm{fl}_{10,3}(y) = 0$.
- For $x = \sqrt{2}$ we have $\mathrm{fl}_{10,3}(x) = 1.41$. ■

The real numbers $\sqrt{2}$ and 1.411 are therefore indistinguishable on the machine with a 2-digit mantissa length. By inputting numbers whose mantissa is longer than the number of significant digits of the machine, *input errors* are already made.

The arithmetic operations on $\mathbb{R}$ are +, -, ·, /. We declare analogous arithmetic operations $\oplus$, $\ominus$, $\odot$, $\oslash$ on the set of floating point numbers. For example, since the (exact) product $x \cdot y$ of floating point numbers x and y need not be a floating point number again, we require for all $x,\ y \in \mathbb{G}_{b,t}$ and all arithmetic operations $* \in \{+,\ -,\ \cdot,\ /\}$:

$$x \circledast y = \mathrm{fl}(x * y)\,.$$

Thus we obtain $x \circledast y$ by calculating $x \circledast y$ and rounding to the nearest floating point number. In this **floating point arithmetic**, which is thus realized on machines, errors are again made. For example, not even the associative law holds.

For machine precision, rounding and floating point arithmetic we have:

The Input, Rounding and Arithmetic Errors

- For every $x \in \mathbb{R}$ there is a $x' \in \mathbb{G}_{b,t}$ with

$$|x - x'| \leq \varepsilon_{b,t}|x|\,.$$

So with the machine numbers we can *small* $x \in \mathbb{R}$ well approximate.
- For every $x \in \mathbb{R}$ there is a $\varepsilon \in \mathbb{R}$ with $|\varepsilon| \leq \varepsilon_{b,t}$ with

$$\mathrm{fl}_{b,t}(x) = x(1 + \varepsilon)\,.$$

So when rounding, only a *small* mistake is made.
- For all $x, y \in \mathbb{G}_{b,t}$ there is a ε with $|\varepsilon| \leq \varepsilon_{b,t}$ with

$$x \circledast y = (x * y)\,(1 + \varepsilon)\,.$$

Thus, only a *small* error is made in floating point arithmetic.

Since basically an error is made in every floating-point operation, it is in principle sensible to keep the number of floating-point operations in a calculation small. In fact, it is often possible to perform one and the same calculation with different numbers of individual operations; we will see examples. The floating point operations are usually counted with **flops** (**flo**ating **p**oint operations).

4.2.3 Loss of Significance

If two nearly equal floating point numbers are subtracted from each other, **loss of significance** occurs. We demonstrate such a loss of significance with MATLAB:

```
» a=(1+1e-15)-1
 a =
   1.1102e-15
 » a*1e15
 ans =
   1.1102
```

Actually we would have expected the result 1.0. This large error can have considerable influence on further calculations.

Such a loss of significance should generally be avoided. Often formulas, where possibly almost equal numbers are subtracted from each other, can be transformed, e.g.

$$\sqrt{x+\varepsilon}-\sqrt{x}=\frac{\varepsilon}{\sqrt{x+\varepsilon}+\sqrt{x}}.$$

4.3 Exercises

4.1 Represent the decimal numbers 2005 and 0.25 in the dual system.

4.2

(a) Represent the decimals 967 and 0.5 in the dual system.
(b) Write the dual number 11001101 as a decimal number.
(c) Determine the product of the following dual numbers $1111 \cdot 11$ as a dual number and do the proof in the decimal system.

4.3 Represent the number 1/11 as a floating point number in the binary system. Use only elementary MATLAB OPERATIONS and loops to do this.

4.4 Write a program that, for a natural number a, outputs the binary representation of a for a natural number a.

4.5 How many normalized machine numbers are there in $\mathbb{M}_{2,4,-3,3}$? Calculate eps, $x_{\min}$ and $x_{\max}$.

4.6 Why does the MATLAB command `realmin/2` does not return `0` or `realmax+realmax/2^60` does not return `inf`?

4.7 Let z_0 and z_1 denote the smallest number of machines that is just greater than 0 or 1, respectively, given the following parameters: $b = 2, t = 24, e_{\min} = -126$ and $e_{\max} = 127$. Enter z_0 and z_1. Which distance is greater: that of z_0 and 0 or that of z_1 and 1?

5 Polynomials

In applied mathematics, one often has to deal with the problem of *decomposing* a polynomial into a product of *linear factors*, if this is possible. We will meet this so fundamental task again and again in different fields of engineering mathematics, e.g. when solving polynomial inequalities, when calculating the eigenvalues of a matrix or also when determining a basis of the solution space of a linear differential equations.

Rational functions are quotients whose numerator and denominator are polynomials. In the *partial fraction decomposition*, rational functions are written as summands of simple rational functions. This decomposition is elementary feasible and is based on the factorization of polynomials. The applications of this decomposition in engineering mathematics are manifold, e.g. when integrating rational functions or also when solving linear differential equations with the help of the Laplace transformation.

5.1 Polynomials: Multiplication and Division

Under a **(real) polynomial** $f = f(x)$ we understand a *formal expression of* the kind

$$f(x) = a_n x^n + \cdots + a_1 x + a_0 = \sum_{k=0}^{n} a_k x^k ,$$

where the **coefficients** $a_0, a_1, \ldots, a_n$ are all from $\mathbb{R}$. If all coefficients are zero, $a_0 = a_1 = \cdots = a_n = 0$, then we have the **zero polynomial** $f = 0$.

Polynomials can be multiplied:

$$\left(\sum_{k=0}^{n} a_k x^k\right)\left(\sum_{k=0}^{m} b_k x^k\right) = \left(\sum_{k=0}^{m+n} c_k x^k\right) \quad \text{with } c_k = \sum_{i+j=k} a_i b_j .$$

C. Karpfinger, *Calculus and Linear Algebra in Recipes*,
https://doi.org/10.1007/978-3-662-65458-3_5

The formula looks complicated, you do not have to remember it, consider the following examples:

Example 5.1

- $3\,(x^2 - 2\,x + 1) = 3\,x^2 - 6\,x + 3$,
- $(x - 1)\,(x^2 - 2) = x^3 - x^2 - 2\,x + 2$,
- $(x - 1)\,(x + 1)\,(x - 2)\,(x + 2) = (x^2 - 1)\,(x^2 - 4) = x^4 - 5\,x^2 + 4$. ■

Note that it is quite easy to multiply polynomials; it will prove much more difficult to *decompose* polynomials into *simple* factors.

MATLAB This multiplication of polynomials is achieved in MATLAB with the command `expand` of the Symbolic Toolbox; here the variable x must be defined as a *symbol* (you can find information on the Internet about the meaning of `syms`), e.g.

```
» syms x; expand((1+x)^2*(1-x)^2)
ans = x^4 - 2*x^2 + 1
```

Terms for Polynomials

We consider a polynomial $f(x) = a_n x^n + \cdots + a_1 x + a_0$ with $a_n \neq 0$. One calls

- a_n the **highest coefficient** of f,
- n the **degree** of f, one writes $n = \deg(f)$,
- $x_0 \in \mathbb{R}$ with $f(x_0) = 0$ **zero of** the polynomial f,
- the polynomials

$$f(x) = a_0, \;\; f(x) = a_0 + a_1 x, \;\; f(x) = a_0 + a_1 x + a_2 x^2,$$

$$f(x) = a_0 + a_1 x + a_2 x^2 + a_3 x^3$$

are called **constant, linear**, **quadratic, cubic**.

Remark One also sets supplementary $\deg 0 = -\infty$ and $-\infty < n$ for all $n \in \mathbb{N}_0$. This *generalizes* some formula, but is not necessary for our purposes.

As is well known, for the zeros of a polynomial we have: f:

- If $\deg f = 0$, then $f(x) = a$ and f has no zero.
- If $\deg f = 1$ then $f(x) = ax + b$ and f has the zero $x_0 = -b/a$.
- If $\deg f = 2$ then $f(x) = ax^2 + bx + c$ and f has
  - two different zeros $x_\pm = \frac{-b \pm \sqrt{b^2 - 4ac}}{2a}$, if $b^2 - 4ac > 0$,

- one zero $x_0 = \frac{-b}{2a}$, if $b^2 - 4ac = 0$,
- not a real zero, if $b^2 - 4ac < 0$.

- If $\deg f \geq 3$, the zeros are often no longer computable by hand. For the cases $\deg p = 3$ or $\deg p = 4$ there are still solution formulas; however, these are too complicated that one can or should remember them. In the examples from school time, one could usually *guess* a zero; however, this is only the case in special cases. In fact, you generally have to rely on the help of a computer to find zeros for polynomials of degree ≥ 3.

Example 5.2

- The polynomial $f(x) = x^2 - 2$ has the highest coefficient 1, the degree 2 and the two zeros $a_1 = \sqrt{2}$ and $a_2 = -\sqrt{2}$.
- The polynomial $f(x) = (x-1)(x+1)(2x-1)(2x^2+1)$ has the highest coefficient 4, the degree 5 and the real zeros $a_1 = 1$, $a_2 = -1$ and $a_3 = 1/2$. ■

We collect some important facts about polynomials:

Some Facts about Polynomials

- Two polynomials $f(x) = a_n x^n + \cdots + a_1 x + a_0$ and $g(x) = b_m x^m + \cdots + b_1 x + b_0$ are **equal** if $a_k = b_k$ for all k.
- **Degree Theorem**: $\deg(f\,g) = \deg(f) + \deg(g)$ for $f \neq 0 \neq g$.
- **Splitting off zeros**: If $x_0 \in \mathbb{R}$ is a zero of a polynomial f of degree n, then there exists a polynomial g of degree $n-1$ with $f(x) = (x - x_0)\,g(x)$, for short:

$$f(x_0) = 0 \Rightarrow f(x) = (x - x_0)g(x) \text{ with } \deg(g) = n - 1\,.$$

- A polynomial $f \neq 0$ of degree n has at most n zeros.
- **Division with remainder**: If f and g are polynomials different from the zero polynomial, then there are polynomials q and r with

$$f = q \cdot g + r \text{ and } \deg r < \deg g\,.$$

In particular:

$$\frac{f(x)}{g(x)} = q(x) + \frac{r(x)}{g(x)} \text{ with } \deg(r) < \deg(g)\,.$$

r is the *remainder of* the division of f by g. In the case $r = 0$ one calls g a **divisor** of f.

Note that for division with remainder, we may assume that $\deg(f) \geq \deg(g)$; otherwise we obtain with $q = 0$ the desired equality.

The equality of polynomials makes the **equating of coefficients** possible which is an effective tool for specifying polynomials with desired properties. We will use this instrument to determine the polynomial g when splitting off a zero $(x - x_0)$ of a polynomial f; from school time this procedure is known under the term **polynomial division**:

Example 5.3 We consider the polynomial p with $p(x) = x^3 - x^2 - x - 2$. It is obviously true $p(2) = 0$. So we make the approach

$$x^3 - x^2 - x - 2 = (x - 2)(ax^2 + bx + c) = ax^3 + (b - 2a)x^2 + (c - 2b)x - 2c\,.$$

Now equating coefficients between the left and right sides yields:

$$a = 1\,,\ b - 2a = -1\,,\ c - 2b = -1\,,\ -2c = -2\,,\ \text{so } a = 1\,,\ b = 1\,,\ c = 1\,.$$

This gives us the decomposition:

$$x^3 - x^2 - x - 2 = (x - 2)(x^2 + x + 1)\,.$$

In school days, this procedure was done with the following arithmetic scheme:

$$\begin{array}{l} x^3 - x^2 - x - 2 = (x - 2)(x^2 + x + 1) \\ \underline{-(x^3 - 2x^2)} \\ \quad x^2 - x - 2 \\ \quad \underline{-(x^2 - 2x)} \\ \qquad\quad x - 2 \end{array}$$

■

You can also perform a polynomial division if the division does not *add up*:
Example 5.4

- With $f(x) = x^2 - 1$ and $g(x) = x + 2$ we get because of

$$x^2 - 1 = (x + 2)(x - 2) + 3$$

the equation

$$\frac{f(x)}{g(x)} = \frac{x^2 - 1}{x + 2} = x - 2 + \frac{3}{x + 2}\,.$$

- Using $f(x) = 4x^5 + 6x^3 + x + 2$ and $g(x) = x^2 + x + 1$ we obtain because of

$$4x^5 + 6x^3 + x + 2 = (x^2 + x + 1)(4x^3 - 4x^2 + 6x - 2) + (-3x + 4)$$

the equation

$$\frac{f(x)}{g(x)} = \frac{4x^5 + 6x^3 + x + 2}{x^2 + x + 1} = 4x^3 - 4x^2 + 6x - 2 + \frac{-3x + 4}{x^2 + x + 1}. \qquad \blacksquare$$

Note that we have performed a reshaping that is important for later purposes: We have a *rational function* $\frac{f(x)}{g(x)}$ whose numerator degree is greater than the denominator degree, is written as the sum of the polynomial and the rational function, where for the latter rational function the numerator degree is less than the denominator degree. We will resort to this decomposition in the integration of rational functions.

5.2 Factorization of Polynomials

Multiplying polynomials is *simple*:

$$(3x^2 - 2x + 1) \cdot (x + 4) = 3x^3 + 10x^2 - 7x + 4\,.$$

In general, it is much more difficult to *factorize* a polynomial $f(x) = a_n x^n + \cdots + a_1 x + a_0$, i.e. to determine a *finest possible* decomposition

$$f = p_1 \cdots p_r$$

with polynomials p_i with $\deg(p_i) \geq 1$. Thereby means *as fine as possible*, that the polynomials $p_1, \ldots, p_r$ cannot be further written as products of non-constant polynomials, e.g.

$$x^3 - 1 = (x - 1)(x^2 + x + 1)\,.$$

A further decomposition of these factors into non-constant polynomials is not possible any more, because the polynomial $x^2 + x + 1$ no longer has a real zero.

If one wants to obtain such a decomposition of a polynomial *as fine as possible*, one does well to first split off all possible zeros; with each zero that one splits off, the degree of the polynomial still to be factorized decreases by one. We have:

Factorizing Polynomials
Every (real) polynomial $f(x) = a_n x^n + \cdots + a_1 x + a_0$ with $\deg(f) \geq 1$ has a decomposition of the form

$$f(x) = a_n(x - x_1)^{r_1} \cdots (x - x_k)^{r_k}(x^2 + p_1 x + q_1)^{s_1} \cdots (x^2 + p_\ell x + q_\ell)^{s_\ell}$$

with $x_i \neq x_j$ for $i \neq j$ and r_i, $s_j \in \mathbb{N}_0$ and further non-decomposable quadratic polynomials $x^2 + p_j x + q_j$.

One calls r_i the **multiplicity** of the zero x_i and says f **splits into linear factors** if $s_j = 0$ for all j; in this case $r_1 + \cdots + r_k = n$.

Example 5.5 We decompose the polynomial f with $f(x) = -x^6 + x^4 - x^2 + 1$. Since f obviously has the zero 1, we get after splitting off this zero

$$f(x) = -(x - 1)(x^5 + x^4 + x + 1).$$

Now obviously the second factor has the zero -1; we get after splitting off this zero

$$f(x) = -(x - 1)(x + 1)(x^4 + 1).$$

The last factor $x^4 + 1$ no longer has a real zero. To decompose it into a product of two necessarily quadratic factors, we make the following approach:

$$\begin{aligned} x^4 + 1 &= (x^2 + ax + b)(x^2 + cx + d) \\ &= x^4 + (a + c)x^3 + (ac + d + b)x^2 + (ad + bc)x + bd. \end{aligned}$$

Equating coefficients yields a system of equations

- $a + c = 0$
- $ac + b + d = 0$
- $ad + bc = 0$
- $bd = 1$.

Because of the penultimate equation we have $b = d$ in the case $a \neq 0$. Then, because of the last equation $b = \pm 1 = d$. We set $d = 1 = b$ and get

$$c = -a \text{ and } a^2 = 2.$$

Thus we obtain $x^4 + 1 = (x^2 + \sqrt{2}x + 1)(x^2 - \sqrt{2}x + 1)$ and thus

$$f(x) = -(x-1)(x+1)(x^2 + \sqrt{2}x + 1)(x^2 - \sqrt{2}x + 1)\,. \qquad \blacksquare$$

MATLAB MATLAB allows the factorization of polynomials with integer zeros with the command `factor`, e.g.

```
» syms x; factor(-x^6+x^4-x^2+1)
ans = -(x^4 + 1)*(x - 1)*(x + 1)
```

Note that `factor` gives no result for non-integer zeros. But MATLAB still offers the functions `solve` and `roots` for determining the zeros of polynomials; here, also the non-real zeros of any quadratic factors are given, e.g.

```
» solve('-x^6+x^4-x^2+1=0','x')     » roots([-1 0 1 0 -1 0 1])
 ans =                   1          ans =-1.0000
                        -1                -0.7071 + 0.7071i
   2^(1/2)*(1/2 + i/2)                    -0.7071 - 0.7071i
  2^(1/2)*(- 1/2 + i/2)                    0.7071 + 0.7071i
   2^(1/2)*(1/2 - i/2)                     0.7071 - 0.7071i
  2^(1/2)*(- 1/2 - i/2)                    1.0000
```

You can see from the results that `solve` calculates symbolically and `roots` numerically.

5.3 Evaluating Polynomials

We want to evaluate a real polynomial f at a point $a \in \mathbb{R}$, i.e. to determine $f(a)$, where

$$f(x) = a_n x^n + a_{n-1}x^{n-1} + \cdots + a_1 x + a_0\,.$$

Two methods are suitable for this purpose:

- In the naive calculation of $f(a)$ for a $a \in \mathbb{R}$ one calculates $a_n a^n$, add to it $a_{n-1}a^{n-1}$, add $a_{n-2}a^{n-2}$ etc.
- At **Horner scheme** one calculates $f(a)$ according to the following pattern:

$$(\cdots((a_n a + a_{n-1})\, a + a_{n-2})\, a + \cdots + a_1)\, a + a_0\,.$$

Naive evaluation requires $3n - 1$ flops (n additions, $n-1$ multiplications for $a^2, \ldots, a^n$ and n multiplications for $a_i a^i$). In the case of the Horner scheme, one gets by with $2\,n$ flops (n additions and n multiplications).

MATLAB MATLAB offers with `polyval` a function for polynomial evaluation. To practice the use of MATLAB, we program this naive polynomial evaluation and the evaluation according to Horner. For this we need the vector $p = [a_n, \ldots, a_1, a_0]$ with the coefficients of the polynomial under consideration and the number a to be predefined:

```
function [y] = polnaiv(p,a)
n=length(p);
y=p(1);
for k=2:n
    y=y+p(k).*a.^(k-1);
end
```

```
function [y] = polhorner(p,a)
n=length(p);
y=p(n);
for k=n-1:-1:1
    y=y.*a+p(k);
end
```

Example 5.6 We consider the polynomial

$$f(x) = (x-2)^9 = x^9 - 18\,x^8 + 144\,x^7 - 672\,x^6 + 2016\,x^5 - 4032\,x^4 + 5376\,x^3 - 4608\,x^2 + 2304\,x - 512\,.$$

We contrast the values obtained with MATLAB in a table:

a	$f(a)$ *naiv*	$f(a)$ *Horner*	$f(a)$ *exakt*
1.97	$0.6366 \cdot 10^{-11}$	$0.2842 \cdot 10^{-11}$	$-0.1968 \cdot 10^{-13}$
1.98	$-0.2046 \cdot 10^{-11}$	$0.7390 \cdot 10^{-11}$	$-0.5120 \cdot 10^{-15}$
1.99	$0.1592 \cdot 10^{-11}$	$0.5343 \cdot 10^{-11}$	$-1 \cdot 10^{-18}$
2.00	$0.0000 \cdot 10^{-11}$	$0.0000 \cdot 10^{-11}$	0
2.01	$-0.6025 \cdot 10^{-11}$	$-0.3752 \cdot 10^{-11}$	$1 \cdot 10^{-13}$

■

5.4 Partial Fraction Decomposition

You add fractions by bringing them to a common denominator:

$$\frac{x}{x^2+1} + \frac{2}{x+1} = \frac{3x^2+x+2}{(x^2+1)(x+1)}\,.$$

But how do you reverse this? That is, how do you find from $\frac{3x^2+x+2}{(x^2+1)(x+1)}$ the *partial fraction decomposition* $\frac{x}{x^2+1} + \frac{2}{x+1}$?

Partial Fraction Decomposition
Any rational function $f(x) = \frac{p(x)}{q(x)}$ with $\deg(p) < \deg(q)$ and

$$q(x) = (x-x_1)^{r_1} \cdots (x-x_k)^{r_k} (x^2+p_1x+q_1)^{s_1} \cdots (x^2+p_\ell x+q_\ell)^{s_\ell}$$

(continued)

with $x_i \neq x_j$ for $i \neq j$ and $r_i, s_j \in \mathbb{N}_0$ and further non-decomposable quadratic polynomials $x^2 + p_j x + q_j$ has a **partial fraction decomposition** of the following form:

$$
\begin{aligned}
f(x) = & \frac{A_1^{(1)}}{x - x_1} + \frac{A_2^{(1)}}{(x - x_1)^2} + \cdots + \frac{A_{r_1}^{(1)}}{(x - x_1)^{r_1}} \\
+ & \qquad \vdots \\
+ & \frac{A_1^{(k)}}{x - x_k} + \frac{A_2^{(k)}}{(x - x_k)^2} + \cdots + \frac{A_{r_k}^{(k)}}{(x - x_k)^{r_k}} \\
+ & \frac{B_1^{(1)} x + C_1^{(1)}}{x^2 + p_1 x + q_1} + \frac{B_2^{(1)} x + C_2^{(1)}}{(x^2 + p_1 x + q_1)^2} + \cdots + \frac{B_{s_1}^{(1)} x + C_{s_1}^{(1)}}{(x^2 + p_1 x + q_1)^{s_1}} \\
+ & \qquad \vdots \\
+ & \frac{B_1^{(\ell)} x + C_1^{(\ell)}}{x^2 + p_\ell x + q_\ell} + \frac{B_2^{(\ell)} x + C_2^{(\ell)}}{(x^2 + p_\ell x + q_\ell)^2} + \cdots + \frac{B_{s_\ell}^{(\ell)} x + C_{s_\ell}^{(\ell)}}{(x^2 + p_\ell x + q_\ell)^{s_\ell}} .
\end{aligned}
$$

The coefficients to be determined $A_j^{(i)}$, $B_j^{(i)}$, $C_j^{(i)}$ are obtained from this approach to partial fraction decomposition according to the following recipe.

Recipe: Determine the Partial Fraction Decomposition

(1) Make the partial fraction decomposition approach:

$$\frac{p(x)}{q(x)} = \frac{A}{(x - x_1)} + \cdots + \frac{Bx + C}{(x^2 + px + q)^s} .$$

(2) Multiply the approach in (1) by $q(x)$ and obtain an equality of polynomials.

(3) Possibly by substituting x_i in (2) some of the coefficients $A_j^{(i)}$, $B_j^{(i)}$, $C_j^{(i)}$ can be determined; possibly also the insertion of special values for x to define a coefficient.

(4) If all the coefficients are not yet determined in (3), obtain the remaining coefficients by equating coefficients of the polynomials in (2).

(5) Once all the coefficients are determined, obtain the partial fraction decomposition from (1) by plotting the coefficients.

Example 5.7

- The partial fraction decomposition of $f(x) = \frac{x}{(x-1)^2(x-2)}$ is obtained as follows:

(1) Approach:

$$\frac{x}{(x-1)^2(x-2)} = \frac{A}{(x-1)} + \frac{B}{(x-1)^2} + \frac{C}{x-2}.$$

(2) Multiplication by $q(x) = (x-1)^2(x-2)$ yields:

$$x = A(x-1)(x-2) + B(x-2) + C(x-1)^2.$$

(3) Choosing $x = 1$ yields $B = -1$, and choosing $x = 2$ yields $C = 2$. Now, since we already know B and C, we can also specify A by any other choice for x. We choose $x = 0$ and get $A = -2$.
(4) is not necessary, since all coefficients are already fixed.
(5) The partial fraction decomposition is

$$f(x) = \frac{x}{(x-1)^2(x-2)} = \frac{-2}{(x-1)} - \frac{1}{(x-1)^2} + \frac{2}{x-2}.$$

- The partial fraction decomposition of $f(x) = \frac{4x^3}{(x-1)(x^2+1)^2}$ we obtain as follows:

(1) Approach:

$$\frac{4x^3}{(x-1)(x^2+1)^2} = \frac{A}{(x-1)} + \frac{Bx+C}{x^2+1} + \frac{Dx+E}{(x^2+1)^2}.$$

(2) Multiplication by $q(x) = (x-1)(x^2+1)^2$ yields:

$$4x^3 = A(x^2+1)^2 + (Bx+C)(x-1)(x^2+1) + (Dx+E)(x-1).$$

(3) Choosing $x = 1$ yields $A = 1$. No other coefficients can be determined in this way.
(4) We substitute $A = 1$ into (2) and after multiplying out the right-hand side we obtain

$$4x^3 = (1+B)x^4 + (C-B)x^3 + (2+D-C+B)x^2 + (C-B+E-D)x + 1 - C - E.$$

Now equating coefficients yields

$$1+B=0,\ C-B=4,\ 2+D-C+B=0,\ C-B+E-D=0,\ 1-C-E=0$$

and we get the following values for the coefficients:

$$B = -1\,,\ C = 3\,,\ D = 2\,,\ E = -2\,.$$

(5) The partial fraction decomposition is

$$f(x) = \frac{4x^3}{(x-1)(x^2+1)^2} = \frac{1}{(x-1)} + \frac{-x+3}{x^2+1} + \frac{2x-2}{(x^2+1)^2}\,.$$

■

Remarks

1. Equating coefficients in step (4) may result in a quite complicated system of equations for the sought coefficients. We will present in Chap. 9 a clear solution method for such systems of equations. In the exercises we will anticipate these solution methods.
2. If the denominator polynomial is not given in factorized form, then a factorization of the denominator polynomial is first necessary before the partial fraction decomposition can begin.
3. A partial fraction decomposition of the given kind exists only if the numerator degree is really smaller than the denominator degree. If this is not the case, first perform a polynomial division.

MATLAB MATLAB provides `residue`, a partial fraction decomposition tool for the case of a denominator polynomial splitting into linear factors with integer zeros: If `z=[a_n ... a_1 a_0]` is the vector with the coefficients of the numerator polynomial and `n=[b_m ... b_1 b_0]` is the vector with the coefficients of the denominator polynomial, then `[a,b]=residue(z,n)` yields two vectors `a` and `b`. The entries in `a` are the numerators of the partial fractions and those in `b` indicate the zeros of the corresponding denominator polynomials; if such a zero occurs more than once, the power of the denominator polynomial always increases by one:

```
» [a,b]=residue([1 0],[1 -4 5 -2])
```

```
a = 2.0000 b = 2.0000
   -2.0000     1.0000
   -1.0000     1.0000
```

means $\frac{x}{x^3-4x^2+5x-2} = \frac{2}{x-2} - \frac{2}{x-1} - \frac{1}{(x-1)^2}$.

5.5 Exercises

5.1 Divide the polynomial $p(x) = x^5 + x^4 - 4\,x^3 + x^2 - x - 2$ by the polynomial

(a) $q(x) = x^2 - x - 1$,
(b) $q(x) = x^2 + x + 1$.

5.2 Factorize the following polynomials:

(a) $p_1(x) = x^3 - 2x - 1$,
(b) $p_2(x) = x^4 - 3x^3 - 3x^2 + 11x - 6$,
(c) $p_3(x) = x^4 - 6x^2 + 7$,
(d) $p_4(x) = 9x^4 + 30x^3 + 16x^2 - 30x - 25$,
(e) $p_5(x) = x^3 - 7x^2 + 4x + 12$,
(f) $p_6(x) = x^4 + x^3 + 2x^2 + x + 1$,
(g) $p_7(x) = x^4 + 4x^3 + 2x^2 - 4x - 3$,
(h) $p_8(x) = x^3 + 1$.

5.3 Determine a partial fraction decomposition for the following expressions:

(a) $\dfrac{x^4 - 4}{x^2(x^2+1)^2}$, (c) $\dfrac{x - 4}{x^3 + x}$, (e) $\dfrac{9x}{2x^3 + 3x + 5}$,

(b) $\dfrac{x}{(1+x)(1+x^2)}$, (d) $\dfrac{x^2}{(x+1)(1-x^2)}$, (f) $\dfrac{4x^2}{(x+1)^2(x^2+1)^2}$.

Trigonometric Functions 6

In this chapter we consider the four **trigonometric functions** sine, cosine, tangent, and cotangent, and their *inverse functions* arc sine, arc cosine, arc tangent, and arc cotangent. In doing so, we summarize the most important properties of these functions and become familiar with their graphs.

We will use these functions in the very next chapter when introducing the complex numbers. In later chapters, we will encounter these functions again in both calculus and linear algebra.

6.1 Sine and Cosine

We consider the unit circle, that is, the circle of radius 1. The circumference of this unit circle is known to be 2π with the **circle number**

$$\pi = 3.141592653589793\ldots.$$

The **radians** gives the angle φ by the length of the arc of the unit circle, which is cut by the angle φ (cf. Figs. 6.1 and 6.2).

We now define functions $\sin : \mathbb{R} \to \mathbb{R}$ and $\cos : \mathbb{R} \to \mathbb{R}$ using this unit circle as follows:

- For $x \in \mathbb{R}_{\geq 0}$ we run through the circle counter clockwise, starting at the point $(1, 0)$, until we have covered the distance x.
- For $x \in \mathbb{R}_{<0}$ we run through the circle clockwise, starting at the point $(1, 0)$, until we have covered the distance $|x|$.

C. Karpfinger, *Calculus and Linear Algebra in Recipes*,
https://doi.org/10.1007/978-3-662-65458-3_6

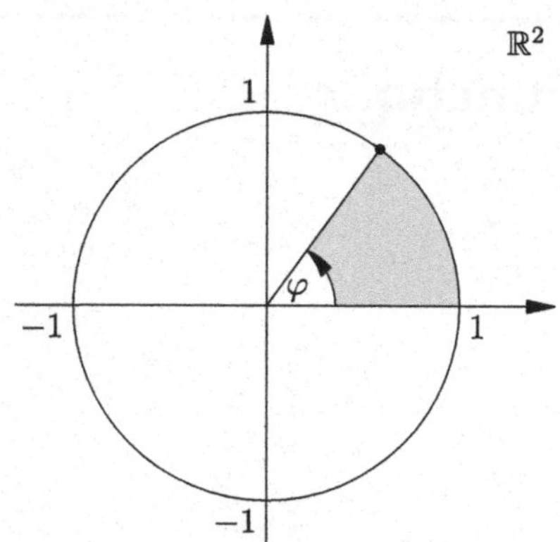

Fig. 6.1 φ in radians

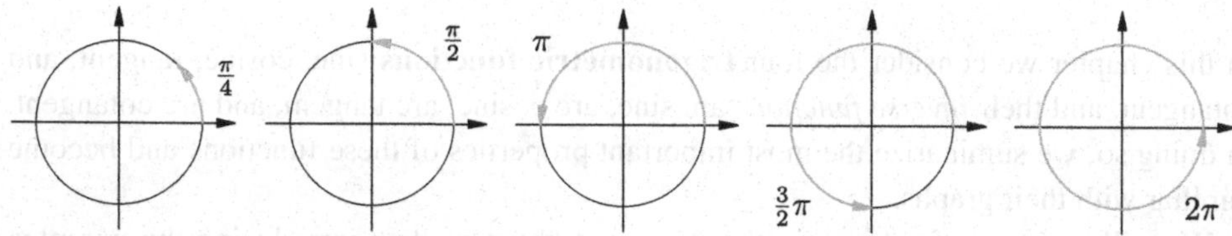

Fig. 6.2 The angles $\frac{\pi}{4}$, $\frac{\pi}{2}$, π, $\frac{3\pi}{2}$, 2π

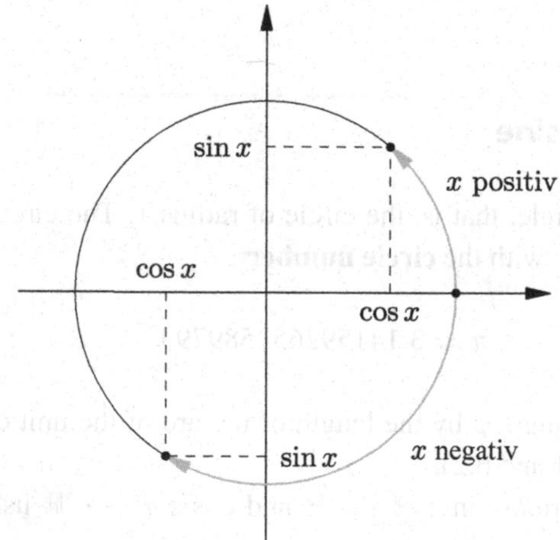

Fig. 6.3 Definition of $\cos(x)$, $\sin(x)$

Whether $x \geq 0$ or $x < 0$, in any case we end up at a point $P = P(x) = (a, b)$ of the unit circle. We declare $\cos(x) = a$ and $\sin(x) = b$ and thus obtain for every $x \in \mathbb{R}$ the numbers $\sin(x)$ and $\cos(x)$. Thus two functions sin and cos are explained: For every $x \in \mathbb{R}$ we get the numbers $\sin(x) \in \mathbb{R}$ and $\cos(x) \in \mathbb{R}$. Cf. Fig. 6.3.

This **sine function** and **cosine function** are written down in detail as follows:

$$\sin : \begin{cases} \mathbb{R} \to & \mathbb{R} \\ x \mapsto & \sin(x) \end{cases}$$

and

$$\cos : \begin{cases} \mathbb{R} \to & \mathbb{R} \\ x \mapsto & \cos(x) \end{cases} .$$

We introduce some values of the functions sin and cos together:

x	0	$\pi/6$	$\pi/4$	$\pi/3$	$\pi/2$	π	$3\pi/2$	2π
sin	0	$1/2$	$1/\sqrt{2}$	$\sqrt{3}/2$	1	0	-1	0
cos	1	$\sqrt{3}/2$	$1/\sqrt{2}$	$1/2$	0	-1	0	1

The following properties of the sine and cosine functions are particularly important:

Properties of sin **and** cos

- The **graphs of** sin and cos (cf. Fig. 6.4):
- The 2π**-periodicity**. For all $x \in \mathbb{R}$ we have:

$$\sin(x) = \sin(x + 2\pi) \quad \text{and} \quad \cos(x) = \cos(x + 2\pi) .$$

- **Zeros**. For all integers $k \in \mathbb{Z}$ we have:

$$\sin(k\pi) = 0 \quad \text{and} \quad \cos(\pi/2 + k\pi) = 0 .$$

It is

$$\pi\mathbb{Z} = \{\pi k \mid k \in \mathbb{Z}\} \text{ and } \pi/2 + \pi\mathbb{Z} = \{\pi/2 + \pi k \mid k \in \mathbb{Z}\}$$

the set of zeros of sin and cos.
- The **symmetry**. For all $x \in \mathbb{R}$ we have:

$$\sin(-x) = -\sin(x) \quad \text{and} \quad \cos(-x) = \cos(x) .$$

The sine is an *odd* function, the cosine an *even* one.

(continued)

- The **boundedness**. For all $x \in \mathbb{R}$ we have:

$$-1 \leq \sin(x),\ \cos(x) \leq 1\,.$$

- According to the **Pythagorean Theorem** we have for all $x \in \mathbb{R}$:

$$\sin^2(x) + \cos^2(x) = 1\,.$$

- The **displacement**. For all $x \in \mathbb{R}$ we have the identities:

$$\sin(x + \pi/2) = \cos(x) \qquad \text{and} \qquad \cos(x - \pi/2) = \sin(x)\,.$$

- The **addition theorems**. For all real numbers $x, y \in \mathbb{R}$ we have:

$$\begin{aligned}\sin(x+y) &= \sin(x)\cos(y) + \cos(x)\sin(y)\\ \cos(x+y) &= \cos(x)\cos(y) - \sin(x)\sin(y)\,.\end{aligned}$$

- For all real numbers $x \in \mathbb{R}$ we have the formulas **angle doubling**:

$$\begin{aligned}\sin(2x) &= 2\sin(x)\cos(x)\,,\\ \cos(2x) &= \cos^2(x) - \sin^2(x) = 2\cos^2(x) - 1\,,\\ \sin^2(x) &= \frac{1-\cos(2x)}{2}\,,\\ \cos^2(x) &= \frac{1+\cos(2x)}{2}\,.\end{aligned}$$

- For all $x \in \mathbb{R}$ and real numbers a and b we have:

$$a\cos(x) + b\sin(x) = R\cos(x - \varphi)$$

with $R = \sqrt{a^2+b^2}$ and $\varphi = \arctan(b/a)$ (see Sect. 6.3).

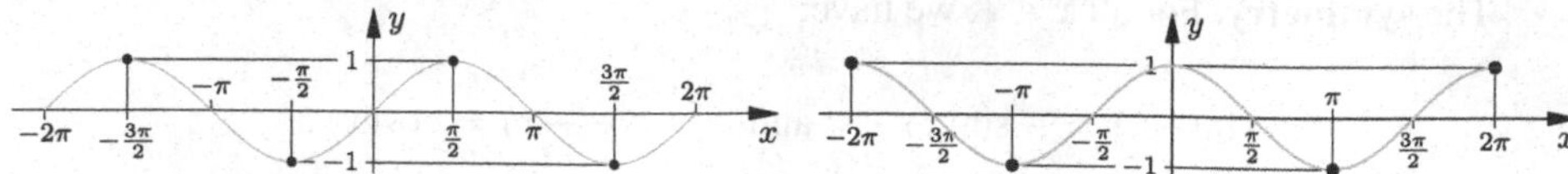

Fig. 6.4 Graphs of sin and cos

Some of these properties are self-evident, for the justifications of the others note the exercises.

6.2 Tangent and Cotangent

Using the functions sine and cosine, we now define two more trigonometric functions, namely **tangent** and **cotangent**:

$$\tan : \begin{cases} \mathbb{R} \setminus (\pi/2 + \pi\mathbb{Z}) \to & \mathbb{R} \\ x & \mapsto \tan(x) = \frac{\sin(x)}{\cos(x)}, \end{cases} \quad \cot : \begin{cases} \mathbb{R} \setminus \pi\mathbb{Z} \to & \mathbb{R} \\ x & \mapsto \cot(x) = \frac{\cos(x)}{\sin(x)}. \end{cases}$$

The restrictions on the definitional domains are of course due to the zeros of the denominators, we write briefly $D_{\tan} = \mathbb{R} \setminus (\pi/2 + \pi\mathbb{Z})$ and $D_{\cot} = \mathbb{R} \setminus \pi\mathbb{Z}$.

We give some values of the functions tan and cot:

x	0	$\pi/6$	$\pi/4$	$\pi/3$	$\pi/2$	π	$3\pi/2$	2π
tan	0	$1/\sqrt{3}$	1	$\sqrt{3}$	n. def.	0	n. def.	0
cot	n. def.	$\sqrt{3}$	1	$1/\sqrt{3}$	0	n. def.	0	n. def.

We state the main properties of tan and cot:

Properties of tan **and** cot

- The **graphs** of tan and cot (cf. Fig. 6.5):
- The π**-periodicity**. For all $x \in D_{\tan}$ and all $x \in D_{\cot}$ we have:

$$\tan(x + \pi) = \tan(x) \quad \text{and} \quad \cot(x + \pi) = \cot(x).$$

- **Zeros**. For all integers $k \in \mathbb{Z}$ we have:

$$\tan(k\pi) = 0 \quad \text{and} \quad \cot(\pi/2 + k\pi) = 0.$$

It is

$$\pi\mathbb{Z} = \{\pi k \mid k \in \mathbb{Z}\} \text{ and } \pi/2 + \pi\mathbb{Z} = \{\pi/2 + \pi k \mid k \in \mathbb{Z}\}$$

the set of zeros of tan and of cot.

(continued)

- **Symmetry**. For all $x \in D_{\tan}$ and all $x \in D_{\cot}$ we have:

$$\tan(-x) = -\tan(x) \qquad \text{and} \qquad \cot(-x) = -\cot(x)\,.$$

Tangent and cotangent are *odd* functions.

- **Unboundedness**. Tangent and cotangent take arbitrarily large and arbitrarily small values.
- The **displacement**. For all admissible $x \in \mathbb{R}$ we have the identities:

$$\tan\left(x + \pi/2\right) = -\cot(x) \qquad \text{and} \qquad \cot\left(x + \pi/2\right) = -\tan(x)\,.$$

- The **addition theorem**. For all admissible real numbers $x, y \in \mathbb{R}$ we have:

$$\tan(x+y) = \frac{\tan(x) + \tan(y)}{1 - \tan(x)\tan(y)}\,.$$

- For every admissible real number $x \in \mathbb{R}$ we have the formulas of **angle doubling**:

$$\tan(2x) = \frac{2\tan(x)}{1-\tan^2(x)} = \frac{2}{\cot(x) - \tan(x)}\,,$$

$$\cot(2x) = \frac{\cot^2(x) - 1}{2\cot(x)} = \frac{\cot(x) - \tan(x)}{2}\,,$$

$$\tan^2(x) = \frac{\sin^2(x)}{1-\sin^2(x)}\,,$$

$$\cot^2(x) = \frac{1-\sin^2(x)}{\sin^2(x)}\,.$$

- For all $x \in (-\pi, \pi)$ we have the formulas:

$$\cos(x) = \frac{1-\tan^2\left(x/2\right)}{1+\tan^2\left(x/2\right)} \qquad \text{and} \qquad \sin(x) = \frac{2\tan\left(x/2\right)}{1+\tan^2\left(x/2\right)}\,.$$

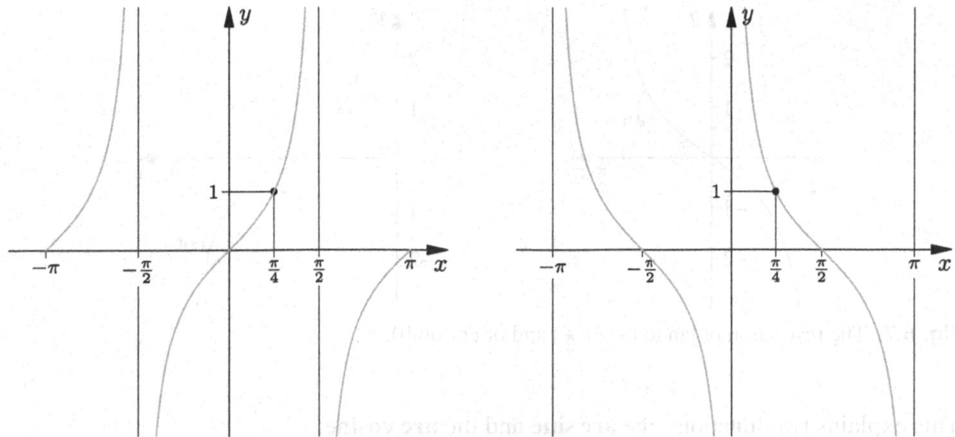

Fig. 6.5 The graphs of tan and cot

All these properties can be derived from the known properties of sin and cos (see exercises).

6.3 The Inverse Functions of the Trigonometric Functions

We consider the sine on the closed interval $\left[-\frac{\pi}{2}, \frac{\pi}{2}\right]$ and the cosine on the closed interval $[0, \pi]$, see Fig. 6.6.

Thereby we find out: For every $y \in [-1, 1]$ there is exactly one

- $x \in \left[-\frac{\pi}{2}, \frac{\pi}{2}\right]$ with $\sin(x) = y$ and
- $x \in [0, \pi]$ with $\cos(x) = y$.

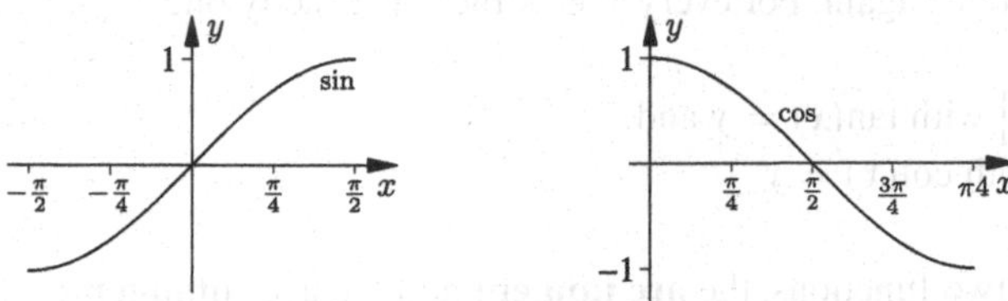

Fig. 6.6 The restriction of sin to $[-\frac{\pi}{2}, \frac{\pi}{2}]$ and of cos to $[0, \pi]$

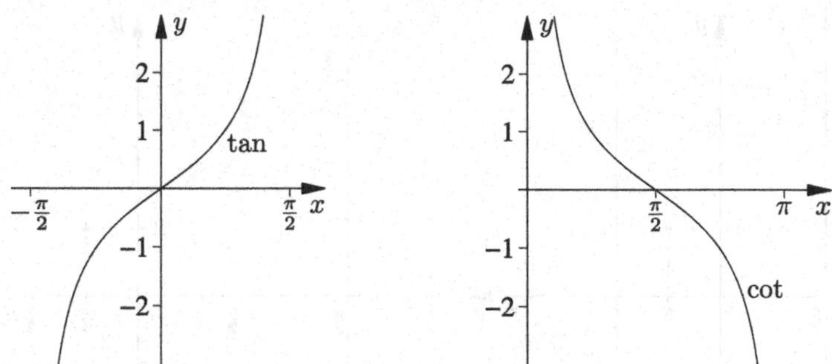

Fig. 6.7 The restriction of tan to $[-\frac{\pi}{2}, \frac{\pi}{2}]$ and of cot on $[0, \pi]$

This explains two functions, the **arc sine** and the **arc cosine**:

$$\arcsin : \begin{cases} [-1, 1] \to & \left[-\frac{\pi}{2}, \frac{\pi}{2}\right] \\ y & \mapsto \text{that } x \text{ with } \sin(x) = y \end{cases}$$

$$\arccos : \begin{cases} [-1, 1] \to & [0, \pi] \\ y & \mapsto \text{that } x \text{ with } \cos(x) = y \end{cases}.$$

Thus, we have for all $x \in \left[-\frac{\pi}{2}, \frac{\pi}{2}\right]$ and for all $x \in [0, \pi]$ and all $y \in [-1, 1]$:

$$\arcsin\left(\sin(x)\right) = x \quad \text{and} \quad \sin\left(\arcsin(y)\right) = y$$
$$\arccos\left(\cos(x)\right) = x \quad \text{and} \quad \cos\left(\arccos(y)\right) = y\,.$$

One calls arcsin and arccos the **inverse function** of sin and cos.

Now we consider analogously the tangent on the closed interval $\left[-\frac{\pi}{2}, \frac{\pi}{2}\right]$ and the cotangent on the closed interval $[0, \pi]$, see Fig. 6.7.

Thereby we notice again: For every $y \in \mathbb{R}$ there is exactly one

- $x \in \left[-\frac{\pi}{2}, \frac{\pi}{2}\right]$ with $\tan(x) = y$ and
- $x \in [0, \pi]$ with $\cot(x) = y$.

This explains two functions, the **arc tangent** and the **arc cotangent**:

$$\arctan : \begin{cases} \mathbb{R} \to & \left[-\frac{\pi}{2}, \frac{\pi}{2}\right] \\ y \mapsto & \text{that } x \text{ with } \tan(x) = y \end{cases}$$

$$\operatorname{arccot} : \begin{cases} \mathbb{R} \to & [0, \pi] \\ y \mapsto & \text{that } x \text{ with } \cot(x) = y \end{cases}.$$

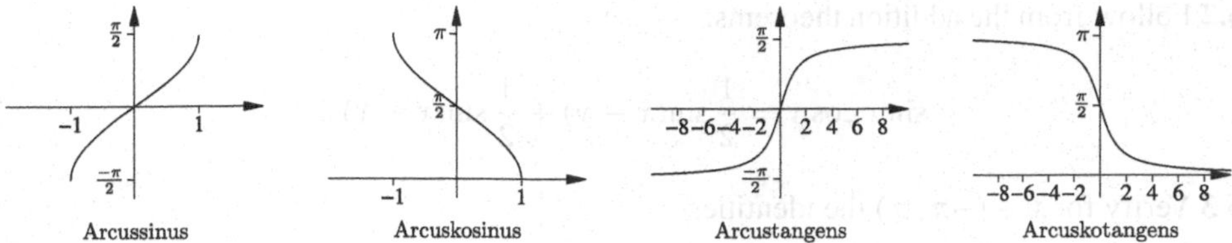

Fig. 6.8 The graphs of the arc functions

It is thus valid for all $x \in \left[-\frac{\pi}{2}, \frac{\pi}{2}\right]$ and for all $x \in [0, \pi]$ and all $y \in \mathbb{R}$:

$$\arctan\big(\tan(x)\big) = x \quad \text{and} \quad \tan\big(\arctan(y)\big) = y$$

$$\operatorname{arccot}\big(\cot(x)\big) = x \quad \text{and} \quad \cot\big(\operatorname{arccot}(y)\big) = y\,.$$

Again one calls arctan and arccot the inverse function of tan and cot.

The graphs of the inverse functions are easily obtained from the graphs of the original function by mirroring at the bisector $y = x$ (cf. Fig. 6.8).

We finally compile a few formulas for the inverse functions, which we will refer to in later chapters:

Some Formulas for the Inverse Functions

For all admissible $x \in \mathbb{R}$ we have:

- $\arcsin(x) = -\arcsin(-x) = \pi/2 - \arccos(x) = \arctan \frac{x}{\sqrt{1-x^2}}$.
- $\arccos(x) = \pi - \arccos(-x) = \pi/2 - \arcsin(x) = \operatorname{arccot} \frac{x}{\sqrt{1-x^2}}$.
- $\arctan(x) = -\arctan(-x) = \pi/2 - \operatorname{arccot}(x) = \arcsin \frac{x}{\sqrt{1+x^2}}$.
- $\operatorname{arccot}(x) = \pi - \operatorname{arccot}(-x) = \pi/2 - \arctan(x) = \arccos \frac{x}{\sqrt{1+x^2}}$.

MATLAB In MATLAB, the trigonometric functions and their inverse functions are `sin`, `cos`, `tan`, `cot`, `asin`, `acos`, `atan`, `acot`.

6.4 Exercises

6.1 Show:

(a) $\cos\big(\arcsin(x)\big) = \sqrt{1-x^2}$ for all $x \in [-1, 1]$.
(b) $\sin(\arctan x) = \frac{x}{\sqrt{1+x^2}}$ for all $x \in \mathbb{R}$.

6.2 Follow from the addition theorems:

$$\sin x \cos y = \frac{1}{2}\sin(x-y) + \frac{1}{2}\sin(x+y).$$

6.3 Verify for $x \in (-\pi, \pi)$ the identities

(a) $\cos x = \dfrac{1-\tan^2(x/2)}{1+\tan^2(x/2)}$,

(b) $\sin x = \dfrac{2\tan(x/2)}{1+\tan^2(x/2)}$,

(c) $\cos^4 x - \sin^4 x = \cos(2x)$.

6.4 Solve the inequality $\sin(2x) \le \sqrt{3}\sin x$ in $\mathbb{R}$.

6.5 Which $x \in \mathbb{R}$ satisfy the equation $5\sin x - 2\cos^2 x = 1$?

6.6 Draw the graphs of $\sin(nx)$ for $x \in [0, 2\pi]$ and $n \in \{1, 2, 3, 4\}$ into a common diagram.

6.7 Show, that for all $x \in \mathbb{R}$ and real numbers a and b we have:

$$a\cos(x) + b\sin(x) = R\cos(x - \varphi)$$

with $R = \sqrt{a^2+b^2}$ and $\varphi = \arctan(b/a)$.

Complex Numbers: Cartesian Coordinates 7

The sets of numbers $\mathbb{N} \subseteq \mathbb{N}_0 \subseteq \mathbb{Z} \subseteq \mathbb{Q} \subseteq \mathbb{R}$ are familiar from school. However, this chain of nested number sets does not stop at $\mathbb{R}$. The *complex* numbers form a number set $\mathbb{C}$ with $\mathbb{R} \subseteq \mathbb{C}$.

When calculating with real numbers, one encounters limitations when taking roots: Since squares of real numbers are always positive, it is not possible in $\mathbb{R}$ to determine roots from negative numbers. This will now be possible in $\mathbb{C}$. It will be shown, that just taking roots in $\mathbb{C}$ becomes a clear matter.

In $\mathbb{C}$ we have the *Fundamental Theorem of algebra*: every polynomial of degree $n \geq 1$ splits over $\mathbb{C}$ in n linear factors. Thus there are over $\mathbb{C}$ no annoying indecomposable quadratic polynomials like $x^2 + 1$ or $x^2 + x + 1$.

In fact, the complex numbers often simplify problems. We will get to know such examples, but for now we will familiarize ourselves with all the essential properties of complex numbers in this and the following chapter.

7.1 Construction of $\mathbb{C}$

We briefly describe the construction of $\mathbb{C}$. To do this, we consider the set

$$\mathbb{R} \times \mathbb{R} = \mathbb{R}^2 = \{(a, b) \mid a, b \in \mathbb{R}\}$$

and explain an addition and a multiplication of elements from $\mathbb{R}^2$ as follows: For two elements $(a, b), (c, d) \in \mathbb{R}^2$ we set:

$$(a, b) + (c, d) = (a + c, b + d) \text{ and } (a, b) \cdot (c, d) = (ac - bd, ad + bc)\,.$$

C. Karpfinger, *Calculus and Linear Algebra in Recipes*,
https://doi.org/10.1007/978-3-662-65458-3_7

Example 7.1

- $(2, 1) + (-1, 7) = (1, 8)$,
- $(2, 1) \cdot (1, 7) = (-5, 15)$,
- $(0, 1) \cdot (0, 1) = (-1, 0)$,
- $(a, 0) \cdot (c, 0) = (ac, 0)$ for all $a, c \in \mathbb{R}$. ■

With these operations + and · we have for all elements $(a, b), (c, d), (e, f) \in \mathbb{R}^2$ the following rules of arithmetic:

- The associative laws: $[(a, b) \dot{+} (c, d)] \dot{+} (e, f) = (a, b) \dot{+} [(c, d) \dot{+} (e, f)]$.
- The commutative laws: $(a, b) \dot{+} (c, d) = (c, d) \dot{+} (a, b)$.
- The distributive law: $(a, b) \cdot [(c, d) + (e, f)] = (a, b) \cdot (c, d) + (a, b) \cdot (e, f)$.
- There is a one-element: $(1, 0)$ satisfies $(1, 0) \cdot (c, d) = (c, d) \quad \forall (c, d) \in \mathbb{R}^2$.
- There is a zero element: $(0, 0)$ satisfies $(0, 0) + (c, d) = (c, d) \quad \forall (c, d) \in \mathbb{R}^2$.
- There are inverse elements: To $(a, b) \in \mathbb{R}^2 \setminus \{(0, 0)\}$ is $\left(\frac{a}{a^2+b^2}, \frac{-b}{a^2+b^2}\right) \in \mathbb{R}^2$ inverse to (a, b) because
$$(a, b) \cdot \left(\frac{a}{a^2 + b^2}, \frac{-b}{a^2 + b^2}\right) = \left(\frac{a^2 + b^2}{a^2 + b^2}, \frac{-ab + ab}{a^2 + b^2}\right) = (1, 0) .$$
One writes $(a, b)^{-1}$ for the inverse of (a, b).
- There are opposite elements: To $(a, b) \in \mathbb{R}^2$ is $(-a, -b) \in \mathbb{R}^2$ opposite to (a, b) because
$$(a, b) + (-a, -b) = (0, 0) .$$
One writes $-(a, b)$ for the opposite of (a, b).

We say, $\mathbb{C} = \mathbb{R}^2$ with these operations + and · is a **field** (like $\mathbb{R}$ or $\mathbb{Q}$). One also speaks of the **field of complex numbers** and calls the elements $(a, b) \in \mathbb{C}$ **complex numbers**.

Example 7.2 Examples of inverse elements are:

$$(2, 1)^{-1} = \left(\frac{2}{2^2 + 1}, \frac{-1}{2^2 + 1}\right) = \left(\frac{2}{5}, \frac{-1}{5}\right) \quad \text{and} \quad (2, 0)^{-1} = \left(\frac{2}{4}, \frac{-0}{4}\right) = \left(\frac{1}{2}, 0\right) .$$ ■

7.2 The Imaginary Unit and Other Terms

The elements $(a, 0) \in \mathbb{C}$ form the real numbers, so for $(a, 0) \in \mathbb{C}$ we can therefore briefly write a. The element $(0, 1)$ will play an excellent role, we set $\mathrm{i} = (0, 1)$ and call this complex number **imaginary unit**. Now with the addition and multiplication of complex numbers introduced above, we get:

$$(a, b) = (a, 0) + (0, b) = (a, 0) + (0, 1) \cdot (b, 0) = a + \mathrm{i}\, b\,.$$

Because $(a, b) = a + \mathrm{i}\, b$ we now have:

$$\mathbb{C} = \{a + \mathrm{i}\, b \mid a, b \in \mathbb{R}\}\,.$$

The addition and multiplication of complex numbers are in this notation:

$$(a + b\,\mathrm{i}) + (c + d\,\mathrm{i}) = a + c + \mathrm{i}(b + d) \text{ and}$$
$$(a + b\,\mathrm{i}) \cdot (c + d\,\mathrm{i}) = a\,c - b\,d + \mathrm{i}(a\,d + b\,c)\,.$$

Because of

$$\mathrm{i}^2 = \mathrm{i} \cdot \mathrm{i} = (0 + 1\,\mathrm{i}) \cdot (0 + 1\,\mathrm{i}) = -1$$

it is easy to remember the rule for multiplication

$$(a + \mathrm{i}\, b) \cdot (c + \mathrm{i}\, d) = a\,c + \mathrm{i}^2\, b\,d + \mathrm{i}\, b\,c + \mathrm{i}\, a\,d\,.$$

You only have to multiply the terms as usual.

Terms and Facts about Complex Numbers
Given a complex number $z = a + \mathrm{i}\, b \in \mathbb{C}$, $a,\ b \in \mathbb{R}$.

- We call
  - $\mathrm{Re}(z) = a \in \mathbb{R}$ the **real part** of z and
  - $\mathrm{Im}(z) = b \in \mathbb{R}$ the **imaginary part** of z.
- The complex numbers $0 + \mathrm{i}\, b$ are called **purely imaginary**.
- Two complex numbers are equal if and only if the real and imaginary parts are equal, i.e.

$$a + \mathrm{i}\, b = c + \mathrm{i}\, d \;\Leftrightarrow\; a = c \text{ and } b = d\,.$$

(continued)

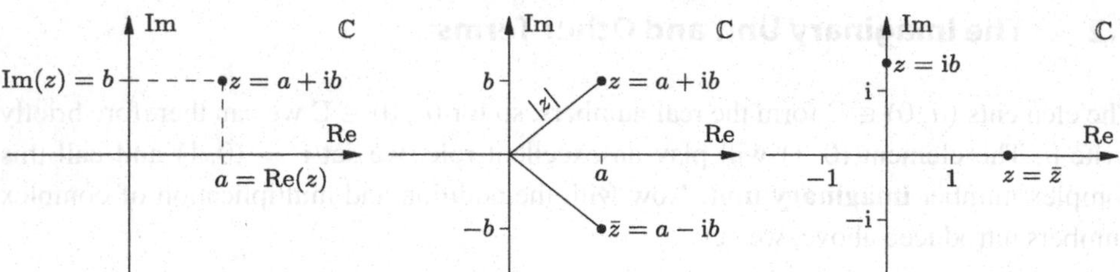

Fig. 7.1 Real and imaginary part of z, length of z and the conjugate of z

- $\overline{z} = a - \mathrm{i}\, b$ is the **conjugate complex number** to z. The number $\overline{z}$ arises from z by reflection on the real axis. In particular, we have:
  - $\overline{z} = z \Leftrightarrow \mathrm{Im}(z) = 0 \Leftrightarrow z \in \mathbb{R}$.
  - $z\overline{z} = (a + \mathrm{i}\, b)(a - \mathrm{i}\, b) = a^2 + b^2 \in \mathbb{R}$.
  - For $z_1,\ z_2 \in \mathbb{C}$ we have:

$$\overline{z_1 + z_2} = \overline{z}_1 + \overline{z}_2 \qquad \text{and} \qquad \overline{z_1 \cdot z_2} = \overline{z}_1 \cdot \overline{z}_2\,.$$

  - $\mathrm{Re}(z) = \frac{1}{2}(z + \overline{z})$.
  - $\mathrm{Im}(z) = \frac{1}{2\mathrm{i}}(z - \overline{z})$.
- One calls the expression

$$|z| = \sqrt{a^2 + b^2} = \sqrt{z\,\overline{z}} \in \mathbb{R}_{\geq 0}$$

the **absolute value** or the **length** from z.
- In $\mathbb{C}$ we have the **triangle inequality**:

$$|z_1 + z_2| \leq |z_1| + |z_2| \text{ for all } z_1,\ z_2 \in \mathbb{C}\,.$$

For some of these terms, see Fig. 7.1.

The division of complex numbers by complex numbers (which is nothing more than multiplication by the inverse $\frac{z}{w} = w^{-1} \cdot z$) is preferably done by expanding with the conjugate complex of the denominator, note the example.

Example 7.3

$$\frac{2+\mathrm{i}}{1-\mathrm{i}} = \frac{2+\mathrm{i}}{1-\mathrm{i}} \cdot \frac{1+\mathrm{i}}{1+\mathrm{i}} = \frac{1+3\mathrm{i}}{2} = \frac{1}{2} + \frac{3}{2}\mathrm{i} \qquad \text{and} \qquad \frac{1}{\mathrm{i}} = \frac{1}{\mathrm{i}} \cdot \frac{(-\mathrm{i})}{(-\mathrm{i})} = -\mathrm{i}\,.$$

■

Remark Because of $\mathrm{i}^2 = -1$ one also likes to write $\mathrm{i} = \sqrt{-1}$ which, however, can lead to confusion, because it is also $-\mathrm{i}$ a root of -1, because it is also true that $(-\mathrm{i})^2 = -1$.

7.3 The Fundamental Theorem of Algebra

The **Fundamental Theorem of Algebra** says that every polynomial $p = a_n x^n + \cdots + a_1 x + a_0$ of degree $n \geq 1$ with complex coefficients $a_0, \ldots, a_n \in \mathbb{C}$ is decomposable into linear factors, i.e., there are different complex numbers $z_1, \ldots, z_k$ and natural numbers $\nu_1, \ldots, \nu_k$ with:

$$p = a_n (x - z_1)^{\nu_1} \cdots (x - z_k)^{\nu_k}.$$

The complex numbers $z_1, \ldots, z_k$ are thereby the **zeros of** p with the **multiplicities** $\nu_1, \ldots, \nu_k$.

In general, it is very difficult to find the zeros of polynomials. We give in the following box a few hints that are useful in finding zeros of polynomials:

Recipe: Tips for Determining Zeros of Polynomials

- **Midnight Formula**: Is $p = ax^2 + bx + c$ is a polynomial with $a, b, c \in \mathbb{C}$, then

$$z_{1/2} = \frac{-b \pm \sqrt{b^2 - 4ac}}{2a} \in \mathbb{C}$$

are the zeros of p, and we have $p = a(x - z_1)(x - z_2)$. Note here:
  - $\sqrt{-r} = \mathrm{i}\sqrt{r}$ for $r \in \mathbb{R}_{>0}$ and
  - $\sqrt{a + b\,\mathrm{i}}$ you get with the following formula: The roots $w_{1,2}$ of $z = a + \mathrm{i}\,b$ are in the case $b \geq 0$:

$$w_{1,2} = \pm \left(\sqrt{\frac{a + \sqrt{a^2 + b^2}}{2}} + \mathrm{i}\sqrt{\frac{-a + \sqrt{a^2 + b^2}}{2}} \right)$$

and in the case $b < 0$:

$$w_{1,2} = \pm \left(\sqrt{\frac{a + \sqrt{a^2 + b^2}}{2}} - \mathrm{i}\sqrt{\frac{-a + \sqrt{a^2 + b^2}}{2}} \right).$$

(continued)

- If $z \in \mathbb{C}$ is a zero of a **real** polynomial $p = a_n x^n + \cdots + a_1 x + a_0$ with $a_0, a_1, \ldots, a_n \in \mathbb{R}$, then $\overline{z} \in \mathbb{C}$ is also a zero of p.
- Can you find a zero $z \in \mathbb{C}$ of the polynomial p of degree n guessed? If so, we can split the polynomial with polynomial division:

$$p(z) = 0 \Leftrightarrow p = (x - z)\, q$$

with a polynomial q of degree $n - 1$.

The formula for the roots $w_{1,2}$ of a complex number need not be remembered. We mention this formula here in order to apply the midnight formula also in the case of a non real *discriminant* $b^2 - 4ac$. Soon we will learn a much simpler formula for determining the roots of a complex number.

Example 7.4

- It is $p = x^2 + 1 = (x + \mathrm{i})(x - \mathrm{i})$ because $\pm\mathrm{i}$ are the zeros of p.
- For the polynomial $p = x^2 + x + 1$ we have because of the midnight formula:

$$p = x^2 + x + 1 = \left(x - \left(\frac{-1}{2} + \frac{\sqrt{3}\,\mathrm{i}}{2}\right)\right)\left(x - \left(\frac{-1}{2} - \frac{\sqrt{3}\,\mathrm{i}}{2}\right)\right).$$

- The polynomial $p = 2x^2 - 8x + 26$ has the zeros according to the midnight formula

$$z_{1/2} = \frac{8 \pm \sqrt{64 - 208}}{4} = 2 \pm \frac{\mathrm{i}\sqrt{144}}{4} = 2 \pm 3\,\mathrm{i}\,.$$

  So it's $p = 2\big(x - (2 + 3\,\mathrm{i})\big)\big(x - (2 - 3\,\mathrm{i})\big)$.
- It is $p = 2\,\mathrm{i}\,x^2 + x + \mathrm{i}$ a quadratic polynomial with complex coefficients. The zeros are:

$$z_{1/2} = \frac{-1 \pm \sqrt{1 + 8}}{4\,\mathrm{i}} = \frac{-1 \pm 3}{4\,\mathrm{i}} \Leftrightarrow z_1 = \frac{1}{2\,\mathrm{i}} = -\frac{\mathrm{i}}{2} \text{ and } z_2 = -\frac{1}{\mathrm{i}} = \mathrm{i}\,.$$

  Accordingly write p as

$$p = 2\,\mathrm{i}\left(x + \frac{\mathrm{i}}{2}\right)(x - \mathrm{i}) = 2\,\mathrm{i}\left(x^2 - \frac{\mathrm{i}}{2}x + \frac{1}{2}\right) = 2\,\mathrm{i}\,x^2 + x + \mathrm{i}\,.$$

- For the polynomial $p = x^2 + 2x + \mathrm{i}$ we obtain the zeros with the midnight formula

$$z_{1/2} = \frac{-2 \pm \sqrt{4 - 4\,\mathrm{i}}}{2}.$$

A root of $z = 4 - 4\,\mathrm{i}$ is

$$w = \sqrt{2 + \sqrt{8}} - \mathrm{i}\sqrt{-2 + \sqrt{8}}\,.$$

This gives us the decomposition $p = (x - (-1 + \frac{w}{2}))(x - (-1 - \frac{w}{2}))$. ■

7.4 Exercises

7.1 Show: If $z \in \mathbb{C}$ is the zero of a real polynomial $p = a_n x^n + \ldots + a_1 x + a_0$ with $a_0, \ldots, a_n \in \mathbb{R}$ then also $\overline{z} \in \mathbb{C}$ is a zero of p.

7.2 Determine real and imaginary parts and the absolute values of:

(a) $(2 - \mathrm{i})(1 + 2\,\mathrm{i})$, (b) $\dfrac{50 - 25\,\mathrm{i}}{-2 + 11\,\mathrm{i}}$, (c) $(1 + \mathrm{i}\sqrt{3})^2$, (d) $\mathrm{i}^{99} + \mathrm{i}^{100} + 2\,\mathrm{i}^{101} - 2$.

7.3 Determine the zeros of $p = z^3 + 4z^2 + 8z$.

7.4 Write every complex number in the form $a + b\,\mathrm{i}$ with $a, b \in \mathbb{R}$:

(a) $(1 + 4\,\mathrm{i}) \cdot (2 - 3\,\mathrm{i})$, (b) $\dfrac{4}{2 + \mathrm{i}}$, (c) $\sum\limits_{n=0}^{2009} \mathrm{i}^n$.

7.5 Sketch the following sets of points in $\mathbb{C}$:

(a) $\{z \mid |z + \mathrm{i}| \leq 3\}$, (b) $\{z \mid \mathrm{Re}(\overline{z} - \mathrm{i}) = z\}$, (c) $\{z \mid |z - 3| = 2|z + 3|\}$.

7.6 Calculate all the complex numbers $z \in \mathbb{C}$ that satisfy the following equations:

(a) $z^2 - 4z + 5 = 0$, (b) $z^2 + (1 - \mathrm{i})z - \mathrm{i} = 0$, (c) $z^2 + 4z + 8 = 0$.

Complex Numbers: Polar Coordinates

8

The complex numbers are the points of $\mathbb{R}^2$. Every complex number $z = a + \mathrm{i}\,b$ with $a, b \in \mathbb{R}$ is uniquely given by the cartesian coordinates $(a, b) \in \mathbb{R}^2$. The plane $\mathbb{R}^2$ can also be thought of as a union of circles around the origin. So every point $z \neq 0$ can be uniquely described by the radius r *of* the circle on which it lies and the angle $\varphi \in (-\pi, \pi]$ which is defined by the positive x-axis and z. We call (r, φ) the *polar coordinates* of z.

With the help of these polar coordinates we can interpret the multiplication of complex numbers very easily, furthermore the exponentiation of complex numbers and the taking of roots from complex numbers becomes clear and simple.

8.1 The Polar Representation

The complex numbers form the so-called **Gaussian number plane** $\mathbb{C}$. Every complex number $z = a + \mathrm{i}\,b$ is uniquely described by its cartesian coordinates (a, b). However, one can describe every point $z \neq 0$ of $\mathbb{C}$ also uniquely by **polar coordinates** (r, φ). Thereby (note Fig. 8.1)

- $r = \sqrt{a^2 + b^2} = |z| \in \mathbb{R}_{>0}$ the length or absolute value of z and
- $\varphi \in (-\pi, \pi]$ the angle formed by z and the positive real axis.

It's called $\varphi \in (-\pi, \pi]$ the **(principal) argument** of z and writes for it $\varphi = \arg(z)$. Thus, for any complex number $z \neq 0$ we have the two possible representations:

$$z = \begin{cases} (a, b) & \text{cartesian coordinates} \\ (r, \varphi) & \text{polar coordinates} \end{cases}.$$

C. Karpfinger, *Calculus and Linear Algebra in Recipes*,
https://doi.org/10.1007/978-3-662-65458-3_8

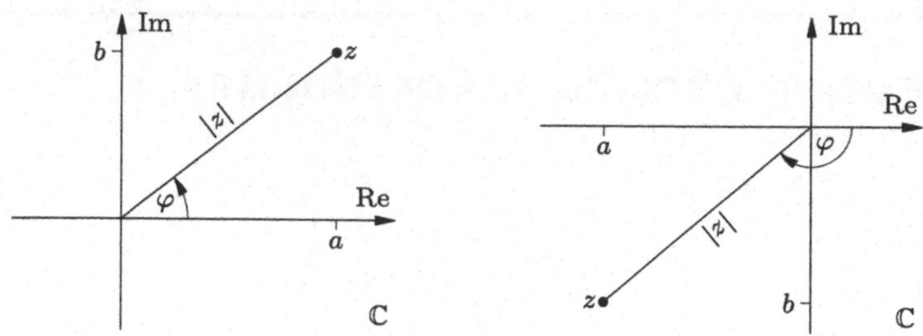

Fig. 8.1 Polar coordinates $(|z|, \varphi)$ of two different complex numbers

Example 8.1 In the following we give for different z the cartesian coordinates (a, b) and polar coordinates (r, φ):

- $z = 1$. Here we have $(a, b) = (1, 0)$ and $(r, \varphi) = (1, 0)$.
- $z = \mathrm{i}$. Here we have $(a, b) = (0, 1)$ and $(r, \varphi) = (1, \pi/2)$.
- $z = -1$. Here we have $(a, b) = (-1, 0)$ and $(r, \varphi) = (1, \pi)$.
- $z = -\mathrm{i}$. Here we have $(a, b) = (0, -1)$ and $(r, \varphi) = (1, -\pi/2)$.
- $z = -1 - \mathrm{i}$. Here we have $(a, b) = (-1, -1)$ and $(r, \varphi) = (\sqrt{2}, -3\pi/4)$.
- $z = -1 + \mathrm{i}$. Here we have $(a, b) = (-1, 1)$ and $(r, \varphi) = (\sqrt{2}, 3\pi/4)$.

■

Because $a = r\cos\varphi$ and $b = r\sin\varphi$ we can write the complex number $z = (a, b) = (r, \varphi)$ using the cartesian coordinates or the polar coordinates as follows:

$$z = a + \mathrm{i}\, b = r\,(\cos\varphi + \mathrm{i}\sin\varphi)\,.$$

This latter notation is also called **polar representation** of the number z. Using the following formulas we can calculate the polar coordinates (r, φ) from the cartesian coordinates (a, b) and vice versa:

Conversion Formulas

- **cartesian → polar:** Given $z = a + \mathrm{i}\, b \neq 0$, $a, b \in \mathbb{R}$. Then:

$$\left.\begin{array}{l} r = \sqrt{a^2 + b^2} \\ \varphi = \begin{cases} \arccos(a/r)\,, & \text{if } b \geq 0 \\ -\arccos(a/r)\,, & \text{if } b < 0 \end{cases} \end{array}\right\} \to z = r\left(\cos(\varphi) + \mathrm{i}\sin(\varphi)\right).$$

(continued)

- **polar → cartesian:** Given $z = r\left(\cos(\varphi) + \mathrm{i}\sin(\varphi)\right)$, $r, \varphi \in \mathbb{R}$. Then:

$$\left.\begin{aligned} a &= r\cos(\varphi) \\ b &= r\sin(\varphi) \end{aligned}\right\} \Rightarrow z = a + \mathrm{i}\, b\,.$$

Example 8.2

- For $z = 1 - \mathrm{i}$ we have $r = \sqrt{1^2 + (-1)^2} = \sqrt{2}$ and $\varphi = -\arccos\left(1/\sqrt{2}\right) = -\pi/4$, so $z = \sqrt{2}\big(\cos(-\pi/4) + \mathrm{i}\sin(-\pi/4)\big)$.
- For $z = 3 + 4\,\mathrm{i}$ we have $r = \sqrt{3^2 + 4^2} = 5$ and $\varphi = \arccos(3/5) = 0.9273$, so $z = 5\big(\cos(0.9273) + \mathrm{i}\sin(0.9273)\big)$.
- For $z = \sqrt{3}(\cos(\pi/6) + \mathrm{i}\sin(\pi/6))$ we have $z = 3/2 + \mathrm{i}\sqrt{3}/2$.

■

MATLAB In MATLAB one obtains with

- `[a,b] = pol2cart(phi,r)` and
- `[phi,r] = cart2pol(a,b)`

the desired coordinates in every case. And if z is a complex number, then

- `real(z)` returns the real part of z,
- `imag(z)` the imaginary part of z,
- `abs(z)` the absolute value of z,
- `conj(z)` the conjugate of z,
- `angle(z)` the argument of z,
- `compass(z)` the pointer plot of z.

8.2 Applications of the Polar Representation

With the help of polar coordinates, multiplication and exponentiation of complex numbers and taking roots of such numbers can be easily represented:

Multiplication, Powers and Roots of Complex Numbers

- If $z_1 = r_1\big(\cos(\varphi_1)+\mathrm{i}\sin(\varphi_1)\big)$ and $z_2 = r_2\big(\cos(\varphi_2)+\mathrm{i}\sin(\varphi_2)\big)$ are two complex numbers not equal to 0, we have the following:

$$z_1 \cdot z_2 = r_1 r_2\big(\cos(\varphi_1+\varphi_2)+\mathrm{i}\sin(\varphi_1+\varphi_2)\big)\,.$$

- For every complex number $z = (\cos(\varphi)+\mathrm{i}\sin(\varphi)) \in \mathbb{C}\setminus\{0\}$ and every natural number $n \in \mathbb{N}$ we have **Moivre's formula**:

$$\big(\cos(\varphi)+\mathrm{i}\sin(\varphi)\big)^n = \cos(n\varphi)+\mathrm{i}\sin(n\varphi)\,.$$

- For every complex number $z = r\big(\cos(\varphi)+\mathrm{i}\sin(\varphi)\big) \in \mathbb{C}\setminus\{0\}$ and each natural number $n \in \mathbb{N}$ we have:

$$z^n = r^n\big(\cos(\varphi)+\mathrm{i}\sin(\varphi)\big)^n = r^n\big(\cos(n\varphi)+\mathrm{i}\sin(n\varphi)\big)\,.$$

- For every complex number $z = r\big(\cos(\varphi)+\mathrm{i}\sin(\varphi)\big) \in \mathbb{C}\setminus\{0\}$ and every $n \in \mathbb{N}$ the n different complex numbers

$$z_k = \sqrt[n]{r}\left(\cos\left(\frac{\varphi+2k\pi}{n}\right)+\mathrm{i}\sin\left(\frac{\varphi+2k\pi}{n}\right)\right),\quad k=0,\,1,\dots,n-1\,,$$

are exactly the n different n-th roots of z, i.e., we have $z_k^n = z$ for all $k \in \{0,\dots,n-1\}$.

So for the product of two complex numbers, the lengths are multiplied and the arguments are added, this formula follows from the addition theorems of sine and cosine. The other formulas follow more or less from this (see Exercise 8.3).

Example 8.3

- The product of $z_1 = \sqrt{2}\,\big(\cos(\pi/4)+\mathrm{i}\sin(\pi/4)\big)$ and $z_2 = \sqrt{3}\,\big(\cos(\pi/2)+\mathrm{i}\sin(\pi/2)\big)$ is:

$$z_1 \cdot z_2 = \sqrt{6}\,\big(\cos(\tfrac{3\pi}{4})+\mathrm{i}\sin(\tfrac{3\pi}{4})\big)\,.$$

- We now determine the first 8 powers of $z = \cos(\pi/4)+\mathrm{i}\sin(\pi/4) = \frac{1}{\sqrt{2}}(1+\mathrm{i})$:
  - $z = \cos(\frac{\pi}{4})+\mathrm{i}\sin(\frac{\pi}{4}) = \frac{1}{\sqrt{2}}(1+\mathrm{i})$.
  - $z^2 = \cos(\frac{\pi}{2})+\mathrm{i}\sin(\frac{\pi}{2}) = \mathrm{i}$.

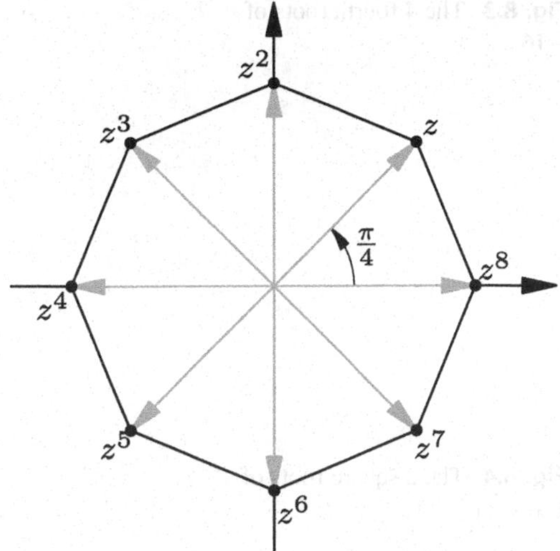

Fig. 8.2 The powers of $z = \frac{1}{\sqrt{2}}(1+\mathrm{i})$

- $z^3 = \cos(\frac{3\pi}{4}) + \mathrm{i}\sin(\frac{3\pi}{4}) = -\frac{1}{\sqrt{2}}(1-\mathrm{i})$.
- $z^4 = \cos(\pi) + \mathrm{i}\sin(\pi) = -1$.
- $z^5 = \cos(\frac{5\pi}{4}) + \mathrm{i}\sin(\frac{5\pi}{4}) = -\frac{1}{\sqrt{2}}(1+\mathrm{i})$.
- $z^6 = \cos(\frac{3\pi}{2}) + \mathrm{i}\sin(\frac{3\pi}{2}) = -\mathrm{i}$.
- $z^7 = \cos(\frac{7\pi}{4}) + \mathrm{i}\sin(\frac{7\pi}{4}) = \frac{1}{\sqrt{2}}(1-\mathrm{i})$.
- $z^8 = \cos(2\pi) + \mathrm{i}\sin(2\pi) = 1$.

Note Fig. 8.2.
In particular $z = \frac{1}{\sqrt{2}}(1+\mathrm{i})$ is a root of i, namely $z^2 = \mathrm{i}$. And analogously $z^3 = -\frac{1}{\sqrt{2}}(1-\mathrm{i})$ is a root of $-\mathrm{i}$ since $(z^3)^2 = z^6 = -\mathrm{i}$.

- The four different 4-th roots of $z = -16$ are found as follows. First, one writes z in polar coordinates: $-16 = 16\big(\cos(\pi) + \mathrm{i}\sin(\pi)\big)$. For every $0 \leq k \leq 3$ is then

$$z_k = \sqrt[4]{16}\Big(\cos\Big(\frac{\pi + 2k\pi}{4}\Big) + \mathrm{i}\sin\Big(\frac{\pi + 2k\pi}{4}\Big)\Big)$$

a 4-th root of -16. Specifically:

$$z_0 = 2\Big(\cos(\tfrac{\pi}{4}) + \mathrm{i}\sin(\tfrac{\pi}{4})\Big) = \sqrt{2} + \sqrt{2}\,\mathrm{i}\,,$$

$$z_1 = 2\Big(\cos(\tfrac{3\pi}{4}) + \mathrm{i}\sin(\tfrac{3\pi}{4})\Big) = -\sqrt{2} + \sqrt{2}\,\mathrm{i}\,,$$

$$z_2 = 2\Big(\cos(\tfrac{5\pi}{4}) + \mathrm{i}\sin(\tfrac{5\pi}{4})\Big) = -\sqrt{2} - \sqrt{2}\,\mathrm{i}\,,$$

$$z_3 = 2\Big(\cos(\tfrac{7\pi}{4}) + \mathrm{i}\sin(\tfrac{7\pi}{4})\Big) = \sqrt{2} - \sqrt{2}\,\mathrm{i}\,.$$

Note Fig. 8.3.

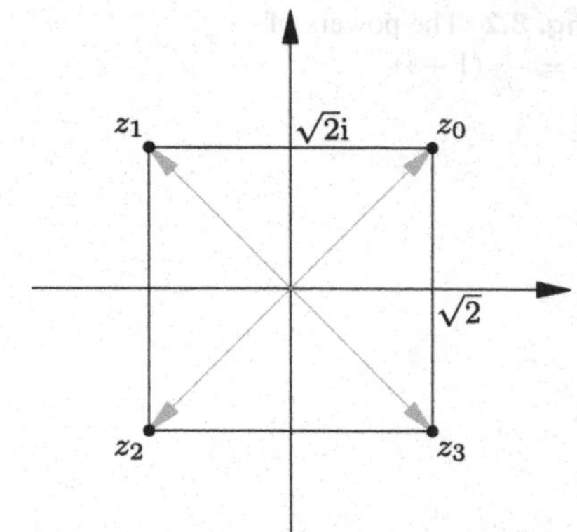

Fig. 8.3 The 4 fourth roots of -16

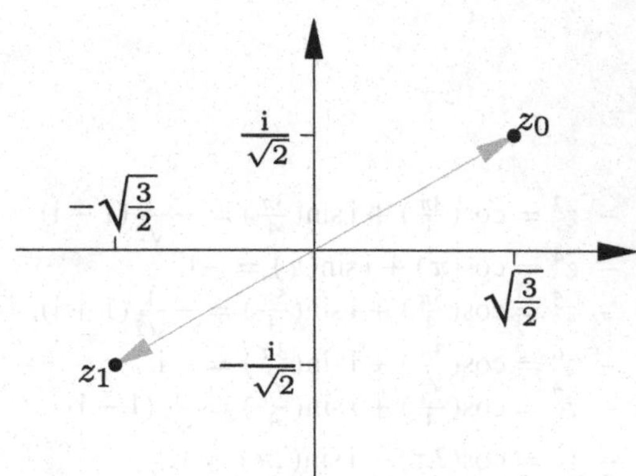

Fig. 8.4 The 2 square roots of $1+\sqrt{3}\,\mathrm{i}$

- Analogously, the square roots of $z = 1+\sqrt{3}\,\mathrm{i}$ can be determined. With the polar representation $z = 2\big(\cos(\pi/3) + \mathrm{i}\sin(\pi/3)\big)$ one obtains the roots:

$$z_0 = \sqrt{2}\Big(\cos\Big(\frac{\pi/3}{2}\Big) + \mathrm{i}\sin\Big(\frac{\pi/3}{2}\Big)\Big) = \sqrt{2}\left(\frac{\sqrt{3}}{2} + \mathrm{i}\,\frac{1}{2}\right) = \sqrt{\frac{3}{2}} + \frac{\mathrm{i}}{\sqrt{2}},$$

$$z_1 = \sqrt{2}\Big(\cos\Big(\frac{7\pi/3}{2}\Big) + \mathrm{i}\sin\Big(\frac{7\pi/3}{2}\Big)\Big) = \sqrt{2}\left(\frac{-\sqrt{3}}{2} + \mathrm{i}\,\frac{-1}{2}\right) = -\sqrt{\frac{3}{2}} - \frac{\mathrm{i}}{\sqrt{2}}.$$

Note that $z_1 = -z_0$. Of course, this must be true, because if z_0 is a square root of z, i.e. $z_0^2 = z$, then of course we also have $(-z_0)^2 = z$.
The roots are plotted in Fig. 8.4.

- The 5-th roots of $z = 1$ (the roots from 1 are also called **roots of unity**) are due to the polar representation $1 = 1\big(\cos(0) + \mathrm{i}\sin(0)\big)$:

$$\begin{aligned} z_0 &= \cos(\tfrac{0\pi}{5}) + \mathrm{i}\sin(\tfrac{0\pi}{5}) = 1\,, \\ z_1 &= \cos(\tfrac{2\pi}{5}) + \mathrm{i}\sin(\tfrac{2\pi}{5})\,, \\ z_2 &= \cos(\tfrac{4\pi}{5}) + \mathrm{i}\sin(\tfrac{4\pi}{5})\,, \\ z_3 &= \cos(\tfrac{6\pi}{5}) + \mathrm{i}\sin(\tfrac{6\pi}{5})\,, \\ z_4 &= \cos(\tfrac{8\pi}{5}) + \mathrm{i}\sin(\tfrac{8\pi}{5})\,. \end{aligned}$$

We can see the roots in Fig. 8.5.

■

Remark The so-called **Euler's formula** (see Chap. 24) reads

$$\mathrm{e}^{\mathrm{i}\varphi} = \cos(\varphi) + \mathrm{i}\sin(\varphi) \text{ for all } \varphi \in \mathbb{R}\,.$$

Thus the polar representation can be summarized even more concisely:

$$z = r\,(\cos(\varphi) + \mathrm{i}\sin(\varphi)) = r\,\mathrm{e}^{\mathrm{i}\varphi}\,.$$

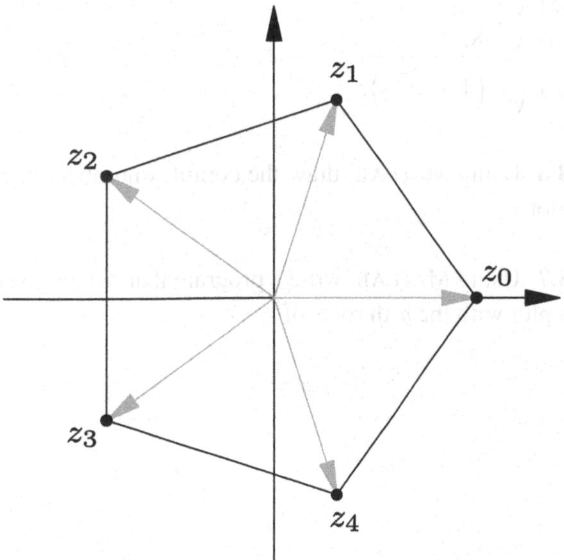

Fig. 8.5 The 5 fifth roots of 1

8.3 Exercises

8.1 Justify why the multiplication, exponentiation and root extraction formulae in Sect. 8.2 apply.

8.2

(a) Give the polar representation for the following complex numbers:

$$z_1 = -2\,\mathrm{i}, \qquad z_2 = \mathrm{i} - 1, \qquad z_3 = \tfrac{1}{2}(-1 + \sqrt{3}\,\mathrm{i}), \qquad z_4 = \tfrac{2}{1-\mathrm{i}}.$$

(b) To the complex numbers with polar coordinates

$$r_1 = 2,\ \varphi_1 = \pi/2, \qquad r_2 = 1,\ \varphi_2 = 3\pi/4, \qquad r_3 = 3,\ \varphi_3 = 5\pi/4,$$
$$r_4 = 4,\ \varphi_4 = 2\pi/3$$

real and imaginary parts are sought.

8.3 Give for $n \in \mathbb{N}$ all solutions of the equation $z^n = 1$ in $\mathbb{C}$ in the polar representation.

8.4 Calculate real and imaginary parts of $(\sqrt{3} + \mathrm{i})^{100}$.

8.5 Calculate the complex roots:

(a) $\sqrt{-2\,\mathrm{i}}$,
(b) $\sqrt[3]{-8}$,
(c) $\sqrt{8\left(1 - \sqrt{3}\,\mathrm{i}\right)}$.

8.6 Using MATLAB, draw the complex numbers $z, z^2, \ldots, z^8$ for $z = \frac{1}{\sqrt{2}}(1 + \mathrm{i})$ into a plot.

8.7 Using MATLAB, write a program that, when given $z = a + b\,\mathrm{i} \in \mathbb{C}$ and $n \in \mathbb{N}$, gives a plot with the n-th roots of z.

Linear Systems of Equations

9

Many problems of linear algebra but also of calculus lead to the task of solving a *linear system of equations*. Such systems of equations can always be solved completely and clearly. This is quite different with the *nonlinear* systems of equations.

The method of choice for solving a system of linear equations is based on the *Gaussian elimination method*. We present this method in detail and also describe the structure of the solution set of such a system.

9.1 The Gaussian Elimination Method

A **linear system of equations**, LGS for short, with m equations in n unknowns $x_1, \ldots, x_n$ can be written in the following form:

$$\begin{array}{ccccc} a_{11}x_1 & + \cdots + & a_{1n}x_n & = & b_1 \\ a_{21}x_1 & + \cdots + & a_{2n}x_n & = & b_2 \\ \vdots & & & & \vdots \\ a_{m1}x_1 & + \cdots + & a_{mn}x_n & = & b_m \end{array} \quad \text{with} \quad a_{ij}, b_i \in \begin{cases} \mathbb{R} & \text{(real LGS)} \\ \mathbb{C} & \text{(complex LGS)} \end{cases}.$$

In order not to make the formulations unnecessarily complicated, we only consider real systems in the following; for complex systems everything works analogously or in the same way.

A ***n*-tuple** $(\ell_1, \ldots, \ell_n) \in \mathbb{R}^n$ is a **solution of** the LGS if for all $i = 1, \ldots, m$ we have:

$$a_{i1}\,\ell_1 + \cdots + a_{in}\,\ell_n = b_i\,.$$

C. Karpfinger, *Calculus and Linear Algebra in Recipes*,
https://doi.org/10.1007/978-3-662-65458-3_9

Example 9.1 In each case, we are looking for the set L of all solutions of the system of linear equations. For the calculation we use the *substitution method*:

- We solve a linear system of equations with two equations in two unknowns:

$$\left.\begin{array}{l} 3x + 2y = 9 \\ 4x - y = 1 \end{array}\right\} \Rightarrow y = 4x - 1 \left.\right\} \Rightarrow 3x + 2(4x - 1) = 9 \left.\right\} \Rightarrow x = 1 .$$

  So there is exactly one solution, and the solution set is $L = \{(1, 3)\}$.
- Now consider an LGS in two unknowns with only one equation:

$$x + y = 1 \Rightarrow x = 1 - y.$$

  We set $y = \lambda \in \mathbb{R}$ arbitrarily and can then determine x as a function of λ as $x = 1 - \lambda$. The solution set is thus $L = \{(1 - \lambda, \lambda) \mid \lambda \in \mathbb{R}\}$, it has infinitely many elements.
- Finally, we consider another LGS with two equations in two unknowns:

$$\left.\begin{array}{l} x + y = 1 \\ 2x + 2y = 5 \end{array}\right\} \Rightarrow y = 1 - x \left.\right\} \Rightarrow 2x + 2(1 - x) = 5 \left.\right\} \Rightarrow 2 = 5 .$$

  The linear system of equations has no solution, we have $L = \emptyset$. ■

In the three examples, we obtained three different types of solution sets. There was either no solution, exactly one solution, or infinitely many solutions. In fact, there are no other possibilities.

The insertion procedure used above is not suitable as soon as one considers more than two equations or unknowns. To motivate the *Gaussian elimination procedure*, we compare the following two systems of linear equations:

$$\begin{array}{rcl} x_1 + x_2 + 2x_3 &=& 2 \\ 2x_1 + 2x_2 - x_3 &=& 1 \\ 3x_1 + 4x_2 + 2x_3 &=& 2 \end{array} \quad \text{and} \quad \begin{array}{rcl} x_1 + x_2 + 2x_3 &=& 2 \\ x_2 - 4x_3 &=& -4 \\ -5x_3 &=& -3 \end{array}$$

In the left-hand system, the solution is not immediately recognizable, whereas in the right-hand system, the last solution component can be ℓ_3 in the last equation. If this is determined, then one receives the penultimate solution component from the equation before last ℓ_2 and thus finally from the first equation the first component ℓ_1—one speaks of **backward substitution** .

The form of the right equation system is called **row step form**. The *Gaussian elimination method* takes advantage of the fact that any linear system of equations can be brought to row step form by so-called *elementary row transformations*. The solution set remains unchanged. Thus, one solves the system on row step form by substituting backwards and obtains the solution set of the original system. The *elementary row transformations* are:

Elementary Row Transformations

- Swap two lines.
- Multiplication of a row by a $\lambda \neq 0$.
- Addition of a λ-fold of a row to another row.

We exemplify how to obtain a row step form using a system with two equations: The system

$$\begin{aligned} ax + by &= c \\ a'x + b'y &= c' \end{aligned}$$

can be brought to row step form in the case $a \neq 0$ by choosing a $\lambda \in \mathbb{R}$ such that $a' + \lambda a = 0$ and then subtracting the λ-fold of the first equation from the second:

$$\begin{array}{ccccc} ax & + & by & = & c \\ \underbrace{(a' + \lambda a)}_{=0}\, x & + & (b' + \lambda b)y & = & c' + \lambda c \end{array}.$$

Example 9.2

- Let (i) denote the first equation and (ii) denote the second equation in the following calculation:

$$\left.\begin{aligned} x + y &= 1 \\ 2x - y &= 5 \end{aligned}\right\} \overset{\text{(ii)}-2\text{(i)}}{\Rightarrow} \left.\begin{aligned} x + y &= 1 \\ 0 - 3y &= 3 \end{aligned}\right\} \Rightarrow \left.\begin{aligned} & \\ y &= -1 \end{aligned}\right\} \Rightarrow \begin{aligned} x &= 2 \\ & \end{aligned}.$$

  So the LGS has the solution set $L = \{(2, -1)\}$.
- With the equation numbering from above we have:

$$\left.\begin{aligned} x + y &= 1 \\ 2x + 2y &= 5 \end{aligned}\right\} \overset{\text{(ii)}-2\text{(i)}}{\Rightarrow} \left.\begin{aligned} x + y &= 1 \\ 0 - 0 &= 3 \end{aligned}\right\} \Rightarrow \begin{aligned} & \\ 0 &= 3 \end{aligned}.$$

  Accordingly, the LGS has no solution, $L = \emptyset$. ■

In the examples it is noticeable that for the row transformations only the coefficients are needed, $x, y, \ldots$ are only *placeholders*. So we can write the whole thing down more economically by omitting the variables and just writing down the coefficients in a *matrix*. We can write any system of linear equations in the following form:

$$\begin{array}{ccc} a_{11}x_1 + \cdots + a_{1n}x_n = b_1 & & \\ \vdots \qquad\qquad \vdots & \longleftrightarrow & \underbrace{\left(\overbrace{\begin{array}{ccc} a_{11} & \cdots & a_{1n} \\ \vdots & & \vdots \\ a_{m1} & \cdots & a_{mn} \end{array}}^{\text{coefficient matrix } A} \left| \begin{array}{c} b_1 \\ \vdots \\ b_m \end{array} \right. \right)}_{\text{ext. coefficient matrix } (A\,|\,b)} . \\ a_{m1}x_1 + \cdots + a_{mn}x_n = b_m & & \end{array}$$

The matrix is called A the **coefficient matrix** and the entire *matrix* $(A \mid b)$ the **extended coefficient matrix**.

With this we can solve linear systems of equations as follows:

Recipe: Solving a System of Linear Equations Using the Gaussian Elimination Method

We obtain the solution set of the linear system of equations (LGS) as follows:

(1) Write down the extended coefficient matrix $(A \mid b)$.
(2) Bring the extended coefficient matrix $(A \mid b)$ to row step form using elementary row transformations.
(3) Create zeros above the *diagonal*:

$$\left(\begin{array}{ccc|c} a_{11} & \dots & a_{1n} & b_1 \\ \vdots & & \vdots & \vdots \\ a_{m1} & \dots & a_{mn} & b_m \end{array} \right) \overset{(2)}{\to} \left(\begin{array}{cccc|c} * & * & * & * & * \\ & * & * & * & \vdots \\ & & * & * & * \\ & & & & \bigstar \end{array} \right) \overset{(3)}{\to} \left(\begin{array}{cccc|c} * & 0 & 0 & * & * \\ & * & 0 & * & \vdots \\ & & * & * & * \\ & & & & \bigstar \end{array} \right)$$

(4) If at the positions ★ there is an entry unequal zero, STOP: The system is not solvable, otherwise:
(5) Obtain the individual components of the solution by backward substitution, where for the x_i, which are not at step edges, a free selectable $\lambda \in \mathbb{R}$ is used.

Remark Step (3) can also be omitted. But the generation of these further zeros *above the diagonal* facilitates the backward substitution in step (4). If you create zeros above the diagonal with (3), you speak of a **reduced row step form**.

Example 9.3

- We solve an LGS with more equations than unknowns:

$$\begin{array}{rl} 2x+\ 4y=2 \\ 3x+\ 6y=3 \\ 5x+10y=5 \end{array} \leftrightarrow \left(\begin{array}{cc|c} 2 & 4 & 2 \\ 3 & 6 & 3 \\ 5 & 10 & 5 \end{array}\right) \rightsquigarrow \left(\begin{array}{cc|c} 1 & 2 & 1 \\ \hline 0 & 0 & 0 \\ 0 & 0 & 0 \end{array}\right).$$

  The solution set is

$$L = \{(1-2\lambda, \lambda) \mid \lambda \in \mathbb{R}\} = \left\{\begin{pmatrix} 1 \\ 0 \end{pmatrix} + \lambda \begin{pmatrix} -2 \\ 1 \end{pmatrix} \mid \lambda \in \mathbb{R}\right\}.$$

- Now consider an LGS with more unknowns than equations:

$$\begin{array}{rl} x+2y+3z=4 \\ 2x+4y+6z=10 \end{array} \leftrightarrow \left(\begin{array}{ccc|c} 1 & 2 & 3 & 4 \\ 2 & 4 & 6 & 10 \end{array}\right) \rightsquigarrow \left(\begin{array}{ccc|c} 1 & 2 & 3 & 4 \\ \hline 0 & 0 & 0 & 2 \end{array}\right).$$

  The LGS obviously has no solution, since $0 \neq 2$. So we have $L = \emptyset$.
- We solve the linear system of equations:

$$\begin{array}{rl} x_1+\ x_2+2x_3=2 \\ 2x_1+2x_2-\ x_3=1 \\ 3x_1+4x_2+2x_3=2 \end{array} \leftrightarrow \left(\begin{array}{ccc|c} 1 & 1 & 2 & 2 \\ 2 & 2 & -1 & 1 \\ 3 & 4 & 2 & 2 \end{array}\right).$$

  Provide elementary row transformations:

$$\left(\begin{array}{ccc|c} 1 & 1 & 2 & 2 \\ 2 & 2 & -1 & 1 \\ 3 & 4 & 2 & 2 \end{array}\right) \rightsquigarrow \left(\begin{array}{ccc|c} 1 & 1 & 2 & 2 \\ 0 & 0 & -5 & -3 \\ 0 & 1 & -4 & -4 \end{array}\right) \rightsquigarrow \left(\begin{array}{ccc|c} 1 & 1 & 2 & 2 \\ 0 & 1 & -4 & -4 \\ 0 & 0 & -5 & -3 \end{array}\right).$$

  By substituting backwards we obtain the solution set $L = \{(12/5,\ -8/5,\ 3/5)\}$.
- Lastly, another LGS with infinitely many solutions:

$$\begin{array}{rl} 3x_1+\ x_2+\ x_3+0x_4=13 \\ x_1+2x_2+2x_3+2x_4=18 \\ x_1+\ x_2+\ x_3+\ x_4=10 \\ 3x_1+\ x_2+\ x_3+5x_4=18 \end{array} \leftrightarrow \left(\begin{array}{cccc|c} 3 & 1 & 1 & 0 & 13 \\ 1 & 2 & 2 & 2 & 18 \\ 1 & 1 & 1 & 1 & 10 \\ 3 & 1 & 1 & 5 & 18 \end{array}\right).$$

Elementary row transformations yield:

$$\rightsquigarrow \left(\begin{array}{cccc|c} 1 & 1 & 1 & 1 & 10 \\ 0 & -2 & -2 & -3 & -17 \\ 0 & 1 & 1 & 1 & 8 \\ 0 & 0 & 0 & 5 & 5 \end{array}\right) \rightsquigarrow \left(\begin{array}{cccc|c} 1 & 0 & 0 & 0 & 2 \\ 0 & 1 & 1 & 1 & 8 \\ 0 & 0 & 0 & -1 & -1 \\ 0 & 0 & 0 & 1 & 1 \end{array}\right) \rightsquigarrow \left(\begin{array}{cccc|c} 1 & 0 & 0 & 0 & 2 \\ 0 & 1 & 1 & 0 & 7 \\ 0 & 0 & 0 & 1 & 1 \\ 0 & 0 & 0 & 0 & 0 \end{array}\right).$$

Given the second line, we choose the third component $\ell_3 = \lambda$ and obtain as solution set

$$L = \{(2,\, 7-\lambda,\, \lambda,\, 1) \mid \lambda \in \mathbb{R}\} = \left\{ \begin{pmatrix} 2 \\ 7 \\ 0 \\ 1 \end{pmatrix} + \lambda \begin{pmatrix} 0 \\ -1 \\ 1 \\ 0 \end{pmatrix} \mid \lambda \in \mathbb{R} \right\}.$$

■

9.2 The Rank of a Matrix

Given a matrix

$$M = \begin{pmatrix} m_{11} & \dots & m_{1t} \\ \vdots & & \vdots \\ m_{s1} & \dots & m_{st} \end{pmatrix}$$

with s rows and t columns. We transform this matrix to row step form with elementary row transformations:

$$M \rightsquigarrow M' = \begin{pmatrix} * & * & * & * \\ & * & * & \vdots \\ & & * & * & * \\ 0 & & & 0 \end{pmatrix} \begin{matrix} \left.\vphantom{\begin{matrix} a \\ b \\ c \end{matrix}}\right\} r \text{ rows not equal } 0 \dots 0 \\ \left.\vphantom{a}\right\} s-r \text{ rows equal } 0 \dots 0 \end{matrix}$$

The number r of the non-zero rows of the row step form of M is called the **rank** of M, in characters $r = \text{rg}(M)$. This term can be used to give a concise solvability criterion for systems of linear equations and, in the case of solvability, also the variety of solutions:

Solvability Criterion for Linear Systems of Equations and the Number of Freely Selectable Parameters

Given is a linear system of equations with m equations and n unknowns with the coefficient matrix A and the extended coefficient matrix $(A \mid b)$.

- The system is solvable if and only if:

$$\operatorname{rg}(A) = \operatorname{rg}(A \mid b) .$$

- If the system is solvable, then:

$$\text{number of free parameters} = n - \operatorname{rg}(A) .$$

- If the system is solvable, then we have the following: The system is uniquely solvable if and only if $\operatorname{rg}(A) = n$.

Example 9.4

- We consider the linear system of equations $x + y = 1$. The extended coefficient matrix

$$\left(\begin{array}{cc|c} 1 & 1 & 1 \end{array}\right)$$

has rank 1. Since $n = 2$ we have $2 - 1 = 1$ parameter, which can be freely chosen. With the choice $\ell_2 = \lambda \in \mathbb{R}$ the solution set is

$$L = \{(1-\lambda,\ \lambda) \mid \lambda \in \mathbb{R}\} = \left\{ \begin{pmatrix} 1 \\ 0 \end{pmatrix} + \lambda \begin{pmatrix} 1 \\ -1 \end{pmatrix} \mid \lambda \in \mathbb{R} \right\} .$$

- If we add the equation $x - y = 1$ to the equation $x + y = 1$, we get the extended coefficient matrix:

$$\left(\begin{array}{cc|c} 1 & 1 & 1 \\ 1 & -1 & 1 \end{array}\right) \rightsquigarrow \left(\begin{array}{cc|c} 1 & 1 & 1 \\ 0 & -2 & 0 \end{array}\right)$$

with rank 2. Thus, $2 - 2 = 0$ parameters are free to be chosen; by the second equation, the LGS is thus uniquely solvable.

- We consider an LGS with the following extended coefficient matrix:

$$\left(\begin{array}{ccccc|c} 0&0&1&3&3&2\\ 1&2&1&4&3&3\\ 1&2&2&7&6&5\\ 2&4&1&5&3&4 \end{array}\right) \rightsquigarrow \left(\begin{array}{ccccc|c} 1&2&1&4&3&3\\ 0&0&1&3&3&2\\ 0&0&1&3&3&2\\ 0&0&-1&-3&-3&-2 \end{array}\right) \rightsquigarrow \left(\begin{array}{ccccc|c} 1&2&0&1&0&1\\ 0&0&1&3&3&2\\ 0&0&0&0&0&0\\ 0&0&0&0&0&0 \end{array}\right).$$

Obviously, we have $\mathrm{rg}(A) = \mathrm{rg}(A\,|\,b) = 2$ so the LGS is solvable. Because of $n = 5$, we have three freely chosen parameters; in particular, there are infinitely many solutions. We choose $\ell_2 = \nu,\ \ell_4 = \mu,\ \ell_5 = \lambda \in \mathbb{R}$. The solution set is then

$$L = \{(1 - 2\nu - \mu,\ \nu,\ 2 - 3\mu - 3\lambda,\ \mu,\ \lambda) \mid \nu, \mu, \lambda \in \mathbb{R}\}$$

$$= \left\{ \begin{pmatrix}1\\0\\2\\0\\0\end{pmatrix} + \lambda \begin{pmatrix}0\\0\\-3\\0\\1\end{pmatrix} + \mu \begin{pmatrix}-1\\0\\-3\\1\\0\end{pmatrix} + \nu \begin{pmatrix}-2\\1\\0\\0\\0\end{pmatrix} \mid \lambda, \mu, \nu \in \mathbb{R} \right\}. \quad \blacksquare$$

9.3 Homogeneous Linear Systems of Equations

A linear system of equations with $b = 0$, i.e., an LGS of the form $(A\,|\,0)$, is called **homogeneous**. If $(A\,|\,b)$ is an arbitrary LGS, then $(A\,|\,0)$ is called the **associated homogeneous linear system of equations**. We record some easily verifiable facts:

On the Structure of the Solution Set of a Homogeneous or Inhomogeneous Linear System of Equations

(a) A homogeneous LGS $(A\,|\,0)$ always has the so-called **trivial solution** $(0, \ldots, 0)$.

(b) If $(k_1, \ldots, k_n)$ and $(\ell_1, \ldots, \ell_n)$ solutions of a homogeneous LGS and $\lambda \in \mathbb{R}$ are arbitrary, then also sum and multiples

$$(k_1, \ldots, k_n) + (\ell_1, \ldots, \ell_n) = (k_1 + \ell_1, \ldots, k_n + \ell_n) \text{ and}$$

$$\lambda \cdot (\ell_1, \ldots, \ell_n) = (\lambda \ell_1, \ldots, \lambda \ell_n)$$

are also solutions of the homogeneous LGS.

(continued)

(c) If $(A \mid b)$ is a solvable LGS with the solution set L and $(A \mid 0)$ is the corresponding homogeneous LGS with the solution set L_h then with a special solution $x = (\ell_1, \dots, \ell_n)$ of the inhomogeneous system $(A \mid b)$:

$$L = x + L_h = \{x + y \mid y \in L_h\}.$$

For justifications of these statements, consider Exercise 9.4.

Consider again all examples in which a system of equations $(A \mid b)$ had infinitely many solutions. We have always given the solution sets in two different notations, e.g.

$$L = \{(2,\ 7-\lambda,\ \lambda,\ 1) \mid \lambda \in \mathbb{R}\} = \left\{ \underbrace{\begin{pmatrix} 2 \\ 7 \\ 0 \\ 1 \end{pmatrix}}_{=:v} + \lambda \underbrace{\begin{pmatrix} 0 \\ -1 \\ 1 \\ 0 \end{pmatrix}}_{=:w} \mid \lambda \in \mathbb{R} \right\}.$$

In this second representation of the solution set, one can clearly see the structure of the solution set: The constant *vector* v is a special solution, the set of *vectors* λw, $\lambda \in \mathbb{R}$ is the solution set of the homogeneous system $(A \mid 0)$.

MATLAB To solve systems of linear equations (and related problems) with MATLAB, consider the following commands:

- The input of a matrix A or a vector b is done exemplarily by:

  `A = [1 2 3 ; 4 5 6 ; 7 8 9]` or `b = [1 ; 2 ; 3]`.

  Entries in rows are separated by a space, rows are separated by a semicolon; `[A b]` produces the extended coefficient matrix.
- `rank(A)` and `rank([A b])` gives the rank of A and of $(A \mid b)$.
- `A\b` gives a (special) solution of $(A \mid b)$ if one exists. If no exact solution exists, then MATLAB may also output a result that is not the exact result, but can be interpreted as the best approximate solution (note the Remark in Sect. 19.3.2). You can tell whether the output solution is exact or an approximate solution by the **residual** `norm(b-A*x)`; if the residual is zero, the result is exact, but note that if the residual is nearly zero, you cannot conclude that the result is nearly exact!

- `null(A,'r')` specifies the vectors that generate the solution set of the homogeneous linear system of equations $(A \mid b)$ (note the Remark above about the solution set).
- `rref(A)` and `rref([A b])` gives a reduced row step form of A and of $(A \mid b)$.

Remark The rank is *sensitive*, i.e., small perturbations in the input matrix A can lead to large changes in $\mathrm{rg}(A)$. For example, in particularly bad cases, the rank of an $n \times n$*-matrix* can fluctuate between 1 and n by just *wiggling* the entries in A a bit. More robust and meaningful statements about rank are provided by the *singular value decomposition of* A; the number of nonzero singular values is the rank of the matrix, see Chap. 42.

9.4 Exercises

9.1 Solve the following systems of linear equations using the Gaussian elimination method:

(a) $$\begin{aligned} 3x_1 - 5x_2 &= 2 \\ -9x_1 + 15x_2 &= -6 \end{aligned}$$

(b) $$\begin{aligned} 2x_1 \phantom{{}+2x_2} + x_3 &= 3 \\ 4x_1 + 2x_2 + x_3 &= 3 \\ -2x_1 + 8x_2 + 2x_3 &= -8 \end{aligned}$$

(c) $$\begin{aligned} -2x_1 + x_2 + 3x_3 - 4x_4 &= -12 \\ -4x_1 + 3x_2 + 6x_3 - 5x_4 &= -21 \\ 2x_1 - 2x_2 - x_3 + 6x_4 &= 10 \\ -6x_1 + 6x_2 + 13x_3 + 10x_4 &= -22 \end{aligned}$$

(d) $$\begin{aligned} x_1 + x_2 + 2x_3 &= 3 \\ 2x_1 + 2x_2 + 5x_3 &= -4 \\ 5x_1 + 5x_2 + 11x_3 &= 6 \end{aligned}$$

(e) $$\begin{aligned} 3x_1 - 5x_2 + x_3 &= -1 \\ -3x_1 + 6x_2 \phantom{{}+x_3} &= 2 \\ 3x_1 - 4x_2 + 2x_3 &= 0 \end{aligned}$$

9.2 Find the solution set of the complex system of equations $(A \mid b)$ with

$$A = \begin{pmatrix} 1-\mathrm{i} & 2\,\mathrm{i} \\ \frac{5}{2+\mathrm{i}} & \frac{(2+4\,\mathrm{i})^2}{1-\mathrm{i}} \end{pmatrix} \quad \text{and} \quad b = \begin{pmatrix} 3+\mathrm{i} \\ 3+16\,\mathrm{i} \end{pmatrix}.$$

9.3 Given the system of linear equations $(A \mid b)$ with

$$A = \begin{pmatrix} 1 & 2 & -1 & -1 \\ 2 & 5 & 1 & 1 \\ 3 & 7 & 2 & 2 \\ -1 & 0 & 1 & \alpha \end{pmatrix} \quad \text{and} \quad b = \begin{pmatrix} 0 \\ 2 \\ \beta \\ 16 \end{pmatrix}.$$

(a) For what values of α and β does this system of equations have
 (i) a unique solution, (ii) no solution, (iii) infinitely many solutions?
(b) Give a solution representation for the case of infinitely many solutions.

9.4 Find the solution set of the following system of complex equations:

$$\begin{aligned} 2x_1 \qquad\quad + \mathrm{i}\,x_3 &= \mathrm{i} \\ x_1 - 3x_2 - \mathrm{i}\,x_3 &= 2\mathrm{i} \\ \mathrm{i}\,x_1 + x_2 + x_3 &= 1 + \mathrm{i} \end{aligned}$$

9.5 Give reasons for the statements in the note box in Sect. 9.3.

9.6 Determine the rank of the following matrices:

(a) $\begin{pmatrix} -2 & -3 \\ 4 & 6 \end{pmatrix}$

(b) $\begin{pmatrix} 3 & 3 & 3 \\ 2 & 2 & 2 \\ -1 & -1 & -1 \end{pmatrix}$

(c) $\begin{pmatrix} 1 & 2 & 3 \\ 2 & 3 & 4 \\ 3 & 4 & 5 \end{pmatrix}$

(d) $\begin{pmatrix} 0 & 0 & 1 \\ 2 & 1 & 0 \\ 1 & 2 & 3 \end{pmatrix}$

9.7 The saddle of a Bonanza bike is to be designed by a cubic polynomial, i.e. by a polynomial function $f(x) = a_3x^3 + a_2x^2 + a_1x + a_0$ where the contour of the saddle is defined by the function values in the range $x \in [-1, 1]$ shall be given. The a_i's are unknown variables resulting from the following design specifications:

(i) At the point $x = 0$ is the seat point. To prevent the rider from sliding $f'(0) = 0$ is demanded.
(ii) At the point $x = -1$, the saddle is said to merge with the bar (which is on the x-axis), therefore $f'(-1) = f(-1) = 0$ is demanded (the driver sits facing left, the backrest is on the right).
(iii) The bike is made to measure for the rider. The desired seat depth b gives the condition $f(0) = -b$.

Set up the linear system of equations that results for the variables $a_3, \ldots, a_0$ and solve it as a function of b. Which polynomial functions correspond to the solutions found?

10 Calculating with Matrices

We have already used matrices to solve systems of linear equations: Here, matrices have been a helpful means of representing systems of linear equations economically and clearly. Matrices also serve as a tool in other, manifold ways. This is one reason to consider matrices in their own right, and to clearly illustrate and practice all kinds of *manipulations* that are possible with them: We will add, multiply with scalars, multiply, exponentiate, transpose, and invert matrices. But everything in order.

10.1 Definition of Matrices and Some Special Matrices

A rectangular number scheme

$$A = \begin{pmatrix} a_{11} & \dots & a_{1n} \\ \vdots & & \vdots \\ a_{m1} & \dots & a_{mn} \end{pmatrix} = \left(a_{ij}\right)_{m,n} = \left(a_{ij}\right)$$

with m rows, n columns and elements $a_{ij} \in \mathbb{R}$ or $a_{ij} \in \mathbb{C}$ we call **real** or **complex** $m \times n$**-matrix**. It is

$$\mathbb{R}^{m \times n} = \left\{(a_{ij})_{m,n} \mid a_{ij} \in \mathbb{R}\right\} \text{ and } \mathbb{C}^{m \times n} = \left\{(a_{ij})_{m,n} \mid a_{ij} \in \mathbb{C}\right\}$$

the set of all real or complex $m \times n$-matrices.

The number a_{ij} in the i-th row and j-th column is called **component**, **entry** or **coefficient** at the **place** (i, j) of the matrix $A = (a_{ij})$. The number i is also called the **row index** and the number j the **column index**.

C. Karpfinger, *Calculus and Linear Algebra in Recipes*,
https://doi.org/10.1007/978-3-662-65458-3_10

Clearly, but worth mentioning: Two matrices $A = (a_{ij})$ and $B = (b_{ij})$ are equal if and only if A and B have the same number of rows and columns and $a_{ij} = b_{ij}$ for all i, j.

In order not to have to constantly mention $\mathbb{R}$ as well as $\mathbb{C}$, we write in the following simplified $\mathbb{K}$ and mean $\mathbb{R}$ or $\mathbb{C}$.

Some Special Matrices

- The $m \times 1$-matrices or $1 \times n$-matrices

$$s = \begin{pmatrix} s_1 \\ \vdots \\ s_m \end{pmatrix} \in \mathbb{K}^{m\times 1} = \mathbb{K}^m \text{ resp. } z = (z_1, \ldots, z_n) \in \mathbb{K}^{1\times n}$$

  are **columns** or **column vectors** or **rows** or **row vectors.**
- The matrix

$$0 = \begin{pmatrix} 0 \ldots 0 \\ \vdots \quad \vdots \\ 0 \ldots 0 \end{pmatrix} \in \mathbb{K}^{m\times n}$$

  is called $m \times n$-**zero matrix**.
- If $m = n$, i.e. $A \in \mathbb{K}^{n\times n}$ so A is called **square matrix**. The most important square matrices are:
  - **Diagonal matrices**:

$$D = \begin{pmatrix} \lambda_1 & & 0 \\ & \ddots & \\ 0 & & \lambda_n \end{pmatrix} = \operatorname{diag}(\lambda_1, \ldots, \lambda_n),$$

  - the $n \times n$-**unit matrix** $E_n = \operatorname{diag}(1, \ldots, 1)$ with the **standard unit vectors** as columns:

$$e_1 = \begin{pmatrix} 1 \\ 0 \\ \vdots \\ 0 \end{pmatrix}, \; e_2 = \begin{pmatrix} 0 \\ 1 \\ \vdots \\ 0 \end{pmatrix}, \ldots, e_n = \begin{pmatrix} 0 \\ 0 \\ \vdots \\ 1 \end{pmatrix} \in \mathbb{K}^n,$$

(continued)

– **upper** and **lower triangular matrices**

$$O = \begin{pmatrix} * & \dots & * \\ & \ddots & \vdots \\ 0 & & * \end{pmatrix} \text{ and } U = \begin{pmatrix} * & & 0 \\ \vdots & \ddots & \\ * & \dots & * \end{pmatrix}.$$

It is often useful to interpret an $m \times n$-matrix $A = (a_{ij}) \in \mathbb{K}^{m \times n}$ as a *collection* of n columns $s_1, \dots, s_n$ and of m rows $z_1, \dots, z_m$:

$$A = \begin{pmatrix} a_{11} & \dots & a_{1n} \\ \vdots & & \vdots \\ a_{m1} & \dots & a_{mn} \end{pmatrix} = (s_1, \dots, s_n) = \begin{pmatrix} z_1 \\ \vdots \\ z_m \end{pmatrix},$$

where

$$s_j = \begin{pmatrix} a_{1j} \\ \vdots \\ a_{mj} \end{pmatrix} \in \mathbb{K}^{m \times 1} \text{ and } z_i = (a_{i1}, \dots, a_{in}) \in \mathbb{K}^{1 \times n}$$

are the columns and rows of A.

10.2 Arithmetic Operations

Transpose, *adding* and *multiplication with scalars* of matrices is a simple story. The *multiplication* of matrices is a bit confusing at first sight. One creates a certain order here, if one first introduces the product row *times* column. If $z \in \mathbb{K}^{1 \times n}$ is a row vector and $s \in \mathbb{K}^{n \times 1}$ a column vector, we multiply row by column as follows:

$$z \cdot s = (a_1, \dots, a_n) \cdot \begin{pmatrix} b_1 \\ \vdots \\ b_n \end{pmatrix} = \sum_{i=1}^{n} a_i b_i .$$

With this product, the multiplication of matrix by matrix can be easily formulated (note the following definition).

Transpose, Addition, Multiplication with Scalars and Multiplication of Matrices

- **Transpose:** To $A = (a_{ij}) \in \mathbb{K}^{m \times n}$ denotes $A^\top = (a_{ji}) \in \mathbb{K}^{n \times m}$ the **to A transposed matrix** or the **transposed** of A:

$$A = \begin{pmatrix} a_{11} & \cdots & a_{1n} \\ \vdots & \ddots & \vdots \\ a_{m1} & \cdots & a_{mn} \end{pmatrix} \xrightarrow{\text{transposing}} A^\top = \begin{pmatrix} a_{11} & \cdots & a_{m1} \\ \vdots & \ddots & \vdots \\ a_{1n} & \cdots & a_{mn} \end{pmatrix}.$$

- **Addition:** For two matrices $A = (a_{ij})$, $B = (b_{ij}) \in \mathbb{K}^{m \times n}$ denotes $A + B = (a_{ij} + b_{ij}) \in \mathbb{K}^{m \times n}$ the **sum**:

$$\begin{pmatrix} a_{11} & \cdots & a_{1n} \\ \vdots & \ddots & \vdots \\ a_{m1} & \cdots & a_{mn} \end{pmatrix} + \begin{pmatrix} b_{11} & \cdots & b_{1n} \\ \vdots & \ddots & \vdots \\ b_{m1} & \cdots & b_{mn} \end{pmatrix} = \begin{pmatrix} a_{11} + b_{11} & \cdots & a_{1n} + b_{1n} \\ \vdots & \ddots & \vdots \\ a_{m1} + b_{m1} & \cdots & a_{mn} + b_{mn} \end{pmatrix}.$$

- **Multiplication with scalars:** For $\lambda \in \mathbb{K}$ and $A = (a_{ij}) \in \mathbb{K}^{m \times n}$ denotes $\lambda \cdot A = (\lambda\, a_{ij}) \in \mathbb{K}^{m \times n}$ the **multiplication by scalar of** A:

$$\lambda \cdot \begin{pmatrix} a_{11} & \cdots & a_{1n} \\ \vdots & \ddots & \vdots \\ a_{m1} & \cdots & a_{mn} \end{pmatrix} = \begin{pmatrix} \lambda\, a_{11} & \cdots & \lambda\, a_{1n} \\ \vdots & \ddots & \vdots \\ \lambda\, a_{m1} & \cdots & \lambda\, a_{mn} \end{pmatrix}.$$

- **Multiplication of matrices:** For two matrices $A = (a_{ij}) \in \mathbb{K}^{m \times n}$ and $B = (b_{jk}) \in \mathbb{K}^{n \times p}$ denotes $A \cdot B = (\sum_{j=1}^{n} a_{ij} b_{jk})_{ik} \in \mathbb{K}^{m \times p}$ the **product**:

$$A \cdot B = \begin{pmatrix} z_1 \\ \vdots \\ z_m \end{pmatrix} \cdot (s_1, \ldots, s_p) = \begin{pmatrix} z_1 \cdot s_1 & \ldots & z_1 \cdot s_p \\ \vdots & & \vdots \\ z_m \cdot s_1 & \ldots & z_m \cdot s_p \end{pmatrix}.$$

- We also set for every $A \in \mathbb{K}^{n \times n}$ and $k \in \mathbb{N}$:

$$A^k = \underbrace{A \cdot A \cdot \ldots \cdot A}_{k \text{ times}} \text{ and } A^0 = E_n .$$

Some remarks:

- When transposing, the ith row becomes the ith column or the ith column becomes the ith row—this is the best way to remember the transpose.
- A square matrix A is called **symmetric**, in case $A^\top = A$, and **skew-symmetric**, if $A^\top = -A$.
- When adding matrices, make sure that only matrices with the same number of rows and columns can be added.
- When multiplying $A \cdot B$ of matrices A and B, make sure that the number of columns of A must be equal to the number of rows of B, and the product has as many rows as A and as many columns as B. We memorably write this down as follows:

$$(m \times n) \cdot (n \times p) = m \times p\,.$$

- The product *matrix times column* gives a column, therefore we get the product *matrix times matrix* also in columns:

$$A \cdot s = \begin{pmatrix} z_1 \\ \vdots \\ z_m \end{pmatrix} \cdot s = \begin{pmatrix} z_1 \cdot s \\ \vdots \\ z_m \cdot s \end{pmatrix} \Rightarrow A \cdot B = A \cdot (s_1, \ldots, s_p) = (A \cdot s_1, \ldots, A \cdot s_p)\,.$$

- When exponentiating, the order does not matter, since multiplication of matrices is associative.
- In addition, we also have the *complex conjugation of* a matrix $A = (a_{ij})_{m,n} \in \mathbb{C}^{m \times n}$: $\overline{A} = (\overline{a}_{ij})_{m,n}$. For this conjugation we have $\overline{A \cdot B} = \overline{A} \cdot \overline{B}$.

For the multiplication $\lambda \cdot A$ with scalars as well as for the multiplication of matrices $A \cdot B$, from now on we omit the point for multiplication; thus we write more simply $\lambda\,A$ and $A\,B$.

Example 10.1

- Transpose turns rows into columns and vice versa:

$$(a_1, \ldots, a_n)^\top = \begin{pmatrix} a_1 \\ \vdots \\ a_n \end{pmatrix}, \begin{pmatrix} a_1 \\ \vdots \\ a_n \end{pmatrix}^\top = (a_1, \ldots, a_n) \text{ and } \begin{pmatrix} 1 & 2 & 3 \\ 4 & 5 & 6 \end{pmatrix}^\top = \begin{pmatrix} 1 & 4 \\ 2 & 5 \\ 3 & 6 \end{pmatrix}.$$

- For $a = \begin{pmatrix} 1 \\ 2 \\ 3 \end{pmatrix}$ and $b = \begin{pmatrix} 3 \\ -2 \\ 2 \end{pmatrix}$ we have

$$a^\top b = (1,\ 2,\ 3) \begin{pmatrix} 3 \\ -2 \\ 2 \end{pmatrix} = 5 \text{ and } a\, b^\top = \begin{pmatrix} 1 \\ 2 \\ 3 \end{pmatrix} (3,\ -2,\ 2) = \begin{pmatrix} 3 & -2 & 2 \\ 6 & -4 & 4 \\ 9 & -6 & 6 \end{pmatrix}.$$

- The matrix $A = \begin{pmatrix} 1 & 2 & 3 \\ 2 & 4 & 5 \\ 3 & 5 & 6 \end{pmatrix}$ is symmetrical and $B = \begin{pmatrix} 0 & 2 & 3 \\ -2 & 0 & 5 \\ -3 & -5 & 0 \end{pmatrix}$ is skew symmetrical.
- Matrices are added component-wise:

$$\begin{pmatrix} 2 & 3 & 9 \\ 5 & 7 & 8 \end{pmatrix} + \begin{pmatrix} 1 & 2 & -1 \\ 2 & 3 & 1 \end{pmatrix} = \begin{pmatrix} 3 & 5 & 8 \\ 7 & 10 & 9 \end{pmatrix} \text{ and } \begin{pmatrix} 1 \\ 2 \end{pmatrix} + \begin{pmatrix} -1 \\ -2 \end{pmatrix} = \begin{pmatrix} 0 \\ 0 \end{pmatrix}.$$

- Multiplication by scalars is also component-wise:

$$2 \begin{pmatrix} 1 & 2 & -1 \\ 2 & 3 & 1 \end{pmatrix} = \begin{pmatrix} 2 & 4 & -2 \\ 4 & 6 & 2 \end{pmatrix} \text{ and } (-2)\,(1,\ 2,\ 3) = (-2,\ -4,\ -6)\,.$$

- When multiplying matrices, the number of columns of the first matrix must be equal to the number of rows of the second matrix:

$$\begin{pmatrix} 2 & 3 & 1 \\ 3 & 5 & 0 \end{pmatrix} \begin{pmatrix} 1 & 2 & 3 & 1 \\ 1 & 0 & 0 & 1 \\ 2 & 5 & 0 & 4 \end{pmatrix} = \begin{pmatrix} 7 & 9 & 6 & 9 \\ 8 & 6 & 9 & 8 \end{pmatrix}.$$

- Multiplication by a diagonal matrix from the left causes the rows to multiply, while multiplication from the right causes the columns to multiply:

$$\begin{pmatrix} 1 & 0 & 0 \\ 0 & 2 & 0 \\ 0 & 0 & 3 \end{pmatrix} \begin{pmatrix} 1 & 2 & 3 \\ 4 & 5 & 6 \\ 7 & 8 & 9 \end{pmatrix} = \begin{pmatrix} 1 & 2 & 3 \\ 8 & 10 & 12 \\ 21 & 24 & 27 \end{pmatrix},\ \begin{pmatrix} 1 & 2 & 3 \\ 4 & 5 & 6 \\ 7 & 8 & 9 \end{pmatrix} \begin{pmatrix} 1 & 0 & 0 \\ 0 & 2 & 0 \\ 0 & 0 & 3 \end{pmatrix} = \begin{pmatrix} 1 & 4 & 9 \\ 4 & 10 & 18 \\ 7 & 16 & 27 \end{pmatrix}.$$

In particular, matrix multiplication is not commutative.

- We consider

$$A = \begin{pmatrix} 1 & \mathrm{i} \\ \mathrm{i} & 1 \end{pmatrix} \text{ and } B = \begin{pmatrix} 2+\mathrm{i} & 1 \\ 0 & 1+\mathrm{i} \end{pmatrix}.$$

We have $(A+B)^2$:

$$\underbrace{\begin{pmatrix} 3+\mathrm{i} & 1+\mathrm{i} \\ \mathrm{i} & 2+\mathrm{i} \end{pmatrix}}_{=A+B} \underbrace{\begin{pmatrix} 3+\mathrm{i} & 1+\mathrm{i} \\ \mathrm{i} & 2+\mathrm{i} \end{pmatrix}}_{=A+B} = \begin{pmatrix} 7+7\mathrm{i} & 3+7\mathrm{i} \\ -2+5\mathrm{i} & 2+5\mathrm{i} \end{pmatrix}.$$

Now we calculate $A^2 + 2AB + B^2$:

$$\underbrace{\begin{pmatrix} 0 & 2\mathrm{i} \\ 2\mathrm{i} & 0 \end{pmatrix}}_{=A^2} + \underbrace{\begin{pmatrix} 4+2\mathrm{i} & 2\mathrm{i} \\ -2+4\mathrm{i} & 2+4\mathrm{i} \end{pmatrix}}_{=2AB} + \underbrace{\begin{pmatrix} 3+4\mathrm{i} & 3+2\mathrm{i} \\ 0 & 2\mathrm{i} \end{pmatrix}}_{=B^2} = \begin{pmatrix} 7+6\mathrm{i} & 3+6\mathrm{i} \\ -2+6\mathrm{i} & 2+6\mathrm{i} \end{pmatrix}.$$

Note: $(A+B)^2 \neq A^2 + 2AB + B^2$.

- The linear system of equations

$$\begin{aligned} a_{11}x_1 + \cdots + a_{1n}x_n &= b_1 \\ \vdots \qquad\quad & \quad\vdots \\ a_{m1}x_1 + \cdots + a_{mn}x_n &= b_m \end{aligned}$$

can be expressed by the coefficient matrix A, the column vector x and the column vector b,

$$A = \begin{pmatrix} a_{11} & \dots & a_{1n} \\ \vdots & \ddots & \vdots \\ a_{m1} & \dots & a_{mn} \end{pmatrix}, \quad x = \begin{pmatrix} x_1 \\ \vdots \\ x_n \end{pmatrix}, \quad b = \begin{pmatrix} b_1 \\ \vdots \\ b_m \end{pmatrix},$$

write briefly as $Ax = b$. ■

We highlight other important types of matrices, viz. *elementary matrices*; the multiplication of an *elementary matrix* from the left onto a matrix A causes an elementary row transformation:

Elementary Matrices

We consider the following **elementary matrices** P_{kl}, $D_k(\lambda)$, and $N_{kl}(\lambda)$:

$$\underbrace{\overset{\begin{matrix} k & l \\ \downarrow & \downarrow \end{matrix}}{\begin{pmatrix} 1 & & & & & \\ & \ddots & & & & \\ & & 0 & & 1 & \\ & & & \ddots & & \\ & & 1 & & 0 & \\ & & & & & \ddots & \\ & & & & & & 1 \end{pmatrix}}}_{=P_{kl}}, \quad \underbrace{\overset{\begin{matrix} k \\ \downarrow \end{matrix}}{\begin{pmatrix} 1 & & & & & \\ & \ddots & & & & \\ & & 1 & & & \\ & & & \lambda & & \\ & & & & 1 & \\ & & & & & \ddots & \\ & & & & & & 1 \end{pmatrix}}}_{=D_k(\lambda)}, \quad \underbrace{\overset{\begin{matrix} k & l \\ \downarrow & \downarrow \end{matrix}}{\begin{pmatrix} 1 & & & & & \\ & \ddots & & & & \\ & & 1 & & \lambda & \\ & & & \ddots & & \\ & & & & 1 & \\ & & & & & \ddots & \\ & & & & & & 1 \end{pmatrix}}}_{=N_{kl}(\lambda)}.$$

The multiplication of an elementary matrix E from the left to a matrix A causes

- the interchange of k-th and l-th row, if $E = P_{kl}$,
- the multiplication of the k-th row by λ, if. $E = D_k(\lambda)$,
- is the addition of the λ-fold of the l-th row to the k-th row, if $E = N_{kl}(\lambda)$,

$$P_{kl}\begin{pmatrix} \vdots \\ z_k \\ \vdots \\ z_l \\ \vdots \end{pmatrix} = \begin{pmatrix} \vdots \\ z_l \\ \vdots \\ z_k \\ \vdots \end{pmatrix}, \quad D_k(\lambda)\begin{pmatrix} \vdots \\ z_k \\ \vdots \\ z_l \\ \vdots \end{pmatrix} = \begin{pmatrix} \vdots \\ \lambda z_k \\ \vdots \\ z_l \\ \vdots \end{pmatrix}, \quad N_{kl}(\lambda)\begin{pmatrix} \vdots \\ z_k \\ \vdots \\ z_l \\ \vdots \end{pmatrix} = \begin{pmatrix} \vdots \\ z_k + \lambda z_l \\ \vdots \\ z_l \\ \vdots \end{pmatrix}.$$

There are many rules for transposing, adding, multiplying and multiplying matrices. We give these calculation rules right after the following section on inverting.

10.3 Inverting Matrices

For some quadratic matrices $A \in \mathbb{K}^{n \times n}$ there is a matrix B with $A\,B = E_n = B\,A$. In this case we write $B = A^{-1}$ and call A^{-1} the **inverse** of A and denote A as **invertible**; we get

A^{-1} as a solution for X of the matrix equation

$$A\,X = E_n\,.$$

Here we determine the matrix $X = A^{-1}$ column by column by solving n systems of equations: With $E_n = (e_1, \ldots, e_n)$ and $X = (s_1, \ldots, s_n)$ the equation $A\,X = E_n$ is in detail:

$$A\,X = (A\,s_1, \ldots, A\,s_n) = (e_1, \ldots, e_n)\,, \quad \text{so } \; A\,s_i = e_i \;\text{ for } i = 1, \ldots, n\,.$$

This n systems of equations we can solve simultaneously:

Recipe: Inverting a Matrix

A matrix $A \in \mathbb{K}^{n\times n}$ is invertible if and only if $\mathrm{rg}(A) = n$.

If $A \in \mathbb{K}^{n\times n}$ is invertible, then we get the inverse A^{-1} by simultaneously solving n systems of equations:

(1) Write down the extended coefficient matrix $(A \mid E_n)$.

(2) Perform elementary row transformations until the reduced row level form E_n is reached:

$$(A \mid E_n) \rightsquigarrow \ldots \rightsquigarrow (E_n \mid B)\,.$$

(3) We then have $B = A^{-1}$.

Remark One can also apply this recipe to a non-invertible matrix A. Since in this case we have $\mathrm{rg}(A) < n$, one will find that it is not possible to make the coefficient matrix to the unit matrix by row transformations, step (2) cannot be completed; i.e., the matrix A is not invertible.

Example 10.2

- For the inverse of the matrix $A = \begin{pmatrix} 2 & 1 \\ 1 & 1 \end{pmatrix}$ we get:

$$\left(\begin{array}{cc|cc} 2 & 1 & 1 & 0 \\ 1 & 1 & 0 & 1 \end{array}\right) \rightsquigarrow \left(\begin{array}{cc|cc} 1 & 1 & 0 & 1 \\ 0 & -1 & 1 & -2 \end{array}\right) \rightsquigarrow \left(\begin{array}{cc|cc} 1 & 0 & 1 & -1 \\ 0 & 1 & -1 & 2 \end{array}\right).$$

Thus $A^{-1} = \begin{pmatrix} 1 & -1 \\ -1 & 2 \end{pmatrix}$.

- For the inverse of the matrix $A = \begin{pmatrix} 6 & 8 & 3 \\ 4 & 7 & 3 \\ 1 & 2 & 1 \end{pmatrix}$ we get:

$$\left(\begin{array}{ccc|ccc} 6 & 8 & 3 & 1 & 0 & 0 \\ 4 & 7 & 3 & 0 & 1 & 0 \\ 1 & 2 & 1 & 0 & 0 & 1 \end{array}\right) \rightsquigarrow \left(\begin{array}{ccc|ccc} 1 & 2 & 1 & 0 & 0 & 1 \\ 0 & -1 & -1 & 0 & 1 & -4 \\ 0 & -4 & -3 & 1 & 0 & -6 \end{array}\right)$$

$$\rightsquigarrow \left(\begin{array}{ccc|ccc} 1 & 0 & -1 & 0 & 2 & -7 \\ 0 & 1 & 1 & 0 & -1 & 4 \\ 0 & 0 & 1 & 1 & -4 & 10 \end{array}\right) \rightsquigarrow \left(\begin{array}{ccc|ccc} 1 & 0 & 0 & 1 & -2 & 3 \\ 0 & 1 & 0 & -1 & 3 & -6 \\ 0 & 0 & 1 & 1 & -4 & 10 \end{array}\right).$$

Thus $A^{-1} = \begin{pmatrix} 1 & -2 & 3 \\ -1 & 3 & -6 \\ 1 & -4 & 10 \end{pmatrix}$.

- We try to find the inverse of the matrix $A = \begin{pmatrix} 1 & 2 & 0 & 4 \\ 1 & 1 & 0 & 2 \\ 0 & 2 & 1 & 0 \\ 2 & 5 & 1 & 6 \end{pmatrix}$:

$$\left(\begin{array}{cccc|cccc} 1 & 2 & 0 & 4 & 1 & 0 & 0 & 0 \\ 1 & 1 & 0 & 2 & 0 & 1 & 0 & 0 \\ 0 & 2 & 1 & 0 & 0 & 0 & 1 & 0 \\ 2 & 5 & 1 & 6 & 0 & 0 & 0 & 1 \end{array}\right) \rightsquigarrow \left(\begin{array}{cccc|cccc} 1 & 2 & 0 & 4 & 1 & 0 & 0 & 0 \\ 0 & -1 & 0 & -2 & -1 & 1 & 0 & 0 \\ 0 & 0 & 1 & -4 & -2 & 2 & 1 & 0 \\ 0 & 1 & 1 & -2 & -2 & 0 & 0 & 1 \end{array}\right)$$

$$\rightsquigarrow \left(\begin{array}{cccc|cccc} 1 & 0 & 0 & 0 & -1 & 2 & 0 & 0 \\ 0 & 1 & 0 & 2 & 1 & -1 & 0 & 0 \\ 0 & 0 & 1 & -4 & -2 & 2 & 1 & 0 \\ 0 & 0 & 1 & -4 & -3 & 1 & 0 & 1 \end{array}\right) \rightsquigarrow \left(\begin{array}{cccc|cccc} 1 & 0 & 0 & 0 & -1 & 2 & 0 & 0 \\ 0 & 1 & 0 & 2 & 1 & -1 & 0 & 0 \\ 0 & 0 & 1 & -4 & -2 & 2 & 1 & 0 \\ 0 & 0 & 0 & 0 & * & * & * & * \end{array}\right).$$

Due to the zero line we have $\text{rg}(A) < 4 = n$, so that A is not invertible. ■

10.4 Calculation Rules

We summarise all calculation rules for calculating with matrices in a clear way and also indicate which rules you might expect but actually do not apply when calculating with matrices.

Calculation Rules for Transposing, Adding, Multiplying and Inverting Matrices

For $A, B, C \in \mathbb{K}^{m\times n}$ and $\lambda, \mu \in \mathbb{K}$ we have:

- The **vector space axioms**:
  - $A + B = B + A$,
  - $(A + B) + C = A + (B + C)$,
  - $A + 0 = A$,
  - $A + (-A) = 0$,
  - $(\lambda\mu)A = \lambda(\mu A)$,
  - $1 \cdot A = A$,
  - $(\lambda + \mu)A = \lambda A + \mu A$,
  - $\lambda(A + B) = \lambda A + \lambda B$.
- Transpose rules:
  - $(A + B)^\top = A^\top + B^\top$,
  - $(\lambda A)^\top = \lambda A^\top$,
  - $(A^\top)^\top = A$.

If $A, B, C \in \mathbb{K}^{n\times n}$ square matrices, then we have:

- Multiplication rules:
  - $(AB)C = A(BC)$,
  - $A(B + C) = AB + AC$,
  - $(A + B)C = AC + BC$,
  - $E_n \cdot A = A = A \cdot E_n$,
  - $(AB)^\top = B^\top A^\top$.
- The rules for inverting: If A and B are invertible, then so is A^{-1} and $A\,B$, and we have:
  - $(A^{-1})^{-1} = A$,
  - $(AB)^{-1} = B^{-1}A^{-1}$.
- The two important rules for the inverses of 2×2 or diagonal matrices:
  - $A = \begin{pmatrix} a & b \\ c & d \end{pmatrix} \Rightarrow A^{-1} = \frac{1}{ad-bc}\begin{pmatrix} d & -b \\ -c & a \end{pmatrix}$, if $ad - bc \neq 0$,
  - $D = \operatorname{diag}(\lambda_1, \ldots, \lambda_n) \Rightarrow D^{-1} = \operatorname{diag}(\lambda_1^{-1}, \ldots, \lambda_n^{-1})$, if $\lambda_1, \ldots, \lambda_n \neq 0$.

The stated rules of arithmetic are easy to prove (with the exception of the associativity of multiplication). However, we dispense with these proofs. It takes a bit of getting used to the fact that some rules one expects we don't have:

- Multiplication of $n \times n$-matrices is **noncommutative** for $n \geq 2$, that is,

$$\text{there are matrices } A,\ B \in \mathbb{K}^{n\times n} \text{ with } A\,B \neq B\,A\,,$$

e.g. in the case $n = 2$:

$$\begin{pmatrix}1&1\\0&0\end{pmatrix}\begin{pmatrix}0&0\\0&1\end{pmatrix} = \begin{pmatrix}0&1\\0&0\end{pmatrix} \neq \begin{pmatrix}0&0\\0&0\end{pmatrix} = \begin{pmatrix}0&0\\0&1\end{pmatrix}\begin{pmatrix}1&1\\0&0\end{pmatrix}.$$

- When multiplying matrices, it might happen $AB = 0$ even though $A \neq 0$ and $B \neq 0$, e.g.

$$\begin{pmatrix}1&0\\0&0\end{pmatrix}\begin{pmatrix}0&0\\1&0\end{pmatrix} = \begin{pmatrix}0&0\\0&0\end{pmatrix}.$$

- *Reducing* is not permissible for matrices: From $AC = BC$ does not necessarily follow $A = B$, e.g.:

$$\begin{pmatrix}1&0\\0&0\end{pmatrix}\begin{pmatrix}0&0\\1&1\end{pmatrix} = \begin{pmatrix}0&0\\1&0\end{pmatrix}\begin{pmatrix}0&0\\1&1\end{pmatrix} = \begin{pmatrix}0&0\\0&0\end{pmatrix}, \text{ but } \begin{pmatrix}1&0\\0&0\end{pmatrix} \neq \begin{pmatrix}0&0\\1&0\end{pmatrix}.$$

Finally, we give how to use MATLAB performs the operations discussed.

MATLAB The operations discussed are obtained with MATLAB as follows:

- With `A'` one obtains the transpose of A.
- With `A*B`, `r*A`, `A+B` and `A^k` we get the product, the multiple with scalars, the sum and the power of A.
- `inv(A)` gives the inverse of A.
- `eye(n)` and `zeros(m,n)` give the $n \times n$ *unit matrix* E_n and the $m \times n$-*zero matrix*.
- `diag([1;2;3])` outputs the 3×3 diagonal matrix with numbers 1, 2, 3 on the diagonal.

10.5 Exercises

10.1 For the matrix $A = \begin{pmatrix}1&2&3\\2&3&4\\3&4&5\end{pmatrix}$ find the expression $A^0 + A + \frac{1}{2}A^2 + \frac{1}{6}A^3$.

10.2 Calculate $\overline{B}^\top B$ and $\overline{B}^\top A\,B$ with the matrix

$$A = \begin{pmatrix} 2 & \mathrm{i} & 0 \\ -\mathrm{i} & 2 & 0 \\ 0 & 0 & 2 \end{pmatrix} \quad \text{and the matrix} \quad B = \begin{pmatrix} 0 & 1/\sqrt{2} & 1/\sqrt{2} \\ 0 & \mathrm{i}/\sqrt{2} & -\mathrm{i}/\sqrt{2} \\ 1 & 0 & 0 \end{pmatrix}.$$

10.3 Form—if possible—with the matrices

$$A = \begin{pmatrix} -2 & 3 \\ 4 & 1 \\ -1 & 5 \end{pmatrix}, \quad B = \begin{pmatrix} 3 & 0 \\ 1 & -7 \end{pmatrix} \quad \text{and} \quad C = \begin{pmatrix} 1 & 4 \\ 0 & -2 \\ 3 & 5 \end{pmatrix}$$

and the vectors

$$x = (1, 0, -4)^\top, \ y = (8, -5)^\top \ \text{and} \ z = (3, 2)^\top$$

the expressions

$$A + C,\ 2B,\ A(y + z),\ C(-4z),\ (A + C)y,\ AB,\ BC,\ AC^\top,\ x^\top A,\ y^\top z,\ yz^\top.$$

10.4 Show:

(a) For every matrix A the matrix $A^\top A$ is symmetric.
(b) For any square matrix A the matrix $A + A^\top$ is symmetric and $A - A^\top$ skew symmetric.
(c) The product of two symmetric matrices A and B is symmetric if and only if $AB = BA$.

10.5 Is the product of quadratic upper and lower triangular matrices again an upper and lower triangular matrix, respectively?

10.6 Given are the matrices

$$A = \begin{pmatrix} 1 & 1 & 1 \\ 2 & 0 & 2 \\ 1 & -2 & 3 \end{pmatrix} \quad \text{and} \quad B = \begin{pmatrix} 0 & 1 & -2 \\ 1 & 1 & 0 \\ 2 & 1 & 1 \end{pmatrix}.$$

(a) Calculate A^{-1}, B^{-1}, $(AB)^{-1}$ and $(2A)^{-1}$.
(b) Is $A + B$ invertible?

10.7 Given are a $n \in \mathbb{N}$ and a matrix $A \in \mathbb{R}^{n\times n}$.

(a) Show by induction according to $m \in \mathbb{N}$:

$$(E_n - A)(E_n + A + A^2 + \cdots + A^{m-1}) = E_n - A^m .$$

(b) Follow from part (a): If $A^m = 0$ for a $m \in \mathbb{N}$, then $E_n - A$ is invertible.

10.8 Determine the solution $X \in \mathbb{R}^{3\times 3}$ of the matrix equation $AX = B$ with

$$A = \begin{pmatrix} 1 & 1 & 1 \\ 0 & 2 & -4 \\ 1 & 0 & 2 \end{pmatrix} \quad \text{and} \quad B = \begin{pmatrix} 1 & 2 & 3 \\ 2 & 3 & 1 \\ 3 & 1 & 2 \end{pmatrix}.$$

10.9

(a) Is the inverse of an invertible symmetric matrix symmetric again?
(b) Does the invertibility of a matrix A always imply the invertibility of $A^\top$?
(c) Is the sum of invertible matrices always invertible?
(d) Is the product of invertible matrices always invertible?

10.10 Invert the following matrices, or show that no inverse exists. In each case, give the rank of the matrix.

(a) $A = \begin{pmatrix} 1 & 4 & -1 \\ -1 & -3 & 5 \\ 5 & 19 & -8 \end{pmatrix}$,

(b) $B = \begin{pmatrix} 1 & 1 & 1 \\ 2021 & -2 & 3 \end{pmatrix}$,

(c) $C = \begin{pmatrix} 0 & 1 & -2 \\ 1 & 1 & 0 \\ 2 & 1 & 1 \end{pmatrix}$,

(d) $D = A + B$,
(e) $E = B + C$,
(f) $F = AB$,
(g) $G = A^\top$.

10.11

(a) Find a 3×3 matrix $A \neq E_3, 0$ with the property $A^2 = A$.
(b) Let $A \in \mathbb{R}^{n\times n}$ where $A^2 = A$. Show that A is invertible if and only if A is the unit matrix $E_n \in \mathbb{R}^{n\times n}$.
(c) Let be $A, B \in \mathbb{R}^{n\times n}$ with $B \neq 0$ and $AB = 0$. Can the matrix A then be invertible?

11 *LR*-Decomposition of a Matrix

We consider the problem to determine to an invertible matrix $A \in \mathbb{R}^{n\times n}$ and a vector $b \in \mathbb{R}^n$ a vector $x \in \mathbb{R}^n$ with $A\,x = b$; in short: we solve the linear system of equations $Ax = b$. Formally, the solution is obtained by $x = A^{-1}b$.

But the calculation of A^{-1} is costly for a large matrix A. Cramer's rule (see recipe in Sect. 12.3 is unsuitable from a numerical point of view for calculating the solution x. In fact, the Gaussian elimination method, which we also discuss in Chap. 9 for the manual solution of an LGS, provides a decomposition of the coefficient matrix A, with the help of which it is possible to calculate a system of equations of the form $A\,x = b$ with invertible A. This so-called *L R-decomposition* is also numerically benign. Systems of equations with up to about 10,000 rows and unknowns can be solved advantageously in this way. For larger systems of equations iterative solution methods are to be preferred (see Chap. 71).

11.1 Motivation

To obtain the mentioned decomposition of a matrix A to solve the LGS $Ax = b$, we first consider the case where A is an upper or lower triangular matrix. The systems of equations

$$\begin{pmatrix} 7 & 0 & 0 \\ -3 & 2 & 0 \\ 4 & -5 & 3 \end{pmatrix} \begin{pmatrix} x_1 \\ x_2 \\ x_3 \end{pmatrix} = \begin{pmatrix} 7 \\ -1 \\ 5 \end{pmatrix} \quad \text{and} \quad \begin{pmatrix} 4 & -5 & 3 \\ 0 & 2 & -3 \\ 0 & 0 & 7 \end{pmatrix} \begin{pmatrix} y_1 \\ y_2 \\ y_3 \end{pmatrix} = \begin{pmatrix} 7 \\ -1 \\ 5 \end{pmatrix}$$

can be solved especially easily because of the triangular shape, you get

$$x_1 = 1\,,\ x_2 = 1\,,\ x_3 = 2 \ \text{ and } \ y_1 = 27/14\,,\ y_2 = 4/7\,,\ y_3 = 5/7\,,$$

C. Karpfinger, *Calculus and Linear Algebra in Recipes*,
https://doi.org/10.1007/978-3-662-65458-3_11

In the first case we speak of a **forward substitution**, in the second case of a **backward substitution**.

This *simple* solvability of systems with triangular matrices can now be used to solve general systems in a simple way: Suppose the matrix $A \in \mathbb{R}^{n \times n}$ can be written as a product of a left lower triangular matrix $L \in \mathbb{R}^{n \times n}$ and a right upper triangular matrix $R \in \mathbb{R}^{n \times n}$ i.e.

$$A = L\,R \text{ where } L = \begin{pmatrix} * & 0 & \dots & 0 \\ * & \ddots & \ddots & \vdots \\ \vdots & \ddots & \ddots & 0 \\ * & \dots & * & * \end{pmatrix} \text{ and } R = \begin{pmatrix} * & * & \dots & * \\ 0 & \ddots & \ddots & \vdots \\ \vdots & \ddots & \ddots & * \\ 0 & \dots & 0 & * \end{pmatrix},$$

we obtain the solution x of the LGS $A\,x = b$ by forward substitution followed by backward substitution:

Recipe: Solve an LGS where $A = LR$

The LGS to solve is $Ax = b$, where $A = LR$ with a left lower triangular matrix L and a right upper triangular matrix R:

(1) First solve the LGS $L\,y = b$ according to y by forward substitution.
(2) Then solve the LGS $R\,x = y$ to x by backward substitution.

Because of $A\,x = b \;\Leftrightarrow\; L\,(R\,x) = b$ the x from (2) is the solution we are looking for.

Although there are two systems of equations to solve with this method, each of them is solvable directly, i.e., without further row transformations. The forward or backward iteration costs in each case about n^2 flops (see Exercise 11.5).

We describe how to obtain a $L\,R$-decomposition of the quadratic coefficient matrix A.

11.2 The $L\,R$-Decomposition: Simplified Variant

You can transform a square matrix A with elementary row transformations to an upper triangular form R; the row transformations correspond to multiplications of elementary matrices from the left (see box in Sect. 10.2):

$$\underbrace{\begin{pmatrix} * & * & * \\ * & * & * \\ * & * & * \end{pmatrix}}_{=A} \xrightarrow{L_1} \underbrace{\begin{pmatrix} * & * & * \\ 0 & * & * \\ 0 & * & * \end{pmatrix}}_{=L_1 A} \xrightarrow{L_2} \underbrace{\begin{pmatrix} * & * & * \\ 0 & * & * \\ 0 & 0 & * \end{pmatrix}}_{=L_2 L_1 A = R}.$$

With $L = L_1^{-1} L_2^{-1}$ we obtain

$$A = L_1^{-1} L_2^{-1} R = L\,R\,.$$

Example 11.1 We decompose the matrix $A = (a_{ij}) = \begin{pmatrix} 2 & 1 & 1 \\ 4 & 3 & 3 \\ 8 & 7 & 9 \end{pmatrix}$; we have:

$$\underbrace{\begin{pmatrix} 1 & 0 & 0 \\ \mathbf{-2} & 1 & 0 \\ \mathbf{-4} & 0 & 1 \end{pmatrix}}_{L_1} \underbrace{\begin{pmatrix} 2 & 1 & 1 \\ 4 & 3 & 3 \\ 8 & 7 & 9 \end{pmatrix}}_{A} = \underbrace{\begin{pmatrix} 2 & 1 & 1 \\ 0 & 1 & 1 \\ 0 & 3 & 5 \end{pmatrix}}_{L_1 A},\ \underbrace{\begin{pmatrix} 1 & 0 & 0 \\ 0 & 1 & 0 \\ 0 & \mathbf{-3} & 1 \end{pmatrix}}_{L_2} \underbrace{\begin{pmatrix} 2 & 1 & 1 \\ 0 & 1 & 1 \\ 0 & 3 & 5 \end{pmatrix}}_{L_1 A} = \underbrace{\begin{pmatrix} 2 & 1 & 1 \\ 0 & 1 & 1 \\ 0 & 0 & 2 \end{pmatrix}}_{L_1 L_2 A = R}.$$

Because

$$L_1^{-1} = \begin{pmatrix} 1 & 0 & 0 \\ 2 & 1 & 0 \\ 4 & 0 & 1 \end{pmatrix} \text{ and } L_2^{-1} = \begin{pmatrix} 1 & 0 & 0 \\ 0 & 1 & 0 \\ 0 & 3 & 1 \end{pmatrix} \text{ we have } L = \begin{pmatrix} 1 & 0 & 0 \\ 2 & 1 & 0 \\ 4 & 3 & 1 \end{pmatrix}.$$

Thus we obtain

$$A = \underbrace{\begin{pmatrix} 1 & 0 & 0 \\ \mathbf{2} & 1 & 0 \\ \mathbf{4} & \mathbf{3} & 1 \end{pmatrix}}_{L} \underbrace{\begin{pmatrix} 2 & 1 & 1 \\ 0 & 1 & 1 \\ 0 & 0 & 2 \end{pmatrix}}_{R}.$$

■

Note how the matrix L results from the row transformations: Below the diagonals of L, the negative values of the elimination factors $\ell_{ik} = -a_{ik}/a_{kk}$, i.e. the numbers $\frac{a_{21}}{a_{11}} = 2$, $\frac{a_{31}}{a_{11}} = 4$ and $\frac{a_{32}}{a_{22}} = 3$—we have marked these numbers in bold in the example. The diagonal of L has only ones as entries, and above the diagonals L has only zeros as entries. This observation suggests a more economical representation of the calculation: We note at the places of the resulting zeros the corresponding entries of L and obtain L as the strict lower triangular matrix by completing the diagonals with ones. We solve the above example again, thus performing elementary row transformations and noting the negative of the elimination factors instead of the resulting zeros, which we separate from the upper part of the matrix by a line for the sake of clarity:

$$\begin{pmatrix} 2 & 1 & 1 \\ 4 & 3 & 3 \\ 8 & 7 & 9 \end{pmatrix} \rightarrow \begin{pmatrix} 2 & 1 & 1 \\ \mathbf{2} & 1 & 1 \\ \mathbf{4} & 3 & 5 \end{pmatrix} \rightarrow \begin{pmatrix} 2 & 1 & 1 \\ \mathbf{2} & 1 & 1 \\ \mathbf{4} & \mathbf{3} & 2 \end{pmatrix}.$$

From this result we get: $L = \begin{pmatrix} 1 & 0 & 0 \\ 2 & 1 & 0 \\ 4 & 3 & 1 \end{pmatrix}$, $R = \begin{pmatrix} 2 & 1 & 1 \\ 0 & 1 & 1 \\ 0 & 0 & 2 \end{pmatrix}$.

We show the procedure with another example:

Example 11.2

$$\begin{pmatrix} 1 & 2 & 3 \\ 4 & 5 & 6 \\ 3 & 9 & 9 \end{pmatrix} \rightarrow \begin{pmatrix} 1 & 2 & 3 \\ 4 & -3 & -6 \\ 3 & 3 & 0 \end{pmatrix} \rightarrow \begin{pmatrix} 1 & 2 & 3 \\ 4 & -3 & -6 \\ 3 & -1 & -6 \end{pmatrix}.$$

That is, we have the following $L\,R$-decomposition:

$$\begin{pmatrix} 1 & 2 & 3 \\ 4 & 5 & 6 \\ 3 & 9 & 9 \end{pmatrix} = \begin{pmatrix} 1 & 0 & 0 \\ 4 & 1 & 0 \\ 3 & -1 & 1 \end{pmatrix} \begin{pmatrix} 1 & 2 & 3 \\ 0 & -3 & -6 \\ 0 & 0 & -6 \end{pmatrix}.$$

■

This procedure always works, unless you have to divide by zero when forming the elimination factors: You can't eliminate non-zero elements with a zero. But we will be able to solve this difficulty quite easily: We simply create a possible zero by swapping rows downwards and thus have even one less elimination to perform. Before we come to this, let us implement our previous procedure in MATLAB.

MATLAB Even with an implementation of this $L\,R$-decomposition on a computer, you will find this overwriting of the entries of A by the entries of L and R, since no additional memory is required in this algorithm. One speaks of **in-situ storage** . The following approach provides the crucial idea for an implementation of the algorithm in MATLAB: Determine w, L_* and R_*so that

$$A = \begin{pmatrix} \alpha & u^\top \\ v & A_* \end{pmatrix} = \begin{pmatrix} 1 & \\ w & L_* \end{pmatrix} \begin{pmatrix} \alpha & u^\top \\ & R_* \end{pmatrix},$$

where $A_*,\ L_*,\ R_* \in \mathbb{R}^{(n-1)\times(n-1)},\ u,\ v,\ w \in \mathbb{R}^{n-1}$ and $\alpha \in \mathbb{R}$. We multiply and get

$$v = \alpha w \text{ and } A_* = wu^\top + L_* R_*\,, \text{ i.e. } w = v/\alpha \text{ and } L_* R_* = A_* - wu^\top\,,$$

In MATLAB, for example, the algorithm is as follows:

```
function [L,R] = LR(A)
[n,n] = size(A);
for j = 1:n-1
  I = j+1:n;
  A(I,j) = A(I,j)/A(j,j);
  A(I,I) = A(I,I)-A(I,j)*A(j,I);
end R = triu(A);
L = eye(n,n) + tril(A,-1);
```

11.3 The LR-Decomposition: General Variant

Our previous method fails as soon as there is a zero on the diagonal during the Gaussian elimination, e.g.

$$\begin{pmatrix} 1 & 2 & 3 \\ 4 & 8 & 6 \\ 3 & 9 & 9 \end{pmatrix} \rightarrow \begin{pmatrix} 1 & 2 & 3 \\ 4 & 0 & -6 \\ 3 & 3 & 0 \end{pmatrix}.$$

In the case of (manual) Gaussian elimination, because of the zero at position (2, 2) we would perform a row swap in order to obtain a non-zeropivot element; in this case, the element $a_{kk} \neq 0$ on the diagonal, eliminating the entries below, is called **pivot element**. Such a swap is implemented by multiplying from the left by a permutation matrix. So one does not decompose A, but PA where P is the product of the permutation matrices realizing the performed row swaps. Thus, if we allow additional row permutations, we obtain a LR-decomposition for any invertible matrix in the following sense:

The *LR*-Decomposition of an Invertible Matrix

For each invertible matrix $A \in \mathbb{R}^{n\times n}$ there exists a lower triangular matrix $L \in \mathbb{R}^{n\times n}$ and an upper triangular matrix $R \in \mathbb{R}^{n\times n}$,

$$L = \begin{pmatrix} 1 & 0 & \dots & 0 \\ * & 1 & \ddots & \vdots \\ \vdots & \ddots & \ddots & 0 \\ * & \dots & * & 1 \end{pmatrix} \quad \text{and} \quad R = \begin{pmatrix} r_{11} & * & \dots & * \\ 0 & r_{22} & \ddots & \vdots \\ \vdots & \ddots & \ddots & * \\ 0 & \dots & 0 & r_{nn} \end{pmatrix},$$

and a matrix $P \in \mathbb{R}^{n\times n}$ which is a product of permutation matrices, with

$$P\,A = L\,R\,.$$

Such a representation of the matrix A is called $L\,R$**-decomposition** of A.

The solution x of $Ax = b$ is obtained by solving $LRx = PAx = Pb$.

Example 11.3 The $L\,R$-decomposition of the matrix $A = \begin{pmatrix} 0 & 1 \\ 1 & 1 \end{pmatrix} \in \mathbb{R}^{2\times 2}$ is

$$\underbrace{\begin{pmatrix} 0 & 1 \\ 1 & 0 \end{pmatrix}}_{=P} \underbrace{\begin{pmatrix} 0 & 1 \\ 1 & 1 \end{pmatrix}}_{=A} = \begin{pmatrix} 1 & 1 \\ 0 & 1 \end{pmatrix} = \underbrace{\begin{pmatrix} 1 & 0 \\ 0 & 1 \end{pmatrix}}_{=L} \underbrace{\begin{pmatrix} 1 & 1 \\ 0 & 1 \end{pmatrix}}_{=R}.$$

■

Since one needs the matrix P of course in the solution of the LGS $Ax = b$ per $L\,R$-decomposition, the question remains how to determine the matrix P. We consider in the following recipe this general $L\,R$-decomposition, here the matrices L, R and P are determined from A:

Recipe: Determine a $L\,R$-Decomposition of A

Given an invertible matrix $A \in \mathbb{R}^{n\times n}$. One obtains matrices L, R and P with $PA = LR$ as follows:

(1) As long as $a_{kk} \neq 0$: Eliminate with a_{kk} the entries below a_{kk}, as described in Sect. 11.2.

(continued)

(2) If $a_{kk} = 0$ swap the row k with a row $l > k$, which makes an elimination possible, over the whole row (i.e. in L and in R) and note the permutation matrix P_{kl} (which emerges from the unit matrix by permuting k-th and l-th row).
(3) Start with (1).
(4) If $k = n$: Obtain $P = P_r \cdots P_1$ where P_1 the first and P_r is the last obtained permutation matrix from step (2) (note the order), L is the lower left part (supplemented by ones on the diagonal), R is the upper right part (with the diagonal).

We show the procedure with an example:

Example 11.4 We determine a LR-decomposition of the following matrix A:

$$\underbrace{\begin{pmatrix} 1 & 1 & 1 & 1 \\ 2 & 2 & -1 & -3 \\ 4 & -1 & -6 & 0 \\ -1 & -2 & -3 & 1 \end{pmatrix}}_{=A} \rightarrow \begin{pmatrix} 1 & 1 & 1 & 1 \\ 2 & 0 & -3 & -5 \\ 4 & -5 & -10 & -4 \\ -1 & -1 & -2 & 2 \end{pmatrix}.$$

Because of $a_{22} = 0$ we have to perform a row swap. We swap rows 2 and 4, noting P_{24} and get

$$\begin{pmatrix} 1 & 1 & 1 & 1 \\ -1 & -1 & -2 & 2 \\ 4 & -5 & -10 & -4 \\ 2 & 0 & -3 & -5 \end{pmatrix} \rightarrow \begin{pmatrix} 1 & 1 & 1 & 1 \\ -1 & -1 & -2 & 2 \\ 4 & 5 & 0 & -14 \\ 2 & 0 & -3 & -5 \end{pmatrix}.$$

Because of $a_{33} = 0$ we have to swap the rows again. We swap lines 3 and 4, we remember P_{34}, and obtain

$$\begin{pmatrix} 1 & 1 & 1 & 1 \\ -1 & -1 & -2 & 2 \\ 2 & 0 & -3 & -5 \\ 4 & 5 & 0 & -14 \end{pmatrix}.$$

Because of

$$P = \begin{pmatrix} 1&0&0&0 \\ 0&0&0&1 \\ 0&1&0&0 \\ 0&0&1&0 \end{pmatrix} = \underbrace{\begin{pmatrix} 1&0&0&0 \\ 0&1&0&0 \\ 0&0&0&1 \\ 0&0&1&0 \end{pmatrix}}_{=P_{34}} \underbrace{\begin{pmatrix} 1&0&0&0 \\ 0&0&0&1 \\ 0&0&1&0 \\ 0&1&0&0 \end{pmatrix}}_{=P_{24}}$$

we get for A the $L\,R$-decomposition:

$$\underbrace{\begin{pmatrix} 1&0&0&0 \\ 0&0&0&1 \\ 0&1&0&0 \\ 0&0&1&0 \end{pmatrix}}_{=P} \underbrace{\begin{pmatrix} 1&1&1&1 \\ 2&2&-1&-3 \\ 4&-1&-6&0 \\ -1&-2&-3&1 \end{pmatrix}}_{=A} = \underbrace{\begin{pmatrix} 1&0&0&0 \\ -1&1&0&0 \\ 2&0&1&0 \\ 4&5&0&1 \end{pmatrix}}_{=L} \underbrace{\begin{pmatrix} 1&1&1&1 \\ 0&-1&-2&2 \\ 0&0&-3&-5 \\ 0&0&0&-14 \end{pmatrix}}_{=R}.$$

■

11.4 The $L\,R$-Decomposition-with Column Pivot Search

In any case, row swaps are necessary whenever division by zero would otherwise result. When swapping, one usually has a wide choice. In exact arithmetic, any nonzero pivot element is equally good. But in floating-point arithmetic on a calculator, the largest pivot element in terms of the absolute value is to be preferred; by this **column pivot search** rounding errors are avoided. The larger the pivot elements are in terms of absolute value, the smaller the absolute values of the elimination factors, so that the elements in the columns still to be processed grow as little as possible.

In the column pivot search, therefore, before the elimination in the s-th column, one exchanges the s-th row with the row that contains the largest of the numbers a_i, $s \leq i \leq n$. Here we assume that the matrix A is invertible.

Example 11.5 We determine the $L\,R$-decomposition with column pivot search of the matrix

$$A = \begin{pmatrix} 0&3&3 \\ -1&3&4 \\ -2&1&5 \end{pmatrix}.$$

Because $|a_{31}| > |a_{21}|, |a_{11}|$ the column pivot search before starting the elimination yields an interchange of the first and the third row; we note P_{13} and eliminate:

$$\begin{pmatrix} 0 & 3 & 3 \\ -1 & 3 & 4 \\ -2 & 1 & 5 \end{pmatrix} \rightarrow \begin{pmatrix} -2 & 1 & 5 \\ -1 & 3 & 4 \\ 0 & 3 & 3 \end{pmatrix} \rightarrow \begin{pmatrix} -2 & 1 & 5 \\ 1/2 & 5/2 & 3/2 \\ 0 & 3 & 3 \end{pmatrix}.$$

Because of $|a_{32}| > |a_{22}|$ the column pivot search returns a permutation of the second and the third row; we note P_{23} and eliminate:

$$\begin{pmatrix} -2 & 1 & 5 \\ 1/2 & 5/2 & 3/2 \\ 0 & 3 & 3 \end{pmatrix} \rightarrow \begin{pmatrix} -2 & 1 & 5 \\ 0 & 3 & 3 \\ 1/2 & 5/2 & 3/2 \end{pmatrix} \rightarrow \begin{pmatrix} -2 & 1 & 5 \\ 0 & 3 & 3 \\ 1/2 & 5/6 & -1 \end{pmatrix}.$$

Thus we obtain with $P = P_{23} P_{13}$ the decomposition:

$$\underbrace{\begin{pmatrix} 0 & 0 & 1 \\ 1 & 0 & 0 \\ 0 & 1 & 0 \end{pmatrix}}_{=P} \underbrace{\begin{pmatrix} 0 & 3 & 3 \\ -1 & 3 & 4 \\ -2 & 1 & 5 \end{pmatrix}}_{=A} = \underbrace{\begin{pmatrix} 1 & 0 & 0 \\ 0 & 1 & 0 \\ 1/2 & 5/6 & 1 \end{pmatrix}}_{=L} \underbrace{\begin{pmatrix} -2 & 1 & 5 \\ 0 & 3 & 3 \\ 0 & 0 & -1 \end{pmatrix}}_{=R}.$$

■

A decomposition $PA = LR$ is all the more desirable when several systems of equations

$$Ax = b_1, \; Ax = b_2, \; \ldots, \; Ax = b_k$$

for one and the same coefficient matrix A are to be solved. Namely, if one has $PA = LR$ already decomposed, this decomposition can be used for each of the systems of equations $Ax = b_i$.

MATLAB With MATLAB you get the matrices L, R and P of the LR-decomposition $PA = LR$ simply by `[L,R,P] = lu(A)`.

11.5 Exercises

11.1

(a) Determine a LR-decomposition of the matrix $A = \begin{pmatrix} -1 & 2 & 3 \\ -2 & 7 & 4 \\ 1 & 4 & -2 \end{pmatrix}$.

(b) Use this LR-decomposition to solve the linear system of equations $Ax = b$ with $b = (12, 24, 3)^\top$.

11.2 Determine a LR-decomposition of the matrix $A = \begin{pmatrix} 1 & 1 & 1 & 1 \\ 2 & 3 & -1 & -3 \\ 4 & -1 & 1 & 1 \\ -1 & -2 & -3 & 1 \end{pmatrix}$.

11.3 Let $A \in \mathbb{R}^{n\times n}$. Determine the computational cost of the LR-decomposition of A (without row swaps) and for the solution of the resulting linear system of equations $Lx = b$ by forward and backward substitution based on the number of floating point operations required.

Note: We have $\sum_{k=1}^{n} k = \frac{n(n+1)}{2}$ as well as $\sum_{k=1}^{n} k^2 = \frac{n(n+\frac{1}{2})(n+1)}{3}$.

11.4 Consider the system of linear equations $Ax = b$ with

$$A = \begin{pmatrix} 1 & 1 \\ 1 & 1+\delta \end{pmatrix}, \quad b = \begin{pmatrix} 1 \\ 1 \end{pmatrix}, \quad \delta = 10^{-15}.$$

(a) Give the exact solution x (without calculation).
(b) Determine A^{-1} as a function of δ.
(c) Compare the following methods for calculating x in MATLAB:
 1. $A^{-1}b$ with A^{-1} from (b)
 2. $A^{-1}b$ with A^{-1} from calculate `inv(A)`
 3. Gauss algorithm (LR-decomposition of A, forward and backward substitution). This can be achieved in MATLAB by `x=A\b`.
(d) Compare the results of 1. and 2. from (c).
(e) Explain the results of (c).

11.5 Determine the LR-decomposition with pivoting of the matrix

$$A = \begin{pmatrix} 1 & 1 & 2 \\ 4 & 0 & 2 \\ 2 & 1 & 1 \end{pmatrix},$$

and solve the LGS with this LR decomposition to solve the LGS $Ax = b$ with $b = (8, 8, 8)^\top$.

11.6 Implement in MATLAB forward or backward substitution to solve an LGS $Ax = b$ with upper and lower triangular matrices, respectively. Test your implementation on examples (use the LR-decomposition `[L,R,P]=lu(A)` of a Matrix A via MATLAB).

The Determinant

12

Each square matrix A has a *determinant* $\det(A)$. With the help of this characteristic of A we can give a decisive invertibility criterion for A: A square matrix A is invertible if and only if $\det(A) \neq 0$. It is this criterion that makes the determinant so useful: we can use it to determine the *eigenvalues* and, in turn, the *principal axis transformation* or *singular value decomposition*—problems, that are so crucial in engineering.

The calculation of the determinant $\det(A)$ for a large matrix A is extremely costly. We give tricks to keep the computation still clear.

In the following $\mathbb{K}$ is always one of the number fields $\mathbb{R}$ or $\mathbb{C}$.

12.1 Definition of the Determinant

We consider a square $n \times n$-matrix $A \in \mathbb{K}^{n \times n}$ and two indices $i, j \in \{1, \dots, n\}$. To this end, we declare the $(n-1) \times (n-1)$-**deletion matrix** A_{ij}, which is derived from A by deleting the i-th row and j-th column.

Example 12.1

$$A = \begin{pmatrix} 1 & 2 & 3 & 4 \\ 5 & 6 & 7 & 8 \\ 4 & 3 & 2 & 1 \\ 8 & 7 & 6 & 5 \end{pmatrix} \quad \Rightarrow \quad A_{23} = \begin{pmatrix} 1 & 2 & 4 \\ 4 & 3 & 1 \\ 8 & 7 & 5 \end{pmatrix} \quad \text{and} \quad A_{32} = \begin{pmatrix} 1 & 3 & 4 \\ 5 & 7 & 8 \\ 8 & 6 & 5 \end{pmatrix}. \quad ■$$

The **determinant** is a mapping that assigns to each quadratic $n \times n$-matrix A with coefficients of $\mathbb{K}$ a number, viz. $\det(A)$. This number $\det(A)$ is obtained recursively as follows:

C. Karpfinger, *Calculus and Linear Algebra in Recipes*,
https://doi.org/10.1007/978-3-662-65458-3_12

In the case $n = 1$: $\det(A) = a_{11}$ and in the case $n \geq 2$:

$$\det(A) = \sum_{i=1}^{n} (-1)^{i+1} a_{i1} \det(A_{i1})$$
$$= a_{11} \det(A_{11}) - a_{21} \det(A_{21}) + \ldots + (-1)^{n+1} a_{n1} \det(A_{n1}) .$$

The determinant of 2×2- and 3×3-matrices.
For $A = (a_{ij}) \in \mathbb{K}^{n \times n}$ we have:

- In the case $n = 2$:

$$\det(A) = a_{11}a_{22} - a_{12}a_{21} .$$

- In the case $n = 3$:

$$\det(A) = a_{11}a_{22}a_{33} + a_{12}a_{23}a_{31} + a_{13}a_{21}a_{32} - \big(a_{13}a_{22}a_{31} + a_{23}a_{32}a_{11} + a_{33}a_{12}a_{21}\big) .$$

One remembers these formulas with the following scheme, which is called in the case $n = 3$ **rule of Sarrus**:

$$\begin{vmatrix} a_{11} & a_{12} \\ a_{21} & a_{22} \end{vmatrix} = \begin{pmatrix} {}^{+}a_{11} & & a_{12}{}^{-} \\ & \times & \\ a_{21} & & a_{22} \end{pmatrix}, \quad \begin{vmatrix} a_{11} & a_{12} & a_{13} \\ a_{21} & a_{22} & a_{23} \\ a_{31} & a_{32} & a_{33} \end{vmatrix} = \begin{matrix} {}^{+}a_{13} \\ a_{23} \\ a_{33} \end{matrix} \begin{pmatrix} {}^{+}a_{11} & {}^{+}a_{12} & a_{13}{}^{-} \\ a_{21} & a_{22} & a_{23} \\ a_{31} & a_{32} & a_{33} \end{pmatrix} \begin{matrix} a_{11}{}^{-} \\ a_{21} \\ a_{31} \end{matrix}$$

In the case $n \geq 4$ there are no such simple mnemonic formulas, see also the following examples. It is common, to write $|A|$ instead of $\det(A)$. We will also use this notation occasionally.

Example 12.2

- We have $\det\begin{pmatrix} 1 & 2 \\ 3 & 4 \end{pmatrix} = 1 \cdot 4 - 3 \cdot 2 = -2$.
- We have $\det\begin{pmatrix} 1 & 2 & 3 \\ 4 & 5 & 6 \\ 7 & 8 & 9 \end{pmatrix} = 1 \cdot 5 \cdot 9 + 2 \cdot 6 \cdot 7 + 3 \cdot 4 \cdot 8 - 3 \cdot 5 \cdot 7 - 6 \cdot 8 \cdot 1 - 9 \cdot 2 \cdot 4 = 0.$

- We have $\begin{vmatrix} 1&2&3&4\\5&6&7&8\\4&3&2&1\\8&7&6&5 \end{vmatrix} = 1\cdot\begin{vmatrix} 6&7&8\\3&2&1\\7&6&5 \end{vmatrix} - 5\cdot\begin{vmatrix} 2&3&4\\3&2&1\\7&6&5 \end{vmatrix} + 4\cdot\begin{vmatrix} 2&3&4\\6&7&8\\7&6&5 \end{vmatrix} - 8\cdot\begin{vmatrix} 2&3&4\\6&7&8\\3&2&1 \end{vmatrix}$.

  Here it becomes clear: computing the determinant of a *large* matrix quickly becomes costly.
- However, if there are only zeros in the first column of a 4×4 matrix below the first element, then:

$$\begin{vmatrix} 1&2&3&4\\0&6&7&8\\0&3&2&1\\0&7&6&5 \end{vmatrix} = 1\cdot\begin{vmatrix} 6&7&8\\3&2&1\\7&6&5 \end{vmatrix}.$$

  In this case, computing the determinant of a 4×4 matrix is essentially computing the determinant of a 3×3-matrix. ■

12.2 Calculation of the Determinant

Numerous calculation rules apply to the determinant. With their help, it becomes possible to generate zeros in a row or column without changing the determinant. By developing according to this row or column, we obtain the determinant of an $n \times n$-matrix by calculating the determinant of a $(n-1) \times (n-1)$-matrix.

Calculation Rules for the Determinant
Given is a square matrix $A = (a_{ij}) \in \mathbb{K}^{n\times n}$. We have:

- **Development according to j-th column:**

$$\forall j \in \{1, \ldots, n\}: \quad \det(A) = \sum_{i=1}^{n} (-1)^{i+j} a_{ij} \det(A_{ij}).$$

- **Development according to i-th row:**

$$\forall i \in \{1, \ldots, n\}: \quad \det(A) = \sum_{j=1}^{n} (-1)^{i+j} a_{ij} \det(A_{ij}).$$

- The determinant does not change when transposed: $\det(A) = \det(A^\top)$.

(continued)

- If A is upper or lower triangular matrix, viz.

$$A = \begin{pmatrix} \lambda_1 & & * \\ & \ddots & \\ 0 & & \lambda_n \end{pmatrix} \quad \text{or} \quad A = \begin{pmatrix} \lambda_1 & & 0 \\ & \ddots & \\ * & & \lambda_n \end{pmatrix},$$

  so for the determinant of A we have:

$$\det(A) = \lambda_1 \cdots \lambda_n .$$

- If A has **block triangle shape**, i.e.

$$A = \begin{pmatrix} B & 0 \\ C & D \end{pmatrix} \quad \text{or} \quad A = \begin{pmatrix} B & C \\ 0 & D \end{pmatrix}$$

  with square matrices B and D and matching matrices 0 and C, then:

$$\det(A) = \det(B)\,\det(D) .$$

- The **Determinant Multiplication Theorem**: If $A = B\,C$ with square matrices B and C, then:

$$\det(A) = \det(B)\,\det(C) .$$

- If A is invertible, $AA^{-1} = E_n$, then we have: $\det(A^{-1}) = \big(\det(A)\big)^{-1}$.
- For $\lambda \in \mathbb{K}$ we have $\det(\lambda\, A) = \lambda^n \det(A)$.
- The determinant under elementary row or column transformations:
  - Swapping two rows or columns changes the sign of the determinant.
  - Multiplying a row or column of A by λ causes the determinant to be multiplied by λ.
  - Addition of the λ-fold of a row or column of A to another row or column does not change the determinant.

$$\det\begin{pmatrix} \vdots \\ z_k \\ \vdots \\ z_l \\ \vdots \end{pmatrix} = -\det\begin{pmatrix} \vdots \\ z_l \\ \vdots \\ z_k \\ \vdots \end{pmatrix}, \ \det\begin{pmatrix} \vdots \\ \lambda z_k \\ \vdots \\ z_l \\ \vdots \end{pmatrix} = \lambda \det\begin{pmatrix} \vdots \\ z_k \\ \vdots \\ z_l \\ \vdots \end{pmatrix}, \ \det\begin{pmatrix} \vdots \\ z_k + \lambda z_l \\ \vdots \\ z_l \\ \vdots \end{pmatrix} = \det\begin{pmatrix} \vdots \\ z_k \\ \vdots \\ z_l \\ \vdots \end{pmatrix}.$$

The last three rules describe the behavior of the determinant under elementary row transformations. Using the last rule, in a column where there is a nonzero entry, one can generate zeros above and below that entry; only one nonzero entry remains in that column, and the determinant of that matrix remains unchanged. By developing according to this column, the determinant is obtained. This is done analogously for a row. The general procedure for calculating the determinant of a matrix $A \in \mathbb{K}^{n\times n}$ with $n \geq 3$ can thus be summarized as:

Recipe: Calculate the Determinant

(1) If A has two equal rows or columns, or two rows or columns that are multiples of each other, then $\det(A) = 0$.

(2) If A has block triangular shape, i.e. $A = \begin{pmatrix} B & 0 \\ C & D \end{pmatrix}$ or $A = \begin{pmatrix} B & C \\ 0 & D \end{pmatrix}$?

If yes: $\det(A) = \det(B)\det(D)$.

If no: Next step.

(3) Is there a row or column with many zeros?

If yes: Develop according to this row or column.

If no: Next step.

(4) Create zeros in a row or column using elementary row or column transformations and develop according to that row or column.

(5) Start over.

Example 12.3

- We compute a determinant by generating many zeros in one column by row operations, and then develop according to it:

$$\det\begin{pmatrix} 4 & 3 & 2 & 1 \\ 3 & 2 & 1 & 4 \\ 2 & 1 & 4 & 3 \\ 1 & 4 & 3 & 2 \end{pmatrix} = \begin{vmatrix} 0 & -13 & -10 & -7 \\ 0 & -10 & -8 & -2 \\ 0 & -7 & -2 & -1 \\ 1 & 4 & 3 & 2 \end{vmatrix} = (-1)^5 \cdot 1 \cdot \begin{vmatrix} -13 & -10 & -7 \\ -10 & -8 & -2 \\ -7 & -2 & -1 \end{vmatrix}$$

$$= \begin{vmatrix} 13 & 10 & 7 \\ 10 & 8 & 2 \\ 7 & 2 & 1 \end{vmatrix} = \begin{vmatrix} -36 & -4 & 0 \\ -4 & 4 & 0 \\ 7 & 2 & 1 \end{vmatrix} = (-1)^6 \cdot 1 \cdot \begin{vmatrix} -36 & -4 \\ -4 & 4 \end{vmatrix}$$

$$= \begin{vmatrix} 36 & 4 \\ 4 & -4 \end{vmatrix} = \begin{vmatrix} 40 & 0 \\ 4 & -4 \end{vmatrix} = -160\,.$$

- Now we compute the determinant of a matrix in block triangular form:

$$\det\begin{pmatrix} 1 & 2 & 0 & 0 & 0 & 0 \\ 1 & 2 & 0 & 0 & 0 & 0 \\ 7 & 8 & 2 & 3 & 3 & 0 \\ 1 & 2 & 0 & 3 & 5 & 0 \\ 2 & 1 & 0 & 1 & 2 & 0 \\ 0 & 0 & 2 & 2 & 3 & 5 \end{pmatrix} = \det\begin{pmatrix} 1 & 2 \\ 1 & 1 \end{pmatrix} \cdot \det\begin{pmatrix} 2 & 3 & 3 \\ 0 & 3 & 5 \\ 0 & 1 & 2 \end{pmatrix} \cdot \det\begin{pmatrix} 5 \end{pmatrix}$$

$$= (-1) \cdot 2 \cdot 1 \cdot 5 = -10\,.$$

- Another way to simplify is to subtract factors from rows or columns:

$$\begin{vmatrix} 2 & 4 \\ 0 & 8 \end{vmatrix} = 2 \cdot \begin{vmatrix} 1 & 2 \\ 0 & 8 \end{vmatrix} = 2 \cdot 8 \cdot \begin{vmatrix} 1 & 2 \\ 0 & 1 \end{vmatrix} \quad \text{or} \quad \begin{vmatrix} 2 & 4 \\ 0 & 8 \end{vmatrix} = 2 \cdot \begin{vmatrix} 1 & 2 \\ 0 & 8 \end{vmatrix} = 2 \cdot 2 \cdot \begin{vmatrix} 1 & 1 \\ 0 & 4 \end{vmatrix}\,.$$

- Yet another example of simplifying and developing:

$$\det\begin{pmatrix} 3 & 1 & 3 & 0 \\ 2 & 4 & 1 & 2 \\ 1 & 0 & 0 & -1 \\ 4 & 2 & -1 & 1 \end{pmatrix} = \begin{vmatrix} 3 & 1 & 3 & 3 \\ 2 & 4 & 1 & 4 \\ 1 & 0 & 0 & 0 \\ 4 & 2 & -1 & 5 \end{vmatrix} = (-1)^4 \cdot 1 \cdot \begin{vmatrix} 1 & 3 & 3 \\ 4 & 1 & 4 \\ 2 & -1 & 5 \end{vmatrix}$$

$$= \begin{vmatrix} -11 & 0 & -9 \\ 4 & 1 & 4 \\ 6 & 0 & 9 \end{vmatrix} = (-1)^4 \cdot 1 \cdot \begin{vmatrix} -11 & -9 \\ 6 & 9 \end{vmatrix} = 3 \cdot \begin{vmatrix} -11 & -9 \\ 2 & 3 \end{vmatrix}$$

$$= 9 \cdot \begin{vmatrix} -11 & -3 \\ 2 & 1 \end{vmatrix} = 9 \cdot \begin{vmatrix} -5 & 0 \\ 2 & 1 \end{vmatrix} = 9 \cdot (-5) = -45\,.$$

- We consider the matrix

$$A = \begin{pmatrix} 1-x & 1 & 1 \\ 1 & 1-x & 1 \\ 1 & 1 & 1-x \end{pmatrix}$$

and determine the numbers $x \in \mathbb{R}$, for which $\det(A) = 0$ is:

$$\begin{vmatrix} 1-x & 1 & 1 \\ 1 & 1-x & 1 \\ 1 & 1 & 1-x \end{vmatrix} = \begin{vmatrix} 1-x & 1 & 1 \\ x & -x & 0 \\ 0 & x & -x \end{vmatrix} = x^2 \cdot \begin{vmatrix} 1-x & 1 & 1 \\ 1 & -1 & 0 \\ 0 & 1 & -1 \end{vmatrix}$$

$$= x^2 \cdot \begin{vmatrix} 1-x & 2-x & 1 \\ 1 & 0 & 0 \\ 0 & 1 & -1 \end{vmatrix} = -x^2 \cdot \begin{vmatrix} 2-x & 1 \\ 1 & -1 \end{vmatrix}$$

$$= -x^2 \cdot \begin{vmatrix} 3-x & 0 \\ 1 & -1 \end{vmatrix} = x^2(3-x)\,.$$

Thus, the determinant is zero exactly when $x = 0$ or $x = 3$. ■

MATLAB In MATLAB, you get the determinant of A by `det(A)`. Here the determinant is given by LR-decomposition $PA = LR$ of a matrix A. For once one has this LR-decomposition, then because of $P^2 = E_n$ and the determinant multiplication theorem $\det(A) = \det(P)\det(L)\det(R) = \pm\det(R)$ where R is an upper triangular matrix whose determinant is simply the product of the diagonal elements. Also the determinant of P is easy to determine, namely $\det(P) = 1$ if P is a product of even numbered permutation matrices and $\det(P) = -1$ else.

12.3 Applications of the Determinant

We know that a matrix $A \in \mathbb{K}^{n\times n}$ is invertible if and only if $\operatorname{rg}(A) = n$. With the determinant we can specify another such **invertibility criterion** (note Exercise 12.1):

Invertibility Criterion for Matrices

A matrix $A \in \mathbb{K}^{n\times n}$ is invertible if and only if $\det(A) \neq 0$.

We show how we can use this criterion: Given a matrix $A \in \mathbb{K}^{n\times n}$. We look for a vector $v \in \mathbb{K}^n$, $v \neq 0$, with

$$A\,v = \lambda\,v \quad \text{for some } \lambda \in \mathbb{K}.$$

This problem finds applications in solving differential equations, finding axes of inertia, etc. We have:

$$\begin{aligned} A\,v = \lambda\,v,\ v \neq 0 \Leftrightarrow\ A\,v - \lambda\,v = 0,\ v \neq 0\ &\Leftrightarrow\ (A - \lambda E_n)\,v = 0,\ v \neq 0 \\ &\Leftrightarrow\ (A - \lambda\,E_n) \text{ has not rank } n \\ &\Leftrightarrow\ (A - \lambda\,E_n) \text{ is not invertible} \\ &\Leftrightarrow\ \det(A - \lambda\,E_n) = 0\,. \end{aligned}$$

Thus, the task is, to determine $\lambda \in \mathbb{K}$ such that $\det(A - \lambda\,E_n) = 0$.

Example 12.4 Let

$$A = \begin{pmatrix} 1 & 2 \\ 2 & 1 \end{pmatrix} \in \mathbb{R}^{2\times 2}\,.$$

We now search $v \in \mathbb{R}^n$ with $A\,v = \lambda\,v$ for a $\lambda \in \mathbb{R}$. According to the above calculation we first determine $\lambda \in \mathbb{R}$ with $\det(A - \lambda\,E_n) = 0$. We have:

$$\begin{aligned} \det(A - \lambda\,E_n) &= \begin{vmatrix} 1-\lambda & 2 \\ 2 & 1-\lambda \end{vmatrix} = (1-\lambda)^2 - 4 \\ &= \lambda^2 - 2\lambda + 1 - 4 = \lambda^2 - 2\lambda - 3 = (\lambda - 3)(\lambda + 1)\,. \end{aligned}$$

Thus it is $\det(A - \lambda\,E_n) = 0$ for $\lambda \in \{-1, 3\}$. We set $\lambda_1 = -1$ and $\lambda_2 = 3$.

So there are $v_1,\, v_2 \in \mathbb{R}^2$ with

$$A\,v_1 = \lambda_1 v_1 \quad \text{and} \quad A\,v_2 = \lambda_2 v_2\,.$$

One finds v_1 and v_2 by solving the linear system of equations $\left(A - \lambda_{1/2} E_2 \middle| 0\right)$:

$$(A - \lambda_1 E_2|0) = \left(\begin{array}{cc|c} 2 & 2 & 0 \\ 2 & 2 & 0 \end{array}\right) \rightsquigarrow \left(\begin{array}{cc|c} 1 & 1 & 0 \\ 0 & 0 & 0 \end{array}\right) \Rightarrow v_1 = \begin{pmatrix} 1 \\ -1 \end{pmatrix}.$$

$$(A - \lambda_2 E_2|0) = \left(\begin{array}{cc|c} -2 & 2 & 0 \\ 2 & -2 & 0 \end{array}\right) \rightsquigarrow \left(\begin{array}{cc|c} 1 & -1 & 0 \\ 0 & 0 & 0 \end{array}\right) \Rightarrow v_2 = \begin{pmatrix} 1 \\ 1 \end{pmatrix}.$$

The sample confirms $A\,v_1 = -v_1$ and $A\,v_2 = 3v_2$. ■

We come back to this problem in Chap. 39 again.

Another application is *Cramer's rule*, this yields the components $\ell_1, \ldots, \ell_n$ of the solution vector x of a uniquely solvable linear system of equations $A\,x = b$ with an invertible matrix $A \in \mathbb{K}^{n\times n}$ and a column vector $b \in \mathbb{K}^n$:

Recipe: Cramer's Rule

Obtain the components ℓ_i of the uniquely determined solution x *of* the LGS $A\,x = b$ with an invertible matrix $A = (s_1, \ldots, s_n) \in \mathbb{K}^{n\times n}$ and a column vector $b \in \mathbb{K}^n$ as follows:

(1) Calculate $\det(A)$.
(2) Replace the i-th column s_i of A by b and get

$$A_i = (s_1, \ldots, s_{i-1}, b, s_{i+1}, \ldots, s_n) \text{ for } i = 1, \ldots, n\,.$$

(3) Calculate $\det(A_i)$ for $i = 1, \ldots, n$.
(4) Get the components ℓ_i of the solution vector x as follows:

$$\ell_i = \frac{\det(A_i)}{\det(A)} \text{ for } i = 1, \ldots, n\,.$$

Example 12.5 We solve the linear system of equations $A\,x = b$ with

$$A = \begin{pmatrix} -1 & 8 & 3 \\ 2 & 4 & -1 \\ -2 & 1 & 2 \end{pmatrix} \quad \text{and} \quad b = \begin{pmatrix} 2 \\ 1 \\ -1 \end{pmatrix}.$$

(1) The matrix A is due to

$$\det(A) = \begin{vmatrix} -1 & 8 & 3 \\ 2 & 4 & -1 \\ -2 & 1 & 2 \end{vmatrix} = \begin{vmatrix} -1 & 8 & 3 \\ 0 & 20 & 5 \\ 0 & 5 & 1 \end{vmatrix} = -\begin{vmatrix} 20 & 5 \\ 5 & 1 \end{vmatrix} = 5 \neq 0$$

invertible.

(2) We obtain the matrices

$$A_1 = \begin{pmatrix} 2 & 8 & 3 \\ 1 & 4 & -1 \\ -1 & 1 & 2 \end{pmatrix}, \quad A_2 = \begin{pmatrix} -1 & 2 & 3 \\ 2 & 1 & -1 \\ -2 & -1 & 2 \end{pmatrix} \quad \text{and} \quad A_3 = \begin{pmatrix} -1 & 8 & 2 \\ 2 & 4 & 1 \\ -2 & 1 & -1 \end{pmatrix}.$$

(3) The determinants of these matrices are calculated as $\det(A_1) = 25$, $\det(A_2) = -5$ and $\det(A_3) = 25$.

(4) Thus

$$\ell_1 = \frac{\det(A_1)}{\det(A)} = \frac{25}{5} = 5, \quad \ell_2 = \frac{-5}{5} = -1 \quad \text{and} \quad \ell_3 = \frac{25}{5} = 5$$

the components of the uniquely determined solution of $A\,x = b$, $x = (5, -1, 5)^\top$. ■

With the Gaussian elimination method we would have obtained the solution more easily. In practice, one does not use Cramer's rule. For a justification of this rule, see the exercises.

12.4 Exercises

12.1 Justify the invertibility criterion for matrices in Sect. 12.3.

12.2 Compute the determinants of the following matrices:

$$\begin{pmatrix} 1 & 2 \\ -2 & -5 \end{pmatrix}, \begin{pmatrix} -1 & 1 & 1 \\ 1 & 0 & -7 \\ 2 & -3 & 5 \end{pmatrix}, \begin{pmatrix} 1 & 3 & -1 & 1 \\ -2 & -5 & 2 & 1 \\ 3 & 4 & 2 & -2 \\ -4 & 2 & -8 & 1 \end{pmatrix}, \begin{pmatrix} 1 & 2 & 0 & 0 & 0 \\ 1 & 1 & 0 & 0 & 0 \\ 7 & 8 & 2 & 3 & 3 \\ 1 & 2 & 0 & 3 & 5 \\ 2 & 1 & 0 & 1 & 2 \end{pmatrix}.$$

12.3 Show with an example that for $A, B, C, D \in \mathbb{R}^{n \times n}$ in general

$$\det \begin{pmatrix} A & B \\ C & D \end{pmatrix} \neq \det A \det D - \det B \det C\,.$$

12.4 Determine the determinant of the following *tridiagonal matrices*.

$$\begin{pmatrix} 1 & \mathrm{i} & 0 & \dots & 0 \\ \mathrm{i} & 1 & \mathrm{i} & \ddots & \vdots \\ 0 & \mathrm{i} & 1 & \ddots & 0 \\ \vdots & \ddots & \ddots & \ddots & \mathrm{i} \\ 0 & \dots & 0 & \mathrm{i} & 1 \end{pmatrix} \in \mathbb{C}^{n \times n}\,.$$

12.5 Write a MATLAB program that calculates the determinant $\det(A)$ after evolution is calculated according to the first column.

12.6 Using Cramer's rule, solve the system of equations $Ax = b$ for

$$A = \begin{pmatrix} 0 & 1 & 3 \\ 2 & 1 & 0 \\ 4 & 1 & 1 \end{pmatrix} \quad \text{and} \quad b = \begin{pmatrix} 4 \\ 3 \\ 6 \end{pmatrix}.$$

12.7 Justify Cramer's rule.

12.8 Write a MATLAB function that solves the system of equations $Ax = b$ using Cramer's rule.

12.9 Find the determinants of the matrices

$$A = \begin{pmatrix} 1 & 1 & 0 & 2 \\ -2 & 2 & 0 & 1 \\ 38 & 7 & -3 & 3 \\ -1 & 2 & 0 & 3 \end{pmatrix}, \quad B = \begin{pmatrix} -3 & 0 & 7^{44} & 0 \\ \frac{22}{23} & 5 & \sqrt{\pi} & 0 \\ 0 & 0 & 6 & 0 \\ -102 & 8^e & e^8 & 10 \end{pmatrix}, \quad C = A^\top B^{-1}.$$

Vector Spaces

13

The notion of *vector space* is a very useful one: many sets of mathematical objects obey one and the same rules and can be grouped under this notion. Whether we consider the solution set of a homogeneous linear system of equations or the set of 2π-periodic functions; these sets form *vector spaces* and their elements thus vectors, which are all subject to the same general rules for vectors.

In this chapter on vector spaces, some ability to abstract is necessary. This is admittedly difficult at the beginning. Perhaps it is a useful tip to suppress intuition: Vector spaces in general elude any intuition, the attempt to imagine something under a function space simply has to fail.

By $\mathbb{K}$ we always denote $\mathbb{R}$ or $\mathbb{C}$.

13.1 Definition and Important Examples

We begin with the definition of a *vector space*. This definition is anything but short and sweet. We want to point out right away that you should not learn this definition by heart, you should just know where to look if you do need to refer back to the definition:

A nonempty set V with an addition $+$ and a multiplication $\cdot$ is called a **vector space over** $\mathbb{K}$ or a $\mathbb{K}$ **-vector space** if we have for all $u,\ v,\ w \in V$ and for all $\lambda, \mu \in \mathbb{K}$:

(1) $v + w \in V, \quad \lambda v \in V,$ (closure)
(2) $u + (v + w) = (u + v) + w,$ (associativity)
(3) There is an element $0 \in V$ with $v + 0 = v$ for all $v \in V$, (zero element)
(4) There is an element $v' \in V : \quad v + v' = 0$ for all $v \in V$, (negative element)
(5) $v + w = w + v$ for all $v, w \in V$, (commutativity)
(6) $\lambda\,(v + w) = \lambda\, v + \lambda\, w,$ (distributivity)

C. Karpfinger, *Calculus and Linear Algebra in Recipes*,
https://doi.org/10.1007/978-3-662-65458-3_13

(7) $(\lambda + \mu)\, v = \lambda\, v + \mu\, v$, (distributivity)
(8) $(\lambda\, \mu)\, v = \lambda\, (\mu\, v)$, (associativity)
(9) $1 \cdot v = v$.

If V a $\mathbb{K}$-vector space, then the elements from V we call **vectors**. The vector 0 from (3) is called **null vector**. The vector v' to v from (4) is called the **opposite vektor** or **inverse vector** or the **negative vector** of v. One writes $-v$ for v' and also $u - v$ instead of $u + (-v)$. In the case $\mathbb{K} = \mathbb{R}$ one also speaks of a **real vector space**, in the case $\mathbb{K} = \mathbb{C}$ of a **complex vector space**. We call + the **vector addition** and $\cdot$ the **multiplication with scalars** or **scalar multiplication**.

This establishes all the notions for the time being, we now consider the four most important examples of $\mathbb{K}$-vector spaces, these are

$$\mathbb{K}^n\,,\ \mathbb{K}^{m\times n}\,,\ \mathbb{K}[x]\,,\ \mathbb{K}^M\,.$$

Example 13.1

- For all natural numbers $n \in \mathbb{N}$ is

$$\mathbb{K}^n = \left\{ \begin{pmatrix} x_1 \\ \vdots \\ x_n \end{pmatrix} \mid x_1, \ldots, x_n \in \mathbb{K} \right\}$$

  with vector addition and multiplication with scalars

$$\begin{pmatrix} x_1 \\ \vdots \\ x_n \end{pmatrix} + \begin{pmatrix} y_1 \\ \vdots \\ y_n \end{pmatrix} = \begin{pmatrix} x_1 + y_1 \\ \vdots \\ x_n + y_n \end{pmatrix} \quad \text{and} \quad \lambda \cdot \begin{pmatrix} x_1 \\ \vdots \\ x_n \end{pmatrix} = \begin{pmatrix} \lambda\, x_1 \\ \vdots \\ \lambda\, x_n \end{pmatrix}$$

  an $\mathbb{K}$-vector space. The zero vector of $\mathbb{K}^n$ is $0 = (0, \ldots, 0)^\top$, and for each vector $v = (v_1, \ldots, v_n)^\top$ is the negative $-v = (-v_1, \ldots, -v_n)^\top$.
- The set

$$\mathbb{K}^{m\times n} = \left\{ \left(a_{ij}\right) \mid a_{ij} \in \mathbb{K} \right\}$$

  of all $m \times n$-matrices over $\mathbb{K}$, $m, n \in \mathbb{N}$, is with the following vector addition and multiplication by scalars

$$\left(a_{ij}\right) + \left(b_{ij}\right) = \left(a_{ij} + b_{ij}\right) \text{ and } \lambda \cdot \left(a_{ij}\right) = \left(\lambda\, a_{ij}\right)$$

an $\mathbb{K}$-vector space with the zero matrix 0 as the zero vector. The negative to the vector $v = (a_{ij}) \in \mathbb{K}^{m \times n}$ is $-v = (-a_{ij})$. The vectors here are matrices.

- The set

$$\mathbb{K}[x] = \{a_0 + a_1 x + \ldots + a_n x^n \mid n \in \mathbb{N}_0,\ a_i \in \mathbb{K}\}$$

of all **polynomials** over $\mathbb{K}$ is with the vector addition and the multiplication with scalars

$$\sum a_i x^i + \sum b_i x^i = \sum (a_i + b_i) x^i \text{ and } \lambda \cdot \sum a_i x^i = \sum (\lambda a_i) x^i$$

an $\mathbb{K}$-vector space with the zero polynomial $\sum 0x^i$ as the zero vector and the negative $-p = \sum (-a_i) x^i$ to the polynomial $p = \sum a_i x^i \in K[x]$. The vectors here are polynomials.

- For each set M is

$$\mathbb{K}^M = \{f \mid f : M \to \mathbb{K} \text{ is a map}\},$$

the set of all mappings from M to $\mathbb{K}$ with vector addition and multiplication with scalars

$$f + g : \begin{cases} M \to & \mathbb{K} \\ x \mapsto & f(x) + g(x) \end{cases} \quad \text{and} \quad \lambda \cdot f : \begin{cases} M \to & \mathbb{K} \\ x \mapsto & \lambda f(x) \end{cases}$$

an $\mathbb{K}$-vector space with zero vector 0, which is the zero mapping

$$0 : \begin{cases} M \to \mathbb{K} \\ x \mapsto 0 \end{cases}$$

and the negative $-f$ to f is:

$$f : \begin{cases} M \to & \mathbb{K} \\ x \mapsto & f(x) \end{cases} \Rightarrow -f : \begin{cases} M \to & \mathbb{K} \\ x \mapsto & -f(x) \end{cases}.$$

The vectors here are mappings. ■

Proving that the axioms (1)–(9) above are fullfield for these four examples $\mathbb{K}^n$, $\mathbb{K}^{m \times n}$, $\mathbb{K}[x]$, $\mathbb{K}^M$ is tedious and boring. We refrain from doing so, but highlight two computational rules that apply to all vectors v of a $\mathbb{K}$-vector space V and all scalars $\lambda \in \mathbb{K}$:

- $0\,v = 0$ and $\lambda\,0 = 0$.

- $\lambda v = 0 \Rightarrow \lambda = 0$ or $v = 0$.

This is justified by the vector space axioms (note Exercise 13.1).

13.2 Subspaces

If one wants to prove, given a set with an addition and a multiplication with scalars from $\mathbb{K}$, that it is a $\mathbb{K}$-vector space, you actually have to justify the nine axioms (1)–(9) mentioned above. This can be laborious. Fortunately, all vector spaces we will ever have to deal with are *subspaces* of one of the four vector spaces $\mathbb{K}^n$, $\mathbb{K}^{m\times n}$, $\mathbb{K}[x]$, $\mathbb{K}^M$. A subset U of a vector space V is called a **subspace** of V if U with the addition $+$ and the multiplication $\cdot$ with scalars of V is again a vector space. And the proof that a subset U is a subspace of a vector space is fortunately quite easy:

Recipe: Proof for Subspace
Let U be a subset of a vector space V over $\mathbb{K}$. If we have

(1) $0 \in U$,
(2) $u,\ v \in U \Rightarrow u + v \in U$,
(3) $u \in U,\ \lambda \in \mathbb{K} \Rightarrow \lambda u \in U$

than the subset U of V is a subspace of V and as such a $\mathbb{K}$-vector space.

So if you want to show that a set is a vector space, use the given recipe to show that this set is a subspace of an appropriate vector space. This makes less work, but gives the same result.

Example 13.2

- The subset $U = \{(x_1, \ldots, x_{n-1}, 0)^\top \in \mathbb{K}^n \mid x_1, \ldots, x_{n-1} \in \mathbb{K}\}$ of the $\mathbb{K}$-vector space $\mathbb{K}^n$ is a subspace of $\mathbb{K}^n$, since we have:

(1) $0 = (0, \ldots, 0, 0)^\top \in U$,
(2)

$$\begin{aligned} u,\ v \in U &\Rightarrow u = (x_1, \ldots, x_{n-1}, 0)^\top,\ v = (y_1, \ldots, y_{n-1}, 0)^\top \\ &\Rightarrow u + v = (x_1 + y_1, \ldots, x_{n-1} + y_{n-1}, 0)^\top \\ &\Rightarrow u + v \in U, \end{aligned}$$

(3) $u \in U, \lambda \in \mathbb{K} \Rightarrow \lambda u = (\lambda x_1, \ldots, \lambda x_{n-1}, 0)^\top \Rightarrow \lambda u \in U$.

- The subset $U = \{\operatorname{diag}(\lambda_1, \ldots, \lambda_n) \in \mathbb{K}^{n\times n} \mid \lambda_1, \ldots, \lambda_n \in \mathbb{K}\}$ of the diagonal matrices of $\mathbb{K}^{n\times n}$ is a subspace of $\mathbb{K}^{n\times n}$, since we have:

(1) $0 = \operatorname{diag}(0, \ldots, 0) \in U$,
(2)

$$\begin{aligned} u, v \in U &\Rightarrow u = \operatorname{diag}(\lambda_1, \ldots, \lambda_n),\ v = \operatorname{diag}(\mu_1, \ldots, \mu_n) \\ &\Rightarrow u + v = \operatorname{diag}(\lambda_1 + \mu_1, \ldots, \lambda_n + \mu_n) \\ &\Rightarrow u + v \in U, \end{aligned}$$

(3) $u \in U, \lambda \in \mathbb{K} \Rightarrow \lambda u = \operatorname{diag}(\lambda \lambda_1, \ldots, \lambda \lambda_n) \Rightarrow \lambda u \in U$.

- The subset $\mathbb{R}[x]_2 = \{a_0 + a_1 x + a_2 x^2 \in \mathbb{R}[x] \mid a_0, a_1, a_2 \in \mathbb{R}\}$ of polynomials of degree less than or equal to 2 is a subspace of $\mathbb{R}[x]$, since we have:

(1) $0 = 0 + 0x + 0x^2 \in \mathbb{R}[x]_2$,
(2)

$$\begin{aligned} p, q \in \mathbb{R}[x]_2 &\Rightarrow p = a_0 + a_1 x + a_2 x^2,\ q = b_0 + b_1 x + b_2 x^2 \\ &\Rightarrow p + q = (a_0 + b_0) + (a_1 + b_1) x + (a_2 + b_2) x^2 \\ &\Rightarrow p + q \in \mathbb{R}[x]_2, \end{aligned}$$

(3) $p \in \mathbb{R}[x]_2, \lambda \in \mathbb{R} \Rightarrow \lambda p = \lambda a_0 + \lambda a_1 x + \lambda a_2 x^2 \Rightarrow \lambda p \in \mathbb{R}[x]_2$.

- The subset $U = \{f \in \mathbb{R}^{\mathbb{R}} \mid f(1) = 0\}$ of all mappings from $\mathbb{R}$ to $\mathbb{R}$, which have value 0 in 1, is a subspace of $\mathbb{R}^{\mathbb{R}}$, since we have:

(1) The zero function $f = 0$ satisfies $f(1) = 0 \Rightarrow f = 0 \in U$,
(2)

$$\begin{aligned} f, g \in U &\Rightarrow f(1) = 0,\ g(1) = 0 \\ &\Rightarrow (f + g)(1) = f(1) + g(1) = 0 + 0 = 0 \\ &\Rightarrow f + g \in U, \end{aligned}$$

(3) $f \in U, \lambda \in \mathbb{R} \Rightarrow (\lambda f)(1) = \lambda f(1) = \lambda 0 = 0 \Rightarrow \lambda f \in U$.

- For each $m \times n$-matrix $A \in \mathbb{K}^{m \times n}$ the solution set $L = \{v \in \mathbb{K}^n \mid A\,v = 0\}$ of the homogeneous linear system of equations $(A \mid 0)$ is a subspace of $\mathbb{K}^n$, since we have:

  (1) $0 = (0, \ldots, 0)^\top \in L$,
  (2) $u, v \in L \Rightarrow A\,u = 0,\ A\,v = 0 \Rightarrow A\,(u + v) = A\,u + A\,v = 0 \Rightarrow u + v \in L$,
  (3) $u \in L,\ \lambda \in \mathbb{K} \Rightarrow A\,u = 0,\ \lambda \in \mathbb{K} \Rightarrow A\,(\lambda\,u) = \lambda\,A\,u = 0 \Rightarrow \lambda\,u \in L$.

  For each vector space V, V itself and $\{0\}$ are subspaces. They are called the **trivial subspaces** of V.
- For two subvector spaces $U_1,\ U_2$ of V also the average $U_1 \cap U_2$ and the sum $U_1 + U_2 = \{u_1 + u_2 \mid u_1 \in U_1,\ u_2 \in U_2\}$ are subspaces of V.
- The set $U = \{f \in \mathbb{R}^\mathbb{R} \mid f(x + 2\pi) = f(x)$ for alle $x \in \mathbb{R}\}$ of all 2π*-periodic functions* $f : \mathbb{R} \to \mathbb{R}$ forms a subspace of $\mathbb{R}^\mathbb{R}$. ■

Since every subspace is in particular a vector space again, we now know numerous examples of vector spaces. We now take this to the extreme by specifying for each subset X of a vector space V a *smallest* subspace U which contains this given subset X, $X \subseteq U \subseteq V$. But for this we first need *linear combinations.*

13.3 Exercises

13.1 Show, that for all vectors v of a $\mathbb{K}$-vector space V and all scalars $\lambda \in \mathbb{K}$ we have:

(a) $0\,v = 0$ and $\lambda\,0 = 0$.
(b) $\lambda\,v = 0 \Rightarrow \lambda = 0$ or $v = 0$.

13.2 For the following sets, decide whether they are subspaces. If the set is not a subspace, give a brief justification.

(a) $U_1 = \{(x, y)^\top \in \mathbb{R}^2 \mid x^2 + y^2 = 0\} \subseteq \mathbb{R}^2$.
(b) $U_2 = \{A \in \mathbb{R}^{4\times4} \mid Ax = 0 \text{ has infinitely many solutions}\} \subseteq \mathbb{R}^{4\times4}$.
(c) $U_3 = \{A \in \mathbb{R}^{2\times2} \mid |\det A| = 1\} \subseteq \mathbb{R}^{2\times2}$.
(d) $U_4 = \{a_0 + a_1x + a_2x^2 \in \mathbb{R}[x]_2 \mid 2a_2 = a_1\} \subseteq \mathbb{R}[x]_2$.

13.3 A function $f : \mathbb{R} \to \mathbb{R}$ is called **even** (or **odd**), if $f(x) = f(-x)$ for all $x \in \mathbb{R}$ (or $f(x) = -f(-x)$ for all $x \in \mathbb{R}$). Let the set of even (or odd) functions be denoted by G (or U). Show that G and U are subspaces of $\mathbb{R}^\mathbb{R}$ and we have $\mathbb{R}^\mathbb{R} = G + U$ and $G \cap U = \{0\}$.
Note: $f(x) = \frac{1}{2}\,(f(x) + f(-x)) + \frac{1}{2}\,(f(x) - f(-x))$ for all $x \in \mathbb{R}$.

13.4 For the following subsets of the vector space $\mathbb{R}^3$ decide whether they are subspaces, and give reasons:

(a) $U_1 = \left\{ \begin{pmatrix} v_1 \\ v_2 \\ v_3 \end{pmatrix} \in \mathbb{R}^3 \mid v_1 + v_2 = 2 \right\}$, (c) $U_3 = \left\{ \begin{pmatrix} v_1 \\ v_2 \\ v_3 \end{pmatrix} \in \mathbb{R}^3 \mid v_1 \, v_2 = v_3 \right\}$,

(b) $U_2 = \left\{ \begin{pmatrix} v_1 \\ v_2 \\ v_3 \end{pmatrix} \in \mathbb{R}^3 \mid v_1 + v_2 = v_3 \right\}$,

13.5

(a) Two motors operate at the same frequency ω but are of different strengths, i.e., they operate at two different amplitudes A_1, A_2. In this case, the motors drive the piston with a harmonic oscillation, i.e., the respective local displacement of the piston as a function of time t is given by

$$y_k(t) = A_k \cos(\omega t + \delta_k) \quad \text{for } k = 1, 2\,.$$

The phase δ_k it is taken into account that the respective piston at the time of switching on ($t = 0$) can already have a certain deflection. By a suitable technical arrangement the motors can be connected together so that the deflections add up, i.e. the total deflection (of a further piston) at time t is

$$y_{\text{ges}}(t) = y_1(t) + y_2(t) = A_1 \cos(\omega t + \delta_1) + A_2 \cos(\omega t + \delta_2)\,.$$

Show that this is also a harmonic oscillation of the same frequency, i.e., there is an amplitude A and a phase δ with $y_{\text{ges}}(t) = A \cos(\omega t + \delta)$.

Hint: The calculation is simplified considerably by a detour through the complex. Write $y_k(t) = \text{Re}(A_k \, \mathrm{e}^{\mathrm{i}(\omega t + \delta_k)})$ and try to find y_{ges} in this form as well.

- Determine A and δ explicitly for $A_1 = 1$, $A_2 = \sqrt{3}$, $\delta_1 = 0$, $\delta_2 = \frac{7}{6}\pi$.
- Show that the set of functions $U = \{y_{A,\delta} \in \mathbb{R}^{\mathbb{R}} \mid A \geq 0,\ \delta \in \mathbb{R}\}$ with $y_{A,\delta}(t) = A \cos(\omega t + \delta)$ is a subspace of the vector space $\mathbb{R}^{\mathbb{R}}$ of all functions from $\mathbb{R}$ to $\mathbb{R}$.

Generating Systems and Linear (In)Dependence 14

Every vector space has a *basis*. A basis is a *linearly independent generating system*. So in order to even know what a basis is, you first have to understand what *linear independence* and *generating system* mean. We will do that in this chapter. Here, a generating system of a vector space is a set that makes it possible to write any vector of the vector space as a sum of multiples of the elements of the generating system. And linear independence guarantees that this representation is unique. But in any case the *representation of* a vector as a sum of multiples of other vectors is the key to everything: One speaks of *linear combinations*.

14.1 Linear Combinations

If a vector space contains two vectors v and w, then it also contains all multiples of v and w and also all sums of all multiples of v and w and of these again all multiples and of these again ... To be able to keep these formulations concise, we introduce the notion of a *linear combination*: If $v_1, \ldots, v_n \in V$ different vectors of a $\mathbb{K}$-vector space V and $\lambda_1, \ldots, \lambda_n \in \mathbb{K}$ the vector is called

$$v = \lambda_1 v_1 + \cdots + \lambda_n v_n = \sum_{i=1}^{n} \lambda_i v_i \in V$$

a **linear combination** of $v_1, \ldots, v_n$ or also from $\{v_1, \ldots, v_n\}$; one also speaks of **a representation** of v by $v_1, \ldots, v_n$. Note that such a representation need by no means be unique, e.g.

$$2\begin{pmatrix}-1\\1\end{pmatrix} + 3\begin{pmatrix}2\\-2\end{pmatrix} = \begin{pmatrix}4\\-4\end{pmatrix} = -4\begin{pmatrix}-1\\1\end{pmatrix} + 0\begin{pmatrix}2\\-2\end{pmatrix}.$$

C. Karpfinger, *Calculus and Linear Algebra in Recipes*,
https://doi.org/10.1007/978-3-662-65458-3_14

Forming linear combinations is a simple story: Choose $v_1, \ldots, v_n \in V$ and $\lambda_1, \ldots, \lambda_n \in \mathbb{K}$, then $v = \lambda_1 v_1 + \cdots + \lambda_n v_n$ is a linear combination of $v_1, \ldots, v_n$. The reverse question, whether a vector is a linear combination of other given vectors, is a bit more interesting:

Recipe: Represent a Vector as a Linear Combination

Given vectors v and $v_1, \ldots, v_n$ of a common $\mathbb{K}$-vector space V. To check whether v is a linear combination of $v_1, \ldots, v_n$ and v is to be written as a linear combination of these vectors, proceed as follows:

(1) Make the approach $\lambda_1 v_1 + \cdots + \lambda_n v_n = v$ in the indefinite $\lambda_1, \ldots, \lambda_n$.

(2) Decide whether the equation in (1) has a solution $\lambda_1, \ldots, \lambda_n$ or not.
 If so, then v is a linear combination of $v_1, \ldots, v_n$, next step.
 If no, then v is not a linear combination of $v_1, \ldots, v_n$.

(3) Determine a solution $\lambda_1, \ldots, \lambda_n$ of the equation in (1) and obtain the required representation for v.

- In the case $V = \mathbb{K}^n$ the approach in (1) yields a linear system of equations.
- In the case $V = \mathbb{K}^{m \times n}$ the approach in (1) yields a coefficient comparison of matrices.
- In the case $V = \mathbb{K}[x]$ the approach in (1) yields a coefficient comparison of polynomials.
- In the case $V = \mathbb{K}^M$ the approach in (1) yields a value comparison of functions.

Example 14.1

- We check whether the vector $v = (0, 1, 1)^\top$ is a linear combination of $v_1 = (1, 2, 3)^\top$, $v_2 = (-1, 1, -2)^\top$, $v_3 = (0, 1, 0)^\top$ and specify such a combination if necessary:

 (1) The approach $\lambda_1 v_1 + \lambda_2 v_2 + \lambda_3 v_3 = v$ yields the linear system of equations with the extended coefficient matrix

$$\left(\begin{array}{ccc|c} 1 & -1 & 0 & 0 \\ 2 & 1 & 1 & 1 \\ 3 & -2 & 0 & 1 \end{array}\right).$$

 (2) We bring the augmented coefficient matrix to row step form and recognize (unique) solvability:

$$\left(\begin{array}{ccc|c} 1 & -1 & 0 & 0 \\ 2 & 1 & 1 & 1 \\ 3 & -2 & 0 & 1 \end{array}\right) \rightsquigarrow \left(\begin{array}{ccc|c} 1 & -1 & 0 & 0 \\ 0 & 3 & 1 & 1 \\ 0 & 1 & 0 & 1 \end{array}\right) \rightsquigarrow \left(\begin{array}{ccc|c} 1 & -1 & 0 & 0 \\ 0 & 1 & 0 & 1 \\ 0 & 0 & 1 & -2 \end{array}\right).$$

(3) The (unique) solution is $(\lambda_1, \lambda_2, \lambda_3) = (1, 1, -2)$, thus

$$v = 1 \cdot \begin{pmatrix} 1 \\ 2 \\ 3 \end{pmatrix} + 1 \cdot \begin{pmatrix} -1 \\ 1 \\ -2 \end{pmatrix} + (-2) \cdot \begin{pmatrix} 0 \\ 1 \\ 0 \end{pmatrix}.$$

- We check that each vector $v = (v_1, \dots, v_n)^\top \in \mathbb{R}^n$ is a linear combination of $e_1, \dots, e_n$ and, if so, specify one:

(1) The approach $\lambda_1 e_1 + \cdots + \lambda_n e_n = v$ yields the linear system of equations with the extended coefficient matrix $(E_n \mid v)$.
(2) Since the system of equations $(E_n \mid v)$ is (uniquely) solvable for every v, every $v \in \mathbb{R}^n$ is a linear combination of $e_1, \dots, e_n$.
(3) The (unique) solution of the LGS is $(\lambda_1, \dots, \lambda_n) = (v_1, \dots, v_n)$, so

$$v = v_1 e_n + \cdots + v_n e_n .$$

- We check whether the polynomial $p = 2x + 1 \in \mathbb{R}[x]$ is a linear combination of $p_1 = x + 1$ and $p_2 = 1$ and specify one if possible:

(1) The approach $\lambda_1 p_1 + \lambda_2 p_2 = p$ yields the following polynomial equation:

$$\lambda_1 x + (\lambda_1 + \lambda_2) = 2x + 1 .$$

(2) Since the equation in (1) is solvable, p is a linear combination of p_1 and p_2.
(3) The (unique) solution of the equation in (2) is $(\lambda_1, \lambda_2) = (2, -1)$, so

$$p = 2p_1 - p_2 .$$

- We check wether the exponential function $\exp \in \mathbb{R}^{\mathbb{R}}$ is a linear combination of $\sin, \cos \in \mathbb{R}^{\mathbb{R}}$ and specify one if possible:

(1) The approach $\lambda_1 \sin + \lambda_2 \cos = \exp$ yields the following equation:

$$\lambda_1 \sin(x) + \lambda_2 \cos(x) = \exp(x) \text{ for all } x \in \mathbb{R} .$$

(2) The equation in (1) is not solvable: with $x = 0$ and $x = \pi/2$ we obtain $\lambda_2 = 1$ and $\lambda_1 = \exp(\pi/2)$ but the equation $\exp(\pi/2)\sin(x) + \cos(x) = \exp(x)$ is not correct for $x = \pi$.

Thus exp is not a linear combination of sin and cos, i.e., there is no $\lambda_1, \lambda_2 \in \mathbb{R}$ such that $\lambda_1 \sin + \lambda_2 \cos = \exp$ is satisfied. ■

14.2 The Span of *X*

The set of all linear combinations of a set X is also called the *linear hull* or the *span* of X:

The Span of *X*

If X is a nonempty subset of a $\mathbb{K}$-vector space V, then the set

$$\langle X \rangle = \left\{ \sum_{i=1}^{n} \lambda_i v_i \mid n \in \mathbb{N},\ \lambda_1, \ldots, \lambda_n \in \mathbb{K},\ v_1, \ldots, v_n \in X \right\} \subseteq V$$

of all linear combinations of X is called the **span** of X or the **linear hull** of X. We have:

- $X \subseteq \langle X \rangle$,
- $\langle X \rangle$ is a subspace of V,
- $\langle X \rangle$ is the smallest subspace U of V with $X \subseteq U$,
- $\langle X \rangle$ is the average of all subvector spaces of V containing X:

$$\langle X \rangle = \bigcap_{\substack{U \text{ subsp. of } V \\ X \subseteq U}} U .$$

- If X is finite, i.e. $X = \{v_1, \ldots, v_n\}$, then we have:

$$\langle X \rangle = \langle v_1, \ldots, v_n \rangle = \left\{ \sum_{i=1}^{n} \lambda_i v_i \mid \lambda_i \in \mathbb{K} \right\} = \mathbb{K} v_1 + \cdots + \mathbb{K} v_n .$$

- One defines complementarily $\langle \emptyset \rangle = \{0\}$.

For $\langle X \rangle = U$ one also says X **generates** the vector space U or X is a **generating system** of U.

We will often be faced with the question whether a subset X of a vector space V generates this vector space. If we can show that every vector v of V is a linear combination of X, then this question must be answered in the affirmative: $\langle X \rangle = V$. However, this typical question is in general not so easy to answer. With the dimension notion, it often becomes easier. More on this in the next chapter, but now a few examples.

Example 14.2

- For every $n \in \mathbb{N}$ is $X = \{e_1, \ldots, e_n\} \subseteq \mathbb{R}^n$ is a generating system of $\mathbb{R}^n$, because:
$$\langle X \rangle = \mathbb{R}\, e_1 + \cdots + \mathbb{R}\, e_n = \mathbb{R}^n .$$
- The set $X = \{(2, 2)^\top, (2, 1)^\top\} \subseteq \mathbb{R}^2$ generates $\mathbb{R}^2$, $\langle X \rangle = \mathbb{R}^2$, since every vector of the $\mathbb{R}^2$ is a linear combination of $(2, 2)^\top$ and $(2, 1)^\top$.
- If one chooses $X = \{(2, 1)^\top\} \subseteq \mathbb{R}^2$ then one obtains for $\langle X \rangle$ a straight line, it is $\langle X \rangle = \mathbb{R}(2, 1)^\top$ the set of all multiples of $(2, 1)^\top$.
- For $X = \mathbb{R}^3$ we have $\langle X \rangle = \mathbb{R}^3$.
- Every vector space has a generating system. In general, there are even many different ones. For example
$$\{1,\ x,\ x^2\} \text{ and } \{2,\ x+1,\ x+2,\ x+3,\ 2x^2\}$$
both are generating systems of the vector space $\mathbb{R}[x]_2$.
- $B = \{1,\ x,\ x^2,\ \ldots\}$ is a generating system of $\mathbb{R}[x]$ since every polynomial $p = a_0 + a_1 x + \cdots + a_n x^n$ is obviously a linear combination of B. ■

14.3 Linear (In)Dependence

We call a set *linearly independent* if fewer vectors also produce *less* space; in linearly independent sets, no element is superfluous in this sense:

Linear Independence and Linear Dependence

Let v be a $\mathbb{K}$-vector space.

- Vectors $v_1, \ldots, v_n \in V$ are called **linearly independent** if for each real subset $T \subsetneq \{v_1, \ldots, v_n\}$ we have:
$$\langle T \rangle \subsetneq \langle v_1, \ldots, v_n \rangle .$$
- Vectors $v_1, \ldots, v_n \in V$ are called **linearly dependent** if they are not linearly independent, i.e.:
$$v_1, \ldots, v_n \text{ lin. dep.} \Leftrightarrow \exists\, T \subsetneq \{v_1, \ldots, v_n\} \text{ with } \langle T \rangle = \langle v_1, \ldots, v_n \rangle .$$

(continued)

- A set $X \subseteq V$ of vectors is called **linearly independent** if the elements $v_1, \ldots, v_n$ of every nonempty finite subset of X are linearly independent (and correspondingly **linearly dependent**, if X is not linearly independent).

Example 14.3 The three vectors

$$v_1 = (1, 0)^\top, \quad v_2 = (0, 1)^\top \text{ and } v_3 = (1, 1)^\top$$

are linearly dependent, because it is

$$\langle v_1, v_2, v_3 \rangle = \mathbb{R}^2 = \langle v_1, v_2 \rangle = \langle v_1, v_3 \rangle = \langle v_2, v_3 \rangle .$$

If we consider only two of the three vectors at a time, e.g. v_1, v_2, but they are linearly independent: For each real subset $T \subsetneq \{v_1, v_2\}$ is $T = \{v_1\}$ or $T = \{v_2\}$ or $T = \emptyset$ and for these subsets

$$\langle \emptyset \rangle = \{0\}, \langle v_1 \rangle = \mathbb{R}v_1, \langle v_2 \rangle = \mathbb{R}v_2 .$$

In each of these three cases we have $\langle T \rangle \subsetneq \mathbb{R}^2 = \langle v_1, v_2 \rangle$. ■

If one wants to prove whether given vectors are linearly dependent or linearly independent, this is not easy with the definition. Almost always the following procedure leads to the goal:

Recipe: Proof of Linear (In)Dependence

Given is a subset $X \subseteq V$ of the $\mathbb{K}$-vector space V with zero vector 0_V. We examine the set X on linear (in)dependence:

(1) Make the approach $\lambda_1 v_1 + \cdots + \lambda_n v_n = 0_V$ with $\lambda_1, \ldots, \lambda_n \in \mathbb{K}$ where

- $X = \{v_1, \ldots, v_n\}$, if X is finite and
- $v_1, \ldots, v_n \in X$ is any finite choice of elements of X if X is not finite.

(2) If the equation in (1) is only possible for $\lambda_1 = \cdots = \lambda_n = 0$ then X is linearly independent.

If the equation in (1) is possible for $\lambda_1, \ldots, \lambda_n$, whereas not all λ_i are zeroes, then X linearly dependent.

Note: The approach $\lambda_1 v_1 + \cdots + \lambda_n v_n = 0_V$ in (1) always has the solution $\lambda_1 = \cdots = \lambda_n = 0$, which is called the *trivial solution*. In (2), the question is whether this is the only solution. If yes: linearly independent, if no: linearly dependent. Therefore, the following way of speaking is also common: If the approach in (1) also has a *nontrivial solution*, then $v_1, \ldots, v_n$ linearly dependent.

In the following examples we write 0 instead of 0_V and always remember that this 0 is the zero vector of the vector space from which the vectors $v_1, \ldots, v_n$ are from.

Example 14.4

- We test the zero vector for linear (in)dependence: (1) $\lambda 0 = 0$. (2) The equation in (1) not only has the solution $\lambda = 0$, but also $\lambda = 1$ satisfies $\lambda 0 = 0$, so 0 is linearly dependent.
- If $v \neq 0$, then v is linearly independent, because the approach (1) $\lambda v = 0$ yields (2) $\lambda = 0$. Note the calculation rule at the end of Sect. 13.1.
- For all natural numbers $n \in \mathbb{N}$ the standard unit vectors $e_1, \ldots, e_n$ are linearly independent, because the approach

$$(1)\ \lambda_1 e_1 + \cdots + \lambda_n e_n = \begin{pmatrix} \lambda_1 \\ \vdots \\ \lambda_n \end{pmatrix} = \begin{pmatrix} 0 \\ \vdots \\ 0 \end{pmatrix} \text{ provides } (2)\ \lambda_1 = \cdots = \lambda_n = 0\,.$$

- On the other hand the three vectors

$$v_1 = \begin{pmatrix} 0 \\ 1 \\ 1 \end{pmatrix}, \quad v_2 = \begin{pmatrix} 1 \\ 1 \\ 1 \end{pmatrix} \quad \text{and} \quad v_3 = \begin{pmatrix} 1 \\ 0 \\ 0 \end{pmatrix}$$

are linearly dependent: The approach $(1)\lambda_1 v_1 + \lambda_2 v_2 + \lambda_3 v_3 = 0$ yields the following LGS, which we immediately put into row step form:

$$\left(\begin{array}{ccc|c} 0 & 1 & 1 & 0 \\ 1 & 1 & 0 & 0 \\ 1 & 1 & 0 & 0 \end{array}\right) \rightsquigarrow \left(\begin{array}{ccc|c} 1 & 0 & -1 & 0 \\ 0 & 1 & 1 & 0 \\ 0 & 0 & 0 & 0 \end{array}\right).$$

(2) This shows that $(\lambda_1, \lambda_2, \lambda_3) = (1, -1, 1)$ is a nontrivial solution: $1v_1 + (-1)v_2 + 1v_3 = 0$.
- The (infinite) set $\{1, x, x^2, x^3, \ldots\} \in \mathbb{R}[x]$ is linearly independent: The approach (1) $\lambda_1 x^{r_1} + \cdots + \lambda_n x^{r_n} = 0$ yields by a coefficient comparison (for the zero polynomial 0, all coefficients are zero) (2): $\lambda_1 = \cdots = \lambda_n = 0$.

- The vectors sin, cos $\in \mathbb{R}^{\mathbb{R}}$ are linearly independent: The approach

$$(1)\ \lambda_1 \cos + \lambda_2 \sin = 0$$

with the zero function $0 : \mathbb{R} \to \mathbb{R}$, $0(x) = 0$ yields the equation:

$$\lambda_1 \cos(x) + \lambda_2 \sin(x) = 0(x) = 0 \text{ for all } x \in \mathbb{R}\,.$$

In particular, this equation yields for $x = 0$ and for $x = \pi/2$:

$$\lambda_1 \cos(0) + \lambda_2 \sin(0) = 0 \text{ and } \lambda_1 \cos(\pi/2) + \lambda_2 \sin(\pi/2) = 0\,.$$

(2) Because of $\sin(0) = 0$ and $\cos(\pi/2) = 0$ this has $\lambda_1 = 0$ and $\lambda_2 = 0$ as a consequence. ■

14.4 Exercises

14.1 For which $r \in \mathbb{R}$ are the following three column vectors from $\mathbb{R}^4$ linearly dependent?

$$\begin{pmatrix} 1 \\ 2 \\ 3 \\ r \end{pmatrix}, \begin{pmatrix} 1 \\ 3 \\ r \\ 0 \end{pmatrix} \text{ and } \begin{pmatrix} 1 \\ r \\ 3 \\ 2 \end{pmatrix}.$$

14.2 Let $A \in \mathbb{R}^{m \times n}$ and vectors $v_1, v_2, \ldots, v_k \in \mathbb{R}^n$ are given. Show:

(a) If $Av_1, Av_2, \ldots, Av_k$ are linearly independent, then this is also true for $v_1, v_2, \ldots, v_k$.
(b) In general, the converse of statement (a) is false.
(c) If $m = n$ and A is invertible, the converse of statement (a) is also true.

14.3 Is the set $\{\cos, \sin, \exp\} \subseteq \mathbb{R}^{\mathbb{R}}$ linearly dependent or linearly independent?

14.4 Prove the following statement or give a counterexample to disprove it: *Given the vectors $x, y, z \in \mathbb{R}^4$. Let the vectors x, y and x, z and y, z be pairwise linearly independent. Then x, y, z are also linearly independent.*

14.5 Are the following sets linearly dependent or linearly independent? Justify your answer. For every of the linearly dependent sets, find as large a linearly independent subset as possible. Also give the linear hull of the sets.

(a) $M_1 = \{(1, 2, 3)^\top, (3, 7, 0)^\top, (1, 3, -6)^\top\}$ in the $\mathbb{R}$-vector space $\mathbb{R}^3$.
(b) $M_2 = \{\mathrm{i}, 1 - \mathrm{i}^2\}$ in the $\mathbb{R}$-vector space $\mathbb{C}$.
(c) $M_3 = \{\mathrm{i}, 1 - \mathrm{i}^2\}$ in the $\mathbb{C}$-vector space $\mathbb{C}$.
(d) $M_4 = \{a_0 + a_1X + a_2X^2 \mid a_0 = a_1 - a_2\}$ in the $\mathbb{R}$-vector space $\mathbb{R}[X]_2$.
(e) $M_5 = \{X^2 - 2, X + 1, X\}$ in the $\mathbb{R}$-vector space $\mathbb{R}[X]_4$.
(f) $M_6 = \{\left(\begin{smallmatrix} 1 & 2 \\ -1 & 5 \end{smallmatrix}\right), \left(\begin{smallmatrix} 2 & 1 \\ 4 & 1 \end{smallmatrix}\right), \left(\begin{smallmatrix} -1 & 2 \\ 2 & -4 \end{smallmatrix}\right), \left(\begin{smallmatrix} 4 & -1 \\ 3 & -3 \end{smallmatrix}\right), \left(\begin{smallmatrix} 3 & 1 \\ 2 & -1 \end{smallmatrix}\right)\}$ in the $\mathbb{R}$-vector space $\mathbb{R}^{2\times 2}$.

14.6 Show that the recipe in Sect. 14.3 gives the correct result.

Bases of Vector Spaces 15

Every vector space V has a *basis* B. A basis is thereby a minimal generating system, in other words a linearly independent generating system, i.e., a basis B generates the vector space, and thereby no element in B is *superfluous*. By specifying a basis, a vector space is completely determined. In this sense, bases will be useful to us: Instead of specifying the vector space, we specify a basis; thus we have the vector space.

A vector space generally has many different bases, but every two bases of a vector space have one thing in common: the number of elements of the bases. This number is called the *dimension of* a vector space. If one knows the dimension of a vector space, much is gained: It can then be quickly decided whether a generating system or a linearly independent set is a basis or not.

As always $\mathbb{K}$ denotes the set of numbers $\mathbb{R}$ or $\mathbb{C}$.

15.1 Bases

The central concept of this chapter is the concept of *bases*:

Basis

A subset B of a $\mathbb{K}$-vector space V is called **basis** of V if

- B is a generating system of V, $\langle B \rangle = V$, and
- B is linearly independent.

C. Karpfinger, *Calculus and Linear Algebra in Recipes*,
https://doi.org/10.1007/978-3-662-65458-3_15

Example 15.1

- For every natural number $n \in \mathbb{N}$ the set $E_n = \{e_1, \ldots, e_n\}$ of the standard unit vectors is a basis of $\mathbb{K}^n$. This basis is called the **standard basis** or also **canonical basis** of $\mathbb{K}^n$.
- For every natural number $n \in \mathbb{N}$ also the set

$$B = \left\{ \begin{pmatrix} 1 \\ 1 \\ \vdots \\ 1 \end{pmatrix}, \begin{pmatrix} 1 \\ \vdots \\ 1 \\ 0 \end{pmatrix}, \ldots, \begin{pmatrix} 1 \\ 0 \\ \vdots \\ 0 \end{pmatrix} \right\}$$

is a basis of $\mathbb{K}^n$ because if we number the vectors as $b_1, \ldots, b_n$ then B is linearly independent, because the approach (1)

$$\lambda_1 b_1 + \cdots + \lambda_n b_n = \lambda_1 \begin{pmatrix} 1 \\ 1 \\ \vdots \\ 1 \end{pmatrix} + \lambda_2 \begin{pmatrix} 1 \\ \vdots \\ 1 \\ 0 \end{pmatrix} + \cdots + \lambda_n \begin{pmatrix} 1 \\ 0 \\ \vdots \\ 0 \end{pmatrix} = \begin{pmatrix} 0 \\ 0 \\ \vdots \\ 0 \end{pmatrix}$$

delivers (2) $\lambda_1 = \lambda_2 = \cdots = \lambda_n = 0$. Furthermore, B is a generating system of the $\mathbb{K}^n$ since the linear system of equations

$$\left(\begin{array}{ccc|c} 1 & \ldots & 1 & v_1 \\ \vdots & \cdot^{\cdot^{\cdot}} & & \vdots \\ 1 & & 0 & v_n \end{array} \right)$$

for all $(v_1, \ldots, v_n)^\top \in \mathbb{R}^n$ is solvable.
- The set $B = \{1, x, x^2, \ldots\}$ is a basis of $\mathbb{K}[x]$, B is namely linearly independent and a generating system of $\mathbb{K}[x]$, note the Examples 14.2 and 14.4.
- If $v_1, \ldots, v_r$ is linearly independent in V, then $B = \{v_1, \ldots, v_r\}$ a basis of the vector space $\langle v_1, \ldots, v_r \rangle$.
- In $\mathbb{K}^{m \times n}$ the set

$$B = \left\{ E_{11}, E_{12}, \ldots, E_{1n}, E_{21}, \ldots, E_{mn} \right\},$$

where E_{ij} except for a one at the position (i, j) contains only zeros as entries, is a basis of $\mathbb{K}^{m \times n}$. The E_{ij} are called **standard unit matrices**. ■

Mnemonic Rules and Important Theorems

- **Existence Theorems:**
  - Any $\mathbb{K}$-vector space V has a basis.
  - Every generating system of V contains a basis of V.
  - Any linearly independent subset of V can be complemented to a basis.
- **Uniqueness:** If B is a basis of V, then every $v \in V$ can be written, except for the order of the summands, in exactly one way in the form

$$v = \lambda_1 b_1 + \cdots + \lambda_r b_r$$

  with $b_1, \ldots, b_r \in B$ and $\lambda_1, \ldots, \lambda_r \in \mathbb{K}$.
- **Equal cardinality of bases:** Two bases B and B' from V have the same number of elements.
- **Dimension:** If B is a basis of the $\mathbb{K}$-vector space V, then $|B|$ is called the **dimension** of V. One writes $\dim(V)$ for the dimension.
- **Useful rules:**
  - If $\dim(V) = n$, so each n linearly independent vectors form a basis.
  - If $\dim(V) = n$ then every generating system with n elements is a basis.
  - If $\dim(V) = n$, then more than n vectors are linearly dependent.
  - If U is a subspace of V with $U \subsetneq V$, then $\dim(U) < \dim(V)$.
  - If U is a subspace of V with $\dim(U) = \dim(V)$, then $U = V$.

Example 15.2

- For every $n \in \mathbb{N}$ we have $\dim\left(\mathbb{K}^n\right) = n$ since the canonical basis $B = \{e_1, \ldots, e_n\}$ contains exactly n elements.
- For all $m, n \in \mathbb{N}$ we have $\dim\left(\mathbb{K}^{m\times n}\right) = m\,n$, since the basis $B = \left\{E_{11}, E_{12}, \ldots, E_{mn}\right\}$ of the standard unit matrices has exactly $m\,n$ elements.
- The vector space $\mathbb{R}[x]$ of the real polynomials has infinite dimension, $\dim\left(\mathbb{R}[x]\right) = \infty$ since the basis $B = \{1, x, x^2, \ldots\}$ is not finite.
- The vector space $U = \langle (0, 1, 0)^\top, (0, 1, 1)^\top, (0, 0, 1)^\top \rangle$ has the dimension 2, $\dim(U) = 2$, because $B = \{(0, 1, 0)^\top, (0, 0, 1)^\top\}$ is a basis of U.
- The vector space $\mathbb{R}[x]_2$ of polynomials of degree less than or equal to 2 has dimension 3, $\dim\left(\mathbb{R}[x]_2\right) = 3$, since $B = \{1, x, x^2\}$ a basis of $\mathbb{R}[x]_2$. ■

The second and the third existence theorems are often formulated as tasks, the solution of which we describe in the following recipe:

Recipe: Shortening a Generating System and Extending a Linearly Independent Set to a Basis

- Let X be a generating system of a vector space V. Then one determines a basis $B \subseteq X$ from V as follows:

  (1) Check whether X is linearly independent.
  If yes: X is a basis.
  If no: Remove from X elements $a_1, \ldots, a_r$, which are linear combinations of other elements from X, and set $\tilde{X} = X \setminus \{a_1, \ldots, a_r\}$.
  (2) Start with $\tilde{X}$ instead of X from the beginning.

- Let X be a linearly independent subset of a vector space V. One then determines a basis $B \supseteq X$ from V as follows:

  (1) Check whether X is a generating system of V.
  If yes: X is a basis.
  If no: Choose from V elements $a_1, \ldots, a_r$ so that $X \cup \{a_1, \ldots, a_r\}$ is linearly independent, and set $\tilde{X} = X \cup \{a_1, \ldots, a_r\}$.
  (2) Start with $\tilde{X}$ instead of X from the beginning.

For column vectors, i.e. $X = \{v_1, \ldots, v_s\}$, $v_i \in \mathbb{K}^n$ these two tasks can be carried out quite simply and with one method: One writes the columns v_i as rows $v_i^\top$ into a matrix M and applies elementary row transformations to get the matrix to row step form:

$$M \rightsquigarrow M' = \begin{pmatrix} * & * & * & * & * \\ & * & * & * & \vdots \\ & & * & * & * \\ 0 & & & & 0 \end{pmatrix} \begin{matrix} \left.\vphantom{\begin{matrix} * \\ * \\ * \end{matrix}}\right\} r \text{ rows not equal } 0 \ldots 0 \\ \left.\vphantom{0}\right\} s-r \text{ rows equal } 0 \ldots 0 \end{matrix}$$

Then we have:

- The transpose of the first r rows of M' form a basis $B = \{b_1, \ldots, b_r\}$ of $\langle X \rangle$, the $s - r$ column vectors from which the last $s - r$ zero rows emerged are linear combinations of B. By choosing the first r rows, one has *shortened* the generating system to a linearly independent generating system of $\langle X \rangle$.

- If you add to the matrix M' by $n-r$ further rows that continue the row step form, then one *extends* the linearly independent set of r vectors to a linearly independent set with n vectors, one thus *extends* the linearly independent set to a basis of the $\mathbb{K}^n$.

Example 15.3

- Given the set

$$E = \left\{ \begin{pmatrix}1\\1\\0\\0\end{pmatrix}, \begin{pmatrix}1\\0\\1\\0\end{pmatrix}, \begin{pmatrix}1\\0\\0\\1\end{pmatrix}, \begin{pmatrix}0\\1\\1\\0\end{pmatrix}, \begin{pmatrix}0\\1\\0\\1\end{pmatrix}, \begin{pmatrix}0\\0\\1\\1\end{pmatrix} \right\} \subseteq \mathbb{R}^4 .$$

We determine a basis of $\langle E\rangle$: First we number the vectors from left to right with $v_1, \ldots, v_6$. We write the columns as rows in a matrix and perform elementary row transformations:

$$\begin{pmatrix}1&1&0&0\\1&0&1&0\\1&0&0&1\\0&1&1&0\\0&1&0&1\\0&0&1&1\end{pmatrix} \rightsquigarrow \begin{pmatrix}1&1&0&0\\0&-1&1&0\\0&-1&0&1\\0&1&1&0\\0&1&0&1\\0&0&1&1\end{pmatrix} \rightsquigarrow \begin{pmatrix}1&1&0&0\\0&1&1&0\\0&0&-1&1\\0&0&2&0\\0&0&1&1\\0&0&1&1\end{pmatrix} \rightsquigarrow \begin{pmatrix}1&1&0&0\\0&1&1&0\\0&0&1&-1\\0&0&0&2\\0&0&0&0\\0&0&0&0\end{pmatrix} .$$

From the row step form, we read: the first four rows are linearly independent, and because of the two zero rows, v_5 and v_6 are linear combinations of $v_1, \ldots, v_4$. Thus, $B = \{v_1, \ldots, v_4\}$ is a basis of $\langle E\rangle$.

- Now consider the set

$$E = \left\{ \begin{pmatrix}1\\-2\\3\\4\end{pmatrix}, \begin{pmatrix}2\\-3\\6\\11\end{pmatrix}, \begin{pmatrix}-1\\3\\-2\\6\end{pmatrix} \right\} \subseteq \mathbb{R}^4 .$$

We determine a basis B of the $\mathbb{R}^4$ with $E \subseteq B$. Again the vectors in E we call $v_1, \ldots, v_3$. We write the columns as rows in a matrix and apply elementary row transformations:

$$\begin{pmatrix}1&-2&3&4\\2&-3&6&11\\-1&3&-2&6\end{pmatrix} \rightsquigarrow \begin{pmatrix}1&-2&3&4\\0&1&0&3\\0&1&1&10\end{pmatrix} \rightsquigarrow \begin{pmatrix}1&-2&3&4\\0&1&0&3\\0&0&1&7\end{pmatrix} .$$

We recognize that $v_1,\ v_2,\ v_3$ are already linearly independent and furthermore: If we add e_4 then we have linearly independent vectors $v_1,\ v_2,\ v_3,\ e_4$ because it also has

$$\begin{pmatrix} 1 & -2 & 3 & 4 \\ 0 & 1 & 0 & 3 \\ 0 & 0 & 1 & 7 \\ 0 & 0 & 0 & 1 \end{pmatrix}$$

a row step form. Thus, $B = \{v_1,\ v_2,\ v_3,\ e_4\}$ is a basis of the $\mathbb{R}^4$. ■

15.2 Applications to Matrices and Systems of Linear Equations

We consider an $m \times n$-matrix $A \in \mathbb{K}^{m\times n}$ with the columns $s_1, \ldots, s_n \in \mathbb{K}^m$ and the rows $z_1, \ldots, z_m \in \mathbb{K}^{1\times n}$:

$$A = (s_1, \ldots, s_n) = \begin{pmatrix} z_1 \\ \vdots \\ z_m \end{pmatrix} = \begin{pmatrix} a_{11} & \ldots & a_{1n} \\ \vdots & & \vdots \\ a_{m1} & \ldots & a_{mn} \end{pmatrix}.$$

Row Rank = Column Rank

One calls

- the span of the columns of A, i.e. $S_A = \langle s_1, \ldots, s_n\rangle \subseteq \mathbb{K}^m$ the **column space** of A and $\dim(S_A)$ the **column rank** of A and
- the span of the rows of A, that is $Z_A = \langle z_1, \ldots, z_m\rangle \subseteq \mathbb{K}^{1\times n}$ the **row space** of A and $\dim(Z_A)$ the **row rank** of A.

We have:

- $S_A = \left\{A\, v \mid v \in \mathbb{K}^n\right\}$ and $Z_A = \left\{v^\top A \mid v \in \mathbb{K}^m\right\}$ and
- rank of A = row rank of A = column rank of A.

How can you determine the row and column rank of A?

Recipe: Determine Row/Column/Space/Rank

Given is the matrix

$$A = \begin{pmatrix} z_1 \\ \vdots \\ z_m \end{pmatrix} = (s_1, \dots, s_n) = \begin{pmatrix} a_{11} & \dots & a_{1n} \\ \vdots & & \vdots \\ a_{m1} & \dots & a_{mn} \end{pmatrix} \in \mathbb{K}^{m \times n}.$$

- Apply elementary row transformations to A and get:

$$A \rightsquigarrow \dots \rightsquigarrow \begin{pmatrix} * & * & * & * & * \\ & * & * & * & \vdots \\ & & * & * & * \\ 0 & & & & 0 \end{pmatrix} = A'.$$

  Then we have: the nonzero rows of A' form a base of the row space of A, the number of these rows is the row rank of A.
- Apply to A *elementary column transformations* (analogous to the elementary row transformations) and get:

$$A \rightsquigarrow \dots \rightsquigarrow \begin{pmatrix} * & & & 0 \\ * & * & & \\ * & * & * & \\ * & \cdots & \cdots * & 0 \end{pmatrix} = A''.$$

  Then we have: the nonzero columns of A' form a base of the column space of A, the number of these columns is the column rank of A.

Alternatively, when determining the column space of A, you can also transpose the matrix A and apply row transformations as usual to $A^\top$.

Example 15.4 We compute row and column space and row and column rank of the following square matrix A by elementary row transformations:

$$A = \begin{pmatrix} 1 & 1 & 1 \\ 1 & 2 & 4 \\ 2 & 3 & 5 \end{pmatrix} \rightsquigarrow \begin{pmatrix} 1 & 1 & 1 \\ 0 & 1 & 3 \\ 0 & 1 & 3 \end{pmatrix} \rightsquigarrow \begin{pmatrix} 1 & 1 & 1 \\ 0 & 1 & 3 \\ 0 & 0 & 0 \end{pmatrix}.$$

Thus, the row space is $Z_A = \langle(1, 1, 1), (0, 1, 3)\rangle$ and the row rank $\dim(Z_A) = 2$. We now perform elementary column transformations and get

$$\begin{pmatrix} 1 & 1 & 1 \\ 1 & 2 & 4 \\ 2 & 3 & 5 \end{pmatrix} \rightsquigarrow \begin{pmatrix} 1 & 0 & 0 \\ 1 & 1 & 3 \\ 2 & 1 & 3 \end{pmatrix} \rightsquigarrow \begin{pmatrix} 1 & 0 & 0 \\ 1 & 1 & 0 \\ 2 & 1 & 0 \end{pmatrix}.$$

Thus, the column space is $S_A = \langle(1, 1, 2)^\top, (0, 1, 1)^\top\rangle$ and the column rank $\dim(S_A) = 2$. ■

The Kernel of a Matrix and Linear Systems of Equations

If $A \in \mathbb{K}^{m\times n}$ then the set

$$\ker(A) = \{v \in \mathbb{K}^n \mid Av = 0\} \subseteq \mathbb{K}^n$$

is called the **kernel** of A. We have:

- The kernel of A is the solution set of the homogeneous linear system of equations $Ax = 0$.
- The kernel of a matrix $A \in \mathbb{K}^{m\times n}$ is a subspace of $\mathbb{K}^n$.
- $\dim(\ker(A)) = n - \mathrm{rg}(A)$.
- For a square matrix $A \in \mathbb{K}^{n\times n}$ we have:

$$\dim(\ker(A)) = \text{number of zero rows in row step form.}$$

- The LGS $Ax = b$ with $A \in \mathbb{K}^{m\times n}$ and $b \in \mathbb{K}^m$ is solvable if and only if $b \in S_A = \{Av \mid v \in \mathbb{K}^n\}$.

To determine the kernel of a matrix $A \in \mathbb{K}^{m\times n}$ the homogeneous linear system of equations with the extended coefficient matrix $(A \mid 0)$ must be solved. To do this, one performs elementary row transformations on the matrix A by. Since the zero column $(\ \mid 0)$ do not change anyway, we omit it from such calculations. Since the kernel of a matrix, i.e. the solution set of $(A \mid 0)$ is a vector space, we can specify it by a basis. The dimension of the kernel is $n - \mathrm{rg}(A)$, in the case of a square matrix A is even exactly equal to the number of zero rows of the row step form of A. Thus, we have a basis of the kernel of a square matrix if we specify as many linearly independent vectors of the kernel as the row step form of A has zero rows. And whether a vector v lies in the kernel of A, the easiest way to tell is that $A'v = 0$ if A' is a (reduced) row step form of A. Note in the following examples:

- The dimension of the kernel is equal to the number of zero rows.
- The column vectors v in the generating system satisfy $A'v = 0$, where A' denotes the (reduced) row level form; of course, the v are chosen in this way.
- Moreover, the column vectors in the generating system have always been chosen so that their linear independence is apparent.

Example 15.5

- $$\ker \begin{pmatrix} 1 & 2 & 3 \\ 4 & 5 & 6 \\ 7 & 8 & 9 \end{pmatrix} = \ker \begin{pmatrix} 1 & 2 & 3 \\ 0 & 3 & 6 \\ 0 & 6 & 12 \end{pmatrix} = \ker \begin{pmatrix} 1 & 0 & -1 \\ 0 & 1 & 2 \\ 0 & 0 & 0 \end{pmatrix} = \langle \begin{pmatrix} 1 \\ -2 \\ 1 \end{pmatrix} \rangle.$$

- $$\ker \begin{pmatrix} -1 & -1 & 2 \\ 1 & 2 & 3 \\ -1 & 0 & 7 \end{pmatrix} = \ker \begin{pmatrix} 1 & 1 & -2 \\ 0 & 1 & 5 \\ 0 & 2 & 10 \end{pmatrix} = \ker \begin{pmatrix} 1 & 0 & -7 \\ 0 & 1 & 5 \\ 0 & 0 & 0 \end{pmatrix} = \langle \begin{pmatrix} 7 \\ -5 \\ 1 \end{pmatrix} \rangle.$$

- $$\ker \begin{pmatrix} 1 & 2 & 3 & 4 \\ 2 & 4 & 6 & 8 \\ 3 & 6 & 9 & 12 \\ 4 & 8 & 12 & 16 \end{pmatrix} = \ker \begin{pmatrix} 1 & 2 & 3 & 4 \\ 0 & 0 & 0 & 0 \\ 0 & 0 & 0 & 0 \\ 0 & 0 & 0 & 0 \end{pmatrix} = \langle \begin{pmatrix} -2 \\ -1 \\ 0 \\ 0 \end{pmatrix} \rangle \langle \begin{pmatrix} 3 \\ 0 \\ -1 \\ 0 \end{pmatrix} \rangle = \langle \begin{pmatrix} 4 \\ 0 \\ 0 \\ -1 \end{pmatrix} \rangle.$$

- $$\ker \begin{pmatrix} 4 & 2 & 2 \\ 2 & 1 & 1 \\ 2 & 1 & 1 \end{pmatrix} = \ker \begin{pmatrix} 2 & 1 & 1 \\ 0 & 0 & 0 \\ 0 & 0 & 0 \end{pmatrix} = \langle \begin{pmatrix} 0 \\ -1 \\ 1 \end{pmatrix}, \begin{pmatrix} 1 \\ -2 \\ 0 \end{pmatrix} \rangle.$$

- $$\ker \begin{pmatrix} -2 & 2 & 2 \\ 2 & -5 & 1 \\ 2 & 1 & -5 \end{pmatrix} = \ker \begin{pmatrix} 1 & -1 & -1 \\ 0 & -3 & 3 \\ 0 & 0 & 0 \end{pmatrix} = \langle \begin{pmatrix} 2 \\ 1 \\ 1 \end{pmatrix} \rangle.$$ ■

15.3 Exercises

15.1 Show that for every $n \in \mathbb{N}$ the set

$$U = \left\{ u = (u_1, \ldots, u_n)^\top \in \mathbb{R}^n \mid u_1 + \cdots + u_n = 0 \right\}$$

forms a vector space. Determine a basis and the dimension of U.

15.2 Determine the dimension of the vector space

$$\langle f_1 : x \mapsto \sin(x),\ f_2 : x \mapsto \sin(2x),\ f_3 : x \mapsto \sin(3x) \rangle \subseteq \mathbb{R}^{\mathbb{R}}.$$

15.3 Let the vectors $u, v \in \mathbb{R}^3$ with $u = (1, -3, 2)^\top$ and $v = (2, -1, 1)^\top$ are given. Check whether $p = (1, 7, -4)^\top$ respectively $q = (2, -5, 4)^\top$ are linear combinations of u and v. If necessary, calculate the representation of p and q with respect to the basis $\{u, v\}$.

15.4 Consider the following homogeneous linear system of equations for $x_1, x_2, x_3, x_4 \in \mathbb{C}$:

$$\begin{aligned} \mathrm{i}x_1 + 4x_2 - (2+\mathrm{i})x_3 - x_4 &= 0 \\ x_1 - \phantom{(2+\mathrm{i})} 5x_3 - 2x_4 &= 0\,. \\ x_1 - \phantom{(2+\mathrm{i})5} x_3 + 2x_4 &= 0 \end{aligned}$$

(a) What is the maximum dimension of the solution space of a system of equations of the above type? What is the minimum size it must be?
(b) Calculate the solution space and give a basis for it.

15.5

(a) Show that $B = \{1, 1-x, (1-x)^2, (1-x)^3\}$ is a basis of the polynomial space $\mathbb{R}[x]_3$.
(b) Give the representation of $p = x^3 - 2x^2 + 7x + 5$ with respect to the basis B.

15.6 A vector space $V \subseteq \mathbb{R}[x]_3$ is generated by the following four polynomials:

$$p = x^3 - 2x^2 + 4x + 1, \qquad r = x^3 + 6x - 5,$$

$$q = 2x^3 - 3x^2 + 9x - 1, \qquad s = 2x^3 - 5x^2 + 7x + 5.$$

Determine $\dim V$ and give a basis of V.

15.7 Determine a basis of the subspace $U = \langle X \rangle$ of $\mathbb{R}^4$ generated by the set

$$X = \left\{ \begin{pmatrix} 0 \\ 1 \\ 0 \\ -1 \end{pmatrix}, \begin{pmatrix} 1 \\ 0 \\ 1 \\ -2 \end{pmatrix}, \begin{pmatrix} -1 \\ -2 \\ 0 \\ 1 \end{pmatrix}, \begin{pmatrix} -1 \\ 0 \\ 1 \\ 0 \end{pmatrix}, \begin{pmatrix} 1 \\ 0 \\ -1 \\ -1 \end{pmatrix}, \begin{pmatrix} 2 \\ 0 \\ -1 \\ 0 \end{pmatrix} \right\}.$$

15.8 Compute the rank and basis each of the kernel, column space, and row space of the following matrices:

$$A = \begin{pmatrix} 1 & 1 & 1 \\ 2 & 1 & 3 \\ 4 & -2 & 1 \end{pmatrix} \quad B = \begin{pmatrix} 2 & 0 & 0 \\ 3 & 0 & 0 \\ 0 & 2 & 0 \end{pmatrix} \quad C = \begin{pmatrix} 1 & 2 \\ 2 & 1 \\ 3 & 2 \\ 2 & 3 \end{pmatrix}$$

15.9 Show that $S_A = \{Av \mid v \in \mathbb{R}^n\}$.

15.10 Prove the statements in the second box in Sect. 15.2.

15.11 We consider the vector space

$$V = \mathbb{R}[x]_2 = \{a_0 + a_1 x + a_2 x^2 | a_0, a_1, a_2 \in \mathbb{R}\}$$

of polynomials of degree ≤ 2. For this, it is well known that $S_0 = \{1, x, x^2\}$ is a basis. Consider further the subsets

$$S_1 = \{x, 2x\},\ S_2 = \{1, x, x+1, x-1\},\ S_3 = \{x^2 + x, x^2 - x\}$$

$$S_4 = \{1, x+1, x^2 + x + 1\},\ S_5 = \{1 + x, x, x^2 + 1, x + 2\},\ S_6 = \{x, x+1, x-1\}$$

of V. Investigate whether the sets $S_1, \ldots, S_6$ are linearly independent or a generating system or a basis of V. Also, for each of the subspaces, determine a basis $B_i \subseteq S_i$ of $\langle S_i \rangle$.

15.12 Complete the linearly independent subset $S \subseteq \mathbb{R}^n$ with vectors from the standard basis $\{e_1, e_2, \ldots, e_n\}$ to a basis of $\mathbb{R}^n$:

(a)

$$S = \left\{ \begin{pmatrix} 1 \\ 2 \\ 0 \end{pmatrix} \right\} \subseteq \mathbb{R}^3,$$

(b)

$$S = \left\{ \begin{pmatrix} -1 \\ 0 \\ 3 \end{pmatrix} \right\} \subseteq \mathbb{R}^3,$$

(c)

$$S = \left\{ \begin{pmatrix} 1 \\ 1 \\ 3 \end{pmatrix} \right\}, \left\{ \begin{pmatrix} 1 \\ 1 \\ 2 \end{pmatrix} \right\} \subseteq \mathbb{R}^3,$$

(d)

$$S = \left\{ \begin{pmatrix} 1 \\ 1 \\ 0 \\ 0 \end{pmatrix} \right\}, \left\{ \begin{pmatrix} 0 \\ 0 \\ 1 \\ 1 \end{pmatrix} \right\} \subseteq \mathbb{R}^4,$$

(e)

$$S = \left\{ \begin{pmatrix} 1 \\ 1 \\ 1 \\ 1 \end{pmatrix} \right\}, \left\{ \begin{pmatrix} 2 \\ 1 \\ 1 \\ 1 \end{pmatrix} \right\}, \left\{ \begin{pmatrix} 1 \\ 1 \\ 1 \\ 2 \end{pmatrix} \right\} \subseteq \mathbb{R}^4.$$

Orthogonality I

16

If a vector space has a *scalar product*, then one can assign a *length* to each vector of this vector space and a *distance* or an intervening *angle* to each two vectors and also ask whether two vectors are *orthogonal*. Thereby a *scalar product* is a product of vectors, where the result is a scalar.

As descriptive as these notions may be, much of the content of the present chapter will not be descriptive: Namely, we also consider vector spaces not equal to the $\mathbb{R}^2$ or $\mathbb{R}^3$, e.g. the vector space of all continuous functions on an interval $[a, b]$. Orthogonality, angles and distances are then not given by the view, but result by evaluating formulas. This step of abstraction, simply applying formulas and suppressing any visualization, is usually difficult for first-year students, even though it sounds so simple. However, this step of abstraction is important, and we will return to the issues raised here in later chapters.

16.1 Scalar Products

We consider a real vector space V. We say a mapping

$$s : \begin{cases} V \times V \to & \mathbb{R} \\ (v, w) \mapsto s(v, w) \end{cases}$$

- is **bilinear** if for all $v,\ v',\ w,\ w' \in V$ and $\lambda \in \mathbb{R}$ we have:
  - $s(\lambda\, v + v', w) = \lambda\, s(v, w) + s(v', w)$. *Linearity in the 1st argument*
  - $s(v, \lambda\, w + w') = \lambda\, s(v, w) + s(v, w')$. *Linearity in the 2nd argument*
- is **symmetric** if for all $v,\ w \in V$ we have:

$$s(v, w) = s(w, v)\,.$$

C. Karpfinger, *Calculus and Linear Algebra in Recipes*,
https://doi.org/10.1007/978-3-662-65458-3_16

- is **positive definite** if for all $v \in V$ we have:

$$s(v, v) \geq 0 \quad \text{and} \quad s(v, v) = 0 \Leftrightarrow v = 0\,.$$

A positive definite symmetric bilinear mapping $s : V \times V \to \mathbb{R}$ is called for short **scalar product**. Instead of $s(v, w)$ one also writes $\langle v\,, w\rangle$, (v, w) or $v \cdot w$. One calls a real vector space V with a scalar product $\langle\ ,\ \rangle$ also **Euclidean vector space**.

Recipe: When Is a Mapping a Scalar Product?
Consider a real vector space V with a mapping

$$\langle\ ,\ \rangle : V \times V \to \mathbb{R},\ (v, w) \mapsto \langle v\,, w\rangle\,.$$

Show that for all $v,\ v',\ w \in V$ and $\lambda \in \mathbb{R}$ we have:

(1) Linearity in the 1st argument: $\langle \lambda\, v + v'\,, w\rangle = \lambda\, \langle v\,, w\rangle + \langle v'\,, w\rangle$.
(2) Symmetry: $\langle v\,, w\rangle = \langle w\,, v\rangle$.
(3) Positive definiteness: $\langle v\,, v\rangle > 0$ for $v \neq 0$.

Then $\langle\ ,\ \rangle$ is a scalar product.

Indeed, because of symmetry, we have linearity in the 2nd argument as well, and because of linearity we have $\langle 0\,, 0\rangle = 0$.

Example 16.1

- The **canonical** or **standard scalar product** is

$$\langle\ ,\ \rangle : \mathbb{R}^n \times \mathbb{R}^n \to \mathbb{R},\ (v, w) \mapsto v^\top w\,.$$

This mapping $\langle\ ,\ \rangle$ is in fact a scalar product, namely for all $v,\ v',\ w \in \mathbb{R}^n$ and $\lambda \in \mathbb{R}$ we have:

(1) Linearity in the 1st argument: $\langle \lambda\, v + v'\,, w\rangle = (\lambda\, v + v')^\top w = (\lambda\, v^\top + v'^\top)\, w = \lambda\, v^\top w + v'^\top w = \lambda\, \langle v\,, w\rangle + \langle v'\,, w\rangle$.
(2) Symmetry: $\langle v\,, w\rangle = v^\top w = w^\top v = \langle w\,, v\rangle$.
(3) Positive definiteness: $\langle v\,, v\rangle = v^\top v = \sum_{i=1}^{n} v_i^2 > 0$ if $v \neq 0$.

- Let $V = \mathbb{R}[x]$ is the vector space of real polynomial functions. Then $\langle\ ,\ \rangle : V \times V \to \mathbb{R}$ with

$$\langle p\,,q\rangle = \int_0^1 p(x)q(x)\,\mathrm{d}x$$

is a scalar product, since for all $p,\ \widetilde{p},\ q \in V$ and $\lambda \in \mathbb{R}$ we have:

(1) Linearity in 1st argument:

$$\begin{aligned}\langle \lambda\, p + \widetilde{p}\,, q\rangle &= \int_0^1 (\lambda\, p + \widetilde{p})(x)q(x)\,\mathrm{d}x = \int_0^1 \lambda\, p(x)q(x) + \widetilde{p}(x)q(x)\,\mathrm{d}x \\ &= \lambda \int_0^1 p(x)q(x)\,\mathrm{d}x + \int_0^1 \widetilde{p}(x)q(x)\,\mathrm{d}x = \lambda\,\langle p\,, q\rangle + \langle \widetilde{p}\,, q\rangle\,.\end{aligned}$$

(2) Symmetry: $\langle p\,, q\rangle = \int_0^1 p(x)q(x)\,\mathrm{d}x = \int_0^1 q(x)p(x)\,\mathrm{d}x = \langle q\,, p\rangle$.

(3) Positive definiteness:

$$\langle p\,, p\rangle = \int_0^1 p(x)^2\,\mathrm{d}x > 0 \text{ for } p \neq 0\,,$$

since the graph of p^2 encloses a positive area with the *x-axis*, namely it is $p(x)^2 \geq 0$ for all $x \in [0, 1]$, and there are $x \in [0, 1]$ with $p(x) > 0$.

We compute, as an example, the scalar product of $p = 1 + x$ and $q = x^2$:

$$\langle p\,, q\rangle = \int_0^1 x^2 + x^3\,\mathrm{d}x = \frac{1}{3}x^3 + \frac{1}{4}x^4\Big|_0^1 = \frac{7}{12}\,.$$

- Similarly, if V is the vector space of all continuous functions on an interval $[a, b]$, then the product $\langle\ ,\ \rangle : V \times V \to \mathbb{R}$ declared as follows is a scalar product:

$$\langle f\,, g\rangle = \int_a^b f(x)g(x)\,\mathrm{d}x\,. \qquad \blacksquare$$

16.2 Length, Distance, Angle and Orthogonality

In Euclidean vector spaces, it is possible to assign a *length* to vectors. This length is explained by means of the scalar product. In $\mathbb{R}^2$ or $\mathbb{R}^3$ this notion of length and the resulting notion of distance, angle and orthogonality corresponds to the descriptive notion of length, provided that the canonical is considered as the scalar product.

Length, Distance, Angle and Orthogonality
If V is a vector space with Euclidean scalar product $\langle\ ,\ \rangle$ then one calls

- the real number $\|v\| = \sqrt{\langle v\,,\, v\rangle}$ the **length** or **norm** of $v \in V$,
- the real number $d(v, w) = \|v - w\| = \|w - v\|$ the **distance** of v and w,
- the real number $\measuredangle(v, w) = \arccos\left(\frac{\langle v\,,w\rangle}{\|v\|\,\|w\|}\right)$ the **angle** between $v \neq 0$ and $w \neq 0$,
- two vectors v and w **perpendicular** or **orthogonal** if $\langle v\,,\, w\rangle = 0$, one writes for it $v \perp w$.

Remark In the exercises we show the **Cauchy-Schwarz inequality**: If $\langle\ ,\ \rangle$ is a scalar product on V, then for all $v, w \in V$:

$$|\langle v\,,\, w\rangle| \leq \|v\|\,\|w\|\,.$$

Due to this inequality we have for all $v,\ w,\ v \neq 0 \neq w$,

$$-1 \leq \frac{\langle v\,,\, w\rangle}{\|v\|\,\|w\|} \leq 1\,,$$

so that $\measuredangle(v, w) \in [0, \pi]$ (see Sect. 6.3) actually exists.

Example 16.2

- We consider the standard scalar product $\langle\ ,\ \rangle$ of the $\mathbb{R}^2$. The vector $(1,\ 1)^\top \in \mathbb{R}^2$ has the length

$$\left\|\begin{pmatrix}1\\1\end{pmatrix}\right\| = \sqrt{(1,1)\begin{pmatrix}1\\1\end{pmatrix}} = \sqrt{2}\,.$$

- We calculate the length of the polynomial $p = 1 + x$ with respect to the scalar product $\langle p , q\rangle = \int_0^1 p(x)q(x)\,\mathrm{d}x$:

$$\|1+x\| = \sqrt{\int_0^1 x^2 + 2x + 1\,\mathrm{d}x} = \sqrt{\frac{1}{3}x^3 + x^2 + x\Big|_0^1} = \sqrt{\frac{7}{3}}\,.$$

- Let $\langle\ ,\ \rangle$ is the standard scalar product of the $\mathbb{R}^2$. We have:

$$\measuredangle\left(\begin{pmatrix}1\\0\end{pmatrix}, \begin{pmatrix}1\\1\end{pmatrix}\right) = \arccos\left(1/\sqrt{2}\right) = \pi/4\,.$$

- In $\mathbb{R}^2$ we have with the standard scalar product $\langle\ ,\ \rangle$

$$\langle\begin{pmatrix}-1\\2\end{pmatrix}, \begin{pmatrix}2\\1\end{pmatrix}\rangle = (-1, 2)\begin{pmatrix}2\\1\end{pmatrix} = 0\,,\ \text{so that}\ \begin{pmatrix}-1\\2\end{pmatrix} \perp \begin{pmatrix}2\\1\end{pmatrix}.$$

- The polynomials $p = x$ and $q = 2 - 3x$ are orthogonal with respect to the scalar product $\langle p , q\rangle = \int_0^1 p(x)q(x)\,\mathrm{d}x$, since

$$\langle p , q\rangle = \int_0^1 2x - 3x^2\,\mathrm{d}x = x^2 - x^3\Big|_0^1 = 0\,,\ \ \text{d.h.}\ p \perp q\,,$$

- The zero vector 0 is orthogonal to all vectors, because of $\langle 0 , v\rangle = 0$, i.e. $0 \perp v$ for all $v \in V$.
- For $v, w \neq 0$ we have:

$$v \perp w \ \Leftrightarrow\ \measuredangle(v, w) = \pi/2\,. \qquad \blacksquare$$

We introduced on the $\mathbb{R}^n$ only the canonical scalar product. There are also other scalar products, not so important for practical applications at first, on the $\mathbb{R}^n$. If you determine the length of a vector $v \in \mathbb{R}^n$ with the canonical scalar product, then one calls $\|v\|$ also the **Euclidean norm** or **Euclidean length**.

16.3 Orthonormal Bases

Each connects to the $\mathbb{R}^2$ respectively $\mathbb{R}^3$ a coordinate system whose axes are orthogonal. This is not by chance, bases of orthogonal vectors have their merits. The axes of the $\mathbb{R}^3$ are generated by the standard unit vectors e_1, e_2, e_3; these are orthogonal vectors of length 1 with respect to the standard scalar product. They form a *orthonormal basis* of $\mathbb{R}^3$. One of our next goals is to give an orthonormal basis to each Euclidean vector space. We succeed in this, provided the dimension of the vector space remains finite.

Because we will not always be dealing with bases, but occasionally only with sets of orthogonal vectors that also do not necessarily have length 1, we need four terms:

Orthogonal/Orthonormal/System/Base
A subset B of a Euclidean vector space V with scalar product $\langle\ ,\ \rangle$ is called

- **orthogonal system of** V, if for all $v, w \in B$ with $v \neq w$ we have $v \perp w$,
- **orthogonal basis of** V if B is an orthogonal system and a basis,
- **orthonormal system of** V if B is an orthogonal system and $\|v\| = 1$ is true for all $v \in B$,
- **orthonormal basis of** V if B is an orthonormal system and a basis.

We abbreviate the most frequently used term *orthonormal basis* with **ONB**.

By **normalizing**, i.e., one replaces a $v \neq 0$ by $\frac{1}{\|v\|}\, v$, one can turn orthogonal systems into orthonormal systems.

Example 16.3

- The following set B is an orthogonal basis of the $\mathbb{R}^3$ with respect to the standard scalar product. By normalizing the elements of B we obtain an orthonormal basis $\tilde{B}$:

$$B = \left\{ \begin{pmatrix} 2 \\ -1 \\ 2 \end{pmatrix}, \begin{pmatrix} 1 \\ 2 \\ 0 \end{pmatrix}, \begin{pmatrix} 2 \\ -1 \\ -5/2 \end{pmatrix} \right\} \to \tilde{B} = \left\{ \frac{1}{3} \begin{pmatrix} 2 \\ -1 \\ 2 \end{pmatrix}, \frac{1}{\sqrt{5}} \begin{pmatrix} 1 \\ 2 \\ 0 \end{pmatrix}, \frac{2}{3\sqrt{5}} \begin{pmatrix} 2 \\ -1 \\ -5/2 \end{pmatrix} \right\}.$$

- For all $n \in \mathbb{N}$ the set $E_n = \{e_1, \ldots, e_n\}$ is an orthonormal basis of the $\mathbb{R}^n$. ■

With the **Kronecker delta**

$$\delta_{ij} = \begin{cases} 1, & \text{if } i = j \\ 0, & \text{if } i \neq j \end{cases}$$

the orthonormality of a basis $B = \{b_1,\ b_2,\ b_3,\ \ldots\}$ can be expressed briefly:

$$B \text{ is ONB } \Leftrightarrow \forall i,\ j:\ \langle b_i\, , b_j \rangle = \delta_{ij}\,.$$

16.4 Orthogonal Decomposition and Linear Combination with Respect to an ONB

We solve the following problems:

- We want a vector $v = u + w$ as a sum of orthogonal vectors $u \perp w$ where the summand u has a given direction.
- We want the coefficients $\lambda_1, \dots, \lambda_n$ of the linear combination $v = \lambda_1 b_1 + \cdots + \lambda_n b_n$ of v with respect to an ONB $B = \{b_1, \dots, b_n\}$.

Both problems can be solved quite easily, we start with the first problem:

Recipe: Orthogonal Decomposition
If $a \neq 0$ is a vector of a Euclidean vector space with the scalar product $\langle \, , \, \rangle$ then any vector $v \in V$ can be represented in the form

$$v = v_a + v_{a^\perp} \quad \text{with } v_a = \lambda\, a \text{ and } v_{a^\perp} \perp a$$

This **orthogonal decomposition of v along a** is obtained as follows:

(1) $v_a = \frac{\langle v, a\rangle}{\langle a, a\rangle}\, a$,
(2) $v_{a^\perp} = v - v_a$.

It's easy to check for that:

$$v_a + v_{a^\perp} = v_a + v - v_a = v\,,\; v_a = \lambda\, a\,,\; \langle v_{a^\perp}, a\rangle = \langle v, a\rangle - \frac{\langle v, a\rangle}{\langle a, a\rangle} \cdot \langle a, a\rangle = 0\,.$$

Example 16.4

- We decompose the vector $v = (1, 2, 3)^\top$ along the vector $a = (1, 0, 1)^\top$. Because of $\langle v, a\rangle = 4$ and $\langle a, a\rangle = 2$ we get:

$$(1)\; v_a = \frac{4}{2}\begin{pmatrix}1\\0\\1\end{pmatrix} = \begin{pmatrix}2\\0\\2\end{pmatrix} \quad \text{and therefore (2) } v_{a^\perp} = \begin{pmatrix}1\\2\\3\end{pmatrix} - \begin{pmatrix}2\\0\\2\end{pmatrix} = \begin{pmatrix}-1\\2\\1\end{pmatrix}.$$

- Let $V = \mathbb{R}[x]$ be the vector space of polynomials over $\mathbb{R}$ with the scalar product

$$\langle p\,,q\rangle = \int_0^1 p(x)q(x)\mathrm{d}x\,.$$

We decompose $p = 1 + x$ along $a = x$ and obtain with the formulas

$$(1) \qquad v_a = \frac{\langle 1+x\,,x\rangle}{\langle x\,,x\rangle}\,x = \frac{\int_0^1 x + x^2\mathrm{d}x}{\int_0^1 x^2\,\mathrm{d}x}\,x = \frac{\frac{1}{2}x^2 + \frac{1}{3}x^3\Big|_0^1}{\frac{1}{3}x^3\Big|_0^1}\,x = \frac{5}{2}\,x\,.$$

Correspondingly we now calculate $v_{a^\perp}$ as

$$(2) \qquad v_{a^\perp} = 1 + x - \frac{5}{2}x = 1 - \frac{3}{2}x\,.$$

We finally check our result. Surely $v = v_a + v_{a^\perp}$ and $v_a = \lambda\,a$ are fulfilled. In addition:

$$\langle v_{a^\perp}\,,a\rangle = \langle 1 - \frac{3}{2}\,x\,,x\rangle = \int_0^1 x - \frac{3}{2}x^2\,\mathrm{d}x = \frac{1}{2}\,x^2 - \frac{1}{2}\,x^3\Big|_0^1 = 0\,. \quad \blacksquare$$

Now for the second problem:

Recipe: Determine the Linear Combination with Respect to an ONB
If $B = \{b_1, \ldots, b_n\}$ is an orthonormal basis of a Euclidean vector space V with respect to the scalar product $\langle\ ,\ \rangle$ then for every $v \in V$ you get the coefficients $\lambda_1, \ldots, \lambda_n$ of the linear combination (uniquely determined up to the order of the summands) $v = \lambda_1 b_1 + \cdots + \lambda_n b_n$ as follows: For $i = 1, \ldots, n$ we have

$$\lambda_i = \langle v\,,b_i\rangle\,.$$

This can be easily verified: For $v = \lambda_1 b_1 + \lambda_2 b_2 + \cdots + \lambda_n b_n$ we have:

$$\langle v\,,b_1\rangle = \lambda_1 \underbrace{\langle b_1\,,b_1\rangle}_{=1} + \lambda_2 \underbrace{\langle b_2\,,b_1\rangle}_{=0} + \cdots + \lambda_n \underbrace{\langle b_n\,,b_1\rangle}_{=0} = \lambda_1$$

$$\langle v\,,b_2\rangle = \lambda_1 \underbrace{\langle b_1\,,b_2\rangle}_{=0} + \lambda_2 \underbrace{\langle b_2\,,b_2\rangle}_{=1} + \cdots + \lambda_n \underbrace{\langle b_n\,,b_2\rangle}_{=0} = \lambda_2 \text{ etc.}$$

Example 16.5 We represent the vector $v = (3,\ 2)^\top \in \mathbb{R}^2$ with respect to the ONB

$$B = \left\{ b_1 = \frac{1}{\sqrt{2}} \begin{pmatrix} 1 \\ 1 \end{pmatrix},\ b_2 = \frac{1}{\sqrt{2}} \begin{pmatrix} 1 \\ -1 \end{pmatrix} \right\}$$

of $\mathbb{R}^2$:

$$\lambda_1 = \langle v\,, b_1 \rangle = \frac{5}{\sqrt{2}}, \quad \lambda_2 = \langle v\,, b_2 \rangle = \frac{1}{\sqrt{2}} \Rightarrow v = \frac{5}{\sqrt{2}} b_1 + \frac{1}{\sqrt{2}} b_2\,.$$ ■

16.5 Orthogonal Matrices

A matrix $A \in \mathbb{R}^{n \times n}$ is called **orthogonal** if $A^\top A = E_n$. Examples of orthogonal matrices are

$$\begin{pmatrix} 0 & -1 & 0 \\ 0 & 0 & -1 \\ -1 & 0 & 0 \end{pmatrix}, \quad \frac{1}{3} \begin{pmatrix} 2 & -1 & 2 \\ 2 & 2 & -1 \\ -1 & 2 & 2 \end{pmatrix} \text{ and } \begin{pmatrix} \cos(\alpha) & \sin(\alpha) \\ \sin(\alpha) & -\cos(\alpha) \end{pmatrix}, \ \alpha \in [0, 2\pi[\,.$$

We can immediately state a number of properties of orthogonal matrices:

Properties of Orthogonal Matrices
For any orthogonal matrix $A \in \mathbb{R}^{n \times n}$ we have:

- A is invertible.
- $A^{-1} = A^\top$.
- The columns of A form an ONB of the $\mathbb{R}^n$.
- The rows of A form an ONB of $\mathbb{R}^n$.
- $\det(A) = \pm 1$.
- A is **length-preserving**, that is $\|Av\| = \|v\|$ for every $v \in \mathbb{R}^n$ (Euclidean norm).
- The product of orthogonal matrices is orthogonal.

The prove is simple: Because of $A^\top A = E_n$ the matrix A is invertible, by multiplying this equation by A^{-1} from the right we get $A^\top = A^{-1}$. If the matrix

$$A = (s_1, \ldots, s_n) = \begin{pmatrix} z_1 \\ \vdots \\ z_n \end{pmatrix}$$

is orthogonal, then

$$A^\top A = \begin{pmatrix} s_1^\top \\ \vdots \\ s_n^\top \end{pmatrix} (s_1, \ldots, s_n) = \begin{pmatrix} s_1^\top s_1 & & s_1^\top s_n \\ & \ddots & \\ s_n^\top s_1 & & s_n^\top s_n \end{pmatrix} = \begin{pmatrix} 1 & & 0 \\ & \ddots & \\ 0 & & 1 \end{pmatrix}$$

and likewise

$$A A^\top = \begin{pmatrix} z_1 \\ \vdots \\ z_n \end{pmatrix} (z_1^\top, \ldots, z_n^\top) = \begin{pmatrix} z_1 z_1^\top & & z_1 z_n^\top \\ & \ddots & \\ z_n z_1^\top & & z_n z_n^\top \end{pmatrix} = \begin{pmatrix} 1 & & 0 \\ & \ddots & \\ 0 & & 1 \end{pmatrix}.$$

So $A^\top A = E_n$ means that the columns of A have length 1 and are perpendicular to each other, and $AA^\top = E_n$ means the same for the rows. Finally, the statement about the determinant follows from the determinant multiplication theorem and from $\det(A^\top) = \det(A)$. From

$$\|Av\|^2 = \langle Av\,, Av\rangle = (Av)^\top (Av) = v^\top A^\top A v = v^\top v = \|v\|^2$$

follows the statement about the conservation of length of A. Finally, for orthogonal matrices A and B:

$$(A\,B)^\top (A\,B) = B^\top A^\top A\,B = B^\top B = E_n\,,$$

so that also the product $A\,B$ is orthogonal.

A whole class of examples of orthogonal matrices are the *reflection matrices*:

Example 16.6 **Reflection matrices:** For every vector $a \in \mathbb{R}^n \setminus \{0\}$ the matrix

$$H_a = E_n - \frac{2}{a^\top a}\, a\, a^\top \in \mathbb{R}^{n \times n}$$

is called **reflection matrix along** $\boldsymbol{a}$. Every such reflection matrix is orthogonal and additionally symmetric, because we have for every $a \in \mathbb{R}^n \setminus \{0\}$:

$$\begin{aligned} H_a^\top H_a &= \left(E_n - \frac{2}{a^\top a}\, a\, a^\top\right)^\top \left(E_n - \frac{2}{a^\top a}\, a\, a^\top\right) \\ &= \left(E_n - \frac{2}{a^\top a}\, a\, a^\top\right) \left(E_n - \frac{2}{a^\top a}\, a\, a^\top\right) \\ &= E_n - \frac{2 \cdot 2}{a^\top a}\, a\, a^\top + \left(\frac{2}{a^\top a}\right)^2 a\, a^\top a\, a^\top = E_n\,. \end{aligned}$$

By $a = (4, 2, 2)^\top$ we obtain, for example, because of $a^\top a = 24$

$$H_a = E_n - \frac{2}{a^\top a} a a^\top = \begin{pmatrix} 1 & 0 & 0 \\ 0 & 1 & 0 \\ 0 & 0 & 1 \end{pmatrix} - \frac{1}{12} \begin{pmatrix} 16 & 8 & 8 \\ 8 & 4 & 4 \\ 8 & 4 & 4 \end{pmatrix} = \begin{pmatrix} -1/3 & -2/3 & -2/3 \\ -2/3 & 2/3 & -1/3 \\ -2/3 & -1/3 & 2/3 \end{pmatrix}. \quad \blacksquare$$

The name *reflection matrix* is explained as follows: Because

$$H_a\, a = a - \frac{2}{a^\top a}\, a\, a^\top a = -a$$

and

$$H_a\, w = w - \frac{2}{a^\top a}\, a\, a^\top w = w \ \text{ for } \ w \perp a$$

the mapping

$$\varphi_{H_a} : \begin{cases} \mathbb{R}^n \to \mathbb{R}^n \\ v \mapsto H_a\, v \end{cases}$$

maps every vector $v = v_a + v_{a^\perp}$ (orthogonal decomposition of v along (a) to $-v_a + v_{a^\perp}$, i.e, v becomes *along a reflected*, note also Fig. 16.1.

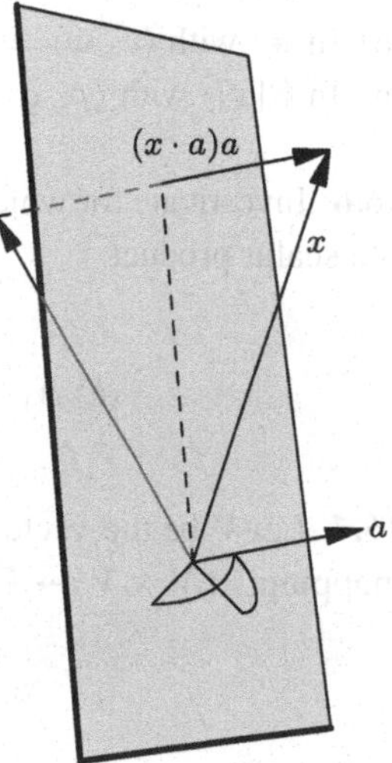

Fig. 16.1 Reflecting along a

16.6 Exercises

16.1 Show the Cauchy-Schwarz inequality: If $\langle \, , \, \rangle$ is a scalar product on V, then for all $v, w \in V$:

$$|\langle v \, , w \rangle| \le \|v\| \, \|w\| \, .$$

16.2 Show that orthogonal vectors other than 0 are linearly independent.

16.3 Write a MATLAB program that decomposes $p = p_a + p_{a^\perp}$, $p \in \mathbb{R}[x]$, along $a \in \mathbb{R}[x]$. Consider hereby the scalar product

$$\langle \, , \, \rangle : \mathbb{R}[x] \times \mathbb{R}[x] \to \mathbb{R} \, , \; (p, q) \mapsto \int_0^1 p(x)q(x)\mathrm{d}x \, .$$

16.4 Let $v = (v_1, v_2)^\top, w = (w_1, w_2)^\top \in \mathbb{R}^2$. Verify that

(a) $\langle v, w \rangle = 4v_1w_1 + 3v_2w_2 + v_1w_2 + v_2w_1$, (b) $\langle v, w \rangle = v_1^2 w_1 + v_2 w_2$

are scalar products in $\mathbb{R}^2$.

16.5 Calculate the angles between the following two vectors. Use the given scalar product for each.

(a) In $\mathbb{R}^3$ with $\langle v \, , w \rangle = v^\top w$: $v = (1, -2, 0)^\top \, , \; w = (2, -1, 1)^\top$.
(b) In $\mathbb{R}[x]_2$ with $\langle p \, , q \rangle = \int_0^1 p(x)q(x)\mathrm{d}x$: $p(x) = x^2 - 2x + 2$, $q(x) = 3x^2 + x - 3$.

16.6 Investigate for which of the following matrices A_1, A_2 the mapping $\langle x \, , y \rangle = x^\top A_i \, y$ is a scalar product:

$$\text{(a) } A_1 = \begin{pmatrix} 1 & 2 \\ 2 & 1 \end{pmatrix}, \qquad \text{(b) } A_2 = \begin{pmatrix} 1 & -1 \\ -1 & 2 \end{pmatrix}.$$

16.7 Let V be the vector space of continuous functions on the interval $[0, 2]$. Check the mapping $s : V \times V \to \mathbb{R}$ given by

$$s(f, g) = f(0)g(0) + f(1)g(1) + f(2)g(2)$$

for bilinearity, symmetry, and positive definiteness. Is s a scalar product?

Orthogonality II

17

We continue with the important topic of *orthogonality*. In doing so, we begin with *Gram and Schmidt's orthonormalization method*, which can be used to construct an orthonormal basis from a basis of an Euclidean vector space. We then consider the *vector product* and *(scalar) triple product*, which are products between vectors in the $\mathbb{R}^3$, and then turn to the orthogonal projection.

17.1 The Orthonormalization Method of Gram and Schmidt

Every finite-dimensional Euclidean vector space has an orthonormal basis. The *orthonormalization method of Gram and Schmidt* is a method that turns an (arbitrary) basis $\{a_1, \ldots, a_n\}$ of an Euclidean vector space V to an orthonormal basis $B = \{b_1, \ldots, b_n\}$:

Recipe: Gram and Schmidt's Orthonormalization Method.
Let $\{a_1, \ldots, a_n\}$ be an arbitrary basis of an Euclidean vector space V with scalar product $\langle\, ,\, \rangle$. Form the vectors $b_1, \ldots, b_n$ as follows:

(1) $b_1 = \frac{1}{\|a_1\|}\, a_1$,
(2) $b_2 = \frac{1}{\|c_2\|}\, c_2$ with $c_2 = a_2 - \langle a_2 , b_1 \rangle\, b_1$,
(3) $b_3 = \frac{1}{\|c_3\|}\, c_3$ with $c_3 = a_3 - \langle a_3 , b_1 \rangle\, b_1 - \langle a_3 , b_2 \rangle\, b_2$,
(4) $\cdots$
(n) $b_n = \frac{1}{\|c_n\|}\, c_n$ with $c_n = a_n - \langle a_n , b_1 \rangle\, b_1 - \cdots - \langle a_n , b_{n-1} \rangle\, b_{n-1}$.

(continued)

C. Karpfinger, *Calculus and Linear Algebra in Recipes*,
https://doi.org/10.1007/978-3-662-65458-3_17

General:

$$b_1 = \frac{1}{\|a_1\|} a_1, \quad b_{k+1} = \frac{1}{\|c_{k+1}\|} c_{k+1} \text{ with } c_{k+1} = a_{k+1} - \sum_{i=1}^{k} \langle a_{k+1}, b_i \rangle \, b_i \, .$$

Then $\{b_1, \ldots, b_n\}$ is an ONB of V.

One simply verifies that the vectors are perpendicular in pairs by taking the scalar product $\langle b_i , b_j \rangle$ to verify that the vectors are perpendicular to each other in pairs.

Example 17.1 We want an orthonormal basis of the $\mathbb{R}^3$ with respect to the standard scalar product. To do this, we start with the base

$$B = \left\{ a_1 = \begin{pmatrix} 1 \\ 0 \\ 0 \end{pmatrix}, \, a_2 = \begin{pmatrix} 1 \\ 1 \\ 0 \end{pmatrix}, \, a_3 = \begin{pmatrix} 1 \\ 1 \\ 1 \end{pmatrix} \right\}$$

and apply to them the Gram-Schmidt orthonormalization procedure:

$$\text{(1) } b_1 = \frac{1}{\|a_1\|} a_1 = \begin{pmatrix} 1 \\ 0 \\ 0 \end{pmatrix},$$

$$\text{(2) } b_2 = \frac{1}{\|c_2\|} c_2 \quad \text{with} \quad c_2 = \begin{pmatrix} 1 \\ 1 \\ 0 \end{pmatrix} - 1 \begin{pmatrix} 1 \\ 0 \\ 0 \end{pmatrix} = \begin{pmatrix} 0 \\ 1 \\ 0 \end{pmatrix} \Rightarrow b_2 = \begin{pmatrix} 0 \\ 1 \\ 0 \end{pmatrix},$$

$$\text{(3) } b_3 = \frac{1}{\|c_3}\|, c_3 \quad \text{with} \quad c_3 = \begin{pmatrix} 1 \\ 1 \\ 1 \end{pmatrix} - 1 \begin{pmatrix} 1 \\ 0 \\ 0 \end{pmatrix} - 1 \begin{pmatrix} 0 \\ 1 \\ 0 \end{pmatrix} = \begin{pmatrix} 0 \\ 0 \\ 1 \end{pmatrix} \Rightarrow b_3 = \begin{pmatrix} 0 \\ 0 \\ 1 \end{pmatrix}.$$

■

The standard basis of $\mathbb{R}^3$, which we received in the example, we could also have easily guessed. Generally one finds in the $\mathbb{R}^2$ and $\mathbb{R}^3$ orthonormal bases by *sharp look* (or with the vector product in the $\mathbb{R}^3$) usually faster than with the Gram-Schmidt method. However, this is not necessarily the case in other vector spaces, as the following example shows:

Example 17.2 We search for an orthonormal basis of the vector space $V = \langle x, x^2 \rangle$ with respect to the scalar product

$$\langle p , q \rangle = \int_0^1 p(x)q(x)\,\mathrm{d}x \, .$$

As a starting basis for this purpose we naturally choose $\{a_1 = x,\ a_2 = x^2\}$. The Gram-Schmidt orthonormalization procedure gives us:

(1) $b_1 = \frac{1}{\|x\|} x = \sqrt{3}\, x\,,\quad$ since $\quad \|x\| = \sqrt{\int_0^1 x^2\, \mathrm{d}x} = \sqrt{\frac{1}{3}},$

(2) $b_2 = \frac{1}{\|c_2\|} c_2 \quad$ with $\quad c_2 = a_2 - \langle a_2\,, b_1\rangle\, b_1.$

We have:

$$c_2 = x^2 - \left(\int_0^1 \sqrt{3}\, x^3\, \mathrm{d}x\right) \sqrt{3}\, x = x^2 - \left(\frac{\sqrt{3}}{4} x^4 \Big|_0^1\right) \sqrt{3}\, x = x^2 - \frac{3}{4} x\,.$$

We can now use this to calculate $\|c_2\|$ as

$$\|c_2\| = \left\| x^2 - \frac{3}{4} x \right\| = \sqrt{\int_0^1 \left(x^2 - \frac{3}{4} x\right)^2 \mathrm{d}x}$$

$$= \sqrt{\int_0^1 x^4 - \frac{3}{2} x^3 + \frac{9}{16} x^2 \mathrm{d}x} = \sqrt{\frac{1}{5} x^5 - \frac{3}{8} x^4 + \frac{3}{16} x^3 \Big|_0^1}$$

$$= \sqrt{\frac{1}{5} - \frac{3}{8} + \frac{3}{16}} = \sqrt{\frac{1}{5} - \frac{3}{16}} = \sqrt{\frac{1}{80}} = \frac{1}{4\sqrt{5}}\,.$$

Thus, an orthonormal basis of V is

$$\left\{ \sqrt{3}\, x,\ 4\sqrt{5}\left(x^2 - \frac{3}{4} x\right) \right\}.$$

■

Remark An implementation of the Gram-Schmidt orthonormalization method is not recommended in this way. Due to rounding errors and cancellation, the vectors obtained by a naive implementation of the method on a computer are generally not orthogonal. Although there is also a numerically *stable* variant of the Gram-Schmidt method, we nevertheless refrain from presenting it, since we use the *Householder transformations* in Chap. 19 as a numerically stable possibility for the construction of an ONB.

17.2 The Vector Product and the (Scalar) Triple Product

In this section we consider the $\mathbb{R}^3$ with its standard scalar product $\langle\ ,\ \rangle$.

In the *vector product* $\times$ two vectors $a, b \in \mathbb{R}^3$ are *multiplied*, the result is again a vector $c = a \times b \in \mathbb{R}^3$. The *scalar triple product* $[\cdot, \cdot, \cdot]$ is a *product of* three vectors $a, b, c \in \mathbb{R}^3$ where the result $[a, b, c] \in \mathbb{R}$ is a scalar. We have:

- $\|a \times b\|$ is the area of the parallelogram spanned by a and b.
- $|[a, b, c]|$ is the volume of the **parallelepiped** spanned by a, b and c.

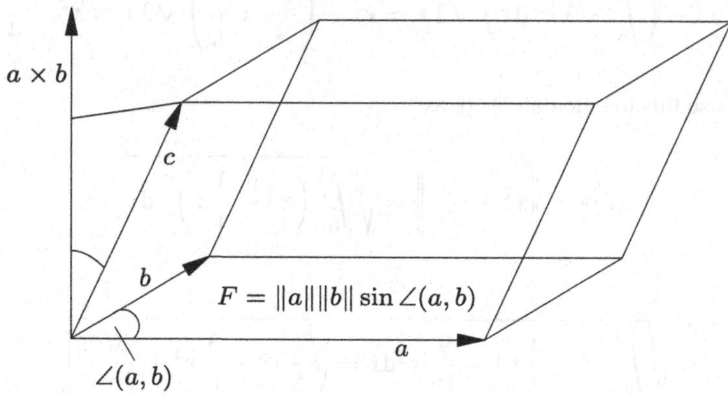

The definition and main characteristics follow:

Definition and Properties of the Vector Product and Triple Product

For vectors

$$a = \begin{pmatrix} a_1 \\ a_2 \\ a_3 \end{pmatrix} \text{ and } b = \begin{pmatrix} b_1 \\ b_2 \\ b_3 \end{pmatrix} \text{ and } c = \begin{pmatrix} c_1 \\ c_2 \\ c_3 \end{pmatrix} \in \mathbb{R}^3$$

we call

$$a \times b = \begin{pmatrix} a_2b_3 - b_2a_3 \\ a_3b_1 - b_3a_1 \\ a_1b_2 - b_1a_2 \end{pmatrix} \in \mathbb{R}^3 \text{ and } [a, b, c] = \langle a \times b\, , c\rangle \in \mathbb{R}$$

the **vector product** of a and b or the **triple product** of a, b and c. We have:

(continued)

1. $\langle x, a \times b \rangle = \det(x, a, b)$ for all $x \in \mathbb{R}^3$.
2. The vector $a \times b$ is perpendicular to a and b, $a \times b \perp a, b$.
3. $\|a \times b\| = \|a\| \, \|b\| \sin \measuredangle(a, b) =$ area of the parallelogram spanned by a, b.
4. If a and b are linearly independent, then $a, b, a \times b$ is a **right handed coordinate system**, i.e. $\det(a, b, a \times b) > 0$.
5. For each vector $a \in \mathbb{R}^3$ we have: $a \times a = 0$.
6. For all $a, b \in \mathbb{R}^3$ we have: $a \times b = -b \times a$.
7. $a \times b = 0 \Leftrightarrow a, b$ are linearly dependent.
8. $\|a \times b\|^2 + |\langle a, b \rangle|^2 = \|a\|^2 \, \|b\|^2$.
9. For all $u, v, w, x \in \mathbb{R}^3$ we have:

   - $u \times (v \times w) = \langle u, w \rangle \, v - \langle u, v \rangle \, w$. (**Grassmann identity**)
   - $\big(u \times (v \times w)\big) + \big(v \times (w \times u)\big) + \big(w \times (u \times v)\big) = 0$. (**Jacobi identity**)
   - $\langle u \times v, w \times x \rangle = \langle u, w \rangle, \langle v, x \rangle - \langle u, x \rangle, \langle v, w \rangle$. (**Lagrangian identity**)

10. $[a, b, c] = \det(a, b, c)$.
11. $|[a, b, c]| =$ volume of the parallelepiped spanned by a, b, c.
12. $[a, b, c] = 0 \Leftrightarrow \{a, b, c\}$ is linearly dependent.
13. $[a, b, c] > 0 \Leftrightarrow a, b, c$ form a right handed coordinate system.

A **right handed coordiante system** is given by the **right hand rule**: $a =$ thumb, $b =$ index finger, $a \times b =$ middle finger; all this, of course, on the right hand.

Example 17.3

- As an application of the vector product, we calculate the area F of the triangle with vertices

$$A = \begin{pmatrix} 1 \\ 2 \\ 3 \end{pmatrix}, \quad B = \begin{pmatrix} -2 \\ 0 \\ 4 \end{pmatrix} \quad \text{and} \quad C = \begin{pmatrix} -1 \\ -1 \\ 2 \end{pmatrix}$$

in $\mathbb{R}^3$. To do this, we take advantage of the fact that, according to property 3, the area of the parallelogram with sides $a = \overline{AB} = B - A$ and $b = \overline{AC} = C - A$ can be calculated. This is, of course, just twice as large as the area of the triangle we are looking for F. Therefore, we have:

$$F = \frac{1}{2} \, \|a\| \, \|b\| \, \sin \measuredangle(a, b) = \frac{1}{2} \, \|a \times b\| = \frac{1}{2} \left\| \begin{pmatrix} -3 \\ -2 \\ 1 \end{pmatrix} \times \begin{pmatrix} -2 \\ -3 \\ -1 \end{pmatrix} \right\| = \frac{5\sqrt{3}}{2}.$$

- As a further application, we use the vector product to determine a normal unit vector u of the plane

$$E = \left\{ x \in \mathbb{R}^3 \mid x = \begin{pmatrix} -3 \\ -2 \\ 1 \end{pmatrix} + t \begin{pmatrix} 1 \\ 2 \\ 2 \end{pmatrix} + s \begin{pmatrix} 0 \\ 1 \\ 1 \end{pmatrix}, \ t, s \in \mathbb{R} \right\}.$$

Because of property 2

$$\widetilde{u} = \begin{pmatrix} 1 \\ 2 \\ 2 \end{pmatrix} \times \begin{pmatrix} 0 \\ 1 \\ 1 \end{pmatrix} = \begin{pmatrix} 0 \\ -1 \\ 1 \end{pmatrix}$$

is perpendicular to the plane E, is therefore normal vector of E. By normalizing we get

$$u = \frac{1}{\sqrt{2}} \begin{pmatrix} 0 \\ -1 \\ 1 \end{pmatrix}.$$

- As an application of the triple product, we calculate the volume of a tetrahedron.

For this we choose as base the triangle with the vertices a_1, a_2 and a_3. we also define $a = a_2 - a_1$, $b = a_3 - a_1$, $c = a_4 - a_1$ and $F_{\square} = \|a \times b\|$ = area of the parallelogram with sides a, b.

$$\begin{aligned} \text{Volume} &= \frac{1}{3} \text{ Floor space} \cdot \text{Height} \\ &= \frac{1}{3} \left(\frac{1}{2} F_{\square} \cdot \text{Height} \right) \\ &= \frac{1}{6} \left(F_{\square} \cdot \text{Height} \right) \\ &= \frac{1}{6} \left| [a, b, c] \right|. \end{aligned}$$

Note Fig. 17.1.

Fig. 17.1 The tetrahedron

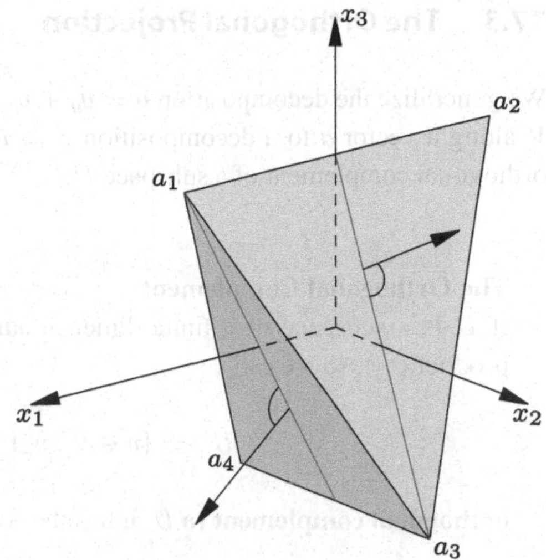

Now we take for a_1, a_2, a_3, a_4 we get the values

$$a_1 = \begin{pmatrix} 2 \\ 0 \\ \sqrt{2} \end{pmatrix}, \quad a_2 = \begin{pmatrix} -2 \\ 0 \\ \sqrt{2} \end{pmatrix}, \quad a_3 = \begin{pmatrix} 0 \\ 2 \\ -\sqrt{2} \end{pmatrix}, \quad a_4 = \begin{pmatrix} 0 \\ -2 \\ -\sqrt{2} \end{pmatrix}$$

then we get for the volume V of the tetrahedron

$$V = \frac{1}{6} \left| \left[\begin{pmatrix} -4 \\ 0 \\ 0 \end{pmatrix}, \begin{pmatrix} -2 \\ 2 \\ -2\sqrt{2} \end{pmatrix}, \begin{pmatrix} -2 \\ -2 \\ -2\sqrt{2} \end{pmatrix} \right] \right| = \frac{16\sqrt{2}}{3}.$$

■

MATLAB In connection with orthogonality we considered the scalar product $a^\top b$ the vector product $a \times b$ and the triple product $[a, b, c]$ with vectors $a, b, c \in \mathbb{R}^3$. In MATLAB, these products are obtained as follows:

```
a'*b, cross(a,b), cross(a,b)'*c
```

17.3 The Orthogonal Projection

We generalize the decomposition $v = v_a + v_{a^\perp}$ of a vector v of an Euclidean vector space V along a vector a to a decomposition $v = u + u^\perp$; for this we need the notion of the orthogonal complement of a subspace U:

The Orthogonal Complement
If U is a subspace of a finite-dimensional Euclidean vector space V with scalar product $\langle\ ,\ \rangle$, so we call

$$U^\perp = \{v \in V \mid v \perp u \text{ for all } u \in U\}$$

orthogonal complement to U, it has the following properties:

- $U^\perp$ is subspace of V.
- $U^\perp \cap U = \{0\}$.
- Every $v \in V$ has exactly one representation of the form

$$v = u + u^\perp \quad \text{with} \quad u \in U \quad \text{and} \quad u^\perp \in U^\perp .$$

- $\dim(V) = n \Rightarrow \dim(U^\perp) = n - \dim(U)$.

To determine the orthogonal complement of a finite-dimensional vector space, proceed as follows:

Recipe: Determine the Orthogonal Complement
If U is a subspace of an Euclidean vector space V with $\dim(V) = n$ and $\dim(U) = r$, then get $U^\perp$ as follows:

(1) Determine a base $\{b_1, \ldots, b_r\}$ of U.
(2) Determine n–r linearly independent vectors $a_1, \ldots, a_{n-r}$ which are orthogonal to all $b_1, \ldots, b_r$.
(3) We then have $U^\perp = \langle a_1, \ldots, a_{n-r}\rangle$.

Example 17.4

- The orthogonal complement to $U = \langle\begin{pmatrix}1\\1\end{pmatrix}\rangle \subseteq \mathbb{R}^2$ is $U^\perp = \langle\begin{pmatrix}1\\-1\end{pmatrix}\rangle$.
- The orthogonal complement to $U = \langle\begin{pmatrix}1\\1\\1\end{pmatrix}\rangle \subseteq \mathbb{R}^3$ is $U^\perp = \langle\begin{pmatrix}1\\0\\-1\end{pmatrix}, \begin{pmatrix}1\\-1\\0\end{pmatrix}\rangle$.

■

If U is a subspace of a finite dimensional Euclidean vector space, then any vector $v \in V$ can represented in exactly one way in the form

$$v = u + u^\perp \text{ with } u \in U \text{ and } u^\perp \in U^\perp .$$

The mapping

$$p_U : \begin{cases} V & \to U \\ v = u + u^\perp \mapsto u \end{cases},$$

which gives to every $v \in V$ the uniquely determined $u \in U$ is called **orthogonal projection** of v onto U.

The decisive hint how to determine $u = p_U(v)$ is given by the following observation: For the vector $u^\perp = v - u$ we have $\|u^\perp\| = \|v - u\| \leq \|v - w\|$ for all $w \in U$ since for all $w \in U$:

$$\|v - w\| = \|\overbrace{v - u}^{=u^\perp} + u - w\| = \sqrt{\langle u^\perp + (u - w)\,, u^\perp + (u - w)\rangle}$$

$$= \sqrt{\|u^\perp\|^2 + \|u - w\|^2 + 2\,\langle u^\perp\,, u - w\rangle} \geq \|u^\perp\| = \|v - u\|\,,$$

because $\langle u^\perp\,, u - w\rangle = 0$, since $u - w \in U$. Therefore one calls $\|u^\perp\| = \|v - u\|$ the **minimum distance** from v to U (Fig. 17.2).

Thus, u is obtained as the solution of the **minimization problem**:

$$\text{determine } u \in U \text{ with } \|v - u\| = \min\,.$$

If $U \subseteq \mathbb{R}^n$, we can formulate this minimization problem as follows: Choose a basis $\{b_1, \ldots, b_r\}$ of U, it is then $u = A\,x$ with $A = (b_1, \ldots, b_r) \in \mathbb{R}^{n\times r}$ for a $x =$

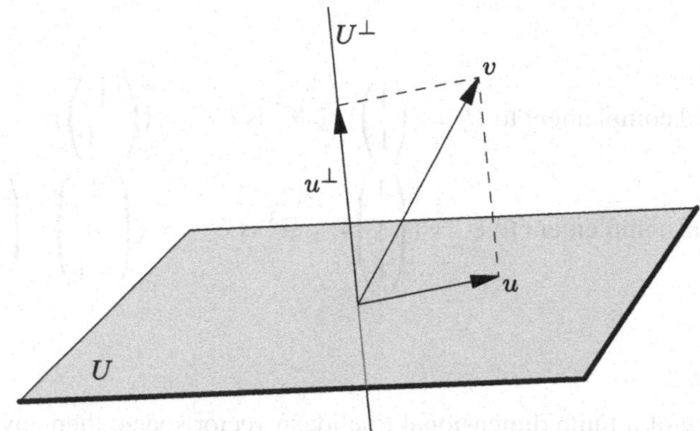

Fig. 17.2 The orthogonal projection

$(\lambda_1, \ldots, \lambda_r)^\top \in \mathbb{R}^r$, the above minimization task is then as follows:

$$\text{determine } x \in \mathbb{R}^r \text{ with } \|v - A\,x\| = \min .$$

This minimization task can be solved in a more general context, we deal with it in the next chapter (see Recipe in Sect. 18.2).

17.4 Exercises

17.1 Prove the properties of the vector product and triple product (see box in Sect. 17.2).

17.2 Let $\mathbb{R}[x]_2$ be the polynomial space with the scalar product $\langle p, q\rangle = \int_0^1 p(x)q(x)\mathrm{d}x$ and $W = \langle 1 + x^2\rangle$ a subspace of $\mathbb{R}[x]_2$.

(a) Determine a basis of $W^\perp$.
(b) Using the Gram-Schmidt method, determine from the base

$$p_1(x) = 1, \quad p_2(x) = x, \quad p_3(x) = x^2$$

of $\mathbb{R}[x]_2$ an orthonormal basis of $\mathbb{R}[x]_2$.

17.3 Given are the vectors

$$p = (3, 0, 4)^\top \text{ and } q = (-1, 2, -2)^\top$$

and the standard scalar product on $\mathbb{R}^3$.

(a) Calculate the angle between p and q.
(b) Give a vector $n \in \mathbb{R}^3$ with $\|n\|_2 = 1$ which is perpendicular on p and q.
(c) Determine $\lambda \in \mathbb{R}$ such that the linear combination $s = p + q + \lambda n$ has the length $\|s\|_2 = \sqrt{13}$.
(d) Determine the area F of the parallelogram spanned by p and q in $\mathbb{R}^3$.
(e) Determine the volume V of the parallelepiped spanned by p, q and n.

17.4 Given the vectors

$$v_1 = (1, 0, 1, 0)^\top, \quad v_2 = (1, 1, 1, 1)^\top, \quad v_3 = (1, 1, 2, 2)^\top \text{ and } v_4 = (0, 1, -1, 0)^\top.$$

Let $W = \langle v_1, v_2, v_3, v_4 \rangle$.

(a) Determine the dimension and a basis of W.
(b) Determine an orthonormal basis of W using the Gram-Schmidt method.

17.5

(a) Calculate the volume of the parallelepiped with edges

$$v_1 = (1, -2, 0)^\top, \quad v_2 = (2, 0, 3)^\top \quad \text{and} \quad v_3 = (3, 1, -1)^\top.$$

(b) Calculate the volume of the parallelepiped with edges

$$w_1 = (1, 2, 3)^\top, \quad w_2 = (-2, 0, 1)^\top \quad \text{and} \quad w_3 = (0, 3, -1)^\top.$$

(c) Compare the results of (a) and (b) and explain the result of the comparison.

17.6

(a) Determine the area F of the parallelogram spanned by the vectors

$$u = (1, 3, 6)^\top \quad \text{and} \quad v = (3, 2, 2)^\top$$

in $\mathbb{R}^3$.
(b) Determine the area D of the triangle in the $\mathbb{R}^3$ with the vertices $(1, 0, -1)^\top$, $(2, 3, 5)^\top$ and $(4, 2, 1)^\top$.
(c) Determine the volume V of the parallelepiped spanned by the vectors

$$u = (1, 3, 6)^\top, \quad v = (3, 2, 2)^\top \quad \text{and} \quad w = (-2, 8, 7)^\top$$

in $\mathbb{R}^3$.

17.7 Determine an orthonormal basis with respect to the standard scalar product of the $\mathbb{R}^4$ of

$$U = \langle \begin{pmatrix} 3 \\ -1 \\ -1 \\ -1 \\ -1 \end{pmatrix}, \begin{pmatrix} -1 \\ 3 \\ -1 \\ -1 \\ -1 \end{pmatrix}, \begin{pmatrix} -1 \\ -1 \\ 3 \\ -1 \end{pmatrix} \rangle \subseteq \mathbb{R}^4.$$

17.8 On the $\mathbb{R}$-vector space $V = \mathbb{R}[x]_3 \subseteq \mathbb{R}[x]$ let the scalar product $\langle \, , \rangle$ by

$$\langle f, g \rangle = \int_{-1}^{1} f(x)g(x)\,\mathrm{d}x$$

for $f, g \in V$ is given.

(a) Determine an orthonormal basis with respect to $\langle \, , \rangle$ of V.
(b) Calculate in V the distance of $f = x + 1$ and $g = x^2 - 1$.

17.9

(a) Write a MATLAB function `[u,w] = orthzer(v,b)` that gives the orthogonal decomposition $v = u + w$, $w \perp b$ of v along b.
(b) Let $\{b_1, \ldots, b_n\}$ be an orthonormal basis of $U \subseteq \mathbb{R}^m$ and $B = (b_1, \ldots, b_n) \in \mathbb{R}^{m \times n}$. Write a MATLAB function `[u,w] = orthzerU(v,B)`, which calculates the orthogonal decomposition $v = u + w$, $u \in U$, $w \in U^\perp$.
(c) Let $\{a_1, \ldots, a_n\}$ a *arbitrary* basis of $V \subseteq \mathbb{R}^m$ and $A = (a_1, \ldots, a_n) \in \mathbb{R}^{m \times n}$. Write a MATLAB function `gramSchmidt(A)`, which uses Gram and Schmidt's method to compute an orthonormal basis of V.

17.10 We consider the vector space $V = \mathbb{R}^3$ with the standard scalar product and the vectors

$$v_1 = \begin{pmatrix} 1/\sqrt{3} \\ 1/\sqrt{3} \\ 1/\sqrt{3} \end{pmatrix}, \; v_2 = \begin{pmatrix} 1/\sqrt{2} \\ -1/\sqrt{2} \\ 0 \end{pmatrix}, \; v_3 = \begin{pmatrix} 1/\sqrt{6} \\ 1/\sqrt{6} \\ -2/\sqrt{6} \end{pmatrix}, \; v = \begin{pmatrix} 2 \\ -2 \\ 1 \end{pmatrix}.$$

(a) Show that $B = \{v_1, v_2, v_3\}$ is an ONB of V.
(b) Write v as a linear combination of B.

The Linear Equalization Problem 18

The *linear equalization problem* is encountered in engineering sciences in various facets, mathematically it is always about one and the same thing: Find an x such that to a vector b and a matrix A the value $\|b - Ax\|$ is minimal. The applications of this are, for example, the *method of least squares, solving over-determined systems of equations* or *determining minimal distances from points to subspaces.*

18.1 The Linear Equalization Problem and Its Solution

We formulate the **linear equalization problem**:

The Linear Equalisation Problem and its Solution

The problem: Let $b \in \mathbb{R}^n$ and a $A \in \mathbb{R}^{n \times r}$ with $n \geq r$. We are looking for a vector $x \in \mathbb{R}^r$ such that

$$\|b - Ax\| = \min .$$

The solution: A vector $x \in \mathbb{R}^r$ is a solution of $\|b - Ax\| = \min$ if and only if x satisfies the following **normal equation**:

$$A^\top A x = A^\top b .$$

(continued)

C. Karpfinger, *Calculus and Linear Algebra in Recipes*,
https://doi.org/10.1007/978-3-662-65458-3_18

The linear equalization problem is uniquely solvable if and only if the rank of A is maximal, i.e., if we have $\mathrm{rg}(A) = r$.

For a justification, see Exercise 18.1.

Thus, to find the solution set of the linear equalization problem $\|b - Ax\| = \min$, one has to determine the solution set of the linear system of equations $A^\top A x = A^\top b$. We can solve this system of equations with the known methods. A numerically stable solution we obtain with the *Q R-decomposition* of A for the case where the system of equations is uniquely solvable (in other words $\mathrm{rg}(A) = r$), which we will discuss in Chap. 19.

We consider the three linear equalization problems in the following three sections:

- Determine to a vector b of a vector space V and a subspace U of V a vector $u \in U$ which has a minimum distance to b.
- Determine a *solution* x of an over-determined, unsolvable linear system of equations $Ax = b$ so that $b - Ax$ has a minimum length.
- Determine a function f whose graph approximates given intercepts $(t_i, y_i) \in \mathbb{R}^2$, $i = 1, \ldots, n$ as well as possible, i.e. that the error $\sum_{i=1}^n (y_i - f(t_i))^2$ becomes minimal.

18.2 The Orthogonal Projection

Let U be a subspace of the Euclidean vector space $\mathbb{R}^n$ with the canonical scalar product $\langle\ ,\ \rangle$ and a vector $b \in \mathbb{R}^n$. Then write b uniquely as $b = u + u^\perp$ with $u \in U$ and $u^\perp \in U^\perp$ (see Sect. 17.3). The searched vector u is the orthogonal projection $u = p_U(b)$ of $b \in \mathbb{R}^n$ on the subspace $U = \langle b_1, \ldots, b_r \rangle \subseteq \mathbb{R}^n$. As such u is a linear combination of the column vectors $b_1, \ldots, b_r \in \mathbb{R}^n$, with $A = (b_1, \ldots, b_r)$ thus we have $u = Ax$ with $x = (\lambda_1, \ldots, \lambda_r)^\top \in \mathbb{R}^r$. We thus know u, if we know x. And x we obtain as the solution of the minimization problem $\|b - Ax\| = \min$. To determine u note the following procedure:

Recipe: Determine the Orthogonal Projection $u = p_U(b)$

The vector u is obtained as follows:

(1) Choose a base $B = \{b_1, \ldots, b_r\}$ of U and set $A = (b_1, \ldots, b_r) \in \mathbb{R}^{n \times r}$.
(2) Solve the uniquely solvable linear system of equations $A^\top A x = A^\top b$ and obtain the solution vector $x = (\lambda_1, \ldots, \lambda_r)^\top \in \mathbb{R}^r$.
(3) It is $u = \lambda_1 b_1 + \cdots + \lambda_r b_r$.

The minimum distance of b to U is then $\|b - u\|$.

Example 18.1 We determine the orthogonal projection of $b = (1, 2, 3)^\top \in \mathbb{R}^3$ onto

$$U = \langle b_1 = (1, 0, 1)^\top, b_2 = (1, 1, 1)^\top \rangle \subseteq \mathbb{R}^3 .$$

(1) $\{b_1, b_2\}$ is a base of U, we set $A = (b_1, b_2) \in \mathbb{R}^{3\times 2}$.
(2) We determine the normal equation:

$$A = \begin{pmatrix} 1 & 1 \\ 0 & 1 \\ 1 & 1 \end{pmatrix} \Rightarrow A^\top A = \begin{pmatrix} 1 & 0 & 1 \\ 1 & 1 & 1 \end{pmatrix} \begin{pmatrix} 1 & 1 \\ 0 & 1 \\ 1 & 1 \end{pmatrix} = \begin{pmatrix} 2 & 2 \\ 2 & 3 \end{pmatrix}, \; A^\top b = \begin{pmatrix} 4 \\ 6 \end{pmatrix}$$

leads to the following system of linear equations, which we immediately put in row-step form:

$$\left(\begin{array}{cc|c} 2 & 2 & 4 \\ 2 & 3 & 6 \end{array}\right) \rightsquigarrow \left(\begin{array}{cc|c} 1 & 1 & 2 \\ 0 & 1 & 2 \end{array}\right) \rightsquigarrow \left(\begin{array}{cc|c} 1 & 0 & 0 \\ 0 & 1 & 2 \end{array}\right).$$

Thus, we have $\lambda_1 = 0$ and $\lambda_2 = 2$.
(3) This leads to $u = 0\,b_1 + 2\,b_2 = (2, 2, 2)^\top$.
We also determine the minimum distance: We have $u^\perp = b - u = (-1, 0, 1)^\top$, consequently $\|u^\perp\| = \sqrt{2}$ is the minimum distance from b *to* U, note further

$$b = u + u^\perp = \begin{pmatrix} 2 \\ 2 \\ 2 \end{pmatrix} + \begin{pmatrix} -1 \\ 0 \\ 1 \end{pmatrix}. \qquad \blacksquare$$

Remark The normal equation becomes particularly simple when $B = \{b_1, \ldots, b_r\}$ an orthonormal basis of U. Namely, it is then $A^\top A = E_r$ the *r-dimensional* unit matrix. The normal equation $A^\top A x = A^\top b$ is in this case $x = A^\top b$.

18.3 Solution of an Over-Determined Linear System of Equations

We consider an **over-determined linear system of equations**, that is, a linear system of equations with more equations than unknowns:

$$A\,x = b \text{ with } A \in \mathbb{R}^{n\times r}, \; n \geq r, \text{ and } b \in \mathbb{R}^n .$$

Inaccuracies in the entries of A and b usually yield equations that cannot be satisfied, so that in general there is no x that $A\,x = b$ is satisfied. It is obvious, as a substitute for the exact solution, to look for a x, so that the **residuum** $b - Ax$ is as small as possible in the sense of the Euclidean norm, i.e., determine an $x \in \mathbb{R}^r$ with $\|b - A\,x\| = \min$. Such a x is called an **optimal solution** of the linear system of equations. If this minimum is equal to zero, then x solves the system of equations even exactly. An optimal solution can be found as follows:

Recipe: Solving an Over-Determined Linear System of Equations.
An optimal solution $x \in \mathbb{R}^r$ of an over determined linear system of equations

$$A\,x = b \text{ with } A \in \mathbb{R}^{n\times r}\,,\ n \geq r\,,\ \text{ and } b \in \mathbb{R}^n$$

is obtained as the solution of the normal equation $A^\top A\,x = A^\top b$.

Example 18.2 We determine an optimal solution of the over determined linear system of equations

$$\begin{aligned} x \quad\quad\quad &= 0.1 \\ x + y \quad\quad &= 6 \\ x + y + z &= 3.1 \\ y \quad\quad &= 1.1 \\ y + z &= 4.2 \end{aligned}$$

The normal equation $A^\top A\,x = A^\top b$ is to be set up, where A is the coefficient matrix of the LGS and b the *right-hand side*:

$$A = \begin{pmatrix} 1\,0\,0 \\ 1\,1\,0 \\ 1\,1\,1 \\ 0\,1\,0 \\ 0\,1\,1 \end{pmatrix} \Rightarrow \begin{cases} A^\top A = \begin{pmatrix} 1\,1\,1\,0\,0 \\ 0\,1\,1\,1\,1 \\ 0\,0\,1\,0\,1 \end{pmatrix} \begin{pmatrix} 1\,0\,0 \\ 1\,1\,0 \\ 1\,1\,1 \\ 0\,1\,0 \\ 0\,1\,1 \end{pmatrix} = \begin{pmatrix} 3\,2\,1 \\ 2\,4\,2 \\ 1\,2\,2 \end{pmatrix} \\ A^\top b = \begin{pmatrix} 1\,1\,1\,0\,0 \\ 0\,1\,1\,1\,1 \\ 0\,0\,1\,0\,1 \end{pmatrix} \begin{pmatrix} 0.1 \\ 6 \\ 3.1 \\ 1.1 \\ 4.2 \end{pmatrix} = \begin{pmatrix} 9.2 \\ 14.4 \\ 7.3 \end{pmatrix} \end{cases}$$

Now only the normal equation $A^\top A\,x = A^\top b$ has to be solved. We give the augmented coefficient matrix and put it in row step form to read off a solution:

$$\left(\begin{array}{ccc|c} 3 & 2 & 1 & 9.2 \\ 2 & 4 & 2 & 14.4 \\ 1 & 2 & 2 & 7.3 \end{array}\right) \rightsquigarrow \left(\begin{array}{ccc|c} 1 & 2 & 2 & 7.3 \\ 0 & -4 & -5 & -12.7 \\ 0 & 0 & -2 & -0.2 \end{array}\right).$$

Thus $x = (1,\ 3.05,\ 0.1)^\top$ is an optimal solution; in fact, it is uniquely determined. ■

18.4 The Method of Least Squares

In an experiment we obtain to n different time points $t_1, \ldots, t_n$ values $y_1, \ldots, y_n \in \mathbb{R}$. We are looking for a function $f : \mathbb{R} \to \mathbb{R}$ which determines the measured values *as closely as possible*, cf. Fig. 18.1.

Here we take f to be a *as good as possible approximation*, if the quantity

$$\left(y_1 - f(t_1)\right)^2 + \cdots + \left(y_n - f(t_n)\right)^2$$

is minimum. This quantity is the sum of the squares of the vertical distances of the graph of the searched function and the given points (t_i, y_i).

The function f is defined by *basis functions* $f_1, \ldots, f_r$, which are given or chosen to fit the problem (see figures above). Then the scalars $\lambda_1, \ldots, \lambda_r$ with $f = \lambda_1 f_1 + \cdots + \lambda_r f_r$ have to be determined.

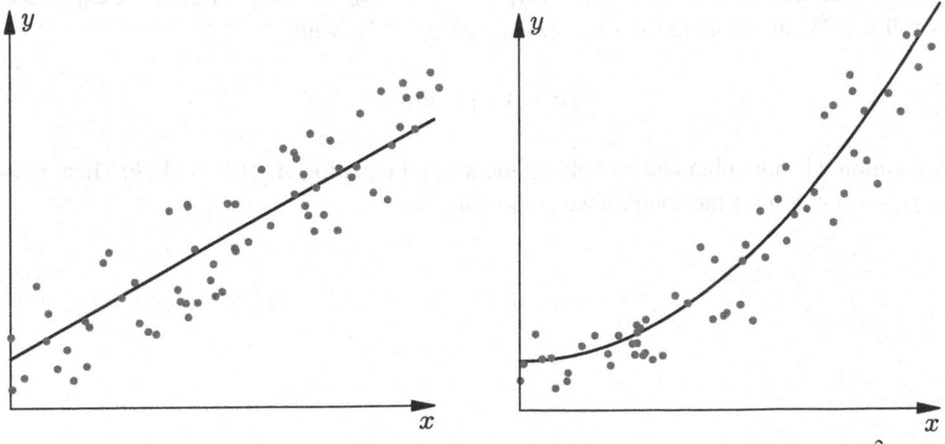

Fig. 18.1 Approximation by a straight line $f = \lambda_1 + \lambda_2 t$ or parabola $f = \lambda_1 + \lambda_2 t + \lambda_3 t^2$

Example 18.3 If one searches for a **best fit straight line** (left image in Fig. 18.1), then one selects

$$\left.\begin{array}{l} f_1 : \mathbb{R} \to \mathbb{R},\ f_1(x) = 1\ \forall x \in \mathbb{R} \\ f_2 : \mathbb{R} \to \mathbb{R},\ f_2(x) = x\ \ \forall x \in \mathbb{R} \end{array}\right\} \Rightarrow f(x) = \lambda_1 + \lambda_2 x\,.$$

If one looks for a **best fit parabola** (right figure in Fig. 18.1), one chooses

$$\left.\begin{array}{l} f_1 : \mathbb{R} \to \mathbb{R},\ f_1(x) = 1\ \forall x \in \mathbb{R} \\ f_2 : \mathbb{R} \to \mathbb{R},\ f_2(x) = x\ \forall x \in \mathbb{R} \\ f_3 : \mathbb{R} \to \mathbb{R},\ f_3(x) = x^2\ \forall x \in \mathbb{R} \end{array}\right\} \Rightarrow f(x) = \lambda_1 + \lambda_2 x + \lambda_3 x^2\,. \qquad \blacksquare$$

If the basis functions $f_1, \ldots, f_r$ are selected or given, you will find the decisive hint how to find the searched values for $\lambda_1, \ldots, \lambda_r \in \mathbb{R}$ with $f = \lambda_1 f_1 + \cdots + \lambda_r f_r$ from the following approach: With the matrix $A \in \mathbb{R}^{n \times r}$, the vector $b \in \mathbb{R}^n$ and the vector $x \in \mathbb{R}^r$

$$A = \begin{pmatrix} f_1(t_1) & \ldots & f_r(t_1) \\ \vdots & & \vdots \\ f_1(t_n) & \ldots & f_r(t_n) \end{pmatrix} \in \mathbb{R}^{n \times r} \text{ and } b = \begin{pmatrix} y_1 \\ \vdots \\ y_n \end{pmatrix} \text{ and } x = \begin{pmatrix} \lambda_1 \\ \vdots \\ \lambda_r \end{pmatrix}$$

we have

$$b - A\,x = \begin{pmatrix} y_1 \\ \vdots \\ y_n \end{pmatrix} - \lambda_1 \begin{pmatrix} f_1(t_1) \\ \vdots \\ f_1(t_n) \end{pmatrix} - \cdots - \lambda_r \begin{pmatrix} f_r(t_1) \\ \vdots \\ f_r(t_n) \end{pmatrix} = \begin{pmatrix} y_1 - f(t_1) \\ \vdots \\ y_n - f(t_n) \end{pmatrix}.$$

So the minimization of the size $\left(y_1 - f(t_1)\right)^2 + \cdots + \left(y_n - f(t_n)\right)^2$ can also be expressed as follows: We are looking for a $x = (\lambda_1, \ldots, \lambda_r)^\top \in \mathbb{R}^r$ with

$$\|b - A\,x\| = \min\,.$$

A solution x is now obtained by solving the normal equation $A^\top A\,x = A^\top b$. Then $f = \lambda_1 f_1 + \cdots + \lambda_r f_r$ is the solution we are looking for.

We formulate the determination of a best fit function to given grid points summarizing as a recipe:

Recipe: Determine a Compensation Function
Given are n grid points $(t_1, y_1), \ldots, (t_n, y_n)$. A best fit function $f = \lambda_1 f_1 + \cdots + \lambda_r f_r$ to given or chosen basis functions $f_1, \ldots, f_r$ is then obtained as follows:

(1) Set $b = (y_1, \ldots, y_n)^\top \in \mathbb{R}^n$ and $A = \begin{pmatrix} f_1(t_1) & \ldots & f_r(t_1) \\ \vdots & & \vdots \\ f_1(t_n) & \ldots & f_r(t_n) \end{pmatrix} \in \mathbb{R}^{n \times r}$.
(2) Solve the normal equation $A^\top A\, x = A^\top b$ and get $x = (\lambda_1, \ldots, \lambda_r)^\top \in \mathbb{R}^r$.
(3) Then $f = \lambda_1 f_1 + \cdots + \lambda_r f_r$ is the best fit function.

Example 18.4 Given are the points

$$(t_1, y_1) = (0, 1),\ (t_2, y_2) = (1, 2),\ (t_3, y_3) = (2, 2),\ (t_4, y_4) = (3, 4),\ (t_5, y_5) = (4, 6)\,.$$

We determine

(a) the best fit straight line $f(x) = \lambda_1 + \lambda_2 x$, i.e., $f_1(x) = 1$ and $f_2(x) = x$ and
(b) the best fit parabola $g(x) = \mu_1 + \mu_2 x + \mu_3 x^2$i.e., $g_1(x) = 1$, $g_2(x) = x$ and $g_3(x) = x^2$.

We start with (a):

(1) It is $b = \begin{pmatrix} 1 \\ 2 \\ 2 \\ 4 \\ 6 \end{pmatrix}$ and $A = \begin{pmatrix} f_1(t_1) & f_2(t_1) \\ f_1(t_2) & f_2(t_2) \\ f_1(t_3) & f_2(t_3) \\ f_1(t_4) & f_2(t_4) \\ f_1(t_5) & f_2(t_5) \end{pmatrix} = \begin{pmatrix} 1 & 0 \\ 1 & 1 \\ 1 & 2 \\ 1 & 3 \\ 1 & 4 \end{pmatrix}$.

(2) Because of

$$A^\top A = \begin{pmatrix} 5 & 10 \\ 10 & 30 \end{pmatrix} \text{ and } A^\top b = \begin{pmatrix} 15 \\ 42 \end{pmatrix}$$

we get as solution of the normal equation:

$$\left(\begin{array}{cc|c} 5 & 10 & 15 \\ 10 & 30 & 42 \end{array}\right) \rightsquigarrow \left(\begin{array}{cc|c} 1 & 2 & 3 \\ 0 & 1 & 1.2 \end{array}\right) \Rightarrow \lambda_2 = 1.2,\ \lambda_1 = 0.6\,.$$

(3) Thus, the best fit straight line we are looking for is

$$f : \mathbb{R} \to \mathbb{R},\ f(x) = 0.6 + 1.2\,x\,.$$

Now (b):

(1) It is $b = \begin{pmatrix} 1 \\ 2 \\ 2 \\ 4 \\ 6 \end{pmatrix}$ and $A = \begin{pmatrix} g_1(t_1) & g_2(t_1) & g_3(t_1) \\ g_1(t_2) & g_2(t_2) & g_3(t_3) \\ g_1(t_3) & g_2(t_3) & g_3(t_3) \\ g_1(t_4) & g_2(t_4) & g_3(t_4) \\ g_1(t_5) & g_2(t_5) & g_3(t_5) \end{pmatrix} = \begin{pmatrix} 1 & 0 & 0 \\ 1 & 1 & 1 \\ 1 & 2 & 4 \\ 1 & 3 & 9 \\ 1 & 4 & 16 \end{pmatrix}$.

(2) Because of

$$A^\top A = \begin{pmatrix} 5 & 10 & 30 \\ 10 & 30 & 100 \\ 30 & 100 & 354 \end{pmatrix} \text{ and } A^\top b = \begin{pmatrix} 15 \\ 42 \\ 142 \end{pmatrix}$$

we get as solution of the normal equation

$$\left(\begin{array}{ccc|c} 5 & 10 & 30 & 15 \\ 10 & 30 & 100 & 42 \\ 30 & 100 & 354 & 142 \end{array}\right) \rightsquigarrow \left(\begin{array}{ccc|c} 1 & 2 & 6 & 3 \\ 0 & 10 & 40 & 12 \\ 0 & 10 & 54 & 16 \end{array}\right) \rightsquigarrow \left(\begin{array}{ccc|c} 1 & 2 & 6 & 3 \\ 0 & 1 & 4 & 1.2 \\ 0 & 0 & 14 & 4 \end{array}\right)$$

and thus $\mu_3 = 2/7,\ \mu_2 = 2/35,\ \mu_1 = 41/35$.

(3) So the best fit parabola g we are looking for is

$$g : \mathbb{R} \to \mathbb{R},\ g(x) = \frac{41}{35} + \frac{2}{35}x + \frac{2}{7}x^2\,.$$

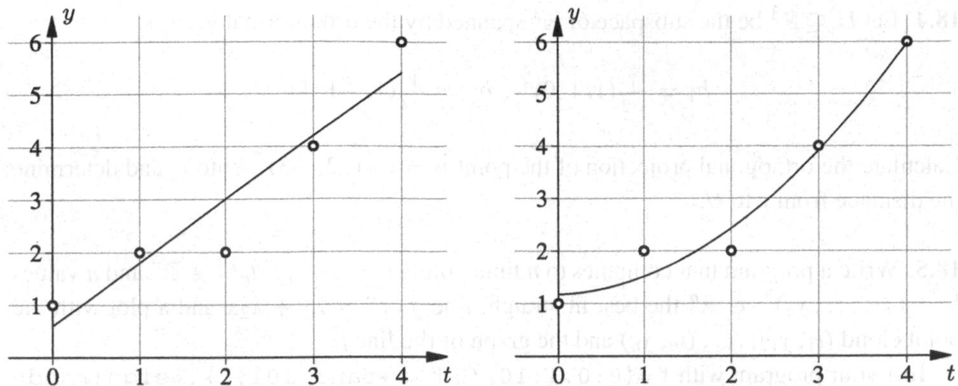

Fig. 18.2 Equalizing line, equalizing parabola and the supporting points

In the figures of Fig. 18.2, the situation is shown. ■

18.5 Exercises

18.1 Let $A \in \mathbb{R}^{n \times r}$ and $b \in \mathbb{R}^n$. Show that the solution sets of the minimization problem $\|b - A\,x\| = \min$ and the normal equation $A^\top A\,x = A^\top b$ are equal. Also show that the solution set has exactly one element if and only if the rank of A is equal to r. Note the box in Sect. 18.1.

18.2 In $\mathbb{R}^4$ let the subspace $U \subseteq \mathbb{R}^4$ be given as

$$U = \langle \begin{pmatrix} 1 \\ -1 \\ 0 \\ 2 \end{pmatrix}, \begin{pmatrix} 0 \\ 2 \\ -2 \\ 1 \end{pmatrix}, \begin{pmatrix} 1 \\ -5 \\ 4 \\ 0 \end{pmatrix} \rangle.$$

Determine the orthogonal projection of the vector $v = (1, 2, -2, -1)^\top$ onto the subspace U. Give a decomposition of $v = u + u^\perp$ with $u \in U$ and $u^\perp \in U^\perp$ and calculate the distance from v to U.

18.3 Let $U = \langle b = (2, 1)^\top \rangle$ and $v = (6, 2)^\top$. Calculate the orthogonal projection of v on the subvector space U. Determine from this an orthogonal decomposition $v = u + u^\perp$ with $u \in U$ and $u^\perp \in U^\perp$. Confirm that this is the orthogonal decomposition of v along b.

18.4 Let $U \subseteq \mathbb{R}^3$ be the subspace of $\mathbb{R}^3$ spanned by the orthonormal vectors

$$b_1 = \frac{1}{\sqrt{2}}(1, 1, 0)^\top, \; b_2 = \frac{1}{\sqrt{3}}(1, -1, 1)^\top.$$

Calculate the orthogonal projection of the point $v = (-1, 2, -3)^\top$ onto U and determine the distance from v to U.

18.5 Write a program that computes to n time points $t = (t_1, \ldots, t_n)^\top \in \mathbb{R}^n$ and n values $b = (y_1, \ldots, y_n)^\top \in \mathbb{R}^n$ the best fit straight line $f(x) = \lambda_1 + \lambda_2 x$ and a plot with the point cloud $(t_1, y_1), \ldots, (t_n, y_n)$ and the graph of the line f.

Test your program with `t=(0:0.1:10)';b=t+rand(101,1).*sign(randn(101,1));`

18.6 **Air resistance:** To reduce the c_w **value of** a car, let it coast at a starting speed v and measure the deceleration a at some speeds v. In an experiment, the following values were obtained:

v [m/s]	10	20	30
a [m/s^2]	0.1225	0.1625	0.2225

Theoretically, the deceleration (negative acceleration) is obtained according to

$$a(v) = r + \frac{\varrho A c_w}{2m} v^2,$$

where the parameter r is due to the velocity-independent rolling friction and the rear term is due to the air friction: A is the area of attack, ϱ the density of the air, m the mass of the car. We have $A = 1\,\text{m}^2$, $\varrho = 1.29\,\text{kg/m}^3$ and $m = 1290$ kg. Estimate with **linear equalization** r and c_w.

18.7

(a) Determine for the measured values

i	1	2	3	4
x_i	−1	0	1	2
y_i	1	0	1	2

a straight line of the form $y(x) = \alpha + \beta x$ so that the deviation

$$\sum_{i=1}^{4} (y(x_i) - y_i)^2$$

becomes minimal. Determine the optimal parameters α and β using the normal equation.

(b) Solve the linear equalization problem $\|b - Ax\| = \min$ for

$$A = \begin{pmatrix} 1 & 1 \\ 1 & 1.0001 \\ 1 & 1.0001 \end{pmatrix} \text{ and } b = \begin{pmatrix} 2 \\ 0.0001 \\ 4.0001 \end{pmatrix}.$$

18.8 Tide prediction: Measurements at a coast give the table

t	0	2	4	6	8	10
h	1.0	1.6	1.4	0.6	0.2	0.8

for the water level h (meters) at daytime t (hours). Estimate under the *natural* assumption that $h(t)$ is due to a harmonic oscillation

$$h(t) = x_1 + x_2 \cos \frac{\pi}{6} t + x_3 \sin \frac{\pi}{6} t$$

described, by means of linear compensation calculation, how large h_{max} and h_{min} are.

18.9 Determine the best fit straight line to the 5 measuring points:

t_i	1	2	3	4	5
y_i	5.0	6.5	9.5	11.5	12.5

Make a drawing.

18.10 In the $\mathbb{R}^3$ are two straight lines

$$G = \{a + \lambda u \mid \lambda \in \mathbb{R}\} \text{ and } H = \{b + \mu v \mid \mu \in \mathbb{R}\}$$

with

$$a = \begin{pmatrix} 1 \\ 3 \\ 2 \end{pmatrix}, \ b = \begin{pmatrix} -2 \\ -2 \\ 1 \end{pmatrix}, \ u = \begin{pmatrix} 1 \\ 1 \\ 2 \end{pmatrix}, \ v = \begin{pmatrix} 2 \\ 3 \\ 4 \end{pmatrix}$$

given. Determine on the straight lines G and H the two points $p = a + \lambda_0 u \in G$ and $q = b + \mu_0 v \in H$ with minimum distance $|p - q|$.

The QR-Decomposition of a Matrix 19

In theory, the linear equilibrium problem is *simple to* solve, requiring *only* the linear system of equations $A^\top A\,x = A^\top b$ to solve. In practical applications, the matrix A usually has a lot of rows, so that solving with pencil and paper is no longer possible. But also the (naive) solving of the normal equation with a computer is not recommended: The calculation of $A^\top A$ and subsequent solving of the LGS $A^\top A\,x = A^\top b$ is unstable and thus leads to inaccurate results. In the numerical solution of the linear compensation problem, the *$Q\,R$-decomposition* of the matrix A is helpful. With the *$Q\,R$ decomposition*, the linear equilibrium problem can be solved in a numerically stable way.

19.1 Full and Reduced $Q\,R$-Decomposition

Every matrix $A \in \mathbb{R}^{n\times r}$ with more rows than columns, i.e. $n \geq r$, can be expressed as the product $A = Q\,R$ with an orthogonal matrix Q and an essentially upper triangular matrix R can be written.

Full and Reduced $Q\,R$-**Decomposition of** A
A decomposition of a matrix $A \in \mathbb{R}^{n\times r}$ with $n \geq r$ of the form $A = Q\,R$ with

- of an orthogonal matrix $Q \in \mathbb{R}^{n\times n}$, i.e.

$$Q^\top Q = E_n \text{ or } Q^{-1} = Q^\top ,$$

- and an *upper* triangular matrix $R \in \mathbb{R}^{n\times r}$ in the following sense

(continued)

C. Karpfinger, *Calculus and Linear Algebra in Recipes*,
https://doi.org/10.1007/978-3-662-65458-3_19

$$R = \begin{pmatrix} \tilde{R} \\ 0 \end{pmatrix} \text{ with } \tilde{R} = \begin{pmatrix} * & \dots & * \\ \vdots & \ddots & \vdots \\ 0 & \dots & * \end{pmatrix} \in \mathbb{R}^{r \times r} \text{ and } 0 \in \mathbb{R}^{(n-r) \times r}$$

is called a **full $Q\,R$-decomposition** of A.

If $A = Q\,R$ is a full $Q\,R$-decomposition of A with an orthogonal matrix $Q \in \mathbb{R}^{n \times n}$ and an *upper* triangular matrix $R = \begin{pmatrix} \tilde{R} \\ 0 \end{pmatrix} \in \mathbb{R}^{n \times r}$, then

$$A = \tilde{Q}\,\tilde{R}\,,$$

where $\tilde{Q} \in \mathbb{R}^{n \times r}$ arises of Q by omitting the last $n - r$ columns, is called a **reduced $Q\,R$-decomposition** of A.

The following pictures visualize the full and reduced $Q\,R$-decomposition:

Notice: If A is quadratic, i.e. $n{=}r$, then the reduced $Q\,R$-decomposition equals the full $Q\,R$-decomposition, in particular in this case $Q = \tilde{Q}$ and $R = \tilde{R}$.

19.2 Construction of the $Q\,R$-Decomposition

We motivate the construction of a full $Q\,R$-decomposition of a matrix $A \in \mathbb{R}^{3\times 3}$. The generalization to a $n \times r$-matrix is then quite simple.

Let $A \in \mathbb{R}^{3\times 3}$. The first step is to determine an orthogonal matrix $H \in \mathbb{R}^{3\times 3}$ with

$$H\,A = \begin{pmatrix} \alpha & * & * \\ 0 & * & * \\ 0 & * & * \end{pmatrix}.$$

We denote the first column of A with s, $A = (s, \ldots)$, and assume that s is not a multiple of e_1, $s \neq \lambda\, e_1$, otherwise we can choose $H = E_3$ and we are done with the first step.

Since now

$$H\,A = H\,(s,\,\ldots) = (H\,s,\,\ldots) = (\alpha\,e_1,\,\ldots)$$

is to be valid, an orthogonal matrix H with $H\,s = \alpha\,e_1$ is needed. Since it is well known that reflection matrices H_a along a (see Example 16.6) are orthogonal, we try it with such a matrix and then set $H = H_a$. We have to determine α and a.

- Since H_a preserves length we have $\|s\| = \|H_a s\| = \|\alpha\,e_1\| = |\alpha|$, i.e. $\alpha = \pm\|s\|$—thus we have α, apart from fixing the sign.
- Because of

$$H_a s = s - \frac{2}{a^\top a} a\,a^\top s = s - \left(\frac{2}{a^\top a} a^\top s\right) a = \alpha\,e_1$$

  we have $a = \lambda\,(s - \alpha\,e_1)$. Since the length of the vector a in the reflection matrix H_a does not play a role, we can choose $a = s - \alpha\,e_1 \neq 0$—this gives us a.
- Now we fix the sign of α, depending on the sign of the first component s_1 of the vector s:

$$\alpha = \begin{cases} +\|s\|\,, & \text{if } s_1 < 0 \\ -\|s\|\,, & \text{if } s_1 \geq 0 \end{cases}.$$

  This choice avoids cancellation: If the first component s_1 is negative, then a positive number is subtracted; if the first component s_1 is positive, then a positive number is added; with this choice, therefore, the phenomenon cannot occur that numbers of approximately equal magnitude are subtracted to approximately zero (see Sect. 4.2.3). If the calculation is correct, this choice does not matter; but if the implementation is done on a computer, this choice of sign has to be considered very well.

With this $\alpha \in \mathbb{R}$ and $a \in \mathbb{R}^3$ one thus obtains $H = H_a = E_3 - \frac{2}{a^\top a} a\,a^\top$, and we have as desired

$$H\,A = \begin{pmatrix} \alpha & * & * \\ 0 & * & * \\ 0 & * & * \end{pmatrix}.$$

One calls the reflection matrices H_a in this context **Householder reflections** or **Householder transformations**.

Example 19.1 For the 3×3-matrix $A = \begin{pmatrix} 1 & 2 & 3 \\ 2 & 5 & 6 \\ 2 & 8 & 9 \end{pmatrix}$ we obtain with $s = \begin{pmatrix} 1 \\ 2 \\ 2 \end{pmatrix}$ and $s_1 = 1$

$$\alpha = -\|s\| = -3 \text{ and } a = s - \alpha\, e_1 = (4, 2, 2)^\top$$

and thus

$$H = H_a = E_3 - \frac{2}{a^\top a} a a^\top = \begin{pmatrix} -1/3 & -2/3 & -2/3 \\ -2/3 & 2/3 & -1/3 \\ -2/3 & -1/3 & 2/3 \end{pmatrix}$$

and

$$H\,A = \begin{pmatrix} -3 & -28/3 & -11 \\ 0 & -2/3 & -1 \\ 0 & 7/3 & 2 \end{pmatrix}. \quad \blacksquare$$

If $H\,A$ is not yet an upper triangular matrix, in a second step we take care of the second column of $H\,A$, i.e., we determine an orthogonal matrix $H' \in \mathbb{R}^{3\times 3}$ with

$$H\,A = \begin{pmatrix} \alpha & * & * \\ 0 & * & * \\ 0 & * & * \end{pmatrix} \Rightarrow H'H\,A = \begin{pmatrix} \alpha & * & * \\ 0 & \alpha' & * \\ 0 & 0 & * \end{pmatrix}, \quad \text{where } H' = \begin{pmatrix} 1 & 0 & 0 \\ 0 & * & * \\ 0 & * & * \end{pmatrix}$$

obviously can be assumed. Then the first row and the first column of $H'H\,A$ and $H\,A$ are equal.

We now obtain H' as in the first step as a Householder transformation H_a with a reflection vector $a = s - \alpha\, e_2$, where we have α and determine s similarly to the first step. Note the following prescription:

Recipe: QR-Decomposition with Householder Transformations.
The full QR-decomposition of a matrix $A \in \mathbb{R}^{n\times r}$ with $n \geq r$, $A = Q\,R$ is obtained after r steps at the latest:

(1) If the first column $s = (s_1, \ldots, s_n)^\top$ of A is not a multiple of e_1:

- Set $\alpha_1 = +\|s\|$ if $s_1 < 0$, and $\alpha_1 = -\|s\|$ if $s_1 \geq 0$.
- Set $a = s - \alpha_1\, e_1 \neq 0$.

(continued)

- With the Householder transformation $H_1 = H_a = E_n - \frac{2}{a^\top a} a a^\top$ we have

$$A_1 = H_1 A = \begin{pmatrix} \alpha_1 & * & \cdots & * \\ 0 & * & \cdots & * \\ \vdots & \vdots & & \vdots \\ 0 & * & \cdots & * \end{pmatrix}.$$

(2) Is $(s_1, \ldots, s_n)^\top$ is the second column of A_1, then set $s = (0, s_2, \ldots, s_n)^\top$. If s is not a multiple of e_2:

- Set $\alpha_2 = +\|s\|$ if $s_2 < 0$, and $\alpha_2 = -\|s\|$ if $s_2 \geq 0$.
- Set $a = s - \alpha_2 e_2 \neq 0$.
- With the Householder transformation $H_2 = H_a = E_n - \frac{2}{a^\top a} a a^\top$ we have

$$A_2 = H_2 H_1 A = \begin{pmatrix} \alpha_1 & * & * & \cdots & * \\ 0 & \alpha_2 & * & \cdots & * \\ \vdots & \vdots & \vdots & & \vdots \\ 0 & 0 & * & \cdots & * \end{pmatrix}.$$

(3) $\cdots$ (r-1)

(r) Is $(s_1, \ldots, s_n)^\top$ the *r-th* column of $A_{r-1} = H_{r-1} \cdots H_1 A$ then set $s = (0, \ldots, 0, s_r, \ldots, s_n)^\top$. If s is not a multiple of e_r:

- Set $\alpha_r = +\|s\|$ if $s_r < 0$, and $\alpha_r = -\|s\|$ if $s_r \geq 0$.
- Set $a = s - \alpha_r e_r \neq 0$.
- With the Householder transformation $H_r = H_a = E_n - \frac{2}{a^\top a} a a^\top$ we have

$$A_r = H_r \cdots H_2 H_1 A = \begin{pmatrix} \tilde{R} \\ 0 \end{pmatrix} \text{ mit } \tilde{R} = \begin{pmatrix} \alpha_1 & \ldots & * \\ \vdots & \ddots & \vdots \\ 0 & \ldots & \alpha_r \end{pmatrix} \text{ and } 0 \in \mathbb{R}^{(n-r)\times r}.$$

(continued)

Because of $H_i^{-1} = H_i$ we get

$$A = Q\,R \text{ with } Q = H_1 \cdots H_r \text{ and } R = A_r\,.$$

Note that the procedure terminates after r steps at the latest, since the matrix A_r already has an upper triangular shape. In the typical tasks in lectures, exercises and exams, one usually considers at most 4×3-matrices, so that one is finished after three steps at the latest. In the following example we consider such a *extreme* case and use for the calculations in some places MATLAB.

Example 19.2 We determine the $Q\,R$-decomposition of the matrix $A = \begin{pmatrix} 2 & 0 & 2 \\ 1 & 0 & 0 \\ 0 & 2 & -1 \\ 2 & 0 & 0 \end{pmatrix}$:

(1) We have $s = (2,\ 1,\ 0,\ 2)^\top \neq \lambda\, e_1$.

- We set $\alpha_1 = -\|s\| = -3$ since $s_1 \geq 0$.
- We set $a = s - \alpha_1\, e_1 = (5,\ 1,\ 0,\ 2)^\top$.
- With the Householder transformation

$$H_1 = E_4 - \frac{2}{a^\top a} a a^\top = \begin{pmatrix} -2/3 & -1/3 & 0 & -2/3 \\ -1/3 & 14/15 & 0 & -2/15 \\ 0 & 0 & 1 & 0 \\ -2/3 & -2/15 & 0 & 11/15 \end{pmatrix}$$

  we have

$$A_1 = H_1 A = \begin{pmatrix} -3 & 0 & -4/3 \\ 0 & 0 & -2/3 \\ 0 & 2 & -1 \\ 0 & 0 & -4/3 \end{pmatrix}.$$

(2) It is $s = (0,\ 0,\ 2,\ 0)^\top \neq \lambda\, e_2$.

- We set $\alpha_2 = -\|s\| = -2$ since $s_2 \geq 0$.
- We set $a = s - \alpha_2\, e_2 = (0,\ 2,\ 2,\ 0)^\top$.

- Using the Householder transformation

$$H_2 = E_4 - \frac{2}{a^\top a} a a^\top = \begin{pmatrix} 1 & 0 & 0 & 0 \\ 0 & 0 & -1 & 0 \\ 0 & -1 & 0 & 0 \\ 0 & 0 & 0 & 1 \end{pmatrix} \text{ delivers } A_2 = H_2 A_1 = \begin{pmatrix} -3 & 0 & -4/3 \\ 0 & -2 & 1 \\ 0 & 0 & 2/3 \\ 0 & 0 & -4/3 \end{pmatrix}.$$

(3) We have $s = (0,\ 0,\ 2/3,\ -4/3)^\top \neq \lambda\, e_3$.

- We set $\alpha_3 = -\|s\| = -\sqrt{20}/3$ since $s_3 \geq 0$.
- We set $a = s - \alpha_3\, e_3 = (0,\ 0,\ (2+\sqrt{20})/3,\ -4/3)^\top$.
- With the Householder transformation $H_3 = E_4 - \frac{2}{a^\top a} a a^\top$,

$$H_3 = \begin{pmatrix} 1 & 0 & 0 & 0 \\ 0 & 1 & 0 & 0 \\ 0 & 0 & -0.4472 & 0.8944 \\ 0 & 0 & 0.8944 & 0.4472 \end{pmatrix}, \quad \text{we get } A_3 = H_3 A_2 = \begin{pmatrix} -3 & 0 & -4/3 \\ 0 & -2 & 1 \\ 0 & 0 & -\sqrt{20}/3 \\ 0 & 0 & 0 \end{pmatrix}.$$

Thus, we have $A = Q\,R$ where $Q = H_1 H_2 H_3$ and $R = A_3$, i.e.

$$Q = \begin{pmatrix} -0.6667 & 0 & -0.7454 & 0 \\ -0.3333 & 0 & 0.2981 & -0.8944 \\ 0 & -1.0000 & 0 & 0 \\ -0.6667 & 0 & 0.5963 & 0.4472 \end{pmatrix} \text{ and } R = \begin{pmatrix} -3 & 0 & -4/3 \\ 0 & -2 & 1 \\ 0 & 0 & -\sqrt{20}/3\,0\,0\,0 \end{pmatrix}.$$

■

If you have the full $Q\,R$-decomposition $A = Q\,R$ determined, one obtains the reduced $Q\,R$-decomposition $A = \tilde{Q}\,\tilde{R}$ quite simply:

- $\tilde{R}$ from R by omitting the lower $n - r$ rows, in MATLAB `R(1:r,:)`, and
- $\tilde{Q}$ from Q by omitting the posterior $n - r$ columns, in MATLAB `Q(:,1:r)`.

MATLAB In MATLAB you get the full $Q\,R$-decomposition of A by `[Q,R]=qr(A)` and the reduced by `[Q,R]=qr(A,0)`.

Remarks

1. Computing the QR-decomposition of a $n \times n$-matrix is generally more stable than computing the LR-decomposition, however, it is also more costly.
2. The Gram-Schmidt orthonormalization method (see recipe in Sect. 17.1 can also be used to construct the QR-decomposition of a matrix A, but it is numerically unstable and therefore not suitable for our purposes.

19.3 Applications of the QR-Decomposition

We discuss two applications of the QR-decomposition of a matrix A.

19.3.1 Solving a System of Linear Equations

Similarly to the LR-decomposition of a square matrix, we can also use the QR-decomposition of a square matrix A to solve a system of linear equations:

Recipe: Solving an LGS with the QR-Decomposition
One obtains a $x \in \mathbb{R}^n$ with $A\,x = b$ for a $b \in \mathbb{R}^m$ as follows:

(1) Determine a QR-decomposition $A = QR$ from A.
(2) Determine x by backward substitution from $Rx = Q^\top b$.

Example 19.3 We solve the linear system of equations $A\,x = b$ with

$$A = \begin{pmatrix} 1 & 2 & 4 \\ 0 & 1 & 0 \\ 1 & 0 & 0 \end{pmatrix} \text{ and } b = \begin{pmatrix} 1 \\ 1 \\ 0 \end{pmatrix}.$$

(1) As QR-decomposition of A we get

$$A = QR = \begin{pmatrix} \frac{-1}{\sqrt{2}} & -1/\sqrt{3} & -1/\sqrt{6} \\ 0 & -1/\sqrt{3} & 2/\sqrt{6} \\ -1/\sqrt{2} & 1/\sqrt{3} & 1/\sqrt{6} \end{pmatrix} \begin{pmatrix} -\sqrt{2} & -\sqrt{2} & -2\sqrt{2} \\ 0 & -\sqrt{3} & -4/3\sqrt{3} \\ 0 & 0 & -2/3\sqrt{6} \end{pmatrix}.$$

(2) By backward substitution we get x from $Rx = Q^\top b$, i.e., from

$$\begin{pmatrix} -\sqrt{2} & -\sqrt{2} & -2\sqrt{2} \\ 0 & -\sqrt{3} & -4/3\sqrt{3} \\ 0 & 0 & -2/3\sqrt{6} \end{pmatrix} \begin{pmatrix} x_1 \\ x_2 \\ x_3 \end{pmatrix} = \begin{pmatrix} -1/\sqrt{2} \\ -2/\sqrt{3} \\ 1/\sqrt{6} \end{pmatrix},$$

thus

$$x_3 = -1/4\,,\ x_2 = 1\ ,\ x_1 = 0\,. \quad \blacksquare$$

19.3.2 Solving the Linear Equalization Problem

The reduced QR-decomposition is used when solving a linear equalization problem or when solving an over determined linear system of equations, i.e., for solving the minimization problem

$$\|b - A\,x\| = \min \quad \text{with } A \in \mathbb{R}^{n\times r} \text{ and } b \in \mathbb{R}^n\,.$$

To determine x in a numerically stable way, proceed as follows:

Recipe: Solve the Linear Equalization Problem with the QR-Decomposition
Let $A \in \mathbb{R}^{n\times r}$ with $n \geq r$ and $\operatorname{rg}(A) = r$ and $b \in \mathbb{R}^n$.
Then one obtains the uniquely determined $x \in \mathbb{R}^r$ with $\|b - A\,x\| = \min$ numerically stable as follows:

(1) Determine a reduced QR-decomposition $A = \tilde{Q}\,\tilde{R}$, $\tilde{Q} \in \mathbb{R}^{n\times r}$, $\tilde{R} \in \mathbb{R}^{r\times r}$.
(2) Solve $\tilde{R}\,x = \tilde{Q}^\top b$.

Example 19.4 We solve the linear equalization problem $\|b - A\,x\| = \min$ with

$$A = \begin{pmatrix} 1 & 1 \\ 2 & 0 \\ 2 & 0 \end{pmatrix}\,,\ b = \begin{pmatrix} 1 \\ 1 \\ 2 \end{pmatrix}$$

using the method described above, with the help of MATLAB: Using `[Q,R] = qr(A,0)` we immediately obtain the matrices $\tilde{Q}$ and $\tilde{R}$ the reduced QR-decomposition:

$$\tilde{Q} = \begin{pmatrix} -0.333 & 0.9428 \\ -0.6667 & -0.2357 \\ -0.6667 & -0.2357 \end{pmatrix} \text{ and } \tilde{R} = \begin{pmatrix} -3.000 & -0.333 \\ 0 & 0.9428 \end{pmatrix}.$$

Using `x=R\ Q'*b` we then obtain the solution $x = \begin{pmatrix} 0.7500 \\ 0.2500 \end{pmatrix}$. ■

MATLAB We recall that a solution x of the linear compensation problem is a best approximation of the possibly unsolvable linear system of equations $Ax = b$: The residual $b - Ax$ is minimized. MATLAB gives this x simply with the command `x = A\b`.

19.4 Exercises

19.1 Compute a QR-decomposition of the matrix $A = \begin{pmatrix} 1 & 1 & 2 \\ 2 & -3 & 0 \\ 2 & 4 & -4 \end{pmatrix}$.

19.2 Given the linear equalization problem defined by

$$A = \begin{pmatrix} 3 & 0 & 0 \\ 4 & 0 & 5 \\ 0 & 3 & -2 \\ 0 & 4 & 4 \end{pmatrix} \quad \text{and} \quad b = \begin{pmatrix} 5 \\ 0 \\ 1 \\ -2 \end{pmatrix}.$$

(a) Determine a QR-decomposition of A.
(b) Give the solution x to the equalization problem. What is the norm of the residual $\|b - Ax\|$?

19.3 Program the QR decomposition using Householder transformations. Test your program using the matrix `A = U S V`, where `U = qr(rand(30));V = qr(rand(30));S = diag(2.\symbol94(-1:-1:-30));`

19.4 Show that the recipe for solving the linear equalization problem with the QR-decomposition in Sect. 19.3.2 works.

19.5 Let $A \in \mathbb{R}^{3\times 2}$ with QR-decomposition

$$Q = \begin{pmatrix} 1/\sqrt{6} & -2/\sqrt{5} & 1/\sqrt{30} \\ 1/\sqrt{6} & 0 & -5/\sqrt{30} \\ 2/\sqrt{6} & 1/\sqrt{5} & 2/\sqrt{30} \end{pmatrix} \quad \text{and} \quad R = \begin{pmatrix} \sqrt{6} & 2\sqrt{6} \\ 0 & -\sqrt{5} \\ 0 & 0 \end{pmatrix},$$

and the vector $b = (-2, 6, 1)^\top$. Solve the linear equalization problem $\|Ax - b\| = \min$ using the reduced QR-decomposition.

Sequences

20

Sequences of real or complex numbers are of fundamental importance for mathematics: With their help, the basic concepts of calculus such as *continuity* and *differentiability* are explained; these concepts can be formulated understandably for an engineer even without the concept of sequences, but in later chapters we will use sequences to explain functions that play a very important role in engineering mathematics, so that we cannot completely dispense with this part of mathematics for engineers either. However, we will keep the presentation concise and only deal with the formulas, rules, properties and criteria that are important for understanding.

20.1 Terms

A **sequence** is a mapping a from $\mathbb{N}_0$ to $\mathbb{R}$ (one speaks then of a **real sequence**) or according to $\mathbb{C}$ (one speaks then of a **complex sequence**); we consider only real sequences for the time being:

$$a : \begin{cases} \mathbb{N}_0 \to \mathbb{R} \\ n \mapsto a(n) \end{cases}.$$

Instead of $a(n)$ one writes a_n and a sequence a is usually given briefly by the **elements of the sequence** a_n, $n \in \mathbb{N}_0$:

$$a = (a_n)_{n \in \mathbb{N}_0} \text{ or shorter } (a_n)_n \text{ or even shorter } (a_n)\,.$$

The elements of a sequence can be given *explicitly* or explained *recursively*, as the following examples show.

C. Karpfinger, *Calculus and Linear Algebra in Recipes*,
https://doi.org/10.1007/978-3-662-65458-3_20

Example 20.1 For the sequences

- $(a_n)_{n\in\mathbb{N}_0}$ with $a_n = 2n$, thus $a_0 = 0,\ a_1 = 2,\ a_2 = 4,\ a_3 = 6,\ a_4 = 8,\ \ldots$ and
- $(a_n)_{n\in\mathbb{N}_0}$ with $a_n = \frac{1}{n^2+1}$, thus $a_0 = 1,\ a_1 = 1/2,\ a_2 = 1/5,\ a_3 = 1/10,\ a_4 = 1/17,\ \ldots$

the elements of the sequences are given **explicit**. In the following examples the elements of the sequences are explained **recursive**:

- $(a_n)_{n\in\mathbb{N}_0}$ with $a_0 = 1$ and $a_1 = 1$ and $a_{i+1} = a_i + a_{i-1}$ for all $i \in \mathbb{N}$. One obtains as first sequence members $a_0 = 1,\ a_1 = 1,\ a_2 = 2,\ a_3 = 3,\ a_4 = 5,\ \ldots$
- $(a_n)_{n\in\mathbb{N}_0}$ with $a_0 = 1$ and $a_{i+1} = 3a_i + 1$ for all $i \in \mathbb{N}$. The first members of the sequence are $a_0 = 1,\ a_1 = 4,\ a_2 = 13,\ a_3 = 40,\ \ldots$

■

In the case of an explicit sequence, one can, for example, specify the 1000th sequence element a_{1000}, whereas for a recursive sequence you first have to specify the 999th sequence element a_{999}, the 998th sequence element a_{998} etc.

Sequences do not have to start with index 0, also $(a_n)_{n\geq 2}$ with $a_n = \frac{1}{n^2-1}$ it makes sense to call a (explicit) sequence. The following are some obvious terms for sequences.

Boundedness and Monotonicity for Sequences

Given a sequence $(a_n)_{n\in\mathbb{N}_0}$. The sequence is called

- **bounded from above** if a $K \in \mathbb{R}$ exists with $a_n \leq K$ for all $n \in \mathbb{N}_0$,
- **bounded from below** if a $K \in \mathbb{R}$ exists with $a_n \geq K$ for all $n \in \mathbb{N}_0$,
- **bounded** if a $K \in \mathbb{R}$ exists with $|a_n| \leq K$ for all $n \in \mathbb{N}_0$,
- **(monotonically) increasing,** if $a_{n+1} \geq a_n$ for all $n \in \mathbb{N}_0$,
- **strictly (monotonically) increasing** if $a_{n+1} > a_n$ for all $n \in \mathbb{N}_0$,
- **(monotonically) decreasing** if $a_{n+1} \leq a_n$ for all $n \in \mathbb{N}_0$,
- **strictly (monotonically) decreasing** if $a_{n+1} < a_n$ for all $n \in \mathbb{N}_0$.

It is clear that a sequence $(a_n)_{n\in\mathbb{N}_0}$ is bounded if and only if $(a_n)_{n\in\mathbb{N}_0}$ is bounded from below and from above. The number K with $a_n \leq K$ and $a_n \geq K$ for all n is called **upper** and **lower bound**.

But how does one decide whether a given sequence is bounded or monotone? The following recipe-like overview shows the essential techniques:

Recipe: Techniques for Proving Boundedness or Monotonicity
Given is a sequence $(a_n)_{n\in\mathbb{N}_0}$.

- Are all members of the sequence positive or negative? If yes: The sequence is bounded by 0 from below or above.
- Often one can conjecture upper and lower bounds after determining the first members of the sequence. This conjecture can often be justified by induction.
- If $a_{n+1} - a_n \geq 0$ or $a_{n+1} - a_n > 0$ for all $n \in \mathbb{N}_0$? If yes: The sequence is increasing or strictly increasing.
- Is $a_{n+1} - a_n \leq 0$ or $a_{n+1} - a_n < 0$ for all $n \in \mathbb{N}_0$? If yes: The sequence is decreasing or strictly decreasing.
- Is $\frac{a_{n+1}}{a_n} \geq 1$ or $\frac{a_{n+1}}{a_n} > 1$ and $a_n > 0$ for all $n \in \mathbb{N}_0$? If yes: The sequence is increasing or strictly increasing.
- Is $\frac{a_{n+1}}{a_n} \leq 1$ or $\frac{a_{n+1}}{a_n} < 1$ and $a_n > 0$ for all $n \in \mathbb{N}_0$? If yes: The sequence is decreasing or strictly decreasing.
- If there is a conjecture that the sequence is (strictly) decreasing or increasing, this can often be justified by induction.

Example 20.2

- The sequence $(a_n)_{n\in\mathbb{N}_0}$ with $a_n = (-1)^n$ is obviously bounded and non-monotonic; where 1 is an upper bound and -1 is a lower bound of $(a_n)_{n\in\mathbb{N}_0}$.
- We now consider the recursive sequence $(a_n)_{n\in\mathbb{N}_0}$ with

$$a_0 = \frac{1}{2} \quad \text{und} \quad a_{n+1} = \frac{1}{2 - a_n} \quad \forall n \in \mathbb{N}_0 .$$

The first members of the sequence are $\frac{1}{2}, \frac{2}{3}, \frac{3}{4}, \frac{4}{5}, \frac{5}{6}, \ldots$. So one suspects $0 < a_n < 1$ for all $n \in \mathbb{N}$. We justify this by induction on$n$:

Base case: For $n = 0$ the statement is correct, because $0 < \frac{1}{2} < 1$.

Induction hypothesis: We assume $0 < a_n < 1$ for a $n \in \mathbb{N}$.

Induction step: We have to show $0 < a_{n+1} < 1$, i.e. because of $a_{n+1} = \frac{1}{2-a_n}$:

$$0 < \frac{1}{2 - a_n} < 1 .$$

And these two inequalities are due to $a_n \in (0, 1)$ (see induction hypothesis) obviously fulfilled, because hereafter $2 - a_n \in (1, 2)$.

Because of the first elements of the sequence we have the conjecture that $(a_n)_{n\in\mathbb{N}_0}$ is strictly monotonically increasing. For example, one confirms this as follows: For all

$n \in \mathbb{N}_0$ we have

$$a_{n+1} - a_n = \frac{1}{2 - a_n} - a_n = \frac{(a_n - 1)^2}{2 - a_n} > 0\,,$$

where, for the last inequality $0 < a_n < 1$ for all $n \in \mathbb{N}_0$ is used, according to which the denominator $2 - a_n$ is always positive.

- The sequence $(a_n)_{n\in\mathbb{N}_0}$ with $a_n = (-2)^n$ is obviously unbounded and not monotone.
- The sequence $(a_n)_{n\geq 1}$ with $a_n = 1 + \frac{1}{n}$ is because of $a_{n+1} - a_n = \frac{1}{n+1} - \frac{1}{n} = -\frac{1}{n(n+1)} < 0$ for all $n \in \mathbb{N}$ strictly monotonically decreasing and bounded from below by 1; bounded from above by 2.
- The sequence $(a_n)_{n\in\mathbb{N}_0}$ with $a_n = 1 + (-1)^n$ is neither monotonically decreasing nor increasing, because:

$$a_0 = 2\,,\ a_1 = 0\,,\ a_2 = 2\,,\ a_3 = 0 \text{ etc.}$$

■

20.2 Convergence and Divergence of Sequences

Sequences *converge* or *diverge*; for a convergent sequence there exists a $a \in \mathbb{R}$ such that in any arbitrarily small neighborhood of a *almost all* sequence elements lie:

Convergence and Limit

A sequence is called (a_n) **convergent** with **limit** $a \in \mathbb{R}$ if for every $\varepsilon > 0$ there is a $N \in \mathbb{N}_0$ with

$$|a_n - a| < \varepsilon \ \text{ for all } \ n \geq N\,.$$

One says then also (a_n) **converges to** a and writes

$$a_n \overset{n\to\infty}{\longrightarrow} a \ \text{ or } \ a_n \longrightarrow a \ \text{ or } \ \lim_{n\to\infty} a_n = a\,.$$

If (a_n) does not converge, then one calls (a_n) **divergent**. And a sequence converging to the limit 0 is called for short **zero sequence**.

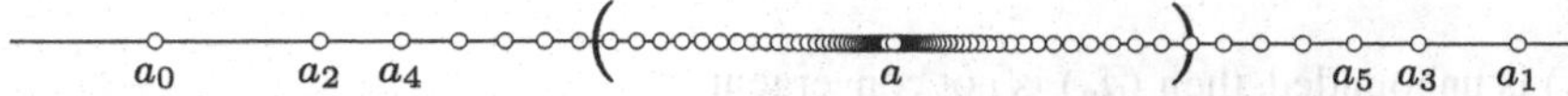

Fig. 20.1 Almost all sequence members lie in the ε-environment of a

Figure 20.1 illustrates the convergence of the sequence (a_n) against the limit a; no matter how small the ε is, there exists a $N \in \mathbb{N}$ so that all sequence elements a_n with $n \geq N$ lie in the ε-environment of a.

One also uses in this sense the way of speaking *almost all* sequence members lie in the ε-environment of a, meaning all but finitely many exceptions.

Example 20.3

- The sequence $(a_n)_{n\geq 1}$ with $a_n = \frac{1}{n}$ converges to 0, because for a $\varepsilon > 0$ we have:

$$|a_n - 0| = \frac{1}{n} < \varepsilon \Leftrightarrow n > \frac{1}{\varepsilon}.$$

If one now sets $N = \lfloor 1/\varepsilon \rfloor + 1$ then for all $n \geq N$ *of* course $|a_n - 0| < \varepsilon$. If for example $\varepsilon = \frac{1}{10}$ then one chooses

$$N = \lfloor 1/\varepsilon \rfloor + 1 = \lfloor 10 \rfloor + 1 = 11 \text{ and gets } |a_n - 0| < \frac{1}{10} \text{ for all } n \geq 11.$$

- The sequence (a_n) with $a_n = (-1)^n$ is divergent: The sequence members are alternately 1 and -1; therefore only ± 1 are possible as limit values, but neither in the $\frac{1}{3}$-environment $(1 - 1/3, 1 + 1/3)$ of 1 nor in the $\frac{1}{3}$-environment $(-1 - 1/3, -1 + 1/3)$ of -1 lie almost all the sequence members.

■

In order to decide whether a sequence converges or not, according to the above definition, one must already know what its limit is. It is often not easy, if not impossible, to guess the limit. Fortunately, there are criteria that are often easy to apply and provide an answer to the question of whether convergence or divergence exists.

Convergence or Divergence Criteria and Other Properties

Given a sequence (a_n).

1. If (a_n) converges, then its limit a is uniquely determined.
2. If (a_n) converges, then (a_n) is bounded.

(continued)

3. If (a_n) is unbounded, then (a_n) is not convergent.
4. **The monotonicity criterion** If (a_n) is bounded and monotonically decreasing or monotonically increasing, then (a_n) is convergent.
5. **The Cauchy criterion** A sequence (a_n) converges if and only if:

$$\forall \varepsilon > 0 \, \exists N \in \mathbb{N}_0 : \quad |a_n - a_m| < \varepsilon \text{ for all } n, m \geq N .$$

Remarks

1. The Cauchy criterion states that sufficiently late members of a sequence are arbitrarily close to each other.
2. The Cauchy and monotonicity criterion allows one to decide convergence even if one does not know the limit.
3. Boundedness alone is not enough for convergence. The consequence (a_n) with $a_n = (-1)^n$ is an example of this.

Example 20.4

- We examine the sequence $(a_n)_{n\geq 1}$ with $a_n = \sum_{k=1}^{n} \frac{1}{k^2}$ for convergence and use the monotonicity criterion. The first elements of the sequence are

$$1, \; 1+1/4, \; 1+1/4+1/9, \; 1+1/4+1/9+1/16, \; \ldots .$$

  - (a_n) is monotonically increasing, because:

$$a_{n+1} - a_n = \frac{1}{(n+1)^2} \geq 0 .$$

  - (a_n) is also limited, because for all $n \in \mathbb{N}$ we have:

$$\begin{aligned} 0 \leq \sum_{k=1}^{n} \frac{1}{k^2} &= 1 + \sum_{k=2}^{n} \frac{1}{k^2} \leq 1 + \sum_{k=2}^{n} \frac{1}{k(k-1)} \leq 1 + \sum_{k=2}^{n} \left(\frac{1}{k-1} - \frac{1}{k} \right) \\ &= 1 + \Bigg(\frac{1}{1} \underbrace{- \frac{1}{2} + \frac{1}{2}}_{=0} + \cdots \underbrace{- \frac{1}{n-1} + \frac{1}{n-1}}_{=0} - \frac{1}{n} \Bigg) \\ &= 1 + \left(1 - \frac{1}{n} \right) < 2 . \end{aligned}$$

The sequence (a_n) is convergent according to the monotonicity criterion. As we will see later (see Example 74.4), we have $a_n \to \pi^2/6$.

- Now we consider the sequence (a_n) with

$$a_0 = 3\,,\ a_1 = 3.1\,,\ a_2 = 3.14\,,\ a_3 = 3.141, \ldots, a_{11} = 3.14159265358 \text{ etc.},$$

which approximates the circular number π. We use the Cauchy criterion to show its convergence. To do this we choose $\varepsilon > 0$. We choose $N \in \mathbb{N}$ large enough so that

$$|a_n - a_m| = 0.0\ldots0x_1\ldots < \varepsilon \ \text{ for all } \ m,\, n \geq N\,.$$

This is possible because from *n-th* sequence element the first n decimal places remain the same.

■

In the case of divergent sequences, several types can be distinguished: There are divergent sequences that are in some sense opposed to $+\infty$ or $-\infty$ *converge*, and those that do not:

Infinite Limits

A sequence (a_n) is said to

- **tend to infinity**, if: $\forall\, K \in \mathbb{R}\, \exists\, N \in \mathbb{N} : \ a_n > K \ \forall n \geq N$,
- **tend to minus infinity**, if: $\forall\, K \in \mathbb{R}\, \exists\, N \in \mathbb{N} : \ a_n < K \ \forall n \geq N$.

One then writes

$$a_n \stackrel{n\to\infty}{\longrightarrow} \pm\infty \ \text{ or } \ \lim_{n\to\infty} a_n = \pm\infty\,.$$

We have:

- $a_n \to +\infty \Rightarrow \frac{1}{a_n} \to 0$,
- $a_n \to -\infty \Rightarrow \frac{1}{a_n} \to 0$,
- $a_n \to 0,\, a_n > 0 \Rightarrow \frac{1}{a_n} \to +\infty$,
- $a_n \to 0,\, a_n < 0 \Rightarrow \frac{1}{a_n} \to -\infty$.

The sequence (a_n) with $a_n = n^2$ tends to $+\infty$, the sequence (a_n) with $a_n = -n$ tends to $-\infty$ and the sequence (a_n) with $a_n = (-1)^n$ diverges.

In the next chapter we show how to determine limits of sequences.

20.3 Exercises

20.1 Given a convergent sequence (a_n) with limit a and a sequence (b_n) with $\lim_{n\to\infty}|b_n - a_n| = 0$. Show

$$\lim_{n\to\infty} b_n = a\,.$$

20.2 Let (a_n) is a sequence of real numbers converging to the limit $a \in \mathbb{R}$, $I = \{i_1, \ldots, i_k\} \subseteq \mathbb{N}$ and $B = \{b_1 \ldots, b_k\} \subseteq \mathbb{R}$. We define the sequence (a'_n) by

$$a'_n = \begin{cases} b_j, & \text{if } n \in I \text{ and } n = i_j, \\ a_n, & \text{if } n \notin I \end{cases}.$$

Show that the sequence (a'_n) converges and determine the limit of the sequence.

20.3 Prove the statements in the box about infinite limits.

Calculation of Limits of Sequences

21

So far we have only asked questions about convergence or divergence and have not yet paid any attention to the calculation of the possibly existing limit. We will do that in this chapter: The methods differ depending on whether one is dealing with an explicit or a recursive sequence.

21.1 Determining Limits of Explicit Sequences

The limit value of an explicit sequence is usually obtained with one of the tools that we compile in a recipe-like manner in the following box.

Recipe: Aids for Calculating Limits of Sequences
If (a_n) and (b_n) are convergent sequences with the limits a and b, i.e. $a_n \to a$ and $b_n \to b$, then we have:

(1) The sequence of sums $(a_n + b_n)$ converges to $a + b$.
(2) The product sequence $(a_n b_n)$ converges to $a\, b$.
(3) The quotient sequence $(a_n/b_n)_{n \geq N}$ converges to a/b, if $b \neq 0$ (there is then a $N \in \mathbb{N}$ with $b_n \neq 0$ for all $n \geq N$).
(4) For all $\lambda \in \mathbb{R}$ converges $(\lambda\, a_n)$ to $\lambda\, a$.
(5) If $a_n \geq 0$ for all n, then $(\sqrt{a_n})$ converges to $\sqrt{a}$.
(6) The sequence of absolute values $(|a_n|)$ converges to $|a|$.
(7) If there exists $N \in \mathbb{N}$ such that $a_n \leq b_n$ for all $n \geq N$, then we have: $a \leq b$.

(continued)

C. Karpfinger, *Calculus and Linear Algebra in Recipes*,
https://doi.org/10.1007/978-3-662-65458-3_21

(8) **Squeeze Theorem**: If $a = b$ and if the sequence (c_n) satisfies the inequality

$$a_n \leq c_n \leq b_n\,,$$

then (c_n) converges to $a = b$.

Complementing (7), we have that from $a_n \to a$, $b_n \to b$ and $a_n < b_n$ for all n does not follow $a < b$. This is shown, for example, by the sequences (a_n) with $a_n = 0$ and (b_n) with $b_n = \frac{1}{n}$: It is $a_n < b_n$ for all n, but nevertheless $a = 0 = b$.

Example 21.1

- For all $q \in \mathbb{R}$ with $0 < |q| < 1$ we have $\lim\limits_{n\to\infty} q^n = 0$. A strict prove of this statement is not so easy, we do without it and content ourselves with an example

$$q = 0.1 \Rightarrow q^2 = 0.01\,,\ q^3 = 0.001\,,\ q^4 = 0.0001\,,\ \ldots$$

- The sequence (a_n) with $a_n = \frac{2}{n} + \left(\frac{1}{3}\right)^n + 7$ converges to 7 due to $\frac{2}{n} \to 0$ and $\left(\frac{1}{3}\right)^n \to 0$.
- The sequence (a_n) with $a_n = \frac{3n^2+7n+8}{5n^2-8n+1}$ converges to $\frac{3}{5}$, because:

$$\lim_{n\to\infty} \frac{3n^2 + 7n + 8}{5n^2 - 8n + 1} = \lim_{n\to\infty} \frac{3 + 7/n + 8/n^2}{5 - 8/n + 1/n^2} = \frac{3+0+0}{5+0+0} = \frac{3}{5}\,.$$

- Using the methods of the last example, one can much more generally state for the sequence (a_n) with $a_n = \frac{a_r n^r + \cdots + a_1 n + a_0}{b_s n^s + \cdots + b_1 n + b_0}$:

$$a_n \to \begin{cases} 0\,, & \text{if } r < s\,, \\ +\infty\,, & \text{if } s < r \text{ and } a_r/b_s \in \mathbb{R}_{>0}\,, \\ -\infty\,, & \text{if } s < r \text{ and } a_r/b_s \in \mathbb{R}_{<0}\,, \\ a_r/b_s\,, & \text{if } s = r\,. \end{cases}$$

- The sequence (a_n) with $a_n = \sqrt{n^2 + 3n + 3} - n$ converges to 3/2, because:

$$a_n = \frac{\left(\sqrt{n^2+3n+3} - n\right)\left(\sqrt{n^2+3n+3} + n\right)}{\sqrt{n^2+3n+3} + n} = \frac{n^2 + 3n + 3 - n^2}{\sqrt{n^2+3n+3} + n}$$

$$= \frac{3n+3}{\sqrt{n^2+3n+3} + n} = \frac{3 + 3/n}{\sqrt{1 + 3/n + 3/n^2} + 1} \overset{n\to\infty}{\longrightarrow} 3/2\,.$$

- We consider the sequence (a_n) with $a_n = \frac{n}{2^n}$. We first have $n^2 \leq 2^n$ for $n \geq 4$, as proved by induction:

$$(n+1)^2 = n^2 + 2n + 1 \leq n^2 + 2n + n = n^2 + 3n \leq n^2 + n \cdot n$$

$$= n^2 + n^2 = 2n^2 \underset{\text{IH}}{\leq} 2 \cdot 2^n = 2^{n+1}$$

With this preliminary consideration, we further have $\frac{1}{2^n} \leq \frac{1}{n^2}$ and with that we get for the sequence (a_n):

$$0 \leq \frac{n}{2^n} \leq \frac{n}{n^2} = \frac{1}{n}.$$

But since the two outer sequences converge to 0, according to the squeeze Theorem $\lim_{n\to\infty} a_n = 0$.
- Now we consider the sequence $(a_n)_{n\geq 1}$ with $a_n = \sqrt[n]{n}$. The first sequence elements are

$$a_1 = 1, \quad a_2 \approx 1.41, \quad a_3 \approx 1.25, \quad a_{100} \approx 1.047, \quad a_{1000} \approx 1.0069.$$

It is reasonable to suspect that (a_n) converges to 1. To show this, we reason that the sequence (b_n) with $b_n = \sqrt[n]{n} - 1$ is a zero sequence:

$$b_n + 1 = \sqrt[n]{n} \Rightarrow n = (1 + b_n)^n = 1 + nb_n + \frac{n(n-1)}{2} b_n^2 + \cdots + b_n^n$$

$$\Rightarrow n \geq 1 + \frac{n(n-1)}{2} b_n^2 \quad \forall n \geq 2 \quad \text{, since } b_n \geq 0$$

$$\Rightarrow 2(n-1) \geq n(n-1) b_n^2 \quad \forall n \geq 2$$

$$\Rightarrow 0 \leq b_n^2 \leq \frac{2}{n} \Rightarrow \lim_{n\to\infty} b_n^2 = 0 \Rightarrow \lim_{n\to\infty} b_n = 0.$$

It is now $\lim_{n\to\infty} a_n = \lim_{n\to\infty} b_n + 1 = 0 + 1 = 1$ and thus $\lim_{n\to\infty} \sqrt[n]{n} = 1$.
- For all real numbers $q \in \mathbb{R}$ the term $n!$ grows faster than q^n, i.e.

$$\frac{q^n}{n!} \overset{n\to\infty}{\longrightarrow} 0.$$

For the proof we choose a $N \in \mathbb{N}$ with $\frac{|q|}{N} \leq \frac{1}{2}$. It is now true for all $n > N$:

$$\frac{|q|^n}{n!} = \frac{|q|}{n} \frac{|q|^{n-1}}{(n-1)!} \leq \frac{1}{2} \frac{|q|^{n-1}}{(n-1)!} \leq \cdots \leq \left(\frac{1}{2}\right)^{n-N} \frac{|q|^N}{N!} = \left(\frac{1}{2}\right)^n \frac{|2q|^N}{N!}.$$

Since N is known, the last fraction is a constant in $\mathbb{R}$. If one now forms the limit, then follows with the squeeze Theorem:

$$0 \leq \left| \frac{q^n}{n!} - 0 \right| \leq \frac{|2q|^N}{N!} \left(\frac{1}{2}\right)^n \stackrel{n\to\infty}{\longrightarrow} 0 \Rightarrow \lim_{n\to\infty} \frac{q^n}{n!} = 0.$$

■

MATLAB We want to mention that MATLAB also offers the possibility to determine limits of explicit sequences with the function `limit`, e.g.

```
» syms n; » limit((n^2+2*n-1)/(2*n^2-2), inf) ans = 1/2
```

or

```
» syms n; » limit(sqrt(n^2+1)-sqrt(n^2-2*n-1), inf) ans =
1
```

21.2 Determining Limits of Recursive Sequences

Now consider a recursively defined sequence (a_n). The elements of the sequence are given by initial values and a **recursion rule**, e.g.

$$\underbrace{a_0 = a\,,\; a_1 = b}_{\text{initial values}} \text{ und } \underbrace{a_{n+1} = \lambda\, a_n - \mu\, a_{n-1}^2}_{\text{recursion rule}}.$$

Suppose that the recursively defined sequence (a_n) converges to a $a \in \mathbb{R}$. Then we have

$$a = \lim_{n\to\infty} a_n = \lim_{n\to\infty} a_{n+1} = \lim_{n\to\infty} a_{n-1}\,,$$

so that in the limit the recursion rule becomes a **fix point equation**, in the above example

$$\underbrace{a = \lambda a - \mu a^2}_{\text{fix point equation}}.$$

If the sequence (a_n) converges to a, then one finds a as the solution of the fixed point equation. Hence, the following recipe for determining the limit of a recursive sequence is obtained:

Recipe: Determine the Limit of a Recursive Sequence
Given is a recursively defined sequence $(a_n)_n$. The limit value a of $(a_n)_n$ is usually determined as follows:

(1) Show that $(a_n)_n$ converges, e.g. by:

- $(a_n)_n$ is bounded and
- $(a_n)_n$ is monotone.

(2) Set up the fixed point equation (replace in the recursion prescription a_{n+1}, a_n, ... by a).
(3) Determine the possible values for a (these are the solutions of the fix point equation).
(4) Consider which values for a are out of the question and which value for a remains.

It is sometimes quite useful to first determine the solutions of the fixed point equation to get a clue as to the value of the lower and upper bounds.

Example 21.2

- We consider the recursive sequence

$$a_0 = 1 \text{ and } a_{n+1} = \sqrt{2\,a_n} \text{ for } n \in \mathbb{N}.$$

(1) The sequence (a_n) converges, since it is bounded and monotone:
We have $0 \leq a_n \leq 2$ for all $n \in \mathbb{N}_0$ since induction shows:

$$0 \leq a_{n+1} = \sqrt{2\,a_n} = \sqrt{2}\sqrt{a_n} \overset{\text{IH}}{\leq} \sqrt{2}\sqrt{2} = 2\,.$$

And we get

$$\frac{a_{n+1}}{a_n} = \frac{\sqrt{2\,a_n}}{a_n} = \sqrt{\frac{2}{a_n}} \geq 1\,.$$

(2) The fix point equation is $a = \sqrt{2\,a}$.
(3) The solutions of the fix point equation are $a = 0$ and $a = 2$.

(4) Since (a_n) is increasing and $a_0 = 1$ is already greater than 0, $a = 0$ cannot be the limit, it must be $a = 2$ the limit we are looking for.

- We study the sequence (a_n) with $a_0 \in (0, 1)$ and $a_{n+1} = 2\,a_n - a_n^2 = a_n(2 - a_n)$.

(1) The sequence (a_n) converges, because it is bounded and monotone:
Boundedness: For all $n \in \mathbb{N}_0$ we have $0 < a_n < 1$ as reasoned by induction:
Base case: For $n = 0$, the statement is true because $0 < a_0 < 1$.
Induction hypothesis: We have $0 < a_n < 1$ for a $n \in \mathbb{N}$.
Induction step: We have to show $0 < a_{n+1} < 1$, i.e. because of $a_{n+1} = a_n(2 - a_n)$:

$$0 < a_n(2 - a_n) < 1 .$$

The first of these two inequalities is due to $a_n \in (0, 1)$ (see induction hypothesis) obviously satisfied, and the second inequality is equivalent to the obviously valid inequality

$$-a_n^2 + 2\,a_n - 1 = -(a_n - 1)^2 < 0 .$$

Monotonicity: Because of

$$a_{n+1} - a_n = a_n - a_n^2 = a_n(1 - a_n) > 0$$

is (a_n) (strictly) monotonically increasing.

(2) The fixed point equation is $a = 2\,a - a^2$, i.e. $a^2 - a = 0$.

(3) The solutions of the fix point equation are $a = 0$ and $a = 1$.

(4) Since (a_n) strictly monotonously increases and $a_0 > 0$, the limit $a = 0$ is not possible and only $a = 1$ remains as limit.

- We consider the so-called **Babylonian method**. For $x \in \mathbb{R}_{>0}$ we declare the sequence (a_n) with

$$a_0 \in \mathbb{R}_{>0} \text{ and } a_{n+1} = \frac{1}{2}\left(a_n + \frac{x}{a_n}\right) \text{ for } n \in \mathbb{N} .$$

We show that this sequence (a_n) converges to $\sqrt{x}$. For example $x = 2$ and $a_0 = 1$ the first members of the sequence are:

$$a_1 = \frac{1}{2}\left(1 + \frac{2}{1}\right) = \frac{3}{2}, \qquad a_2 = \frac{1}{2}\left(\frac{3}{2} + \frac{2}{3/2}\right) = 1.41\overline{6}\ldots .$$

(1) With the possible exception of a_0 the sequence (a_n) is bounded from below by $\sqrt{x}$, thus $a_{n+1} \geq \sqrt{x}$ for all $n \in \mathbb{N}_0$: From the obviously correct inequality

$$0 \leq \left(a_n - \sqrt{x}\right)^2 = a_n^2 - 2a_n\sqrt{x} + x$$

follows namely

$$a_n^2 + x \geq 2a_n\sqrt{x} \text{ and though } \underbrace{\frac{1}{2}\left(a_n + \frac{x}{a_n}\right)}_{a_{n+1}} \geq \sqrt{x}\,.$$

With the possible exception of a_0 the sequence (a_n) is monotonically decreasing, since for all $n \in \mathbb{N}$:

$$a_{n+1} - a_n = \frac{1}{2}\left(a_n + \frac{x}{a_n}\right) - a_n = \frac{1}{2}\left(\frac{-a_n^2 + x}{a_n}\right) \leq 0\,.$$

Since (a_n) is monotonically decreasing for $n \geq 1$ and bounded from below, (a_n) is also bounded.

(2) The fix point equation is $a = \frac{1}{2}\left(a + \frac{x}{a}\right)$ i.e. $a^2 - x = 0$.

(3) As solutions of the fix point equations we obtain $a = \pm\sqrt{x}$.

(4) Since $a_n \geq 0$ for all $n \in \mathbb{N}_0$, the limit value $a = -\sqrt{x}$ is excluded, so there remains only $\lim_{n\to\infty} a_n = \sqrt{x}$.

21.3 Exercises

21.1 For convergent sequences (a_n) and (b_n) we have $\lim_{n\to\infty}(a_n + b_n) = \lim_{n\to\infty} a_n + \lim_{n\to\infty} b_n$. Give examples with

$$\lim_{n\to\infty} c_n = +\infty \quad \text{und} \quad \lim_{n\to\infty} d_n = -\infty$$

for which the above statement is false. In particular, for $e \in \mathbb{R}$ should be true:

(a) $\lim\limits_{n\to\infty} (c_n + d_n) = +\infty$, (b) $\lim\limits_{n\to\infty} (c_n + d_n) = -\infty$, (c) $\lim\limits_{n\to\infty} (c_n + d_n) = e$.

21.2 Investigate whether the following sequences converge and determine their limits, if possible:

(a) $a_n = \frac{(2n+3)(n-1)}{n^2+n-4}$,

(b) $b_n = \sqrt{n+\sqrt{n}} - \sqrt{n-\sqrt{n}}$,

(c) $c_n = \prod_{k=2}^{n}\left(1-\frac{1}{k^2}\right)$,

(d) $d_n = \binom{2n}{n}2^{-n}$,

(e) $e_n = \sqrt{n+4} - \sqrt{n+2}$,

(f) $f_n = \left(\frac{5n}{2n+1}\right)^4$,

(g) $g_n = \frac{n^2-1}{n+3} - \frac{n^3+1}{n^2+1}$,

(h) $h_n = \sqrt{n(n+3)} - n$,

(i) $i_n = \frac{(4n+3)(n-2)}{n^2+n-2}$,

(j) $j_n = \sqrt{n+\sqrt{2n}} - \sqrt{n-\sqrt{2n}}$,

(k) $k_n = \frac{(4n^2+3n-2)(4n-2)}{(4n-2)(2n+1)(n-4)}$,

(l) $l_n = \sqrt{n^2+2n} - n$.

21.3 Examine the following recursively defined sequences for convergence and determine the limits if possible:

(a) $a_1 = 0$, $a_{n+1} = \frac{1}{4}(a_n - 3)$ for $n \geq 1$,

(b) $b_1 = 0$, $b_{n+1} = \sqrt{2+b_n}$ for $n \geq 1$,

(c) $c_1 = 2$, $c_{n+1} = \frac{3}{4-c_n}$ for $n \geq 1$,

(d) $d_1 = 0$, $d_{n+1} = 3d_n + 2$ for $n \geq 1$.

21.4 Using the squeeze Theorem, show that for any $\alpha > 0$ we have $\lim_{n\to\infty} \sqrt[n]{\alpha} = 1$.
Note: Use $a_n = \sqrt[n]{\alpha} - 1$ and Bernoulli's inequality from Exercise 2.1.

21.5 For $x, a_0 \in \mathbb{R}_{>0}$ the sequence (a_n) with $a_{n+1} = \frac{1}{2}\left(a_n + \frac{x}{a_n}\right)$, $n \in \mathbb{N}$, converges to $\sqrt{x}$.

Write a MATLAB-function `[an, n] = root(x, a0, tol)` that returns the value and index of the first member of the sequence a_n for which $|a_n^2 - x| < $ `tol`.

Series

22

With the help of *series* we will explain important functions. But this is future-talk, more details in Chap. 24 on *power series*. But we want to make clear already here, that the notion of a series is fundamental for our purposes.

With sequences we have already laid the essential foundation, since series are special sequences. But unlike sequences, it is usually very difficult to determine the limit for series. But that doesn't matter much, because for series there are some tools available, which allow to decide about convergence or divergence of the series. And this knowledge alone is generally sufficient.

22.1 Definition and Examples

For a series the on the one hand suggestive, on the other hand confusing notation $\sum_{k=0}^{\infty} a_k$ is common. We describe briefly, how this notation is used, so that the mistake, that here infinite summands are added, is eliminated once and for all: Given is a real sequence $(a_k)_{k\in\mathbb{N}_0}$. For this sequence we consider another sequence $(s_n)_{n\in\mathbb{N}_0}$ where the sequence elements s_n are formed with the help of the sequence elements a_k in the following way:

$$s_0 = a_0\,,\ s_1 = a_0 + a_1\,,\ s_2 = a_0 + a_1 + a_2, \ldots,\ s_n = \sum_{k=0}^{n} a_k\,,\ \ldots$$

The sequence elements s_n are called **partial sums** of $(a_k)_{k\in\mathbb{N}_0}$ and the sequence $(s_n)_{n\in\mathbb{N}_0}$ of the partial sums of $(a_k)_{k\in\mathbb{N}_0}$ is called **series** with the **series elements** a_k. Instead of

C. Karpfinger, *Calculus and Linear Algebra in Recipes*,
https://doi.org/10.1007/978-3-662-65458-3_22

$(s_n)_{n\in\mathbb{N}_0}$ one also writes $\sum_{k=0}^{\infty} a_k$ this shorthand notation is obvious:

$$(s_n)_n = \left(\sum_{k=0}^{n} a_k\right)_{n\in\mathbb{N}_0} = \sum_{k=0}^{\infty} a_k\,.$$

Note that a series does not have an infinite number of summands added together (you can't, you wouldn't be able to handle it), a series is a sequence (s_n) which can therefore converge or diverge.

Example 22.1

- The **harmonic series**

$$\sum_{k=1}^{\infty} \frac{1}{k}$$

  is the sequence $(s_n)_{n\in\mathbb{N}_0}$ of the partial sums $s_n = 1 + \frac{1}{2} + \frac{1}{3} + \cdots + \frac{1}{n}$.
- The **geometric series**

$$\sum_{k=0}^{\infty} q^k$$

  is the sequence $(s_n)_{n\in\mathbb{N}_0}$ of the partial sums $s_n = 1 + q + q^2 + \cdots + q^n$.

Since series are sequences, we can examine them for convergence and divergence; here we can make a refinement in the notion of convergence:

Convergence and Absolute Convergence of Series

Given is a series $\sum_{k=0}^{\infty} a_k$. We say the series

- **converges to** $a \in \mathbb{R}$ if the sequence $(s_n)_{n\in\mathbb{N}_0} = \left(\sum_{k=0}^{n} a_k\right)_{n\in\mathbb{N}_0}$ of partial sums converges to a.
  One calls a in this case the **value of the series** and denotes it also by $\sum_{k=0}^{\infty} a_k$.
- **converges absolutely** if the sequence $(t_n)_{n\in\mathbb{N}_0} = \left(\sum_{k=0}^{n} |a_k|\right)_{n\in\mathbb{N}_0}$ of the absolute values of the series elements converges.

Every absolutely convergent series also converges.

So absolute convergence is *better* than convergence per se: dealing with absolutely convergent series is much easier. Convergent series that do not converge absolutely are also called **conditionally convergent** .

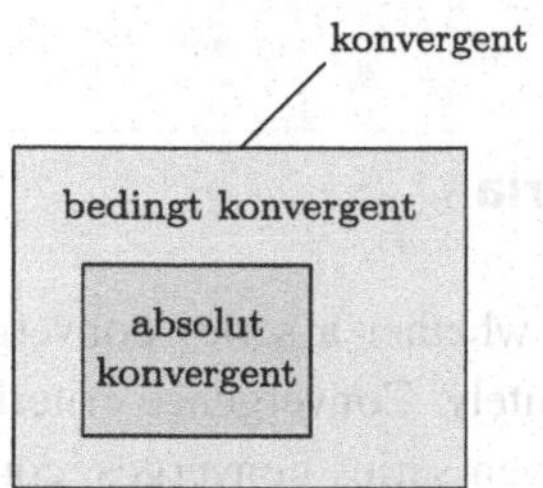

Whether a series is convergent or even absolutely convergent can often be determined by looking at the *n-th* partial sum s_n; note the following examples:

Example 22.2

- The geometric series $\sum_{k=0}^{\infty} q^k$ converges in the case $|q| < 1$, because for the *n-th* partial sum we have:

$$s_n = 1 + q + q^2 + \cdots + q^n = \frac{1 - q^{n+1}}{1 - q} \overset{n \to \infty}{\longrightarrow} \frac{1}{1 - q}.$$

  So $\sum_{k=0}^{\infty} q^k = \frac{1}{1-q}$ is the value of the geometric series for $|q| < 1$. The geometric series also converges absolutely, which can be seen by replacing q^k with $|q^k|$.
- The series $\sum_{k=1}^{\infty} \frac{1}{k(k+1)}$ also converges, we have for the *n-th* partial sum:

$$s_n = \sum_{k=1}^{n} \frac{1}{k(k+1)} = \sum_{k=1}^{n} \left(\frac{1}{k} - \frac{1}{k+1} \right) = 1 - \frac{1}{n+1} \overset{n \to \infty}{\longrightarrow} 1.$$

  Therefore $\sum_{k=1}^{\infty} \frac{1}{k(k+1)} = 1$ is the value of the series. Since all series members are positive, this series also converges absolutely.
- The series $\sum_{k=0}^{\infty} k^2$ on the other hand diverges, we have for the *n-th* partial sum:

$$s_n = 1 + 4 + 9 + \cdots + n^2 \overset{n \to \infty}{\longrightarrow} \infty.$$

- The harmonic series $\sum_{k=1}^{\infty} \frac{1}{k}$ diverges (note exercise 22.3).
- The **alternating harmonic series**

$$\sum_{k=1}^{\infty} \frac{(-1)^{k+1}}{k}$$

converges (we will prove this later), but it does not converge absolutely, because the series $\sum_{k=1}^{\infty} \frac{1}{k}$ of the sums, which is the harmonic series, does not converge. So the alternating harmonic series is conditionally convergent.

22.2 Convergence Criteria

So far, it is not easy to decide whether a series converges or not, and if it converges, whether it also converges absolutely. Convergence criteria are statements that can often be used to decide whether a given series converges, converges absolutely, or diverges. Unfortunately, these criteria do not generally tell us what the value of the series is in the case of convergence. But then we can often live with that.

Convergence and Divergence Criteria

Given is a series $\sum\limits_{k=0}^{\infty} a_k$.

- **The zero sequence criterion** The series $\sum_{k=0}^{\infty} a_k$ diverges, if $(a_k)_k$ is not a zero sequence.
- **The Leibniz criterion** The alternating series $\sum_{k=0}^{\infty}(-1)^k a_k$ converges if $(a_k)_k$ is a monotonically decreasing zero sequence (which implies $a_k \geq 0$). For the value S of the series, we have the estimation

$$\left| S - \sum_{k=0}^{n}(-1)^k a_k \right| \leq a_{n+1} .$$

- **The comparison test (major)**: The series $\sum_{k=0}^{\infty} a_k$ converges absolutely if a **convergent majorant** exists, i.e.

$$\exists \text{ convergent series } \sum_{k=0}^{\infty} b_k : \quad |a_k| \leq b_k \quad \forall k \geq N \in \mathbb{N} .$$

- **The comparison test (minor)**: The series $\sum_{k=0}^{\infty} a_k$ diverges if a **divergent minor** exists, i.e.

$$\exists \text{ divergent series } \sum_{k=0}^{\infty} b_k : \quad 0 \leq b_k \leq a_k \quad \forall k \geq N \in \mathbb{N} .$$

(continued)

- **The ratio test** If there exists $r = \lim\limits_{k\to\infty} \left|\frac{a_{k+1}}{a_k}\right|$ then we have:

$$\text{In the case} \begin{cases} r < 1 & \text{converges the series absolutely.} \\ r > 1 & \text{diverges the series.} \\ r = 1 & \text{is all possible.} \end{cases}$$

- **The root test**: Exists $r = \lim\limits_{k\to\infty} \sqrt[k]{|a_k|}$ then we have:

$$\text{In the case} \begin{cases} r < 1 & \text{converges the series absolutely.} \\ r > 1 & \text{diverges the series.} \\ r = 1 & \text{is all possible.} \end{cases}$$

Note carefully the formulation of the zero sequence criterion: *If* (a_k) is not a zero sequence, *then* $\sum a_k$ *divergent.* If (a_k) is a zero sequence, then $\sum a_k$ does not necessarily converge; the simplest counterexample is the harmonic series:

$$\left(\frac{1}{k}\right)_k \text{ is a zero sequence, but } \sum_{k=1}^{\infty} \frac{1}{k} \text{ diverges.}$$

Example 22.3

- The series $\sum_{k=0}^{\infty} \frac{k^2+7}{5k^2+1}$ diverges, since $\left(\frac{k^2+7}{5k^2+1}\right)_k$ *is not a zero sequence.*
- The series $\sum_{k=1}^{\infty} \frac{(-1)^{k+1}}{k}$ is called **alternating harmonic series**. It converges according to the Leibniz criterion, since $\left(\frac{1}{k}\right)$ is a monotonically decreasing zero sequence.
- The series $\sum_{k=1}^{\infty} \frac{1}{k^2-k+1}$ converges, since $\sum_{k=2}^{\infty} \frac{1}{(k-1)^2} = \sum_{k=1}^{\infty} \frac{1}{k^2}$ is a convergent majorant (note Example 20.4):

$$\forall k \in \mathbb{N}: \quad k^2 - k + 1 = (k-1)^2 + k \geq (k-1)^2 \overset{\forall k \geq 2}{\Longrightarrow} \frac{1}{k^2-k+1} \leq \frac{1}{(k-1)^2}.$$

- The series $\sum_{k=1}^{\infty} \frac{1}{\sqrt{k}}$ diverges, since $\sum_{k=1}^{\infty} \frac{1}{k}$ is a divergent minor:

$$\forall k \in \mathbb{N}: \quad \sqrt{k} \leq k \Rightarrow 0 \leq \frac{1}{k} \leq \frac{1}{\sqrt{k}}.$$

- The **general harmonic series** $\sum_{k=1}^{\infty} \frac{1}{k^{\alpha}}$, $\alpha > 0$, converges for $\alpha > 1$ and diverges for $\alpha \leq 1$ (see Exercise 22.3).
- The series $\sum_{k=1}^{\infty} \frac{(-1)^k}{k^2}$ converges absolutely, since $\sum_{k=1}^{\infty} \frac{1}{k^2}$ is a convergent majorant.
- The series $\sum_{k=1}^{\infty} \frac{(-1)^k}{k}$ converges according to the Leibniz criterion, but does not converge absolutely.
- The series $\sum_{k=1}^{\infty} (-1)^k \frac{\cos(\sqrt{k})}{k^{5/2}}$ converges absolutely because of the comparison test, we have:

$$\left| (-1)^k \frac{\cos(\sqrt{k})}{k^{5/2}} \right| \leq \frac{1}{k^{5/2}} \leq \frac{1}{k^2} \quad \forall k \in \mathbb{N}.$$

Therefore $\sum_{k=1}^{\infty} \frac{1}{k^2}$ is a convergent majorant.

- The series $\sum_{k=0}^{\infty} \frac{2^k}{k!}$ converges absolutely because of the ratio test:

$$r = \lim_{k \to \infty} \left| \frac{2^{k+1}}{(k+1)!} \frac{k!}{2^k} \right| = \lim_{k \to \infty} \frac{2}{k+1} = 0 < 1.$$

- The series $\sum_{k=1}^{\infty} \left(\frac{1}{3} - \frac{1}{\sqrt{k}} \right)^k$ converges absolutely by the root test because:

$$r = \lim_{k \to \infty} \sqrt[k]{\left| \left(\frac{1}{3} - \frac{1}{\sqrt{k}} \right)^k \right|} = \lim_{k \to \infty} \left| \frac{1}{3} - \frac{1}{\sqrt{k}} \right| = \frac{1}{3} < 1.$$

The following two examples prove that both convergence and divergence can exist in the ratio test in the case $r = 1$:

- The harmonic series $\sum_{k=1}^{\infty} \frac{1}{k}$ is known to diverge. Using the ratio test, it follows:

$$r = \lim_{k \to \infty} \frac{1}{k+1} k = 1.$$

- The series $\sum\limits_{k=1}^{\infty} \frac{1}{k^2}$ is known to converge. By the ratio test, it follows:

$$r = \lim_{k \to \infty} \frac{1}{(k+1)^2} k^2 = 1.$$

■

We calculate two more concrete values of series:

Example 22.4

- We consider the series $\sum_{k=1}^{\infty} \frac{3}{\pi^k}$. Its convergence can be easily proved by the ratio test. To determine its value, we use the limit formula of the geometric series. We have:

$$\sum_{k=1}^{\infty} \frac{3}{\pi^k} = \sum_{k=0}^{\infty} \frac{3}{\pi^k} - 3 = 3\left(\sum_{k=0}^{\infty} \frac{1}{\pi^k} - 1\right) = 3\left(\frac{1}{1-\frac{1}{\pi}} - 1\right)$$

$$= 3\left(\frac{1}{\frac{\pi-1}{\pi}} - 1\right) = 3\left(\frac{\pi}{\pi-1} - \frac{\pi-1}{\pi-1}\right) = \frac{3}{\pi-1}.$$

- Now we consider the alternating series $\sum_{k=1}^{\infty}(-1)^k \frac{4^k+2}{5^k}$. First, we prove their convergence using the Leibniz criterion. This follows if we can show that $\left(\frac{4^k+2}{5^k}\right)$ is a monotonically decreasing zero sequence. Now we have:

$$\frac{4^k+2}{5^k} = \left(\frac{4}{5}\right)^k + \frac{2}{5^k} \overset{k\to\infty}{\longrightarrow} 0,$$

so that $\left(\frac{4^k+2}{5^k}\right)$ is a zero sequence, and because of

$$\frac{a_{k+1}}{a_k} = \frac{4^{k+1}+2}{5^{k+1}} \frac{5^k}{4^k+2} = \frac{4\cdot 4^k + 1\cdot 2}{5\cdot 4^k + 5\cdot 2} \le 1$$

the sequence is $\left(\frac{4^k+2}{5^k}\right)$ is monotonically decreasing. According to the Leibniz criterion, the series converges.

Having shown convergence, we now determine the value of the series. To do this, we use the formula for the geometric series:

$$\sum_{k=1}^{\infty}(-1)^k \frac{4^k+2}{5^k} = \sum_{k=1}^{\infty}\left(-\frac{4}{5}\right)^k + 2\sum_{k=1}^{\infty}\left(-\frac{1}{5}\right)^k$$

$$= \sum_{k=0}^{\infty}\left(-\frac{4}{5}\right)^k + 2\sum_{k=0}^{\infty}\left(-\frac{1}{5}\right)^k - 1 - 2$$

$$= \frac{1}{1+4/5} + \frac{2}{1+1/5} - 3 = -\frac{7}{9}.$$

Since the series is a sum of two convergent geometric series, the sum is naturally convergent. So the above proof of convergence was ultimately superfluous, but certainly useful for practice. ■

Special rules apply to absolutely convergent series, we record them:

Rules for Absolutely Convergent Series
Let $\sum_{k=0}^{\infty} a_k$ and $\sum_{k=0}^{\infty} b_k$ are absolutely convergent series.

- We have the triangle inequality:

$$\left|\sum_{k=0}^{\infty} a_k\right| \leq \sum_{k=0}^{\infty} |a_k| .$$

- Arbitrary rearrangements of the summation yield the same series value.
- **The Cauchy product** We define the Cauchy product

$$\sum_{k=0}^{\infty} c_k \quad \text{mit} \quad c_k = \sum_{l=0}^{k} a_l b_{k-l}$$

of the series $\sum_{k=0}^{\infty} a_k$ and $\sum_{k=0}^{\infty} b_k$. If these are absolutely convergent, then their Cauchy product converges to $\sum_{k=0}^{\infty} c_k$ and we have

$$\sum_{k=0}^{\infty} c_k = \sum_{k=0}^{\infty} a_k \cdot \sum_{k=0}^{\infty} b_k .$$

MATLAB It is possible to calculate one or the other value of a series with MATLAB using the function symsum; unfortunately MATLAB cannot determine all values, nor can it always decide on convergence or divergence (note the examples in the exercises), for example we get the value $\pi^2/6$ of the (convergent) series $\sum_{k=1}^{\infty} 1/k^2$ or the value ln(2) of the (convergent) series $\sum_{k=1}^{\infty} (-1)^{k+1}/k$ as follows:

```
» syms k; » symsum(1/k^2,1,inf) ans = pi^2/6  » syms k; » symsum((-1)^(k+1)/k,1,inf)
ans = log(2)
```

22.3 Exercises

22.1 Prove that the harmonic series $\sum_{k=1}^{\infty} \frac{1}{k}$ diverges.

22.2 Prove that the general harmonic series $\sum_{k=1}^{\infty} \frac{1}{k^\alpha}$, $\alpha > 0$, converges for $\alpha > 1$ and diverges for $\alpha \leq 1$.

22.3 Examine the following series for convergence or divergence, determine the value of the series if possible.

(a) $\sum\limits_{k=1}^{\infty} \frac{2k^2+k+7}{(k+2)(k-7)}$,

(b) $\sum\limits_{k=1}^{\infty} \frac{k!}{k^k}$,

(c) $\sum\limits_{k=1}^{\infty} \frac{k+4}{k^2-3k+1}$,

(d) $\sum\limits_{k=1}^{\infty} \frac{(k+1)^{k-1}}{(-k)^k}$,

(e) $\sum\limits_{k=1}^{\infty} \frac{1}{5^k}$,

(f) $\sum\limits_{k=1}^{\infty} \frac{4k}{3k^2+5}$,

(g) $\sum\limits_{k=1}^{\infty} \frac{4k}{4k^2+8}$,

(h) $\frac{1}{2} + \frac{2}{3} + \frac{3}{4} + \frac{4}{5} + \ldots$,

(i) $\sum\limits_{k=1}^{\infty} \frac{2}{3^k}$,

(j) $\sum\limits_{k=1}^{\infty} \frac{2k}{k!}$,

(k) $\sum\limits_{k=1}^{\infty} \frac{1}{100k}$,

(l) $\sum\limits_{k=1}^{\infty} \frac{(k+1)^{k-1}}{(-k)^k}$,

(m) $\sum\limits_{k=1}^{\infty} \left(1 - \frac{1}{k}\right)$,

(n) $\sum\limits_{k=3}^{\infty} \frac{k+1}{k^2-k-2}$,

(o) $\sum\limits_{k=0}^{\infty} \frac{k^3}{4^k}$,

(p) $\sum\limits_{k=1}^{\infty} \left(\frac{-9k-10}{10k}\right)^k$,

(q) $\sum_{k=1}^{\infty} \frac{1}{k^k}$,

(r) $\sum_{k=0}^{\infty} \frac{k^2}{2^k}$.

22.4

(a) Show that the series $\sum_{k=0}^{\infty} \frac{2+3\cdot(-1)^k}{k+1}$ is alternating and that $\lim_{k\to\infty} \frac{2+3\cdot(-1)^k}{k+1} = 0$. Why is the Leibniz criterion not applicable?

(b) Why does the series converge $\sum_{k=0}^{\infty} \frac{(-1)^k}{k+2} \cdot \frac{k+1}{k+3}$?

22.5 Using MATLAB, calculate the following values:

(a) $\sum_{k=1}^{\infty} \frac{1}{(4k-1)(4k+1)}$.

(b) $\sum_{k=0}^{\infty} (1/2)^k$, $\sum_{k=0}^{\infty} (1/10)^k$, $\sum_{k=m}^{\infty} (1/10)^k$.

(c) $\sum_{k=0}^{\infty} (-1)^k \frac{1}{2k+1}$.

22.6 Write a program that approximates the value of a Leibniz convergent alternating series.

22.7 Examine the following series for convergence and divergence:

(a) $\sum_{k=1}^{\infty} (-1)^k \sin(\frac{20}{k})$,

(b) $\sum_{k=0}^{\infty} (\frac{2k^2+3k+1}{5k^2+k+3})^k$,

(c) $\sum_{k=1}^{\infty} \cos(\frac{3}{k})$,

(d) $\sum_{k=0}^{\infty} \frac{2k+3}{3k^2+5}$,

(e) $\sum_{k=0}^{\infty} \frac{4k+5}{3k^3+1}$.

22.8 Show that the following series converge and determine the limits.

(a) $\sum_{k=1}^{\infty} (\cos(\frac{5}{k}) - \cos(\frac{5}{k+1}))$,

(b) $\sum_{k=0}^{\infty} \frac{2^k - 3^{2k}}{10^k}$,

(c) $\sum_{k=1}^{\infty} \frac{1}{(2k)^2}$,

(d) $\sum_{k=0}^{\infty} \frac{1}{(2k+1)^2}$.

For (c),(d) you may use the series value $\sum_{k=1}^{\infty} \frac{1}{k^2} = \frac{\pi^2}{6}$.

Mappings

23

We already had first contacts with *functions*, more general *mappings*. In addition, we are familiar with the *concept of functions* from our school days. In this chapter, we will look at general properties of mappings or functions, which will help us to correctly understand many properties that have so far been formulated only *vaguely*, e.g. the *invertibility of* mappings.

23.1 Terms and Examples

A **mapping** f from the set D into the set W *assigns* to each element $x \in D$ exactly one element $y \in W$; we write $f(x) = y$. This *coherence of* f, D, W and $y = f(x)$ is expressed by a mapping f as follows:

$$f : \begin{cases} D \to & W \\ x \mapsto & f(x) = y \end{cases} \quad \text{or } f : D \to W,\ x \mapsto f(x) = y \text{ or } f : D \to W,\ f(x) = y.$$

If D and W are subsets of $\mathbb{R}$, $\mathbb{C}$, $\mathbb{R}^n$ or $\mathbb{C}^n$, we also speak of a **function** instead of a mapping.

If $f : D \to W,\ f(x) = y$ is a mapping, it is called

- D the **domain**,
- W is the **codomain**,
- $f(x) = y$ the **mapping rule**,
- $f(D) = \{f(x) \mid x \in D\} \subseteq W$ the **range** of f,
- $\text{graph}(f) = \{(x, f(x)) \mid x \in D\} \subseteq D \times W$ the **graph** of f.

C. Karpfinger, *Calculus and Linear Algebra in Recipes*,
https://doi.org/10.1007/978-3-662-65458-3_23

From our school days we are used to mappings being given *explicitly*, i.e., the mapping rule is given concretely, e.g. $f : \mathbb{R} \to \mathbb{R}$, $f(x) = 2x^2 + \sin(x)$. But we can deal with a mapping even if no explicit mapping rule is given, e.g. $f : \mathbb{R} \to \mathbb{R}$, $x \mapsto y$ where x and y satisfy the equation $e^y + y^3 = x$; one speaks then of an **implicit function**. We deal with implicit functions in Chap. 52; in the following we consider only explicit functions.

Example 23.1

- The function $f : [-\sqrt{3}, \sqrt{3}] \to \mathbb{R}$, $f(x) = x^2$ has the range $f([-\sqrt{3}, \sqrt{3}]) = [0, 3] \subseteq \mathbb{R}$ The graph is shown in Fig. 23.1 below.
- We give a function *section-wise*:

$$f : [-1, 1] \to \mathbb{R},\ f(x) = \begin{cases} 2x, & -1 \le x < 0 \\ x^2, & x = 0 \\ 1, & 0 < x \le 1 \end{cases} .$$

  The range of f is $f([-1, 1]) = [-2, 0] \cup \{1\} \subseteq \mathbb{R}$. Note Fig. 23.1.
- We consider a function in two variables x and y:

$$f : \begin{cases} \mathbb{R}^2 & \to \quad \mathbb{R} \\ (x, y) & \mapsto \ f(x, y) = x\, y\, e^{-2(x^2+y^2)} \end{cases} .$$

  Here, each point of the plane $\mathbb{R}^2$, which is the domain, is mapped to the real number $x\, y\, e^{-2(x^2+y^2)}$. The graph of this function is the set $\text{Graph}(f) = \{(x, y, x\, y\, e^{-2(x^2+y^2)}) \mid (x, y) \in \mathbb{R}^2\} \subseteq \mathbb{R}^3$, this *surface* is shown in Fig. 23.2. ■

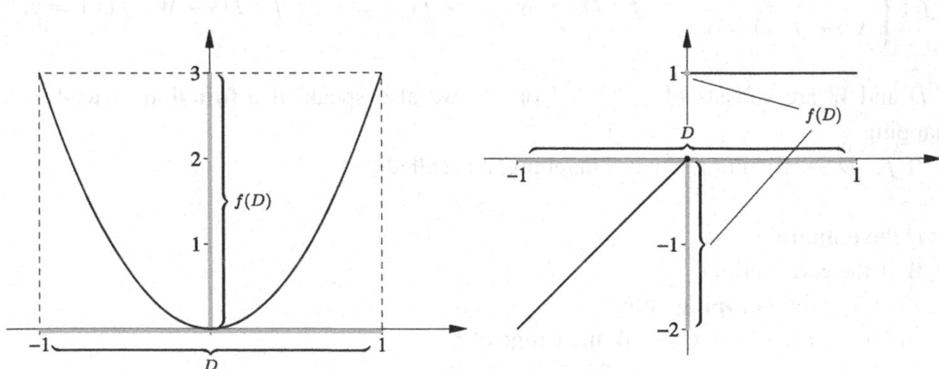

Fig. 23.1 The domains D, the images $f(D)$, the graphs of the functions under consideration

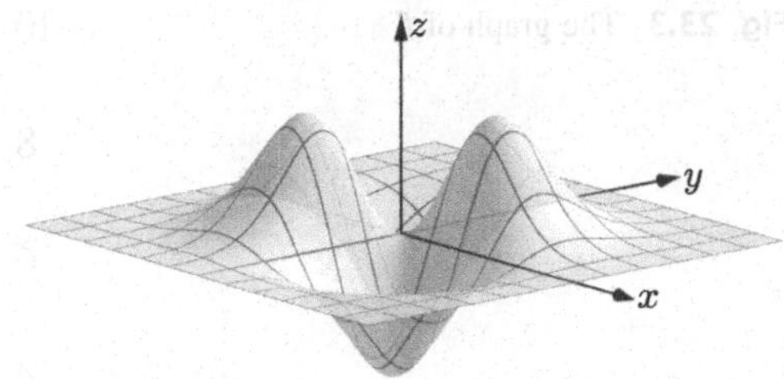

Fig. 23.2 The graph of a function in two (real) variables is a surface in the $\mathbb{R}^3$

MATLAB MATLAB offers several ways to define a function and plot its graph. Please refer to the internet or the MATLAB script for the terms `fplot`, `ezplot` and `ezsurf`. The graph of the above function $f : \mathbb{R}^2 \to \mathbb{R}$ with $f(x, y) = x\, y\, \mathrm{e}^{-2(x^2+y^2)}$ is obtained with

```
ezsurf('x*y*exp(-2*(x^2+y^2))',[-2,2,-2,2])
```

The function specification must be enclosed in inverted commas, the numbers in `[-2,2,-2,2]` stand in sequence for the restricted definition range chosen in the representation $(x, y) \in [-2, 2] \times [-2, 2]$.

Even a section-wise explained function is easily representable with MATLAB, e.g. by means of a function file:

```
function [ y ] = f( x )
if x<=1
  y=x^2;
else
  y=-(x-1)^2+1;
end
```

The graph of this function can then be obtained with `fplot('f',[-3,3])`, for example, where we have also specified `grid on` to get the *grid* for better orientation (see Fig. 23.3).

23.2 Composition, Injective, Surjective, Bijective

One can possibly *compose* mappings: Are $f : D \to W$ and $g : D' \to W'$ mappings, where $f(D) \subseteq D'$, then one calls

$$g \circ f : \begin{cases} D \to & W' \\ x \mapsto (g \circ f)(x) = g\big(f(x)\big) \end{cases}$$

the **composition** of f with g. Note that $f(D)$ must lie in the domain of g.

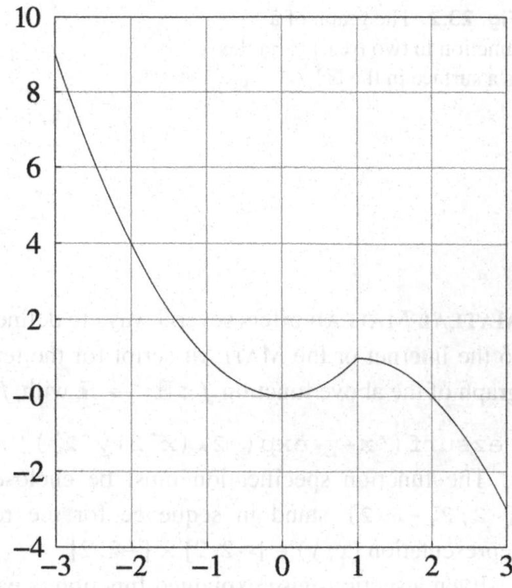

Fig. 23.3 The graph of f

Example 23.2

- We consider the functions $f : \mathbb{R} \to \mathbb{R},\ f(x) = x^2$ and $g : \mathbb{R} \to \mathbb{R},\ g(x) = \mathrm{e}^x$. Then:

$$g \circ f : \mathbb{R} \to \mathbb{R},\ g \circ f(x) = g\big(f(x)\big) = \mathrm{e}^{x^2}$$

$$f \circ g : \mathbb{R} \to \mathbb{R}, f \circ g(x) = f\big(g(x)\big) = \mathrm{e}^{2x} .$$

  Note $f \circ g \neq g \circ f$, because $g \circ f(1) = \mathrm{e} \neq \mathrm{e}^2 = f \circ g(1)$.
- Now we consider $f : (0, \infty) \to \mathbb{R},\ f(x) = \frac{-1}{x}$ and $g : (0, \infty) \to \mathbb{R},\ g(x) = \sqrt{x}$. Then:

$$f \circ g : (0, \infty) \to \mathbb{R},\ f \circ g(x) = f\big(g(x)\big) = f(\sqrt{x}) = \frac{-1}{\sqrt{x}} .$$

  On the other hand $g \circ f$ cannot be formed at all, since $f(0, \infty) = (-\infty, 0)$ is not in the domain of definition of g.
- Now let matrices $B \in \mathbb{R}^{m \times n}$ and $A \in \mathbb{R}^{p \times m}$ are given. For this we consider the mappings

$$f_B : \begin{cases} \mathbb{R}^n \to \mathbb{R}^m \\ v \mapsto B\,v \end{cases} \text{ and } f_A : \begin{cases} \mathbb{R}^m \to \mathbb{R}^p \\ v \mapsto A\,v \end{cases} .$$

We compose f_A with f_B and get $f_A \circ f_B = f_{A\,B}$ since

$$f_A \circ f_B(v) = f_A(B\,v) = A\,(B\,v) = (A\,B)\,v = f_{A\,B}(v) \quad \text{for all } v \in \mathbb{R}^n .$$

■

Remarks

1. This composition of mappings is interesting for the following reason: One can often easily prove properties for *simple* mappings. If one then knows that these properties are inherited by composition one can prove these properties also for quite complicated mappings, namely for compositions of simple mappings, an example of such a property is the *differentiability*: If f and g are *differentiable*, then also $g \circ f$.
2. the **elementary functions** are: Polynomial functions, rational functions, root functions, power functions, trigonometric functions and their inverse functions, the *hyperbolic* functions and their inverse functions, exponential function and its inverse function, the logarithm function, and compostions of these functions. This is to be seen more as an agreement than a definition, in fact there is some disagreement on the description of elementary functions, we definitely include all the functions listed among these elementary functions. By the fact that compositons of elementary functions are elementary again, the variety of elementary functions is guaranteed.

Injectivity, Surjectivity and Bijectivity

A mapping $f : D \to W$ is called

- **injective** if from $f(x_1) = f(x_2)$, $x_1, x_2 \in D$, follows $x_1 = x_2$.
- **surjective** if for each $y \in W$ there exists a $x \in D$ with $f(x) = y$.
- **bijective** if it is injective and surjective.

We give further, sometimes quite descriptive, descriptions of these terms:

- A mapping f is injective if and only if from $x_1 \neq x_2$ also $f(x_1) \neq f(x_2)$ follows.
- A mapping f is injective if and only if f takes on all different values.
- A mapping $f : D \to W$ is surjective if and only if $f(D) = W$.
- A mapping $f : D \to W$ is surjective if and only if every element in W is a value of the function f.
- The mapping $f : D \to W$ is bijective if and only if for every $y \in W$ there is exactly one $x \in D$ with $f(x) = y$.

- The mapping $f : D \to W$ is bijective if and only if every element in W is taken exactly once as the value of the function.

Recipe: Injectivity, Surjectivity, Bijectivity of a Real Function
If $f : D \to W$ is a real function, i.e. $D, W \subseteq \mathbb{R}$, whose graph is known, then we have:

- f is injective if and only if every horizontal line intersects the graph in at most one point.
- f is surjective if and only if every horizontal line passing through a $y \in W$ intersects the graph at least once.
- f is bijective if and only if every horizontal line passing through a $y \in W$ intersects the graph exactly once.

Example 23.3

- For any set X, the mapping $\mathrm{Id}_X : X \to X, \mathrm{Id}(x) = x$ is bijective. One calls the mapping Id_X the **identity**.
- We now consider four mappings with different domains and codomains, but each with the same mapping rule $f(x) = x^2$:
  The function $f_1 : \mathbb{R} \to \mathbb{R},\ f(x) = x^2$ is
  - not surjective, because $\nexists x \in \mathbb{R} : \quad f(x) = -1$.
  - not injective, because $f(1) = 1 = f(-1)$ and $1 \neq -1$.

  The function $f_2 : \mathbb{R} \to \mathbb{R}_{\geq 0},\ f(x) = x^2$ is
  - surjective, because $\forall\, y \in \mathbb{R}_{\geq 0} : \exists \sqrt{y} \in \mathbb{R} : \quad f(\sqrt{y}) = y$.
  - not injective, because $f(1) = 1 = f(-1)$ and $1 \neq -1$.

  The function $f_3 : \mathbb{R}_{\geq 0} \to \mathbb{R},\ f(x) = x^2$ is
  - not surjective, because $\nexists x \in \mathbb{R} : \quad f(x) = -1$.
  - injective, because $x_1^2 = x_2^2 \Rightarrow (x_1 - x_2)(x_1 + x_2) = 0 \Rightarrow x_1 = x_2$.

  The function $f_4 : \mathbb{R}_{\geq 0} \to \mathbb{R}_{\geq 0},\ f(x) = x^2$ is
  - surjective, because $\forall\, y \in \mathbb{R}_{\geq 0} : \exists \sqrt{y} \in \mathbb{R} : \quad f(\sqrt{y}) = y$.
  - injective, because $x_1^2 = x_2^2 \Rightarrow (x_1 - x_2)(x_1 + x_2) = 0 \Rightarrow x_1 = x_2$.

    Thus f_4 is bijective.

  The graphs of the four functions follow; note that all the properties can be easily read off from the graph using the above recipe, see Fig. 23.4.
- Let now be

$$f : \mathbb{R}_{>0} \to \mathbb{R}_{>1/2},\ f(x) = \frac{1}{2}\sqrt{1 + \frac{1}{x^2}}\,.$$

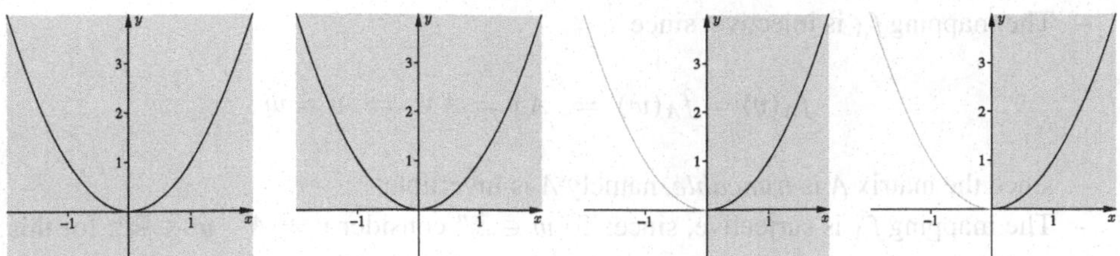

Fig. 23.4 The function f with $f(x) = x^2$ with different domains and codomains

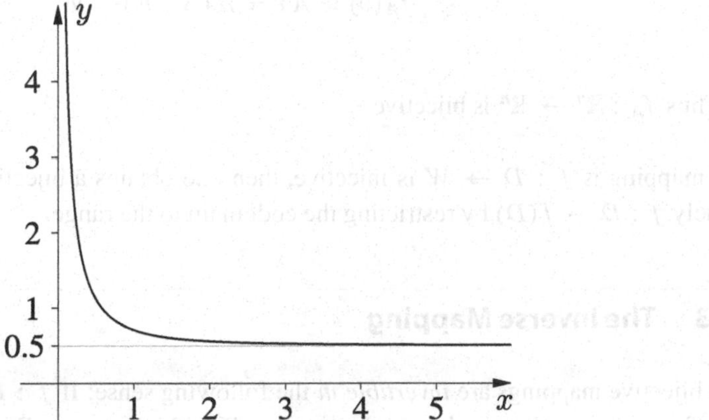

Fig. 23.5 $f : \mathbb{R}_{>0} \to \mathbb{R}_{>1/2}$ is bijective

This function is bijective because:

– f is injective: namely, if $x_1, x_2 \in \mathbb{R}_{>0}$ with $f(x_1) = f(x_2)$, we have:

$$\frac{1}{2}\sqrt{1+\frac{1}{x_1^2}} = \frac{1}{2}\sqrt{1+\frac{1}{x_2^2}} \implies x_1^2 = x_2^2 \overset{x_1,x_2>0}{\implies} x_1 = x_2\,.$$

– f is surjective: To $y \in \mathbb{R}_{>1/2}$ consider $x = \frac{1}{\sqrt{4y^2-1}} \in \mathbb{R}_{>0}$, we have:

$$f(x) = f\left(\frac{1}{\sqrt{4y^2-1}}\right) = \frac{1}{2}\sqrt{1+\frac{1}{1/(4y^2-1)}} = \frac{1}{2}\sqrt{4y^2} = y\,.$$

Since f is injective and surjective, the function f is bijective.

Figure 23.5 shows the graph of the function under consideration: any horizontal straight line through a point from the interval $(1/2, \infty)$ of the y-axis intersects the graph at exactly one point. Therefore, the function is bijective.

- We consider an invertible matrix $A \in \mathbb{R}^{n\times n}$ and for this the mapping $f_A : \mathbb{R}^n \to \mathbb{R}^n$, $f(v) = A\,v$.

- The mapping f_A is injective, since

$$f_A(v) = f_A(w) \Rightarrow A\,v = A\,w \Rightarrow v = w\,,$$

  since the matrix A is *truncable*, namely A is invertible.
- The mapping f_A is surjective, since: To $w \in \mathbb{R}^n$ consider $v = A^{-1}w \in \mathbb{R}^n$; for this v we have:

$$f_A(v) = A\,v = A\,(A^{-1}w) = w\,.$$

Thus $f_A : \mathbb{R}^n \to \mathbb{R}^n$ is bijective. ■

If a mapping is $f : D \to W$ is injective, then one obtains a bijective mapping from it, namely $f : D \to f(D)$ by restricting the codomain to the range.

23.3 The Inverse Mapping

The bijective mappings are *invertible in* the following sense: If $f : D \to W$ is bijective, then for each $y \in W$ there is exactly one $x \in D$ with $f(x) = y$. So if we assign to each $y \in W$ this uniquely determined $x \in D$, we get a mapping $g : W \to D$. This mapping g has the characteristic properties:

$$g \circ f : D \to D,\ g \circ f(x) = g\big(f(x)\big) = g(y) = x\,,\ \text{i.e. } g \circ f = \mathrm{Id}_D\,,$$

$$f \circ g : W \to W,\ f \circ g(y) = f\big(g(y)\big) = f(x) = y\,,\ \text{i.e. } f \circ g = \mathrm{Id}_W\,.$$

Instead of g we write f^{-1} and call f^{-1} the **inverse mapping** of f or f also **invertible**.

The Inverse

Given is a mapping $f : D \to W$. We have:

- f is invertible if and only if f is bijective.
- f is invertible if and only if there is a mapping $g : W \to D$ with

$$g \circ f = \mathrm{Id}_D \quad\text{and}\quad f \circ g = \mathrm{Id}_W\,;$$

  in this case $f^{-1} = g$.
- If f^{-1} is bijective, then f is the inverse function of f^{-1}.

We look at some known affiliations $f \leftrightarrow f^{-1}$.

Example 23.4

- The inverse function of the bijective function $f : \mathbb{R}_{\geq 0} \to \mathbb{R}_{\geq 0},\ f(x) = x^2$ is

$$f^{-1} : \mathbb{R}_{\geq 0} \to \mathbb{R}_{\geq 0},\ f^{-1}(x) = \sqrt{x}\,,$$

because $f^{-1} \circ f(x) = \sqrt{x^2} = x$ for all $x \in \mathbb{R}_{\geq 0}$ and $f \circ f^{-1}(x) = \left(\sqrt{x}\right)^2 = x$ for all $x \in \mathbb{R}_{\geq 0}$.

- The exponential function $\exp : \mathbb{R} \to \mathbb{R}_{>0}$ is also bijective. Its inverse function is the natural logarithm function

$$\ln : \mathbb{R}_{>0} \to \mathbb{R},\ x \mapsto \ln(x)\,,$$

Therefore

$$\ln \circ \exp(x) = \ln\left(\mathrm{e}^x\right) = x\ \forall x \in \mathbb{R} \ \text{ and } \ \exp \circ \ln(x) = \mathrm{e}^{\ln(x)} = x\ \forall x \in \mathbb{R}_{>0}\,.$$

- The bijective function $\cos : [0, \pi] \to [-1, 1]$ has the inverse function

$$\arccos : [-1, 1] \to [0, \pi],\ x \to \arccos(x)\,.$$

■

In these examples we *knew* or we *defined* the inverse mappings accordingly. But how else to find $f^{-1} : W \to D$ to a reversible mapping $f : D \to W$? This is often not possible at all, sometimes the following recipe helps:

Recipe: Determine the Inverse Mapping
If $f : D \to W$ is a invertible mapping, you get $f^{-1} : W \to D$ possibly as follows:

(1) Solve (if possible) the equation $f(x) = y$ to x, thus $x = g(y)$.
(2) Set $y = x$ and $g = f^{-1}$.

Example 23.5

- The mapping $f : \mathbb{R}_{\geq 0} \to \mathbb{R}_{\geq 0},\ f(x) = x^2$ is bijective. We determine the inverse function:

  (1) Solving the equation $f(x) = y$ to $x = g(y)$ yields:

  $$f(x) = y \Leftrightarrow x^2 = y \Leftrightarrow x = \sqrt{y} = g(y)\,.$$

  (2) Now we replace y by x and g by f^{-1}. We get the inverse function $f^{-1}(x) = \sqrt{x}$.

- We consider again the bijective function

  $$f : \mathbb{R}_{>0} \to \mathbb{R}_{>\frac{1}{2}},\ f(x) = \frac{1}{2}\sqrt{1 + \frac{1}{x^2}}\,.$$

  We determine the inverse function f^{-1}:

  (1) First we solve the equation $f(x) = y$ to $x = g(y)$:

  $$\begin{aligned} f(x) = y &\Rightarrow \frac{1}{2}\sqrt{1 + \frac{1}{x^2}} = y \\ &\Rightarrow 4y^2 = 1 + \frac{1}{x^2} \Rightarrow x = \frac{1}{\sqrt{4y^2 - 1}} = g(y)\,. \end{aligned}$$

  (2) Now we replace y by x and g by f^{-1}. We get the inverse function

  $$f^{-1} : \mathbb{R}_{>\frac{1}{2}} \to \mathbb{R}_{>0},\ f^{-1}(x) = \frac{1}{\sqrt{4x^2 - 1}}\,.$$

- Finally, we consider the function $f : [0, \pi] \to [1, \pi - 1],\ f(x) = \cos(x) + x$. This function is also bijective, as you can figure out by looking at the graph. Nevertheless, the inverse function of f is not so easy to state, since it is not clear how to solve the equation $f(x) = y$ to $x = g(y)$. ■

23.4 Bounded and Monotone Functions

Boundedness and *monotonicity* are explained for functions analogous to the corresponding terms for sequences:

Bounded and Monotone Functions

A function $f : D \to W$ is called

- **bounded from above** if $\exists\, K \in \mathbb{R} : \ f(x) \le K \quad \forall x \in D$.
- **bounded from below** if $\exists\, K \in \mathbb{R} : \ K \le f(x) \quad \forall x \in D$.
- **bounded** if $\exists\, K \in \mathbb{R} : \ |f(x)| \le K \quad \forall x \in D$.
- **monotonically increasing**, if $\forall x, y \in D : \ x < y \Rightarrow f(x) \le f(y)$.
- **strictly monotonically increasing**, if $\forall x, y \in D : \ x < y \Rightarrow f(x) < f(y)$.
- **monotonically decreasing** if $\forall x, y \in D : \ x < y \Rightarrow f(x) \ge f(y)$.
- **strictly monotonically decreasing** if $\forall x, y \in D : \ x < y \Rightarrow f(x) > f(y)$.

If f is bounded from above, below or at all, this means the graph of f is below, above, or between horizontal lines, see Fig. 23.6.

Example 23.6

- $f : \mathbb{R} \to \mathbb{R}, \ f(x) = 2$ is monotonically increasing and monotonically decreasing.
- $f : \mathbb{R}_{\ge 0} \to \mathbb{R}, \ f(x) = x^2$ is strictly monotonically increasing.
- $f : \mathbb{R} \to \mathbb{R}, \ f(x) = \lfloor x \rfloor = \max\{z \in \mathbb{Z} \,|\, z \le x\}$ is monotonically increasing. ■

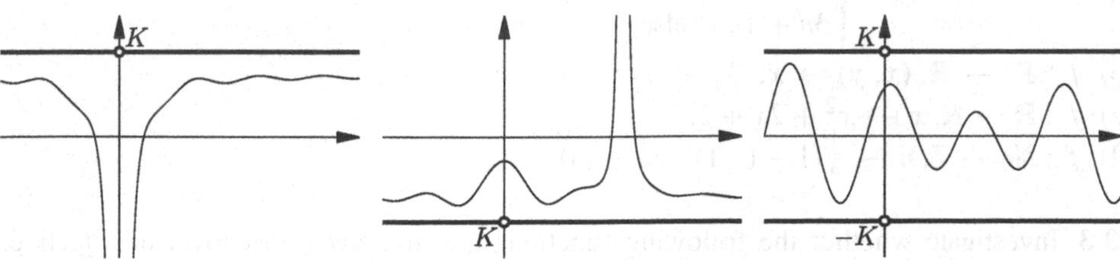

Fig. 23.6 A function bounded from above, one from below, and one bounded at all

We explain the important connections to injectivity and bijectivity, respectively:

Strict Monotonicity and Injectivity or Bijectivity

- If $f : D \to W$ is strictly monotone (increasing or decreasing), then f injective.
- If $f : D \to W$ is strictly monotonic (increasing or decreasing), then $f : D \to f(D)$ bijective, i.e., there is an inverse function f^{-1}.

Remark In principle, it is difficult to decide whether a function is monotonic or strictly monotonic (decreasing or increasing) on a domain D. The essential tool for this is the *derivative*, which of course we know from our school days. We give this tool in a clear representation in Sect. 27.1 and do without further details here.

23.5 Exercises

23.1 Give two mappings each from $\mathbb{N}$ to $\mathbb{N}$ which are

(a) injective, but not surjective,
(b) surjective, but not injective,
(c) are injective and surjective.

23.2 Examine the following mappings for injectivity, surjectivity, and bijectivity:

(a) $f : \mathbb{N} \to \mathbb{N},\ n \mapsto \begin{cases} n/2\,, & \text{if } n \text{ even} \\ 3n+1\,, & \text{else} \end{cases}$,
(b) $f : \mathbb{R}^2 \to \mathbb{R},\ (x, y) \mapsto x$,
(c) $f : \mathbb{R} \to \mathbb{R},\ x \mapsto x^2 + 2x + 2$,
(d) $f : \mathbb{N}_0 \to \mathbb{Z},\ n \mapsto \frac{1}{4}(1 - (-1)^n(2n + 1))$.

23.3 Investigate whether the following functions are injective, surjective, or bijective. Also, give the range and (if existent) the inverse function of f.

(a) $f : [-3, 1] \to [-4, 0],\ f(x) = x - 1$,
(b) $f : [-2, 3] \to \mathbb{R},\ f(x) = 3x + 2$,
(c) $f : \mathbb{R} \to \mathbb{R},\ f(x) = 3x + 2$.

23.4 Show that the mapping

$$f : \begin{cases} (-1,1) & \to & \mathbb{R} \\ x & \mapsto & \frac{x}{1-x^2} \end{cases}$$

is bijective.

23.5 Which of the following functions have an inverse function? State them if possible.

(a) $f : \mathbb{R} \setminus \{0\} \to \mathbb{R} \setminus \{0\}$ with $f(x) = \frac{1}{x^2}$,
(b) $f : \mathbb{R} \setminus \{0\} \to \mathbb{R} \setminus \{0\}$ with $f(x) = \frac{1}{x^3}$.

Power Series

24

Power series are series in an *indeterminate x*. For some values of x the power series may converge, for others it may diverge. The range of all those x for which a power series converges is the *domain of convergence* of the power series. The task for *power series* is usually to determine the domain of convergence K for this series. The set K plays an important role: Each *power series* yields a function on K; in this way we obtain important functions.

Even functions, which at first sight have nothing to do with series, can often be understood as such; one can often *develop* functions *in series*. Whether we consider the *Taylor series* or the *Fourier series*, it is always a matter of understanding a *complicated* function as a *sum of* simple functions.

24.1 The Domain of Convergence of Real Power Series

A **(real) power series** f is a series of the form

$$f(x) = \sum_{k=0}^{\infty} a_k\,(x-a)^k \quad \text{with} \quad a,\, a_0,\, a_1,\, \ldots \in \mathbb{R}\,.$$

We call $a \in \mathbb{R}$ the **development point**, x the **indeterminate** (which can be *substituted* by real numbers) and the a_k the **coefficients** of the power series f.

If you insert a real number into the power series f for x, the result is a series as we know it.

Example 24.1 We consider the power series $f(x) = \sum_{k=0}^{\infty} x^k$ around the development point $a = 0$. Let us set $x = 1$ and $x = 1/2$, we obtain for

C. Karpfinger, *Calculus and Linear Algebra in Recipes*,
https://doi.org/10.1007/978-3-662-65458-3_24

- $x = 1$ the divergent series $f(1) = \sum\limits_{k=0}^{\infty} 1$ and for
- $x = 1/2$ the convergent series $f(1/2) = \sum\limits_{k=0}^{\infty} (1/2)^k$. ■

We are concerned with the task of finding for a given power series f with $f(x) = \sum_{k=0}^{\infty} a_k(x-a)^k$ the **domain of convergence**, which is the set

$$K(f) = \{x \in \mathbb{R} \mid f(x) \text{ converges}\}$$

of all x for which f converges. To do this, we observe in advance that every power series converges at least at its development point, viz.

$$f(x) = \sum_{k=0}^{\infty} a_k(x-a)^k \;\Rightarrow\; f(a) = a_0 \in \mathbb{R}\,.$$

With the help of the following result we can say a bit more precisely that the domain of convergence is $K(f)$ of a (real) power series f with development point a is an interval with the center a, i.e. of the form

$$(a-R, a+R) \text{ or } [a-R, a+R] \text{ or } (a-R, a+R] \text{ or } [a-R, a+R)\,,$$

whereas it is usually the easier task to determine this radius R of convergence, $R \in \mathbb{R}_{\geq 0} \cup \{\infty\}$; more complicated is in general the decision whether the boundary points are $a \pm R$ still belong to the convergence domain or not.

The Radius of Convergence of a Power Series

For each power series $f(x) = \sum_{k=0}^{\infty} a_k(x-a)^k$ there is a $R \in \mathbb{R}_{\geq 0} \cup \{\infty\}$ with

$$f(x) \begin{cases} \text{converges absolute} & \forall\, x \text{ with } |x-a| < R \\ \text{diverges} & \forall\, x \text{ with } |x-a| > R \\ \text{no general statement possible} & \forall\, x \text{ with } |x-a| = R \end{cases}$$

Here we make the following agreements:

- In the case $R = 0$ $K(f) = \{a\}$, and
- in the case $R = \infty$ we have $K(f) = \mathbb{R}$.

(continued)

One calls R the **radius of convergence**. The two most important formulas for calculating R for a power series $f(x) = \sum_{k=0}^{\infty} a_k (x-a)^k$ are:

$$R = \lim_{k\to\infty} \left|\frac{a_k}{a_{k+1}}\right| \quad \text{and} \quad R = \lim_{k\to\infty} \frac{1}{\sqrt[k]{|a_k|}},$$

if exists $\lim_{k\to\infty} \left|\frac{a_k}{a_{k+1}}\right| \in \mathbb{R}_{\geq 0} \cup \{\infty\}$ resp. $\lim_{k\to\infty} \frac{1}{\sqrt[k]{|a_k|}} \in \mathbb{R}_{\geq 0} \cup \{\infty\}$.

We explain the first formula for calculating the radius of convergence: To do this, we agree on $1/0 = \infty$ and $1/\infty = 0$, we then have due to $\lim_{k\to\infty} \left|\frac{a_k}{a_{k+1}}\right| \in \mathbb{R}_{\geq 0} \cup \{\infty\}$:

$$\lim_{k\to\infty} \left|\frac{a_{k+1}}{a_k}\right| = q \in \mathbb{R}_{\geq 0} \cup \{\infty\}.$$

Now, using the ratio test for series, we obtain:

$$\left|\frac{a_{k+1}(x-a)^{k+1}}{a_k(x-a)^k}\right| = \left|\frac{a_{k+1}}{a_k}\right| |x-a| \stackrel{k\to\infty}{\longrightarrow} q\,|x-a| \begin{cases} < 1 & \Rightarrow \text{abs. convergent} \\ > 1 & \Rightarrow \text{divergent} \\ = 1 & \Rightarrow \text{no statement} \end{cases}$$

Thus, we have:

for all x with $|x-a| < 1/q$ converges $f(x)$ absolute.

for all x with $|x-a| > 1/q$ diverges $f(x)$.

for all x with $|x-a| = 1/q$ no statement is possible.

The set of all x with $|x-a| < 1/q$ is an open interval around a with *radius*

$$\lim_{k\to\infty} \left|\frac{a_k}{a_{k+1}}\right| = 1/q = R.$$

Recipe: Determine the Domain of Convergence of a Power Series

To determine the domain of convergence of the power series $f(x) = \sum_{k=0}^{\infty} a_k(x-a)^k$ proceed as follows:

(1) Determine the radius of convergence R with the known formulas

$$R = \lim_{k\to\infty} \left|\frac{a_k}{a_{k+1}}\right| \quad \text{or} \quad R = \lim_{k\to\infty} \frac{1}{\sqrt[k]{|a_k|}} .$$

(2) If $R = 0$ or $R = \infty$ then set $K(f) = \{a\}$ or $K(f) = \mathbb{R}$, DONE.
Otherwise, examine the power series f on the boundaries $a - R$ and $a + R$: Consider the series

$$f(a-R) = \sum_{k=0}^{\infty} a_k(-R)^k \text{ and } f(a+R) = \sum_{k=0}^{\infty} a_k R^k$$

and apply one of the well-known convergence or divergence criteria from Sect. 22.2 (but not the ratio or root test).

Note in applying this recipe that the power series f really has the form $f(x) = \sum_{k=0}^{\infty} a_k(x-a)^k$. The recipe does not necessarily work for, say, a power series of the form $f(x) = \sum_{k=0}^{\infty} a_k(x-a)^{2k}$. For such series one should use the ratio or root test for series, cf. Exercise 24.5.

Example 24.2

- We first consider the **geometric series** $f(x) = \sum_{k=0}^{\infty} x^k$.
  (1) Because of

  $$\left|\frac{a_k}{a_{k+1}}\right| = 1 \overset{k\to\infty}{\longrightarrow} R = 1$$

  the geometric series has radius of convergence $R = 1$.
  (2) In the boundary points -1 and 1 we have

  $$f(-1) = \sum_{k=0}^{\infty}(-1)^k \text{ and } f(1) = \sum_{k=0}^{\infty} 1^k ,$$

note, that the sequences of the series elements are not zero sequences; therefore, according to the zero sequence criterion, both series diverge. The domain of convergence is therefore $K(f) = (-1, 1)$.

- We consider the power series $f(x) = \sum_{k=1}^{\infty} \frac{2k!+1}{k!}(x+2)^k$.
  (1) We obtain

$$\left|\frac{a_k}{a_{k+1}}\right| = \frac{2k!+1}{k!}\,\frac{(k+1)!}{2(k+1)!+1} = \frac{2(k+1)!+(k+1)}{2(k+1)!+1}$$
$$= \frac{2+1/k!}{2+1/(k+1)!} \xrightarrow{k\to\infty} R = 1\,.$$

  Thus the radius of convergence is $R = 1$.
  (2) For the right boundary point $x = -1 = a + R$ (the development point is $a = -2$) of $K(f)$ we have:

$$f(-1) = \sum_{k=1}^{\infty} \frac{2k!+1}{k!} = \sum_{k=1}^{\infty} 2 + \frac{1}{k!}\,.$$

  Since the sequence of series elements is not a zero sequence, we have divergence at the boundary point $x = 2$ according to the zero sequence criterion.
  For the left boundary point $x = -3 = a - R$ of $K(f)$ we have analogously: The series

$$f(-3) = \sum_{k=1}^{\infty} (-1)^k \left(2 + \frac{1}{k!}\right)$$

  diverges. Altogether, then, we obtain the domain of convergence $K(f) = (-3, -1)$.
- Now we consider the power series $f(x) = \sum_{k=1}^{\infty} \frac{1}{k} x^k$.
  (1) We calculate:

$$\left|\frac{a_k}{a_{k+1}}\right| = \frac{k+1}{k} \xrightarrow{k\to\infty} R = 1\,.$$

  Thus, the radius of convergence of f is $R = 1$.
  (2) At the right and left boundary points $x = 1$ and $x = -1$, respectively, we have:

$$f(1) = \sum_{k=1}^{\infty} \frac{1}{k} \quad \text{and} \quad f(-1) = \sum_{k=1}^{\infty} (-1)^k \frac{1}{k}\,.$$

  This is the divergent harmonic series or convergent alternating harmonic series. Thus, we obtain the domain of convergence $K(f) = [-1, 1)$.

- Also the power series $f(x) = \sum_{k=0}^{\infty} \frac{(x-2)^k}{k^2+1}$ around the development point $a = 2$ has the radius of convergence $R = 1$, because

$$\frac{(k+1)^2+1}{k^2+1} \overset{k\to\infty}{\longrightarrow} R = 1 .$$

For the boundary points $x = 1$ and $x = 3$ of the convergence interval we have:

$$f(1) = \sum_{k=0}^{\infty} \frac{(-1)^k}{k^2+1} \text{ converges and } f(3) = \sum_{k=0}^{\infty} \frac{1}{k^2+1} \text{ converges.}$$

The domain of convergence of f is therefore $K(f) = [1, 3]$. ■

A power series f defines a function $f : K(f) \to \mathbb{R}$, $f(x) = \sum_{k=0}^{\infty} a_k (x-a)^k$: Each x from the convergence domain is assigned the value of the series $f(x)$. This function is in the inner points $(a - R, a + R)$ of the domain of convergence $K(f)$ differentiable and integrable (more on this in Chaps. 26 and 30).

24.2 The Domain of Convergence of Complex Power Series

Besides the real power series also the **complex power series** plays an important role, i.e. the series of the form

$$f(z) = \sum_{k=0}^{\infty} a_k (z-a)^k \text{ with } a, a_0, a_1, \ldots \in \mathbb{C} .$$

Fortunately we don't have to start from the beginning now, everything is still valid for $z \in \mathbb{C}$; only the convergence interval $|x - a| < R$ with center $a \in \mathbb{R}$ now becomes a circle of convergence $|z - a| < R$ with center $a \in \mathbb{C}$ (Fig. 24.1):

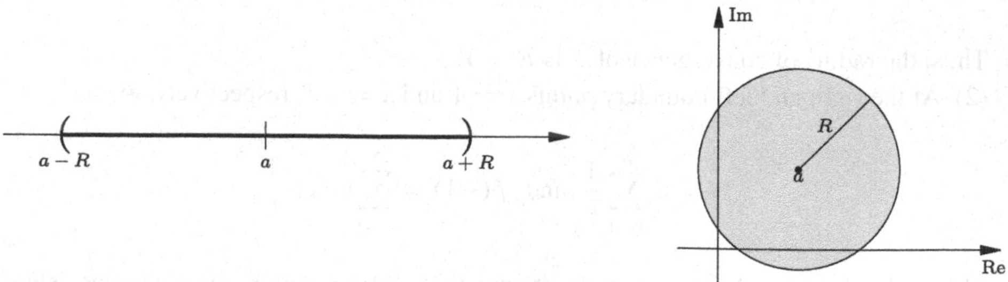

Fig. 24.1 The radius of convergence of a real or complex power series

To determine the radius R, we can again use the familiar formulas:

$$R = \lim_{n\to\infty} \left|\frac{a_k}{a_{k+1}}\right| \quad \text{and} \quad R = \lim_{k\to\infty} \frac{1}{\sqrt[k]{|a_k|}}.$$

Example 24.3

- The power series $f(z) = \sum_{k=0}^{\infty} \frac{1}{k!} z^k$ converges for all $z \in \mathbb{C}$, because

$$\left|\frac{(k+1)!}{k!}\right| = k+1 \overset{k\to\infty}{\longrightarrow} R = \infty.$$

So the circle of convergence in this case is quite $\mathbb{C}$.
- The power series $f(z) = \sum_{k=0}^{\infty} 2^k (z - \mathrm{i})^k$ converges in the circle with center $a = \mathrm{i}$ and radius $R = 1/2$:

$$\frac{1}{\sqrt[k]{2^k}} = \frac{1}{2} \overset{k\to\infty}{\longrightarrow} R = \frac{1}{2}.$$

- The power series $f(z) = \sum_{k=0}^{\infty} k^k (z-1)^k$ converges only for $z = 1$, since due to

$$\frac{1}{\sqrt[k]{k^k}} = \frac{1}{k} \overset{k\to\infty}{\longrightarrow} R = 0$$

their radius of convergence is $R = 0$. ■

For later purposes, we record one more important theorem:

The Identity Theorem—Equating Coefficients

If we have for two power series $f(z) = \sum_{k=0}^{\infty} a_k (z-a)^k$ and $g(z) = \sum_{k=0}^{\infty} b_k (z-a)^k$ with the same development point a and a radius $r > 0$ the equality

$$\sum_{k=0}^{\infty} a_k (z-a)^k = \sum_{k=0}^{\infty} b_k (z-a)^k$$

for all z with $|z - a| < r$ then it follows $a_k = b_k$ for all $k \in \mathbb{N}_0$.

This identity theorem allows equating coefficients analogous to equating coefficients for polynomials (see Sect. 5.1), with the help of which we will be able to derive many nontrivial identities (see Sect. 28.2).

24.3 The Exponential and the Logarithmic Function

The power series $\sum_{k=0}^{\infty} \frac{1}{k!} z^k$ is called **exponential series**. According to example 24.3 the exponential series converges for each $z \in \mathbb{C}$. We thus obtain the important **(complex) exponential function**

$$\exp : \begin{cases} \mathbb{C} \to & \mathbb{C} \\ z \mapsto & \sum\limits_{k=0}^{\infty} \frac{1}{k!} z^k \end{cases}.$$

We deal with this complex version of the exponential function in the chapters on the complex analysis from Chap. 80 on. The only properties of this complex function that we will use in what follows are:

$$|\exp(\mathrm{i}\,x)| = 1 \text{ and } \exp(\mathrm{i}\,x) = \cos(x) + \mathrm{i}\sin(x) \text{ for all } x \in \mathbb{R}.$$

We summarize the most important properties of the exponential function for the time being:

The Exponential Function

The **(real) exponential function** is given by

$$\exp : \begin{cases} \mathbb{R} \to & \mathbb{R} \\ x \mapsto & \sum\limits_{k=0}^{\infty} \frac{1}{k!} x^k \end{cases}.$$

It is reasonable and usual to write e^x instead of $\exp(x)$. We have:

- $\mathrm{e}^x\,\mathrm{e}^y = \mathrm{e}^{x+y}$ for all $x,\ y \in \mathbb{R}$ (**functional equation**).
- $\mathrm{e}^{-x} = \frac{1}{\mathrm{e}^x}$ for all $x \in \mathbb{R}$.
- $|\,\mathrm{e}^{\mathrm{i}\,x}\,| = 1$ for all $x \in \mathbb{R}$.
- $\mathrm{e}^{\mathrm{i}\,x} = \cos(x) + \mathrm{i}\sin(x)$ for all $x \in \mathbb{R}$ (**Euler's formula**).
- $\mathrm{e}^x > 0$ for all $x \in \mathbb{R}$.
- $\exp : \mathbb{R} \to \mathbb{R}_{>0}$ is bijective.

Remarks

1. Using Euler's formula, we can find the polar representation $z = r(\cos(\varphi) + \mathrm{i}\sin(\varphi))$ with $r = |z|$ and $\varphi = \arg(z)$ of a complex number $z \neq 0$ succinctly:

$$z = r\left(\cos(\varphi) + \mathrm{i}\sin(\varphi)\right) = r\,\mathrm{e}^{\mathrm{i}\varphi}\,.$$

2. Moreover, it follows from Euler's formula:

$$\cos(x) = \mathrm{Re}\left(\mathrm{e}^{\mathrm{i}x}\right) \qquad \text{and} \qquad \sin(x) = \mathrm{Im}\left(\mathrm{e}^{\mathrm{i}x}\right)\,.$$

This is an alternative (and equivalent) definition of sine and cosine. We obtain from it power series representations of sin and cos, we have for all $x \in \mathbb{R}$:

- $\cos(x) = \sum\limits_{k=0}^{\infty}(-1)^k \frac{x^{2k}}{(2k)!} = 1 - \frac{x^2}{2} + \frac{x^4}{4!} - + \ldots,$
- $\sin(x) = \sum\limits_{k=0}^{\infty}(-1)^k \frac{x^{2k+1}}{(2k+1)!} = x - \frac{x^3}{3!} + \frac{x^5}{5!} - + \ldots.$

Both series converge absolutely for all $x \in \mathbb{R}$. Their representation follows from:

$$\mathrm{e}^{\mathrm{i}x} = \sum_{k=0}^{\infty}\frac{(\mathrm{i}x)^k}{k!} = \sum_{k=0}^{\infty}(-1)^k\frac{x^{2k}}{(2k)!} + \mathrm{i}\sum_{k=0}^{\infty}(-1)^k\frac{x^{2k+1}}{(2k+1)!}\,,$$

since $\mathrm{i}^{2k} = (-1)^k$ and $\mathrm{i}^{2k+1} = (-1)^k\,\mathrm{i}$.

Since the exponential function $\exp : \mathbb{R} \to \mathbb{R}_{>0}$ is bijective, there exists an inverse function to exp which is the **natural logarithm** ln. Figure 24.2 shows the graph of the exponential and logarithm functions.

We summarize the main properties of the logarithm.

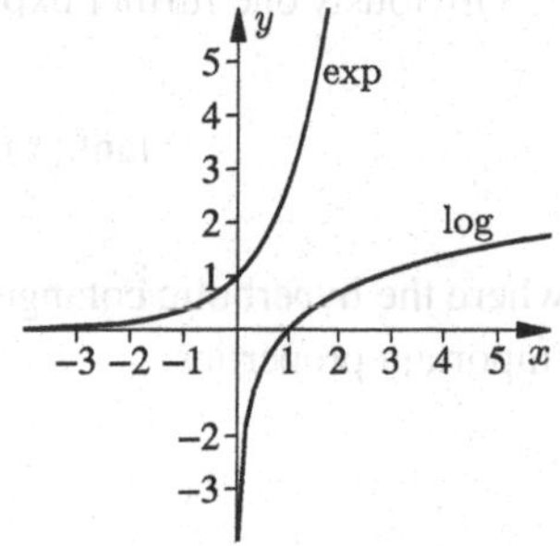

Fig. 24.2 The graphs of exp and ln

The Logarithm Function
The **logarithm function**

$$\ln : \begin{cases} \mathbb{R}_{>0} \to & \mathbb{R} \\ x & \mapsto \ln(x) \end{cases}$$

has the properties:

- $\exp(\ln(x)) = x$ for all $x \in \mathbb{R}_{>0}$.
- $\ln(\exp(x)) = x$ for all $x \in \mathbb{R}$.
- $\ln(x\,y) = \ln(x) + \ln(y)$ for all $x,\ y \in \mathbb{R}_{>0}$.
- $\ln(x^r) = r\,\ln(x)$ for all $x \in \mathbb{R}_{>0}$ and $r \in \mathbb{Q}$.
- $\ln(x/y) = \ln(x) - \ln(y)$ for all $x,\ y \in \mathbb{R}_{>0}$.
- $\ln(1) = 0$ and $\ln(\mathrm{e}) = 1$.

Remark The general power function is $a^x = \mathrm{e}^{x\ln(a)}$ for all $a \in \mathbb{R}_{>0}$.

24.4 The Hyperbolic Functions

Finally, using the exponential function, we explain the following two functions which are important for engineering mathematics, namely, the **hyperbolic cosine** and the **hyperbolic sine**:

$$\cosh : \begin{cases} \mathbb{R} \to & \mathbb{R} \\ x \mapsto & \cosh(x) = \frac{\mathrm{e}^x + \mathrm{e}^{-x}}{2} \end{cases} \quad \text{and} \quad \sinh : \begin{cases} \mathbb{R} \to & \mathbb{R} \\ x \mapsto & \sinh(x) = \frac{\mathrm{e}^x - \mathrm{e}^{-x}}{2} \end{cases}.$$

The graphs of these functions are shown in Fig. 24.3.

Obviously one further explains the **hyperbolic tangent** and the **hyperbolic cotangent**

$$\tanh(x) = \frac{\sinh(x)}{\cosh(x)} \quad \text{and} \quad \coth(x) = \frac{\cosh(x)}{\sinh(x)},$$

where the hyperbolic cotangent for $x = 0$ is not explained. We again summarize the most important properties:

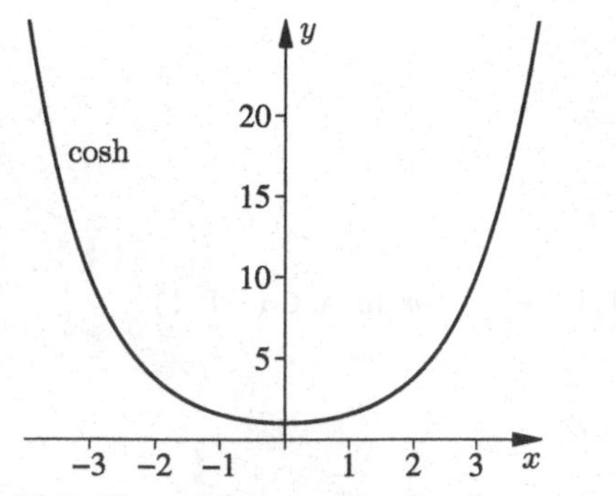

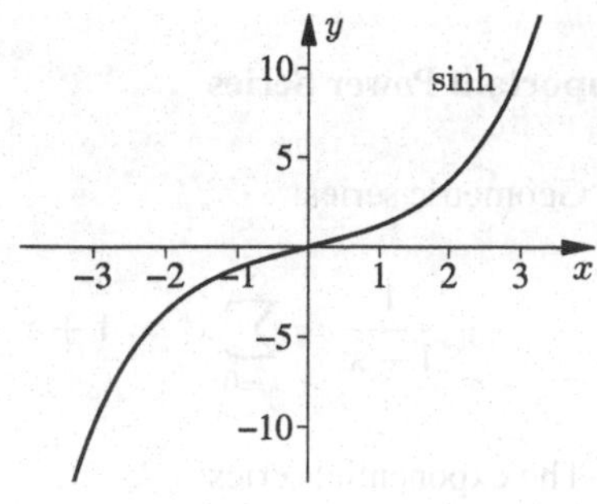

Fig. 24.3 The graphs of cosh and sinh

The Hyperbolic Functions
The most important properties of hyperbolic functions are:

- $\cosh(-x) = \cosh(x)$ for all $x \in \mathbb{R}$.
- $\sinh(-x) = -\sinh(x)$ for all $x \in \mathbb{R}$.
- $\cosh^2(x) - \sinh^2(x) = 1$ for all $x \in \mathbb{R}$.
- $\cosh : \mathbb{R}_{\geq 0} \to \mathbb{R}_{\geq 1}$ is bijective, for the inverse function **inverse hyperbolic cosine** arcosh we have

$$\operatorname{arcosh}(x) = \ln(x + \sqrt{x^2 - 1}) \text{ for all } x \in \mathbb{R}_{\geq 1}.$$

- $\sinh : \mathbb{R} \to \mathbb{R}$ is bijective, for the inverse function **inverse hyperbolic sine** arsinh we have

$$\operatorname{arsinh}(x) = \ln(x + \sqrt{x^2 + 1}) \text{ for all } x \in \mathbb{R}.$$

With the exponential series, we easily obtain power series representations for the hyperbolic cosine and the hyperbolic sine, respectively:

$$\cosh(x) = \sum_{k=0}^{\infty} \frac{1}{(2k)!} x^{2k} \text{ and } \sinh(x) = \sum_{k=0}^{\infty} \frac{1}{(2k+1)!} x^{2k+1} \text{ for all } x \in \mathbb{R}.$$

There are only a few power series that you absolutely have to know by heart, but among these few are the following series, which we summarize again here:

Important Power Series

- Geometric series:

$$\frac{1}{1-x} = \sum_{k=0}^{\infty} x^k = 1 + x + x^2 + x^3 + \cdots \text{ for all } x \in (-1, 1)\,.$$

- The exponential series:

$$\exp(x) = \sum_{k=0}^{\infty} \frac{1}{k!} x^k = 1 + x + \frac{x^2}{2!} + \frac{x^3}{3!} + \cdots \text{ for all } x \in \mathbb{R}\,.$$

- Cosine series:

$$\cos(x) = \sum_{k=0}^{\infty} (-1)^k \frac{x^{2k}}{(2k)!} = 1 - \frac{x^2}{2} + \frac{x^4}{4!} - + \ldots \text{ für alle } x \in \mathbb{R}\,.$$

- The sine series:

$$\sin(x) = \sum_{k=0}^{\infty} (-1)^k \frac{x^{2k+1}}{(2k+1)!} = x - \frac{x^3}{3!} + \frac{x^5}{5!} - + \ldots \text{ für alle } x \in \mathbb{R}\,.$$

24.5 Exercises

24.1 Determine the radius of convergence of the following power series:

(a) $\sum\limits_{k=1}^{\infty} (k^4 - 4k^3) z^k$,

(b) $\sum\limits_{k=0}^{\infty} 2^k z^{2k}$.

24.2 Determine the domain of convergence $K(f)$ of the following power series

(a) $f(x) = \sum\limits_{k=1}^{\infty} \frac{1}{k 2^k} (x-1)^k$,

(b) $f(x) = \sum\limits_{k=0}^{\infty} \frac{k^4}{4^k} x^k$,

(c) $f(x) = \sum\limits_{k=2}^{\infty} \frac{1}{\sqrt{k^2-1}} x^k$,

(d) $f(x) = \sum\limits_{k=0}^{\infty} (k^2 + 2^k)x^k$.

(e) $f(x) = \sum\limits_{k=1}^{\infty} \frac{1}{3^k k^2} x^k$.

(f) $f(x) = \sum\limits_{k=5}^{\infty} 5^k k^5 x^k$.

(g) $f(x) = \sum\limits_{k=0}^{\infty} k! x^k$.

(h) $f(x) = \sum\limits_{k=1}^{\infty} \frac{1}{k^k} x^k$.

24.3 Show the identity

$$\cosh^2 x - \sinh^2 x = 1\,.$$

24.4 Show the addition theorem

$$\cosh(x + y) = \cosh x \cosh y + \sinh x \sinh y\,.$$

Limits and Continuity

25

The notion of *limit* does not only play a role for sequences, also a function $f : D \to W$ has limits at the places $a \in D$ and at the *boundary points a* of D. We form these limits using sequences and thus obtain an idea about the *behavior of* the function $f : D \to W$ for x near a.

The *continuity of* a function f can again be formulated in terms of the notion of limit. The notion of a continuous function is quite simple as long as we consider bounded functions on bounded intervals: A function is *continuous* if the graph of the function f can be drawn *without interruption*. The *continuity* of a function has far-reaching consequences concerning the existence of zeros and extrema. We state these theorems and describe a procedure to determine zeros of *continuous* functions at least approximately even if this is not possible analytically.

25.1 Limits of Functions

The concept of the *limit of* a function f is explained with the help of sequences. We start with the following observation: Let $f : D \to W$ be a function and $(a_n)_n$ a sequence with $a_n \in D$ for all n, then so is $\big(f(a_n)\big)_n$ a sequence:

$$(a_n)_n \text{ is a sequence } \Rightarrow \big(f(a_n)\big)_n \text{ is a sequence}.$$

It is essential that the elements of the sequence a_n are in the domain of f, so that it is at all possible to determine the sequence members $f(a_n)$.

C. Karpfinger, *Calculus and Linear Algebra in Recipes*,
https://doi.org/10.1007/978-3-662-65458-3_25

Example 25.1 We consider the sequence $(a_n)_n$ with $a_n = \frac{1}{n^2+1}$ and the function $f : \mathbb{R} \to \mathbb{R},\ f(x) = e^{\sin(x)}$. The a_n can be expressed by f, and we thus obtain the new sequence

$$\left(f(a_n)\right)_n \quad \text{mit} \quad f(a_n) = e^{\sin\left(\frac{1}{n^2+1}\right)} .$$

■

The Limit of *f* in *a*

Given are a function $f : D \to W$ and two elements $a,\ c \in \mathbb{R} \cup \{\pm\infty\}$. We say: The function $f : D \to W$ has in a **the limit** c, if for every sequence $(a_n)_n$ in D with $a_n \neq a$ which converges to a, the sequence $\left(f(a_n)\right)_n$ converges to c. For this we write:

$$\lim_{x\to a} f(x) = c .$$

One expresses this fact briefly as follows:

$$\lim_{x\to a} f(x) = c \Leftrightarrow \forall\, (a_n)_n \text{ with } \lim_{n\to\infty} a_n = a \text{ we have } \lim_{n\to\infty} f(a_n) = c .$$

The concept of limit is not very simple. But fortunately we do not often have to go back to this definition. If we temporarily assume the notion of continuity, or if we remember it from our school days, we have the following: In the domain D, the determination of the limit is almost always completely unproblematic: If f is *continuous* in D, then the limit of f in $a \in D$ always equals $f(a)$, i.e., the value of f in the point a; in this case, only the *boundary points of* the domain of definition are of special interest. For *discontinuous* functions, the discontinuity points must also be considered.

Example 25.2

- We determine the limit of $f : \mathbb{R} \to \mathbb{R},\ f(x) = x^2 + 1$ in $a = 0$ and $a = \infty$:
  $a = 0$: For any zero sequence $(a_n)_n$ converges $\left(f(a_n)\right)_n = \left(a_n^2 + 1\right)_n$ towards 1, i.e.

  $$\lim_{x\to 0} f(x) = 1 .$$

  $a = \infty$: For every sequence $(a_n)_n$ with $a_n \to \infty$ diverges $\left(f(a_n)\right)_n = \left(a_n^2 + 1\right)_n$ against ∞, i.e.

  $$\lim_{x\to\infty} f(x) = \infty .$$

We determine the limit of $f : (0, 1) \to \mathbb{R},\ f(x) = 1/x$ in $a = 1$ and $a = 0$:

- $a = 1$: For each sequence $(a_n)_n$ with $a_n \to 1$ converges $\big(f(a_n)\big)_n = (1/a_n)_n$ against 1, i.e.

$$\lim_{x\to 1} f(x) = 1\,.$$

$a = 0$: For any zero sequence $(a_n)_n$ in $(0, 1)$ diverges $\big(f(a_n)\big)_n = (1/a_n)_n$ against ∞, i.e.

$$\lim_{x\to 0} f(x) = \infty\,.$$

- We try to find the limit of the step function $f : \mathbb{R} \to \mathbb{R},\ f(x) = \lfloor x \rfloor = \max\{z \in \mathbb{Z} \mid z \le x\}$ in $a = 0$:
For all zero sequences $(a_n)_n$ with $a_n > 0$ we have:

$$f(a_n) = 0 \text{ and hence } \lim_{n\to\infty} f(a_n) = 0\,.$$

For all zero sequences $(a_n)_n$ with $a_n < 0$ we have:

$$f(a_n) = -1 \text{ and } \lim_{n\to\infty} f(a_n) = -1\,.$$

Thus $\lim_{x\to 0} f(x)$ does not exist. There is no $c \in \mathbb{R} \cup \{\pm\infty\}$ with $\lim_{x\to 0} f(x) = c$.

- We consider the so-called **cardinal sine**

$$f : \mathbb{R} \setminus \{0\} \to \mathbb{R},\ f(x) = \frac{\sin(x)}{x}\,.$$

Does f in $a = 0$ has a limit, i.e., exists $\lim_{x\to 0} \frac{\sin(x)}{x}$?
To calculate this and similar limits, use *L'Hospital's rule* or the power series expansion of sin. Note the following recipe. ■

In fact, to compute limits of functions one more or less never has to go back to the definition, we give in a recipe-like manner the main methods for determining limits, these are essentially based on the following computational rules for determining limits : if $f, g : D \to \mathbb{R}$ are functions with the limits $\lim_{x\to a} f(x) = c$ and $\lim_{x\to a} g(x) = d$, then we have:

- $\lim_{x\to a} \big(\lambda\, f(x) + \mu\, g(x)\big) = \lambda\, c + \mu\, d$ for all $\lambda,\ \mu \in \mathbb{R}$.
- $\lim_{x\to a} \big(f(x)\, g(x)\big) = c\, d$.
- $\lim_{x\to a} f(x)/g(x) = c/d$, if $d \neq 0$.

Recipe: Determine Limits

To determine the limit $\lim_{x\to a} f(x) = c$ of a function $f : D \to \mathbb{R}$ note:

- If $a \in D$ and f is an elementary function, then $\lim_{x\to a} f(x) = f(a)$.
- If a is a zero of the denominator of $f(x) = g(x)/h(x)$ then one can possibly *reduce* the fraction by $x - a$:

  If g and h are polynomials with the common zero a, the result after reducing by $(x - a)$ is a rational function $\tilde{f}(x) = \tilde{g}(x)/\tilde{h}(x)$ where a is no longer a zero of the denominator; then we have: $\lim_{x\to a} f(x) = \lim_{x\to a} \tilde{f}(x) = \tilde{f}(a)$.

  Occasionally, clever expansion also helps, so that the numerator simplifies significantly and reducing becomes possible after this simplification (this *coincidence* often occurs in academic examples).

  If g and h are power series with $h(a) = 0$ then after reducing $(x - a)$ possibly we get a function $\tilde{f}(x) = \tilde{g}(x)/\tilde{h}(x)$ where a is no longer a zero of the denominator. We then have:

$$\lim_{x\to a} f(x) = \lim_{x\to a} \tilde{f}(x) = \tilde{f}(a) .$$

- If $a = \pm\infty$ and $f(x) = g(x)/h(x)$ a rational function with $\deg(g) = r$ and highest coefficient a_r and $\deg(h) = s$ with highest coefficient b_s then

$$\lim_{x\to\pm\infty} = \begin{cases} a_r/b_s\,, & \text{if } r = s \\ \infty\,, & \text{if } r > s \text{ and } a_r/b_s > 0 \\ -\infty\,, & \text{if } r > s \text{ and } a_r/b_s < 0 \\ 0\,, & \text{if } r < s \end{cases} .$$

L'Hospital's rule for determining the limit of a function uses the derivative and is therefore not presented until Sect. 27.5. If you have mastered derivation, you can skip ahead and apply the rule to the following examples.

Example 25.3

- $\lim_{x\to 2} \frac{2x^2+x+3}{x^2+x+12} = \frac{13}{18}$ and $\lim_{x\to\infty} \frac{2x^2+x+3}{x^2+x+12} = 2$,
- $\lim_{x\to 1} \frac{x^3-1}{x-1} = \lim_{x\to 1} \frac{(x-1)(x^2+x+1)}{x-1} = \lim_{x\to 1} x^2 + x + 1 = 3$,
- $\lim_{x\to 0} \frac{\sqrt{1+x}-1}{x} = \lim_{x\to 0} \frac{(\sqrt{1+x}-1)(\sqrt{1+x}+1)}{x(\sqrt{1+x}+1)} = \lim_{x\to 0} \frac{1+x-1}{x(\sqrt{1+x}+1)} = \lim_{x\to 0} \frac{1}{\sqrt{1+x}+1} = \frac{1}{2}$,
- $\lim_{x\to 0} \frac{\sin(x)}{x} = \lim_{x\to 0} \frac{x-x^3/3!+x^5/5!-+\cdots}{x} = \lim_{x\to 0} \frac{1-x^2/3!+x^4/5!-+\cdots}{1} = 1$. ■

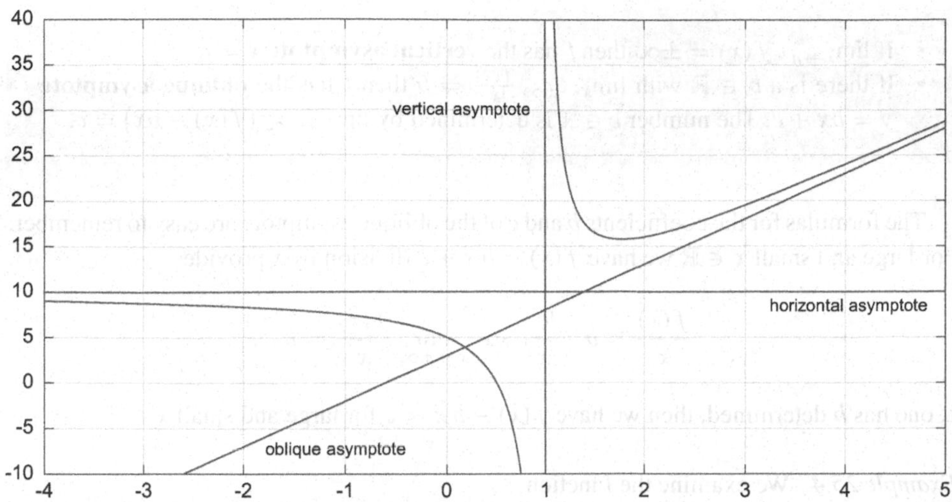

Fig. 25.1 The various asymptotes

25.2 Asymptotes of Functions

A typical application of the limits of functions is the determination of *asymptotes*; these are horizontal, vertical or oblique straight lines to which the graph of a function is *close to*, see Fig. 25.1.

For a vertical asymptote at a point $a \in \mathbb{R}$ one has to distinguish by the *approximation* $x \to a$ between *approximation from the left* and *approximation from the right.* We express this by the following symbols:

$$\lim_{x \to a^+} f(x) \text{ and } \lim_{x \to a^-} f(x)$$

and by this we mean that for the limit process $x \to a^+$ consider only sequences whose sequence elements are on the right of a, correspondingly, for $x \to a^-$ only sequences whose sequence elements are on the left of a. We have already seen this in example 25.2 for the step function $f(x) = \lfloor x \rfloor$ in $a = 0$: we got $\lim_{x \to 0^+} \lfloor x \rfloor = 0$ and $\lim_{x \to 0^-} \lfloor x \rfloor = -1$.

We find the asymptotes using the following methods:

Recipe: Determine Asymptotes
Given the function $f : D \setminus \{a\} \to \mathbb{R}$.

- If there is a $c \in \mathbb{R}$ with $\lim_{x \to \pm\infty} f(x) = c$ then f has the **horizontal asymptote** $y = c$.

(continued)

- If $\lim_{x\to a^\pm} f(x) = \pm\infty$ then f has the **vertical asymptote** $x = a$.
- If there is a $b \in \mathbb{R}$ with $\lim_{x\to\pm\infty} \frac{f(x)}{x} = b$ then f has the **oblique asymptote** $y = bx + c$. The number $c \in \mathbb{R}$ is determined by $\lim_{x\to\pm\infty} \big(f(x) - bx\big) = c$.

The formulas for the coefficients b and c of the oblique asymptote are easy to remember: For large and small $x \in \mathbb{R}$ we have $f(x) \approx bx + c$ division by x provides

$$\frac{f(x)}{x} \approx b + \frac{c}{x}, \text{ so } \lim_{x\to\pm\infty} \frac{f(x)}{x} = b\,.$$

If one has b determined, then we have $f(x) - bx \approx c$ for large and small $x \in \mathbb{R}$.

Example 25.4 We examine the function

$$f : \mathbb{R} \setminus \{1\} \to \mathbb{R},\ f(x) = \frac{x^3}{x^2 - 2x + 1}$$

for horizontal, vertical and oblique asymptotes:

- Horizontal asymptotes $y = c$: It is

$$\lim_{x\to\pm\infty} f(x) = \lim_{x\to\pm\infty} \frac{x^3}{x^2 - 2x + 1} = \lim_{x\to\pm\infty} \frac{x}{1 - 2/x + 1/x^2} = \pm\infty\,.$$

  Therefore there are no horizontal asymptotes.
- Vertical asymptotes $x = a$: The denominator has a zero at $a = 1$, and since

$$\lim_{x\to 1^\pm} f(x) = \lim_{x\to 1^\pm} \frac{x^3}{(x-1)^2} = \infty\,,$$

  there is a vertical asymptote $x = 1$.
- Oblique asymptotes $y = bx + c$: We calculate

$$\lim_{x\to\pm\infty} \frac{f(x)}{x} = \lim_{x\to\pm\infty} \frac{x^2}{x^2 - 2x + 1} = 1$$

  and get b= 1. Now we calculate

$$\lim_{x\to\pm\infty} \big(f(x) - 1\cdot x\big) = \lim_{x\to\pm\infty} \left(\frac{x^3}{(x-1)^2} - \frac{x(x-1)^2}{(x-1)^2}\right) = \lim_{x\to\pm\infty} \frac{2x^2 - x}{(x-1)^2} = 2\,.$$

  It is thus $c = 2$. Hence $y = x + 2$ is an oblique asymptote.

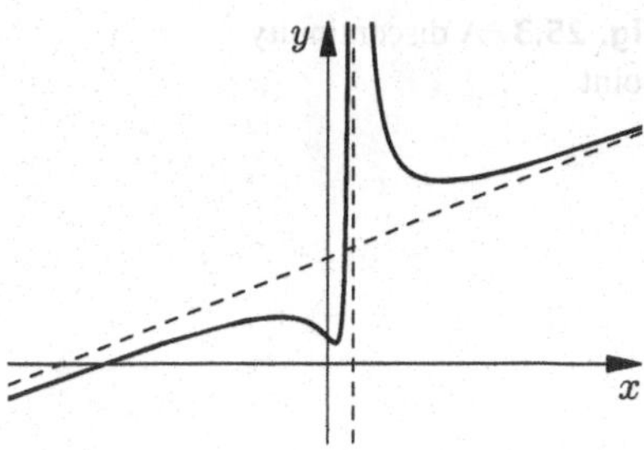

Fig. 25.2 The asymptotes of f

In Fig. 25.2, we see our results confirmed; the MATLAB code to generate the graph reads:

```
» ezplot('x^3/(x^2 - 2*x + 1)', [-6,6]) » grid on » hold
on » ezplot('x+2', [-6,6])
```

■

25.3 Continuity

Most functions one deals with in engineering mathematics are continuous. Here, *continuity* vividly states that the graph of the function has no *jumps*. We will be more precise: a function $f : D \subseteq \mathbb{R} \to \mathbb{R}$ is

- **continuous in** $a \in D$, if $\lim_{x \to a} f(x) = f(a)$, and
- **continuous in** D if f is continuous in every $a \in D$.

The function f is therefore continuous in a, if the limit $\lim_{x \to a} f(x)$ exists and is equal to $f(a)$. This continuity in a can be formally expressed as

$$\lim_{x \to a} f(x) = f\left(\lim_{x \to a} x\right).$$

Is f is **not** continuous in a, so there is a sequence $(a_n)_n$ with $a_n \to a$ and $f(a_n) \not\to f(a)$. That is at a *jump point* given, note Fig. 25.3.

The rules for the computation of limits show:

Continuous Functions
If $f, g : D \subseteq \mathbb{R} \to \mathbb{R}$ are continuous functions, so are

$$\lambda f + \mu g, \quad f g, \quad f/g, \quad f \circ g,$$

where $\lambda,\ \mu \in \mathbb{R}$ are real numbers and for the quotient f/g assume that $g(x) \neq 0$ for all $x \in D$.

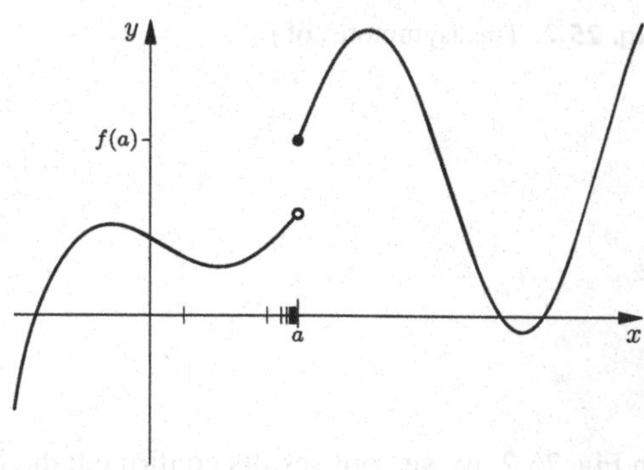

Fig. 25.3 A discontinuity point

Example 25.5

- Continuous are: Polynomial functions, rational functions, power series functions in the interior of their domain of convergence, exp, ln, sin, cos etc.
- Non-continuous are: The step function $\lfloor\cdot\rfloor : \mathbb{R} \to \mathbb{R}, \lfloor x \rfloor = \max\{z \in \mathbb{Z} \mid z \le x\}$ and in general every function with a *jump*. ■

It may be that a function is not defined at a certain point, in particular it is not continuous at this point. In certain cases, however, it can be made continuous by specifying a value at this point. E.g. the function

$$f : \mathbb{R} \setminus \{1\} \to \mathbb{R},\ f(x) = \frac{(x-1)(x+3)}{x-1}$$

is continuous in $\mathbb{R} \setminus \{1\}$, and we have $\lim_{x\to 1} f(x) = 4$. The function $g : \mathbb{R} \to \mathbb{R},\ g(x) = x + 3$ is continuous in $\mathbb{R}$ and it is $g = f$ in $\mathbb{R} \setminus \{1\}$. We thus have f **continuously extended**.

The opinion that *continuity* means that you can draw the graph of the function f without interruption with the pencil is not too reliable. This is because the following continuous function has a graph that cannot be drawn without setting it down:

$$f : \mathbb{R} \setminus \{0\} \to \mathbb{R},\ f(x) = \frac{1}{x}.$$

25.4 Important Theorems about Continuous Functions

Continuity has numerous consequences, we keep them collected: ■

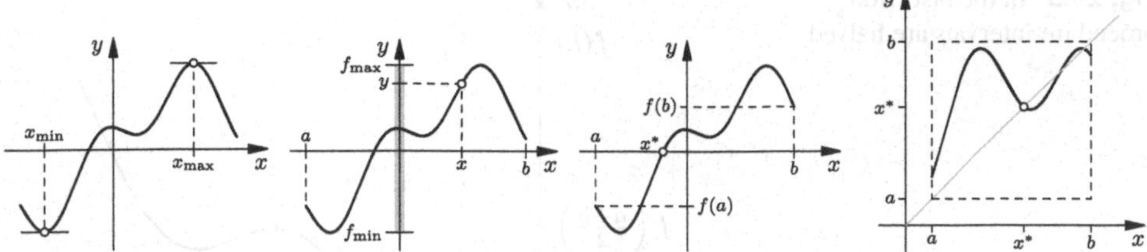

Fig. 25.4 Pictures of the important theorems of continuous functions

Theorems on Continuous Functions

Given are a bounded and closed interval $D = [a, b] \subseteq \mathbb{R}$ and a continuous function $f : [a, b] \to \mathbb{R}$. Then also $f([a, b])$ is also a bounded and closed interval, and we have:

(1) **Theorem of maximum and minimum:** There are places $x_{\max}$, $x_{\min} \in [a, b]$ with

$$f_{\min} = f(x_{\min}) \leq f(x) \leq f(x_{\max}) = f_{\max} \quad \forall x \in [a, b].$$

One calls $x_{\min}$ **minimum point** and $x_{\max}$ **maximum point.**

(2) **Intermediate value Theorem**: For every $y \in [f_{\min}, f_{\max}]$ there is a $x^* \in [a, b]$ with $f(x^*) = y$.

(3) **Bolzano's Theorem**: Is $f(a) < 0$ and $f(b) > 0$ (or vice versa), then there is a $x^* \in [a, b]$ with $f(x^*) = 0$.

(4) **Fixed point Theorem**: If $f : [a, b] \to [a, b]$ is continuous, then there exists a $x^* \in [a, b]$ with $f(x^*) = x^*$.

The four pictures in Fig. 25.4 illustrate these four theorems in the order given:

Note that the propositions are so-called existence theorems: There are $x_{\min}$ and $x_{\max}$ and x^*. But the propositions do not give any hints how to find these points. With the help of the differential calculus we will find $x_{\min}$ and $x_{\max}$ for *differentiable* functions, and with the help of the bisection method we can determine x^* at least approximately (note the following section).

25.5 The Bisection Method

The *bisection method* is a very simple and stable method to approximate a zero of a continuous scalar function f approximately. It gets by with one function evaluation at each iteration.

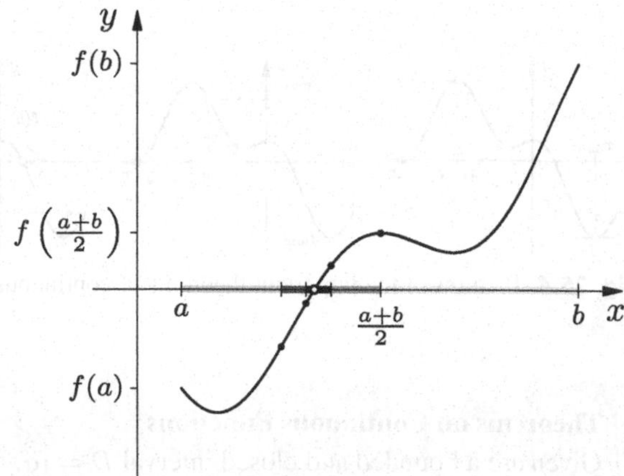

Fig. 25.5 In the bisection procedure intervals are halved

To do this, consider a continuous function $f : [a, b] \to \mathbb{R}$, which has in a a negative value and in b a positive value, $f(a) < 0$ and $f(b) > 0$. Because of continuity, the function f has according to Bolzano's theorem (see Sect. 25.4 a zero $x^* \in [a, b]$. We now determine the sign of f at the center $c = \frac{a+b}{2}$ of this interval:

- If $f(c) > 0$, then a point x^* we are looking for is in $[a, c]$, in this case start from the beginning with $a = a$ and $b = c$.
- If $f(c) < 0$ then a point x^* we are looking for is in $[c, b]$, in this case start from the beginning with $a = c$ and $b = b$.
- If $f(c) = 0$, finished.

Starting from the interval $I_0 = [a, b]$ a sequence $(I_k)_k$ of nested intervals is determined, i.e. $I_{k+1} \subseteq I_k$ for all k, where the zero x^* we are looking for lies in all these intervals. Since the interval length is halved at each step, we thus obtain x^* arbitrarily exact. Note Fig. 25.5.

Somewhat more formally and generally, the procedure sounds as follows:

Recipe: The Bisection Method
The function $f : [a_0, b_0] \to \mathbb{R}$ be continuous with $f(a_0)\, f(b_0) < 0$. A zero x^* of f is obtained approximately by the **bisection method**: For $k = 0, 1, 2, \ldots$ calculate

- $x_k = \frac{1}{2}(a_k + b_k)$ and $f(x_k)$.
- Set

$$a_{k+1} = a_k \text{ and } b_{k+1} = x_k\,, \text{ if } f(a_k)\, f(x_k) < 0\,,$$

$$a_{k+1} = x_k \text{ and } b_{k+1} = b_k\,, \text{ if } f(x_k)\, f(b_k) < 0\,.$$

(continued)

- Termination of the iteration, if $b_k - a_k < \text{tol}$ where $\text{tol} \in \mathbb{R}_{>0}$

is given.
Furthermore we have

$$|x_k - x^*| \leq b_k - a_k \leq \frac{1}{2^k}|b_0 - a_0| .$$

Therefore, the bisection method *converges* in any case, i.e. approximates a zero with arbitrary accuracy.

The bisection method is also called **interval halving method**.

MATHLAB The bisection method can be easily implemented in MATLAB, we have formulated this in Exercise 25.6. In the following examples, we use such a program.

Example 25.6

- After 27 iterations, the correct eight decimal places of a zero point x^* of the function $f(x) = \cos(x) - x$ in the interval $[0, 1]$ are:

$$x^* = 0.73908513\ldots$$

■

- After 28 iterations, the correct eight decimal places of a zero of the function x^* of the function $f(x) = x^6 - x - 2$ in the interval $[0, 2]$ are:

$$x^* = 1.21486232\ldots$$

25.6 Exercises

25.1 Show that every polynomial of odd degree has a zero in $\mathbb{R}$.

25.2 Prove the fixedpoint Theorem from Sect. 25.2.

25.3 Using the bisection method, calculate the value of $\sqrt{3}$ to two decimal places.

25.4 For $n \in \mathbb{N}$ let the function $g_n : \mathbb{R} \to \mathbb{R}$ is defined by

$$g_n(x) = \frac{nx}{1 + |nx|}.$$

(a) Show that for every $n \in \mathbb{N}$ the function g_n is continuous.
(b) Show that for every $x \in \mathbb{R}$

$$g(x) = \lim_{n \to \infty} g_n(x)$$

exists, and investigate in which points $x \in \mathbb{R}$ the function $g : \mathbb{R} \to \mathbb{R}$ is continuous.

25.5 Investigate where the following functions are defined and where they are continuous. Can the functions be extended continuously into points $x \in \mathbb{R}$ that do not belong to their domain?

(a) $f(x) = \frac{\frac{x-2}{x^2+3x-10}}{1 - \frac{4}{x+3}}$,
(b) $g(x) = \frac{\sqrt{2+x} - \sqrt{3x-3}}{\sqrt{3x-2} - \sqrt{11-5x}}$.

25.6 Let $\alpha, \beta \in \mathbb{R}$ with $\alpha < \beta$. Show that the equation

$$\frac{x^2+1}{x-\alpha} + \frac{x^6+1}{x-\beta} = 0$$

has a solution $x \in \mathbb{R}$ with $\alpha < x < \beta$.

25.7 Find all the asymptotes of the function

$$f : \mathbb{R} \setminus \{\pm 1\} \to \mathbb{R}, \quad f(x) = \frac{x^3-x^2+2x}{x^2-1}.$$

25.8 Determine the following limits:

(a) $\lim_{x \to 0} \frac{\sin x \cos x}{x \cos x - x^2 - 3x}$,
(b) $\lim_{x \to 2} \frac{x^4-2x^3-7x^2+20x-12}{x^4-6x^3+9x^2+4x-12}$,
(c) $\lim_{x \to \infty} \frac{2x-3}{x-1}$,
(d) $\lim_{x \to \infty} \left(\sqrt{x+1} - \sqrt{x}\right)$,
(e) $\lim_{x \to 0} \left(\frac{1}{x} - \frac{1}{x^2}\right)$,
(f) $\lim_{x \to 1} \frac{x^n-1}{x^m-1}$ $(n, m \in \mathbb{N})$,

(g) $\lim_{x\to 0}\left(\sqrt{x}\,\sin\frac{1}{x}\right)$,
(h) $\lim_{x\to 1}\frac{x^3-2x+1}{x-1}$,
(i) $\lim_{x\to\infty} 2x-\sqrt{4x^2-x}$,
(j) $\lim_{x\to\frac{\pi}{2}}\left(\tan^2 x-\frac{1}{\cos^2 x}\right)$,
(k) $\lim_{x\to 0}\frac{1-\sqrt{x+1}}{x}$.

25.9 Write a MATLAB function that implements the bisection method.

25.10 Determine to at least eight correct decimal places each of the real zeros of

(a) $f(x)=\sin(x)+x^2-1$,
(b) $f(x)=\mathrm{e}^x-3x^2$.

25.11 In the following a function $f : D\to\mathbb{R}$ is given. Decide whether f is continuous, and otherwise state all points of discontinuity of f.

(a) $f:\mathbb{R}\to\mathbb{R},\ f\mapsto\begin{cases}-x^2+2 & \text{für } x\le 1\\ \frac{2}{x+1} & \text{für } x>1.\end{cases}$
(b) $f:\mathbb{R}\to\mathbb{R},\ f\mapsto\begin{cases}-x^2+2 & \text{für } x\le 1\\ \frac{3}{x+1} & \text{für } x>1.\end{cases}$
(c) $f:\mathbb{R}\setminus\{1\}\to\mathbb{R},\ f\mapsto\frac{x}{(1-x)^2}$.
(d) $f:\mathbb{R}\to\mathbb{R},\ f\mapsto\begin{cases}\frac{1}{(1-x)^2} & \text{für } x\ne 1\\ 0 & \text{für } x=1.\end{cases}$
(e) $f:[-1,1)\to\mathbb{R},\ f\mapsto\begin{cases}\exp(\frac{x}{1-x^2}) & \text{für } x\in(-1,1)\\ 0 & \text{für } x=-1.\end{cases}$

Differentiation

26

With *differentiation* we now meet the core of calculus. Most functions of engineering mathematics are not only continuous, they are even *differentiable*. With this *differentiation* now the possibility opens up to determine extrema of such functions. This is the essential application of this theory. But also the monotonic behaviour of functions can be evaluated with this theory, and last but not least we can often determine the zeros of *differentiable* functions with an efficient method.

But before we get to these numerous applications of *differentiation*, we need to briefly explain how to think of it and what the rules are for *differentiation*. Many of these rules will be familiar from school days, but some will be new. We will give an overview of these rules and round off this chapter with numerous examples, some of which are sure to amaze.

26.1 The Derivative and the Derivative Function

We consider a continuous function $f : D \subseteq \mathbb{R} \to \mathbb{R}$ and its graph graph(f). To two numbers x_0 and $x_0 + h$ from D with $h > 0$ we consider the two points $\big(x_0, f(x_0)\big)$ and $\big(x_0 + h, f(x_0 + h)\big)$ of the graph through which we draw a secant, see Fig. 26.1.

With the rule *slope = opposite cathetus by adjacent cathetus* we get for the slope of the secant

$$\frac{f(x_0 + h) - f(x_0)}{h}.$$

We now let h go towards 0, i.e. we form a limit.

C. Karpfinger, *Calculus and Linear Algebra in Recipes*,
https://doi.org/10.1007/978-3-662-65458-3_26

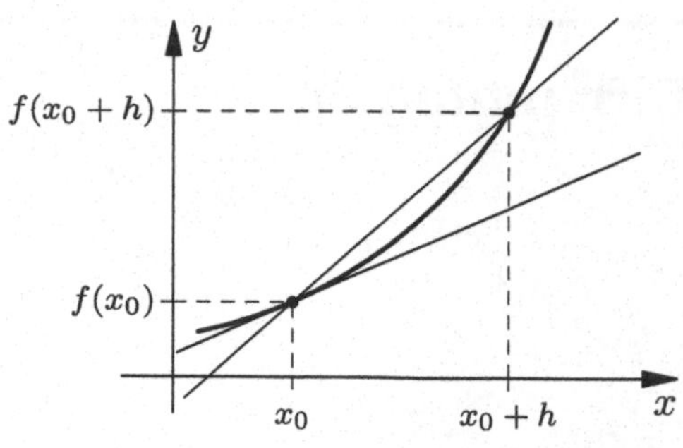

Fig. 26.1 The secant becomes a tangent

If the graph of the function f is only *smooth* enough (as in the figure), then this limit also exists, and it becomes for $h \to 0$ from the secant a tangent to the point $\big(x_0, f(x_0)\big)$ of the graph. The slope of this tangent is then

$$\lim_{h \to 0} \frac{f(x_0 + h) - f(x_0)}{h} .$$

Differentiability
A function $f : D \subseteq \mathbb{R} \to \mathbb{R}$ is called

- **differentiable in** $x_0 \in D$ if the limit

$$c = \lim_{h \to 0} \frac{f(x_0 + h) - f(x_0)}{h}$$

exists. In this case this limit c is called the **derivative** from f at the point x_0 and one writes

$$f'(x_0) = \frac{\mathrm{d}f}{\mathrm{d}x}(x_0) = c .$$

- **differentiable on** D, if f in all $x_0 \in D$ is differentiable. In this case

$$f' : D \to \mathbb{R}, \ x \mapsto f'(x)$$

is a (new) function. It is called the **derivative function** to f.

The idea conveys: A function is differentiable if a tangent can be formed at every point of the graph in an unambiguous way. For example, a function is not differentiable at a point if it has a jump or a kink there. We will support this idea once with some precise examples. We will then quickly obtain rules with which we can determine the derivative of a function more easily.

Example 26.1

- The function $f : \mathbb{R} \to \mathbb{R},\ f(x) = ax + b$ with $a, b \in \mathbb{R}$ is differentiable on $\mathbb{R}$. For all $x_0 \in \mathbb{R}$ namely, we have:

$$\lim_{h\to 0} \frac{a(x_0+h)+b-(ax_0+b)}{h} = \lim_{h\to 0} \frac{ah}{h} = a\,.$$

  The derivative of f in each x_0 is thus

$$f'(x_0) = a \quad \forall\, x_0 \in \mathbb{R}\,.$$

  The derivative function is thus $f' : \mathbb{R} \to \mathbb{R},\ f'(x) = a$. In the case $a = 0$ and $b \in \mathbb{R}$ we thus obtain for f with $f(x) = b$ the derivative function f' with $f'(x) = 0$.
- Also the function $f : \mathbb{R} \to \mathbb{R},\ f(x) = x^2$ is on $\mathbb{R}$ differentiable. For all $x_0 \in \mathbb{R}$ namely, we have:

$$\lim_{h\to 0} \frac{(x_0+h)^2 - x_0^2}{h} = \lim_{h\to 0} \frac{x_0^2 + 2hx_0 + h^2 - x_0^2}{h} = \lim_{h\to 0} 2\,x_0 + h = 2\,x_0\,.$$

  The derivative function f' of f is thus given by $f'(x) = 2x$.
- Now we consider the function $f : \mathbb{R} \to \mathbb{R},\ f(x) = |x|$ whose graph has a kink in (0, 0) has. The function f is not differentiable in 0, because the limit value $\lim_{h\to 0} \frac{f(0+h)-f(0)}{h}$ does not exist:

$$\text{If } (x_n) \text{ is a zero sequence with } x_n > 0, \text{ then follows } \lim_{n\to\infty} \frac{|x_n|}{x_n} = 1\,,$$

$$\text{but if } (x_n) \text{ is a zero sequence with } x_n < 0, \text{ then follows } \lim_{n\to\infty} \frac{|x_n|}{x_n} = -1\,.$$

  However, the magnitude function is differentiable on any interval that does not contain 0 (cf. example above). ■

According to the last example, the (everywhere) continuous function is not (everywhere) differentiable; conversely, however, it can be shown that continuity is a prerequisite for differentiability. We conclude these theoretical considerations for now with a summary of these facts:

Differentiability and Continuity or Linear Approximation

- If $f : D \subseteq \mathbb{R} \to \mathbb{R}$ is in $x_0 \in D$ differentiable, then f is continuous in x_0.
- If $f : D \subseteq \mathbb{R} \to \mathbb{R}$ is in $x_0 \in D$ continuous, then f need not be differentiable in x_0.
- If $f : D \subseteq \mathbb{R} \to \mathbb{R}$ is differentiable in $x_0 \in D$, the polynomial function g

$$g : \mathbb{R} \to \mathbb{R}, \; g(x) = f(x_0) + f'(x_0)(x - x_0)$$

of degree 1 is called **linear approximation** of f in x_0. The graph of g is the tangent at the point $(x_0, f(x_0))$ to the graph of f; the tangent equation is

$$y = f(x_0) + f'(x_0)(x - x_0)\,.$$

26.2 Derivation Rules

In this section we give clearly the derivation rules, some of which are familiar from school days. With their help one can usually find the derivative function f' for a differentiable function f. These rules are based on the fact that sums, multiples, products, quotients and compositions of differentiable functions are differentiable again. Power series are also differentiable in the interior of their convergence domain:

Derivation Rules

- If f, g are differentiable in x, then for all $\lambda, \mu \in \mathbb{R}$:

$$[\lambda f(x) + \mu g(x)]' = \lambda f'(x) + \mu g'(x)\,.$$

- If f, g are differentiable in x, then we have the **product rule**:

$$[f(x)\, g(x)]' = f'(x)\, g(x) + f(x)\, g'(x)\,.$$

- If f, g are differentiable in x and $g(x) \neq 0$, then we have the **quotient rule**:

(continued)

$$\left[\frac{f(x)}{g(x)}\right]' = \frac{g(x)f'(x) - f(x)g'(x)}{g(x)^2}\,.$$

- If g in x and f in $g(x)$ are differentiable, then we have the **chain rule**:

$$\left[f(g(x))\right]' = f'\big(g(x)\big)\,g'(x)\,.$$

- If f is invertible and differentiable in $x = f^{-1}(y)$ and $f'(x) \neq 0$, then we have:

$$\left[f^{-1}(y)\right]' = \frac{1}{f'\big(f^{-1}(y)\big)}\,.$$

- If f is a power series function, $f(x) = \sum_{k=0}^{\infty} a_k(x-a)^k$ with radius of convergence R, then f on $(a-R, a+R)$ is differentiable, and we have:

$$f' : (a-R, a+R) \to \mathbb{R} \text{ with } f'(x) = \sum_{k=1}^{\infty} k\,a_k(x-a)^{k-1}\,.$$

The power series f' has again the radius of convergence R.

We now give numerous examples in which the derivation rules are used. Thus we obtain a whole list of partly known identities.

Example 26.2

- Using the exponential series and the derivative rule for power series, we can differentiate exp:

$$(\mathrm{e}^x)' = \left(\sum_{n=0}^{\infty} \frac{x^n}{n!}\right)' = \sum_{n=1}^{\infty} n\,\frac{x^{n-1}}{n!} = 1 + x + \frac{x^2}{2!} + \cdots = \sum_{n=0}^{\infty} \frac{x^n}{n!} = \mathrm{e}^x\,.$$

- Similarly, we also find the derivative function of the sine:

$$\big(\sin(x)\big)' = \left(\sum_{n=0}^{\infty} (-1)^n \frac{x^{2n+1}}{(2n+1)!}\right)' = \sum_{n=0}^{\infty} (-1)^n \frac{x^{2n}}{(2n)!} = \cos(x)\,.$$

- The derivative of the cosine is then calculated as $\big(\cos(x)\big)' = -\sin(x)$.

- Knowing this, we can now use the quotient rule to differentiate the tangent:

$$\begin{aligned}\big(\tan(x)\big)' &= \left(\frac{\sin(x)}{\cos(x)}\right)' = \frac{\cos(x)\cos(x) - \sin(x)\,(-\sin(x))}{\cos^2(x)} \\ &= 1 + \tan^2(x) = \frac{1}{\cos^2(x)}\,.\end{aligned}$$

- Similarly, we obtain as the derivative of the cotangent : $\cot'(x) = -\frac{1}{\sin^2(x)}$.
- In differentiating the natural logarithm, we take advantage of the fact that $\ln = \exp^{-1}$:

$$\big(\ln(x)\big)' = \frac{1}{\exp'(\ln(x))} = \frac{1}{\exp(\ln(x))} = \frac{1}{x}\,.$$

- Similarly, we can differentiate arctan via the inverse function tan:

$$\arctan'(x) = \frac{1}{\tan'(\arctan(x))} = \frac{1}{1 + \tan^2(\arctan(x))} = \frac{1}{1 + x^2}\,.$$

- One also differentiate the inverse functions of the other trigonometric functions in this way:

$$\arccos'(x) = \frac{-1}{\sqrt{1 - x^2}}\,, \quad \operatorname{arccot}'(x) = \frac{-1}{1 + x^2}\,, \quad \arcsin'(x) = \frac{1}{\sqrt{1 - x^2}}\,.$$

- According to the rule for power series, the function $f(x) = x^n$ with $n \in \mathbb{N}$ has the derivative $f'(x) = nx^{n-1}$. According to the derivative rule for inverse functions, we can thus determine the derivative of $g(x) = \sqrt[n]{x} = x^{1/n}$:

$$\big(\sqrt[n]{x}\big)' = \big(x^{1/n}\big)' = g'(x) = \frac{1}{f'(g(x))} = \frac{1}{nx^{(n-1)/n}} = \frac{1}{nx^{1-1/n}} = \frac{1}{n}\,x^{\frac{1}{n}-1}\,.$$

Using the chain rule, we are now able to differentiate x^r for rational numbers $r \in \mathbb{Q}$. For this let $r = \frac{p}{q}$ with $p, q \in \mathbb{N}$, as well as $g(x) = x^p$ and $f(x) = x^{\frac{1}{q}} = \sqrt[q]{x}$, then we have:

$$\begin{aligned}\big(x^r\big)' &= \Big(x^{\frac{p}{q}}\Big)' = \Big(\big(x^p\big)^{\frac{1}{q}}\Big)' = \Big(f\big(g(x)\big)\Big)' = f'\big(g(x)\big)\,g'(x) \\ &= \frac{1}{q}\big(x^p\big)^{\frac{1}{q}-1}\,p\,x^{p-1} = \frac{p}{q}\,x^{\frac{p}{q}-p+(p-1)} = r x^{r-1}\,.\end{aligned}$$

- We use our derivation rule for rational exponents and get:

$$\left(\sqrt{x}\right)' = \frac{1}{2}x^{-\frac{1}{2}} = \frac{1}{2\sqrt{x}} \quad \text{and} \quad \left(\frac{1}{x^2}\right)' = -\frac{2}{x^3}\,.$$

- The chain rule makes it easy to derive powers:

$$\left(\left(x^3+1\right)^7\right)' = 7\,(x^3+1)^6\,3x^2\,.$$

- Other examples of the application of the chain rule are:

$$\left(\cos\left(\sin(x)\right)\right)' = -\sin\left(\sin(x)\right)\cos(x)\,,$$

$$\left(\operatorname{arccot}\left(\cos(ax)\right)\right)' = -\frac{1}{1+\cos^2(ax)}\left(-\sin(ax)\right)a = \frac{a\sin(ax)}{1+\cos^2(ax)}\,,$$

$$\left(\left(\sin(x^4-x)\right)^5\right)' = 5\left(\sin(x^4-x)\right)^4\cos(x^4-x)\,(4x^3-1)\,.$$

- For the derivation of x^x we first cleverly rewrite this expression as $\exp\left(\ln(x^x)\right)$ and use the logarithm rule $\ln(a^b) = b\ln(a)$:

$$\left(x^x\right)' = \left(\mathrm{e}^{\ln(x^x)}\right)' = \left(\mathrm{e}^{x\ln(x)}\right)' = \mathrm{e}^{x\ln(x)}\left(1\cdot\ln(x) + x\,\frac{1}{x}\right) = x^x\left(\ln(x)+1\right).$$

■

The derivative function f' of a differentiable function can in turn be continuous or even differentiable. If it is even differentiable, we can determine the derivative function $f'' = (f')'$ of f', i.e. form the *second derivative of f*. Now we can ask again whether this is continuous or even differentiable, etc. We write $f^{(k)}$ for the k-th derivative, $k \in \mathbb{N}_0$ (in particular $f^{(0)} = f$) if it exists, and write or say for each set $D \subseteq \mathbb{R}$:

- $C^0(D) = \{f : D \to \mathbb{R} \mid f \text{ is continuous}\}$—the set of continuous functions defined on D.
- $C^k(D) = \{f : D \to \mathbb{R} \mid f^{(k)} \text{ exists and is continuous}\}$—the set of **$k$-times continuously differentiable** functions defined on D, $k \in \mathbb{N}_0$.
- $C^\infty(D) = \{f : D \to \mathbb{R} \mid f \text{ is differentiable any number of times}\}$—the set of arbitrarily often differentiable functions on D.

Example 26.3 exp, sin, cos, polynomial functions and power series functions are examples of arbitrarily often differentiable functions on $\mathbb{R}$, i.e. they are elements of $C^\infty(\mathbb{R})$.

We end this section with the *mean value Theorem* of the differential calculus and two of its main consequences.

The Mean Value Theorem of Differential Calculus and First Consequences
If $f : [a, b] \subseteq \mathbb{R} \to \mathbb{R}$ is a continuous function which is on (a, b) differentiable, then there exists a $x_0 \in (a, b)$ with

$$f'(x_0) = \frac{f(b) - f(a)}{b - a}.$$

It follows:

- If $f'(x) = 0$ for all x of an interval D, then $f : D \to \mathbb{R}$ is constant, i.e. $f(x) = c$ for a $c \in \mathbb{R}$.
- Is $f'(x) = g'(x)$ for all x of an interval D for two functions $f, g : D \to \mathbb{R}$, then $f - g : D \to \mathbb{R}$ is constant, so $f(x) = g(x) + c$ for a $c \in \mathbb{R}$.

The mean value Theorem is best illustrated by a sketch (see Fig. 26.2): It states that the secant passes through the points $(a, f(a))$ and $(b, f(b))$ is parallel to a tangent to a point $(x_0, f(x_0))$ of the graph for a x_0 with $a < x_0 < b$.

MATLAB Of course, we can also obtain the derivative of a function *f* with MATLAB, for this we explain in advance a *symbol* x and then form with `diff` the derivative function of a function *f*, e.g.

```
» syms x; » f(x) = exp(sin(x)); » df(x)=diff(f(x)) df(x)
= exp(sin(x))*cos(x)
```

Now `df(x)` is the derivative function, for example `df(0)=1`.

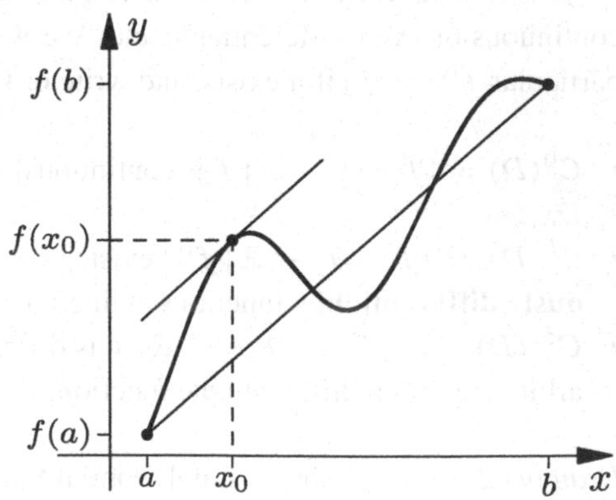

Fig. 26.2 The secant is parallel to a tangent line

26.3 Numerical Differentiation

It is often difficult, if not impossible, to determine the derivative function of a function. This is the case, for example, when the function f itself is not given, but is only known, for example, which values $y_i = f(x_i)$ it has in discrete points $x_0, \ldots, x_n$. Numerical differentiation provides a way to determine approximations for the values $f'(x_i),\ f''(x_i),\ \ldots$ of the derivative functions in such situations.

Note: It is not the derivative function that is approximated, but values of the derivative function.

We obtain such approximations by a simple approach: We determine an interpolation polynomial p by interpolation points

$$(x_0, f(x_0)),\ (x_1, f(x_1)),\ \ldots$$

where in general the x-values are equidistant, i.e. $x_k = x_0 + k\,h$ for all $k = 1, 2, \ldots$ with the *step size* h. The (simple) interpolation polynomial p approximates the (complicated) function f, therefore it is only obvious to use as approximation for the derivative of f between the interpolation points the derivative of the interpolation polynomial p in these points.

We do not determine the interpolating polynomial explicitly, but give the formulas for the approximation of the values of the first derivatives of f by the procedure we have described:

Formulas of Numerical Differentiation
If $f : D \to \mathbb{R}$ is a sufficiently often differentiable function and if $x_0, \ldots, x_3 \in D$ are equidistant points with the step size $h = x_{k+1} - x_k$ for $k = 0, 1, 2$, then approximate values of the derivative function are obtained as follows:

- $f'(x) \approx \frac{f(x_1)-f(x_0)}{h}$ for $x \in [x_0, x_1]$,
- $f''(x) \approx \frac{f(x_2)-2f(x_1)+f(x_0)}{h^2}$ for $x \in [x_0, x_2]$,
- $f'''(x) \approx \frac{f(x_3)-3f(x_2)+3f(x_1)-f(x_0)}{h^3}$ for $x \in [x_0, x_3]$.

Two kinds of errors happen in numerical differentiation:

- **Discretization error:** The computation of the derivative is done by a limit process $h \to 0$, this limit process is in a sense replaced in numerical differentiation by instead using a *small* h. The discretization error therefore becomes smaller with the choice of a smaller h.

- **Rounding error:** The subtraction of numbers of approximately the same size in the numerator of the approximating quotients leads to cancellation and thus to rounding errors. This error becomes larger if h is chosen smaller.

This interaction of discretization and rounding error has now the consequence that the step size must be chosen with care, in fact there is in general an optimal step size h, in the following example this is about 10^{-5}:

Example 26.4 We compute numerically the first and second derivatives of the function $f : \mathbb{R} \to \mathbb{R}$, $f(x) = e^{x^2} \sin(x)$ at the point $x_0 = 1$ and obtain the following values, underlining the correct digits:

h	$f'(x_0)$ numerisch	$f''(x_0)$ numerisch
10^{-1}	6.067189794992709	17.478644617728918
10^{-2}	6.043641979728198	17.313215006673488
10^{-3}	6.043406888888825	17.311568821121170
10^{-4}	6.043404538029762	17.311552280574460
10^{-5}	6.043404514555207	17.311552191756622
10^{-6}	6.043404514599615	17.310153310745591
10^{-7}	6.043404510158723	17.275070263167439
10^{-8}	6.043404399136421	4.440892098500625

■

Remark The importance of numerical differentiation cannot be underestimated. Since one can also numerically differentiate functions in several variables in an analogous way, we will be able to use numerical differentiation in the one-dimensional and multidimensional Newton method for determining zeros of functions or in the gradient method for numerically determining extrema of functions in several variables.

26.4 Exercises

26.1 Show that an in $x_0 \in D$ differentiable function $f : D \to \mathbb{R}$ in x_0 is continuous.

26.2 Let $f : \mathbb{R} \to \mathbb{R}$ given by $f(x) = x|x|$. Show that f is on $\mathbb{R}$ continuous and differentiable.

26.3 Calculate for $2 + \frac{1}{x} > 0$ the first derivative of $f(x) = \left(2 + \frac{1}{x}\right)^x$.

26.4 A function $f : \mathbb{R} \to \mathbb{R}$ is called **even** if $f(x) = f(-x)$ for all $x \in \mathbb{R}$, f is called **odd** if $f(x) = -f(-x)$ for all $x \in \mathbb{R}$.

Check which of the following functions are even or odd. State this also for the corresponding derivatives:

(a) $f(x) = x^2 + \cos x$,
(b) $g(x) = x^2 \tan x$,
(c) $h(x) = x \sin x$,
(d) $k(x) = \mathrm{e}^x$.

Show in general that the derivative of a differentiable even (odd) function is odd (even).

26.5 Calculate the first derivatives of the following functions:

(a) $f_1(x) = 3 \arctan x + \frac{3x}{x^2+1} + \frac{2x}{(x^2+1)^2}$,
(b) $f_2(x) = \ln \frac{x^2+1}{x^2-1}$,
(c) $f_3(x) = \frac{\sqrt{2x+3}}{\sqrt{4x+5}}$,
(d) $f_4(x) = \arccos \frac{1}{x}$,
(e) $f_5(x) = \ln(\sin x) - x \cot x$,
(f) $f_6(x) = -8(x + \frac{2}{x}) + 4\ln(x+3)$,
(g) $f_7(x) = \frac{\sqrt{2x^2+3}}{\sqrt{4x+2}}$,
(h) $f_8(x) = \ln(\sin x) - x \cos x$,
(i) $f_9(x) = x^2 \tan x$,
(j) $f_{10}(x) = \tan\left(\frac{\sin x^2 + \cos x}{\ln \frac{1}{x^2} + 2}\right)$,
(k) $f_{11}(x) = \frac{2x^3 - x^2 + 7x + 1}{(2x^2-1)^2}$,
(l) $f_{12}(x) = 2x + \frac{1}{2x^2}$,
(m) $f_{13}(x) = x^2 \cos\left(2x^2 + \sin \frac{1}{x}\right)$,
(n) $f_{14}(x) = x^{\cos x}$.

26.6 Let $f : \mathbb{R}_+ \to \mathbb{R}$ where $f(x) = \frac{\sin x}{\sqrt{x}}$. Show: For all $x \in \mathbb{R}_{>0}$ we have

$$f''(x) + \frac{1}{x} f'(x) + \left(1 - \frac{1}{4x^2}\right) f(x) = 0 .$$

26.7 Write a MATLAB script that returns the values in Example 26.4.

26.8 For the derivative of a function f we have the approximation

$$f'(x_0) \approx \big(f(x_0 + h) - f(x_0)\big)/h .$$

(a) Calculate in MATLAB the above approximation of $f'(\pi/4)$ for $f(x) = \sin(x)$ and $h = 10^{-4}$.
(b) Write a MATLAB script that performs this calculation for various $h = 10^{-k}$ where k has the values $1, 2, \ldots, 10$.

(c) Compare the values from (b) with the exact derivative. For which h does the best approximation result?

26.9 We consider the function

$$f : \mathbb{R} \to \mathbb{R}, \quad x \mapsto \begin{cases} x^2 \sin(\frac{1}{x}) & \text{für } x \neq 0 \\ 0 & \text{für } x = 0 \end{cases}.$$

Show that f is differentiable and determine f'.

Applications of Differential Calculus I 27

Differential calculus has many applications in engineering mathematics. Among these many applications, some are familiar from school days, such as assessing monotonicity and convexity or concavity, or determining local extrema. We also discuss a method for determining limit values, namely *L'Hospital's rule*.

27.1 Monotonicity

We refer to the four suggestive notions of *(strictly) monotonically increasing* or *(strictly) monotonically decreasing* (see Sect. 23.4) and tie in with the remark in Sect. 23.4:

Criterion for Monotonicity
If $f : [a, b] \to \mathbb{R}$ is continuous and (a, b) differentiable, then

- f is monotonically increasing if and only if $f'(x) \geq 0 \quad \forall x \in (a, b)$.
- f is monotonically decreasing if and only if $f'(x) \leq 0 \quad \forall x \in (a, b)$.
- f is strictly monotonically increasing if $f'(x) > 0 \quad \forall x \in (a, b)$.
- f is strictly monotonically decreasing if $f'(x) < 0 \quad \forall x \in (a, b)$.

The converse of the last two statements are in general not true. Here are brief examples:

- The function $f : [-1, 1] \to \mathbb{R},\ f(x) = x^3$ is strictly monotonically increasing (and differentiable on $(-1, 1)$), but $f' : \mathbb{R} \to \mathbb{R},\ f'(x) = 3x^2$ does not satisfy $f'(x) > 0$ for all $x \in (-1, 1)$, because $f'(0) = 0$.

C. Karpfinger, *Calculus and Linear Algebra in Recipes*,
https://doi.org/10.1007/978-3-662-65458-3_27

- The function $f:[-1,1]\to\mathbb{R},\ f(x)=-x^3$ is strictly monotonically decreasing (and differentiable on $(-1,1)$), but $f':\mathbb{R}\to\mathbb{R},\ f'(x)=-3x^2$ does not satisfy $f'(x)<0$ for all $x\in(-1,1)$, because $f'(0)=0$.

An essential application of this monotonicity test is discussed in the following section: With the help of the monotonicity test it is often possible to decide whether a differentiable function f has an extrema in a place x_0 of the domain of f.

27.2 Local and Global Extrema

We summarize *maxima* and *minima* under the term *extrema* and distinguish *local* and *global extrema*:

Extrema and Extremal Places
We consider a function $f: D\subseteq\mathbb{R}\to\mathbb{R}$. One calls a $x_0\in D$ **point of a**

- **global maximum** if $f(x_0)\geq f(x)\quad\forall x\in D$.
  One then calls $f(x_0)$ **the global maximum.**
- **global minimum** if $f(x_0)\leq f(x)\quad\forall x\in D$.
  One then calls $f(x_0)$ **the global minimum.**
- **local maximum** if $\exists\,\varepsilon>0:\ f(x_0)\geq f(x)\quad\forall x\in(x_0-\varepsilon,x_0+\varepsilon)$.
  One then calls $f(x_0)$ **a local maximum.**
- **local minimum** if $\exists\,\varepsilon>0:\ f(x_0)\leq f(x)\quad\forall x\in(x_0-\varepsilon,x_0+\varepsilon)$.
  One then calls $f(x_0)$ **a local minimum.**

If we have even $>$ instead of $\geq$ or $<$ instead of $\leq$, we speak of **strict** local or global extrema.

The global maximum is the largest overall value that the function assumes on its domain, the global minimum is correspondingly the smallest overall value. The value is unique, but it may well be assumed at different points.

A point of a local extrema always lie in the *interior* of the domain of f, an ε-environment of such an extremal place within the domain is required. If one restricts the function to such a (possibly very small) environment U of a place x_0 of a local extremum, then this place x_0 is the point of a global extremum of the function restricted to U. Note Fig. 27.1.

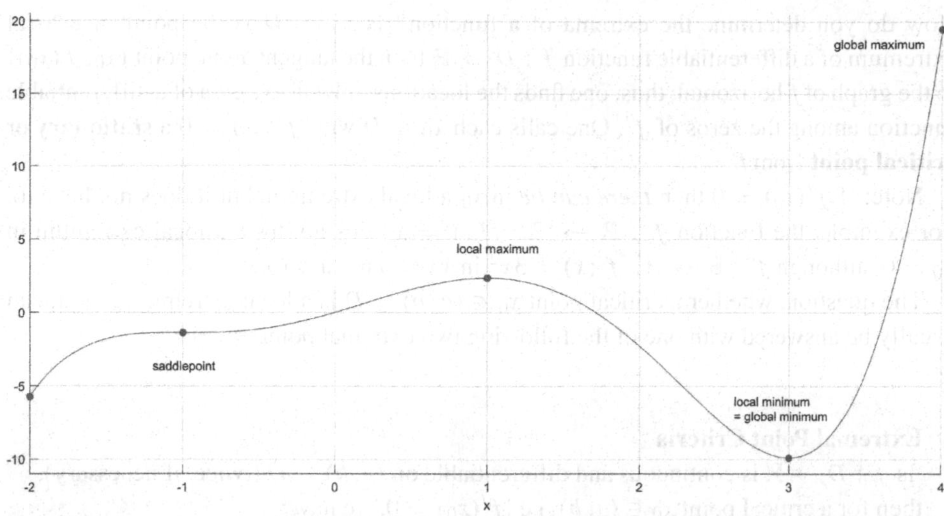

Fig. 27.1 Local and global extrema

Example 27.1

- For the function $f : [-2, 2] \to \mathbb{R}$, $f(x) = 2$ every $x \in [-2, 2]$ is a point of a global and local minimum and maximum with the value 2. There are no strict extrema.
- As extremal points or extremal values of the polynomial function $f : [0, 3] \to \mathbb{R}$, $f(x) = 2x^3 - 9x^2 + 12x$ we have

  a global minimum with value 0 at position 0,
  a global maximum with value 9 at place 3,
  a local minimum with value 4 at place 2,
  a local maximum with value 5 at location 1.

Note the graph of the function in Fig. 27.2. ■

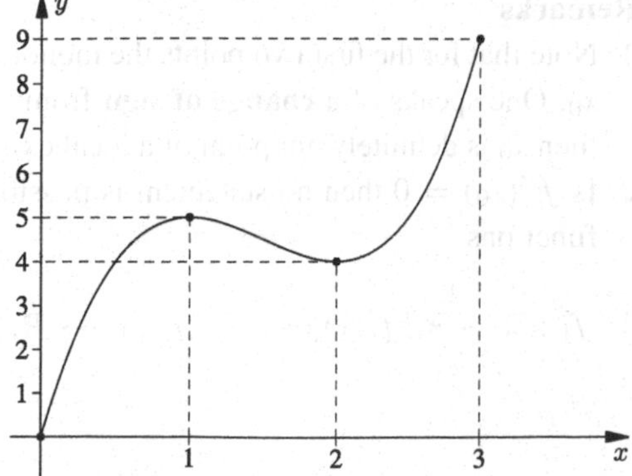

Fig. 27.2 Extrema of f

How do you determine the extrema of a function? Is $x_0 \in D$ is the point of a local extremum of a differentiable function $f : D \to \mathbb{R}$ then the tangent at the point $(x_0, f(x_0))$ to the graph of f horizontal; thus, one finds the locations of local extrema of a differentiable function among the zeros of f'. One calls each $x_0 \in D$ with $f'(x_0) = 0$ a **stationary** or **critical point** from f.

Note: If $f'(x_0) = 0$ then *there can be* in x_0 a local extremum, but it does not have to. For example, the function $f : \mathbb{R} \to \mathbb{R},\ f(x) = x^3$ dos not have a local extremum in $x_0 = 0$, although $f' : \mathbb{R} \to \mathbb{R},\ f'(x) = 3x^2$ in $x_0 = 0$ has a zero.

The question, whether a critical point $x_0 \in (a, b) \subseteq D$ is a local extremum or not, can usually be answered with one of the following two extremal point criteria:

Extremal Point Criteria

Is $f : D \to \mathbb{R}$ is continuous and differentiable on $(a, b) \subseteq D$ (twice, if necessary), then for a critical point $x_0 \in (a, b)$, i.e. $f'(x_0) = 0$, we have:

- x_0 is a point of a local minimum if a $\varepsilon > 0$ exists with:

$$f'(x) < 0 \quad \forall x \in (x_0 - \varepsilon, x_0) \text{ and } f'(x) > 0 \quad \forall x \in (x_0, x_0 + \varepsilon)\,.$$

- x_0 is a point of a local maximum if a $\varepsilon > 0$ exists with:

$$f'(x) > 0 \quad \forall x \in (x_0 - \varepsilon, x_0) \text{ and } f'(x) < 0 \quad \forall x \in (x_0, x_0 + \varepsilon)\,.$$

- x_0 is a point of a local maximum if $f''(x_0) < 0$.
- x_0 is a point of a local minimum if $f''(x_0) > 0$.

If $f(x_0)$ isn't extremal (but $f'(x_0) = 0$), we call $(x_0, f(x_0))$ a **saddlepoint**.

Remarks

1. Note that for the first two points the monotonicity of the function f changes at the point x_0. One speaks of a **change of sign** from f' in x_0. If there is no change of sign in x_0, then x_0 is definitely not point of a local extremum.
2. Is $f''(x_0) = 0$ then no statement is possible. Anything can happen: For example, the functions

$$f_1 : \mathbb{R} \to \mathbb{R},\ f_1(x) = x^4, \quad f_2 : \mathbb{R} \to \mathbb{R},\ f_2(x) = -x^4, \quad f_3 : \mathbb{R} \to \mathbb{R},\ f_3(x) = x^3$$

all have a stationary point in $x_0 = 0$ and it is also

$$f_1''(0) = f_2''(0) = f_3''(0) = 0\,.$$

But nevertheless f_1 has a local minimum in 0, f_2 has a local maximum in 0 and f_3 has neither a minimum nor a maximum in 0.

3. If f is in x_0 not differentiable and if we have in x_0 a change of sign of f', then in x_0 nevertheless is a local extremum. For example, $f : \mathbb{R} \to \mathbb{R},\ f(x) = |x|$ has in $x_0 = 0$ a local minimum.

27.3 Determination of Extrema and Extremal Points

Determining the local extrema is usually unproblematic. One determines the zeros of f', i.e., the stationary points, and decide whether each stationary point is an extrema using one of the above criteria.

The theorem of maximum and minimum in Sect. 25.4 states that a differentiable function f has a global maximum and minimum in any case if the domain D of f is a closed and bounded interval, i.e. of the form $D = [a, b]$ with real numbers $a < b$.

If the point x_0 of a global extremum $f(x_0)$ is in the interior (a, b) of D, then this global extremum is also a local extremum. If the point x_0 of the global extremum is not in the interior of (a, b) then it lies in a boundary point a or b. So in the case $D = [a, b]$ the global extremal points are found under the points of the local extrema or the boundary points.

But also in the cases $D = [a, b)$ or $D = (a, b]$ or $D = (a, b)$ with $a, b \in \mathbb{R} \cup \{\pm\infty\}$ there can be global extrema, but there need not be. In any case, there is no global maximum if $\lim_{x\to a} f(x) = \infty$ or $\lim_{x\to b} f(x) = \infty$; there is no greatest function value in this situations, because the set of values is unbounded from above. Similarly, there is no global minimum if $\lim_{x\to a} f(x) = -\infty$ or $\lim_{x\to b} f(x) = -\infty$. We summarize and add:

Recipe: Determine the Extremal Places

The extremal places of a function which may be differentiated twice

$$f : D \subseteq \mathbb{R} \to \mathbb{R},\ x \mapsto f(x)$$

can be found as follows:

(1) Determine f'.
(2) Determine the critical points of f, i.e., the zeros $a_1, \ldots, a_n \in D$ of f'.

(continued)

(3) Get the points of local extrema: Decide with one of the extremum criteria whether there is a local maximum or minimum in $a_1, \ldots, a_n$.
(4) Obtain the local extrema: Determine the values of $f(a_i)$ if in a_i there is a local extremum.
(5) Determine the following *values* at the boundary points of D:

- if $D = [a, b], a, b \in \mathbb{R}$, then determine $f(a), f(b)$,
- if $D = (a, b), a, b \in \mathbb{R} \cup \{\infty\}$, then determine $\lim_{x\to a} f(x)$, $\lim_{x\to b} f(x)$,
- if $D = [a, b), a \in \mathbb{R}, b \in \mathbb{R} \cup \{\infty\}$ then determine $f(a)$, $\lim_{x\to b} f(x)$,
- if $D = (a, b], a \in \mathbb{R} \cup \{\infty\}, b \in \mathbb{R}$, then determine $\lim_{x\to a} f(x)$, $f(b)$.

If $D = D_1 \cup \cdots \cup D_r$ is a union of disjoint intervals, treat each subinterval separately.
(6) Consider the values in (4) and (5):

- Does a smallest real value $y_{\min}$ exist? If yes, then $y_{\min}$ is the global minimum, all x_i with $f(x_i) = y_{\min}$ are points of the global minimum.
- Does a greatest real value $y_{\max}$ exist? If yes, then $y_{\max}$ is the global maximum, all x_i with $f(x_i) = y_{\max}$ are points of the global maximum.
- Otherwise there is no global extremum and therefore no points of global extrema.

Example 27.2

- We determine the extrema of the (differentiable) function

$$f : [-1, 1] \to \mathbb{R}, \ f(x) = x^2\sqrt{1 - x^2}.$$

(1) We obtain for the derivative

$$f'(x) = 2x\sqrt{1 - x^2} + x^2 \frac{1}{2} \frac{(-2x)}{\sqrt{1 - x^2}} = \frac{2x(1 - x^2) - x^3}{\sqrt{1 - x^2}} = \frac{-3x^3 + 2x}{\sqrt{1 - x^2}}.$$

(2) We calculate the zeros of the derivative:

$$f'(x) = 0 \Leftrightarrow -3x^3 + 2x = 0 \Leftrightarrow x(2 - 3x^2) = 0$$

$$\Leftrightarrow x = 0 \lor x = \sqrt{\frac{2}{3}} \lor x = -\sqrt{\frac{2}{3}}.$$

This gives us the stationary points $x_1 = 0$, $x_2 = \sqrt{\frac{2}{3}}$ and $x_3 = -\sqrt{\frac{2}{3}}$.

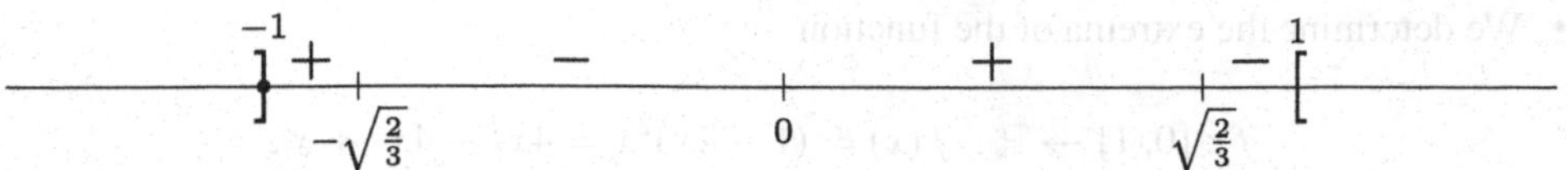

Fig. 27.3 Changes of signs of f'

(3) Now we consider the sign of

$$f'(x) = \frac{-3\left(x+\sqrt{2/3}\right)x\left(x-\sqrt{2/3}\right)}{\sqrt{1-x^2}},$$

where we factorized the polynomial in the numerator, note Fig. 27.3.

(4) Thus, by the sign change criterion, we have:

$$\text{a local maximum in } x_1 = -\sqrt{\frac{2}{3}} \text{ with value } \frac{2}{3\sqrt{3}},$$

$$\text{a local minimum in } x_2 = 0 \text{ with value } 0$$

$$\text{and a local maximum in } x_3 = \sqrt{\frac{2}{3}} \text{ with value } \frac{2}{3\sqrt{3}}.$$

(5) To find the global extrema, we determine the values at the boundaries of the domain of f and obtain:

$$f(-1) = 0 \quad \text{und} \quad f(1) = 0.$$

(6) We thus have global maxima in $\pm\sqrt{\frac{2}{3}}$ with value $\frac{2}{3\sqrt{3}}$ and global minima in $\pm 1,\ 0$ with value 0.

The whole fact is also illustrated by the graph of the function in Fig. 27.4.

Fig. 27.4 The extrema of f

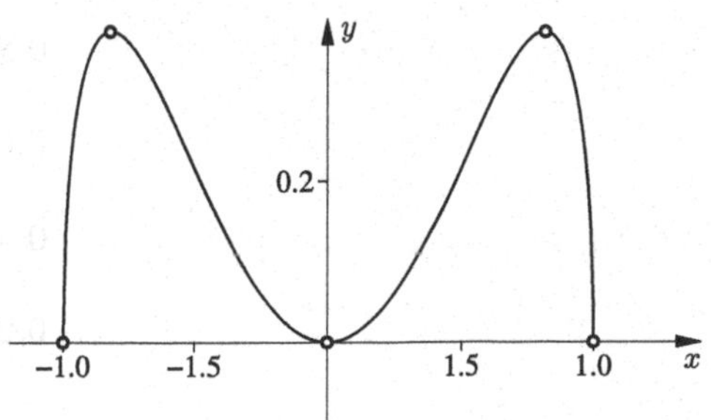

- We determine the extrema of the function

$$f : [0, 1] \to \mathbb{R}\,,\ f(x) = (1 - 2x)^2 x = 4x^3 - 4x^2 + x\,.$$

(1) We have $f'(x) = 12x^2 - 8x + 1$.
(2) The zeros of f' are

$$a_1 = \frac{8 + \sqrt{64 - 48}}{24} = \frac{1}{2} \text{ and } a_2 = \frac{8 - \sqrt{64 - 48}}{24} = \frac{1}{6}\,.$$

(3) Because of $f''(x) = 24x - 8$ we have

$$f'(a_1) > 0 \text{ and } f'(a_2) < 0\,.$$

Therefore, at the point a_1 there is a local minimum and at the point a_2 there is a local maximum.
(4) The local minimum has the value $f(a_1) = 0$, the local maximum has the value $f(a_2) = 2/27$.
(5) To find the global extrema, we determine the values at the boundery points of the domain and obtain:

$$f(0) = 0 \quad \text{and} \quad f(1) = 1\,.$$

(6) The global maximum is 1, it is at the point $b = 1$. The global minimum is 0, it is at the points $a = 0$ and $a_1 = 1/2$

Figure 27.5 shows the graph of the function; we have indicated the extrema. ■

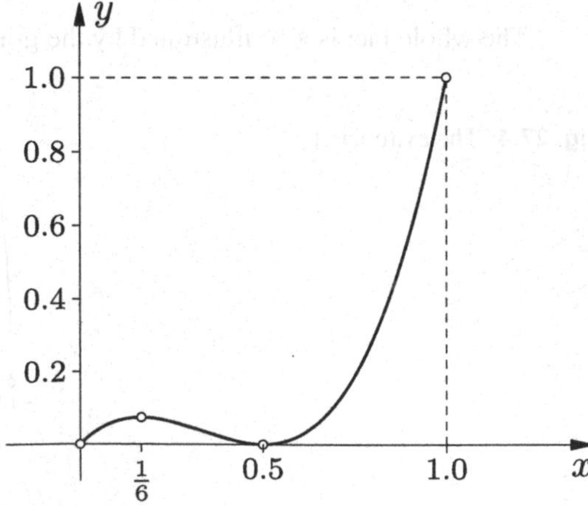

Fig. 27.5 The extrema of f

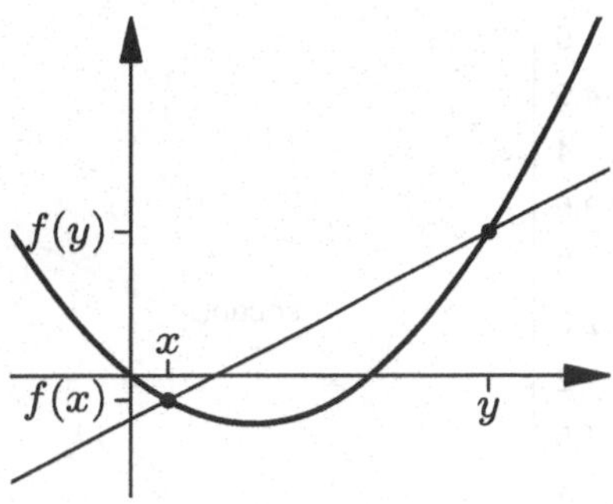

Fig. 27.6 The secant

27.4 Convexity

We consider a (twice differentiable) function $f : [a, b] \subseteq \mathbb{R} \to \mathbb{R}$. To two points $x < y$, $x, y \in [a, b]$ we consider the straight line through the points $(x, f(x))$ and $(y, f(y))$, see Fig. 27.6.

The equation of this straight line is

$$g(z) = \frac{f(y) - f(x)}{y - x}(z - x) + f(x).$$

We have $g(x) = f(x)$ and $g(y) = f(y)$ and the points in the interval $[x, y]$ are given by

$$z = x + t(y - x), \qquad t \in [0, 1].$$

Thus, the equation of the straight line is:

$$g\big(z(t)\big) = \big(f(y) - f(x)\big)t + f(x) = (1 - t)f(x) + tf(y).$$

Convexity and Concavity
A function $f : [a, b] \to \mathbb{R}$ is called

- **convex** on $[a, b]$ if for all $x, y \in [a, b]$ where $x \neq y$ we have:

$$f\big(x + t(y - x)\big) = f\big((1 - t)x + ty\big) \leq (1 - t)f(x) + tf(y) \; \forall t \in [0, 1].$$

- **strictly convex** on $[a, b]$ if for all $x, y \in [a, b]$ with $x \neq y$:

$$f\big(x + t(y - x)\big) = f\big((1 - t)x + ty\big) < (1 - t)f(x) + tf(y) \; \forall t \in (0, 1).$$

(continued)

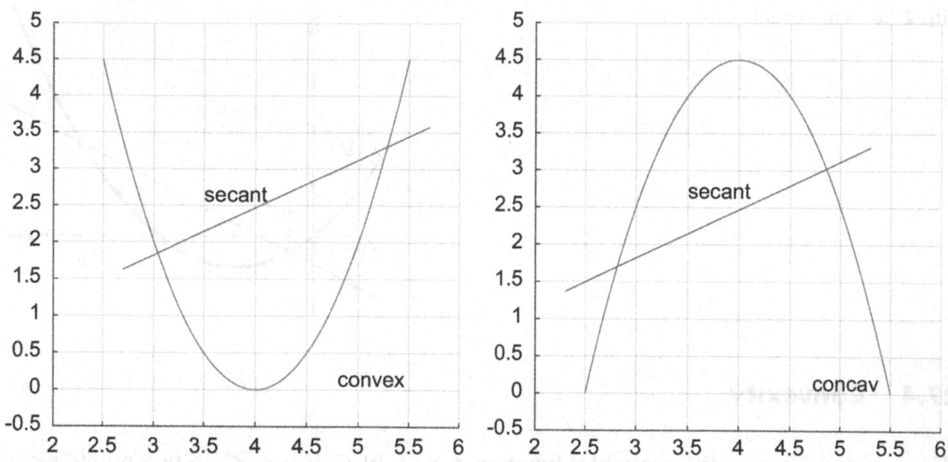

Fig. 27.7 Convex = graph lies below the secant, Concave = graph lies above the secant

- **concave** on $[a, b]$ if for all $x, y \in [a, b]$ with $x \neq y$:

$$f\big(x + t(y - x)\big) = f\big((1 - t)x + ty\big) \geq (1 - t)f(x) + tf(y) \;\forall t \in [0, 1]\,.$$

- **strictly concave** on $[a, b]$ if for all $x, y \in [a, b]$ with $x \neq y$:

$$f\big(x + t(y - x)\big) = f\big((1 - t)x + ty\big) > (1 - t)f(x) + tf(y) \;\forall t \in (0, 1)\,.$$

Convex means graphically that the graph of f lies below the secant, concave means that it lies above the secant, see Fig. 27.7.

The convexity or concavity of a function f can often be determined by the second derivative f'':

Criterion for Convexity and Concavity

If $f : [a, b] \to \mathbb{R}$ is twice continuously differentiable, then we have:

- f is convex on $[a, b]$ if $f''(x) \geq 0 \;\forall x \in [a, b]$.
- f is concave on $[a, b]$ when $f''(x) \leq 0 \;\forall x \in [a, b]$.
- If $f''(x) > 0 \;\forall x \in [a, b]$ so f is strictly convex.
- If $f''(x) < 0 \;\forall x \in [a, b]$ so f is strictly concave.

The two statements below are not reversible, as the example of the function $f : [-1, 1] \to \mathbb{R},\ f(x) = x^4$ shows. f is strictly convex, but $f''(0) = 0$. Analogous is $f(x) = -x^4$ an example of a strictly concave function with $f''(0) = 0$.

Example 27.3 The functions $f : \mathbb{R} \to \mathbb{R},\ f(x) = x^2$ and $\exp : \mathbb{R} \to \mathbb{R}$ are strictly convex. ■

27.5 The Rule of L'Hospital

L'Hospital's rule helps in determining limits (cf. the prescription in Sect. 25.1 with the remark that follows):

Rule of L'Hospital
Given two differentiable functions $f, g : (a, b) \to \mathbb{R}$ where $-\infty \le a < b \le \infty$. Let further:

$$\lim_{x \to b} f(x) = \lim_{x \to b} g(x) = 0 \quad \text{or} \quad \lim_{x \to b} f(x) = \lim_{x \to b} g(x) = \infty .$$

If $\lim_{x \to b} \frac{f'(x)}{g'(x)} \in \mathbb{R} \cup \{\pm\infty\}$, then

$$\lim_{x \to b} \frac{f(x)}{g(x)} = \lim_{x \to b} \frac{f'(x)}{g'(x)} .$$

This applies analogously to limit values $x \to a$.

We note the sign $\frac{0}{0}$ and $\frac{\infty}{\infty}$ above the equal sign when we use L'Hospital's rule.

Example 27.4

- We have $\lim_{x \to 0} \cos(x) - 1 = 0$ and $\lim_{x \to 0} \tan(x) = 0$. So we can calculate the following limit:

$$\lim_{x \to 0} \frac{\cos(x) - 1}{\tan(x)} \overset{\frac{0}{0}}{=} \lim_{x \to 0} \frac{-\sin(x)}{\frac{1}{\cos^2(x)}} = 0 .$$

- Another example is the limit against 0 of the cardinal sine:

$$\lim_{x\to 0}\frac{\sin(x)}{x}\overset{\frac{0}{0}}{=}\lim_{x\to 0}\frac{\cos(x)}{1}=1\,.$$

- In the next case, we use L'Hospital's rule several times:

$$\lim_{x\to 0}\left(\frac{1}{\sin(x)}-\frac{1}{x}\right)=\lim_{x\to 0}\frac{x-\sin(x)}{x\sin(x)}\overset{\frac{0}{0}}{=}\lim_{x\to 0}\frac{1-\cos(x)}{\sin(x)+x\cos(x)}$$
$$\overset{\frac{0}{0}}{=}\lim_{x\to 0}\frac{\sin(x)}{2\cos(x)-x\sin(x)}=0\,.$$

- Now consider limits $x\to\infty$:

$$\lim_{x\to\infty}\frac{\sqrt{x}}{\ln(x)}\overset{\frac{\infty}{\infty}}{=}\lim_{x\to\infty}\frac{1}{2\sqrt{x}}x=\infty\,.$$

- Last, an example where L'Hospital's rule does not work. It is

$$\lim_{x\to\infty}\sin(x)+2x=\lim_{x\to\infty}\cos(x)+2x=\infty\,,$$

but still the limit

$$\lim_{x\to\infty}\frac{\sin(x)+2x}{\cos(x)+2x}$$

of the quotient cannot be calculated with L'Hospital's rule, because the limit value

$$\lim_{x\to\infty}\frac{\cos(x)+2}{-\sin(x)+2}$$

from the quotient of the derivatives does not exist. However, there are other ways to determine the limit:

$$\lim_{x\to\infty}\frac{\sin(x)+2x}{\cos(x)+2x}=\lim_{x\to\infty}\frac{\cos(x)-\cos(x)+\sin(x)+2x}{\cos(x)+2x}$$
$$=\lim_{x\to\infty}1+\frac{\sin(x)-\cos(x)}{\cos(x)+2x}=1\,.$$

■

27.6 Exercises

27.1 A beverage manufacturer wants to save costs in the production of beverage cans. A beverage can should always have a volume of $V_0 = 0.41$ and be cylindrical (we assume in this problem that it is in fact exactly a circular cylinder). How must the height and radius of the cylinder be chosen if as little material as possible is to be used for production?

27.2 Let the function $f : \left[-\frac{3}{2}; \frac{3}{2}\right] \to \mathbb{R}$ with $f(x) = x^2 - 2\,|x| + 1$.

Determine the zeros of this function and its symmetry behavior, calculate (where possible) the derivative, all asymptotes, monotonic behavior, local and global maxima and minima, and indicate where the function is convex or concave. Then sketch the graph of the function f and fill in the information in the graph.

27.3 Show:

(a) $\ln(1 + x) \le \arctan x$ for $x \in [0, 1]$,
(b) $\arctan y + \arctan(1/y) = \pi/2$ for $y > 0$.

27.4 You have bought a can of Coke that has a perfect cylindrical shape. The mass M of the can (without contents) is uniformly distributed over the whole can, the can has height H and volume V. You want the can to be as stable as possible, so the center of gravity of the can (including contents) should be as low as possible. To simplify matters, we assume that cola has a density of 1. How much cola (filling height in percent of the can height) do you have to drink so that the center of gravity reaches its lowest level?

27.5 Let $f : \mathbb{R} \to \mathbb{R}$ defined by $f(x) = 2\cos x - x$. Show that the function f has infinitely many local maxima and minima, but no global extrema.

27.6 Determine the following limits:

(a) $\lim_{x\to 0} \frac{e^x - 1}{x}$,
(b) $\lim_{x\to 0} \frac{\sqrt{\cos(ax)} - \sqrt{\cos(bx)}}{x^2}$,
(c) $\lim_{x\to 0} \frac{\ln^2(1+3x) - 2\sin^2 x}{1 - e^{-x^2}}$,
(d) $\lim_{x\to -1^+} (x + 1)\tan\frac{\pi x}{2}$,
(e) $\lim_{x\to\infty} \frac{\cosh x}{e^x}$,
(f) $\lim_{x\to 0} \frac{\cos x - 1}{x}$,
(g) $\lim_{x\to 1^-} \frac{\pi/2 - \arcsin x}{\sqrt{1-x}}$,
(h) $\lim_{x\to 1} \sin(\pi x) \cdot \ln|1 - x|$,

(i) $\lim_{x\to 0}(1+\arctan x)^{\frac{1}{x}}$,
(j) $\lim_{x\to 0}\frac{\sin x+\cos x}{x}$,
(k) $\lim_{x\to\infty}\frac{\ln(\ln x)}{\ln x}$,
(l) $\lim_{x\to 0}\frac{1}{e^x-1}-\frac{1}{x}$,
(m) $\lim_{x\to 0}\cot(x)(\arcsin(x))$.

Applications of Differential Calculus II

28

We discuss further applications of differentiation, such as *Newton's method* for approximating zeros of functions and *Taylor expansion* for approximating functions by polynomials or representing functions by power series.

28.1 The Newton Method

The *Newton method* is a method for the approximate determination of a solution of an equation of the type $f(x) = 0$ for a differentiable function f. Given is a (differentiable) function $f : I \to \mathbb{R}$, the search is for a x^* with $f(x^*) = 0$.

To approximate x^*, we choose a $x_0 \in I$ which is close to the searched point x^* and determine the intersection x_1 of the tangent $y = f(x_0) + f'(x_0)(x - x_0)$ to the point $(x_0, f(x_0))$ of the graph of f with the x-axis:

$$0 = f(x_0) + f'(x_0)(x_1 - x_0) \Rightarrow x_1 = x_0 - \frac{f(x_0)}{f'(x_0)}.$$

Often x_1 is a better approximation to the searched point x^* as x_0. Note Fig. 28.1.

Now one carries out this construction with x_1 instead of x_0, i.e., one forms

$$x_2 = x_1 - \frac{f(x_1)}{f'(x_1)}.$$

This **Newton method**, that is, the generation of the recursive sequence (x_n) with

$$x_0 \in I \text{ and } x_{n+1} = x_n - \frac{f(x_n)}{f'(x_n)} \text{ for } n \in \mathbb{N}_0$$

C. Karpfinger, *Calculus and Linear Algebra in Recipes*,
https://doi.org/10.1007/978-3-662-65458-3_28

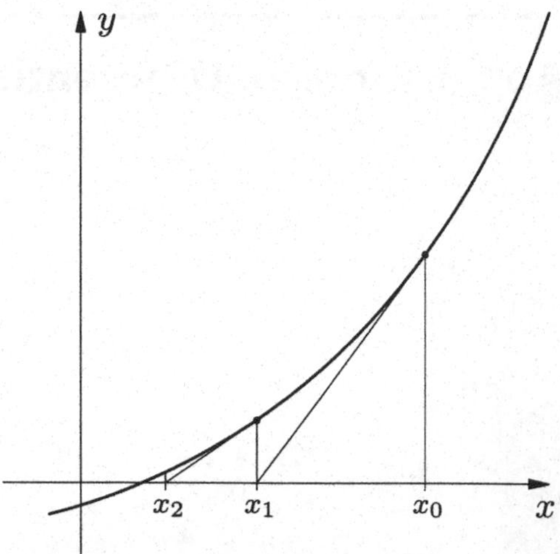

Fig. 28.1 Calculate $x_1, x_2, \ldots$

need not converge, i.e. that the sequence (x_n) of the iterates not necessarily converges to a zero point x^* of f (see below). However, if the method converges, the convergence is often *quadratic*, i.e., the number of correct digits doubles with each iteration. So if you start with one correct digit, after three iterations you have eight correct digits in favorable cases.

There are two reasons to abort the iteration:

- Termination, if a searched zero x^* is sufficiently well approximated and
- Termination if no convergence is expected.

The first reason to abort provides the abort criterion: STOP if

$$|x_n - x^*| < \text{tol for a given tol } > 0 .$$

Since in practice x^* isn't known, one stops the iteration if two successive iterations do not differ more than tol, i.e. if

$$|x_{n+1} - x_n| < \text{tol for a given tol } > 0 .$$

Indeed, one can reason that for *large n* we have

$$|x_n - x^*| \leq |x_{n+1} - x_n| .$$

The second reason to terminate is given by the termination criterion: STOP, if

$$|f(x_{n+1})| > |f(x_n)| .$$

We describe the procedure in a recipe-like way:

Recipe: The (One-Dimensional) Newton Method
Given is a twice continuously differentiable function $f : I \to \mathbb{R}$. For the approximate determination of a zero $x^* \in I$ proceed after choice of a tolerance limit tol > 0 as follows:

(1) Choose a $x_0 \in I$ in the neighborhood of x^*.
(2) As long as $|x_{n+1} - x_n| \geq$ tol and $|f(x_{n+1})| \leq |f(x_n)|$ determine

$$x_{n+1} = x_n - \frac{f(x_n)}{f'(x_n)}.$$

If

$$f(x^*) = 0 \text{ and } f'(x^*) \neq 0,$$

then there also exists a neighborhood U of x^* so that the iteration

$$x_0 \in U \text{ and } x_{k+1} = x_k - \frac{f(x_k)}{f'(x_k)}, \quad k = 0, 1, 2, \ldots$$

converges for every x_0 of U to the zero x^*. The convergence is **quadratic**, i.e.

$$x_{n+1} - x^* = C\,(x_n - x^*)^2 \text{ for a } C \in \mathbb{R}.$$

The Newton method is easy to program, note Exercise 28.5. In the following example we have used this program.

Example 28.1 We determine approximately the two zeros of the function

$$f : \mathbb{R} \to \mathbb{R}, \; f(x) = \mathrm{e}^{x^2} - 4\,x^2 \sin(x).$$

As the graph of this function shows, see Fig. 28.2, the two zeros lie in the neighborhood of $x_0 = 1$ and $x_0 = 1.5$. Therefore, we choose these two numbers as initial values and obtain the following iterates, underlining the correct digits in each case:

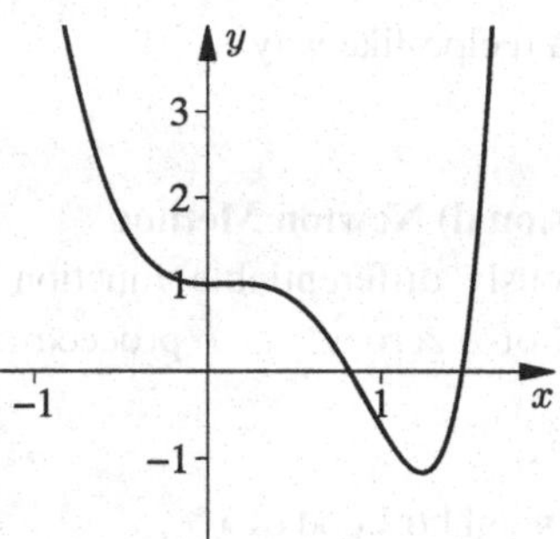

Fig. 28.2 Determine the zeros of f

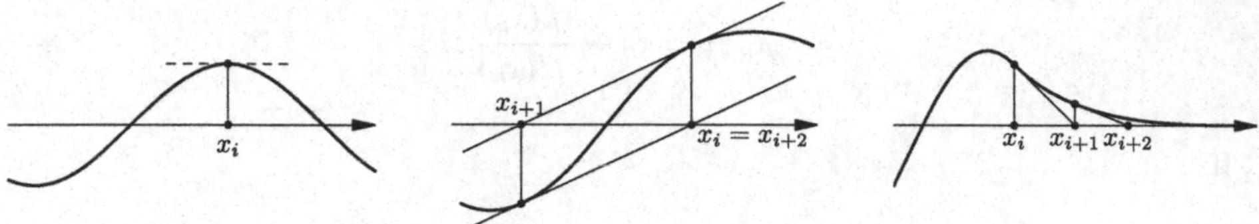

Fig. 28.3 In these situations the Newton method does not converge

n	x_n	x_n
0	1.000000000000000	1.500000000000000
1	0.812637544357997	1.467819084821214
2	0.817801747254039	1.463803347440465
3	0.817786886188853	1.463745290635512
4	0.817786886068805	1.463745278642304

■

Figure 28.3 shows three situations in which Newton's method does not converge.

MATLAB provides with `fzero` a function for the numerical determination of an approximate solution x^* of an equation $f(x) = 0$. Here a starting value x_0 can be specified, in the neighborhood of which a zero point x^* of f is assumed. Alternatively, an interval $[a, b]$ can be specified, in which a zero x^* is searched for, e.g.

```
» fzero('x.^3-2*x-5',2) ans = 2.094551481542327
```

or

```
» fzero('exp(2*x)-sin(x)-x.^2',[-2,0]) ans = -0.986474879875717
```

28.2 Taylor Expansion

Let $f : I \to \mathbb{R}$ be a m-times differentiable function and $a \in I$. We want to approximate this function by a polynomial. To do this, we consider the following polynomial obtained from f by successive differentiation:

$$T_{m,f,a}(x) = f(a) + f'(a)(x-a) + \frac{f''(a)}{2!}(x-a)^2 + \cdots + \frac{f^{(m)}(a)}{m!}(x-a)^m$$
$$= \sum_{k=0}^{m} \frac{f^{(k)}(a)}{k!}(x-a)^k,$$

where $f^{(0)} = f$. We then have:

$$f(a) = T_{m,f,a}(a),\ f'(a) = T'_{m,f,a}(a),\ f''(a) = T''_{m,f,a}(a), \ldots, f^{(m)}(a) = T^{(m)}_{m,f,a}(a).$$

Thus the two functions f and $T_{m,f,a}$ have many similarities: In the point a the function values as well as the values of the first m derivatives of f and $T_{m,f,a}$ are equal.

Figure 28.4 shows the graph of the function $f : [-\pi, \pi] \to \mathbb{R}$, $f(x) = x\cos(x)$ (bold line) and the polynomials (thin lines):

$$T_{3,f,0}(x) = x - \frac{x^3}{2},\quad T_{5,f,0}(x) = x - \frac{x^3}{2} + \frac{x^5}{24},\quad T_{7,f,0}(x) = x - \frac{x^3}{2} + \frac{x^5}{24} - \frac{x^7}{720}.$$

The graphs of the polynomials coincide more and more with increasing degree near the development point $a = 0$ with the graph of the function f. This behavior does not come by chance:

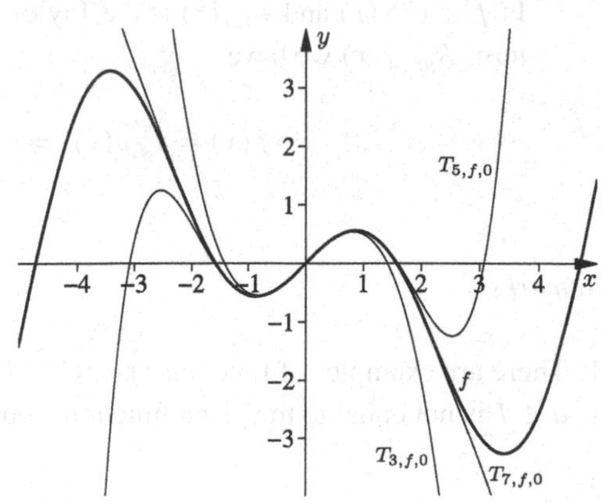

Fig. 28.4 The function f and some Taylor polynomials

Taylor Polynomial, Taylor Series
Let $I \subseteq \mathbb{R}$ and $a \in I$.

- If $f : I \to \mathbb{R}$ is a m-times differentiable function, the polynomial

$$T_{m,f,a}(x) = \sum_{k=0}^{m} \frac{f^{(k)}(a)}{k!}(x-a)^k$$

is called the **m-th Taylor polynomial to f** in the **development point** a with the **remainder**

$$R_{m+1}(x) = f(x) - T_{m,f,a}(x) .$$

- Are $f \in C^{m+1}(I)$ and $T_{m,f,a}(x)$ the m-th Taylor polynomial of f in a, then the remainder term $R_{m+1}(x)$ has the two representations

$$R_{m+1}(x) = \frac{f^{(m+1)}(\xi)}{(m+1)!}(x-a)^{m+1} = \frac{1}{m!}\int_a^x (x-t)^m f^{(m+1)}(t)\,\mathrm{d}t$$

with a ξ between a and x.
- Is $f \in C^\infty(I)$, then one calls

$$T_{f,a}(x) = \sum_{k=0}^{\infty} \frac{f^{(k)}(a)}{k!}(x-a)^k$$

the **Taylor series of f in** a.
- Is $f \in C^\infty(I)$ and $T_{f,a}(x)$ is the Taylor series of f in a, then with the remainder term $R_{m+1}(x)$ we have

$$f(x) = T_{f,a}(x) \Leftrightarrow \lim_{m\to\infty} R_{m+1}(x) = 0 .$$

Remarks

1. There are examples of functions $f \in C^\infty(I)$ so that the Taylor series $T_{f,a}(x)$ from f in $a \in I$ is not equal to the given function f on I. For example, the function

$$f : \mathbb{R} \to \mathbb{R}, \ f(x) = \begin{cases} \mathrm{e}^{-1/x^2}, & x \neq 0 \\ 0, & x = 0 \end{cases}$$

has for the development point $a = 0$ the zero series as Taylor series (this is not quite easy to see). However, such functions are rather exceptional.

2. The formula $R_{m+1}(x) = \frac{f^{(m+1)}(\xi)}{(m+1)!}(x-a)^{m+1}$ suggests that the error is small if the following conditions are satisfied:

$$\left.\begin{array}{l} \bullet\ m \text{ big} \\ \bullet\ x \text{ is close } a \\ \bullet\ f^{(m+1)} \text{ is bounded} \end{array}\right\} \Rightarrow R_{m+1}(x) \approx 0\,.$$

In essence, this is true. One can estimate the residual $R_{m+1}(x)$, if one sets a bound C for $f^{(m+1)}$ on I, i.e., if one knows a C with $|f^{(m+1)}(x)| \le C\ \forall x \in I$.

Example 28.2

- Let $f : \mathbb{R} \to \mathbb{R},\ f(x) = 1 + 3x + 2x^2$ and $a = 1$. The first three Taylor polynomials have the form

$$T_{0,f,1}(x) = 6\,, \quad T_{1,f,1}(x) = 6 + 7(x-1)\,, \quad T_{2,f,1} = 6 + 7(x-1) + 2(x-1)^2\,.$$

- We now consider the exponential function $\exp : \mathbb{R} \to \mathbb{R},\ \exp(x) = \mathrm{e}^x$ with $a = 0$. The first three Taylor polynomials are as follows:

$$T_{0,\exp,0}(x) = 1\,, \quad T_{1,\exp,0}(x) = 1 + x\,, \quad T_{2,\exp,0}(x) = 1 + x + \frac{1}{2!}x^2\,.$$

Since the m-th derivative of $\exp(x)$ is again $\exp(x)$ is, we can generally state the m-th Taylor polynomial in 0, it reads:

$$T_{m,\exp,0}(x) = 1 + x + \frac{x^2}{2} + \cdots + \frac{x^m}{m!} = \sum_{k=0}^{m} \frac{x^k}{k!}\,.$$

and the Taylor series of exp in $a = 0$ is the well-known power series representation of the exponential function

$$\exp(x) = \sum_{k=0}^{\infty} \frac{x^k}{k!}\,.$$

- For sine and cosine, we obtain the Taylor series in $a = 0$:

$$T_{\sin,0}(x) = x - \frac{x^3}{3!} + \frac{x^5}{5!} - \frac{x^7}{7!} + - \cdots$$

and

$$T_{\cos,0}(x) = 1 - \frac{x^2}{2!} + \frac{x^4}{4!} - \frac{x^6}{6!} + - \cdots$$

- If f is a polynomial of degree m, then $f^{(m+1)} = 0$. The error R_{m+1} is then 0, in particular $T_{f,a} = f$.
- For the exponential function $\exp : \mathbb{R} \to \mathbb{R}$ we have:

$$\exp(x) = \sum_{k=0}^{m} \frac{x^k}{k!} + \underbrace{\frac{e^\xi}{(m+1)!} x^{m+1}}_{R_{m+1}(x)}.$$

Here ξ is between 0 and x. For example, for $x = 1$

$$\left| e - \sum_{k=0}^{m} \frac{1}{k!} \right| \leq \frac{3}{(m+1)!},$$

in particular $|\, e - \sum_{k=0}^{10} \frac{1}{k!} | < 7.5 \cdot 10^{-8}$.
- If f is given by a power series, i.e. $f(x) = \sum_{n=0}^{\infty} a_n (x-a)^n$ then this is the Taylor series for f at the development point a with $f(a) = a_0$, $f'(a) = a_1$, $f''(a) = 2\,a_2$, $f'''(a) = 3!\,a_3, \ldots$, so $T_{f,a} = f$. ■

MATLAB provides with `taylortool` a tool, which outputs not only the Taylor polynomial of any degree but also the graph of the Taylor polynomial under consideration.

28.3 Remainder Estimates

According to the box in Sect. 28.2 we have the following equation for the Taylor polynomial $T_{m,f,a}(x)$ with the development point a for a sufficiently often differentiable function f:

$$f(x) - T_{m,f,a}(x) = \frac{f^{(m+1)}(\xi)}{(m+1)!}(x-a)^{m+1} = R_{m+1}(x),$$

with ξ between a and x. Here we have chosen the *Lagrangian representation* of the remainder term.

Typical questions in this context are:

- What is the worst case error when the function f is given by the Taylor polynomial $T_{m,f,a}$ in a neighborhood U of a or in a point x_0?

Here we have to estimate $|R_{m+1}(x)|$ for a given m and U or x_0.

- How large must we choose m so that the function f and the Taylor polynomial $T_{m,f,a}$ differ by at most a error ε in an environment U of a or in a point x_0?

  Here m is to be determined in such a way that $|R_{m+1}(x)| \leq \varepsilon$ for all $x \in U$ or for $x = x_0$ for a given ε and U or x_0.

We consider examples to these questions:

Example 28.3

- We consider the following function f, the development point a and the environment U of a:

$$f(x) = \frac{1}{\sqrt[3]{(x+1)^2}}, \quad a = 0, \quad U = \left[-\frac{1}{2}, \frac{1}{2}\right].$$

We determine what is the worst-case error of f by substituting f by $T_{3,f,a}$ on U. To do this, we first determine the fourth derivative of f:

$$f'(x) = -\frac{2}{3}(x+1)^{-\frac{5}{3}}, \quad f''(x) = \frac{10}{9}(x+1)^{-\frac{8}{3}},$$
$$f'''(x) = -\frac{80}{27}(x+1)^{-\frac{11}{3}}, \quad f''''(x) = \frac{880}{81}(x+1)^{-\frac{14}{3}}.$$

It is not necessary, but we take the opportunity to determine the third Taylor polynomial, which, because of

$$f(0) = 1, \quad f'(0) = -\frac{2}{3}, \quad f''(0) = \frac{10}{9}, \quad f'''(-1) = -\frac{80}{27}$$

is as follows:

$$T_{3,f,0}(x) = 1 - \frac{2}{3}x + \frac{5}{9}x^2 - \frac{40}{81}x^3.$$

For the remainder term we now obtain:

$$R_4(x) = \frac{f''''(\xi)}{4!}x^4 = \frac{880}{81 \cdot 4!}(\xi+1)^{-\frac{14}{3}}x^4$$

with a ξ between x and $a = 0$, i.e. $\xi \in [-\frac{1}{2}, \frac{1}{2}]$. Because of

$$(\xi+1)^{-\frac{14}{3}} \leq \left(\frac{1}{2}\right)^{-\frac{14}{3}} \quad \text{and} \quad |x| \leq \frac{1}{2}$$

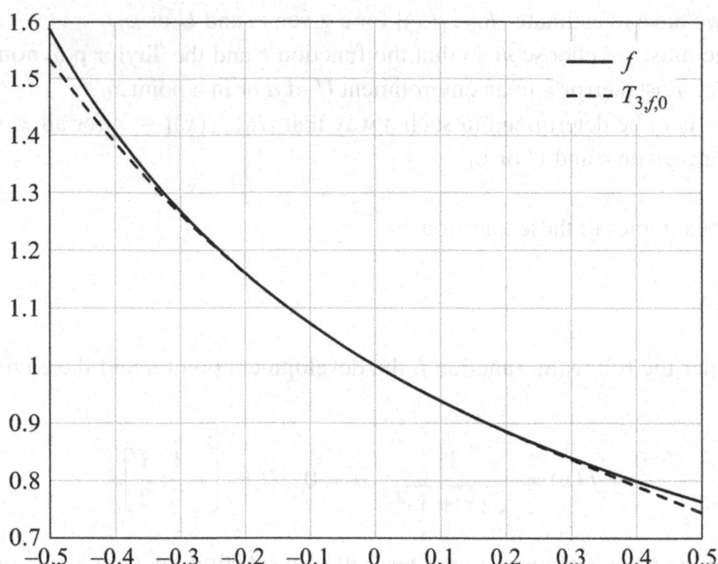

Fig. 28.5 The function f adjacent to its third Taylor polynomial on the interval under consideration. U

for all $\xi,\ x \in]-\frac{1}{2}, \frac{1}{2}[$ we now have:

$$|R_4(x)| = \left|\frac{880}{81 \cdot 4!}(\xi+1)^{-\frac{14}{3}} x^4\right| \le \frac{880}{81 \cdot 4!}\left(\frac{1}{2}\right)^{-\frac{14}{3}}\left(\frac{1}{2}\right)^4 \approx 0.7186\,.$$

Note Fig. 28.5.

- We consider the following function f, the development point a and the environment U of a:

$$f(x) = \ln\frac{(3+x)^2}{4}\,,\ a = -1\,,\ U = \left[-\frac{3}{2}, -\frac{1}{2}\right].$$

We determine $m \in \mathbb{N}$ so that $|R_{m+1}(x)| \le 0.1$. To do this, we first determine the n-th derivative of f. Because

$$f'(x) = \frac{2}{(3+x)},\ f''(x) = \frac{-2}{(3+x)^2},\ f'''(x) = \frac{4}{(3+x)^3},\ f''''(x) = \frac{-12}{(3+x)^4}$$

we assume for all $n \in \mathbb{N}$:

$$f^{(n)}(x) = (-1)^{n+1}\frac{2(n-1)!}{(3+x)^n}\,.$$

This can easily be proved by induction to n. While it is not necessary, we take the opportunity to determine the (arbitrarily chosen) third Taylor polynomial. Because of

$$f(-1) = 0\,,\ f'(-1) = 1\,,\ f''(-1) = -\frac{1}{2}\,, f'''(-1) = \frac{1}{2}$$

it reads as follows:

$$T_{3,f,-1}(x) = (x+1) - \frac{1}{4}(x+1)^2 + \frac{1}{12}(x+1)^3\,.$$

For the remainder term (but now, of course, for a general $m \in \mathbb{N}$) we now obtain:

$$R_{m+1}(x) = \frac{f^{(m+1)}(\xi)}{(m+1)!}(x+1)^{m+1} = \frac{(-1)^{m+2}\frac{2m!}{(3+\xi)^{m+1}}}{(m+1)!}(x+1)^{m+1}$$

with a ξ between x and $a = -1$, i.e. $\xi \in \left[-\frac{3}{2}, -\frac{1}{2}\right]$. Because

$$\frac{1}{(3+\xi)^{m+1}} \leq \frac{1}{(3/2)^{m+1}} \text{ and } |x+1| \leq \left(\frac{1}{2}\right)^{m+1}$$

for all $\xi,\ x \in \left[-\frac{3}{2}, -\frac{1}{2}\right]$ we now have:

$$|R_{m+1}(x)| = \left|\frac{(-1)^{m+2}\frac{2m!}{(3+\xi)^{m+1}}}{(m+1)!}(x+1)^{m+1}\right| = \frac{2}{(3+\xi)^{m+1}(m+1)}|x+1|^{m+1}$$

$$\leq \frac{2}{(3/2)^{m+1}(m+1)}\left(\frac{1}{2}\right)^{m+1} = \frac{2}{3^{m+1}(m+1)} < 0.1\,,$$

Where this last inequality obviously is already for $m = 2$ satisfied. Thus, the second Taylor polynomial approximates $T_{2,f,-1}$ the given function f in the interval $U = \left[-\frac{3}{2}, -\frac{1}{2}\right]$ with an error of at most 0.1. Note Fig. 28.6.

- We take up the last example again, but do not consider a neighborhood U of a, but a point x_0 not far from the development point a:

$$f(x) = \ln\frac{(3+x)^2}{4}\,,\ a = -1\,,\ x_0 = 0\,,$$

and determine an upper limit for the error $f(0) - T_{2,f,-1}(0)$:

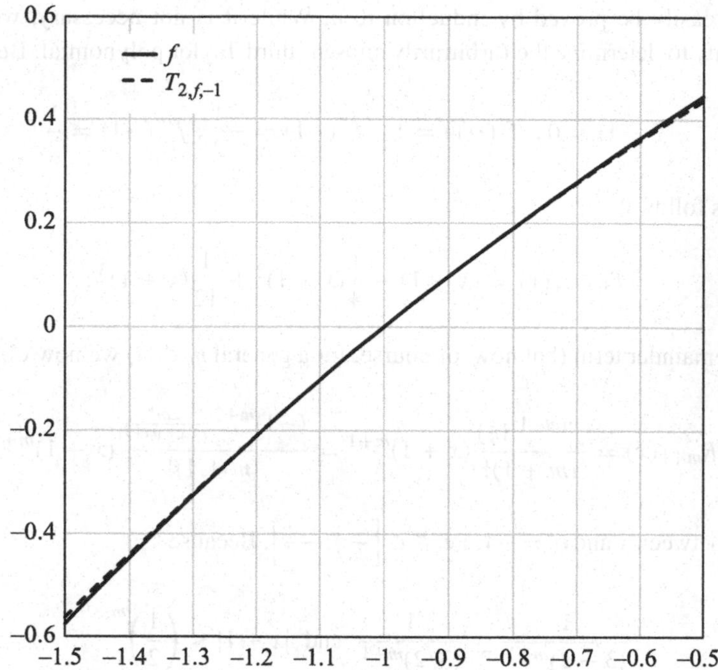

Fig. 28.6 The function f adjacent to its second Taylor polynomial on the interval under consideration. *U*

For the remainder element we get with $m = 2$ (see above):

$$R_3(0) = \frac{2}{(3+\xi)^3 \cdot 3} |0+1|^3 \leq \frac{2}{2^3 \cdot 3} = \frac{1}{12}$$

because of

$$\frac{1}{(3+\xi)^{m+1}} \leq \frac{1}{2^{m+1}}$$

for all $\xi \in [-1, 0]$—note that ξ is between $x_0 = 0$ and $a = -1$. ■

28.4 Determination of Taylor Series

The Taylor series $T_{f,a}(x)$ of a function f in a is a power series with radius of convergence $R \in \mathbb{R}_{\geq 0} \cup \{\infty\}$ (see Sect. 24.1). As already noted, usually $f(x) = T_{f,a}(x)$ for all $x \in (a-R, a+R)$. So we can represent a complicated function f by the simple power series $T_{f,a}(x)$ at least in the interior of the convergence region of the power series $T_{f,a}(x)$. We use the way of speaking for this: The Taylor series represents the function or the Taylor series is a power series representation of the function f and identify the function f with its Taylor series.

To determine the Taylor series of a function, heed the following recipe-like recommendations:

Recipe: Determine the Taylor Series of a Function f

If f is an infinitely often differentiable function on an interval I and $a \in I$, one obtains the Taylor series $T_{f,a}(x)$ from f in a as follows:

- **Using Taylor's formula**: To be calculated is the k-th derivative $f^{(k)}$ at the place a for all $k \in \mathbb{N}_0$. This procedure is often very tedious or not possible at all.
- **Differentiate or integrate known series**: If the function f is the derivative function $f = g'$ or a antiderivative function $f = G$ of a function g whose Taylor series $T_{g,a}(x)$ is known? If yes, then by differentiation or integration of the Taylor series of g one gets the Taylor series of f:
  - $f = g'$ and $g(x) = T_{g,a}(x) \Rightarrow T_{f,a}(x) = T'_{g,a}(x)$
  - $f = G$ and $g(x) = T_{g,a}(x) \Rightarrow T_{f,a}(x) = \int T_{g,a}(x)$.
- **Insert known series**: Is the function f the sum, product, or quotient of functions whose Taylor series are known? If yes, you get the Taylor series of f by substituting these known Taylor series.
- **Equating coefficients**: The (undetermined) approach $f(x) = \sum_{k=0}^{\infty} a_k(x - a)^k$ leads to a coefficient comparison, from which the coefficients a_k can be determined.

We recall some important power series in Sect. 24.4.

Example 28.4

- An example of integrating and differentiating known series:

$$\arctan'(x) = \frac{1}{1+x^2} = \sum_{k=0}^{\infty}(-x^2)^k \Rightarrow \arctan(x) = \sum_{k=0}^{\infty} \frac{(-1)^k}{2k+1} x^{2k+1} .$$

And:

$$\frac{1}{1-x} = \sum_{k=0}^{\infty} x^k \Rightarrow \left(\frac{1}{1-x}\right)' = \frac{1}{(1-x)^2} = \sum_{k=1}^{\infty} k x^{k-1} .$$

- An example of substituting known series:

$$\cosh(x) = \frac{1}{2}\left(e^x + e^{-x}\right) \Rightarrow \cosh(x) = \frac{1}{2}\left(\sum_{k=0}^{\infty}\frac{x^k}{k!} + \sum_{k=0}^{\infty}\frac{(-1)^k x^k}{k!}\right) = \sum_{k=0}^{\infty}\frac{x^{2k}}{(2k)!}$$

$$\sinh(x) = \frac{1}{2}\left(e^x - e^{-x}\right) \Rightarrow \sinh(x) = \frac{1}{2}\left(\sum_{k=0}^{\infty}\frac{x^k}{k!} - \sum_{k=0}^{\infty}\frac{(-1)^k x^k}{k!}\right) = \sum_{k=0}^{\infty}\frac{x^{2k+1}}{(2k+1)!}.$$

- An example of equating coefficients:

$$\begin{aligned}\frac{x}{1-x} = \sum_{k=0}^{\infty} a_k x^k \Rightarrow x = \left(\sum_{k=0}^{\infty} a_k x^k\right)(1-x) &= \sum_{k=0}^{\infty} a_k x^k - \sum_{k=0}^{\infty} a_k x^{k+1} \\ &= (a_0 + a_1 x + a_2 x^2 + \ldots) - (a_0 x + a_1 x^2 + \ldots) \\ &= a_0 + \sum_{k=1}^{\infty}(a_k - a_{k-1})x^k .\end{aligned}$$

It follows $a_0 = 0$, $a_1 - a_0 = 1$, so $a_1 = 1$, $a_2 - a_1 = 0$, so $a_2 = 1$, etc. Thus we get:

$$\frac{x}{1-x} = \sum_{k=1}^{\infty} x^k .$$

■

These methods of obtaining a power series representation of a function work the same way for complex functions, note the following example:

Example 28.5 We determine a power series representation for $f(z) = \frac{1+z^2}{1-z}$ with $|z| < 1$ around the development point $a = 0$, so

$$f(z) = \frac{1+z^2}{1-z} = \sum_{k=0}^{\infty} a_k z^k .$$

Because of $\frac{1}{1-z} = \sum_{k=0}^{\infty} z^k$ we obtain :

$$\sum_{k=0}^{\infty} a_k z^k = (1+z^2)\sum_{k=0}^{\infty} z^k = \sum_{k=0}^{\infty} z^k + \sum_{k=0}^{\infty} z^{k+2} = \sum_{k=0}^{\infty} z^k + \sum_{k=2}^{\infty} z^k = 1 + z + \sum_{k=2}^{\infty} 2z^k .$$

Thus we obtain

$$f(z) = 1 + z + \sum_{k=2}^{\infty} 2z^k .$$

■

Remark Such power series representations of functions provide interesting representations of important numbers in passing, for example, from $\arctan(x) = \sum_{k=0}^{\infty} \frac{(-1)^k}{2k+1} x^{2k+1}$ by substituting $x = 1$ we get because of $\arctan(1) = \pi/4$:

$$\sum_{k=0}^{\infty} \frac{(-1)^k}{2k+1} = \frac{\pi}{4} .$$

28.5 Exercises

28.1 Let $f(x) = \mathrm{e}^{2x} - 3\pi$. Determine approximately a zero of f using Newton's method. Start with the initial value $x_0 = 1.1$ and use the termination criterion $|f(x_k)| \leq 10^{-5}$.

28.2 Let $f(x) = (x-1)^2 - 4$. Determine approximately a zero of f using Newton's method. One uses the initial values $x_0 = 1.1$ and $x_0 = 0.9$ and the termination criterion $|f(x_k)| \leq 10^{-5}$.

28.3 Determine all the derivatives of the following functions at the point 0 and, using the Taylor formula, give the Taylor polynomials T_n around the development point 0:

(a) $f(x) = \frac{1}{1-x}$,
(b) $g(x) = \frac{1}{(1-x)^2}$.

28.4 Calculate the Taylor series of the following functions at their respective development points a:

(a) $f(x) = 2^x \,, \quad a = 0,$

(b) $f(x) = \begin{cases} \frac{\sin x - x}{x^3}, & \text{if } x \neq 0 \\ -\frac{1}{6}, & \text{if } x = 0 \end{cases} , \quad a = 0,$

(c) $f(x) = \frac{1}{2+3x} \,, \quad a = 2,$

(d) $f(x) = -\frac{3}{(2+3x)^2} \,, \quad a = 2.$

Also determine the radii of convergence $R \geq 0$ and investigate for which points $x \in (a-R, a+R)$ the Taylor series coincides with the respective function.

28.5 Give the Taylor polynomial T_{10} of sin and cos around the development point $x = 0$. Use `taylortool` to get an idea about the approximation.

28.6 Develop the tangent at the point 0 to the fifth order, each using

(a) the Taylor formula,
(b) of the known series expansions of sin and cos,
(c) its inverse function, the $\arctan x = x - \frac{x^3}{3} + \frac{x^5}{5} - \frac{x^7}{7} \pm \cdots$ and equating coefficients.

28.7 Determine for the following functions $f : \mathbb{R} \to \mathbb{R}$ the Taylor polynomials $T_{m,f,a}$ for $m = 0, 1, 2$ for the following functions and consider the graphs of these polynomials. To do this, use `taylortool` from MATLAB.

(a) $f(x) = x^2 - 4x + 4$ with $a = 1$,
(b) $f(x) = \frac{1}{1+x}$ with $a = 0$,
(c) $f(x) = x \sin(x)$ with $a = 0$.

28.8 Program Newton's method.

Polynomial and Spline Interpolation

29

We determine for given interpolation points (x_i, y_i) a polynomial p with $p(x_i) = y_i$. We find this polynomial by evaluating *Lagrange's interpolation formula*. As impressively simple as it is to determine this *interpolation polynomial*, as effective is this tool: We will apply this *polynomial interpolation* several times in later chapters, for instance for the numerical approximation of certain integrals or solutions of initial value problems.

In addition to polynomial interpolation, we also consider *spline interpolation* to given grid points. The goal here is not to specify a closed function that interpolates the grid points, but rather to specify a section-wise defined function whose graph passes through the given grid points as *smoothly* as possible.

29.1 Polynomial Interpolation

Polynomial **interpolation** is the determination of a polynomial whose graph passes through given *interpolation* points; it is called *interpolation* because the discrete interpolation points are connected by a *continuous* function (Fig. 29.1).

It is surprisingly easy to specify an interpolation polynomial to given grid points:

Lagrange's Interpolation Formula
Given are $n+1$ interpolation points

$$(x_0, y_0),\ (x_1, y_1), \ldots, (x_n, y_n) \in \mathbb{R} \times \mathbb{R}.$$

(continued)

C. Karpfinger, *Calculus and Linear Algebra in Recipes*,
https://doi.org/10.1007/978-3-662-65458-3_29

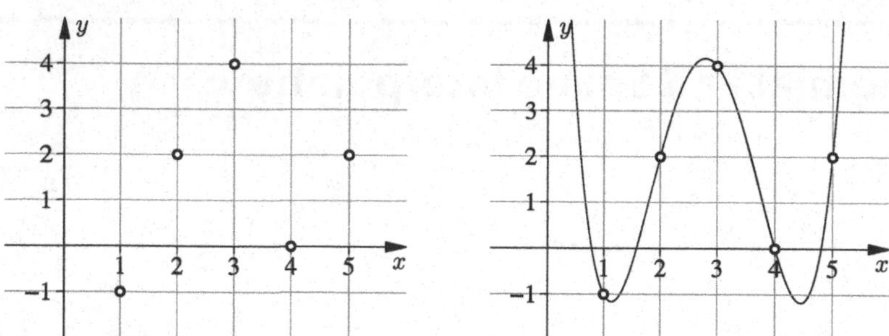

Fig. 29.1 For interpolation points (x_i, y_i) an interpolating polynomial is searched

Then there is exactly one polynomial f of degree $\leq n$ with $f(x_0) = y_0, \ldots, f(x_n) = y_n$. The polynomial f is given by **Lagrange's interpolation formula**

$$f(x) = \sum_{i=0}^{n} y_i \prod_{\substack{j=0 \\ j \neq i}}^{n} \frac{x - x_j}{x_i - x_j},$$

written out is this formula for the cases $n = 1$ and $n = 2$:

$$f(x) = y_0 \frac{(x - x_1)}{(x_0 - x_1)} + y_1 \frac{(x - x_0)}{(x_1 - x_0)} \quad \text{and}$$

$$f(x) = y_0 \frac{(x - x_1)(x - x_2)}{(x_0 - x_1)(x_0 - x_2)} + y_1 \frac{(x - x_0)(x - x_2)}{(x_1 - x_0)(x_1 - x_2)} + y_2 \frac{(x - x_0)(x - x_1)}{(x_2 - x_0)(x_2 - x_1)}.$$

One can easily convince oneself that we have $\deg(f) \leq n$ and f interpolates the grid points, i.e. $f(x_0) = y_0, \ldots, f(x_n) = y_n$ is fulfilled.

Example 29.1 We determine the interpolation polynomial to the interpolation points with the given formula

$$(1, 2), \ (2, 3), \ (3, 6).$$

By inserting the values $x_0 = 1, x_1 = 2, x_2 = 3$ and $y_0 = 2, y_1 = 3, y_2 = 6$ we get

$$f(x) = 2\frac{(x-2)(x-3)}{(1-2)(1-3)} + 3\frac{(x-1)(x-3)}{(2-1)(2-3)} + 6\frac{(x-1)(x-2)}{(3-1)(3-2)} = x^2 - 2x + 3.$$

■

The explicit specification of the polynomial by Lagrangian is for the cases $n \geq 5$ by hand, but numerically stable even for very large n. Another way to specify the uniquely determined coefficients of the interpolation polynomial of degree $\leq n$ for $n+1$ grid points is provided by the following procedure of Newton, which is very clear for calculating by pencil for small n, but is numerically unstable for large n:

Recipe: Determining the Newton Interpolation Polynomial
Given are $n+1$ interpolation points

$$(x_0, y_0)\,,\ (x_1, y_1), \ldots, (x_n, y_n) \in \mathbb{R} \times \mathbb{R}\,.$$

Then one obtains the uniquely determined interpolation polynomial $f(x) = a_n x^n + \cdots + a_1 x + a_0$ as follows:

(1) Make the approach

$$\begin{aligned} f(x) =& \lambda_0 + \lambda_1(x - x_0) + \lambda_2(x - x_0)(x - x_1) + \cdots \\ &+ \lambda_n(x - x_0)(x - x_1) \cdots (x - x_{n-1})\,. \end{aligned}$$

(2) Determine successively $\lambda_0, \lambda_1, \ldots, \lambda_n$ by evaluating f at the locations $x_0, x_1, \ldots, x_n$ taking into account $f(x_i) = y_i$:

$$\begin{aligned} y_0 &= f(x_0) = \lambda_0 \\ y_1 &= f(x_1) = \lambda_0 + \lambda_1(x_1 - x_0) \\ &\ \vdots \\ y_n &= f(x_n) = \lambda_0 + \lambda_1(x_n - x_0) + \cdots + \lambda_n(x_n - x_0) \cdots (x_n - x_{n-1})\,. \end{aligned}$$

(3) Substituting the λ_i from (2) into the approach in (1), after multiplying out the parentheses, we obtain the coefficients $a_n, \ldots, a_1, a_0$.

This interpolation according to Newton has the advantage over the one according to Lagrange that further interpolation points can be added; we show this in the following example.

Example 29.2 We use the given procedure to determine the interpolation polynomial to the grid points

$$(1, 2)\,,\ (2, 3)\,,\ (3, 6)\,.$$

(1) We make the approach $f(x) = \lambda_0 + \lambda_1(x-1) + \lambda_2(x-1)(x-2)$.
(2) We determine the coefficients λ_0, λ_1, λ_2:

$$2 = f(1) = \lambda_0 \Rightarrow \lambda_0 = 2$$

$$3 = f(2) = 2 + \lambda_1 \Rightarrow \lambda_1 = 1$$

$$6 = f(3) = 2 + 1 \cdot 2 + \lambda_2 \cdot 2 \Rightarrow \lambda_2 = 1\,.$$

(3) From (1) and (2) we get:

$$f(x) = 2 + 1\,(x-1) + 1\,(x-1)(x-2) = x^2 - 2x + 3\,.$$

Now we add the further interpolation point $(x_3, y_3) = (4, 5)$ and obtain the corresponding λ_3 from the approach $f(x) = x^2 - 2x + 3 + \lambda_3(x-1)(x-2)(x-3)$ by the requirement

$$5 = f(4) = 11 + \lambda_3(4-1)(4-2)(4-3) \Rightarrow \lambda_3 = -1\,.$$

Thus we obtain

$$f(x) = x^2 - 2x + 3 - (x-1)(x-2)(x-3) = -x^3 + 7x^2 - 13x + 9\,.$$

■

MATLAB MATLAB offers the function `polyfit`, with which the coefficients of the uniquely determined interpolation polynomial can be determined. With the vectors `x=[x_0 x_1 ...x_n]` and `y=[y_0 y_1 ...y_n]` to the interpolation points $(x_0, y_0), \ldots, (x_n, y_n)$ one then obtains the coefficients $a_n, \ldots, a_1, a_0$ in this order as entries of f:

```
» f = polyfit(x,y,n)
```

In fact, the `polyfit` function can be used much more universally; you can find more information about this at `help polyfit`.

For MATLAB there is also the toolbox `chebfun`, which takes up the idea of Lagrange interpolation and approximates functions by a sufficiently large number of interpolation points up to machine precision.

Finally, we note that an equidistant or nearly equidistant distribution of the interpolation points (as we have done so far) at large n leads to a *oscillation of* the interpolation polynomial near the boundaries of the interpolation interval. This oscillation near the boundaries can be seen well in Fig. 29.2. One can easily remedy this phenomenon by choosing the interpolation points appropriately; near the boundaries of the interpolation interval, the interpolation points must be denser.

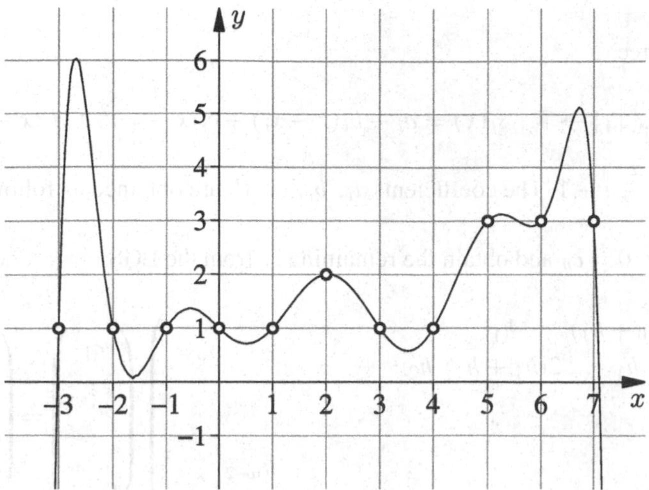

Fig. 29.2 The polynomial oscillates near the boundaries

29.2 Construction of Cubic Splines

We consider $n+1$ grid points

$$(x_0, y_0)\,,\ (x_1, y_1), \ldots, (x_n, y_n) \in \mathbb{R} \times \mathbb{R} \text{ with } x_0 < x_1 < \cdots < x_n$$

with the distances $h_i = x_{i+1} - x_i$ for $i = 0, \ldots, n-1$. The places $x_0, x_1, \ldots, x_n$ are called in this context also **node**. The task of finding a function s with $s(x_i) = y_i$ for all i, we have solved with polynomial interpolation in Sect. 29.1. In the following, however, we make additional demands on the interpolating function s. These additional requirements determine the function s uniquely and make it possible to determine it:

Recipe: Determine the Cubic Spline Function
There is exactly one function s with the properties

- $s(x_i) = y_i$ for all $i = 0, \ldots, n$,
- s is on each subinterval $[x_i, x_{i+1}]$ is a polynomial of at most the third degree,
- s is twice continuously differentiable on $[x_0, x_n]$,
- the following **natural boundary conditions** $s''(x_0) = 0 = s''(x_n)$.

One calls s is the **cubic spline function** to the interpolation points $(x_0, y_0), \ldots, (x_n, y_n)$. This function s is given by n polynomials $s_0, \ldots, s_{n-1}$ at most

(continued)

of third degree,

$$s_i : [x_i, x_{i+1}] \to \mathbb{R}\,,\; s_i(x) = a_i + b_i(x - x_i) + c_i(x - x_i)^2 + d_i(x - x_i)^3\,,$$

for $i = 0, \ldots, n-1$. The coefficients $a_i,\ b_i,\ c_i,\ d_i$ are obtained as follows:

(1) Set $c_0 = 0 = c_n$ and obtain the remaining c_i from the LGS

$$\begin{pmatrix} 2(h_0+h_1) & h_1 & & & \\ h_1 & 2(h_1+h_2) & h_2 & & \\ & \ddots & \ddots & \ddots & \\ & & \ddots & \ddots & h_{n-2} \\ & & & h_{n-2} & 2(h_{n-2}+h_{n-1}) \end{pmatrix} \begin{pmatrix} c_1 \\ \vdots \\ \vdots \\ c_{n-1} \end{pmatrix} = \begin{pmatrix} r_1 \\ \vdots \\ \vdots \\ r_{n-1} \end{pmatrix},$$

where $r_i = 3\left(\frac{y_{i+1}-y_i}{h_i} - \frac{y_i - y_{i-1}}{h_{i-1}}\right)$ for $i = 1, \ldots, n-1$.

(2) Finally set for $i = 0, \ldots, n-1$:

$$a_i = y_i\,,\quad b_i = \frac{y_{i+1}-y_i}{h_i} - \frac{2c_i + c_{i+1}}{3} h_i\,,\quad d_i = \frac{c_{i+1} - c_i}{3h_i}\,.$$

We have the boundary conditions $s''(x_0) = 0 = s''(x_n)$, in other words $c_0 = 0 = c_n$ are chosen. Besides these natural boundary conditions, in practice also the

- **complete boundary conditions** $s'(x_0) = y_0'$ and $s'(x_n) = y_n'$ or the
- **not-a-knot boundary conditions** $s_0'''(x_1) = s_1'''(x_1)$ and $s_{n-2}'''(x_{n-1}) = s_{n-1}'''(x_{n-1})$

play an important role. Choosing other boundary conditions changes the LGS slightly in the above box. But any kind of boundary conditions determines a unique cubic spline s.

Example 29.3 We determine the cubic spline function s to the grid points

$$(x_0, y_0) = (1, 2)\,,\ (x_1, y_1) = (2, 3)\,,\ (x_2, y_2) = (3, 2)\,,\ (x_3, y_3) = (4, 1)\,.$$

Because $h_i = x_{i+1} - x_i = 1$ for all $i = 0,\ 1,\ 2$ the formulas simplify considerably.

(1) We determine the coefficients $c_0, \ldots, c_3$. For this we first determine r_1 and r_2:

$$r_1 = -6\,,\ r_2 = 0\,.$$

Now we get from the LGS

$$\begin{pmatrix} 4 & 1 \\ 1 & 4 \end{pmatrix} \begin{pmatrix} c_1 \\ c_2 \end{pmatrix} = \begin{pmatrix} -6 \\ 0 \end{pmatrix}$$

the values for c_1 and c_2 in addition to the already known values for c_0 and c_3:

$$c_0 = 0\,,\ c_1 = -8/5\,,\ c_2 = 2/5\,,\ c_3 = 0\,.$$

(2) The values for the coefficients a_i are given by the numbers y_i:

$$a_0 = 2\,,\ a_1 = 3\,,\ a_2 = 2\,.$$

And finally, using the numbers c_i we obtain the values of the remaining coefficients b_i and d_i:

$$b_0 = 23/15\,,\ b_1 = -1/15\,,\ b_2 = -19/15\,,\ d_0 = -8/15\,,\ d_1 = 2/3\,,\ d_2 = -2/15\,.$$

Thus, we obtain the spline function s by the three polynomials of degree 3, each explained on the intervals given:

$$s_0 : [1,2] \to \mathbb{R}\,,\ s_0(x) = 2 + \frac{23}{15}(x-1) - \frac{8}{15}(x-1)^3\,,$$

$$s_1 : [2,3] \to \mathbb{R}\,,\ s_1(x) = 3 - \frac{1}{15}(x-2) - \frac{8}{5}(x-2)^2 + \frac{2}{3}(x-2)^3\,,$$

$$s_2 : [3,4] \to \mathbb{R}\,,\ s_2(x) = 2 - \frac{19}{15}(x-3) + \frac{2}{5}(x-3)^2 - \frac{2}{15}(x-3)^3\,.$$

In Fig. 29.3, we have the graph of the spline function s, i.e. the graphs of the polynomial functions s_0, s_1 and s_2 are plotted. Note how smoothly the graph interpolates the grid points. ■

MATLAB As the number of knots increases, the calculations quickly become very time-consuming. Of course, it makes sense to leave the construction of cubic splines to MATLAB. MATLAB already has a corresponding function preinstalled, our example above can be solved in MATLAB by simply entering

```
» x = 1:4; y = [2 3 2 1]; » cs = spline(x,[0 y 0]); » xx
= linspace(1,4,101); » plot(x,y,'o',xx,ppval(cs,xx),'-');
```

The data of the spline function are in `cs`; the zeros in `[0 y 0]` are the natural boundary conditions, and in `ppval(cs,xx)` the spline function is applied to the vector

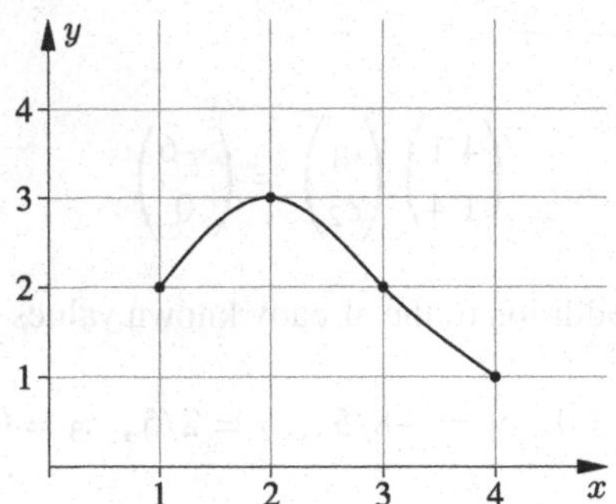

Fig. 29.3 The spline function s

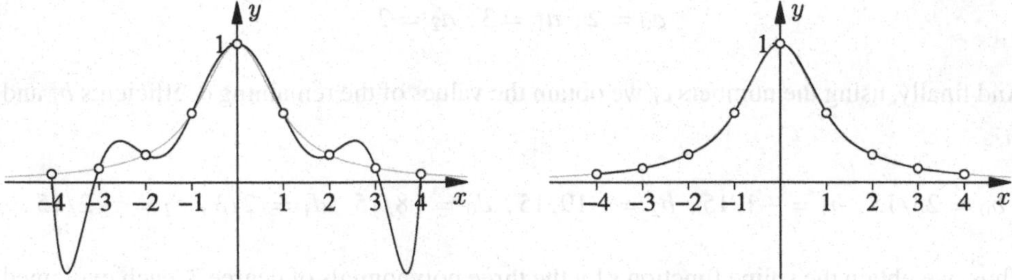

Fig. 29.4 A comparison: polynomial and spline interpolation through the same interpolation points

xx. Even more convenient is the `spline tool` that MATLAB offers: By entering the vectors x and y, you get the plot of the spline function directly and can edit it further.

In Fig. 29.4 we again compare polynomial interpolation and spline interpolation to the nine grid points

$$(k, f(k)) \text{ für } k = -4, \ldots, 4$$

with the function $f(x) = \frac{1}{1+x^2}$. The polynomial in polynomial interpolation has degree 8. Because of the equidistance of the grid points, the interpolating polynomial oscillates strongly near the boundaries. In the case of the spline function, each two adjacent grid points are connected by a polynomial of degree at most 3.

These images are generated with MATLAB as follows:

```
» x = -4:4; y = 1./(1+x.^2); » polyfit(x,y,8) »
plot(x,y,'o',xx,polyval(p,xx)) » xlim([-4.5,4.5]) »
ylim([0,1.05]) » grid on » x=-4:4; y=1./(1+x.^2); » cs =
spline(x,[0 y 0]); » xx = linspace(-4,4,101); »
plot(x,y,'o',xx,ppval(cs,xx),'-'); » xlim([-4.5,4.5]) »
ylim([0,1.05]) » grid on
```

If s is the spline function to the interpolation points $(x_i, f(x_i))$, $i = 0, \ldots, n$, for a function f which is continuously differentiable at least four times, then we have for every $x \in [x_0, x_n]$ the following error estimate:

$$|s(x) - f(x)| \leq M\, K\, \Delta^4\,,$$

where $\Delta = \max\{|x_{i+1} - x_i| \mid i = 0, \ldots, n-1\}$, $K = \Delta / \min_{i=0,\ldots,n-1}\{|x_{i+1} - x_i|\}$ and $M = \max\{f(x) \mid x \in [x_0, x_n]\}$.

29.3 Exercises

29.1 Determine the interpolation polynomial of degree 4 for the 5 interpolation points

$$(-2, 1)\,, (-1, 1)\,,\ (0, 2)\,,\ (1, 1)\,,\ (2, 1)\,.$$

29.2 Determine the cubic spline function s at the grid points

$$(x_0, y_0) = (0, 1)\,,\ (x_1, y_1) = (1, 0)\,,\ (x_2, y_2) = (3, 0)\,,\ (x_3, y_3) = (6, 0)\,.$$

29.3 Write a MATLAB function that functions on vectors x and y with $x = (x_i)$ and $y = y_i$ outputs a plot of the interpolation points (x_i, y_i) and the corresponding interpolation polynomial.

29.4 Write a MATLAB function that outputs the cubic spline function to grid points $(x_0, y_0), \ldots, (x_n, y_n)$ to interpolation points.

29.5 The sketch below shows the graph of a spline function $s = (s_0, s_1, s_2)$ that interpolates the grid points (1, 2), (2, 3), (3, 1), (4, 2). The following three polynomials are the cubic polynomials s_0, s_1, s_2, but not necessarily in that order. Assign these polynomials to the cubic polynomials s_0, s_1, s_2 correctly. Briefly justify your choice.

$$f(x) = 1 - (x-3) + 3\,(x-3)^2 - (x-3)^3,$$
$$g(x) = -x^3 + 3\,x^2 - x + 1,$$
$$h(x) = 2\,x^3 - 15\,x^2 + 35\,x - 23\,.$$

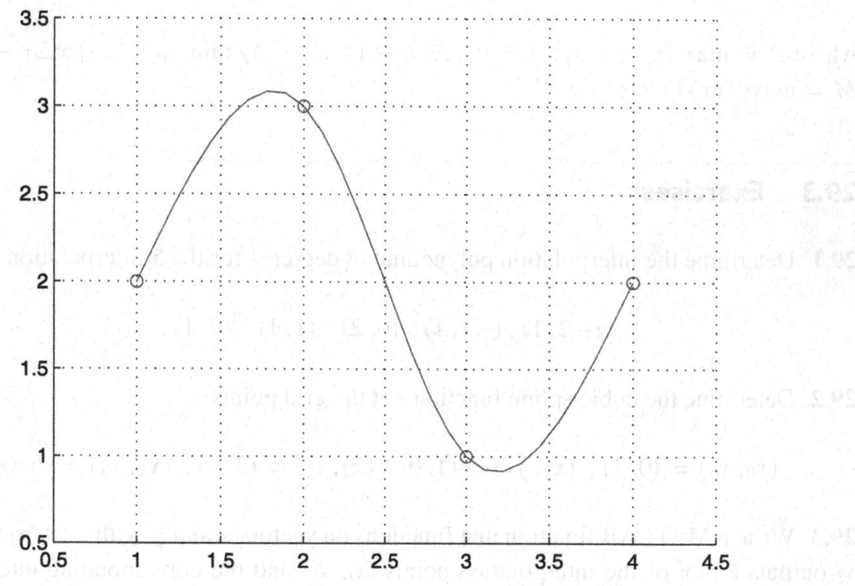

Integration I

30

There are two types of *integration of* a function f: With the *definite integration*, an area is determined that lies between the graph of f and x-axis, with the *indeterminate integration* a *antiderivative* F of f is determined, i.e. a function F with $F' = f$. The connection between these two types is very close and is clarified in the Fundamental Theorem of Calculus.

Along with differential calculus, integral calculus is at the heart of calculus. Just as there are derivation rules, there are also integration rules. We present the most important ones in this chapter. While the derivation is rather easy, the integration often requires tricks to determine an *integral*.

30.1 The Definite Integral

We consider the graph of a function

$$f : [a, b] \to \mathbb{R}$$

and want to find the area A enclosed between the graph and the *x-axis*. Note the accompanying Fig. 30.1.

To do this, we approximate this area A by rectangles, in two ways: We consider a *decomposition* $Z = \{x_0, x_1, x_2, \ldots, x_n\}$ of $[a, b]$ in n subintervals $[x_0, x_1], [x_1, x_2], \ldots, [x_{n-1}, x_n]$, thus

$$a = x_0 < x_1 < x_2 < \cdots < x_n = b\,,$$

C. Karpfinger, *Calculus and Linear Algebra in Recipes*,
https://doi.org/10.1007/978-3-662-65458-3_30

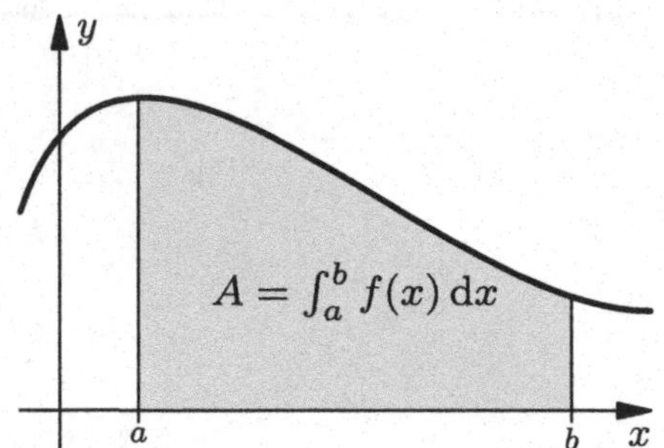

Fig. 30.1 The area we are looking for

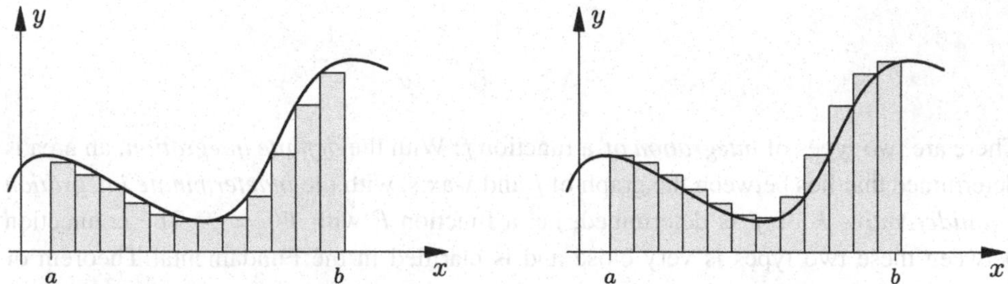

Fig. 30.2 Upper sum and lower sum are approximations for the searched surface area

and determine for each open subinterval (x_{i-1}, x_i) the numbers

$$m_i = \inf\left\{f(x) \mid x \in [x_{i-1}, x_i]\right\} \text{ und } M_i = \sup\left\{f(x) \mid x \in [x_{i-1}, x_i]\right\}.$$

We now calculate the **lower sum** $U_Z(f)$ and the **upper sum** $O_Z(f)$ (see also Fig. 30.2):

$$U_Z(f) = \sum_{i=1}^{n} m_i(x_i - x_{i-1}) \quad \text{and} \quad O_Z(f) = \sum_{i=1}^{n} M_i(x_i - x_{i-1}).$$

The lower sum is the area that is enclosed by the *smaller* rectangles, the upper sum is that enclosed by the *larger* rectangles. Of course, the upper sum is larger than the lower sum, and the sought area A enclosed by the graph is in between:

$$U_Z(f) \leq A \leq O_Z(f).$$

Now we set

$$U(f) = \sup\left\{U_Z(f) \mid Z \text{ is decomposition of } [a, b]\right\}$$

$$O(f) = \inf\left\{O_Z(f) \mid Z \text{ is decomposition of } [a, b]\right\}$$

and arrive at the essential notion: a function is called $f : [a, b] \to \mathbb{R}$ **(Riemann) integrable** if $U(f) = O(f)$. The number $U(f) = O(f) \in \mathbb{R}$ is then called the **definite (Riemann) integral**, one writes:

$$\int_a^b f(x)\,\mathrm{d}x = U(f) = O(f)\,.$$

Example 30.1

- Every constant function $f : [a, b] \to \mathbb{R},\ f(x) = c$ is integrable with the definite integral

$$\int_a^b c\,\mathrm{d}x = c\,(b - a)\,.$$

- Every staircase function $f : [a, b] \to \mathbb{R}$ is integrable.
- The function

$$f : [0, 1] \to \mathbb{R},\ f(x) = \begin{cases} 1\,, & \text{if } x \in \mathbb{Q} \\ 0\,, & \text{if } x \notin \mathbb{Q} \end{cases}$$

is not integrable, because for each decomposition Z contains $[x_{i-1}, x_i]$ points from $\mathbb{Q}$ and from $\mathbb{R} \setminus \mathbb{Q}$. So it is always

$$m_i = \inf\{f(x) \mid x \in [x_{i-1}, x_i]\} = 0 \quad \text{and} \quad M_i = \sup\{f(x) \mid x \in [x_{i-1}, x_i]\} = 1\,.$$

Therefore it follows $U_Z(f) = 0$, $O_Z(f) = 1$ for all decompositions Z and thus

$$U(f) = 0 \neq 1 = O(f)\,.$$

■

For the sake of simplicity, we have so far only considered functions whose values were always positive. But of course all previous and all further observations apply just as well to functions which also take negative values: The area of the approximating *rectangles* below the x-axis is reduced for negative function values because of $f(x) < 0$ for such x is evaluated negatively in the integral.

For the determination of the value

$$\int_a^b f(x)\,\mathrm{d}x$$

fortunately we do not have to use this definition by upper and lower sums. Rather, we will get to know a method to determine this value with the *indefinite integral*. Finally, we will note the essential properties of the definite integral:

Important Properties and Statements about Integrable Functions
For functions $f, g : [a, b] \to \mathbb{R}$ we have:

- If f is continuous or monotone, then f is integrable.
- If f is integrable, so is its absolute value $|f| : [a, b] \to \mathbb{R}$, $|f|(x) = |f(x)|$.
- If f is integrable, then:

$$\left|\int_a^b f(x)\,\mathrm{d}x\right| \leq \int_a^b |f(x)|\,\mathrm{d}x\,.$$

- If f and g are integrable, then also $\lambda f + g$, $\lambda \in \mathbb{R}$, and we have:

$$\int_a^b \big(\lambda f(x) + g(x)\big)\,\mathrm{d}x = \lambda \int_a^b f(x)\,\mathrm{d}x + \int_a^b g(x)\,\mathrm{d}x\,.$$

- If f and g are integrable and $f(x) \leq g(x)$ for all $x \in [a, b]$, then we have:

$$\int_a^b f(x)\,\mathrm{d}x \leq \int_a^b g(x)\,\mathrm{d}x\,.$$

- If f is integrable, then one sets

$$\int_a^b f(x)\,\mathrm{d}x = -\int_b^a f(x)\,\mathrm{d}x\,, \quad \text{so} \quad \int_a^a f(x)\,\mathrm{d}x = 0\,.$$

- If f is integrable, then for each $c \in [a, b]$:

(continued)

$$\int_a^b f(x)\,\mathrm{d}x = \int_a^c f(x)\,\mathrm{d}x + \int_c^b f(x)\,\mathrm{d}x .$$

- If f is integrable, then we have the following for the function

$$\tilde{f} : [a, b] \to \mathbb{R},\ \tilde{f}(x) = \begin{cases} f(x) & , x \neq x_0 \\ \omega & , x = x_0 \end{cases},$$

with $x_0 \in [a, b]$ and $\omega \in \mathbb{R}$:

$$\int_a^b f(x)\,\mathrm{d}x = \int_a^b \tilde{f}(x)\,\mathrm{d}x .$$

- **The Mean Value Theorem:** If $f : [a, b] \to \mathbb{R}$ is continuous, then there exists a $\xi \in [a, b]$ so that

$$\int_a^b f(x)\,\mathrm{d}x = f(\xi)(b - a) .$$

The Mean Value Theorem of the integral calculus will prove very useful; its statement is clear from the accompanying Fig. 30.3: The area between the graph of f and x-axis is the interval length $b - a$ multiplied by $f(\xi)$ for a ξ between a and b.

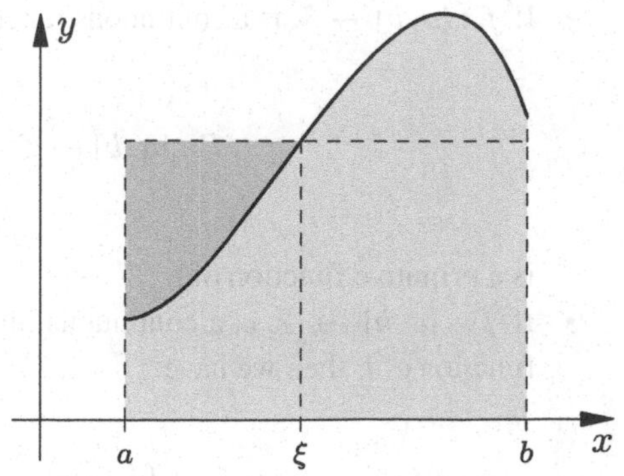

Fig. 30.3 The rectangular area is the integral

30.2 The Indefinite Integral

We consider a function $f : [a, b] \to \mathbb{R}$. A differentiable function $F : [a, b] \to \mathbb{R}$ is called **antiderivative** or **primitive funktion** of f, if $F' = f$, i.e. $F'(x) = f(x)$ for all $x \in [a, b]$. If F is an antiderivative of f, then one can show (see Exercise 30.3), that $\{F + c \mid c \in \mathbb{R}\}$ is the set of all primitive functions of f. This set is called the **indefinite integral** of f, one writes

$$\int f(x)\,\mathrm{d}x = \{F + c \mid c \in \mathbb{R}\}.$$

Often one writes the indefinite integral not as a set, but more carelessly as

$$\int f(x)\,\mathrm{d}x = F + c \quad \text{or even} \quad \int f(x)\,\mathrm{d}x = F.$$

Example 30.2 The function $f : \mathbb{R} \to \mathbb{R}$, $f(x) = 2x$ obviously has the primitive function $F : \mathbb{R} \to \mathbb{R}$, $F(x) = x^2$, the following notations are common for the indefinite integral:

$$\int 2x\,\mathrm{d}x = \{x^2 + c \mid c \in \mathbb{R}\} \quad \text{or} \quad \int 2x\,\mathrm{d}x = x^2 + c \quad \text{or} \quad \int 2x\,\mathrm{d}x = x^2.$$

■

The following Theorem puts together what belongs together, the definite and indefinite integral, the differential and the integral calculus:

The Fundamental Theorem of Calculus

- If $f : [a, b] \to \mathbb{R}$ is a continuous function, then

$$F : [a, b] \to \mathbb{R},\ F(x) = \int_a^x f(t)\,\mathrm{d}t$$

  is a primitive function of f.
- If $f : [a, b] \to \mathbb{R}$ is a continuous function and $F : [a, b] \to \mathbb{R}$ is a primitive function of f, then we have:

$$\int_a^b f(x)\,\mathrm{d}x = F(b) - F(a).$$

We have given the proof of this Theorem as Exercise 30.3. Note that in the first part of this Fundamental Theorem, a primitive function is given for every continuous function. However, this is not given in *closed form*. It is often not even possible to give a closed representation of an (existing) primitive function, for example there is no closed representation of

$$\int \frac{\sin(x)}{x} \, dx \text{ and } \int e^{-x^2} \, dx \, .$$

The second part of the Fundamental Theorem provides the essential method of calculating certain integrals:

Recipe: Calculate the Definite Integral

To determine the integral $\int\limits_a^b f(x)\,dx$ proceed as follows:

(1) Calculate an antiderivative $F : [a, b] \to \mathbb{R}$ of f.

(2) Calculate $F(b) - F(a) = \int\limits_a^b f(x)\,dx$.

Example 30.3 We calculate the definite Riemann integral $\int\limits_1^3 2x\,dx$ according to the above procedure:

(1) $F : [1, 3] \to \mathbb{R},\ F(x) = x^2$ is an antiderivative of $f : [1, 3] \to \mathbb{R},\ f(x) = 2x$.

(2) It is $\int\limits_1^3 2x\,dx = F(3) - F(1) = 9 - 1 = 8$. ■

Now we only need to know how to find an antiderivative for a function f. This is called **to integrate**. In this sense, the function f is often called **integrand** and F the **integral**. From Chap. 26 we already know some integrals F to integrands f:

f	x^n	$\cos(x)$	$\sin(x)$	$\exp(x)$	$\frac{1}{x}$	$\frac{1}{\cos^2(x)}$	$\frac{1}{\sin^2(x)}$
F	$\frac{1}{n+1}x^{n+1}$	$\sin(x)$	$-\cos(x)$	$\exp(x)$	$\ln(\lvert x\rvert)$	$\tan(x)$	$-\cot(x)$

f	0	$\frac{1}{1+x^2}$	$\frac{1}{\sqrt{1-x^2}}$	$-\frac{1}{\sqrt{1-x^2}}$	$-\frac{1}{1+x^2}$
F	c	$\arctan(x)$	$\arcsin(x)$	$\arccos(x)$	$\operatorname{arccot}(x)$

.

For the determination of further integrals the following rules are useful:

Integration Rules

- **Linearity:** For all $\lambda, \mu \in \mathbb{R}$:

$$\int \lambda f(x) + \mu g(x) \mathrm{d}x = \lambda \int f(x) \mathrm{d}x + \mu \int g(x) \mathrm{d}x .$$

- **Integration by parts:**

$$\int u(x) v'(x) \mathrm{d}x = u(x) v(x) - \int u'(x) v(x) \mathrm{d}x .$$

- **Integration by substitution:**

$$\int f\big(\underbrace{g(x)}_{t}\big) \underbrace{g'(x) \mathrm{d}x}_{\mathrm{d}t} = \int f(t) \mathrm{d}t .$$

- **Logarithmic integration:**

$$\int \frac{g'(x)}{g(x)} \mathrm{d}x = \ln\left(|g(x)|\right) .$$

- **Integration of power series:** : The power series function

$$f(x) = \sum_{k=0}^{\infty} a_k (x-a)^k$$

has the antiderivative

$$F(x) = \sum_{k=0}^{\infty} \frac{a_k}{k+1} (x-a)^{k+1} .$$

If f has the radius of convergence R, so also F.

These rules justify the following recipe-like recommendations for integration:

Recipe: Recommendations for Integrating
When determining an antiderivative $F(x) = \int f(x)\mathrm{d}x$ consider the following recommendations:

- Because of linearity, only *normalized* summands are to be integrated.
- If the integrand is $f = uv'$ is a product of two functions u and v'which is not integrable at first go, but so that $u'v$ is integrable, then choose integration by parts.
- If the integrand contains a function in x as a factor that reappears as a derivative, choose the integration by substitution.
- If the integrand is a quotient such that the numerator (up to possibly a multiple) is the derivative of the denominator, choose logarithmic integration.

Example 30.4

- Using integration by parts, we get:

$$\int x\cos(x)\mathrm{d}x = \begin{vmatrix} u=x & u'=1 \\ v'=\cos(x) & v=\sin(x) \end{vmatrix}$$
$$= x\sin(x) - \int \sin(x)\mathrm{d}x = x\sin(x) + \cos(x).$$

- With a trick also $\ln(x)$ can be integrated partially:

$$\int \ln(x)\,\mathrm{d}x = \begin{vmatrix} u=\ln(x) & u'=\frac{1}{x} \\ v'=1 & v=x \end{vmatrix} = x\ln(x) - \int \frac{1}{x}x\,\mathrm{d}x = x\big(\ln(x)-1\big).$$

- For the function $\mathrm{e}^x\sin(x)$ we have to integrate twice:

$$\int \mathrm{e}^x\sin(x)\,\mathrm{d}x = \begin{vmatrix} u=\sin(x) & u'=\cos(x) \\ v'=\mathrm{e}^x & v=\mathrm{e}^x \end{vmatrix} = \mathrm{e}^x\sin(x) - \int \mathrm{e}^x\cos(x)\,\mathrm{d}x$$
$$= \begin{vmatrix} u=\cos(x) & u'=-\sin(x) \\ v'=\mathrm{e}^x & v=\mathrm{e}^x \end{vmatrix}$$
$$= \mathrm{e}^x\sin(x) - \left(\mathrm{e}^x\cos(x) + \int \mathrm{e}^x\sin(x)\,\mathrm{d}x\right)$$
$$= \mathrm{e}^x\big(\sin(x)-\cos(x)\big) - \int \mathrm{e}^x\sin(x)\,\mathrm{d}x\,.$$

By rearranging we now get:

$$2\int e^x \sin(x)\,dx = e^x\left(\sin(x)-\cos(x)\right) \Rightarrow \int e^x \sin(x)\,dx = \frac{e^x}{2}\left(\sin(x)-\cos(x)\right).$$

•

$$\int \cos\left(e^x\right) e^x \; dx = \left| \begin{array}{l} t = e^x \\ dt = e^x \; dx \end{array} \right| = \int \cos(t)\,dt = \sin(t) = \sin\left(e^x\right).$$

•

$$\int \tan(x)\,dx = -\int \frac{1}{\cos(x)}\left(-\sin(x)\right)dx = \left| \begin{array}{l} t = \cos(x) \\ dt = -\sin(x)\,dx \end{array} \right|$$
$$= -\int \frac{1}{t}\,dt = -\ln\left(|t|\right) = -\ln\left(|\cos(x)|\right).$$

•

$$\int \frac{e^x}{(1+e^x)^2}\,dx = \left| \begin{array}{l} t = 1+e^x \\ dt = e^x \; dx \end{array} \right| = \int \frac{1}{t^2}\,dt = -\frac{1}{t} = -\frac{1}{1+e^x}.$$

• Sometimes you have to combine integration by parts and integration by substitution:

$$\int \frac{x}{\sin^2(x)}\,dx = \left| \begin{array}{ll} u = x & u' = 1 \\ v' = \frac{1}{\sin^2(x)} & v = -\frac{\cos(x)}{\sin(x)} \end{array} \right| = -x\frac{\cos(x)}{\sin(x)} + \int \frac{\cos(x)}{\sin(x)}\,dx$$
$$= \left| \begin{array}{l} t = \sin(x) \\ dt = \cos(x)\,dx \end{array} \right| = -x\cot(x) + \int \frac{1}{t}\,dt = -x\cot(x) + \ln\left(|\sin(x)|\right).$$

•

$$\int \frac{3x}{x^2+1}\,dx = \frac{3}{2}\int \frac{2x}{x^2+1}\,dx = \frac{3}{2}\ln(x^2+1).$$

•

$$\int \frac{\cos(x)}{-2+\sin(x)}\,dx = \ln\left(|-2+\sin(x)|\right).$$

•

$$\int \frac{x+2}{x^2+4x+9}\,dx = \frac{1}{2}\int \frac{2x+4}{x^2+4x+9}\,dx = \frac{1}{2}\ln\left(|x^2+4x+9|\right).$$

- We consider the function $f : \mathbb{R}_{>-1} \to \mathbb{R},\ f(x) = \ln(x+1)$. Its derivation is

$$f'(x) = \frac{1}{1+x}.$$

Is $|x| < 1$, this expression can also be represented as a power series, we have:

$$\frac{1}{1+x} = \sum_{k=0}^{\infty}(-x)^k.$$

For $|x| < 1$ therefore we have $f'(x) = \sum_{k=0}^{\infty}(-x)^k$, and integration provides:

$$\begin{aligned} f(x) = \ln(x+1) &= \int f'(x)\,\mathrm{d}x = \int \sum_{k=0}^{\infty}(-x)^k\,\mathrm{d}x = \sum_{k=0}^{\infty}\frac{(-1)^k}{k+1}x^{k+1} \\ &= \sum_{k=1}^{\infty}\frac{(-1)^{k+1}}{k}x^k = x - \frac{x^2}{2} + \frac{x^3}{3} - \frac{x^4}{4} + - \ldots \end{aligned}$$

The constant c in the indefinite integral is 0, since also $\ln(x+1)$ does not contain a constant. This power series has the convergence range $(-1, 1]$, because for $|x| < 1$ it converges according to the above theorem and for the boundery points we have:

$$x = -1: \quad \sum_{k=1}^{\infty}\frac{(-1)^{k+1}}{k}(-1)^k = \sum_{k=1}^{\infty}\frac{(-1)^{2k+1}}{k} = -\sum_{k=1}^{\infty}\frac{1}{k} \quad \text{diverges;}$$

$$x = 1: \quad \sum_{k=1}^{\infty}\frac{(-1)^{k+1}}{k}\cdot 1^k = \sum_{k=1}^{\infty}\frac{(-1)^{k+1}}{k} \quad \text{converges.}$$

In particular, in the case $x = 1$:

$$\sum_{k=1}^{\infty}\frac{(-1)^{k+1}}{k} = \ln(2).$$

We have thus determined the value of the alternating harmonic series.

- The function $f : \mathbb{R} \to \mathbb{R},\ f(x) = \mathrm{e}^{-x^2}$ has no elementary antiderivative. But with the exponential series we have:

$$\mathrm{e}^x = \sum_{k=0}^{\infty}\frac{x^k}{k!} \Rightarrow \mathrm{e}^{-x^2} = \sum_{k=0}^{\infty}\frac{(-1)^k}{k!}x^{2k}.$$

We can now integrate and obtain:

$$F : \mathbb{R} \to \mathbb{R},\ F(x) = \sum_{k=0}^{\infty} \frac{(-1)^k}{(2k+1)\,k!} x^{2k+1}.$$

This is a primitive function of *f*. ■

You can also use the integration by substitution the *other way* round, by not substituting an expression in x by t, but by replacing x by an expression in t:

Example 30.5

- $$\int \frac{1}{\mathrm{e}^x + 1}\,\mathrm{d}x = \left| \begin{array}{l} x = \ln(t) \\ \mathrm{d}x = \frac{1}{t}\,\mathrm{d}t \end{array} \right| = \int \frac{1}{t+1} \cdot \frac{1}{t}\,\mathrm{d}t = \int \frac{1}{t} - \frac{1}{t+1}\,\mathrm{d}t$$
$$= \ln\left(|t|\right) - \ln\left(|t+1|\right) = \ln\left(\left|\frac{t}{t+1}\right|\right) = \ln\left(\frac{\mathrm{e}^x}{\mathrm{e}^x + 1}\right).$$

- $$\int \arcsin(x)\,\mathrm{d}x = \left| \begin{array}{l} x = \sin(t) \\ \mathrm{d}x = \cos(t)\,\mathrm{d}t \end{array} \right| = \int \arcsin\left(\sin(t)\right)\cos(t)\,\mathrm{d}t$$
$$= \int t\cos(t)\,\mathrm{d}t = \left| \begin{array}{ll} u = t & u' = 1 \\ v' \cos(t) & v = \sin(t) \end{array} \right|$$
$$= t\sin(t) - \int \sin(t)\,\mathrm{d}t = t\sin(t) + \cos(t)$$
$$= t\sin(t) + \sqrt{1 - \sin^2(t)} = x\arcsin(x) + \sqrt{1 - x^2}.$$

■

In the second recipe in Sect. 30.2 we proposed to determine a definite integral by determining an antiderivative and then evaluating this antiderivative at the boundaries. In integration by parts or integration by substitution, there is an alternative that avoids the need to determine the antiderivative:

Recipe: Calculate a Definite Integral by Integration by Parts or Integration by Substitution.

The definite integral is obtained as follows:

$$\int_a^b uv' = uv\Big|_a^b - \int_a^b u'v \quad \text{and} \quad \int_a^b f\big(g(x)\big)\, g'(x)\,\mathrm{d}x = \int_{g(a)}^{g(b)} f(t)\,\mathrm{d}t\,.$$

Example 30.6

- $$\int_0^{\mathrm{e}} x\,\mathrm{e}^x\,\mathrm{d}x = \left|\begin{array}{ll} u = x & u' = 1 \\ v' = \mathrm{e}^x & v = \mathrm{e}^x \end{array}\right| = x\,\mathrm{e}^x\Big|_0^{\mathrm{e}} - \int_0^{\mathrm{e}} \mathrm{e}^x\,\mathrm{d}x = \mathrm{e}^{\mathrm{e}+1} - \mathrm{e}^{\mathrm{e}} + 1\,.$$
- $$\int_0^{\ln(2)} \frac{\mathrm{e}^x}{(1+\mathrm{e}^x)^2}\,\mathrm{d}x = \left|\begin{array}{l} t = \mathrm{e}^x \\ \mathrm{d}t = \mathrm{e}^x\,\mathrm{d}x \end{array}\right| = \int_1^2 \frac{1}{(1+t)^2}\,\mathrm{d}t = -\frac{1}{1+t}\Big|_1^2 = -\frac{1}{3} + \frac{1}{2}\,.$$

■

In this chapter, we have presented some integration techniques that can be used to determine an antiderivative for many different integrands. In the following chapter, we will discuss further integration techniques in order to be able to deal with other integrands, such as rational functions.

30.3 Exercises

30.1 Show: If F is a primitive function of f, then $\{F + c \mid c \in \mathbb{R}\}$ is the set of all primitive functions of f.

30.2 Prove the Fundamtental Theorem of Differential and Integral Calculus of Sect. 30.2.

30.3 Calculate the following definite integrals:

(a) $\int_0^{\pi/9} \sin(\pi/3 - 3x)\,dx$, (g) $\int_0^1 r^2\sqrt{1-r}\,dr$, (m) $\int_0^{\pi} \sqrt{1+\sin x}\,dx$,

(b) $\int_0^{\ln 2} e^x\sqrt{e^x - 1}\,dx$, (h) $\int_0^1 \frac{e^x}{(1+e^x)^2}\,dx$, (n) $\int_0^{\pi} \frac{x^2+2\sqrt{x}+\sin x}{\tan(\sqrt{x})}\,dt$,

(c) $\int_0^1 2x(1-e^x)\,dx$, (i) $\int_1^2 \frac{2x^3-3x^2+4x-5}{x}\,dx$, (o) $\int_0^{\pi/2} \frac{\sin(2x)}{\cos x+\cos^2 x}\,dx$,

(d) $\int_e^{e^2} (2x - \ln x)\,dx$, (j) $\int_0^{\pi} 1 - 2\sin^2 x\,dx$, (p) $\int_0^{\pi/2} \frac{\cos^3 x}{1-\sin x}\,dx$,

(e) $\int_{-\sqrt{3}}^{-1} \frac{1}{(1+x^2)\arctan x}\,dx$, (k) $\int_0^1 \frac{2x-2}{1+x^2}\,dx$, (q) $\int_0^1 \frac{1}{1+\sqrt{1+x}}\,dx$.

(f) $\int_{\pi/6}^{\pi/2} \frac{x}{\sin^2 x}\,dx$, (l) $\int_0^1 2^{x+1}\,dx$,

30.4 Calculate the following indefinite integrals:

(a) $\int x\,\sin x\,dx$, (h) $3\int \frac{x^3}{\sqrt{x^2+1}}\,dx$, (o) $\int x^2 e^{-x}\,dx$,

(b) $\int \frac{x}{\cosh^2 x}\,dx$, (i) $\int x^2\cos x\,dx$, (p) $4\int \frac{x\arcsin(x^2)}{\sqrt{1-x^4}}\,dx$,

(c) $\int \frac{\ln(x^2)}{x^2}\,dx$, (j) $\int \frac{\sin^2 x\cos^2 x}{(\cos^3 x+\sin^3 x)^2}\,dx$, (q) $15\int \sqrt{x}(x-1)\,dx$,

(d) $\int \frac{x}{a^2x^2+c^2}\,dx$, (k) $26\int e^{-x}\cos(5x)\,dx$, (r) $\int \frac{\cos(\ln x)}{x}\,dx$,

(e) $\int x\sqrt{1+4x^2}\,dx$, (l) $\int \sin(\ln x)\,dx$, (s) $\int \frac{e^x}{e^x+e^{-x}}\,dx$,

(f) $\int \frac{1-x^2}{1+x^2}\,dx$, (m) $\int x\sin(x^2)\cos^2(x^2)\,dx$, (t) $\int \frac{\ln^2 x}{x}\,dx$.

(g) $4\int (x^2+1)e^{2x}\,dx$, (n) $2\int x\sin(x^2)\,dx$,

30.5 Let $f : \mathbb{R} \to \mathbb{R}$ be a continuous and $g : \mathbb{R} \to \mathbb{R}$ be a differentiable function. The function $F : \mathbb{R} \to \mathbb{R}$ is defined by

$$F(x) = \int_0^{g(x)} f(t)\,dt.$$

Show that F is differentiable on $\mathbb{R}$. What is the derivative $F'(x)$?

30.6 Let $a,\ b > 0$. Calculate the area of the ellipse explained by $\frac{x^2}{a^2} + \frac{y^2}{b^2} = 1$.

Integration II

31

An antiderivative can be determined for every rational function. The procedure is clear, but computationally expensive and thus error-prone. We give a description of this procedure in a recipe. By a standard substitution, integrands which are *rational functions* in sine and cosine functions can always be converted into *true* rational functions. Thus we are able to determine primitive functions also to such integrands.

The applications of integration are essentially the determination of area contents; but with a little interpretive skill we can also determine surfaces and volumes enclosed by *rotating* graphs.

(Definite) integrals often cannot be determined exactly analytically. The remedy here is *numerical integration*; here a definite integral is calculated approximately.

31.1 Integration of Rational Functions

Integrating polynomials is a trivial matter, integrating rational functions is a case for higher mathematics, although it is also *easy*: Indeed, we only need to know what the integrals of a few simple rational functions are. We can reduce all other rational functions to these few simple integrals thanks to polynomial division and partial fraction decomposition. In advance, we give the integrals of these few simple rational functions. That these are in each case integrals of the given rational functions can be easily verified by derivation:

C. Karpfinger, *Calculus and Linear Algebra in Recipes*,
https://doi.org/10.1007/978-3-662-65458-3_31

Integrals of the Basic Rational Functions

We have for all $m \in \mathbb{N}$ and polynomials $x^2 + p\,x + q$ with $p^2 - 4\,q < 0$:

- $\int \frac{dx}{(x-x_k)^m} = \begin{cases} \ln|x - x_k| & \text{for } m = 1 \\ -\frac{1}{(m-1)\,(x-x_k)^{m-1}} & \text{for } m \geq 2 \end{cases}$
- $\int \frac{Bx+C}{x^2+px+q}\,dx = \frac{B}{2}\ln\left(x^2+px+q\right) + \left(C - \frac{Bp}{2}\right)\int \frac{dx}{x^2+px+q}$
- $\int \frac{dx}{x^2+px+q} = \frac{2}{\sqrt{4q-p^2}}\arctan\frac{2x+p}{\sqrt{4q-p^2}}$

For $m \geq 2$ we further have

- $\int \frac{Bx+C}{(x^2+px+q)^m}\,dx = -\frac{B}{2\,(m-1)\,(x^2+px+q)^{m-1}} + \left(C - \frac{Bp}{2}\right)\int \frac{dx}{(x^2+px+q)^{m-1}}$
- $\int \frac{dx}{(x^2+px+q)^m} = \frac{2x+p}{(m-1)\,(4q-p^2)\,(x^2+px+q)^{m-1}} + \frac{2\,(2m-3)}{(m-1)\,(4q-p^2)}\int \frac{dx}{(x^2+px+q)^{m-1}}$

Attention: These formulas are really only valid for the case $p^2 < 4q$.

Now we can integrate any rational function $\frac{A(x)}{Q(x)}$. Proceed as described in the following recipe:

Recipe: Integration of Rational Functions

To determine the integral of a rational function, i.e. an integral of the form

$$\int \frac{A(x)}{Q(x)}\,dx$$

with polynomials $A(x)$ and $Q(x)$, proceed as follows:

(1) If $\deg A \geq \deg Q$ then use polynomial division and get

$$\frac{A(x)}{Q(x)} = P(x) + \frac{B(x)}{Q(x)}$$

with a polynomial $P(x)$ and $\deg B < \deg Q$ (note the box in Sec. 5.1).

(2) Decompose the polynomial Q into indecomposable factors:

$$Q(x) = (x - a_1)^{r_1} \cdots (x - a_n)^{r_n} (x^2 + p_1 x + q_1)^{s_1} \cdots (x^2 + p_m x + q_m)^{s_m}$$

In this case $p_i^2 - 4\,q_i < 0$ for all $i = 1, \ldots, m$ (note the box in Sect. 5.2).

(continued)

(3) Get a partial fraction decomposition of $\frac{B(x)}{Q(x)}$ by:

$$\frac{B(x)}{Q(x)} = \frac{P_1}{(x-a_1)} + \cdots + \frac{P_l}{(x^2+p_m x+q_m)^{s_m}} .$$

In this case $\deg P_i \leq 1$ for all $i = 1, \ldots, l$ (note the box in Sec. 5.4).

(4) Integrate the individual *summands* using the formulas above:

$$\int \frac{A(x)}{Q(x)}\,\mathrm{d}x = \int P(x)\,\mathrm{d}x + \int \frac{P_1}{(x-a_1)}\,\mathrm{d}x + \cdots + \int \frac{P_l}{(x^2+p_m x+q_m)^{s_m}}\,\mathrm{d}x .$$

Example 31.1

- We determine the integral of

$$\frac{A(x)}{Q(x)} = \frac{2x^4+x^3+4x^2+1}{(x-1)(x^2+1)^2} .$$

(1) Because of $\deg A < \deg Q$ we do not need the polynomial division.
(2) The decomposition of the denominator into factors that cannot be further decomposed has already been done.
(3) The partial fraction decomposition is:

$$\frac{2x^4+x^3+4x^2+1}{(x-1)(x^2+1)^2} = \frac{2}{(x-1)} + \frac{1}{(x^2+1)} + \frac{x}{(x^2+1)^2} .$$

(4) Because of

$$\int \frac{x}{(x^2+1)^2}\mathrm{d}x = -\frac{1}{2(x^2+1)}$$

we get

$$\int \frac{2x^4+x^3+4x^2+1}{(x-1)(x^2+1)^2}\mathrm{d}x = \int \frac{2}{(x-1)}\mathrm{d}x + \int \frac{1}{(x^2+1)}\mathrm{d}x + \int \frac{x}{(x^2+1)^2}\mathrm{d}x$$
$$= 2\ln|x-1| + \arctan(x) - \frac{1}{2(x^2+1)} .$$

- We determine the integral of

$$\frac{A(x)}{Q(x)} = \frac{4x^5 + 6x^3 + x + 2}{x^2 + x + 1}.$$

(1) Because of $\deg A \geq \deg Q$ we have to use polynomial division. We have already done this in Example 5.1:

$$\frac{4x^5 + 6x^3 + x + 2}{x^2 + x + 1} = 4x^3 - 4x^2 + 6x - 2 + \frac{-3x + 4}{x^2 + x + 1}.$$

(2) The decomposition of the denominator into factors that cannot be decomposed further has already been done.
(3) Partial fraction decomposition is no longer necessary.
(4) Because of

$$\begin{aligned}\int \frac{-3x + 4}{x^2 + x + 1} \mathrm{d}x &= \frac{-3}{2} \ln(x^2 + x + 1) + \left(4 - \frac{-3}{2}\right) \int \frac{\mathrm{d}x}{x^2 + x + 1} \\ &= \frac{-3}{2} \ln(x^2 + x + 1) + \frac{11}{2} \frac{2}{\sqrt{3}} \arctan\left(\frac{2x + 1}{\sqrt{3}}\right)\end{aligned}$$

we get:

$$\begin{aligned}&\int \frac{4x^5 + 6x^3 + x + 2}{x^2 + x + 1} \mathrm{d}x \\ &= \int 4x^3 - 4x^2 + 6x - 2 \mathrm{d}x + \int \frac{-3x + 4}{x^2 + x + 1} \mathrm{d}x \\ &= x^4 - \frac{4}{3} x^3 + 3x^2 - 2x + \frac{-3}{2} \ln(x^2 + x + 1) + \frac{11}{\sqrt{3}} \arctan\left(\frac{2x + 1}{\sqrt{3}}\right).\end{aligned}$$

■

31.2 Rational Functions in Sine and Cosine

By a **rational function in sine and cosine** we understand a quotient $r(x)$ whose numerator and denominator are *polynomials* in $\sin(x)$ and $\cos(x)$, e.g.

$$r(x) = \frac{1}{\sin(x)} \quad \text{or} \quad r(x) = \frac{\sin^2(x)}{1 + \sin(x)}.$$

We give a method of finding an antiderivative to any such rational function. The following substitution plays the key role:

$$t = \tan(x/2) \text{ implies } \frac{\mathrm{d}t}{\mathrm{d}x} = \frac{1}{2}\left(1 + \tan^2(x/2)\right).$$

Thus we obtain

$$\mathrm{d}x = \frac{2}{t^2+1}\,\mathrm{d}t\,,\ \sin(x) = \frac{2\tan(x/2)}{1+\tan^2(x/2)} = \frac{2t}{t^2+1}\,,$$
$$\cos(x) = \frac{1-\tan^2(x/2)}{1+\tan^2(x/2)} = \frac{1-t^2}{t^2+1}\,.$$

Recipe: Integration of Rational Functions in Sine and Cosine

If $r(x) = \frac{A(x)}{Q(x)}$ is a rational function in sine and cosine functions, then one finds an antiderivative $R(x)$ of $r(x)$ as follows:

(1) Is the numerator a multiple of the derivative of the denominator, i.e. $A(x) = \lambda\, Q'(x)$? If yes, then $R(x) = \lambda \ln(|Q(x)|)$ is an antiderivative of $r(x)$. If no, next step.
(2) Use the substitution $t = \tan(x/2)$ i.e., substitute

$$\mathrm{d}x = \frac{2}{t^2+1}\,\mathrm{d}t\,,\ \sin(x) = \frac{2t}{t^2+1}\,,\ \cos(x) = \frac{1-t^2}{t^2+1}$$

and get

$$\int \frac{A(x)}{Q(x)}\mathrm{d}x = \int \frac{\tilde{A}(t)}{\tilde{Q}(t)}\mathrm{d}t$$

with a rational function $\tilde{r}(t) = \frac{\tilde{A}(t)}{\tilde{Q}(t)}$.
(3) Using the recipe in Sect. 31.1 determine an antiderivative $\tilde{R}(t)$ of the rational function $\tilde{r}(t) = \frac{\tilde{A}(t)}{\tilde{Q}(t)}$.
(4) Back substitution yields an antiderivative $R(x)$ of $r(x)$.

Example 31.2 We determine an antiderivative of $r(x) = \frac{\sin^2(x)}{1+\sin(x)}$:

(1) The numerator is not a multiple of the derivative of the denominator.
(2) We substitute $t = \tan(x/2)$:

$$\int \frac{\sin^2(x)}{1+\sin(x)}\,\mathrm{d}x = \int \frac{\left(\frac{2t}{t^2+1}\right)^2}{1+\frac{2t}{t^2+1}} \frac{2}{1+t^2}\,\mathrm{d}t = \int \frac{8t^2}{(t^2+1)^2(1+t^2+2t)}\,\mathrm{d}t\,.$$

(3) We determine a primitive function $\tilde{R}(t)$ to $\tilde{r}(t) = \frac{8t^2}{(t^2+1)^2(1+t^2+2t)}$ (see exercises):

$$\tilde{R}(t) = 2\int \frac{4t^2}{(1+t^2)^2(1+t)^2}\,\mathrm{d}t = 2\left(-\frac{1}{1+t} - \frac{1}{1+t^2} - \arctan(t)\right).$$

(4) Back substitution yields

$$R(x) = -\frac{2}{1+\tan(x/2)} - \frac{2}{1+\tan^2(x/2)} - x\,.$$

■

We could continue the topic of *integration with special integrands* and present more schemes that can be used to determine antiderivatives of special integrands, but we break off with this topic here. In fact the *manual* integration does not play such a fundamental role in practice as it might seem at first, especially since a computer can also do the integration. In the following we will show how to use MATLAB to calculate definite and indefinite integrals:

MATLAB can be used to calculate both indefinite and definite integrals.

An indefinite integral $\int f(x)\mathrm{d}x$:

`syms x; int(f(x),x)` e.g. » `int(x^2*sin(x),x) ans = 2*x*sin(x) - cos(x)*(x^2 - 2)`

A definite integral $\int_a^b f(x)\mathrm{d}x$:

`syms x; int(f(x),x,a,b)` e.g. » `int(x^2*sin(x),x,0,2) ans = 4*sin(2) - 4*cos(1)^2` » `double(ans) ans = 2.4695`

Note:

- For indefinite integration, MATLAB does not output an integration constant.
- For rational functions, MATLAB may not output a result. In this case, a manually performed partial fraction decomposition can help.

31.3 Numerical Integration

Numerical integration is the approximate calculation of a definite integral $\int_a^b f(x)\mathrm{d}x$. The procedure can be described as follows:

Recipe: Numerical Integration Strategy

Proceed as follows to get an approximate value for $\int_a^b f(x)\mathrm{d}x$:

(1) Divide the interval $[a, b]$ into subintervals $[x_i, x_{i+1}]$, $i = 0, 1, \ldots, n-1$.
(2) Substitute the integrand f on each subinterval $[x_i, x_{i+1}]$ by a simple integrable function p_i for each $i = 0, 1, \ldots, n-1$.
(3) Obtain the approximate value

$$\int_a^b f(x)\mathrm{d}x \approx \sum_{i=0}^{n-1} \int_{x_i}^{x_{i+1}} p_i(x)\mathrm{d}x .$$

We consider only the simplest cases:

- The subintervals are equidistant, i. e. $x_i = a + ih$ with $i = 0, 1, \ldots, n$ and $h = \frac{b-a}{n}$.
- The easy to integrate functions p_i on the subintervals $[x_i, x_{i+1}]$ are polynomials of degree 1 or 2.

Note Fig. 31.1.

And now the best: We don't even need to know what the coefficients of the polynomials are (these can be determined, of course), we can also give the sum in (3) directly in the above recipe:

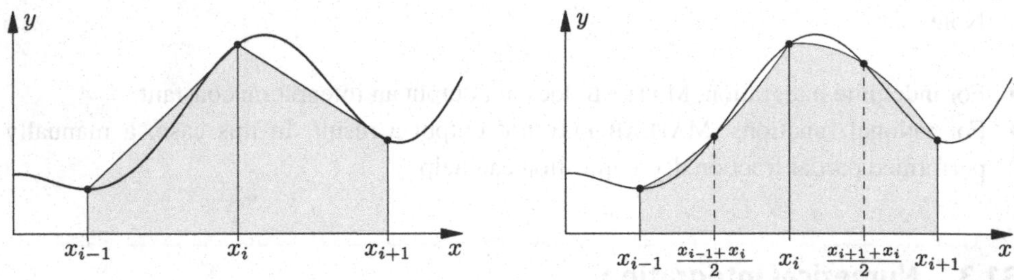

Fig. 31.1 Approximation of the area under the graph by linear or quadratic polynomials

Newton-Cotes Formulas

To calculate the integral, one approximates $\int_a^b f(x)\mathrm{d}x$ the integrand $f(x)$ on each of the equidistant subintervals $[x_i, x_{i+1}]$, $i = 0, \ldots, n-1$ by linear or quadratic polynomials, one obtains with $h = \frac{b-a}{n}$:

- **Trapezoidal rule:** Approximation by linear polynomials, $n \in \mathbb{N}$:

$$T(h) = h\left(\frac{1}{2}f(x_0) + f(x_1) + \cdots + f(x_{n-1}) + \frac{1}{2}f(x_n)\right).$$

- **Simpson's rule:** Approximation by quadratic polynomials, $n \in \mathbb{N}$:

$$S(h) = \frac{h}{6}\left(f(x_0) + 4f\left(\frac{x_0+x_1}{2}\right) + 2f(x_1) + 4f\left(\frac{x_1+x_2}{2}\right) + \cdots \right.$$
$$\left. \cdots + 2f(x_{n-1}) + 4f\left(\frac{x_{n-1}+x_n}{2}\right) + f(x_n)\right).$$

By approximation with polynomials of higher degree one obtains further formulas, we omit the presentation of these formulas. For the effort or for the error estimation note:

- The trapezoidal rule requires $n+1$ function evaluations, the error can be estimated by.

$$\left|\int_a^b f(x)\mathrm{d}x - T(h)\right| \leq \frac{b-a}{12}h^2 \max_{a\leq x\leq b} |f''(x)|.$$

- Simpson's rule requires $2n + 1$ function evaluations, the error can be estimated by.

$$\left| \int_a^b f(x)\mathrm{d}x - S(h) \right| \leq \frac{b-a}{180} h^4 \max_{a \leq x \leq b} |f^{(4)}(x)| .$$

MATLAB It is very easy to program the trapezoidal rule and Simpson's rule in MATLAB, we have formulated this as Exercise 31.3. But in fact these rules are already implemented in MATLAB: The integral $\int_a^b f(x)\mathrm{d}x$ is calculated using the trapezoidal rule by entering `trapz(x,y)`. Here the vectors x and y are `x =[a, a+h ..., b]` and `y =[f(a), f(a+h) ..., f(b)]`.

Simpson's rule is available with `quad('f(x)',a,b)` in an *adaptive* variant, i.e., *step-width control* takes place.

Example 31.3 We calculate the definite integral $\int_0^\pi x \sin(x)\mathrm{d}x$:

- With exact integration we get $\int_0^\pi x \sin(x)\mathrm{d}x = \pi = 3.14159265358979\ldots$.
- With $h = 0.01$ we obtain with `x=0:0.01:pi` and `y=x.*sin(x)`:

$$\texttt{trapz(x, y)} = 3.141562517136044.$$

- With $h = 0.001$ we obtain with `x=0:0.001:pi` and `y=x.*sin(x)`:

$$\texttt{trapz(x, y)} = 3.141591840234830.$$

- Using `quad('x.*sin(x)',0,pi)` we get

```
quad('x.*sin(x)',0,pi) = 3.141592657032484".
```

■

31.4 Volumes and Surfaces of Solids of Revolution

We let the graph of a continuous function $f : [a, b] \to \mathbb{R}$ around x-axis rotate and obtain a **solid of revolution**, see Fig. 31.2.

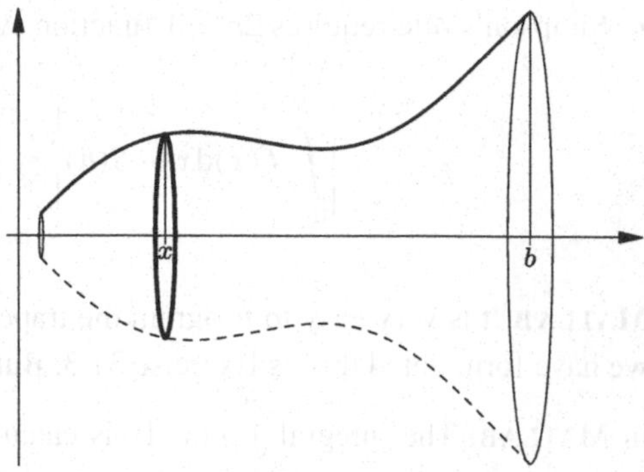

Fig. 31.2 A solid of revolution is obtained by rotating a graph around the x-axis

We obtain the volume and surface area of this solid of revolution as follows:

Volume and Surface Area of a Solid of Revolution
By the continuous function $f : [a, b] \to \mathbb{R}$ we get a solid of revolution by rotating the graph of f around the x-axis. Then

- $V = \pi \int\limits_a^b (f(x))^2 \mathrm{d}x$ is the volume and
- $O = 2\pi \int\limits_a^b f(x)\sqrt{1 + (f'(x))^2} \mathrm{d}x$ is the surface area

of the solid of revolution.

Example 31.4 For any $r > 0$, the graph of $f : [-r, r] \to \mathbb{R}$, $f(x) = \sqrt{r^2 - x^2}$ is a semicircle with radius r. If we let this rotate around the x-axis, we get a sphere of radius r.

We can now calculate the volume of this sphere, it is:

$$V_{\text{sphere}} = \pi \int\limits_{-r}^{r} r^2 - x^2 \mathrm{d}x = \pi \left(r^2 x - \frac{1}{3} x^3 \Big|_{-r}^{r} \right)$$
$$= \pi \left(r^3 - \frac{1}{3} r^3 + r^3 - \frac{1}{3} r^3 \right) = \frac{4}{3} \pi r^3 .$$

And we obtain the surface area of this sphere because of

$$f'(x) = -\frac{x}{\sqrt{r^2 - x^2}}$$

by:

$$O_{\text{sphere}} = 2\pi \int_{-r}^{r} \sqrt{r^2 - x^2}\sqrt{1 + \frac{x^2}{r^2 - x^2}}\,\mathrm{d}x = 2\pi \int_{-r}^{r} \sqrt{r^2 - x^2 + x^2}\,\mathrm{d}x$$

$$= 2\pi \left(rx \Big|_{-r}^{r} \right) = 4\pi r^2 .$$

■

These formulas, after simply renaming the variables, of course also work for solids of revolution, which are formed by rotating graphs around the *y-axis*.

31.5 Exercises

31.1 Determine an antiderivative of

(a) $\frac{1}{\sin x}$,
(b) $\frac{x}{(1+x)(1+x^2)}$,
(c) $\frac{\tan x}{1+\tan x}$,
(d) $\frac{x-4}{x^3+x}$,
(e) $\frac{x^2}{(x+1)(1-x^2)}$,
(f) $\frac{9x}{2x^3+3x+5}$,
(g) $\frac{4x^2}{(x+1)^2(x^2+1)^2}$,
(h) $\frac{\sin^2 x}{1+\sin x}$,
(i) $\frac{x^7-28x^3}{(x^2-8)(x^4+8x^2+16)}$,
(j) $\frac{x^3+1}{x(x-1)^3}$,
(k) $\frac{\sin^2 x \cos^2 x}{(\cos^3 x+\sin^3 x)^2}$,
(l) $\frac{x^2+1}{x^3+1}$,
(m) $\frac{3\,\mathrm{e}^x+4\,\mathrm{e}^{-x}+2}{1-\mathrm{e}^{2x}}$.

31.2 Determine the following integrals:

(a) $\displaystyle\int_1^{\sqrt{3}} \frac{x^4 - 4}{x^2(x^2 + 1)^2}\mathrm{d}x$,

(b) $\int\limits_0^{\pi/3} \frac{1}{\cos x} \mathrm{d}x$.

31.3 Program the trapezoidal rule and Simpson's rule. Test both functions for $I = \int\limits_0^{\pi} \sin(x)\, \mathrm{d}x$ with $n = 5$ and $n = 50$. Can you confirm the error estimate of Sec. 31.3?

31.4 An onion of height h arises as a solid of revolution of the graph of the function

$$f_{a,h} : [0, h] \to \mathbb{R}, \quad x \mapsto f_{a,h}(x) := \frac{ax}{h}\sqrt{\frac{x}{h} - (\frac{x}{h})^2}$$

to find the x-axis. Here a is another parameter, which is half the width b (the maximum of the function f) of the onion.

(a) Plot the graph of the function $f_{1,1}$ to get an idea of the onion.
(b) Determine the half-width b of the onion (as a function of the parameters a, h).
(c) Calculate the volume of the onion as a function of a, h. Show that there is a constant σ such that the onion volume is given by the formula $V = \sigma b^2 h$ is given. It is called σ the *onion number*.
(d) Determine the surface area of the onion for $a = h = 1$ (approximately) using MATLAB.

Improper Integrals

32

We now determine integrals over unbounded intervals or unbounded functions. Such integrals are the basis for *integral transformations* like the Laplace or Fourier transform. The essential tool for the determination of such improper integrals is the notion of limit: We namely determine a fictitious limit d and calculate a definite integral $I = I(d)$ as a function of d and then consider whether, for example, the limit $\lim_{d \to \pm\infty} I(d)$ exists.

32.1 Calculation of Improper Integrals

A **improper integral** is an integral over an unbounded interval or over an unbounded function (see Fig. 32.1).

We summarize all essential terms in this context in one box:

Improper Integrals

- **Unbounded intervals:** For $a, b \in \mathbb{R}$ and $f : [a, \infty) \to \mathbb{R}$ or $f : (-\infty, b] \to \mathbb{R}$ or $f : (-\infty, \infty) \to \mathbb{R}$ (if the respective limits exist):
  - $\displaystyle\int_a^\infty f(x)\,dx = \lim_{b \to \infty} \int_a^b f(x)\,dx,$
  - $\displaystyle\int_{-\infty}^b f(x)\,dx = \lim_{a \to -\infty} \int_a^b f(x)\,dx,$
  - $\displaystyle\int_{-\infty}^\infty f(x)\,dx = \int_{-\infty}^c f(x)\,dx + \int_c^\infty f(x)\,dx$, where $c \in \mathbb{R}$.

(continued)

C. Karpfinger, *Calculus and Linear Algebra in Recipes*,
https://doi.org/10.1007/978-3-662-65458-3_32

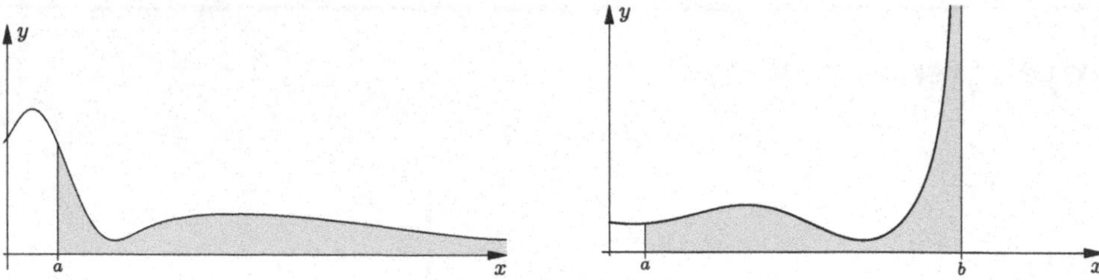

Fig. 32.1 An integral over an unbounded interval or unbounded function

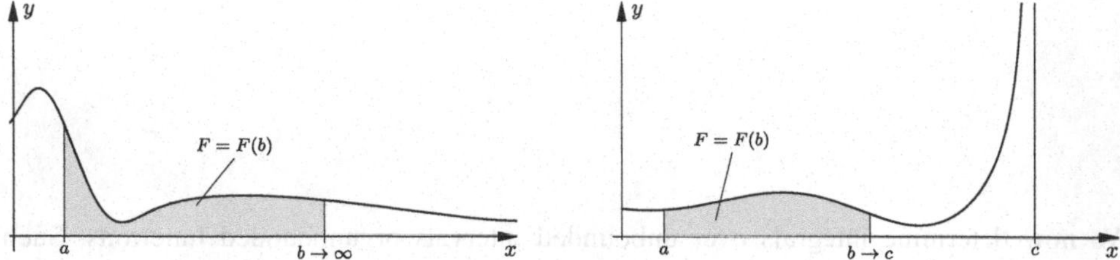

Fig. 32.2 In the case of an improper integral, a *common* integral from a to b is determined, then b goes against ∞ or c

- CHW $\int_{-\infty}^{\infty} f(x)\,\mathrm{d}x = \lim_{a\to\infty} \int_{-a}^{a} f(x)\,\mathrm{d}x$ is called the **Cauchy principal value** from f if this limit exists.

- **Unbounded functions:** For $a, b \in \mathbb{R}$ and $f : [a, c) \to \mathbb{R}$ or $f : (c, b] \to \mathbb{R}$ (if the respective limit values exist):
  - $\int_a^c f(x)\,\mathrm{d}x = \lim_{b\to c^-} \int_a^b f(x)\,\mathrm{d}x$,
  - $\int_c^b f(x)\,\mathrm{d}x = \lim_{b\to c^+} \int_a^b f(x)\,\mathrm{d}x$.

If the respective limit exists and is finite, then the improper integral is said to **exist** or **converge**, otherwise it is called **nonexistent** or **divergent**.

The following sketches in Fig. 32.2 show the determination of two improper integrals:

Example 32.1

•

$$\int_1^{\mathrm{e}} \frac{1}{x\ln(x)}\,\mathrm{d}x = \lim_{a\to 1^+} \int_a^{\mathrm{e}} \frac{1}{x\ln(x)}\,\mathrm{d}x = \left| \begin{matrix} t = \ln(x) & \mathrm{e} \to 1 \\ \mathrm{d}t = \frac{1}{x}\,\mathrm{d}x & a \to \ln(a) \end{matrix} \right|$$

$$= \lim_{a\to 1^+} \int_{\ln(a)}^{1} \frac{1}{t}\,\mathrm{d}t = \lim_{a\to 1^+} \ln(t)\Big|_{\ln(a)}^{1} = \lim_{a\to 1^+} -\ln\big(\ln(a)\big)$$

$$= -\lim_{b\to 0} \ln(b) = \infty\,.$$

The improper integral is therefore divergent or does not exist.

- Now an example where the improper integral exists:

$$\int_1^\infty \frac{\ln(x)}{x^2}\,\mathrm{d}x = \lim_{b\to\infty} \int_1^b \frac{\ln(x)}{x^2}\,\mathrm{d}x = \left| \begin{array}{ll} u = \ln(x) & u' = \frac{1}{x} \\ v' = \frac{1}{x^2} & v = -\frac{1}{x} \end{array} \right|$$

$$= \lim_{b\to\infty} \left(-\frac{\ln(x)}{x}\Big|_1^b + \int_1^b \frac{1}{x^2}\,\mathrm{d}x \right) = \lim_{b\to\infty} -\frac{\ln(b)}{b} - \frac{1}{b} + 1 = 1\,.$$

- The improper integral $\int\limits_{-\infty}^{\infty} \frac{1}{1+x^2}\,\mathrm{d}x$ exists, namely

$$\int_{-\infty}^{\infty} \frac{1}{1+x^2}\,\mathrm{d}x = \int_{-\infty}^{0} \frac{1}{1+x^2}\,\mathrm{d}x + \int_{0}^{\infty} \frac{1}{1+x^2}\,\mathrm{d}x$$

$$= \lim_{a\to-\infty} \int_a^0 \frac{1}{1+x^2}\,\mathrm{d}x + \lim_{b\to\infty} \int_0^b \frac{1}{1+x^2}\,\mathrm{d}x$$

$$= \lim_{a\to-\infty} -\arctan(a) + \lim_{b\to\infty} \arctan(b) = \frac{\pi}{2} + \frac{\pi}{2} = \pi\,.$$

- The improper integral $\int\limits_{-\infty}^{\infty} x\,\mathrm{d}x$ does not exist, because the improper integrals $\int\limits_{-\infty}^{c} x\,\mathrm{d}x$ and $\int\limits_{c}^{\infty} x\,\mathrm{d}x$ do not exist.
- The Cauchy principal value of the function $f(x) = x$ is 0, because we have:

$$\mathrm{CHW} \int_{-\infty}^{\infty} x\,\mathrm{d}x = \lim_{a\to\infty} \int_{-a}^{a} x\mathrm{d}x = \lim_{a\to\infty} \frac{1}{2}x^2\Big|_{-a}^{a} = \lim_{a\to\infty} \left(\frac{a^2}{2} - \frac{a^2}{2}\right) = 0\,.$$

- $$\int_1^{\infty} \frac{1}{x}\,\mathrm{d}x = \lim_{b\to\infty} \int_1^{b} \frac{1}{x}\,\mathrm{d}x = \lim_{b\to\infty} \ln(b) = \infty\,.$$

- $$\int_0^{1} \frac{1}{x}\,\mathrm{d}x = \lim_{a\to 0+} \int_a^{1} \frac{1}{x}\,\mathrm{d}x = \lim_{a\to 0+} -\ln(a) = \infty\,.$$

■

MATLAB MATLAB can also calculate improper integrals, where you set `Inf` or `-Inf` for the limits $\pm\infty$ for unbounded intervals. But also for unbounded functions the calculation with MATLAB is possible, e.g.

```
syms x; int(1/sqrt(x), 0,1) ans = 2  syms x; int(1/x,
0,1) ans = Inf  syms x; int(1/x^2,1,Inf) ans = 1
```

32.2 The Comparison Test for Improper Integrals

We will often be faced with the problem of deciding whether an improper integral exists or not. The actual value is then often not interesting at all. The *comparison test* provides such a method:

The Comparison Test
Are $f, g : [a, \infty) \to \mathbb{R}$ functions which are integrable on any bounded interval $[a, b]$, then we have: If $|f(x)| \leq g(x)$ for all $x \in [a, \infty)$ and exists $\int_a^{\infty} g(x)\,\mathrm{d}x$ then the following improper integral also exists:

$$\int_a^{\infty} f(x)\mathrm{d}x\,.$$

The criterion applies analogously to functions $f, g : (-\infty, b] \to \mathbb{R}$.

Example 32.2

- The improper integral $\int_1^\infty \frac{1}{x+x^2}\,\mathrm{d}x$ exists, since for all $x \geq 1$ we have:

$$\frac{1}{x+x^2} \leq \frac{1}{x^2} \quad \text{and} \quad \int_1^\infty \frac{1}{x^2}\mathrm{d}x \text{ exists,}$$

note Exercise 32.1.
- The improper integral $\int_0^1 \frac{1+\cos(x)}{x}\,\mathrm{d}x$ on the other hand does not exist, because for all $x \in [0, 1]$ we have $\cos(x) \geq 0$, thus:

$$\frac{1+\cos(x)}{x} \geq \frac{1}{x}.$$

But since $\int_0^1 \frac{1}{x}\mathrm{d}x$ does not exist, also $\int_0^1 \frac{1+\cos(x)}{x}\,\mathrm{d}x$ does not exist.
- The improper integral $\int_0^\infty \mathrm{e}^{-x^2}\,\mathrm{d}x$ exists, because we can decompose it as

$$\int_0^\infty \mathrm{e}^{-x^2}\,\mathrm{d}x = \int_0^1 \mathrm{e}^{-x^2}\,\mathrm{d}x + \int_1^\infty \mathrm{e}^{-x^2}\,\mathrm{d}x\,,$$

where the integral $\int_0^1 \mathrm{e}^{-x^2}\,\mathrm{d}x$ exists as a proper integral and the improper integral $\int_1^\infty \mathrm{e}^{-x^2}\,\mathrm{d}x$ exists because we have for $x \geq 1$:

$$x^2 \geq x \Rightarrow \mathrm{e}^{x^2} \geq \mathrm{e}^x \Rightarrow \mathrm{e}^{-x^2} \leq \mathrm{e}^{-x} \quad \text{and} \quad \int_1^\infty \mathrm{e}^{-x}\,\mathrm{d}x = \lim_{b\to\infty} -\mathrm{e}^{-b} + \mathrm{e} = \mathrm{e}\,.$$

- Also the improper integral $\int_0^\infty \frac{\sin(x)}{x}\,\mathrm{d}x$ exists. Again we decompose

$$\int_0^\infty \frac{\sin(x)}{x}\mathrm{d}x = \int_0^1 \frac{\sin(x)}{x}\mathrm{d}x + \int_1^\infty \frac{\sin(x)}{x}\mathrm{d}x\,.$$

The integral $\int_0^1 \frac{\sin(x)}{x}\,\mathrm{d}x$ exists, since $\frac{\sin(x)}{x}$ on $(0, 1]$ limited and $\lim_{x\to 0^+} \frac{\sin(x)}{x} = 1$ is. With the integral $\int_1^\infty \frac{\sin(x)}{x}\,\mathrm{d}x$ lies the estimation $\left|\frac{\sin(x)}{x}\right| \leq \frac{1}{x}$ close. But unfortunately this does not help, because the improper integral $\int_1^\infty \frac{1}{x}\,\mathrm{d}x$ does not exist. Instead, we make do with a trick and integrate by parts first:

$$\int_1^b \frac{\sin(x)}{x}\,\mathrm{d}x = \left| \begin{array}{ll} u = \frac{1}{x} & u' = -\frac{1}{x^2} \\ v' = \sin(x) & v = -\cos(x) \end{array} \right| = \underbrace{-\frac{\cos(x)}{x}\Big|_1^b}_{\stackrel{b\to\infty}{\longrightarrow}\cos(1)} - \int_1^b \frac{\cos(x)}{x^2}\,\mathrm{d}x\,.$$

The improper integral $\int_1^\infty \frac{\cos(x)}{x^2}\,\mathrm{d}x$ is now convergent, since for all $x \in [1,\infty)$ we have:

$$\left|\frac{\cos(x)}{x^2}\right| \leq \frac{1}{x^2} \quad \text{and} \quad \int_1^\infty \frac{1}{x^2}\mathrm{d}x \text{ exists.}$$

Thus also exists $\int_1^\infty \frac{\sin(x)}{x}\mathrm{d}x$ and thus also $\int_0^\infty \frac{\sin(x)}{x}\mathrm{d}x$.

As we will see later, indeed

$$\int_0^\infty \frac{\sin(x)}{x}\mathrm{d}x = \frac{\pi}{2}\,.$$

■

Note: If the improper integral $\int_{-\infty}^\infty f(x)\mathrm{d}x$ exists, it is equal to its Cauchy principal value, i.e.

$$\int_{-\infty}^\infty f(x)\,\mathrm{d}x \text{ exists} \quad \Rightarrow \quad \int_{-\infty}^\infty f(x)\,\mathrm{d}x = \mathrm{CHW}\int_{-\infty}^\infty f(x)\,\mathrm{d}x\,.$$

32.3 Exercises

32.1 Show:

$$\int_0^1 \frac{\mathrm{d}x}{x^\alpha} = \begin{cases} \infty, & \text{if } \alpha \geq 1 \\ \frac{1}{1-\alpha}, & \text{if } \alpha < 1 \end{cases} \quad \text{and} \quad \int_1^\infty \frac{\mathrm{d}x}{x^\alpha} = \begin{cases} \frac{1}{\alpha-1}, & \text{if } \alpha > 1 \\ \infty, & \text{if } \alpha \leq 1 \end{cases}\,.$$

32.2 Examine the following improper integrals for convergence and give their value if possible:

(a) $\int_0^{e} |\ln x| \, dx$,

(b) $\int_0^{2} \frac{1}{\sqrt{|1-x^2|}} \, dx$,

(c) $\int_1^{2} \frac{dx}{\ln x}$,

(d) $\int_1^{\infty} x^x \, e^{-x^2} \, dx$,

(e) $\int_0^{1} \ln x \, dx$,

(f) $\int_3^{\infty} \frac{4x}{x^2-4} \, dx$.

(g) $\int_{-\infty}^{\infty} |x| \, e^{-x^2} \, dx$,

(h) $\int_0^{e} \ln x \, dx$.

Separable and Linear Differential Equations of First Order

33

The topic of *differential equations* is one of the most important topics in engineering and scientific mathematics. Differential equations describe movements, flows, bends, models, ...

Therefore, in engineering and science, one is usually confronted with differential equations very early in one's studies, especially in physics. Some types of differential equations can be solved using the methods developed so far. We cover some such types in this and the next chapters and show how to solve them in a recipe-like manner.

In fact, the examples in this chapter are not representative of real-world examples. In practice, one has much more complicated differential equations where a solution function $x(t)$ cannot usually be given analytically; one then uses numerical methods to obtain approximately the value $x(t)$ at certain points t of the solution x. We will also deal with these topics (see Sect. 36). But in order to understand where the problems lie in finding solutions of differential equations, one should also consider a few simple solvable equations.

33.1 First Differential Equations

We start with a remark about the notation: So far we mostly considered functions f in the variable x, i.e. $f(x)$. Sometimes we also had a function y in the variable x, so $y(x)$. Since differential equations typically have *location functions* in the variable t such as *time*, it is common and useful to consider functions x in the variable t, i.e. $x(t)$. Furthermore, it is common in physics to use a dot for the derivative; therefore, we should not opt out here and use this notation as well, so we write $\dot{x}$ instead of x', $\ddot{x}$ instead of x'' etc.

C. Karpfinger, *Calculus and Linear Algebra in Recipes*,
https://doi.org/10.1007/978-3-662-65458-3_33

We now consider equations such as

$$\dot{x}(t) = -2\,t\,(x(t))^2\,.$$

We are looking for the set of all functions $x = x(t)$, which satisfy such an equation. A solution to this equation is something like $x(t) = \frac{1}{t^2+1}$.

An equation of this kind, in which functions $x = x(t)$ are searched for in **one** variable t, is called **ordinary differential equation**. Other examples are

$$\dot{x}(t) = 2\,x(t)\,,\ t\,x(t) = \ddot{x}(t)\,,\ \ddot{x}(t) = t\ \mathrm{e}^{x(t)}\,.$$

We consider only ordinary differential equations in this section, from now on we omit the adjective *ordinary*. Instead of *(ordinary) differential equation* we write *ODE* for short.

Differential equations come from the natural sciences and mathematical modelling: The attempt to formulate laws of nature, laws of formation, models, ... mathematically ends with an ODE and associated boundary or initial value conditions. An ODE expresses a dependence between the variable t, the function x and the derivative $\dot{x}$ of this function. Here, an ODE describes the changing behavior of these quantities with respect to each other. As an example we consider the **radioactive decay**:

Example 33.1 There is a quantity Q_0 radioactive material at the time $t_0 = 0$ is given. We are looking for a function $Q = Q(t)$ that indicates what amount of radioactive material is still present at the time t. From physical observations and theoretical assumptions, we know that the rate at which the radioactive material decays is directly proportional to the amount of material still present. This gives rise to the following ODE:

$$\frac{\mathrm{d}\,Q}{\mathrm{d}\,t}(t) = \dot{Q}(t) = -r\,Q(t)\,.$$

Furthermore we know $Q(t_0) = Q_0$ (one speaks of *initial condition*). The proportionality constant r, $r > 0$, is the decay rate, different for each radioactive material; this number r is known. The function we are looking for is $Q = Q(t)$.

Before dealing with first simple types of analytically solvable ODEs, we introduce a few suggestive notions:

- The **order** of an ODE is the highest derivative of the searched function $x = x(t)$ which occurs in the ODE.
- An ODE is called **linear** if x and all derivatives of x occur in the ODE in the first power and not in sin-, exp-, ... functions.

Example 33.2

- The ODE $\mathrm{e}^{\dot{x}} = \sin(x) + x^2$ is a nonlinear 1st order ODE.
- The ODE $\dddot{x} + 2\ddot{x} + 14 = 0$ is a linear ODE of 3rd order. ■

We treat two types of ODEs for which one can give a simple solution scheme: separable ODEs and 1st order linear ODEs.

33.2 Separable Differential Equations

An ODE is called **separable** if it can be written in the form

$$\dot{x} = f(t)\, g(x)\,.$$

So for separable ODEs you can write $\dot{x}$ on one side, on the other side there is a product of two functions f and g, where f is a function in t and g is a function in x:

$$\dot{x} = \underbrace{2}_{=f(t)}\ \underbrace{x}_{=g(x)}\,,\ \dot{x} = \underbrace{t}_{=f(t)}\ \underbrace{x}_{=g(x)}\,,\ \dot{x} = \underbrace{-2t}_{=f(t)}\ \underbrace{x^2}_{=g(x)}\,.$$

33.2.1 The Procedure for Solving a Separable Differential Equation

A separable ODE is solved according to the following scheme:

Recipe: Solving Separable Differential Equations

Given is a separable ODE $\dot{x} = f(t)\, g(x)$.

(1) Separation of variables: Write $\dot{x} = \frac{\mathrm{d}x}{\mathrm{d}t}$ and push out everything that starts with t to one side of the equation and everything to do with x to the other:

$$\frac{\mathrm{d}x}{\mathrm{d}t} = f(t)\, g(x) \Rightarrow \frac{1}{g(x)}\, \mathrm{d}x = f(t)\, \mathrm{d}t\,.$$

(2) Integrate on both sides, the integration constants c_l and c_r, which are obtained on the left and the right side, shift to the right and put $c = c_r - c_l$:

$$\int \frac{1}{g(x)}\, \mathrm{d}x = \int f(t)\, \mathrm{d}t + c\,.$$

(continued)

(3) Solve the equation $\int \frac{1}{g(x)}\, \mathrm{d}x = \int f(t)\, \mathrm{d}t + c$ obtained in (2) to $x = x(t)$. For each admissible $c \in \mathbb{R}$ you then have a solution.
(4) Give additionally the solutions $x = x(t)$ with $g(x) = 0$ (this case is excluded in (1)).

When solving separable ODEs, we typically obtain an equation of the form in (2).

$$\ln(|x(t)|) = h(t) + c \text{ with a function } h \text{ and a constant } c.$$

We solve this equation by applying the exponential function to $x = x(t)$:

$$\begin{aligned} \ln(|x(t)|) = h(t) + c \;&\Rightarrow\; |x(t)| = \mathrm{e}^{h(t)+c} = \mathrm{e}^{c}\, \mathrm{e}^{h(t)} \\ &\Rightarrow\; x(t) = \pm \tilde{c}\, \mathrm{e}^{h(t)} \\ &\Rightarrow\; x(t) = c\, \mathrm{e}^{h(t)} \end{aligned}$$

with a *new* $c \in \mathbb{R} \setminus \{0\}$. Note that c may also be negative, while e^c is always positive. We will use this in the following examples.

Example 33.3

- We solve the ODE $\dot{x} = 2x$.

(1) Separation of variables: $\frac{\mathrm{d}x}{\mathrm{d}t} = 2x \;\Rightarrow\; \frac{1}{x}\, \mathrm{d}x = 2\, \mathrm{d}t$.
(2) We integrate on both sides:

$$\int \frac{1}{x}\, \mathrm{d}x = \int 2\, \mathrm{d}t \;\Rightarrow\; \ln|x| = 2t + c\,.$$

(3) We solve for $x = x(t)$ by applying the exponential function on both sides, and obtain for each $c \in \mathbb{R} \setminus \{0\}$ we obtain the solution

$$x(t) = c\, \mathrm{e}^{2t}\,.$$

(4) It is also x with $x(t) = 0$ a solution of the ODE, in summary our solutions are $x(t) = c\, \mathrm{e}^{2t}$ with $c \in \mathbb{R}$.

- We solve the ODE $t\,x = \dot{x}$.

(1) Separation of variables: $\frac{\mathrm{d}x}{\mathrm{d}t} = t\,x \;\Rightarrow\; \frac{1}{x}\, \mathrm{d}x = t\, \mathrm{d}t$.

(2) We integrate on both sides:

$$\int \frac{1}{x}\,\mathrm{d}x = \int t\,\mathrm{d}t \Rightarrow \ln x = \frac{1}{2}t^2 + c\,.$$

(3) We solve for $x = x(t)$ by applying the exponential function on both sides, and obtain the solution for each $c \in \mathbb{R} \setminus \{0\}$ we obtain the solution

$$x(t) = c\,\mathrm{e}^{\frac{1}{2}t^2}\,.$$

(4) It is also x with $x(t) = 0$ a solution of the ODE, in summary our solutions are $x(t) = c\,\mathrm{e}^{\frac{1}{2}t^2}$ with $c \in \mathbb{R}$.

- We solve the ODE $\dot{x} = -2\,t\,x^2$.

(1) Separation of variables: $\frac{\mathrm{d}x}{\mathrm{d}t} = -2\,t\,x^2 \Rightarrow \frac{1}{x^2}\,\mathrm{d}x = -2t\,\mathrm{d}t$.
(2) We integrate on both sides:

$$\int \frac{1}{x^2}\,\mathrm{d}x = \int -2t\,\mathrm{d}t \Rightarrow -\frac{1}{x} = -t^2 + c\,.$$

(3) We solve for $x = x(t)$ by inverting the equation, and for each $c \in \mathbb{R}$ we get the solution

$$x(t) = \frac{1}{t^2 - c}\,.$$

(4) It is also x with $x(t) = 0$ is a solution of the ODE. ■

The c is determined by an initial condition, we will take care of this now.

33.2.2 Initial Value Problems

In general, an ODE describes a *motion*. The solution $x(t)$ can be interpreted as follows:

- At time t a particle is at the location $x(t)$, or
- at time t the *quantum* $Q(t)$ of radioactive material is present (note the above example of radioactive decay).

The separable ODE in radioactive decay can be solved easily, we get.

$$\frac{\mathrm{d}Q}{\mathrm{d}t} = -r\,Q(t) \Rightarrow Q(t) = c\,\mathrm{e}^{-rt}\,,\ c \in \mathbb{R}\,.$$

The c is now determined with an **initial condition** as for example $Q(0) = Q_0$, i.e., at time $t = 0$, i.e. at the beginning of the observation, the quantum Q_0 of radioactive material is present. We now put this condition into our solution manifold $Q(t) = c\,\mathrm{e}^{-rt}$, $c \in \mathbb{R}$, and thereby nail down the constant c fixed:

$$Q_0 = Q(0) = c\,\mathrm{e}^{-r0} = c\,.$$

Thus we now obtain the uniquely determined solution

$$Q(t) = Q_0\,\mathrm{e}^{-rt}\,.$$

This function satisfies the ODE and the initial condition:

$$\frac{\mathrm{d}\,Q}{\mathrm{d}\,t} = -r\,Q(t) \ \text{ and } \ Q(0) = Q_0\,.$$

One speaks of an **initial value problem**, in short **IVP** if we treat it like an ODE **and** an initial condition. To solve an IVP with a separable ODE, consider the following recipe:

Recipe: Solving an IVP with Separable ODE

The solution of the IVP

$$\dot{x} = f(t)\,g(x)\,,\ x(t_0) = x_0$$

is obtained as follows:

(1) Determine the general solution $x = x(t)$ of the ODE $\dot{x} = f(t)\,g(x)$ (with the integration constant c with the recipe in Sect. 33.3.
(2) Determine c from the equation $x(t_0) = x_0$ with $x(t)$ from (1).

Example 33.4 We consider the IVP $\dot{x} = -2\,t\,x^2\,,\ x(1) = 1/2\,.$

(1) According to the above example, the general solution of the separable ODE is

$$x(t) = \frac{1}{t^2 - c} \ \text{ with } \ c \in \mathbb{R}\,.$$

(2) We compute c from the following equation:

$$\frac{1}{2} = x(1) = \frac{1}{1-c} \Rightarrow c = -1 .$$

Thus $x(t) = \frac{1}{t^2+1}$ is the (uniquely determined) solution of the IVP. ■

33.3 The Linear Differential Equation of First Order

We consider 1st order linear ODEs in this section. The general appearance of a 1st order linear ODE is as follows

$$b(t)\,\dot{x}(t) + a(t)\,x(t) = s(t) .$$

By dividing the equation by $b(t)$, we can assume the following appearance of a 1st order linear ODE:

$$\dot{x}(t) + a(t)\,x(t) = s(t) .$$

The solution set of this ODE can be given in a formula, we have:

The Solution Formula for a Linear Differential Equation of 1st Order
The solution set L of the linear ODE of 1st order $\dot{x}(t) + a(t)\,x(t) = s(t)$ is

$$L = \left\{ \mathrm{e}^{-\int a(t)\mathrm{d}t} \left(\int \mathrm{e}^{\int a(t)\mathrm{d}t}\, s(t)\mathrm{d}t + c \right) \mid c \in \mathbb{R} \right\} .$$

So solving a linear ODE of 1st order is done by determining the given integral. But you can't easily remember this formula. In fact, you don't have to. We obtain the solution by a simple approach, which we now introduce, since we will need it again in the next chapter. We consider the 1st order linear ODE and the **corresponding homogeneous** ODE

$$\dot{x}(t) + a(t)\,x(t) = s(t) \longrightarrow \dot{x}(t) + a(t)\,x(t) = 0 .$$

The homogeneous ODE thus arises from the original one by replacing the **perturbation function** $s(t)$ is replaced by the zero function 0.

This homogeneous ODE is separable, viz.

$$\frac{1}{x}\,\mathrm{d}x = -a(t)\,\mathrm{d}t\,,$$

such that $x(t) = c\,\mathrm{e}^{-\int a(t)\mathrm{d}t}$ with $x \in \mathbb{R}$ is the general solution of the homogeneous ODE, so

$$L_h = \{c\,\mathrm{e}^{-\int a(t)\mathrm{d}t} \mid c \in \mathbb{R}\}$$

is the solution set of the homogeneous ODE.

We now determine the general solution set of the inhomogeneous ODE. This consists of a **particulate** solution x_p, which is *a* solution of the inhomogeneous ODE, and the general solution set L_h of the homogeneous ODE (see Exercise 33.4):

$$L = x_p + L_h = \{x_p + x_h \mid x_h \in L_h\}\,.$$

Thus, to determine L, we need, in addition to L_h a particular solution x_p. We find such a solution by **variation of the parameter** c of the solution $x(t) = c\,\mathrm{e}^{-\int a(t)\mathrm{d}t}$ of the homogeneous ODE, i.e., one sets

$$x_p(t) = c(t)\,\mathrm{e}^{-\int a(t)\mathrm{d}t}$$

with a *function* $c(t)$—in this sense the constant c is varied. With this approach we enter the inhomogeneous ODE and thereby determine the unknown function $c(t)$ in order to obtain the special solution $x_p(t) = c(t)\,\mathrm{e}^{-\int a(t)\mathrm{d}t}$; we have because of $\dot{x}_p = \dot{c}(t)\,\mathrm{e}^{-\int a(t)\mathrm{d}t} - a(t)\,c(t)\,\mathrm{e}^{-\int a(t)\mathrm{d}t}$:

$$\begin{aligned}\dot{x}_p + a(t)x_p &= \dot{c}(t)\,\mathrm{e}^{-\int a(t)\mathrm{d}t} - a(t)\,c(t)\,\mathrm{e}^{-\int a(t)\mathrm{d}t} + a(t)\,c(t)\,\mathrm{e}^{-\int a(t)\mathrm{d}t}\\ &= \dot{c}(t)\,\mathrm{e}^{-\int a(t)\mathrm{d}t} = s(t)\,,\end{aligned}$$

thus

$$\dot{c}(t) = \mathrm{e}^{\int a(t)\mathrm{d}t}\,s(t)\,,\ \text{ i.e. }\ c(t) = \int \mathrm{e}^{\int a(t)\mathrm{d}t}\,s(t)\mathrm{d}t$$

and thus

$$x_p(t) = c(t)\,\mathrm{e}^{-\int a(t)\mathrm{d}t} = \int \mathrm{e}^{\int a(t)\mathrm{d}t}\,s(t)\mathrm{d}t\,\mathrm{e}^{-\int a(t)\mathrm{d}t}\,.$$

We summarize the procedure in a recipe-like manner, but we also note that one does not usually give the solution set of an ODE; rather, one gives the **general solution**, one writes briefly

$$x_a(t) = x_p(t) + c\,\mathrm{e}^{-\int a(t)\mathrm{d}t},\ c \in \mathbb{R} \text{ instead of } L = \{x_p(t) + c\,\mathrm{e}^{-\int a(t)\mathrm{d}t} \mid c \in \mathbb{R}\}.$$

Recipe: Solve a Linear Differential Equation of 1st Order

We obtain as follows the general solution x_a of the ODE

$$\dot{x}(t) + a(t)\,x(t) = s(t)\,.$$

(1) Determine the general solution $x_h(t) = c\,\mathrm{e}^{-\int a(t)\mathrm{d}t}$ of the separable homogeneous ODE $\dot{x}(t) + a(t)\,x(t) = 0$.
(2) Determine a particulate solution by varying parameters $x_p(t) = c(t)\,\mathrm{e}^{-\int a(t)\mathrm{d}t}$: Substitute this x_p into the inhomogeneous ODE and obtain $c(t)$ and thus $x_p(t)$.
(3) Get the general solution $x_a(t) = x_p(t) + c\,\mathrm{e}^{-\int a(t)\mathrm{d}t}$, $c \in \mathbb{R}$.

Example 33.5

- We determine the general solution of the following 1st order linear ODE.

$$\dot{x} - \frac{1}{t}x = 3t\,.$$

(1) Solution of the homogeneous ODE: The homogeneous ODE is $\dot{x} - \frac{1}{t}x = 0$. Separation returns $x_h(t) = c\,t$, $c \in \mathbb{R}$.
(2) Variation of parameters: We set $x_p(t) = c(t)\,t$ into the inhomogeneous ODE:

$$\dot{x}_p(t) - \frac{1}{t}x_p = \dot{c}(t)\,t + c(t) - \frac{1}{t}c(t)\,t = \dot{c}(t)\,t = 3t\,.$$

This gives us $\dot{c}(t) = 3$ we choose $c(t) = 3\,t$ and obtain the particulate solution

$$x_p(t) = 3\,t^2\,.$$

(3) The general solution is $x_a(t) = 3\,t^2 + c\,t$, $c \in \mathbb{R}$. ■

Remarks

1. An initial condition $x(t_0) = x_0$ determines the constant c.
2. Sometimes a particulate solution can be guessed or found by a suitable approach (cf. the *right-hand side type* approach in the next chapter in Sect. 34.2). In this case, the (usually quite laborious) step (2) is omitted.
3. For the sake of clarity we state again: For certain types of differential equations there is a solution method. *Most* differential equations, however, cannot be solved analytically; one then has to rely on numerical solution methods.

33.4 Exercises

33.1 Give all the solutions of the following ODEs:

(a) $\dot{x}\,t = 2\,x$,
(b) $\dot{x} = \frac{2t}{t^2+1}\,x$,
(c) $x\,(1-t)\,\dot{x} = 1 - x^2$,
(d) $\dot{x}\,(x+1)^2 + t^3 = 0$.

33.2 Solve the following initial value problems with separable ODEs:

(a) $\dot{x} = -\frac{t}{x}\,,\ x(1) = 1$,
(b) $\dot{x} = \mathrm{e}^x \sin t,\ x(0) = 0$,
(c) $t^2 x = (1+t)\dot{x},\ x(0) = 1$.

33.3 Determine the solutions of the following initial value problems. Use variation of parameters to determine a particulate solution.

(a) $(1+t^2)\,\dot{x} - t\,x = \sqrt{1+t^2}\,,\ x(t_0) = x_0$,
(b) $\mathrm{e}^{-t}\,\dot{x} + 2\,\mathrm{e}^t\,x = \mathrm{e}^t\,,\ x(0) = \frac{1}{2} + \frac{1}{\mathrm{e}}$.

33.4 Show: If x_p is a particulate solution of a linear ODE of the 1st order and L_h is the solution set of the corresponding homogeneous ODE, then $L = x_p + L_h$ is the solution set of the original ODE.

33.5 Just now Xaver was served his Maß beer, but it was poured so hastily that it consisted entirely of foam. Approximately, beer foam decays exponentially with a certain half-life T_0, i.e., the beer foam volume $V(t)$ at time t is given by $V(t) = V_0 \cdot (\frac{1}{2})^{t/T_0}$. A realistic value is $T_0 = 50$s, and the foam volume at the beginning is $V_0 = 1$l. Xaver now starts drinking immediately and begins to drink thirstily, without stopping. Thereby he constantly drinks $a = 20\,\text{ml/s}$ foam away.

(a) Show that the beer foam volume without Xaver's intervention of an ODE $\dot{V} = -kV$ and determine k as a function of T_0.
(b) Set up an ODE for the beer foam volume V that accounts for Xaver's thirst, and solve it.
(c) Determine the time T_e at which the foam has completely disappeared.

34 Linear Differential Equations with Constant Coefficients

In the linear differential equations we can distinguish two types: There are those in which all the coefficients are constant, and those in which this is not the case, that is, in which some of the coefficients are functions in *t*. One immediately suspects that finding solutions for those with non-constant coefficients is in general more difficult. In fact, there is already no general method for finding a solution if only the order is greater than or equal to 2. This makes it all the more surprising that all linear differential equations with constant coefficients can in general be solved by a clear scheme (provided the perturbation function does not interfere too much). We deal with this in the present chapter.

The general form of an *nth*-order linear differential equation with constant coefficients is

$$a_n\, x^{(n)}(t) + a_{n-1}\, x^{(n-1)}(t) + \cdots + a_1\, \dot{x}(t) + a_0\, x(t) = s(t)$$

with $a_n, \ldots, a_0 \in \mathbb{R}$ and $a_n \neq 0$. If the **perturbation function** $s = s(t)$ is the zero function, the differential equation is called **homogeneous**, otherwise **inhomogeneous**.

34.1 Homogeneous Linear Differential Equations with Constant Coefficients

The general form of a homogeneous linear ODE with constant coefficients is

$$a_n\, x^{(n)}(t) + a_{n-1}\, x^{(n-1)}(t) + \cdots + a_1\, \dot{x}(t) + a_0 x(t) = 0$$

with $a_n, \ldots, a_0 \in \mathbb{R}$ and $a_n \neq 0$. It is not difficult to show that sum and scalar multiples of solutions are solutions again (see Exercise 34.4), more generally we have:

C. Karpfinger, *Calculus and Linear Algebra in Recipes*,
https://doi.org/10.1007/978-3-662-65458-3_34

The Solution Space of a Homogeneous Linear Differential Equation
The set of all solutions of a homogeneous linear ODE of *nth* order is an n-dimensional subspace L_h of $\mathbb{R}^{\mathbb{R}}$.
If $x_1, \ldots, x_n$ are linearly independent solutions of $\mathbb{R}^{\mathbb{R}}$ then

$$L_h = \{c_1 x_1 + \cdots + c_n x_n \mid c_1, \ldots, c_n \in \mathbb{R}\}.$$

Then one calls

$$x_h(t) = c_1 x_1(t) + \cdots + c_n x_n(t) \text{ with } c_1, \ldots, c_n \in \mathbb{R}$$

the **general solution of** the homogeneous ODE. A basis $\{x_1, \ldots, x_n\}$ of L_h is also called a **(real) fundamental system**.

Thus, we have determined all solutions if we succeed in specifying n linearly independent solutions, that is, a real fundamental system. In order to find any solutions at all, we make the approach $x(t) = \mathrm{e}^{\lambda t}$. Substituting this $x(t)$ into the ODE

$$a_n x^{(n)}(t) + a_{n-1} x^{(n-1)}(t) + \cdots + a_1 \dot{x}(t) + a_0 x(t) = 0$$

one obtains because of $\dot{x} = \lambda \mathrm{e}^{\lambda t}$, $\ddot{x} = \lambda^2 \mathrm{e}^{\lambda t}, \ldots, x^{(n)} = \lambda^n \mathrm{e}^{\lambda t}$ the equation

$$a_n \lambda^n \mathrm{e}^{\lambda t} + a_{n-1} \lambda^{n-1} \mathrm{e}^{\lambda t} + \cdots + a_1 \lambda \mathrm{e}^{\lambda t} + a_0 \mathrm{e}^{\lambda t} = 0.$$

Factoring out and shortening $\mathrm{e}^{\lambda t} \neq 0$ yields the **characteristic equation** $p(\lambda) = 0$ or the **characteristic polynomial** $p(\lambda)$:

$$p(\lambda) = a_n \lambda^n + a_{n-1} \lambda^{n-1} + \cdots + a_1 \lambda + a_0 = 0.$$

For every solution λ of the characteristic equation, that is, for every zero of the characteristic polynomial p is $x(t) = \mathrm{e}^{\lambda t}$ a solution of the ODE. And now comes the best: If $\lambda_1, \ldots, \lambda_r$ different solutions, then the functions $x_1 = \mathrm{e}^{\lambda_1 t}, \ldots, x_r = \mathrm{e}^{\lambda_r t}$ are linearly independent. So we don't have to worry about the linear independence, we get it for free.

Example 34.1 We consider the homogeneous linear ODE

$$\ddot{x} + \dot{x} - 6x = 0 \text{ with characteristic equation } \lambda^2 + \lambda - 6 = 0.$$

The characteristic equation has the two solutions $\lambda_1 = 2$ and $\lambda_2 = -3$. Thus we have $x_1(t) = \mathrm{e}^{2t}$ and $x_2(t) = \mathrm{e}^{-3t}$ are two linearly independent solutions of the ODE, i.e., $\{\mathrm{e}^{2t}, \mathrm{e}^{-3t}\}$ is a real fundamental system of the ODE. The general solution is

$$x_a(t) = c_1 \mathrm{e}^{2t} + c_2 \mathrm{e}^{-3t} \quad \text{with } c_1, c_2 \in \mathbb{R}. \quad \blacksquare$$

If the characteristic polynomial has a multiple zero, there seems to be a problem at first, because the number of different zeros is then really smaller than the dimension of the searched solution space. However, we can easily find enough linearly independent solutions, because if λ is a zero multiplicity m of the characteristic polynomial $p(\lambda) = \sum a_k \lambda^k$ of the homogeneous linear ODE $\sum a_k x^{(k)} = 0$, then the functions

$$x_1 = \mathrm{e}^{\lambda t}, \; x_2 = t\,\mathrm{e}^{\lambda t}, \ldots, x_m = t^{m-1}\,\mathrm{e}^{\lambda t}$$

are exactly m linearly independent solutions.

Example 34.2 We consider the homogeneous linear ODE

$$\ddddot{x} + 2\,\dddot{x} - 2\,\dot{x} - x = 0 \text{ with characteristic equation } \lambda^4 + 2\,\lambda^3 - 2\,\lambda - 1 = 0.$$

The characteristic equation has the two solutions $\lambda_1 = 1$ (single, i.e. of multiplicity 1) and $\lambda_2 = -1$ (of multiplicity 3), namely

$$\lambda^4 + 2\,\lambda^3 - 2\,\lambda - 1 = (\lambda - 1)(\lambda + 1)^3.$$

Thus,

$$x_1(t) = \mathrm{e}^t, \; x_2(t) = \mathrm{e}^{-t}, \; x_3(t) = t\,\mathrm{e}^{-t}, \; x_4(t) = t^2\,\mathrm{e}^{-t}$$

are four linearly independent solutions of the ODE, i.e., $\{\mathrm{e}^t, \mathrm{e}^{-t}, t\,\mathrm{e}^{-t}, t^2\,\mathrm{e}^{-t}\}$ is a real fundamental system of the ODE. The general solution is

$$x_a(t) = c_1 \mathrm{e}^t + c_2 \mathrm{e}^{-t} + c_3\, t\,\mathrm{e}^{-t} + c_4\, t^2\,\mathrm{e}^{-t} \quad \text{mit } c_1, c_2, c_3, c_4 \in \mathbb{R}. \quad \blacksquare$$

If a zero λ of the characteristic polynomial is not real, $\lambda = a + \mathrm{i}\,b$ where $b \neq 0$, then the conjugate complex $\overline{\lambda} = a - \mathrm{i}\,b$ is also a zero of the characteristic polynomial p. Thus, for each such zero λ, one obtains a pair of complex solutions $x_1(t) = \mathrm{e}^{\lambda t}$ and $x_2(t) = \mathrm{e}^{\overline{\lambda} t}$. But since one is interested in real solutions, one now proceeds as follows:

One chooses one of the two complex solutions $x = x_1$ or $x = x_2$ and discards the second complex solution. Decompose the chosen solution x into real and imaginary parts:

$$x(t) = \mathrm{e}^{\lambda t} = \mathrm{e}^{(a+\mathrm{i}\,b)t} = \mathrm{e}^{at}\,\mathrm{e}^{\mathrm{i}\,bt} = \mathrm{e}^{at}\,(\cos(b\,t) + \mathrm{i}\sin(b\,t))$$
$$= \mathrm{e}^{at}\cos(b\,t) + \mathrm{i}\,\mathrm{e}^{at}\sin(b\,t)\,.$$

Then we have

$$\mathrm{Re}(x(t)) = \mathrm{e}^{at}\cos(b\,t) \text{ and } \mathrm{Im}(x(t)) = \mathrm{e}^{at}\sin(b\,t)\,.$$

One can show that the real and imaginary parts of a complex solution of a homogeneous linear ODE are two linearly independent real solutions of that ODE. Altogether the balance is correct again: Each pair λ, $\overline{\lambda}$ conjugate complex zeros of the characteristic polynomial yields two real linearly independent solutions.

By the way, it is also clear now that it doesn't matter which of the two complex solutions you take: Each complex solution yields a pair of real solutions, and the two pairs generate the same real solution space.

Example 34.3 We consider the homogeneous linear ODE

$$\ddot{x} - 4\,\dot{x} + 13\,x = 0 \text{ with characteristic equation } \lambda^2 - 4\,\lambda + 13 = 0\,.$$

The characteristic equation has the two solutions $\lambda_1 = 2 + 3\,\mathrm{i}$ and $\lambda_2 = 2 - 3\,\mathrm{i}$. Thus there are

$$x_1(t) = \mathrm{e}^{2t}\cos(3t) \text{ and } x_2(t) = \mathrm{e}^{2t}\sin(3t)$$

two linearly independent real solutions of the ODE, i.e., $\{\mathrm{e}^{2t}\cos(3t),\ \mathrm{e}^{2t}\sin(3t)\}$ is a real fundamental system of the ODE. The general solution is

$$x_a(t) = c_1\,\mathrm{e}^{2t}\cos(3t) + c_2\,\mathrm{e}^{2t}\sin(3t) \text{ mit } c_1,\ c_2 \in \mathbb{R}\,. \qquad \blacksquare$$

We describe the general procedure for solving a homogeneous linear ODE with constant coefficients:

Recipe: Solving a Homogeneous Linear ODE with Constant Coefficients
Find the general solution of the following ODE as follows:

$$a_n\,x^{(n)} + a_{n-1}\,x^{(n-1)} + \cdots + a_1\,\dot{x} + a_0\,x = 0 \text{ mit } a_0, \ldots, a_n \in \mathbb{R}\,.$$

(continued)

(1) Get the characteristic equation $p(\lambda) = \sum_{k=0}^{n} a_k \lambda^k = 0$.
(2) Determine all solutions of $p(\lambda) = 0$, i.e., decompose $p(\lambda)$ into the form

$$p(\lambda) = (\lambda - \lambda_1)^{m_1} \cdots (\lambda - \lambda_r)^{m_r} = 0 \text{ mit } \lambda_1, \ldots, \lambda_r \in \mathbb{C}.$$

(3) Give n linearly independent solutions $x_1, \ldots, x_n$ of the solution space L_h of the ODE as follows:

- If $\lambda = \lambda_i \in \mathbb{R}$ with $m = m_i \in \mathbb{N}$, then choose

$$e^{\lambda t}, \, t\, e^{\lambda t}, \ldots, t^{m-1} e^{\lambda t}.$$

- If $\lambda = a + \mathrm{i}\, b = \lambda_i \in \mathbb{C} \setminus \mathbb{R}$ with $m = m_i \in \mathbb{N}$: Delete $\overline{\lambda}_i$ and choose

$$e^{at} \cos(b\, t), \, t\, e^{at} \cos(b\, t), \ldots, t^{m-1} e^{at} \cos(b\, t)$$

$$e^{at} \sin(b\, t), \, t\, e^{at} \sin(b\, t), \ldots, t^{m-1} e^{at} \sin(b\, t).$$

This gives a total of n linearly independent real solutions $x_1, \ldots, x_n$. It is then $L_h = \{c_1 x_1 + \cdots + c_n x_n \mid c_1, \ldots, c_n \in \mathbb{R}\}$ the solution space of the ODE and

$$x_h(t) = c_1 x_1(t) + \cdots + c_n x_n(t) \text{ with } c_1, \ldots, c_n \in \mathbb{R}$$

the general solution of the homogeneous ODE.

The question whether given n solutions $x_1, \ldots, x_n : I \to \mathbb{R}$ of a homogeneous linear ODE are linearly independent, i.e. they form a fundamental system of the ODE, can be decided with the **Wronski determinant**

$$W(t) = \det \begin{pmatrix} x_1(t) & \ldots & x_n(t) \\ \dot{x}_1(t) & \ldots & \dot{x}_n(t) \\ \vdots & & \vdots \\ x_1^{(n-1)}(t) & \ldots & x_n^{(n-1)}(t) \end{pmatrix}$$

namely we have:

The Wronski Determinant
The solutions $x_1, \ldots, x_n : I \to \mathbb{R}$ of

$$a_n(t)\, x^{(n)} + \cdots + a_1(t)\, \dot{x} + a_0(t)\, x = 0$$

form a fundamental system if and only if $W(t) \neq 0$ for at least one $t \in I$. We then have $W(t) \neq 0$ for all $t \in I$.

Example 34.4 The ODE $\ddot{x} + x = 0$ has the two solutions sin and cos. These form because of

$$W(t) = \det \begin{pmatrix} \sin(t) & \cos(t) \\ \cos(t) & -\sin(t) \end{pmatrix} = -1$$

a fundamental system of the ODE. ■

34.2 Inhomogeneous Linear Differential Equations with Constant Coefficients

An inhomogeneous linear ODE with constant coefficients has the form

$$a_n\, x^{(n)} + a_{n-1}\, x^{(n-1)} + \cdots + a_1\, \dot{x} + a_0\, x = s(t)$$

with $s(t) \neq 0$. We are looking for the solution set L of this ODE. We denote by L_h the solution space of the **the homogeneous** ODE which follows from the ODE by substituting $s(t)$ by the zero function 0. The solution space L_h is determined by the method described in Sect. 34.1 For the solution set L *of* the inhomogeneous ODE we have:

The Solution Space of an Inhomogeneous Linear Differential Equation with Constant Coefficients
For the set L of all solutions of a linear ODE *of nth* order with constant coefficients we have

$$L = x_p(t) + L_h = \{x_p(t) + x_h(t) \,|\, x_h \in L_h\},$$

(continued)

where L_h is the solution space of the homogeneous ODE belonging to the ODE and $x_p(t)$ is a particulate solution of the inhomogeneous ODE. The general solution x_a has the form

$$x_a(t) = x_p(t) + x_h(t) .$$

Thus, to determine the general solution of an inhomogeneous linear ODE with constant coefficients, we need the general solution of the associated homogeneous ODE and a particulate solution. Such a particulate solution x_p can be found by means of one of the two approaches:

- Variation of parameters,
- approach of the right-hand side type.

34.2.1 Variation of Parameters

If $x_h(t) = c_1 x_1 + \cdots + c_n x_n$ is the general solution of the homogeneous ODE *n-th* order, then one makes with the **variation of parameters** the approach

$$x_p(t) = c_1(t)\, x_1 + \cdots + c_n(t)\, x_n .$$

One *varies* the constants $c_1, \ldots, c_n$, by expressing them as functions in the variable t, and thus enters the inhomogeneous ODE. One determines coefficient functions in this approach $c_1(t), \ldots, c_n(t)$ such that x_p is a solution of the inhomogeneous ODE. If one uses this approach in the inhomogeneous ODE of order n, it is important to note that the product rule must be applied to the derivation, since the c_i are functions in t. Thus, for the first derivative, we obtain the already abundantly complicated expression

$$\dot{x}_p = (c_1 \dot{x}_1 + \cdots + c_n \dot{x}_n) + (\dot{c}_1 x_1 + \cdots + \dot{c}_n x_n) .$$

To keep this approach clear, we now simply set $\dot{c}_1 x_1 + \cdots + \dot{c}_n x_n = 0$; for $\ddot{x}_p$ we then get

$$\ddot{x}_p = (c_1 \ddot{x}_1 + \cdots + c_n \ddot{x}_n) + (\dot{c}_1 \dot{x}_1 + \cdots + \dot{c}_n \dot{x}_n) .$$

If higher derivatives of x_p are necessary (i.e. if $n > 2$), then set the second expression equal to zero, $\dot{c}_1 \dot{x}_1 + \cdots + \dot{c}_n \dot{x}_n = 0$ and continue with this principle; for practical cases we come up with $n = 2$. We therefore break off at this point. In the case $n = 2$ we obtain with the general approach $x_p(t) = c_1(t)\,x_1 + c_2(t)\,x_2$ and the requirement $\dot{c}_1 x_1 + \cdots + \dot{c}_n x_n = 0$:

$$\begin{aligned} s(t) &= a_2\ddot{x}_p + a_1\dot{x}_p + a_0 x_p \\ &= a_2[(c_1\ddot{x}_1 + c_2\ddot{x}_2) + (\dot{c}_1\dot{x}_1 + \dot{c}_2\dot{x}_2)] + a_1(c_1\dot{x}_1 + c_2\dot{x}_2) + a_0(c_1\,x_1 + c_2\,x_2) \\ &= c_1\,(a_2\ddot{x}_1 + a_1\dot{x}_1 + a_0 x_1) + c_2\,(a_2\ddot{x}_2 + a_1\dot{x}_2 + a_0 x_2) + a_2\,(\dot{c}_1\dot{x}_1 + \dot{c}_2\dot{x}_2)\,. \end{aligned}$$

Since $a_2\ddot{x}_1 + a_1\dot{x}_1 + a_0 x_1 = 0$ and $a_2\ddot{x}_2 + a_1\dot{x}_2 + a_0 x_2 = 0$, we obtain the functions $c_1(t)$ and $c_2(t)$ as follows:

Recipe: Determine a Particulate Solution with Variation of the Constants
If $x_h = c_1 x_1 + c_2 x_2$ is the general solution of the homogeneous ODE $a_2\,\ddot{x} + a_1\,\dot{x} + a_0 x = 0$ then one obtains a partial solution x_p of the inhomogeneous ODE

$$a_2\,\ddot{x} + a_1\,\dot{x} + a_0 x = s(t)$$

by the approach $x_p = c_1(t)\,x_1 + c_2(t)\,x_2$.
The functions $c_1(t)$ and $c_2(t)$ are obtained by solving the system

$$\begin{aligned} \dot{c}_1 x_1 + \dot{c}_2 x_2 &= 0 \\ \dot{c}_1\dot{x}_1 + \dot{c}_2\dot{x}_2 &= s(t)/a_2 \end{aligned}$$

and indefinite integration of the solutions $\dot{c}_1(t)$ and $\dot{c}_2(t)$.

So although we have more or less arbitrarily set $\dot{c}_1 x_1 + \dot{c}_2 x_2 = 0$ this method yields a solution (provided that the given system is solvable and that primitive functions of the solutions $\dot{c}_1(t)$ and $\dot{c}_2(t)$ can be given). The method works analogously for higher orders, in the case $n = 3$ one obtains $\dot{c}_1(t),\ \dot{c}_2(t),\ \dot{c}_3(t)$ from the system

$$\begin{aligned} \dot{c}_1 x_1 + \dot{c}_2 x_2 + \dot{c}_3 x_3 &= 0 \\ \dot{c}_1\dot{x}_1 + \dot{c}_2\dot{x}_2 + \dot{c}_3\dot{x}_3 &= 0 \\ \dot{c}_1\ddot{x}_1 + \dot{c}_2\ddot{x}_2 + \dot{c}_3\ddot{x}_3 &= s(t)/a_3\,. \end{aligned}$$

We solve an example in the case $n = 2$.

Example 34.5 We determine a partial solution of the inhomogeneous ODE

$$\ddot{x} - 2\,\dot{x} + x = (1 + t)\,\mathrm{e}^t\ .$$

Because $p(\lambda) = (\lambda - 1)^2$ the general solution of the corresponding homogeneous ODE is

$$x(t) = c_1 \, \mathrm{e}^t + c_2 \, t \, \mathrm{e}^t \, ,$$

in particular $x_1 = \mathrm{e}^t$ and $x_2 = t \, \mathrm{e}^t$. We vary the parameters, i.e., we set

$$x_p(t) = c_1(t) \, \mathrm{e}^t + c_2(t) \, t \, \mathrm{e}^t \, .$$

We obtain $\dot{c}_1$ and $\dot{c}_2$ as solutions of the system

$$\begin{aligned} \dot{c}_1 \, \mathrm{e}^t + \dot{c}_2 \, t \, \mathrm{e}^t &= 0 \\ \dot{c}_1 \, \mathrm{e}^t + \dot{c}_2 \, (1+t) \, \mathrm{e}^t &= (1+t) \, \mathrm{e}^t \, . \end{aligned}$$

A short calculation yields

$$\dot{c}_1 = -\dot{c}_2 t \text{ and } \dot{c}_2 = 1 + t \, .$$

Thus we find

$$c_1(t) = \frac{-t^2}{2} - \frac{t^3}{3} \text{ and } c_2(t) = t + \frac{t^2}{2} \, .$$

As a special solution of the ODE we thus obtain

$$x_p(t) = \mathrm{e}^t \left(\frac{t^2}{2} + \frac{t^3}{6} \right) . \qquad \blacksquare$$

The variation of the constants always leads to a particulate solution. Unfortunately, the effort to determine the coefficient functions $c_i(t)$ is a lot of work if the order is $n \geq 2$. Here one likes to take every possible shortcut that presents itself. And such a shortcut exists in any case whenever the perturbation function is of special construction. Here then the *approach of the right-hand side type* offers itself.

34.2.2 Approach of the Right-Hand Side Type

In the right-hand side type approach, one assumes that a particulate solution $x_p(t)$ is of the same type as the perturbation function $s(t)$. We consider an inhomogeneous linear ODE with constant coefficients of the form:

$$a_n \, x^{(n)} + a_{n-1} \, x^{(n-1)} + \cdots + a_1 \, \dot{x} + a_0 x = s(t) \, .$$

$s(t)$	$x_p(t)$
$b(t)$	$A(t)$, if $p(0) \neq 0$, $t^r A(t)$, if 0 is a z. of mult. r of p.
$b(t)\,\mathrm{e}^{at}$	$A(t)\,\mathrm{e}^{at}$, if $p(a) \neq 0$, $t^r A(t)\,\mathrm{e}^{at}$, if a is a z. of mult. r of p.
$b(t)\cos(b\,t)$	$A(t)\cos(b\,t) + B(t)\sin(b\,t)$, if $p(\mathrm{i}\,b) \neq 0$, $t^r[A(t)\cos(b\,t) + B(t)\sin(b\,t)]$, if $\mathrm{i}\,b$ is a z. of mult. r of p.
$b(t)\sin(b\,t)$	$A(t)\cos(b\,t) + B(t)\sin(b\,t)$, if $p(\mathrm{i}\,b) \neq 0$, $t^r[A(t)\cos(b\,t) + B(t)\sin(b\,t)]$, if $\mathrm{i}\,b$ is a z. of mult. r of p.
$b(t)\,\mathrm{e}^{at}\cos(b\,t)$	$[A(t)\cos(b\,t) + B(t)\sin(b\,t)]\,\mathrm{e}^{at}$, if $p(a+\mathrm{i}\,b) \neq 0$, $t^r[A(t)\cos(b\,t) + B(t)\sin(b\,t)]\,\mathrm{e}^{at}$, if $a+\mathrm{i}\,b$ is a z. of mult. r of p.
$b(t)\,\mathrm{e}^{at}\sin(b\,t)$	$[A(t)\cos(b\,t) + B(t)\sin(b\,t)]\,\mathrm{e}^{at}$, if $p(a+\mathrm{i}\,b) \neq 0$, $t^r[A(t)\cos(b\,t) + B(t)\sin(b\,t)]\,\mathrm{e}^{at}$, if $a+\mathrm{i}\,b$ is a z. of mult. r of p.

Approach of the Right-Hand Side Type
If the perturbation function $s(t)$ of an inhomogeneous linear ODE with the characteristic polynomial $p(\lambda) = a_n\lambda^n + \cdots + a_1\lambda + a_0$ is of the form

$$s(t) = (b_0 + b_1 t + \cdots + b_m t^m)\,\mathrm{e}^{at}\cos(b\,t) \text{ or}$$

$$s(t) = (b_0 + b_1 t + \cdots + b_m t^m)\,\mathrm{e}^{at}\sin(b\,t),$$

then set in the case $p(a+\mathrm{i}\,b) \neq 0$

$$x_p(t) = [(A_0 + A_1 t + \cdots + A_m t^m)\cos(b\,t) + \\ + (B_0 + B_1 t + \cdots + B_m t^m)\sin(b\,t)]\,\mathrm{e}^{at}$$

and in the case $a+\mathrm{i}\,b$ is a zero of multiplicity r of p set

$$x_p(t) = t^r[(A_0 + A_1 t + \cdots + A_m t^m)\cos(b\,t) + \\ + (B_0 + B_1 t + \cdots + B_m t^m)\sin(b\,t)]\,\mathrm{e}^{at}\,.$$

We describe this general approach of the right-hand side type again for the special cases $a = 0$ and/or $b = 0$ in the perturbation function $s(t)$ in the following table. Here we set $b(t) = b_0 + b_1 t + \cdots + b_m t^m$, $A(t) = A_0 + A_1 t + \cdots + A_m t^m$ and $B(t) = B_0 + B_1 t + \cdots + B_m t^m$, moreover we abbreviate *zero of multiplicity* r by *z. of mult.* r.

If $s(t)$ is of the given form, then one enters the inhomogeneous ODE with the corresponding approach and thus obtains an equation in which the numbers $A_0, \ldots, A_m$ and $B_0, \ldots, B_m$ are to be determined. This can be done by equating coefficients.

Example 34.6 We determine a particulate solution x_p of the inhomogeneous ODE

$$\ddot{x} - 2\dot{x} + x = (1+t)\, \mathrm{e}^t \,.$$

The perturbation function $s(t) = (1+t)\, \mathrm{e}^t$ is of the form

$$s(t) = (b_0 + b_1 t + \cdots + b_m t^m)\, \mathrm{e}^{at} \cos(bt)\,,$$

where $b_0 = 1 = b_1$, $m = 1$, $a = 1$ and $b = 0$. Since $\lambda = a + \mathrm{i}\, b = 1$ is a double zero of $p = \lambda^2 - 2\lambda + \lambda = (\lambda - 1)^2$ we make the approach

$$x_p(t) = t^2 (A_0 + A_1 t)\, \mathrm{e}^t \,.$$

We enter the inhomogeneous ODE with this approach and obtain, because of

$$\dot{x}_p(t) = [2\, A_0 t + (A_0 + 3\, A_1)\, t^2 + A_1 t^3]\, \mathrm{e}^t \quad \text{and}$$
$$\ddot{x}_p(t) = [2\, A_0 + (4\, A_0 + 6\, A_1)\, t + (A_0 + 6\, A_1)\, t^2 + A_1 t^3]\, \mathrm{e}^t$$

we obtain the equation

$$\ddot{x}_p - 2\dot{x}_p + x_p(t) = (2\, A_0 + 6\, A_1 t)\, \mathrm{e}^t = (1+t)\, \mathrm{e}^t = s(t)\,.$$

Equating coefficients yields $A_0 = 1/2$ and $A_1 = 1/6$, so that

$$x_p(t) = \left(\frac{t^2}{2} + \frac{t^3}{6}\right) \mathrm{e}^t$$

is a particulate solution. Compare this with the Example 34.5. ■

Thus, the following procedure for finding the solution set of an (inhomogeneous) linear ODE with constant coefficients is appropriate:

Recipe: Solving a Linear ODE with Constant Coefficients
The solution set L of the linear ODE with constant coefficients

$$a_n x^{(n)} + a_{n-1} x^{(n-1)} + \cdots + a_1 \dot{x} + a_0 x = s(t)$$

is obtained as follows:

(1) Determine the solution set L_h of the corresponding homogeneous differential equation (set $s(t) = 0$).
(2) Determine a particulate solution x_p by the *right-hand side type approach* or by *varying parameters*.
(3) Obtain L by $L = x_p + L_h$.

By specifying n initial conditions

$$x(t_0) = x_0, \ldots, x^{(n-1)}(t_0) = x_{n-1}$$

the constants $c_1, \ldots, c_n$ are determined. Note that the number of initial conditions is just the order of the ODE. Clearly, the order of the ODE is also the dimension of the solution space of the homogeneous ODE and thus equal to the number of free constants $c_1, \ldots, c_n$; and around this n numbers, you also need n conditions.

MATLAB With MATLAB we solve a ODE or an IVP using the function `dsolve`. We show this with examples, the general procedure is then clear:

```
» dsolve('Dx=2*x, x(0)=1')
ans =
exp(2*t)
» dsolve('D2x=2*x')
ans =
C8*exp(2^(1/2)*t) + C9*exp(-2^(1/2)*t)
» dsolve('D2x=2*t, x(0)=1, Dx(0)=1')
ans =
t^3/3 + t + 1
```

We conclude this chapter with a useful tool: If one has to solve a linear ODE with a perturbation function $s(t)$, which is a sum of two functions $s_1(t)$ and $s_2(t)$ then the *superposition principle* is helpful, according to which one can only use for each summand $s_i(t)$ one solution each $x_i(t)$ has to be determined. The solution for the more complicated perturbation function $s(t)$ is then obtained by *superposition*, more precisely:

The Superposition Principle
If x_1 is a solution of the linear ODE

$$x^{(n)} + a_{n-1}(t)\, x^{(n-1)} + \cdots + a_1(t)\, \dot{x} + a_0(t)\, x = s_1(t)$$

and x_2 a solution of the linear ODE

$$x^{(n)} + a_{n-1}(t)\, x^{(n-1)} + \cdots + a_1(t)\, \dot{x} + a_0(t)\, x = s_2(t)\,,$$

then $\alpha x_1 + \beta x_2$ with $\alpha, \beta \in \mathbb{R}$ is a solution of the linear ODE

$$x^{(n)} + a_{n-1}(t)\, x^{(n-1)} + \cdots + a_1(t)\, \dot{x} + a_0(t)\, x = \alpha s_1(t) + \beta s_2(t)\,.$$

In particular, with every two solutions of a homogeneous linear ODE, every linear combination is always a solution again.

34.3 Exercises

34.1 Show that sum and scalar multiples of solutions of a homogeneous linear ODE are again solutions of this linear ODE.

34.2 Determine a continuous function $x : \mathbb{R} \to \mathbb{R}$ which for all $t \in \mathbb{R}$ satisfies the following equation:

$$x(t) + \int_0^t x(\tau)\mathrm{d}\tau = \frac{1}{2}t^2 + 3t + 1\,.$$

Proceed as follows:

(a) First rewrite the integral equation into a ODE with initial condition.
(b) What is a general solution x_h for the associated homogeneous ODE?
(c) To determine a particulate solution, use $x_p(t)$ the right-hand side type approach and give the general solution of the ODE from (a).
(d) Find the solution of the IVP from (a) and hence a solution to the integral equation.

34.3 Solve the IVPs

(a) $x^{(4)} - x = t^3$, $x(0) = 2$, $\dot{x}(0) = 0$, $\ddot{x}(0) = 2$, $\dddot{x}(0) = -6$.
(b) $\ddot{x} + 2\dot{x} - 3x = \mathrm{e}^t + \sin t$, $x(0) = \dot{x}(0) = 0$.

(c) $\dddot{x} + \ddot{x} - 5\dot{x} + 3x = 6 \sinh 2t$, $x(0) = \dot{x}(0) = 0$, $\ddot{x}(0) = 4$.

34.4 Given the ODE $\ddot{x} - 7\dot{x} + 6x = \sin t$.

(a) Determine the general solution.
(b) For which initial values $x(0)$, $\dot{x}(0)$ is the solution periodic?

34.5 Determine a real fundamental system for each of the following linear ODEs:

(a) $\ddot{x} + 4\dot{x} - 77x = 0$.
(b) $\ddot{x} + 8\dot{x} + 16x = 0$.
(c) $\ddot{x} + 10\dot{x} + 29x = 0$.
(d) $\ddot{x} + 2\dot{x} = 0$.
(e) $\dddot{x} = 0$.

34.6 Investigate with the help of the oscillation equation

$$m\ddot{x} + b\dot{x} + cx = 0$$

the motion of a mass of $m = 50\,\text{kg}$, which is connected to an elastic spring of spring constant $c = 10200\,\text{N/m}$ if the system has the damping factor $b = 2000\,\text{kg/s}$. Let the mass at the beginning of the motion ($t = 0$) in the equilibrium position with velocity $v_0 = 2.8$ m/s is pushed ($x(0) = 0\,\text{m}$, $\dot{x}(0) = 2.8\,\text{m/s}$). Sketch the course of the movement.

34.7 Determine all functions $w(t)$, $t \geq 0$, with

$$w^{(4)} + 4a^4 w = 1\,, \quad a > 0\,, \quad w(0) = w'(0) = 0\,, \quad \lim_{t\to\infty} |w(t)| < \infty\,.$$

(Bending line of a one-sided infinitely long rail in a ballast bed with free support at $t = 0$.)

Some Special Types of Differential Equations 35

For a few types of differential equations, a solution procedure can be given for the analytical solution. We have already treated the separable, the linear 1st order and the linear *nth* order differential equations with constant coefficients. In this chapter, we consider some other types of differential equations that can be solved using a special approach.

To be sure that we get all solutions in each case, we recall the result in Sect. 34.1. A homogeneous linear ODE *n-th* order has an *n-dimensional* solution space. Thus we have always determined all solutions of a homogeneous linear ODE of *nth* order if n linearly independent solutions can be given.

35.1 The Homogeneous Differential Equation

A **homogeneous differential equation** is one of the form

$$\dot{x} = \varphi\left(\frac{x}{t}\right) \text{ with continuous } \varphi : I \to \mathbb{R}\,.$$

Example 35.1 An example of a homogeneous ODE, which is not immediately obvious, is as follows

$$t^2\,\dot{x} = t\,x + x^2\,.$$

C. Karpfinger, *Calculus and Linear Algebra in Recipes*,
https://doi.org/10.1007/978-3-662-65458-3_35

A division by t^2 provides

$$\dot{x} = \frac{x}{t} + \left(\frac{x}{t}\right)^2,$$

where we now have to consider $t \neq 0$. The function φ is given by $\varphi(z) = z + z^2$. ■

By substitution $z = \frac{x}{t}$, a homogeneous ODE becomes a separable differential equation, viz.

$$x = t\,z, \quad \text{so } \dot{x} = z + t\,\dot{z},$$

thus $\dot{x} = \varphi(\frac{x}{t})$ becomes the separable ODE

$$z + t\,\dot{z} = \varphi(z), \quad \text{so } \dot{z} = \frac{1}{t}\,(\varphi(z) - z).$$

To solve a homogeneous ODE, proceed as follows:

Recipe: Solving a Homogeneous Differential Equation
We solve the homogeneous ODE $\dot{x} = \varphi(\frac{x}{t})$:

(1) By the substitution $z = \frac{x}{t}$ in the given homogeneous ODE we get the separable ODE for $z(t)$:

$$\dot{z} = \frac{1}{t}\,(\varphi(z) - z).$$

(2) Solve the separable ODE by solving the integral and then solving for $z(t)$:

$$\int \frac{\mathrm{d}z}{\varphi(z) - z} = \ln|t| + c.$$

Note that $t \neq 0$ and $\varphi(z) \neq z$.

(3) One obtains the solution x by back substitution: Replace z by $\frac{x}{t}$.

Example 35.2 We continue the above example and start from the homogeneous ODE $\dot{x} = \frac{x}{t} + (\frac{x}{t})^2$.

(1) With the substitution $z = \frac{x}{t}$ the given ODE is transformed into $z + t\,\dot{z} = z + z^2$, i. e. into the separable ODE

$$\dot{z} = \frac{1}{t}\,z^2\,.$$

(2) We solve the separable ODE

$$\int \frac{\mathrm{d}z}{z^2} = \frac{-1}{z} = \ln|t| + c\,, \text{ so } z(t) = \frac{-1}{\ln|t| + c} \text{ and } z(t) = 0\,.$$

(3) Back substitution $z = \frac{x}{t}$ yields

$$x(t) = \frac{-t}{\ln|t| + c} \text{ and } x(t) = 0\,. \quad \blacksquare$$

35.2 The Euler Differential Equation

An **Euler differential equation** is a linear ODE of *nth* order with non-constant coefficients $a_k(t) = a_k\,t^k$ for all $k = 0, \ldots, n$, in detail

$$a_n\,t^n x^{(n)} + \cdots + a_1\,t\,\dot{x} + a_0\,x = s(t) \text{ with } a_k \in \mathbb{R}$$

and a perturbation function $s(t)$. The general solution L_a of such a ODE is again obtained by the general solution L_h of the corresponding homogeneous ODE

$$a_n\,t^n x^{(n)} + \cdots + a_1\,t\,\dot{x} + a_0\,x = 0$$

and a particular solution x_p of the inhomogeneous ODE; thus we have

$$L_a = x_p + L_h\,,$$

where the solution space L_h is a is an n-dimensional vector space, since the Eulerian ODE is linear. The number n is the order of the ODE. The specification of n linearly independent solutions $x_1, \ldots, x_n$ thus yields the solution $x = c_1x_1 + \cdots + c_nx_n$ of the homogeneous Euler ODE.

A particulate solution can be found with the by now familiar variation of parameters (see Sect. 34.2). The problem remains to determine the solution of the homogeneous ODE.

For this one makes the approach $x(t) = t^\alpha$, $t > 0$. One enters with this approach into the homogeneous ODE belonging to Euler's and obtains because of $(t^\alpha)^{(k)} = \alpha\,(\alpha -$

$1) \cdots (\alpha - (k-1))\, t^{\alpha-k}$ an equation of degree n for α:

$$\begin{aligned} 0 &= a_n t^n \alpha\,(\alpha-1)\cdots(\alpha-(n-1))\, t^{\alpha-n} + \cdots + a_1 t\, \alpha\, t^{\alpha-1} + a_0 t^\alpha \\ &= a_n \alpha\,(\alpha-1)\cdots(\alpha-(n-1))\, t^\alpha + \cdots + a_1 \alpha\, t^\alpha + a_0 t^\alpha \\ &= (a_n \alpha\,(\alpha-1)\cdots(\alpha-(n-1)) + \cdots + a_1\alpha + a_0)\, t^\alpha\,. \end{aligned}$$

For $t \neq 0$ α must be a solution of the polynomial equation

$$a_n \alpha\,(\alpha-1)\cdots(\alpha-(n-1)) + \cdots + a_1\alpha + a_0 = 0$$

of degree n in α. To solve an Eulerian ODE, proceed as in the case of a linear ODE with constant coefficients:

Recipe: Solve an Euler Differential Equation
Given is the Euler ODE

$$a_n t^n x^{(n)} + \cdots + a_1 t\, \dot{x} + a_0 x = s(t) \text{ with } a_k \in \mathbb{R} \text{ and a perturbation function } s(t)\,.$$

The general solution x_a is obtained as follows:

(1) Get the characteristic equation $p(\alpha) = 0$:

$$p(\alpha) = a_n \alpha\,(\alpha-1)\cdots(\alpha-(n-1)) + \cdots + a_1\alpha + a_0 = 0\,.$$

(2) Determine all solutions of $p(\alpha) = 0$, i.e., decompose $p(\alpha) = 0$ into the form

$$p(\alpha) = (\alpha-\alpha_1)^{m_1}\cdots(\alpha-\alpha_r)^{m_r} = 0 \text{ mit } \alpha_1, \ldots, \alpha_r \in \mathbb{C}\,.$$

(3) Give n linearly independent solutions $x_1, \ldots, x_n$ of the solution space U of the homogeneous ODE as follows:

- If $\alpha = \alpha_i \in \mathbb{R}$ with $m = m_i \in \mathbb{N}$ then select

$$t^\alpha,\ t^\alpha \ln(t), \ldots, t^\alpha (\ln(t))^{m-1}\,.$$

(continued)

- If $\alpha = a + \mathrm{i}\, b = \alpha_i \in \mathbb{C} \setminus \mathbb{R}$ with $m = m_i \in \mathbb{N}$: Strike $\overline{\alpha}_i$ and select

$$t^a \sin(b\ln(t)),\ t^a \sin(b\ln(t))\ln(t), \ldots, t^a \sin(b\ln(t))(\ln(t))^{m-1},$$

$$t^a \cos(b\ln(t)),\ t^a \cos(b\ln(t))\ln(t), \ldots, t^a \cos(b\ln(t))(\ln(t))^{m-1}.$$

This gives a total of n linearly independent real solutions $x_1, \ldots, x_n$. It is then $U = \{c_1x_1 + \cdots + c_nx_n \mid c_1, \ldots, c_n \in \mathbb{R}\}$ the solution space of the homogeneous ODE and

$$x_h(t) = c_1x_1(t) + \cdots + c_nx_n(t) \text{ with } c_1, \ldots, c_n \in \mathbb{R}$$

the (general) solution of the homogeneous ODE.

(4) Determine by variation of parameters a particulate solution x_p of the inhomogeneous ODE.

(5) The general solution is $x_a = x_p + x_h$.

Example 35.3

- We solve the homogeneous Eulerian ODE

$$t^2\ddot{x} + t\,\dot{x} - n^2x = 0 \text{ with } n \in \mathbb{N}_0.$$

With the approach $x(t) = t^\alpha$ we obtain the characteristic equation

$$\alpha\,(\alpha - 1) + \alpha - n^2 = 0 \Leftrightarrow \alpha^2 = n^2 \Leftrightarrow \alpha = \pm n.$$

This gives us the general solution

$$x(t) = \begin{cases} c_1t^n + c_2t^{-n}, & \text{if } n \neq 0 \\ c_1 + c_2\ln(t), & \text{if } n = 0 \end{cases}. \qquad \blacksquare$$

35.3 Bernoulli's Differential Equation

A nonlinear ODE of the form

$$\dot{x}(t) = a(t)\,x(t) + b(t)\,x^\alpha(t) \text{ with } \alpha \in \mathbb{R} \setminus \{0,\ 1\}$$

is called **Bernoulli's ODE**. The substitution $z(t) = (x(t))^{1-\alpha}$, i.e. $x(t) = (z(t))^{\frac{1}{1-\alpha}}$, turns the nonlinear Bernoulli's ODE for x into a linear ODE for z. For the proof we set $\dot{x}(t) = \frac{1}{1-\alpha}\dot{z}(t)x^{\alpha}(t)$, $x(t)$ and x^{α} into the Bernoulli's ODE and divide the resulting equation by $x^{\alpha}(t)$:

$$\tfrac{1}{1-\alpha}\,\dot{z}(t)\,x^{\alpha}(t) = a(t)\,x(t) + b(t)\,x^{\alpha}(t) \;\Leftrightarrow\; \tfrac{1}{1-\alpha}\,\dot{z}(t) = a(t)\,z(t) + b(t)\,.$$

Therefore, the following procedure for solving a Bernoulli's ODE results:

Recipe: Solving a Bernoulli's ODE
To solve the Bernoulli's ODE

$$\dot{x}(t) = a(t)\,x(t) + b(t)\,x^{\alpha}(t) \text{ with } \alpha \in \mathbb{R} \setminus \{0, 1\}$$

proceed as follows:

(1) Determine the general solution $z = z(t)$ of the linear ODE of 1st order

$$\frac{1}{1-\alpha}\dot{z}(t) = a(t)\,z(t) + b(t$$

using the recipe in Sect. 33.3

(2) Obtain the solution by back substitution $x(t) = (z(t))^{\frac{1}{1-\alpha}}$.

(3) Any initial condition specifies the constant c.

Example 35.4 The **logistic ODE** is

$$\dot{x}(t) = a\,x(t) - b\,x^2(t)$$

and is thus a Bernoulli's ODE (with $\alpha = 2$). We solve an IVP with a logistic ODE, viz.

$$\dot{x}(t) = x(t) - x^2(t) \text{ with } x(0) = 2\,.$$

(1) We solve the linear ODE $\dot{z}(t) = -z(t) + 1$:

The solution of the homogeneous ODE is $z_h(t) = c\,\mathrm{e}^{-t}$, a particulate solution is $z_p(t) = 1$ (*Variation of parameters* or *approach of the right-hand side type*). Thus

$$z_a(t) = 1 + c\,\mathrm{e}^{-t} \text{ with } c \in \mathbb{R}$$

is the general solution of the linear ODE.

(2) We obtain the general solution

$$x_a(t) = \frac{1}{1 + c\,\mathrm{e}^{-t}} \text{ with } c \in \mathbb{R}.$$

(3) The initial condition $x(0) = 2$ gives $2 = \frac{1}{1+c}$, thus $c = -1/2$; we obtain the solution

$$x(t) = \frac{1}{1 - \mathrm{e}^{-t}/2}. \quad \blacksquare$$

35.4 The Riccati Differential Equation

A **Riccati's ODE** is a 1st order nonlinear ODE of the form.

$$\dot{x}(t) = a(t)\,x^2(t) + b(t)\,x(t) + r(t).$$

There is no general solution procedure for this ODE. However, if one knows a particulate solution $x_p = x_p(t)$ of this ODE, e.g. by trying, one can determine all solutions of this ODE:

Recipe: Solving a Riccati's ODE
If a solution $x_p = x_p(t)$ of the Riccati's ODE

$$\dot{x}(t) = a(t)x^2(t) + b(t)\,x(t) + r(t)$$

is known, then all solutions of this ODE are obtained as follows:

(1) Using the recipe in Sect. 35.3 determine the general solution $z_a = z_a(t)$ of Bernoulli's ODE

$$\dot{z}(t) = a(t)\,z(t)^2 + (2\,x_p(t)a(t) + b(t))z(t).$$

(2) Give the general solution $x_a = x_a(t)$ of the Riccati ODE:

$$x_a(t) = x_p(t) + z_a(t).$$

(3) Any initial condition specifies the constant c.

Example 35.5
We solve the Riccati's ODE

$$\dot{x}(t) = -\frac{1}{t^2 - t} x^2(t) + \frac{1 + 2t}{t^2 - t} x(t) - \frac{2t}{t^2 - t}.$$

Apparently $x_p = x_p(t) = 1$ is a solution of this ODE. We apply the above prescription:

(1) We use the recipe in Sect. 35.3 to determine the general solution $z_a = z_a(t)$ of the following Bernoulli's ODE with $\alpha = 2$:

$$\dot{z}(t) = \frac{-1 + 2t}{t^2 - t} z(t) + \frac{-1}{t^2 - t} z^2(t).$$

(1) We use the recipe in Sect. 33.3 the general solution $y = y(t)$ of the linear ODE of 1st order

$$\dot{y}(t) = \frac{1 - 2t}{t^2 - t} y(t) + \frac{1}{t^2 - t}.$$

(1) It is $y_h(t) = \frac{c}{t^2 - t}$ (separation of variables) the general solution of the homogeneous linear ODE.
(2) It is $y_p(t) = \frac{1}{t-1}$ (variation of parameters) a particulate solution of the linear ODE.
(3) It is $y_a(t) = \frac{t+c}{t^2-t}$ the general solution of the linear ODE.

(2) By back substitution $z_a(t) = (y_a(t))^{\frac{1}{1-2}}$ we obtain the general solution of Bernoulli's ODE:

$$z_a(t) = \frac{t^2 - t}{t + c}.$$

(2) This gives us the general solution of Riccati's ODE:

$$x_a(t) = 1 + \frac{t^2 - t}{t + c} = \frac{t^2 + c}{t + c}. \quad \blacksquare$$

There are many other special types of ODEs that can be solved by suitable approaches. We refrain from presenting these types. For one thing, this brief excursion into the subject of differential equations should leave a first and positive impression on the reader: Yes, we can solve many ODEs using a scheme. On the other hand, we do not want to hide the reality: The ODEs one often has to deal with in practice are often not solvable analytically. If one nevertheless wants to have *solutions*, one has to rely on numerical methods. We will talk about these methods as well. But first we will give a last method, with which one can give solutions of ODEs (possibly only Taylor polynomials of these). This method of solution springs, as it were, from an act of desperation, but it is extremely fruitful.

35.5 The Power Series Approach

We consider a linear ODE with coefficient functions that are not necessarily constant.

$$x^{(n)} + a_{n-1}(t)\, x^{(n-1)} + \cdots + a_1(t)\, \dot{x} + a_0(t)x = s(t)\,.$$

Our previous solution methods all fail if this does not happen to be an Eulerian, Bernoulliian, Riccatiian ODE or one with constant coefficients. An idea to nevertheless get a solution or at least a Taylor polynomial of a solution comes from the following observation: If all functions $a_0(t), \ldots, a_{n-1}(t),\ s(t)$ in a Taylor series around a point a can be developed (cf. Chap. 28), then there exists a solution x of this ODE, which can also be represented as a power series with development point a, i.e., we have the following

$$x(t) = \sum_{k=0}^{\infty} c_k (t-a)^k$$

in a neighborhood of a. The ODE by means of a **power series approach** to solve now means to determine the coefficients c_k. If you can determine all of them, then you have found a solution function x in power series notation. If you can only the first $n+1$ coefficients $c_0, \ldots, c_n$ determine then you have the Taylor polynomial $T_{n,x,a}(x)$ of a solution function x found. The solution method is as follows:

Recipe: Solve a ODE Using Power Series Approach
To solve a linear ODE

$$x^{(n)} + a_{n-1}(t)\, x^{(n-1)} + \cdots + a_1(t)\, \dot{x} + a_0(t)x = s(t)$$

by means of a power series approach around the development point a proceed as follows:

(1) Develop all functions $a_0(t), \ldots, a_{n-1}(t),\ s(t)$ into Taylor series a and obtain the following representation of the ODE:

$$x^{(n)} + \sum_{k=0}^{\infty} c_k^{(a_{n-1})} (t-a)^k\, x^{(n-1)} + \cdots + \sum_{k=0}^{\infty} c_k^{(a_0)} (t-a)^k x = \sum_{k=0}^{\infty} c_k^{(s)} (t-a)^k\,.$$

(continued)

(2) Set $x(t) = \sum_{k=0}^{\infty} c_k(t-a)^k$ with the unknown coefficients c_k, $k \in \mathbb{N}_0$ into the ODE in (1) and obtain a new representation of the ODE.
(3) Summarize the coefficients for equal powers of $t-a$ on the left side of the ODE from (2) as compactly as possible; here index shifts are often necessary, one obtains:

$$d_0 + d_1(t-a) + d_2(t-a)^2 + \ldots = \sum_{k=0}^{\infty} c_k^{(s)}(t-a)^k\,,$$

where the d_l depend on c_k.
(4) Equating coefficients, $d_k = c_k^{(s)}$, often leads to recursion formulas, with which the coefficients $c_n, c_{n+1}, \ldots$ can be determined by the coefficients $c_0, \ldots, c_{n-1}$. If initial conditions $x(a) = x_0, \ldots, x^{(n-1)}(a) = x_{n-1}$ are given, then

$$c_0 = x_0\,,\ c_1 = x_1\,,\ c_2 = \frac{1}{2}x_2, \ldots, c_{n-1} = \frac{1}{(n-1)!}x_{n-1}\,.$$

This should be taken into account right away in the recursion formulas.

In general, because of the increasing computational effort, one has to stop the successive determination of the coefficients $c_0, c_1, \ldots$ and thus only a Taylor polynomial is obtained as approximate solution.

Example 35.6 We apply the power series approach to the following IVP:

$$\ddot{x} + (t-1)\,x = \mathrm{e}^t\,,\ x(1) = 2\,,\ \dot{x}(1) = -1\,.$$

Since the initial conditions in $t_0 = 1$ are given, we choose as development point $a = 1$:

(1) The development of the coefficient function $a_0(t) = t-1$ around the development point $a = 1$ is already done. What remains is the development of the perturbation function $s(t) = \mathrm{e}^t$ around the point $a = 1$, which is due to $\mathrm{e}^t = \mathrm{e}\,\mathrm{e}^{t-1}$ apparently $s(t) = \sum_{k=0}^{\infty} \frac{\mathrm{e}}{k!}(t-1)^k$, thus we obtain

$$\ddot{x} + (t-1)\,x = \sum_{k=0}^{\infty} \frac{\mathrm{e}}{k!}(t-1)^k\,.$$

(2) We set $x(t) = \sum_{k=0}^{\infty} c_k (t-1)^k$ with the unknown coefficients c_k, $k \in \mathbb{N}_0$, into the differential equation; because of $\dot{x}(t) = \sum_{k=1}^{\infty} k\, c_k\, (t-1)^{k-1}$ and $\ddot{x}(t) = \sum_{k=2}^{\infty} (k-1)\, k\, c_k\, (t-1)^{k-2}$ we obtain:

$$\sum_{k=2}^{\infty} (k-1)\, k\, c_k\, (t-1)^{k-2} + \sum_{k=0}^{\infty} c_k\, (t-1)^{k+1} = \sum_{k=0}^{\infty} \frac{\mathrm{e}}{k!} (t-1)^k .$$

(3) To summarize the coefficients for equal powers of $t-1$ on the left-hand side of the ODE from (2), we reindex to obtain the representation:

$$2\, c_2 (t-1)^0 + \sum_{k=1}^{\infty} ((k+1)\,(k+2)\, c_{k+2} + c_{k-1})\,(t-1)^k = \sum_{k=0}^{\infty} \frac{\mathrm{e}}{k!} (t-1)^k .$$

(4) Substituting the initial conditions, we obtain:

$$2 = x(1) = c_0 \text{ and } -1 = \dot{x}(1) = c_1 ,$$

so that $c_0 = 2$ and $c_1 = -1$ have already been determined. We now compare the coefficients to the left and right of the equal sign in front of the same powers $(t-1)^k$ and consider $c_0 = 2$ and $c_1 = -1$:

$$\begin{aligned}
&(t-1)^0 : c_2 = \frac{\mathrm{e}}{2} \\
&(t-1)^1 : 6\, c_3 + c_0 = \mathrm{e} && \Rightarrow c_3 = \frac{1}{6}(\mathrm{e}-2) \\
&(t-1)^2 : 12\, c_4 + c_1 = \frac{\mathrm{e}}{2} && \Rightarrow c_4 = \frac{1}{12}\left(\frac{\mathrm{e}}{2} + 1\right) \\
&(t-1)^3 : 20\, c_5 + c_2 = \frac{\mathrm{e}}{6} && \Rightarrow c_5 = \frac{1}{20}\left(\frac{\mathrm{e}}{6} - \frac{\mathrm{e}}{2}\right) \\
&\qquad \vdots && \qquad \vdots
\end{aligned}$$

Thus the first members of the Taylor expansion of the solution are $x(t)$

$$x(t) = 2 - (t-1) + \frac{\mathrm{e}}{2}(t-1)^2 + \frac{1}{6}(\mathrm{e}-2)\,(t-1)^3 + \frac{1}{12}\left(\frac{\mathrm{e}}{2} + 1\right)(t-1)^4 \ldots$$

Figure 35.1 shows, next to the graph of the correct solution, the graph of this approximate solution (dashed line) in a neighborhood of 1. In a neighborhood of 1, our solution approximates the exact solution. ■

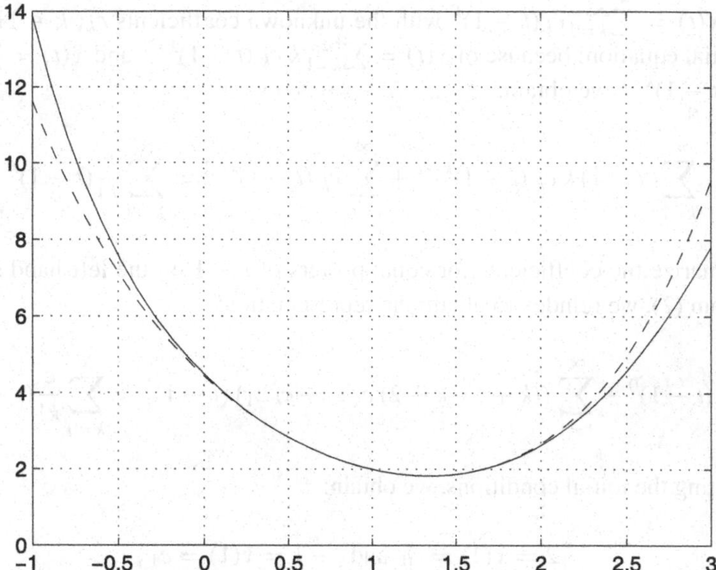

Fig. 35.1 Solution and approximation

The requirement that the ODE under consideration is linear can be waived: The power series approach can also be used to solve nonlinear ODEs. Only then, e. g., powers of series have to be calculated; equating coefficients then quickly becomes an inscrutable story in general. For the practical applications this solution method is omitted, therefore we do not consider examples for it.

35.6 Exercises

35.1 Solve the IVPe:

(a) $\dot{x}\,t - 2\,x - t = 0,\, t>0,\, x(2) = 6,$
(b) $\dot{x} - \frac{1}{2}\cot t\, x + \cos t\, x^3 = 0,\, 0<t<\pi,\, x(\pi/2) = 1.$

35.2 Solve the following ODEs:

(a) $4x\dot{x} - x^2 + 1 + t^2 = 0,$
(b) $t(t-1)\dot{x} - (1+2t)x + x^2 + 2t = 0,$
(c) $t\dot{x} = x + \sqrt{x^2 - t^2},$
(d) $t^2\ddot{x} + t\dot{x} - x = 0,$
(e) $t^2 x^{(4)} + 3\dddot{x} - \frac{7}{t}\dot{x} + \frac{8}{t^2}x = 0,$
(f) $t^2\ddot{x} - t\dot{x} + 2x = 0.$

35.3 Given is the IVP $\dot{x} = tx + t\,, \quad x(0) = 0$.

(a) Represent the solution as a power series, find the first five non vanishing members, and use them to calculate an approximation of $x(2)$.
(b) By repeatedly differentiating the ODE, determine the Taylor expansion of $x(t)$ at the point $t_0 = 0$.
(c) Find the solution $x(t)$ explicitly in closed form, derive from it the power series expansion at $t_0 = 0$ and calculate $x(2)$.

35.4 Using the power series approach, solve the IVP for $(1 + t^2)\,\ddot{x} + t\dot{x} - x = 0$ with the initial values $x(0) = 0$, $\dot{x}(0) = 1$ or $x(0) = \dot{x}(0) = 1$.

35.5 Determine the general solution of the following ODEs using a power series approach:
(a) $\ddot{x} - tx = 0$. (b) $\dot{x} + tx = 0$. (c) $\dot{x} - tx = 1 + t$.

35.6 Let x be the solution of the IVP $\ddot{x} + tx + \mathrm{e}^{t^2} = 0$ with initial values $x(0) = \dot{x}(0) = 1$. Using a (truncated) power series approach, determine the Taylor polynomial $T_{4,x,0}$ from x of degree 4 around the development point 0.

Numerics of Ordinary Differential Equations I 36

Differential equations and hence initial value problems occupy a role in engineering and science which should not be underestimated. We have devoted numerous chapters to this problem, which is so fundamental. In Chaps. 33, 34 and 35 we dealt with the (exact) analytical solution of differential equations and initial value problems, respectively. In the mentioned chapters we also mentioned several times that initial value problems are analytically solvable only in rare cases. In most cases one has to be content with approximate solutions. Thereby one does not determine the searched function $x = x(t)$ by approximation, but in general the values $x(t_i)$ of the unknown function x at discrete points $t_0, \ldots, t_n$.

36.1 First Procedure

Given is an initial value problem (IVP)

$$\dot{x} = f(t, x) \text{ with } x(t_0) = x_0$$

with a differential equation (ODE) of 1st order $\dot{x} = f(t, x)$. We assume that the IVP is uniquely solvable. Furthermore, no explicit solution method is to be known, so that one must be content with approximations.

A first approximation method for the solution *x*(*t*) of the IVP is obtained as follows: We subdivide the interval $[t_0, t]$ for an $t > t_0$ equidistant with the **step size** $h = \frac{t-t_0}{n}$ by the **interpolation points**

$$t_k = t_0 + k\,h \text{ with } k = 0, 1, \ldots, n,$$

C. Karpfinger, *Calculus and Linear Algebra in Recipes*,
https://doi.org/10.1007/978-3-662-65458-3_36

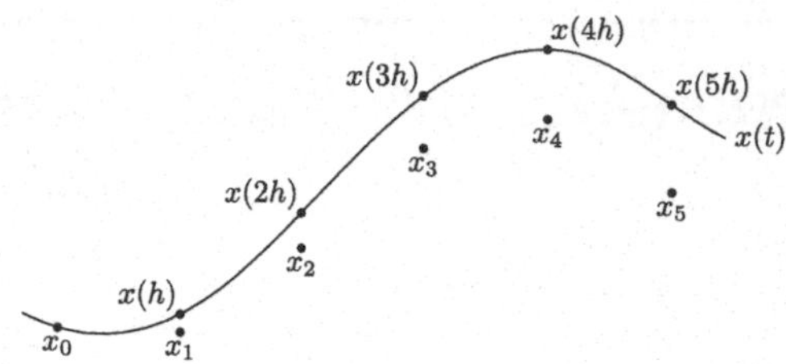

Fig. 36.1 Exact values and approximate values

i.e. $t_n = t$ and determine from the initial value $x_0 = x(t_0)$ one after the other approximations $x_1, \ldots, x_n$ for the exact (and unknown) function values $x(t_1), \ldots, x(t_n)$. One speaks of a **discretization** of the IVP.

We approximate the exact values $x(t_k)$ of the solution x at the points $t_k = t_0 + k\,h$ by numbers x_k (in the adjacent Fig. 36.1 we choose $t_0 = 0$). The various methods differ in the way in which x_{k+1} is calculated from $x_0, \ldots, x_k$. One distinguishes in principle:

- **Single-step method**, where x_{k+1} only from x_k and
- **Multistep method** where x_{k+1} from $x_{k-j+1}, \ldots, x_k$ for a $j \in \{2, \ldots, k+1\}$ is calculated.

Simple Single-Step Methods
With the following methods we obtain approximate solutions x_k for the exact solution $x(t_k)$ of the IVP

$$\dot{x} = f(t, x) \text{ with } x(t_0) = x_0$$

at the points $t_k = t_0 + k\,h$ with $k = 0, 1, \ldots, n$ and $h = \frac{t - t_0}{n}$ for a $n \in \mathbb{N}$.

- In the **explicit Euler method** the approximation points x_k for $x(t_k)$ are recursively from x_0 determined by

$$x_{k+1} = x_k + h\,f(t_k, x_k)\,,\; k = 0, 1, 2 \ldots .$$

- With the **midpoint rule** one determines the x_k recursively according to

$$x_{k+1} = x_k + h\,f\left(\frac{t_k + t_{k+1}}{2}, \frac{x_k + x_{k+1}}{2}\right)\,,\; k = 0, 1, 2 \ldots .$$

- In the **implicit Euler method** one determines the x_k recursively according to

$$x_{k+1} = x_k + h\,f\,(t_{k+1}, x_{k+1})\,,\; k = 0, 1, 2 \ldots .$$

Note: With the midpoint rule and the implicit Euler method, the values x_{k+1} are noch explicitly given, they also appear in

$$f\left(\frac{t_k + t_{k+1}}{2}, \frac{x_k + x_{k+1}}{2}\right) \quad \text{and} \quad f(t_{k+1}, x_{k+1}),$$

they are only implicitly given. To calculate these values x_{k+1}, a linear or nonlinear equation (depending on f) has to be solved. In general, the Newton method is suitable for this.

In all three methods, however, the calculated values x_k are then approximations to the exact values $x(t_k)$ of the solution of the IVP.

Example 7 We consider the IVP

$$\dot{x} = t - x\,, \; x(0) = 1\,.$$

with the exact solution

$$x(t) = 2\,\mathrm{e}^{-t} + t - 1\,.$$

We now determine approximate solutions with the three given methods x_k for $x(t_k)$ for the discrete points $t_1, \ldots, t_n$. We choose

$$t = 0.6 \text{ and } n = 10\,, \text{ so } h = 0.06 \text{ and } t_k = k \cdot 0.06 \text{ with } k = 0,\, 1,\, ,\, \ldots,\, 10\,.$$

Using MATLAB we get the results in the following table:

t_k	x_k exakt	x_k Euler (expl.)	x_k Mittelpunktsregel	x_k Euler (impl.)
0.00	1.0000	1.0000	1.0000	1.0000
0.06	0.9435	0.9400	0.9435	0.9468
0.12	0.8938	0.8872	0.8938	0.9000
0.18	0.8505	0.8412	0.8505	0.8592
0.24	0.8133	0.8015	0.8131	0.8242
0.30	0.7816	0.7678	0.7815	0.7945
0.36	0.7554	0.7397	0.7552	0.7699
0.42	0.7341	0.7170	0.7339	0.7501
0.48	0.7176	0.6991	0.7174	0.7348
0.54	0.7055	0.6860	0.7053	0.7238
0.60	0.6976	0.6772	0.6974	0.7168

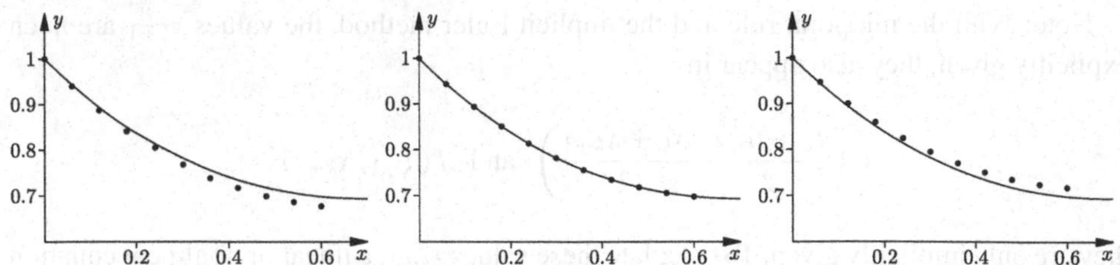

Fig. 36.2 Exact solution and approximations using explicit Euler, the midpoint rule, implicit Euler

In the following Fig. 36.2, the obtained approximate values are plotted next to the graph of the exact solution: ■

In the examples of single-step methods discussed so far, the intervals between $t_0, t_1, \ldots, t_n$ are equal, of course this does not have to be so, more generally one speaks of a **time grid**

$$\Delta = \{t_0, t_1, \ldots, t_n\} \subseteq \mathbb{R}$$

with the **step sizes** $h_j = t_{j+1} - t_j$ for $j = 0, \ldots, n-1$ and the **maximum step size**

$$h_\Delta = \max\{h_j \mid j = 0, \ldots, n-1\}.$$

In the approximate solution of an IVP, one determines a **lattice function**

$$x_\Delta : \Delta \to \mathbb{R}$$

with $x_\Delta(t_j) \approx x(t_j)$ for all $j = 0, \ldots, n-1$.

In a single-step method, successive

$$x_\Delta(t_1),\ x_\Delta(t_2), \ldots, x_\Delta(t_n)$$

are calculated, whereby in the calculation of $x_\Delta(t_{k+1})$ only $x_\Delta(t_k)$ is included. This is suggestively abbreviated with the following notation:

$$x_\Delta(t_0) = x_0\,,\ x_\Delta(t_{k+1}) = \Psi(x_\Delta(t_k),\ t_k,\ h_k)$$

and also briefly speaks of the single-step method Ψ.

Remark In Chap. 73 we will discuss the terms *consistency order* and *convergence order* of singe-step methods in order to obtain an estimate of the goodness of these methods. The consistency order and the convergence order of a method usually coincide. In the case of *consistency*, a local error is considered as a function of a step size h. With the *convergence*, on the other hand, one obtains a global estimate of the quality of a single-step method as a function of a *time grid* $\{t_0, t_1, \ldots, t_n\}$. Thus, a high consistency order locally ensures that the error made by the single-step method Ψ when the step size is reduced is *h quickly* becomes vanishingly small. So, in this view, it is desirable to have single-step methods with high consistency order at hand. We anticipate:

- The explicit Euler method has consistency order $p = 1$.
- The midpoint rule has consistency order $p = 2$.
- The classical Runge-Kutta procedure discussed in the next section has consistency order $p = 4$.

36.2 Runge-Kutta Method

For the calculation of the approximate solutions x_k for the exact solution $x(t_k)$ of the IVP

$$\dot{x} = f(t, x) \text{ with } x(t_0) = x_0$$

at the points $t_k = t_0 + k\,h$ with $k = 0, 1, \ldots, n$ and $h = \frac{t-t_0}{n}$ for a $n \in \mathbb{N}$ we have given three single-step methods in Sect. 36.1, where we get x_{k+1} by the following expression:

$$x_k + h\,f(t_k, x_k) \text{ resp. } x_k + h\,f\left(\frac{t_k + t_{k+1}}{2}, \frac{x_k + x_{k+1}}{2}\right) \text{ resp. } x_k + h\,f\,(t_{k+1}, x_{k+1})$$

for $k = 0, 1, 2 \ldots$. The *2-stage Runge-Kutta method* is a single-step method in which the approximate value x_{k+1} is obtained in the following manner:

$$x_{k+1} = x_k + h\,f\left(t_k + \frac{h}{2}, x_k + \frac{h}{2} f(t_k, x_k)\right).$$

The *form* is the same, and of course all this can be generalized to so-called *s-stage Runge-Kutta methods*, which seem correspondingly more complicated, but still carry the same basic principle.

Remark The idea of Runge-Kutta methods originates from the Taylor expansion of the exact solution x in t_0. From the integral representation of the IVP we obtain:

$$\begin{aligned} x(t_0 + h) &= x_0 + \int_{t_0}^{t_0+h} f(s, x(s))\mathrm{d}s \\ &= x_0 + h\,f\left(t_0 + \frac{h}{2}, x\left(t_0 + \frac{h}{2}\right)\right) + \cdots \\ &= x_0 + h\,f\left(t_0 + \frac{h}{2}, x_0 + \frac{h}{2} f(t_0, x_0)\right) + \cdots . \end{aligned}$$

The General and the Classical Runge-Kutta Method
A *s*-**stage Runge-Kutta method** is

$$x_{k+1} = x_k + h \sum_{i=1}^{s} b_i k_i$$

with

$$k_i = f\left(t + c_i h, x + h \sum_{j=1}^{s} a_{ij} k_j\right)$$

for $i = 1, \dots, s$. Here one calls k_i the **steps**, the number s the **number of stages** and the coefficients b_i **weights**.
If $a_{ij} = 0$ for $j \geq i$, then the Runge-Kutta method is explicit, otherwise it is implicit.
The **classical Runge-Kutta method** is a 4th order method, it is

$$x_{k+1} = x_k + \frac{h}{6}(k_1 + 2k_2 + 2k_3 + k_4)$$

with

$$k_1 = f(t_k, x_k), \ k_2 = f\left(t_k + \frac{h}{2}, x_k + \frac{h}{2} k_1\right),$$

$$k_3 = f\left(t_k + \frac{h}{2}, x_k + \frac{h}{2} k_2\right), \ k_4 = f(t_k + h, x_k + h\, k_3).$$

A Runge-Kutta process can be described uniquely by the quantities

$$c = (c_1, \dots, c_s)^\top \in \mathbb{R}^s, \ b = (b_1, \dots, b_s)^\top \in \mathbb{R}^s \text{ and } A = (a_{ij})_{ij} \in \mathbb{R}^{s \times s}$$

unambiguously. One notes this clearly in the **Butcher scheme**

$$\begin{array}{c|c} c & A \\ \hline & b \end{array}$$

Example 8 With the classical Runge-Kutta method, we have the Butcher scheme.

$$\begin{array}{c|cccc}
0 & 0 & 0 & 0 & \\
1/2 & 1/2 & 0 & 0 & \\
1/2 & 0 & 1/2 & 0 & 0 \\
1 & 0 & 0 & 1 & 0 \\
\hline
 & 1/6 & 1/3 & 1/3 & 1/6
\end{array}$$

We now consider the example 36.1 again and obtain the approximate solutions using the classical Runge-Kutta method:

t_k	$x(t_k)$ exakt	x_k Runge-Kuttaverfahren
0.00	1.0000	1.0000
0.06	0.943529067168497	0.943529080000000
0.12	0.893840873434315	0.893840897602823
0.18	0.850540422822544	0.850540456964110
0.24	0.813255722133107	0.813255765004195
0.30	0.781636441363436	0.781636491831524
0.36	0.755352652142062	0.755352709176929
0.42	0.734093639630113	0.734093702295764
0.48	0.717566783612282	0.717566851059467
0.54	0.705496504747979	0.705496576207267
0.60	0.697623272188053	0.697623346963412

In Fig. 36.3, the obtained approximate values are plotted next to the graph of the exact solution. ■

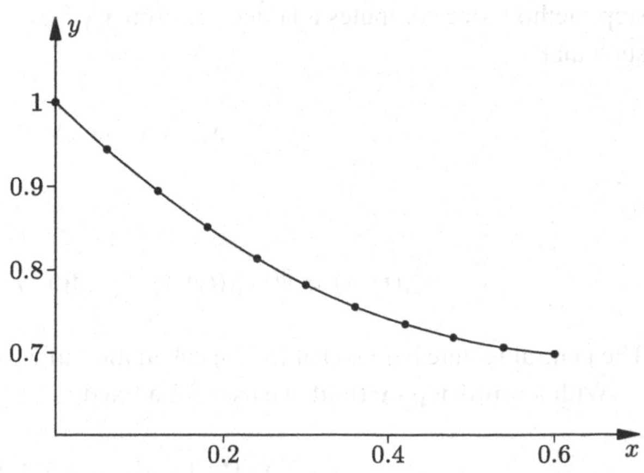

Fig. 36.3 Exact solution and approximations using classical Runge-Kutta method

Remarks 1. Up to now, we have always selected a fixed step size h. The step size may vary from step to step. With the so-called step size *control*, the step size is designed in such a way that, on the one hand, the computational effort is low (the step size should be large here), but, on the other hand, the approximate solution approximates the exact solution well (the step size should be small here). We do not describe the step-size control in detail, but we want to emphasize that the possibility of adaptive step-size control is the main advantage of the one-step methods compared to the multi-step methods.

2. The s-stage Runge-Kutta methods become more complicated with large s; the advantages of the high order are quickly can celled out by the many necessary function evaluations. In practice one uses *s-stage* methods with $4 \leq s \leq 7$.

MATLAB MATLAB implements methods for the numerical solution of IVPs. Typical functions are `ode45` or `ode113`; note the descriptions of these functions by calling e.g. `doc ode45`. All methods are called in a similar way, the exact conventions can be found in the description.

`ode113` is a *multi-step procedure* with *variable order and step-size control.*

36.3 Multistep Methods

In this concluding section on the numerics of ordinary differential equations, we briefly discuss *multistep methods*. In these methods, the information from the previously calculated support points is used to calculate an approximate value. One is carried thereby by the hope to obtain a high order with few f-evaluations.

We consider again the IVP:

$$\dot{x} = f(t, x) \text{ with } x(t_0) = x_0 .$$

Let the exact (unknown) solution $x : [t_0, t] \to \mathbb{R}$ is continuously differentiable. In a single-step method, one computes a lattice function $x_\Delta : \Delta \to \mathbb{R}$ on a lattice $\Delta = \{t_0, \ldots, t_n\}$ such that

$$x_\Delta \approx x \text{ on } \Delta ,$$

by

$$x_\Delta(t_{j+1}) = \Psi(x_\Delta(t_j), t_j, h_j) \text{ for } j = 0, 1, 2, \ldots .$$

The central feature here is that for the calculation of $x_\Delta(t_{j+1})$ only $x_\Delta(t_j)$ is used.

With a **multistep method** one uses for a fixed $k \in \mathbb{N}$ the approximate values

$$x_\Delta(t_{j-k+1}), \ldots, x_\Delta(t_j) ,$$

to calculate the approximate value $x_\Delta(t_{j+1})$. Thereby two problems arise:

- For the first step in a k-step procedure one needs additionally $k-1$ initial values.
- It seems that an equidistant grid is necessary.

The first problem is solved by using a single-step method to obtain the additional values. The second problem can also be circumvented: There is a step-size control for multistep methods, but it is much more complicated than for single-step methods. We will not discuss this problem and will only consider multistep methods with constant step size h.

Multistep Methods
Given is an equidistant lattice $\Delta = \{t_0, t_1, \ldots, t_n\}$ with the step size $h = t_{j+1} - t_j$ for all j.

- The **explicit midpoint rule**:

$$x_{j+1} = x_{j-1} + 2h\, f(t_j, x_j)$$

is a 2-step method of consistency order $p=2$.
- The **Adams-Bashforth method**:

$$x_{j+k} = x_{j+k-1} + \tau \sum_{i=0}^{k-1} \beta_i f(t_{j+i}, x_{j+i})$$

is an explicit k-step method of the consistency order $p = k$.
- The **Adams-Moulton method**:

$$x_{j+k} = x_{j+k-1} + \tau \sum_{i=0}^{k} \beta_i f(t_{j+i}, x_{j+i})$$

is an implicit k-step method of the consistency order $p = k + 1$.

The multistep methods arise from the common idea of representing in

$$x_{j+k} = x_{j+k-1} + \int_{t_{j+k-1}}^{t_{j+k}} f(t, x(t))\mathrm{d}t$$

the integrand $f(t, x(t))$ by a polynomial $q(t)$, i.e.

$$q(t_{j+i}) = f(t_{j+i}, x_{j+i}) \text{ for } j = 0, \ldots, k-1 \text{ or } k\,.$$

Thus one obtains

$$x_{j+k} = x_{j+k-1} + \int_{t_{j+k-1}}^{t_{j+k}} q(t)\,\mathrm{d}t\,.$$

By integrating this polynomial, one obtains the weights β_i of the above methods. Consider the following examples.

Example 36.3 For $k = 3, 4$ we give the formulas (with an index shift) with the weights:

- The Adams-Bashforth method for $k = 3$:

$$x_{j+1} = x_j + \frac{h}{12}\big(23 f(t_j, x_j) - 16 f(t_{j-1}, x_{j-1}) + 5 f(t_{j-2}, x_{j-2})\big)\,.$$

- The Adams-Bashforth method for $k = 4$:

$$\begin{aligned} x_{j+1} = x_j + \frac{h}{24}\big(&55 f(t_j, x_j) - 59 f(t_{j-1}, x_{j-1}) \\ &+ 37 f(t_{j-2}, x_{j-2}) - 9 f(t_{j-3}, x_{j-3})\big)\,. \end{aligned}$$

- The Adams-Moulton method for $k = 1$:

$$x_{j+1} = x_j + \frac{h}{2}\big(f(t_{j+1}, x_{j+1}) + f(t_j, x_j)\big)\,.$$

- The Adams-Moulton method for $k = 2$:

$$x_{j+1} = x_j + \frac{h}{12}\big(5 f(t_{j+1}, x_{j+1}) + 8 f(t_j, x_j) - f(t_{j-1}, x_{j-1})\big)\,.$$

- The Adams-Moulton method for $k = 3$:

$$x_{j+1} = x_j + \frac{h}{24}\big(9 f(t_{j+1}, x_{j+1}) + 19 f(t_j, x_j) - 5 f(t_{j-1}, x_{j-1}) + f(t_{j-2}, x_{j-2})\big)\,.$$

- The Adams-Moulton method for $k = 4$:

$$\begin{aligned} x_{j+1} = {} & x_j + \frac{h}{720}\big(251 f(t_{j+1}, x_{j+1}) + 646 f(t_j, x_j) \\ & - 264 f(t_{j-1}, x_{j-1}) + 106 f(t_{j-2}, x_{j-2}) - 19 f(t_{j-3}, x_{j-3})\big)\,. \end{aligned}$$

■

36.4 Exercises

36.1 Program the explicit and implicit Euler method as well as the midpoint rule.

36.2 We consider the IVP

$$\dot{x} = 1 + (x - t)^2, \quad x(0) = 1/2 .$$

Choose as step size $h = 1/2$ and calculate the value $x(3/2)$ using the

(a) explicit Euler method,
(b) classical Runge-Kutta method.

Compare your results to the exact solution $x(t) = t + \frac{1}{2-t}$.

36.3 Implement the classic Runge-Kutta method. Choose as step sizes $h = 0.1; 0.01; 0.001$ and use it to calculate the value $x(1.8)$ for the IVP from Exercise 36.2 Compare your results with the exact solution.

36.4 Solve the IVP from Exercise 36.2 using the Adams-Moulton method. In the first step use $k = 1$ and then $k = 2$. Compare your result for the value $x(1)$ with the results from Exercise 36.4.

Note: When solving the quadratic equation for x_{j+1} choose in each case the value closest to x_j.

36.5 For the 2-stage Runge-Kutta method, determine the coefficients of $A \in \mathbb{R}^{2\times 2}, b, c \in \mathbb{R}^2$ of the Butcher scheme.

36.6 Use the Euler method and the Runge-Kutta method to solve the IVP $\dot{x} = 2t \sin(t^2) x,\ x(0) = 1$ on the interval $t \in [0, \sqrt{2\pi}]$ with $N = 30$ steps and plot both results. The exact solution is $x(t) = \mathrm{e}^{1-\cos(t^2)}$. Without plotting the exact solution, can you tell which of the approximations is more accurate?

36.7 A rocket car with a mass of 1000kg has filled up with an additional 1000kg of fuel. At time 0 (car is stationary) the fuel is ignited, which now burns explosively, so that the amount of fuel remaining at time t until the end of the burning time is $t_e = 10s$ given by $1000kg - 10\frac{kg}{s^2}t^2$ is given. At the same time the vehicle is subjected to $t < t_e$ the driving force $1000N \cdot (t^2/s^2)$. Due to the air resistance, the car is also subjected to the force $0.7\frac{Ns^2}{m^2}v^2$ against the driving direction.

(a) Determine the differential equation for the velocity v of the car at time t using Newton's 3rd axiom (force = mass times acceleration).
(b) Determine the velocity of the car at time t_e using the MATLAB command ode45. Also plot the velocity curve. How many grid points did MATLAB choose? Compare the results with the self-implemented Runge-Kutta method from the previous task.
(c) Use MATLAB to determine how far the car has traveled until the time t_e.

Linear Mappings and Transformation Matrices 37

A *linear mapping* is a mapping $f : V \to W$ between $\mathbb{K}$-vector spaces V and W with the property $f(\lambda v + w) = \lambda f(v) + f(w)$ for all $\lambda \in \mathbb{K}$ and $v, w \in V$. Such a mapping is thus *compatible* with vector addition and multiplication by scalars; one also speaks of a *structure-preserving* mapping. What interests us most about such mappings is the possibility of determining a *transformation matrix* M to the linear mapping f after choosing bases B and C in the vector spaces V and W. Applying the linear mapping f onto a vector v thus becomes a multiplication of the representing matrix M on the coordinate vector of v.

As in earlier chapters on linear algebra, $\mathbb{K}$ again denotes one of the two number fields $\mathbb{R}$ or $\mathbb{C}$.

37.1 Definitions and Examples

We consider two $\mathbb{K}$-vector spaces V and W. A mapping $f : V \to W$ is **linear** or a **homomorphism** if for all $\lambda \in \mathbb{K}$ and all $v,\ w \in V$ we have

$$f(\lambda v + w) = \lambda f(v) + f(w).$$

We conclude:

Properties of Linear Mappings
Let U, V and W be vector spaces over a field $\mathbb{K}$.

(continued)

C. Karpfinger, *Calculus and Linear Algebra in Recipes*,
https://doi.org/10.1007/978-3-662-65458-3_37

- If $f : V \to W$ is a linear mapping, then $f(0_V) = 0_W$ for the zero vectors 0_V of V and 0_W of W.
- If $f : V \to W$ and $g : W \to U$ are linear mappings, then the composition $g \circ f : V \to U$ is also a linear mapping.
- If $f : V \to W$ is a bijective linear mapping, then the inverse mapping $f^{-1} : W \to V$ is also a bijective linear mapping.

We have posed the proofs of these facts as Exercise 37.1. Note that the first property provides a good way to expose a mapping as nonlinear; the proof of whether an f is linear or nonlinear is as follows:

Recipe: Test Whether f is Linear or Not

Given is a mapping $f : V \to W$ between $\mathbb{K}$-vector spaces V and W. To check whether f is linear or not, proceed as follows:

(1) Check, if $f(0) = 0$.

- If no, then f is not linear.
- If yes, continue to the next step.

(2) Choose $\lambda \in \mathbb{K}$ and $v, w \in V$ and try to prove one of the two equivalent conditions:

- $f(\lambda v + w) = \lambda f(v) + f(w)$.
- $f(v + w) = f(v) + f(w)$ and $f(\lambda v) = \lambda f(v)$.

If the proof fails, continue to the next step.

(3) Search for numbers for λ or vectors v and w such that one of the equalities in (2) is not satisfied.

If powers or nonlinear functions appear in the mapping prescription of f such as cos, exp or products of coefficients of v or w, this is an indication that f is not linear. It is then best to start with step (3).

Because of its great importance, we highlight the first example in a box (cf. also the Examples 23.2 and 23.3), further examples follow immediately afterwards:

The Linear Mapping $f_A : \mathbb{K}^n \to \mathbb{K}^m$, $f_A(v) = A\,v$
Given are matrices $A \in \mathbb{K}^{m\times n}$ and $B \in \mathbb{K}^{r\times m}$. Then we have:

- The mapping $f_A : \mathbb{K}^n \to \mathbb{K}^m$ with $f_A(v) = A\,v$ is linear.
- It is $f_B \circ f_A = f_{BA}$ a linear mapping from $\mathbb{K}^n$ in $\mathbb{K}^r$.
- It is $f_A : \mathbb{K}^n \to \mathbb{K}^m$ invertible if and only if $m = n$ and A is invertible. In this case $f_A^{-1} = f_{A^{-1}}$.

The first statement here follows from:

$$f_A(\lambda\,v + w) = A\,(\lambda\,v + w) = \lambda\,A\,v + A\,w = \lambda\,f_A(v) + f_A(w)$$

for all $\lambda \in \mathbb{K}$ and $v,\ w \in \mathbb{K}^n$.

Example 37.1

- Let $n \in \mathbb{N}$. For any invertible matrix $A \in \mathbb{K}^{n\times n}$ the mapping $f : \mathbb{K}^{n\times n} \to \mathbb{K}^{n\times n}$ with $f(X) = A\,X\,A^{-1}$ satisfies the equation $f(0) = 0$. The mapping is linear, namely, we have:

$$f_A(\lambda\,X + Y) = A\,(\lambda\,X + Y)\,A^{-1} = \lambda\,A\,X\,A^{-1} + A\,Y\,A^{-1} = \lambda\,f(X) + f(Y)\,.$$

- Let $V = W = \mathbb{R}[x]_n = \{a_0 + a_1x + a_2x^2 + \cdots + a_nx^n \mid a_i \in \mathbb{R}\}$ for $n \geq 1$. The mapping

$$f : \mathbb{R}[x]_n \to \mathbb{R}[x]_n,\ f(p) = p'$$

satisfies $f(0) = 0$. The mapping is linear, namely

$$f(\lambda p + q) = (\lambda p + q)' = \lambda p' + q' = \lambda f(p) + f(q)\,.$$

- The mapping

$$f : \mathbb{R}^2 \to \mathbb{R},\ f\left((v_1, v_2)^\top\right) = v_1\,v_2$$

satisfies $f(0) = 0$ but $f(v + w) = f(v) + f(w)$ is obviously not satisfied, for example

$$1 = f\left((1,1)^\top\right) = f\left((1,0)^\top + (0,1)^\top\right) \neq f\left((1,0)^\top\right) + f\left((0,1)^\top\right) = 0 + 0 = 0\,.$$

- Let $V = C(I)$ be the vector space of continuous mappings on the interval $I = [-\pi, \pi]$. The mapping $f : V \to \mathbb{R}$, $f(g) = \int_{-\pi}^{\pi} g(x)\mathrm{d}x$ is linear. Namely, we have

$$\begin{aligned} f(\lambda g + h) &= \int_{-\pi}^{\pi} \lambda g(x) + h(x)\mathrm{d}x = \lambda \int_{-\pi}^{\pi} g(x)\mathrm{d}x + \int_{-\pi}^{\pi} h(x)\mathrm{d}x \\ &= \lambda f(g) + f(h) . \end{aligned}$$ ■

Remark One calls a mapping $f : V \to W$ between $\mathbb{K}$-vector spaces V and W**affine-linear** if

$$f(v) = g(v) + a$$

with a linear mapping $g : V \to W$ and a constant vector $a \in W$. Thus an affine-linear mapping is of an additive constant a apart from a linear mapping. With $a = 0$, we obtain that every linear mapping is also affine-linear.

37.2 Image, Kernel and the Dimensional Formula

Image, Kernel and the Dimensional Formula
If $f : V \to W$ is linear, then

- $\ker(f) = f^{-1}(\{0\}) = \{v \in V \mid f(v) = 0\} \subseteq V$—the **kernel of** f and
- $\mathrm{im}(f) = f(V) = \{f(v) \mid v \in V\} \subseteq W$—the **range of** f

are subspaces of V and W. The dimension

- of the kernel is called the **defect** of f and one writes $\mathrm{def}(f) = \dim(\ker(f))$,
- of the range of f is called the **rank** of f and one writes $\mathrm{rg}(f) = \dim(\mathrm{im}(f))$.

We further have:

- **The dimension formula for linear mappings:**

$$\dim(V) = \dim(\ker(f)) + \dim(\mathrm{im}(f)) \text{ i.e. } \dim(V) = \mathrm{def}(f) + \mathrm{rg}(f) .$$

(continued)

- The mapping f is injective if and only if $\ker(f) = \{0\}$, i.e. $\operatorname{def}(f) = 0$.
- If V and W are finite dimensional, then in the case $\dim(V) = \dim(W)$:

$$f \text{ is surjective } \Leftrightarrow f \text{ is injective } \Leftrightarrow f \text{ is bijective} .$$

Figure 37.1 shows the kernel and range of a linear mapping $f : V \to W$.
The assertions listed in the box above are easily verifiable except for the dimensional formula. Consider Exercise 37.2.

Example 37.2

- For the linear mapping $f : \mathbb{R} \to \mathbb{R}$, $f(v) = a\, v$ with $a \in \mathbb{R}$ we have:
  - If $a = 0$, then $\ker(f) = \mathbb{R}$, so $\operatorname{def}(f) = 1$, and $\operatorname{im}(f) = \{0\}$, so $\operatorname{rg}(f) = 0$.
  - If $a \neq 0$, then $\ker(f) = \{0\}$, so $\operatorname{def}(f) = 0$ and $\operatorname{im}(f) = \mathbb{R}$, so $\operatorname{rg}(f) = 1$.

  In the case $a = 0$ the mapping f is neither injective nor surjective nor bijective, but in the case $a \neq 0$ injective and surjective and bijective.
- For every matrix $A \in \mathbb{R}^{m \times n}$ is $f_A : \mathbb{R}^n \to \mathbb{R}^m$ with $f_A(v) = A\, v$ linear. We have:
  - $\ker(f_A) = \ker(A)$ and $\operatorname{def}(f_A) = n - \operatorname{rg}(A)$.
  - $\operatorname{im}(f_A) = S_A =$ column space of A and $\operatorname{rg}(f_A) = \operatorname{rg}(A)$.

  The mapping f_A is injective if and only if $\operatorname{rg}(A) = n$ and surjective if and only if $\operatorname{rg}(A) = m$ and bijective if and only if $m = n$ and $\operatorname{rg}(A) = n$.
- For the linear mapping $f : \mathbb{R}[x]_n \to \mathbb{R}[x]_n$ with $f(p) = p'$ we have:
  - $\ker(f) = \{p \in \mathbb{R}[x]_n \mid \deg(p) \leq 0\} = \mathbb{R}$, so $\operatorname{def}(f) = 1$.
  - $\operatorname{im}(f) = \{p \in \mathbb{R}[x]_n \mid \deg(p) \leq n - 1\} = \mathbb{R}[x]_{n-1}$, so $\operatorname{rg}(f) = n$. ■

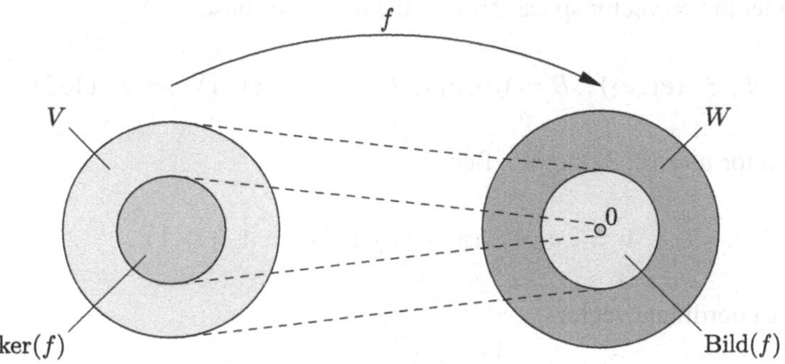

Fig. 37.1 The kernel lies in V, the image lies in W, Bild(f) means the range of f

37.3 Coordinate Vectors

According to the Theorem for unambiguous representability in the box in Sect. 15.1 every vector v of an n-dimensional $\mathbb{K}$-vector space V with basis $B = \{b_1, \ldots, b_n\}$ is, apart from the order of the summands, in exactly one way of the form $v = \lambda_1 b_1 + \cdots + \lambda_n b_n$ with $\lambda_1, \ldots, \lambda_n \in \mathbb{K}$. We achieve uniqueness if we fix the order of the basic elements $b_1, \ldots, b_n$:

Coordinate Vector

If $B = (b_1, \ldots, b_n)$ is a **ordered basis** of the $\mathbb{K}$-vector space V, then any vector $v \in V$ has a unique representation as linear combination of B:

$$v = \lambda_1 b_1 + \cdots + \lambda_n b_n \,, \quad \text{where } \lambda_1, \ldots, \lambda_n \in \mathbb{K} \,.$$

One calls ${}_B v = (\lambda_1, \ldots, \lambda_n)^\top \in \mathbb{K}^n$ the **coordinate vector** of v with respect to B. And we have for all $\lambda \in \mathbb{K}$ and $v,\ w \in V$

$${}_B(\lambda\, v + w) = \lambda\, {}_B v + {}_B w \,,$$

i.e., the bijective mapping ${}_B : V \to \mathbb{K}^n,\ v \mapsto {}_B v$ is linear.

Note: For $B = \{b_1, \ldots, b_n\}$ the order of the elements does not matter, we have $\{b_1, \ldots, b_n\} = \{b_n, \ldots, b_1\}$, whereas for $B = (b_1, \ldots, b_n)$ on the other hand, the order does play a role, we have $(b_1, \ldots, b_n) \neq (b_n, \ldots, b_1)$.

Example 37.3

- We consider the $\mathbb{R}$-vector space $\mathbb{R}^2$ with the (ordered) bases

$$E_2 = (e_1, e_2)\,,\ B = (e_2, e_1)\,,\ C = (c_1 = (1, 1)^\top, c_2 = (1, 2)^\top)$$

and the vector $v = (2, 3)^\top \in \mathbb{R}^2$. Because of

$$v = 2\, e_1 + 3\, e_2 = 3\, e_2 + 2\, e_1 = 1\, c_1 + 1\, c_2$$

we get the coordinate vectors

$${}_{E_2} v = \begin{pmatrix} 2 \\ 3 \end{pmatrix}\,,\ {}_B v = \begin{pmatrix} 3 \\ 2 \end{pmatrix}\,,\ {}_C v = \begin{pmatrix} 1 \\ 1 \end{pmatrix}\,.$$

- The $\mathbb{R}$-vector space $\mathbb{R}[x]_2$ has the (ordered) bases

$$B = (b_1 = 1, b_2 = x, b_3 = x^2) \text{ and } C = (c_1 = x^2 + x + 1, c_2 = x + 1, c_3 = 1)\,.$$

The polynomial $p = 2\,x^2 + 11$ has the representations

$$p = 11\,b_1 + 0\,b_2 + 2\,b_3 = 2c_1 - 2c_2 + 11c_3$$

and so the coordinate vectors

$${}_B p = \begin{pmatrix} 11 \\ 0 \\ 2 \end{pmatrix} \quad \text{or} \quad {}_C p = \begin{pmatrix} 2 \\ -2 \\ 11 \end{pmatrix}. \qquad \blacksquare$$

In the given examples it was always easy to determine the coordinate vector by *trying*, in general the determination of the coefficients $\lambda_1, \ldots, \lambda_n \in \mathbb{K}$ in the equation $v = \lambda_1 b_1 + \cdots + \lambda_n b_n$ amounts to solving a system of linear equations. This linear system of equations is uniquely solvable, since the representation of a vector v as a linear combination with respect to an ordered basis $B = (b_1, \ldots, b_n)$ of an n-dimensional $\mathbb{K}$-vector space V is unique. Compare the prescription in Sect. 14.1 where there $v_1, \ldots, v_n$ are to be replaced by $b_1, \ldots, b_n$ and in the step (2) of the recipe there the positive answer is valid in any case. Compare also the exercises.

37.4 Transformation Matrices

Now, we consider a linear mapping $f : V \to W$ between finite-dimensional $\mathbb{K}$-vector spaces V and W:

Transformation Matrix
Let V and W finite dimensional $\mathbb{K}$-vector spaces with

- $\dim(V) = n$ and $B = (b_1, \ldots, b_n)$ a base of V,
- $\dim(W) = m$ and $C = (c_1, \ldots, c_m)$ a base of W.

If $f : V \to W$ linear, then the $m \times n$-matrix

$${}_C M(f)_B = \big({}_C f(b_1), \ldots, {}_C f(b_n)\big) \in \mathbb{K}^{m \times n}$$

(continued)

is called the **transformation matrix of f with respect to B and C**.
The i-th column of the transformation matrix contains the coordinate vector of the image of the i-th base vector.
The properties *injective*, *surjective* and *bijective* of the mapping f can be found in one and therefore every transformation matrix $A = {}_C M(f)_B$:

- f is injective if and only if $\ker(A) = \{0\}$.
- f is surjective if and only if $\operatorname{rg}(A) = m$.
- f is bijective if and only if A is invertible.

We clarify how the matrix ${}_C M(f)_B$ *represents* the linear mapping f: To do this, we consider a vector $v \in V$ and represent it with respect to the ordered basis $B = (b_1, \ldots, b_n)$ of V, $v = \lambda_1 b_1 + \cdots + \lambda_n b_n$. Now we apply the linear mapping f on v and represent the image $f(v) \in W$ with respect to the ordered basis C of W:

$$f(v) = \lambda_1 f(b_1) + \cdots + \lambda_n f(b_n), \text{ also } {}_C f(v) = \lambda_1 {}_C f(b_1) + \cdots + \lambda_n {}_C f(b_n).$$

Now, on the other hand, we multiply the transformation matrix ${}_C M(f)_B$ by the coordinate vector ${}_B v$ and we get:

$${}_C M(f)_B \, {}_B v = ({}_C f(b_1), \ldots, {}_C f(b_n)) \begin{pmatrix} \lambda_1 \\ \vdots \\ \lambda_n \end{pmatrix} = \lambda_1 {}_C f(b_1) + \cdots + \lambda_n {}_C f(b_n).$$

Thus

$${}_C M(f)_B \, {}_B v = {}_C f(v).$$

Thus, applying f on v is ultimately the multiplication of the transformation matrix ${}_C M(f)_B$ of f with the coordinate vector of v.

The special case $V = \mathbb{K}^n$ with the standard base $B = E_n$ and $W = \mathbb{K}^m$ with the standard basis $C = E_m$ thus yields because of ${}_{E_m} f(e_i) = f(e_i)$:

The Linear Mappings of $\mathbb{K}^n$ in $\mathbb{K}^m$
For each linear mapping $f : \mathbb{K}^n \to \mathbb{K}^m$ there is a matrix $A \in \mathbb{K}^{m \times n}$ with $f = f_A : \mathbb{K}^n \to \mathbb{K}^m$, $f_A(v) = A\, v$. The matrix A is obtained column-wise as follows

$$A = {}_{E_m} M(f)_{E_n} = (f(e_1), \ldots, f(e_n)).$$

Now, at the latest, it should also be clear why we have the meaning of the example $f_A : \mathbb{K}^n \to \mathbb{K}^m$, $f_A(v) = A\,v$ in Sect. 37.1 in such a way: Ultimately, any linear mapping of $\mathbb{K}^n$ in the $\mathbb{K}^m$ is of this form, more generally even any linear mapping between finite-dimensional vector spaces, if one identifies the vectors with their coordinate vectors.

We bring more examples of transformation matrices.

Example 37.4

- The transformation matrix ${}_{E_2}M(f_A)_{E_3}$ of the linear mapping $f_A : \mathbb{R}^3 \to \mathbb{R}^2$, $f(v) = A\,v$ for $A = (a_1, a_2, a_3) \in \mathbb{R}^{2\times 3}$ with respect to the standard bases E_3 of $\mathbb{R}^3$ and E_2 of $\mathbb{R}^2$ is because of

$$f_A(e_1) = (a_1, a_2, a_3)\,e_1 = a_1\,,$$

$$f_A(e_2) = (a_1, a_2, a_3)\,e_2 = a_2\,,$$

$$f_A(e_3) = (a_1, a_2, a_3)\,e_3 = a_3$$

thus the matrix A, i.e. $A = {}_{E_2}M(f_A)_{E_3}$. More generally, for any linear mapping $f_A : \mathbb{K}^n \to \mathbb{K}^m$, $f_A(v) = A\,v$ with $A \in \mathbb{K}^{m\times n}$

$${}_{E_m}M(f_A)_{E_n} = A\,.$$

- Now let $V = \mathbb{R}[x]_2 = W$ with the bases $B = (1, x, x^2) = C$. The mapping $f : V \to W$, $p \mapsto xp'$ is linear and has the transformation matrix

$${}_C M(f)_B = \begin{pmatrix} 0 & 0 & 0 \\ 0 & 1 & 0 \\ 0 & 0 & 2 \end{pmatrix} \in \mathbb{R}^{3\times 3},$$

since

$$f(1) = 0 = 0\cdot 1 + 0\cdot x + 0\cdot x^2 \Rightarrow {}_C f(1) = (0,\,0,\,0)^\top\,,$$

$$f(x) = x = 0\cdot 1 + 1\cdot x + 0\cdot x^2 \Rightarrow {}_C f(x) = (0,\,1,\,0)^\top\,,$$

$$f(x^2) = 2x^2 = 0\cdot 1 + 0\cdot x + 2\cdot x^2 \Rightarrow {}_C f(x^2) = (0,\,0,\,2)^\top\,.$$

With $p = 2x^2+3x+4$ we have, for example: $f(p) = f(2x^2+3x+4) = x(4x+3) = 4x^2 + 3x$ and

$$_C M(f)_B \, _B p = \begin{pmatrix} 0 & 0 & 0 \\ 0 & 1 & 0 \\ 0 & 0 & 2 \end{pmatrix} \begin{pmatrix} 4 \\ 3 \\ 2 \end{pmatrix} = \begin{pmatrix} 0 \\ 3 \\ 4 \end{pmatrix} = {}_C f(p) \, . \qquad \blacksquare$$

Remark We describe in Sect. 38.3 the procedure for determining the transformation matrix of a linear mapping of the form $f : V \to V$ as a recipe. By then we will have a second way to determine such a transformation matrix.

37.5 Exercises

37.1 Prove the properties of linear mappings in the box in Sect. 37.1

37.2 Show the claims in the box in Sect. 37.2 (except for the dimensional formula).

37.3 Determine the kernel and image of each of the linear mappings:

(a) $f : \mathbb{R}^2 \to \mathbb{R}^2, (v_1, v_2) \mapsto (v_2, v_1)$,
(b) $f : V \to W, v \mapsto 0$,
(c) $f : \mathbb{R}^3 \to \mathbb{R}^2, (v_1, v_2, v_3) \mapsto (v_1 + v_2, v_2)$,
(d) $\frac{\mathrm{d}}{\mathrm{d}x} : \mathbb{R}[x] \to \mathbb{R}[x], p \mapsto \frac{\mathrm{d}}{\mathrm{d}x}(p)$.

37.4 Given is the linear mapping $f_A : \mathbb{R}^2 \to \mathbb{R}^3, v \mapsto A\,v$ with $A = \begin{pmatrix} 1 & 3 \\ 4 & 2 \\ -1 & 0 \end{pmatrix}$.

(a) Determine the image and kernel of f_A.
(b) Is f_A injective, surjective, bijective?

37.5 Is there a linear mapping f from $\mathbb{R}^2$ in the $\mathbb{R}^2$ with $\ker f = \operatorname{image} f$?

37.6 Which of the following mappings are linear? Give a brief justification or counterexample for each!

(a) $f_1 : \mathbb{R} \to \mathbb{R}^3$ with $f_1(v) = (v + 1, 2v, v - 3)^\top$,
(b) $f_2 : \mathbb{R}^4 \to \mathbb{R}^2$ with $f_2(v_1, v_2, v_3, v_4) = (v_1 + v_2, v_1 + v_2 + v_3 + v_4)^\top$,

(c) $f_3\colon \mathbb{R}^4 \to \mathbb{R}^2$ with $f_3(v_1, v_2, v_3, v_4) = (v_1 v_2, v_3 v_4)^\top$,
(d) $f_4\colon \mathbb{R}^n \to \mathbb{R}^2$ with $f_4(v) = ((1, 0, 0, \ldots, 0)v) \cdot (1, 2)^\top$,
(e) $f_5\colon \mathbb{R}[x]_3 \to \mathbb{R}[x]_4$ with
$f_5(a_0 + a_1 x + a_2 x^2 + a_3 x^3) = (a_0 + a_1) + 2(a_1 + a_2)x + 3(a_2 + a_3)x^2 + 4(a_3 + a_0)x^3 + 5x^4$.

37.7 Let the linear mappings $f_1, f_2\colon \mathbb{R}^3 \to \mathbb{R}^3$ given by

$$f_1(x, y, z) = (3x, x - y, 2x + y + z)^\top \quad \text{and} \quad f_2(x, y, z) = (x - y, 2x + z, 0)^\top.$$

(a) Determine bases of $\operatorname{im}(f_i)$, $\operatorname{im}(f_1 \circ f_2)$, $\ker(f_i)$, $\ker(f_1 \circ f_2)$, $i = 1,2$.
(b) Are the mappings f_1 and f_2 injective or surjective?

37.8 Consider for $n \geq 1$ the mapping. $f\colon \mathbb{R}[x]_{n-1} \to \mathbb{R}[x]_n$, defined by

$$(f(p))(x) = \int_0^x p(t)\mathrm{d}t.$$

(a) Show that f is a linear mapping.
(b) Determine the transformation matrix A of this linear mapping with respect to the bases $(1, x, \ldots, x^{n-1})$ of $\mathbb{R}[x]_{n-1}$ and $(1, x, \ldots, x^n)$ of $\mathbb{R}[x]_n$.
(c) Is the mapping f injective? Is it surjective?

37.9 Let $a = (a_1, a_2, a_3)^\top \in \mathbb{R}^3$ given by $\|a\| = 1$. The linear mapping $f : \mathbb{R}^3 \to \mathbb{R}^3$, $f(x) = x - 2(x^\top a)\, a$ is a reflection on a perpendicular plane through the origin.

(a) Use a sketch to illustrate the mapping f.
(b) Calculate $f \circ f$.
(c) What is the transformation matrix A of f with respect to the canonical basis?
(d) Find a basis $B = (b_1, b_2, b_3)$ with $f(b_1) = -b_1,\ f(b_2) = b_2,\ f(b_3) = b_3$. Give the transformation matrix $\tilde{A}$ of f with respect to B.

Base Transformation

38

38.1 The Tansformation Matrix of the Composition of Linear Mappings

Every linear mapping f between finite-dimensional $\mathbb{K}$-vector spaces V and W can be represented by choosing bases of V and W by a matrix. If now g is another linear mapping from W into a vector space U, for which there is of course also a transformation matrix, then the transformation matrix of the composition $g \circ f$ is just the product of the transformation matrices of f and g, if only the choice of bases is made appropriately. We are more accurate:

The Transformation Matrix of the Composition of Linear Mappings
We consider three $\mathbb{K}$-vector spaces

- V with $\dim(V) = n$ and ordered base $B = (b_1, \ldots, b_n)$,
- W with $\dim(W) = m$ and ordered base $C = (c_1, \ldots, c_m)$,
- U with $\dim(U) = r$ and ordered base $D = (d_1, \ldots, d_r)$

and linear mappings $f : V \to W$ and $g : W \to U$ with the transformation matrices ${}_C M(f)_B$ and ${}_D M(g)_C$. Then for the transformation matrix ${}_D M(g \circ f)_B$ of $g \circ f$ with respect to the bases B and D we have:

$$ {}_D M(g \circ f)_B = {}_D M(g)_C \, {}_C M(f)_B \, . $$

C. Karpfinger, *Calculus and Linear Algebra in Recipes*,
https://doi.org/10.1007/978-3-662-65458-3_38

In fact, this formula is the motivation for matrix multiplication: we explained the multiplication of matrices so that we have this formula. This formula has far-reaching consequences; the *basic transformation formula* follows from this. Before we get into this, let's verify the formula with an example:

Example 38.1 We consider the two linear mappings

$$f : \mathbb{R}[x]_2 \to \mathbb{R}[x]_1 ,\ f(p) = p' \text{ and } g : \mathbb{R}[x]_1 \to \mathbb{R}[x]_3 ,\ g(p) = x^2 p .$$

The composition of these two linear mappings is the linear mapping

$$g \circ f : \mathbb{R}[x]_2 \to \mathbb{R}[x]_3 ,\ g \circ f(p) = g(f(p)) = x^2 p' .$$

We choose the canonical bases

$$B = (1, x, x^2) \text{ of } \mathbb{R}[x]_2 \text{ and } C = (1, x) \text{ of } \mathbb{R}[x]_1 \text{ and } D = (1, x, x^2, x^3) \text{ of } \mathbb{R}[x]_3$$

and obtain the representation matrices

$${}_CM(f)_B = \begin{pmatrix} 0 & 1 & 0 \\ 0 & 0 & 2 \end{pmatrix} \text{ and } {}_DM(g)_C = \begin{pmatrix} 0 & 0 \\ 0 & 0 \\ 1 & 0 \\ 0 & 1 \end{pmatrix} \text{ and } {}_DM(g \circ f)_B = \begin{pmatrix} 0 & 0 & 0 \\ 0 & 0 & 0 \\ 0 & 1 & 0 \\ 0 & 0 & 2 \end{pmatrix} .$$

And we also have

$${}_DM(g \circ f)_B = {}_DM(g)_C \, {}_CM(f)_B . \qquad \blacksquare$$

38.2 Base Transformation

In general, a vector space has many different bases. And the transformation matrices of one and the same linear mapping with respect to different bases naturally have different appearances. The *base transformation formula* allows the calculation of the transformation matrix of a linear mapping with respect to bases B' and C' if the transformation matrix of this linear mapping with respect to bases is B and C is already known:

The Base Transformation Formula

Let $f : V \to W$ is a linear mapping and furthermore

- $B = (b_1, \ldots, b_n)$ and $B' = (b'_1, \ldots, b'_n)$ two bases of V and
- $C = (c_1, \ldots, c_m)$ and $C' = (c'_1, \ldots, c'_m)$ two bases of W.

Then we have the **base transformation formula**:

$$_{C'}M(f)_{B'} = {_{C'}M(\mathrm{Id}_W)_C}\, {_C M(f)_B}\, {_B M(\mathrm{Id}_V)_{B'}} .$$

Here

$$_{C'}M(\mathrm{Id})_C = ({_{C'}c_1}, \ldots, {_{C'}c_m}) \in \mathbb{K}^{m \times m} \text{ and}$$
$$_B M(\mathrm{Id})_{B'} = ({_B b'_1}, \ldots, {_B b'_n}) \in \mathbb{K}^{n \times n}$$

are the transfromation matrices of the identities with respect to the given bases,

$$\mathrm{Id} : W \to W , \ \mathrm{Id}(w) = w \ \text{ and } \ \mathrm{Id} : V \to V , \ \mathrm{Id}(v) = v .$$

In particular, for the linear mapping $f : \mathbb{K}^n \to \mathbb{K}^n$, $f(v) = A\,v$ with $A \in \mathbb{K}^{n \times n}$ and a base or invertible matrix $B \in \mathbb{K}^{n \times n}$ we have

$$_B M(f)_B = B^{-1} A\, B .$$

This is easy to prove, because with the statement in the box in Sect. 38.1, we have:

$$_{C'}M(f)_{B'} = {_{C'}M(\mathrm{Id} \circ f \circ \mathrm{Id})_{B'}} = {_{C'}M(\mathrm{Id})_C}\,{_C M(f)_B}\,{_B M(\mathrm{Id})_{B'}} .$$

The statement in the special case $V = \mathbb{K}^n = W$, $f : \mathbb{K}^n \to \mathbb{K}^n$, $f(v) = A\,v$, $A \in \mathbb{K}^{n \times n}$ results just as simply from $B = {_{E_n}M(\mathrm{Id})_B}$ and

$$_B M(\mathrm{Id})_{E_n}\, {_{E_n}M(\mathrm{Id})_B} = {_B M(\mathrm{Id})_B} = E_n ,$$

namely, it follows from this

$$_B M(\mathrm{Id})_{E_n} = {_{E_n}M(\mathrm{Id})_B^{-1}} = B^{-1} .$$

Important for our purposes is the following interpretation of the base transformation formula: If B is a base of the n-dimensional vector space $\mathbb{K}^n$ and $f : \mathbb{K}^n \to \mathbb{K}^n$ is a linear mapping with $f(v) = A\,v$, $A \in \mathbb{K}^{n\times n}$, then we have

$$_B M(f)_B = B^{-1} A\, B$$

with the invertible matrix B. In general, two $n\times n$-matrices A and M are called **similarly** if there exists an invertible matrix S such that $M = S^{-1} A\, S$. Thus, any two transformation matrices of a linear mapping are similar to each other. Similar matrices have many common properties, the most important of which we have collected in Sect. 39.3.

38.3 The Two Methods for Determining Transformation Matrices

With the basis transformation formula we have found another way to determine the transformation matrix of a linear mapping with respect to the chosen basis. In applications, it is mainly essential to find a transformation matrix of the form $f : V \to V$, i.e., a linear mapping from a vector space into itself. A further *simplification* we get by the fact that one usually uses only one base B, i.e. B=C. We formulate the possibilities of having a transformation matrix $_B M(f)_B$ to determine, as a recipe:

Recipe: Determine the Transformation Matrix $_B M(f)_B$
If $f : V \to V$ is a linear mapping, then we get the transformation matrix $A = {}_B M(f)_B$ with respect to the ordered basis $B = (b_1, \ldots, b_n)$ in one of the following ways:

- In the case $V = \mathbb{K}^n$ and $B = E_n$ we have

$$A = (f(e_1), \ldots, f(e_n)) .$$

- Determine for each $i = 1, \ldots, n$ the coordinate vector $_B f(b_i) = (\lambda_1, \ldots, \lambda_n)^\top$ from

$$f(b_i) = \lambda_1 b_1 + \cdots + \lambda_n b_n$$

and get

$$A = ({}_B f(b_1), \ldots, {}_B f(b_n)) .$$

(continued)

- If a transformation matrix ${}_C M(f)_C$ is known, then

$$A = S^{-1} {}_C M(f)_C \, S \text{ with } S = {}_C M(\mathrm{Id})_B \, .$$

Example 38.2

- Let $f : \mathbb{R}^3 \to \mathbb{R}^3,\ v = (v_1, v_2, v_3)^\top \mapsto f(v) = (v_1 + v_3,\ v_2 - v_1,\ v_1 + v_3)^\top$. We consider the base

$$B = \left(\begin{pmatrix} 1 \\ 0 \\ 0 \end{pmatrix}, \begin{pmatrix} 1 \\ 2 \\ 0 \end{pmatrix}, \begin{pmatrix} 1 \\ 0 \\ 1 \end{pmatrix} \right)$$

of $\mathbb{R}^3$. The transformation matrix ${}_B M(f)_B$ can now be calculated in the following two different ways.

1st option: We represent the images of the basis vectors with respect to the basis B:

- $$f\left(\begin{pmatrix} 1 \\ 0 \\ 0 \end{pmatrix} \right) = \begin{pmatrix} 1 \\ -1 \\ 1 \end{pmatrix} = 1/2 \begin{pmatrix} 1 \\ 0 \\ 0 \end{pmatrix} + (-1/2) \begin{pmatrix} 1 \\ 2 \\ 0 \end{pmatrix} + 1 \begin{pmatrix} 1 \\ 0 \\ 1 \end{pmatrix},$$

- $$f\left(\begin{pmatrix} 1 \\ 2 \\ 0 \end{pmatrix} \right) = \begin{pmatrix} 1 \\ 1 \\ 1 \end{pmatrix} = (-1/2) \begin{pmatrix} 1 \\ 0 \\ 0 \end{pmatrix} + 1/2 \begin{pmatrix} 1 \\ 2 \\ 0 \end{pmatrix} + 1 \begin{pmatrix} 1 \\ 0 \\ 1 \end{pmatrix},$$

- $$f\left(\begin{pmatrix} 1 \\ 0 \\ 1 \end{pmatrix} \right) = \begin{pmatrix} 2 \\ -1 \\ 2 \end{pmatrix} = 1/2 \begin{pmatrix} 1 \\ 0 \\ 0 \end{pmatrix} + (-1/2) \begin{pmatrix} 1 \\ 2 \\ 0 \end{pmatrix} + 2 \begin{pmatrix} 1 \\ 0 \\ 1 \end{pmatrix}.$$

The columns of the transformation matrix are now the calculated coefficients:

$${}_B M(f)_B = \begin{pmatrix} 1/2 & -1/2 & 1/2 \\ -1/2 & 1/2 & -1/2 \\ 1 & 1 & 2 \end{pmatrix}.$$

2nd option: we use the base transformation formula:

$$_BM(f)_B = {}_BM(\mathrm{Id})_{E_3}\ {}_{E_3}M(f)_{E_3}\ {}_{E_3}M(\mathrm{Id})_B\,.$$

The transformation matrix concerning the standard base E_3 reads

$$_{E_3}M(f)_{E_3} = \begin{pmatrix} 1 & 0 & 1 \\ -1 & 1 & 0 \\ 1 & 0 & 1 \end{pmatrix}.$$

The matrix $S = {}_{E_3}M(\mathrm{Id})_B$ contains the elements of B as columns and $_BM(\mathrm{Id})_{E_3}$ is S^{-1}:

$$_{E_3}M(\mathrm{Id})_B = \begin{pmatrix} 1 & 1 & 1 \\ 0 & 2 & 0 \\ 0 & 0 & 1 \end{pmatrix} \quad \text{and} \quad {}_BM(\mathrm{Id})_{E_3} = \begin{pmatrix} 1 & -1/2 & -1 \\ 0 & 1/2 & 0 \\ 0 & 0 & 1 \end{pmatrix}.$$

Thus we have:

$$_BM(f)_B = 1/2 \begin{pmatrix} 2 & -1 & -2 \\ 0 & 1 & 0 \\ 0 & 0 & 2 \end{pmatrix} \begin{pmatrix} 1 & 0 & 1 \\ -1 & 1 & 0 \\ 1 & 0 & 1 \end{pmatrix} \begin{pmatrix} 1 & 1 & 1 \\ 0 & 2 & 0 \\ 0 & 0 & 1 \end{pmatrix} = 1/2 \begin{pmatrix} 1 & -1 & 1 \\ -1 & 1 & -1 \\ 2 & 2 & 4 \end{pmatrix}.$$

- We consider the linear mapping

$$f : \mathbb{R}[x]_2 \to \mathbb{R}[x]_2,\ p \mapsto x^2 p'' + p'(1)\,.$$

Let further be the ordered bases $B = (1, x, x^2)$ and $C = (x^2 + x + 1,\ x + 1,\ 1)$ of the $\mathbb{R}$-vector space $\mathbb{R}[x]_2$ of all polynomials of degree less than or equal to 2 are given. By

$$f(x^2 + x + 1) = 2x^2 + 3 = 2\,(x^2 + x + 1) - 2\,(x + 1) + 3 \cdot 1$$

$$f(x + 1) = 1 = 0\,(x^2 + x + 1) + 0\,(x + 1) + 1 \cdot 1$$

$$f(1) = 0 = 0\,(x^2 + x + 1) + 0\,(x + 1) + 0 \cdot 1$$

are

$$ {}_BM(f)_B = \begin{pmatrix} 0 & 1 & 2 \\ 0 & 0 & 0 \\ 0 & 0 & 2 \end{pmatrix} \quad \text{and} \quad {}_CM(f)_C = \begin{pmatrix} 2 & 0 & 0 \\ -2 & 0 & 0 \\ 3 & 1 & 0 \end{pmatrix} $$

the transformation matrices of f with regard to B and C. Again we can now calculate ${}_CM(f)_C$ by using the base transformation formula. There are

$$ S = {}_BM(\mathrm{Id})_C = \begin{pmatrix} 1 & 1 & 1 \\ 1 & 1 & 0 \\ 1 & 0 & 0 \end{pmatrix} \quad \text{and} \quad {}_CM(\mathrm{Id})_B = S^{-1} = \begin{pmatrix} 0 & 0 & 1 \\ 0 & 1 & -1 \\ 1 & -1 & 0 \end{pmatrix} $$

and thus

$$ {}_CM(f)_C = \begin{pmatrix} 0 & 0 & 1 \\ 0 & 1 & -1 \\ 1 & -1 & 0 \end{pmatrix} \begin{pmatrix} 0 & 1 & 2 \\ 0 & 0 & 0 \\ 0 & 0 & 2 \end{pmatrix} \begin{pmatrix} 1 & 1 & 1 \\ 1 & 1 & 0 \\ 1 & 0 & 0 \end{pmatrix} = \begin{pmatrix} 2 & 0 & 0 \\ -2 & 0 & 0 \\ 3 & 1 & 0 \end{pmatrix}. \blacksquare $$

But what is the point of all this now? The question is, whether one can find a basis B for which the transformation matrix ${}_BM(f)_B$ has a diagonal shape. If so, there are crucial advantages associated with it. We will deal with this problem in the next section.

38.4 Exercises

38.1 Given are two ordered bases A and B of $\mathbb{R}^3$,

$$ A = \left(\begin{pmatrix} 8 \\ -6 \\ 7 \end{pmatrix}, \begin{pmatrix} -16 \\ 7 \\ -13 \end{pmatrix}, \begin{pmatrix} 9 \\ -3 \\ 7 \end{pmatrix} \right), \quad B = \left(\begin{pmatrix} 1 \\ -2 \\ 1 \end{pmatrix}, \begin{pmatrix} 3 \\ -1 \\ 2 \end{pmatrix}, \begin{pmatrix} 2 \\ 1 \\ 2 \end{pmatrix} \right), $$

and a linear mapping $f : \mathbb{R}^3 \to \mathbb{R}^3$ which is with respect to the basis A has the following transformation matrix

$$ {}_AM(f)_A = \begin{pmatrix} 1 & -18 & 15 \\ -1 & -22 & 15 \\ 1 & -25 & 22 \end{pmatrix}. $$

Determine the transformation matrix ${}_BM(f)_B$ from f with respect to the base B.

38.2 Given a linear mapping $f : \mathbb{R}^3 \to \mathbb{R}^3$. The transformation matrix of f with respect to the ordered standard basis $E_3 = (e_1, e_2, e_3)$ of $\mathbb{R}^3$ is:

$$_{E_3}M(f)_{E_3} = \begin{pmatrix} 4 & 0 & -2 \\ 1 & 3 & -2 \\ 1 & 2 & -1 \end{pmatrix} \in \mathbb{R}^{3\times 3}.$$

(a) Show: $B = \left(\begin{pmatrix} 2 \\ 2 \\ 3 \end{pmatrix}, \begin{pmatrix} 1 \\ 1 \\ 1 \end{pmatrix}, \begin{pmatrix} 2 \\ 1 \\ 1 \end{pmatrix} \right)$ is an ordered base of $\mathbb{R}^3$.

(b) Determine the transformation matrix $_BM(f)_B$ and the transformation matrix S with $_BM(f)_B = S^{-1}\, {}_{E_3}M(f)_{E_3}\, S$.

38.3 Let $f : \mathbb{R}^3 \to \mathbb{R}^2$ and $g : \mathbb{R}^2 \to \mathbb{R}^4$ with

$$f(v_1, v_2, v_3) = \begin{pmatrix} v_1 + v_2 \\ 2v_1 + v_2 + v_3 \end{pmatrix} \text{ and } g(v_1, v_2) = \begin{pmatrix} 2v_1 + v_2 \\ 2v_1 + v_2 \\ v_2 \\ v_1 + v_2 \end{pmatrix}$$

linear mappings, $B = E_3$, $C = E_2$ and $D = E_4$ the standard bases of $\mathbb{R}^3$, $\mathbb{R}^2$ and $\mathbb{R}^4$. Determine the transformation matrices of f with respect to B and C and g with respect to C and D and of $g \circ f$ with respect to B and D.

38.4 Given are the ordered standard basis $E_2 = \left(\begin{pmatrix} 1 \\ 0 \end{pmatrix}, \begin{pmatrix} 0 \\ 1 \end{pmatrix} \right)$ of $\mathbb{R}^2$, $B = \left(\begin{pmatrix} 1 \\ 1 \\ 1 \end{pmatrix}, \begin{pmatrix} 1 \\ 1 \\ 0 \end{pmatrix}, \begin{pmatrix} 1 \\ 0 \\ 0 \end{pmatrix} \right)$ of $\mathbb{R}^3$ and $C = \left(\begin{pmatrix} 1 \\ 1 \\ 1 \\ 1 \end{pmatrix}, \begin{pmatrix} 1 \\ 1 \\ 1 \\ 0 \end{pmatrix}, \begin{pmatrix} 1 \\ 1 \\ 0 \\ 0 \end{pmatrix}, \begin{pmatrix} 1 \\ 0 \\ 0 \\ 0 \end{pmatrix} \right)$ of $\mathbb{R}^4$.

Now consider two linear mappings $f : \mathbb{R}^2 \to \mathbb{R}^3$ and $g : \mathbb{R}^3 \to \mathbb{R}^4$ defined by

$$f\left(\begin{pmatrix} v_1 \\ v_2 \end{pmatrix} \right) = \begin{pmatrix} v_1 - v_2 \\ 0 \\ 2\,v_1 - v_2 \end{pmatrix} \text{ and } g\left(\begin{pmatrix} v_1 \\ v_2 \\ v_3 \end{pmatrix} \right) = \begin{pmatrix} v_1 + 2\,v_3 \\ v_2 - v_3 \\ v_1 + v_2 \\ 2\,v_1 + 3\,v_3 \end{pmatrix}.$$

Determine the transformation matrices $_BM(f)_{E_2}$, $_CM(g)_B$ and $_CM(g \circ f)_{E_2}$.

38.5 Let the linear mapping $f: \mathbb{R}^3 \to \mathbb{R}^3$ is given by

$$f(e_1) = 3e_3, \quad f(e_2) = e_1 - e_2 - 9e_3 \text{ and } f(e_3) = 2e_2 + 7e_3.$$

Give the transformation matrices of f with respect to the standard basis $E = (e_1, e_2, e_3)$ and with respect to the following basis $B = (b_1, b_2, b_3)$ from $\mathbb{R}^3$:

$$b_1 = (1, 1, 1)^\top, \quad b_2 = (1, 2, 3)^\top \text{ and } b_3 = (1, 3, 6)^\top.$$

38.6 Let the linear mapping $f: \mathbb{R}^2 \to \mathbb{R}^3$ given by

$$f(x, y) = (y, 2x - 2y, 3x)^\top.$$

(a) Give the transformation matrix of f with respect to the standard bases of $\mathbb{R}^2$, $\mathbb{R}^3$.
(b) Determine the transformation matrix of f with respect to the bases $B = (b_1, b_2)$ of $\mathbb{R}^2$ and $C = (c_1, c_2, c_3)$ of $\mathbb{R}^3$ with

$$b_1 = (1, 1)^\top, \quad b_2 = (5, 3)^\top$$

and

$$c_1 = (1, 2, 2)^\top, \quad c_2 = (1, 3, 4)^\top, \quad c_3 = (2, 4, 5)^\top.$$

(c) Let there be $x = 2e_1 - 4e_2$. What are the coordinates of $f(x)$ with respect to the basis (c_1, c_2, c_3)?

Diagonalization: Eigenvalues and Eigenvectors 39

With *diagonalizing* matrices we have reached the center of linear algebra. The key to diagonalizing are vectors v not equal to the zero vector with $A\,v = \lambda\,v$ for a $\lambda \in \mathbb{K}$—one calls v *eigenvector* and λ *Eigenvalue*. When diagonalizing a matrix. $A \in \mathbb{K}^{n\times n}$ one determines all eigenvalues of A and a basis of $\mathbb{K}^n$ from eigenvectors.

The applications of diagonalizing matrices are many, topics such as principal axis transformation, singular value decomposition and matrix exponential function for solving systems of differential equations are based on diagonalizing.

As often before $\mathbb{K}$ one of the two number fields $\mathbb{R}$ or $\mathbb{C}$.

39.1 Eigenvalues and Eigenvectors of Matrices

We start with the central concepts:

Eigenvalue, Eigenvector, Eigenspace
Given is a square matrix $A \in \mathbb{K}^{n\times n}$. If

$$A\,v = \lambda\,v \text{ with } v \neq 0 \text{ and } \lambda \in \mathbb{K},$$

then we call

- $v \in V \setminus \{0\}$ an **eigenvector** of A corresponding to the **eigenvalue** $\lambda \in \mathbb{K}$ and
- $\lambda \in \mathbb{K}$ has an **eigenvalue** of A with **eigenvector** $v \in V \setminus \{0\}$.

(continued)

C. Karpfinger, *Calculus and Linear Algebra in Recipes*,
https://doi.org/10.1007/978-3-662-65458-3_39

If λ is an eigenvalue of A, the subspace

- $\text{Eig}_A(\lambda) = \{v \in \mathbb{K}^n \mid A\, v = \lambda\, v\}$ is called the **eigenspace** of A to the eigenvalue λ and
- $\dim(\text{Eig}_A(\lambda))$ the **geometric multiplicity** of the eigenvalue λ, one writes

$$\text{geo}(\lambda) = \dim(\text{Eig}_A(\lambda))\,.$$

The elements $v \neq 0$ of the eigenspace $\text{Eig}_A(\lambda)$ are exactly the eigenvectors of A to the eigenvalue λ; in particular, every linear combination of eigenvectors is either an eigenvector again or the null vector.

The fact that $\text{Eig}_A(\lambda)$ is a subspace of $\mathbb{K}^n$ follows quite simply with the prescription in Sect. 13.2:

(1) We have $0 \in \text{Eig}_A(\lambda)$, since $A\,0 = \lambda\,0$.
(2) If $u,\ v \in \text{Eig}_A(\lambda)$, then $A\,u = \lambda\,u$ and $A\,v = \lambda\,v$. It follows

$$A\,(u+v) = A\,u + A\,v = \lambda\,u + \lambda\,v = \lambda\,(u+v)\,,$$

so that $u + v \in \text{Eig}_A(\lambda)$.
(3) If $u \in \text{Eig}_A(\lambda)$ and $\mu \in \mathbb{K}$, then $A\,u = \lambda\,u$. It follows

$$\mu\,A\,u = \mu\,\lambda\,u\,,\ \text{i.e.,}\ A\,(\mu\,u) = \lambda\,(\mu\,u)\,,$$

so that $\mu\,u \in \text{Eig}_A(\lambda)$.

We could have had this even simpler: Because of

$$A\,v = \lambda\,v \;\Leftrightarrow\; (A - \lambda\,E_n)\,v = 0$$

we have

$$\text{Eig}_A(\lambda) = \ker(A - \lambda\,E_n)\,.$$

And as the kernel of a matrix $\text{Eig}_A(\lambda)$ is of course a subspace of $\mathbb{K}^n$.

Example 39.1 We consider the matrix $A = \begin{pmatrix} -4 & 6 \\ -3 & 5 \end{pmatrix}$. Because of

$$\begin{pmatrix} -4 & 6 \\ -3 & 5 \end{pmatrix} \begin{pmatrix} 1 \\ 1 \end{pmatrix} = 2 \begin{pmatrix} 1 \\ 1 \end{pmatrix} \quad \text{and} \quad \begin{pmatrix} -4 & 6 \\ -3 & 5 \end{pmatrix} \begin{pmatrix} 2 \\ 1 \end{pmatrix} = (-1) \begin{pmatrix} 2 \\ 1 \end{pmatrix}$$

the matrix has A at least the two different eigenvalues $\lambda_1 = 2$ and $\lambda_2 = -1$ with corresponding eigenvectors $v_1 = (1, 1)^\top$ and $v_2 = (2, 1)^\top$. ■

39.2 Diagonalizing Matrices

The central Theorem is: *A matrix $A \in \mathbb{K}^{n\times n}$ is diagonalizable if and only if there is a basis of $\mathbb{K}^n$ existing of eigenvectors of* A, more precisely:

Diagonalizing Matrices

A matrix $A \in \mathbb{K}^{n\times n}$ is called **diagonalizable** if there is an invertible matrix $B \in \mathbb{K}^{n\times n}$ such that

$$D = B^{-1} A\, B$$

is a diagonal matrix. In this case it is called

- D a **diagonal form** of A,
- B a **the matrix A diagonalizing matrix** and
- determining D and B **diagonalizing A**.

The matrix $A \in \mathbb{K}^{n\times n}$ is diagonalizable if and only if there is a basis of $\mathbb{K}^n$ consisting of eigenvectors of A, i.e., there exists an ordered basis $B = (b_1, \ldots, b_n)$ of $\mathbb{K}^n$ and $\lambda_1, \ldots, \lambda_n \in \mathbb{K}$ which are not necessarily all different, with

$$A\, b_1 = \lambda_1 b_1, \ldots, A\, b_n = \lambda_n b_n\,.$$

In this case

- the matrix $D = \operatorname{diag}(\lambda_1, \ldots, \lambda_n)$ is a diagonal form of A and
- the matrix $B = (b_1, \ldots, b_n)$ a A diagonalizing matrix.

The proof is simple: If $A \in \mathbb{K}^{n\times n}$ is diagonalizable, then there exists a diagonal matrix $D = \operatorname{diag}(\lambda_1, \ldots, \lambda_n)$ and an invertible matrix $B = (b_1, \ldots, b_n)$ with

$$\begin{aligned} D = B^{-1} A B \;&\Leftrightarrow\; A B = B D \\ &\Leftrightarrow\; (A b_1, \ldots, A b_n) = (\lambda_1 b_1, \ldots, \lambda_n b_n) \\ &\Leftrightarrow\; A b_1 = \lambda_1 b_1, \ldots, A b_n = \lambda_n b_n \,. \end{aligned}$$

Since the matrix B is invertible, the columns of B also form a basis of $\mathbb{K}^n$. Thus everything is shown.

Diagonalizing a matrix $A \in \mathbb{K}^{n\times n}$ means to determine

- the not necessarily different eigenvalues $\lambda_1, \ldots, \lambda_n$ and
- an ordered basis $B = (b_1, \ldots, b_n)$ of $\mathbb{K}^n$ consisting of eigenvectors of A.

From this we obtain the diagonal form $D = \operatorname{diag}(\lambda_1, \ldots, \lambda_n)$ and the diagonalizing matrix $B = (b_1, \ldots, b_n)$.

Not every matrix is diagonalizable, examples will follow later. Now first an example of a diagonalizable matrix:

Example 39.2 The matrix $A = \begin{pmatrix} -4 & 6 \\ -3 & 5 \end{pmatrix} \in \mathbb{R}^{2\times 2}$ from Example 39.1 is diagonalizable since

- $b_1 = (1, 1)^\top$ and $b_2 = (2, 1)^\top$ are eigenvectors of A: We have

$$A b_1 = \lambda_1 b_1 \text{ and } A b_2 = \lambda_2 b_2 \text{ with } \lambda_1 = 2 \text{ and } \lambda_2 = -1 \,.$$

- $B = (b_1, b_2)$ is an ordered basis of $\mathbb{R}^2$ because of the linear independence of b_1 and b_2.

So we have

$$D = \begin{pmatrix} 2 & 0 \\ 0 & -1 \end{pmatrix} = \begin{pmatrix} 1 & 2 \\ 1 & 1 \end{pmatrix}^{-1} A \begin{pmatrix} 1 & 2 \\ 1 & 1 \end{pmatrix} \text{ and } D = \begin{pmatrix} -1 & 0 \\ 0 & 2 \end{pmatrix} = \begin{pmatrix} 2 & 1 \\ 1 & 1 \end{pmatrix}^{-1} A \begin{pmatrix} 2 & 1 \\ 1 & 1 \end{pmatrix} .$$

Note that this equality is not to be verified, it must be correct (which you are welcome to verify). ■

For the understanding of later facts the following interpretation of the *diagonalizing* is helpful: To diagonalize a matrix A means to determine a diagonal matrix D which is similar to A. The matrices A and D are transformation matrices of the linear mapping $f_A : \mathbb{K}^n \to \mathbb{K}^n$ with $f_A(v) = A\,v$. The columns of the matrix B form a basis of eigenvectors $b_1, \ldots, b_n$ of A to the eigenvalues $\lambda_1, \ldots, \lambda_n$, therefore

$$A\,b_1 = \lambda_1 b_1, \ldots, A\,b_n = \lambda_n b_n .$$

With the mnemonic rule *in the i-th column of the transformation matrix contains the coordinate vector of the image of the i-th basis vector* we get the transformation matrix ${}_B M(f_A)_B$ of f_A with respect to B and the base transformation formula:

$${}_B M(f_A)_B = \begin{pmatrix} \lambda_1 & & \\ & \ddots & \\ & & \lambda_n \end{pmatrix} \quad \text{and} \quad {}_B M(f_A)_B = B^{-1} A\,B .$$

But how do you determine the eigenvalues or eigenvectors of a matrix? The essential tool is the *characteristic polynomial* of A.

39.3 The Characteristic Polynomial of a Matrix

We are looking for the eigenvalues of $A \in \mathbb{K}^{n \times n}$, i.e. the numbers $\lambda \in \mathbb{K}$ with $A\,b = \lambda\,b$ where $b \in \mathbb{K}^n$ and $b \neq 0$. Because of

$$A\,b = \lambda\,b \text{ with } b \neq 0 \Leftrightarrow (A - \lambda\,E_n)\,b = 0 \text{ with } b \neq 0 \Leftrightarrow \det(A - \lambda\,E_n) = 0$$

the eigenvalues λ are thus found to be zeros of the following polynomial

$$\chi_A = \det(A - x\,E_n) .$$

The Characteristic Polynomial of a Matrix A
To $A \in \mathbb{K}^{n \times n}$ consider the **characteristic polynomial**

$$\chi_A = \det(A - x\,E_n) .$$

We assume in what follows that χ_A via $\mathbb{K}$ splits into linear factors, i.e.

$$\chi_A = (\lambda_1 - x)^{\nu_1} \cdots (\lambda_r - x)^{\nu_r}$$

(continued)

with different $\lambda_1, \ldots, \lambda_r \in \mathbb{K}$. We then have:

- $\lambda_1, \ldots, \lambda_r$ are the different eigenvalues of A, there are no others.
- χ_A has the degree n, correspondingly $\nu_1 + \cdots + \nu_r = n$.
- A has at most n different eigenvalues (if $r = n$).
- One calls the power ν_i the **algebraic multiplicity** of the eigenvalue λ_i, one writes $\text{alg}(\lambda_i) = \nu_i$.
- For $1 \le i \le r$ we have: $1 \le \text{geo}(\lambda_i) \le \text{alg}(\lambda_i)$.

Note that $\text{geo}(\lambda)$ is the dimension of the eigenspace to the eigenvalue λ of $A \in \mathbb{K}^{n \times n}$. Since eigenvectors to different eigenvalues are linearly independent (cf. Exercise 39.5), the union of bases of eigenspaces is again a linearly independent set. Now, if each eigenspace is as *large* as possible, i.e. $\text{geo}(\lambda) = \text{alg}(\lambda)$ for any eigenvalue λ, then the union of the bases of all eigenspaces is a n-element linearly independent subset of the n-dimensional vector space $\mathbb{K}^n$; in this situation we have a basis of $\mathbb{K}^n$ consisting of eigenvectors of A. Thus, the following criterion for diagonalizability and the subsequent recipe for diagonalizing a matrix are plausible:

Criterion for Diagonalizability
A matrix $A \in \mathbb{K}^{n \times n}$ is diagonalizable if and only if the characteristic polynomial χ_A splits into linear factors via $\mathbb{K}$ and $\text{alg}(\lambda) = \text{geo}(\lambda)$ for each eigenvalue λ of A.

In particular, every matrix $A \in \mathbb{K}^{n \times n}$ with n different eigenvalues can be diagonalized.

If you want to diagonalize a matrix A, proceed as follows:

Recipe: Diagonalize a Matrix A
Given is a square matrix $A \in \mathbb{K}^{n \times n}$ which we want to diagonalize.

(1) Determine the characteristic polynomial χ_A and decompose it into linear factors, if possible:

$$\chi_A = (\lambda_1 - x)^{\nu_1} \cdots (\lambda_r - x)^{\nu_r}.$$

(continued)

- We have $\nu_1 + \cdots + \nu_r = n$.
- $\lambda_1, \ldots, \lambda_r$ are the different eigenvalues with the respective algebraic multiplicity $\text{alg}(\lambda_i) = \nu_i$.

If χ_A does not completely split into linear factors, STOP: A is not diagonalizable, else:

(2) For each eigenvalue λ_i determine the eigenspace $\text{Eig}_A(\lambda_i)$,

$$\text{Eig}_A(\lambda_i) = \ker(A - \lambda_i E_n) = \langle B_i \rangle \,,$$

by specifying a basis B_i of $\text{Eig}_A(\lambda_i)$. We have $|B_i| = \text{geo}(\lambda_i)$.

If $\text{geo}(\lambda_i) \neq \text{alg}(\lambda_i)$ for one i, STOP: A is not diagonalizable, else:

(3) $B = B_1 \cup \cdots \cup B_r$ is a basis of $\mathbb{K}^n$ consisting of eigenvectors of A. Order the basis $B = (b_1, \ldots, b_n)$ and obtain in the case $A\,b_1 = \lambda_1 b_1, \ldots, A\,b_n = \lambda_n b_n$:

$$\text{diag}(\lambda_1, \ldots, \lambda_n) = B^{-1} A\, B \,.$$

Example 39.3

- We diagonalize the matrix $A = \begin{pmatrix} -2 & -8 & -12 \\ 1 & 4 & 4 \\ 0 & 0 & 1 \end{pmatrix} \in \mathbb{R}^{3\times 3}$.

(1) A has the characteristic polynomial

$$\chi_A = \det \begin{pmatrix} -2-x & -8 & -12 \\ 1 & 4-x & 4 \\ 0 & 0 & 1-x \end{pmatrix} = (1-x)\det \begin{pmatrix} -2-x & -8 \\ 1 & 4-x \end{pmatrix}$$
$$= (1-x)\big((-2-x)(4-x) - (-8)\cdot 1\big) = (1-x)\,x\,(x-2)\,.$$

Thus A has the three eigenvalues $\lambda_1 = 0,\ \lambda_2 = 1$ and $\lambda_3 = 2$.

(2) We obtain the eigenspaces:

$$\mathrm{Eig}_A(0) = \ker(A - 0 \cdot E_3) = \ker \begin{pmatrix} -2 & -8 & -12 \\ 1 & 4 & 4 \\ 0 & 0 & 1 \end{pmatrix} = \ker \begin{pmatrix} 1 & 4 & 4 \\ 0 & 0 & 1 \\ 0 & 0 & 0 \end{pmatrix} = \langle \begin{pmatrix} 4 \\ -1 \\ 0 \end{pmatrix} \rangle,$$

$$\mathrm{Eig}_A(1) = \ker(A - 1 \cdot E_3) = \ker \begin{pmatrix} -3 & -8 & -12 \\ 1 & 3 & 4 \\ 0 & 0 & 0 \end{pmatrix} = \ker \begin{pmatrix} 1 & 3 & 4 \\ 0 & 1 & 0 \\ 0 & 0 & 0 \end{pmatrix} = \langle \begin{pmatrix} -4 \\ 0 \\ 1 \end{pmatrix} \rangle,$$

$$\mathrm{Eig}_A(2) = \ker(A - 2 \cdot E_3) = \ker \begin{pmatrix} -4 & -8 & -12 \\ 1 & 2 & 4 \\ 0 & 0 & -1 \end{pmatrix} = \ker \begin{pmatrix} 1 & 2 & 4 \\ 0 & 0 & 1 \\ 0 & 0 & 0 \end{pmatrix} = \langle \begin{pmatrix} -2 \\ 1 \\ 0 \end{pmatrix} \rangle.$$

(3) It is $B = (b_1, b_2, b_2)$ with

$$b_1 = \begin{pmatrix} 4 \\ -1 \\ 0 \end{pmatrix}, \; b_2 = \begin{pmatrix} -4 \\ 0 \\ 1 \end{pmatrix}, \; b_3 = \begin{pmatrix} -2 \\ 1 \\ 0 \end{pmatrix}$$

an ordered basis of $\mathbb{R}^3$ consisting of eigenvectors of A. Because of $A\,b_1 = 0\,b_1$, $A\,b_2 = 1\,b_2$ and $A\,b_3 = 2\,b_3$ we have

$$D = \begin{pmatrix} 0 & 0 & 0 \\ 0 & 1 & 0 \\ 0 & 0 & 2 \end{pmatrix} = B^{-1} A\, B \text{ with } B = (b_1, b_2, b_3).$$

Note: The order of the basis vectors is important here. The basis $C = (b_3, b_1, b_2)$ provides, for example, the diagonal matrix

$$\tilde{D} = \begin{pmatrix} 2 & 0 & 0 \\ 0 & 0 & 0 \\ 0 & 0 & 1 \end{pmatrix} = C^{-1} A\, C \text{ with } C = (b_3, b_1, b_2).$$

- Now we diagonalize the matrix $A = \begin{pmatrix} 1 & -4 \\ 1 & 1 \end{pmatrix} \in \mathbb{C}^{2\times 2}$.

(1) A has the characteristic polynomial

$$\chi_A(x) = \det(A - x\,E_2) = \det\begin{pmatrix} 1-x & -4 \\ 1 & 1-x \end{pmatrix}$$

$$= (1-x)^2 + 4 = x^2 - 2x + 5 = \big(x - (1-2\,\mathrm{i})\big)\big(x - (1+2\,\mathrm{i})\big)\,.$$

The (complex) eigenvalues are thus $\lambda_1 = 1 - 2\,\mathrm{i}$ and $\lambda_2 = 1 + 2\,\mathrm{i}$.

(2) The eigenspaces are computed as

$$\mathrm{Eig}_A(1-2\,\mathrm{i}) = \ker\begin{pmatrix} 2\,\mathrm{i} & -4 \\ 1 & 2\,\mathrm{i} \end{pmatrix} = \ker\begin{pmatrix} 1 & 2\,\mathrm{i} \\ 0 & 0 \end{pmatrix} = \langle\begin{pmatrix} 2\,\mathrm{i} \\ 1 \end{pmatrix}\rangle\,,$$

$$\mathrm{Eig}_A(1+2\,\mathrm{i}) = \ker\begin{pmatrix} -2\,\mathrm{i} & -4 \\ 1 & -2\,\mathrm{i} \end{pmatrix} = \ker\begin{pmatrix} 1 & -2\,\mathrm{i} \\ 0 & 0 \end{pmatrix} = \langle\begin{pmatrix} -2\,\mathrm{i} \\ -1 \end{pmatrix}\rangle\,.$$

(3) It is $B = (b_1, b_2)$ with

$$b_1 = \begin{pmatrix} 2\,\mathrm{i} \\ -1 \end{pmatrix}\,,\; b_2 = \begin{pmatrix} -2\,\mathrm{i} \\ -1 \end{pmatrix}$$

an ordered basis of $\mathbb{C}^2$ consisting of eigenvectors of A. Because $A\,b_1 = (1-2\,\mathrm{i})\,b_1$ and $A\,b_2 = (1+2\,\mathrm{i})\,b_2$ we have

$$\begin{pmatrix} 1-2\,\mathrm{i} & 0 \\ 0 & 1+2\,\mathrm{i} \end{pmatrix} = B^{-1} A\,B \text{ with } B = (b_1, b_2)\,. \qquad \blacksquare$$

The **trace** of a matrix $S = (s_{ij})$ is defined as the sum of all diagonal elements:

$$\mathrm{trace}(S) = \sum_{i=1}^{n} s_{ii}\,.$$

It is a trifle to compute the trace of a matrix. Remarkably, the trace is also the sum of the eigenvalues of a matrix (even if the matrix is not diagonalizable). Therefore, the trace provides a wonderful way to check the calculation: If the trace of A is not the sum of the calculated eigenvalues, then one has certainly miscalculated. Also the determinant of A is closely related to the eigenvalues of A:

det(A) = Product of the Eigenvalues, trace(A) = Sum of the Eigenvalues
If $\lambda_1, \ldots, \lambda_n$ are the not necessarily different eigenvalues of a matrix $A \in \mathbb{K}^{n\times n}$ then we have:

$$\det(A) = \lambda_1 \cdots \lambda_n \qquad \text{and} \qquad \operatorname{trace}(A) = \lambda_1 + \cdots + \lambda_n\,,$$

i.e., the determinant of A is the product of the eigenvalues, the trace of A is the sum of the eigenvalues.

Check these facts with the above examples.

We recall the notion of *similarity of* matrices: Two $n \times n$-matrices A and B are called similar if there is an invertible matrix S with $B = S^{-1} A\, S$. The main common properties of similar matrices are:

Common Properties of Similar Matrices
If $A,\ B \in \mathbb{K}^{n\times n}$ are similar, then we have:

- $\chi_A = \chi_B$.
- A and B have the same eigenvalues.
- The eigenvalues of A and B have the same algebraic and geometric multiplicities.

Remark A matrix $A \in \mathbb{K}^{n\times n}$ is not diagonalizable, for example, if

- χ_A via $\mathbb{K}$ does not split into linear factors, e.g. $A = \begin{pmatrix} 0 & -1 \\ 1 & 0 \end{pmatrix} \in \mathbb{R}^{2\times 2}$, or
- $\operatorname{geo}(\lambda) < \operatorname{alg}(\lambda)$ for an eigenvalue λ of A, e.g. $A = \begin{pmatrix} 2 & 1 \\ 0 & 2 \end{pmatrix} \in \mathbb{R}^{2\times 2}$; here $\chi_A(x) = (2-x)^2$, e.g. $\operatorname{alg}(2) = 2$, and $\operatorname{Eig}_A(2) = \langle (1,0)^\top \rangle$, thus $\operatorname{geo}(2) = 1$.

MATLAB You get the eigenvalues with MATLAB by `eig(A)`. If one also wants a diagonalizing matrix B, one enters `[B,D]=eig(A)`. If one is only interested in the characteristic polynomial χ_A of a matrix A, then `poly(A)` gives the coefficients $a_n, \ldots, a_0$ (in this order) of χ_A.

39.4 Diagonalization of Real Symmetric Matrices

In general, it is difficult to decide whether a matrix $A \in \mathbb{K}^{n \times n}$ is diagonalizable. One has to check whether the characteristic polynomial splits into linear factors and whether algebraic and geometric multiplicities coincide for each eigenvalue. This is quite different for a real symmetric matrix. For such a matrix we have:

Diagonalizability of a Real Symmetric Matrix
If A is a real symmetric matrix, $A \in \mathbb{R}^{n \times n}$, $A^\top = A$, then we have:

- A is diagonalizable.
- All eigenvalues of A are real.
- Eigenvectors to different eigenvalues are perpendicular to each other (with respect to the standard scalar product $\langle v\,,\,w\rangle = v^\top w$).
- The matrix A diagonalizing matrix B can be chosen to be orthogonal, i.e. $B^{-1} = B^\top$.

Since a matrix $B \in \mathbb{R}^{n \times n}$ is orthogonal if and only if the columns $b_1, \ldots, b_n$ form an orthonormal basis of $\mathbb{R}^n$ the last point therefore states that the eigenvectors $b_1, \ldots, b_n$ of A in the case of a real symmetric matrix A form an ONB of $\mathbb{R}^n$. According to the penultimate point, eigenvectors are orthogonal to different eigenvectors per se. So when determining an ONB of $\mathbb{R}^n$ consisting of eigenvectors of A only orthogonal basis vectors within the eigenspaces have to be determined; we can therefore still use our recipe in Sect. 39.3 for the diagonalization of a real symmetric matrix A, in which we only add ONBs B_i in the eigenspaces.

Example 39.4

- We consider the real and symmetric matrix

$$A = \begin{pmatrix} 1 & 3 \\ 3 & 1 \end{pmatrix} \in \mathbb{R}^{2 \times 2}.$$

(1) A has the characteristic polynomial $\chi_A(x) = (1-x)^2 - 9 = (x+2)(x-4)$, i.e. the eigenvalues $\lambda_1 = 4$ and $\lambda_2 = -2$.

(2) We obtain as eigenspaces:

$$\operatorname{Eig}_A(4) = \ker \begin{pmatrix} -3 & 3 \\ 3 & -3 \end{pmatrix} = \langle \begin{pmatrix} 1 \\ 1 \end{pmatrix} \rangle \text{ and } \operatorname{Eig}_A(-2) = \langle \begin{pmatrix} 1 \\ -1 \end{pmatrix} \rangle,$$

whereby we have determined $\text{Eig}_A(-2)$ simply by specifying the $(1, 1)^\top$orthogonal vector $(1, -1)^\top$.

(3) If one now chooses

$$b_1 = \tfrac{1}{\sqrt{2}}(1, 1)^\top \text{ and } b_2 = \tfrac{1}{\sqrt{2}}(1, -1)^\top,$$

then $B = (b_1, b_2)$ is an ONB of the $\mathbb{R}^2$ of eigenvectors and thus

$$B^{-1} A\, B = B^\top A\, B = \begin{pmatrix} 4 & 0 \\ 0 & -2 \end{pmatrix}.$$

- Now we consider the matrix

$$A = \begin{pmatrix} 1 & 1 & 1 \\ 1 & 1 & 1 \\ 1 & 1 & 1 \end{pmatrix} \in \mathbb{R}^{3\times 3}.$$

(1) The characteristic polynomial of A is

$$\begin{aligned}
\chi_A(x) &= \det \begin{pmatrix} 1-x & 1 & 1 \\ 1 & 1-x & 1 \\ 1 & 1 & 1-x \end{pmatrix} = \det \begin{pmatrix} 1-x & 1 & 1 \\ 1 & 1-x & 1 \\ 0 & x & -x \end{pmatrix} \\
&= x \cdot \det \begin{pmatrix} 1-x & 2 & 1 \\ 1 & 2-x & 1 \\ 0 & 0 & -1 \end{pmatrix} = -x \cdot \det \begin{pmatrix} 1-x & 2 \\ 1 & 2-x \end{pmatrix} \\
&= -x\big((1-x)(2-x)-2\big) = -x(x^2-3x) = -x^2(x-3)\,.
\end{aligned}$$

The matrix A therefore has the two eigenvalues $\lambda_1 = 0$ and $\lambda_2 = 3$ with the algebraic multiplicity $\text{alg}(0) = 2$ and $\text{alg}(3) = 1$.

(2) The associated eigenspaces are

$$\text{Eig}_A(0) = \ker \begin{pmatrix} 1 & 1 & 1 \\ 1 & 1 & 1 \\ 1 & 1 & 1 \end{pmatrix} = \ker \begin{pmatrix} 1 & 1 & 1 \\ 0 & 0 & 0 \\ 0 & 0 & 0 \end{pmatrix} = \langle \begin{pmatrix} 1 \\ -1 \\ 0 \end{pmatrix}, \begin{pmatrix} 1 \\ 0 \\ -1 \end{pmatrix} \rangle,$$

$$\text{Eig}_A(3) = \ker \begin{pmatrix} -2 & 1 & 1 \\ 1 & -2 & 1 \\ 1 & 1 & -2 \end{pmatrix} = \ker \begin{pmatrix} 1 & 1 & -2 \\ 0 & -3 & 3 \\ 0 & 0 & 0 \end{pmatrix} = \langle \begin{pmatrix} 1 \\ 1 \\ 1 \end{pmatrix} \rangle.$$

(3) With the choice

$$b_1 = \frac{1}{\sqrt{2}}(1, -1, 0)^\top, \quad b_2 = \frac{1}{\sqrt{6}}(1, 1, -2)^\top, \quad b_3 = \frac{1}{\sqrt{3}}(1, 1, 1)^\top$$

you get with $B = (b_1, b_2, b_3)$ an ONB of the $\mathbb{R}^3$ and thus

$$B^{-1} A\, B = B^\top A\, B = \begin{pmatrix} 0 & 0 & 0 \\ 0 & 0 & 0 \\ 0 & 0 & 3 \end{pmatrix}. \quad \blacksquare$$

39.5 Exercises

39.1 Give the eigenvalues and eigenvectors of the following complex matrices:

$$\text{(a) } A = \begin{pmatrix} 3 & -1 \\ 1 & 1 \end{pmatrix}, \quad \text{(b) } B = \begin{pmatrix} 0 & 1 \\ 1 & 0 \end{pmatrix}, \quad \text{(c) } C = \begin{pmatrix} 0 & -1 \\ 1 & 0 \end{pmatrix}.$$

39.2 Show that eigenvectors to different eigenvalues are linearly independent.

39.3 Diagonalize, if possible, the following real matrices:

$$\text{(a) } A = \begin{pmatrix} 1 & 0 & 0 \\ 0 & 2 & 0 \\ 0 & 0 & 3 \end{pmatrix}, \quad \text{(c) } C = \begin{pmatrix} 1 & 3 & 6 \\ -3 & -5 & -6 \\ 3 & 3 & 4 \end{pmatrix}, \quad \text{(e) } F = \begin{pmatrix} -3 & 1 & -1 \\ -7 & 5 & -1 \\ -6 & 6 & -2 \end{pmatrix},$$

$$\text{(b) } B = \begin{pmatrix} 2 & 1 & 0 \\ 0 & 2 & 0 \\ 0 & 0 & 3 \end{pmatrix}, \quad \text{(d) } D = \begin{pmatrix} 1 & -3 & 3 \\ 3 & -5 & 3 \\ 6 & -6 & 4 \end{pmatrix}, \quad \text{(f) } G = \begin{pmatrix} 1 & -1 & 1 \\ 0 & 3 & 0 \\ -1 & 0 & 3 \end{pmatrix}.$$

39.4 Calculate all the eigenvalues and corresponding eigenvectors of the following matrices:

$$\text{(a) } A = \begin{pmatrix} 0 & 1 & -1 \\ 0 & 1 & 0 \\ 1 & 0 & 0 \end{pmatrix},$$

$$\text{(b) } B = \begin{pmatrix} 1 & 2 & 0 \\ 0 & 1 & 0 \\ -1 & 2 & -2 \end{pmatrix},$$

(c) $C = \begin{pmatrix} 1 & 0 & 0 \\ 0 & \cos\alpha & -\sin\alpha \\ 0 & \sin\alpha & \cos\alpha \end{pmatrix}$,

(d) $D = \begin{pmatrix} 2 & -2 & 2 \\ -2 & 2 & -2 \\ -2 & 2 & -2 \end{pmatrix}$.

39.5 Let $A \in \mathbb{R}^{n\times n}$ an orthogonal matrix and $\lambda \in \mathbb{C}$ an eigenvalue of A. Show that $|\lambda| = 1$.

39.6

(a) Show the following statement: Is $A \in \mathbb{R}^{n\times n}$ is a symmetric matrix and v_1 and v_2 are two eigenvectors of A at eigenvalues λ_1 and λ_2, where $\lambda_1 \neq \lambda_2$ then v_1 and v_2 are orthogonal to each other.
(b) Given the matrix $A = \begin{pmatrix} 0 & -1 & -2 \\ -1 & 0 & -2 \\ -2 & -2 & -3 \end{pmatrix}$.
Determine all the eigenvalues of A and give a basis of the eigenspaces.
(c) Further determine an orthogonal matrix U such that $U^\top A\, U$ has a diagonal form.

39.7 Let v be an eigenvector to the eigenvalue λ of a matrix A.

(a) Is v also an eigenvector of A^2? To which eigenvalue?
(b) Moreover, if A is invertible, is v also an eigenvector to A^{-1}? To which eigenvalue?

39.8 Give a basis of eigenvectors for each of the following matrices.

(a) $A = \begin{pmatrix} 1 & 2 \\ 2 & 1 \end{pmatrix}$ (b) $A = \begin{pmatrix} 2 & 0 & -1 & -4 \\ -3 & 1 & 3 & 0 \\ 2 & 0 & -1 & -2 \\ 1 & 0 & -1 & -3 \end{pmatrix}$ (c) $A = \begin{pmatrix} 1+\mathrm{i} & 0 \\ -6 & 1-\mathrm{i} \end{pmatrix}$.

39.9 Let the matrix $A \in \mathbb{R}^{3\times 3}$ be given as $A = \begin{pmatrix} 1 & 4 & 0 \\ 2 & 3 & 0 \\ 0 & 0 & 1 \end{pmatrix}$.

(a) Show that the vectors $v_1 = (1, 1, 0)^\top$ and $v_2 = (0, 0, 1)^\top$ are eigenvectors of A. Determine the corresponding eigenvalues.

(b) Does A have any other eigenvalues? If necessary, calculate these eigenvalues and associated eigenvectors.
(c) Show that $\mathbb{R}^3$ has a basis B which consists of eigenvectors of A. Determine the representation matrix of the linear mapping $f: x \mapsto Ax$ with respect to the basis B.
(d) Use the previous results to calculate as simply as possible the matrix A^5.

39.10 The Fibonacci numbers $F_0, F_1, F_2, \ldots$ are recursively defined by the rule

$$F_0 = 0, \qquad F_1 = 1, \qquad F_n = F_{n-1} + F_{n-2} \text{ for } n \geq 2.$$

(a) Determine a matrix $A \in \mathbb{R}^{2\times 2}$ which satisfies the equation $(F_n, F_{n-1})^\top = A\,(F_{n-1}, F_{n-2})^\top$.
(b) How must $k \in \mathbb{N}$ be chosen so that $(F_n, F_{n-1})^\top = A^k\,(F_1, F_0)^\top$?
(c) Calculate all the eigenvalues and eigenvectors of the matrix A.
(d) Compute an invertible matrix T and a diagonal matrix D with the property $D = T^{-1} A\, T$.
(e) Use the representation of A from (d) to represent A^k for the number determined in (b).
(f) Use the previous partial results to construct an explicit representation for the Fibonacci numbers F_n (without recursion).

39.11 Let $A \in \mathbb{R}^{n\times n}$. Show:

(a) If $\lambda \in \mathbb{C}$ an eigenvalue of A, so is $\overline{\lambda} \in \mathbb{C}$ such a value.
(b) If $v \in \mathbb{C}^n$ is an eigenvector of A, then so is $\overline{v} \in \mathbb{C}^n$ is such a vector (where $\overline{v} = (\overline{v}_i)$ for $v = (v_i) \in \mathbb{C}^n$).

Give the complex eigenvalues and eigenvectors of the matrix $\begin{pmatrix} 0 & 1 \\ -1 & 0 \end{pmatrix}$.

40 Numerical Calculation of Eigenvalues and Eigenvectors

40.1 Gerschgorin Circles

A first, rough but often useful localization of the eigenvalues of a square matrix $A = (a_{ij}) \in \mathbb{C}^{n\times n}$ is obtained with the *Gerschgorin circles*:

The Gershgorin Theorem
The n eigenvalues of the complex matrix $A = (a_{ij}) \in \mathbb{C}^{n\times n}$ lie in the union $\bigcup_{i=1}^{n} K_i$ of n **Gershgorin circles**

$$K_i = \{z \in \mathbb{C} \mid |z - a_{ii}| \le \sum_{\substack{j=1\\ j\neq i}}^{n} |a_{ij}|\}, \quad i = 1, \ldots, n\,.$$

Are $M_1, \ldots, M_r$ different circular disks of $\{K_1, \ldots, K_n\}$ and $M_{r+1}, \ldots, M_n$ the rest of the n circular disks and do we have

$$\left(\bigcup_{i=1}^{r} M_i\right) \cap \left(\bigcup_{i=r+1}^{n} M_i\right) = \emptyset\,,$$

so $\bigcup_{i=1}^{r} M_i$ contains exactly r eigenvalues and $\bigcup_{i=r+1}^{n} M_i$ exactly $n-r$ eigenvalues.

Note that this theorem does not guarantee that in every Gerschgorin circle there is also an eigenvalue, unless the circle is disjoint to all other circles.

C. Karpfinger, *Calculus and Linear Algebra in Recipes*,
https://doi.org/10.1007/978-3-662-65458-3_40

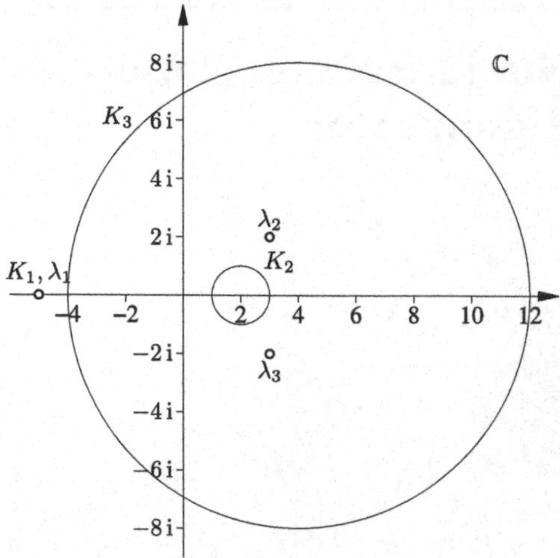

Fig. 40.1 The three Gerschgorin circles

Example 40.1 The matrix $A = \begin{pmatrix} -5 & 0 & 0 \\ 0 & 2 & 1 \\ 3 & -5 & 4 \end{pmatrix} \in \mathbb{C}^{3\times 3}$ has the three Gerschgorin circles

$$K_1 = \{-5\},$$
$$K_2 = \{z \in \mathbb{C} \mid |z-2| \le 1\},$$
$$K_3 = \{z \in \mathbb{C} \mid |z-4| \le 8\}.$$

These circles are plotted in Fig. 40.1. The circle K_1 contains because of

$$K_1 \cap (K_2 \cup K_3) = \emptyset$$

exactly one eigenvalue. This can only be $\lambda_1 = -5$ be. The circle K_2 does not contain an eigenvalue, nor is it disjoint from the other circles. In fact, the matrix A has the eigenvalues $\lambda_1 = -5$, $\lambda_2 = 3 + 2\,\mathrm{i}$ and $\lambda_3 = 3 - 2\,\mathrm{i}$. ■

For a proof of the first part of Gerschgorin's Theorem, see Exercise 40.5.

One knows the eigenvalues according to Gerschgorin's Theorem more precisely the smaller these circular disks are. In the extreme case of a diagonal matrix, Gerschgorin's Theorem even gives the eigenvalues exactly. Otherwise, Gerschgorin's Theorem gives a *rather good* approximation of the eigenvalues if the matrix A is **strictly diagonally dominant**, i.e., the absolute value of its diagonal elements a_{ii} are each greater than the

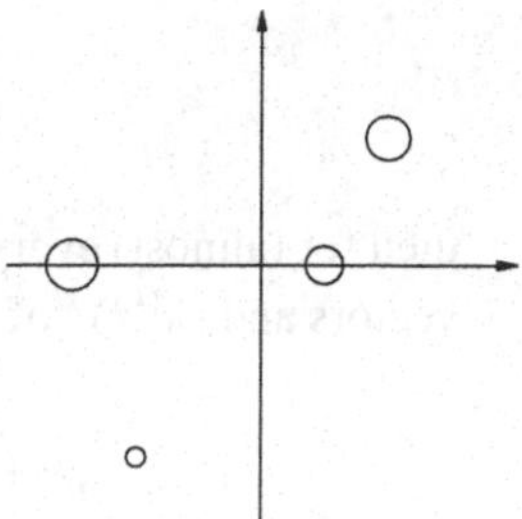

Fig. 40.2 Strictly diagonal dominant matrix

sum of the absolute values of the remaining respective line entries a_{ij}:

$$\sum_{\substack{j=1 \\ j \neq i}}^{n} |a_{ij}| < |a_{ii}| \text{ for all } i = 1, \ldots, n\,.$$

The Gerschgorin circles of a strictly diagonal dominant matrix have a *small* radius, they are in this case rather *small*, see Fig. 40.2.

Gerschgorin circles are often used as follows: *If no Gerschgorin circle of a matrix A contains the zero point, then A is invertible.*

In this case, $\lambda = 0$ cannot be an eigenvalue of A; the determinant, which is the product of the eigenvalues, is not zero in this case. Thus A is invertible.

40.2 Vector Iteration

The largest eigenvalue (more accurate its absolute value) of a complex matrix often plays an important role. One calls the number

$$\max\{|\lambda| \mid \lambda \text{ is an eigenvalue of } A \in \mathbb{C}^{n \times n}\}$$

the **spectral radius** of the matrix A. This number plays an important role in various numerical methods. The *vector iteration* is a simple method which determines for a diagonalizable matrix A the largest eigenvalue of A with an associated eigenvector. It is not necessary to determine all eigenvalues of A. We get more specific:

Vector Iteration or von Mises Iteration

If $A \in \mathbb{C}^{n \times n}$ is a diagonalizable matrix with the eigenvalues $\lambda_1, \ldots, \lambda_n$ and the largest eigenvalue λ_1,

(continued)

$$|\lambda_1| > |\lambda_2| \geq \cdots \geq |\lambda_n|\,,$$

then for (almost) every starting vector $v^{(0)} \in \mathbb{C}^n$ of length 1 the sequences $(v^{(k)})_k$ of vectors and $(\lambda^{(k)})_k$ of complex numbers with

$$v^{(k+1)} = \frac{A\,v^{(k)}}{\|A\,v^{(k)}\|} \quad \text{and} \quad \lambda^{(k+1)} = \frac{(v^{(k)})^\top A\,v^{(k)}}{(v^{(k)})^\top v^{(k)}}$$

converge to an eigenvector v, $v^{(k)} \to v$, and to the eigenvalue λ_1, $\lambda^{(k)} \to \lambda_1$ from A.

The essential idea for a justification of this statement is in the following considerations: If $(v_1, \ldots, v_n)$ a basis of the $\mathbb{C}^n$ from eigenvectors of the matrix A, we can use the normalized starting vector $v^{(0)}$ as a linear combination of this basis:

$$v^{(0)} = \mu_1 v_1 + \mu_2 v_2 + \cdots + \mu_n v_n\,.$$

We multiply $v^{(0)}$ with A and get $w^{(1)} = A\,v^{(0)}$ and more generally

$$w^{(k)} = A^k v^{(0)} = \mu_1\,\lambda_1^k v_1 + \cdots + \mu_n\,\lambda_n^k v_n\,.$$

It follows

$$\frac{1}{\lambda_1^k} w^{(k)} = \mu_1 v_1 + \mu_2 \left(\frac{\lambda_2}{\lambda_1}\right)^k v_2 + \cdots + \mu_n \left(\frac{\lambda_n}{\lambda_1}\right)^k v_n\,.$$

Because λ_1 is greater (its absolute value) than the other eigenvalues $\lambda_2, \ldots, \lambda_n$ for sufficiently large $k \in \mathbb{N}$ we get

$$\frac{1}{\lambda_1^k} w^{(k)} \approx \mu_1 v_1\,,$$

so that because of

$$A\,w^{(k)} = w^{(k+1)} \approx \lambda_1^{k+1}\,\mu_1 v_1 \approx \lambda_1 w^{(k)}$$

an approximate value for the eigenvalue λ_1 as well as a vector $w^{(k)}$ which, in the event of $\mu_1 \neq 0$, can be approximated as an eigenvector to the eigenvalue λ_1. In order to obtain convergence, a normalization must be performed in each step.

Remark Theoretically, the starting vector must not be chosen arbitrarily, we must have $\mu_1 \neq 0$. But in practice one knows the eigenvector v_1 not at all, so it must be left to chance whether $\mu_1 \neq 0$. However, this is not problematic at all, since the rounding errors occurring in the calculations usually ensure that the calculations lead to an approximate solution even without this precondition.

The *rate of convergence* depends on the value $|\lambda_2/\lambda_1| < 1$. The smaller this quotient is –we then speak of eigenvalues that are *well separated*– the faster the method converges.

Example 40.2 We determine approximately the largest eigenvalue (absolute value) and an associated eigenvector of the matrix $A = \begin{pmatrix} 1 & 1 & 0 \\ 3 & -1 & 2 \\ 2 & -1 & 3 \end{pmatrix}$. As starting vector we choose $v^{(0)} = (1, 0, 0)^\top$ and obtain:

k	$x^{(k)}$	$\lambda^{(k)}$	k	$x^{(k)}$	$\lambda^{(k)}$
0	$(1.0000, 0.0000, 0.0000)^\top$		5	$(0.2970, 0.6109, 0.7339)^\top$	2.7303
1	$(0.2673, 0.8018, 0.5345)^\top$	1.0000	6	$(0.3086, 0.5942, 0.7427)^\top$	2.9408
2	$(0.5298, 0.5298, 0.6623)^\top$	1.8571	7	$(0.2979, 0.5996, 0.7428)^\top$	3.0306
3	$(0.2923, 0.6577, 0.6942)^\top$	3.4912	8	$(0.3005, 0.5958, 0.7448)^\top$	2.9869
4	$(0.3463, 0.5860, 0.7326)^\top$	2.7303	9	$(0.2981, 0.5970, 0.7448)^\top$	3.0068

With MATLAB we obtain for comparison the eigenvector $v = (0.2981, 0.5963, 0.7454)^\top$ to the eigenvalue $\lambda = 3$. ■

A termination criterion for vector iteration is: STOP, if the difference of two successive approximations for λ falls below a tolerance limit.

Remark By an *inverse* vector iteration one can also determine a possibly existing smallest positive eigenvalue λ with associated eigenvector v. For this one essentially uses the fact that λ^{-1} is then a largest eigenvalue of A^{-1} (with the same eigenvector v). We do not treat this method, but turn to methods which determine the totality of all eigenvalues of a matrix numerically.

40.3 The Jacobian Method

The *Jacobian method* transfers a real symmetric matrix $A \in \mathbb{R}^{n \times n}$ by successive multiplication with particularly simple orthogonal matrices $S_1, \ldots, S_r$ or $S_1^\top, \ldots, S_r^\top$ to an approximate diagonal form,

$$A \mapsto A^{(1)} = S_1^\top A\, S_1,\ A^{(1)} \mapsto A^{(2)} = S_2^\top A^{(1)}\, S_2,\ \ldots$$

In each step $A^{(k)} \to A^{(k+1)}$ the sum of the squares of the non-diagonal components of the matrices $A^{(k)} = (a_{ij}^{(k)})$ is reduced, which is the number

$$N(A^{(k)}) = \sum_{\substack{i,j=1 \\ i \neq j}}^{n} (a_{ij}^{(k)})^2.$$

If you perform such transformations until the sum of the non-diagonal components of a matrix $A^{(r)} = D$ is zero, you have finally reached diagonal form. The eigenvalues form the diagonal entries of D, and the columns of the then orthogonal matrix

$$S = S_1 \cdots S_r$$

contains then because of $D = (S_1 \cdots S_r)^\top A\,(S_1 \cdots S_r)$ the searched eigenvectors.

In fact, in practice, one will specify an error bound ε and stop the iteration as soon as the sum of squares of the off-diagonal elements of a matrix $A^{(r)}$ falls below the error bound ε.

We now give explicitly the transforming orthogonal matrices $S_1, S_2, \ldots$ which deliver a sequence of matrices

$$A \to A^{(1)} \to A^{(2)} \cdots$$

which ultimately *converges to* a diagonal matrix.

In the following we assume that the symmetric matrix $A = (a_{ij}) \in \mathbb{R}^{n \times n}$ does not already have diagonal form. Then there are p, q with $p < q$ and $a_{pq} \neq 0$. We now choose the transforming matrix in such a way that the two non-zero entries $a_{pq} = a_{qp}$ disappear by this transformation.

Using the real numbers a_{pq}, a_{pp} and a_{qq} we can form the following three numbers:

$$D = \frac{a_{pp} - a_{qq}}{\sqrt{(a_{pp} - a_{qq})^2 + 4a_{pq}^2}}, \quad c = \sqrt{\frac{1+D}{2}}, \quad s = \begin{cases} \sqrt{\frac{1-D}{2}}, & \text{if } a_{pq} > 0 \\ -\sqrt{\frac{1-D}{2}}, & \text{if } a_{pq} < 0 \end{cases}.$$

We have $c^2 + s^2 = 1$, therefore there is a $\alpha \in [0, 2\pi[$ with $c = \cos\alpha$ and $s = \sin\alpha$.

With the help of the numbers c and s we now form the matrix

$$S = \begin{pmatrix} 1 & & & & & & \\ & \ddots & & & & & \\ & & c & & -s & & \\ & & & \ddots & & & \\ & & s & & c & & \\ & & & & & \ddots & \\ & & & & & & 1 \end{pmatrix} \begin{matrix} \\ \\ \leftarrow p \\ \\ \leftarrow q \\ \\ \\ \end{matrix} .$$

This matrix S is obviously orthogonal, since $S^\top S = E_n$. In fact, the matrix S describes a rotation by angle α about zero in the plane containing the *p-th* and *q-th* coordinate axis. Hence the following term:

Jacobi Rotation
If $A \in \mathbb{R}^{n \times n}$ is a symmetric matrix and S as described above, then the transformation $A \mapsto \tilde{A} = S^\top A\, S$ changes at most the components of *p-th* and *q-th* row and column. We have

$$N(\tilde{A}) = N(A) - 2\, a_{pq}^2 ,$$

so that the sum of squares of the non diagonal components after performing such a **Jacobi rotation** becomes genuinely smaller. Furthermore $\tilde{A}$ is again symmetric, so that another Jacobian rotation can be performed.

In general, one reaches the goal faster if one always chooses p and q in such a way that the element a_{pq} has a large absolute value. The search of this element is for large n costly. One then *eliminates* row by row at the places

$$(1, 2),\ (1, 3), \ldots, (1, n),\ (2, 3), \ldots, (n - 1, n) ,$$

where a pair (p, q) is omitted if $|a_{pq}|$ is smaller than a given error bound.

Note, that at the places (p, q) where zeros have already been generated, numbers unequal to zero can arise again in the following iterations. This is no problem: The sum of squares of the non-diagonal components decreases. To get below a given tolerance level, several runs may be necessary.

Example 40.3 Usually the Jacobian method is used for **tridiagonal matrices**, i.e., for matrices with three diagonals like the following symmetric matrix.

$$A = \begin{pmatrix} -2 & 1 & 0 & 0 \\ 1 & -2 & 1 & 0 \\ 0 & 1 & -2 & 1 \\ 0 & 0 & 1 & -2 \end{pmatrix} \in \mathbb{R}^{4\times 4}.$$

We have $N(A) = 6$; by successive elimination at the points $(1,2),\ (2,3),\ (3,4)$ we obtain in turn:

$$A = \begin{pmatrix} -2 & 1 & 0 & 0 \\ 1 & -2 & 1 & 0 \\ 0 & 1 & -2 & 1 \\ 0 & 0 & 1 & -2 \end{pmatrix} \xrightarrow{c=s=\sqrt{1/2}} A^{(1)} = \begin{pmatrix} -1.000 & 0 & 0.7071 & 0 \\ 0 & -3.000 & 0.7071 & 0 \\ 0.7071 & 0.7071 & -2.0000 & 1.0000 \\ 0 & 0 & 1.0000 & -2.0000 \end{pmatrix}$$

$$\xrightarrow{c=0.4597\ s=0.8881} A^{(2)} = \begin{pmatrix} -1.000 & 0.6280 & 0.3251 & 0 \\ 0.6280 & -1.6340 & 0 & 0.8881 \\ 0.3251 & 0 & -3.3660 & 0.4597 \\ 0 & 0.8881 & 0.4597 & -2.0000 \end{pmatrix}$$

$$\xrightarrow{c=0.3810\ s=0.9246} A^{(3)} = \begin{pmatrix} -1.000 & 0.6280 & 0.0949 & -0.3109 \\ 0.6280 & -1.6340 & 0.8494 & 0.2592 \\ 0.0949 & 0.8494 & -1.8597 & 0 \\ -0.3109 & 0.2592 & 0 & -3.5063 \end{pmatrix}.$$

After another such *sweep*, i.e., a sweep through all nonzero non diagonal components, we get

$$A^{(9)} = \begin{pmatrix} -0.4165 & 0.1519 & -0.1421 & -0.0587 \\ 0.1519 & -1.3582 & -0.0337 & -0.0085 \\ -0.1421 & -0.0337 & -2.6084 & 0 \\ -0.0587 & -0.0085 & 0 & -3.6169 \end{pmatrix}$$

with $N(A^{(9)}) = 0.0959$. Another *sweep* returns the already *near diagonal matrix*

$$A^{(15)} = \begin{pmatrix} -0.3820 & 0.0007 & 0.0001 & 0.0000 \\ 0.0007 & -1.3820 & -0.0000 & -0.0000 \\ 0.0001 & -0.0000 & -2.6180 & -0.0000 \\ 0.0000 & -0.0000 & -0.0000 & -3.6180 \end{pmatrix}$$

with $N(A^{(15)}) = 1.0393 \cdot 10^{-6}$. And the matrix $S_1 \cdots S_{15}$ with the approximations of the eigenvectors is

$$\begin{pmatrix} 0.3722 & -0.6012 & 0.6015 & -0.3717 \\ 0.6018 & -0.3713 & -0.3717 & 0.6015 \\ 0.6012 & 0.3722 & -0.3717 & -0.6015 \\ 0.3713 & 0.6018 & 0.6015 & 0.3717 \end{pmatrix}.$$

The *exact* eigenvalues of A are

$$\lambda_1 = -3.6180\,,\ \lambda_2 = -2.6180\,,\ \lambda_3 = -1.3820\,,\ \lambda_4 = -0.3820\,. \quad \blacksquare$$

The Jacobian method is simple, transparent, numerically stable and easy to implement. However, it is only suitable for symmetric matrices. There are also methods for the approximate determination of eigenvalues and eigenvectors of arbitrary matrices. The most important method is the so-called QR-method, which we will discuss in the next section.

40.4 The QR-Method

With the QR-method, the eigenvalues of an arbitrary (square) matrix are determined approximately. It is a frequently used method. We describe the basic procedure and remind of the QR-decomposition $A = QR$ of a (square) matrix $A \in \mathbb{R}^{n\times n}$ with an orthogonal matrix $Q \in \mathbb{R}^{n\times n}$ and an upper triangular matrix $R \in \mathbb{R}^{n\times n}$ (see Chap. 19). With the help of the QR-decomposition, we generate a sequence $(A_k)_k$ of $n \times n$-matrices starting with a square matrix $A \in \mathbb{R}^{n\times n}$ in the following way, where we set $A_0 = A$:

- We determine the QR-decomposition of A_0, i.e. $A_0 = Q_0R_0$ and set $A_1 = R_0Q_0$.
- We determine the QR-decomposition of A_1, i.e. $A_1 = Q_1R_1$ and set $A_2 = R_1Q_1$
- ...
- General: Decompose $A_k = Q_kR_k$ and set $A_{k+1} = R_kQ_k$.

Note that because of $A_{k+1} = Q_k^\top A_k Q_k$ for each k the matrix A_{k+1} is similar to A_k and thus the matrices $A, A_1, A_2, \ldots$ all have the same eigenvalues with the same multiplicities.

We thus obtain a sequence (A_k) of square matrices which converges under suitable conditions against an upper triangular matrix, more precisely:

The QR-Method

The sequence (A_k) of matrices A_k converges to a matrix of the form

$$A_\infty = \begin{pmatrix} A_{11} & * & * & * \\ 0 & A_{22} & * & * \\ \vdots & \ddots & \ddots & * \\ 0 & \ldots & 0 & A_{ss} \end{pmatrix}$$

with 1×1- or 2×2-matrices $A_{11}, \ldots, A_{ss}$. We have:

- The eigenvalues of A are the eigenvalues of the matrices $A_{11}, \ldots, A_{ss}$.
- If A_{jj} is a 1×1-matrix, the eigenvalue of A_{jj} is real.
- If A_{jj} is a 2×2-matrix, then the two eigenvalues of A_{jj} are complex conjugated.
- If $|\lambda_i| \neq |\lambda_j|$ for the n eigenvalues of A, then all boxes $A_{11}, \ldots, A_{ss}$ are single-row, i.e., A_∞ is an upper triangular matrix.
- If A is symmetric and $|\lambda_i| \neq |\lambda_j|$ for the n eigenvalues of A, then the sequence $(P_k)_k$ with $P_k = Q_0 Q_1 \cdots Q_k$ converges against an orthogonal matrix whose columns are an ONB of the $\mathbb{R}^n$ consisting of eigenvectors of A and A_∞ is a diagonal matrix with the eigenvalues of A as diagonal elements.

The QR-method is easy to program, if one refers to the in MATLAB implemented QR-decomposition with `[Q,R]=qr(A)`. Note Exercise 40.5.

Example 40.4 We consider the matrix $A = \begin{pmatrix} 1 & 2 & 3 \\ 2 & 4 & 5 \\ 3 & 5 & 6 \end{pmatrix}$. Using `[V,D]=eig(A)` we obtain an *exact* transformation matrix (i.e. eigenvectors of A) and the *exact* eigenvalues

$$V = \begin{pmatrix} 0.7370 & 0.5910 & 0.3280 \\ 0.3280 & -0.7370 & 0.5910 \\ -0.5910 & 0.3280 & 0.7370 \end{pmatrix} \text{ and } D = \begin{pmatrix} -0.5157 & 0 & 0 \\ 0 & 0.1709 & 0 \\ 0 & 0 & 11.3448 \end{pmatrix}.$$

we use MATLAB to calculate the first iterates A_1, A_2, A_3, ... of the matrix $A = \begin{pmatrix} 1 & 2 & 3 \\ 2 & 4 & 5 \\ 3 & 5 & 6 \end{pmatrix}$:

k	A_k	P_k
0	$\begin{pmatrix} 1.0000 & 2.0000 & 3.0000 \\ 2.0000 & 4.0000 & 5.0000 \\ 3.0000 & 5.0000 & 6.0000 \end{pmatrix}$	$\begin{pmatrix} -0.2673 & -0.3586 & -0.8944 \\ -0.5345 & -0.7171 & 0.4472 \\ -0.8018 & 0.5976 & -0.0000 \end{pmatrix}$
1	$\begin{pmatrix} 11.2143 & 1.1819 & 0.3586 \\ 1.1819 & -0.2143 & -0.2673 \\ 0.3586 & -0.2673 & 0.0000 \end{pmatrix}$	$\begin{pmatrix} 0.3316 & 0.8362 & -0.4369 \\ 0.5922 & 0.1759 & 0.7863 \\ 0.7344 & -0.5195 & -0.4369 \end{pmatrix}$
2	$\begin{pmatrix} 11.3446 & -0.0549 & -0.0062 \\ -0.0549 & -0.4896 & -0.1307 \\ -0.0062 & -0.1307 & 0.1450 \end{pmatrix}$	$\begin{pmatrix} -0.3278 & -0.6968 & -0.6380 \\ -0.5909 & -0.3756 & 0.7139 \\ -0.7371 & 0.6111 & -0.2886 \end{pmatrix}$
3	$\begin{pmatrix} 11.3448 & 0.0025 & 0.0001 \\ 0.0025 & -0.5128 & -0.0448 \\ 0.0001 & -0.0448 & 0.1680 \end{pmatrix}$	$\begin{pmatrix} 0.3280 & 0.7496 & -0.5749 \\ 0.5910 & 0.3119 & 0.7439 \\ 0.7370 & -0.5837 & -0.3408 \end{pmatrix}$
4	$\begin{pmatrix} 11.3448 & -0.0001 & -0.0000 \\ -0.0001 & -0.5154 & -0.0149 \\ -0.0000 & -0.0149 & 0.1706 \end{pmatrix}$	$\begin{pmatrix} -0.3280 & -0.7327 & -0.5963 \\ -0.5910 & -0.3333 & 0.7346 \\ -0.7370 & 0.5934 & -0.3237 \end{pmatrix}$
5	$\begin{pmatrix} 11.3448 & 0.0000 & 0.0000 \\ 0.0000 & -0.5157 & -0.0049 \\ 0.0000 & -0.0049 & 0.1709 \end{pmatrix}$	$\begin{pmatrix} 0.3280 & 0.7384 & -0.5892 \\ 0.5910 & 0.3262 & 0.7378 \\ 0.7370 & -0.5902 & -0.3294 \end{pmatrix}$
6	$\begin{pmatrix} 11.3448 & -0.0000 & -0.0000 \\ -0.0000 & -0.5157 & -0.0016 \\ -0.0000 & -0.0016 & 0.1709 \end{pmatrix}$	

■

Remark In practice, the matrix A is first brought to a so-called *Hessenberg form*, i.e. to an upper triangular form, where it is allowed that there are non-zero entries in the first lower secondary diagonal. Furthermore, in order to accelerate the convergence, one performs a so-called *shift*, i.e., one does not decompose the matrix A_k, but the matrix $A_k - \sigma_k E_n$ for a to-be-chosen $\sigma_k \in \mathbb{R}$. Thus, the method generally yields the eigenvalues of even very large matrices very quickly.

40.5 Exercises

40.1 Prove that the set of eigenvalues of a matrix $A \in \mathbb{C}^{n \times n}$ lie in the union of the n Gerschgorin circles of this matrix.

40.2 Determine the Gerschgorin circles of the following matrices:

$$\text{(a) } A = \begin{pmatrix} 4 & 1 & 0 \\ 1 & 4 & 1 \\ 0 & 1 & 4 \end{pmatrix}, \quad \text{(b) } B = \begin{pmatrix} 2 & 1 & 0.5 \\ 0.2 & 5 & 0.7 \\ 1 & 0 & 6 \end{pmatrix}, \quad \text{(c) } C = \begin{pmatrix} 3 & 0.1 & 0.1 \\ 0.1 & 7 & 1 \\ 0.1 & 1 & 5 \end{pmatrix}.$$

40.3 Program the $Q\,R$-method.

40.4 Program the vector iteration. Let the iteration terminate when the distance between two iterations $\lambda^{(k+1)}$ and $\lambda^{(k)}$ is below a given tolerance tol. Test your program on examples.

Quadrics

41

41.1 Terms and First Examples

Let $A \in \mathbb{R}^{n \times n}$ with $A^\top = A$, $b \in \mathbb{R}^n$ and $c \in \mathbb{R}$. The function $q : \mathbb{R}^n \to \mathbb{R}$ with

$$q(x) = x^\top A\, x + b^\top x + c$$

is called a **quadratic polynomial** in the indefinite $x = (x_1, \ldots, x_n)^\top$ with **quadratic part** $x^\top A\, x$ and **linear part** $b^\top x$. These designations are quite obvious, in the cases (most important to us) $n = 2$ and $n = 3$ we have with $A = (a_{ij})$, $b = (b_i)$ and $x = (x_i)$:

$$q(x) = a_{11}x_1^2 + 2a_{12}x_1x_2 + a_{22}x_2^2 + b_1x_1 + b_2x_2 + c \text{ in case } n = 2 \text{ and in case } n = 3$$

$$q(x) = a_{11}x_1^2 + 2a_{12}x_1x_2 + 2a_{13}x_1x_3 + a_{22}x_2^2 + 2a_{23}x_2x_3 + a_{33}x_3^2 + b_1x_1 + b_2x_2 + b_3x_3 + c.$$

Quadrics

The set Q of all $x \in \mathbb{R}^n$ which solve an equation of the form

$$x^\top A\, x + b^\top x + c = 0 \text{ with } A \in \mathbb{R}^{n \times n},\ A^\top = A,\ b \in \mathbb{R}^n,\ c \in \mathbb{R}$$

is called a **quadric**.

C. Karpfinger, *Calculus and Linear Algebra in Recipes*,
https://doi.org/10.1007/978-3-662-65458-3_41

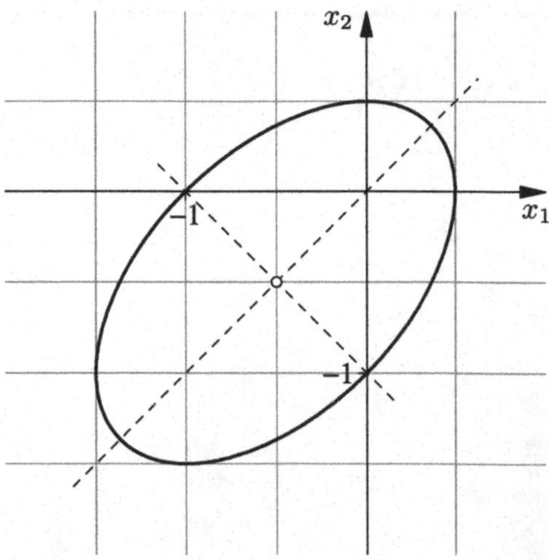

Fig. 41.1 Main axes of the ellipse

Example 41.1 We obtain with $A = \begin{pmatrix} 2 & 0 \\ 0 & -3 \end{pmatrix}$ and $b = 0$ and $c = -6$ the equation

$$2\,x_1^2 - 3\,x_2^2 - 6 = 0\,, \quad \text{i.e.} \quad \left(\frac{x_1}{\sqrt{3}}\right)^2 - \left(\frac{x_2}{\sqrt{2}}\right)^2 = 1\,,$$

so that $Q = \{x = (x_1, x_2)^\top \in \mathbb{R}^2 \mid 2\,x_1^2 - 3\,x_2^2 - 6 = 0\}$ is a hyperbola. ■

In this example, it was very easy to see which curve this quadric is. But which curve is given, for example, by the equation

$$2\,x_1^2 - 2x_1x_2 + 2\,x_2^2 + x_1 + x_2 - 1 = 0\,?$$

We can plot this and other solution sets with MATLAB:

MATLAB With MATLAB it is easy to get a picture of the described area in the $\mathbb{R}^2$. You get the zero set of $q(x) = 2\,x_1^2 - 2x_1x_2 + 2\,x_2^2 + x_1 + x_2 - 1$ in $\mathbb{R}^2$ e.g. as follows

```
» ezplot('2*x.^2-2*x.*y+2*y.^2+x+y-1', [-2,1,-2,1])
» grid on
» title('')
```

Note Fig. 41.1.

In fact, there are only a few substantially different types of quadrics in $\mathbb{R}^2$ or in $\mathbb{R}^3$. The type of the quadric is to be decided by means of the eigenvalues of A. To do this, one performs a *principal axis transformation* and, if necessary, a *translation*: The goal here is to describe the quadric with respect to another basis, so that the type of the quadric can be easily identified.

We first show this with an example before formulating the general procedure:

Example 41.2 We determine the *normal form* of the quadric Q, which is given as the zero set of the equation

$$2\,x_1^2 - 2x_1x_2 + 2\,x_2^2 + x_1 + x_2 - 1 = 0\,.$$

In matrix notation, this equation is

$$x^\top A\,x + b^\top x + c = 0 \text{ where } A = \begin{pmatrix} 2 & -1 \\ -1 & 2 \end{pmatrix},\ b = \begin{pmatrix} 1 \\ 1 \end{pmatrix},\ c = -1\,.$$

(1) We determine an ONB of the $\mathbb{R}^2$ consisting of eigenvectors of A:

$$\chi_A = (2-x)^2 - 1 = (x-1)\,(x-3)\,, \text{ so that } \lambda_1 = 1\,,\ \lambda_2 = 3$$

are the eigenvalues of A. Because

$$\mathrm{Eig}_A(1) = \ker(A - 1\,E_2) = \ker\begin{pmatrix} 1 & -1 \\ -1 & 1 \end{pmatrix} = \langle \begin{pmatrix} 1 \\ 1 \end{pmatrix} \rangle \text{ and}$$

$$\mathrm{Eig}_A(3) = \ker(A - 3\,E_2) = \ker\begin{pmatrix} -1 & -1 \\ -1 & -1 \end{pmatrix} = \langle \begin{pmatrix} 1 \\ -1 \end{pmatrix} \rangle$$

we obtain the ONB or orthogonal matrix

$$B = \begin{pmatrix} \frac{1}{\sqrt{2}} & \frac{1}{\sqrt{2}} \\ \frac{1}{\sqrt{2}} & \frac{-1}{\sqrt{2}} \end{pmatrix}.$$

We now set $y = B^\top x$ and thus obtain a new coordinate system (y_1, y_2) which is again rectangular because of the orthogonality of the matrix B as the coordinate system (x_1, x_2). Because of $B^\top = B^{-1}$ we now have $x = B\,y$. We substitute this into the equation $x^\top A\,x + b^\top x + c = 0$ and get because of

$$D = B^\top A\,B = \begin{pmatrix} 1 & 0 \\ 0 & 3 \end{pmatrix} \text{ and } d^\top = b^\top B = (\sqrt{2}, 0)$$

the following equation of the quadric with respect to the Cartesian coordinate system $y = (y_1, y_2)^\top$:

$$0 = x^\top A\, x + b^\top x + c = y^\top B^\top A\, B\, y + b^\top B\, y + c$$

$$= y^\top \begin{pmatrix} 1 & 0 \\ 0 & 3 \end{pmatrix} y + \begin{pmatrix} \sqrt{2} \\ 0 \end{pmatrix}^\top y - 1 = y_1^2 + 3\, y_2^2 + \sqrt{2}\, y_1 - 1 .$$

In this new representation of the quadric with respect to the Cartesian coordinate system (y_1, y_2) there is no longer a mixed term $y_1 y_2$.

(2) In a second step we eliminate the linear part by completing the square: Because of

$$y_1^2 + \sqrt{2}\, y_1 = \left(y_1 + \frac{1}{\sqrt{2}}\right)^2 - \frac{1}{2}$$

we now set

$$z_1 = y_1 + \frac{1}{\sqrt{2}} \text{ and } z_2 = y_2$$

and get

$$y_1^2 + 3\, y_2^2 + \sqrt{2}\, y_1 - 1 = 0 \Leftrightarrow z_1^2 + 3\, z_2^2 - \frac{3}{2} = 0 .$$

(4) In a last step, we divide by 3/2 and obtain the following equation of the quadric in the Cartesian coordinate system (z_1, z_2):

$$\left(\frac{z_1}{\sqrt{3/2}}\right)^2 + \left(\frac{z_2}{\sqrt{1/2}}\right)^2 = 1 .$$

From this last representation, we can see that it is an ellipse with the semi-axes $a = \sqrt{3/2}$ and $b = \sqrt{1/2}$. ■

We note: The first step eliminates the mixed terms by diagonalizing the matrix A, and in the second step, completing the square eliminates the linear part. Note that this example is a particularly lucky case: Namely, if an eigenvalue is 0, then the elimination of any linear component by completing the square is not possible. We will see this in the next example, but we have already recorded the basic procedure for this case in a recipe.

41.2 Transformation to Normal Form

We obtain the *normal form of* a quadric Q as described in the following recipe:

Recipe: Determine the Normal Form of a Quadric—Principal Axis Theorem
Given is a quadric $Q = \{x \in \mathbb{R}^n \mid q(x) = 0\}$ where

$$q(x) = x^\top A\,x + b^\top x + c \text{ with } A \in \mathbb{R}^{n\times n},\ A^\top = A,\ b \in \mathbb{R}^n,\ c \in \mathbb{R}.$$

(1) **Principal axis transformation to eliminate the mixed terms:** Determine an ONB $B = (b_1, \ldots, b_n)$ of the $\mathbb{R}^n$ from eigenvectors of A, $A\,b_1 = \lambda_1 b_1, \ldots, A\,b_n = \lambda_n b_n$ where

$$\lambda_1, \ldots, \lambda_r \neq 0 \text{ and } \lambda_{r+1}, \ldots, \lambda_n = 0 \text{ for } r \leq n.$$

Set $y = B^\top x$, i.e. $x = B\,y$. By substituting x in the equation $x^\top A\,x + b^\top x + c = 0$ we obtain the equation

$$y^\top D\,y + d^\top y + c = 0$$

with $D = \operatorname{diag}(\lambda_1, \ldots, \lambda_n) \in \mathbb{R}^{n\times n}$ and $d = (d_i)$ where $d^\top = b^\top B \in \mathbb{R}^n$, i.e.

$$(*)\quad \lambda_1 y_1^2 + \cdots + \lambda_r y_r^2 + d_1 y_1 + \cdots + d_n y_n + c = 0.$$

(2) **Translation to eliminate the linear component (completing the square):**

- In the case $\lambda_i \neq 0$ and $d_i \neq 0$ set $z_i = y_i + \frac{d_i}{2\lambda_i}$.
- In the case $\lambda_i = 0$ or $d_i = 0$ set $z_i = y_i$.

One thus obtains from $(*)$ the equation

$$(**)\quad \lambda_1 z_1^2 + \cdots + \lambda_r z_r^2 + d_{r+1} z_{r+1} + \cdots + d_n z_n + e = 0 \text{ with an } e \in \mathbb{R},$$

where e arises from c by adding the correction terms (completing the squares).

(continued)

(3) **Translation to eliminate the constant part:** If $d_k \neq 0$ for a $k > r$, then set $\tilde{z}_k = z_k + \frac{e}{d_k}$ for this k and $\tilde{z}_i = z_i$ otherwise. In this case one obtains from (**)

$$(***)\quad \lambda_1 \tilde{z}_1^2 + \cdots + \lambda_r \tilde{z}_r^2 + d_{r+1}\tilde{z}_{r+1} + d_{r+2}\tilde{z}_{r+2} + \cdots + d_n \tilde{z}_n = 0\,.$$

(4) **Normal form:** By renaming back and possibly swapping the variables and multiplying by a suitable real number $\neq 0$ you get from (**) or $(***)$ in the cases $n = 2$ or $n = 3$ one of the **normal forms** for Q given in the table in Sect. 41.2.

Remark More generally, one can show that for any $n \in \mathbb{N}$ in the case $Q \neq \emptyset$ one of the following **normal forms** is achievable:

$$\left(\left(\frac{x_1}{\alpha_1}\right)^2 + \cdots + \left(\frac{x_p}{\alpha_p}\right)^2\right) - \left(\left(\frac{x_{p+1}}{\alpha_{p+1}}\right)^2 + \cdots + \left(\frac{x_r}{\alpha_r}\right)^2\right) = 0$$

$$\left(\left(\frac{x_1}{\alpha_1}\right)^2 + \cdots + \left(\frac{x_p}{\alpha_p}\right)^2\right) - \left(\left(\frac{x_{p+1}}{\alpha_{p+1}}\right)^2 + \cdots + \left(\frac{x_r}{\alpha_r}\right)^2\right) = 1$$

$$\left(\left(\frac{x_1}{\alpha_1}\right)^2 + \cdots + \left(\frac{x_p}{\alpha_p}\right)^2\right) - \left(\left(\frac{x_{p+1}}{\alpha_{p+1}}\right)^2 + \cdots + \left(\frac{x_r}{\alpha_r}\right)^2\right) - 2x_{r+1} = 0\,,$$

where in the first two cases $1 \leq p \leq r \leq n$ and in the last case $1 \leq p \leq r \leq n-1$. Conceptually, one distinguishes these types (in order) with **cone**, **midpoint quadrics** and **paraboloid**.

Note that in the translation to eliminate the linear component in case $\lambda_i \neq 0$ and $d_i \neq 0$ by this substitution just the linear term $d_i y_i$ is eliminated by completing the square, namely we have:

$$\lambda_i y_i^2 + d_i y_i = \lambda_i \left(y_i^2 + \frac{d_i}{\lambda_i} y_i\right) = \lambda_i \left(y_i + \frac{d_i}{2\lambda_i}\right)^2 - \frac{d_i^2}{4\lambda_i}\,.$$

In exercises, the quadratic equation is often given in the detailed representation. However, there is nothing wrong with transforming this verbose representation into a matrix representation; one just has to always remember that one has to substitute the mixed terms $x_i x_j$ at the position (i, j) of the matrix A must be weighted by a factor of 1/2 (this is owed to the symmetry of the matrix A). Note the following examples:

Example 41.3

- We determine the normal form of the quadric Q, which is given by

$$x_1^2 + 4x_2^2 - 4x_1x_2 + 2x_1 + x_2 - 1 = 0\,.$$

In matrix notation this equation reads

$$x^\top A\,x + b^\top x + c = 0 \text{ where } A = \begin{pmatrix} 1 & -2 \\ -2 & 4 \end{pmatrix},\ b = \begin{pmatrix} 2 \\ 1 \end{pmatrix},\ c = -1\,.$$

(1) **Principal axis transformation:** We determine an ONB of $\mathbb{R}^2$ consisting of eigenvectors of A:

$$\chi_A = (1-x)\,(4-x) - 4 = x\,(x-5)\,, \text{ so that } \lambda_1 = 5\,,\ \lambda_2 = 0$$

the eigenvalues of A are. Because

$$\mathrm{Eig}_A(5) = \ker(A - 5E_2) = \ker \begin{pmatrix} -4 & -2 \\ -2 & -1 \end{pmatrix} = \langle \frac{1}{\sqrt{5}} \begin{pmatrix} 1 \\ -2 \end{pmatrix} \rangle \text{ and}$$

$$\mathrm{Eig}_A(0) = \ker(A) = \ker \begin{pmatrix} 1 & -2 \\ -2 & 4 \end{pmatrix} = \langle \frac{1}{\sqrt{5}} \begin{pmatrix} 2 \\ 1 \end{pmatrix} \rangle$$

we obtain the ONB or orthogonal matrix

$$B = \frac{1}{\sqrt{5}} \begin{pmatrix} 1 & 2 \\ -2 & 1 \end{pmatrix}.$$

We obtain with $y = B^\top x$ and $d^\top = b^\top B = (0, \sqrt{5})$ the equation

$$(*) \quad 5y_1^2 + \sqrt{5}y_2 - 1 = 0\,.$$

(2) **Translation to eliminate the linear component:** Completing the square is omitted, we set $z_1 = y_1$ and $z_2 = y_2$ and obtain

$$(**) \quad 5z_1^2 + \sqrt{5}z_2 - 1 = 0\,.$$

(3) **Translation to eliminate the constant part:** Because $d_2 = \sqrt{5} \neq 0$ we set $\tilde{z}_2 = z_2 - \frac{1}{\sqrt{5}}$ and $\tilde{z}_1 = z_1$ and get

$$(***) \quad 5\tilde{z}_1^2 + \sqrt{5}\tilde{z}_2 = 0\,.$$

(4) **Normal form:** By renaming back $x_k = \tilde{z}_k$, dividing by $\sqrt{5}$ and multiplying the equation by 2 we get the equation of a parabola:

$$\left(\frac{x_1}{1/\sqrt[4]{20}}\right)^2 + 2x_2 = 0\,.$$

- We determine the normal form of the quadric Q, which is given by

$$7x_1^2 - 2x_2^2 + 7\,x_3^2 + 8\,x_1x_2 - 10\,x_1x_3 + 8x_2x_3 + \sqrt{6}x_1 - \sqrt{6}x_2 + 1 = 0\,.$$

In matrix notation this equation is

$$x^\top A\,x + b^\top x + c = 0 \text{ where } A = \begin{pmatrix} 7 & 4 & -5 \\ 4 & -2 & 4 \\ -5 & 4 & 7 \end{pmatrix},\ b = \begin{pmatrix} \sqrt{6} \\ -\sqrt{6} \\ 0 \end{pmatrix},\ c = 1\,.$$

(1) **Principal axis transformation:** We determine an ONB of $\mathbb{R}^2$ consisting of eigenvectors of A:

$$\chi_A = (-6 - x)\,(6 - x)\,(12 - x)\,,\ \text{so that } \lambda_1 = -6\,,\ \lambda_2 = 6\,,\ \lambda_3 = 12$$

are the eigenvalues of A. Because of

$$\mathrm{Eig}_A(-6) = \ker(A + 6E_3) = \langle \tfrac{1}{6}\,(1, -2, 1)^\top \rangle \text{ and}$$

$$\mathrm{Eig}_A(6) = \ker(A - 6E_3) = \langle \tfrac{1}{3}(1, 1, 1)^\top \rangle \text{ and}$$

$$\mathrm{Eig}_A(12) = \ker(A - 12E_3) = \langle \tfrac{1}{2}\,(1, 0, -1)^\top \rangle$$

we obtain the ONB or orthogonal matrix after normalizing the given eigenvectors:

$$B = \begin{pmatrix} \frac{1}{\sqrt{6}} & \frac{1}{\sqrt{3}} & \frac{1}{\sqrt{2}} \\ \frac{-2}{\sqrt{6}} & \frac{1}{\sqrt{3}} & 0 \\ \frac{1}{\sqrt{6}} & \frac{1}{\sqrt{3}} & \frac{-1}{\sqrt{2}} \end{pmatrix}.$$

We get with $y = B^\top x$ and $d^\top = b^\top B = (3, 0, \sqrt{3})$ the equation

$$(*) \quad -6y_1^2 + 6y_2^2 + 12y_3^2 + 3y_1 + \sqrt{3}y_3 + 1 = 0.$$

(2) **Translation to eliminate the linear component:** We set $z_1 = y_1 - \frac{3}{12}$, $z_2 = y_2$ and $z_3 = y_3 + \frac{\sqrt{3}}{24}$ and receive

$$(**) \quad -6z_1^2 + 6z_2^2 + 12z_3^2 + \frac{21}{16} = 0.$$

(3) **Translation to eliminate the constant fraction:** N/A.
(4) **Normal form:** By renaming back $x_k = z_k$, multiplying by -1 and dividing by 21/16 we get the equation.

$$\frac{x_1^2}{\alpha_1^2} - \frac{x_2^2}{\alpha_2^2} - \frac{x_3^2}{\alpha_3^2} = 1 \text{ where } \alpha_1 = \sqrt{\frac{21}{16 \cdot 6}}, \ \alpha_2 = \sqrt{\frac{21}{16 \cdot 6}}, \ \alpha_3 = \sqrt{\frac{21}{16 \cdot 12}}.$$

This is the equation of a two-shell hyperboloid. ■

$\frac{x_1^2}{a_1^2} - \frac{x_2^2}{a_2^2} = 0$	Intersecting straight lines	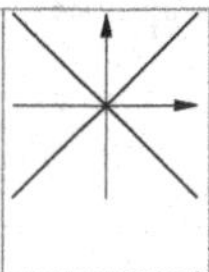	$\frac{x_1^2}{a_1^2} + \frac{x_2^2}{a_2^2} = 1$	Ellipse	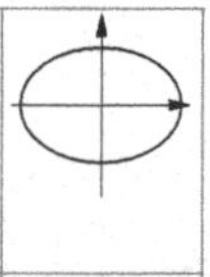
$\frac{x_1^2}{a_1^2} - \frac{x_2^2}{a_2^2} = 1$	Hyperbola	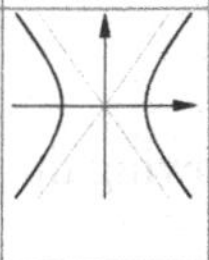	$\frac{x_1^2}{a_1^2} + \frac{x_2^2}{a_2^2} = 0$	Point	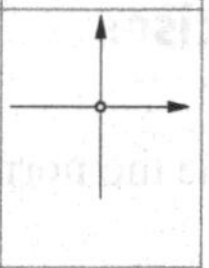
$\frac{x_1^2}{a_1^2} = 1$	Parallel straight lines	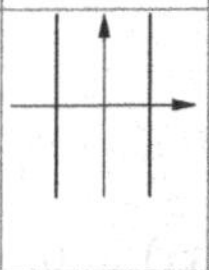	$\frac{x_1^2}{a_1^2} - 2x_2 = 0$	Parabola	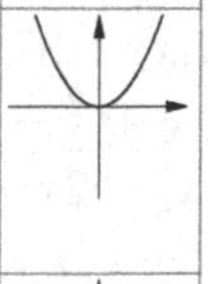
$\frac{x_1^2}{a_1^2} = 0$	Straight line	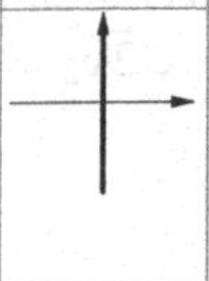	$-\frac{x_1^2}{a_1^2} = 1$ $-\frac{x_1^2}{a_1^2} - \frac{x_2^2}{a_2^2} = 1$	Empty set	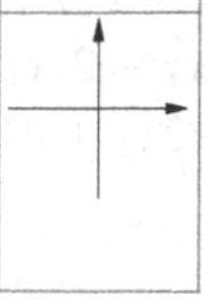

$\frac{x_1^2}{a_1^2}+\frac{x_2^2}{a_2^2}+\frac{x_3^2}{a_3^2}=1$	Ellipsoid		$\frac{x_1^2}{a_1^2}+\frac{x_2^2}{a_2^2}-\frac{x_3^2}{a_3^2}=1$	Single-shell hyperboloid	
$\frac{x_1^2}{a_1^2}+\frac{x_2^2}{a_2^2}+\frac{x_3^2}{a_3^2}=0$	Point		$\frac{x_1^2}{a_1^2}+\frac{x_2^2}{a_2^2}-\frac{x_3^2}{a_3^2}=0$	Square cone	
$\frac{x_1^2}{a_1^2}-\frac{x_2^2}{a_2^2}-\frac{x_3^2}{a_3^2}=1$	Double-shell hyperboloid		$-\frac{x_1^2}{a_1^2}-\frac{x_2^2}{a_2^2}-\frac{x_3^2}{a_3^2}=1$	Empty set	
$\frac{x_1^2}{a_1^2}+\frac{x_2^2}{a_2^2}-2x_3=0$	Elliptical paraboloid		$\frac{x_1^2}{a_1^2}-\frac{x_2^2}{a_2^2}-2x_3=0$	Hyperbolical paraboloid	
$\frac{x_1^2}{a_1^2}+\frac{x_2^2}{a_2^2}=0$	Straight line		$\frac{x_1^2}{a_1^2}-2x_3=0$	Parabolical cylinder	
$\frac{x_1^2}{a_1^2}+\frac{x_2^2}{a_2^2}=1$	Elliptical cylinder		$\frac{x_1^2}{a_1^2}-\frac{x_2^2}{a_2^2}=1$	Hyperbolical cylinder	
$\frac{x_1^2}{a_1^2}-\frac{x_2^2}{a_2^2}=0$	Intersecting planes		$x_1^2=0$	Plane	
$\frac{x_1^2}{a_1^2}=1$	Parallel planes		$-\frac{x_1^2}{a_1^2}=1$ $-\frac{x_1^2}{a_1^2}-\frac{x_2^2}{a_2^2}=1$	Empty set	

41.3 Exercises

41.1 Determine the normal forms of the following quadrics Q given by:

(a) $13x_1^2-32x_1x_2+37x_2^2=45$,
(b) $x_1^2-4x_1x_2+4x_2^2-6x_1+12x_2+8=0$,
(c) $7x_2^2+24x_1x_2-2x_2+24=0$,
(d) $-2(x_1^2+x_2^2+x_3^2)+2(x_1x_2+x_1x_3+x_2x_3)=0$,
(e) $x_1^2+x_2^2+x_3^2-2(x_1x_2+x_1x_3+x_2x_3)+\sqrt{2}x_2=1$,
(f) $x_1^2+2x_2^2+2x_1+8x_2+x_3+3=0$.

41.2 For $c \geq 0$ let Q be the quadric given by

$$c(x_1^2 + x_2^2 + x_3^2) + 6x_1x_2 - 8x_2x_3 + 8x_1 + 6x_3 = 0\,.$$

(a) Write Q in the form $x^\top Ax + a^\top x + \alpha = 0$ with $A^\top = A$ and confirm that one of the principal axis directions of Q is perpendicular to the plane E: $4x_1 + 3x_3 - 5 = 0$.
(b) Determine a principal axis system (n_1, n_2, n_3). What is the equation of Q in the coordinates transformed to principal axes (case distinction!)?

41.3 If $r_i = (x_i, y_i, z_i) \in \mathbb{R}^3$ $(i = 1, \ldots, n)$ are the points of rigidly connected mass points (let the connections be massless) with masses m_i $(i = 1, \ldots, n)$, then the inertia tensor of this rigid body is

$$J = \sum_i m_i \begin{pmatrix} y_i^2 + z_i^2 & -x_i y_i & -x_i z_i \\ -y_i x_i & x_i^2 + z_i^2 & -y_i z_i \\ -z_i x_i & -z_i y_i & x_i^2 + y_i^2 \end{pmatrix}.$$

(a) Let the inertia tensor J for the cube with corners be $r_i = (\pm 1, \text{Â}\pm 1, \pm 1), i = 1, \ldots, 8$ (length unit m), where in all corners the mass is 1 kg and only in $(-1, -1, -1)$ the mass 2 kg.
(b) Compute the principal moments of inertia and axes of the given cube (i.e., the eigenvalues and an ONB of eigenvectors). *Note:* The eigenvalues of J are 19 and 16.
(c) Determine all $\omega \in \mathbb{R}^3$ for which $T_0 = \frac{1}{2}\,\omega^\top J\,\omega = 1.5\,\frac{\text{kg m}^2}{\text{s}^2}$.

42 Schur Decomposition and Singular Value Decomposition

Matrix factorizations such as LR-decomposition, $A = PLR$, the QR-decomposition, $A = QR$, the diagonalization $A = BDB^{-1}$, are advantageous in a wide variety of applications in engineering mathematics. We discuss in this chapter further factorizations, viz. *Schur decomposition* and the *singular value decomposition of* a matrix A. These decompositions find applications in numerical mathematics, but also in signal and image processing. Both methods pick up on old familiar concepts and therefore also repeat many concepts developed in earlier chapters on linear algebra. We formulate these factorizations in a recipe-like manner, drawing on earlier recipes.

42.1 The Schur Decomposition

We remember: A matrix $A \in \mathbb{K}^{n\times n}$ is diagonalizable if there exists an invertible matrix $S \in \mathbb{K}^{n\times n}$ such that $S^{-1}AS$ is a diagonal matrix. It is only obvious to call a matrix **triagonalizable** if there is an invertible matrix $S \in \mathbb{K}^{n\times n}$ such that $S^{-1}AS$ is an upper triangular matrix,

$$S^{-1}AS = R = \begin{pmatrix} \lambda_1 & \cdots & * \\ & \ddots & \vdots \\ 0 & & \lambda_n \end{pmatrix}.$$

We remember further: A matrix A is diagonalizable if and only if the characteristic polynomial χ_A decomposes into linear factors and for each eigenvalue of A coincides the geometric and algebraic multiplicities. Since the triangulation is *weaker* than diagonalizing, one rightly expects that triagonalizing a matrix is also possible under weaker conditions. And so it is: the second condition for equality of multiplicities is not necessary:

C. Karpfinger, *Calculus and Linear Algebra in Recipes*,
https://doi.org/10.1007/978-3-662-65458-3_42

A square matrix A is triagonalizable if and only if the characteristic polynomial χ_A decomposes into linear factors. It gets even better: The matrix transforming to triagonal form S can then even be chosen orthogonally, this is stated by the following theorem:

Schur Decomposition Theorem

For each matrix $A \in \mathbb{R}^{n\times n}$ with characteristic polynomial splitting into linear factors χ_A there is an orthogonal matrix $Q \in \mathbb{R}^{n\times n}$, $Q^{-1} = Q^\top$, with

$$Q^\top A\, Q = R = \begin{pmatrix} \lambda_1 & \cdots & * \\ & \ddots & \vdots \\ 0 & & \lambda_n \end{pmatrix}.$$

This representation is called the **Schur decomposition** of A.

Since the matrices $Q^\top A\, Q = R$ and A are similar and the (not necessarily different) eigenvalues of R are the diagonal entries $\lambda_1, \ldots, \lambda_n$, these numbers are also the eigenvalues of A, i.e., $\lambda_1, \ldots, \lambda_n$ are the zeros of the characteristic polynomial χ_A. However, it is still completely open how the matrix Q of the Schur decomposition is determined. We come to this in the following section.

Remark What can be done if χ_A does not split into linear factors? Then the problem can be solved by *complexification*: One takes the matrix A to be a complex matrix $A \in \mathbb{C}^{n\times n}$; the characteristic polynomial $\chi_A \in \mathbb{C}[x]$ splits over $\mathbb{C}$ into linear factors; one is then dealing with non real eigenvalues and eigenvectors. In this case, there exists a very similar *shorthand decomposition of* A, namely there is then a matrix $Q \in \mathbb{C}^{n\times n}$ with $\overline{Q}^\top Q = E_n$, i.e. $\overline{Q}^\top = Q^{-1}$ so that $\overline{Q}^\top A\, Q = R$ is an upper triangular matrix. One calls a matrix $Q \in \mathbb{C}^{n\times n}$ with $\overline{Q}^\top = Q^{-1}$ **unitary** , which is the *complex version of* orthogonal. We treat only the real Schur decomposition in the following.

42.2 Calculation of the Schur Decomposition

We present a recipe to determine the Schur decomposition of a matrix $A \in \mathbb{R}^{n\times n}$. To do this, we take the columns of the matrix A in turn. We start with the first column and determine an orthogonal matrix Q_1 such that the first column of $Q_1^\top A\, Q_1$ at most in the first position has a nonzero entry. In a second step we provide with an orthogonal matrix Q_2 for then the second column of $Q_2^\top A\, Q_2$ has at most non-zero entries in the first two places, and so on:

$$\underbrace{\begin{pmatrix} * & * & * \\ * & * & * \\ * & * & * \end{pmatrix}}_{=A} \longrightarrow \underbrace{\begin{pmatrix} * & * & * \\ 0 & * & * \\ 0 & * & * \end{pmatrix}}_{=Q_1^\top A\, Q_1} \longrightarrow \underbrace{\begin{pmatrix} * & * & * \\ 0 & * & * \\ 0 & 0 & * \end{pmatrix}}_{=Q_2^\top A\, Q_2} .$$

In doing so, we determine a normalized eigenvector in each step, which we complete to an ONB. The computational effort is kept within limits, since the characteristic polynomial only applies to the initial matrix A is to be calculated. Here one obtains the n possibly multiple occurring eigenvalues $\lambda_1, \ldots, \lambda_n$ with which we bring column by column to the desired form. We describe the procedure in a recipe that basically gives a constructive proof of the Schur decomposition theorem. We only assume that the characteristic polynomial decomposes into linear factors:

Recipe: Determine the Schur Decomposition of a Matrix

The Schur decomposition $R = Q^\top A\, Q$ with an upper triangular matrix R and an orthogonal matrix Q of a matrix $A \in \mathbb{R}^{n\times n}$ with the characteristic polynomial splitting into linear factors

$$\chi_A = (\lambda_1 - x) \cdots (\lambda_n - x)$$

is obtained at the latest $n - 1$ steps:

(1) If the first column $s = (s_1, \ldots, s_n)^\top$ from $A_1 = A$ is not a multiple of e_1:

- Determine an eigenvector v to the eigenvalue λ_1 of A_1 and complete it to an ONB of the $\mathbb{R}^n$, i.e. to an orthogonal matrix $B_1 = (v, v_2, \ldots, v_n)$.
- Calculate

$$B_1^\top A_1\, B_1 = \left(\begin{array}{c|c} \lambda_1 & * \cdots * \\ \hline 0 & \\ \vdots & A_2 \\ 0 & \end{array}\right) \quad \text{with } A_2 \in \mathbb{R}^{(n-1)\times(n-1)} .$$

- Set $Q_1 = B_1$.

(2) If the first column $s = (s_1, \ldots, s_{n-1})^\top$ of A_2 is not a multiple of e_1:

(continued)

- Determine an eigenvector v to the eigenvalue λ_2 of A_2 and complete it to an ONB of the $\mathbb{R}^{n-1}$, i.e. to an orthogonal matrix $B_2 = (v, v_2, , \dots, v_{n-1})$.
- Calculate

$$B_2^\top A_2 B_2 = \left(\begin{array}{c|ccc} \lambda_2 & * & \cdots & * \\ \hline 0 & & & \\ \vdots & & A_3 & \\ 0 & & & \end{array}\right) \quad \text{with } A_3 \in \mathbb{R}^{(n-2)\times(n-2)} .$$

- Set $Q_2 = Q_1 \left(\begin{array}{c|c} 1 & 0 \\ \hline 0 & B_2 \end{array}\right)$.

(3) $\cdots (n-1)$.

Finally set $Q = Q_{n-1}$. We have $Q^{-1} = Q^\top$ and the Schur decomposition of A is

$$Q^\top A\, Q = \begin{pmatrix} \lambda_1 & \cdots & * \\ & \ddots & \vdots \\ 0 & & \lambda_n \end{pmatrix} = R .$$

Example 42.1 We determine the Schur decomposition of $A = \begin{pmatrix} -3 & -4 & 0 \\ 4 & 5 & 0 \\ 3 & 5 & 1 \end{pmatrix} \in \mathbb{R}^{3\times 3}$ with $\chi_A = (1-x)^3$.

(1) The first column $s = (-3, 4, 3)^\top$ from $A_1 = A$ is not a multiple of e_1:

- An eigenvector v to the eigenvalue $\lambda_1 = 1$ of A_1 is $v = (0, 0, 1)^\top$; we complete this to an ONB, i.e. to an orthogonal matrix $B_1 = (v, v_2, v_3)$ of $\mathbb{R}^3$:

$$B_1 = \begin{pmatrix} 0 & 1 & 0 \\ 0 & 0 & 1 \\ 1 & 0 & 0 \end{pmatrix} .$$

- We compute

$$B_1^\top A_1\, B_1 = \left(\begin{array}{c|cc} 1 & 3 & 5 \\ \hline 0 & -3 & -4 \\ 0 & 4 & 5 \end{array}\right) \quad \text{with } A_2 = \begin{pmatrix} -3 & -4 \\ 4 & 5 \end{pmatrix} \in \mathbb{R}^{2\times 2}.$$

- We set $Q_1 = \begin{pmatrix} 0 & 1 & 0 \\ 0 & 0 & 1 \\ 1 & 0 & 0 \end{pmatrix}$.

(2) The first column $s = (-3, 4)^\top$ of A_2 is not a multiple of e_1:

- An eigenvector v to the eigenvalue $\lambda_1 = 1$ of A_2 is $v = (1, -1)^\top$; we complete this to an ONB, i.e. to an orthogonal matrix $B_1 = (v, v_2)$ of $\mathbb{R}^2$:

$$B_2 = \frac{1}{\sqrt{2}} \begin{pmatrix} 1 & 1 \\ -1 & 1 \end{pmatrix}.$$

- We calculate

$$B_2^\top A_2\, B_2 = \left(\begin{array}{c|c} 1 & -8 \\ \hline 0 & 1 \end{array}\right) \quad \text{with } A_3 = (1) \in \mathbb{R}^{1\times 1}.$$

- We set

$$Q_2 = \begin{pmatrix} 0 & 1 & 0 \\ 0 & 0 & 1 \\ 1 & 0 & 0 \end{pmatrix} \left(\begin{array}{c|cc} 1 & 0 & 0 \\ \hline 0 & 1/\sqrt{2} & 1/\sqrt{2} \\ 0 & -1/\sqrt{2} & 1/\sqrt{2} \end{array}\right) = \begin{pmatrix} 0 & 1/\sqrt{2} & 1/\sqrt{2} \\ 0 & -1/\sqrt{2} & 1/\sqrt{2} \\ 1 & 0 & 0 \end{pmatrix}.$$

By $Q = Q_2$ we now obtain the Schur decomposition $Q^\top A\, Q = R$ of A:

$$\underbrace{\begin{pmatrix} 0 & 0 & 1 \\ 1/\sqrt{2} & -1/\sqrt{2} & 0 \\ 1/\sqrt{2} & 1/\sqrt{2} & 0 \end{pmatrix}}_{=Q^\top} \underbrace{\begin{pmatrix} -3 & -4 & 0 \\ 4 & 5 & 0 \\ 3 & 5 & 1 \end{pmatrix}}_{=A} \underbrace{\begin{pmatrix} 0 & 1/\sqrt{2} & 1/\sqrt{2} \\ 0 & -1/\sqrt{2} & 1/\sqrt{2} \\ 1 & 0 & 0 \end{pmatrix}}_{=Q} = \underbrace{\begin{pmatrix} 1 & -\sqrt{2} & 4\sqrt{2} \\ 0 & 1 & -8 \\ 0 & 0 & 1 \end{pmatrix}}_{=R}. \qquad \blacksquare$$

MATLAB With MATLAB you can get a Schur decomposition of a matrix A by the input of `[Q,R] = schur(A)`.

42.3 Singular Value Decomposition

In the *singular value decomposition*, an arbitrary matrix $A \in \mathbb{R}^{m \times n}$ is written as a product of three matrices U, Σ and $V^\top$,

$$A = U\,\Sigma\,V^\top \text{ with } U \in \mathbb{R}^{m \times m}\,,\ \Sigma \in \mathbb{R}^{m \times n}\,,\ V \in \mathbb{R}^{n \times n}\,,$$

where U and V are orthogonal and Σ is a *diagonal matrix* (Fig. 42.1).

Singular Value Decomposition Theorem
For each matrix $A \in \mathbb{R}^{m \times n}$ there are two orthogonal matrices $U \in \mathbb{R}^{m \times m}$ and $V \in \mathbb{R}^{n \times n}$ and a *diagonal matrix* $\Sigma \in \mathbb{R}^{m \times n}$ with

$$\Sigma = \underbrace{\begin{pmatrix} \sigma_1 & & & 0 & 0 & \dots & 0 \\ & \ddots & & & \vdots & & \vdots \\ 0 & & & \sigma_m & 0 & \dots & 0 \end{pmatrix}}_{\text{in case } m \leq n} \quad \text{or} \quad \Sigma = \underbrace{\begin{pmatrix} \sigma_1 & & 0 \\ & \ddots & \\ 0 & & \sigma_n \\ 0 & \dots & 0 \\ \vdots & & \vdots \\ 0 & \dots & 0 \end{pmatrix}}_{\text{in case } n \leq m},$$

where $\sigma_1 \geq \sigma_2 \geq \dots \geq \sigma_k \geq 0$, $k = \min\{m, n\}$ and

$$A = U\,\Sigma\,V^\top .$$

The numbers $\sigma_1, \dots, \sigma_k$ are called the **singular values** from A and the representation $A = U\,\Sigma\,V^\top$ a **singular value decomposition** of A.

In the next section we show how to determine a singular value decomposition of a matrix A.

Remark Often, one just calls the σ_i with $\sigma_i > 0$ singular values. We also allow $\sigma_i = 0$.

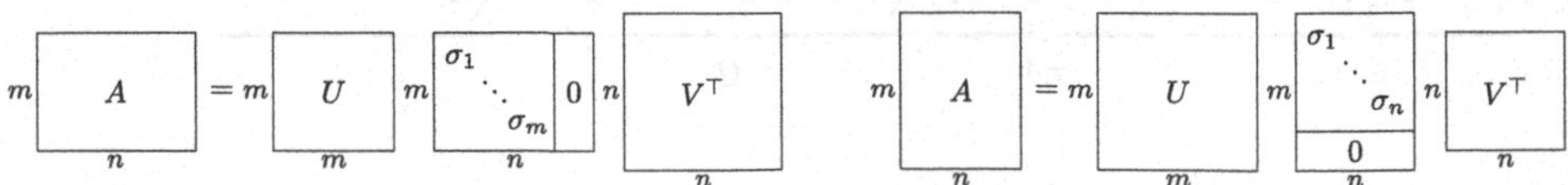

Fig. 42.1 Dimensions of the matrices A, U, Σ, $V^\top$ in the *singular value decomposition*

42.4 Determination of the Singular Value Decomposition

The following considerations motivate the construction of the matrices V, Σ, U of a singular value decomposition $A = U\,\Sigma\,V^\top$ of $A \in \mathbb{R}^{m\times n}$ with orthogonal matrices U, V:

We consider the linear mapping

$$f_A : \mathbb{R}^n \to \mathbb{R}^m \text{ with } f_A(v) = A\,v\,.$$

The representation matrix with respect to the canonical bases E_n and E_m is A:

$$A = {}_{E_m}M(f_A)_{E_n}\,.$$

Now, according to the base transformation formula (see Sect. 38.2) with the bases U of $\mathbb{R}^m$ and V of $\mathbb{R}^n$:

$${}_U M(f_A)_V = {}_U M(\mathrm{Id})_{E_m}\,{}_{E_m}M(f_A)_{E_n}\,{}_{E_n}M(\mathrm{Id})_V\,.$$

Because $V = {}_{E_n}M(\mathrm{Id})_V$ and $U^\top = U^{-1} = {}_U M(\mathrm{Id})_{E_m}$ and $A = {}_{E_m}M(f_A)_{E_n}$ this last equation is

$${}_U M(f_A)_V = U^\top A\,V = \Sigma\,,$$

so that Σ is the representation matrix of the linear mapping f_A with respect to the bases $V = (v_1, \ldots, v_n)$ and $U = (u_1, \ldots, u_m)$. Since in the *i-th* column of the representation matrix is the coordinate vector of the image of the *i-th* basis vector, we have the following

$$(*) \qquad A\,v_i = \sigma_i\,u_i \text{ for all } i = 1, \ldots, \min\{m, n\}\,.$$

Because of $A^\top = V\,\Sigma^\top U^\top$ the matrix $A^\top$ has the same singular values as A, it follows:

$$(**) \qquad A^\top u_i = \sigma_i\,v_i \text{ for all } i = 1, \ldots, \min\{m, n\}$$

and thus:

$$A^\top A\,v_i \overset{(*)}{=} \sigma_i\,A^\top u_i \overset{(**)}{=} \sigma_i^2\,v_i \text{ for all } i = 1, \ldots, n\,.$$

So we get the singular value decomposition as follows:

Recipe: Determine the Singular Value Decomposition of a Matrix

To determine the singular value decomposition $A = U\Sigma V^\top$ of a matrix $A \in \mathbb{R}^{m\times n}$ proceed as follows:

(1) Determine the eigenvalues $\lambda_1, \dots, \lambda_n$ of $A^\top A \in \mathbb{R}^{n\times n}$ and order them:

$$\lambda_1 \geq \lambda_2 \geq \cdots \geq \lambda_r > \lambda_{r+1} = \cdots = \lambda_n = 0 \text{ with } 1 \leq r \leq n\,.$$

Determine an ONB $(v_1, \dots, v_n)$ of $\mathbb{R}^n$ consisting of eigenvectors of $A^\top A$, $A^\top A v_i = \lambda_i v_i$ and obtain the orthogonal matrix $V = (v_1, \dots, v_n) \in \mathbb{R}^{n\times n}$.

(2) Set

$$\Sigma = \underbrace{\begin{pmatrix} \sigma_1 & & & 0\ 0 \dots 0 \\ & \ddots & & \vdots \quad \vdots \\ 0 & & \sigma_m & 0 \dots 0 \end{pmatrix}}_{\text{in case } m\leq n} \quad \text{or} \quad \Sigma = \underbrace{\begin{pmatrix} \sigma_1 & & 0 \\ & \ddots & \\ 0 & & \sigma_n \\ 0 & \dots & 0 \\ \vdots & & \vdots \\ 0 & \dots & 0 \end{pmatrix}}_{\text{in case } n\leq m}$$

with $\sigma_i = \sqrt{\lambda_i}$ for all $i = 1, \dots, \min\{m, n\}$.

(3) Determine $u_1, \dots, u_r$ from

$$u_i = \frac{1}{\sigma_i} A\, v_i \text{ for } i = 1, \dots, r$$

and complete in the case $r < m$ the vectors $u_1, \dots, u_r$ to an ONB, i. e. to an orthogonal matrix $U = (u_1, \dots, u_m)$.

Step (1) yields V, step (2) yields Σ, and finally step (3) yields U.

Remarks

1. Because $A^\top A$ is positive semidefinite (cf. Chap. 45), the λ_i are all greater than or equal to 0.
2. For the number r in step (1), the following is true $r = \text{rg}(A)$.

3. Because of

$$\langle u_i\,, u_j\rangle = u_i^\top u_j = \frac{1}{\sigma_i}\frac{1}{\sigma_j} v_i^\top A^\top A v_j = \frac{1}{\sigma_i}\frac{1}{\sigma_j}\lambda_j v_i^\top v_j = \delta_{ij} = \begin{cases} 1, & i = j \\ 0, & i \neq j \end{cases}$$

the vectors u_i obtained in the third step are elements of an ONB without further action.

Example 42.2 We determine the singular value decomposition of the matrix $A = \begin{pmatrix} -1 & 1 & 0 \\ -1 & -1 & 1 \end{pmatrix} \in \mathbb{R}^{2\times 3}$.

(1) We first calculate the eigenvalues and eigenvectors of the product

$$A^\top A = \begin{pmatrix} 2 & 0 & -1 \\ 0 & 2 & -1 \\ -1 & -1 & 1 \end{pmatrix}.$$

The characteristic polynomial $\chi_{A^\top A} = x\,(2-x)\,(3-x)$ yields the eigenvalues (ordered by magnitude) $\lambda_1 = 3,\ \lambda_2 = 2$ and $\lambda_3 = 0$. The eigenspaces are

$$\mathrm{Eig}_{A^\top A}(3) = \langle \begin{pmatrix} 1 \\ 1 \\ -1 \end{pmatrix} \rangle, \quad \mathrm{Eig}_{A^\top A}(2) = \langle \begin{pmatrix} 1 \\ -1 \\ 0 \end{pmatrix} \rangle, \quad \mathrm{Eig}_{A^\top A}(0) = \langle \begin{pmatrix} 1 \\ 1 \\ 2 \end{pmatrix} \rangle.$$

Thus we obtain the orthogonal matrix

$$V = \frac{1}{\sqrt{6}} \begin{pmatrix} \sqrt{2} & \sqrt{3} & 1 \\ \sqrt{2} & -\sqrt{3} & 1 \\ -\sqrt{2} & 0 & 2 \end{pmatrix}.$$

(2) With the singular values $\sigma_1 = \sqrt{3}$ and $\sigma_2 = \sqrt{2}$ and $m = 2 < 3 = n$ we get

$$\Sigma = \begin{pmatrix} \sqrt{3} & 0 & 0 \\ 0 & \sqrt{2} & 0 \end{pmatrix}.$$

(3) We now determine the orthonormal basis $U = (u_1, u_2)$ as

$$u_1 = \frac{1}{\sigma_1} A v_1 = \frac{1}{\sqrt{3}} \begin{pmatrix} -1 & 1 & 0 \\ -1 & -1 & 1 \end{pmatrix} \frac{1}{\sqrt{3}} \begin{pmatrix} 1 \\ 1 \\ -1 \end{pmatrix} = \begin{pmatrix} 0 \\ -1 \end{pmatrix},$$

$$u_2 = \frac{1}{\sigma_2} A v_2 = \frac{1}{\sqrt{2}} \begin{pmatrix} -1 & 1 & 0 \\ -1 & -1 & 1 \end{pmatrix} \frac{1}{\sqrt{2}} \begin{pmatrix} 1 \\ -1 \\ 0 \end{pmatrix} = \begin{pmatrix} -1 \\ 0 \end{pmatrix}.$$

The matrix U is therefore

$$U = \begin{pmatrix} 0 & -1 \\ -1 & 0 \end{pmatrix}.$$

The singular value decomposition of A is thus:

$$A = \begin{pmatrix} 0 & -1 \\ -1 & 0 \end{pmatrix} \begin{pmatrix} \sqrt{3} & 0 & 0 \\ 0 & \sqrt{2} & 0 \end{pmatrix} \frac{1}{\sqrt{6}} \begin{pmatrix} \sqrt{2} & \sqrt{2} & -\sqrt{2} \\ \sqrt{3} & -\sqrt{3} & 0 \\ 1 & 1 & 2 \end{pmatrix}.$$

■

Remark An important application of singular value decomposition is image compression. A (digital) image with $m\,n$ pixels can be represented by a $m \times n$-matrix. For many images, the sequence of singular values (σ_i) has a considerable drop, i.e., above a certain (small) s the values σ_i for $i > s$ are small in relation to the σ_i with $i \leq s$. If one sets in the singular value decomposition $A = U \Sigma V^\top$ all σ_i with $i > s$ equal zero, then a new matrix is obtained $\tilde{\Sigma}$, so that the transition from A to $\tilde{A} = U \tilde{\Sigma} V^\top$ does represent a loss of data, but it is barely visible in the image. The gain is that for evaluating $\tilde{A}$ only the first s columns of $U\, \tilde{\Sigma}$ and from V must store, so in total $s\,(n + m)$ instead of $m\,n$ entries. This can result in significant data compression.

MATLAB Of course MATLAB also offers a function for the singular value decomposition of a matrix A. By entering `[U,S,V] = svd(A)` you get matrices U, V and $S = \Sigma$.

42.5 Exercises

42.1 Given is the matrix $A = \begin{pmatrix} 3 & 1 & 2 \\ 0 & 2 & -1 \\ 0 & 1 & 4 \end{pmatrix}$.

(a) Show that $v = (1, 0, 0)^\top$ is an eigenvector of A and give the associated eigenvalue λ.
(b) Determine the Schur decomposition $R = Q^\top A Q$ of A such that $(1, 0, 0)^\top$ is the first column of Q.

42.2 Determine the Schur decompositions of the matrices

$$\text{(a)} \quad A = \begin{pmatrix} 1 & -1 & 1 \\ 0 & 3 & 0 \\ -1 & 0 & 3 \end{pmatrix}, \qquad \text{(b)} \quad A = \begin{pmatrix} 2 & 0 & 0 & 0 \\ 2 & 2 & 0 & 0 \\ 1 & -1 & 2 & -1 \\ 0 & 1 & 0 & 2 \end{pmatrix}.$$

42.3 Determine the singular value decompositions of the matrices

$$\text{(a) } A = \begin{pmatrix} 1 & 1 & 3 \\ 1 & 1 & -3 \end{pmatrix}, \quad \text{(c) } A^\top = \begin{pmatrix} 1 & 1 \\ 1 & 1 \\ 3 & -3 \end{pmatrix}, \quad \text{(e) } C = \begin{pmatrix} 8 & -4 & 0 & 0 \\ -1 & -7 & 0 & 0 \\ 0 & 0 & 1 & -1 \end{pmatrix},$$

$$\text{(b) } B = \begin{pmatrix} 2 \\ 2 \\ 1 \end{pmatrix}, \quad \text{(d) } B^\top = (2, 2, 1), \quad \text{(f) } D = \begin{pmatrix} 2 & 1 & 2 \\ 1 & -2 & 1 \end{pmatrix}.$$

42.4 A monochrome image in a 3×3 grid is stored by a real 3×3-matrix whose entries correspond to the gray level values at each pixel. The image of a cross hair is thus represented by the matrix $A = \begin{pmatrix} 0 & 1 & 0 \\ 1 & 1 & 1 \\ 0 & 1 & 0 \end{pmatrix} \in \mathbb{R}^{3\times 3}$ matrix. Perform the singular value decomposition, and compress the data by replacing the smallest singular value with 0. What gray scale image results after data compression?

42.5 The table below shows the relationship between current and voltage in a simple electrical circuit.

(a) Use MATLAB to calculate a singular value decomposition of the matrix consisting of the values in the tables (Let **U** be the 1st row and **I** be the 2nd row of the matrix).
(b) Make (also using MATLAB) a plot of the data points.
(c) Plot the results of the singular value decomposition (where reasonably possible) on the plot. Interpret the result.

U	I
−1.03	−2.23
0.74	1.61
−0.02	−0.02
0.51	0.88
−1.31	−2.39
0.99	2.02
0.69	1.62
−0.12	−0.35
−0.72	−1.67
1.11	2.46

The Jordan Normal Form I

43

Not every square matrix $A \in \mathbb{R}^{n \times n}$ is diagonalizable. But if the characteristic polynomial χ_A splits into linear factors, then at least a Schur decomposition exists (see Chap. 42). The *Jordan normal form* is, in a sense, an improvement of a Schur decomposition: it exists under the same conditions as a Schur decomposition does and is a particularly simple upper triangular matrix: it has diagonal shape except for some ones on the upper secondary diagonal. The essential thing is that every complex matrix A has such a *Jordan normal form* J. The determination of the A to Jordan normal form J transforming matrix S, which is the matrix S with $J = S^{-1} A\, S$, is somewhat involved: The first step to do this is to determine the generalized eigenspaces. We do that in the present chapter, in the next chapter we show how to get S from this.

43.1 Existence of the Jordan Normal Form

A matrix $(a_{ij}) \in \mathbb{C}^{s \times s}$ is called a **Jordan box** to $\lambda \in \mathbb{C}$ if

$$a_{11} = \cdots = a_{ss} = \lambda\,,\ a_{12} = \cdots = a_{s-1,s} = 1 \text{ and } a_{ij} = 0 \text{ else}\,,$$

i.e. (we leave out the zeros here),

$$(a_{ij}) = \begin{pmatrix} \lambda & 1 & & \\ & \ddots & \ddots & \\ & & \ddots & 1 \\ & & & \lambda \end{pmatrix}.$$

C. Karpfinger, *Calculus and Linear Algebra in Recipes*,
https://doi.org/10.1007/978-3-662-65458-3_43

A Jordan box is therefore, apart from the ones in the upper secondary diagonal, a diagonal matrix. Examples of Jordan boxes are

$$\begin{pmatrix} \lambda \end{pmatrix}, \begin{pmatrix} \lambda & 1 \\ 0 & \lambda \end{pmatrix}, \begin{pmatrix} \lambda & 1 & 0 \\ 0 & \lambda & 1 \\ 0 & 0 & \lambda \end{pmatrix}, \begin{pmatrix} \lambda & 1 & 0 & 0 \\ 0 & \lambda & 1 & 0 \\ 0 & 0 & \lambda & 1 \\ 0 & 0 & 0 & \lambda \end{pmatrix}$$

A matrix $J \in \mathbb{C}^{n \times n}$ is called **Jordan matrix** if

$$J = \begin{pmatrix} J_1 & & \\ & \ddots & \\ & & J_l \end{pmatrix}$$

where the matrices $J_1, \ldots, J_l$ are Jordan boxes on the diagonal. Here, the diagonal entries λ_i of the J_i need not be different and of course 1×1 Jordan boxes are allowed.

Example 43.1

- Jordan matrices with one Jordan box:

$$\begin{pmatrix} \boxed{1} \end{pmatrix}, \begin{pmatrix} 1 & 1 \\ 0 & 1 \end{pmatrix}, \begin{pmatrix} 0 & 1 \\ 0 & 0 \end{pmatrix}, \begin{pmatrix} 2 & 1 & 0 \\ 0 & 2 & 1 \\ 0 & 0 & 2 \end{pmatrix}.$$

- Jordan matrices with two Jordan boxes:

$$\begin{pmatrix} \boxed{1} & \\ & \boxed{2} \end{pmatrix}, \begin{pmatrix} \boxed{0} & \\ & \boxed{0} \end{pmatrix}, \begin{pmatrix} \boxed{3} & & \\ & 2 & 1 \\ & 0 & 2 \end{pmatrix}, \begin{pmatrix} 0 & 1 & \\ 0 & 0 & \\ & & \boxed{-1} \end{pmatrix}.$$

- Jordan matrices with three Jordan boxes:

$$\begin{pmatrix} \boxed{2} & & \\ & \boxed{2} & \\ & & \boxed{2} \end{pmatrix}, \begin{pmatrix} 2 & 1 & & \\ 0 & 2 & & \\ & & \boxed{3} & \\ & & & \boxed{-1} \end{pmatrix}, \begin{pmatrix} \boxed{1} & & & \\ & 0 & 1 & \\ & 0 & 0 & \\ & & & \boxed{1} \end{pmatrix}. \quad \blacksquare$$

Jordan Basis and Jordan Normal Form

For every matrix $A \in \mathbb{K}^{n \times n}$ with splitting characteristic polynomial χ_A there exists an ordered basis $B = (b_1, \ldots, b_n)$ of $\mathbb{K}^n$ and a Jordan matrix $J \in \mathbb{K}^{n \times n}$ with Jordan boxes $J_1, \ldots, J_l$ such that

$$J = \begin{pmatrix} J_1 & & \\ & \ddots & \\ & & J_l \end{pmatrix} = B^{-1} A B .$$

We call any such basis B a **Jordan base** of the $\mathbb{K}^n$ to A and the matrix J a **Jordan normal form** to A.

In particular, for every complex matrix there is a Jordan basis and a Jordan normal form.

Remark In general, a Jordan normal form is not unique. If the boxes are swapped, a Jordan normal form is created again, in the Jordan base the corresponding Jordan base vectors are swapped as well.

A Jordan normal form differs from a diagonal form at most by the fact that it has some ones in the first upper secondary diagonal.

Example 43.2 Each diagonal matrix has Jordan normal form. And also the matrices

$$\begin{pmatrix} 1 & 1 & & & & & \\ 0 & 1 & & & & & \\ & & 2 & 1 & & & \\ & & 0 & 2 & & & \\ & & & & 3 & 1 & 0 \\ & & & & 0 & 3 & 1 \\ & & & & 0 & 0 & 3 \end{pmatrix}, \quad \begin{pmatrix} 0 & & & & & & \\ & 0 & 1 & 0 & & & \\ & 0 & 0 & 1 & & & \\ & 0 & 0 & 0 & & & \\ & & & & 0 & 1 & 0 \\ & & & & 0 & 0 & 1 \\ & & & & 0 & 0 & 0 \end{pmatrix}$$

have Jordan normal form. On the other hand.

$$A = \begin{pmatrix} 1 & 1 \\ 0 & -1 \end{pmatrix}$$

is not Jordan normal form, because only equal entries may be in the Jordan boxes on the diagonal. But there are

$$J = \begin{pmatrix} \boxed{1} & \\ & \boxed{-1} \end{pmatrix} \text{ and } J' = \begin{pmatrix} \boxed{-1} & \\ & \boxed{1} \end{pmatrix}$$

the two different Jordan normal forms of this matrix A, since A is diagonalizable. Jordan bases in this example are bases of eigenvectors of A. ■

43.2 Generalized Eigenspaces

An essential step in determining a Jordan basis B to a matrix $A \in \mathbb{K}^{n\times n}$ is the determination of *generalized eigenspaces* to each eigenvalue λ of A. This step is not complicated, but is computationally expensive for large A. These generalized eigenspaces are *nested* vector spaces $\ker(A - \lambda E_n)^i \subseteq \ker(A - \lambda E_n)^{i+1}$:

Generalized Eigenspaces

For an eigenvalue λ of a matrix $A \in \mathbb{K}^{n\times n}$ consider the matrix $N = A - \lambda E_n$. For this matrix $N \in \mathbb{K}^{n\times n}$ there exists a $r \in \mathbb{N}$ with

$$\{0\} \subsetneq \ker N \subsetneq \ker N^2 \subsetneq \cdots \subsetneq \ker N^r = \ker N^{r+1}.$$

One calls the spaces $\ker N^i$ **generalized eigenspaces** to the eigenvalue λ and the *largest* among them, i.e. $\ker N^r$, also **hauptspace** to the eigenvalue λ. We have:

- The first generalized eigenspace $\ker N$ is the eigenspace of A to the eigenvalue λ.
- The dimension of $\ker N$ is the geometric multiplicity of the eigenvalue λ, $\dim \ker N = \text{geo}(\lambda)$.
- The dimension of $\ker N^r$ is the algebraic multiplicity of the eigenvalue λ, $\dim \ker N^r = \text{alg}(\lambda)$.

The number r is the smallest natural number with $\ker N^r = \ker N^{r+1}$, it is said that the chain becomes *stationary*. So far the number r, apart from the fact that it exists, plays no role. In fact, this number plays a very crucial role in the construction of the Jordan basis. But more on that later.

As vector spaces, the generalized eigenspaces are already completely determined by specifying a basis. And since we have always $\ker(A - \lambda E_n)^i \subseteq \ker(A - \lambda E_n)^{i+1}$, we obtain a basis of a *larger* generalized eigenspace $\ker(A - \lambda E_n)^{i+1}$ by corresponding addition of an already determined basis of $\ker(A - \lambda\, E_n)^i$. Therefore, to determine the generalized eigenspaces, proceed as follows:

Recipe: Determine the Generalized Eigenspaces

If λ is an eigenvalue of geometric multiplicity s, $s = \text{geo}(\lambda)$, and algebraic multiplicity t, $t = \text{alg}(\lambda)$, of a matrix $A \in \mathbb{K}^{n\times n}$ and $N = A - \lambda\, E_n$, then one obtains the *chain*

$$\{0\} \subsetneq \ker N \subsetneq \ker N^2 \subsetneq \cdots \subsetneq \ker N^r = \ker N^{r+1}$$

of generalized eigenspaces as follows:

(1) Determine an ordered basis B_1 of the eigenspace $\ker N$. If $|B_1| = t$: STOP, else:
(2) Calculate N^2 and add the base B_1 from (1) to a base B_2 of the generalized eigenspace $\ker N^2$. If $|B_2| = t$ STOP, else:
(3) Calculate N^3 and add the base B_2 from (2) to a base B_3 of the generalized eigenspace $\ker N^3$. If $|B_3| = t$: STOP, else:
(4) ...

We are with the step r finished, if the basis $B_r = (b_1, \ldots, b_t)$ exactly has t elements, where t is the algebraic multiplicity of the eigenvalue λ.

Example 43.3

- We consider the matrix $A = \begin{pmatrix} 1 & 1 & 1 \\ 0 & 1 & 1 \\ 0 & 0 & 1 \end{pmatrix}$ with $\chi_A = (1 - x)^3$.

  The matrix A has the only eigenvalue $\lambda = 1$ with $t = \text{alg}(1) = 3$. We set

$$N = A - 1\, E_3 = \begin{pmatrix} 0 & 1 & 1 \\ 0 & 0 & 1 \\ 0 & 0 & 0 \end{pmatrix}.$$

  (1) Then we have

$$\ker N = \ker \begin{pmatrix} 0 & 1 & 1 \\ 0 & 0 & 1 \\ 0 & 0 & 0 \end{pmatrix} = \langle \begin{pmatrix} 1 \\ 0 \\ 0 \end{pmatrix} \rangle.$$

  Thus, $B_1 = ((1, 0, 0)^\top)$.

(2) We have

$$\ker N^2 = \ker \begin{pmatrix} 0&0&1 \\ 0&0&0 \\ 0&0&0 \end{pmatrix} = \langle \begin{pmatrix} 1 \\ 0 \\ 0 \end{pmatrix}, \begin{pmatrix} 0 \\ 1 \\ 0 \end{pmatrix} \rangle .$$

So we get $B_2 = ((1, 0, 0)^\top, (0, 1, 0)^\top)$.

(3) We have

$$\ker N^3 = \ker \begin{pmatrix} 0&0&0 \\ 0&0&0 \\ 0&0&0 \end{pmatrix} = \langle \begin{pmatrix} 1 \\ 0 \\ 0 \end{pmatrix}, \begin{pmatrix} 0 \\ 1 \\ 0 \end{pmatrix}, \begin{pmatrix} 0 \\ 0 \\ 1 \end{pmatrix} \rangle .$$

We set $B_3 = ((1, 0, 0)^\top, (0, 1, 0)^\top, (0, 0, 1)^\top)$. Because of $|B_3| = 3 = \text{alg}(1)$ we're done.

In summary, we get the chain $\{0\} \subsetneq \ker N \subsetneq \ker N^2 \subsetneq \ker N^3$ with $r = 3$:

$$\{0\} \subsetneq \langle \begin{pmatrix} 1 \\ 0 \\ 0 \end{pmatrix} \rangle \subsetneq \langle \begin{pmatrix} 1 \\ 0 \\ 0 \end{pmatrix}, \begin{pmatrix} 0 \\ 1 \\ 0 \end{pmatrix} \rangle \subsetneq \langle \begin{pmatrix} 1 \\ 0 \\ 0 \end{pmatrix}, \begin{pmatrix} 0 \\ 1 \\ 0 \end{pmatrix}, \begin{pmatrix} 0 \\ 0 \\ 1 \end{pmatrix} \rangle .$$

- We consider the matrix $A = \begin{pmatrix} 3&1&0&0 \\ -1&1&0&0 \\ 1&1&3&1 \\ -1&-1&-1&1 \end{pmatrix}$ with $\chi_A = (2 - x)^4$.

  The matrix A has the only eigenvalue $\lambda = 2$ with $t = \text{alg}(1) = 4$. We set

$$N = A - 2\,E_4 = \begin{pmatrix} 1&1&0&0 \\ -1&-1&0&0 \\ 1&1&1&1 \\ -1&-1&-1&-1 \end{pmatrix} .$$

(1) We have

$$\ker N = \ker \begin{pmatrix} 1&1&0&0 \\ -1&-1&0&0 \\ 1&1&1&1 \\ -1&-1&-1&-1 \end{pmatrix} = \langle \begin{pmatrix} 1 \\ -1 \\ 0 \\ 0 \end{pmatrix}, \begin{pmatrix} 0 \\ 0 \\ 1 \\ -1 \end{pmatrix} \rangle .$$

Thus, $B_1 = ((1, -1, 0, 0)^\top, (0, 0, 1, -1)^\top)$.

(2) We have

$$\ker N^2 = \ker 0 = \langle \begin{pmatrix} 1 \\ -1 \\ 0 \\ 0 \end{pmatrix}, \begin{pmatrix} 0 \\ 0 \\ 1 \\ -1 \end{pmatrix}, \begin{pmatrix} 0 \\ 1 \\ 0 \\ 0 \\ 0 \\ 0 \end{pmatrix}, \begin{pmatrix} 0 \\ 0 \\ 0 \\ 1 \end{pmatrix} \rangle.$$

It follows $B_2 = ((1,-1,0,0)^\top, (0,0,1,-1)^\top, (0,1,0,0)^\top, (0,0,0,1)^\top)$. Because of $|B_2| = 4 = \text{alg}(2)$ we're done.

In summary, we get the chain $\{0\} \subsetneq \ker N \subsetneq \ker N^2$ with $r=2$:

$$\{0\} \subsetneq \langle \begin{pmatrix} 1 \\ -1 \\ 0 \\ 0 \end{pmatrix}, \begin{pmatrix} 0 \\ 0 \\ 1 \\ -1 \end{pmatrix} \rangle \subsetneq \langle \begin{pmatrix} 1 \\ -1 \\ 0 \\ 0 \end{pmatrix}, \begin{pmatrix} 0 \\ 0 \\ 1 \\ -1 \end{pmatrix}, \begin{pmatrix} 0 \\ 1 \\ 0 \\ 0 \end{pmatrix}, \begin{pmatrix} 0 \\ 0 \\ 0 \\ 1 \end{pmatrix} \rangle.$$

- We consider the matrix $A = \begin{pmatrix} -3 & -1 & 4 & -3 & -1 \\ 1 & 1 & -1 & 1 & 0 \\ -1 & 0 & 2 & 0 & 0 \\ 4 & 1 & -4 & 5 & 1 \\ -2 & 0 & 2 & -2 & 1 \end{pmatrix}$ with $\chi_A = (1-x)^4(2-x)$.

  The matrix A has the two eigenvalues 1 and 2.
  Eigenvalue $\lambda = 1$: It is $t = \text{alg}(1) = 4$. We set

$$N = A - 1\,E_5 = \begin{pmatrix} -4 & -1 & 4 & -3 & -1 \\ 1 & 0 & -1 & 1 & 0 \\ -1 & 0 & 1 & 0 & 0 \\ 4 & 1 & -4 & 4 & 1 \\ -2 & 0 & 2 & -2 & 0 \end{pmatrix}.$$

(1) We have

$$\ker N = \ker \begin{pmatrix} -4 & -1 & 4 & -3 & -1 \\ 1 & 0 & -1 & 1 & 0 \\ -1 & 0 & 1 & 0 & 0 \\ 4 & 1 & -4 & 4 & 1 \\ -2 & 0 & 2 & -2 & 0 \end{pmatrix} = \langle \begin{pmatrix} 1 \\ 0 \\ 1 \\ 0 \\ 0 \end{pmatrix}, \begin{pmatrix} 0 \\ -1 \\ 0 \\ 0 \\ 1 \end{pmatrix} \rangle.$$

Thus, $B_1 = ((1,0,1,0,0)^\top, (0,-1,0,0,1)^\top)$.

(2) We have

$$\ker N^2 = \ker \begin{pmatrix} 1 & 1 & -1 & 1 & 1 \\ 1 & 0 & -1 & 1 & 0 \\ 3 & 1 & -3 & 3 & 1 \\ 3 & 0 & -3 & 0 & \\ -2 & 0 & 2 & -2 & 0 \end{pmatrix} = \langle \begin{pmatrix} 1 \\ 0 \\ 1 \\ 0 \\ 0 \end{pmatrix}, \begin{pmatrix} 0 \\ -1 \\ 0 \\ 0 \\ 1 \end{pmatrix}, \begin{pmatrix} -1 \\ 0 \\ 0 \\ 1 \\ 0 \end{pmatrix} \rangle .$$

Set $B_2 = ((1, 0, 1, 0, 0)^\top, (0, -1, 0, 0, 1)^\top, (-1, 0, 0, 1, 0)^\top)$.

(3) We have

$$\ker N^3 = \ker \begin{pmatrix} 0 & 0 & 0 & 0 & 0 \\ 1 & 0 & -1 & 1 & 0 \\ 2 & 0 & -2 & 2 & 0 \\ 3 & 0 & -3 & 3 & 0 \\ -2 & 0 & 2 & -2 & 0 \end{pmatrix} = \langle \begin{pmatrix} 1 \\ 0 \\ 1 \\ 0 \\ 0 \\ 0 \end{pmatrix}, \begin{pmatrix} 0 \\ -1 \\ 0 \\ 0 \\ 1 \end{pmatrix}, \begin{pmatrix} -1 \\ 0 \\ 0 \\ 1 \\ 0 \end{pmatrix}, \begin{pmatrix} 0 \\ 0 \\ 0 \\ 0 \\ 1 \end{pmatrix} \rangle .$$

It follows $B_3 = ((1, 0, 1, 0, 0)^\top, (0, -1, 0, 0, 1)^\top, (-1, 0, 0, 1, 0)^\top, (0, 0, 0, 0, 1)^\top)$. Because of $|B_3| = 4 = \mathrm{alg}(1)$ we're done.

In summary, we have the chain $\{0\} \subsetneq \ker N \subsetneq \ker N^2 \subsetneq \ker N^3$ with $r=3$:

$$\{0\} \subsetneq \langle \begin{pmatrix} 1 \\ 0 \\ 1 \\ 0 \\ 0 \\ 0 \end{pmatrix}, \begin{pmatrix} 0 \\ -1 \\ 0 \\ 0 \\ 1 \end{pmatrix} \rangle \subsetneq \langle \begin{pmatrix} 1 \\ 0 \\ 1 \\ 0 \\ 0 \end{pmatrix}, \begin{pmatrix} 0 \\ -1 \\ 0 \\ 1 \end{pmatrix}, \begin{pmatrix} -1 \\ 0 \\ 0 \\ 1 \\ 0 \end{pmatrix} \rangle \subsetneq \langle \begin{pmatrix} 1 \\ 0 \\ 1 \\ 0 \\ 0 \\ 0 \end{pmatrix}, \begin{pmatrix} 0 \\ -1 \\ 0 \\ 0 \\ 1 \end{pmatrix}, \begin{pmatrix} -1 \\ 0 \\ 0 \\ 1 \\ 0 \end{pmatrix}, \begin{pmatrix} 0 \\ 0 \\ 0 \\ 0 \\ 1 \end{pmatrix} \rangle .$$

Eigenvalue $\lambda = 2$: It is $t = \mathrm{alg}(2) = 1$. We set

$$N = A - 2\,E_5 = \begin{pmatrix} -5 & -1 & 4 & -3 & -1 \\ 1 & -1 & -1 & 1 & 0 \\ -1 & 0 & 0 & 0 & 0 \\ 4 & 1 & -4 & 3 & 1 \\ -2 & 0 & 2 & -2 & -1 \end{pmatrix} .$$

(1) We have

$$\ker N = \ker \begin{pmatrix} -5 & -1 & 4 & -3 & -1 \\ 1 & -1 & -1 & 1 & 0 \\ -1 & 0 & 0 & 0 & 0 \\ 4 & 1 & -4 & 3 & 1 \\ -2 & 0 & 2 & -2 & -1 \end{pmatrix} = \langle \begin{pmatrix} 0 \\ 1 \\ 2 \\ 3 \\ -2 \end{pmatrix} \rangle .$$

Set $B_1 = ((0, 1, 2, 3, -2)^\top)$. Because of $|B_1| = 1 = \mathrm{alg}(2)$ we're done.

We have the chain $\{0\} \subsetneq \ker N$ with $r = 1$:

$$\{0\} \subsetneq \langle (0, 1, 2, 3, -2)^\top \rangle . \quad \blacksquare$$

With this determination of the generalized eigenspaces, we have anticipated the computationally intensive part of determining a Jordan basis of a matrix A in advance. We turn to the construction of Jordan bases in the following chapter. Exercises for the determination of generalized eigenspaces can be found in the next chapter.

43.3 Exercises

43.1 Justify why for any matrix $A \in \mathbb{K}^{n \times n}$ with the eigenvalue $\lambda \in \mathbb{K}$ we have

$$\ker(A - \lambda E_n)^i \subseteq \ker(A - \lambda E_n)^{i+1} .$$

43.2 Let $J = \begin{pmatrix} \boxed{\lambda} & & \\ & \lambda & 1 \\ & 0 & \lambda \end{pmatrix}$ be a Jordan normal form of a matrix A with Jordan basis $B = (b_1, b_2, b_3)$. Show that $\tilde{J} = \begin{pmatrix} \lambda & 1 & \\ 0 & \lambda & \\ & & \boxed{\lambda} \end{pmatrix}$ is a Jordan normal form of A with Jordan basis $\tilde{B} = (b_2, b_3, b_1)$.

The Jordan Normal Form II

44

For each square complex matrix A there is a Jordan normal form J, i.e., there exists an invertible matrix $B \in \mathbb{C}^{n \times n}$ with $J = B^{-1} A B$. The columns of B form an associated Jordan basis. We obtain such a matrix or Jordan basis B by *successively traversing* the generalized eigenspaces. The key role is played by the matrices $N = A - \lambda E_n$ for the eigenvalues λ of A.

44.1 Construction of a Jordan Base

Let us briefly explain the idea that drives us to construct a Jordan basis in the given way: To do this, we consider, as an example, a matrix $A \in \mathbb{R}^{6 \times 6}$ with a Jordan basis $B = (b_1, \ldots, b_6)$ and the corresponding Jordan normal form

$$J = \begin{pmatrix} \boxed{\mu} & & & & & \\ & \lambda & 1 & & & \\ & 0 & \lambda & & & \\ & & & \lambda & 1 & 0 \\ & & & 0 & \lambda & 1 \\ & & & 0 & 0 & \lambda \end{pmatrix} = B^{-1} A B \ \text{ with } \lambda \neq \mu .$$

Our question is: How to find J and B to A? The diagonal elements of J are obtained quite simply. Since A and J are similar, then A and J also have the same characteristic polynomial

C. Karpfinger, *Calculus and Linear Algebra in Recipes*,
https://doi.org/10.1007/978-3-662-65458-3_44

and also the same eigenvalues:

$$\chi_A = (\mu - x)\,(\lambda - x)^5\,.$$

The matrix A has eigenvalues μ and λ with $\text{alg}(\mu) = 1$ and $\text{alg}(\lambda) = 5$.

Moreover, since J is the transformation matrix of the linear mapping $f : \mathbb{R}^6 \to \mathbb{R}^6$ with $f(v) = A\,v$ with respect to the base B (note the base transformation formula in Sect. 38.2), we have the following equalities:

- $A\,b_1 = \mu\,b_1$,
- $A\,b_2 = \lambda\,b_2$,
- $A\,b_3 = 1\,b_2 + \lambda\,b_3$,
- $A\,b_4 = \lambda\,b_4$,
- $A\,b_5 = 1\,b_4 + \lambda\,b_5$,
- $A\,b_6 = 1\,b_5 + \lambda\,b_6$.

From these equations we can see that the basic elements b_1, b_2 and b_4 are eigenvectors of A. We reformulate the other three equations to:

$$A\,b_3 - \lambda\,b_3 = b_2\,,\ A\,b_5 - \lambda\,b_5 = b_4\,,\ A\,b_6 - \lambda\,b_6 = b_5\,.$$

With the abbreviation $N = A - \lambda\,E_6$ these equations are

$$N\,b_3 = b_2,\ N\,b_5 = b_4\,,\ N\,b_6 = b_5\,.$$

Notice now:

- Since b_2 lies in the eigenspace of A, we have $N\,b_2 = 0$, i.e. $b_2 \in \ker N$. It follows

$$b_3 \in \ker N^2 \setminus \ker N\,.$$

- Since b_4 lies in the eigenspace of A, we have $N\,b_4 = 0$, i.e. $b_4 \in \ker N$. It follows

$$b_5 \in \ker N^2 \setminus \ker N\,.$$

- Since b_5 lies in the kernel of N^2, $N^2\,b_5 = 0$, it follows

$$b_6 \in \ker N^3 \setminus \ker N^2\,.$$

Here we see how the generalized eigenspaces to an eigenvalue λ come into play: The chain of generalized eigenspaces to an eigenvalue λ of the matrix A under consideration is as follows.

$$\begin{array}{cccccc} \{0\} & \subsetneq \ker N & \subsetneq & \ker N^2 & \subsetneq & \ker N^3 \\ & b_2, b_4 & & b_2, b_4, b_3, b_5 & & b_2, b_4, b_3, b_5, b_6 \end{array}$$

with the dimensions

$$\dim\ker N = 2\,, \quad \dim\ker N^2 = 4\,, \quad \dim\ker N^3 = 5\,.$$

To determine the vectors $b_1, \ldots, b_6$ of a Jordan basis one proceeds as follows: One determines the chain of generalized eigenspaces as described in the recipe in Sect. 43.2 and determine the vectors $b_1, \ldots, b_6$ as follows by **successively traversing** this chain of generalized eigenspaces:

- Choose $b_6 \in \ker N^3 \setminus \ker N^2$.
- Set $b_5 = N\, b_6$. We then have $b_5 \in \ker N^2 \setminus \ker N$.
- Set $b_4 = N\, b_5$. We then have $b_4 \in \ker N \setminus \{0\}$.

As b_4 is an eigenvector to the eigenvalue λ, a Jordan box closes here. We obtain a 3×3-Jordan box after this *back-to-front run through the chain*, the number 3 is coming from $r = 3$. With respect to the constructed basis $B = (\ldots, b_4, b_5, b_6)$ the Jordan matrix J is

$$J = \begin{pmatrix} \ddots & & & \\ & \lambda & 1 & 0 \\ & 0 & \lambda & 1 \\ & 0 & 0 & \lambda \end{pmatrix} = B^{-1} A\, B\,.$$

Because of $\dim\ker N^2 = \dim\ker N + 2$ there is in $\ker N^2 \setminus \ker N$ another vector b_3, which is to the already constructed b_5 linearly independent:

- Choose $b_3 \in \ker N^2 \setminus \ker N$, linearly independent to b_5.
- Set $b_2 = N\, b_3$. We then have $b_2 \in \ker N \setminus \{0\}$.

As b_2 is an eigenvector to the eigenvalue λ, a Jordan box closes here again. We obtain a 2×2 Jordan box after this *run through the chain*. With respect to the basis constructed so far $B = (\ldots, b_2, b_3, b_4, b_5, b_6)$ the Jordan matrix J is

$$J = \begin{pmatrix} \ddots & & \\ & \boxed{\begin{matrix} \lambda & 1 \\ 0 & \lambda \end{matrix}} & \\ & & \boxed{\begin{matrix} \lambda & 1 & 0 \\ 0 & \lambda & 1 \\ 0 & 0 & \lambda \end{matrix}} \end{pmatrix} = B^{-1} A\, B\,.$$

Since λ has algebraic multiplicity 5 and we have already constructed 5 basis vectors to the eigenvalue λ, we now turn to the next eigenvalue μ. Here the situation is particularly simple:

- Choose $b_1 \in \ker(A - \mu\, E_6) \setminus \{0\}$.

It is then $B = (b_1, \ldots, b_6)$ a Jordan basis with the Jordan normal form as originally given J.

We notice quite generally: If $b \in \ker N^{i+1} \setminus \ker N^i$, then $N\, b \in \ker N^i \setminus \ker N^{i-1}$ (see Exercise 44.3). By successively multiplying a chosen vector $b \in \ker N^r \setminus \ker N^{r-1}$ with N we *traverse* the chain of generalized eigenspaces from hauptspace to eigenspace. One obtains a maximal Jordan box to the eigenvalue λ with r rows and r columns. Each further traversal leads to Jordan boxes of at most the same length, but in general shorter. Consequently, in this successive traversal of the chain of generalized eigenspaces to an eigenvalue λ, we obtain progressively smaller Jordan boxes, which continue our large $r \times r$-Jordan box towards the upper left in the Jordan matrix. The following procedure is based on these considerations, by means of which one can easily determine a Jordan basis and, of course, an associated Jordan normal form in most cases.

Recipe: Determine a Jordan Basis and a Jordan Normal Form

A Jordan basis $B = (b_1, \ldots, b_n)$ and a Jordan normal form J of a matrix $A \in \mathbb{K}^{n \times n}$ are obtained in general as follows:

(1) Determine the characteristic polynomial $\chi_A \in \mathbb{K}[x]$:

- If χ_A does not split into linear factors: STOP, there is no Jordan normal form to A, else:

(continued)

- Obtain the decomposition $\chi_A = (\lambda_1 - x)^{\nu_1} \cdots (\lambda_k - x)^{\nu_k}$ and thus the different eigenvalues $\lambda_1, \ldots, \lambda_k$ with their algebraic multiplicities $\nu_1, \ldots, \nu_r$.

(2) Choose an eigenvalue λ of A and determine for $N = A - \lambda E_n$ the following chain of nested subspaces as in the recipe in Sect. 43.2:

$$\{0\} \subsetneq \ker N \subsetneq \ker N^2 \subsetneq \cdots \subsetneq \ker N^r = \ker N^{r+1} .$$

(3) **Successively traversing the chain of generalized eigenspaces:**

- Choose $b_r \in \ker N^r \setminus \ker N^{r-1}$ and set

$$b_{r-1} = N\, b_r\,,\ b_{r-2} = N\, b_{r-1}, \ldots, b_1 = N\, b_2 \in \ker N .$$

It is then $(b_1, \ldots, b_r)$ the *last part* to the eigenvalue λ of the Jordan basis to be determined $B = (\ldots, b_1, \ldots, b_r)$ to A with a $r \times r$-Jordan box to the eigenvalue λ. STOP if $n{=}r$, then B is a Jordan basis to A, otherwise:

- Choose in the largest generalized eigenspace with $\dim \ker N^s \geq \dim \ker N^{s-1} + 2$ a $a_s \in \ker N^s \setminus \ker N^{s-1}$ which is linearly independent to the already determined $b_s \in \ker N^s$ from the previous step and set

$$a_{s-1} = N\, a_s\,,\ a_{s-2} = N\, a_{s-1}, \ldots, a_1 = N\, a_2 \in \ker N .$$

It is then $(a_1, \ldots, a_s)$ the *penultimate* part to the eigenvalue λ of the Jordan basis to be determined $B = (\ldots, a_1, \ldots, a_s, b_1, \ldots, b_r)$ to A with a $s \times s$-Jordan box to the eigenvalue λ. STOP if $n = r + s$, then B is a Jordan basis to A, otherwise:

- ...

(4) Check whether the vectors from equal generalized eigenspaces are linearly independent. If yes, then all vectors are linearly independent, if no, then a different *starting vector* in the largest generalized eigenspace must be chosen on the run that generated the vector that led to the linear dependence. (This is rarely the case).

(5) Choose the next eigenvalue λ and start from (2).

In typical exercises to the Jordan normal form, one usually has the case $n \leq 6$. For these *short* bases, it is useful to number the basis vectors $b_1, \ldots, b_n$ as in the introductory example right from the start and to use b_n instead of b_r to begin with. We will do the same in the following examples, completing the Examples 43.3 by the construction of a Jordan base:

Example 44.1

- We consider the matrix $A = \begin{pmatrix} 1 & 1 & 1 \\ 0 & 1 & 1 \\ 0 & 0 & 1 \end{pmatrix}$.

  (1) We have $\chi_A = (1-x)^3$.
  (2) Using $N = A - 1\,E_3$ we have the following chain of nested generalized eigenspaces:
  $$\{0\} \subsetneq \langle \begin{pmatrix} 1 \\ 0 \\ 0 \end{pmatrix} \rangle \subsetneq \langle \begin{pmatrix} 1 \\ 0 \\ 0 \end{pmatrix}, \begin{pmatrix} 0 \\ 1 \\ 0 \end{pmatrix} \rangle \subsetneq \langle \begin{pmatrix} 1 \\ 0 \\ 0 \end{pmatrix}, \begin{pmatrix} 0 \\ 1 \\ 0 \end{pmatrix}, \begin{pmatrix} 0 \\ 0 \\ 1 \end{pmatrix} \rangle .$$
  (3) We choose $b_3 = (0, 0, 1)^\top \in \ker N^3 \setminus \ker N^2$ and set
  $$b_2 = N\,b_3 = \begin{pmatrix} 0 & 1 & 1 \\ 0 & 0 & 1 \\ 0 & 0 & 0 \end{pmatrix} \begin{pmatrix} 0 \\ 0 \\ 1 \end{pmatrix} = \begin{pmatrix} 1 \\ 1 \\ 0 \end{pmatrix}$$
  and
  $$b_1 = N\,b_2 = \begin{pmatrix} 0 & 1 & 1 \\ 0 & 0 & 1 \\ 0 & 0 & 0 \end{pmatrix} \begin{pmatrix} 1 \\ 1 \\ 0 \end{pmatrix} = \begin{pmatrix} 1 \\ 0 \\ 0 \end{pmatrix} \in \ker N\,.$$
  (4) and (5) are omitted.
  Thus $B = (b_1,\ b_2,\ b_3)$ is a Jordan basis to A with the Jordan normal form
  $$J = \begin{pmatrix} \boxed{\begin{matrix} 1 & 1 & 0 \\ 0 & 1 & 1 \\ 0 & 0 & 1 \end{matrix}} \end{pmatrix} = B^{-1} A\,B\,.$$

- We consider the matrix $A = \begin{pmatrix} 3 & 1 & 0 & 0 \\ -1 & 1 & 0 & 0 \\ 1 & 1 & 3 & 1 \\ -1 & -1 & -1 & 1 \end{pmatrix}$.

(1) We have $\chi_A = (2-x)^4$.
(2) Using $N = A - 2\,E_4$ we have the following chain of nested generalized eigenspaces:

$$\{0\} \subsetneq \langle \begin{pmatrix}1\\-1\\0\\0\end{pmatrix}, \begin{pmatrix}0\\0\\1\\-1\end{pmatrix} \rangle \subsetneq \langle \begin{pmatrix}1\\-1\\0\\0\end{pmatrix}, \begin{pmatrix}0\\0\\1\\-1\end{pmatrix}, \begin{pmatrix}0\\1\\0\\0\end{pmatrix}, \begin{pmatrix}0\\0\\0\\1\end{pmatrix} \rangle .$$

(3) We choose $b_4 = (0,0,0,1)^\top \in \ker N^2 \setminus \ker N$ and set

$$b_3 = N\,b_4 = \begin{pmatrix}1&1&0&0\\-1&-1&0&0\\1&1&1&1\\-1&-1&-1&-1\end{pmatrix}\begin{pmatrix}0\\0\\0\\1\end{pmatrix} = \begin{pmatrix}0\\0\\1\\-1\end{pmatrix} \in \ker N\,.$$

We choose $b_2 = (0,1,0,0)^\top \in \ker N^2 \setminus \ker N$ and set

$$b_1 = N\,b_2 = \begin{pmatrix}1&1&0&0\\-1&-1&0&0\\1&1&1&1\\-1&-1&-1&-1\end{pmatrix}\begin{pmatrix}0\\1\\0\\0\end{pmatrix} = \begin{pmatrix}1\\-1\\1\\-1\end{pmatrix} \in \ker N\,.$$

(4) Clearly, b_1 and b_3 are linearly independent.
(5) Dropped.

Thus $B = (b_1, b_2, b_3, b_4)$ is a Jordan basis to A with the Jordan normal form

$$J = \begin{pmatrix}2&1&&\\&2&&\\&&2&1\\&&&2\end{pmatrix} = B^{-1}A\,B\,.$$

- We consider the matrix $A = \begin{pmatrix}-3&-1&4&-3&-1\\1&1&-1&1&0\\-1&0&2&0&0\\4&1&-4&5&1\\-2&0&2&-2&1\end{pmatrix}$.

(1) We have $\chi_A = (1-x)^4(2-x)$.
(2) We choose the eigenvalue $\lambda = 1$ and have with $N = A - 1\,E_5$ the following chain of nested generalized eigenspaces:

$$\{0\} \subsetneq \langle \begin{pmatrix}1\\0\\1\\0\\0\end{pmatrix}, \begin{pmatrix}0\\-1\\0\\0\\1\end{pmatrix} \rangle \subsetneq \langle \begin{pmatrix}1\\0\\1\\0\\0\end{pmatrix}, \begin{pmatrix}0\\-1\\0\\0\\1\end{pmatrix}, \begin{pmatrix}-1\\0\\0\\1\\0\end{pmatrix} \rangle \subsetneq \langle \begin{pmatrix}1\\0\\1\\0\\0\end{pmatrix}, \begin{pmatrix}0\\-1\\0\\0\\1\end{pmatrix}, \begin{pmatrix}-1\\0\\0\\1\\0\end{pmatrix}, \begin{pmatrix}0\\0\\0\\0\\1\end{pmatrix} \rangle.$$

(3) We choose $b_5 = (0,0,0,1)^\top \in \ker N^3 \setminus \ker N^2$ and set

$$b_4 = N\,b_5 = \begin{pmatrix}-4 & -1 & 4 & -3 & -1\\ 1 & 0 & -1 & 1 & 0\\ -1 & 0 & 1 & 0 & 0\\ 4 & 1 & -4 & 4 & 1\\ -2 & 0 & 2 & -2 & 0\end{pmatrix}\begin{pmatrix}0\\0\\0\\0\\1\end{pmatrix} = \begin{pmatrix}-1\\0\\0\\1\\0\end{pmatrix} \text{ and}$$

$$b_3 = N\,b_4 = \begin{pmatrix}-4 & -1 & 4 & -3 & -1\\ 1 & 0 & -1 & 1 & 0\\ -1 & 0 & 1 & 0 & 0\\ 4 & 1 & -4 & 4 & 1\\ -2 & 0 & 2 & -2 & 0\end{pmatrix}\begin{pmatrix}-1\\0\\0\\1\\0\end{pmatrix} = \begin{pmatrix}1\\0\\1\\0\\0\end{pmatrix} \in \ker N\,.$$

We choose $b_2 = (0,-1,0,0,1)^\top \in \ker N \setminus \{0\}$.
(4) N/A.
(5) We choose the eigenvalue $\lambda = 2$ and have with $N = A - 2\,E_5$ the following chain of nested generalized eigenspaces:

$$\{0\} \subsetneq \langle (0,1,2,3,-2)^\top \rangle\,.$$

(3) We choose $(0, 1, 2, 3, -2)^\top$.
(4) and (5) are omitted.

So $B = (b_1, b_2, b_3, b_4, b_5)$ is a Jordan base to A with the Jordan normal form

$$J = \begin{pmatrix} \boxed{2} & & & & \\ & \boxed{1} & & & \\ & & 1 & 1 & 0 \\ & & 0 & 1 & 1 \\ & & 0 & 0 & 1 \end{pmatrix} = B^{-1} A\, B\,.$$

■

44.2 Number and Size of Jordan Boxes

Often one is only interested in the Jordan normal form J and can do without knowing a Jordan base. Then, fortunately, one is not always dependent on determining a Jordan base, a Jordan normal form J of $A \in \mathbb{K}^{n \times n}$ is namely for small $n \leq 6$ already known by the following numbers:

On the Number and Size of Jordan Boxes

Let $A \in \mathbb{K}^{n \times n}$ be a matrix with splitting characteristic polynomial χ_A. If $\lambda \in \mathbb{K}$ is an eigenvalue of A with the chain of generalized eigenspaces.

$$\underbrace{\ker N}_{\dim = \text{geo}(\lambda)} \subsetneq \ker N^2 \subsetneq \cdots \subsetneq \underbrace{\ker N^r}_{\dim = \text{alg}(\lambda)} = \ker N^{r+1}\,,$$

where $N = A - \lambda\, E_n$, then we have:

1. The dimension of the eigenspace $\ker N$ is the number of Jordan boxes to the eigenvalue λ.
2. The number r is the row number of the largest Jordan box to the eigenvalue λ.

Example 44.2

- If $A \in \mathbb{C}^{3\times 3}$ is a matrix with the (only) eigenvalue 2 with $\text{alg}(2) = 3$ and $\text{geo}(2) = 2$, then a Jordan normal form of A must be one of the following types:

$$\begin{pmatrix} 2 & & \\ & 2 & 1 \\ & 0 & 2 \end{pmatrix}, \quad \begin{pmatrix} 2 & 1 & \\ 0 & 2 & \\ & & 2 \end{pmatrix}.$$

- If $A \in \mathbb{C}^{4\times 4}$ a matrix with the (only) eigenvalue 2 with $\text{alg}(2) = 4$ and $\text{geo}(2) = 2$, then a Jordan normal form of A

$$\text{in case } r = 2: \begin{pmatrix} 2 & 1 & & \\ & 2 & & \\ & & 2 & 1 \\ & & & 2 \end{pmatrix} \quad \text{and in case } r = 3: \begin{pmatrix} 2 & & & \\ & 2 & 1 & \\ & & 2 & 1 \\ & & & 2 \end{pmatrix}.$$

■

MATLAB With `[B,J]=jordan(A)` one obtains with MATLAB a Jordan normal form J and a corresponding *transforming* matrix B or Jordan basis B.

44.3 Exercises

44.1 Show: If $b \in \ker N^{i+1} \setminus \ker N^i$ then $Nb \in \ker N^i \setminus \ker N^{i-1}$.

44.2 Determine Jordan normal forms and associated Jordan bases of the following matrices A, viz. Jordan matrices J and Jordan bases B with $J = B^{-1}AB$:

(a) $A = \begin{pmatrix} 7 & -1 \\ 4 & 3 \end{pmatrix}$,

(b) $A = \begin{pmatrix} -1 & 0 & 0 \\ 5 & -1 & 3 \\ 2 & 0 & -1 \end{pmatrix}$,

(c) $A = \begin{pmatrix} 2 & 1 & 1 \\ 0 & 2 & 4 \\ 0 & 0 & 3 \end{pmatrix}$,

(d) $A = \begin{pmatrix} 2 & 2 & 0 & 0 \\ 0 & 2 & 0 & 0 \\ 1 & 2 & 2 & 1 \\ 3 & 4 & 0 & 2 \end{pmatrix}$,

(e) $A = \begin{pmatrix} 3 & 1 & 0 & 0 \\ -1 & 1 & 0 & 0 \\ 1 & 1 & 3 & 1 \\ -1 & -1 & -1 & 1 \end{pmatrix}$.

44.3 Let $A \in \mathbb{C}^{n \times n}$ with characteristic polynomial $\chi_A(x) = (\lambda - x)^n$. Further let $s = \dim \mathrm{Eig}_A(\lambda)$ is the geometric multiplicity of the eigenvalue λ and r is the smallest natural number with $(A - \lambda E_n)^r = 0$. Determine the possible Jordan normal forms of the matrix A for the following triples (n, s, r):

$$(5, 3, 1)\,,\ (5, 3, 2)\,,\ (5, 3, 3)\,,\ (5, 1, 4)\,,\ (6, 2, 3)\,,\ (6, 1, 2)\,.$$

Note: Not every triple is possible!

44.4 We consider the sequence $(g_n)_{n \in \mathbb{N}_0}$ with

$$g_0 = 0, \quad g_1 = 1, \quad g_{n+1} = -4g_{n-1} + 4g_n \quad \text{for } n \geq 1\,.$$

Determine the sequence element g_{20}. Do this as follows:

(a) Describe the recursion by a matrix A.
(b) Determine a Jordan normal form J of A and the transformation matrix S with $S^{-1}AS = J$.
(c) Write J as the sum $D+N$ with a diagonal matrix D and a matrix N with $N^2 = 0$.
(d) Use the binomial formula $(D + N)^k = \sum_{i=0}^{k} \binom{k}{i} D^{k-i} N^i$ to calculate J^{19}.
(e) Now determine A^{19}.

Definiteness and Matrix Norms 45

A real number is positive or negative or zero. For symmetric matrices a similar distinction is possible by means of *definiteness*. The definiteness will play a decisive role in the evaluation of extremal places of a function of several variables. The definiteness of a symmetric matrix can be judged by its (real) eigenvalues.

It is often not only useful to distinguish matrices into *positive* or *negative*, one can also assign matrices a *length* or *norm* to matrices. There are different possibilities. One important norm is the *spectral norm* of a matrix A. It is calculated by means of the eigenvalues of $A^\top A$.

45.1 Definiteness of Matrices

In the determination of the extreme value of a function in several variables, but also in the applications of linear algebra, the *definiteness of* matrices plays an important role. It is quickly explained what *positive* or *negative definite* means for a matrix. But deciding whether a given matrix is positive or negative definite is unfortunately not always easy, especially for larger matrices.We learn about criteria that requires computing the eigenvalues or several determinants; this is an undertaking that requires considerable effort for larger matrices. We start with the terms:

Definiteness of Symmetric Matrices

We call a real symmetric $n \times n$-matrix A

(continued)

C. Karpfinger, *Calculus and Linear Algebra in Recipes*,
https://doi.org/10.1007/978-3-662-65458-3_45

- **positive definite** if $v^\top A\, v > 0$ for all $v \in \mathbb{R}^n \setminus \{0\}$,
- **negative definite** if $v^\top A\, v < 0$ for all $v \in \mathbb{R}^n \setminus \{0\}$,
- **positive semi definite** if $v^\top A\, v \geq 0$ for all $v \in \mathbb{R}^n \setminus \{0\}$,
- **negative semi definite** if $v^\top A\, v \leq 0$ for all $v \in \mathbb{R}^n \setminus \{0\}$,
- **indefinite** if there are vectors $v,\ w \in \mathbb{R}^n$ with $v^\top A\, v > 0$ and $w^\top A\, w < 0$.

Note that symmetry is inherent in the notion of *definiteness*: positive definite matrices are symmetric. Further, a positive definite matrix is also positive semi definite. Using the following examples, we can also easily specify positive semi definite matrices that are not positive definite. The same is true for negative semi definite matrices.

Example 45.1

- For a diagonal matrix $D = \operatorname{diag}(\lambda_1, \ldots, \lambda_n) \in \mathbb{R}^{n\times n}$ we obviously have:

$$D \text{ is positive definite } \Leftrightarrow \lambda_1, \ldots, \lambda_n > 0\,,$$

$$D \text{ is negative definite } \Leftrightarrow \lambda_1, \ldots, \lambda_n < 0\,,$$

$$D \text{ is positive semi definite } \Leftrightarrow \lambda_1, \ldots, \lambda_n \geq 0\,,$$

$$D \text{ is negative semi definite } \Leftrightarrow \lambda_1, \ldots, \lambda_n \leq 0\,,$$

$$D \text{ is indefinite } \Leftrightarrow \exists\, i,\ j \text{ with } \lambda_i < 0,\ \lambda_j > 0\,.$$

- The square matrix

$$A = \begin{pmatrix} 2 & 1 \\ 1 & 1 \end{pmatrix} \in \mathbb{R}^{2\times 2}$$

is positive definite: A is symmetric and for all $v = (v_1, v_2)^\top \in \mathbb{R}^2 \setminus \{0\}$ we have:

$$v^\top A v = (v_1, v_2)\begin{pmatrix} 2 & 1 \\ 1 & 1 \end{pmatrix}\begin{pmatrix} v_1 \\ v_2 \end{pmatrix} = 2v_1^2 + 2v_1v_2 + 1v_2^2 = (v_1 + v_2)^2 + v_1^2 > 0\,.$$

■

But how do you decide for a larger symmetric matrix A, which does not have an even diagonal shape, whether A is positive or negative (semi) definite or indefinite? We give two criteria, for this we introduce a new term:

For each real $n \times n$-matrix $A = (a_{ij})_{n,n}$ and each number $k \in \{1, \ldots, n\}$ the determinant of the upper left $k \times k$-submatrix $(a_{ij})_{k,k}$ of A is called **principal minor**.

The n principal minors of a $n \times n$-matrix $A = (a_{ij})_{n,n}$ are given by:

$$\left|a_{11}\right| , \begin{vmatrix} a_{11} & a_{12} \\ a_{21} & a_{22} \end{vmatrix} , \begin{vmatrix} a_{11} & a_{12} & a_{13} \\ a_{21} & a_{22} & a_{23} \\ a_{31} & a_{32} & a_{33} \end{vmatrix} , , \ldots , \begin{vmatrix} a_{11} & \cdots & a_{1n} \\ \vdots & & \vdots \\ a_{n1} & \cdots & a_{nn} \end{vmatrix} .$$

Criteria for Determining Definiteness

- **The eigenvalue criterion:** A real symmetric $n \times n$-matrix $A \in \mathbb{R}^{n \times n}$ is
  - positive definite if and only if all eigenvalues of A are positive,
  - negative definite if and only if all eigenvalues of A are negative,
  - positive semi definite if and only if all eigenvalues of A are positive or zero,
  - negative semi definite if and only if all eigenvalues of A are negative or zero,
  - indefinite if and only if A has positive and negative eigenvalues.
- **Principal minors criterion:** A real symmetric $n \times n$-matrix $A \in \mathbb{R}^{n \times n}$ is
  - positive definite if and only if all n principal minors are positive,
  - negative definite if and only if the n principal minors are alternating as follows:

$$\det(a_{ij})_{11} < 0 , \det(a_{ij})_{22} > 0 , \det(a_{ij})_{33} < 0 , \ldots$$

The eigenvalues of a symmetric real matrix A are, according to a result in the box in Sect. 39.4 always real, so that it is also reasonable to aks for $\lambda \geq 0$ or $\lambda \leq 0$ for the eigenvalues λ of A.

Example 45.2

- We consider the symmetric matrix $A = \begin{pmatrix} 2 & 1 \\ 1 & 1 \end{pmatrix} \in \mathbb{R}^{2 \times 2}$:
  - Eigenvalue criterion: The eigenvalues of the matrix A are the two positive numbers. $\frac{3 \pm \sqrt{5}}{2}$. Thus the matrix A positive definite.
  - Principal minor criterion: The principal minors are

$$|2| = 2 > 0 \quad \text{and} \quad \begin{vmatrix} 2 & 1 \\ 1 & 1 \end{vmatrix} = 1 > 0 .$$

  Thus the matrix is positive definite.

- We consider the matrix $B = \begin{pmatrix} -1 & 0 & 2 \\ 0 & -1 & 0 \\ 2 & 0 & -8 \end{pmatrix} \in \mathbb{R}^{3\times 3}$:
  - Eigenvalue criterion: The eigenvalues of the matrix A are the three negative numbers. $-1, \frac{-9\pm\sqrt{65}}{2}$. Thus the matrix is A negative definite.
  - Principal minor criterion: The principal minors are

$$\begin{vmatrix} -1 \end{vmatrix} = -1 < 0, \quad \begin{vmatrix} -1 & 0 \\ 0 & -1 \end{vmatrix} = 1 > 0, \quad \begin{vmatrix} -1 & 0 & 2 \\ 0 & -1 & 0 \\ 2 & 0 & -8 \end{vmatrix} = -\begin{vmatrix} -1 & 2 \\ 2 & -8 \end{vmatrix} = -4 < 0.$$

Thus the matrix is A negative definite.

■

In the extreme determination of functions of several variables, we will often be faced with the problem of deciding about the definiteness of a symmetric matrix $A \in \mathbb{R}^{2\times 2}$. This is possible because of the box in Sect. ?? only with the determinant $\det(A)$ and the trace $\text{trace}(A)$ The determinant is namely the product of the eigenvalues and the trace is the sum of the eigenvalues. Whether the two eigenvalues of a symmetric 2×2-matrix have the same or different sign or one of the two is zero can be seen from the determinant; with the trace we can see whether both are negative or positive in the case that both signs are the same. We record the procedure in a recipe-like manner:

Recipe: Definiteness of 2×2-Matrices

To determine whether a symmetric 2×2-matrix $A = \begin{pmatrix} a & b \\ b & c \end{pmatrix}$ is positive or negative (semi) definite or indefinite, proceed as follows:

(1) Determine $\det(A) = a\,c - b^2$ and $\text{trace}(A) = a + c$.
(2)
- If $\det(A) < 0$, then A is indefinite.
- If $\det(A) = 0$ and $\text{trace}(A) \geq 0$, then A positive semi definite.
- If $\det(A) = 0$ and $\text{trace}(A) \leq 0$, then A negative semi definite.
- If $\det(A) > 0$ and $\text{trace}(A) > 0$, then A is positive definite.
- If $\det(A) > 0$ and $\text{trace}(A) < 0$, then A is negative definite.

MATLAB For larger matrices, the numerical methods for the approximate determination of the eigenvalues from Chap. 40 can be used. MATLAB gives with the function `eig` a function to determine the eigenvalues of the matrix A.

45.2 Matrix Norms

In Chap. 16, more specifically in Sect. 16.2, we have explained the term *length* or *norm* of a vector in a Euclidean vector space with the help of an Euclidean scalar product $\langle \cdot , \cdot \rangle$.

45.2.1 Norms

We now introduce such a term *length* on an arbitrary $\mathbb{R}$- or $\mathbb{C}$-vector space, independent of a scalar product. We write $\mathbb{K}$ and mean $\mathbb{R}$ or $\mathbb{C}$:

Norm

A mapping $N : V \to \mathbb{R}$ of a $\mathbb{K}$-vector space V is called a **norm** on V if we have:

(N_1) $N(v) \geq 0$ for all $v \in V$ and $N(v) = 0 \Leftrightarrow v = 0$.
(N_2) $N(\lambda v) = |\lambda|\, N(v)$ for all $\lambda \in \mathbb{K},\ v \in V$.
(N_3) $N(v + w) \leq N(v) + N(w)$ for all $v,\ w \in V$ (triangle inequality).

A vector space with a norm is also called a **normed vector space**.

These conditions (N_1)–(N_3) are quite natural, if we only consider that we want to have a *length* with this norm. Note that no scalar product appears in the definition. But if $\langle \cdot , \cdot \rangle$ is a scalar product of a real vector space V, then the mapping $\|\cdot\| : V \to \mathbb{R}$ with $\|v\| = \sqrt{\langle v , v \rangle}$ is a norm in the sense explained here (note the Exercise 45.3).

In the following examples we mainly consider norms on the vector space $\mathbb{R}^n$ but now remember that we are interested in *matrix norms*, i.e., in norms on the vector space $\mathbb{R}^{n\times n}$ of quadratic matrices; such we will then explain with the help of the following norms on the $\mathbb{R}^n$.

Example 45.3

- We consider the following mappings N_1, N_2 and N_∞ from $V = \mathbb{R}^n$ to $\mathbb{R}$ which are given by
  - $N_1((v_1, \ldots , v_n)^\top) = \sum_{i=1}^{n} |v_i|$,
  - $N_2((v_1, \ldots , v_n)^\top) = \sqrt{\sum_{i=1}^{n} |v_i|^2}$,
  - $N_\infty((v_1, \ldots , v_n)^\top) = \max\{|v_i| \mid i = 1, \ldots , n\}$.

  The mappings $N_1,\ N_2,\ N_\infty$ are norms of the $\mathbb{R}^n$. Figure 45.1 shows in each case the vectors of $\mathbb{R}^2$, which have length 1 with respect to these three different norms.

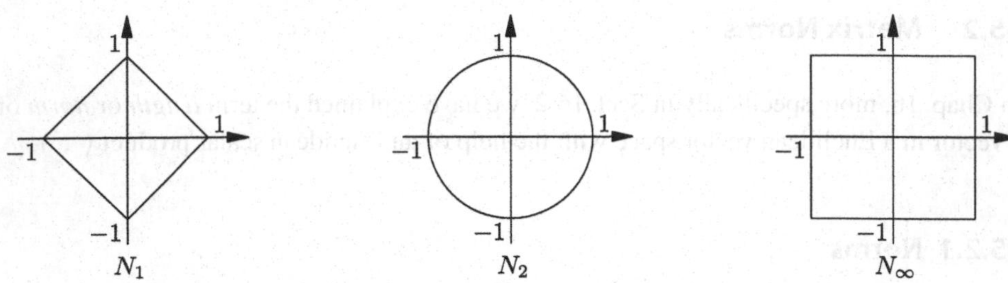

Fig. 45.1 Unit circles with respect to the norms N_1, N_2 and N_∞

We call

- N_1 also **1-norm** and write ℓ^1**-norm**,
- N_2 also **Euclidean norm** and write ℓ^2**-norm**,
- N_∞ also **maximum norm** and also write ℓ^∞**-norm.**

The ℓ^2-norm of course is just the known Euclidean length $\|v\| = \sqrt{\langle v\,,\,v\rangle}$ where $\langle\cdot\,,\cdot\rangle$ is the canonical scalar product of $\mathbb{R}^n$.

- Completely analogous to N_2 from the above example, we can prove that the mapping

$$N : \mathbb{R}^{m\times n} \to \mathbb{R} \text{ with } N(A) = \sqrt{\sum_{i=1}^{m}\sum_{j=1}^{n} a_{ij}^2}$$

on the vector space $V = \mathbb{R}^{m\times n}$ of $m \times n$-matrices is a norm; this norm is called the **Frobenius norm.** ■

MATLAB With `norm(v,1)`, `norm(v,2)`, `norm(v,inf)` and `norm(A,'fro')` one obtains in MATLAB according to our notation the ℓ^1-, ℓ^2-, ℓ^∞-norm of a vector v or the Frobenius norm of a matrix A.

45.2.2 Induced Matrix Norm

We're interested in **matrix norms** of $\mathbb{R}^{n\times n}$. In doing so, we want to have norms that have additional properties. They should be *submultiplicative* and *compatible* with a vector norm:

Submultiplicity and Compatibility

We call a matrix norm $\|\cdot\|$ at $\mathbb{R}^{n\times n}$

(continued)

- **submultiplicative** if for all $A,\ B \in \mathbb{R}^{n\times n}$ we have:

$$\|A\,B\| \le \|A\|\,\|B\|\,,$$

- **compatible** with a vector norm $\|\cdot\|_V$ if for all $A \in \mathbb{R}^{n\times n}$ and $v \in \mathbb{R}^n$:

$$\|Av\|_V \le \|A\|\,\|v\|_V\,.$$

Now we come to the essential *construction* of matrix norms using vector norms. Namely, by the process described as follows, we obtain from each vector norm $\|\cdot\|_V$ on the vector space $V = \mathbb{R}^n$ a matrix norm $\|\cdot\|$ on $\mathbb{R}^{n\times n}$; and now comes the best: the matrix norm constructed in this way is automatically submultiplicative and compatible with the vector norm $\|\cdot\|_V$ with the help of which it arises, more precisely:

Induced or Natural Matrix Norm

Each vector norm $\|\cdot\|_V$ of the $\mathbb{R}^n$ defines a matrix norm $\|\cdot\|$ of $\mathbb{R}^{n\times n}$. To do this, one sets for $A \in \mathbb{R}^{n\times n}$:

$$\|A\| = \sup_{v\in\mathbb{R}^n,\ \|v\|_V=1} \|Av\|_V\,.$$

One calls $\|\cdot\|$ the from $\|\cdot\|_V$ **induced** or **natural matrix norm**. For the by a vector norm $\|\cdot\|_V$ induced matrix norm $\|\cdot\|$ we have:

- The matrix norm $\|\cdot\|$ is compatible with the vector norm $\|\cdot\|_V$.
- The matrix norm $\|\cdot\|$ is submultiplicative.

The calculation of $\|A\|$ using the given supremum is not very practical. Nor will we do it often: We consider which matrix norms on $\mathbb{R}^{n\times n}$ arise in this way from the vector norms ℓ^1, ℓ^2 and ℓ^∞ of $\mathbb{R}^n$. For the proofs see Exercise 45.3.

Before we now state the main matrix norms, we note two essential facts:

Important Properties of Natural Matrix Norms

For each natural matrix norm $\|\cdot\|$ of $\mathbb{R}^{n\times n}$ we have

- $\|E_n\| = 1$.
- $|\lambda| \le \|A\|$ for each eigenvalue λ of A.

The proofs are easy to follow: We have

$$\|E_n\| = \sup_{v \in \mathbb{R}^n,\ \|v\|_V = 1} \|E_n v\|_V = \sup_{v \in \mathbb{R}^n,\ \|v\|_V = 1} \|v\|_V = 1\,.$$

And if λ is an eigenvalue of A with an associated normalized eigenvector $v \in \mathbb{R}^n$, $\|v\|_V = 1$, then

$$\|Av\|_V = \|\lambda v\|_V = |\lambda|\,\|v\|_V = |\lambda| \text{ and } \|Av\|_V \le \|A\|\|v\|_V = \|A\|\,.$$

From this we obtain the second assertion.

The from ℓ^1-, ℓ^2- and ℓ^∞-Norm Induced Matrix Norms

In the following let $A = (a_{ij}) \in \mathbb{R}^{n \times n}$ be a square matrix.

- The ℓ^1-norm on $\mathbb{R}^n$ induces on $\mathbb{R}^{n \times n}$ the matrix norm $\|\cdot\|_1$ with

$$\|A\|_1 = \max\{|a_{1i}| + \cdots + |a_{ni}| \mid 1 \le i \le n\}\,,$$

so $\|A\|_1$ is the *maximum column sum (in absolute values).*

- The ℓ^∞-norm on $\mathbb{R}^n$ induces on $\mathbb{R}^{n \times n}$ the matrix norm $\|\cdot\|_\infty$ with

$$\|A\|_\infty = \max\{|a_{i1}| + \cdots + |a_{in}| \mid 1 \le i \le n\}\,,$$

so it is $\|A\|_\infty$ the *maximum row sum (in absolute values).*

- The ℓ^2-norm on $\mathbb{R}^n$ induces on $\mathbb{R}^{n \times n}$ the matrix norm $\|\cdot\|_2$ with

$$\|A\|_2 = \max\{\sqrt{\mu} \mid \mu \text{ is eigenvalue of } A^\top A\}\,,$$

so it is $\|A\|_2$ is the *square root of the largest eigenvalue of* $A^\top A$. One calls $\|\cdot\|_2$ the **spectral norm**.

If $A \in \mathbb{R}^{n \times n}$ symmetric, $A^\top = A$, then we have for the spectral norm:

$$\|A\|_2 = \max\{|\mu| \mid \mu \text{ is eigenvalue of } A\}\,.$$

Moreover, for any matrix $A \in \mathbb{R}^{n \times n}$ and any matrix norm induced by a vector norm $\|\cdot\|$ we have:

$$\|A\|_2 \le \|A\|\,.$$

Note that because of the positive semi definiteness of the symmetric matrix $A^\top A$ in fact all eigenvalues of $A^\top A$ are also greater than or equal to 0, so that the root at $\|A\|_2$ also makes sense.

We show the claims for the ℓ^∞- and ℓ^2-norm for symmetric A. The assertion for ℓ^1-norm is shown to be similar to that for ℓ^∞:

- ℓ^∞ induces $\|\cdot\|_\infty$: For the maximum norm $\|\cdot\|_\infty$ of $\mathbb{R}^n$ we have $\|v\|_\infty = \max\{|v_i| \mid 1 \leq i \leq n\}$. For a matrix $A \in \mathbb{R}^{n\times n}$ with the rows $z_1, \ldots, z_n$ we get:

$$\begin{aligned}\|A\|_\infty = \sup_{\|v\|_\infty=1} \|Av\|_\infty &= \sup_{\|v\|_\infty=1} \|(z_1 v, \ldots, z_n v)^\top\|_\infty \\ &= \sup_{\|v\|_\infty=1} \max\{|z_i v| \mid 1 \leq i \leq n\} \\ &= \sup_{\|v\|_\infty=1} \max\{|z_{i1}v_1 + \cdots + z_{in}v_n| \mid 1 \leq i \leq n\} \\ &= \max\{|z_{i1}| + \cdots + |z_{in}| \mid 1 \leq i \leq n\}\,.\end{aligned}$$

  Thus $\|A\|_\infty$ is therefore the maximum row sum of A.
- ℓ^2 induces $\|\cdot\|_2$ (for symmetric A): If $A \in \mathbb{R}^{n\times n}$ is a real symmetric matrix, then there exists an ONB $B = (b_1, \ldots, b_n)$ of the $\mathbb{R}^n$ consisting of eigenvectors of A, i.e. a base B of $\mathbb{R}^n$ with

$$A\,b_1 = \lambda_1 b_1, \ldots, A\,b_n = \lambda_n b_n \text{ and } \langle b_i\,, b_j\rangle = \delta_{ij}\,,$$

  where we write $\lambda_{\max}$ for the largest of the real eigenvalues $\lambda_1, \ldots, \lambda_n$. We represent a $v \in \mathbb{R}^n \setminus \{0\}$ with respect to B,

$$v = \mu_1 b_1 + \cdots + \mu_n b_n \text{ with } \mu_1, \ldots, \mu_n \in \mathbb{R}\,,$$

  and calculate the ℓ^2-norm of v:

$$\|v\|_2 = \sqrt{\langle v\,, v\rangle} = \sqrt{\mu_1^2 + \cdots + \mu_n^2}\,.$$

  In order to determine now $\|A\|_2$ for the matrix $\|\cdot\|_2$ induced by the vector norm $\|\cdot\|_2$ we calculate:

$$\begin{aligned}\|Av\|_2 = \|\mu_1\lambda_1 b_1 + \cdots + \mu_n\lambda_n b_n\|_2 &= \sqrt{(\mu_1\lambda_1)^2 + \cdots + (\mu_n\lambda_n)^2} \\ &\leq |\lambda_{\max}|\sqrt{\mu_1^2 + \cdots + \mu_n^2} = |\lambda_{\max}|\,\|v\|_2\,.\end{aligned}$$

Thus we obtain:

$$\|A\|_2 = \sup_{v \in \mathbb{R}^n \setminus \{0\}} \frac{\|Av\|_2}{\|v\|_2} \leq |\lambda_{\max}| \,.$$

On the other hand, since $|\lambda| \leq \|A\|_2$ for all eigenvalues λ of A, it follows together

$$|\lambda_{\max}| = \|A\|_2 \,.$$

In particular, these results imply that for any matrix norm induced by a vector norm there is an upper bound for all eigenvalues of A. Thus, all eigenvalues of A can be constrained by the maximum row sum or maximum column sum:

$$|\lambda| \leq \|A\|_\infty = \max_{1 \leq i \leq n} \sum_{j=1}^{n} |a_{ij}| \text{ and } |\lambda| \leq \|A\|_1 = \max_{1 \leq j \leq n} \sum_{i=1}^{n} |a_{ij}|.$$

Example 45.4

- For the matrix $A = \begin{pmatrix} 1 & -1 & 5 \\ 0 & 2 & 0 \\ 0 & 0 & 1 \end{pmatrix}$ we have

$$\|A\|_1 = 6 \text{ and } \|A\|_\infty = 7 \text{ and thus } |\lambda| \leq 6$$

for each eigenvalue λ of A, which is a very rough estimate, since 1, 2 and 1 are the exact eigenvalues of A.
- We compute the spectral norms of two matrices.
  - The matrix $A = \begin{pmatrix} 1 & 1 \\ 1 & 1 \end{pmatrix}$ is real and symmetric and has the eigenvalues 0 and 2. Thus we have

$$\|A\|_2 = 2.$$

  - The matrix $A = \begin{pmatrix} 1 & 1 \\ 0 & 1 \end{pmatrix}$ is not symmetric, so we have to determine the eigenvalues of $A^\top A$: The eigenvalues of

$$A^\top A = \begin{pmatrix} 1 & 1 \\ 1 & 2 \end{pmatrix} \quad \text{are } \lambda_1 = \frac{3+\sqrt{5}}{2} \text{ and } \lambda_2 = \frac{3-\sqrt{5}}{2} \,,$$

so that

$$\|A\|_2 = \sqrt{\frac{3+\sqrt{5}}{2}}\,.$$

■

Finally, we note that the set of eigenvalues of a matrix is also called the **spectrum** of A and for this we write $\sigma(A)$, i.e.

$$\sigma(A) = \{\lambda \in \mathbb{C} \mid \lambda \text{ is eigenvalue of } A \in \mathbb{C}^{n \times n}\}\,.$$

Further, one calls

$$\rho(A) = \max\{|\lambda| \mid \lambda \in \sigma(A)\}$$

the **spectral radius of**A. The term is suggestive: a circle of radius $\rho(A)$ around 0 in $\mathbb{C}$ contains all eigenvalues of A; it is the smallest circle with this property. If $A \in \mathbb{R}^{n \times n}$ is symmetric, then $\|A\|_2$ is just the spectral radius.

45.3 Exercises

45.1 A matrix $M \in \mathbb{R}^{n \times n}$ is called positive semi definite if $v^\top M v \geq 0$ for all $v \in \mathbb{R}^n$.

(a) Show that a positive semi definite matrix has only nonnegative eigenvalues.
(b) Follow from problem part (a) that for $A \in \mathbb{R}^{m \times n}$ the matrix $A^\top A$ has only nonnegative eigenvalues.

45.2 Compute the spectral norms of the following matrices:

$$A = \begin{pmatrix} 0 & -1 & -2 \\ -1 & 0 & -2 \\ -2 & -2 & -3 \end{pmatrix}, \quad B = \begin{pmatrix} 3 & 0 & -1 \\ 0 & 2 & 0 \\ -1 & 0 & 3 \end{pmatrix}.$$

45.3 Prove the eigenvalue criterion for establishing the definiteness of a real symmetric matrix.

45.4 Show that the length of vectors of a Euclidean vector space is a norm.

45.5 Prove the statements in the box on induced matrix norms in Sect. 45.2.2.

45.6

(a) Show that the Frobenius norm is a norm on $\mathbb{R}^{n\times n}$.
(b) Show that the Frobenius norm is compatible with the Euclidean vector norm $\|\cdot\|_2$ and is submultiplicative.
(c) Why is the Frobenius norm for $n > 1$ not induced by any vector norm?

45.7 Calculate $\|A\|_1$ and $\|A\|_\infty$ for the matrix $A = \begin{pmatrix} 1 & 2 & 3 \\ 2 & -3 & 4 \\ 2 & 4 & -5 \end{pmatrix} \in \mathbb{R}^{3\times 3}$.

46 Functions of Several Variables

We now turn to the analysis of functions of several variables. Thus we consider functions $f : D \to W$ with $D \subseteq \mathbb{R}^n$ and $W \subseteq \mathbb{R}^m$ for natural numbers m and n. To this end, we first show by means of quite a few examples what kinds of such functions can still be illustrated at all. Finally, we generalize open and closed intervals, sequences, and limits of sequences to the vector space $\mathbb{R}^n$ and finally explain the continuity of functions of several variables analogous to the case of a function of one variable.

With these concepts some new phenomena appear, but also much of one-dimensional analysis remains in its basic features.

46.1 The Functions and Their Representations

So far, our functions under consideration have always had the form $f : D \subseteq \mathbb{R} \to \mathbb{R}$, $x \mapsto f(x)$, so in each case a real number x of a domain D is a real value $f(x)$ assigned. This is the special case $n = 1$ and $m = 1$ of a **vector-valued function in n variables,** i.e. a function of the form

$$f : D \subseteq \mathbb{R}^n \to \mathbb{R}^m \,,\; x = \begin{pmatrix} x_1 \\ \vdots \\ x_n \end{pmatrix} \mapsto f(x) = \begin{pmatrix} f_1(x_1, \ldots, x_n) \\ \vdots \\ f_m(x_1, \ldots, x_n) \end{pmatrix}.$$

In $\mathbb{R}^2$ and $\mathbb{R}^3$ we write instead of $x_1,\ x_2,\ x_3$ simple $x,\ y,\ z$. But with this we would run the risk of confusing the vector x with the variable x. Therefore, from now on we write vectors in bold, we write $\boldsymbol{x}$ or $\boldsymbol{a}$, if $\boldsymbol{x} \in \mathbb{R}^n$ or $\boldsymbol{a} \in \mathbb{R}^n$. For a function we write from now on $\boldsymbol{x} \mapsto f(\boldsymbol{x})$.

C. Karpfinger, *Calculus and Linear Algebra in Recipes*,
https://doi.org/10.1007/978-3-662-65458-3_46

It is common to distinguish the following special types of vector-valued functions in n variables:

- **Curve**: $n = 1$ and $m \in \mathbb{N}$, specifically:
  - **Plane curves**: $\gamma : D \subseteq \mathbb{R} \to \mathbb{R}^2$, thus $n = 1$ and $m = 2$,
  - **Space curves**: $\gamma : D \subseteq \mathbb{R} \to \mathbb{R}^3$, so $n = 1$ and $m = 3$.
- **Surfaces**: $\phi : D \subseteq \mathbb{R}^2 \to \mathbb{R}^3$, so $n = 2$ and $m = 3$.
- **Scalar fields**: $f : D \subseteq \mathbb{R}^n \to \mathbb{R}$, so $n \in \mathbb{N}$ and $m = 1$.
- **Vector fields**: $\boldsymbol{v} : D \subseteq \mathbb{R}^n \to \mathbb{R}^n$, so $n = m$.

In one-dimensional analysis, we were nearly almost able to represent a function $f : D \to \mathbb{R}$ by the *graph* $\text{Graph}(f) = \{(x, f(x)) \mid x \in D\} \subseteq \mathbb{R} \times \mathbb{R}$. Unfortunately, in multidimensional calculus this only works in special cases anymore. We can represent in general:

- the range $\gamma(D)$ of a plane curve $\gamma : D \subseteq \mathbb{R} \to \mathbb{R}^2$ or space curve $\gamma : D \subseteq \mathbb{R} \to \mathbb{R}^3$,
- the range $\phi(D)$ of a surface $\phi : D \subseteq \mathbb{R}^2 \to \mathbb{R}^3$,
- the graph $\text{Graph}(f) = \{(x, y, f(x, y)) \mid (x, y) \in D\}$ of a scalar field $f : D \subseteq \mathbb{R}^2 \to \mathbb{R}$,
- the range of a vector field $\boldsymbol{v} : D \subseteq \mathbb{R}^2 \to \mathbb{R}^2$ and $\boldsymbol{v} : D \subseteq \mathbb{R}^3 \to \mathbb{R}^3$.

The ranges or graphs in the above table are from the following examples:

Example 46.1

- Curves: It is
  - $\gamma : [0, 2\pi] \to \mathbb{R}^2,\ \gamma(t) = \begin{pmatrix} 2\cos(t) \\ \sin(t) \end{pmatrix}$ a plane curve,
  - $\gamma : [0, 8\pi] \to \mathbb{R}^3,\ \gamma(t) = \begin{pmatrix} e^{-0.1t}\cos(t) \\ e^{-0.1t}\sin(t) \\ t \end{pmatrix}$ a space curve.
- Surfaces: It is
  - $\phi : [0, 2\pi] \times [0, 2] \to \mathbb{R}^3,\ \phi(u, v) = \begin{pmatrix} \cos(u) \\ \sin(u) \\ v \end{pmatrix}$ a surface,
  - $\phi : [0, 2\pi] \times [0, 2\pi] \to \mathbb{R}^3,\ \phi(u, v) = \begin{pmatrix} (2 + \cos(u))\cos(v) \\ (2 + \cos(u))\sin(v) \\ \sin(u) \end{pmatrix}$ a surface.
- Scalar fields:

  It is
  - $f : [-4, 4] \times [-5, 5] \to \mathbb{R},\ f(x, y) = 2\,(x^2 + y^2)$ a scalar field,
  - $f : \mathbb{R}^2 \to \mathbb{R},\ f(x, y) = x\,y\,e^{-(x^2+y^2)}$ a scalar field.

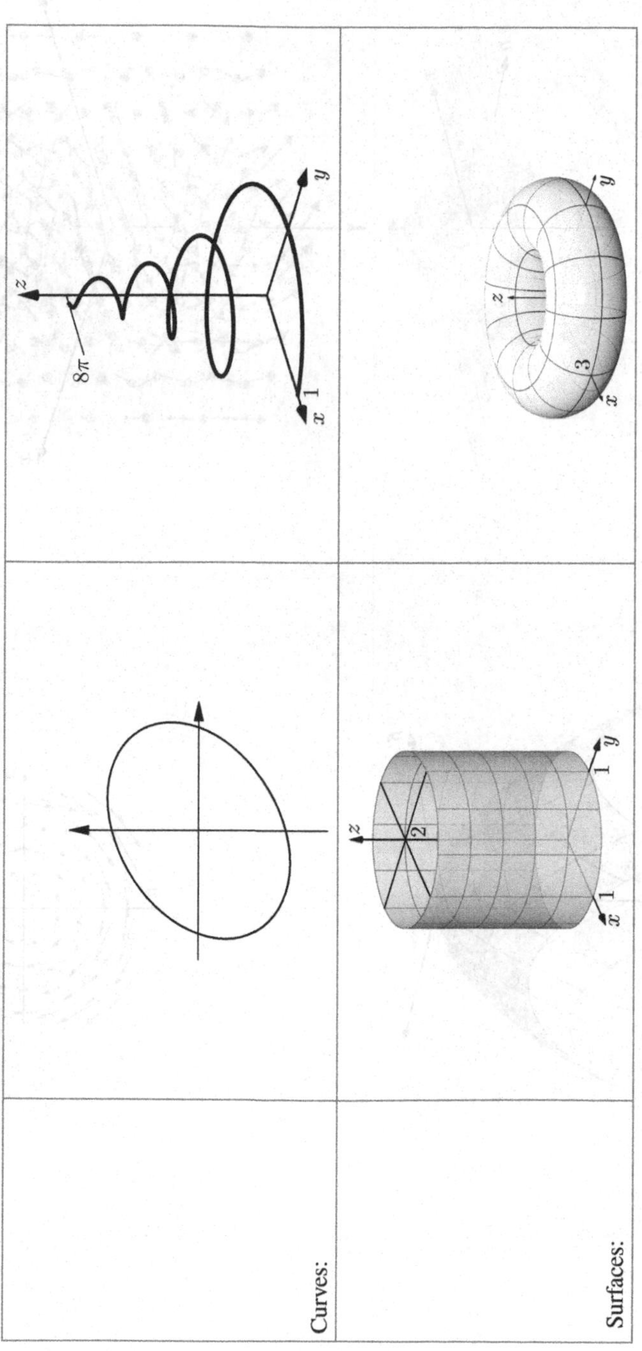
Curves:
Surfaces:
z
8π
x
1
y
2
3

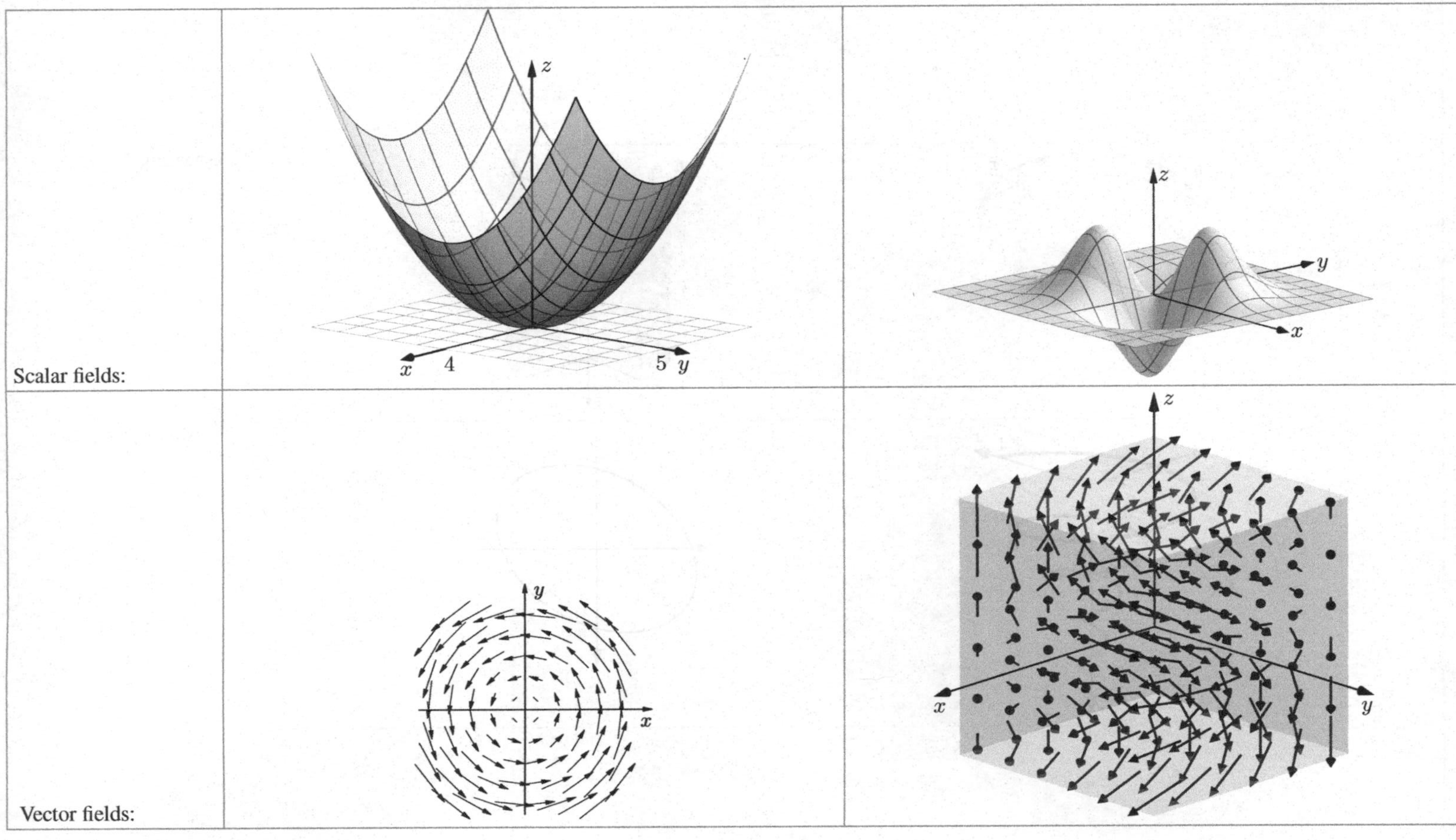

Scalar fields:
z
x
4
5
y
z
y
x
Vector fields:
y
x
z
x
y

- Vector fields:
  It is
  - $v : \mathbb{R}^2 \to \mathbb{R}^2$, $v(x, y) = \begin{pmatrix} -y \\ x \end{pmatrix}$ a vector field,
  - $v : \mathbb{R}^3 \to \mathbb{R}^3$, $v(x, y, z) = \begin{pmatrix} z \\ y \\ -x \end{pmatrix}$ a vector field. ■

MATLAB The above images (and of course many other images in an analogous way) can be obtained with MATLAB, e.g. as follows

```
» ezplot('2*cos(t)','sin(t)',[0,2*pi])
» ezplot3('exp(-0.1*t)*cos(t)','exp(-0.1*t)*sin(t)','t',
[0,8*pi])
» ezsurf('cos(u)','sin(u)','v',[0.2*pi,0.2])
» ezsurf('(2+cos(u))*cos(v)','
      (2+cos(u))*sin(v)','sin(u)',[0,2*pi,0,2*pi])
» ezmesh('2*(x^2+y^2)',[-4,4,-5,5])
» ezmesh('x*y*exp(-(x^2+y^2))')
» [x,y] = meshgrid(-2:0.5:2); quiver(x,y,-y,x)
» [x,y,z] = meshgrid(-2:0.5:2); quiver3(x,y,z,z,y,-x)
```

46.2 Some Topological Terms

In the analysis of one variable, the domains of definition were *open*, *half-open* or *closed* intervals, i.e. (a, b), $(a, b]$, $[a, b)$ or $[a, b]$, we called the points a and b suggestively *boundary points*. Similar notions we now meet again in the multidimensional.

For each subset $D \subseteq \mathbb{R}^n$ we denote by $D^c = \mathbb{R}^n \setminus D$ the complement of D in $\mathbb{R}^n$, and let $\|\cdot\|$ be the Euclidean norm on the $\mathbb{R}^n$, thus

$$\|(x_1, \ldots, x_n)^\top\| = \sqrt{x_1^2 + \cdots + x_n^2}\,.$$

- A point $\boldsymbol{x}_0 \in D$ is called **interior point** of D, if a $\varepsilon > 0$ exists, so that the **ε-ball**

$$B_\varepsilon(\boldsymbol{x}_0) = \left\{\boldsymbol{x} \in \mathbb{R}^n \mid, \|\boldsymbol{x} - \boldsymbol{x}_0\| < \varepsilon\right\} \subseteq D\,,$$

  i.e., there is an (open) ball around $\boldsymbol{x}_0$, which is completely contained in D.
- The set of all interior points of D is called **the interior** of D and is denoted by $\mathring{D}$.

- The set D is called **open** if $\overset{\circ}{D} = D$, i.e., if every point of D is an interior point.
- A point $\boldsymbol{x}_0 \in \mathbb{R}^n$ is called **boundary point** of D, if for all $\varepsilon > 0$ we have:

$$B_\varepsilon(\boldsymbol{x}_0) \cap D \neq \emptyset \quad \text{and} \quad B_\varepsilon(\boldsymbol{x}_0) \cap D^c \neq \emptyset,$$

  i.e., every ε-ball around $\boldsymbol{x}_0$ contains points of D as well as of the complement of D.
- The set of all boundary points of D, is called the **boundary** of D. It is denoted by ∂D.
- The set $\overline{D} = D \cup \partial D$ is called the **closure** of D.
- The set D is called **closed**, if $\partial D \subseteq D$, then $\overline{D} = D$.
- The set D is called **bounded** if there is a real $K \in \mathbb{R}$ such that $\|\boldsymbol{x}\| < K$ for all $\boldsymbol{x} \in D$.
- The set D is called **compact** if D is closed and bounded.
- The set D is called **convex** if for all $\boldsymbol{x},\ \boldsymbol{y} \in D$ and for all $\lambda \in \mathbb{R}$ with $0 \leq \lambda \leq 1$ we have $\lambda\,\boldsymbol{x} + (1-\lambda)\,\boldsymbol{y} \in D$. This says that the connecting line between $\boldsymbol{x}$ and $\boldsymbol{y}$ runs completely in D. See Fig. 46.1.

Boundary points and inner points are also exactly what one imagines by it, this shows exemplarily Fig. 46.2 with a subset D in $\mathbb{R}^2$.

Example 46.2

- The set $D = [0, 1] \subseteq \mathbb{R}$ is bounded (we have $|x| < 2$ for all $x \in D$) and closed (the boundary $\partial D = \{0, 1\}$ is in D included). Therefore D compact.

Fig. 46.1 A convex (= konvex) and a nonconvex (= nicht konvex) set

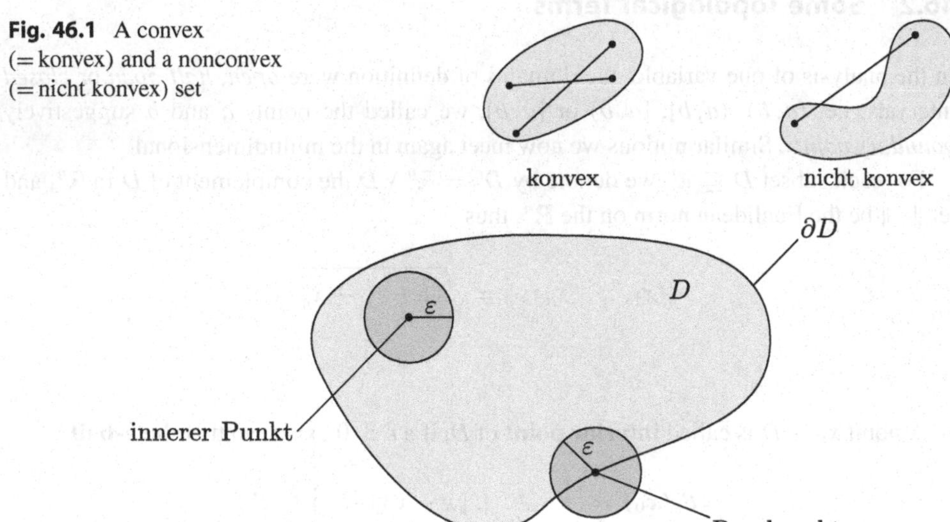

Fig. 46.2 Interior point (= innerer Punkt) and boundary point (= Randpunkt)

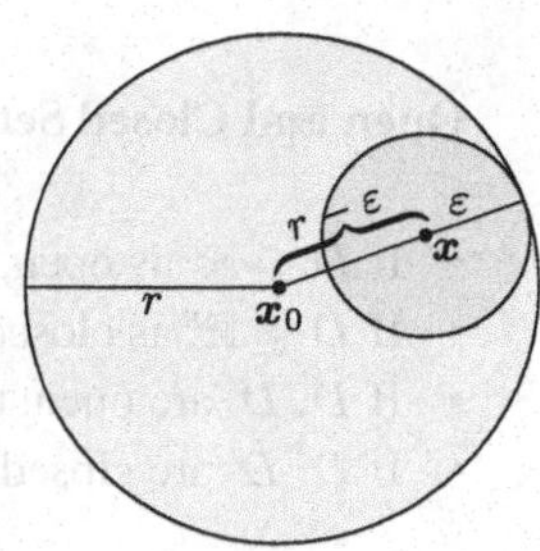

Fig. 46.3 x is an interior point

- The set $D = [0, 1) \subseteq \mathbb{R}$ is bounded (see above) and not closed (the boundary point 1 is not in D included). Therefore D is not compact. However, D is not open either, $0 \in D$ is not an interior point.
- For each $\boldsymbol{x}_0 \in \mathbb{R}^n$ and $r > 0$ the ball $B_r(\boldsymbol{x}_0) = \{\boldsymbol{x} \in \mathbb{R}^n \mid \|\boldsymbol{x} - \boldsymbol{x}_0\| < r\}$ is open: Each point is an interior point, note also the accompanying Fig. 46.3. Thus we have that $\overset{\circ}{B_r}(\boldsymbol{x}_0) = B_r(\boldsymbol{x}_0)$. Moreover it seems to be true

$$\partial B_r(\boldsymbol{x}_0) = \left\{\boldsymbol{x} \in \mathbb{R}^n \mid \|\boldsymbol{x} - \boldsymbol{x}_0\| = r\right\} .$$

So the conclusion is

$$\overline{B_r(\boldsymbol{x}_0)} = K_r(\boldsymbol{x}_0) = \left\{\boldsymbol{x} \in \mathbb{R}^n \mid \|\boldsymbol{x} - \boldsymbol{x}_0\| \leq r\right\} .$$

For $K_r(\boldsymbol{x}_0)$ thus we have

$$\overline{K_r(\boldsymbol{x}_0)} = K_r(\boldsymbol{x}_0) \text{ and } \overset{\circ}{K_r}(\boldsymbol{x}_0) = B_r(\boldsymbol{x}_0) .$$

- For $D = \mathbb{R}^n$ we have:
  - $\mathbb{R}^n$ is closed, since $\partial\, \mathbb{R}^n = \emptyset$ and $\emptyset \subseteq \mathbb{R}^n$, and
  - $\mathbb{R}^n$ is open, since every point of $\mathbb{R}^n$ is an interior point.
- For $D = \emptyset$ we have:
  - $\emptyset$ is closed, since $\partial\, \emptyset = \emptyset$ and $\emptyset \subseteq \emptyset$, and
  - $\emptyset$ is open, since every point of the empty set is an interior point. ■

The complement of an open set is closed, and the complement of a closed set is open: With this result, the assertions of the last example follow quite simply from those of the example before last. We record these and other results in a box:

Open and Closed Sets and Their Complements

- If $D \subseteq \mathbb{R}^n$ is open, then the complement $D^c = \mathbb{R}^n \setminus D$ is closed.
- If $D \subseteq \mathbb{R}^n$ is closed, then D^c is open.
- If D, D' are open, then also $D \cup D'$ and $D \cap D'$ are open.
- If D, D' are closed, then also $D \cup D'$ and $D \cap D'$ are closed.

46.3 Consequences, Limits, Continuity

As in the one-dimensional we explain the *continuity of* a function with the help of *sequences*. Instead of the sequences $(x_k)_{k\in\mathbb{N}_0}$ of real numbers in the $\mathbb{R}^1$ we now consider **sequences** $(\boldsymbol{x}^{(k)})_{k\in\mathbb{N}_0}$ of vectors in $\mathbb{R}^n$, i.e. $\boldsymbol{x}^{(k)} = (x_1^{(k)}, \ldots, x_n^{(k)})^\top \in \mathbb{R}^n$ for each $k \in \mathbb{N}_0$.

We say the sequence $(\boldsymbol{x}^{(k)})_{k\in\mathbb{N}_0}$ **converges to the limit** x if

$$\lim_{k\to\infty} \|\boldsymbol{x}^{(k)} - \boldsymbol{x}\| = 0\,.$$

That is, a sequence of vectors $(\boldsymbol{x}^{(k)})$ converges to $\boldsymbol{x}$, if the real sequence $(\|\boldsymbol{x}^{(k)} - \boldsymbol{x}\|)$ of distances converges to zero. We write in this case

$$\boldsymbol{x}^{(k)} \stackrel{k\to\infty}{\longrightarrow} \boldsymbol{x} \text{ or } \boldsymbol{x}^{(k)} \to \boldsymbol{x} \text{ or } \lim_{k\to\infty} \boldsymbol{x}^{(k)} = \boldsymbol{x}\,.$$

It is easy to consider:

Convergence = Component-Wise Convergence

We have

$$\boldsymbol{x}^{(k)} = \begin{pmatrix} x_1^{(k)} \\ \vdots \\ x_n^{(k)} \end{pmatrix} \to \boldsymbol{x} = \begin{pmatrix} x_1 \\ \vdots \\ x_n \end{pmatrix} \Leftrightarrow x_1^{(k)} \to x_1, \ldots, x_n^{(k)} \to x_n\,.$$

Example 46.3 We consider the sequence $(\boldsymbol{x}^{(k)})_{k\in\mathbb{N}}$ with

$$\boldsymbol{x}^{(k)} = (2 + (-1)^k/k, 2/k^2)^\top \in \mathbb{R}^2\,.$$

Because of $2 + (-1)^k/k \to 2$ and $2/k^2 \to 0$ we have

$$x^{(k)} = \begin{pmatrix} 2 + \frac{(-1)^k}{k} \\ \frac{2}{k^2} \end{pmatrix} \to \begin{pmatrix} 2 \\ 0 \end{pmatrix} . \quad \blacksquare$$

We now come to the central notion of *continuity* in this first chapter on functions in several variables. Here we explain this notion completely analogous to the one-dimensional case, avoiding now the detour over the limit values of functions (which are interesting in the one-dimensional case by themselves, but have much less importance in the multidimensional case):

Continuity of Functions of Several Variables

Given is a vector-valued function $f : D \subseteq \mathbb{R}^n \to \mathbb{R}^m$ in n variables. We say that the function f is

- **continuous in $a \in D$** if for each sequence $(x^{(k)})_{k\in\mathbb{N}_0}$ in D with $x^{(k)} \to a$ the sequence $(f(x^{(k)}))_{k\in\mathbb{N}_0}$ in $\mathbb{R}^m$ converges to $f(a)$ and
- **continuous on D** if f in each $a \in D$ is continuous.

Continuity = component-wise continuity: A vector-valued function

$$f : D \subseteq \mathbb{R}^n \to \mathbb{R}^m, \quad \begin{pmatrix} x_1 \\ \vdots \\ x_n \end{pmatrix} \mapsto \begin{pmatrix} f_1(x_1, \ldots, x_n) \\ \vdots \\ f_m(x_1, \ldots, x_n) \end{pmatrix}$$

is continuous in a or on D, if each component function

$$f_i : D \to \mathbb{R}, \ (x_1, \ldots, x_n)^\top \to f_i(x_1, \ldots, x_n), \ i \in \{1, \ldots, n\}$$

is continuous in a or on D.

Thanks to this last theorem, we only need to consider when scalar fields are continuous. And therefore we have as in the one-dimensional case:

Continuous Functions

If $f, g : D \subseteq \mathbb{R}^n \to \mathbb{R}$ continuous functions, so are

$$\lambda f + \mu g, \quad f g, \quad \frac{f}{g},$$

where $\lambda,\ \mu \in \mathbb{R}$ are real numbers (and we assume for the quotient f/g that $g(\boldsymbol{x}) \neq 0$ for all $\boldsymbol{x} \in D$).

Furthermore, if $g : D \subseteq \mathbb{R}^n \to W \subseteq \mathbb{R}^m$ and $f : W \subseteq \mathbb{R}^m \to \mathbb{R}^p$ are continuous mappings, so also

$$f \circ g.$$

Example 46.4

- All functions in Example 46.1 are continuous on their respective domains.
- We consider the function

$$f : \mathbb{R}^2 \to \mathbb{R},\ f(x, y) = \begin{cases} \frac{2xy}{x^2+y^2} & (x, y) \neq (0, 0) \\ 0 & (x, y) = (0, 0) \end{cases}.$$

Note that the function f on $\mathbb{R}^2 \setminus \{(0, 0)\}$ is continuous as a rational function. Only the point $(0, 0)$ is critical. On the coordinate axes $y = 0$ or $x = 0$ the function is outside the zero point $(0, 0)$ constant zero. But still: The function f is not continuous in $(0, 0)$.

For let us consider the sequence $(\boldsymbol{x}^{(k)})$ with $\boldsymbol{x}^{(k)} = (x^{(k)}, y^{(k)}) = (1/k, 1/k)$ which converges to $(0, 0)$ then we have the following for the sequence of images

$$f(\boldsymbol{x}^{(k)}) = \frac{2/k^2}{1/k^2 + 1/k^2} = 1 \quad \forall k \in \mathbb{N}_0 .$$

So we have

$$f(\boldsymbol{x}^{(k)}) \to 1 \neq 0 = f(0, 0);$$

and this says that the function f in $(0, 0)$ is not continuous (note Fig. 46.4). ■

A continuous function $f : [a, b] \to \mathbb{R}$ on the compact interval $[a, b] \subset \mathbb{R}$ has a maximum and a minimum (see Sect. 25.4). This theorem of maximum and minimum we also have for scalar fields:

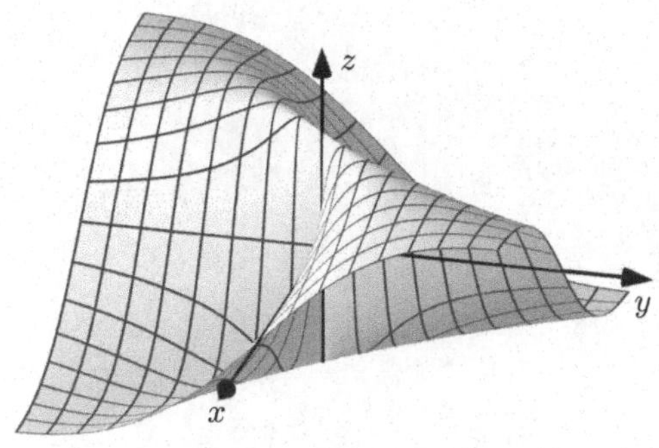

Fig. 46.4 An in (0, 0) discontinuous function

Theorem of Maximum and Minimum

If $D \subseteq \mathbb{R}^n$ is compact and $f : D \to \mathbb{R}$ is a continuous scalar field, then f has on D a maximum and a minimum, i.e., there are

$$\boldsymbol{x}_{\min}, \boldsymbol{x}_{\max} \in D \text{ with } f(\boldsymbol{x}_{\min}) \leq f(\boldsymbol{x}) \leq f(\boldsymbol{x}_{\max}) \text{ for all } \boldsymbol{x} \in D.$$

The answer to the question of how these $x_{\min}$ and $x_{\max}$ can be determined concretely, as in the one-dimensional case, is provided by differential calculus.

46.4 Exercises

46.1 Let $D \subseteq \mathbb{R}^n$. By $D^c = \mathbb{R}^n \setminus D$ we denote the complement of D. Show:

(a) Every point of D is either an interior point of D or a boundary point of D.
(b) If D is open, then $D \cap \partial D = \emptyset$.
(c) If D is open, then D^c closed.
(d) If D is closed, then D^c is open.

46.2 Examine the subsets of $\mathbb{R}^2$, $M_1 =]-1, 1[^2$, $M_2 =]-1, 1]^2$, $M_3 = [-1, 1]^2$, $\mathbb{R}^2$ and $\emptyset$ for interior points and boundary points. Which of the sets are open or closed?

Partial Differentiation: Gradient, Hessian Matrix, Jacobian Matrix 47

When differentiating a function f of a variable x, one examines the changing behavior of f *in direction* x. For a scalar field f in the n variables $x_1, \ldots, x_n$ there are many *directions* in which the function can change. The *partial derivatives* give this change behavior in the directions of the axes, the *directional* derivative much more generally in any direction. Fortunately, this *partial derivation* (and also forming the *directional* derivative) does not introduce any new difficulties: one simply derives according to the variable under consideration, as one is used to doing in the one-dimensional case, *freezing* all other variables. In this way we easily obtain the *gradient* as a collection of the first partial derivatives, and the *Hessian matrix* as a collection of the second partial derivatives of a scalar field f and the Jacobian matrix as a collection of the first partial derivatives of a vector-valued function in several variables.

47.1 The Gradient

The graph of a continuous scalar field $f : D \subseteq \mathbb{R}^2 \to \mathbb{R}$, D open, in two variables x and y is a surface in $\mathbb{R}^3$. We consider as an example such a surface in Fig. 47.1.

Three straight line segments run through the bold marked point of the definition area in the directions of the x-axis, y-axis and $\boldsymbol{v}$. The images of these straight line segments are marked on the graph. We can see from these lines the change behavior of the function in these directions and determine the slope of the tangent at this point in the direction under consideration as a measure of this change behavior.

We know how this works from one-dimensional calculus, since after restricting the multidimensional function to one direction we have $\boldsymbol{e}_1 = (1, 0)^\top$ or $\boldsymbol{e}_2 = (0, 1)^\top$ or $\boldsymbol{v} = (v_1, v_2)^\top$ we essentially have only one function in one variable. We explain more generally for a function in n variables:

C. Karpfinger, *Calculus and Linear Algebra in Recipes*,
https://doi.org/10.1007/978-3-662-65458-3_47

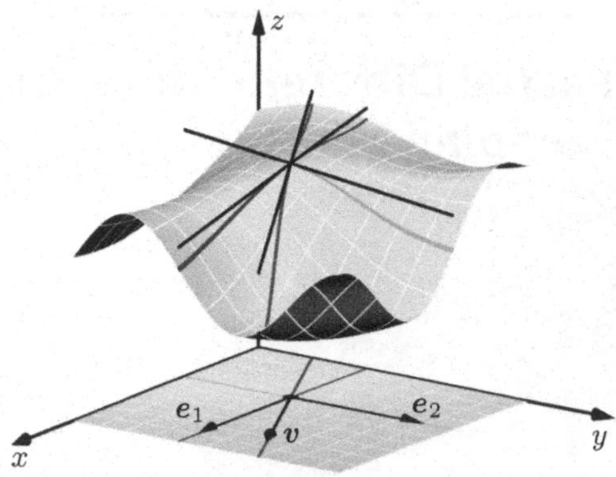

Fig. 47.1 Slope in different directions

Directional Derivative, Partial Derivative, Gradient

Given is a scalar field

$$f : D \subseteq \mathbb{R}^n \to \mathbb{R},\ \boldsymbol{x} = (x_1, \ldots, x_n)^\top \mapsto f(\boldsymbol{x}) = f(x_1, \ldots, x_n)\,.$$

- For any vector $\boldsymbol{v}$ with $\|\boldsymbol{v}\| = 1$ denotes for $\boldsymbol{a} \in D$ the number

$$\frac{\partial f}{\partial \boldsymbol{v}}(\boldsymbol{a}) = \partial_{\boldsymbol{v}} f(\boldsymbol{a}) = f_{\boldsymbol{v}}(\boldsymbol{a}) = \lim_{h \to 0} \frac{f(\boldsymbol{a} + h\,\boldsymbol{v}) - f(\boldsymbol{a})}{h}$$

  the **directional derivative** of f in $\boldsymbol{a}$ in direction $\boldsymbol{v}$, if this limit exists.
- The directional derivative

$$\frac{\partial f}{\partial x_i}(\boldsymbol{a}) = \partial_i f(\boldsymbol{a}) = f_{x_i}(\boldsymbol{a}) = \lim_{h \to 0} \frac{f(\boldsymbol{a} + h\,\boldsymbol{e}_i) - f(\boldsymbol{a})}{h}$$

  in $\boldsymbol{a}$ in the direction of the coordinate axes $\boldsymbol{e}_1, \ldots, \boldsymbol{e}_n$ is called the **partial derivative** of f in $\boldsymbol{a}$ with respect to x_i if this limit exists.
- If the n partial derivatives $f_{x_1}, \ldots, f_{x_n}$ exist in $\boldsymbol{a} \in D$, then one calls f **partial differentiable** in $\boldsymbol{a}$ and the vector

$$\nabla f(\boldsymbol{a}) = \begin{pmatrix} f_{x_1}(\boldsymbol{a}) \\ \vdots \\ f_{x_n}(\boldsymbol{a}) \end{pmatrix}$$

(continued)

the **gradient** of f in $\boldsymbol{a}$. The symbol ∇ is also called **nabla operator** and also writes grad $f(\boldsymbol{a}) = \nabla f(\boldsymbol{a})$.

- If the functions $f_{x_1}, \ldots, f_{x_n}$ are continuous, then the directional derivate $\frac{\partial f}{\partial \boldsymbol{v}}(\boldsymbol{a})$ of f for $\boldsymbol{v}$ with $\|\boldsymbol{v}\| = 1$ and $\boldsymbol{a} \in D$ can be calculated as

$$\frac{\partial f}{\partial \boldsymbol{v}}(\boldsymbol{a}) = \langle \nabla f(\boldsymbol{a}), \boldsymbol{v} \rangle .$$

Written in detail, the partial derivative of f in $\boldsymbol{x}$ to x_i is as follows:

$$\frac{\partial f}{\partial x_i}(\boldsymbol{x}) = \lim_{h \to 0} \frac{f(x_1, \ldots, x_{i-1}, x_i + h, x_{i+1}, \ldots, x_n) - f(x_1, \ldots, x_n)}{h} .$$

From this illustration, one can see that the partial derivative to x_i is as in the one-dimensional case: We derivate to x_i and all other variables are considered as constants. So nothing happens during partial differentiation *new*, and the directional derivative of a function f at a point $\boldsymbol{x}$ in the direction of a normalized vector $\boldsymbol{v}$ is obtained simply by taking the scalar product of $\boldsymbol{v}$ with the gradient of f at the point under consideration.

Example 47.1

- For $f : \mathbb{R}^2 \to \mathbb{R},\ f(x, y) = x^2 y^3 + x$ the partial derivatives are

$$f_x(x, y) = 2xy^3 + 1\,,\ f_y(x, y) = 3x^2y^2\,,\ \text{so}\ \nabla f(x, y) = \begin{pmatrix} 2xy^3 + 1 \\ 3x^2y^2 \end{pmatrix} .$$

- For $f : \mathbb{R}^3 \to \mathbb{R},\ f(x, y, z) = x^2 + e^{yz}$ and $v = \frac{1}{\sqrt{3}}(1, 1, 1)^\top$ the partial derivatives are

$$f_x(x, y, z) = 2x\,,\ f_y(x, y, z) = z\,e^{yz}\,,\ f_z(x, y, z) = y\,e^{yz}\,,$$

so

$$\nabla f(x, y, z) = \begin{pmatrix} 2x \\ z\,e^{yz} \\ y\,e^{yz} \end{pmatrix} \ \text{and e. g.}\ \frac{\partial f}{\partial v}(1, 1, 0) = \langle \begin{pmatrix} 2 \\ 0 \\ 1 \end{pmatrix}, \frac{1}{\sqrt{3}} \begin{pmatrix} 1 \\ 1 \\ 1 \end{pmatrix} \rangle = \sqrt{3} .$$

- For $f : \mathbb{R}^n \to \mathbb{R},\ f(x) = a^\top x$, where $a = (a_1, \dots, a_n)^\top \in \mathbb{R}^n$, we have:
$$\frac{\partial f}{\partial x_i}(x) = a_i \text{ for each } i = 1, \dots, n\,, \quad \text{so } \nabla f(x) = a\,.$$
- The partial derivatives and hence the gradient of the scalar field $f : \mathbb{R}^n \to \mathbb{R}$ with $f(x) = x^\top A\, x$ with a matrix $A \in \mathbb{R}^{n \times n}$, can be obtained most easily by determining the limit value $\lim_{h\to 0} \frac{f(x+he_i)-f(x)}{h}$. For this we first consider the difference quotient

$$\begin{aligned}\frac{f(x + h\, e_i) - f(x)}{h} &= \frac{1}{h}\left(x^\top A\, x + h\, x^\top A\, e_i + h\, e_i^\top A\, x + h^2 e_i^\top A\, e_i - x^\top A\, x\right)\\ &= x^\top A\, e_i + e_i^\top A\, x + h\, e_i^\top A\, e_i\,.\end{aligned}$$

Now we let h go towards 0 and get

$$\begin{aligned}\lim_{h\to 0}\frac{f(x + h\, e_i) - f(x)}{h} &= x^\top A\, e_i + e_i^\top A\, x = e_i^\top A^\top x + e_i^\top A\, x\\ &= e_i^\top (A^\top + A)\, x\,.\end{aligned}$$

Thus

$$\nabla f(x) = (A^\top + A)\, x \text{ and } \nabla f(x) = 2\, A\, x\,, \quad \text{if } A^\top = A\,. \qquad \blacksquare$$

The significance of the gradient or directional derivative lies in the following interpretation, which, for clarity, we reproduce very graphically:

We summarize the graph of a function f in two variables x and y as a *mountainous landscape* and we are in this mountainous *landscape* at the point P: If we now go in the direction of the positive x-axis, we are going downhill, so the directional derivative in this direction is negative. If, on the other hand, we walk in the direction of the negative x-axis, we go uphill, the directional derivative in this direction is positive, see Fig. 47.2.

Note that when walking in the direction of v, one obviously gets downhill much faster, the directional derivative is negative and thereby significantly larger in absolute values than that in the direction of the positive x-axis. The remarkable thing is that we can determine the direction of the steepest ascent or descent without knowing the graph. These directions are given by the gradient of the function f at the point under consideration:

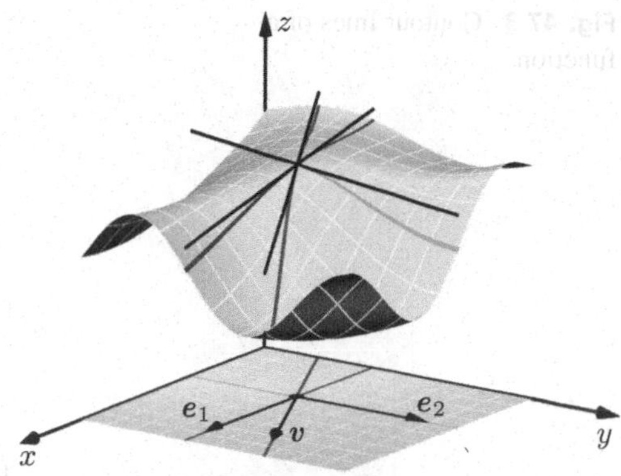

Fig. 47.2 Directional derivative in different directions

The Gradient Indicates the Direction of the Steepest Slope
If $f : D \subseteq \mathbb{R}^n \to \mathbb{R}$, D open, a partially differentiable scalar field with continuous derivatives $f_{x_1}, \ldots, f_{x_n}$ and $\boldsymbol{a} \in D$, then in the case $\nabla f(\boldsymbol{a}) \neq 0$:

- f grows most steeply in the direction $\nabla f(\boldsymbol{a})$ and
- f falls most steeply in the direction $-\nabla f(\boldsymbol{a})$.

Further, the gradient is perpendicular to the **contour lines** or **levels** $N_c = \{\boldsymbol{x} \in D \mid f(\boldsymbol{x}) = c\}$ with $c \in f(D) \subseteq \mathbb{R}$.

The statement about the orthogonality of the gradient with the contour lines can be justified with the help of the *chain rule*. We have given this reasoning as well as the reasoning for the statement about the extremal growth in the direction of the gradient or in the opposite direction as Exercise 47.4.

Here we leave it with an illustrative explanation in the case $n = 2$ to this orthogonality and note in advance that the *perpendicularity* of the gradient $\nabla f(\boldsymbol{x})$ on the contour line N_c naturally refers to the perpendicularity of the gradient on the tangent to the contour line. One can use the contour lines of f can be interpreted as contour lines of a (mountain) landscape (think of a topographic map), see Fig. 47.3.

If we now look at two closely *adjacent contour* lines at level c and $c + \varepsilon$ then locally the shortest path from a point of the contour line N_c to the contour line $N_{c+\varepsilon}$ is found by walking at right angles to the approximately parallel contour lines. This just says that the gradient is perpendicular to the contour line.

Example 47.2 We determine the direction in which the graph of the scalar field $f : \mathbb{R}^2 \to \mathbb{R}$, $f(x, y) = \frac{1}{2}(x^2 + y^2)$ at the point $\boldsymbol{a} = (1, 1)$ grows or falls most strongly: Because

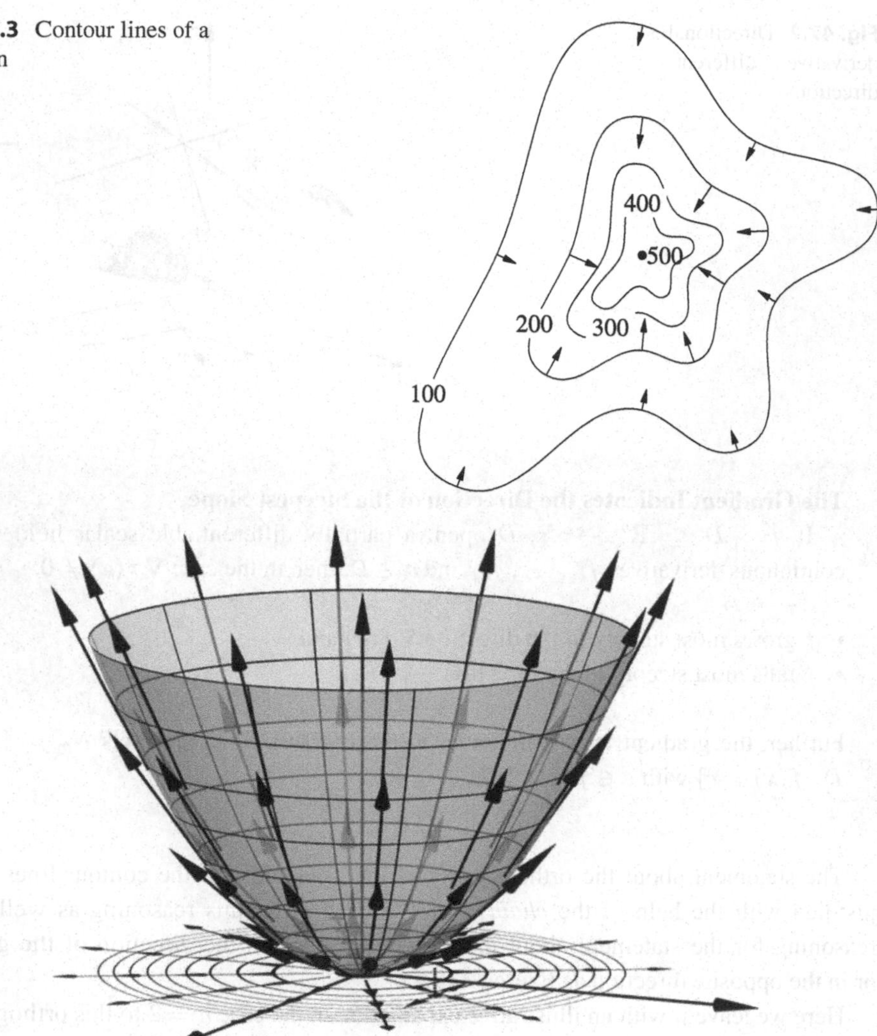

Fig. 47.3 Contour lines of a function

Fig. 47.4 The gradient and the direction of strongest growth

of $\nabla f(x, y) = (x, y)^\top$ the function grows most strongly in the direction $(1, 1)^\top$ and falls most strongly in the direction $(-1, -1)^\top$, see Fig. 47.4. ■

Remark The descent *method* or *gradient method* for the determination of local minima of a function in several variables is based on the above illustration of the direction of descent when *walking in mountains*. We will discuss this method in Chap. 72.

47.2 The Hessian Matrix

The partial derivatives $f_{x_1}, \ldots, f_{x_n}$ of a scalar field $f : D \subseteq \mathbb{R}^n \to \mathbb{R}$, D open, are themselves scalar fields again, $f_{x_1}, , \ldots, f_{x_n} : D \subseteq \mathbb{R}^n \to \mathbb{R}$. If these scalar fields are continuous, we call f **continuous partial differentiable**. And if these partial derivatives are in turn partially differentiable, then we can determine the **second partial derivatives of** f, for which we write

$$\partial_{x_j}\partial_{x_i} f(x) = \frac{\partial^2 f}{\partial x_j \partial x_i}(x) = \partial_j \partial_i f(x) = f_{x_i x_j}(x).$$

We will only very rarely need *higher* partial derivatives. But it is quite obvious, how to proceed now, or what is meant with the **k-th partial** derivatives and in general under **higher partial derivatives** of f. If the k-th partial derivatives are all continuous, then we call f understandably **k-times continuous partial differentiable** or briefly a **C^k-function**, i. e, $f \in C^k(D)$ with $k \in \mathbb{N}_0 \cup \{\infty\}$, where:

- $C^0(D) = \{f : D \subseteq \mathbb{R}^n \to \mathbb{R} \mid f$ is continuous$\}$,
- $C^k(D) = \{f : D \subseteq \mathbb{R}^n \to \mathbb{R} \mid f$ is k-times continuously partial differentiable$\}$, $k \in \mathbb{N}$,
- $C^\infty(D) = \{f : D \subseteq \mathbb{R}^n \to \mathbb{R} \mid f$ is continuously partially differentiable arbitrarily many times$\}$.

Example 47.3 The scalar field $f : \mathbb{R}^2 \to \mathbb{R},\ f(x, y) = 3x^2y^2$ has the first and second partial derivatives

$$f_x(x, y) = 6xy^2, \quad f_y(x, y) = 6x^2y$$

$$f_{xx}(x, y) = 6y^2, \quad f_{xy}(x, y) = 12xy, \quad f_{yx}(x, y) = 12xy, \quad f_{yy}(x, y) = 6x^2. \quad \blacksquare$$

In this example, the mixed second partial derivatives are equal: $f_{xy}(x, y) = 12xy = f_{yx}(x, y)$. This is not a coincidence, but the content of the following important theorem:

Schwarz's Theorem

If $f : D \subseteq \mathbb{R}^n \to \mathbb{R}$ is a scalar field and $f \in C^2(D)$, then for all $i,\ j = 1, \ldots, n$:

$$f_{x_i x_j} = f_{x_j x_i}.$$

It is not at all easy to find an example of a function f with various mixed second partial derivatives. According to Schwarz's Theorem, these second partial derivatives of such a function should not be continuous. Such a function is for example

$$f : \mathbb{R}^2 \to \mathbb{R},\ f(x, y) = \begin{cases} \frac{xy(x^2-y^2)}{x^2+y^2} & (x, y) \neq (0, 0) \\ 0 & (x, y) = (0, 0) \end{cases}$$

With a little effort, you can prove that $f_{xy}(0, 0) = -1$ and $f_{yx}(0, 0) = 1$ is valid for this f. But such examples are rather *exceptions*, in all further examples we will have to deal with the conditions of Schwarz's Theorem will be fulfilled.

The Hessian Matrix

If $f : D \subseteq \mathbb{R}^n \to \mathbb{R}$ is twice partially differentiable, then we call the $n \times n$-matrix

$$H_f(\boldsymbol{x}) = \begin{pmatrix} f_{x_1x_1}(\boldsymbol{x}) & \dots & f_{x_1x_n}(\boldsymbol{x}) \\ \vdots & & \vdots \\ f_{x_nx_1}(\boldsymbol{x}) & \dots & f_{x_nx_n}(\boldsymbol{x}) \end{pmatrix}$$

the **Hessian matrix** of f in $\boldsymbol{x}$. The Hessian matrix is symmetric if $f \in C^2(D)$.

Example 47.4

- For each $\boldsymbol{a} \in \mathbb{R}^n$ the scalar field $f : \mathbb{R}^n \to \mathbb{R},\ f(\boldsymbol{x}) = \boldsymbol{a}^\top \boldsymbol{x}$ has the gradient $\nabla f(\boldsymbol{x}) = \boldsymbol{a}$ and thus the Hessian matrix

$$H_f(\boldsymbol{x}) = 0\,.$$

- For each $A \in \mathbb{R}^{n\times n}$ the scalar field $f : \mathbb{R}^n \to \mathbb{R},\ f(\boldsymbol{x}) = \boldsymbol{x}^\top A\,\boldsymbol{x}$ has the gradient $\nabla f(\boldsymbol{x}) = (A^\top + A)\,\boldsymbol{x}$ and thus the Hessian matrix

$$H_f(\boldsymbol{x}) = A^\top + A \ \text{ or } \ 2\,A\,, \ \text{ if } A \text{ is symmetric.}$$

- If $f : \mathbb{R}^2 \to \mathbb{R},\ f(x, y) = x^2 + y\sin(x)$, then we have:

$$\nabla f(x, y) = \begin{pmatrix} 2x + y\cos(x) \\ \sin(x) \end{pmatrix} \ \text{ and thus } \ H_f(x, y) = \begin{pmatrix} 2 - y\sin(x) & \cos(x) \\ \cos(x) & 0 \end{pmatrix}.$$

- For $f : \mathbb{R}^n \setminus \{0\} \to \mathbb{R}$, $f(\boldsymbol{x}) = \|\boldsymbol{x}\|$ we obtain:

$$\nabla f(\boldsymbol{x}) = \frac{1}{\|\boldsymbol{x}\|}\boldsymbol{x} \text{ and hence } H_f(\boldsymbol{x}) = \frac{1}{\|\boldsymbol{x}\|}E_n - \frac{1}{\|\boldsymbol{x}\|^3}\boldsymbol{x}\boldsymbol{x}^\top,$$

where E_n denotes the $n \times n$-unit matrix. ■

We conclude this section with three remarks that point to future problems.

Remarks

1. Gradient and Hessian matrix of a scalar field will play the roles of the first and second derivatives of a real function in a variable: The zeros of the gradient will be the candidates for extremal sites; using the Hessian matrix, we will be able to decide in many cases whether the candidates are indeed extremal sites.
2. The partial derivative of a scalar field is not the *derivative* of the scalar field. Unfortunately, in the case of functions of several variables, the notion of derivative is not quite as simple as in the one-dimensional case. Here one considers not only the partial differentiability but also the *total differentiability*. More about this in Chap. 51.
3. A **partial differential equation** is an equation in which, in addition to the function f sought in several variables, partial derivatives of f and the variables of the function f can appear, e.g. the **Laplace equation** $f_{xx} + f_{yy} = 0$ is such a partial differential equation. While determining solutions of such partial differential equations is usually a very difficult undertaking, it is very easy to check whether a given function is a solution of this partial differential equation. It is for instance $f(x, y) = x^2 - y^2$ a solution of Laplace's equation. Note the exercises for more examples.

47.3 The Jacobian Matrix

So far we have explained the partial differentiability only for scalar fields $f : D \subseteq \mathbb{R}^n \to \mathbb{R}$. We now simply transfer the (*k-times* continuous) partial differentiability *component-wise* on vector-valued functions in multiple variables. We call a function

$$f : D \subseteq \mathbb{R}^n \to \mathbb{R}^m \text{ with } \boldsymbol{x} = (x_1, \ldots, x_n) \mapsto \begin{pmatrix} f_1(x_1, \ldots, x_n) \\ \vdots \\ f_m(x_1, \ldots, x_n) \end{pmatrix}$$

in $\boldsymbol{a} \in D$ or on D **partially differentiable** or **k-times continuously partially differentiable** for $k \in \mathbb{N}_0 \cup \{\infty\}$ if all component functions $f_1, \ldots, f_m$ are in $\boldsymbol{a} \in D$ or on

D partially differentiable or k-times continuously differentiable; this is then also briefly referred to as a C^k-**function** and one writes $f \in C^k(D)$.

If $f : D \subseteq \mathbb{R}^n \to \mathbb{R}^m$ is on D partially differentiable, then there exist the $m\,n$ partial derivatives $\frac{\partial f_i}{\partial x_j}$ for $1 \leq i \leq m,\ 1 \leq j \leq n$. These partial derivatives are summarized in the **Jacobian matrix** or **functional matrix** $Df(\boldsymbol{x})$ which is the $m \times n$-matrix

$$Df(\boldsymbol{x}) = \left(\frac{\partial f_i}{\partial x_j}(\boldsymbol{x})\right)_{ij} = \begin{pmatrix} \frac{\partial f_1}{\partial x_1}(\boldsymbol{x}) & \dots & \frac{\partial f_1}{\partial x_n}(\boldsymbol{x}) \\ \vdots & & \vdots \\ \frac{\partial f_m}{\partial x_1}(\boldsymbol{x}) & \dots & \frac{\partial f_m}{\partial x_n}(\boldsymbol{x}) \end{pmatrix} = \begin{pmatrix} \nabla f_1(\boldsymbol{x})^\top \\ \vdots \\ \nabla f_m(\boldsymbol{x})^\top \end{pmatrix}.$$

The lines of $Df(\boldsymbol{x})$ are, in order, the transposed gradients of the scalar fields $f_1, \dots, f_m$. In particular, in the case $m = 1$

$$Df(\boldsymbol{x}) = \nabla f(\boldsymbol{x})^\top.$$

Besides the notation $Df(\boldsymbol{x})$ the notation $J_f(\boldsymbol{x})$ for the Jacobian matrix is also common.

Example 47.5

- For $f : \mathbb{R}^2 \to \mathbb{R}^3,\ f(x, y) = (x^2 y,\ \sin(y),\ x^2 + y^2)^\top$ the Jacobian matrix is

$$Df(x, y) = \begin{pmatrix} 2xy & x^2 \\ 0 & \cos(y) \\ 2x & 2y \end{pmatrix}.$$

- For $f : \mathbb{R}^n \to \mathbb{R}^m,\ f(x) = A\,\boldsymbol{x}$ with $A \in \mathbb{R}^{m \times n}$ the Jacobian matrix is $Df(\boldsymbol{x}) = A$.
- Is $g : D \subseteq \mathbb{R}^n \to \mathbb{R},\ g \in C^2(D)$ is a twice continuously differentiable scalar field, then $f = \nabla g : D \subseteq \mathbb{R}^n \to \mathbb{R}^n$ with

$$f(x_1, \dots, x_n) = \nabla g(x_1, \dots, x_n) = \begin{pmatrix} g_{x_1}(x_1, \dots, x_n) \\ \vdots \\ g_{x_n}(x_1, \dots, x_n) \end{pmatrix}$$

a vector field with the Jacobian matrix

$$Df(\boldsymbol{x}) = \begin{pmatrix} g_{x_1 x_1}(\boldsymbol{x}) & \dots & g_{x_1 x_n}(\boldsymbol{x}) \\ \vdots & & \vdots \\ g_{x_n x_1}(\boldsymbol{x}) & \dots & g_{x_n x_n}(\boldsymbol{x}) \end{pmatrix}.$$

$Df(\boldsymbol{x}) = H_g(\boldsymbol{x})$ is the Hessian matrix of g. ■

We conclude by noting the following computational rules for the Jacobian matrix:

Calculation Rules for the Jacobian Matrix
If $f, g : D \subseteq \mathbb{R}^n \to \mathbb{R}^m$ are partially differentiable, then for all $\boldsymbol{x} \in D$:

- $D(f+g)(\boldsymbol{x}) = Df(\boldsymbol{x}) + Dg(\boldsymbol{x})$, **(additivity)**
- $D(\lambda f)(\boldsymbol{x}) = \lambda\, Df(\boldsymbol{x})$ for all $\lambda \in \mathbb{R}$ **(homogeneity)**
- $D(f(\boldsymbol{x})^\top g(\boldsymbol{x})) = f(\boldsymbol{x})^\top Dg(\boldsymbol{x}) + g(\boldsymbol{x})^\top Df(\boldsymbol{x})$. **(product rule)**

If $f : D \subseteq \mathbb{R}^n \to \mathbb{R}^m$ and $g : D' \subseteq \mathbb{R}^l \to \mathbb{R}^n$ with $g(D') \subseteq D$ are partially differentiable, then we also have:

- $D(f \circ g)(\boldsymbol{x}) = Df(g(\boldsymbol{x}))\, Dg(\boldsymbol{x})$. **(chain rule)**

This chain rule is component-wise for $h = f \circ g$:

$$\frac{\partial h_i}{\partial x_j}(\boldsymbol{x}) = \sum_{k=1}^{n} \frac{\partial f_i}{\partial x_k}(g(\boldsymbol{x}))\, \frac{\partial g_k}{\partial x_j}(\boldsymbol{x}) \ \text{ for } 1 \le i \le m,\ 1 \le j \le l\,.$$

We still formulate the chain rule for the following special cases, which are of great importance in applications:

- $n = 1$ and $m = 1$: If $f : D \subseteq \mathbb{R} \to \mathbb{R}$ is a function and $g : D' \subseteq \mathbb{R}^l \to \mathbb{R}$ a scalar field, then $h = f \circ g : D' \subseteq \mathbb{R}^l \to \mathbb{R}$ is a scalar field, and we have

$$\nabla(f \circ g)(\boldsymbol{x}) = f'(g(\boldsymbol{x}))\, \nabla g(\boldsymbol{x})\,, \text{ i. e. } \frac{\partial h}{\partial x_j}(\boldsymbol{x}) = f'(g(\boldsymbol{x}))\, \frac{\partial g}{\partial x_j}(\boldsymbol{x})\,,\ j = 1, \ldots, l\,.$$

- $l = 1$ and $m = 1$: If $f : D \subseteq \mathbb{R}^n \to \mathbb{R}$ is a scalar field and $g : [a, b] \to \mathbb{R}^n$ a curve, then the composition $h = f \circ g : [a, b] \to \mathbb{R}$ is a function in one variable t, and we have

$$\dot{h}(t) = \nabla(f(g(t))^\top \dot{g}(t) = \sum_{k=1}^{n} \frac{\partial f}{\partial x_k}(g(t))\, \dot{g}_k(t)\,.$$

Example 47.6

- We consider the functions $f : \mathbb{R}_{>0} \to \mathbb{R}$ with $f(x) = \ln(x)$ and $g : \mathbb{R}^l \setminus \{0\} \to \mathbb{R}$ with $g(\boldsymbol{x}) = \|\boldsymbol{x}\| = \sqrt{\sum_{i=1}^{l} x_i^2}$. Then

$$h = f \circ g : \mathbb{R}^l \setminus \{0\} \to \mathbb{R},\ h(\boldsymbol{x}) = \ln(\|\boldsymbol{x}\|)$$

is a scalar field. Because of $\frac{\partial g}{\partial x_i}(\boldsymbol{x}) = \frac{x_i}{\|\boldsymbol{x}\|}$ we obtain the gradient $\nabla g(\boldsymbol{x}) = \frac{\boldsymbol{x}}{\|\boldsymbol{x}\|}$. So the chain rule provides:

$$\nabla h(\boldsymbol{x}) = f'(g(\boldsymbol{x}))\, \nabla g(\boldsymbol{x}) = \frac{1}{\|\boldsymbol{x}\|} \frac{\boldsymbol{x}}{\|\boldsymbol{x}\|} = \frac{\boldsymbol{x}}{\|\boldsymbol{x}\|^2}.$$

The function h is **radially symmetric** i.e. from $\|\boldsymbol{x}\| = \|\boldsymbol{y}\|$ follows $h(\boldsymbol{x}) = h(\boldsymbol{y})$.
- Let $g : [0, 2\pi] \to \mathbb{R}^2,\ g(t) = (\cos(t), \sin(t))^\top$ be a plane curve and furthermore $f : \mathbb{R}^2 \to \mathbb{R},\ f(x, y) = x^2 + xy + y^2$ a scalar field. Then

$$h = f \circ g : [0, 2\pi] \to \mathbb{R},\ h(t) = 1 + \cos(t)\sin(t)$$

is a real function of one variable. Because of $\nabla f(x, y) = (2x + y, 2y + x)^\top$ and $\dot{g}(t) = (-\sin(t), \cos(t))^\top$, we have by the chain rule:

$$\dot{h}(t) = \nabla f(g(t))^\top \dot{g}(t) = \begin{pmatrix} 2\cos(t) + \sin(t) \\ 2\sin(t) + \cos(t) \end{pmatrix}^\top \begin{pmatrix} -\sin(t) \\ \cos(t) \end{pmatrix}$$

$$= -2\sin(t)\cos(t) - \sin^2(t) + 2\sin(t)\cos(t) + \cos^2(t) = \cos^2(t) - \sin^2(t).$$

■

47.4 Exercises

47.1 Prove the statements about the gradient in the note box in Sect. 47.1.

47.2 Calculate the first and second partial derivatives, the gradient, and the directional derivative in the direction of $(1, -1)^\top$ of the following functions from $\mathbb{R}^2$ to $\mathbb{R}$:

(a) $f(x, y) = 2x^2 + 3xy + y$,
(b) $g(x, y) = xy^2 + y\,\mathrm{e}^{-xy}$,
(c) $h(x, y) = x \sin y$.

47.3 Show that the function $f\colon \mathbb{R}^2 \to \mathbb{R}$ given by $f(x, y) = xy + x\ln(y/x)$ for $x, y > 0$ satisfies the following equation:

$$x\partial_x f + y\partial_y f = xy + f\,.$$

47.4 Calculate for the following functions $f\colon \mathbb{R}^2 \to \mathbb{R}$ the gradient:

(a) $f(x, y) = 2x + 3y$,
(b) $f(x, y) = \sqrt{x^2 + y^2}$,
(c) $f(x, y) = \ln(1 + x^2y^4)$,
(d) $f(x, y) = 8 - 3x \sin y$.

47.5 For the composition of the following functions, calculate the gradient or first derivative using the chain rule. Check the result by deriving the composition directly:

(a) $f(x, y) = f_2(f_1(x, y))$ with $f_1(x, y) = xy$ and $f_2(t) = \mathrm{e}^t$,
(b) $h(t) = h_2(h_1(t))$ with $h_1(t) = (\cos t, \sin t)$ and $h_2(x, y) = x^2 + y^2$.

47.6 Justify why there cannot be a twice continuously differentiable function $f : \mathbb{R}^2 \to \mathbb{R}$ which is $\frac{\partial}{\partial x} f(x, y) = x^2y$ and $\frac{\partial}{\partial y} f(x, y) = x^3$.

47.7 Verify that for $f\colon \mathbb{R}^2 \setminus \{(0, 0)\} \to \mathbb{R}$ with $f(x, y) = \frac{x-y}{x^2+y^2}$ we have the identity $\partial_x\partial_y f = \partial_y\partial_x f$.

47.8 Calculate the Jacobian matrices of the functions

(a) $f(x, y, z) = \begin{pmatrix} x + y \\ x^2z \end{pmatrix}$, (c) $f(x, y, z) = \begin{pmatrix} \mathrm{e}^{xy} + \cos^2 z \\ xyz - \mathrm{e}^{-z} \\ \sinh(xz) + y^2 \end{pmatrix}$,

(b) $f(x, y, z) = \begin{pmatrix} z + x^2 \\ xy \\ 2y \end{pmatrix}$, (d) $f(x, y, z) = x^2 + yz + 2$.

47.9 Let it be $f\colon \mathbb{R}^3 \to \mathbb{R}^3$ defined by

$$f(x, y, z) = (y + \exp z, z + \exp x, x + \exp y)^\top.$$

Determine the Jacobian matrix of the inverse function of f at the point $(1 + \mathrm{e}, 2, \mathrm{e})^\top$.

47.10 Let $f : \mathbb{R}^2 \to \mathbb{R}^2$ defined by

$$f(x, y) = (x + 2y^2, y - \sin x)^\top .$$

Determine the Jacobian matrix of the inverse function of f at the point $(2 + \frac{\pi}{2}, 0)^\top$.

47.11 Calculate the Jacobian matrix for $g = g_2 \circ g_1$ using the chain rule and check the result by deriving directly. Let $g_1 : \mathbb{R}^3 \to \mathbb{R}^2$ and $g_2 : \mathbb{R}^2 \to \mathbb{R}^3$ are given by

$$g_1(x, y, z) = (x + y, y + z)^\top \quad \text{and} \quad g_2(u, v) = (uv, u + v, \sin(u + v))^\top .$$

47.12 Let $\boldsymbol{v}_1 : \mathbb{R}^3 \to \mathbb{R}^2$ and $\boldsymbol{v}_2 : \mathbb{R}^2 \to \mathbb{R}^3$ defined by

$$\boldsymbol{v}_1(x, y, z) = (x + y, y + z)^\top \text{ and } \boldsymbol{v}_2(x, y) = (xy, x + y, \sin(x + y))^\top .$$

Calculate the Jacobian matrix of the vector field $\boldsymbol{v} = \boldsymbol{v}_2 \circ \boldsymbol{v}_1$ at the position $\boldsymbol{x} = (x, y, z)$.

47.13 Let $f_1 : \mathbb{R}^2 \to \mathbb{R}^4$ defined by

$$f_1(x_1, x_2) = (x_1 x_2, x_2^2, \sin^2 x_2^3, x_1)^\top$$

and $f_2 : \mathbb{R}^4 \to \mathbb{R}^3$ by

$$f_2(x_1, x_2, x_3, x_4) = (\arctan(x_1 x_2), 5 \cos x_4, \ln(1 + x_1^2 + x_2^2))^\top .$$

Calculate the Jacobian matrix of the vector field $f = f_2 \circ f_1$ at the position $(0, 0)$.

47.14 Show that the function

$$u(x, y, t) = \frac{1}{t} \exp\left(-\frac{x^2 + y^2}{4t}\right)$$

for $(x, y) \in \mathbb{R}^2$ and $t > 0$ is a solution of the partial DGL

$$u_t = u_{xx} + u_{yy}$$

(the *heat conduction equation*).

47.15 Determine $\frac{du}{dt}$ for $u(x, y, z) = e^{xy^2} + z$ with $x = \cos t$, $y = \sin t$, $z = t^2$.

48 Applications of Partial Derivatives

In Chap. 28 we mentioned applications of the differentiation of one variable. We now do the same with the (partial) differentiation of functions of several variables: We describe the *(multidimensional) Newton method* for the determination of zeros of vector fields and the *Taylor expansion* for scalar fields, in order to approximate given scalar fields locally by a *tangential plane* or a *parabola.* For this purpose, we do not have to learn anything new in terms of content, but only compile previously created knowledge.

48.1 The (Multidimensional) Newton Method

We briefly recall Newton's (one-dimensional) method for the approximate determination of a zero point x^* of a twice continuously differentiable function $f : I \subseteq \mathbb{R} \to \mathbb{R}$. Starting from an initial value x_0, one forms the sequence members

$$x_{k+1} = x_k - \frac{f(x_k)}{f'(x_k)} \quad \text{for } k = 0, 1, 2, \ldots .$$

According to the recipe in Sect. 28.1 we know that this sequence *quickly* against the sought zero x^* converges if only the initial value x_0 is in the near of x^*. The immediate generalization to the multidimensional case is now:

If $f : D \subseteq \mathbb{R}^n \to \mathbb{R}^n$ is a vector field and $\boldsymbol{x}^* \in D$ a **zero** of f, i.e. $f(\boldsymbol{x}^*) = 0$ then, starting from a starting vector $\boldsymbol{x}_0 \in D$ form further sequence members by the iteration

$$\boldsymbol{x}_{k+1} = \boldsymbol{x}_k - (Df(\boldsymbol{x}_k))^{-1} f(\boldsymbol{x}_k) \quad \text{for } k = 0, 1, 2, \ldots ,$$

C. Karpfinger, *Calculus and Linear Algebra in Recipes*,
https://doi.org/10.1007/978-3-662-65458-3_48

where $Df(\boldsymbol{x}_k)$ is the Jacobian matrix of the function f at the point $\boldsymbol{x}_k$. Thereby one is carried by the hope that with growing k the sequence members $\boldsymbol{x}_k$ approximate a sought zero $\boldsymbol{x}^*$.

Before describing this Newton method in general, we draw attention to two problematic issues:

- **Avoid the explicit computation of** $(Df(\boldsymbol{x}_k))^{-1}$**:** At each iteration (theoretically) the inverse $(Df(\boldsymbol{x}_k))^{-1}$ of the Jacobian matrix has to be determined. Inverting matrices, however, is costly and should be avoided, especially when implemented on a computer. Fortunately, this can also be done quite easily. We decompose the iteration rule into two parts: We first compute Δ_k as the solution of the linear system of equations $Df(\boldsymbol{x}_k)\Delta\boldsymbol{x}_k = f(\boldsymbol{x}_k)$ with the extended coefficient matrix $(Df(\boldsymbol{x}_k) \mid f(\boldsymbol{x}_k))$ (e.g. with the numerically stable LR-decomposition) and then obtain $\boldsymbol{x}_{k+1}$ as the difference of $\boldsymbol{x}_k$ with this $\Delta\boldsymbol{x}_k$, in short:

$$\boldsymbol{x}_{k+1} = \boldsymbol{x}_k - \Delta\boldsymbol{x}_k\,, \text{ where } Df(\boldsymbol{x}_k)\Delta\boldsymbol{x}_k = f(\boldsymbol{x}_k) \text{ for } k = 0, 1, 2, \ldots .$$

- **Termination criteria:** As in the one-dimensional case, one should terminate the iteration if a sought zero is $\boldsymbol{x}^*$ sufficiently well approximated or if no convergence is expected. While as in the one-dimensional case, a null is then sufficiently well approximated, i.e. $\|\boldsymbol{x}_k - \boldsymbol{x}^*\| < \text{tol}$ for a given tol > 0, if two successive iterates $\boldsymbol{x}_k$ and $\boldsymbol{x}_{k+1}$ are sufficiently close to each other, i.e. $\|\boldsymbol{x}_k - \boldsymbol{x}_{k+1}\| < \text{tol}$, unfortunately the second termination criterion cannot be salvaged from the one-dimensional case: rather, one terminates the iteration if the **natural monotonicity test** fails, i.e. STOP if

$$\|Df(\boldsymbol{x}_k)^{-1}f(\boldsymbol{x}_{k+1})\| > \|Df(\boldsymbol{x}_k)^{-1}f(\boldsymbol{x}_k)\| \;(= \|\Delta\boldsymbol{x}_k\|)\,.$$

We record this procedure in the form of a recipe:

Recipe: The (Multidimensional) Newton method

Given is a C^2-function $f : D \subseteq \mathbb{R}^n \to \mathbb{R}^n$, D open and convex. For the approximate determination of a zero $\boldsymbol{x}^* \in D$ of f proceed as follows after choosing a tolerance limit tol > 0:

(1) Choose a $\boldsymbol{x}_0 \in D$ in the near of $\boldsymbol{x}^*$.
(2) As long as $\|\boldsymbol{x}_{k+1} - \boldsymbol{x}_k\| \geq \text{tol}$ and $\|Df(\boldsymbol{x}_k)^{-1}f(\boldsymbol{x}_{k+1})\| \leq \|Df(\boldsymbol{x}_k)^{-1}f(\boldsymbol{x}_k)\|$ determine

$$\boldsymbol{x}_{k+1} = \boldsymbol{x}_k - \Delta\boldsymbol{x}_k \text{ mit } Df(\boldsymbol{x}_k)\Delta\boldsymbol{x}_k = f(\boldsymbol{x}_k)\,,\ k = 0, 1, 2, \ldots .$$

(continued)

If det $Df(\boldsymbol{x}^*) \neq 0$ then there exists a neighborhood U of $\boldsymbol{x}^*$ such that the iteration

$$\boldsymbol{x}_0 \in U \text{ and } \boldsymbol{x}_{k+1} = \boldsymbol{x}_k - (Df(\boldsymbol{x}_k))^{-1} f(\boldsymbol{x}_k)\,,\ k = 0,\,1,\,2,\,\ldots$$

converges for each $\boldsymbol{x}_0$ from U against the zero $\boldsymbol{x}^*$. The convergence is **quadratic**, i.e.

$$\|\boldsymbol{x}_{k+1} - \boldsymbol{x}^*\| = C\,\|\boldsymbol{x}_k - \boldsymbol{x}^*\|^2 \text{ for a } C \in \mathbb{R}\,.$$

It is not difficult to generalize the MATLAB function for the (one-dimensional) Newton method from Exercise 28.8 to the multidimensional method. You should definitely do this for practice and check the following example as well as the examples in the exercises.

Remarks

1. When calculating with pencil and paper, the calculation effort for large n for the calculation of $(Df(\boldsymbol{x}_k))^{-1}$ is very high. It is recommended to proceed from a step k on, with firm $(Df(\boldsymbol{x}_k))^{-1}$ or to update it after a few steps. However, the convergence is then no longer quadratic. This is called the *simplified Newton method.*
2. For the choice of the starting vector $\boldsymbol{x}_0$ there is no systematic procedure. The starting vector should be close to the searched solution. Here one often has to rely on estimations and a lot of background information of the given problem.

Example 48.1 We use Newton's method to determine an approximate solution to the nonlinear system of equations

$$x = 0.1\,x^2 + \sin(y)$$

$$y = \cos(x) + 0.1\,y^2\,.$$

The solutions are the zeros of the function

$$f(x, y) = \begin{pmatrix} 0.1\,x^2 + \sin(y) - x \\ \cos(x) + 0.1\,y^2 - y \end{pmatrix} = \begin{pmatrix} 0 \\ 0 \end{pmatrix}.$$

We have

$$Df(x, y) = \begin{pmatrix} 0.2\,x - 1 & \cos(y) \\ -\sin(x) & 0.2\,y - 1 \end{pmatrix},$$

so

$$(Df(x,y))^{-1} = \frac{1}{\Delta(x,y)} \begin{pmatrix} 0.2\,y-1 & -\cos(y) \\ \sin(x) & 0.2\,x-1 \end{pmatrix}$$

with $\Delta(x, y) = (0.2x - 1)(0.2y - 1) + \sin(x)\cos(y)$. So the general step in Newton's method is

$$\begin{pmatrix} x_{k+1} \\ y_{k+1} \end{pmatrix} = \begin{pmatrix} x_k \\ y_k \end{pmatrix} - \frac{1}{\Delta(x_k, y_k)} \begin{pmatrix} 0.2\,y_k-1 & -\cos(y_k) \\ \sin(x_k) & 0.2\,x_k-1 \end{pmatrix} \begin{pmatrix} 0.1\,x_k^2+\sin(y_k)-x_k \\ \cos(x_k)+0.1\,y_k^2-y_k \end{pmatrix}.$$

With the initial values $x_0 = 0.8$ and $y_0 = 0.8$ we obtain:

k	x_k	y_k
0	0.80000000000	0.80000000000
1	0.764296288278366	0.783713076688571
2	0.764070576897057	0.783396762842286
3	0.764070550812738	0.783396774300478
4	0.764070550812738	0.783396774300478

■

48.2 Taylor Development

As in the one-dimensional case, we now approximate a function by a *Taylor polynomial.* Because we consider functions in several variables, the corresponding Taylor polynomials are also polynomials in several variables. We already know such polynomials from the Chap. 41 on quadrics.

48.2.1 The Zeroth, First and Second Taylor Polynomial

The *m-th* Taylor polynomial in the development point $a \in I$ of a (m-times differentiable) function $f : I \subseteq \mathbb{R} \to \mathbb{R},\ x \mapsto f(x)$ in a variable is according to Sect. 28.2:

$$T_{m,f,a}(x) = f(a) + f'(a)(x-a) + \frac{1}{2}f''(a)(x-a)^2 + \cdots + \frac{1}{m!}f^{(m)}(a)(x-a)^m\,.$$

The role of the first and second derivatives f' and f'' of a scalar field $f : D \subseteq \mathbb{R}^n \to \mathbb{R}$ is assumed by the gradient ∇f and the Hessian matrix H_f. This gives us the first *Taylor polynomials* for a scalar field:

Zeroth, First and Second Taylor Polynomials of a Scalar Field
Given is an open and convex set $D \subseteq \mathbb{R}^n$ and a point $\boldsymbol{a} \in D$. If $f : D \subseteq \mathbb{R}^n \to \mathbb{R}$ is a twice continuously differentiable scalar field, then we call

- $T_{0,f,\boldsymbol{a}}(\boldsymbol{x}) = f(\boldsymbol{a})$ the **zeroth** and
- $T_{1,f,\boldsymbol{a}}(\boldsymbol{x}) = f(\boldsymbol{a}) + \nabla f(\boldsymbol{a})^\top(\boldsymbol{x} - \boldsymbol{a})$ the **first** and
- $T_{2,f,\boldsymbol{a}}(\boldsymbol{x}) = f(\boldsymbol{a}) + \nabla f(\boldsymbol{a})^\top(\boldsymbol{x} - \boldsymbol{a}) + \frac{1}{2}(\boldsymbol{x} - \boldsymbol{a})^\top H_f(\boldsymbol{a})(\boldsymbol{x} - \boldsymbol{a})$ the **second Taylor polynomial** at the **development point** $\boldsymbol{a} \in D$.

Note that, as in the one-dimensional case, we have

$$f(\boldsymbol{a}) = T_{2,f,\boldsymbol{a}}(\boldsymbol{a}), \ \nabla f(\boldsymbol{a}) = \nabla T_{2,f,\boldsymbol{a}}(\boldsymbol{a}), \ H_f(\boldsymbol{a}) = H_{T_{2,f,\boldsymbol{a}}}(\boldsymbol{a}),$$

so that we can write the Taylor polynomial again as *approximation* of the scalar field f around the development point $\boldsymbol{a}$. Of course, we can also define for $m > 2$ a *m-th* Taylor polynomial to scalar fields, the formalism is just a bit more complicated. We do this following the examples and explanations below.

Example 48.2

- We determine the first three Taylor polynomials of the scalar field $f : \mathbb{R}^2 \to \mathbb{R}$ with $f(x, y) = x^2 - y + xy$ at the development point $\boldsymbol{a} = (1, 0)^\top \in \mathbb{R}^2$. For this we determine in advance all partial derivatives of 1st and 2nd order:

$$f_x(x, y) = 2x + y, \ f_y(x, y) = -1 + x,$$

$$f_{xx}(x, y) = 2, \ f_{xy}(x, y) = 1 = f_{yx}(x, y), \ f_{yy}(x, y) = 0.$$

Thus the gradient and Hessian matrix are in $\boldsymbol{a} = (1, 0)^\top$:

$$\nabla f(1, 0) = \begin{pmatrix} 2 \\ 0 \end{pmatrix} \text{ and } H_f(1, 0) = \begin{pmatrix} 2 & 1 \\ 1 & 0 \end{pmatrix}.$$

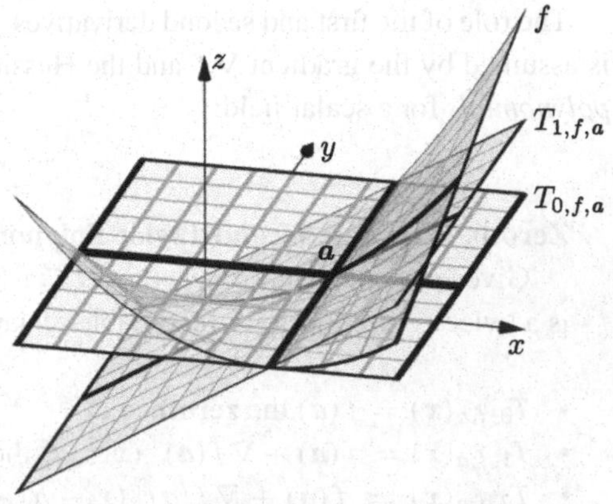

Fig. 48.1 The function and Taylor polynomials

Thus, we obtain for the first three Taylor polynomials:

$$T_{0,f,a}(x,y) = f(1,0) = 1\,,$$

$$T_{1,f,a}(x,y) = f(1,0) + \nabla f(1,0)^{\top}\begin{pmatrix} x-1 \\ y-0 \end{pmatrix} = 1 + 2(x-1) + 0(y-0) = 2x-1\,,$$

$$T_{2,f,a}(x,y) = f(1,0) + \nabla f(1,0)^{\top}\begin{pmatrix} x-1 \\ y-0 \end{pmatrix} + \frac{1}{2}(x-1, y-0)H_f(1,0)\begin{pmatrix} x-1 \\ y-0 \end{pmatrix}$$

$$= 2x - 1 + \frac{1}{2}(2(x-1)^2 + 2(x-1)y) = x^2 + xy - y\,.$$

We have the graphs of the three Taylor polynomials $T_{0,f,a}$, $T_{1,f,a}$ and $T_{2,f,a}$ in Fig. 48.1 drawn in.

- We determine the first three Taylor polynomials of the scalar field $f : \mathbb{R}^2 \to \mathbb{R}$ with $f(x,y) = x^3\,\mathrm{e}^y$ at the development point $\boldsymbol{a} = (1,0)^{\top} \in \mathbb{R}^2$. For this purpose we determine in advance all partial derivatives of 1st and 2nd order:

$$f_x(x,y) = 3\,x^2\,\mathrm{e}^y\,,\quad f_y(x,y) = x^3\,\mathrm{e}^y\,,$$

$$f_{xx}(x,y) = 6\,x\,\mathrm{e}^y\,,\quad f_{xy}(x,y) = 3\,x^2\,\mathrm{e}^y = f_{yx}(x,y)\,,\quad f_{yy}(x,y) = x^3\,\mathrm{e}^y\,.$$

Thus the gradient and Hessian matrix are in $\boldsymbol{a} = (1,0)^{\top}$:

$$\nabla f(1,0) = \begin{pmatrix} 3 \\ 1 \end{pmatrix} \quad \text{and} \quad H_f(\boldsymbol{a}) = \begin{pmatrix} 6 & 3 \\ 3 & 1 \end{pmatrix}.$$

So we get for the first three Taylor polynomials:

$$T_{0,f,\boldsymbol{a}}(x, y) = f(1, 0) = 1\,,$$
$$T_{1,f,\boldsymbol{a}}(x, y) = 1 + 3(x - 1) + 1(y - 0) = 3x + y - 2\,,$$
$$T_{2,f,\boldsymbol{a}}(x, y) = 1 + 3(x - 1) + 1(y - 0) + \frac{1}{2}(6(x - 1)^2 + 6(x - 1)y + y^2)$$
$$= 1 + 3(x - 1) + y + 3(x - 1)^2 + \frac{1}{2}y^2 + 3(x - 1)y\,. \quad \blacksquare$$

The graph of $T_{1,f,\boldsymbol{a}}$ is the **tangent plane** in $\boldsymbol{a}$ of f. An idea of this is possible in the case $n = 2$: The image of the function $T_{1,f,\boldsymbol{a}}$ is the set $\{z = f(\boldsymbol{a}) + \nabla f(\boldsymbol{a})^\top(\boldsymbol{x} - \boldsymbol{a}) \in \mathbb{R}^3 \mid \boldsymbol{x} \in D\}$. The set of all these points is given by a plane equation of the form

$$r\,x + s\,y - z = t\,.$$

In the first example, the plane equation of the tangent plane to f at the point is $\boldsymbol{a} = (1, 0)^\top$ so that $2x - z = 1$ and in the second example we obtain the equation for the tangent plane $3x + y - z = 2$.

In Fig. 48.2 we see the graph of the scalar field $f : \mathbb{R}^2 \to \mathbb{R}$ with $f(x, y) = x^3\,\mathrm{e}^y$, the tangent plane at the point $(1, 0)^\top$ and the quadric at the point $(1, 0)^\top$.

If you create the plot with MATLAB, you can rotate the graph with its Taylor approximations in all directions.

48.2.2 The General Taylor polynomial

Sometimes one needs Taylor polynomials higher than 2nd order. In order to provide a *Taylor formula* for this situation as well, we briefly recall the motivation of the Taylor expansion: The Taylor polynomial *m-th* order is supposed to be the polynomial with the same *k-th* partial derivatives at the development point a as the function f for all $k \le m$. Moreover, the Taylor polynomial of order m is of course also a polynomial of degree m. Such a general polynomial of degree m in the development point $\boldsymbol{a} = (a_1, \ldots, a_n)^\top \in \mathbb{R}^n$ reads:

$$p_m(x_1, \ldots, x_n) = \sum_{k_1+\ldots+k_n \le m} a_{k_1,\ldots,k_n}(x_1 - a_1)^{k_1} \ldots (x_n - a_n)^{k_n}\,.$$

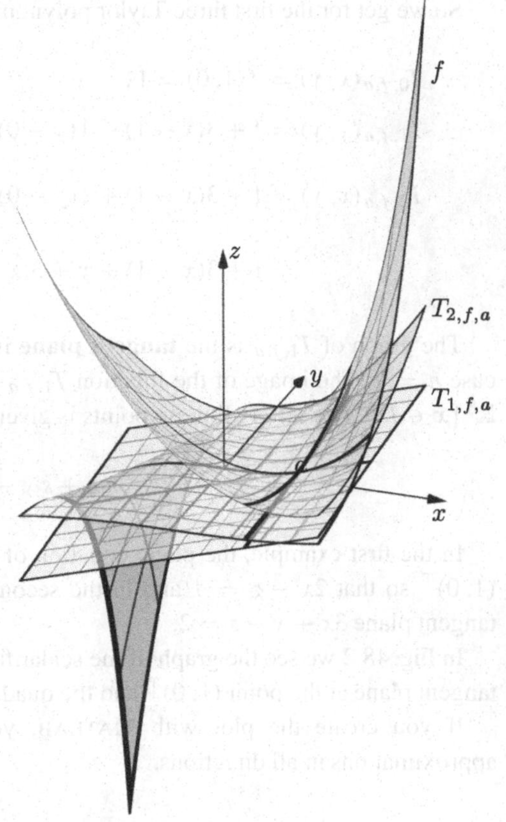

Fig. 48.2 The function and Taylor polynomials

Example 48.3 The general polynomial of degree 3 in the variables x and y and development point $\boldsymbol{a} = (1, 0)^\top$ is

$$\begin{aligned} p_3(x, y) = {} & a_{0,0} \\ & + a_{1,0}(x-1) + a_{0,1}y \\ & + a_{2,0}(x-1)^2 + a_{1,1}(x-1)y + a_{0,2}y^2 \\ & + a_{3,0}(x-1)^3 + a_{2,1}(x-1)^2y + a_{1,2}(x-1)y^2 + a_{0,3}y^3 \\ = {} & \sum_{k_1+k_2 \leq 3} a_{k_1,k_2}(x-1)^{k_1}y^{k_2}\,. \end{aligned}$$

■

The condition that the polynomial p_m has the same *k-th* partial derivatives in a as a given function f, yields a determining equation for the coefficients $a_{k_1,\dots,k_n}$, this one reads:

The *m*-th Taylor Polynomial

If $f : D \subseteq \mathbb{R}^n \to \mathbb{R}$ is a m-times continuously partially differentiable scalar field, then the *m-th* Taylor polynomial in a development point $\boldsymbol{a} = (a_1, \dots, a_n)^\top \in D$ is:

$$T_{m,f,\boldsymbol{a}}(\boldsymbol{x}) = \sum_{k_1+\ldots+k_n \le m} a_{k_1,\ldots,k_n} (x_1 - a_1)^{k_1} \ldots (x_n - a_n)^{k_n}$$

with

$$a_{k_1,\ldots,k_n} = \frac{1}{k_1! \ldots k_n!} \frac{\partial^{k_1+\ldots+k_n}}{\partial x_1^{k_1} \ldots \partial x_n^{k_n}} f(\boldsymbol{a}) .$$

Example 48.4 We determine the third Taylor polynomial $T_{3,f,\boldsymbol{a}}$ of the scalar field $f : \mathbb{R}^2 \to \mathbb{R}$ with $f(x, y) = x^3 \mathrm{e}^y$ at the development point $\boldsymbol{a} = (1, 0)^\top \in \mathbb{R}^2$ (cf. the Example 48.2). To do this, we determine in advance all the 1st, 2nd and 3rd order partial derivatives at the point $\boldsymbol{a}$, where for the polynomial of degree 3 according to Eq. 48.3, we only need the partial derivatives in bold type:

- $\boldsymbol{f_x(a) = 3x^2 e^y |_a = 3}$,
- $\boldsymbol{f_y(a) = x^3 e^y |_a = 1}$,
- $\boldsymbol{f_{xx}(a) = 6x e^y |_a = 6}$,
- $\boldsymbol{f_{xy}(a) = 3x^2 e^y |_a = 3}$,
- $f_{yx}(\boldsymbol{a}) = 3x^2 \mathrm{e}^y |_a = 3$,
- $\boldsymbol{f_{yy}(a) = x^3 e^y |_a = 1}$
- $\boldsymbol{f_{xx}(a) = 6 e^y |_a = 6}$,
- $\boldsymbol{f_{xxy}(a) = 6x e^y |_a = 6}$,
- $f_{xyx}(\boldsymbol{a}) = 6x \mathrm{e}^y |_a = 6$,
- $\boldsymbol{f_{xy}(a) = 3x^2 e^y |_a = 3}$,
- $f_{yxx}(\boldsymbol{a}) = 6x \mathrm{e}^y |_a = 6$,
- $f_{yxy}(\boldsymbol{a}) = 3x^2 \mathrm{e}^y |_a = 3$,
- $f_{yyx}(\boldsymbol{a}) = 3x^2 \mathrm{e}^y |_a = 3$,
- $\boldsymbol{f_{yy}(a) = x^3 e^y |_a = 1}$.

With $\boldsymbol{f(a) = 1}$ we thus obtain the Taylor polynomial of the third degree:

$$T_{3,f,\boldsymbol{a}}(x, y) = 1 + 3(x-1) + 1y + 3(x-1)^2 + 3(x-1)y + \frac{1}{2}y^2 + \\ + 1(x-1)^3 + 3(x-1)^2 y + \frac{3}{2}(x-1)y^2 + \frac{1}{6}y^3 .$$

■

The general formulation of the *Taylor formula* for a scalar field $f : D \subseteq \mathbb{R}^n \to \mathbb{R}$ in a development point $\boldsymbol{a} \in \mathbb{R}^n$ can be done succinctly and elegantly with the help of the following formalism: We substitute $\boldsymbol{x}$ by $\boldsymbol{a}+\boldsymbol{h}$ with $\boldsymbol{h} = (h_1, \ldots, h_n)^\top \in \mathbb{R}^n$. This turns $\boldsymbol{x}$ *lies in the near of* $\boldsymbol{a}$ to $\boldsymbol{h}$ *lies in the near of* 0. We now formally compute with the Nabla operator as follows:

$$\boldsymbol{h} \cdot \nabla = h_1 \frac{\partial}{\partial x_1} + \cdots + h_n \frac{\partial}{\partial x_n} = \sum_{i=1}^{n} h_i \frac{\partial}{\partial x_i}$$

and thus

$$\boldsymbol{h} \cdot \nabla f(\boldsymbol{x}) = h_1 \frac{\partial}{\partial x_1} f(\boldsymbol{x}) + \cdots + h_n \frac{\partial}{\partial x_n} f(\boldsymbol{x}) = \sum_{i=1}^{n} h_i \frac{\partial}{\partial x_i} f(\boldsymbol{x})$$

and analogously with powers of $\boldsymbol{h} \cdot \nabla$:

$$(\boldsymbol{h} \cdot \nabla)^k = \sum_{i_1,\ldots,i_k=1}^{n} h_{i_1} \cdots h_{i_k} \frac{\partial^k}{\partial x_{i_1} \cdots \partial x_{i_k}}$$

and thus

$$(\boldsymbol{h} \cdot \nabla)^k f(\boldsymbol{x}) = \sum_{i_1,\ldots,i_k=1}^{n} h_{i_1} \cdots h_{i_k} \frac{\partial^k}{\partial x_{i_1} \cdots \partial x_{i_k}} f(\boldsymbol{x}) .$$

This is simply remembered with $(\boldsymbol{h} \cdot \nabla)^k = \left(h_1 \frac{\partial}{\partial x_1} + \cdots + h_n \frac{\partial}{\partial x_n} \right)^k$. So now the Taylor formula is quite simple:

Taylor's Theorem—the Taylor Formula

If $f : D \subseteq \mathbb{R}^n \to \mathbb{R}$ is a $(m+1)$-time continuously partial differentiable scalar field, D is open and convex, then for $\boldsymbol{a} \in D$ and $\boldsymbol{h} \in \mathbb{R}^n$ with $\boldsymbol{a} + \boldsymbol{h} \in D$ we have the **Taylor formula**

$$f(\boldsymbol{a}+\boldsymbol{h}) = f(\boldsymbol{a}) + (\boldsymbol{h} \cdot \nabla) f(\boldsymbol{a}) + \frac{1}{2!} (\boldsymbol{h} \cdot \nabla)^2 f(\boldsymbol{a}) + \cdots + \frac{1}{m!} (\boldsymbol{h} \cdot \nabla)^m f(\boldsymbol{a}) + R_{m+1}(\boldsymbol{a}, \boldsymbol{h})$$

with a remainder

$$R_{m+1}(\boldsymbol{a}, \boldsymbol{h}) = \frac{1}{(m+1)!} (\boldsymbol{h} \cdot \nabla)^{m+1} f(\boldsymbol{a} + \xi \boldsymbol{h}) \text{ for } \xi \in (0, 1) .$$

48.3 Exercises

48.1 We are looking for a solution of the nonlinear system of equations

$$10x - \cos y = 0\,,$$
$$-\cos x + 5y = 0\,.$$

We choose the starting vector $\boldsymbol{x}_0 = (0, 0)^\top$.
(Note: It is $\cos 0.1 = 0.995$, $\cos 0.2 = 0.98$.)

(a) Using Newton's method, calculate $\boldsymbol{x}_1$.
(b) Calculate with the simplified Newton method $\boldsymbol{x}_1, \boldsymbol{x}_2$.

Note: The simplified Newton method calculates only in every *k-th* step, for a fixed $k \geq 2$, $Df(\boldsymbol{x}_n)$ new, in between we use the old derivative matrix in each case.

48.2 Let $f(x, y) = \begin{pmatrix} x^2 + y^2 - 2 \\ y - 1/x \end{pmatrix}$.

(a) Determine the Jacobian matrix of f.
(b) Formulate Newton's method for solving the equation $f(x, y) = 0$.
(c) Calculate an iterate to the initial value of $(x_0, y_0) = (5/4, 5/4)$.

48.3 Write a MATLAB function

```
function[ x,xvec,deltax ] = newtonverf( f,Df,x,TOL ),
```

which function handle `f` and `Df`, an initial value `x` and a desired precision `TOL` as input. This function is then to compute the Newton method to the desired precision or until the natural monotonicity test is violated. As a return the approximation `x` of the zero, the sequence of iterates and the precision at the last iteration shall be returned.

48.4 Enter the expressions $f(\boldsymbol{a}) + (\boldsymbol{h} \cdot \nabla) f(\boldsymbol{a})$ and $f(\boldsymbol{a}) + (\boldsymbol{h} \cdot \nabla) f(\boldsymbol{a}) + \frac{1}{2!}(\boldsymbol{h} \cdot \nabla)^2 f(\boldsymbol{a})$ from the Taylor formula for the case $n = 2$ and compare this with the Taylor polynomials T_1 and T_2.

48.5 Determine at the point $\boldsymbol{p} = (1, 0, f(1, 0))^\top$ the tangent plane to the graph of the function $f(x, y) = x^2 - y - x\,\mathrm{e}^y$.

48.6 Let $f : \mathbb{R}^2 \to \mathbb{R}$ given by

$$f(x, y) = 2x^3 - 5x^2 + 3xy - 2y^2 + 9x - 9y - 9.$$

Calculate the third degree Taylor polynomial of f at the position $\boldsymbol{a} = (1, -1)^\top$.

48.7 Develop the function $f : \mathbb{R}^2 \to \mathbb{R}$, to $\boldsymbol{a} = (1/\mathrm{e}, -1)^\top$ into a second order Taylor polynomial with

$$f(x, y) = y \ln x + x\,\mathrm{e}^{y+2}\,.$$

48.8 Develop $f(x, y) = x^y$ at the point $\boldsymbol{a} = (1,\ 1)^\top$ into a 2nd order Taylor polynomial and thus calculate approximately $\sqrt[10]{(1.05)^9}$.

48.9 Let $f : \mathbb{R}^2 \to \mathbb{R}$ given by

$$f(x, y) = \exp(x^2 + y^3) + xy(x + y).$$

Calculate the third degree Taylor polynomial of f at the position $\boldsymbol{a} = (0, 0)^\top$.

48.10 Given the function $f : D \subseteq \mathbb{R}^2 \to \mathbb{R}, \quad f(x, y) = \ln |x^{\frac{3}{2}} \cos(y)|$.

(a) Calculate the value of the function and its partial derivatives up to and including the 2nd order at the point $\boldsymbol{a} = (1, 0)^\top \in D$.
(b) Give the 2nd degree Taylor polynomial of the function at the point $\boldsymbol{a} = (1, 0)^\top$.
(c) The equation $f(x, y) = 0$ can be approximated by a quadric equation using the Taylor polynomial, state it.
(d) Determine the normal form of the quadric from (c) an.

48.11 Determine for $f(x, y) = \ln(x - y)$ the 2nd order Taylor polynomial where only powers of x and $y + 1$ occur in the development.

Extreme Value Determination

49

The value set of a scalar field $f : D \subseteq \mathbb{R}^n \to \mathbb{R}$ lies in $\mathbb{R}$. Thus it is possible to distinguish the values of a scalar field f by magnitude and to ask whether local or global extreme values are assumed by f. Fortunately, this search for *extremal places* and *extrema* can be treated analogously to the one-dimensional case: One determines the candidates as zeros of the gradient (the counterpart of the first derivative) and then checks with the Hessian matrix (the counterpart of the second derivative) whether the candidates are indeed extremal. In the search for global extrema, the *boundary* of the domain of f has to be taken into account.

49.1 Local and Global Extrema

According to the theorem of maximum and minimum in Sect. 46.3 a continuous scalar field $f : D \subseteq \mathbb{R}^n \to \mathbb{R}$ with a compact D has a (global) maximum and (global) minimum. As in the one-dimensional case, such *global extrema* can of course exist alongside other *local extrema* for non-compact D, we define (cf. Sect. 27.2):

Extrema and Extremal Places
We consider a scalar field $f : D \subseteq \mathbb{R}^n \to \mathbb{R}$. One calls a $\boldsymbol{x}_0 \in D$ **place of a**

- **global maximum** if $f(\boldsymbol{x}_0) \geq f(\boldsymbol{x}) \quad \forall \boldsymbol{x} \in D$,
  one then calls $f(\boldsymbol{x}_0)$ **the global maximum** of f,
- **global minimum** if $f(\boldsymbol{x}_0) \leq f(\boldsymbol{x}) \quad \forall \boldsymbol{x} \in D$,
  one then calls $f(\boldsymbol{x}_0)$ **the global minimum** of f,

(continued)

C. Karpfinger, *Calculus and Linear Algebra in Recipes*,
https://doi.org/10.1007/978-3-662-65458-3_49

- **local maximum** if $\exists\, \varepsilon > 0: \; f(\boldsymbol{x}_0) \geq f(\boldsymbol{x}) \quad \forall\, \boldsymbol{x} \in B_\varepsilon(\boldsymbol{x}_0)$,
  one then calls $f(\boldsymbol{x}_0)$ **a local maximum** of f,
- **local minimum** if $\exists\, \varepsilon > 0: \; f(\boldsymbol{x}_0) \leq f(\boldsymbol{x}) \quad \forall\, \boldsymbol{x} \in B_\varepsilon(\boldsymbol{x}_0)$,
  one then calls $f(\boldsymbol{x}_0)$ **a local minimum** of f.

If even $>$ instead of $\geq$ or $<$ instead of $\leq$ applies, one speaks of **strict** local or global extrema.

As usual $B_\varepsilon(\boldsymbol{x}_0) = \{\boldsymbol{x} \in \mathbb{R}^n \mid \|\boldsymbol{x} - \boldsymbol{x}_0\| < \varepsilon\}$ is the ε-neighborhood of $\boldsymbol{x}_0$.

In the case $n = 2$, the graph of a continuous $f : D \subseteq \mathbb{R}^2 \to \mathbb{R}$ is a surface in the $\mathbb{R}^3$, the extrema, local and global, correspond exactly to the notion or known case of a function of one variable: Global extrema are the total largest and smallest values of the function on D, local extrema are local, that is, for the restriction of the function f to a possibly very small neighborhood U of the considered place, largest and smallest value of f on U, see Fig. 49.1.

How do you determine the extrema of a scalar field? If $\boldsymbol{x}_0 \in D$ is a point of a local extremum of the continuously partially differentiable scalar field $f : D \subseteq \mathbb{R}^n \to \mathbb{R}$ then the tangent plane at the point $(\boldsymbol{x}_0, f(\boldsymbol{x}_0))$ to the graph of f is horizontal. So one finds the places of local extrema of a partially differentiable function among the **zeros** of $\nabla f(\boldsymbol{x})$: One calls each $\boldsymbol{x}_0 \in D$ with $\nabla f(\boldsymbol{x}_0) = \boldsymbol{0}$ a **stationary** or **critical point** of f.

Note that if $\nabla f(\boldsymbol{x}_0) = \boldsymbol{0}$, then in $\boldsymbol{x}_0$ *can be* a local extremum, it does not *have to be*. For example, the function $f : \mathbb{R}^2 \to \mathbb{R},\; f(x, y) = x^3$ doesn't have in $\boldsymbol{x}_0 = (0, 0)$ a local extremum, although $\nabla f(0, 0) = (0, 0)^\top$. Every stationary point which is neither a point of a local minimum nor a point of a local maximum is called **saddle point** of f.

The question whether in a critical point $\boldsymbol{x}_0 \in D$ is a local extremum or saddle point can often be answered by the following extreme point criterion:

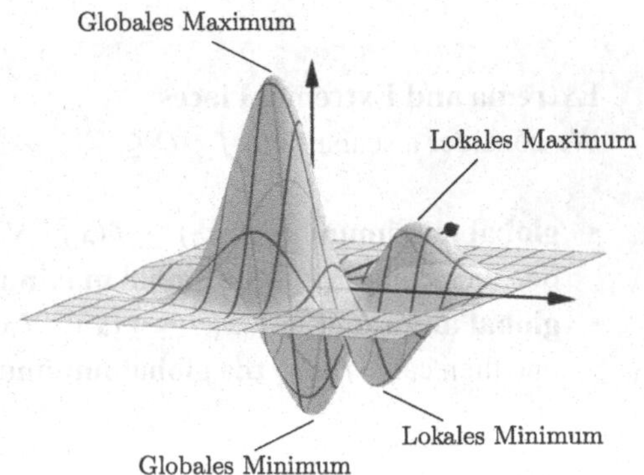

Fig. 49.1 Local (= Lokales) and global (= Globales) maximum and minimum

Extreme Point Criterion

If $f : U \subseteq \mathbb{R}^n \to \mathbb{R}$, U open, is a twice continuously differentiable scalar field with the (symmetric) Hessian matrix H_f, then for a critical point $\boldsymbol{x}_0 \in U$:

- If $H_f(\boldsymbol{x}_0)$ is negative definite, then $\boldsymbol{x}_0$ is the place of a local maximum.
- If $H_f(\boldsymbol{x}_0)$ is positive definite, then $\boldsymbol{x}_0$ is the place of a local minimum.
- If $H_f(\boldsymbol{x}_0)$ indefinite, then $\boldsymbol{x}_0$ place of a saddle point.
- If $H_f(\boldsymbol{x}_0)$ is semi definite (and not definite), no general statement is possible.

Whether a matrix A is positive, negative, semi-definite, or indefinite is decided using the definiteness criteria from Chap. 45.

Remarks

1. Sometimes the term *saddle point* is defined differently, namely, as a point $\boldsymbol{x}_0 \in D$ such that in each neighborhood of $\boldsymbol{x}_0$ there are points $\boldsymbol{a}$ and $\boldsymbol{b}$ with $f(\boldsymbol{a}) < f(\boldsymbol{x}_0) < f(\boldsymbol{b})$. The above criterion is also valid for this definition.
2. If $H_f(\boldsymbol{x}_0)$ is semi definite (and not definite), then anything can happen: For example, the following scalar fields $f_i : \mathbb{R}^2 \to \mathbb{R}$ with

$$f_1(x, y) = x^4 + y^2, \quad f_2(x, y) = -(x^4 + y^2), \quad f_3(x, y) = x^3 - y^2$$

all have a stationary point in $\boldsymbol{x}_0 = (0, 0)^\top$. And their respective Hessian matrices $H_{f_1}(0, 0)$, $H_{f_2}(0, 0)$, $H_{f_3}(0, 0)$ are all in $\boldsymbol{x}_0$ semidefinite (and not definite). But nevertheless f_1 has in $(0, 0)$ a local minimum, f_2 has in $(0, 0)$ a local maximum and f_3 has neither a minimum nor a maximum in $(0, 0)$ (in this last example we set $y = 0$ to see that the function for each positive x has positive values and for each negative x has negative values).

Example 49.1 To have a better idea of a *saddle*, let's consider the **monkey saddle**, which is the graph of the scalar field.

$$f : \mathbb{R}^2 \to \mathbb{R} \text{ with } f(x, y) = x^3 - 3\,x\,y^2 .$$

We determine the gradient $\nabla f(x, y)$ and the Hessian matrix $H_f(x, y)$:

$$\nabla f(x, y) = \begin{pmatrix} 3x^2 - 3y^2 \\ -6xy \end{pmatrix} \text{ and } H_f(x, y) = \begin{pmatrix} 6x & -6y \\ -6y & -6x \end{pmatrix} .$$

We obtain the zeros of the gradient in this case quite simply: For the second component $-6xy$ has to be zero, we must have $x = 0$ or $y = 0$. In either of these two cases, the first component $3x^2 - 3y^2$ is zero if and only if the other variable is zero: This means that $\boldsymbol{x}_0 = (0, 0)$ is the only stationary point of this scalar field. The Hessian matrix in the stationary point $\boldsymbol{x}_0 = (0, 0)^\top$ is the zero matrix, $H_f(0, 0) = 0$, and thus semidefinite (and not definite). Thus our extreme point criterion is not applicable. The typical procedure to induce a decision nevertheless is as follows:

We restrict the scalar field $f(x, y) = x^3 - 3\,x\,y^2$ to a straight line. For this we choose the x-axis, because then $y = 0$ and the function is $f(x, 0) = x^3$; in each neighborhood of the point $\boldsymbol{x}_0 = (0, 0)$ there are points $\boldsymbol{a}$ and $\boldsymbol{b}$ with $f(\boldsymbol{a}) < f(\boldsymbol{x}_0) < f(\boldsymbol{b})$ namely for $\boldsymbol{a}$ on the negative x-axis and $\boldsymbol{b}$ on the positive x-axis.

This restriction of a scalar field to a straight line, as here to the x-axis, is an extremely useful and versatile method for obtaining further information about the local or global behavior of the function f. For example, one can see that $\boldsymbol{x}_0 = (0, 0)$ is the place of a saddle point, but we also see that because of

$$\lim_{x\to\infty} f(x, 0) = \infty \text{ and } \lim_{x\to\infty} f(x, 0) = -\infty$$

the function f is unbounded above and below, so in particular it also has no global extrema.

The graph of f, the monkey saddle, is shown in the accompanying Fig. 49.2: It has three sinks. Two sinks are for the monkey's legs, as in any other saddle, and a third sink is for the tail. If you plot the graph with MATLAB, you can rotate it in all directions and clearly see the saddle point $(0, 0)$. ■

The last example shows the typical procedure when examining a stationary point with semi definite (and not definite) Hessian matrix. A similar procedure should be known in the case $f''(x_0) = 0$ from one-dimensional analysis.

49.2 Determination of Extrema and Extremal Points

The determination of the local extrema of $f : D \subseteq \mathbb{R}^n \to \mathbb{R}$ is in principle unproblematic: One determines the zeros of ∇f (this is often computationally tedious, but if necessary can be solved by computer, e.g. with Newton's method) and decide with the above criterion whether the individual stationary points are extrema or saddle points.

The global extrema of $f : D \subseteq \mathbb{R}^n \to \mathbb{R}$ can be found as in the one-dimensional case among the local extrema and the values of f on the boundary of D, provided that f is defined there. This was mostly unproblematic in the one-dimensional case, since the boundary consists of numbers a and b or $\pm\infty$. In the multidimensional, unfortunately, the boundary can get very complicated. In the typical exercises to this topic we usually

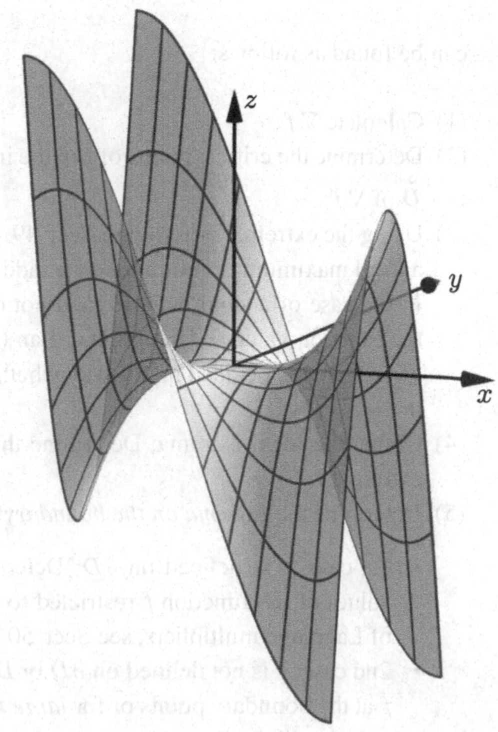

Fig. 49.2 The monkey saddle

have the case $n = 2$. The boundary of D is thereby typically the boundary of a circle or rectangle, or the domain D is unbounded.

In any case, there is no global maximum if the function f is unbounded from above; in this situation there is no largest function value, and analogously there is no global minimum if f is unbounded from below. Such a decision is usually made as follows: One fixes all but one variable x_i of f and then consider $\lim_{x_i \to a} f(\ldots, x_i, \ldots)$ for a $a \in \mathbb{R} \cup \{\infty\}$, in short: one constrains f onto a straight line (see above example on the monkey saddle).

We summarize and add:

Recipe: Determine the Extremal Points
The extremal points of a twice continuously differentiable scalar field

$$f : D \subseteq \mathbb{R}^n \to \mathbb{R}, \; \boldsymbol{x} \mapsto f(\boldsymbol{x})$$

(continued)

can be found as follows:

(1) Calculate ∇f.
(2) Determine the critical points of f in the interior of D i.e., the zeros. $\boldsymbol{a}_1, \ldots, \boldsymbol{a}_n \in \overset{\circ}{D}$ of ∇f.
(3) Using the extremal criterion in Sect. 49.1, decide whether in $\boldsymbol{a}_1, \ldots, \boldsymbol{a}_n$ there is a local maximum or minimum or a saddle point.
In the case of a semi definite (and not definite) Hessian matrix in $\boldsymbol{a}_k$ consider the function values of the points of an (arbitrarily small) neighborhood of $\boldsymbol{a}_k$; often it can be decided in this way whether an extremum or a saddle point exists in $\boldsymbol{a}_k$.
(4) Obtain the local extrema: Determine the values $f(\boldsymbol{a}_i)$ if in $\boldsymbol{a}_i$ there is a local extremum.
(5) Determine the *extrema on the boundary*:

- 1st case: f is defined on ∂D: Determine the extremal points and extremal values of the function f restricted to the boundary (possibly use the method of Lagrange multipliers, see Sect. 50.3).
- 2nd case: f is not defined on ∂D or D is unbounded: Determine the limits of f at the boundary points or for *large* $\boldsymbol{x} \in D$.

(6) By comparing the values from (4) and (5), decide whether and, if so, at which points the global maximum and global minimum are assumed:

- Exists a smallest real value $y_{\min}$? If yes, then $y_{\min}$ is the global minimum, all $\boldsymbol{a}_i$ with $f(\boldsymbol{a}_i) = y_{\min}$ are places of the global minimum.
- Does a greatest real value exist $y_{\max}$? If yes, then $y_{\max}$ is the global maximum, all $\boldsymbol{a}_i$ with $f(\boldsymbol{a}_i) = y_{\max}$ are places of the global maximum.
- Otherwise, there is no global extremum and hence no places of global extrema.

Example 49.2

- We determine the extremal places and extrema of the scalar field

$$f : \mathbb{R}^2 \to \mathbb{R}, \ f(x, y) = x^2 + 3y^4 .$$

(1) The gradient is $\nabla f(x, y) = (2x, 12y^3)^\top$.
(2) Because of

$$\nabla f(x, y) = \begin{pmatrix} 2x \\ 12y^3 \end{pmatrix} = \begin{pmatrix} 0 \\ 0 \end{pmatrix} \Leftrightarrow \begin{pmatrix} x \\ y \end{pmatrix} = \begin{pmatrix} 0 \\ 0 \end{pmatrix}$$

the point $\boldsymbol{a} = (0, 0)^\top$ is the only stationary point of f.
(3) We determine $H_f(\boldsymbol{a})$:

$$H_f(x, y) = \begin{pmatrix} 2 & 0 \\ 0 & 36y^2 \end{pmatrix} \Rightarrow H_f(0, 0) = \begin{pmatrix} 2 & 0 \\ 0 & 0 \end{pmatrix}.$$

The matrix $H_f(0, 0)$ is positive semi definite. Therefore we consider an (arbitrarily small) ε-neighborhood of $\boldsymbol{a} = (0, 0)$. We have for all $(x, y) \in B_\varepsilon(0, 0)$:

$$f(x, y) \geq 0 \text{ and } f(x, y) = 0 \Leftrightarrow (x, y) = (0, 0).$$

The place $\boldsymbol{a} = (0, 0)$ is therefore the place of a local minimum of f.
(4) The local minimum has the value $f(0, 0) = 0$.
(5) Because of $D = \mathbb{R}^2$ we consider $f(x, y)$ for *large* x and y, respectively. Obviously, for each $K \in \mathbb{R}$ real numbers x and y exists with $f(x, y) > K$ so that f is unbounded from above. More precisely: If one sets $y = 0$, then we have

$$\lim_{x \to \infty} f(x, 0) = \lim_{x \to \infty} x^2 = \infty,$$

so the function f does not have a global maximum. This is also true for any other straight line in $\mathbb{R}^2$.
(6) It is $f_{\min} = 0$ the smallest function value, i.e. the global minimum. And $\boldsymbol{a} = (0, 0)$ is the (only) place of the global minimum. A global maximum does not exist.

- We determine the extremal points and extrema of the scalar field

$$f : \mathbb{R}^2 \to \mathbb{R} \text{ with } f(x, y) = (x - 1)^2 - y^4 - 4y.$$

(1) The gradient is $\nabla f(x, y) = (2x - 2, -4y^3 - 4)^\top$.
(2) Because of

$$\nabla f(x, y) = \begin{pmatrix} 2x - 2 \\ -4y^3 - 4 \end{pmatrix} = \begin{pmatrix} 0 \\ 0 \end{pmatrix} \Leftrightarrow \begin{pmatrix} x \\ y \end{pmatrix} = \begin{pmatrix} 1 \\ -1 \end{pmatrix}$$

the point $\boldsymbol{a} = (1, -1)^\top$ is the only stationary point of f.

(3) We determine $H_f(\boldsymbol{a})$:

$$H_f(x,y)=\begin{pmatrix}2 & 0\\ 0 & -12y^2\end{pmatrix} \Rightarrow H_f(1,-1)=\begin{pmatrix}2 & 0\\ 0 & -12\end{pmatrix}.$$

The Hessian Matrix $H_f(\boldsymbol{a})$ is indefinite, so $\boldsymbol{a}$ is the place of a saddle point.

(4) not applicable.

(5) Because of $D=\mathbb{R}^2$ we consider $f(x,y)$ for *large* x and y. Obviously, for every $K\in\mathbb{R}$ exists real numbers x and y with $f(x,y)>K$ resp. $f(x,y)<K$ such that f is unbounded from above and below. More precisely: If one sets $x=1$ and $y=0$, then

$$\lim_{y\to\infty} f(1,y)=-\infty \text{ resp. } \lim_{x\to\infty} f(x,0)=\infty.$$

(6) The function has neither a global maximum nor a global minimum.

- We determine the extremal points and extrema of the scalar field

$$f:[-2,2]\times[-3,3]\to\mathbb{R} \text{ with } f(x,y)=\frac{x^3-3x}{1+y^2}.$$

(1) The gradient is $\nabla f(x,y)=\left(\frac{3(x^2-1)}{1+y^2}, -\frac{x^3-3x}{(1+y^2)^2}\cdot 2y\right)^{\top}$.

(2) Because of

$$\nabla f(x,y)=\begin{pmatrix}\frac{3(x^2-1)}{1+y^2}\\ -\frac{x^3-3x}{(1+y^2)^2}\cdot 2y\end{pmatrix}=\begin{pmatrix}0\\0\end{pmatrix} \Leftrightarrow \begin{pmatrix}x\\y\end{pmatrix}=\begin{pmatrix}1\\0\end{pmatrix} \text{ or } \begin{pmatrix}x\\y\end{pmatrix}=\begin{pmatrix}-1\\0\end{pmatrix}$$

the points $\boldsymbol{a}_1=(1,0)^{\top}$ and $\boldsymbol{a}_2=(-1,0)^{\top}$ are the only stationary points of f.

(3) We determine $H_f(\boldsymbol{a}_k)$, for this we need first of all the Hessian matrix

$$H_f(x,y)=\begin{pmatrix}\frac{6x}{1+y^2} & -\frac{6y(x^2-1)}{(1+y^2)^2}\\ -\frac{6y(x^2-1)}{(1+y^2)^2} & \frac{(x^3-3x)(6y^4+4y^2-2)}{(1+y^2)^4}\end{pmatrix}.$$

By substituting the stationary points we get

$$H_f(1,0)=\begin{pmatrix}6 & 0\\ 0 & 4\end{pmatrix} \text{ and } H_f(-1,0)=\begin{pmatrix}-6 & 0\\ 0 & -4\end{pmatrix}.$$

The matrix $H_f(1, 0)$ is positive definite, therefore $\boldsymbol{a}_1 = (1, 0)^\top$ place of a local minimum, and the matrix $H_f(-1, 0)$ is negative definite, so that $\boldsymbol{a}_2 = (-1, 0)^\top$ is the place of a local maximum.

(4) As values for the local minimum and local maximum we obtain

$$f(1, 0) = -2 \text{ and } f(-1, 0) = 2\,.$$

(5) Because of $D = [-2, 2] \times [-3, 3]$ we consider $f(x, y)$ on the four parts of the boundary:

$$D_1 = \{(x, y)^\top \,|\, x = 2\,,\ y \in [-3, 3]\}\,,$$

$$D_2 = \{(x, y)^\top \,|\, x = -2\,,\ y \in [-3, 3]\}\,,$$

$$D_3 = \{(x, y)^\top \,|\, x \in [-2, 2]\,,\ y = 3\}\,,$$

$$D_4 = \{(x, y)^\top \,|\, x \in [-2, 2]\,,\ y = -3\}\,.$$

See Fig. 49.3.

We now examine the scalar field f for extrema on these parts:

D_1: $f : [-3, 3] \to \mathbb{R}$, $f(2, y) = \frac{2}{1+y^2}$ This function has a local maximum in $y = 0$ with the value $f(2, 0) = 2$.

D_2: $f : [-3, 3] \to \mathbb{R}$, $f(-2, y) = \frac{-2}{1+y^2}$ This function has a local minimum in $y = 0$ with the value $f(-2, 0) = -2$.

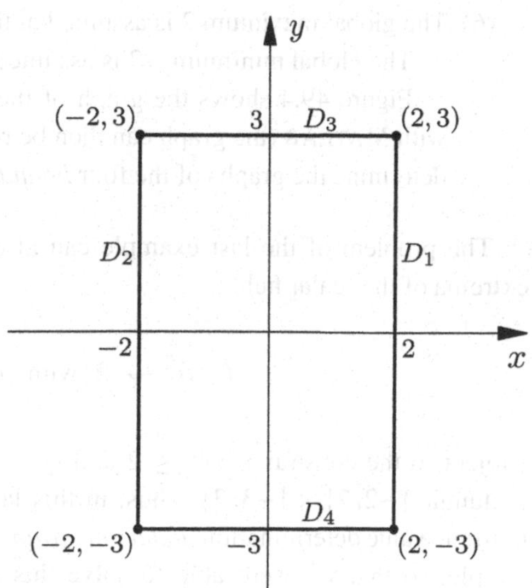

Fig. 49.3 The parts of the boundary of D

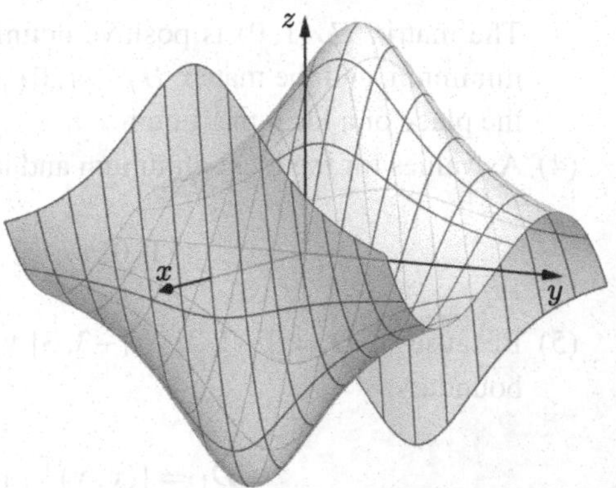

Fig. 49.4 The function on D

D_3: $f : [-2, 2] \to \mathbb{R}$, $f(x, 3) = \frac{1}{10}(x^3 - 3x)$: This function has a local minimum in $x = 1$ with the value $f(1, 3) = \frac{-2}{10}$ and a local maximum in $x = -1$ with value $f(-1, 3) = \frac{2}{10}$.

D_4: $f : [-2, 2] \to \mathbb{R}$, $f(x, -3) = \frac{1}{10}(x^3 - 3x)$: This function has a local minimum in $x = 1$ with the value $f(1, -3) = \frac{-2}{10}$ and a local maximum in $x = -1$ with value $f(-1, -3) = \frac{2}{10}$.

It remains to determine the values in the boundary points of the boundary, that is, in the corners, we have

$$f(2, 3) = \frac{2}{10}, \; f(2, -3) = \frac{2}{10}, \; f(-2, 3) = \frac{-2}{10}, \; f(-2, -3) = \frac{-2}{10}.$$

(6) The global maximum 2 is assumed at the points $(2, 0)$ and $(-1, 0)$.
The global minimum -2 is assumed at the points $(-2, 0)$ and $(1, 0)$.
Figure 49.4 shows the graph of the function f. Do not miss to plot this graph with MATLAB (the graph can then be rotated in all directions) and, in particular, to determine the graphs of the four *boundary functions* in the picture. ■

The problem of the last example can also be formulated differently: Determine the extrema of the scalar field

$$f : \mathbb{R}^2 \to \mathbb{R} \text{ with } f(x, y) = \frac{x^3 - 3x}{1 + y^2}$$

subject to the constraints $|x| \leq 2$ and $|y| \leq 3$ which are exactly those extrema in the rectangle $[-2, 2] \times [-3, 3]$. Thus, in this last example, we have already performed an extreme value determination *under constraints*. In this case, the constraint was particularly simple, so that we were able to solve this task using the methods determined so far. However, if the constraint is *more complicated*, we have to use other methods to solve the extreme value problem. We will do that in the next chapter.

49.3 Exercises

49.1 Let $f: \mathbb{R}^2 \to \mathbb{R}$ given by

$$f(x, y) = x^3 + y^3 - 3xy.$$

Determine all local and global extrema of f.

49.2 Given the function $f: \mathbb{R}^2 \to \mathbb{R}$ defined by

$$f(x, y) = 2x^2 - 3xy^2 + y^4 .$$

(a) Show that f along any straight line through the zero point $(0, 0)^\top$ has a local minimum in $(0, 0)^\top$.
(b) Show that f has no local minimum in $(0, 0)^\top$.

49.3 Given a cuboid with side lengths $x, y, z > 0$. Let the sum of the sides be given by the condition $x + y + z = 1$.

(a) Represent the surface area of the cuboid as a function $f: (0, 1) \times (0, 1) \to \mathbb{R}$.
(b) Determine the critical points of f.
(c) Give the maximum surface area of the cuboid, and show that this is indeed a local maximum.

49.4 Show that $(0, 0, 0)^\top$ is a stationary point of the function $f: \mathbb{R}^3 \to \mathbb{R}$ with

$$f(x, y, z) = \cos^2 z - x^2 - y^2 - \exp(xy)$$

is. Investigate whether f has a local maximum or minimum at this point.

49.5 Compute all the stationary points of the following functions and classify them:

(a) $f(x, y) = xy + x - 2y - 2$,
(b) $f(x, y) = x^2y^2 + 4x^2y - 2xy^2 + 4x^2 - 8xy + y^2 - 8x + 4y + 4$,
(c) $f(x, y) = 4\,\mathrm{e}^{x^2+y^2} - x^2 - y^2$,
(d) $f(x, y, z) = -\ln(x^2 + y^2 + z^2 + 1)$.

49.6 Determine for the function $f(x, y) = y^4 - 3xy^2 + x^3$

(a) local and global extrema and saddle points,
(b) maximum and minimum for $(x, y) \in [-\frac{5}{2}, \frac{5}{2}] \times [-2, 2]$.

49.7 Determine the location and type of the local extrema of

$$f(x, y) = (x^2 + y^2)\,\mathrm{e}^{-(x+y)}\,.$$

49.8 Given the function $f : \mathbb{R}^2 \to \mathbb{R}$, $f(x, y) = x^3 + y^3 - 3\,a\,x\,y$ with $a \in \mathbb{R}$. Determine

(a) the stationary points of f,
(b) local and global extrema and saddle points of f

as a function of a.

Extreme Value Determination Under Constraints 50

In practice, extrema of scalar fields in n variables are usually to be determined $x_1, \ldots, x_n$ *under constraints*. Such constraints can often be defined as sets of zeros of partially differentiable functions in the variables $x_1, \ldots, x_n$. There are then essentially two methods of determining the extremal places and extrema sought, the *substitution method* and the *method of Lagrange multipliers*. Nothing new happens with the substitution method, we have already considered a first example in the last chapter; only the substitution method is not as universally applicable as the method of Lagrange multipliers, which in turn has the disadvantage that it often leads to nonlinear systems of equations that are difficult to solve.

50.1 Extrema Under Constraints

If $f : D \subseteq \mathbb{R}^n \to \mathbb{R}$ is a partially differentiable scalar field, we can in principle use the methods of Chap. 49 to determine the local and global extrema of this scalar field on D. If now D_z is a subset of D, $D_z \subseteq D$, the global extrema of f on D_z will in general be different from those on D; we consider f only on a possibly very small subset D_z from D.

Example 50.1 The scalar field $f : \mathbb{R}^2 \to \mathbb{R}$, $f(x, y) = x + y + 5$ has on $D = \mathbb{R}^2$ neither local nor global extrema. But if one restricts the domain D of f on the set $D_z = \{(x, y) \in \mathbb{R}^2 \mid x^2 - y = 0\}$ then f on D_z very well has a global minimum (but not a maximum), note Fig. 50.1. ■

The task of finding the extrema of the function f on D_z in the last example can also be formulated as follows: Determine the extrema of the function

$$f : \mathbb{R}^2 \to \mathbb{R}\,,\ f(x, y) = x + y + 5 \ \text{ under the constraint } \ g(x, y) = x^2 - y = 0\,.$$

C. Karpfinger, *Calculus and Linear Algebra in Recipes*,
https://doi.org/10.1007/978-3-662-65458-3_50

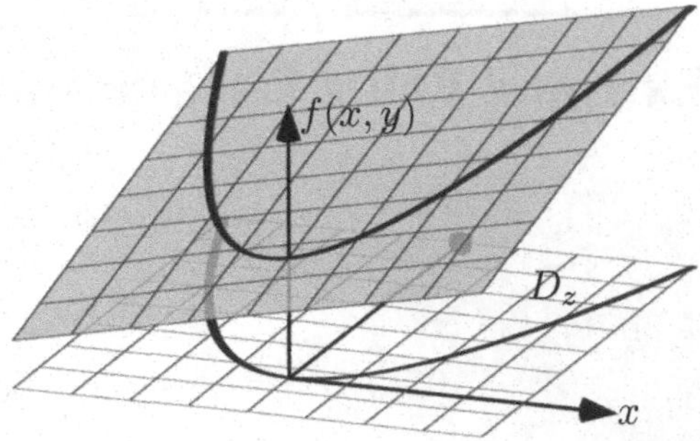

Fig. 50.1 D_z and $f(D_z)$

We generalize the procedure to n variables:

Extreme Value Problems With Constraints
Under an **extreme value problem with constraints** we understand the task of finding the extremal points and extrema of a scalar field $f : D \subseteq \mathbb{R}^n \to \mathbb{R}$ on a subset $D_z \subseteq D$. The set D_z of **admissible points** is usually defined as the set of zeros of one or more scalar fields $g_1, \ldots, g_m$ with $m < n$ given, i.e.

$$D_z = \{\boldsymbol{x} \in D \mid g_1(\boldsymbol{x}) = \cdots = g_m(\boldsymbol{x}) = 0\}.$$

We briefly describe such a problem with: *Determine the extrema of* $f : D \subseteq \mathbb{R}^n \to \mathbb{R}$ *under the constraints* $g_1(\boldsymbol{x}) = \cdots = g_m(\boldsymbol{x}) = 0$.
There are two main solution methods for this problem:

- The **substitution method**: Especially useful in the case $m = 1$ and $n = 2$ (one constraint and two variables x and y): Solve, if possible, the constraint $g(x, y) = 0$ according to x or y and insert this constraint into the scalar field f. You then get a function in one variable whose extremal points and extrema are determinable.
- The **method of Lagrange multipliers**: Offers especially in the case where resolving the constraint after a variable is not possible, or $n \geq 3$, or $m \geq 2$: One sets up the *Lagrangian function L*, form its gradient, determine the zeros of this gradient, and thus obtain the candidates for the extremal points.

The requirement that the number of constraints m is smaller than the number of variables is necessary, but also *natural*: if there are more constraints than variables, the set D_z of admissible points is empty in general.

In the following two sections, we discuss the substitution method and the method of Lagrange multipliers for a constraint $g(\boldsymbol{x}) = 0$. Then, in another section, we discuss the method of Lagrange multipliers for multiple constraints.

50.2 The Substitution Method

The operation of the substitution method is easy to make clear: If $f : D \subseteq \mathbb{R}^2 \to \mathbb{R}$ a scalar field in two variables x and y and let the constraint $g(x, y) = 0$ be solvable for one variable x or y, for example $y = h(x)$ then the function $\tilde{f}(x) = f(x, h(x))$ is a function in the single variable x. It is then $\tilde{f}$ the constraint of the function f on the set of admissible points. Therefore, the following prescription is obtained:

Recipe: Solve the Extremal Problem Under a Constraint Using the Substitution Method

The extrema of $f : D \subseteq \mathbb{R}^2 \to \mathbb{R}$ under the constraint $g(x, y) = 0$ are obtained with a solvable constraint as follows:

(1) Solve the constraint $g(x, y) = 0$ for x or y and get $y = h(x)$ or $x = h(y)$ (possibly a case distinction is necessary).
(2) Replace x or y into $f(x, y)$ by $h(y)$ or $h(x)$ and get $\tilde{f}(y)$ or $\tilde{f}(x)$.
(3) Determine the extremal points $a_1, \ldots, a_n$ and extrema $\tilde{f}(a_1), \ldots, \tilde{f}(a_n)$ of $\tilde{f}$.
(4) Give the extremal places $(a_i, h(a_i))$ or $(h(a_i), a_i)$ and extrema $\tilde{f}(a_i)$ for all i of f under the constraint $g(x, y) = 0$.

Example 50.2

- Compare Example 50.1 above: We solve the extreme value problem with the substitution method

$$f : \mathbb{R}^2 \to \mathbb{R},\ f(x, y) = x + y + 5 \text{ under the constraint } g(x, y) = x^2 - y = 0\,.$$

(1) We solve $g(x, y) = 0$ for y:

$$y = x^2\,.$$

(2) Substituting $y = x^2$ into the scalar field yields the function

$$\tilde{f} : \mathbb{R} \to \mathbb{R} \text{ with } \tilde{f}(x) = x + x^2 + 5\,.$$

(3) As the place of a local and global minimum of $\tilde{f}$ we get $a = -1/2$ with $\tilde{f}(a) = 19/4$.

(4) Thus it is

$$\left(\frac{-1}{2}, \frac{1}{4}\right) \text{ and } f\left(\frac{-1}{2}, \frac{1}{4}\right) = \frac{19}{4}$$

place and value of the global minimum of the function f under the constraint $g(x, y) = 0$.

- We solve the extreme value problem with the substitution method

$$f : \mathbb{R}^2 \to \mathbb{R},\ f(x, y) = x^3 y^3 \text{ under the constraint } g(x, y) = x^2 + 2\,y^2 - 1 = 0\,.$$

(1) We solve $g(x, y) = 0$ for one variable:

$$\text{1st case: } x = \sqrt{1 - 2y^2} \text{ and 2nd case: } x = -\sqrt{1 - 2y^2}\,.$$

(2) Case 1: Inserting $x = \sqrt{1 - 2y^2}$ into the scalar field yields the function

$$\tilde{f}_1 : \left[\tfrac{-1}{\sqrt{2}}, \tfrac{1}{\sqrt{2}}\right] \to \mathbb{R} \text{ with } \tilde{f}_1(y) = (1 - 2\,y^2)^{3/2} y^3\,.$$

2nd case: Inserting $x = -\sqrt{1 - 2y^2}$ into the scalar field yields the function

$$\tilde{f}_2 : \left[\tfrac{-1}{\sqrt{2}}, \tfrac{1}{\sqrt{2}}\right] \to \mathbb{R} \text{ with } \tilde{f}_2(y) = -(1 - 2\,y^2)^{3/2} y^3\,.$$

(3) Case 1: As candidates of extremal places we get because of $\tilde{f}_1'(y) = -3\,y^2\sqrt{1 - 2y^2}(4y^2 - 1)$ the places

$$a_1 = \frac{-1}{\sqrt{2}},\ a_2 = \frac{-1}{2},\ a_3 = 0,\ a_4 = \frac{1}{2},\ a_5 = \frac{1}{\sqrt{2}}\,.$$

From $\tilde{f}_1''(y) = \frac{6y(20y^4 - 11y^2 + 1)}{\sqrt{1 - 2y^2}}$ and a sign consideration of $\tilde{f}_1'$ in an environment of $a_3 = 0$ and determining the values at the boundary points a_1 and a_5 we obtain:

- $\tilde{f}_1$ has a local minimum in $a_2 = -1/2$ with the value $\tilde{f}_1(a_2) = -\sqrt{2}/32$.
- $\tilde{f}_1$ has no local extremum in $a_3 = 0$.
- $\tilde{f}_1$ has a local maximum in $a_4 = 1/2$ with the value $\tilde{f}_1(a_4) = \sqrt{2}/32$.
- $\tilde{f}_1$ has no extremum in $a_1 = -1/\sqrt{2}$ and $a_5 = 1/\sqrt{2}$ since $\tilde{f}_1(a_1) = 0 = \tilde{f}_1(a_5)$.

Case 2: For symmetry reasons we obtain for $\tilde{f}_2$:

- $\tilde{f}_2$ has a local maximum in $a_2 = -1/2$ with the value $\tilde{f}_2(a_2) = \sqrt{2}/32$.
- $\tilde{f}_2$ has no local extremum in $a_3 = 0$.
- $\tilde{f}_2$ has a local minimum in $a_4 = 1/2$ with the value $\tilde{f}_2(a_4) = -\sqrt{2}/32$.

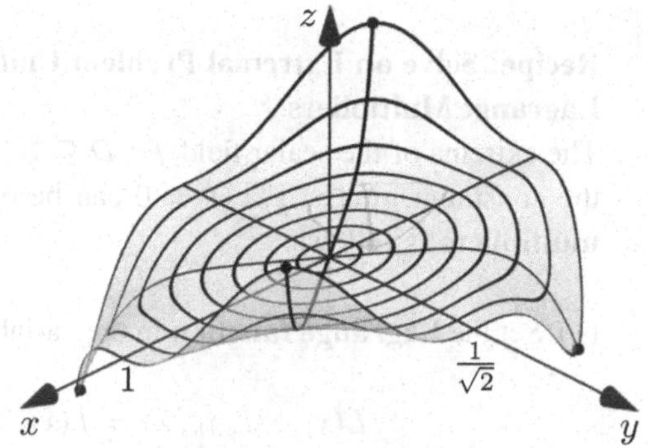

Fig. 50.2 The graph of f

- $\tilde{f}_2$ has no extremum in $a_1 = -1/\sqrt{2}$ and $a_5 = 1/\sqrt{2}$ since $\tilde{f}_2(a_1) = 0 = \tilde{f}_2(a_5)$.

(4) Thus there are
- $(1/\sqrt{2}, -1/2)$ place of a local minimum with $f(1/\sqrt{2}, -1/2) = -\sqrt{2}/32$.
- $(-1/\sqrt{2}, -1/2)$ place of a local maximum with $f(-1/\sqrt{2}, -1/2) = \sqrt{2}/32$.
- $(1/\sqrt{2}, 1/2)$ place of a local maximum with $f(1/\sqrt{2}, -1/2) = \sqrt{2}/32$.
- $(-1/\sqrt{2}, 1/2)$ place of a local minimum with $f(-1/\sqrt{2}, -1/2) = -\sqrt{2}/32$.

Every local extremum is also a global one. The above Fig. 50.2 shows the graph $\{(x, y, f(x, y)) \mid x^2 + 2y^2 \leq 1\}$ of the scalar field $f(x, y) = x^3 y^3$. One recognizes at the boundary of this surface the symmetrically lying extrema.

With MATLAB you get the image of the ellipse $x^2 + 2y^2 = 1$ at f with

```
ezplot3('cos(t)','sin(t)/sqrt(2)','(cos(t)).^3.
*(sin(t)/sqrt(2)).^3')
```

This is the curve whose extrema we have determined.

■

50.3 The Method of Lagrange Multipliers

The *method of Lagrange multipliers* is a theoretically very simple method to solve an extremal problem with a scalar field in n variables and one (or more—but only in the next section) constraint(s), which can not be solved for one variable: One sets up the *Lagrange function*, determines the gradient and its zeros, and searches among them for the extrema, more precisely:

Recipe: Solve an Extremal Problem Under a Constraint Using the Method of Lagrange Multipliers

The extrema of the scalar field $f : D \subseteq \mathbb{R}^n \to \mathbb{R}$ in the variables $x_1, \ldots, x_n$ under the constraint $g(x_1, \ldots, x_n) = 0$ can be obtained with the **method of Lagrange multipliers** as follows:

(1) Set the **Lagrange function** in the variables $x_1, \ldots, x_n, \lambda$:

$$L(x_1, \ldots, x_n, \lambda) = f(x_1, \ldots, x_n) + \lambda\, g(x_1, \ldots, x_n).$$

(2) Determine the zeros $\boldsymbol{a}_1, \ldots, \boldsymbol{a}_k \in \mathbb{R}^n$ of the gradient

$$\nabla L(x_1, \ldots, x_n, \lambda) = \begin{pmatrix} f_{x_1}(x_1, \ldots, x_n) + \lambda\, g_{x_1}(x_1, \ldots, x_n) \\ \vdots \\ f_{x_n}(x_1, \ldots, x_n) + \lambda\, g_{x_n}(x_1, \ldots, x_n) \\ g(x_1, \ldots, x_n) \end{pmatrix}.$$

These zeros $\boldsymbol{a}_1, \ldots, \boldsymbol{a}_k \in \mathbb{R}^n$ are **candidates for extremal places.**

(3) Further determine the zeros $\boldsymbol{b}_1, \ldots, \boldsymbol{b}_l \in \mathbb{R}^n$ of the gradient

$$\nabla g(x_1, \ldots, x_n) = \begin{pmatrix} g_{x_1}(x_1, \ldots, x_n) \\ \vdots \\ g_{x_n}(x_1, \ldots, x_n) \end{pmatrix}.$$

Add all $\boldsymbol{b}_i$ with $g(\boldsymbol{b}_i) = 0$ to the candidates for extrema.

(4) Determine the values $f(\boldsymbol{a}_i)$ and $f(\boldsymbol{b}_i)$ for all candidates for extremal places and decide where the global extrema are. If any exist, they are among the candidates.

Remarks

1. The values for λ are not of interest at all, nevertheless it is often useful to first determine the possible values for λ in order to get the possible values for $\boldsymbol{a}_i$.
2. The candidates $\boldsymbol{b}_i$ are rare (we show an example which teaches not to forget them).
3. The biggest problem lies in the determination of the zeros of the gradient $\nabla L(x_1, \ldots, x_n, \lambda)$. If one does not find a solution, then a solution by Newton's method is a good idea.

We try the procedure on the two examples above from Example 50.2 and compare the effort:

Example 50.3

- We solve the extreme value problem with the method of Lagrange multipliers:

$$f : \mathbb{R}^2 \to \mathbb{R}\,,\; f(x, y) = x + y + 5 \text{ under the constraint } g(x, y) = x^2 - y = 0\,.$$

(1) The Lagrange function is:

$$L(x, y, \lambda) = x + y + 5 + \lambda\,(x^2 - y)\,.$$

(2) We determine the zeros of the gradient $\nabla L(x, y, \lambda)$:

$$\nabla L(x, y, \lambda) = \begin{pmatrix} 1 + 2\,\lambda\,x \\ 1 - \lambda \\ x^2 - y \end{pmatrix} = 0 \;\Leftrightarrow\; \lambda = 1\,,\; x = -1/2\,,\; y = 1/4\,.$$

This gives us the candidate $(x, y) = (-1/2, 1/4)$.

(3) Zeros of the gradient of g do not exist, namely we have

$$\nabla g(x, y) = \begin{pmatrix} 2x \\ -1 \end{pmatrix} \neq 0 \text{ for all } (x, y)\,.$$

Thus there are no further candidates to be added.

(4) In the point $(-1/2, 1/4)$ the function f has the global minimum $f(-1/2, 1/4) = 19/4$.

- We solve the following extreme value problem with the method of Lagrange multipliers:

$$f : \mathbb{R}^2 \to \mathbb{R},\; f(x, y) = x^3 y^3 \text{ under the constraint } g(x, y) = x^2 + 2\,y^2 - 1 = 0\,.$$

(1) The Lagrange function is:

$$L(x, y, \lambda) = x^3 y^3 + \lambda(x^2 + 2y^2 - 1)\,.$$

(2) We determine the zeros of the gradient $\nabla L(x, y, \lambda)$:

$$\nabla L(x, y, \lambda) = \begin{pmatrix} 3x^2y^3 + 2\lambda x \\ 3x^3y^2 + 4\lambda y \\ x^2 + 2y^2 - .1 \end{pmatrix} = 0 \Leftrightarrow \begin{array}{c} 3x^2y^3 + 2\lambda x = 0 \\ 3x^3y^2 + 4\lambda y = 0 \\ x^2 + 2y^2 - 1 = 0 \end{array} .$$

By subtracting the y-fold of the second equation from the x-fold of the first, we get:

$$\begin{aligned} xL_x(x, y, \lambda) - yL_y(x, y, \lambda) &= 3x^3y^3 + 2\lambda x^2 - 3x^3y^3 - 4\lambda y^2 \\ &= 2\lambda(x^2 - 2y^2) = 0 \\ &\Leftrightarrow \lambda = 0 \vee x = \pm\sqrt{2}y. \end{aligned}$$

We now distinguish several cases and subcases to find all solutions:
Case 1. $\lambda = 0$: It follows from the first equation. $x = 0 \vee y = 0$.
  Case 1a. $x= 0$: It is then $y = \pm\sqrt{1/2}$.
  Case 1b. $y= 0$: It is then $x = \pm 1$.
In this case the candidates for extremal places are

$$\boldsymbol{a}_1 = (0, \tfrac{1}{\sqrt{2}}),\ \boldsymbol{a}_2 = (0, \tfrac{-1}{\sqrt{2}}),\ \boldsymbol{a}_3 = (1, 0) \text{ and } \boldsymbol{a}_4 = (-1, 0)\,.$$

Case 2. $x = \pm\sqrt{2}y$: From the third equation then follows $y = \pm 1/2,\ x = \pm 1/\sqrt{2}$.
Here we get the further candidates for extremal places

$$\boldsymbol{a}_5 = \left(\tfrac{1}{\sqrt{2}}, \tfrac{1}{2}\right),\ \boldsymbol{a}_6 = \left(\tfrac{1}{\sqrt{2}}, \tfrac{-1}{2}\right),\ \boldsymbol{a}_7 = \left(\tfrac{-1}{\sqrt{2}}, \tfrac{1}{2}\right) \text{ and } \boldsymbol{a}_8 = \left(\tfrac{-1}{\sqrt{2}}, \tfrac{-1}{2}\right).$$

(3) Because of $g(0, 0) \neq 0$ is the only zero $(0, 0)$ of the gradient is not another candidate.
(4) We now determine the function values at the points $\boldsymbol{a}_1, \ldots, \boldsymbol{a}_8$ and preserved:

$$\begin{aligned} &f(0, \tfrac{\pm 1}{\sqrt{2}}) = f(\pm 1, 0) = 0, \\ &f\left(\tfrac{1}{\sqrt{2}}, \tfrac{1}{2}\right) = f\left(\tfrac{-1}{\sqrt{2}}, \tfrac{-1}{2}\right) = \tfrac{\sqrt{2}}{32}, \\ &f\left(\tfrac{-1}{\sqrt{2}}, \tfrac{1}{2}\right) = f\left(\tfrac{1}{\sqrt{2}}, \tfrac{-1}{2}\right) = \tfrac{-\sqrt{2}}{32}. \end{aligned}$$

The function f therefore has the global minimum $-\sqrt{2}/32$ in the two places $(-1/\sqrt{2}, 1/2)$, $(1/\sqrt{2}, -1/2)$ and the global maximum $\sqrt{2}/32$ at the two places $(1/\sqrt{2}, 1/2)$, $(-1/\sqrt{2}, -1/2)$. ■

We explain the idea on which Lagrange's multiplier rule is based:

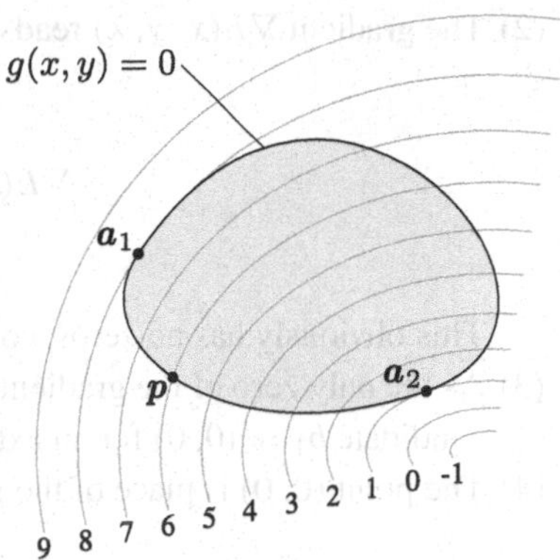

Fig. 50.3 The curve $g = 0$ and contour lines of f

For this purpose, we consider Fig. 50.3, which shows, on the one hand, the *curve of* the points (x, y) with $g(x, y) = 0$ and on the other hand some contour lines of the function f whose extrema under the constraint $g(x, y) = 0$ are searched. The searched extremal places can now be characterized as follows:

At the point $\boldsymbol{p}$ there is certainly no extremum, since the function values increase when moving clockwise on the curve, but decrease when moving counterclockwise (this is indicated by the contour lines of the function). At the point $\boldsymbol{a}_1$ there is a maximum: Both when moving clockwise and counterclockwise, the function values fall. At point $\boldsymbol{a}_2$ the situation is exactly the opposite; there is a minimum in $\boldsymbol{a}_2$. These points $\boldsymbol{a}_1$ and $\boldsymbol{a}_2$ can be easily determined by f and g: In $\boldsymbol{a}_1$ and $\boldsymbol{a}_2$ the gradients of f and g are linearly dependent, and the gradients of f and g are perpendicular to the contour lines of f and g. Thus it follows: There is a $\lambda \in \mathbb{R}$ with $\nabla f(\boldsymbol{a}_1) + \lambda \nabla g(\boldsymbol{a}_1) = 0$, analogously for $\boldsymbol{a}_2$. but this condition just says that we find the extema under the zeros of the gradient of the Lagrange function.

Finally, we show with an example that one does well to consider the candidates $\boldsymbol{b}_i$ from step (3) of the above recipe:

Example 50.4

- We solve the extreme value problem with Lagrange's multiplier rule

$$f : \mathbb{R}^2 \to \mathbb{R}\,, \ f(x, y) = y \text{ under the constraint } g(x, y) = x^2 - y^3 = 0\,.$$

(1) The Lagrange function is:

$$L(x, y, \lambda) = y + \lambda(x^2 - y^3)\,.$$

(2) The gradient $\nabla L(x, y, \lambda)$ reads:

$$\nabla L(x, y, \lambda) = \begin{pmatrix} 2\lambda x \\ 1 - 3\lambda y^2 \\ x^2 - y^3 \end{pmatrix}.$$

This obviously has no zeros, so there are no candidates $\boldsymbol{a}_i$ exist.

(3) As the only zero of the gradient ∇g we get $(0, 0)$. Because of $g(0, 0) = 0$ we get the candidate $\boldsymbol{b}_1 = (0, 0)$ for an extremal.

(4) The point $(0, 0)$ is place of the global minimum $f(0, 0) = 0$. ■

Often the constraint is also given by an inequality, e.g.

Determine the extrema of $f(x, y)$ *under the constraint* $g(x, y) = x^2 + y^2 \leq 1$.

One solves such a problem quite simply and pragmatically:

(1) Determine the local extrema of f inside the circle.
(2) Determine the extrema of f on the boundary, i.e. under the constraint $g(x, y) = x^2 + y^2 - 1 = 0$ using the substitution method or Lagrange's multiplier rule.

50.4 Extrema Under Multiple Constraints

The above procedure can be easily applied to the extreme value determination of a function $f : D \subseteq \mathbb{R}^n \to \mathbb{R}$ in n variables under $m < n$ constraints, we generalize the procedure into a recipe:

Recipe: Solve an Extremal Problem Under Multiple Constraints Using the Method of Lagrange Multipliers

The extrema of the scalar field $f : D \subseteq \mathbb{R}^n \to \mathbb{R}$ under $m < n$ constraints

$$g_1(\boldsymbol{x}) = 0, \ldots, g_m(\boldsymbol{x}) = 0$$

can be obtained with the **method of Lagrange multipliers** as follows:

(1) Set the **Lagrange function** in the variables $\boldsymbol{x} = (x_1, \ldots, x_n), \lambda_1, \ldots, \lambda_m$:

$$L(\boldsymbol{x}, \lambda_1, \ldots, \lambda_m) = f(\boldsymbol{x}) + \lambda_1 g_1(\boldsymbol{x}) + \cdots + \lambda_m g_m(\boldsymbol{x}).$$

(continued)

(2) Determine the zeros $\boldsymbol{a}_1, \ldots, \boldsymbol{a}_k$ of the gradient

$$\nabla L(x_1, \ldots, x_n, \lambda_1, \ldots, \lambda_m).$$

These zeros $\boldsymbol{a}_1, \ldots, \boldsymbol{a}_k \in \mathbb{R}^n$ are **candidates for extremal places.**

(3) Further determine the vectors $\boldsymbol{b}_1, \ldots, \boldsymbol{b}_l \in \mathbb{R}^n$ with $\operatorname{rg} Dg(\boldsymbol{b}_i) < m$ for the Jacobian matrix

$$Dg(x_1, \ldots, x_n) = \left(\frac{\partial g_i}{\partial x_j}(x_1, \ldots, x_n)\right)_{i,j} \in \mathbb{R}^{m \times n}$$

of the function

$$g : \mathbb{R}^n \to \mathbb{R}^m, (x_1, \ldots, x_n) \mapsto \left(g_1(x_1, \ldots, x_n), \ldots, g_m(x_1, \ldots, x_n)\right)^\top.$$

Add all $\boldsymbol{b}_i$ with $g_1(\boldsymbol{b}_i) = \cdots = g_m(\boldsymbol{b}_i) = 0$ to the candidates for extremal places.

(4) Determine the values $f(\boldsymbol{a}_i)$ and $f(\boldsymbol{b}_i)$ for all candidates for extremal places and decide where the global extrema are. If any exist, they are among the candidates.

Example 50.5 We are looking for the extrema of the function

$$f : \mathbb{R}^3 \to \mathbb{R}, \ f(x, y, z) = x^2$$

under the constraints

$$g_1(x, y, z) = x^2 + y^2 + z^2 - 1 = 0 \quad \text{and} \quad g_2(x, y, z) = x - z = 0.$$

(1) The Lagrange function is:

$$L(x, y, z, \lambda, \mu) = x^2 + \lambda(x^2 + y^2 + z^2 - 1) + \mu(x - z).$$

(2) We determine the zeros of the gradient $\nabla L(x, y, z\lambda, \mu)$:

$$\nabla L(x, y, z, \lambda, \mu) = 0 \iff \begin{aligned} 2x + 2\lambda x + \mu &= 0 \\ 2\lambda y &= 0 \\ 2\lambda z - \mu &= 0\,. \\ x^2 + y^2 + z^2 - 1 &= 0 \\ x - z &= 0 \end{aligned}$$

We distinguish two cases:
Case 1. $\lambda = 0$: Then in turn we have $\mu = 0, x = 0, z = 0, y = \pm 1$. We thus obtain the candidates:

$$(0,\,1,\,0) \quad \text{and} \quad (0,\,-1,\,0)\,.$$

Case 2. $\lambda \neq 0$: It follows directly $y = 0$ and $x = z$ and thus the equation $2x^2 = 1$. Here we find as candidates therefore the places

$$(\tfrac{1}{\sqrt{2}},\,0,\,\tfrac{1}{\sqrt{2}}) \quad \text{and} \quad (\tfrac{-1}{\sqrt{2}},\,0,\,\tfrac{-1}{\sqrt{2}})\,.$$

(3) We now determine the Jacobian matrix Dg of the function $g : \mathbb{R}^3 \to \mathbb{R}^2$, $g(x, y, z) = \big(g_1(x, y, z),\, g_2(x, y, z)\big)^\top$:

$$Dg(x, y, z) = \begin{pmatrix} 2x & 2y & 2z \\ 1 & 0 & -1 \end{pmatrix}.$$

This has a rank unequal 2 if and only if $y = 0$ and $x = -z$. So that the constraint $g_2(x, y, z) = 0$ is fulfilled, then $x = y = z = 0$ must be, and thus $g_1(x, y, z) = -1 \neq 0$. So there are no other candidates for extremal places.

(4) We determine the function values at the places from (2):

$$f(0,\,1,\,0) = f(0,\,-1,\,0) = 0\,,\ f(\tfrac{1}{\sqrt{2}},\,0,\,\tfrac{1}{\sqrt{2}}) = f(\tfrac{-1}{\sqrt{2}},\,0,\,\tfrac{-1}{\sqrt{2}}) = \tfrac{1}{2}$$

and thus obtain minimal places at $(0, 1, 0)$ and $(0, -1, 0)$ and maximum places at $(1/\sqrt{2}, 0, 1/\sqrt{2})$ and $(-1/\sqrt{2}, 0, -1/\sqrt{2})$. ■

50.5 Exercise

50.1 Let $D = \{(x, y) \in \mathbb{R}^2 \mid x^2 + y^2 \leq 1\}$ be the unit circle. Determine the extremal points and extrema of the function.

$$f : D \to \mathbb{R}^2 \text{ with } f(x, y) = x^2 - xy + y^2 - x\,.$$

50.2 Determine the extrema of $f : \mathbb{R}^2 \to \mathbb{R}$, $f(x, y) = x^2 + y^2$ under the constraint $g : \mathbb{R}^2 \to \mathbb{R}$, $g(x, y) = y - x^2 + 3 = 0$.

50.3 Determine the maxima and minima of the polynomial

$$f(x, y) = 4x^2 - 3xy$$

on the closed circular disk

$$K = \{(x, y)^\top \in \mathbb{R}^2 \,|\, x^2 + y^2 \le 1\}.$$

Note: Consider f in the interior and on the boundary of K and use the approach with Lagrange multipliers to study the function on the boundary of K.

50.4 Using the method of Lagrange multipliers, determine those points on the boundary of the circle

$$x^2 + y^2 - 2x + 2y + 1 = 0,$$

which have the smallest or largest distance from the point $(-1, 1)$ and give the distances between them.

50.5 Determine the local and global extrema of the function

$$f : E \to \mathbb{R},\; f(x, y) = x^2 - \tfrac{xy}{2} + \tfrac{y^2}{4} - x\,,$$

where

$$E = \left\{(x, y)^\top \in \mathbb{R}^2 \,|\, x^2 + \tfrac{y^2}{4} \le 1\right\}.$$

Note: To study the function on the boundary of E use the approach with Lagrange multipliers.

50.6 Determine the global extrema of the function $f(x, y) = x^2 - xy + y^2 - x$ on the set $S = \{(x, y) \in \mathbb{R}^2 \,|\, x^2 + y^2 \le 1\}$.

50.7 Given are the point $P = (0, 1)$ and the hyperbola $x^2 - y^2 = 2$. Determine the points on the hyperbola which have the shortest distance to P. Do this as follows:

(a) Give the corresponding Lagrange function $L(x, y, \lambda)$.
(b) Determine the local extrema.
(c) Which points of the hyperbola have the smallest distance from P and what is its magnitude.

50.8 Which points of the ellipse $4x^2 + y^2 - 4 = 0$ have from the point $(2, 0)$ extremal distance?

51 Total Differentiation, Differential Operators

So far we have *only* considered partial derivatives or directional derivatives for functions in several variables. In addition to these *special* derivation terms, there is also the *total derivation*. This total derivative is explained as a local approximation of a function f by a linear function and finally leads to the *total differential*, which allows a linearized error estimate.

Finally, we present a clear overview of the differential operators *gradient*, *Laplace*, *divergence* and *rotation*, which are important in the following chapters, and partly their interpretations.

51.1 Total Differentiability

While the directional derivatives, and thus in particular the partial derivatives, describe the behavior or slope of a function in an isolated direction, the *total* derivative approximates the graph of the function f by a *tangent hyperplane*; in particular, it assesses the behavior of the function in each direction. This *total differentiability* is much stronger than partial differentiability: if a function is totally differentiable, it is also partially differentiable, but there are partially differentiable functions that are not totally differentiable.

C. Karpfinger, *Calculus and Linear Algebra in Recipes*,
https://doi.org/10.1007/978-3-662-65458-3_51

Total Differentiability
A (vector-valued) function $f : D \subseteq \mathbb{R}^n \to \mathbb{R}^m$, D open, in n variables is called **totally differentiable**

- **in $\boldsymbol{a} \in D$** if there is a linear mapping $L : \mathbb{R}^n \to \mathbb{R}^m$ with

$$(*) \qquad \lim_{\boldsymbol{h} \to \boldsymbol{0}} \frac{f(\boldsymbol{a}+\boldsymbol{h}) - f(\boldsymbol{a}) - L(\boldsymbol{h})}{\|\boldsymbol{h}\|} = \boldsymbol{0},$$

- on $\boldsymbol{D}$, if f is in every $\boldsymbol{a} \in D$ (total) differentiable.

The (in general, on $\boldsymbol{a} \in D$ dependent) linear mapping L is called the **total differential** of f in $\boldsymbol{a}$ and we write for it $\mathrm{d}f(\boldsymbol{a})$. We have:

- **Representation matrix:** The Jacobian matrix $Df(\boldsymbol{a})$ is the representation matrix of $\mathrm{d}f(\boldsymbol{a})$ with respect to the canonical bases E_n and E_m, $Df(\boldsymbol{a}) = {}_{E_m}M(\mathrm{d}f(\boldsymbol{a}))_{E_n}$.
- **Chain rule:** If $g : D \subseteq \mathbb{R}^n \to \mathbb{R}^m$ and $f : D' \subseteq \mathbb{R}^m \to \mathbb{R}^l$ with $g(D) \subseteq D'$ the composition

$$f \circ g : D \subseteq \mathbb{R}^n \to \mathbb{R}^l$$

can be formed. If g is in $a \in D$ and f in $g(\boldsymbol{a}) \in D'$ are totally differentiable, then we have:

$$\mathrm{d}(f \circ g)(\boldsymbol{a}) = \mathrm{d}f(g(\boldsymbol{a})) \circ \mathrm{d}g(\boldsymbol{a}).$$

Remark In the special case $n = m = 1$, the condition $(*)$ implies just the existence of the limit $f'(a) = \lim_{h \to 0} \frac{f(a+h)-f(a)}{h}$ so that we get back the differentiability notion for a real function of a variable from Chap. 26.

Example 51.1 The function $f : \mathbb{R}^n \to \mathbb{R}^m$ with $f(\boldsymbol{x}) = A\,\boldsymbol{x} + \boldsymbol{b}$, where $A \in \mathbb{R}^{m \times n}$ and $\boldsymbol{b} \in \mathbb{R}^m$, is totally differentiable. With the linear mapping $L : \mathbb{R}^n \to \mathbb{R}^m$, $L(\boldsymbol{x}) = A\boldsymbol{x}$ namely $f(\boldsymbol{a}+\boldsymbol{h}) = f(\boldsymbol{a}) + L(\boldsymbol{h})$ for every $\boldsymbol{a} \in \mathbb{R}^n$, so that independently of $\boldsymbol{h}$ we have:

$$\frac{f(\boldsymbol{a}+\boldsymbol{h}) - f(\boldsymbol{a}) - L(\boldsymbol{h})}{\|\boldsymbol{h}\|} = \boldsymbol{0}. \qquad \blacksquare$$

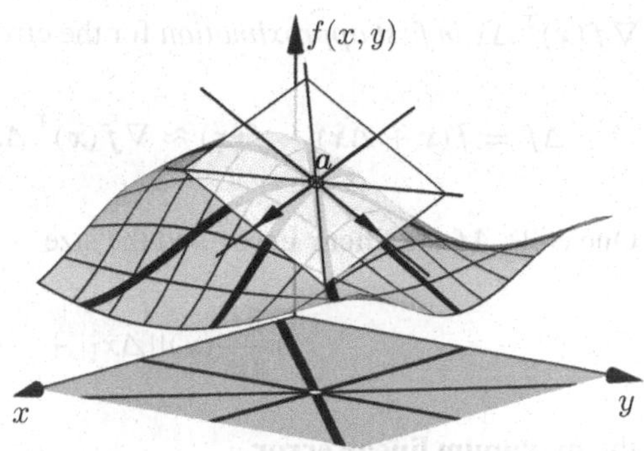

Fig. 51.1 A tangent plane

In the future, we will no longer resort to the definition of total differentiability when deciding whether a function is totally differentiable. It is usually much easier to apply the following criterion:

Criterion for Total Differentiability
If $f : D \subseteq \mathbb{R}^n \to \mathbb{R}^m$ is continuously partial differentiable, then f is totally differentiable.

Thus, any partially differentiable function whose Jacobian matrix contains continuous functions as entries is totally differentiable.

The connection between partial differentiability, the existence of all directional derivatives, and total differentiability is not straightforward. We list some facts for a scalar field $f : D \subseteq \mathbb{R}^n \to \mathbb{R}$:

- If an f is totally differentiable, then f is partially differentiable and there exist at every point $\boldsymbol{a} \in D$ also all directional derivatives. The totality of the tangents in a point $\boldsymbol{a}$ gives the *Tangent hyperplane*, in $\mathbb{R}^3$ i.e. the tangent plane (Fig. 51.1).
- If f is partially differentiable, then f is not necessarily also totally differentiable.
- If f is totally differentiable, then f is also continuous. There are non-continuous functions that are partially differentiable.
- If there exist for f in every $\boldsymbol{a} \in D$ all directional derivatives, then f is not necessarily also totally differentiable.

51.2 The Total Differential

We consider a totally differentiable scalar field $f : D \subseteq \mathbb{R}^n \to \mathbb{R}$ with the Jacobian matrix $\nabla f(\boldsymbol{x})^\top$. For a small $\Delta \boldsymbol{x} = (\Delta x_1, \ldots, \Delta x_n)^\top$ one then obtains because of $L(\Delta \boldsymbol{x}) =$

$\nabla f(\boldsymbol{x})^\top \Delta x$ *in first approximation* for the error

$$\Delta f = f(\boldsymbol{x} + \Delta \boldsymbol{x}) - f(\boldsymbol{x}) \approx \nabla f(\boldsymbol{x})^\top \Delta \boldsymbol{x} = \frac{\partial f}{\partial x_1}(\boldsymbol{x})\Delta x_1 + \cdots + \frac{\partial f}{\partial x_n}(\boldsymbol{x})\Delta x_n .$$

One calls Δf the **linear error** and the size

$$|\frac{\partial f}{\partial x_1}(\boldsymbol{x})||\Delta x_1| + \cdots + |\frac{\partial f}{\partial x_n}(\boldsymbol{x})||\Delta x_n|$$

the **maximum linear error**.

Example 51.2 We consider the scalar field $f : \mathbb{R}^3_{>0} \to \mathbb{R}$ with $f(x, y, z) = x\,y\,z$, i.e., f assigns the *edge lengths* x, y and z the volume $x\,y\,z$ of a cuboid with the edge lengths x, y and z.

Measuring the length of such a cuboid we get $x = l$, $y = b$ and $z = h$ with a measurement accuracy of $|\Delta x|, |\Delta y|, |\Delta z| \le 0.1$. Because of $\nabla f(x, y, z) = (y\,z, x\,z, x\,y)^\top$ we obtain for the maximum linear error

$$b\,h\,|\Delta x| + l\,h\,|\Delta y| + l\,b\,|\Delta z| \le \frac{1}{10}(b\,h + l\,h + l\,b) . \qquad ■$$

In natural sciences and engineering the following conception of the total differential is common: In the above consideration, i.e. the *first approximation*

$$\Delta f \approx \frac{\partial f}{\partial x_1}\Delta x_1 + \cdots + \frac{\partial f}{\partial x_n}\Delta x_n ,$$

we consider *infinitesimally* small Δx_i and write for it $\mathrm{d}x_i$. Here, *infinitesimally* small means the smallest measurable quantity of the particular problem under consideration. We obtain the following representation of the total differential:

The Total Differential

If f is totally differentiable, then the **total differential** has the representation

$$\mathrm{d}f = \frac{\partial f}{\partial x_1}\mathrm{d}x_1 + \cdots + \frac{\partial f}{\partial x_n}\mathrm{d}x_n \text{ or } \mathrm{d}f(\boldsymbol{x}) = \frac{\partial f}{\partial x_1}(\boldsymbol{x})\mathrm{d}x_1 + \cdots + \frac{\partial f}{\partial x_n}(\boldsymbol{x})\mathrm{d}x_n .$$

One calls $\mathrm{d}x_1, \ldots, \mathrm{d}x_n$ also **differentials** of the coordinates $x_1, \ldots, x_n$. In this representation the total differential has the interpretation: If f is a (totally differentiable) function in the variables $x_1, \ldots, x_n$, then small changes $\mathrm{d}x_1, \ldots, \mathrm{d}x_n$ in the variables result in the change $\mathrm{d}f$ as a result.

Note that because of the infinitesimal smallness of the dx_i the equal sign = is justified instead of ≈.

Example 51.3 We consider the volume $V = V(T, P)$ of an ideal gas at temperature T and under pressure P:

$$V(T, P) = \frac{nRT}{P}$$

with constants n and R. If temperature and pressure change simultaneously, the change in volume is obtained as a total differential

$$\mathrm{d}V = \frac{\partial V}{\partial T}\mathrm{d}T + \frac{\partial V}{\partial P}\mathrm{d}P = \frac{nR}{P}\mathrm{d}T - \frac{nRT}{P^2}\mathrm{d}P\,.$$

In particular, we obtain the volume change

$$\mathrm{d}V = \frac{nR}{P}\mathrm{d}T \text{ or } \mathrm{d}V = -\frac{nRT}{P^2}\mathrm{d}P\,,$$

if you only change the temperature and keep the pressure ($\mathrm{d}P = 0$) or only the pressure is changed and the temperature is left unchanged ($\mathrm{d}T = 0$). ■

51.3 Differential Operators

A **differential operator** is a mapping that maps a function to another, with the derivative playing a fundamental role. An important differential operator in the analysis of several variables is the gradient $\nabla : f \to \nabla f$. There are other important operators, which we collect in the box below. Here we use the notation which is advantageous for this case $\partial_i f$ for f_{x_i} and accordingly $\partial_j \partial_i f$ for $f_{x_i x_j}$ or $\partial_i^2 f$ for $\partial_i \partial_i f$:

Differential Operators—Gradient, Laplace, Divergence, and Rotation
In the following formulas, all partial derivatives should exist and be continuous:

- **Gradient**: The gradient ∇ maps a scalar field $f : D \subseteq \mathbb{R}^n \to \mathbb{R}$ to the vector field ∇f:

$$\nabla f = \begin{pmatrix} \partial_1 f \\ \vdots \\ \partial_n f \end{pmatrix}.$$

(continued)

- **Laplace operator**: The Laplace operator Δ maps a scalar field $f : D \subseteq \mathbb{R}^n \to \mathbb{R}$ to the scalar field Δf:

$$\Delta f = \sum_{i=1}^{n} \partial_i^2 f = \partial_1^2 f + \cdots + \partial_n^2 f .$$

- **Divergence**. The divergence div maps a vector field $\boldsymbol{v} = (v_1, \ldots, v_n)^\top : D \subseteq \mathbb{R}^n \to \mathbb{R}^n$ to the scalar field div $\boldsymbol{v}$:

$$\operatorname{div} \boldsymbol{v} = \sum_{i=1}^{n} \partial_i v_i = \partial_1 v_1 + \cdots + \partial_n v_n .$$

- **Rotation**: The rotation rot maps a vector field $\boldsymbol{v} = (v_1, v_2, v_3)^\top : D \subseteq \mathbb{R}^3 \to \mathbb{R}^3$ to the vector field rot $\boldsymbol{v}$:

$$\operatorname{rot} \boldsymbol{v} = \begin{pmatrix} \partial_2 v_3 - \partial_3 v_2 \\ \partial_3 v_1 - \partial_1 v_3 \\ \partial_1 v_2 - \partial_2 v_1 \end{pmatrix} .$$

Note that the divergence and rotation can be formally written as the scalar product and vector product, respectively, of ∇ with $\boldsymbol{v}$:

$$\operatorname{div} \boldsymbol{v} = \langle \nabla , \boldsymbol{v} \rangle = \langle \begin{pmatrix} \partial_1 \\ \vdots \\ \partial_n \end{pmatrix} , \begin{pmatrix} v_1 \\ \vdots \\ v_n \end{pmatrix} \rangle \text{ and } \operatorname{rot} \boldsymbol{v} = \nabla \times \boldsymbol{v} = \begin{pmatrix} \partial_1 \\ \partial_2 \\ \partial_3 \end{pmatrix} \times \begin{pmatrix} v_1 \\ v_2 \\ v_3 \end{pmatrix} .$$

Example 51.4

- The vector field $\boldsymbol{v} : \mathbb{R}^3 \to \mathbb{R}^3,\ \boldsymbol{v}(x, y, z) = (x^2, z, \sin(y))^\top$ has the divergence $\operatorname{div} \boldsymbol{v}(x, y, z) = 2x$.
- The vector field $\boldsymbol{v} : \mathbb{R}^3 \to \mathbb{R}^3, \boldsymbol{v}(x, y, z) = (xy, y^2, xz)^\top$ has the rotation $\operatorname{rot} \boldsymbol{v}(x, y, z) = (0, -z, -x)^\top$. ■

There are several useful identities for the differential operators considered, we summarize them clearly and refer to the exercises for proofs of some of these formulas:

Formulas for Gradient, Laplace, Divergence and Rotation

If $\boldsymbol{v}$ and $\boldsymbol{u}$ are twice continuously differentiable vector fields and g is a twice continuously differentiable scalar field, then (with the agreement $\Delta \boldsymbol{v} = (\Delta v_1, \Delta v_2, \Delta v_3)^\top$) we have:

- $\operatorname{div}(\operatorname{rot}(\boldsymbol{v})) = 0$.
- $\operatorname{rot}(\nabla f) = \boldsymbol{0}$.
- $\operatorname{div}(\nabla f) = \Delta f$.
- $\nabla(\operatorname{div} \boldsymbol{v}) = \operatorname{rot}\operatorname{rot}(\boldsymbol{v}) + \Delta \boldsymbol{v}$.
- $\operatorname{rot}(g\,\boldsymbol{v}) = g \operatorname{rot} \boldsymbol{v} - \boldsymbol{v} \times \nabla g$.
- $\operatorname{rot}(g\,\nabla g) = \boldsymbol{0}$.
- $\operatorname{rot}(\boldsymbol{v} \times \boldsymbol{u}) = (\boldsymbol{u}\nabla)\boldsymbol{v} - (\boldsymbol{v}\nabla)\boldsymbol{u} + \boldsymbol{v} \operatorname{div} \boldsymbol{u} - \boldsymbol{u} \operatorname{div} \boldsymbol{v}$.

We interpret the divergence and the rotation:

Divergence: We consider a small cuboid (or a small rectangle in the $\mathbb{R}^2$) in a *flow* given by a differentiable velocity field $\boldsymbol{v} = (v_1, v_2, v_3)^\top$. Fluid flows into and possibly out of the cuboid. We determine how much fluid leaves the cuboid net, i.e. how much more flows out than flows in. To do this, we first consider only the *x direction*, see Fig. 51.2.

The liquid volume (in the x-direction) is

$$[v_1(x + \mathrm{d}x, y, z) - v_1(x, y, z)]\,\mathrm{d}y\,\mathrm{d}z = \frac{v_1(x + \mathrm{d}x, y, z) - v_1(x, y, z)}{\mathrm{d}x}\,\mathrm{d}x\mathrm{d}y\mathrm{d}z\,.$$

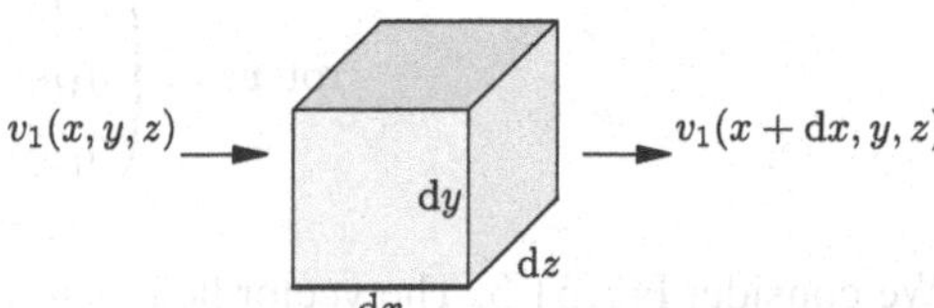

Fig. 51.2 A cuboid is defined by v *flowing* through it, here in the x-direction

Analogous expressions are obtained for y- and z-directions. This gives us the total volume gain

$$\left(\frac{v_1(x+\mathrm{d}x, y, z) - v_1(x, y, z)}{\mathrm{d}x} + \frac{v_2(x, y+\mathrm{d}y, z) - v_2(x, y, z)}{\mathrm{d}y} + \frac{v_3(x, y, z+\mathrm{d}z) - v_3(x, y, z)}{\mathrm{d}z}\right) \mathrm{d}x\mathrm{d}y\mathrm{d}z.$$

division by the volume element $\mathrm{d}x\mathrm{d}y\mathrm{d}z$ yields the volume gain per volume element; in the limit transition one finally obtains the divergence

$$\operatorname{div}(\boldsymbol{v}) = \frac{\partial v_1}{\partial x} + \frac{\partial v_2}{\partial y} + \frac{\partial v_3}{\partial z}.$$

The divergence thus indicates whether at a point $(x, y, z) \in D$ fluid *arises* or is *lost* or whether *equilibrium*:

$\operatorname{div}(\boldsymbol{v}) > 0 \Rightarrow$ More flows out than in: **source**
$\operatorname{div}(\boldsymbol{v}) < 0 \Rightarrow$ More flows in than out: **sink**
$\operatorname{div}(\boldsymbol{v}) = 0 \Rightarrow$ There is as much inflow as outflow: **source – free**

One calls $\operatorname{div}(\boldsymbol{v})$ therefore also **source density of$\boldsymbol{v}$**.

Example 51.5 The rotation of a vector field $\boldsymbol{v} = (v_1,\ v_2,\ v_3)^\top$ is defined as

$$\operatorname{rot}(\boldsymbol{v}) = \begin{pmatrix} \partial_2 v_3 - \partial_3 v_2 \\ \partial_3 v_1 - \partial_1 v_3 \\ \partial_1 v_2 - \partial_2 v_1 \end{pmatrix} \in \mathbb{R}^3.$$

We consider Fig. 51.3. The vector field here is of the form $\boldsymbol{v}(x, y, z) = (-y,\ x,\ 0)^\top$. The drawn square will rotate (rotate) counterclockwise around its own axis in this flow, the axis of rotation points out of the drawing plane.

The axis of rotation is

$$\operatorname{rot}(\boldsymbol{v}) = \begin{pmatrix} 0 & - & 0 \\ 0 & - & 0 \\ \partial_1 v_2 & - & \partial_2 v_1 \end{pmatrix} = \begin{pmatrix} 0 \\ 0 \\ 2 \end{pmatrix}.$$

For the components of rotation, the change in the i-th component in j-direction compared to the change of the j-th component in i-direction. The rotation is also referred to as the

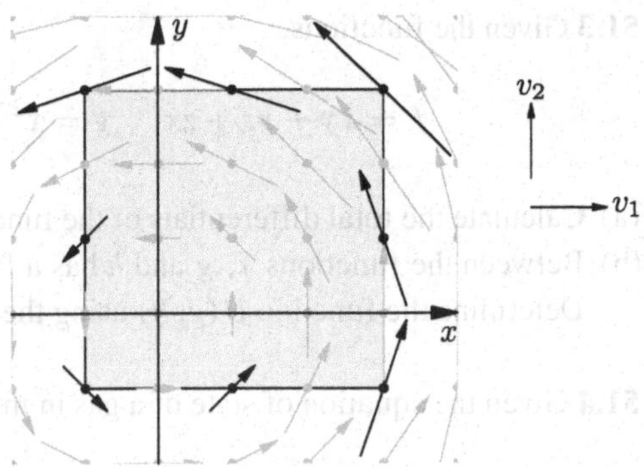

Fig. 51.3 The vector field makes the square rotate

vortex density of v. And one calls a vector field $\boldsymbol{v} : D \subseteq \mathbb{R}^3 \to \mathbb{R}^3$ **vortex-free,** if $\operatorname{rot}(\boldsymbol{v}) = \boldsymbol{0}$ to D.

51.4 Exercises

51.1 Show the following statements:

(a) For a twice continuously differentiable vector field $\boldsymbol{v} \colon D \subseteq \mathbb{R}^3 \to \mathbb{R}^3$ we have for all $\boldsymbol{x} \in D$

$$\operatorname{div}(\operatorname{rot} \boldsymbol{v}(\boldsymbol{x})) = 0.$$

(b) For a twice continuously differentiable scalar field $f \colon D \subseteq \mathbb{R}^3 \to \mathbb{R}$ we have for all $\boldsymbol{x} \in D$

$$\operatorname{rot}(\nabla f(\boldsymbol{x})) = \boldsymbol{0}\,.$$

(c) For continuously differentiable functions $g \colon D \subseteq \mathbb{R}^3 \to \mathbb{R}$ and $\boldsymbol{v} \colon D \subseteq \mathbb{R}^3 \to \mathbb{R}^3$ we have for all $\boldsymbol{x} \in D$

$$\operatorname{div}(g(\boldsymbol{x})\boldsymbol{v}(\boldsymbol{x})) = g(\boldsymbol{x}) \operatorname{div} \boldsymbol{v}(\boldsymbol{x}) + \boldsymbol{v}(\boldsymbol{x})^\top \nabla g(\boldsymbol{x}).$$

51.2 Show

$$\operatorname{rot}(\operatorname{rot} \boldsymbol{v}) = -\Delta \boldsymbol{v} + \operatorname{grad}(\operatorname{div} \boldsymbol{v})\,,$$

where the components of $\boldsymbol{v}$ shall be twice continuously differentiable.

51.3 Given the functions

$$f = xy + yz + zx, \quad g = x^2 + y^2 + z^2, \quad h = x + y + z.$$

(a) Calculate the total differentials of the functions.
(b) Between the functions f, g and h has a functional relation of the form $f = U(g, h)$. Determine the function $U(g, h)$ using the total differentials.

51.4 Given the equation of state of a gas in implicit form

$$f(P, V, T) = 0,$$

where P the pressure, V the volume and T the temperature of the gas. Let it be shown that the partial derivatives are $\frac{\partial P}{\partial T}\big|_V$, $\frac{\partial T}{\partial V}\big|_P$ and $\frac{\partial P}{\partial V}\big|_T$ where an index indicates the variable held constant, satisfy the following equation

$$\left.\frac{\partial P}{\partial T}\right|_V \left.\frac{\partial T}{\partial V}\right|_P = -\left.\frac{\partial P}{\partial V}\right|_T.$$

51.5

(a) Show that the function

$$u(x, y) = \cosh x \sin y - x^2$$

solves the two-dimensional Poisson equation $-\Delta u = 2$.
(b) Show that the one-dimensional heat conduction equation $u_t - k\Delta u = 0$ is solved using $k > 0$ from the function

$$u(x, t) = \mathrm{e}^{-t} \sin(x/\sqrt{k})$$

is solved.
(c) Show that with $r = \sqrt{x^2 + y^2 + z^2}$ and $(x, y, z) \neq (0, 0, 0)$ the function

$$u(r, t) = \frac{1}{r} \sin(r - ct)$$

solves the three-dimensional wave equation $u_{tt} - c^2 \Delta u = 0$.

51.6 Calculate for the vector field $\boldsymbol{v}(x, y, z) = \begin{pmatrix} x & -y & xz \end{pmatrix}$:

(a) $\operatorname{div} \boldsymbol{v} = \nabla \cdot \boldsymbol{v}$
(b) $\operatorname{rot} \boldsymbol{v} = \nabla \times \boldsymbol{v}$
(c) $\operatorname{div} \operatorname{rot} \boldsymbol{v} = \nabla \cdot (\nabla \times \boldsymbol{v})$
(d) $\operatorname{rot}(\operatorname{rot} \boldsymbol{v}) = \nabla \times (\nabla \times \boldsymbol{v})$
(e) $\nabla(\operatorname{div} \boldsymbol{v}) = \nabla(\nabla \cdot \boldsymbol{v})$

Implicit Functions

52

We have already discussed in Chap. 23 the topic *implicit functions*: In practice, functions are often not given by explicit specification of the mapping rule, but are implicitly determined by an equation. We encountered this problem several times in the solutions of differential equations. In this chapter we provide a method how to handle implicit functions nevertheless. For example, under certain conditions it will be possible to determine the derivative of an implicit function at a point x without knowing an explicit mapping rule of this function.

52.1 Implicit Functions: The Simple Case

A function $y : I \to \mathbb{R}$ is often given by an equation of the form $F(x, y) = 0$ which may not be explicitly solvable to $y = y(x)$.

We consider such an equation $F(x, y) = 0$ in the variables x and y and the zero set $N_0 = \{(x, y) \in \mathbb{R}^2 \mid F(x, y) = 0\}$. The set N_0 is the *level set of* the function F to the level 0, i.e. the intersection of the graph of F with the x-y-plane.

Example 52.1

- We consider the function $F(x, y) = x^2(1 - x^2) - y^2$. The graph of this function can be seen with some level lines in the following figure. The equation $F(x, y) = 0$ yields the set of zeros shown in Fig. 52.1—this curve is called **lemniscate**.

C. Karpfinger, *Calculus and Linear Algebra in Recipes*,
https://doi.org/10.1007/978-3-662-65458-3_52

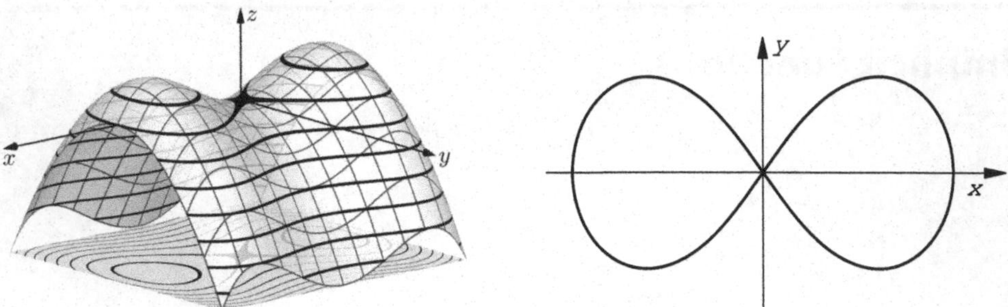

Fig. 52.1 The lemniscate is the level line to level 0 of the function F

This set is certainly not the graph of a function $f : \mathbb{R} \to \mathbb{R}$. However, if we restrict the domain of definition of F on $]0, 1[\times]0, \infty[$ we can solve the equation $F(x, y) = 0$ to y:

$$x^2(1 - x^2) - y^2 = 0 \Leftrightarrow y = \sqrt{x^2(1 - x^2)}\,.$$

The graph of the function

$$f :]0, 1[\to \mathbb{R} \text{ with } f(x) = \sqrt{x^2(1 - x^2)}$$

is the upper right part of the lemniscate (see Fig. 52.2).
With a little effort, we can also solve the equation for x : To do this, we restrict F to $]1/\sqrt{2}, \infty[\times]-1/2, 1/2[$ and get:

$$x^2\,(1 - x^2) - y^2 = 0 \Leftrightarrow x = \sqrt{1/2 + 1/2\sqrt{1 - 4y^2}}\,.$$

The graph of the function

$$g :]-1/2, 1/2[\to \mathbb{R} \text{ with } g(y) = \sqrt{1/2 + 1/2\sqrt{1 - 4y^2}}$$

is the right-hand part of the lemniscate (see Fig. 52.2).
Note that the images of f and g are each again not graphs of the other variable; graphs of real-valued functions of a real-valued variable have no vertical tangents.

- We consider the function $F(x, y) = \mathrm{e}^y + y^3 - x$. The equation $F(x, y) = 0$ yields the zero set shown in Fig. 52.3. This level set is the graph of a function $y = y(x)$. But this function y is not to be given explicitly, because the equation

$$\mathrm{e}^y + y^3 - x = 0$$

is not solvable to y. ■

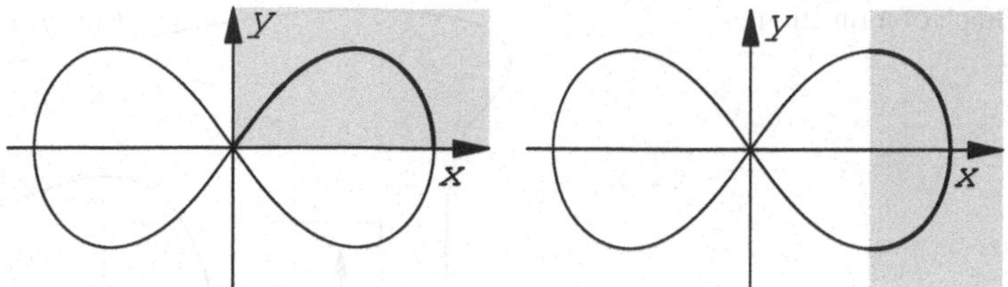

Fig. 52.2 The *shaded* part of the lemniscate is the graph of a function in x or y

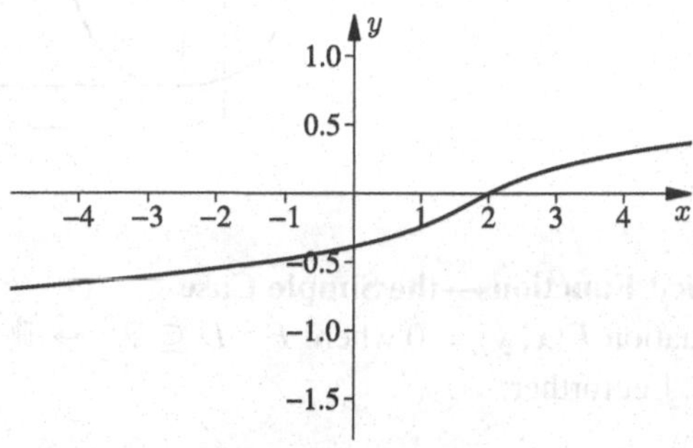

Fig. 52.3 Graph of a function y

In each of these examples, a function is *implicitly* given by an equation $F(x, y) = 0$, more precisely:

> **Implicit Function**
> We consider an equation $F(x, y) = 0$ where $F : D \subseteq \mathbb{R}^2 \to \mathbb{R}$ is a function in the variables x and y. This equation $F(x, y) = 0$ declares an **implicit function** $f : I \to J$ if $I \times J \subseteq D$ and for each $x \in I$ there is exactly one $y = f(x) \in J$, so that $F(x, f(x)) = 0$ is satisfied.
> Here also the roles of x and y may be interchanged.

One also uses the suggestive way of speaking *the zero set* $F(x, y) = 0$ *is locally graph of a function* f in the sense that a subset of the zero set of $F : D \subseteq \mathbb{R}^2 \to \mathbb{R}$ is graph of a function $f : I \to J$ of a variable (see Fig. 52.4).

Now the question is how to find out that by the equation $F(x, y) = 0$ an implicit function f is explained. From the above example of the lemniscate we already know that a *vertical tangent* speaks against a local solvability. The following theorem about implicit functions gives information about this and even allows us to differentiate the implicit function f at a point x with the help of the function F:

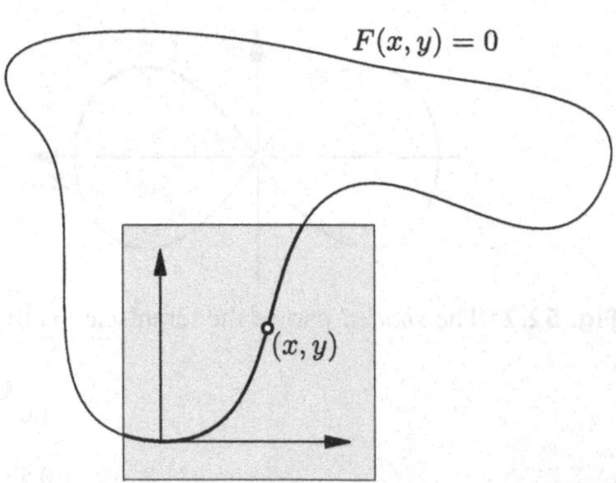

Fig. 52.4 Graph of a function f

Theorem on Implicit Functions—the Simple Case
We consider an equation $F(x, y) = 0$ where $F : D \subseteq \mathbb{R}^2 \to \mathbb{R}$ is a function in the variables x and y is. Let further:

- $D \subseteq \mathbb{R}^2$ is open.
- $F : D \to \mathbb{R}$ is continuously partially differentiable on D.
- There is $(x_0, y_0) \in D$ with $F(x_0, y_0) = 0$.
- $F_y(x_0, y_0) \neq 0$.

Then there are open intervals $I \subseteq \mathbb{R}$ with $x_0 \in I$ and $J \subseteq \mathbb{R}$ with $y_0 \in J$ with

$$I \times J \subseteq D \text{ and } F_y(x, y) \neq 0 \text{ for all } (x, y) \in I \times J .$$

There is also an implicit function $f : I \to J$ with $F(x, f(x)) = 0$ for all $x \in I$ where for all $x \in I$ we have:

$$f'(x) = \frac{-F_x(x, f(x))}{F_y(x, f(x))} \text{ and}$$

$$f''(x) = \frac{-(F_{xx}(x, f(x)) + 2F_{xy}(x, f(x))f'(x) + F_{yy}(x, f(x))(f'(x))^2)}{F_y(x, f(x))},$$

if $F \in C^2(D)$.

It is remarkable that one can determine the derivative of f with this theorem, even if one does not know f itself. This fact results from the chain rule: First we rewrite $F(x, y)$, we have:

$$F(x, y) = F \circ \gamma(x) \text{ with } \gamma(x) = \begin{pmatrix} x \\ f(x) \end{pmatrix}.$$

Now we derive the equation $F(x, y) = 0$ according to the chain rule and obtain:

$$0 = (F \circ \gamma)'(x) = \nabla F(\gamma(x))^\top \dot{\gamma}(x) = \left(F_x(x, f(x)), F_y(x, f(x))\right) \begin{pmatrix} 1 \\ f'(x) \end{pmatrix}$$

$$= F_x(x, f(x)) + F_y(x, f(x))\, f'(x).$$

From this follows the formula for f', by deriving it again we then also obtain the formula for f''.

Example 52.2 We consider the function $F : \mathbb{R}^2 \to \mathbb{R}$ with $F(x, y) = x^2(1 - x^2) - y^2$ (see above Example 52.1). We have:

- $D = \mathbb{R}^2$ is open.
- F is continuously partially differentiable on $D = \mathbb{R}^2$.
- $(\frac{1}{\sqrt{2}}, \frac{1}{2})^\top \in D$ satisfies $F(\frac{1}{\sqrt{2}}, \frac{1}{2}) = 0$.
- $F_y(\frac{1}{\sqrt{2}}, \frac{1}{2}) = -1 \neq 0$.

Therefore, by the theorem on implicit functions, it follows that there exist two intervals $I \subseteq \mathbb{R}$ with $1/\sqrt{2} \in I$ and $J \subseteq \mathbb{R}$ with $1/2 \in J$ as well as a function $f : I \to J$ with

$$F(x, f(x)) = 0 \text{ for all } x \in I.$$

We determine the derivative f', we have:

$$f'(x) = -F_x(x, f(x))/F_y(x, f(x)) = \frac{x - 2x^3}{y}.$$

By $x = 1/\sqrt{2}$ we obtain a candidate extremum of f, although we have f has not been explicitly stated. ■

In the situation of the above Theorem about implicit functions, we have: If $f'(x) = 0$, thus x is a candidate for an extremal point of the implicit function f, than the second derivative is $f''(x) = -F_{xx}(x, f(x)/F_y(x, f(x))$ so that we can use the following recipe to decide about extremal points of an implicit function, if necessary:

Recipe: Determine Extrema of an Implicit Function
Extrema of an implicit function $f : I \to J$ which is given by an equation $F(x, y) = 0$ with a C^2-function F are found as follows:

(1) Determine places (x_0, y_0) with $x_0 \in I$ and $F(x_0, y_0) = 0$ and $F_x(x_0, y_0) = 0$.
(2)

- If $F_{xx}(x_0, y_0)/F_y(x_0, y_0) < 0$, then x_0 is a place of a local minimum of f.
- If $F_{xx}(x_0, y_0)/F_y(x_0, y_0) > 0$ then x_0 is a place of a local maximum of f.

Example 52.3 We consider again the function $F : \mathbb{R}^2 \to \mathbb{R}$ with $F(x, y) = x^2(1 - x^2) - y^2$.

(1) According to the above example, the point $(x_0, y_0) = (1/\sqrt{2}, 1/2)$ satisfies both $x \in I$ as well as $F(x, y) = 0$ and $F_x(x, y) = 0$.
(2) Because of $F_{xx}(x, y)/F_y(x, y) = (6x^2 - 1)/y$ we have $F_{xx}(x_0, y_0)/F_y(x_0, y_0) > 0$ so that the implicit function f in $x_0 = 1/\sqrt{2}$ has a local maximum. ■

52.2 Implicit Functions: The General Case

So far we have *curves* in the $\mathbb{R}^2$ represented locally as graphs of implicit functions, which are given by an equation of the form $F(x, y) = 0$ for a scalar field F in the two variables x and y. Now we ask much more generally about the *solvability of* an equation of the form $F(x_1, \dots, x_n) = \mathbf{0}$ where F is not only a scalar field but also a vector-valued function in the n variables $x_1, \dots, x_n$:

$$F : D \subseteq \mathbb{R}^n \to \mathbb{R}^m \quad \text{with} \quad F(x_1, \dots, x_n) = \begin{pmatrix} F_1(x_1, \dots, x_n) \\ \vdots \\ F_m(x_1, \dots, x_n) \end{pmatrix} \quad \text{and } m < n\,.$$

The condition $m < n$ is quite natural, since this makes the system of equations $F(x_1, \dots, x_n) = 0$ *solvable*: It has in general a non-empty set of solutions. It follows

$$n = k + m \text{ for } k \in \mathbb{N}\,.$$

Thus F is a function of $\mathbb{R}^{k+m}$ to $\mathbb{R}^m$, and each $z \in \mathbb{R}^{k+m}$ has a decomposition of the form:

$$z = (x, y) \text{ for } x \in \mathbb{R}^k \text{ and } y \in \mathbb{R}^m .$$

For $k = m = 1$ this corresponds to the *simple case* we considered in the last section.

- In the *simple case*, we have an equation $F(x, y) = 0$, $x \in \mathbb{R}$, $y \in \mathbb{R}$ and $(x_0, y_0) \in \mathbb{R}^2$ with $F(x_0, y_0) = 0$ and search for a function $f : I \subseteq \mathbb{R} \to J \subseteq \mathbb{R}$ with $F(x, f(x)) = 0$ where $x_0 \in I$ and $y_0 \in J$.
- In the *general case*, we have an equation $F(\boldsymbol{x}, \boldsymbol{y}) = \boldsymbol{0}$, $\boldsymbol{x} \in \mathbb{R}^k$, $\boldsymbol{y} \in \mathbb{R}^m$ and $(\boldsymbol{x}_0, \boldsymbol{y}_0) \in \mathbb{R}^{k+m}$ with $F(\boldsymbol{x}_0, \boldsymbol{y}_0) = \boldsymbol{0}$ and search for a function $f : I \subseteq \mathbb{R}^k \to J \subseteq \mathbb{R}^m$ whith $F(\boldsymbol{x}, f(\boldsymbol{x})) = \boldsymbol{0}$ where $\boldsymbol{x}_0 \in I$ and $\boldsymbol{y}_0 \in J$.

As already indicated here, we now always solve for the *part* $\boldsymbol{y} \in \mathbb{R}^m$. After a renaming of the variables and a subsequent reordering, this can always be achieved.

Theorem on Implicit Functions—the General Case
We consider an equation $F(\boldsymbol{x}, \boldsymbol{y}) = \boldsymbol{0}$, $\boldsymbol{x} \in \mathbb{R}^k$ and $\boldsymbol{y} \in \mathbb{R}^m$ where $F : D \subseteq \mathbb{R}^k \times \mathbb{R}^m \to \mathbb{R}^m$ is a function in $n = k + m$ variables $\boldsymbol{x} = (x_1, \ldots, x_k)$ and $\boldsymbol{y} = (y_1, \ldots, y_m)$ is. We assume:

- $D \subseteq \mathbb{R}^m \times \mathbb{R}^k$ is open.
- $F : D \to \mathbb{R}^m$ is continuously differentiable on D.
- There are $(\boldsymbol{x}_0, \boldsymbol{y}_0) \in D$, $\boldsymbol{x}_0 \in \mathbb{R}^k$ and $\boldsymbol{y}_0 \in \mathbb{R}^m$, with $F(\boldsymbol{x}_0, \boldsymbol{y}_0) = \boldsymbol{0}$.
- The submatrix

$$DF_{\boldsymbol{y}}(\boldsymbol{x}_0, \boldsymbol{y}_0) = \left(\frac{\partial F_i}{\partial y_j}(\boldsymbol{x}_0, \boldsymbol{y}_0)\right)_{\substack{i=1,\ldots,m \\ j=1,\ldots,m}}$$

  of the Jacobian matrix $DF(\boldsymbol{x}_0, \boldsymbol{y}_0)$ is invertible.

Then there are open sets $I \subseteq \mathbb{R}^k$ with $\boldsymbol{x}_0 \in I$ and $J \subseteq \mathbb{R}^m$ with $\boldsymbol{y}_0 \in J$ with

$$I \times J \subseteq D \text{ and } DF_{\boldsymbol{y}}(\boldsymbol{x}, \boldsymbol{y}) \text{ is invertible for all } (\boldsymbol{x}, \boldsymbol{y}) \in I \times J .$$

There is also an implicit function $f : I \to J$ with $F(\boldsymbol{x}, f(\boldsymbol{x})) = \boldsymbol{0}$ for all $\boldsymbol{x} \in I$ where for all $\boldsymbol{x} \in I$ we have:

$$Df(\boldsymbol{x}) = -(DF_{\boldsymbol{y}}(\boldsymbol{x}, f(\boldsymbol{x}))^{-1} DF_{\boldsymbol{x}}(\boldsymbol{x}, f(\boldsymbol{x})) .$$

Example 52.4

- We consider the function

$$F : \mathbb{R}^3 \to \mathbb{R} \text{ with } F(x, y, z) = \sin(z + y - x^2) - \frac{1}{\sqrt{2}}.$$

So we are in the case $n = 3$, $m = 1$, $k = 2$; searched is a function $f : I \subseteq \mathbb{R}^2 \to \mathbb{R}$.

The point $(x_0, y_0, z_0) = (0, 0, \pi/4)^\top$ fulfills $F(x_0, y_0, z_0) = 0$. The Jacobian matrix DF is

$$DF(x, y, z) = \left(-2x\cos(z + y - x^2),\ \cos(z + y - x^2),\ \cos(z + y - x^2)\right), \text{ so}$$

$$DF(0, 0, \tfrac{\pi}{4}) = \left(0, \tfrac{1}{\sqrt{2}}, \tfrac{1}{\sqrt{2}}\right).$$

Thus F can be solved for y or z but not to x, since the first component is 0. We solve to z, so we are in the situation $\boldsymbol{x} = (x, y) \in \mathbb{R}^2$ and $\boldsymbol{y} = z$, where $\boldsymbol{x}_0 = (0, 0)$, $\boldsymbol{y}_0 = \pi/4$, since $x_0 = y_0 = 0$ and $z_0 = \pi/4$.

According to the Theorem on implicit functions there are open sets $I = B_\varepsilon((0, 0)) \subseteq \mathbb{R}^2$ and $J = B_\delta(\pi/4)$ and a function

$$f : I \to J,\ (x, y) \mapsto z \text{ with } F(x, y, f(x, y)) = 0 \text{ for all } (x, y) \in B_\varepsilon((0, 0)).$$

The Jacobian matrix of the function f is according to the Theorem on implicit functions:

$$\begin{aligned} Df(x, y) &= -(DF_{\boldsymbol{y}}(\boldsymbol{x}, f(\boldsymbol{x})))^{-1} DF_{\boldsymbol{x}}(\boldsymbol{x}, f(\boldsymbol{x})) \\ &= -(\cos(z + y - x^2))^{-1}(-2x\cos(z + y - x^2), \cos(z + y - x^2)) \\ &= (2x, -1). \end{aligned}$$

In fact, the function f can even be specified concretely in this case:

$$\sin(z + y - x^2) = 1/\sqrt{2} \Leftrightarrow z = f(x, y) = \pi/4 + x^2 - y.$$

Because $f_x = 2x$ and $f_y = -1$ we find our calculations confirmed.

- We consider the function

$$F : \mathbb{R}^3 \to \mathbb{R}^2 \text{ with } F(x, y, z) = \begin{pmatrix} x^2 + y^2 - z^2 - 8 \\ \sin(\pi x) + \sin(\pi y) + \sin(\pi z) \end{pmatrix}.$$

So we are in the case $n = 3$, $m = 2$, $k = 1$; searched is a function $f : I \subseteq \mathbb{R} \to \mathbb{R}^2$.

The point $(x_0, y_0, z_0) = (2, 2, 0)^\top$ fulfills $F(x_0, y_0, z_0) = \mathbf{0}$. The Jacobian matrix DF is

$$DF(x, y, z) = \begin{pmatrix} 2x & 2y & -2z \\ \pi\cos(\pi x) & \pi\cos(\pi y) & \pi\cos(\pi z) \end{pmatrix},$$

so

$$DF(2, 2, 0) = \begin{pmatrix} 4 & 4 & 0 \\ \pi & \pi & \pi \end{pmatrix}.$$

Thus F can be resolved to y and z or to x and z but not to x and y, because the matrix consisting of the first two columns is not invertible. We resolve to y and z, so we are in the situation $\boldsymbol{x} = x \in \mathbb{R}$ and $\boldsymbol{y} = (y, z) \in \mathbb{R}^2$, whereby $\boldsymbol{x}_0 = 2$, $\boldsymbol{y}_0 = (2, 0)$, since $x_0 = y_0 = 2$ and $z_0 = 0$.

According to the theorem on implicit functions, there are open sets $I = B_\varepsilon(2) \subseteq \mathbb{R}$ and $J = B_\delta(2, 0)$ and a function

$$f : I \to J,\ x \mapsto (y, z) \text{ with } F(x, f(x)) = \mathbf{0} \text{ for all } x \in B_\varepsilon(2).$$

The Jacobian matrix of the function f by the theorem on implicit functions is:

$$\begin{aligned} Df(x) &= -(DF_{\boldsymbol{y}}(\boldsymbol{x}, f(\boldsymbol{x})))^{-1} DF_{\boldsymbol{x}}(\boldsymbol{x}, f(\boldsymbol{x})) \\ &= -\begin{pmatrix} 2y & -2z \\ \pi\cos\pi y & \pi\cos\pi z \end{pmatrix}^{-1} \begin{pmatrix} 2x \\ \pi\cos\pi x \end{pmatrix}. \end{aligned}$$ ■

52.3 Exercises

52.1 Let $F : D \subseteq \mathbb{R}^2 \to \mathbb{R}$, D open, a C^1-function. The level lines $N_c = \{(x, y) \mid F(x, y) = c\} \neq \emptyset$ define (implicitly) curves (possibly degenerate to a point). Show:

(a) If $F_x(x, y) = 0$ and $\nabla F(x, y) \neq 0$, then N_c has a horizontal tangent there.
(b) If $F_y(x, y) = 0$ and $\nabla F(x, y) \neq 0$, then N_c has a vertical tangent there.

52.2 Let $F(x, y) = x^2 - xy + y^2$.

(a) Where does the curve defined by $F(x, y) = 2$ have horizontal and vertical tangents?
(b) Why can the curve in a neighborhood of $(\sqrt{2}, 0)$ can be represented as a graph of a C^1-function $y = f(x)$?
(c) Calculate $f'(\sqrt{2})$.

52.3 Given the nonlinear system of equations

$$F(x, y, z) = \begin{pmatrix} x^2 + y^2 + z^2 - 6\sqrt{x^2 + y^2} + 8 \\ x^2 + y^2 + z^2 - 2x - 6y + 8 \end{pmatrix} = \begin{pmatrix} 0 \\ 0 \end{pmatrix}.$$

(a) Show that $F(0, 3, 1) = 0$.
(b) Using the Theorem on implicit functions, check that the equation $F(x, y, z) = 0$ at the point $(0, 3, 1)$ locally can be solved to x and y or to x and z or to y and z and, if necessary, perform this solution.

52.4 Let it be reasoned that $F(x, y, z) = z^3 + 4z - x^2 + xy^2 + 8y - 7 = 0$ in the neighborhood of each $(x, y) \in \mathbb{R}^2$ can be represented as a graph of a function $z = f(x, y)$ in a neighborhood. Calculate there the gradient of *f*.

52.5 Let $F(x, y) = x^3 + y^3 - 3xy$. Where does the curve defined by $F(x, y) = 0$ implicitly has horizontal and vertical tangents, where singular points (that is, points (x_0, y_0) with $F_x(x_0, y_0) = 0 = F_y(x_0, y_0)$? Why can the curve in every point with $x < 0$ be represented in each neighborhood as a graph of a C^1-function $y = f(x)$? One computes there $f'(x)$.

52.6 Investigate whether the nonlinear systems of equations

$$\begin{aligned} x + y - \sin z &= 0 \\ \exp z - x - y^3 &= 1 \end{aligned} \quad \text{and} \quad \begin{aligned} x + y - \sin z &= 0 \\ \exp x - x^2 + y &= 1 \end{aligned}$$

in a neighborhood of $(0, 0, 0)^\top$ to (y, z) can be solved.

52.7 Let $F(x, y, z) = z^2 - x^2 - y^2 - 2xz - 2yz - 2xy - 1$ and $N_0 = \{(x, y, z) \mid F(x, y, z) = 0\}$.

(a) Show: For every $(x, y) \in \mathbb{R}^2$ there is a neighborhood U, in which N_0 as the graph of a function $z = f(x, y)$ can be represented.
(b) Calculate the gradient $\nabla f(x, y)$.

52.8 Given the function

$$F(x, y) = x^3 - 3xy^2 + 16.$$

Show: For $x, y > 0$ the curve given by $F(x, y) = 0$ can be represented as the graph of a function $y = f(x)$. Determine the local extrema of $f(x)$.

52.9 Given the nonlinear system of equations

$$F(x, y, z) = \begin{pmatrix} 2\sin x + z\,e^y \\ \sqrt{x^2 + y^2 + 3} \end{pmatrix} = \begin{pmatrix} 0 \\ 2 \end{pmatrix}$$

(a) Calculate the Jacobian matrix of the function F.
(b) Check whether the system of equations at the point $(0, 1, 0)$ is locally solvable to x and y or to x and z or to y and z.

52.10

(a) Prove the local solvability of

$$x^2 - 2xy - y^2 - 2x + 2y + 2 = 0$$

to x in a neighborhood of $(x_0, y_0) = (3, 1)$ and calculate $h'(1)$ and $h'(1)$ for the implicitly defined function $x = h(y)$.
(b) Explicitly calculate the function $h(y)$ from (a), give its domain and confirm the calculated derivative values.

52.11 Let $F : \mathbb{R}^2 \to \mathbb{R}$ defined by

$$F(x, y) = x\,e^{-xy} - 4y.$$

(a) Show that by $F(x, y) = 0$ a function $y = f(x)$ for all $x \in \mathbb{R}$ is implicitly defined.
(b) Calculate the values $f(0)$ and $f'(0)$.
(c) Determine the stationary points of f.

53 Coordinate Transformations

We have learned in $\mathbb{C} = \mathbb{R}^2$ two ways to represent each element $z \neq 0$ uniquely: $z = (a, b)$ with the Cartesian coordinates a and b or $z = (r, \varphi)$ with the polar coordinates r and φ. Behind this representation of elements with respect to different *coordinate systems* lies a coordinate transformation $(r, \varphi) \to (a, b)$. In the $\mathbb{R}^3$ several such transformations are of special interest, especially cylindrical and spherical coordinates play a fundamental role in multidimensional engineering analysis, because many problems of engineering mathematics can be described and solved much easier in special coordinates.

53.1 Transformations and Transformation Matrices

Every point in the $\mathbb{R}^3$ is uniquely determined by its Cartesian coordinates x, y, and z. This uniqueness of the representation of each element is expected from each coordinate system. Thus, if one has two coordinate systems, one expects that there is a bijection ϕ from the set of all points with respect to one coordinate system K_1 to the set of all points with respect to the other coordinate system K_2, $\phi : x_{K_1} \to x_{K_2}$. This abstract notion of a *coordinate system change* leads to the following definition:

Coordinate Transformation and Transformation Matrix

If B and D are open subsets of the $\mathbb{R}^n$ a continuously differentiable bijection $\phi : B \to D$ is called a **coordinate transformation** if ϕ^{-1} is also continuously differentiable.

If $\phi : B \to D$ is a coordinate transformation, the quadratic Jacobian matrix $D\phi$ is called **transformation matrix** and the determinant $\det D\phi$ the **functional determinant** or **Jacobian determinant**.

C. Karpfinger, *Calculus and Linear Algebra in Recipes*,
https://doi.org/10.1007/978-3-662-65458-3_53

In addition to Cartesian coordinates, the most important coordinates of $\mathbb{R}^3$ are cylindrical and spherical coordinates:

$$\underbrace{(x,\, y,\, z)}_{\text{cartesian}} \qquad \underbrace{(r,\, \varphi,\, z)}_{\text{cylinder}} \qquad \underbrace{(r,\, \varphi,\, \vartheta)}_{\text{sphere}} .$$

In $\mathbb{R}^2$ polar coordinates play a fundamental role. In the following section we consider these coordinates and the associated transformation matrices or Jacobian determinants.

53.2 Polar, Cylindrical and Spherical Coordinates

Polar coordinates form a coordinate system in $\mathbb{R}^2$, *cylinder* coordinates complement the polar coordinates by the z-coordinate to a coordinate system of $\mathbb{R}^3$. In addition to the cylinder coordinates *spherical* coordinates a frequently used coordinate system of $\mathbb{R}^3$.

Polar, Cylindrical and Spherical Coordinates

- **Polar coordinates.** It is

$$\phi : \begin{cases} \mathbb{R}_{>0} \times [0, 2\pi[\to & \mathbb{R}^2 \setminus \{0\} \\ \begin{pmatrix} r \\ \varphi \end{pmatrix} & \mapsto \begin{pmatrix} x\ y \end{pmatrix} = \begin{pmatrix} r \cos\varphi \\ r \sin\varphi \end{pmatrix} \end{cases}$$

A coordinate transformation with the transformation matrix and functional determinant.

$$D\phi(r, \varphi) = \begin{pmatrix} \cos\varphi & -r \sin\varphi \\ \sin\varphi & r \cos\varphi \end{pmatrix} \quad \text{and} \quad \det D\phi(r, \varphi) = r .$$

One calls r, φ **polar coordinates.**

- **Cylinder coordinates.** It is

$$\phi : \begin{cases} \mathbb{R}_{>0} \times [0, 2\pi[\times \mathbb{R} \to & \mathbb{R}^3 \setminus z\text{-axis} \\ \begin{pmatrix} r \\ \varphi \\ z \end{pmatrix} & \mapsto \begin{pmatrix} x \\ y \\ z \end{pmatrix} = \begin{pmatrix} r \cos\varphi \\ r \sin\varphi \\ z \end{pmatrix} \end{cases}$$

(continued)

a coordinate transformation with the transformation matrix and functional determinant

$$D\phi(r,\varphi,z)=\begin{pmatrix}\cos\varphi & -r\sin\varphi & 0\\ \sin\varphi & r\cos\varphi & 0\\ 0 & 0 & 1\end{pmatrix}\ \text{and}\ \det D\phi(r,\varphi,z)=r\,.$$

One calls r, φ, z **cylinder coordinates.**

- **Spherical coordinates.** It is

$$\phi:\begin{cases}\mathbb{R}_{>0}\times[0,2\pi[\times]0,\pi[\to & \mathbb{R}^3\setminus z\text{-axis}\\ \begin{pmatrix}r\\ \varphi\\ \vartheta\end{pmatrix} \mapsto & \begin{pmatrix}x\\ y\\ z\end{pmatrix}=\begin{pmatrix}r\cos\varphi\sin\vartheta\\ r\sin\varphi\sin\vartheta\\ r\cos\vartheta\end{pmatrix}\end{cases}$$

a coordinate transformation with the transformation matrix and functional determinant

$$D\phi(r,\varphi,\vartheta)=\begin{pmatrix}\cos\varphi\sin\vartheta & -r\sin\varphi\sin\vartheta & r\cos\varphi\cos\vartheta\\ \sin\varphi\sin\vartheta & r\cos\varphi\sin\vartheta & r\sin\varphi\cos\vartheta\\ \cos\vartheta & 0 & -r\sin\vartheta\end{pmatrix}\ \text{and}$$

$$\det D\phi(r,\varphi,\vartheta)=-r^2\sin\vartheta\,.$$

One calls r, φ, ϑ **spherical coordinates.**

Figure 53.1 shows the representation of an (arbitrary) point P of $\mathbb{R}^2$ or $\mathbb{R}^3$ with respect to polar, cylindrical and spherical coordinates.

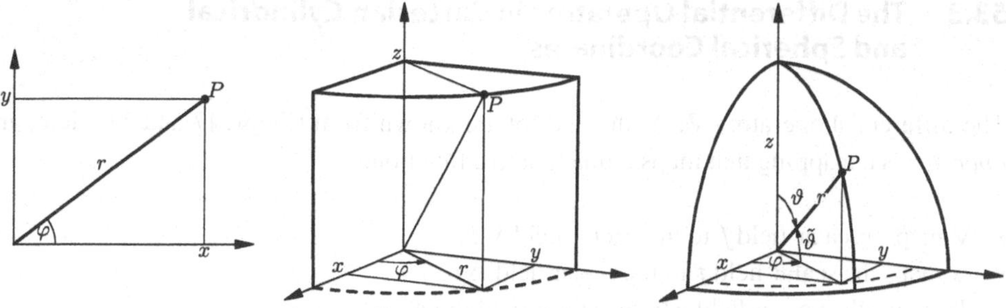

Fig. 53.1 Polyr coordinates in the $\mathbb{R}^2$ and cylindrical and spherical coordinates in $\mathbb{R}^3$

In some books, the spherical coordinates use the angle $\tilde{\vartheta}$ instead of ϑ (see 3rd picture in Fig. 53.1). Because of $\tilde{\vartheta} = \pi/2 - \vartheta$ are then $]0, \pi[$ through $]-\pi/2, \pi/2[$ and $\sin\vartheta$ through $\cos\tilde{\vartheta}$ and $\cos\vartheta$ by $\sin\tilde{\vartheta}$ to replace.

Polar coordinates are especially useful for point-symmetric problems in the $\mathbb{R}^2$ favorable. Cylindrical coordinates are advantageously used in axisymmetric problems. And spherical coordinates are especially advantageous for point-symmetric problems.

Occasionally one needs other coordinate systems, we state for completeness:

- **Affine coordinates:**

$$\phi : \begin{cases} \mathbb{R}^n & \to & \mathbb{R}^n \\ \begin{pmatrix} u_1 \\ \vdots \\ u_n \end{pmatrix} & \mapsto & \begin{pmatrix} x_1 \\ \vdots \\ x_n \end{pmatrix} = A \begin{pmatrix} u_1 \\ \vdots \\ u_n \end{pmatrix} + b \end{cases} \quad \text{with inv. } A \in \mathbb{R}^{n \times n} \text{ and } b \in \mathbb{R}^n .$$

- **Elliptic coordinates:**

$$\phi : \begin{cases} \mathbb{R}_{>0} \times [0, 2\pi[& \to & \mathbb{R}^2 \\ \begin{pmatrix} \eta \\ \varphi \end{pmatrix} & \mapsto & \begin{pmatrix} x \\ y \end{pmatrix} = \begin{pmatrix} c \cosh\eta \cos\varphi \\ c \sinh\eta \sin\varphi \end{pmatrix} \end{cases} \quad \text{with } c \in \mathbb{R}_{>0} .$$

- **Parabolic coordinates:**

$$\phi : \begin{cases} \mathbb{R}_{>0} \times \mathbb{R}_{>0} \times [0, 2\pi[& \to & \mathbb{R}^3 \\ \begin{pmatrix} \xi \\ \eta \\ \varphi \end{pmatrix} & \mapsto & \begin{pmatrix} x \\ y \\ z \end{pmatrix} = \begin{pmatrix} \xi\,\eta \cos\varphi \\ \xi\,\eta \sin\varphi \\ 1/2\,(\xi^2 - \eta^2) \end{pmatrix} \end{cases} .$$

53.3 The Differential Operators in Cartesian Cylindrical and Spherical Coordinates

The differential operators ∇, Δ, div and rot are known from Chaps. 47 and 51. Here, an *operator* is a mapping that maps a function to a function:

- ∇ maps a scalar field f to the vector field ∇f,
- Δ maps the scalar field f to the scalar field Δf,
- div maps the vector field $\boldsymbol{v}$ to the scalar field div $\boldsymbol{v}$,
- rot maps the vector field $\boldsymbol{v}$ to the vector field rot $\boldsymbol{v}$.

So far we have considered these differential operators only in Cartesian coordinates. If, however, we have given a scalar field f or vector field $\boldsymbol{v}$ in cylindrical or spherical coordinates, i.e. $f = f(r, \varphi, z)$ or $\boldsymbol{v} = \boldsymbol{v}(r, \varphi, z)$ or $f = f(r, \varphi, \vartheta)$ or $\boldsymbol{v} = \boldsymbol{v}(r, \varphi, \vartheta)$, one naturally expects that these operators also have their individual representations in the respective coordinate systems. The *transformation* of the operators are generally very tedious and make heavy use of the chain rule. In the following box one finds a clear representation of the operators mentioned in Cartesian, cylindrical and spherical coordinates (for the transformation see the exercises):

The Differential Operators in Cartesian, Cylindrical and Spherical Coordinates

The differential operators ∇, Δ, div and rot have the following representations with respect to the different coordinate systems:

- **Cartesian coordinates.** Given is a scalar field $f = f(x, y, z)$ or a vector field $\boldsymbol{v} = (v_1(x, y, z), v_2(x, y, z), v_3(x, y, z))^\top$:
  - Gradient ∇:

$$\nabla f(x, y, z) = \begin{pmatrix} \frac{\partial f}{\partial x} \\ \frac{\partial f}{\partial y} \\ \frac{\partial f}{\partial z} \end{pmatrix}.$$

  - Laplace Δ:

$$\Delta f(x, y, z) = \frac{\partial^2 f}{\partial x \partial x} + \frac{\partial^2 f}{\partial y \partial y} + \frac{\partial^2 f}{\partial z \partial z}.$$

  - Divergence div:

$$\operatorname{div} \boldsymbol{v}(x, y, z) = \frac{\partial v_1}{\partial x} + \frac{\partial v_2}{\partial y} + \frac{\partial v_3}{\partial z}.$$

  - Rotation rot:

$$\operatorname{rot} \boldsymbol{v}(x, y, z) = \begin{pmatrix} \frac{\partial v_3}{\partial y} - \frac{\partial v_2}{\partial z} \\ \frac{\partial v_1}{\partial z} - \frac{\partial v_3}{\partial x} \\ \frac{\partial v_2}{\partial x} - \frac{\partial v_1}{\partial y} \end{pmatrix}.$$

(continued)

- **Polar coordinates.** Given is a scalar field $f = f(r, \varphi)$:
  - Gradient ∇:

$$\nabla f(r, \varphi) = \begin{pmatrix} \frac{\partial f}{\partial r} \\ \frac{1}{r}\frac{\partial f}{\partial \varphi} \end{pmatrix}.$$

  - Laplace Δ:

$$\Delta f(r, \varphi) = \frac{\partial^2 f}{\partial r \partial r} + \frac{1}{r}\frac{\partial f}{\partial r} + \frac{1}{r^2}\frac{\partial^2 f}{\partial \varphi \partial \varphi}.$$

- **Cylinder coordinates.** Given is a scalar field $f = f(r, \varphi, z)$ or a vector field $\boldsymbol{v} = (v_1(r, \varphi, z), v_2(r, \varphi, z), v_3(r, \varphi, z))^\top$:
  - Gradient ∇:

$$\nabla f(r, \varphi, z) = \begin{pmatrix} \frac{\partial f}{\partial r} \\ \frac{1}{r}\frac{\partial f}{\partial \varphi} \\ \frac{\partial f}{\partial z} \end{pmatrix}.$$

  - Laplace Δ:

$$\Delta f(r, \varphi, z) = \frac{\partial^2 f}{\partial r \partial r} + \frac{1}{r}\frac{\partial f}{\partial r} + \frac{1}{r^2}\frac{\partial^2 f}{\partial \varphi \partial \varphi} + \frac{\partial^2 f}{\partial z \partial z}.$$

  - Divergence div:

$$\operatorname{div} \boldsymbol{v}(r, \varphi, z) = \frac{1}{r}\frac{\partial (r v_1)}{\partial r} + \frac{1}{r}\frac{\partial v_2}{\partial \varphi} + \frac{\partial v_3}{\partial z}.$$

  - Rotation rot:

$$\operatorname{rot} \boldsymbol{v}(r, \varphi, z) = \begin{pmatrix} \frac{1}{r}\frac{\partial v_3}{\partial \varphi} - \frac{\partial v_2}{\partial z} \\ \frac{\partial v_1}{\partial z} - \frac{\partial v_3}{\partial r} \\ \frac{1}{r}\frac{\partial (r v_2)}{\partial r} - \frac{1}{r}\frac{\partial v_1}{\partial \varphi} \end{pmatrix}.$$

- **Spherical coordinates.** Given is a scalar field $f = f(r, \varphi, \vartheta)$ or a vector field $\boldsymbol{v} = (v_1(r, \varphi, \vartheta), v_2(r, \varphi, \vartheta), v_3(r, \varphi, \vartheta))^\top$:

(continued)

– Gradient ∇:

$$\nabla f(r,\varphi,\vartheta) = \begin{pmatrix} \frac{\partial f}{\partial r} \\ \frac{1}{r\sin\vartheta}\frac{\partial f}{\partial \varphi} \\ \frac{1}{r}\frac{\partial f}{\partial \vartheta} \end{pmatrix}.$$

– Laplace Δ:

$$\Delta f(r,\varphi,\vartheta) = \frac{\partial^2 f}{\partial r \partial r} + \frac{2}{r}\frac{\partial f}{\partial r} + \frac{1}{r^2\sin^2\vartheta}\frac{\partial^2 f}{\partial\varphi\partial\varphi} + \frac{\cos\vartheta}{r^2\sin\vartheta}\frac{\partial f}{\partial\vartheta} + \frac{1}{r^2}\frac{\partial^2 f}{\partial\vartheta\partial\vartheta}.$$

– Divergence div:

$$\operatorname{div} \boldsymbol{v}(r,\varphi,\vartheta) = \frac{1}{r^2}\frac{\partial(r^2 v_1)}{\partial r} + \frac{1}{r\sin\vartheta}\frac{\partial v_2}{\partial\varphi} + \frac{1}{r\sin\vartheta}\frac{\partial(v_3\sin\vartheta)}{\partial\vartheta}.$$

– Rotation rot:

$$\operatorname{rot} \boldsymbol{v}(r,\varphi,\vartheta) = \begin{pmatrix} \frac{1}{r\sin\vartheta}\frac{\partial(v_2\sin\vartheta)}{\partial\vartheta} - \frac{1}{r\sin\vartheta}\frac{\partial v_3}{\partial\varphi} \\ \frac{1}{r}\frac{\partial(rv_3)}{\partial r} - \frac{1}{r}\frac{\partial v_1}{\partial\vartheta} \\ \frac{1}{r\sin\vartheta}\frac{\partial v_1}{\partial\varphi} - \frac{1}{r}\frac{\partial(rv_2)}{\partial r} \end{pmatrix}.$$

Here, of course, one must think of the vectors in Cartesian, cylindrical, and spherical coordinates, respectively, as vectors with respect to the corresponding bases, that is, as coordinate vectors with respect to the bases $\boldsymbol{e}_x, \boldsymbol{e}_y, \boldsymbol{e}_z$ or $\boldsymbol{e}_r, \boldsymbol{e}_\varphi, \boldsymbol{e}_z$ or $\boldsymbol{e}_r, \boldsymbol{e}_\varphi, \boldsymbol{e}_\vartheta$.

Example 53.1

- For the vector field $\boldsymbol{v}(r,\varphi,z) = (z,\ 0,\ r)^\top$ in cylindrical coordinates we have

$$\operatorname{div} \boldsymbol{v}(r,\varphi,z) = \frac{z}{r} \quad \text{and} \quad \operatorname{rot} \boldsymbol{v}(r,\varphi,z) = \begin{pmatrix} 0 \\ 0 \\ 0 \end{pmatrix}.$$

- For the scalar field $f(r,\varphi,z) = r^2 + z^3$ in cylindrical coordinates we have:

$$\nabla f(r,\varphi,z) = \begin{pmatrix} 2r \\ 0 \\ 3z^2 \end{pmatrix} \quad \text{and} \quad \Delta f(r,\varphi,z) = 4 + 6z.$$

- Considering $f(x, y, z) = x^2 + y^2 + z^3$ one obtains

$$\nabla f(x, y, z) = \begin{pmatrix} 2x \\ 2y \\ 3z^2 \end{pmatrix} \quad \text{and} \quad \Delta f(x, y, z) = 2 + 2 + 6z = 4 + 6z\,.$$

- For the scalar field $f(r, \varphi, \vartheta) = r^2 + r$ in spherical coordinates we have:

$$\nabla f(r, \varphi, \vartheta) = \begin{pmatrix} 2r + 1 \\ 0 \\ 0 \end{pmatrix} \quad \text{and} \quad \Delta f(r, \varphi, \vartheta) = 6 + \frac{2}{r}\,.$$

- For the vector field $\boldsymbol{v}(r, \varphi, \vartheta) = \frac{1}{r}\left(\cos^2 \vartheta,\ \sin \vartheta,\ -\sin \vartheta \cos \vartheta\right)^\top$ in spherical coordinates we have:

$$\begin{aligned} \operatorname{div} \boldsymbol{v}(r, \varphi, \vartheta) &= \frac{1}{r^2} \frac{\partial}{\partial r}\left(r \cos^2 \vartheta\right) + \frac{1}{r \sin \vartheta} \frac{\partial}{\partial \varphi}\left(\frac{\sin \vartheta}{r}\right) \\ &\quad + \frac{1}{r \sin \vartheta} \frac{\partial}{\partial \vartheta}\left(-\frac{\sin^2 \vartheta \cos \vartheta}{r}\right) \\ &= \frac{\cos^2 \vartheta}{r^2} + \frac{1}{r^2 \sin \vartheta}\left(-2 \sin \vartheta \cos^2 \vartheta + \sin^3 \vartheta\right) \\ &= \frac{\sin^2 \vartheta - \cos^2 \vartheta}{r^2}\,. \end{aligned}$$

■

53.4 Conversion of Vector Fields and Scalar Fields

It is often desirable to represent a given vector field or scalar field in Cartesian coordinates as a vector field in cylindrical or spherical coordinates (or vice versa), for example to exploit certain symmetries. Thus, integrating or applying differential operators in suitable coordinate systems often becomes much easier.

Example 53.2

- The following two scalar fields

$$f_{\text{cart}}(x, y, z) = \sqrt{x^2 + y^2 + z^2} \quad \text{and} \quad f_{\text{sph}}(r, \varphi, \vartheta) = r$$

assign to each point of the $\mathbb{R}^3$ its distance from the origin. We get exemplarily

$$\nabla f_{\mathrm{cart}}(x, y, z) = \frac{1}{\sqrt{x^2 + y^2 + z^2}} \begin{pmatrix} x \\ y \\ z \end{pmatrix} \quad \text{and} \quad \nabla f_{\mathrm{sph}}(r, \varphi, \vartheta) = \begin{pmatrix} 1 \\ 0 \\ 0 \end{pmatrix}.$$

- The following two vector fields

$$\boldsymbol{v}_{\mathrm{cart}}(x, y, z) = \frac{1}{\sqrt{x^2 + y^2 + z^2}} \begin{pmatrix} x \\ y \\ z \end{pmatrix} \quad \text{and} \quad \boldsymbol{v}_{\mathrm{sph}}(r, \varphi, \vartheta) = \begin{pmatrix} 1 \\ 0 \\ 0 \end{pmatrix}$$

normalize each vector from $\mathbb{R}^3 \setminus \{0\}$. We obtain exemplarily

$$\operatorname{div} \boldsymbol{v}_{\mathrm{cart}}(x, y, z) = \frac{2}{\sqrt{x^2 + y^2 + z^2}} \quad \text{and} \quad \operatorname{div} \boldsymbol{v}_{\mathrm{sph}}(r, \varphi, \vartheta) = \frac{2}{r}. \qquad \blacksquare$$

With the help of the respective transformation matrix it is easy to convert a scalar field or a vector field in Cartesian coordinates into a scalar field or vector field in cylindrical or spherical coordinates. We give this conversion in a recipe-like manner:

Recipe: Convert Scalar and Vector Fields to Cylindrical and Spherical Coordinates, Respectively

In the following we consider a scalar field $f : D \subseteq \mathbb{R}^3 \to \mathbb{R}$ or a vector field $\boldsymbol{v} : D \subseteq \mathbb{R}^3 \to \mathbb{R}^3$ where

$$f_{\mathrm{cart}} \text{ or } f_{\mathrm{cyl}} \text{ or } f_{\mathrm{sph}} \text{ and } \boldsymbol{v}_{\mathrm{cart}} \text{ or } \boldsymbol{v}_{\mathrm{cyl}} \text{ or } \boldsymbol{v}_{\mathrm{sph}}$$

denote the representations of this scalar field and vector field in Cartesian and cylindrical and spherical coordinates, respectively. We obtain the other representation by means of the orthogonal matrix S_{cyl} and S_{sph} (note $S_{\mathrm{cyl}}^{-1} = S_{\mathrm{cyl}}^{\top}$ and $S_{\mathrm{sph}}^{-1} = S_{\mathrm{sph}}^{\top}$),

$$S_{\mathrm{cyl}} = \begin{pmatrix} \cos\varphi & -\sin\varphi & 0 \\ \sin\varphi & \cos\varphi & 0 \\ 0 & 0 & 1 \end{pmatrix} \quad \text{or} \quad S_{\mathrm{sph}} = \begin{pmatrix} \cos\varphi\sin\vartheta & -\sin\varphi & \cos\varphi\cos\vartheta \\ \sin\varphi\sin\vartheta & \cos\varphi & \sin\varphi\cos\vartheta \\ \cos\vartheta & 0 & -\sin\vartheta \end{pmatrix},$$

as follows:

(continued)

- Given $f_{\text{cart}} = f_{\text{cart}}(x, y, z)$:
  - Get $f_{\text{cyl}} = f_{\text{cyl}}(r, \varphi, z)$ by substituting $x = r\cos\varphi$, $y = r\sin\varphi$, $z = z$ in f_{cart}:

$$f_{\text{cyl}}(r, \varphi, z) = f_{\text{cart}}(r\cos\varphi, r\sin\varphi, z)\,.$$

  - Get $f_{\text{sph}} = f_{\text{sph}}(r, \varphi, \vartheta)$ by substituting $x = r\cos\varphi\sin\vartheta$, $y = r\sin\varphi\sin\vartheta$, $z = r\cos\vartheta$ to f_{cart}:

$$f_{\text{sph}}(r, \varphi, \vartheta) = f_{\text{cart}}(r\cos\varphi\sin\vartheta, r\sin\varphi\sin\vartheta, r\cos\vartheta)\,.$$

- Given is $\boldsymbol{v}_{\text{cart}} = \boldsymbol{v}_{\text{cart}}(x, y, z)$:
  - Get $\boldsymbol{v}_{\text{cyl}} = \boldsymbol{v}_{\text{cyl}}(r, \varphi, z)$ by substituting $x = r\cos\varphi$, $y = r\sin\varphi$, $z = z$ in $\boldsymbol{v}_{\text{cart}}$ and multiplication by S_{cyl}^{-1}:

$$\boldsymbol{v}_{\text{cyl}}(r, \varphi, z) = S_{\text{cyl}}^{-1}\,\boldsymbol{v}_{\text{cart}}(r\cos\varphi, r\sin\varphi, z)\,.$$

  - Get $\boldsymbol{v}_{\text{sph}} = \boldsymbol{v}_{\text{sph}}(r, \varphi, \vartheta)$ by substituting $x = r\cos\varphi\sin\vartheta$, $y = r\sin\varphi\sin\vartheta$, $z = r\cos\vartheta$ to $\boldsymbol{v}_{cart}$ and multiplication by S_{sph}^{-1}:

$$\boldsymbol{v}_{\text{sph}}(r, \varphi, \vartheta) = S_{\text{sph}}^{-1}\,\boldsymbol{v}_{\text{cart}}(r\cos\varphi\sin\vartheta, r\sin\varphi\sin\vartheta, r\cos\vartheta)\,.$$

Example 53.3

- The representation of the scalar field $f_{\text{cart}}(x, y, z) = x^2 + y^2 + z^3$ in cylindrical coordinates is as follows:

$$f_{\text{cyl}}(r, \varphi, z) = r^2\cos^2\varphi + r^2\sin^2\varphi + z^3 = r^2 + z^3\,.$$

- The representation of the scalar field $f_{\text{cart}}(x, y, z) = x^2 + y^2 + z^2 + \sqrt{x^2 + y^2 + z^2}$ in spherical coordinates is as follows:

$$f_{\text{sph}}(r, \varphi, \vartheta) = r^2 + r\,.$$

- The representation of the vector field $\boldsymbol{v}_{\text{cart}}(x, y, z) = \frac{1}{\sqrt{x^2+y^2}}(xz, yz, x^2 + y^2)^\top$ in cylindrical coordinates is as follows:

$$\boldsymbol{v}_{\text{cyl}}(r, \varphi, z) = \begin{pmatrix} \cos\varphi & \sin\varphi & 0 \\ -\sin\varphi & \cos\varphi & 0 \\ 0 & 0 & 1 \end{pmatrix} \frac{1}{r} \begin{pmatrix} r\cos\varphi\, z \\ r\sin\varphi\, z \\ r^2 \end{pmatrix} = \begin{pmatrix} z \\ 0 \\ r \end{pmatrix}.$$

- The representation of the vector field $\boldsymbol{v}_{\text{cart}}(x, y, z) = \frac{1}{x^2+y^2+z^2}(-y, x, z)^\top$ in spherical coordinates is as follows:

$$\boldsymbol{v}_{\text{sph}}(r, \varphi, \vartheta) = \frac{1}{r^2} \begin{pmatrix} \cos\varphi\sin\vartheta & \sin\varphi\sin\vartheta & \cos\vartheta \\ -\sin\varphi & \cos\varphi & 0 \\ \cos\varphi\cos\vartheta & \sin\varphi\cos\vartheta & -\sin\vartheta \end{pmatrix} \begin{pmatrix} -r\sin\varphi\sin\vartheta \\ r\cos\varphi\sin\vartheta \\ r\cos\vartheta \end{pmatrix}$$

$$= \frac{1}{r^2} \begin{pmatrix} r\cos^2\vartheta \\ r\sin\vartheta \\ -r\sin\vartheta\cos\vartheta \end{pmatrix}. \qquad \blacksquare$$

The above formulas can all be derived with the help of the chain rule. Besides the mentioned cylindrical and spherical coordinates, there are other coordinate systems, generally called *curvilinear coordinate systems*. When investigating differential equations it is usual and useful to represent them in different coordinate systems.

53.5 Exercises

53.1 Calculate $\operatorname{rot}\boldsymbol{v}$ and $\operatorname{div}\boldsymbol{v}$ in Cartesian coordinates and in cylindrical coordinates, where:

$$\boldsymbol{v}(x, y, z) = \frac{1}{\sqrt{x^2+y^2}} \begin{pmatrix} xz \\ yz \\ x^2+y^2 \end{pmatrix}.$$

53.2 Given is the scalar field $f(x, y, z) = (x^2 + y^2 + z^2)^2$. Calculate ∇f and Δf in Cartesian coordinates and in spherical coordinates.

53.3 Given is the vector field $\boldsymbol{v}$ on $\mathbb{R}^3 \setminus z$-axis with

$$\boldsymbol{v}(x, y, z) = \frac{1}{x^2 + y^2} \begin{pmatrix} -y \\ x \\ z \end{pmatrix} .$$

Set the vector field $\boldsymbol{v}$ in spherical coordinates and calculate $\operatorname{rot} \boldsymbol{v}$ and $\operatorname{div} \boldsymbol{v}$ in Cartesian coordinates and in spherical coordinates.

53.4 Given is the scalar field $f(x, y, z) = x^2 + y^3 + z^2 + xz$. Calculate ∇f and Δf in Cartesian coordinates and in cylindrical coordinates.

53.5 Show the representation of the Laplace operator in cylindrical coordinates:

$$\Delta = \frac{1}{r} \frac{\partial}{\partial r} \left(r \frac{\partial}{\partial r} \right) + \frac{1}{r^2} \frac{\partial^2}{\partial \varphi^2} + \frac{\partial^2}{\partial z^2} .$$

Curves I

54

One can fill books with the topic of *curves*. One can debate for hours about the definition of the term *curve*. We do not do that. We are also only interested in *plane* curves and *space curves*, we do not consider pathological exceptions. Plane curves and space curves have a beginning and an end, a length and a curvature, they can intersect or be represented differently. These are suggestive terms and facts, which actually mean exactly what one imagines them to mean.

Curves often appear in the applications of mathematics: Wires with a charge density, trajectories of particles, spiral components with a density—with the help of curve integrals we will be able to determine the total charge or mass of such *curves*.

54.1 Terms

The term *curve* already appears in Chap. 46.1 on: We explained a *curve* γ as a mapping from $D \subseteq \mathbb{R}$ in $\mathbb{R}^n$. This definition is much too general for our purposes. We want to have *smooth lines* that are in $\mathbb{R}^2$ or $\mathbb{R}^3$ run and represent wires or paths of particles. In order to have more freedom, we also want to *seamless compositions of* such *smooth lines* are again called curves (see Fig. 54.1).

We achieve *smoothness* by requiring continuous differentiability, and *seamless composition by* requiring piecewise continuity:

C. Karpfinger, *Calculus and Linear Algebra in Recipes*,
https://doi.org/10.1007/978-3-662-65458-3_54

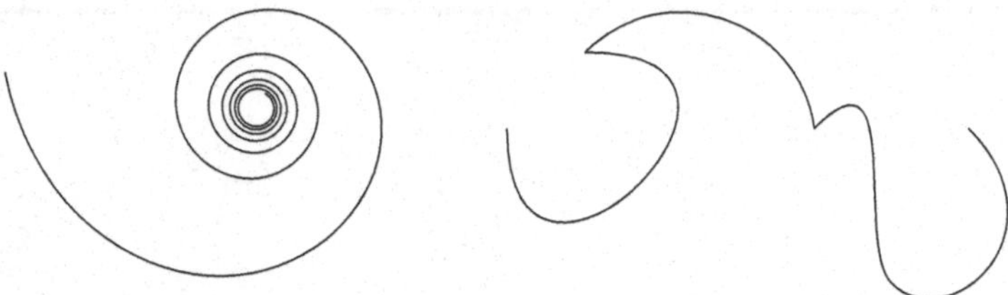

Fig. 54.1 Examples of curves, more precisely: traces of curves

Curves or Piecewise Continuously Differentiable Curves

A **curve** is a mapping

$$\gamma : I \subseteq \mathbb{R} \to \mathbb{R}^n \text{ with } \gamma(t) = \begin{pmatrix} x_1(t) \\ \vdots \\ x_n(t) \end{pmatrix}$$

of an interval I with continuous component functions $x_1, \ldots, x_n : I \to \mathbb{R}$.

- In the case $n = 2$ one speaks of a **plane curve**.
- In the case $n = 3$ one speaks of a **space curve**.
- The image $\gamma(I) = \{(x_1(t), \ldots, x_n(t))^\top \mid t \in I\}$ is called the **trace of** γ.
- In the case $I = [a, b]$ the point $\gamma(a)$ is called the **starting point** and $\gamma(b)$ the **end point** of γ. One calls γ **closed** if $\gamma(a) = \gamma(b)$.

Further, one calls γ

- **continuously differentiable** or short C^1**-curve** if the component functions $x_1, \ldots, x_n$ are continuously differentiable, and
- **twice continuously differentiable** or short C^2**-curve** if the component functions $x_1, \ldots, x_n$ are twice continuously differentiable, and
- **piecewise continuously differentiable** or briefly **piecewise** C^1**-curve** if the interval I can be divided in such a way that $x_1, \ldots, x_n$ are continuously differentiable on each subinterval.

It is customary to dispense with the fine distinction between *curve* (= image) and *trace of the* curve (= the range of the mapping) and to equate the curve with the trace. Note that a (trace of a) curve is not only a set of points, but quite essentially also has a *direction of passage*. This *direction* cannot be seen from the trace, it results from the mapping rule. It will always be useful to reverse the direction of a curve, fortunately this can be done quite easily if I is a closed interval: If $\gamma : [a, b] \to \mathbb{R}^n$, $t \to \gamma(t)$, is a curve, then the curve

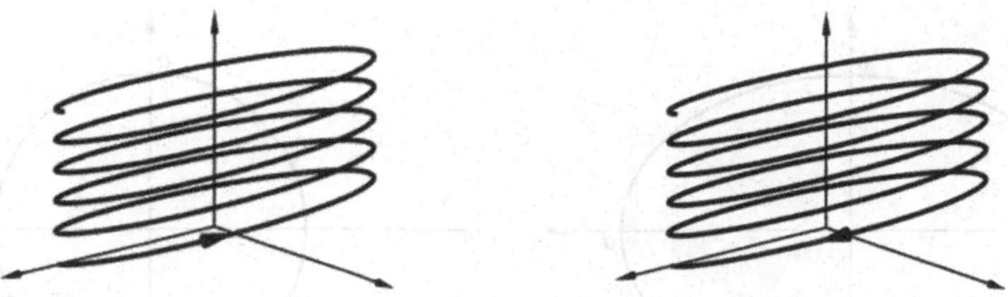

Fig. 54.2 The tracks are the same, the directions of passage are different

passes through $\tilde{\gamma} : [a, b] \to \mathbb{R}$, $\tilde{\gamma}(t) = \gamma(a+b-t)$ traverses the same trace in the reverse direction.

Example 54.1 The trace of the curve $\gamma : [0, 10\pi] \to \mathbb{R}^3$ with $\gamma(t) = \begin{pmatrix} 4\cos(t) \\ \sin(t) \\ 0.1\,t \end{pmatrix}$ is a *helical curve*. The curve

$$\tilde{\gamma} : [0, 10\pi] \to \mathbb{R}^3 \text{ with } \tilde{\gamma}(t) = \begin{pmatrix} 4\cos(10\pi - t) \\ \sin(10\pi - t) \\ 0.1\,(10\pi - t) \end{pmatrix}$$

passes through the same track; but from *top to bottom* (see Fig. 54.2). ■

The curves we will be dealing with will always be piecewise continuously differentiable. A piecewise continuously differentiable curve can also be obtained by putting curves together: If $\gamma_1 : I_1 \to \mathbb{R}^n, \dots, \gamma_k : I_k \to \mathbb{R}^n$ are curves whose traces are *connected*, i.e., the endpoint of γ_i is the starting point of γ_{i+1}, so we write $\gamma_1 + \cdots + \gamma_k$ for the *total curve* γ which is piecewise continuously differentiable if the individual curves are $\gamma_1, \dots, \gamma_k$ are continuously differentiable.

In the following examples all curves are piecewise continuously diferentiable.

Example 54.2

- $\gamma : [0, 2\pi] \to \mathbb{R}^2$, $\gamma(t) = \begin{pmatrix} a\cos(t) \\ b\sin(t) \end{pmatrix}$ has as its trace an ellipse with the semi axes a and b, in the case $a = b$ we get a circle of radius a (see Fig. 54.3).
- The trace of the curve $\gamma : [0, 4] \to \mathbb{R}^2$, $\gamma(t) = (t, \sqrt{t})^\top$ is the graph of the function $f : [0, 4] \to \mathbb{R}_{>0}$ with $f(x) = \sqrt{x}$. In general, for each function $f : [a, b] \to \mathbb{R}$ the curve

$$\gamma : [a, b] \to \mathbb{R}^2, \gamma(t) = \begin{pmatrix} t \\ f(t) \end{pmatrix}$$

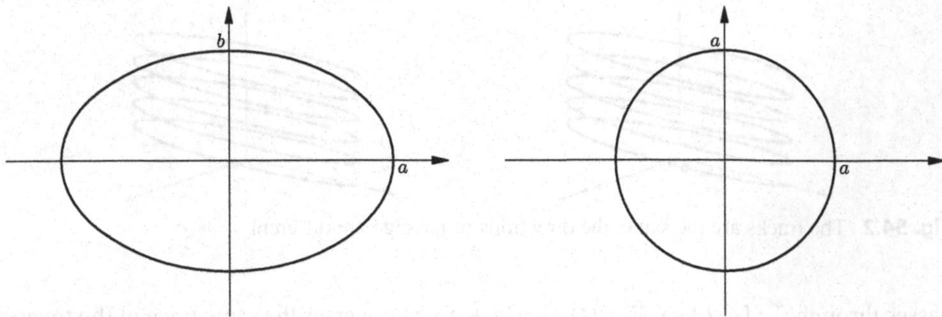

Fig. 54.3 A circle is an ellipse with equal semi-axes

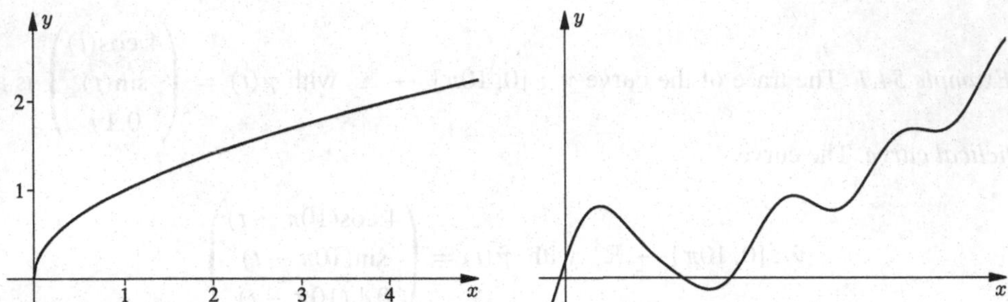

Fig. 54.4 Graphs of functions are traces of special curves

has as trace the graph of f (see Fig. 54.4).

- The curve

$$\gamma : [0, 2\pi n] \to \mathbb{R}^3,\ \gamma(t) = \begin{pmatrix} r\cos(t) \\ r\sin(t) \\ ht \end{pmatrix}$$

with $r,\ h \in \mathbb{R}_{>0}$ has as trace a **helical** with diameter 2r, **height** $2\pi h$ and **number of turns** n (see Fig. 54.5).

- If one sets the following curves $\gamma_1 : [0, \pi] \to \mathbb{R}^2$, $\gamma_2 : [0, 1] \to \mathbb{R}^2$, $\gamma_3 : [\pi/2, 3\pi/2] \to \mathbb{R}^2$ with

$$\gamma_1(t) = \begin{pmatrix} -\cos(t) \\ \sin(t) \end{pmatrix},\ \gamma_2(t) = \begin{pmatrix} 1+t \\ t \end{pmatrix},\ \gamma_3(t) = \begin{pmatrix} 2-\cos(t) \\ \sin(t) \end{pmatrix}$$

one obtains the piecewise continuously differentiable curve $\gamma = \gamma_1 + \gamma_2 + \gamma_3$. We have the trace of this curve in Fig. 54.6 illustrated. ■

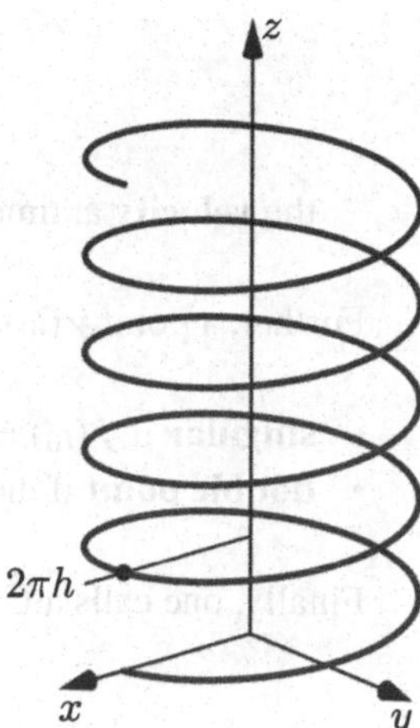

Fig. 54.5 The pitch is $2\pi h$

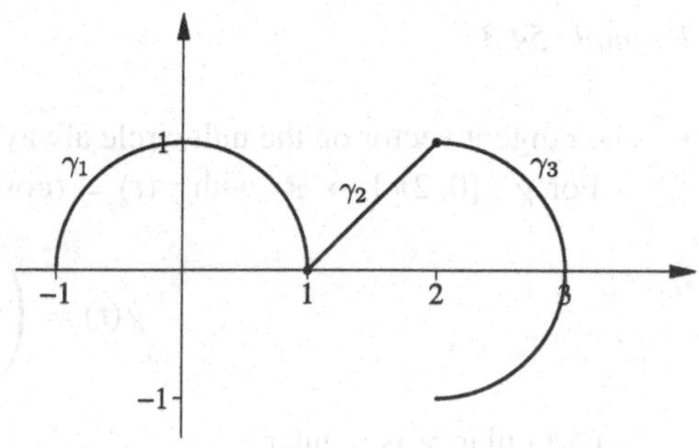

Fig. 54.6 A compound curve

What follows is a list of terms, all of which are highly suggestive:

Terms for Curves
Given is a piecewise C^1-curve,

$$\gamma : I \to \mathbb{R}^n \text{ with } \gamma(t) = (x_1(t), \ldots, x_n(t))^\top .$$

- One then calls

$$\dot{\gamma}(t) = (\dot{x}_1(t), \ldots, \dot{x}_n(t))^\top$$

the **tangent vector** or **velocity vector** of γ at *time* t and

(continued)

$$\|\dot{\gamma}(t)\| = \sqrt{\dot{x}_1(t)^2 + \cdots + \dot{x}_n(t)^2}$$

the **velocity** at time t.

Further, a point $\gamma(t_0)$ of a curve is called

- **singular** if $\dot{\gamma}(t_0) = 0$ and
- **double point** if there are $t_0,\ t_1 \in I$, $t_0 \neq t_1$, with $\gamma(t_0) = \gamma(t_1)$.

Finally, one calls a C^1-curve without singular points **regular**.

Example 54.3

- The tangent vector on the unit circle always has length 1 (see Fig. 54.7):
  For $\gamma : [0, 2\pi] \to \mathbb{R}^2$ with $\gamma(t) = (\cos(t), \sin(t))^\top$ we have namely

$$\dot{\gamma}(t) = \begin{pmatrix} -\sin(t) \\ \cos(t) \end{pmatrix}.$$

  In particular, γ is regular.
  For $t = 0$ and $t = \pi$, we obtain the vertical tangents $(0, 1)^\top$ and $(0, -1)^\top$. For $t = \pi/2$ and $t = 3\pi/2$ one obtains the horizontal tangents $(-1, 0)^\top$ and $(1, 0)^\top$.
- We consider the curve

$$\gamma : [0, 1] \to \mathbb{R}^2 \text{ with } \gamma(t) = \begin{pmatrix} t^2 - t \\ t^3 - t \end{pmatrix},$$

  whose trace is shown in Fig. 54.8. We examine the curve for singular points.

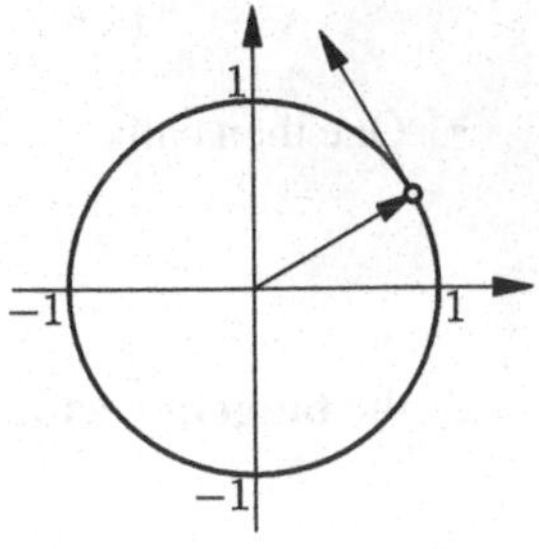

Fig. 54.7 Tangent vector at the circle

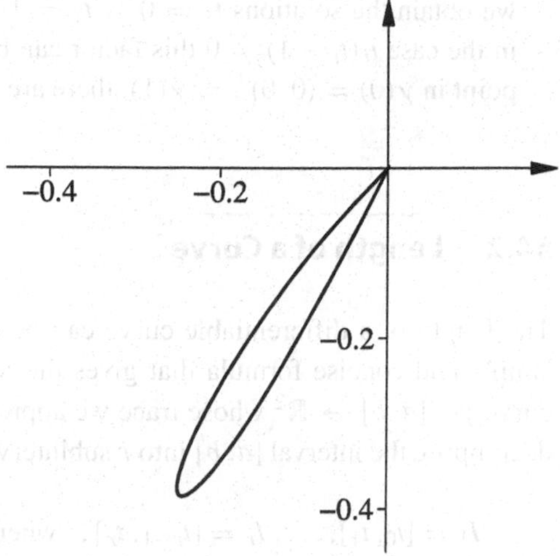

Fig. 54.8 A regular curve

The tangent vector is

$$\dot{\gamma}(t) = \begin{pmatrix} 2t - 1 \\ 3t^2 - 1 \end{pmatrix}.$$

Because of

$$2t - 1 = 0 \Leftrightarrow t = \frac{1}{2}$$

and

$$3t^2 - 1 = 0 \Leftrightarrow t = \pm\frac{1}{\sqrt{3}}$$

we obtain:

- There are no singular points, the curve γ is regular.
- At the curve point $\gamma(1/2)$ the curve has the vertical tangent $(0, -1/4)^\top$.
- At the curve point $\gamma(1/\sqrt{3})$ the curve has the horizontal tangent $((2-\sqrt{3})/\sqrt{3}, 0)^\top$.

To determine the double points, we make the approach:

$$t_1^2 - t_1 = t_2^2 - t_2 \text{ and } t_1^3 - t_1 = t_2^3 - t_2 .$$

Because of

$$t_i^2 - t_i = t_i(t_i - 1) \text{ and } t_i^3 - t_i = t_i(t_i - 1)(t_i + 1)$$

we obtain the solutions $t_1 = 0 \wedge t_2 = 1$ and $t_1 = 1 \wedge t_2 = 0$ and $t_1 = t_2$ (note that in the case $t_i(t_i - 1) \neq 0$ this factor can be truncated). Thus the curve γ has a double point in $\gamma(0) = (0, 0)^\top = \gamma(1)$, there are no further double points. ■

54.2 Length of a Curve

The length of a differentiable curve can be determined by an integral. We motivate the simple and concise formula that gives the length of a curve. To do this, we consider a curve $\gamma : [a, b] \to \mathbb{R}^2$ whose trace we approximate by a polygonal curve as follows: We decompose the interval $[a, b]$ into r subintervals,

$$I_1 = [t_0, t_1], \ldots, I_r = [t_{r-1}, t_r], \text{ where } a = t_0 < t_1 < \cdots < t_{r-1} < t_r = b,$$

and consider the composite curve $\tilde{\gamma} = \gamma_1 + \cdots + \gamma_r$, where γ_i connects the curve points $\gamma(t_{i-1})$ and $\gamma(t_i)$ by a line segment, see Fig. 54.9. The length of the curve $\tilde{\gamma}$ is

$$L(\tilde{\gamma}) = \sum_{i=1}^{r} \|\gamma(t_i) - \gamma(t_{i-1})\|.$$

Now we apply the mean value theorem of differential calculus:

This states that in each subintervall there are $t_i^{(x)}, t_i^{(y)} \in I_i$ with

$$\gamma(t_i) - \gamma(t_{i-1}) = \begin{pmatrix} x(t_i) - x(t_{i-1}) \\ y(t_i) - y(t_{i-1}) \end{pmatrix} = (t_i - t_{i-1}) \begin{pmatrix} \dot{x}(t_i^{(x)}) \\ \dot{y}(t_i^{(y)}) \end{pmatrix}.$$

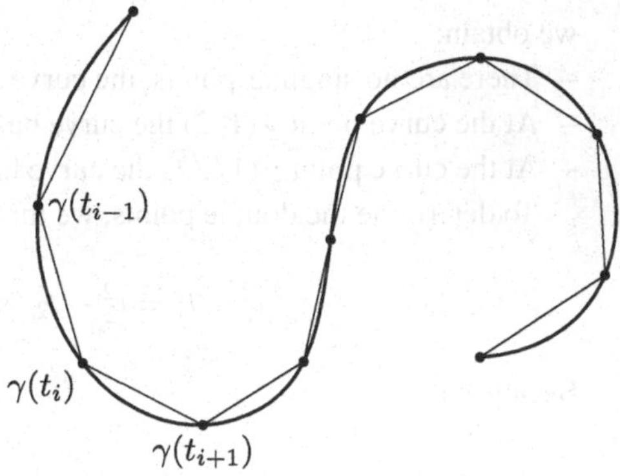

Fig. 54.9 Approximation of the length by line segments

By *refining* the decomposition (the maximum interval length approaches 0) we obtain an integral in the length formula for the sum:

The Length of a Curve
Let $\gamma : [a, b] \to \mathbb{R}^n$ a C^1-curve. We call

$$L(\gamma) = \int_a^b \|\dot{\gamma}(t)\| \mathrm{d}t$$

the **arc length** or **length** of the curve γ. The **arc length function**

$$s : [a, b] \to [0, L(\gamma)],\ s(t) = \int_a^t \|\dot{\gamma}(\tau)\| \mathrm{d}\tau$$

gives to each $t \in [a, b]$ the length of the curve from $\gamma(a)$ to $\gamma(t)$.

In the case

- of a connected curve $\gamma = \gamma_1 + \cdots + \gamma_r$ we have the formula

$$L(\gamma) = L(\gamma_1) + \cdots + L(\gamma_r)\,,$$

- of a curve with unbounded *time interval* $I = [a, \infty)$ we have for $\gamma : [a, \infty) \to \mathbb{R}^n$:

$$L(\gamma) = \int_a^\infty \|\dot{\gamma}(t)\| \mathrm{d}t\,,$$

- of a curve which parametrizes the graph of a differentiable function $f : [a, b] \to \mathbb{R}$, $\gamma : [a, b] \to \mathbb{R}^2$ with $\gamma(t) = (t, f(t))^\top$, we obtain for the length of the graph of f because of $\|\dot{\gamma}(t)\| = \sqrt{1 + (f'(t))^2}$ the formula

$$L(\mathrm{Graph}(f)) = L(\gamma) = \int_a^b \sqrt{1 + (f'(t))^2} \mathrm{d}t\,.$$

Example 54.4

- The trace of

$$\gamma : [0, 2\pi] \to \mathbb{R}^2,\ \gamma(t) = \begin{pmatrix} r\cos(t) \\ r\sin(t) \end{pmatrix} \quad \text{with } \dot{\gamma}(t) = \begin{pmatrix} -r\sin(t) \\ r\cos(t) \end{pmatrix}$$

is a circle of radius r. The circumference is the length of the curve:

$$L(\gamma) = \int_0^{2\pi} \sqrt{r^2\sin^2(t) + r^2\cos^2(t)}\,\mathrm{d}t = \int_0^{2\pi} r\,\mathrm{d}t = rt\Big|_0^{2\pi} = 2\pi r\,.$$

- We determine the arc length function for the **logarithmic spiral**

$$\gamma : [0, \infty) \to \mathbb{R}^2,\ \gamma(t) = \begin{pmatrix} \mathrm{e}^{-t}\cos(t) \\ \mathrm{e}^{-t}\sin(t) \end{pmatrix}.$$

Note the Fig. 54.10.
For the arc length function we need the velocity vector, which is

$$\dot{\gamma}(t) = \begin{pmatrix} -\mathrm{e}^{-t}(\cos(t) + \sin(t)) \\ -\mathrm{e}^{-t}(\sin(t) - \cos(t)) \end{pmatrix}.$$

With this we now get the arc length function

$$s(t) = \int_0^t \sqrt{2\,\mathrm{e}^{-2\tau}\left(\cos^2(\tau) + \sin^2(\tau)\right)}\mathrm{d}\tau = \sqrt{2}\int_0^t \mathrm{e}^{-\tau}\,\mathrm{d}\tau = \sqrt{2}(1 - \mathrm{e}^{-t})\,.$$

Because $\lim_{t\to\infty}\sqrt{2}(1 - \mathrm{e}^{-t}) = \sqrt{2}$ we have $L(\gamma) = \sqrt{2}$.

Fig. 54.10 Logarithmic spiral

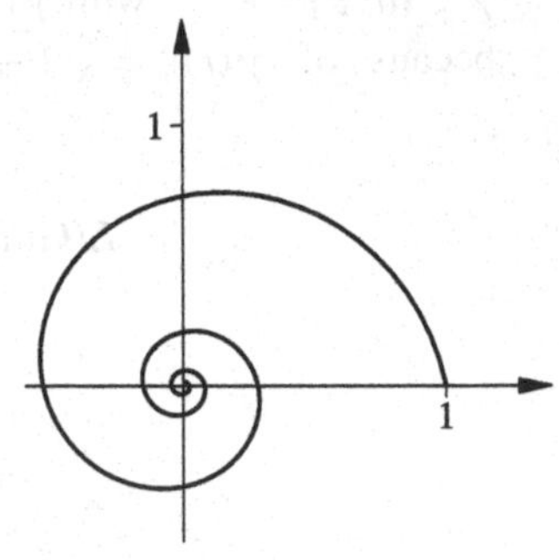

- The length of the graph of $f : [0, 2] \to \mathbb{R}$ with $f(t) = t^2$ is

$$L(f) = \int_0^2 \sqrt{1 + 4t^2}\mathrm{d}t = \frac{1}{2}t\sqrt{4t^2 + 1} + \frac{1}{4}\ln(2t + \sqrt{4t^2 + 1})\Big|_0^2$$

$$= \sqrt{17} + \frac{\ln(4 + \sqrt{17})}{4},$$

where we obtained the given primitive function using MATLAB. ■

54.3 Exercises

54.1 Given the curve $\gamma(t) = (t^2, t + 2)^\top$, $t \in \mathbb{R}_{\geq 0}$.

(a) Determine singular points and horizontal and vertical tangents.
(b) Calculate the arc length function.

54.2 Calculate the arc length of the following curve:

$$\gamma: [0, a] \to \mathbb{R}^3 \text{ with } \gamma(t) = t\,(\cos t, \sin t, 1)^\top .$$

54.3 We consider the cone given by $x^2 + y^2 = (z - 2)^2$ with $0 \leq z \leq 2$. Let an ant climb this cone on a path that (i) starts at the point $(2, 0, 0)$, (ii) ends in the cone apex, (iii) goes around the cone three times, and (iv) increases linearly in height.

(a) Find a parametrization $\gamma : [0, 1] \to \mathbb{R}^3$ of this path and convince yourself that this path is really **on the** cone.
(b) Calculate the velocity vector $\dot{\gamma}(t)$ and the acceleration vector $\ddot{\gamma}(t)$.
(c) Construct an expression for the length of the curve γ and evaluate it using MATLAB.
(d) On the way back, the ant takes the same path. Give a parametrization of the form $\tilde{\gamma} : [0, 2\pi] \to \mathbb{R}^3$ for its return path.

Curves II

55

Now that we know numerous examples of plane curves and space curves and can also calculate the length of curves, we turn to other special properties of curves: Curves can be parameterized in many ways. Among these many ways, parameterization *by arc length* plays a prominent role. We introduce this parameterization. Further, curve points generally have an *accompanying tripod*, *curvature*, and *torsion*. These vectors or quantities are easy to determine. *Leibniz's sector formula* allows the computation of area enclosed by curves or, more generally, the area swept by a *driving ray*.

55.1 Reparameterization of a Curve

The following curves each have the unit circle as their trace:

$$\gamma_1 : [0, 2\pi] \to \mathbb{R}^2,\ \gamma_1(t) = \begin{pmatrix} \cos(t) \\ \sin(t) \end{pmatrix} \text{ and } \gamma_2 : [0, 1] \to \mathbb{R}^2,\ \gamma_2(t) = \begin{pmatrix} \cos(2\pi t^2) \\ \sin(2\pi t^2) \end{pmatrix}.$$

The velocities $\|\dot{\gamma}_i(t)\|$ with which the unit circle is traversed are, however, different:

$$\|\dot{\gamma}_1(t)\| = 1 \text{ and } \|\dot{\gamma}_2(t)\| = 4\pi t\,.$$

The curves γ_1 and γ_2 can be obtained by **reparameterization**. We have:

$$\gamma_2(t) = \gamma_1(2\pi t^2)\,.$$

C. Karpfinger, *Calculus and Linear Algebra in Recipes*,
https://doi.org/10.1007/978-3-662-65458-3_55

Reparameterization of a Curve
If $\gamma : [a, b] \to \mathbb{R}^n$ is a curve and $h : [c, d] \to [a, b]$ is a strictly monotonically increasing function with $h(c) = a$ and $h(d) = b$, then

$$\tilde{\gamma} : [c, d] \to \mathbb{R}^n \text{ with } \tilde{\gamma}(t) = \gamma(h(t))$$

is a curve with the same trace as γ, i.e. $\tilde{\gamma}([c, d]) = \gamma([a, b])$.
We say, $\tilde{\gamma}$ arises from γ by **reparameterization**.

Among the many possible parameterizations of a regular curve, one is excellent; it is called the *natural parameterization* or the *parameterization by arc length*. This is obtained as follows:

Recipe: Parametrization by Arc length
Given is a regular curve $\gamma : [a, b] \to \mathbb{R}^n$. The **parameterization by arc length** or **natural parameterization** $\tilde{\gamma}$ of γ is obtained as follows:

(1) Determine the arc length function $s(t)$, in particular the length $L(\gamma)$.
(2) Determine the inverse function $s^{-1}(t)$, $t \in [0, L(\gamma)]$; we have

$$s^{-1}(0) = a \text{ and } s^{-1}(L(\gamma)) = b\,.$$

(3) Obtain the natural parametrization $\tilde{\gamma}$

$$\tilde{\gamma} : [0, L(\gamma)] \to \mathbb{R}^n \text{ with } \tilde{\gamma}(t) = \gamma\big(s^{-1}(t)\big)\,.$$

We have $\|\dot{\tilde{\gamma}}(t)\| = 1$ for all $t \in [0, L(\gamma)]$.

Up to now we have always considered curves on closed intervals $[a, b]$ since the formalism for this is somewhat simpler, but of course we can also consider curves on unbounded intervals like $[a, \infty)$. We'll do that right away in the following example.

Example 55.1 We determine the natural parametrization of the logarithmic spiral

$$\gamma : [0, \infty) \to \mathbb{R}^2 \text{ with } \gamma(t) = \begin{pmatrix} e^{-t}\cos(t) \\ e^{-t}\sin(t) \end{pmatrix}.$$

(1) According to Example 54.4 we have:

$$s(t) = \int_0^t \sqrt{2}\,e^{-\tau}\,d\tau = \sqrt{2}(1 - e^{-t}) \text{ and } L(\gamma) = \sqrt{2}.$$

(2) To determine s^{-1}, we solve the term $s = \sqrt{2}(1 - e^{-t})$ for t:

$$s = \sqrt{2} - \sqrt{2}\,e^{-t} \Leftrightarrow e^{-t} = \frac{\sqrt{2} - s}{\sqrt{2}} \Leftrightarrow -t = \ln\left(\frac{\sqrt{2} - s}{\sqrt{2}}\right)$$

$$\Leftrightarrow t = \ln\left(\frac{\sqrt{2}}{\sqrt{2} - s}\right).$$

Thus the reparameterization function is

$$s^{-1} : [0, \sqrt{2}) \to [0, \infty),\ s^{-1}(t) = \ln\left(\frac{\sqrt{2}}{\sqrt{2} - t}\right).$$

(3) We now obtain the natural parameterization $\tilde{\gamma} : [0, \sqrt{2}) \to \mathbb{R}^2$ with

$$\tilde{\gamma}(t) = \gamma\left(s^{-1}(t)\right) = \begin{pmatrix} e^{-\ln\left(\frac{\sqrt{2}}{\sqrt{2}-t}\right)} \cos\left(\ln\left(\frac{\sqrt{2}}{\sqrt{2}-t}\right)\right) \\ e^{-\ln\left(\frac{\sqrt{2}}{\sqrt{2}-t}\right)} \sin\left(\ln\left(\frac{\sqrt{2}}{\sqrt{2}-t}\right)\right) \end{pmatrix} = \begin{pmatrix} \frac{\sqrt{2}-t}{\sqrt{2}} \cos\left(\ln\left(\frac{\sqrt{2}}{\sqrt{2}-t}\right)\right) \\ \frac{\sqrt{2}-t}{\sqrt{2}} \sin\left(\ln\left(\frac{\sqrt{2}}{\sqrt{2}-t}\right)\right) \end{pmatrix}.$$

■

The natural parameterization of a curve has many advantages, for example, many formulas become much simpler with this parameterization. In the next section we will introduce numerous formula for curves. Note how much these formulas simplify when γ is parameterized by arc length, namely, we have then $\|\dot{\gamma}(t)\| = 1$ for all t.

55.2 Frenet–Serret Frame, Curvature and Torsion

In each curve point $\gamma(t)$ of a (space) curve γ, we have a *Frenet–Serret frame* and a *curvature* and a *torsion*:

- The Frenet–Serret frame is an orthonormal basis consisting of the tangent unit vector $T(t)$, the binormal unit vector $B(t)$ and the principal normal unit vector $N(t)$ in every curve point $\gamma(t)$. Here $T(t)$ and $N(t)$ generate the *osculating plane* E, which is the plane to which the curve in $\gamma(t)$ is attached to.
- Curvature is a measure of the deviation from straight line motion. If the curvature is zero, the motion is straight. Curvature is explained as the rate of change of the tangent unit vector with respect to the arc length.
- Torsion is a measure of the deviation of the curve from the plane path. If the torsion is zero, the motion is in a plane. The torsion is explained as the rate of change of the binormal unit vector, related to the arc length.

We summarize the essential formulas.

Frenet–Serret Frame, Osculating Plane, Curvature, Torsion
Given is a three times differentiable curve

$$\gamma : I \to \mathbb{R}^n \text{ with } \gamma(t) = (x_1(t), \ldots, x_n(t))^\top \text{ and } \dot{\gamma}(t) \times \ddot{\gamma}(t) \neq \mathbf{0}.$$

- Case $n = 2$: The tangent vector $\dot{\gamma}(t) = (\dot{x}_1(t), \dot{x}_2(t))^\top$ and the **normal vector** $n(t) = (-\dot{x}_2(t), \dot{x}_1(t))^\top$ are in the curve point $\gamma(t)$ perpendicular.
- Case $n = 3$: The **Frenet–Serret frame** is the right hand system $(T(t), N(t), B(t))$ of the three normalized and orthogonal vectors

$$T(t) = \frac{\dot{\gamma}(t)}{\|\dot{\gamma}(t)\|}, \quad B(t) = \frac{(\dot{\gamma}(t) \times \ddot{\gamma}(t))}{\|\dot{\gamma}(t) \times \ddot{\gamma}(t)\|}, \quad N(t) = B(t) \times T(t).$$

One calls
- $T(t)$ the **tangent unit vector**,
- $B(t)$ the **binormal unit vector** and
- $N(t)$ the **principal normal unit vector**.

(continued)

The vectors T and N span the **osculating plane**. This osculating plane E has at the curve point $\gamma(t)$ the presentation:

$$E : \boldsymbol{x} = \gamma(t) + \lambda\, T(t) + \mu\, N(t) \text{ with } \lambda,\ \mu \in \mathbb{R}\,.$$

- The **curvature** of a plane curve ($n = 2$) is

$$\kappa(t) = \frac{|\det(\dot{\gamma}(t),\ \ddot{\gamma}(t))|}{\|\dot{\gamma}(t)\|^3} = \frac{|\dot{x}_1(t)\,\ddot{x}_2(t) - \dot{x}_2(t)\,\ddot{x}_1(t)|}{(\dot{x}_1(t)^2 + \dot{x}_2(t)^2)^{3/2}}\,.$$

- The **curvature** of a space curve ($n = 3$) is

$$\kappa(t) = \frac{\|\dot{\gamma}(t) \times \ddot{\gamma}(t)\|}{\|\dot{\gamma}(t)\|^3}\,.$$

- The **torsion** of a space curve ($n = 3$) is

$$\tau(t) = \frac{|\det(\dot{\gamma}(t), \ddot{\gamma}(t), \dddot{\gamma}(t))|}{\|\dot{\gamma}(t) \times \ddot{\gamma}(t)\|^2}\,.$$

Figure 55.1 shows a plane curve with tangent and normal vector as well as a space curve with the Frenet–Serret frame and osculating plane in a curve point.

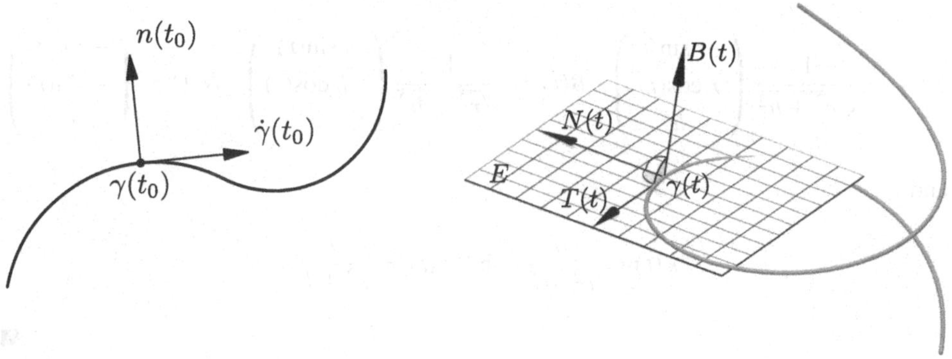

Fig. 55.1 Tangent and normal vector as well as the Frenet–Serret frame and osculating plane

Example 55.2 We consider the helix

$$\gamma : [0, 2\pi] \to \mathbb{R}^3 \text{ with } \gamma(t) = \begin{pmatrix} r\cos(t) \\ r\sin(t) \\ h\,t \end{pmatrix} \text{ and } r,\, h \in \mathbb{R}_{>0}\,.$$

First we calculate the first three derivatives:

$$\dot\gamma(t) = \begin{pmatrix} -r\sin(t) \\ r\cos(t) \\ h \end{pmatrix}, \quad \ddot\gamma(t) = \begin{pmatrix} -r\cos(t) \\ -r\sin(t) \\ 0 \end{pmatrix}, \quad \dddot\gamma(t) = \begin{pmatrix} r\sin(t) \\ -r\cos(t) \\ 0 \end{pmatrix}$$

and thus the quantities

$$\|\dot\gamma(t)\| = \sqrt{r^2+h^2}\,, \quad \dot\gamma(t) \times \ddot\gamma(t) = \begin{pmatrix} rh\sin(t) \\ -rh\cos(t) \\ r^2 \end{pmatrix}, \quad \|\dot\gamma(t) \times \ddot\gamma(t)\| = r\sqrt{r^2+h^2}$$

and

$$\det(\dot\gamma(t), \ddot\gamma(t), \dddot\gamma(t)) = \begin{vmatrix} -r\sin(t) & -r\cos(t) & r\sin(t) \\ r\cos(t) & -r\sin(t) & -r\cos(t) \\ h & 0 & 0 \end{vmatrix} = h\,r^2\,.$$

After this preliminary work, we only need to state the results:

$$T(t) = \frac{1}{\sqrt{r^2+h^2}} \begin{pmatrix} -r\sin(t) \\ r\cos(t) \\ h \end{pmatrix}, \quad B(t) = \frac{1}{\sqrt{r^2+h^2}} \begin{pmatrix} h\sin(t) \\ -h\cos(t) \\ r \end{pmatrix}, \quad N(t) = \begin{pmatrix} -\cos(t) \\ -\sin(t) \\ 0 \end{pmatrix}$$

and

$$\kappa(t) = \frac{r}{r^2+h^2} \text{ and } \tau(t) = \frac{h}{r^2+h^2}\,.$$

■

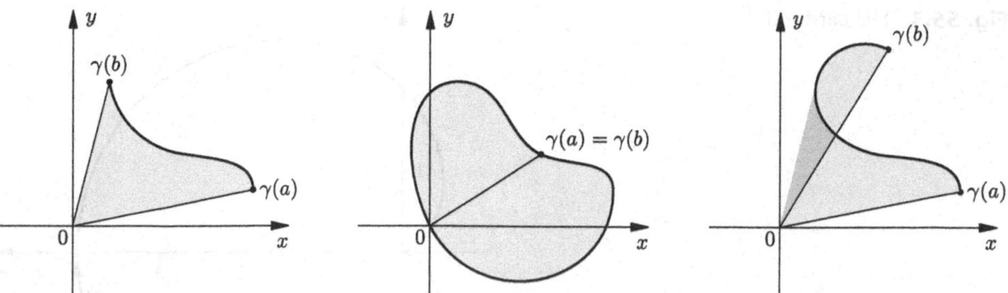

Fig. 55.2 Leibniz's sector formula gives the area swept by the ray

55.3 The Leibniz Sector Formula

With Leibniz's sector formula one calculates the (oriented) area, which a driving ray of a double point free curve section sweeps. In particular, one can determine the area of a region enclosed by a curve.

The driving ray drawn in the following Fig. 55.2 sweeps an area that lies between the initial ray $\overline{0\,\gamma(a)}$ and the final ray $\overline{0\,\gamma(b)}$. Parts of the area may be swept more than once.

The following Leibniz sector formula is used to calculate the area F in each case. Areas that are swept more than once are *shortened* because an *oriented* area is calculated, i.e. areas that are swept when the direction of the driving ray is reversed are calculated negatively. If, however, one wants to explicitly determine the total area swept by the driving beam (independent of the orientation), the sector formula also offers the possibility for this:

Leibniz's Sector Formula

The area covered by the driving ray of a plane double point free C^1-curve $\gamma : [a, b] \to \mathbb{R}^2$ with $\gamma(t) = (x(t), y(t))^\top$ is

$$F(\gamma) = \frac{1}{2}\left|\int_a^b x(t)\dot{y}(t) - \dot{x}(t)y(t)\mathrm{d}t\right| .$$

It is

$$\tilde{F}(\gamma) = \frac{1}{2}\int_a^b |x(t)\dot{y}(t) - \dot{x}(t)y(t)|\,\mathrm{d}t$$

the total area swept by the driving ray.

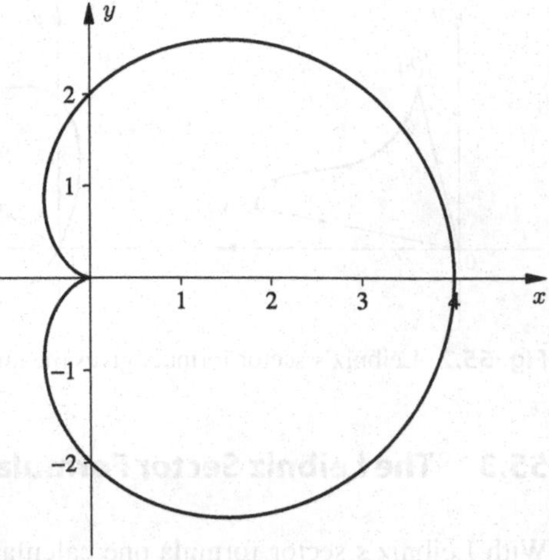

Fig. 55.3 The cardioid

Example 55.3 We calculate the area $F = F(\gamma)$ which is included by the **cardioid**

$$\gamma : [0, 2\pi] \to \mathbb{R}^2 \text{ with } \gamma(t) = \begin{pmatrix} a\cos t\,(1+\cos t) \\ a\sin t\,(1+\cos t) \end{pmatrix}$$

with $a > 0$. In Fig. 55.3 we have chosen $a = 1$.

We have:

$$F(\gamma) = \frac{1}{2}\left|\int_0^{2\pi} x(t)\dot{y}(t) - \dot{x}(t)y(t)\mathrm{d}t\right|$$

$$= \frac{a^2}{2}\left|\int_0^{2\pi} 1 + 2\cos^3 t + \cos^4 t + 2\cos t\sin^2 t + \cos^2 t\sin^2 t\mathrm{d}t\right| = \frac{3a^2\pi}{2},$$

To calculate this integral, we used MATLAB. ■

55.4 Exercises

55.1 Show that the recipe for determining the natural parametrization in Sect. 55.1 works, in particular that we have $\|\dot{\tilde{\gamma}}(t)\| = 1$ for all t.

55.2 Parameterize the following curve by arc length:

$$\gamma(t) = (x(t), y(t)) = \left(\ln\sqrt{1+t^2}, \arctan t\right)^\top, \quad t \in [0, 2].$$

55.3 Let a,b > 0. Given the curve $\gamma(t) = (x(t), y(t))^\top, t \in [0, 2\pi]$ with

$$x(t) = a\cos t \quad \text{and} \quad y(t) = b\sin t,$$

which passes through an ellipse.

(a) Determine the points (x, y) of the curve where the curvature is maximum.
(b) Using Leibniz's sector formula, calculate the area of the ellipse.

55.4 Using Leibniz's sector formula, calculate the area of the region enclosed by the two curves

$$\gamma_1(t) = \begin{pmatrix} 2 - t^2 \\ t \end{pmatrix}, \qquad \gamma_2(t) = \begin{pmatrix} 1 \\ t \end{pmatrix}, \qquad t \in \mathbb{R}$$

enclosed by the two curves.

55.5 A point P on the tread of a rolling wheel describes a periodic curve, which is called a cycloid (see figure).

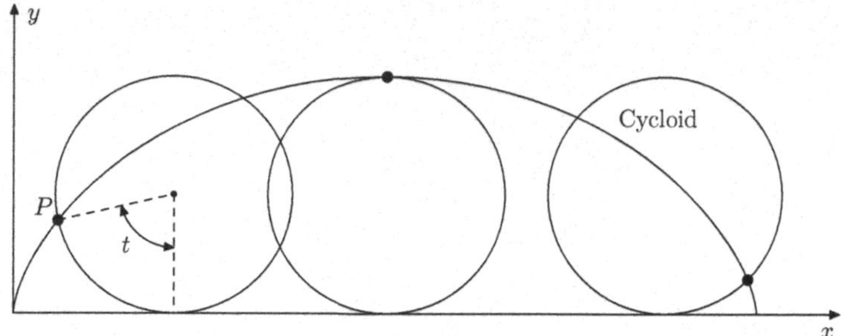

(a) Give a parameter representation for the cycloid. To do this, use as a parameter the angle t shown in the figure.
(b) Calculate the area under a cycloid arc using Leibniz's sector formula.
(c) Calculate the curvature of the cycloid for $0 < t < 2\pi$.

55.6 Calculate the arc length of the following curve and its reparameterization by arc length:

$$\gamma(t) = (x(t), y(t)) = (t, \cosh(t/2) - 1)^\top, \quad |t| \leq 5.$$

55.7 Determine the curvature of the curve γ given by

$$\gamma(t) = (t, 1 - \cos t)^\top , \quad t \in [0, 1].$$

55.8 Calculate the arc length of the curve γ defined by

$$\gamma(t) = \mathrm{e}^{-2t}(\cos t, \sin t)^\top , \quad t \in [0, \infty) ,$$

and determine the reparameterization of γ according to the arc length.

Line Integrals

56

We distinguish between two types of line integrals: *scalar* line integrals and *vector* line integrals. In a scalar line integral a scalar field is integrated along a curve, in a vectorial one a vector field. In applications, scalar integrals determine a mass or charge of the described curve, while vector integrals determine a work done when a particle moves along the curve.

56.1 Scalar and Vector Line Integrals

We want to integrate a scalar field or a vector field along a curve. It is clear what we need for this: In the scalar case a curve and a scalar field, in the vector case a curve and a vector field. We start with the definition and examples and explain the background or ideas afterwards.

The Scalar and the Vectorial Line Integral
Given is a curve $\gamma : [a, b] \to \mathbb{R}^n$ with $\gamma(t) = \big(x_1(t), \ldots, x_n(t)\big)^\top$.

- For a scalar field $f : D \subseteq \mathbb{R}^n \to \mathbb{R}$ with $\gamma([a, b]) \subseteq D$ the integral

$$\int_\gamma f \, \mathrm{d}s = \int_a^b f\big(\gamma(t)\big) \, \|\dot{\gamma}(t)\| \, \mathrm{d}t$$

  is called the **scalar line integral** of f along γ.
- For a vector field $\boldsymbol{v} : D \subseteq \mathbb{R}^n \to \mathbb{R}^n$ with $\gamma([a, b]) \subseteq D$ the integral

(continued)

C. Karpfinger, *Calculus and Linear Algebra in Recipes*,
https://doi.org/10.1007/978-3-662-65458-3_56

$$\int_{\gamma} \boldsymbol{v} \cdot \mathrm{d}s = \int_a^b \boldsymbol{v}(\gamma(t))^{\top} \dot{\gamma}(t)\,\mathrm{d}t$$

is called the **vectorial line integral** of $\boldsymbol{v}$ along γ.

Note the distinction in the notation: The multiplication sign in front of the ds in the case of the vectorial line integral is meant to remind us of the scalar product, which is used to form this line integral. We will consistently use this sign to distinguish between scalar and vector line integrals.

Before we continue with theoretical considerations, we show with two examples how *easy* it is to set up the line integrals, if only the ingredients curve and scalar field or vector field are available. The only *hurdle* to overcome is the acceptance of the simple fact that

$$f(\gamma(t)) \text{ or } \boldsymbol{v}(\gamma(t))$$

only means that the variable x_i in f or $\boldsymbol{v}$ by the i-th component $x_i(t)$ of the curve γ is to be replaced.

Example 56.1 Given is the curve $\gamma : [0, 2\pi] \to \mathbb{R}^2$ with $\gamma(t) = \big(2\cos(t), \sin(t)\big)^{\top}$ and the scalar field f or vector field $\boldsymbol{v}$:

$$f : \mathbb{R}^2 \to \mathbb{R}\,,\ f(x, y) = x\,y^2 \ \text{ or } \ \boldsymbol{v} : \mathbb{R}^2 \to \mathbb{R}^2\,,\ \boldsymbol{v}(x, y) = \begin{pmatrix} -y \\ x^2 \end{pmatrix}.$$

- The scalar line integral is due to $f(\gamma(t)) = 2\cos(t)\sin^2(t)$ and $\|\dot{\gamma}(t)\| = \sqrt{4\sin^2(t) + \cos^2(t)}$:

$$\int_{\gamma} f \mathrm{d}s = \int_a^{2\pi} f\big(\gamma(t)\big)\, \|\dot{\gamma}(t)\|\ \mathrm{d}t = \int_0^{2\pi} 2\cos(t)\sin^2(t)\sqrt{4\sin^2(t) + \cos^2(t)}\mathrm{d}t\,.$$

- The vectorial line integral is due to $\boldsymbol{v}(\gamma(t)) = \big(-\sin(t), 4\cos^2(t)\big)^{\top}$ and $\dot{\gamma}(t) = (-2\sin(t), \cos(t))^{\top}$:

$$\begin{aligned}\int_{\gamma} \boldsymbol{v} \cdot \mathrm{d}s &= \int_0^{2\pi} \boldsymbol{v}(\gamma(t))^{\top} \dot{\gamma}(t)\,\mathrm{d}t = \int_a^b \begin{pmatrix} -\sin t \\ 4\cos^2 t \end{pmatrix}^{\top} \begin{pmatrix} -2\sin t \\ \cos t \end{pmatrix} \mathrm{d}t \\ &= \int_0^{2\pi} 2\sin^2(t) + 4\cos^3(t)\mathrm{d}t\,.\end{aligned}$$

We should not care about determining these integrals now; we just wanted to convince ourselves that setting up the integrals is a snap if only the ingredients are at hand. ■

If $\gamma(a) = \gamma(b)$, the curve γ is called closed. For the line integral along a closed curve γ, one also writes

$$\oint_\gamma f\,\mathrm{d}s = \int_\gamma f\,\mathrm{d}s \text{ and } \oint_\gamma \boldsymbol{v}\cdot \mathrm{d}\boldsymbol{s} = \int_\gamma \boldsymbol{v}\cdot \mathrm{d}\boldsymbol{s}\,.$$

The vectorial line integral along a closed curve is also called **circulation** of $\boldsymbol{v}$ along γ.

The following calculation rules for scalar or vectorial line integrals are immediately clear:

Calculation Rules for Line Integrals

For a curve $\gamma : [a, b] \to \mathbb{R}^n$ we have:

- For all $\lambda, \mu \in \mathbb{R}$ and scalar fields $f, g : D \subseteq \mathbb{R}^n \to \mathbb{R}$ or vector fields $\boldsymbol{v}, \boldsymbol{w} : D \subseteq \mathbb{R}^n \to \mathbb{R}^n$ we have:

$$\int_\gamma (\lambda f + \mu g)\mathrm{d}s = \lambda \int_\gamma f\mathrm{d}s + \mu \int_\gamma g\mathrm{d}s \text{ and } \int_\gamma (\lambda \boldsymbol{v} + \mu \boldsymbol{w})\cdot \mathrm{d}\boldsymbol{s} = \lambda \int_\gamma \boldsymbol{v}\cdot \mathrm{d}\boldsymbol{s} + \mu \int_\gamma \boldsymbol{w}\cdot \mathrm{d}\boldsymbol{s}.$$

- If $\gamma = \gamma_1 + \ldots + \gamma_r$ is composite, then we have:

$$\int_\gamma f\mathrm{d}s = \int_{\gamma_1} f\mathrm{d}s + \ldots + \int_{\gamma_r} f\mathrm{d}s \text{ and } \int_\gamma \boldsymbol{v}\cdot \mathrm{d}\boldsymbol{s} = \int_{\gamma_1} \boldsymbol{v}\cdot \mathrm{d}\boldsymbol{s} + \ldots + \int_{\gamma_r} \boldsymbol{v}\cdot \mathrm{d}\boldsymbol{s}\,.$$

- If one reverses the orientation of γ, i.e. $\gamma \to -\gamma$, then we have:

$$\int_{-\gamma} f\mathrm{d}s = \int_\gamma f\mathrm{d}s \text{ and } \int_{-\gamma} \boldsymbol{v}\cdot \mathrm{d}\boldsymbol{s} = -\int_\gamma \boldsymbol{v}\cdot \mathrm{d}\boldsymbol{s}\,.$$

In the typical tasks on this topic or applications, the curves along which integration takes place are given by sketches. The trick is then to find a parameterization of the individual curve pieces. The rest can be worked off according to the following recipe:

Recipe: Calculate a Line Integral

To calculate a scalar line integral $\int_\gamma f\mathrm{d}s$ or vectorial line integral $\int_\gamma \boldsymbol{v}\cdot \mathrm{d}\boldsymbol{s}$ proceed as follows:

(continued)

(1) Determine a parametrization of the curve $\gamma = \gamma_1 + \cdots + \gamma_r$, i.e.

$$\gamma_i : [a_i, b_i] \to \mathbb{R}^n \text{ with } \gamma_i(t) = (x_1^{(i)}(t), \ldots, x_n^{(i)}(t))^\top \text{ for all } i = 1, \ldots, r .$$

(2) Determine for each $i = 1, \ldots, r$ the integral:

- In the case of a scalar line integral:

$$I_i = \int_{\gamma_i} f \mathrm{d}s = \int_{a_i}^{b_i} f(\gamma_i(t)) \, \|\dot{\gamma}_i(t)\| \mathrm{d}t ,$$

- In the case of a vectorial line integral:

$$I_i = \int_{\gamma_i} \boldsymbol{v} \cdot \mathrm{d}s = \int_a^b \boldsymbol{v}(\gamma_i(t))^\top \dot{\gamma}_i(t) \mathrm{d}t .$$

(3) Get $I = \int_\gamma f \mathrm{d}s$ or $I = \int_\gamma \boldsymbol{v} \cdot \mathrm{d}s$ by $I = I_1 + \cdots + I_r$.

The determination of the *ordinary* integrals in step (2) is occasionally problematic; especially the root of the scalar line integrals quickly leads to integrals which can no longer be determined elementarily. For the solution of these integrals one resorts to the methods from Chaps. 30 and 31 or to integral tables or MATLAB.

Example 56.2

- Given the curve $\gamma : [0, 2\pi] \to \mathbb{R}^2$ with $\gamma(t) = \left(\cos(t), \sin(t)\right)^\top$ and the scalar field $f : \mathbb{R}^2 \to \mathbb{R}$ with $f(x, y) = 2$. The line integral of f along γ is:

$$\int_\gamma f \, \mathrm{d}s = \int_0^{2\pi} f(\gamma(t)) \, \|\dot{\gamma}(t)\| \, \mathrm{d}t = \int_0^{2\pi} 2 \cdot 1 \, \mathrm{d}t = 4\pi .$$

  This is the surface area of a cylinder of height 2 of radius 1.
- Given is the curve $\gamma : [0, 2] \to \mathbb{R}^3$, $\gamma(t) = \left(t, t^3, 3\right)^\top$ and the vector field $\boldsymbol{v} : \mathbb{R}^3 \to \mathbb{R}^3$, $\boldsymbol{v}(x, y, z) = \left(xy, x - z, xz\right)^\top$. The vectorial line integral of $\boldsymbol{v}$ along γ has the value

$$\int_\gamma \boldsymbol{v} \cdot \mathrm{d}s = \int_0^2 \boldsymbol{v}(\gamma(t))^\top \dot{\gamma}(t) \mathrm{d}t = \int_0^2 \begin{pmatrix} t^4 \\ t - 3 \\ 3t \end{pmatrix}^\top \begin{pmatrix} 1 \\ 3t^2 \\ 0 \end{pmatrix} \mathrm{d}t = \int_0^2 t^4 + 3t^3 - 9t^2 \mathrm{d}t = -\frac{28}{5}.$$

- Given is the closed curve $\gamma : [0, 2\pi] \to \mathbb{R}^2,\ \gamma(t) = \big(\cos(t),\ \sin(t)\big)^\top$ and the vector field $\boldsymbol{v} : \mathbb{R}^2 \to \mathbb{R}^2,\ \boldsymbol{v}(x, y) = \big(-y,\ x\big)^\top$. The vectorial line integral of $\boldsymbol{v}$ along γ is then

$$\oint_\gamma \boldsymbol{v} \cdot \mathrm{d}s = \int_0^{2\pi} \begin{pmatrix} -\sin(t) \\ \cos(t) \end{pmatrix}^\top \begin{pmatrix} -\sin(t) \\ \cos(t) \end{pmatrix} \mathrm{d}t = 2\pi .$$

- If we again choose the closed curve γ from the last example and the vector field $\boldsymbol{v} : \mathbb{R}^2 \to \mathbb{R}^2,\ \boldsymbol{v}(x, y) = \big(3x, 0\big)^\top$, then we have:

$$\oint_\gamma \boldsymbol{v} \cdot \mathrm{d}s = \int_0^{2\pi} \begin{pmatrix} 3\cos(t) \\ 0 \end{pmatrix}^\top \begin{pmatrix} -\sin(t) \\ \cos(t) \end{pmatrix} \mathrm{d}t = \int_0^{2\pi} -3\cos(t)\sin(t)\mathrm{d}t = \frac{3}{2}\cos^2(t)\Big|_0^{2\pi} = 0.$$

■

An interpretation of the scalar line integral in the case $n = 2$ is simple: The trace of the curve γ runs in the x-y-plane, the scalar field f has at the curve point $\gamma(t)$ the value $f(\gamma(t))$. In the scalar line integral

$$\int_\gamma f \mathrm{d}s = \int_a^b f(\gamma(t))\, \|\dot{\gamma}(t)\| \mathrm{d}t$$

the area included by the graph of f with the trace of γ is determined (see Fig. 56.1). In this case, the *arc element* $\mathrm{d}s = \|\dot{\gamma}(t)\|\mathrm{d}t$ takes the tangential vector $\dot{\gamma}$ of the trace of γ at the curve point under consideration is taken into account.

The line integral over the scalar field $f = 1$ thus yields the arc length of the curve (see Sect. 54.2).

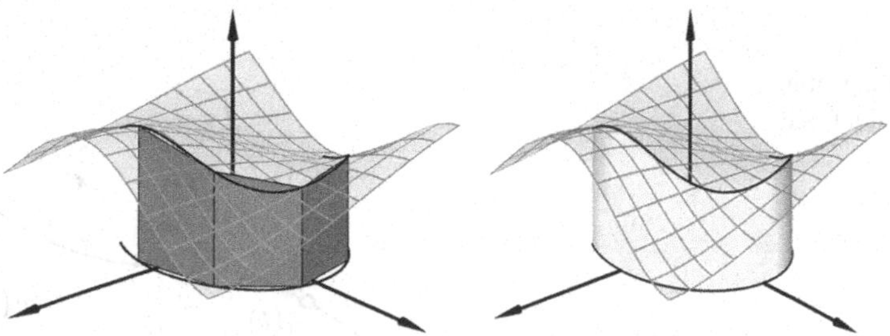

Fig. 56.1 In the scalar line integral, the area between γ and $f(\gamma)$ is determined

56.2 Applications of the Line Integrals

Line integrals have many applications in science and engineering. This is due to the following interpretations:

- **Scalar line integral:** The scalar $f(\gamma(t))$ can be taken as *assignment* of the curve point $\gamma(t)$: this scalar field f can be a mass density or a charge density. The integral of the scalar field f along this curve then gives the total mass or total charge of γ.
- **Vectorial line integral:** The vector $\boldsymbol{v}(\gamma(t))$ can be understood as acting *force* at the point $\gamma(t)$ on a particle located there and moving along the curve γ. By the scalar product $\boldsymbol{v}(\gamma(t))^\top \dot{\gamma}(t)$ the tangential component of this force along γ is calculated (see Fig. 56.2). The integral of the vector field $\boldsymbol{v}$ along this curve then provides the work that must be done to move the particle along γ.

We consider a typical application of the scalar line integral in the following box:

Total Mass, Total Charge, Centre of Gravity, Geometric Centre of Gravity
If $\rho(x, y)$ or $\rho(x, y, z)$ is a **mass density** or a **charge density of** a curve γ (e.g. a wire), then the **total mass** or **total charge** is:

$$M(\gamma) = \int_\gamma \rho \, \mathrm{d}s \, .$$

And the **centre of gravity** $S = (s_1, s_2)$ or $S = (s_1, s_2, s_3)$ we obtain by

$$s_1 = \frac{1}{M(\gamma)} \int_\gamma x\rho \, \mathrm{d}s \, , \; s_2 = \frac{1}{M(\gamma)} \int_\gamma y\rho \, \mathrm{d}s \, , \; s_3 = \frac{1}{M(\gamma)} \int_\gamma z\rho \, \mathrm{d}s \, .$$

With $\rho = 1$, we obtain the **geometric centre**, we then have $M(\gamma) = L(\gamma)$.

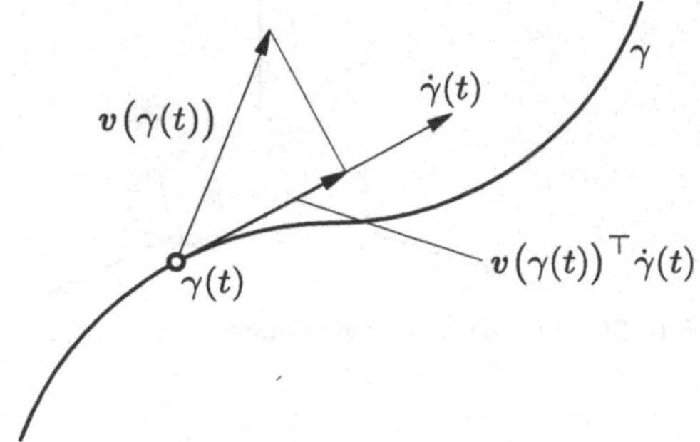

Fig. 56.2 Tangential component of a force

Example 56.3 Let be given the mass density $\rho : \mathbb{R}^2 \to \mathbb{R},\ \rho(x, y) = x$ and the curve $\gamma = \gamma_1 + \gamma_2$, where

$$\gamma_1 : [0, 1] \to \mathbb{R}^2,\ \gamma_1(t) = \begin{pmatrix} t \\ t^2 \end{pmatrix} \quad \text{and} \quad \gamma_2 : [0, 1] \to \mathbb{R}^2,\ \gamma_2(t) = \begin{pmatrix} 1-t \\ 1 \end{pmatrix}.$$

For the total mass of γ we then have:

$$\begin{aligned} M(\gamma) &= \int_\gamma \rho\,\mathrm{d}s = \int_{\gamma_1} \rho\,\mathrm{d}s + \int_{\gamma_2} \rho\,\mathrm{d}s = \int_0^1 t\sqrt{1+4t^2}\,\mathrm{d}t + \int_0^1 (1-t)\,\mathrm{d}t \\ &= \frac{1}{12}(1+4t^2)^{3/2}\Big|_0^1 + t\Big|_0^1 - \frac{1}{2}t^2\Big|_0^1 = \frac{1}{12}(5^{3/2}-1) + \frac{1}{2} = 1.3484\,. \end{aligned}$$

For the coordinates s_1, s_2 of the center of gravity $S = (s_1, s_2)$ we obtain (with MATLAB):

$$s_1 = \frac{1}{M(\gamma)}\int_\gamma x\rho\mathrm{d}s = \frac{1}{M(\gamma)}\left(\int_0^1 t^2\sqrt{1+4t^2}\,\mathrm{d}t + \int_0^1 (1-t)^2\,\mathrm{d}t\right) = 0.6969\,,$$

$$s_2 = \frac{1}{M(\gamma)}\int_\gamma y\rho\mathrm{d}s = \frac{1}{M(\gamma)}\left(\int_0^1 t^3\sqrt{1+4t^2}\,\mathrm{d}t + \int_0^1 (1-t)\,\mathrm{d}t\right) = 0.7225\,.$$

We also determine the geometric center of gravity $\tilde{S} = (\tilde{s}_1, \tilde{s}_2)$, where for the calculations using MATLAB. To do this, we first need the length of the curve γ:

$$L(\gamma) = \int_0^1 \sqrt{1+4t^2}\,\mathrm{d}t + \int_0^1 1\,\mathrm{d}t = 2.4789\,.$$

Now we obtain the coordinates $\tilde{s}_1$ and $\tilde{s}_2$ of the geometric center of gravity:

$$\begin{aligned} \tilde{s}_1 &= \frac{1}{L(\gamma)}\left(\int_0^1 t\sqrt{1+4t^2}\,\mathrm{d}t + \int_0^1 (1-t)\,\mathrm{d}t\right) = 0.5439\,, \\ \tilde{s}_2 &= \frac{1}{L(\gamma)}\left(\int_0^1 t^2\sqrt{1+4t^2}\,\mathrm{d}t + \int_0^1 1\,\mathrm{d}t\right) = 0.6480\,. \end{aligned}$$

Figure 56.3 shows the curve $\gamma = \gamma_1 + \gamma_2$ with the center of gravity S with the given density ρ and geometric centroid $\tilde{S}$. ■

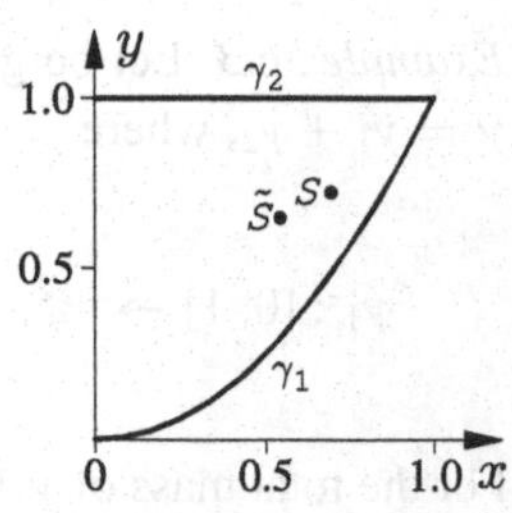

Fig. 56.3 The centroids S and $\tilde{S}$

56.3 Exercises

56.1 Determine the following scalar and vector line integrals, respectively:

(a) $\gamma : [0, 2\pi] \to \mathbb{R}^3$, $\gamma(t) = (\cos t, \sin t, t)^\top$ and $f(x, y, z) = x^2 + y\,z$.
(b) γ is the connecting distance from $(0, 0)^\top$ to $(1, 1)^\top$ and $\boldsymbol{v}(x, y) = (2y, \mathrm{e}^x)^\top$.

56.2 A helical spring is defined by the curve $\gamma : [0, 4\pi] \to \mathbb{R}^3$ with $\gamma(t) = (\cos t, \sin t, \frac{1}{2}t)^\top$ with the line mass density $\rho(x, y, z) = z$. Calculate the mass and the center of gravity of the helical spring.
56.3 Let be given the vector fields $\boldsymbol{v}\colon \mathbb{R}^2 \to \mathbb{R}^2$ and $\boldsymbol{w}\colon \mathbb{R}^2 \to \mathbb{R}^2$ by

$$\boldsymbol{v}(x, y) = \begin{pmatrix} x^2 - y \\ x + y^2 \end{pmatrix} \quad \text{and} \quad \boldsymbol{w}(x, y) = \begin{pmatrix} x + y^2 \\ 2xy \end{pmatrix}.$$

Calculate both for $\boldsymbol{v}$ as well as for $\boldsymbol{w}$ in each case the line integral of $A = (0, 1)^\top$ according to $B = (1, 2)^\top$:

(a) Along the straight line connecting the two points,
(b) along the line consisting of the lines from A to $(1, 1)^\top$ and from $(1, 1)^\top$ to B,
(c) along the parabola $y = x^2 + 1$.

56.4 Let $G \subseteq \mathbb{R}^2$ be the bounded domain bounded by the two graphs of the functions $x = 1 - \frac{1}{4}y^2$ and $x = \frac{1}{2}(y^2 - 1)$. Moreover, let be a vector field $\boldsymbol{v}$ defined by $\boldsymbol{v}(x, y) = (xy, y^2)^\top$.

(a) Parameterize the G bounding curve.
(b) Calculate $\displaystyle\oint_{\partial G} \boldsymbol{v} \cdot \mathrm{d}\boldsymbol{s}$.

Gradient Fields

57

Most of the vector fields one deals with in technology and natural sciences are *force fields*. In mathematics, these and other fields are referred to as *gradient fields*. The calculation of vectorial curve integrals is generally much simpler in such fields: One determines a *primitive function* of the field and obtains the value of the vectorial curve integral by substituting the start and end point of the curve into the primitive function; the difference of these values is the value of the vectorial curve integral. In particular, the value is not dependent on the path of the curve.

57.1 Definitions

In this section, D is always a **domain**, i.e.

- D is open, and
- to two points each $A, B \in D$ there is a C^1-curve $\gamma : [a, b] \to D$ with $\gamma(a) = A$, $\gamma(b) = B$.

Note that the trace of the curve connecting two points of D must also be completely in D.

We come to the crucial notion of *gradient field*. This is a vector field to which there is a *primitive function*:

C. Karpfinger, *Calculus and Linear Algebra in Recipes*,
https://doi.org/10.1007/978-3-662-65458-3_57

Gradient Field, Primitive Function
A continuous vector field $\boldsymbol{v} : D \subseteq \mathbb{R}^n \to \mathbb{R}^n$ is called a **gradient field** if there is a scalar field $f : D \subseteq \mathbb{R}^n \to \mathbb{R}$ with

$$\nabla f = \boldsymbol{v}\,, \quad \text{i.e. } f_{x_1} = v_1, \ldots, f_{x_n} = v_n\,.$$

In this case we call f a **primitive function** of $\boldsymbol{v}$.

A gradient field $\boldsymbol{v}$ is also called a **potential field** or **conservative field**. And if f is a primitive function of $\boldsymbol{v}$, then $-f$ is also called **potential** of $\boldsymbol{v}$.

Example 57.1

- The vector field $\boldsymbol{v} : \mathbb{R}^2 \to \mathbb{R}^2$, $\boldsymbol{v}(x, y) = \left(x, y\right)^\top$ is a gradient field, it has the primitive function

$$f : \mathbb{R}^2 \to \mathbb{R},\ f(x, y) = \frac{1}{2}(x^2 + y^2)\,.$$

- The vector field $\boldsymbol{v} : \mathbb{R}^2 \to \mathbb{R}^2$, $\boldsymbol{v}(x, y) = \left(-y, x\right)^\top$ is not a gradient field, it does not have a primitive function: namely, if $f = f(x, y)$ is a primitive function, then we have $f_x = -y$ and $f_y = x$ and therefore

$$-xy + g(y) = f(x, y) = xy + h(x)\,.$$

This is not satisfiable, for it would then have to be true: $2xy = g(y) - h(x)$. ■

We clearly summarize the essential properties of gradient fields, in particular the connection with vectorial curve integrals:

Vectorial Line Integrals in Gradient Fields
If $\boldsymbol{v} : D \subseteq \mathbb{R}^n \to \mathbb{R}^n$ is a continuous gradient field with primitive function f, then for each piecewise C^1-curve $\gamma : [a, b] \to D$:

$$\int_\gamma \boldsymbol{v} \cdot \mathrm{d}\boldsymbol{s} = f\big(\gamma(b)\big) - f\big(\gamma(a)\big).$$

(continued)

In particular, we have:

- $\oint_\gamma \boldsymbol{v} \cdot \mathrm{d}s = 0$ for every closed curve γ.
- $\int_\gamma \boldsymbol{v} \cdot \mathrm{d}s$ is **path independent**, i.e.

$$\int_\gamma \boldsymbol{v} \cdot \mathrm{d}s = \int_{\tilde{\gamma}} \boldsymbol{v} \cdot \mathrm{d}s$$

for each two piecewise C^1-curves γ and $\tilde{\gamma}$ with the same starting and ending points.

If g is another primitive function of $\boldsymbol{v}$, then $g = f + c$ with $c \in \mathbb{R}$.

We summarize: A gradient field is a vector field $\boldsymbol{v}$ with a primitive function f. If f is a primitive function of $\boldsymbol{v}$, it does not take much effort to determine a vector curve integral for this $\boldsymbol{v}$. To exploit this, we should solve the following two problems:

- How can you tell from a vector field $\boldsymbol{v} : D \subseteq \mathbb{R}^n \to \mathbb{R}^n$ whether it is a gradient field?
- How do you determine a primitive function f for a gradient field $\boldsymbol{v}$?

We solve these two problems in turn in the next two sections.

57.2 Existence of a Primitive Function

A domain $D \subseteq \mathbb{R}^n$ is **simply connected** if every closed curve in D without double points is within D continuous contractible to a point of D.

Example 57.2

- Simply connected domains in the $\mathbb{R}^2$ are: $\mathbb{R}^2$, any half-plane, any circle, any convex region, ...
- Not simply connected domains in the $\mathbb{R}^2$ are: $\mathbb{R}^2 \setminus \{0\}$, each dotted circle, ...
- Simply connected domains in the $\mathbb{R}^3$ are: $\mathbb{R}^3$, any half-space, any sphere, any dotted half-space, any dotted sphere, any convex region, ...
- Not simply connected domains in the $\mathbb{R}^3$ are: $\mathbb{R}^3 \setminus x$-axis, any torus, ... ■

With this notion we now get a solution to our first problem:

Existence of a Primitive Function
A continuously differentiable vector field $\boldsymbol{v} : D \subseteq \mathbb{R}^n \to \mathbb{R}^n$, $\boldsymbol{v} = (v_1, \ldots, v_n)^\top$ is a gradient field if:

- D is simply connected, and
- we have the **integrability condition** $\frac{\partial v_i}{\partial x_k} = \frac{\partial v_k}{\partial x_i}$ for all $1 \leq i,\ k \leq n$.

In particular, then $\boldsymbol{v}$ has a primitive function f.

Since we only consider plane curves or space curves, we are only interested in the cases $n = 2$ and $n = 3$:

- In the case $n = 2$, the condition is $\frac{\partial v_1}{\partial y} = \frac{\partial v_2}{\partial x}$.
- In the case $n = 3$, it reads $\frac{\partial v_1}{\partial y} = \frac{\partial v_2}{\partial x},\ \frac{\partial v_2}{\partial z} = \frac{\partial v_3}{\partial y},\ \frac{\partial v_3}{\partial x} = \frac{\partial v_1}{\partial z}$.

With the help of rotation both cases can be reconciled: We can formulate the integrability condition briefly as

$$\operatorname{rot} \boldsymbol{v} = \boldsymbol{0}\,.$$

Thereby one understands a two-dimensional vector field $v = (v_1, v_2)^\top$ as a plane three-dimensional vector field as follows

$$\boldsymbol{v} = \begin{pmatrix} v_1 \\ v_2 \end{pmatrix} \to \boldsymbol{v} = \begin{pmatrix} v_1 \\ v_2 \\ 0 \end{pmatrix} \Rightarrow \operatorname{rot} \boldsymbol{v} = \begin{pmatrix} \frac{\partial v_3}{\partial y} - \frac{\partial v_2}{\partial z} \\ \frac{\partial v_1}{\partial z} - \frac{\partial v_3}{\partial x} \\ \frac{\partial v_2}{\partial x} - \frac{\partial v_1}{\partial y} \end{pmatrix} = \begin{pmatrix} 0 \\ 0 \\ \frac{\partial v_2}{\partial x} - \frac{\partial v_1}{\partial y} \end{pmatrix},$$

since such a plane vector field of z is independent and we have $v_3 = 0$.

Example 57.3

- The vector field

$$\boldsymbol{v} : \mathbb{R}^3 \to \mathbb{R}^3,\ \boldsymbol{v}(x, y, z) = \begin{pmatrix} e^x y + 1 \\ e^x + z \\ y \end{pmatrix}$$

is a gradient field, because $\boldsymbol{v}$ is continuously differentiable, $\mathbb{R}^3$ is simply continuous, and we have

$$\operatorname{rot} \boldsymbol{v} = \begin{pmatrix} 1-1 \\ 0-0 \\ \mathrm{e}^x - \mathrm{e}^x \end{pmatrix} = \begin{pmatrix} 0 \\ 0 \\ 0 \end{pmatrix}.$$

- For the vector field

$$\boldsymbol{v} : \mathbb{R}^2 \setminus \{0\} \to \mathbb{R}^2, \ \boldsymbol{v}(x, y) = \frac{1}{x^2+y^2} \begin{pmatrix} -y \\ x \end{pmatrix}$$

on the other hand, the integrability condition is valid, $\operatorname{rot} \boldsymbol{v} = \boldsymbol{0}$, since

$$\frac{\partial v_1}{\partial y} = \frac{-(x^2+y^2) + y \cdot 2y}{(x^2+y^2)^2} = \frac{(x^2+y^2) - x \cdot 2x}{(x^2+y^2)^2} = \frac{\partial v_2}{\partial x},$$

but $\mathbb{R}^2 \setminus \{0\}$ is not simply connected.

In this case, a vectorial curve integral over a closed curve need not yield zero. We test this with the closed curve $\gamma : [0, 2\pi] \to \mathbb{R}^2, \ \gamma(t) = \big(\cos(t), \ \sin(t)\big)^\top$. Indeed we have:

$$\oint_\gamma \boldsymbol{v} \cdot \mathrm{d}s = \int_0^{2\pi} \begin{pmatrix} -\sin(t) \\ \cos(t) \end{pmatrix}^\top \begin{pmatrix} -\sin(t) \\ \cos(t) \end{pmatrix} \mathrm{d}t = 2\pi \neq 0.$$

However, if one shifts the curve in such a way that it no longer goes around the origin, for example, if one considers the curve

$$\widetilde{\gamma} : [0, 2\pi] \to \mathbb{R}^2, \ \widetilde{\gamma}(t) = \big(3 + \cos(t) \ 4 + \sin(t)\big),$$

then its trace lies in the simply connected area $D = \big\{(x, y) \mid x, y > 0\big\}$ and so the following is valid

$$\oint_{\widetilde{\gamma}} \boldsymbol{v} \cdot \mathrm{d}s = 0. \qquad \blacksquare$$

57.3 Determination of a Primitive Function

Now we take care of the second problem: If $\boldsymbol{v} : D \subseteq \mathbb{R}^n \to \mathbb{R}^n$ with $n = 2$ or $n = 3$ is a gradient field, $\boldsymbol{v} = (v_1, v_2)^\top$ or $\boldsymbol{v} = (v_1, v_2, v_3)^\top$, then there is a primitive function f to $\boldsymbol{v}$, i.e., there is a scalar field $f : D \subseteq \mathbb{R}^n \to \mathbb{R}$ with

- $n = 2$: $f_x = v_1$, $f_y = v_2$ or
- $n = 3$: $f_x = v_1$, $f_y = v_2$, $f_z = v_3$.

It is therefore obvious, to determine f by successive integration as described in the following recipe:

Recipe: Determine a Primitive Function of a Gradient Field

- Case $n = 2$: If $\boldsymbol{v} : D \subseteq \mathbb{R}^2 \to \mathbb{R}^2$ is a gradient field, $\boldsymbol{v} = (v_1, v_2)^\top$ then one finds a primitive function f of $\boldsymbol{v}$ as follows:
  (1) Integration of v_1 to x:
  $$f(x, y) = \int v_1(x, y)\mathrm{d}x + g(y)\,.$$
  (2) Derive f from (1) to y and equating with v_2 yields an equation for $g_y(y)$:
  $$f_y(x, y) = v_2(x, y) \;\Rightarrow\; g_y(y)\,.$$
  (3) Integration of $g_y(y)$ to y with the constant $c = 0$ yields $g(y)$.
  (4) Set $g(y)$ from (3) into f from (1) and obtain a primitive function f.

- Case $n = 3$: If $\boldsymbol{v} : D \subseteq \mathbb{R}^3 \to \mathbb{R}^3$ is a gradient field, $\boldsymbol{v} = (v_1, v_2, v_3)^\top$, then one finds a primitive function f of $\boldsymbol{v}$ as follows:
  (1) Integration of v_1 to x:
  $$f(x, y, z) = \int v_1(x, y, z)\mathrm{d}x + g(y, z)\,.$$

(continued)

(2) Derive f from (1) to y and equating with v_2 yields an equation for $g_y(y, z)$:

$$f_y(x, y, z) = v_2(x, y, z) \Rightarrow g_y(y, z).$$

(3) Integration of $g_y(y, z)$ to y provides:

$$g(y, z) = \int g_y(y, z)\mathrm{d}y + h(z).$$

This $g(y, z)$ is entered into the f from (1) and thus obtains f up to the indefinite summand $h(z)$.

(4) Derive f from (3) to z and equating with v_3 yields an equation for $h_z(z)$:

$$f_z(x, y, z) = v_3(x, y, z) \Rightarrow h_z(z).$$

(5) Integration of $h_z(z)$ to z with the constant $c = 0$ yields $h(z)$.

(6) Set $h(z)$ from (5) into f from (3) and get a primitive function f.

The order of the variables, after which successive integration takes place, may of course be interchanged at will.

Example 57.4 We have already established that the vector field

$$\boldsymbol{v} : \mathbb{R}^3 \to \mathbb{R}^3,\ \boldsymbol{v}(x, y, z) = \begin{pmatrix} v_1 \\ v_2 \\ v_3 \end{pmatrix} = \begin{pmatrix} \mathrm{e}^x\, y + 1 \\ \mathrm{e}^x + z \\ y \end{pmatrix}$$

is a gradient field. We now compute a primitive function $f : \mathbb{R}^3 \to \mathbb{R}$ of $\boldsymbol{v}$.

(1) From $f_x(x, y, z) = v_1(x, y, z) = \mathrm{e}^x\, y + 1$ follows by integration to x:

$$f(x, y, z) = \mathrm{e}^x\, y + x + g(y, z).$$

(2) If we derive f from (1) to y, then by equating with v_2:

$$f_y(x, y, z) = \mathrm{e}^x + g_y(y, z) = \mathrm{e}^x + z$$

and thus $g_y(y, z) = z$.

(3) We integrate $g_y(y, z) = z$ to y and obtain $g(y, z) = yz + h(z)$ and thus with (1)

$$f(x, y, z) = e^x y + x + yz + h(z).$$

(4) Derive f from (3) to z and equating the result with v_3 yields:

$$f_z(x, y, z) = y + h_z(z) = y.$$

(5) So we have $h_z(z) = 0$ and thus $h(z) = 0$.

(6) Using h from (5), we obtain from (3) the primitive function $f(x, y, z) = e^x y + x + yz$.

57.4 Exercises

57.1 Determine the value of the vectorial curve integral $\int_\gamma \boldsymbol{v} \cdot d\boldsymbol{s}$ where

$$\boldsymbol{v} : \mathbb{R}^3 \to \mathbb{R}^3,\ \boldsymbol{v}(x, y, z) = \begin{pmatrix} 2x + y \\ x + 2yz \\ y^2 + 2z \end{pmatrix}$$

and

$$\gamma : [0, 2\pi] \to \mathbb{R}^3,\ \gamma(t) = \begin{pmatrix} \sin^2(t) + t \\ \cos(t)\sin(t) + \cos^2(t) \\ \sin(t) \end{pmatrix}.$$

57.2 Given the vector field

$$\boldsymbol{v}(x, y) = \begin{pmatrix} x^2 - y \\ y^2 + x \end{pmatrix}.$$

(a) Is there a primitive function $f(x, y)$ for $\boldsymbol{v}$?

(b) Calculate the integral of the curve $\int_\gamma \boldsymbol{v} \cdot d\boldsymbol{s}$ along

(i) of a straight line from $(0, 1)$ to $(1, 2)$,

(ii) the parabola $y = x^2 + 1$ from $(0, 1)$ to $(1, 2)$.

57.3 Given the vector field

$$\boldsymbol{v}(x, y, z) = \begin{pmatrix} 2xz^3 + 6y \\ 6x - 2yz \\ 3x^2z^2 - y^2 \end{pmatrix}.$$

(a) Calculate rot $\boldsymbol{v}$.
(b) Calculate the curve integral $\int_\gamma \boldsymbol{v} \cdot \mathrm{d}s$ along the spiral γ with parameterization $\gamma(t) = (\cos t, \sin t, t)^\top$ for $t \in [0, 2\pi]$.

57.4 We consider the following vector field $\boldsymbol{v}$ and curve γ:

$$\boldsymbol{v}(\boldsymbol{x}) = \begin{pmatrix} z^3 + y^2 \cos x \\ -4 + 2y \sin x \\ 2 + 3xz^2 \end{pmatrix}, \quad \boldsymbol{x} \in \mathbb{R}^3, \quad \gamma(t) = \begin{pmatrix} \tan t \\ \tan^2 t \\ \tan^3 t \end{pmatrix}, \quad t \in [0, \tfrac{\pi}{4}].$$

Show that $\boldsymbol{v}$ is a gradient field, and compute a primitive function of $\boldsymbol{v}$ and hence the curve integral $\int_\gamma \boldsymbol{v} \cdot \mathrm{d}s$.

57.5 Given the vector field $\boldsymbol{v}$ and the curve γ.

$$\boldsymbol{v}(\boldsymbol{x}) = \begin{pmatrix} 3x^2y \\ x^3 + 2z^3 \\ 6yz^2 \end{pmatrix}, \quad \boldsymbol{x} \in \mathbb{R}^3, \quad \gamma(t) = \begin{pmatrix} \sin t \\ -\cos t \\ t \end{pmatrix}, \quad t \in [0, \pi].$$

Calculate rot $\boldsymbol{v}$ and $\int_\gamma \boldsymbol{v} \cdot \mathrm{d}s$.

57.6 Given the plane vector field

$$\boldsymbol{v} \colon \mathbb{R}^2 \to \mathbb{R}^2, \quad \boldsymbol{x} = \begin{pmatrix} x \\ y \end{pmatrix} \mapsto \boldsymbol{v}(\boldsymbol{x}) = \begin{pmatrix} 1 + x + y^2 \\ y + 2xy \end{pmatrix}$$

(a) Check that $\boldsymbol{v}$ is a potential field.
(b) Calculate the curve integral $\int_C \boldsymbol{v}(\boldsymbol{x})\mathrm{d}\boldsymbol{x}$, where C is the positively oriented part of the unit circle located in the upper half plane.

Multiple Integrals

58

The definite integral $\int_a^b f(x)\,\mathrm{d}x$ yields the area that lies between $[a, b] \subseteq \mathbb{R}$ and the graph of f. This notion can be easily generalized: The *multiple integral* $\int_D f(x_1, \dots, x_n)\,\mathrm{d}x_1 \cdots \mathrm{d}x_n$ yields the *volume* that lies between the domain $D \subseteq \mathbb{R}^n$ and the graph of f. If D is a subset of $\mathbb{R}^2$, then this is a (three-dimensional) volume.

58.1 Integration Over Rectangles or Cuboids

We consider a (continuous) scalar field $f : D = [a, b] \times [c, d] \subseteq \mathbb{R}^2 \to \mathbb{R}$ in two variables. The domain D is the rectangle $[a, b] \times [c, d]$ in $\mathbb{R}^2$, and the graph of f encloses a volume with this rectangle (Fig. 58.1 is drawn a cross section of this volume), which we will now compute.

If we consider a fixed number $y_0 \in [c, d]$ the integral

$$F(y_0) = \int_{x=a}^{b} f(x, y_0)\,\mathrm{d}x$$

gives the area of the cross section $\{(x, y_0, f(x, y_0)) \mid x \in [a, b]\}$ of the enclosed volume V (see Fig. 58.1). By an integration of $F(y)$, $y \in [c, d]$ over the interval $[c, d]$ we get the enclosed volume V,

$$V = \int_{y=c}^{d} F(y)\mathrm{d}y = \int_{y=c}^{d} \left(\int_{x=a}^{b} f(x, y)\mathrm{d}x \right) \mathrm{d}y\,.$$

Here it does not matter whether one first goes to x and then to y or vice versa, as long as the function f is continuous on the rectangle $[a, b] \times [c, d]$. This can be done analogously for three-dimensional *rectangles*, i.e. cuboids $[a, b] \times [c, d] \times [e, f] \subseteq \mathbb{R}^3$ for functions

C. Karpfinger, *Calculus and Linear Algebra in Recipes*,
https://doi.org/10.1007/978-3-662-65458-3_58

Fig. 58.1 A cross section of the volume to be determined

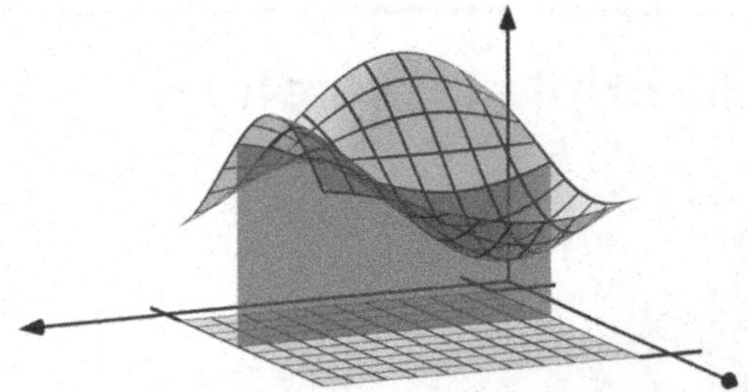

$f = f(x, y, z)$ in three variables. Thus we get the simplest **multiple integrals** as integrals over rectangles and cuboids. We summarize:

Integration Over a Rectangle or a Cuboid
If $f : D = [a, b] \times [c, d] \to \mathbb{R}$ or $f : D = [a, b] \times [c, d] \times [e, f] \to \mathbb{R}$ is a scalar field, then we declare the **iterated integrals**

$$\int\int_D f(x, y)\mathrm{d}x\mathrm{d}y = \int_c^d \int_a^b f(x, y)\mathrm{d}x\mathrm{d}y = \int_c^d \left(\int_a^b f(x, y)\mathrm{d}x\right)\mathrm{d}y$$

and

$$\int\int_D f(x, y)\mathrm{d}y\mathrm{d}x = \int_a^b \int_c^d f(x, y)\mathrm{d}y\mathrm{d}x = \int_a^b \left(\int_c^d f(x, y)\mathrm{d}y\right)\mathrm{d}x\,,$$

analogously for triple integrals, e.g.:

$$\int\int\int_D f(x, y, z)\mathrm{d}x\mathrm{d}y\mathrm{d}z = \int_e^f \int_c^d \int_a^b f(x, y, z)\mathrm{d}x\mathrm{d}y\mathrm{d}z = \int_e^f \left(\int_c^d \left(\int_a^b f(x, y)\mathrm{d}x\right)\mathrm{d}y\right)\mathrm{d}z.$$

Theorem of Fubini: Is If f is continuous, then the iterated integrals have the same value, e.g.

$$\int_c^d \int_a^b f(x, y)\,\mathrm{d}x\,\mathrm{d}y = \int_a^b \int_c^d f(x, y)\,\mathrm{d}y\,\mathrm{d}x\,.$$

Example 58.1

- We determine the multiple integral over the domain $D = [a, b] \times [c, d]$ for the scalar field $f : [a, b] \times [c, d] \to \mathbb{R}$ with $f(x, y) = 1$:

$$\int_{y=c}^d \int_{x=a}^b 1\mathrm{d}x\mathrm{d}y = \int_{y=c}^d x\Big|_a^b \mathrm{d}y = \int_{y=c}^d b - a\mathrm{d}y = (b-a)y\Big|_{y=c}^d = (b-a)(d-c)\,.$$

- Analog for $D = [a, b] \times [c, d] \times [e, f]$ and $f(x, y, z) = 1$:

$$\int_{z=e}^{f}\int_{y=c}^{d}\int_{x=a}^{b} 1\mathrm{d}x\mathrm{d}y\mathrm{d}z = \int_{z=e}^{f}\int_{y=c}^{d} (b-a)\mathrm{d}y\mathrm{d}z = \int_{z=e}^{f} (b-a)(d-c)\mathrm{d}z$$
$$= (b-a)(d-c)(f-e).$$

- We determine the integral of the scalar field $f : [0, 1] \times [1, 2] \times [2, 3] \to \mathbb{R}$ with $f(x, y, z) = x + y + z$:

$$\int_{z=2}^{3}\int_{y=1}^{2}\int_{x=0}^{1} x+y+z\,\mathrm{d}x\mathrm{d}y\mathrm{d}z = \int_{z=2}^{3}\int_{y=1}^{2} \frac{1}{2}x^2 + yx + zx\Big|_{x=0}^{1} \mathrm{d}y\mathrm{d}z$$
$$= \int_{z=2}^{3}\int_{y=1}^{2} \frac{1}{2} + y + z\,\mathrm{d}y\,\mathrm{d}z = \int_{z=2}^{3} \frac{1}{2}y + \frac{1}{2}y^2 + zy\Big|_{y=1}^{2} \mathrm{d}z$$
$$= \int_{z=2}^{3} \frac{1}{2} + \frac{3}{2} + z\,\mathrm{d}z = 2z + \frac{1}{2}z^2\Big|_{z=2}^{3} = \frac{9}{2}. \qquad \blacksquare$$

As the first examples show, by integrating the function f with $f(x, y) = 1$ or $f(x, y, z) = 1$ over the domain D one gets the area of D: Namely one determines the *volume* V with height $h = 1$ above the domain D. This is the area of the domain D in terms of size. This enables us to determine areas and volumes of *more complicated* domains in the next section.

Sometimes one wants to guard against too many integral signs, in order to keep the view on the essential things free, and then writes e.g.

$$\int_D f = \int_D f\mathrm{d}F = \int\int_D f\mathrm{d}x\mathrm{d}y \text{ or } \int_D f = \int_D f\mathrm{d}V = \int\int\int_D f\mathrm{d}x\mathrm{d}y\mathrm{d}z$$

for a domain $D \subseteq \mathbb{R}^2$ or $D \subseteq \mathbb{R}^3$.

MATLAB With MATLAB you get iterated integrals for example with

```
» syms x y z
» int(int(int(x+y+z,x,0,1),y,1,2),z,2,3)
ans = 9/2
```

58.2 Normal Domains

More general than rectangles in the $\mathbb{R}^2$ are the areas of $\mathbb{R}^2$ of the form

$$D = \{(x, y) \,|\, a \leq x \leq b, u(x) \leq y \leq o(x)\} \quad \text{but}$$

$$D = \{(x, y) \,|\, c \leq y \leq d, u(y) \leq x \leq o(y)\},$$

where u and o are functions in a variable, u stands for *lower bound*, o for *upper* bound. So with such a domain, one variable can be restricted between real numbers, the other variable takes values between a *lower function* u and an *upper function* o. One speaks of a **normal domain** D, if D can be written in this form, note Fig. 58.2.

Every rectangle $D = [a, b] \times [c, d]$ is a normal area, namely with $o(x) = d$ and $u(x) = c$ or $o(y) = b$ and $u(y) = a$ or

$$D = \{(x, y) \,|\, a \leq x \leq b,\ c \leq y \leq d\}.$$

We now declare accordingly **normal domains** D in $\mathbb{R}^3$ such a parallelepiped has the form

$$\begin{aligned}
D &= \{(x, y, z) \,|\, a \leq x \leq b,\ u(x) \leq y \leq o(x),\ \tilde{u}(x, y) \leq z \leq \tilde{o}(x, y) \\
\text{or } D &= \{(x, y, z) \,|\, c \leq y \leq d,\ u(y) \leq x \leq o(y),\ \tilde{u}(x, y) \leq z \leq \tilde{o}(x, y) \\
\text{or } D &= \{(x, y, z) \,|\, e \leq z \leq f,\ u(z) \leq x \leq o(z),\ \tilde{u}(x, z) \leq y \leq \tilde{o}(x, z) \\
\text{or } D &= \{(x, y, z) \,|\, a \leq x \leq b,\ u(x) \leq z \leq o(x),\ \tilde{u}(x, z) \leq y \leq \tilde{o}(x, z) \\
\text{or } D &= \{(x, y, z) \,|\, c \leq y \leq d,\ u(y) \leq z \leq o(y),\ \tilde{u}(y, z) \leq x \leq \tilde{o}(y, z) \\
\text{or } D &= \{(x, y, z) \,|\, e \leq z \leq f,\ u(z) \leq y \leq o(z),\ \tilde{u}(y, z) \leq x \leq \tilde{o}(y, z)\}.
\end{aligned}$$

Every cuboid in the $\mathbb{R}^3$ is a normal domain with in each case constant lower and upper functions u, o, $\tilde{u}$, $\tilde{o}$. A somewhat more complicated normal domain of the type $a \leq x \leq b$, $u(x) \leq y \leq o(x)$, $\tilde{u}(x, y) \leq z \leq \tilde{o}(x, y)$ can be seen in Fig. 58.3.

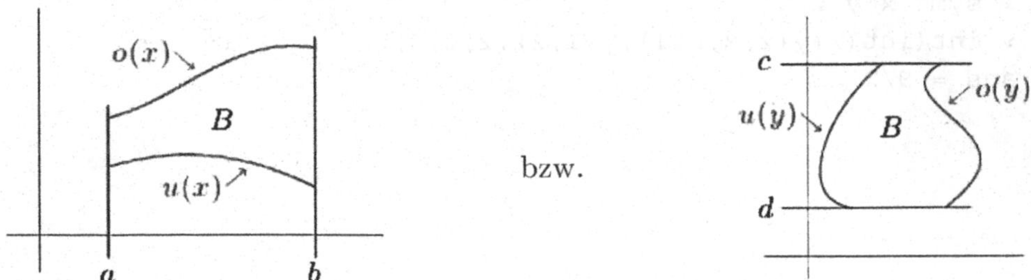

Fig. 58.2 A normal domain generalizes a rectangle: two sides remain straight and parallel, the other two sides are graphs of functions u and o

Fig. 58.3 A normal domain in the $\mathbb{R}^3$

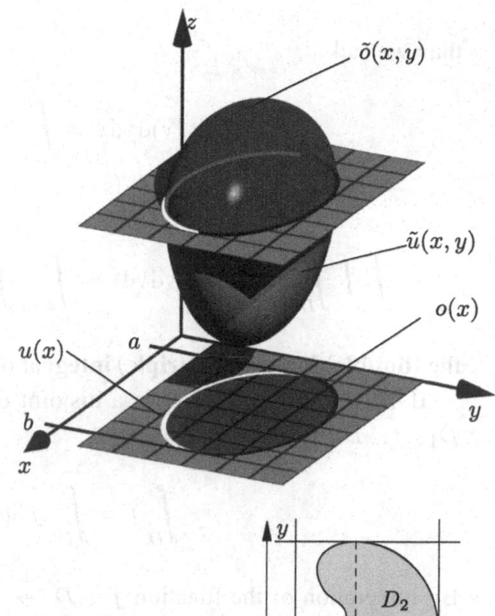

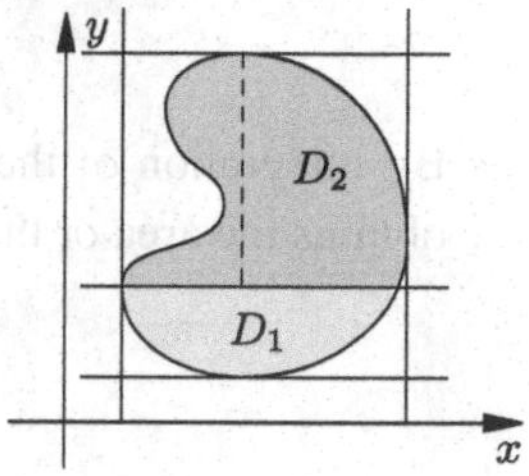

Fig. 58.4 D_1 and D_2 are normal domains

If D is not a normal domain, then the set D can in general be easily disjointly decomposed into subdomains, $D = D_1 \cup \cdots \cup D_r$ with $D_i \cap D_j = \emptyset$ for $i \neq j$ where the individual parts $D_1, \ldots, D_r$ are then normal domains (see Fig. 58.4).

58.3 Integration Over Normal Domains

We now explain an integration of functions f over normal domains D: The normal domain D is the domain of the function f, and in this integration the volume is determined that lies between the normal domain D and the graph of f.

Integration Over Normal Domains
If D is a normal domain, e.g.

$$D = \{(x, y) \mid a \leq x \leq b,\ u(x) \leq y \leq o(x)\} \text{ or}$$

$$D = \{(x, y, z) \mid a \leq x \leq b,\ u(x) \leq y \leq o(x),\ \tilde{u}(x, y) \leq z \leq \tilde{o}(x, y)\},$$

(continued)

than we call

$$\int\int_D f(x,y)\mathrm{d}y\mathrm{d}x = \int_{x=a}^{b}\int_{y=u(x)}^{o(x)} f(x,y)\mathrm{d}y\mathrm{d}x \quad \text{or}$$

$$\int\int\int_D f(x,y,z)\mathrm{d}z\mathrm{d}y\mathrm{d}x = \int_{x=a}^{b}\int_{y=u(x)}^{o(x)}\int_{z=\tilde{u}(x,y)}^{\tilde{o}(x,y)} f(x,y,z)\mathrm{d}z\mathrm{d}y\mathrm{d}x$$

the **(double) integral** or **(triple) integral** over the normal domain D.

If $D = D_1 \cup \cdots \cup D_r$ is a disjoint decomposition of D in normal domains $D_1, \ldots, D_r$ then

$$\int_D f = \int_{D_1} f + \cdots + \int_{D_r} f\,.$$

By integration of the function $f : D \to \mathbb{R}$, $f(x,y) = 1$ or $f(x,y,z) = 1$ one obtains the area or the volume of D:

$$\int_D 1 = \text{area or volume of } D.$$

If D is not a rectangle, the order of integration must not be reversed, and in fact it cannot be reversed at all: In this integration over a normal domain, a variable is successively *consumed* with each integration, and a real number remains at the end. If you were to swap the order, this would no longer work.

Example 58.2

- We consider the area bounded by the graphs of two real functions $u = u(x)$ and $o = o(x)$ a variable x over the interval $[a, b]$ which is the normal domain (see also Fig. 58.5):

$$B = \{(x,y) \mid a \le x \le b\,,\ u(x) \le y \le o(x)\}\,.$$

By integrating the unity function we get the area of B:

$$\int_B f = \int_{x=a}^{b}\int_{y=u(x)}^{o(x)} 1\,\mathrm{d}y\,\mathrm{d}x = \int_{x=a}^{b} o(x) - u(x)\,\mathrm{d}x = \int_{x=a}^{b} o(x)\,\mathrm{d}x - \int_{x=a}^{b} u(x)\,\mathrm{d}x\,.$$

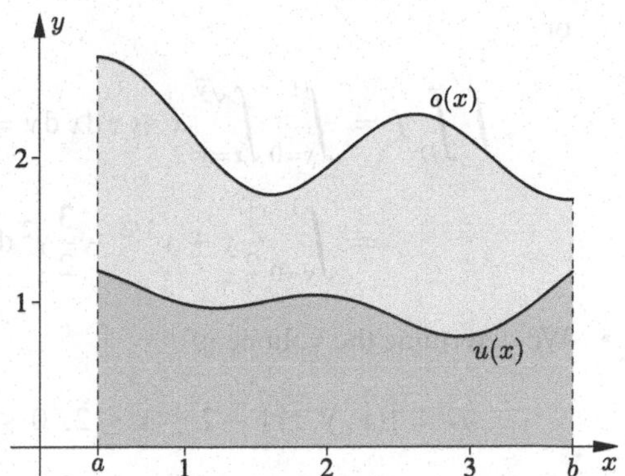

Fig. 58.5 The normal domain between $o(x)$ and $u(x)$

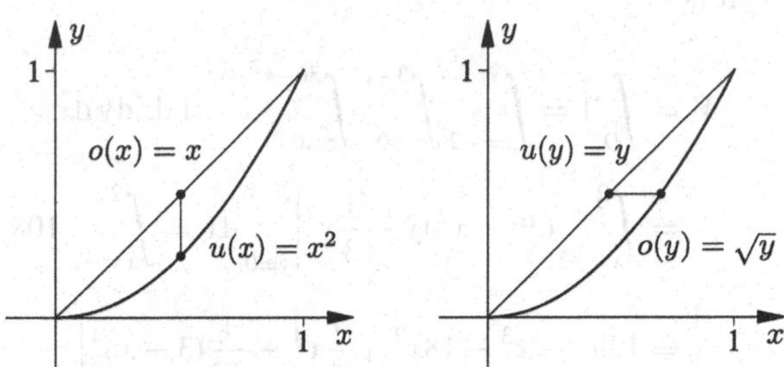

Fig. 58.6 A normal domain, *left* with respect to x, on the *right* with respect to y

This should not surprise anyone: Of course, the area we are looking for is just the difference of the integrals over $o(x)$ and $u(x)$.

- We integrate the function $f : D \subseteq \mathbb{R}^2 \to \mathbb{R}$ with $f(x, y) = x + y$ for the following normal domain D, which we can take to be one in two ways:

$$D = \{(x, y) \mid 0 \leq x \leq 1,\ x^2 \leq y \leq x\} \quad \text{but}$$
$$D = \{(x, y) \mid 0 \leq y \leq 1,\ y \leq x \leq \sqrt{y}\},$$

cf. the two pictures in Fig. 58.6:

Now we determine the double integrals over the two representations of the normal domain:

$$\int\int_D f = \int_{x=0}^{1} \int_{y=x^2}^{x} x + y \,\mathrm{d}y\mathrm{d}x = \int_{x=0}^{1} xy + \frac{1}{2}y^2 \Big|_{y=x^2}^{x} \mathrm{d}x$$
$$= \int_{x=0}^{1} \frac{3}{2}x^2 - x^3 - \frac{1}{2}x^4 \,\mathrm{d}x = \frac{1}{2}x^3 - \frac{1}{4}x^4 - \frac{1}{10}x^5 \Big|_{x=0}^{1} = \frac{3}{20}$$

or

$$\iint_D f = \int_{y=0}^{1} \int_{x=y}^{\sqrt{y}} x + y \,\mathrm{d}x\,\mathrm{d}y = \int_{y=0}^{1} \frac{1}{2}x^2 + xy\Big|_{x=y}^{\sqrt{y}} \mathrm{d}y$$
$$= \int_{y=0}^{1} \frac{1}{2}y + y^{3/2} - \frac{3}{2}y^2 \,\mathrm{d}y = \frac{1}{4}y^2 + \frac{2}{5}y^{5/2} - \frac{1}{2}y^3\Big|_{y=0}^{1} = \frac{3}{20}.$$

- We determine the volume of

$$D = \{(x, y, z) \mid -2 \le x \le 2,\ 0 \le y \le 3 - x,\ 0 \le z \le 36 - x^2 - y^2\}.$$

It is

$$V = \int_D 1 = \int_{x=-2}^{2} \int_{y=0}^{3-x} \int_{z=0}^{36-x^2-y^2} 1 \,\mathrm{d}z\,\mathrm{d}y\,\mathrm{d}x = \int_{x=-2}^{2} \int_{y=0}^{3-x} 36 - x^2 - y^2 \,\mathrm{d}y\,\mathrm{d}x$$
$$= \int_{x=-2}^{2} (36 - x^2)y - \frac{1}{3}y^3\Big|_{y=0}^{3-x} \mathrm{d}x = \int_{x=-2}^{2} 108 - 3x^2 - 36x + x^3 - \frac{1}{3}(3 - x)^3 \,\mathrm{d}x$$
$$= 108x - x^3 - 18x^2 + \frac{1}{4}x^4 + \frac{1}{12}(3 - x)^4\Big|_{x=-2}^{2}$$
$$= 216 - 8 - 72 + 4 + \frac{1}{12} + 216 - 8 - 4 + 72 - \frac{625}{12} = 364\,.$$

- Finally, we determine the volume of a cylinder D with radius R and height h. For this we write D as a normal domain, e.g.

$$D = \{(x, y, z) \mid -R \le x \le R\,,\ -\sqrt{R^2 - x^2} \le y \le \sqrt{R^2 - x^2}\,,\ 0 \le z \le h\}\,.$$

This gives us the volume

$$V_{\mathrm{Zyl}} = \int_D 1 = \int_{x=-R}^{R} \int_{y=-\sqrt{R^2-x^2}}^{\sqrt{R^2-x^2}} \int_{z=0}^{h} 1 \,\mathrm{d}z\mathrm{d}y\mathrm{d}x$$
$$= \int_{x=-R}^{R} \int_{y=-\sqrt{R^2-x^2}}^{\sqrt{R^2-x^2}} h \,\mathrm{d}y\mathrm{d}x = \int_{x=-R}^{R} 2h\sqrt{R^2 - x^2}\,\mathrm{d}x$$
$$= h\left[x\sqrt{R^2 - x^2} + R^2 \arcsin(x/R)\right]_{x=-R}^{R} = \pi R^2 h\,,$$

where we use MATLAB obtained the following primitive function:

$$\int \sqrt{R^2 - x^2}\mathrm{d}x = \frac{1}{2}\left(x\sqrt{R^2 - x^2} + R^2 \arcsin(x/R)\right)\,. \qquad \blacksquare$$

The cylinder D in the last example is in cylinder coordinates (see Chap. 53)

$$D = \{(r, \varphi, z) \mid 0 \leq r \leq R\,,\ 0 \leq \varphi \leq 2\pi\,,\ 0 \leq z \leq h\}\,.$$

Thus D is a *cuboid* in cylindrical coordinates. It is to be expected that integration in cylindrical coordinates is much simpler. Unfortunately, we must not simply change the coordinates, we must also change the *infinitesimally small* volume element $\mathrm{d}x\mathrm{d}y\mathrm{d}z$ in cylindrical coordinates. The *transformation formula*, which is the content of the next chapter, shows how this is done.

MATLAB With MATLAB, you can obtain integrals over normal domains as follows, for example

```
» syms x y z
» int(int(int(1,z,0,36-x^2-y^2),y,0,3-x),x,-2,2)
ans = 364
```

58.4 Exercises

58.1 Let D denote the from $x = y^2$ and $y = \frac{1}{2}x - \frac{3}{2}$ enclosed area in the $\mathbb{R}^2$. Calculate the double integral $\iint_B (x+y)\mathrm{d}F$ in two ways, by calculating on the one hand $\mathrm{d}F = \mathrm{d}x\mathrm{d}y$ and on the other $\mathrm{d}F = \mathrm{d}y\mathrm{d}x$ on the other.

58.2 Calculate the integral of the function $f(x, y) = x^2 y$ over the domain

$$D = \{(x, y) \in \mathbb{R}^2 \mid 0 \leq y \leq \tfrac{1}{2}x,\ x \leq y^2 + 1,\ x \leq 2\}$$

for both orders of integration.

58.3 Determine the area of the surface enclosed by the curves

$$y^2 + x - 1 = 0\ y^2 - x - 1 = 0$$

by the curves.

58.4 Given is the double integral

$$\int_{-1}^{1} \int_{x^2}^{1} f(x, y)\mathrm{d}y\mathrm{d}x$$

(a) Sketch the domain of integration.
(b) State the double integral with the order of integration reversed.
(c) Calculate the integral for $f(x, y) = 2x \sin x^2$.

58.5 Show that for $f: \mathbb{R}^2 \to \mathbb{R}$, $f(x, y) = (2 - xy)xy\,\mathrm{e}^{-xy}$ we have:

$$\int_0^1 \int_0^\infty f(x, y)\mathrm{d}y\mathrm{d}x \neq \int_0^\infty \int_0^1 f(x, y)\mathrm{d}x\mathrm{d}y$$

What can be concluded from this?

58.6 Calculate the following integrals:

(a) $\iiint_D xyz\, \mathrm{d}x\mathrm{d}y\mathrm{d}z$ where $D \subseteq \mathbb{R}^3$ denotes the volume that remains when one subtracts from the parallelepiped with vertices $P_1 = (0, 0, 0)$, $P_2 = (3, 0, 0)$, $P_3 = (3, 0, 1)$, $P_4 = (0, 0, 1)$, $P_5 = (0, 2, 0)$, $P_6 = (3, 2, 0)$, $P_7 = (3, 2, 1)$, $P_8 = (0, 2, 1)$ cuts off the tetrahedron P_6, P_7, P_8, P_3 defined tetrahedron.
(b) $\iiiint_D x_2\, \mathrm{d}x_1\mathrm{d}x_2\mathrm{d}x_3\mathrm{d}x_4$ where $D \subseteq \mathbb{R}^4$ is defined by $D: 0 < x_1 < 1, 0 < x_2 < 2, 0 < x_3 < x_1, 0 < x_4 < x_2x_3$.
(c) $Q = \sqrt{2g} \iint_D \sqrt{y}\, \mathrm{d}x\mathrm{d}y$ where D which is given by the four points $(-b/2, c)$, $(b/2, c)$, $(-a/2, c + d)$ and $(a/2, c + d)$ is defined trapezoid. The numerical values are: $a = 8$, cm, $b = 3$ cm, $c = 4$ cm, $d = 5$ cm and $g = 981$ cm/s^2 (gravitational acceleration). According to Torricelli, Q is the outflow velocity through the orifice D.

Substitution for Multiple Variables

59

When *integrating* in the multi-dimensional, i.e. over x, y and z, everything remains the same: One integrates successively over the individual variables. This multidimensional integration is at least theoretically unproblematic. Computational difficulties usually arise when the domain D over which the integral extends can only be represented with complications concerning Cartesian coordinates. Often the representation of the domain D is simpler by choosing, for example, polar, cylindrical or spherical coordinates. One integrates then better over these coordinates, whereby the *substitution for multiple variables* indicates which *correction factor* one must consider when changing the coordinate system.

59.1 Integration via Polar, Cylindrical, Spherical and Other Coordinates

Often integrals are to be formed over normal domains $D \subseteq \mathbb{R}^n$ which, with respect to the given Cartesian coordinates, lead to functions which are difficult or impossible to integrate elementarily, since the lower and upper limits, e.g. $u(x)$, $\tilde{u}(x, y)$ and $o(x)$, $\tilde{o}(x, y)$ lead to complicated integrals after the insertion into the primitive functions of even simple integrands. A way out of this difficult situation offers the possibility to change to other coordinate systems, if possible in such a way, that these areas D can be described as rectangles or cuboids with respect to these coordinate systems. For example

- circular rings or circles in polar coordinates as rectangles,
- cylinders in cylindrical coordinates as cuboids,
- spheres or hemispheres in spherical coordinates as cuboids.

C. Karpfinger, *Calculus and Linear Algebra in Recipes*,
https://doi.org/10.1007/978-3-662-65458-3_59

Example 59.1

- The points of the circular ring $D = \{(x, y) \mid R_1 \leq \sqrt{x^2 + y^2} \leq R_2\}$ for $R_1, R_2 \in \mathbb{R}_{>0}$ have the polar coordinates

$$(r, \varphi) \text{ with } r \in [R_1, R_2] \text{ and } \varphi \in [0, 2\pi[\,.$$

See Fig. 59.1.

- The points of the cylinder $D = \{(x, y, z) \mid \sqrt{x^2 + y^2} \leq R,\ 0 \leq z \leq h\}$ for $R \in \mathbb{R}_{>0}$ have the polar coordinates

$$(r, \varphi, z) \text{ with } r \in [0, R]\,,\ \varphi \in [0, 2\pi[\text{ and } z \in [0, h]\,.$$

See Fig. 59.2.

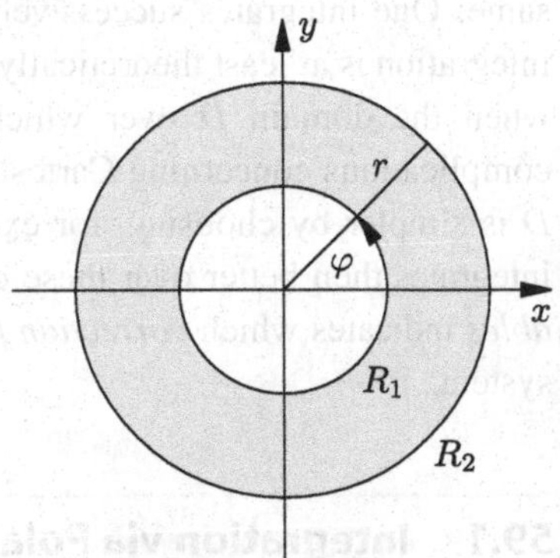

Fig. 59.1 Polar coordinates

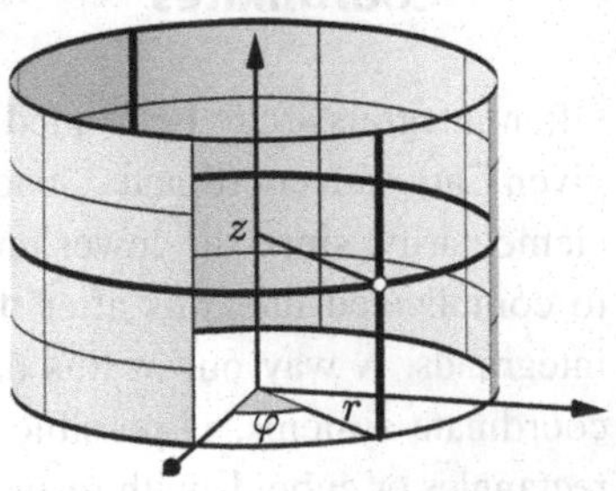

Fig. 59.2 Cylinder coordinates

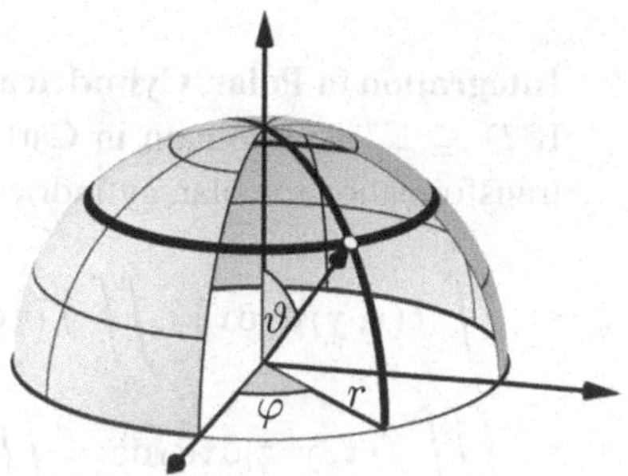

Fig. 59.3 Spherical coordinates

- The points of the *northern hemisphere* $D = \{(x, y, z) \mid \sqrt{x^2 + y^2 + z^2} \leq R,\ 0 \leq z \leq \sqrt{R}\}$ for $R \in \mathbb{R}_{>0}$ have the spherical coordinates

$$(r, \varphi, \vartheta) \text{ with } r \in [0, R]\,,\ \varphi \in [0, 2\pi[\text{ and } \vartheta \in [0, \pi/2]\,.$$

See Fig. 59.3. ■

We consider a coordinate transformation $\phi : B \to D$, $B, D \subseteq \mathbb{R}^n$. Here ϕ is a continuously differentiable bijection whose inverse is again continuously differentiable, see Chap. 53. If now $f : D \subseteq \mathbb{R}^n \to \mathbb{R}$ is a scalar field, then $f \circ \phi : B \to \mathbb{R}$ is such a field which can be differentiated from f differs only by the bijection ϕ. The respective integrals over the domain differ *only* by the factor $|\det D\phi|$, more precisely:

Substitution for Multiple Variables
If B and D are open subsets of the $\mathbb{R}^n$ and $\phi : B \to D$ is a coordinate transformation, then we have:

$$\int_D f(x_1, \ldots, x_n)\, \mathrm{d}x_1 \cdots \mathrm{d}x_n$$
$$= \int_B f(\phi(y_1, \ldots, y_n)) \big|\det D\phi(y_1, \ldots, y_n)\big| \mathrm{d}y_1 \cdots \mathrm{d}y_n.$$

So to get an integral $\int_D f$ over a *complicated* area D, one chooses a clever coordinate transformation $\phi : B \to D$ so that the integral $\int_B f |\det D\phi|$ is easy to determine. Typically B will be a rectangle if $n = 2$, or a cuboid if $n = 3$, and typically we are concerned with polar, cylindrical, or spherical coordinates; we give the transformation formulas explicitly for these most important cases, and the determinants of the Jacobian matrices are well known after Chap. 53:

Integration in Polar, Cylindrical and Spherical Coordinates
If $D \subseteq \mathbb{R}^n$ is a domain in Cartesian coordinates and $\phi : B \to D$ a coordinate transformation to polar, cylindrical, or spherical coordinates, then

- $$\iint_D f(x, y)\,\mathrm{d}y\mathrm{d}x = \iint_B f(r\cos\varphi, r\sin\varphi)\,\boldsymbol{r}\,\mathrm{d}r\mathrm{d}\varphi,$$
- $$\iiint_D f(x, y, z)\mathrm{d}x\mathrm{d}y\mathrm{d}z = \iiint_B f(r\cos\varphi,\ r\sin\varphi,\ z)\,\boldsymbol{r}\,\mathrm{d}r\,\mathrm{d}\varphi\,\mathrm{d}z,$$
- $$\iiint_D f(x, y, z)\,\mathrm{d}x\mathrm{d}y\mathrm{d}z$$
  $$= \iiint_B f(r\cos\varphi\sin\vartheta, r\sin\varphi\sin\vartheta, r\cos\vartheta)\,\boldsymbol{r^2\sin\vartheta}\,\mathrm{d}r\mathrm{d}\varphi\mathrm{d}\vartheta.$$

A common mistake is that the functional determinant $|D\phi|$ is forgotten in the integrand; we have therefore highlighted these functional determinants in bold in the above formula.

Example 59.2

- The volume of the cylinder from the top is thus calculated as.

$$V = \int_{r=0}^{R}\int_{\varphi=0}^{2\pi}\int_{z=0}^{h} 1\cdot r\,\mathrm{d}z\,\mathrm{d}\varphi\,\mathrm{d}r = \pi R^2 h.$$

- Let $D = \{(x, y) \in \mathbb{R}^2 \mid x > 0,\ 1 \leq x^2 + y^2 \leq 4\}$ be a half circular ring with outer circle radius 2 and inner circle radius 1. We use polar coordinates in order to calculate the function $f(x, y) = x(x^2 + y^2)$ via D to integrate:

$$\iint_D x(x^2 + y^2)\,\mathrm{d}x\,\mathrm{d}y = \int_{\varphi=-\pi/2}^{\pi/2}\int_{r=1}^{2} r\cos\varphi(r^2\cos^2\varphi + r^2\sin^2\varphi)r\,\mathrm{d}r\mathrm{d}\varphi$$
$$= \int_{\varphi=-\pi/2}^{\pi/2}\int_{r=1}^{2} r^4\cos\varphi\,\mathrm{d}r\mathrm{d}\varphi$$

$$= \int_{\varphi=-\pi/2}^{\pi/2} \frac{r^5}{5} \cos\varphi \Big|_1^2 \,\mathrm{d}\varphi$$

$$= \int_{\varphi=-\pi/2}^{\pi/2} \frac{31}{5} \cos\varphi \,\mathrm{d}\varphi = \frac{31}{5} \sin\varphi \Big|_{-\pi/2}^{\pi/2} = \frac{62}{5}.$$

- We now determine the following integral

$$\int_{x=0}^{1} \int_{y=0}^{1-x} \mathrm{e}^{\frac{y}{x+y}} \,\mathrm{d}y\,\mathrm{d}x$$

over the triangle D with the corners $(0, 0)$, $(0, 1)$ and $(1, 0)$ with the coordinate transformation given as follows:

$$\phi : (0, 1) \times (0, 1) \to D \text{ with } \phi(u, v) = \begin{pmatrix} u(1-v) \\ uv \end{pmatrix}.$$

The Jacobi determinant $\det D\phi$ we obtain with the Jacobian matrix:

$$D\phi(u, v) = \begin{pmatrix} 1-v & -u \\ v & u \end{pmatrix} \Rightarrow |\det(D\phi)| = u.$$

Thus, using the transformation formula:

$$\int_{x=0}^{1} \int_{y=0}^{1-x} \mathrm{e}^{\frac{y}{x+y}} \,\mathrm{d}y\,\mathrm{d}x = \int_{u=0}^{1} \int_{v=0}^{1} \mathrm{e}^{v}\, u \,\mathrm{d}v\,\mathrm{d}u = \int_{0}^{1} \frac{1}{2} \mathrm{e}^{v} \,\mathrm{d}v = \frac{1}{2}(\mathrm{e}-1).$$

Note: In this example the coordinate transformation was given. In general, the difficulty lies in finding a suitable coordinate transformation, which has to be chosen in such a way that, on the one hand, the area B is as rectangular as possible (or a cuboid in the three-dimensional case) and, on the other hand, the integrand in the new coordinates together with the Jacobian determinant can be integrated without any problems. ■

MATLAB With MATLAB it is possible to evaluate multiple integrals with respect to arbitrary coordinate systems. For example, we obtain the value of the first example above with

```
» syms r phi
» int(int(r^4*cos(phi),r,1,2),phi,-pi/2,pi/2)
ans = 62/5
```

The integral over this *right semicircular ring* D we could have determined also in cartesian coordinates, especially since with MATLAB the integration can be done without problems. The half circular ring D can be written as a disjoint union of three normal domains, $D = D_1 \cup D_2 \cup D_3$, where

$$D_1 = \{(x, y) \mid -2 \leq y \leq -1\,,\ 0 \leq x \leq \sqrt{4 - y^2}\,,$$

$$D_2 = \{(x, y) \mid -1 \leq y \leq 1\,,\ \sqrt{1 - y^2} \leq x \leq \sqrt{4 - y^2}\,,$$

$$D_3 = \{(x, y) \mid 1 \leq y \leq 2\,,\ 0 \leq x \leq \sqrt{4 - y^2}\}\,.$$

We obtain with MATLAB

```
» syms x y
» I1=int(int(x*(x^2 + y^2),x,0,sqrt(4-y^2)),y,-2,-1);
» I2=int(int(x*(x^2 + y^2),x,sqrt(1-y^2),sqrt(4-y^2)),y,-1,1);
» I3=int(int(x*(x^2 + y^2),x,0,sqrt(4-y^2)),y,1,2);
» I1+I2+I3
ans = 62/5
```

59.2 Application: Mass and Center of Gravity

A typical application of multiple integration is the determination of masses or charges and centroids of two- or three-dimensional domains D, where a mass or charge density $\rho = \rho(x, y)$ or $\rho = \rho(x, y, z)$ is given. In practice, to describe the domains D mostly polar, cylindrical or spherical coordinates are to prefer. Then one should absolutely determine these integrals by means of these coordinates, one considers for this above substitution for multiple variables:

Mass, Centre of Gravity, Geometric Centre of Gravity
If $D \subseteq \mathbb{R}^2$ or $D \subseteq \mathbb{R}^3$ is a domain with the density $\rho(x, y)$ or $\rho(x, y, z)$, then

$$M = \int_D \rho\, \mathrm{d}x\mathrm{d}y \qquad \text{or} \qquad M = \int_D \rho\, \mathrm{d}x\mathrm{d}y\mathrm{d}z$$

(continued)

is the **mass** of D, and it is $S = (s_1, s_2)^\top$ or $S = (s_1, s_2, s_3)^\top$ with

$$s_1 = \frac{1}{M}\int_D \rho\, x\,, \quad s_2 = \frac{1}{M}\int_D \rho\, y\,, \quad s_3 = \frac{1}{M}\int_D \rho\, z$$

the **centre of gravity**. With $\rho = 1$ one obtains the **geometric centre of gravity**.

Example 59.3 We determine the center of gravity of the northern hemisphere

$$H = \{(x, y, z) \mid -1 \leq x \leq 1,\ -\sqrt{1-x^2} \leq y \leq \sqrt{1-x^2},\ 0 \leq z \leq \sqrt{1-x^2-y^2}\},$$

where $\rho(x, y, z) = z$. First, we need the mass in the process. We use spherical coordinates and get:

$$M = \int_{\vartheta=0}^{\pi/2}\int_{\varphi=0}^{2\pi}\int_{r=0}^{1} r\cos\vartheta\, r^2 \sin\vartheta\, \mathrm{d}r\, \mathrm{d}\varphi\, \mathrm{d}\vartheta = \int_{\vartheta=0}^{\pi/2} \frac{\pi}{2}\cos\vartheta\sin\vartheta\, \mathrm{d}\vartheta = \frac{\pi}{2}\cdot\frac{1}{2}\sin^2\vartheta\Big|_0^{\pi/2} = \frac{\pi}{4}.$$

For symmetry reasons $s_1 = 0 = s_2$. Also

$$\begin{aligned} s_3 &= \frac{4}{\pi}\int_{\vartheta=0}^{\pi/2}\int_{\varphi=0}^{2\pi}\int_{r=0}^{1} r^2\cos^2\vartheta\, r^2\sin\vartheta\, \mathrm{d}r\, \mathrm{d}\varphi\, \mathrm{d}\vartheta = 8\int_{\vartheta=0}^{\pi/2}\frac{1}{5}\cos^2\vartheta\, \sin\vartheta\, \mathrm{d}\vartheta \\ &= \frac{-8}{15}\cos^3\vartheta\Big|_{\vartheta=0}^{\pi/2} = \frac{8}{15}. \end{aligned}$$

■

59.3 Exercises

59.1 Determine the multiple integral

$$\int_D \arctan\tfrac{x-y}{x+y}\, \mathrm{d}x\, \mathrm{d}y\,, \quad \text{where } D = \{(x, y)^\top \mid x^2 + y^2 \leq 2\}.$$

(a) Use the coordinate transformation

$$x = s\,(\cos t + \sin t), \quad y = s\,(\cos t - \sin t) \quad \text{with} \quad s \in [0, \infty[, \quad t \in [0, 2\pi[$$

in the given integral and give the multiple integral in the new coordinates.

(b) Calculate the multiple integral.

59.2 Calculate the multiple integral

$$\int_D e^{(x+y)/(x-y)}\,\mathrm{d}x\mathrm{d}y,$$

where D is the trapezoidal area with the vertices $(1, 0)$, $(2, 0)$, $(0, -2)$ and $(0, -1)$. *Note:* Use the coordinate transformation $s=x+y$, $t=x-y$.

59.3 The error integral,

$$\mathcal{I} = \int_0^\infty e^{-x^2}\,\mathrm{d}x\,,$$

which is important for many applications, cannot be determined via an analytically calculated primitive function. In order to nevertheless determine its value exactly, the following detour can be taken via multiple integrals:

(a) First compute the multiple integral

$$\mathcal{K}_R = \int_{D_R} e^{-x^2}\,e^{-y^2}\,\mathrm{d}x\mathrm{d}y$$

with respect to the first quadrant of a circular disk D_R of radius R, i.e. $D_R = \{(x, y) \in \mathbb{R}^2 \mid \sqrt{x^2+y^2} \le R,\ x, y \ge 0\}$.

(b) As can be easily convinced, we have the following for the improper integral

$$\mathcal{I}^2 = \lim_{A\to\infty} \int_0^A e^{-x^2}\,\mathrm{d}x \int_0^A e^{-y^2}\,\mathrm{d}y = \lim_{A\to\infty} \int_0^A \int_0^A e^{-x^2}\,e^{-y^2}\,\mathrm{d}x\mathrm{d}y\,.$$

Estimate the integral

$$\mathcal{I}_A^2 = \int_0^A \int_0^A e^{-x^2}\,e^{-y^2}\,\mathrm{d}x\mathrm{d}y$$

from above and below by multiple integrals $\mathcal{K}_R$ and calculate from this the error integral per $A \to \infty$.

59.4 Let R and α be positive. The circular plate $B = \{(x, y) \in \mathbb{R}^2 \mid x^2 + y^2 \le R^2\}$ of a capacitor is charged by electrons, which are distributed according to the surface charge density $\varrho(x, y) = -\alpha(R^2 - x^2 - y^2)$ of B.

(a) Calculate the total charge $Q = \iint_B \varrho \, \mathrm{d}F$ of the plate directly.
(b) Use polar coordinates to simplify the calculation.

59.5 Let $D = \{(x, y) \in \mathbb{R}^2 \mid 4 \geq x^2 + y^2 \geq 1, x \geq y \geq 0\}$. Using polar coordinates, calculate the integral $\iint_D \frac{y}{x^4} \mathrm{d}x\mathrm{d}y$.

59.6 Determine the center of gravity of the northern hemisphere $D = \{(x, y, z) \mid x^2 + y^2 + z^2 \leq R^2$ with $z \geq 0\}$ with the density $\rho(x, y, z) = z$.

59.7 Consider the cone K in the $\mathbb{R}^3$ with the apex $(0, 0, 3)^\top$ and the base $x^2 + y^2 \leq 1$ in the plane $z= 0$. Let the (inhomogeneous) mass density ρ of K is given by $\rho(x, y, z) = 1 - \sqrt{x^2 + y^2}$.

(a) Using cylindrical coordinates, determine the volume V and the total mass M of K.
(b) Determine the center of mass of the cone.

59.8 Let D be the triangle with vertices $(0, 0)$, $(1, 0)$ and $(1, 1)$ and $K = \{(x, y) \in \mathbb{R}^2 \mid x^2 + y^2 \leq a^2\}$ with $a> 0$. Calculate:

(a) $\displaystyle\iint_D x\, y\mathrm{d}x\mathrm{d}y$

(d) $\displaystyle\iint_K \mathrm{e}^{-x^2-y^2}\, \mathrm{d}x\mathrm{d}y$

(b) $\displaystyle\iint_D \frac{2y}{x+1}\mathrm{d}x\mathrm{d}y$

(e) $\displaystyle\iint_K \frac{\mathrm{d}x\mathrm{d}y}{1+x^2+y^2}$

(c) $\displaystyle\iint_D \mathrm{e}^{-x^2}\, \mathrm{d}x\mathrm{d}y$

(f) $\displaystyle\iint_K \sin(x^2+y^2)\mathrm{d}x\mathrm{d}y$

Surfaces and Surface Integrals 60

We have already discussed the notion of a *surface* in Chap. 46: Whereas a space curve is a function in a parameter t, a surface is a function in two parameters u and v. The best thing is: A surface is also exactly what you imagine it to be. Important are surfaces of simple bodies like spheres, cylinders, tori, cones, but also graphs of scalar fields $f : D \subseteq \mathbb{R}^2 \to \mathbb{R}$.

Analogous to the scalar and vectorial line integrals we will introduce scalar and vectorial *surface integrals*. These integrals for surfaces have a similar descriptive interpretation as those for curves.

60.1 Regular Surfaces

With curves we first distinguished between curve and range of the curve: the curve was the mapping, the range of the curve was then what one imagines under the curve. This will now be quite similar with *surfaces*.

In applications, *closed* surfaces will play an important role, i.e. spherical, cubic or cylindrical surfaces. In order to be able to include such surfaces under the term *surface*, a fineness must be taken into account: A spherical surface consists of one piece, so to speak, whereas a cylindrical surface consists of a shell surface, bottom and top. This surface is thus composed of surface pieces, as in the case of the cube. We can reconcile such *surfaces* by considering piecewise continuously differentiable surfaces, thus we also admit surfaces like the cylinder surface—the *seams* of mantle, bottom and lid are the boundaries of continuously differentiable surface pieces.

C. Karpfinger, *Calculus and Linear Algebra in Recipes*,
https://doi.org/10.1007/978-3-662-65458-3_60

Surfaces, Regular Surfaces

A continuous, piecewise continuously differentiable function

$$\phi : B \subseteq \mathbb{R}^2 \to \mathbb{R}^3 \text{ with } \phi(u, v) = \begin{pmatrix} x(u, v) \\ y(u, v) \\ z(u, v) \end{pmatrix}$$

is called a **surface**. If $\phi_u(u, v) \times \phi_v(u, v) \neq 0$ in all (u, v) with possibly finitely many exceptions, then the surface ϕ is called **regular**.

The range of a surface is a surface in space. In the following we will no longer distinguish so meticulously between the mapping *surface* and the *surface* as range of the mapping and we will also refer again and again to a **parameterization** ϕ of a surface, by which we mean a function ϕ whose range represents the given surface.

Note that since ϕ is partially differentiable, we can determine the **tangent vectors**

$$\frac{\partial \phi}{\partial u}(u, v) = \phi_u(u, v) \text{ and } \frac{\partial \phi}{\partial v}(u, v) = \phi_v(u, v)$$

at each point (u, v). The surface ϕ is regular if the two tangent vectors $\frac{\partial \phi}{\partial u}$ and $\frac{\partial \phi}{\partial v}$ in all (u, v) (with possibly finitely many exceptions) are linearly independent. In this case the vector $\phi_u(u, v) \times \phi_v(u, v)$ that is orthogonal to the surface ϕ is not the zero vector (see Fig. 60.1).

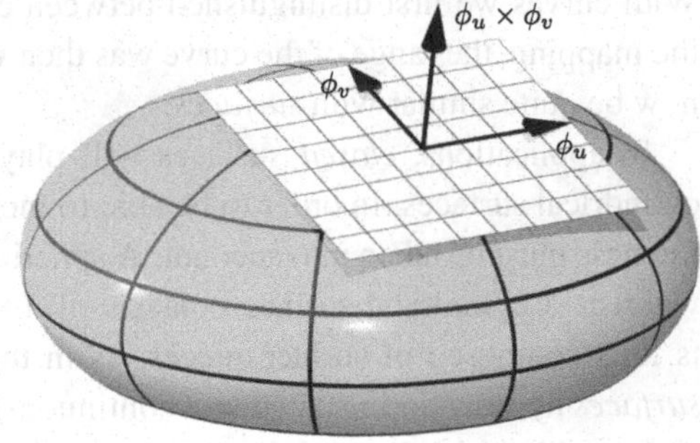

Fig. 60.1 A regular surface with linearly independent tangent vectors ϕ_u, ϕ_v and a normal vector perpendicular to $\phi_u \times \phi_v$

Example 60.1

- The graph of a continuous mapping $f : [a, b] \times [c, d] \to \mathbb{R}$ is a surface, a parametrization of this surface is:

$$\phi : [a, b] \times [c, d] \to \mathbb{R}^3,\ \phi(u, v) = \begin{pmatrix} u \\ v \\ f(u, v) \end{pmatrix}.$$

- The lateral surface of a cylinder of radius $R \in \mathbb{R}_{>0}$ and height h has the parametrization:

$$\phi : [0, 2\pi] \times [0, h] \to \mathbb{R}^3,\ \phi(u, v) = \begin{pmatrix} R\cos u \\ R\sin u \\ v \end{pmatrix}.$$

- A spherical surface of radius $R \in \mathbb{R}_{>0}$ has the parameterization

$$\phi : [0, \pi] \times [0, 2\pi] \to \mathbb{R}^3,\ \phi(\vartheta, \varphi) = \begin{pmatrix} R\cos\varphi\sin\vartheta \\ R\sin\varphi\sin\vartheta \\ R\cos\vartheta \end{pmatrix}.$$

  We have:

$$\phi_\vartheta \times \phi_\varphi = \begin{pmatrix} R\cos\varphi\cos\vartheta \\ R\sin\varphi\cos\vartheta \\ -R\sin\vartheta \end{pmatrix} \times \begin{pmatrix} -R\sin\varphi\sin\vartheta \\ R\cos\varphi\sin\vartheta \\ 0 \end{pmatrix} = R^2\sin\vartheta \begin{pmatrix} \cos\varphi\sin\vartheta \\ \sin\varphi\sin\vartheta \\ \cos\vartheta \end{pmatrix},$$

  that's easy to remember: $\phi_\vartheta \times \phi_\varphi$ is the $R^2\sin\vartheta$-fold of the *spherical coordinates*. We will often refer to this result in the next pages and chapters. We also determine the length of $\phi_\vartheta \times \phi_\varphi$:

$$\left\|\phi_\vartheta \times \phi_\varphi\right\| = R^2|\sin\vartheta|\sqrt{\sin^2\vartheta\left(\cos^2\varphi + \sin^2\varphi\right) + \cos^2\vartheta} = R^2\sin\vartheta\,,$$

  where we used in the last equals sign that $\vartheta \in [0, \pi]$.
- A **helical surface** is obtained by the following parametrization:

$$\phi : [0, a] \times [c, d] \to \mathbb{R}^3 \text{ with } \phi(u, v) = \begin{pmatrix} u\cos v \\ u\sin v \\ b\,v \end{pmatrix}$$

  with $b \in \mathbb{R}_{>0}$ (see Fig. 60.2).

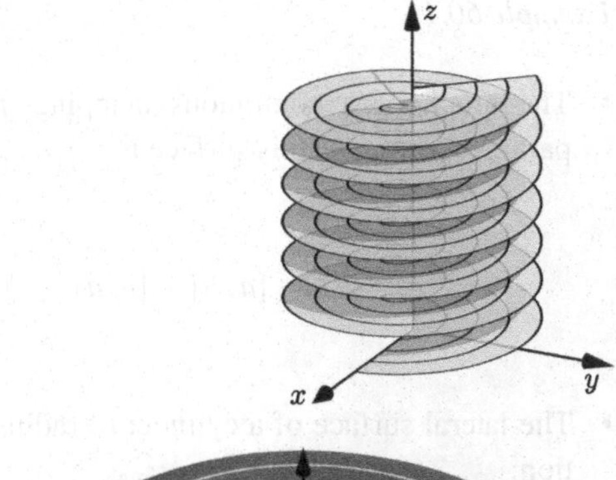

Fig. 60.2 A helical surface

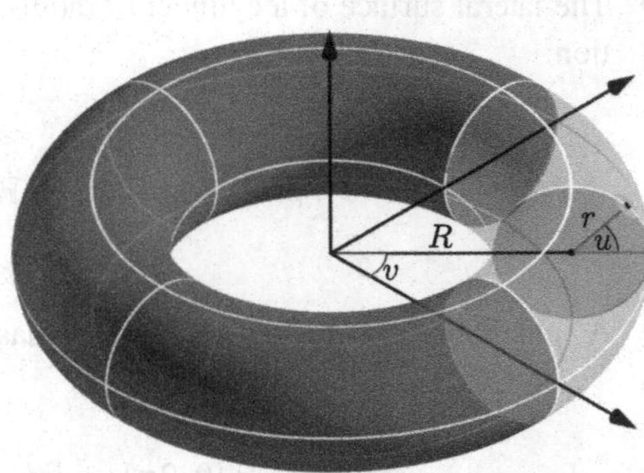

Fig. 60.3 A torus

- A **torus** with *inner radius* r and *outer radius* R, $r < R$, is formed by rotation of a circle of radius r in the x-z-plane with center $(R, 0)^\top$ around the z-axis. The result is a *donut*, see Fig. 60.3 This donut has the following parameterization:

$$\phi : [0, 2\pi] \times [0, 2\pi] \to \mathbb{R}^3, \quad \phi(u, v) = \begin{pmatrix} (R + r\cos u)\cos v \\ (R + r\cos u)\sin v \\ r\sin u \end{pmatrix}.$$

■

60.2 Surface Integrals

The role of $\dot{\gamma}(t)$ in the case of line integrals is taken over in the case of surface integrals by $\phi_u(u, v) \times \phi_v(u, v)$. If one exchanges here the two parameters u and v, then (because of $\phi_v(v, u) \times \phi_u(v, u) = -\phi_u(u, v) \times \phi_v(u, v)$) we get the negative of the to ϕ_u and ϕ_v orthogonal vector, which of course is also orthogonal to ϕ_u and ϕ_v. This swapping of parameters is thus roughly equivalent to reorienting a curve, $\gamma \to -\gamma$. In the following vectorial surface integral, thus *along the curve* γ of the vectorial line integral has to be replaced by *in the direction* $\phi_u \times \phi_v$:

The Scalar and the Vector Surface Integral
Given is a regular surface

$$\phi : B \subseteq \mathbb{R}^2 \to \mathbb{R}^3 \text{ with } \phi(u, v) = \big(x(u, v), y(u, v), z(u, v)\big)^\top .$$

- For a scalar field $f : D \subseteq \mathbb{R}^3 \to \mathbb{R}$ with $\phi(B) \subseteq D$ the integral

$$\iint_\phi f \mathrm{d}s = \iint_B f\big(\phi(u, v)\big) \,\|\phi_u(u, v) \times \phi_v(u, v)\| \mathrm{d}u\mathrm{d}v$$

  is called the **scalar surface integral** of f over ϕ.
- For a vector field $\boldsymbol{v} : D \subseteq \mathbb{R}^3 \to \mathbb{R}^3$ with $\phi(B) \subseteq D$ the integral

$$\iint_\phi \boldsymbol{v} \cdot \mathrm{d}s = \iint_B \boldsymbol{v}(\phi(u, v))^\top \big(\phi_u(u, v) \times \phi_v(u, v)\big)\mathrm{d}u\mathrm{d}v$$

  is called the **vectorial surface integral** or the **flux** of $\boldsymbol{v}$ through ϕ **in direction** $\phi_u(u, v) \times \phi_v(u, v)$.

The value of the scalar surface integral with the 1-*function* $f(x, y, z) = 1$ is the area $O(\phi)$ of ϕ, more precisely

$$O(\phi) = \iint_\phi \mathrm{d}s = \iint_B \|\phi_u \times \phi_v\| \mathrm{d}u\mathrm{d}v .$$

Of course, like the multiple integral and also the line integrals, the surface integrals are also *additive*, i.e., for each composite surface $\phi = \phi_1 + \cdots + \phi_r$ we have:

$$\iint_\phi = \iint_{\phi_1} + \cdots + \iint_{\phi_r} .$$

Example 60.2

- We consider the function $f : \mathbb{R}^3 \to \mathbb{R},\ f(x, y, z) = z^2$ and in addition the *upper* hemisphere of radius $R = 1$,

$$\phi : [0, \pi/2] \times [0, 2\pi] \to \mathbb{R}^3,\ \phi(\vartheta, \varphi) = \begin{pmatrix} \cos\varphi \sin\vartheta \\ \sin\varphi \sin\vartheta \\ \cos\vartheta \end{pmatrix}.$$

The scalar surface integral of f over ϕ has because of $\|\phi_\vartheta \times \phi_\varphi\| = \sin\vartheta$ (see Example 60.1) the value:

$$\iint\limits_{\phi} f\,\mathrm{d}s = \int\limits_{\vartheta=0}^{\pi/2} \int\limits_{\varphi=0}^{2\pi} \cos^2\vartheta \sin\vartheta \,\mathrm{d}\varphi\,\mathrm{d}\vartheta = 2\pi \frac{-1}{3} \cos^3\vartheta \Big|_{\vartheta=0}^{\pi/2} = \frac{2\pi}{3}.$$

- We consider the vector field $\boldsymbol{v} : \mathbb{R}^3 \to \mathbb{R}^3,\ \boldsymbol{v}(x, y, z) = (y, x, z)^\top$ and the surface

$$\phi : [0, \pi/2] \times [0, 2\pi] \to \mathbb{R}^3,\ \phi(\vartheta, \varphi) = \begin{pmatrix} \cos\varphi \sin\vartheta \\ \sin\varphi \sin\vartheta \\ \cos\vartheta \end{pmatrix}.$$

From Example 60.1 we know:

$$\phi_\vartheta(\vartheta, \varphi) \times \phi_\varphi(\vartheta, \varphi) = \sin\vartheta \begin{pmatrix} \cos\varphi \sin\vartheta \\ \sin\varphi \sin\vartheta \\ \cos\vartheta \end{pmatrix}.$$

Thus, for the vectorial surface integral:

$$\begin{aligned} \iint\limits_{\phi} \boldsymbol{v} \cdot \mathrm{d}\boldsymbol{s} &= \int\limits_{\vartheta=0}^{\pi/2} \int\limits_{\varphi=0}^{2\pi} \begin{pmatrix} \sin\varphi \sin\vartheta \\ \cos\varphi \sin\vartheta \\ \cos\vartheta \end{pmatrix}^\top \begin{pmatrix} \sin^2\vartheta \cos\varphi \\ \sin^2\vartheta \sin\varphi \\ \sin\vartheta \cos\vartheta \end{pmatrix} \mathrm{d}\varphi\,\mathrm{d}\vartheta \\ &= \int\limits_{\vartheta=0}^{\pi/2} \int\limits_{\varphi=0}^{2\pi} 2\sin^3\vartheta \sin\varphi \cos\varphi + \sin\vartheta \cos^2\vartheta \,\mathrm{d}\varphi\,\mathrm{d}\vartheta = 2\pi/3. \end{aligned}$$

- As the area of the sphere of radius R we get with $\|\phi_\vartheta \times \phi_\varphi\| = R^2 \sin\vartheta$:

$$O(\phi) = \int_{\vartheta=0}^{\pi} \int_{\varphi=0}^{2\pi} R^2 \sin\vartheta \,\mathrm{d}\varphi\,\mathrm{d}\vartheta = 2\pi R^2 \big(-\cos\vartheta\big)\Big|_{\vartheta=0}^{\pi} = 4\pi R^2 .$$

- We have seen that we can parameterize the graph of a function $f : B \subseteq \mathbb{R}^2 \to \mathbb{R}$ as a surface by

$$\phi(u, v) = \begin{pmatrix} u \\ v \\ f(u, v) \end{pmatrix} .$$

Therefore $\phi_u(u, v) = \big(1,\ 0,\ f_u(u, v)\big)^\top$ and $\phi_v(u, v) = \big(0,\ 1,\ f_v(u, v)\big)^\top$, and for the area of this surface we get:

$$O(f) = \iint_B \|\phi_u \times \phi_v\| \,\mathrm{d}u\,\mathrm{d}v = \int_B \sqrt{1 + f_u^2 + f_v^2}\,\mathrm{d}u\,\mathrm{d}v .$$

■

Flux: A vector field $\boldsymbol{v} : D \subseteq \mathbb{R}^3 \to \mathbb{R}^3$ can be understood as the velocity field of a flowing fluid. We consider a surface $\phi : B \subseteq \mathbb{R}^2 \to \mathbb{R}^3$ with $\phi(B) \subseteq D$, i.e., ϕ is given by $\boldsymbol{v}$ *flows through*. Then $\iint_\phi \boldsymbol{v} \cdot \mathrm{d}s$ the amount of fluid that per unit time flows through the surface $\phi(B)$ (see Fig. 60.4).

This is explained as follows: It corresponds to $\|\phi_u \times \phi_v\|$ the area of the mesh (parallelogram), on which $\phi_u \times \phi_v$ is perpendicular. Per time unit a parallelepiped flows through this mesh. The volume of this parallelepiped is $v\big(\phi(u, v)\big)^\top (\phi_u \times \phi_v)$. Summing up all infinitesimally small meshes leads to

$$\iint_\phi \boldsymbol{v} \cdot \mathrm{d}s = \iint_B \boldsymbol{v}\big(\phi(u, v)\big)^\top (\phi_u \times \phi_v)\,\mathrm{d}u\,\mathrm{d}v .$$

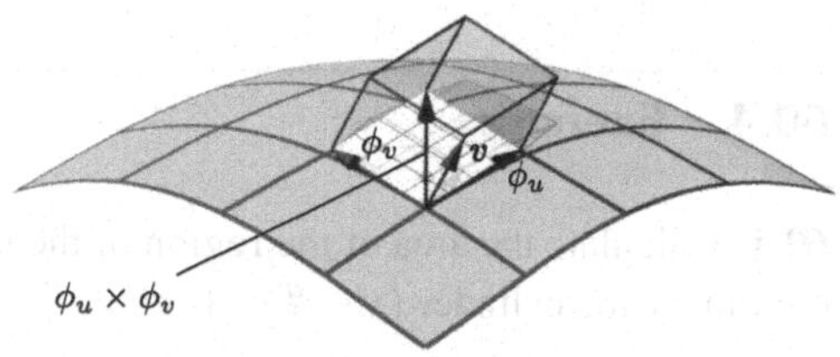

Fig. 60.4 A spat penetrates through a mesh

60.3 Overview of the Integrals

We give an overview of the different integrals and the meaning of their values, since this is important for understanding the *integral theorems* in the next chapter.

Overview of the Integrals

- **Multiple integrals:** Given is a scalar field $f : D \subseteq \mathbb{R}^n \to \mathbb{R}$.
  - $n = 2$: $\iint_D f \, dxdy$ is the volume enclosed between the graph of f and the domain D. If f is a mass density, the mass of D is determined.
  - $n = 3$: $\iiint_D f \, dxdydz$ is the mass of D if f is a mass density.

  To calculate the integral $\iint_D f$ or $\iiint_D f$ it is often advantageous to use a different coordinate system (note the substitution for multiple variables).
- **Line integrals:** Given is a curve $\gamma : [a, b] \subseteq \mathbb{R} \to \mathbb{R}^n$ and a scalar field $f : D \subseteq \mathbb{R}^n \to \mathbb{R}$ with $\gamma([a, b]) \subseteq D$ or a vector field $\boldsymbol{v} : D \subseteq \mathbb{R}^n \to \mathbb{R}^n$ with $\gamma([a, b]) \subseteq D$:
  - scalar line integral: $\int_\gamma f \, ds = \int_a^b f(\gamma(t)) \, \|\dot{\gamma}(t)\| \, dt$.

    Calculate the area between $\gamma(t)$ and $f(\gamma(t))$. If $f = 1$, then the curve length of γ is determined. If f is a mass density, then the mass of γ is determined.
  - vectorial line integral: $\int_\gamma \boldsymbol{v} \cdot ds = \int_a^b \boldsymbol{v}(\gamma(t))^\top \dot{\gamma}(t) \, dt$.

    Calculated is the work that has to be done to move a particle from $\gamma(a)$ to $\gamma(b)$.

  Surface integrals: Given is a surface $\phi : B \subseteq \mathbb{R}^2 \to \mathbb{R}^3$ and a scalar field $f : D \subseteq \mathbb{R}^3 \to \mathbb{R}$ with $\phi(B) \subseteq D$ or a vector field $\boldsymbol{v} : D \subseteq \mathbb{R}^3 \to \mathbb{R}^3$ with $\phi(B) \subseteq D$:
- - scalar surface integral: $\iint_\phi f \, ds = \iint_B f(\phi(u, v)) \, \|\phi_u \times \phi_v\| \, du \, dv$.

    This is used to calculate the three-dimensional volume that lies between $\phi(u, v)$ and $f(\phi(u, v))$. If $f = 1$, then the area of ϕ is determined.
  - vectorial surface integral: $\iint_\phi \boldsymbol{v} \cdot ds = \iint_B \boldsymbol{v}(\phi(u, v))^\top (\phi_u \times \phi_v) \, du \, dv$.

    Here, the flux of $\boldsymbol{v}$ through ϕ in the direction $\phi_u \times \phi_v$is calculated.

60.4 Exercises

60.1 Calculate the area of the region of the surface $\{(x, y) \mid x^2 + y^2 + z^2 \leq a^2\}$ which is cut out by the cylinder $(x - \frac{a}{2})^2 + y^2 = \frac{a^2}{4}$. See the figure below.

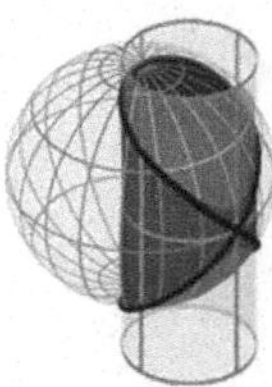

60.2 Calculate the area of the surface of the intersection of the two cylinders $x^2+z^2 \leq a^2$ and $y^2+z^2 \leq a^2$. See the figure above.

60.3 Let us calculate the area of the helical surface

$$\phi(r, \varphi) = \begin{pmatrix} r\cos\varphi \\ r\sin\varphi \\ \varphi \end{pmatrix}, \quad r \in [0, 1],\ \varphi \in [0, 2\pi].$$

60.4 Given is the cylinder Z of height $h > 0$ above circle around the origin in the x-y-plane with radius $R > 0$.

(a) Describe the cylindrical shell of Z in appropriate coordinates.
(b) Calculate the flux of the vector field $\boldsymbol{v}$ through the cylindrical surface of Z from the inside to the outside, where

$$\boldsymbol{v} : \mathbb{R}^3 \to \mathbb{R}^3, \quad (x, y, z)^\top \mapsto (xz + y, yz - x, z)^\top.$$

60.5 Let a pencil have the base of a regular hexagon with edge length $a = 4$ mm. Calculate the surface area of the pencil tip assuming that the side face makes an angle of 10° with the pencil lead.
Note: Use the substitution for multiple variables and the fact that the pencil point, the subset of a certain circular cone of the form $(mz)^2 = x^2 + y^2$ can be written as a graph over the plane hexagon.

Integral Theorems I

61

The *Integral Theorems of Green, Gauss* and *Stokes* form the core of *vector analysis*. What these theorems have in common is that two different integrals for a bounded domain B are equal if only the integrands are closely related.

In this first part on Integral Theorems, we cover the *plane theorems* of Green and Gauss. They are not plane theorems, of course; no, they are so called because they take place in the plane $\mathbb{R}^2$.

61.1 Green's Theorem

We start with a simple version of Green's Theorem. If one understands this simple version, the general version does not cause any difficulties.

To formulate Green's Theorem, we need the following notion: If B is a (bounded) domain in the $\mathbb{R}^2$ with a (closed) boundary ∂B as in Fig. 61.1, we say that the boundary ∂B of B is **positively oriented** or ∂B is **positively parametrized** if, given a chosen parametrization $\gamma : [a, b] \to \mathbb{R}^2$ of the boundary ∂B, this boundary is traversed in such a way that B on the passage to the left of ∂B lies.

Remark It is quite useful at this point to take a look at Chap. 54 at this point: If $\gamma : [a, b] \to \mathbb{R}^2$ is a parameterization of ∂B, then $\tilde{\gamma} : [a, b] \to \mathbb{R}^2$ with $\tilde{\gamma}(t) = \gamma(a + b - t)$ is also a parameterization of ∂B, where the direction of passage is reversed. If we have found a parametrization of a boundary, we will also find a positive parametrization (if necessary, we have to change the direction of passage, which is a trivial matter). By the way, only the sign of the vectorial curve integral is changed by reversing the direction of passage, see Sect. 56.1.

C. Karpfinger, *Calculus and Linear Algebra in Recipes*,
https://doi.org/10.1007/978-3-662-65458-3_61

Fig. 61.1 ∂B is positively oriented

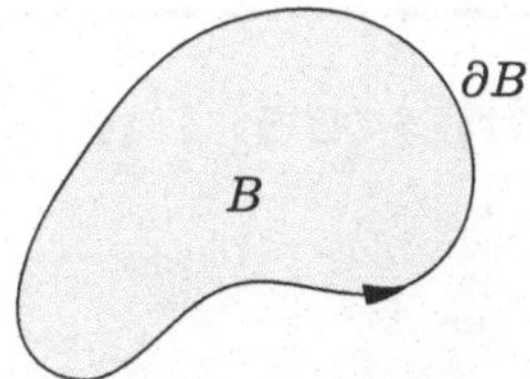

With this we are already in a position to formulate the first Integral Theorem (in a simple version):

Green's Theorem—Simple Version
If B is a domain in $\mathbb{R}^2$ with a (closed) positive parameterized boundary ∂B, then for any continuously differentiable vector field $\boldsymbol{v} : D \subseteq \mathbb{R}^2 \to \mathbb{R}^2$ with $\boldsymbol{v}(x, y) = \left(v_1(x, y), v_2(x, y)\right)^\top$ and $B \subseteq D$ we have:

$$\iint\limits_B \frac{\partial v_2}{\partial x} - \frac{\partial v_1}{\partial y}\,\mathrm{d}x\mathrm{d}y = \int\limits_{\partial B} \boldsymbol{v} \cdot \mathrm{d}\boldsymbol{s}\,.$$

Note: Green's Theorem states that the multiple integral over the domain B for the scalar field $\frac{\partial v_2}{\partial x} - \frac{\partial v_1}{\partial y}$ has the same value as the vectorial curve integral over the positively oriented boundary of B for the vector field $\boldsymbol{v}$. Thus, one obtains the value of one of these integrals by computing the value of the other integral. For example, if it is complicated to calculate the vectorial curve integral $\int_{\partial B} \boldsymbol{v} \cdot \mathrm{d}\boldsymbol{s}$ then determine the perhaps much more easily computable multiple integral $\iint_B \frac{\partial v_2}{\partial x} - \frac{\partial v_1}{\partial y}\mathrm{d}x\mathrm{d}y$.

It makes sense to verify Green's Theorem with an example, i.e. we calculate for a chosen domain B and vector field $\boldsymbol{v}$ and convince ourselves that both times the same result is obtained:

Example 61.1 We consider the domain

$$B = \left\{(x, y) \in \mathbb{R}^2 \mid \frac{x^2}{a^2} + \frac{y^2}{b^2} \le 1\right\}.$$

This is an ellipse with the semi-axes a and b, $a,\ b > 0$. A positive parametrization of ∂B is

$$\gamma : [0, 2\pi] \to \mathbb{R}^2,\ \gamma(t) = \begin{pmatrix} a\cos(t) \\ b\sin(t) \end{pmatrix}.$$

We now verify Green's Theorem for the vector field $v : \mathbb{R}^2 \to \mathbb{R}^2,\ v(x, y) = (-y, x)^\top$ by computing both integrals.

- For the multiple integral we obtain:

$$\iint_B \frac{\partial v_2}{\partial x} - \frac{\partial v_1}{\partial y} \,\mathrm{d}x\,\mathrm{d}y = \int_{x=-a}^{a} \int_{y=-b\sqrt{1-\frac{x^2}{a^2}}}^{b\sqrt{1-\frac{x^2}{a^2}}} 2\,\mathrm{d}y\,\mathrm{d}x = \frac{4b}{a}\int_{x=-a}^{a} \sqrt{a^2-x^2}\,\mathrm{d}x$$

$$= \frac{4b}{a}\left[\frac{1}{2}\left(x\sqrt{a^2-x^2} + a^2 \arcsin\left(\frac{x}{a}\right)\right)\right]_{x=-a}^{a}$$

$$= \frac{2b}{a}\left(a^2\frac{\pi}{2} + a^2\frac{\pi}{2}\right) = 2\,a\,b\,\pi\,,$$

where we used MATLAB to determine the given primitive function.

- For the vectorial curve integral we obtain:

$$\int_{\partial B} v \cdot \mathrm{d}s = \int_0^{2\pi} \begin{pmatrix} -b\sin(t) \\ a\cos(t) \end{pmatrix}^\top \begin{pmatrix} -a\sin(t) \\ b\cos(t) \end{pmatrix} \mathrm{d}t = ab\int_0^{2\pi} 1\,\mathrm{d}t = 2\,a\,b\,\pi\,.$$

So, in fact, both integrals give the same result, as predicted by Green's Theorem. ■

A useful application of Green's Theorem is the following formula for calculating the area of B by means of a vectorial curve integral. Namely, if one chooses the (continuously differentiable) vector field $v(x, y) = (-y, x)^\top$, then by Green's Theorem we have:

$$\int_{\partial B} v \cdot \mathrm{d}s = \iint_B \frac{\partial x}{\partial x} + \frac{\partial y}{\partial y}\mathrm{d}x\mathrm{d}y = 2\iint_B 1\mathrm{d}x\mathrm{d}y\,,$$

so that we have the area $F(B) = \iint_B 1\mathrm{d}x\mathrm{d}y$ of B by integration of v along the positively oriented boundary of ∂B more precisely:

Area Calculation with Green's Theorem

If B is a (bounded) domain in the $\mathbb{R}^2$ with a (closed) positively parametrized boundary ∂B then one obtains the area $F(B)$ from B by:

$$F(B) = \frac{1}{2}\int_{\partial B} \begin{pmatrix} -y \\ x \end{pmatrix} \cdot \mathrm{d}s\,.$$

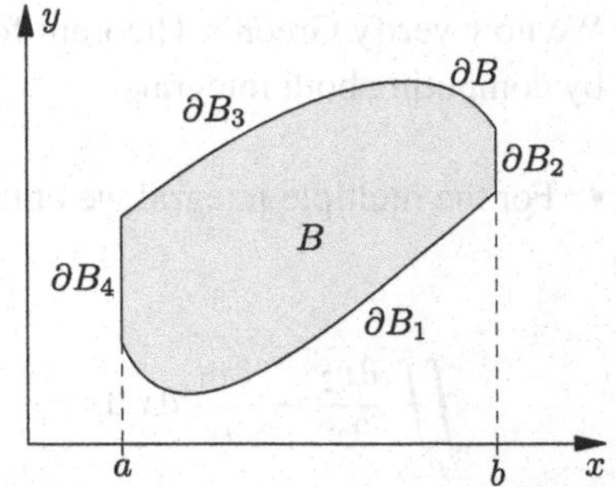

Fig. 61.2 Composite boundary

We already know this formula: This is the Leibniz sector formula for bounded domains (see Sect. 55.3).

Green's Theorem can also be formulated more generally for more complicated domains B. Here we consider domains whose boundaries consist of composite curves. Consider Fig. 61.2. Here the boundary pieces must then each be parameterized in such a way that altogether the boundary ∂B is positively oriented, i.e., that B to the left of ∂B lies.

Green's Theorem—General Version
Let there be $B \subseteq \mathbb{R}^2$ a bounded domain whose closed boundary ∂B is the union of finitely many, piecewise continuously differentiable, positively parameterized curves $\gamma_1, \ldots, \gamma_k$.
For each continuously differentiable vector field $\boldsymbol{v} : D \subseteq \mathbb{R}^2 \to \mathbb{R}^2$ with $\boldsymbol{v}(x, y) = \big(v_1(x, y), v_2(x, y)\big)^\top$ and $B \subseteq D$ then we have:

$$\iint\limits_B \frac{\partial v_2}{\partial x} - \frac{\partial v_1}{\partial y} \mathrm{d}x\mathrm{d}y = \int\limits_{\partial B} \boldsymbol{v} \cdot \mathrm{d}\boldsymbol{s} = \sum_{i=1}^{k} \int\limits_{\gamma_i} \boldsymbol{v} \cdot \mathrm{d}\boldsymbol{s} .$$

61.2 The Plane Theorem of Gauss

We proceed as with Green's Theorem and begin with a simple variant of the theorem:

Again we consider a domain B with positive parameterized boundary ∂B (i.e. ∂B is traversed in such a way that B is on the left). In addition, we consider the normal vector $\boldsymbol{n}$ on ∂B, which is at each point of ∂B points *outward*, see Fig. 61.3.

This vector $\boldsymbol{n}$ has length 1, so it is normalized, and is orthogonal to the tangent vector $\dot{\gamma}(t)$ if ∂B is given by γ. Therefore we obtain $\boldsymbol{n}$ simply by

$$\dot{\gamma}(t) = \begin{pmatrix} a \\ b \end{pmatrix} \Rightarrow \boldsymbol{n} = \pm \frac{1}{\sqrt{a^2 + b^2}} \begin{pmatrix} b \\ -a \end{pmatrix} ,$$

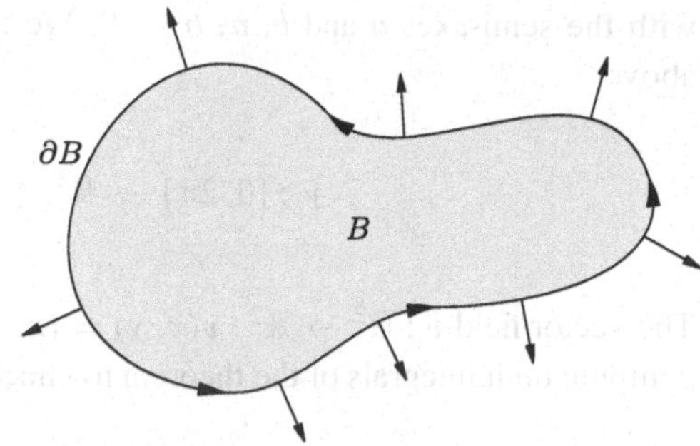

Fig. 61.3 The boundary is positively oriented and the normal vector points outward

where the sign is still to be chosen so that $\boldsymbol{n}$ points outwards.

With this we can formulate the second Integral Theorem (in a simple version):

The Plane Theorem of Gauss—Simple Version

If B is a domain in $\mathbb{R}^2$ with a (closed) positive parameterized boundary ∂B and an outward pointing normal vector $\boldsymbol{n}$, then for any continuously differentiable vector field $\boldsymbol{v} : D \subseteq \mathbb{R}^2 \to \mathbb{R}^2$ with $\boldsymbol{v}(x, y) = \big(v_1(x, y), v_2(x, y)\big)^\top$ and $B \subseteq D$:

$$\iint_B \operatorname{div} \boldsymbol{v} \, \mathrm{d}x\mathrm{d}y = \int_{\partial B} \boldsymbol{v}^\top \boldsymbol{n} \, \mathrm{d}s .$$

Note that the notation $\boldsymbol{v}^\top \boldsymbol{n}$ is common, but takes some getting used to: actually $\boldsymbol{v}$ is a vector with component functions in x and y and $\boldsymbol{n}$ a vector with component functions in t; this does not fit together. What is meant by $\boldsymbol{v}^\top \boldsymbol{n}$ is actually $(\boldsymbol{v} \circ \gamma)^\top \boldsymbol{n}$ with the positive parametrization γ of the boundary ∂B.

Gauss's Theorem states that the scalar curve integral over the positively oriented boundary of B for the scalar field $\boldsymbol{v}^\top \boldsymbol{n}$ has the same value as the multiple integral over the domain B for the scalar field $\operatorname{div} \boldsymbol{v}$. So you get the value of one of these integrals by calculating the value of the other integral. For example, if it is complicated to determine the multiple integral $\iint_B \operatorname{div} \boldsymbol{v} \, \mathrm{d}x\mathrm{d}y$ then determine the scalar line integral $\int_{\partial B} \boldsymbol{v}^\top \boldsymbol{n} \, \mathrm{d}s$ which is perhaps much easier to calculate.

We will also verify this Integral Theorem with an example:

Example 61.2 We consider again the ellipse

$$B = \left\{ (x, y) \in \mathbb{R}^2 \;\middle|\; \frac{x^2}{a^2} + \frac{y^2}{b^2} \le 1 \right\}$$

with the semi-axes a and b, $a, b > 0$. We also choose the positive parametrization as above:

$$\gamma : [0, 2\pi] \to \mathbb{R}^2, \ \gamma(t) = \begin{pmatrix} a\cos(t) \\ b\sin(t) \end{pmatrix}.$$

The vector field $\boldsymbol{v} : \mathbb{R}^2 \to \mathbb{R}^2$, $\boldsymbol{v}(x, y) = (x, y)^\top$ has the divergence $\operatorname{div} \boldsymbol{v} = 2$. We again compute both integrals of the theorem to check their equality.

- For the area integral, we get (using the calculation from above):

$$\iint_B \operatorname{div} \boldsymbol{v} \, \mathrm{d}x\mathrm{d}y = \iint_B 2 \, \mathrm{d}x \, \mathrm{d}y = 2\, a\, b\, \pi \, .$$

- To solve the curve integral, we first determine the normal vector:

$$\dot{\gamma}(t) = \begin{pmatrix} -a\sin(t) \\ b\cos(t) \end{pmatrix} \Rightarrow \boldsymbol{n} = \frac{1}{\sqrt{a^2\sin^2(t) + b^2\cos^2(t)}} \begin{pmatrix} b\cos(t) \\ a\sin(t) \end{pmatrix}.$$

  As can easily be seen for $t=0$ ($\gamma(0) = (a, 0)^\top$), the sign must be chosen as indicated so that $\boldsymbol{n}$ points outward. We now compute the integral:

$$\begin{aligned}
\int_{\partial B} \boldsymbol{v}^\top \boldsymbol{n} \mathrm{d}s &= \int_0^{2\pi} \boldsymbol{v}\big(\gamma(t)\big)^\top \boldsymbol{n} \, \|\dot{\gamma}(t)\| \, \mathrm{d}t \\
&= \int_0^{2\pi} \frac{1}{\sqrt{a^2\sin^2(t) + b^2\cos^2(t)}} \begin{pmatrix} a\cos(t) \\ b\sin(t) \end{pmatrix}^\top \begin{pmatrix} b\cos(t) \\ a\sin(t) \end{pmatrix} \left\| \begin{pmatrix} -a\sin(t) \\ b\cos(t) \end{pmatrix} \right\| \mathrm{d}t \\
&= \int_0^{2\pi} \frac{1}{\sqrt{a^2\sin^2(t) + b^2\cos^2(t)}} \big(ab\cos^2(t) + ab\sin^2(t)\big) \sqrt{a^2\sin^2(t) + b^2\cos^2(t)} \, \mathrm{d}t \\
&= 2\, a\, b\, \pi \, .
\end{aligned}$$

In fact, we again get the same result for both integrals. ■

Like Green's Theorem, Gauss's plane Theorem can be generalized accordingly to more complicated domains B which are bounded by composite curve pieces.

The Plane Theorem of Gauss—General Version

Let $B \subseteq \mathbb{R}^2$ be a bounded domain whose closed boundary ∂B is a union of finitely many, piecewise continuously differentiable, positively parameterized curves $\gamma_1, \ldots, \gamma_k$ with the respective outward pointing normal vectors $\boldsymbol{n}_1, \ldots, \boldsymbol{n}_k$. For each continuously differentiable vector field $\boldsymbol{v} : D \subseteq \mathbb{R}^2 \to \mathbb{R}^2$ with $B \subseteq D$ then we have:

$$\iint_B \operatorname{div} \boldsymbol{v} \, \mathrm{d}x\mathrm{d}y = \int_{\partial B} \boldsymbol{v}^\top \boldsymbol{n} \, \mathrm{d}s = \sum_{i=1}^{k} \int_{\gamma_i} \boldsymbol{v}^\top \boldsymbol{n}_i \, \mathrm{d}s \, .$$

The Theorems of Green and Gauss take place in the $\mathbb{R}^2$. In each case, a multiple integral over a domain B in $\mathbb{R}^2$ was equal to a line integral over the plane curve ∂B. We will now treat two further integral theorems which have their existence in the $\mathbb{R}^3$: So we will consider bounded domains B in $\mathbb{R}^3$ and its boundaries ∂B, which are surfaces. In principle there are two possibilities:

- The domain B is *closed* such as a cube, the boundary is then the cube surface, in particular a surface ($\to$ *Gauss's Divergence Theorem*).
- The area B is *open* such as a piece of a surface, the boundary is then the boundary of the piece of surface, in particular a space curve ($\to$ Theorem of *Stokes*).

With this we can already guess which integrals will be the same: In the first case a (three-dimensional) multiple integral and a surface integral, in the second case a surface integral and a line integral.

61.3 Exercises

61.1 Verify Green's Theorem for the vector field $\boldsymbol{v}(x, y) = (2xy - x^2, x + y^2)^\top$ and the domain B, which bounded by $y = x^2$ and $y^2 = x$.

61.2 Verify Gauss' Theorem in the plane for the function $u(x, y) = (x^2 + y^2)^{5/2}$:

$$\iint_B \Delta u \, \mathrm{d}x\mathrm{d}y = \oint_{\partial B} \langle \nabla u, n \rangle \, \mathrm{d}s \, ,$$

where B is a circular disk of radius R.

61.3 Let P be the parallelogram with vertices (0, 0), (1, 0), (1, 1) and (2, 1).

(a) Give the limits of the multiple integral

$$\iint_P f(x, y)\,\mathrm{d}x\mathrm{d}y$$

in the order of integration given.

(b) Calculate the multiple integral for the function $f(x, y) = y\,\mathrm{e}^{-y^2}$.

(c) Calculate the integral of the curve $\int_{\partial P} v \cdot \mathrm{d}s$ over the boundary ∂P of P (positive sense of orbit) for the vector field $v(x, y) = (-y^3, xy^2)^\top$. Use Green's Theorem.

Integral Theorems II

62

The *Divergence Theorem of Gauss* and the *Theorem of Stokes* are *spatial* Integral Theorems, namely they take place in the $\mathbb{R}^3$. We continue our tradition: we first describe simple versions of these Theorems, verify them by examples, and then give more general versions.

The true usefulness of these Theorems becomes apparent only in hydrodynamics and electricity. With the help of these Integral Theorems, elegant notations of physical relationships are possible, such as the representation of Maxwell's equations in differential or integral form.

62.1 The Divergence Theorem of Gauss

As before, we first consider a simple version of the Theorem. To do this, let B be an area from Fig. 62.1 with the respective surface $\phi = \partial B$ and their respective outward normal vectors $\phi_u \times \phi_v$, which (because we will need them) we enter immediately on each surface piece:

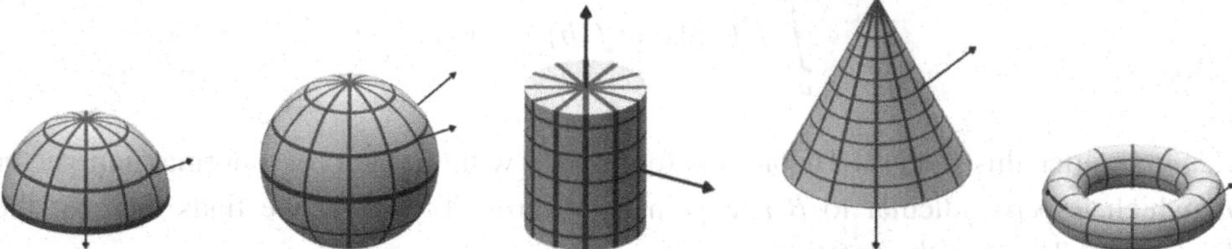

Fig. 62.1 We consider in advance only hemispheres, spheres, cylinder, cones, tori

C. Karpfinger, *Calculus and Linear Algebra in Recipes*,
https://doi.org/10.1007/978-3-662-65458-3_62

For simplicity, we consider for now only these special spatial domains, i.e. hemisphere, sphere, cylinder, cone, torus or cuboid, which have the following great advantage: By eventual choice of special coordinates, we can define both the domains B (i.e., the volumes) and the surfaces ∂B (i.e., the surface pieces) in a simple way. We will turn to more general *spatial domains* with *more complicated surfaces* later and will leave it at the theory.

The Divergence Theorem of Gauss—Simple Version
Let B be a hemisphere, sphere, cylinder, cone, torus or cuboid with the surface (boundary) $\phi = \partial B$ where $\phi_u \times \phi_v$ points outward, then for any continuously differentiable vector field $\boldsymbol{v} : G \subseteq \mathbb{R}^3 \to \mathbb{R}^3$ with $B \subseteq G$:

$$\iiint_B \operatorname{div} \boldsymbol{v} \, \mathrm{d}x \, \mathrm{d}y \, \mathrm{d}z = \iint_{\partial B} \boldsymbol{v} \cdot \mathrm{d}s \,,$$

here the flux is calculated in the direction $\phi_u \times \phi_v$.

Gauss' Divergence Theorem states that the multiple integral over the domain B for the scalar field div $\boldsymbol{v}$ has the same value as the vectorial surface integral over $\phi = \partial B$ in the direction of $\phi_u \times \phi_v$. That is the flux through the surface *outward* for the vector field v. Thus, one obtains the value of one of these integrals by calculating the value of the other integral. For example, if it is complicated to calculate the vectorial surface integral $\iint_{\partial B} \boldsymbol{v} \cdot \mathrm{d}s$, then one determines the perhaps clearly more simply computable multiple integral $\iiint_B \operatorname{div} \boldsymbol{v} \mathrm{d}x \mathrm{d}y \mathrm{d}z$.

Nor shall we fail to give the typical interpretation of this proposition: That which flows through the surface from the inside to the outside engenders in a source in the inside. Analogously, as much disappears in a sink inside as the negative flow from inside to outside indicates. The one-dimensional analogue of this divergence theorem is well known: For $B = [a, b]$ we have $\partial B = \{a, b\}$ and $\operatorname{div} f = f'$:

$$\int_a^b f'(x) \mathrm{d}x = f(b) - f(a) \,.$$

Remark Often this Integral Theorem is formulated with the help of a normal unit vector $\boldsymbol{n}$, which is perpendicular to B and points outwards. Typically one finds then for the Divergence Theorem the notation

$$\iiint_B \operatorname{div} \boldsymbol{v} \, \mathrm{d}x \, \mathrm{d}y \, \mathrm{d}z = \iint_{\partial B} \boldsymbol{v}^\top \boldsymbol{n} \, \mathrm{d}s \,,$$

where $\boldsymbol{v}^\top \boldsymbol{n}$ is again to be understood in the sense of $(\boldsymbol{v} \circ \boldsymbol{\phi})^\top \boldsymbol{n}$ (see the comment on Gauss' plane Theorem in Sect. 61.2). In fact, however, this integral is equal to the integral we have given, because we have:

$$\iint_{\partial B} \boldsymbol{v}^\top \boldsymbol{n}\, \mathrm{d}s = \iint_{\partial B} \boldsymbol{v}(\boldsymbol{\phi}(u,v))^\top \frac{\pm 1}{\|\boldsymbol{\phi}_u \times \boldsymbol{\phi}_v\|} (\boldsymbol{\phi}_u \times \boldsymbol{\phi}_v)\, \|\boldsymbol{\phi}_u \times \boldsymbol{\phi}_v\| \mathrm{d}u \mathrm{d}v = \iint_{\partial B} \boldsymbol{v} \cdot \mathrm{d}\boldsymbol{s}\,.$$

We verify Gauss's Divergence Theorem with an example.

Example 62.1 We choose as B the upper hemisphere of radius $R > 0$,

$$B = \left\{(x, y, z)^\top \in \mathbb{R}^3 \mid x^2 + y^2 \le R^2,\ 0 \le z \le \sqrt{R^2 - x^2 - y^2}\right\}.$$

Further let be the following vector field $\boldsymbol{v}$ given:

$$\boldsymbol{v} : \mathbb{R}^3 \to \mathbb{R}^3,\ \boldsymbol{v}(x, y, z) = \begin{pmatrix} xz^2 \\ x^2 y \\ y^2 z \end{pmatrix}.$$

We parameterize the surface ∂B from B, which consists of two surface pieces: $\partial B = \boldsymbol{\phi}^{(1)} + \boldsymbol{\phi}^{(2)}$; the *roof* $\boldsymbol{\phi}^{(1)}$ and the *bottom* $\boldsymbol{\phi}^{(2)}$:

$$\boldsymbol{\phi}^{(1)} : (\vartheta, \varphi) \mapsto \begin{pmatrix} R\cos\varphi \sin\vartheta \\ R\sin\varphi \sin\vartheta \\ R\cos\vartheta \end{pmatrix} \quad \text{with} \quad \vartheta \in [0, \pi/2],\ \varphi \in [0, 2\pi]\,,$$

$$\boldsymbol{\phi}^{(2)} : (r, \varphi) \mapsto \begin{pmatrix} r\cos\varphi \\ r\sin\varphi \\ 0 \end{pmatrix} \quad \text{with} \quad r \in [0, R],\ \varphi \in [0, 2\pi]\,.$$

The normal vector $\boldsymbol{\phi}_\vartheta^{(1)} \times \boldsymbol{\phi}_\varphi^{(1)}$ of the roof we have already determined in the Example 60.1 and for the normal vector $\boldsymbol{\phi}_\varphi^{(2)} \times \boldsymbol{\phi}_r^{(2)}$ of the floor we get:

$$\boldsymbol{\phi}_\vartheta^{(1)} \times \boldsymbol{\phi}_\varphi^{(1)} = R^2 \sin\vartheta \begin{pmatrix} \cos\varphi \sin\vartheta \\ \sin\varphi \sin\vartheta \\ \cos\vartheta \end{pmatrix} \text{ and } \boldsymbol{\phi}_\varphi^{(2)} \times \boldsymbol{\phi}_r^{(2)} = \begin{pmatrix} -r\sin\varphi \\ r\cos\varphi\ 0 \end{pmatrix} \times \begin{pmatrix} \cos\varphi \\ \sin\varphi \\ 0 \end{pmatrix} = \begin{pmatrix} 0 \\ 0 \\ -r \end{pmatrix}.$$

Because $\cos\vartheta \ge 0$ for $\vartheta \in [0, \pi/2]$ and $r > 0$, both normal vectors point outward. (If this were not so, one could reverse the orientation by swapping the factors in the vector product).

Now we have all the ingredients to be able to calculate the two integrals in the Divergence Theorem:

- For the multiple integral, we obtain using spherical coordinates:

$$\iiint\limits_B \operatorname{div} \boldsymbol{v} \mathrm{d}x\mathrm{d}y\mathrm{d}z = \iiint\limits_B x^2 + y^2 + z^2 \mathrm{d}x\mathrm{d}y\mathrm{d}z$$

$$= \int\limits_{r=0}^{R} \int\limits_{\varphi=0}^{2\pi} \int\limits_{\vartheta=0}^{\pi/2} r^2 r^2 \sin\vartheta \,\mathrm{d}\vartheta\, \mathrm{d}\varphi \mathrm{d}r = 2\pi \frac{1}{5} R^5 = \frac{2}{5}\pi R^5 .$$

- And for the surface integral, we obtain with the abbreviations s_ϑ for $\sin\vartheta$, c_ϑ for $\cos\vartheta$ etc:

$$\iint\limits_{\partial B} \boldsymbol{v}\cdot \mathrm{d}\boldsymbol{s} = \iint\limits_{\partial\phi_1} \boldsymbol{v}\cdot \mathrm{d}\boldsymbol{s} + \iint\limits_{\partial\phi_2} \boldsymbol{v}\cdot \mathrm{d}\boldsymbol{s}$$

$$= \int\limits_{\vartheta=0}^{\pi/2} \int\limits_{\varphi=0}^{2\pi} \begin{pmatrix} R^3 c_\vartheta^2 c_\varphi s_\vartheta \\ R^3 s_\vartheta^3 c_\varphi^2 s_\varphi \\ R^3 s_\vartheta^2 s_\varphi^2 c_\vartheta \end{pmatrix}^{\mathrm{T}} R^2 s_\vartheta \begin{pmatrix} c_\varphi s_\vartheta \\ s_\varphi s_\vartheta \\ c_\vartheta \end{pmatrix} \mathrm{d}\varphi \mathrm{d}\vartheta + \int\limits_{r=0}^{R} \int\limits_{\varphi=0}^{2\pi} \begin{pmatrix} 0 \\ r^3 c_\varphi^2 s_\varphi \\ 0 \end{pmatrix}^{\mathrm{T}} \begin{pmatrix} 0 \\ 0 \\ -r \end{pmatrix} \mathrm{d}\varphi \mathrm{d}r$$

$$= \int\limits_{\vartheta=0}^{\pi/2} \int\limits_{\varphi=0}^{2\pi} R^5 c_\vartheta^2 c_\varphi^2 s_\vartheta^3 + R^5 s_\vartheta^5 c_\varphi^2 s_\varphi^2 + R^5 s_\vartheta^3 s_\varphi^2 c_\vartheta^2 \mathrm{d}\varphi \mathrm{d}\vartheta$$

$$= R^5 \int\limits_{\vartheta=0}^{\pi/2} \int\limits_{\varphi=0}^{2\pi} s_\vartheta^5 c_\varphi^2 s_\varphi^2 + s_\vartheta^3 c_\vartheta^2 (s_\varphi^2 + c_\varphi^2) \mathrm{d}\varphi \mathrm{d}\vartheta$$

$$= R^5 \int\limits_{\vartheta=0}^{\pi/2} \int\limits_{\varphi=0}^{2\pi} \sin^5\vartheta \cos^2\varphi \sin^2\varphi + \sin^3\vartheta \cos^2\vartheta \mathrm{d}\varphi \mathrm{d}\vartheta$$

$$= R^5 \int\limits_{\vartheta=0}^{\pi/2} 2\pi \cos^2\vartheta \sin^3\vartheta + \sin^5\vartheta \frac{1}{4}\pi \mathrm{d}\vartheta$$

$$= R^5 \left(2\pi \frac{2}{15} + \frac{1}{4}\pi \frac{8}{15}\right) = R^5 \frac{6\pi}{15} = \frac{2}{5}\pi R^5 . \qquad \blacksquare$$

A generalization of the Divergence Theorem of Gauss is possible: Essential is, that the considered domain B is analogous to the spheres, hemispheres, cylinders, ... considered above, restricted with a closed surface. An exact mathematical description of this simple

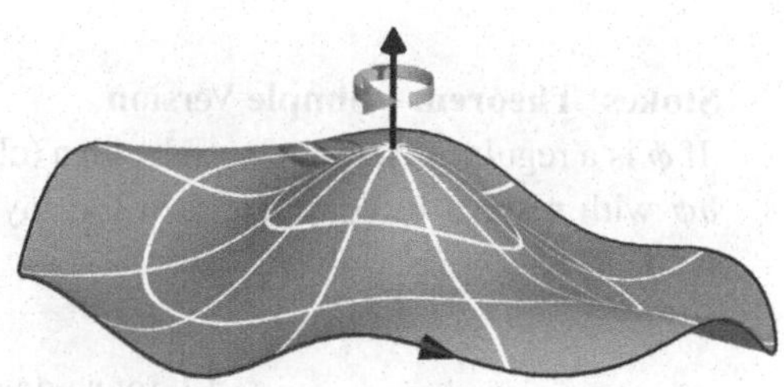

Fig. 62.2 A positively oriented boundary of a surface

notion is abundantly difficult; thus this Divergence Theorem with its so simply imaginable statement quickly sounds terribly complicated, but behind it is only: What flows outward through the surface must arise inside.

Gauss' Divergence Theorem—General Version
If B is a domain of $\mathbb{R}^3$ with piecewise continuously differentiable surface ∂B, then for any continuously differentiable vector field $v : D \subseteq \mathbb{R}^3 \to \mathbb{R}^3$ with $B \subseteq D$:

$$\iiint_B \operatorname{div} v \, \mathrm{d}x\mathrm{d}y\mathrm{d}z = \iint_{\partial B} v \cdot \mathrm{d}s \,,$$

where, in the case of the surface integral, the parametrization ϕ for each partial surface must be chosen so that $\phi_u \times \phi_v$ points outwards.

62.2 Stokes' Theorem

We now turn to the fourth and final Integral Theorem, *Stokes' Theorem*. Again, we first consider a simple version of this Theorem and assume a (bounded) surface ϕ with a closed boundary $\partial\phi$. Here we assume that the surface ϕ has no *pathologies*, i.e., has two sides, is regular, etc. The boundary of this surface is a space curve which we can traverse in two different directions. We can also declare on both sides at each point a normal vector n. Graphs of continuously differentiable scalar fields in two variables on a bounded domain of definition have such properties, see Fig. 62.2.

We specify the direction of passage of the boundary $\partial\phi$ by choosing a surface side: we call the space curve $\partial\phi$ **positively oriented** or **positively parameterized** with respect to $\phi_u \times \phi_v$ if, when grasping the vector $\phi_u \times \phi_v$ with the right hand, where the vector runs in the direction of the thumb stretched upwards, the direction of passage of the curve thereby points in the direction of the curved fingers (see Fig. 62.2).

Thus we can already formulate Stokes' Theorem in a simple version:

Stokes' Theorem—Simple Version

If ϕ is a regular surface in the $\mathbb{R}^3$ with a (closed) positively parameterized boundary $\partial\phi$ with respect to $\phi_u \times \phi_v$, then for any continuously differentiable vector field $\boldsymbol{v} : D \subseteq \mathbb{R}^3 \to \mathbb{R}^3$ with $\phi \subseteq D$:

$$\iint\limits_{\phi} \operatorname{rot} \boldsymbol{v} \cdot \mathrm{d}s = \int\limits_{\partial\phi} \boldsymbol{v} \cdot \mathrm{d}s\,.$$

Stokes' Theorem states that the vectorial line integral over the positively oriented boundary $\partial\phi$ of ϕ for the vector field $\boldsymbol{v}$ has the same value as the surface integral over the surface ϕ in the direction of $\phi_u \times \phi_v$ for the vector field $\operatorname{rot} \boldsymbol{v}$, which is the flux of $\operatorname{rot} \boldsymbol{v}$ through ϕ in the direction $\phi_u \times \phi_v$. Thus, one obtains the value of one of these integrals by calculating the value of the other integral. For example, if it is complicated to determine the vectorial surface integral $\iint_\phi \operatorname{rot} \boldsymbol{v} \cdot \mathrm{d}s$ let us determine the vectorial line integral $\int_{\partial\phi} \boldsymbol{v} \cdot \mathrm{d}s$, which is perhaps much easier to compute.

By the way, while $\iint_\phi \boldsymbol{v} \cdot \mathrm{d}s$ is called the flux of $\boldsymbol{v}$ by ϕ in the direction $\phi_u \times \phi_v$ the surface integral $\iint_\phi \operatorname{rot} \boldsymbol{v} \cdot \mathrm{d}s$ is called **vortex flux** of $\boldsymbol{v}$ through ϕ in the direction $\phi_u \times \phi_v$. Thus the statement of Stokes' Theorem can be summarized vividly: *The circulation of* $\boldsymbol{v}$ *along* $\partial\phi$ *is equal to the vortex flux of* $\boldsymbol{v}$ *through* ϕ.

Remark Also for this Theorem, a different notation using a normal unit vector $\boldsymbol{n}$ perpendicular to the surface ϕ is common in the literature. Typically, one then finds the notation:

$$\iint\limits_{\phi} \operatorname{rot} \boldsymbol{v}^\top \boldsymbol{n}\, \mathrm{d}s = \oint_{\partial\phi} \boldsymbol{v} \cdot \mathrm{d}s$$

for Stokes' Theorem, which, as already stated in Sect. 62.1), is just a reformulation of our vectorial surface integral as a scalar surface integral.

We will also verify this Integral Theorem with an example:

Example 62.2 We consider the surface ϕ of the upper hemisphere of radius 2,

$$\phi = \left\{(x, y, z)^\top \in \mathbb{R}^3 \,|\, x^2 + y^2 \leq 4,\ z = \sqrt{4 - x^2 - y^2}\right\},$$

with the boundary $\partial\phi = \{(x, y, 0) \mid x^2 + y^2 = 4\}$. We check Stokes' Theorem for the vector field

$$\boldsymbol{v} : \mathbb{R}^3 \to \mathbb{R}^3,\ \boldsymbol{v}(x, y, z) = \begin{pmatrix} -y \\ x \\ 1 \end{pmatrix} \quad \text{with} \quad \operatorname{rot} \boldsymbol{v} = \begin{pmatrix} \partial_2 v_3 - \partial_3 v_2 \\ \partial_3 v_1 - \partial_1 v_3 \\ \partial_1 v_2 - \partial_2 v_1 \end{pmatrix} = \begin{pmatrix} 0 \\ 0 \\ 2 \end{pmatrix}.$$

We parameterize ϕ by

$$\phi : (\vartheta, \varphi) \mapsto \begin{pmatrix} 2\cos\varphi \sin\vartheta \\ 2\sin\varphi \sin\vartheta \\ 2\cos\vartheta \end{pmatrix} \quad \text{with} \quad \vartheta \in [0, \pi/2],\ \varphi \in [0, 2\pi].$$

Consequently, then

$$\partial\phi : t \mapsto \begin{pmatrix} 2\cos t \\ 2\sin t \\ 0 \end{pmatrix},\ t \in [0, 2\pi] \quad \text{and} \quad \phi_\vartheta \times \phi_\varphi = 4\sin\vartheta \begin{pmatrix} \cos\varphi \sin\vartheta \\ \sin\varphi \sin\vartheta \\ \cos\vartheta \end{pmatrix}.$$

We can now compute both integrals:

- For the vectorial surface integral we get:

$$\iint_\varphi \operatorname{rot} \boldsymbol{v} \cdot \mathrm{d}\boldsymbol{s} = \int_{\varphi=0}^{2\pi} \int_{\vartheta=0}^{\pi/2} \begin{pmatrix} 0 \\ 0 \\ 2 \end{pmatrix}^{\top} 4\sin\vartheta \begin{pmatrix} \sin\vartheta\cos\varphi \\ \sin\vartheta\sin\varphi \\ \cos\vartheta \end{pmatrix} \mathrm{d}\vartheta\, \mathrm{d}\varphi$$

$$= 4 \int_{\varphi=0}^{2\pi} \int_{\vartheta=0}^{\pi/2} 2\sin\vartheta\cos\vartheta\, \mathrm{d}\vartheta\, \mathrm{d}\varphi$$

$$= 4 \int_{\varphi=0}^{2\pi} \left[\sin^2\vartheta\right]_{\vartheta=0}^{\pi/2} \mathrm{d}\varphi = 4 \int_{\varphi=0}^{2\pi} 1\, \mathrm{d}\varphi = 8\pi.$$

- For the vectorial line integral we obtain:

$$\int_{\partial\phi} \boldsymbol{v} \cdot \mathrm{d}\boldsymbol{s} = \int_0^{2\pi} \begin{pmatrix} -2\sin t \\ 2\cos t \\ 1 \end{pmatrix}^{\top} \begin{pmatrix} -2\sin t \\ 2\cos t \\ 0 \end{pmatrix} \mathrm{d}t = \int_0^{2\pi} 4\, \mathrm{d}t = 8\pi. \qquad \blacksquare$$

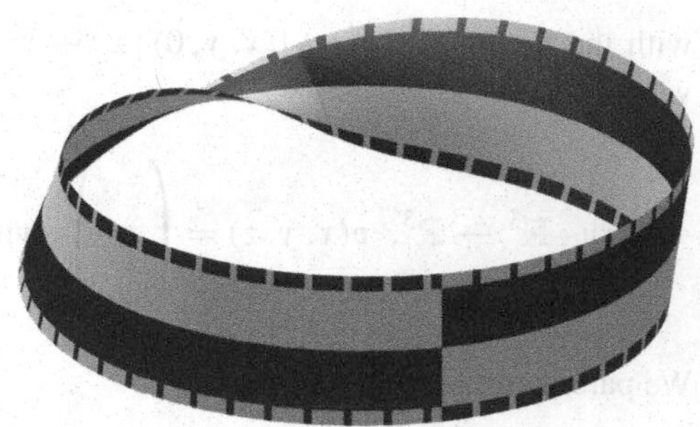

Fig. 62.3 The Möbius strip is not bilaterally

Stokes' Theorem can also be stated more generally:

Stokes' Theorem—General Version

If ϕ is a *two-sided* piecewise regular surface with double-point-free closed boundary $\partial\phi$, then for any vector field $\boldsymbol{v} : D \to \mathbb{R}^3$ with $\phi \subseteq D$:

$$\iint\limits_{\phi} \operatorname{rot} \boldsymbol{v} \cdot \mathrm{d}s = \int\limits_{\partial\phi} \boldsymbol{v} \cdot \mathrm{d}s\,,$$

if $\partial\phi$ is positively oriented traversed.

In particular, if ϕ_1 and ϕ_2 surfaces with the same boundary $\partial\phi_1 = \partial\phi_2$, then we have:

$$\iint\limits_{\phi_1} \operatorname{rot} \boldsymbol{v} \cdot \mathrm{d}s = \iint\limits_{\phi_2} \operatorname{rot} \boldsymbol{v} \cdot \mathrm{d}s\,.$$

By *two-sided* we mean that two sides can be distinguished. Regular surfaces are always two-sided: just choose the sign of $\pm(\phi_u \times \phi_v)$ uniform for all points on ϕ. An example of a non-bipartite surface is the Möbius strip shown in Fig. 62.3.

Finally, we give the following formulas which are used to show the uniqueness of solutions of partial differential equations. These formulas follow from the Divergence Theorem of Gauss:

Green's Integral Formulas
If f and g are scalar fields (twice differentiable) and B a bounded domain with the boundary ∂B, then we have:

$$\iiint\limits_{B} f\,\Delta g + \langle \nabla f\,,\nabla g\rangle\, dxdydz = \iint\limits_{\partial B} f\,\nabla g \cdot ds$$

and
$$\iiint\limits_{B} f\,\Delta g - g\,\Delta f\, dxdydz = \iint\limits_{\partial B} f\,\nabla g - g\,\nabla f \cdot ds\,.$$

In particular, for $f = 1$:

$$\iiint\limits_{B} \Delta g\, dxdydz = \iint\limits_{\partial B} \nabla g \cdot ds\,.$$

62.3 Exercises

62.1 Using the Integral Theorems, derive from the following differential form of the **Maxwell's equations** to produce the integral representation:

- $\text{rot}(H) - \dot{D} = j$,
- $\text{rot}(E) + \dot{B} = 0$,
- $\text{div}(D) = \rho$,
- $\text{div}(B) = 0$.

62.2 Show that Green's Theorem is a special case of Stokes' Theorem.

62.3 Let us confirm Stokes' Theorem

$$\iint\limits_{\phi} \text{rot}\,\boldsymbol{v} \cdot ds = \oint\limits_{\partial\phi} \boldsymbol{v} \cdot ds \quad \text{for the vector field} \quad \boldsymbol{v}(x, y, z) = \begin{pmatrix} 3y \\ -xz \\ yz^2 \end{pmatrix},$$

where ϕ is the surface given by the paraboloid $2z = x^2 + y^2$ and bounded by the plane $z = 2$. For the surface normal vector choose the one with negative z-component.

62.4 Given are the vector field $\boldsymbol{v}$ and the surface ϕ:

$$\boldsymbol{v}(\boldsymbol{x}) = \begin{pmatrix} 1 \\ xz \\ xy \end{pmatrix}, \ \boldsymbol{x} \in \mathbb{R}^3 \text{ and } \phi(\varphi, \vartheta) = \begin{pmatrix} \cos\varphi \sin\vartheta \\ \sin\varphi \sin\vartheta \\ \cos\vartheta \end{pmatrix}, \ \varphi \in [0, 2\pi], \ \vartheta \in \left[0, \tfrac{\pi}{4}\right].$$

Using Stokes' Theorem, calculate the surface integral $\iint_\phi \operatorname{rot} \boldsymbol{v} \cdot \mathrm{d}s$.

62.5 Using Gauss's Divergence Theorem, compute the multiple integral

$$\iint_\phi \boldsymbol{v} \cdot \mathrm{d}s \quad \text{for the vector field} \quad \boldsymbol{v}(x, y, z) = \begin{pmatrix} z^2 - x \\ -xy \\ 3z \end{pmatrix},$$

where ϕ is the surface of the domain B which is bounded by $z = 4 - y^2$ and the three planes $x = 0$, $x = 3$, $z = 0$.

62.6 For fixed $R > 0$, we introduce Torus coordinates:

$$T : [0, R] \times [0, 2\pi] \times [0, 2\pi] \to \mathbb{R}^3, \quad \begin{pmatrix} \varrho \\ \varphi \\ \vartheta \end{pmatrix} \mapsto \begin{pmatrix} (R + \varrho\cos\vartheta)\cos\varphi \\ (R + \varrho\cos\vartheta)\sin\varphi \\ \varrho\sin\vartheta \end{pmatrix}.$$

Determine

(a) the points of $\mathbb{R}^3$ which can be represented,
(b) the area of the torus $T_R^r = T([0, r] \times [0, 2\pi] \times [0, 2\pi])$ with $r \in [0, R]$,
(c) the flux of the vector field $\boldsymbol{v} : \mathbb{R}^3 \to \mathbb{R}^3$, $\boldsymbol{v}(\boldsymbol{x}) = \boldsymbol{x}$, through the surface of T_R^r,
(d) the flux of the vector field $\boldsymbol{v} : \mathbb{R}^3 \to \mathbb{R}^3$, $\boldsymbol{v}(\boldsymbol{x}) = \boldsymbol{x}$, through the surface of T_R^r with the help of the Theorem of Gauss.

62.7 We consider the torus T given by the parametrization

$$\phi : [0, 1] \times [0, 2\pi) \times [0, 2\pi) \longrightarrow \mathbb{R}^3 \quad \text{mit} \quad \phi(r, \theta, \varphi) = \begin{pmatrix} (2 + r\cos\theta)\cos\varphi \\ (2 + r\cos\theta)\sin\varphi \\ r\sin\theta \end{pmatrix}.$$

(a) The torus T intersects the half-plane $\{(x, y, z) \in \mathbb{R}^3 | y = 0, x \geq 0\}$. Give a parametrization of the intercept surface S.

(b) Calculate the flux $\int_S \operatorname{rot} \boldsymbol{v} \cdot \mathrm{d}F$ for the vector field

$$\boldsymbol{v} = \begin{pmatrix} y\,\mathrm{e}^z \\ z\,\mathrm{e}^x \\ x\,\mathrm{e}^y \end{pmatrix},$$

where the normal vector of the intersection S with $\boldsymbol{n} = (0, 1, 0)^\top$ is given.

(c) Give a parametrization of the curve, which one must go through in order to apply Stokes' Theorem to calculate the surface integral.

62.8 Given the scalar field

$$u : \mathbb{R}^3 \setminus \{0\} \to \mathbb{R} \quad \text{with} \quad u(x, y, z) = \frac{1}{\sqrt{x^2 + y^2 + z^2}}.$$

(a) Calculate ∇u and $\operatorname{div}(\nabla u)$.

(b) Determine the flux of the vector field ∇u through the surface of the sphere ∂K (from the inside to the outside), where the sphere is K is given by

$$(x-2)^2 + y^2 + z^2 \leq 1.$$

General Information on Differential Equations 63

We continue our discussion of (ordinary) differential equations (ODEs for short) in Chap. 36 with the numerics of differential equations. We start with the consideration of the *direction field of* a 1st order ODE and provide information about the conditions under which an initial value problem (IVP for short) has exactly one solution. With the help of matrices and the theory of functions of several variables we are able to trace back every explicit differential equation of *n*th order to a system of differential equations of the first order. Therefore, we direct our attention to 1st order differential equation systems (ODE systems for short) in the remaining chapters on differential equations.

63.1 The Directional Field

We consider an **explicit** 1st order ODE, i.e. a ODE of the form

$$\dot{x} = f(t, x)$$

with a function $f : G \subseteq \mathbb{R}^2 \to \mathbb{R}$ (here, *explicit* means that the ODE—as stated—can be solved for $\dot{x}$).

By this ODE each point $(t, x) \in G$ is assigned the *direction* $(1, f(t, x))^\top$.

For illustration one draws in $G \subseteq \mathbb{R}^2$ a short line passing through the point $(t, x)^\top$ with the gradient $\dot{x} = f(t, x)$. One speaks of a **line element**. The totality of all line elements in $G \subseteq \mathbb{R}^2$ is called the **directional field** (note Fig. 63.1).

A solution $x = x(t)$ of the ODE $\dot{x} = f(t, x)$ is a function whose slope at every point on its graph coincides with the slope of the associated line element. Thus, constructing the graph of a solution means constructing the curves that fit into the constructed direction field. This does not give an analytical formula for the solution, but at least an overview of

C. Karpfinger, *Calculus and Linear Algebra in Recipes*,
https://doi.org/10.1007/978-3-662-65458-3_63

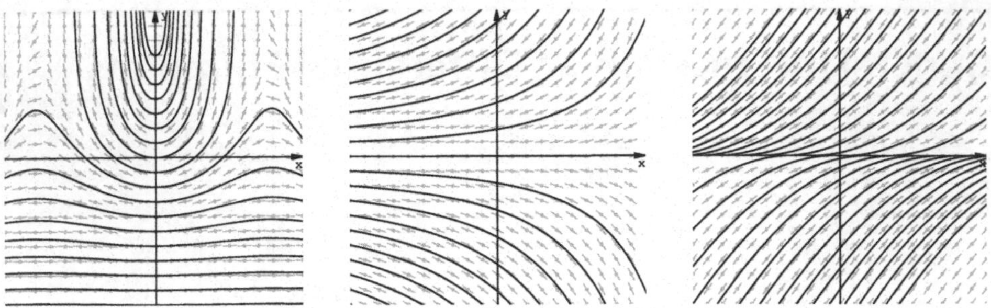

Fig. 63.1 Direction fields of the three given differential equations

the course of the solution. More precisely expressed, this means: A function $x = x(t)$ is a solution of the ODE $\dot{x} = f(t, x)$ if the graph of x is a **field line** i.e. if the tangent to the graph of x at each point $(t, x(t))$ contains the line element through $(t, x) \in G$ is contained. Additionally one can determine **isoclines** which are the curves on which $\dot{x} = f(t, x) = c$ is constant. On the isoclines, the line elements have the same slope.

The three pictures in Fig. 63.1 show the directional field and some field lines of the ODEs

$$\dot{x} = \mathrm{e}^x \sin(t) \text{ and } \dot{x} = x\, \mathrm{e}^t \text{ and } \dot{x} = \sqrt{|x|}\,.$$

63.2 Existence and Uniqueness of Solutions

For an initial value problems so far, we have always found a unique solution. This is no coincidence, there is a profound existence and uniqueness Theorem behind it, which says that under certain conditions exactly one solution of the initial value problem exists. Mathematicians try to formulate these conditions as weakly as possible, so that this Theorem can be applied to as many initial value problems as possible. We make it a bit easier for ourselves and assume more than absolutely necessary:

Existence and Uniqueness Theorem

Given is the initial value problem (IVP)

$$x^{(n)} = f(t, x, \ldots, x^{(n-1)}),\ x(t_0) = x_0, \ldots, x^{(n-1)}(t_0) = x_{n-1}$$

(continued)

with an explicit ODE. Is there a neighborhood $G \subseteq \mathbb{R}^{n+1}$ of $(t_0, x_0, \ldots, x_{n-1})$ such that

- the function f is continuous on G and
- the function f is continuously partially differentiable with respect to the $2nd, 3rd., \ldots, (n+1)$-th variable on G,

then there exists exactly one solution $x = x(t)$ of the IVP which is given on an interval $I \subseteq \mathbb{R}$ with $t_0 \in I$.

We can immediately infer an interesting property of different solutions of a ODE if each IVP with this ODE is uniquely solvable: the graphs of different solutions x_1 and x_2 of such a ODE cannot intersect, i.e., either two solutions completely agree or they do not agree at any time.

If the interval I is chosen to be maximal, the corresponding solution is called $x : I \to \mathbb{R}$ a **maximum solution** of the IVP. It is often very difficult to determine the maximum solution.

Example 63.1

- The IVP

$$\dddot{x} = 0 \text{ with } x(1) = 0\,,\ \dot{x}(1) = 1\,,\ \ddot{x}(1) = 2$$

has exactly one solution, since the function $f(t, x, \dot{x}, \ddot{x}) = 0$ continuous in any neighborhood $G \subseteq \mathbb{R}^4$ from $(1, 0, 1, 2)^\top$ and f is obviously continuously partial differentiable after the 2nd, 3rd and 4th variable.

The uniquely determined solution is

$$x(t) = t^2 - t\,,$$

it's on $I = \mathbb{R}$ explained.

- The IVP

$$\dot{x} = \frac{\sqrt{1 - x^2}}{t^2} \quad \text{with } x(1) = 0$$

has exactly one solution, since the function $f(t, x) = \frac{\sqrt{1-x^2}}{t^2}$ in an neighborhood of $(1, 0)^\top \in \mathbb{R}^2$, e.g. in $G = (0, \infty) \times (-1, 1)$, is continuous and in this neighborhood also partially differentiable with respect to x. With a separation of the variables we obtain the solution

$$x(t) = \sin(c - 1/t).$$

As a maximum interval I we get because of the initial condition $I = (0, \infty)$.

- The IVP

$$\dot{x} = 3\,x^{2/3} \text{ with } x(0) = 0$$

has the two different solutions

$$x(t) = t^3 \text{ and } x(t) = 0.$$

Since the function $f(t, x) = 3\,x^{2/3}$ in every neighborhood G of $(0, 0)^\top \in \mathbb{R}^2$ is not continuously partial differentiable with respect to the 2nd variable, the existence and uniqueness Theorem is not applicable either. ■

Remark We consider again the IVP

$$x^{(n)} = f(t, x, , \dots, x^{(n-1)}),\ x(t_0) = x_0, \dots, x^{(n-1)}(t_0) = x_{n-1}.$$

In practice, the initial conditions or the function f have uncertainties due to measurement limitations; the **stability theorem** states that in the case of a continuously differentiable function f small errors in f or in the initial conditions also entail small errors in the solution.

63.3 Transformation to 1st Order Systems

We consider an explicit linear ODE of nth order

$$x^{(n)} + a_{n-1}(t)\,x^{(n-1)} + \cdots + a_1(t)\,\dot{x} + a_0(t)\,x = s(t).$$

We are looking for the set of all functions x, which satisfy this ODE. We now want to show that the ODE can be reduced to a **differential equation system**, in short ODE system, of 1st order, i.e. can be traced back to a ODE of the form

$$\dot{z} = A\,z + s$$

with a vector $z = z(t) = (z_1(t), \dots, z_n(t))^\top$ a square matrix $A = A(t)$ with n rows and a vector $s = s(t) = (s_1(t), \dots, s_n(t))^\top$.

We introduce auxiliary functions for this, we set

$$z_0 = x^{(0)}\,,\; z_1 = x^{(1)} = \dot z_0\,,\; z_2 = x^{(2)} = \dot z_1\,,\ldots,\; z_{n-1} = x^{(n-1)} = \dot z_{n-2}\,.$$

We now declare the vector-valued function

$$z(x) = \begin{pmatrix} z_0(t) \\ z_1(t) \\ \vdots \\ z_{n-1}(t) \end{pmatrix} \quad \text{with the derivative } \dot z(t) = \begin{pmatrix} \dot z_0(t) \\ \dot z_1(t) \\ \vdots \\ \dot z_{n-1}(t) \end{pmatrix}.$$

Since two vectors are equal if and only if they agree componentwise, and since

$$\dot z_{n-1} = x^{(n)} = -a_{n-1}\,x^{(n-1)} - \cdots - a_1\,\dot x - a_0\,x + s(t)\,,$$

we obtain with the above explained auxiliary functions

$$\underbrace{\begin{pmatrix} 0 & 1 & 0 & \cdots & 0 \\ 0 & 0 & 1 & \cdots & 0 \\ \vdots & & & \ddots & \vdots \\ 0 & \cdots & & 0 & 1 \\ -a_0 & -a_1 & \cdots & \cdots & -a_{n-1} \end{pmatrix}}_{=:A} \begin{pmatrix} z_0 \\ z_1 \\ \vdots \\ z_{n-2} \\ z_{n-1} \end{pmatrix} + \underbrace{\begin{pmatrix} 0 \\ 0 \\ \vdots \\ 0 \\ s(t) \end{pmatrix}}_{=:s} = \begin{pmatrix} \dot z_0 \\ \dot z_1 \\ \vdots \\ \dot z_{n-2} \\ \dot z_{n-1} \end{pmatrix}.$$

With the vector-valued functions z, $\dot z$ and s and the matrix-valued function A, we obtain the 1st order ODE system

$$\dot z = A\,z + s\,.$$

Reduction to 1st Order Systems

If one reduces an nth-order ODE to a 1st-order ODE system, then we have:

- If x is a solution of the nth-order ODE, then $(x, \dot x, \ldots, x^{(n-1)})^\top$ is a solution of the 1st order ODE system.
- If $(x_1, x_2, \ldots, x_n)^\top$ is a solution of the 1st order ODE system, then x_1 is a solution of the ODE of nth order.

(continued)

If, in addition, a solution x of the ODE nth order fulfills the initial conditions

$$x(t_0) = x_0, \ldots, x^{(n-1)}(t_0) = x_{n-1},$$

then the corresponding solution $(x_1, \ldots, x_n)^\top$ of the 1st order ODE system satisfies the initial condition

$$x_1(t_0) = x_0, \ldots, x_n(t_0) = x_{n-1}.$$

This result can be formulated even more generally for arbitrary explicit ODEs and thus has a far-reaching meaning: it suffices to consider 1st order ODE systems; all other ODEs can be reduced to such systems.

Example 63.2

- The IVP

$$\ddot{x} - \dot{x} - x = 0 \text{ with } x(0) = 1 \text{ and } \dot{x}(0) = 2$$

with a 2nd order ODE is reduced to the 1st order ODE system with the initial condition

$$\dot{z} = \begin{pmatrix} \dot{z}_0 \\ \dot{z}_1 \end{pmatrix} = \begin{pmatrix} 0 & 1 \\ 1 & 1 \end{pmatrix} \begin{pmatrix} z_0 \\ z_1 \end{pmatrix} = A\,z \text{ with } z(0) = \begin{pmatrix} 1 \\ 2 \end{pmatrix}.$$

- The ODE

$$\overset{\cdots\cdot}{x} - 3\dddot{x} - \sin(t)\,\dot{x} + t^2 x = t\,\mathrm{e}^t$$

of 4th order is reduced to the ODE system

$$\dot{z} = \begin{pmatrix} \dot{z}_0 \\ \dot{z}_1 \\ \dot{z}_2 \\ \dot{z}_3 \end{pmatrix} = \begin{pmatrix} 0 & 1 & 0 & 0 \\ 0 & 0 & 1 & 0 \\ 0 & 0 & 0 & 1 \\ -t^2 & \sin(t) & 0 & 3 \end{pmatrix} \begin{pmatrix} z_0 \\ z_1 \\ z_2 \\ z_3 \end{pmatrix} + \begin{pmatrix} 0 \\ 0 \\ 0 \\ t\,\mathrm{e}^t \end{pmatrix} = A\,z + s$$

1st order. ■

63.4 Exercises

63.1 For the ODE $\dot{x} = \sqrt{1 - tx}$ sketch the direction field and draw some solution curves on which curves are the points where the solution curves have $x = x(t)$ have vanishing or extreme slope?

63.2 Given the ODE $\dot{x} = |t + x|$, $(t, x) \in \mathbb{R}^2$.

(a) Determine the isoclines $\dot{x} = c$ with $c \in \{0, 1, 2, 3\}$ and sketch the direction field of the ODE with their help.
(b) Sketch the approximate course of the solution curves through the points $(1, 0)^\top$ and $(-1, 0)^\top$.

63.3 Is the IVP $\dot{x} = \mathrm{e}^x\, t$ with $x(0) = x_0$ uniquely solvable? Determine the maximum solution.

63.4 Determine the maximum solution of the IVP

(a) $\dot{x} = t\,x^2$, $x(0) = 1$.
(b) $\dot{x} = t\,x^2$, $x(0) = -1$.

63.4 Formulate the ODE $\ddot{x} = -x$ with the initial condition $x(t_0) = x_0$, $\dot{x}(t_0) = x_1$ as a 1st order ODE system with the corresponding initial condition.

The Exact Differential Equation

64

Not every 1st order differential equation has a solution method. However, if it is *exact*, we find a clear procedure for finding a solution of such an ODE with the method of determining a primitive function to a gradient field. The best thing is: Even if a ODE is not exact, it can often be made exact by multiplication with a *multiplier*.

64.1 Definition of Exact ODEs

If $f(t, x)$ and $g(t, x)$ are continuously differentiable functions, then a ODE of the form

$$f(t, x) + g(t, x)\,\dot{x} = 0$$

is an **exact differential equation**, if $f_x = g_t$.

Instead of $\dot{x}$ we can also write $\frac{\mathrm{d}x}{\mathrm{d}t}$ instead. If we substitute this into the above ODE and multiply by $\mathrm{d}t$ we get the usual notation for an exact ODE:

$$f(t, x)\,\mathrm{d}t + g(t, x)\,\mathrm{d}x = 0 \text{ with } f_x = g_t\,.$$

Example 64.1 The following ODEs are exact:

- $(-2\,t - x\,\sin(t)) + (2\,x + \cos t)\,\dot{x} = 0$.
- $3\,t^2 - 2\,a\,t + a\,x - 3\,x^2\dot{x} + a\,t\,\dot{x} = 0$ for every $a \in \mathbb{R}$.
- Any separable ODE $\dot{x} = f(t)\,g(x)$ with continuously differentiable f and g and $g(x) \neq 0$ is exact, since it can be expressed in the form

$$-f(t)\,\mathrm{d}t + \frac{1}{g(x)}\,\mathrm{d}x = 0\,.$$

C. Karpfinger, *Calculus and Linear Algebra in Recipes*,
https://doi.org/10.1007/978-3-662-65458-3_64

Now notice that $f_x = 0 = g_t$.

- The ODE

$$\dot{x} + x - t\,x^2 = 0$$

is not exact, but if we multiply it by the factor $\mu(t, x) = \frac{e^{-t}}{x^2}$, $x \neq 0$, we get the exact ODE

$$\frac{e^{-t}}{x^2}\,\dot{x} + \frac{e^{-t}}{x}\,(1 - tx) = 0\,.$$

■

64.2 The Solution Procedure

To determine the solution $x = x(t)$ of an exact ODE, we resort to the well-known scheme for the determination of a primitive function of a gradient field (cf. recipe in Sect. 57.3): Namely, if $F(t, x)$ is a primitive function of the gradient field (note that $f_x = g_t$ is just the integrability condition)

$$\boldsymbol{v}(t, x) = \begin{pmatrix} f(t, x) \\ g(t, x) \end{pmatrix},$$

then $F_t = f(t, x)$ and $F_x = g(t, x)$. Therefore, if one solves for $c \in \mathbb{R}$ the equation $F(t, x) = c$ to $x = x(t)$ then one obtains with this a function x in t. Because of the chain rule and two-sided differentiation of the equation $F(t, x) = c$ we get the equation

$$\frac{\mathrm{d}}{\mathrm{d}t} F(t, x) = F_t + F_x\,\dot{x} = f(t, x) + g(t, x)\,\dot{x} = 0\,.$$

Thus $x = x(t)$ is a solution of the given exact ODE. To solve an exact ODE, proceed as follows:

Recipe: Solve an Exact Differential Equation
Given is the exact ODE

$$f(t, x) + g(t, x)\,\dot{x} = 0 \;\; \text{(we have } f_x = g_t\text{)}\,.$$

(continued)

One obtains the general solution $x(t)$ (possibly only in implicit form $F(t,x) = c$) as follows:

(1) Determine a primitive function $F(t,x)$ of $\boldsymbol{v} = \begin{pmatrix} f \\ g \end{pmatrix}$ by successive integration:

- $F(t,x) = \int f \mathrm{d}t + G(x)$.
- Determine $G_x(x)$ from $F_x = \frac{\partial}{\partial x}(\int f \mathrm{d}t) + G_x(x) = g$.
- Determine $G(x)$ from $G_x(x)$ by integration.
- Get $F(t,x)$ from the first step.

(2) Solve the equation $F(t,x) = c$ with $c \in \mathbb{R}$ to $x = x(t)$ (if possible).
(3) The solution $x = x(t)$ depending on c is the general solution of the exact ODE.
(4) The c is determined by any initial condition that may exist.

The step (2) is often not feasible. The solution $x(t)$ is then implicitly given by the implicit equation $F(t,x) = c$. The c can nevertheless be determined by an initial condition.

Example 64.2 We solve the exact ODE

$$(-2t - x\sin(t)) + (2x + \cos t)\,\dot{x} = 0\,.$$

The ODE is exact because

$$\frac{\partial}{\partial t}(2x + \cos t) = -\sin t \text{ and } \frac{\partial}{\partial x}(-2t - x\sin t) = -\sin t\,.$$

(1) By integration, we obtain a primitive function,

$$F(t,x) = \int -2t - x\sin(t)\mathrm{d}t + G(x) = -t^2 + x\cos(t) + G(x)\,,$$

$$F_x(t,x) = \cos(t) + G_x(x) = 2x + \cos(t)\,,$$

So $G_x(x) = 2x$, i.e. $G(x) = x^2$. Thus we obtain $F(t,x) = -t^2 + x\cos(t) + x^2$.

(2) The equation $F(t,x) = c$ is

$$x^2 + \cos(t)x - (t^2 + c) = 0\,.$$

Solve to x returns for x the two possibilities

$$x_{1/2}(t) = \frac{-\cos(t) \pm \sqrt{\cos^2(t) + 4(t^2 + c)}}{2} .$$

(3) The general solution is

$$x(t) = \begin{cases} \frac{-\cos(t)+\sqrt{\cos^2(t)+4(t^2+c)}}{2} & \text{or} \\ \frac{-\cos(t)-\sqrt{\cos^2(t)+4(t^2+c)}}{2} \end{cases} ,$$

depending on the initial condition to be satisfied. Note that, for example, for $c \geq 0$ the graph of the first solution (with the + sign) in the upper half-plane ($x(t) \geq 0$ for all $t \in \mathbb{R}$), whereas the graph of the second solution (with the − sign) runs in the lower half plane ($x(t) \leq 0$ for all $t \in \mathbb{R}$).

(4) For the initial condition $x(0) = 0$ we get because of

$$x_{1/2}(0) = \frac{-1 \pm \sqrt{1 + 4c}}{2} = 0$$

immediately $c = 0$, and choosing the + sign in the general solution, i.e.

$$x(t) = \frac{-\cos(t) + \sqrt{\cos^2(t) + 4t^2}}{2} ,$$

the solution of the IVP with the given exact ODE and the initial condition is $x(0) = 0$.

■

64.2.1 Integrating Factors: Euler's Multiplier

The last considered ODE in Example 64.1 was not exact at first, after multiplication by the factor $\mu(t, x) \neq 0$ but it is. In this case one speaks of an **integrating factor** or **Euler's multiplier** $\mu(t, x)$: By multiplying the equation by this factor, an initially non-exact ODE becomes exact:

$$f(t, x)\mathrm{d}t + g(t, x)\,\mathrm{d}x = 0 \longrightarrow \underbrace{\mu(t, x)\, f(t, x)}_{=:\tilde{f}(t,x)}\mathrm{d}t + \underbrace{\mu(t, x)\, g(t, x)}_{=:\tilde{g}(t,x)}\,\mathrm{d}x = 0 ,$$

Where $\tilde{f}_x(t, x) = \tilde{g}_t(t, x)$, i.e., $\mu(t, x)$ satisfies the equation

$$\mu_x f + \mu\, f_x = \mu_t g + \mu\, g_t .$$

If $x = x(t)$ is the general solution of this exact ODE, then $x(t)$ is also the general solution of the original non-exact ODE (see Exercise 64.3).

We summarize:

Solution of a Non Exact Differential Equation with an Integrating Factor

If μ is an integrating factor that makes the non exact ODE $f + g\dot{x} = 0$ exactly, then the solution of the exact ODE is $\mu f + \mu g y\dot{x} = 0$ is the sought solution of the ODE $f + g\dot{x} = 0$.

The solution of the exact ODE $\mu f + \mu g\dot{x} = 0$ is obtained by the method described in Sect. 64.2).

Now there is only one problem: How to find an integrating factor $\mu = \mu(t, x)$? The decisive hint is given by the equation

$$\mu_x f + \mu f_x = \mu_t g + \mu g_t .$$

If $\mu = \mu(t, x)$ solves this equation for given f and g, then μ is an integrating factor. This equation is a partial ODE whose solutions are generally difficult to determine. But we don't need to determine all the solutions; certain ones, as simple as possible, will suffice.

It is possible to describe systematically the procedure for solving such a problem in special cases. Nevertheless, we will refrain from doing so. In fact, this description will be very complicated, and this effort is not justifiable. In practical examples, this method of solving a non-exact ODE by means of an integrating factor hardly plays a role; solutions are only found in very *artificial* examples. In practice, numerical methods are used to solve such non-exact ODEs. We therefore dispense with the systematics and show by examples how to guess an integrating factor in simple special cases.

The simplest case is that μ depends only on one of the two variables t or x. Then $\mu_x = 0$ or $\mu_t = 0$, and the partial ODE becomes much simpler:

Example 64.3 For a (non-exact) ODE of the form

$$f(t, x) + \dot{x} = 0$$

with $f_x \neq 0$ and $g(t, x) = 1$ the partial ODE for the integrating factor is $\mu = \mu(t, x)$

$$\mu_x f + \mu f_x = \mu_t .$$

Specifically for $f(t, x) = t^2 + x \sin(t)$ we obtain the non-exact ODE with the corresponding partial ODE

$$t^2 + x \sin(t) + \dot{x} = 0 \text{ with } \mu_x(t^2 + x \sin(t)) + \mu \sin(t) = \mu_t .$$

We consider a μ that satisfies $\mu_x = 0$ because then we are left with the much simpler ODE $\mu \sin(t) = \mu_t$ which obviously has the solution $\mu(t, x) = e^{-\cos(t)} \mu$. Thus we have found an integrating factor. We obtain the exact ODE with this μ

$$e^{-\cos(t)} t^2 + e^{-\cos(t)} x \sin(t) + e^{-\cos(t)} \dot{x} = 0 .$$

■

Somewhat more generally than in this example, one can reason:

Integrating Factors in Special Cases
Given the non exact ODE $f + g\dot{x} = 0$.

- If

$$\frac{f_x - g_t}{g} = u(t)$$

depends only on t, then $\mu = \mu(t, x) = e^{\int u(t) dt}$ is an integrating factor.

- If

$$\frac{g_t - f_x}{f} = u(x)$$

depends only on x, then $\mu = \mu(t, x) = e^{\int u(x) dx}$ is an integrating factor.

Remark It may be useful to perform a coordinate transformation to polar coordinates: An ODE in the (Cartesian) coordinates t and x is transformed into a ODE in polar coordinates r and φ. This new ODE may well be easier to deal with than the original ODE. Note the example in the problems.

64.3 Exercises

64.1 Show: If $x(t)$ is the general solution of the exact ODE given by an Eulerian multiplier $\mu(t, x)$ from a non-exact $f + g\,\dot{x} = 0$ ODE, then $x(t)$ also solves the original non exact ODE.

64.2 Determine the solutions of the following ODEs, if necessary determine an integrating factor:

(a) $t + x - (x - t)\,\dot{x} = 0.$
(b) $(1 + tx)\,\dot{x} + x^2 = 0.$
(c) $\frac{1-\cosh x\,\cos t}{(\cosh x-\cos t)^2}\,\dot{x} - \frac{\sinh x\,\sin t}{(\cosh x-\cos t)^2} = 0.$

64.3 Solve the following initial value problems, if necessary determine an integrating factor:

(a) $\dot{x} = \frac{x-t^2}{t}$ with $x(1) = 2$.
(b) $2t + 3\cos(x) + (2x - 3t\sin(x))\dot{x} = 0$ with $x(1) = 0$.
(c) $tx^2 - 1 + (t^2x - 1)\dot{x} = 0$ with $x(0) = 0$.

64.4 Solve the following ODE:

$$x - \left[(t^2 + x^2)^{3/2} + t\right]\dot{x} = 0\,.$$

Note: Introduce polar coordinates.

Linear Differential Equation Systems I

65

We consider 1st order explicit linear differential equation systems,

$$\dot{\boldsymbol{x}}(t) = A(t)\,\boldsymbol{x}(t) + \boldsymbol{s}(t)\,.$$

Here $A(t)$ is a $n \times n$-matrix with the component functions $a_{ij}(t)$ and $\boldsymbol{s}(t)$ is a curve, i.e. a $n \times 1$-vector, the **perturbation function**. We are looking for the set of all solution curves $\boldsymbol{x} = \boldsymbol{x}(t)$.

However, the ODE system cannot be solved in this generality. Only in special cases it is possible to state explicitly the totality of all solutions.

In this first chapter on linear ODE systems we consider the case of a constant and diagonalizable matrix $A \in \mathbb{R}^{n\times n}$ and a non-perturbative perturbation function, namely $\boldsymbol{s}(t) = \boldsymbol{0}$.

With the help of the eigenvalues and eigenvectors of A it will be easy to solve this case completely. The essential tool for this is the exponential function for matrices, which will also be the basis for the more general case, which we will discuss in the next chapter.

65.1 The Exponential Function for Matrices

We recall the series representation of the exponential function, we have for any real number b:

$$\mathrm{e}^b = \sum_{k=0}^{\infty} \frac{b^k}{k!} = 1 + b + \frac{b^2}{2!} + \cdots \in \mathbb{R}\,.$$

C. Karpfinger, *Calculus and Linear Algebra in Recipes*,
https://doi.org/10.1007/978-3-662-65458-3_65

Therefore, for every square matrix $B \in \mathbb{R}^{n\times n}$ we define:

$$\mathrm{e}^B = \sum_{k=0}^{\infty} \frac{1}{k!} B^k = E_n + B + \frac{1}{2!} B^2 + \cdots \in \mathbb{R}^{n\times n}.$$

Example 65.1

- We compute e^B for the matrix $B = \begin{pmatrix} 1 & 1 \\ 0 & 0 \end{pmatrix}$. Because of $B^k = B$ for $k \geq 1$ we have:

$$\mathrm{e}^B = \begin{pmatrix} 1 & 0 \\ 0 & 1 \end{pmatrix} + \begin{pmatrix} 1 & 1 \\ 0 & 0 \end{pmatrix} + \frac{1}{2}\begin{pmatrix} 1 & 1 \\ 0 & 0 \end{pmatrix} + \cdots = \begin{pmatrix} \mathrm{e} & \mathrm{e}-1 \\ 0 & 1 \end{pmatrix}.$$

- For the matrix $E_{12} = \begin{pmatrix} 0 & 1 \\ 0 & 0 \end{pmatrix} \in \mathbb{R}^{2\times 2}$ we have because of $E_{12}^k = \begin{pmatrix} 0 & 0 \\ 0 & 0 \end{pmatrix}$ For $k \geq 2$:

$$\mathrm{e}^{E_{12}} = \begin{pmatrix} 1 & 0 \\ 0 & 1 \end{pmatrix} + \begin{pmatrix} 0 & 1 \\ 0 & 0 \end{pmatrix} + \frac{1}{2}\begin{pmatrix} 0 & 0 \\ 0 & 0 \end{pmatrix} + \cdots = \begin{pmatrix} 1 & 1 \\ 0 & 1 \end{pmatrix}.$$

- For diagonal matrices, the rule is:

$$\exp\begin{pmatrix} \lambda_1 & \cdots & 0 \\ \vdots & \ddots & \vdots \\ 0 & \cdots & \lambda_n \end{pmatrix} = \begin{pmatrix} 1 & \cdots & 0 \\ \vdots & \ddots & \vdots \\ 0 & \cdots & 1 \end{pmatrix} + \begin{pmatrix} \lambda_1 & \cdots & 0 \\ \vdots & \ddots & \vdots \\ 0 & \cdots & \lambda_n \end{pmatrix} + \begin{pmatrix} \frac{1}{2}\lambda_1^2 & \cdots & 0 \\ \vdots & \ddots & \vdots \\ 0 & \cdots & \frac{1}{2}\lambda_n^2 \end{pmatrix} + \cdots$$

$$= \begin{pmatrix} \mathrm{e}^{\lambda_1} & \cdots & 0 \\ \vdots & \ddots & \vdots \\ 0 & \cdots & \mathrm{e}^{\lambda_n} \end{pmatrix}.$$

■

In the last examples we could determine e^B, because B^k was obvious for all natural numbers. Of course, this is not the case in the general case. But with the help of the following calculation rules we can e^B explicitly determine for all diagonalizable matrices.

Calculation Rules for the Exponential Function for Matrices

(a) For all $A,\ B \in \mathbb{R}^{n\times n}$ with $A\,B = B\,A$ we have:

$$\mathrm{e}^{A+B} = \mathrm{e}^{A}\,\mathrm{e}^{B}\,.$$

(b) For any invertible matrix $S \in \mathbb{R}^{n\times n}$ we have:

$$S^{-1}\,\mathrm{e}^{A}\,S = \mathrm{e}^{S^{-1}\,A\,S}\,.$$

In particular, for any diagonalizable matrix $A \in \mathbb{R}^{n\times n}$ with the eigenvalues $\lambda_1, \ldots, \lambda_n$ and eigenvectors $\boldsymbol{v}_1, \ldots, \boldsymbol{v}_n$ for $S = (\boldsymbol{v}_1, \ldots, \boldsymbol{v}_n)$:

$$\mathrm{e}^{A} = S \begin{pmatrix} \mathrm{e}^{\lambda_1} & \cdots & 0 \\ \vdots & \ddots & \vdots \\ 0 & \cdots & \mathrm{e}^{\lambda_n} \end{pmatrix} S^{-1}\,.$$

We consider that the addition is correct: If $A \in \mathbb{R}^{n\times n}$ is a diagonalizable matrix, then there exist an invertible matrix $S \in \mathbb{C}^{n\times n}$ and complex numbers $\lambda_1, \ldots, \lambda_n$ with the property

$$S^{-1}\,A\,S = \begin{pmatrix} \lambda_1 & \cdots & 0 \\ \vdots & \ddots & \vdots \\ 0 & \cdots & \lambda_n \end{pmatrix} = D\,,\ \text{i.e., } A = S\,D\,S^{-1}\,.$$

Now, using the rules of arithmetic, we obtain:

$$\mathrm{e}^{A} = \mathrm{e}^{S\,D\,S^{-1}} = S\,\mathrm{e}^{D}\,S^{-1}\,.$$

Thus we can e^A determine for diagonalizable matrices:

Recipe: Determine e^A for a Diagonalizable A
If $A \in \mathbb{R}^{n \times n}$ is a diagonalizable matrix, then obtain e^A as follows:

(1) Determine the eigenvalues $\lambda_1, \ldots, \lambda_n \in \mathbb{R}$ and an associated basis $\boldsymbol{v}_1, \ldots, \boldsymbol{v}_n$ of $\mathbb{R}^n$ from eigenvectors of A.
(2) Set $S = (\boldsymbol{v}_1, \ldots, \boldsymbol{v}_n)$ and $D = \operatorname{diag}(\lambda_1, \ldots, \lambda_n)$.
(3) Determine S^{-1}.
(4) Form the product $e^A = S \operatorname{diag}(e^{\lambda_1}, \ldots, e^{\lambda_n}) S^{-1}$.

In MATLAB, you call the exponential function for matrices with `expm(A)` for a previously declared A.

Example 65.2 We determine e^A for the diagonalizable matrix

$$A = \begin{pmatrix} 1 & 1 & 1 \\ 1 & 1 & 1 \\ 1 & 1 & 1 \end{pmatrix}.$$

(1) The eigenvalues are obviously $\lambda_1 = 0$, $\lambda_2 = 0$ and $\lambda_3 = 3$. A basis of eigenvectors is easily guessed:

$$\boldsymbol{v}_1 = \begin{pmatrix} 1 \\ -1 \\ 0 \end{pmatrix}, \ \boldsymbol{v}_2 = \begin{pmatrix} 1 \\ 0 \\ -1 \end{pmatrix}, \ \boldsymbol{v}_3 = \begin{pmatrix} 1 \\ 1 \\ 1 \end{pmatrix}.$$

(2) We obtain the matrices ;

$$S = \begin{pmatrix} 1 & 1 & 1 \\ -1 & 0 & 1 \\ 0 & -1 & 1 \end{pmatrix} \text{ and } e^D = \begin{pmatrix} 1 & 0 & 0 \\ 0 & 1 & 0 \\ 0 & 0 & e^3 \end{pmatrix}.$$

(3) As the inverse of S is calculated:

$$S^{-1} = \begin{pmatrix} 1/3 & -2/3 & 1/3 \\ 1/3 & 1/3 & -2/3 \\ 1/3 & 1/3 & 1/3 \end{pmatrix}.$$

(4) Finally, we compute the product $S\,\mathrm{e}^D\,S^{-1}$ and obtain

$$\mathrm{e}^A = \begin{pmatrix} 1 & 1 & 1 \\ -1 & 0 & 1 \\ 0 & -1 & 1 \end{pmatrix} \begin{pmatrix} 1 & 0 & 0 \\ 0 & 1 & 0 \\ 0 & 0 & \mathrm{e}^3 \end{pmatrix} \frac{1}{3} \begin{pmatrix} 1 & -2 & 1 \\ 1 & 1 & -2 \\ 1 & 1 & 1 \end{pmatrix} = \frac{1}{3} \begin{pmatrix} 2+\mathrm{e}^3 & -1+\mathrm{e}^3 & -1+\mathrm{e}^3 \\ -1+\mathrm{e}^3 & 2+\mathrm{e}^3 & -1+\mathrm{e}^3 \\ -1+\mathrm{e}^3 & -1+\mathrm{e}^3 & 2+\mathrm{e}^3 \end{pmatrix}.$$

■

65.2 The Exponential Function as a Solution of Linear ODE Systems

Using the exponential function for matrices, we can now solve ODE systems: We recall the simple case $n = 1$: we consider for $a \in \mathbb{R}$

$$\text{the ODE } \dot{x} = a\,x \quad \text{or the IVP } \dot{x} = a\,x \text{ and } x(t_0) = x_0\,.$$

Then

$$x(t) = c\,\mathrm{e}^{at} \text{ with } c \in \mathbb{R} \text{ or } x(t) = x_0\,\mathrm{e}^{a(t-t_0)}$$

is the general solution of the ODE or the uniquely determined solution of the IVP.

As one can easily convince oneself with the rules for the derivation of differentiable functions in several variables, completely analogous to the case $n = 1$, we have the following for an arbitrary $n \in \mathbb{N}$:

The Solution of a Homogeneous Linear ODE System or an IVP
Given

$$\text{the ODE system } \dot{\boldsymbol{x}} = A\boldsymbol{x} \text{ or the IVP } \dot{\boldsymbol{x}} = A\,\boldsymbol{x} \text{ with } \boldsymbol{x}(t_0) = \boldsymbol{x}_0$$

with $A \in \mathbb{R}^{n\times n}$ and $\boldsymbol{x}_0 \in \mathbb{R}^n$. Then the function

$$\boldsymbol{x} = \boldsymbol{x}(t) = \mathrm{e}^{tA}\,\boldsymbol{c} \text{ with a } \boldsymbol{c} \in \mathbb{R}^n \text{ or } \boldsymbol{x} = \boldsymbol{x}(t) = \mathrm{e}^{(t-t_0)A}\,\boldsymbol{x}_0$$

is the general solution of the ODE system or the unique solution of the IVP.

These formulas are easy to remember, everything is analogous to the one-dimensional case. In order to give the solution now still concretely, is only $\mathrm{e}^{tA}\,\boldsymbol{c}$ or $\mathrm{e}^{(t-t_0)A}\,\boldsymbol{x}_0$ to calculate. But for this *only* the matrix e^{tA} or $\mathrm{e}^{(t-t_0)A}$ and then multiply it by the (undetermined) vector $\boldsymbol{c}$ or by the (determined) vector $\boldsymbol{x}_0$ respectively. The t or the $t - t_0$ doesn't bother much, it's a number that can be *passed by* any matrix. Thus, the following solution path is seemingly pre-drawn:

Recipe: Solving a ODE System or an IVP with a Diagonalizable A—1st Version
Given is the ODE system or the IVP with diagonalizable matrix $A \in \mathbb{R}^{n \times n}$:

$$\dot{\boldsymbol{x}} = A\,\boldsymbol{x} \text{ or } \dot{\boldsymbol{x}} = A\,\boldsymbol{x} \text{ with } \boldsymbol{x}(0) = \boldsymbol{x}_0\,.$$

To determine the solution $\boldsymbol{x}_a$ or $\boldsymbol{x}$ proceed as follows:

(1) Determine the not necessarily different eigenvalues $\lambda_1, \ldots, \lambda_n$ with corresponding eigenvectors $\boldsymbol{v}_1, \ldots, \boldsymbol{v}_n$ and obtain $S = (\boldsymbol{v}_1, \ldots, \boldsymbol{v}_n)$ and $D = \operatorname{diag}(\lambda_1, \ldots, \lambda_n)$.
(2) The general (complex) solution of the ODE system is:

$$\boldsymbol{x}_a(t) = \mathrm{e}^{tA}\,\boldsymbol{c} = S\,\mathrm{e}^{tD}\,S^{-1}\boldsymbol{c} \text{ with } \boldsymbol{c} = (c_1, \ldots, c_n)^\top \in \mathbb{C}^n\,,$$

and the uniquely determined (complex) solution of the IVP is:

$$\boldsymbol{x}(t) = \mathrm{e}^{(t-t_0)A}\,\boldsymbol{x}_0 = S\,\mathrm{e}^{(t-t_0)D}\,S^{-1}\boldsymbol{x}_0\,.$$

This method of solving the ODE system or IVP sounds very simple, but is actually very computationally expensive; one must determine a basis S of eigenvectors, invert the matrix S, and finally compute the product of three matrices, then multiply this matrix by $\boldsymbol{c}$ or $\boldsymbol{x}_0$. In fact, one proceeds differently; we show how this can be done in the following section.

65.3 The Solution for a Diagonalizable A

It is $\boldsymbol{x}(t) = \mathrm{e}^{tA}\,\boldsymbol{c}$ with $\boldsymbol{c} \in \mathbb{R}^n$ the general solution, i.e., if $S = (\boldsymbol{v}_1, \ldots, \boldsymbol{v}_n)$ is a basis of $\mathbb{R}^n$ consisting of eigenvectors of the diagonalizable matrix A and $D = \operatorname{diag}(\lambda_1, \ldots, \lambda_n)$ is the diagonal matrix with the eigenvalues $\lambda_1, \ldots, \lambda_n$ on the diagonal, then by the abbreviation $\tilde{c} = S^{-1}c$, $\tilde{c} = (c_1, \ldots, c_n)^\top$, we have

$$\begin{aligned}\boldsymbol{x}_a(t) = \mathrm{e}^{tA}\,\boldsymbol{c} = S\,\mathrm{e}^{tD}\,S^{-1}\boldsymbol{c} = (\boldsymbol{v}_1, \ldots, \boldsymbol{v}_n) &\begin{pmatrix} \mathrm{e}^{\lambda_1 t} & & 0 \\ & \ddots & \\ 0 & & \mathrm{e}^{\lambda_n t} \end{pmatrix} \begin{pmatrix} c_1 \\ \vdots \\ c_n \end{pmatrix} \\ = c_1\,\mathrm{e}^{\lambda_1 t}\,\boldsymbol{v}_1 + \cdots + c_n\,\mathrm{e}^{\lambda_n t}\,\boldsymbol{v}_n\,.&\end{aligned}$$

The solution $\boldsymbol{x}(t) = \mathrm{e}^{(t-t_0)A}\,\boldsymbol{x}_0$ of the IVP $\dot{\boldsymbol{x}} = A\,\boldsymbol{x}$ with $\boldsymbol{x}(t_0) = \boldsymbol{x}_0$ is then obtained from this general solution $\boldsymbol{x}_a(t)$ of the ODE system by solving the system of equations

$$\boldsymbol{x}_a(t_0) = c_1\,\mathrm{e}^{\lambda_1 t_0}\,\boldsymbol{v}_1 + \cdots + c_n\,\mathrm{e}^{\lambda_n t_0}\,\boldsymbol{v}_n = \boldsymbol{x}_0$$

and thereby obtaining the numbers $c_1, \ldots, c_n$. With this the solution $\boldsymbol{x}(t)$ is found.

We have so far ignored only one problem: If the ODE system is real (and we always assume it is), then it is possible that some eigenvalues are non real after all. The solution we construct is then also non real, and you don't want to have something like that. The problem can be solved: If $\lambda = a + \mathrm{i}\,b \in \mathbb{C} \setminus \mathbb{R}$ is an eigenvalue of $A \in \mathbb{R}^{n\times n}$ with associated eigenvector $\boldsymbol{v} \in \mathbb{C}^n$ then also $\overline{\lambda} = a - \mathrm{i}\,b$ is an eigenvalue of A with associated eigenvector $\overline{\boldsymbol{v}}$. Using our previous system, we obtain the two non real solutions

$$\mathrm{e}^{\lambda t}\,\boldsymbol{v} \text{ and } \mathrm{e}^{\overline{\lambda} t}\,\overline{\boldsymbol{v}}\,.$$

Now we delete one of the two solutions, e.g. $\mathrm{e}^{\overline{\lambda} t}\,\overline{\boldsymbol{v}}$, and choose the real part and imaginary part of the remaining solution $\mathrm{e}^{\lambda t}\,\boldsymbol{v}$; these give two real linearly independent solutions

$$\begin{aligned}\mathrm{Re}(\mathrm{e}^{\lambda t}\,\boldsymbol{v}) &= \mathrm{e}^{at}(\cos(b\,t)\,\mathrm{Re}(\boldsymbol{v}) - \sin(b\,t)\,\mathrm{Im}(\boldsymbol{v})) \text{ and}\\ \mathrm{Im}(\mathrm{e}^{\lambda t}\,\boldsymbol{v}) &= \mathrm{e}^{at}(\sin(b\,t)\,\mathrm{Re}(\boldsymbol{v}) + \cos(b\,t)\,\mathrm{Im}(\boldsymbol{v}))\,.\end{aligned}$$

Thus we proceed to the solution as described below:

Recipe: Solving an ODE System or an IVP with Diagonalizable *A*—2nd Version
Given the ODE system or the IVP with diagonalizable matrix $A \in \mathbb{R}^{n\times n}$:

$$\dot{\boldsymbol{x}} = A\,\boldsymbol{x} \text{ or } \dot{\boldsymbol{x}} = A\,\boldsymbol{x} \text{ with } \boldsymbol{x}(0) = \boldsymbol{x}_0\,.$$

To determine the solution $\boldsymbol{x}_a$ or $\boldsymbol{x}$ proceed as follows:

(1) Determine the not necessary different eigenvalues $\lambda_1, \ldots, \lambda_n$ with corresponding eigenvectors $\boldsymbol{v}_1, \ldots, \boldsymbol{v}_n$.
(2) The general complex solution of the ODE system is:

$$\boldsymbol{x}_a(t) = c_1\,\mathrm{e}^{\lambda_1 t}\,\boldsymbol{v}_1 + \cdots + c_n\,\mathrm{e}^{\lambda_n t}\,\boldsymbol{v}_n \text{ with } c_1, \ldots, c_n \in \mathbb{C}\,.$$

(3) The general real solution is obtained from the general complex solution: If $\lambda, \overline{\lambda} \in \mathbb{C}\setminus\mathbb{R}$ with $\lambda = a + \mathrm{i}\,b$ are complex eigenvalues of A with the eigenvectors

(continued)

$\boldsymbol{v}, \overline{\boldsymbol{v}}$ then delete in the complex solution the summand $d\,\mathrm{e}^{\overline{\lambda} t}\,\overline{\boldsymbol{v}}$ and replace the summand $c\,\mathrm{e}^{\lambda t}\,\boldsymbol{v}$ by

$$d_1\,\mathrm{e}^{at}\,(\cos(b\,t)\,\mathrm{Re}(\boldsymbol{v}) - \sin(b\,t)\,\mathrm{Im}(\boldsymbol{v})) + d_2\,\mathrm{e}^{at}\,(\sin(b\,t)\,\mathrm{Re}(\boldsymbol{v}) + \cos(b\,t)\,\mathrm{Im}(\boldsymbol{v}))$$

with $d_1,\ d_2 \in \mathbb{R}$.

(4) Determine with the $\boldsymbol{x}_a(t)$ from (3) the numbers $c_1, \ldots, c_n$ from the LGS

$$\boldsymbol{x}_a(t_0) = c_1\,\mathrm{e}^{\lambda_1 t_0}\,\boldsymbol{v}_1 + \cdots + c_n\,\mathrm{e}^{\lambda_n t_0}\,\boldsymbol{v}_n = \boldsymbol{x}_0$$

and obtain the uniquely determined solution of the IVP:

$$\boldsymbol{x}(t) = c_1\,\mathrm{e}^{\lambda_1 t}\,\boldsymbol{v}_1 + \cdots + c_n\,\mathrm{e}^{\lambda_n t}\,\boldsymbol{v}_n\,.$$

Note that the exponential function does not appear at all in the solution.

Example 65.3

- Compare Example 65.2: We determine the general solution of the ODE-system and the IVP

$$\dot{\boldsymbol{x}} = A\,\boldsymbol{x} \ \text{ and } \ \dot{\boldsymbol{x}} = A\,\boldsymbol{x} \ \text{ with } \ \boldsymbol{x}(1) = \begin{pmatrix} 1 \\ 0 \\ 1 \end{pmatrix} \ \text{ and } \ A = \begin{pmatrix} 1 & 1 & 1 \\ 1 & 1 & 1 \\ 1 & 1 & 1 \end{pmatrix}.$$

(1) As eigenvalues with corresponding eigenvectors we obtain $\lambda_1 = 0$, $\lambda_2 = 0$ and $\lambda_3 = 3$ with

$$\boldsymbol{v}_1 = \begin{pmatrix} 1 \\ -1 \\ 0 \end{pmatrix},\ \boldsymbol{v}_2 = \begin{pmatrix} 1 \\ 0 \\ -1 \end{pmatrix},\ \boldsymbol{v}_3 = \begin{pmatrix} 1 \\ 1 \\ 1 \end{pmatrix},$$

(2) We obtain the general solution $\boldsymbol{x}_a(t)$ of the ODE system:

$$\boldsymbol{x}_a(t) = c_1 \begin{pmatrix} 1 \\ -1 \\ 0 \end{pmatrix} + c_2 \begin{pmatrix} 1 \\ 0 \\ -1 \end{pmatrix} + c_3\,\mathrm{e}^{3t} \begin{pmatrix} 1 \\ 1 \\ 1 \end{pmatrix},\ c_1,\ c_2,\ c_3 \in \mathbb{C}.$$

(3) like (2) (write $\mathbb{R}$ instead of $\mathbb{C}$).

(4) To solve the IVP, solve the following system of equations:

$$x_a(1) = c_1 \begin{pmatrix} 1 \\ -1 \\ 0 \end{pmatrix} + c_2 \begin{pmatrix} 1 \\ 0 \\ -1 \end{pmatrix} + c_3\,\mathrm{e}^3 \begin{pmatrix} 1 \\ 1 \\ 1 \end{pmatrix} = \begin{pmatrix} 1 \\ 0 \\ 1 \end{pmatrix}.$$

A solution is obviously $c_1 = \frac{2}{3}$, $c_2 = \frac{-1}{3}$ and $c_3 = \frac{2}{3\,\mathrm{e}^3}$; thus the uniquely determined solution of the IVP is

$$x(t) = \begin{pmatrix} 1/3 \\ -2/3 \\ 1/3 \end{pmatrix} + \mathrm{e}^{3(t-1)} \begin{pmatrix} 2/3 \\ 2/3 \\ 2/3 \end{pmatrix}.$$

We solve the IVP $\dot{x} = A\,x$, extensively

$$\dot{x} = \begin{pmatrix} \dot{x}_1 \\ \dot{x}_2 \end{pmatrix} = \begin{pmatrix} -x_2 \\ x_1 \end{pmatrix} \text{ mit } x(0) = \begin{pmatrix} 1 \\ 1 \end{pmatrix}.$$

- (1) The matrix $A = \begin{pmatrix} 0 & -1 \\ 1 & 0 \end{pmatrix}$ has the eigenvalues $\lambda_1 = \mathrm{i}$ and $\lambda_2 = -\mathrm{i}$ with the corresponding eigenvectors

$$v_1 = \begin{pmatrix} 1 \\ -\mathrm{i} \end{pmatrix} \text{ and } v_2 = \begin{pmatrix} 1 \\ \mathrm{i} \end{pmatrix}.$$

(2) We obtain the general solution $x_a(t)$ of the ODE system:

$$x_a(t) = c_1\,\mathrm{e}^{\mathrm{i}t} \begin{pmatrix} 1 \\ -\mathrm{i} \end{pmatrix} + c_2\,\mathrm{e}^{-\mathrm{i}t} \begin{pmatrix} 1 \\ \mathrm{i} \end{pmatrix}, \quad c_1, c_2, \in \mathbb{C}.$$

(3) Delete $c_2\,\mathrm{e}^{-\mathrm{i}t}\,v_2$ and decompose $c_1\,\mathrm{e}^{\mathrm{i}t}\,v_1$ into real and imaginary parts:

$$\mathrm{e}^{\mathrm{i}t} \begin{pmatrix} 1 \\ -\mathrm{i} \end{pmatrix} = (\cos(t) + \mathrm{i}\sin(t)) \left(\begin{pmatrix} 1 \\ 0 \end{pmatrix} + \mathrm{i} \begin{pmatrix} 0 \\ -1 \end{pmatrix} \right) = \begin{pmatrix} \cos(t) \\ \sin(t) \end{pmatrix} + \mathrm{i} \begin{pmatrix} \sin(t) \\ -\cos(t) \end{pmatrix}.$$

Thus, the general real solution of the ODE system is

$$x_a(t) = c_1 \begin{pmatrix} \cos(t) \\ \sin(t) \end{pmatrix} + c_2 \begin{pmatrix} \sin(t) \\ -\cos(t)\,. \end{pmatrix}, \quad c_1, c_2, \in \mathbb{R}.$$

(4) By solving the system of equations $x_a(0) = (1, 1)^\top$ we obtain the uniquely determined solution of the IVP

$$x(t) = \begin{pmatrix} \cos(t) - \sin(t) \\ \sin(t) + \cos(t) \end{pmatrix}.$$

■

65.4 Exercises

65.1 Give an example for $e^{A+B} \neq e^A e^B$.

65.2 Determine the solutions of the following IVPs:

(a) $\dot{x} = Ax, \quad x(0) = \begin{pmatrix} 0 \\ 2 \\ 2 \end{pmatrix}$ where $A = \begin{pmatrix} 5 & 0 & 4 \\ 0 & 4 & 0 \\ 4 & 0 & 5 \end{pmatrix}$,

(b) $\dot{x} = Ax, \quad x(0) = \begin{pmatrix} 1 \\ 1 \end{pmatrix}$ where $A = \begin{pmatrix} -1 & 1 \\ -1 & -1 \end{pmatrix}$.

65.3 We consider a population of wild boars (W) and snails (S) whose population is described by real functions $W, S : \mathbb{R}_{\geq 0} \to \mathbb{R}_{\geq 0}$. These are supposed to satisfy the following ODEs:

$$\dot{W}(t) = -W(t) + S(t) - 2 \quad \text{and} \quad \dot{S}(t) = S(t) - 2W(t) \qquad (*).$$

(a) Find a pair $(w_0, s_0) \in \mathbb{R}^2$ so that $W(t) = w_0$, $S(t) = s_0$ for $t \in \mathbb{R}$ describes a constant solution of $(*)$.

(b) Find a solution of $(*)$ with $W(0) = 3$, $S(0) = 6$, by taking the approach

$$W(t) = w(t) + w_0\,, \quad S(t) = s(t) + s_0 \quad (t \in \mathbb{R})$$

and solving the resulting ODE for $w, s : \mathbb{R} \to \mathbb{R}$.

(c) Sketch the solution $t \mapsto (W(t), S(t))$.

Linear Differential Equation Systems II

66

We continue to consider 1st order explicit linear differential equation systems,

$$\dot{\boldsymbol{x}}(t) = A(t)\,\boldsymbol{x}(t) + \boldsymbol{s}(t)\,,$$

where in this second chapter on this topic we still consider a constant, but not necessarily diagonalizable matrix $A \in \mathbb{R}^{n\times n}$ and again consider $s = 0$.

With the help of the Jordan normal form of A it will be possible to solve this case completely as well. Once again, the key to the goal lies in the exponential function for matrices.

66.1 The Exponential Function as a Solution of Linear ODE Systems

Let there be a real matrix $A \in \mathbb{R}^{n\times n}$ is given. According to the mnemonic box in Sect. 65.2: the solution of the ODE system is $\dot{\boldsymbol{x}} = A\,\boldsymbol{x}$ or the IVP $\dot{\boldsymbol{x}} = A\,\boldsymbol{x}$ with $\boldsymbol{x}(0) = \boldsymbol{x}_0$ in any case

$$\boldsymbol{x}(t) = \mathrm{e}^{tA}\,\boldsymbol{c} \;\text{ with a }\; \boldsymbol{c} \in \mathbb{R}^n \;\text{ or }\; \boldsymbol{x} = \boldsymbol{x}(t) = \mathrm{e}^{(t-t_0)A}\,\boldsymbol{x}_0\,.$$

The only question is how to determine these matrices. If A is diagonalizable, then the solution algorithm in Sect. 65.3 gives the solution. But what to do if A is not diagonalizable? The matrix A is diagonalizable if and only if

- the characteristic polynomial χ_A splits and
- for each eigenvalue λ of A we have $\mathrm{alg}(\lambda) = \mathrm{geo}(\lambda)$.

C. Karpfinger, *Calculus and Linear Algebra in Recipes*,
https://doi.org/10.1007/978-3-662-65458-3_66

If this is not fulfilled, the ODE system or the IVP can still be solved:

- If the characteristic polynomial χ_A does not split, we take the matrix A to be a complex matrix, over $\mathbb{C}$ splits χ_A.
- If $\mathrm{alg}(\lambda) \neq \mathrm{geo}(\lambda)$ for an eigenvalue λ of A, then A is not diagonalizable, but there exists a Jordan normal form of A.

We have:

The Exponential Function for a Non-Diagonalizable Matrix
For each matrix $A \in \mathbb{R}^{n\times n}$ there are (possibly non-real) matrices J and S from $\mathbb{C}^{n\times n}$ so that

$$J = S^{-1} A\, S = \begin{pmatrix} \lambda_1 & \varepsilon_1 & & \\ & \ddots & \ddots & \\ & & \ddots & \varepsilon_{n-1} \\ & & & \lambda_n \end{pmatrix} = \underbrace{\begin{pmatrix} \lambda_1 & & & \\ & \ddots & & \\ & & \ddots & \\ & & & \lambda_n \end{pmatrix}}_{=:D} + \underbrace{\begin{pmatrix} 0 & \varepsilon_1 & & \\ & \ddots & \ddots & \\ & & \ddots & \varepsilon_{n-1} \\ & & & 0 \end{pmatrix}}_{=:N}$$

with $\varepsilon_i \in \{0,\ 1\}$ and we have

$$\mathrm{e}^A = S \,\mathrm{diag}\left(\mathrm{e}^{\lambda_1}, \ldots, \mathrm{e}^{\lambda_n}\right) \left(E_n + N + \frac{1}{2} N^2 + \cdots \right) S^{-1}.$$

Because of $D\,N = N\,D$ we have according to the calculation rule (a) in Sect. 65.1:

$$\begin{aligned} \mathrm{e}^A &= \mathrm{e}^{SJS^{-1}} = S\,\mathrm{e}^J\,S^{-1} = S\,\mathrm{e}^{D+N}\,S^{-1} = S\,\mathrm{e}^D\,\mathrm{e}^N\,S^{-1} \\ &= S\ \mathrm{diag}\left(\mathrm{e}^{\lambda_1}, \ldots, \mathrm{e}^{\lambda_n}\right) \left(E_n + N + \frac{1}{2} N^2 + \cdots \right) S^{-1}. \end{aligned}$$

Since for N, because of its special shape, there is a natural number k with $N^k = 0$, $k < n$, this sum breaks $E_n + N + \frac{1}{2}N^2 + \cdots$ rapidly. We show this exemplarily for $n = 4$ in the *worst* case $\varepsilon_i = 1$ for all i:

$$\begin{pmatrix} 0&1&0&0 \\ 0&0&1&0 \\ 0&0&0&1 \\ 0&0&0&0 \end{pmatrix} \overset{N\cdot}{\rightarrow} \begin{pmatrix} 0&0&1&0 \\ 0&0&0&1 \\ 0&0&0&0 \\ 0&0&0&0 \end{pmatrix} \overset{N\cdot}{\rightarrow} \begin{pmatrix} 0&0&0&1 \\ 0&0&0&0 \\ 0&0&0&0 \\ 0&0&0&0 \end{pmatrix} \overset{N\cdot}{\rightarrow} \begin{pmatrix} 0&0&0&0 \\ 0&0&0&0 \\ 0&0&0&0 \\ 0&0&0&0 \end{pmatrix},$$

for any multiplication the diagonal with the ones *slides* up one row. So this gives us a possibility to compute e^A for any constant matrix A.

Recipe: Determine e^A for a Non-Diagonalizable A

Given a matrix $A \in \mathbb{R}^{n \times n}$. To compute e^A proceed as follows:

(1) Determine a Jordan normal form with associated Jordan basis S,

$$J = S^{-1} A\, S\,.$$

(2) Write $J = D + N$ with a diagonal matrix D and *nilpotent* matrix N.

(3) Get $e^A = S \operatorname{diag}\left(e^{\lambda_1}, \ldots, e^{\lambda_n}\right) \left(E_n + N + \frac{1}{2}N^2 + \cdots\right) S^{-1}$.

Example 66.1

- We consider the matrix $A = \begin{pmatrix} -1 & 1 & 0 \\ 0 & -1 & 1 \\ 0 & 0 & -1 \end{pmatrix}$.

  (1) The matrix A is already in Jordan normal form, $J = A$, we have $S = E_3$.

  (2) We decompose J into a sum of a diagonal matrix D with a nilpotent matrix N:

$$J = D + N \text{ with } D = \begin{pmatrix} -1 & 0 & 0 \\ 0 & -1 & 0 \\ 0 & 0 & -1 \end{pmatrix} \text{ and } N = \begin{pmatrix} 0 & 1 & 0 \\ 0 & 0 & 1 \\ 0 & 0 & 0 \end{pmatrix}.$$

  (3) Because of $N^3 = 0$ we get

$$e^A = E_3 \operatorname{diag}(e^{-1}, e^{-1}, e^{-1}) \left(E_3 + N + \tfrac{1}{2}N^2\right) E_3^{-1}$$

$$= \begin{pmatrix} e^{-1} & 0 & 0 \\ 0 & e^{-1} & 0 \\ 0 & 0 & e^{-1} \end{pmatrix} \begin{pmatrix} 1 & 1 & 1/2 \\ 0 & 1 & 1 \\ 0 & 0 & 1 \end{pmatrix} = \begin{pmatrix} e^{-1} & e^{-1} & \frac{1}{2}e^{-1} \\ 0 & e^{-1} & e^{-1} \\ 0 & 0 & e^{-1} \end{pmatrix}.$$

We consider the matrix $A = \begin{pmatrix} 3 & 1 & 0 & 0 \\ -1 & 1 & 0 & 0 \\ 1 & 1 & 3 & 1 \\ -1 & -1 & -1 & 1 \end{pmatrix}$.

- (1) According to Example 44.1 we have for the matrix

$$J = \begin{pmatrix} 2 & 1 & 0 & 0 \\ 0 & 2 & 0 & 0 \\ 0 & 0 & 2 & 1 \\ 0 & 0 & 0 & 2 \end{pmatrix} = S^{-1} A\, S \text{ where } S = \begin{pmatrix} 1 & 0 & 0 & 0 \\ -1 & 1 & 0 & 0 \\ 1 & 0 & 1 & 0 \\ -1 & 0 & -1 & 1 \end{pmatrix}.$$

  It is J a Jordan normal form and S a Jordan basis to A.
  (2) We decompose J into a sum of a diagonal matrix D with a nilpotent matrix N:

$$J = D + N \text{ with } D = \begin{pmatrix} 2 & 0 & 0 & 0 \\ 0 & 2 & 0 & 0 \\ 0 & 0 & 2 & 0 \\ 0 & 0 & 0 & 2 \end{pmatrix} \text{ and } N = \begin{pmatrix} 0 & 1 & 0 & 0 \\ 0 & 0 & 0 & 0 \\ 0 & 0 & 0 & 1 \\ 0 & 0 & 0 & 0 \end{pmatrix}.$$

- Because of $N^2 = 0$ we get

$$\mathrm{e}^A = S \operatorname{diag}(\mathrm{e}^2, \mathrm{e}^2, \mathrm{e}^2, \mathrm{e}^2)\, (E_4 + N)\, S^{-1}$$

$$= \begin{pmatrix} 1 & 0 & 0 & 0 \\ -1 & 1 & 0 & 0 \\ 1 & 0 & 1 & 0 \\ -1 & 0 & -1 & 1 \end{pmatrix} \begin{pmatrix} \mathrm{e}^2 & 0 & 0 & 0 \\ 0 & \mathrm{e}^2 & 0 & 0 \\ 0 & 0 & \mathrm{e}^2 & 0 \\ 0 & 0 & 0 & \mathrm{e}^2 \end{pmatrix} \begin{pmatrix} 1 & 1 & 0 & 0 \\ 0 & 1 & 0 & 0 \\ 0 & 0 & 1 & 1 \\ 0 & 0 & 0 & 1 \end{pmatrix} \begin{pmatrix} 1 & 0 & 0 & 0 \\ 1 & 1 & 0 & 0 \\ -1 & 0 & 1 & 0 \\ 0 & 0 & 1 & 1 \end{pmatrix}$$

$$= \begin{pmatrix} 2\,\mathrm{e}^2 & \mathrm{e}^2 & 0 & 0 \\ -\mathrm{e}^2 & 0 & 0 & 0 \\ \mathrm{e}^2 & \mathrm{e}^2 & 2\,\mathrm{e}^2 & \mathrm{e}^2 \\ -\mathrm{e}^2 & -\mathrm{e}^2 & -\mathrm{e}^2 & 0 \end{pmatrix}.$$

■

66.2 The Solution for a Non-Diagonalizable A

Now that we know how to compute e^J for a Jordan matrix J, it is easy to determine the solution of a ODE system or an IVP with a nondiagonalizable matrix A:

Recipe: Solving an ODE System or an IVP with Non-Diagonalizable A
Given is

$$\text{the ODE system } \dot{x} = A\,x \quad \text{or the IVP } \dot{x} = A\,x \text{ with } x(t_0) = x_0$$

with a nondiagonalizable matrix $A \in \mathbb{R}^{n\times n}$ and $x_0 \in \mathbb{R}^n$.

(1) Determine a Jordan basis S of the $\mathbb{C}^n$ and the corresponding Jordan normal form J of A, i.e. $S^{-1} A\, S = J$.
(2) Calculate $e^{tA} = S\, e^{tJ}\, S^{-1}$ or $e^{(t-t_0)A} = S\, e^{(t-t_0)J}\, S^{-1}$.
(3) Obtain the general solution of the ODE system as $x(t) = e^{tA}\, c$, $c \in \mathbb{R}^n$ or the unique solution of the IVP as $x(t) = e^{(t-t_0)A}\, x_0$.

In fact, the recipe is also applicable to a diagonalizable matrix: The Jordan normal form is then a diagonal form.

Example 66.2 We solve the IVP $\dot{x} = A\,x$ with $A = \begin{pmatrix} 3 & 1 \\ -1 & 1 \end{pmatrix}$ and $x(0) = \begin{pmatrix} 1 \\ 1 \end{pmatrix}$.

(1) Obviously, the columns v_1 and v_2 of $S = \begin{pmatrix} 1 & 1 \\ -1 & 0 \end{pmatrix}$ form a Jordan basis to A, and we have

$$J = \begin{pmatrix} 2 & 1 \\ 0 & 2 \end{pmatrix} = S^{-1} A\, S\,.$$

(2) We calculate $e^{tA} = S\, e^{tJ}\, S^{-1}$ and set $J = D + N$ with $D = \begin{pmatrix} 2 & 0 \\ 0 & 2 \end{pmatrix}$ and $N = \begin{pmatrix} 0 & 1 \\ 0 & 0 \end{pmatrix}$; thus we obtain for the time being

$$e^{tJ} = e^{t(D+N)} = e^{tD}\, e^{tN} = \begin{pmatrix} e^{2t} & 0 \\ 0 & e^{2t} \end{pmatrix} \cdot \begin{pmatrix} 1 & t \\ 0 & 1 \end{pmatrix} = \begin{pmatrix} e^{2t} & t\, e^{2t} \\ 0 & e^{2t} \end{pmatrix}$$

and finally

$$e^{tA} = S\,e^{tJ}\,S^{-1} = \begin{pmatrix} 1 & 1 \\ -1 & 0 \end{pmatrix} \begin{pmatrix} e^{2t} & t\,e^{2t} \\ 0 & e^{2t} \end{pmatrix} \begin{pmatrix} 0 & -1 \\ 1 & 1 \end{pmatrix} = \begin{pmatrix} (t+1)\,e^{2t} & t\,e^{2t} \\ -t\,e^{2t} & (1-t)\,e^{2t} \end{pmatrix}.$$

(3) We obtain as unique solution of the IVP

$$\boldsymbol{x}(t) = e^{tA}\,\boldsymbol{x}_0 = \begin{pmatrix} e^{2t}\,(2t+1) \\ (1-2t)\,e^{2t} \end{pmatrix}.$$

■

There is one problem we have not mentioned so far: It can now also be the case that an eigenvalue is complex: If one wants to have a real solution, the real and imaginary parts of the complex contribution still have to be determined; the contribution of the conjugate complex eigenvalue has to be discarded, real and imaginary parts of the one complex contribution form two real linearly independent contributions—the balance is then correct again.

We conclude by stating how the matrix is e^{tJ} looks like in general for a Jordan box $J \in \mathbb{R}^{k\times k}$ for the eigenvalue λ:

$$\begin{aligned} e^{tJ} &= e^{t(D+N)} = e^{tD}\,e^{tN} = e^{tD}\left(E_k + t\,N + \frac{1}{2}t^2N^2 + \frac{1}{3!}t^3N^3 + \frac{1}{4!}t^4N^4 + \cdots\right) \\ &= \begin{pmatrix} e^{\lambda t} & & & \\ & \ddots & & \\ & & \ddots & \\ & & & e^{\lambda t} \end{pmatrix} \begin{pmatrix} 1 & t & \frac{1}{2}t^2 & \ldots & \frac{1}{(k-1)!}t^{k-1} \\ & \ddots & \ddots & \ddots & \vdots \\ & & \ddots & \ddots & \frac{1}{2}t^2 \\ & & & \ddots & t \\ & & & & 1 \end{pmatrix}. \end{aligned}$$

66.3 Exercises

66.1 Determine the solution of the following IVP:

$$\dot{\boldsymbol{x}} = A\boldsymbol{x}, \quad \boldsymbol{x}(0) = \begin{pmatrix} 2 \\ 1 \\ 1 \end{pmatrix} \quad \text{where } A = \begin{pmatrix} -3 & 0 & 1 \\ -2 & -2 & 2 \\ -1 & 0 & -1 \end{pmatrix}.$$

66.2 Find functions $x, y : \mathbb{R} \to \mathbb{R}$ that correspond to the ODE system

$$\ddot{x}(t) = y(t), \quad \dot{y}(t) = -\dot{x}(t) + 2y(t) \quad (t \in \mathbb{R})$$

and satisfy $x(0) = \dot{x}(0) = 0$, $y(0) = 1$.
Note: Set $\boldsymbol{u}(t) = (x(t), \dot{x}(t), y(t))^\top$ and find a matrix $A \in \mathbb{R}^{3\times 3}$ that solves the equation $\dot{\boldsymbol{u}} = A\boldsymbol{u}$. Solve this system using the Jordan normal form.

66.3 Find a solution $x : \mathbb{R} \to \mathbb{R}$ of the ODE

$$\ddot{x}(t) = 2\dot{x}(t) - x(t)\,, \quad x(0) = 1\,,\ \dot{x}(0) = 2 \quad (t \in \mathbb{R})\,.$$

66.3 Solve the following IVP:

$$\begin{aligned} \dot{x}_1(t) &= -5x_1(t) + x_2(t) \\ \dot{x}_2(t) &= -5x_2(t) + x_3(t) \quad \text{with} \quad x(0) = (x_1(0), x_2(0), x_3(0))^\top = (2, 0, 1)^\top\,. \\ \dot{x}_3(t) &= -5x_3(t) \end{aligned}$$

Linear Differential Equation Systems III 67

We further consider 1st order explicit linear differential equation systems,

$$\dot{\boldsymbol{x}}(t) = A(t)\,\boldsymbol{x}(t) + \boldsymbol{s}(t)\,,$$

where in this third chapter on the subject we consider the general case.

The solution set of such a system is composed of the solution set of the homogeneous system and a particulate solution. In general, it is not possible to determine the solution set of the homogeneous system. But if you do have this set (e.g. by trial and error or guessing), you can obtain a particular solution by varying the parameters, and thus the complete solution.

We also discuss some points about *stability*; this involves studying the behavior of solutions to a ODE system near *equilibrium points*.

67.1 Solving ODE Systems

Given is a linear ODE system with an n-row square matrix $A(t)$,

$$\dot{\boldsymbol{x}} = A(t)\,\boldsymbol{x} + \boldsymbol{s}(t) \text{ with } A(t) = (a_{ij}(t)) \text{ and } \boldsymbol{s}(t) = (s_i(t))\,.$$

We are looking for the set L of all solutions $\boldsymbol{x} = (x_1(t), \ldots, x_n(t))^\top$ of this system. If L_h is the solution set of the corresponding homogeneous system and $\boldsymbol{x}_p$ is a partial solution of the inhomogeneous system, then the set L is given by

$$L = \boldsymbol{x}_p + L_h\,.$$

C. Karpfinger, *Calculus and Linear Algebra in Recipes*,
https://doi.org/10.1007/978-3-662-65458-3_67

If A is a constant matrix, we can use the methods from the last two chapters to determine a basis of L_h. Namely, we have

$$L_h = \{\mathrm{e}^{tA}\, \boldsymbol{c} \mid \boldsymbol{c} \in \mathbb{R}^n\}.$$

However, if A is not a constant matrix (and this is the case of interest in the present chapter), then there is in fact no general method of calculating the solution set L_h of the homogeneous system. We will give a set of solutions in the examples and exercises if necessary. One can then use the Wronski determinant to decide whether such a system of given solutions is complete: L_h is a vector space of dimension n (where n is the number of rows of $A(t)$); the Wronski determinant is a tool to decide whether a system of n solutions is linearly independent. Once such a system of n linearly independent solutions has been found (we speak then of a *fundamental system*), it is possible with the variation of the constants to determine a sought particular solution.

We compile the main results that help to find the solution set of the system

$$\dot{\boldsymbol{x}} = A(t)\,\boldsymbol{x} + \boldsymbol{s}(t).$$

Compilation of Results

Given is the ODE system $\dot{\boldsymbol{x}} = A(t)\,\boldsymbol{x} + \boldsymbol{s}(t)$ where the coefficient functions $a_{ij}(t)$ and $s_i(t)$ of $A(t)$ and $\boldsymbol{s}(t)$ are continuous on a common interval I. Then we have:

- The solution set L of the ODE system is of the form

$$L = \boldsymbol{x}_p + L_h\,, \quad \text{one writes } \boldsymbol{x}_a(t) = \boldsymbol{x}_p(t) + \boldsymbol{x}_h(t)\,.$$

  Here are
  - L_h is the solution set of the homogeneous system $\dot{\boldsymbol{x}} = A(t)\,\boldsymbol{x}$,
  - $\boldsymbol{x}_h(t)$ the general solution of the homogeneous system $\dot{\boldsymbol{x}} = A(t)\,\boldsymbol{x}$,
  - $\boldsymbol{x}_p(t)$ a particulate solution of the system $\dot{\boldsymbol{x}} = A(t)\,\boldsymbol{x} + \boldsymbol{s}(t)$,
  - $\boldsymbol{x}_a(t)$ the general solution of the system $\dot{\boldsymbol{x}} = A(t)\,\boldsymbol{x} + \boldsymbol{s}(t)$.
- The homogeneous ODE system $\dot{\boldsymbol{x}} = A(t)\,\boldsymbol{x}$ has n linearly independent solutions $\boldsymbol{x}_1, \ldots, \boldsymbol{x}_n$, the general solution of the homogeneous system has the form $\boldsymbol{x}_h(t) = c_1\boldsymbol{x}_1(t) + \cdots + c_n\boldsymbol{x}_n(t)$. One writes also briefly $\boldsymbol{x}_h(t) = X(t)\,\boldsymbol{c}$ with $X(t) = (\boldsymbol{x}_1, \ldots, \boldsymbol{x}_n)$ and $\boldsymbol{c} \in \mathbb{R}^n$.
- For each n linearly independent solutions $\boldsymbol{x}_1, \ldots, \boldsymbol{x}_n$ of the homogeneous system is called a **fundamental system** of solutions.

(continued)

- If $\boldsymbol{x}_1, \ldots, \boldsymbol{x}_n$ solutions of the homogeneous system, then they are a fundamental system of solutions exactly if the so-called **Wronski determinant**

$$W(t) = \det(\boldsymbol{x}_1(t), \ldots, \boldsymbol{x}_n(t)) \neq 0$$

for one (and therefore for all) $t \in I$ is not equal to zero.
- A partial solution $\boldsymbol{x}_p$ of the inhomogeneous system is found by **variation of parameters**: If $\boldsymbol{x}_h(t) = X(t)\,\boldsymbol{c}$ with $X(t) = (\boldsymbol{x}_1, \ldots, \boldsymbol{x}_n)$ and $\boldsymbol{c} \in \mathbb{R}^n$ is the general solution of the homogeneous ODE system, then set $\boldsymbol{x}_p(t) = X(t)\,\boldsymbol{c}(t)$ and obtain $\boldsymbol{c}(t)$ by

$$\boldsymbol{c}(t) = \int X^{-1}(t)\,\boldsymbol{s}(t)\mathrm{d}t\,,$$

where this integral over the vector $X^{-1}(t)\,\boldsymbol{s}(t)$ is understood componentwise.

Thus, the following procedure for determining the solution of such a ODE system is obvious:

Recipe: Solving an ODE System
To determine the general solution $\boldsymbol{x}_a(t)$ of the ODE system $\dot{\boldsymbol{x}} = A(t)\boldsymbol{x} + \boldsymbol{s}(t)$ with continuous coefficient functions proceed as follows:

(1) Determine n distinct solutions $\boldsymbol{x}_1, \ldots, \boldsymbol{x}_n$ of the homogeneous system $\dot{\boldsymbol{x}} = A(t)\boldsymbol{x}$:

- If A is constant, then determine the matrix by diagonalization or via the Jordan normal form $\mathrm{e}^{tA} = X(t) = (\boldsymbol{x}_1, \ldots, \boldsymbol{x}_n)$.
- If A is not constant, you may find by trial and error or guessing $\boldsymbol{x}_1, \ldots, \boldsymbol{x}_n$.
- In the typical exercises $\boldsymbol{x}_1, \ldots, \boldsymbol{x}_n$ are usually given.

(2) Test the solutions $\boldsymbol{x}_1, \ldots, \boldsymbol{x}_n$ with the Wronski determinant for linear independence:

- If $\det(\boldsymbol{x}_1, \ldots, \boldsymbol{x}_n) \neq 0$ then $\boldsymbol{x}_1, \ldots, \boldsymbol{x}_n$ are linearly independent → (3).
- If $\det(\boldsymbol{x}_1, \ldots, \boldsymbol{x}_n) = 0$ then $\boldsymbol{x}_1, \ldots, \boldsymbol{x}_n$ are linearly dependent → (1).

(continued)

(3) Determine a particulate solution by varying the parameters

$$\boldsymbol{x}_p = X(t)\boldsymbol{c}(t)\,, \text{ where } X(t) = (\boldsymbol{x}_1, \ldots, \boldsymbol{x}_n) \text{ and } \boldsymbol{c}(t) = \int X^{-1}(t)\,\boldsymbol{s}(t)\,\mathrm{d}t\,.$$

(4) Obtain the general solution $\boldsymbol{x}_a(t) = \boldsymbol{x}_p(t) + c_1\boldsymbol{x}_1(t) + \cdots + c_n\boldsymbol{x}_n(t)$ with $c_1, \ldots, c_n \in \mathbb{R}$.

(5) A possible initial condition $\boldsymbol{x}_a(t_0) = \boldsymbol{x}_0$ yields an LGS by which the coefficients $c_1, \ldots, c_n$ are determined.

Example 67.1 We determine the general solution of the following ODE system:

$$\dot{\boldsymbol{x}} = \begin{pmatrix}\dot{x}_1\\ \dot{x}_2\end{pmatrix} = \begin{pmatrix}1 & 1\\ 0 & \frac{2}{t}\end{pmatrix}\begin{pmatrix}x_1\\ x_2\end{pmatrix} + \begin{pmatrix}-2\,\mathrm{e}^t\\ t^2\,\mathrm{e}^t\end{pmatrix} \quad \text{with } t > 0\,.$$

(1) We consider the two solutions

$$\boldsymbol{x}_1 = \begin{pmatrix}\mathrm{e}^t\\ 0\end{pmatrix} \quad \text{and} \quad \boldsymbol{x}_2 = \begin{pmatrix}-t^2 - 2t - 2\\ t^2\end{pmatrix}$$

of the homogeneous system found by trial and error.

(2) We test the two solutions $\boldsymbol{x}_1$ and $\boldsymbol{x}_2$ from (1) with the Wronski determinant on linear independence; we have

$$W(t) = \det\begin{pmatrix}\mathrm{e}^t & -t^2 - 2t - 2\\ 0 & t^2\end{pmatrix} = \mathrm{e}^t\,t^2 \neq 0$$

for all t > 0. Thus, we have a fundamental system; we set

$$X(t) = \begin{pmatrix}\mathrm{e}^t & -t^2 - 2t - 2\\ 0 & t^2\end{pmatrix}.$$

(3) We find a partial solution by varying the parameters, i.e., we make the approach $\boldsymbol{x}_p = X(t)\,\boldsymbol{c}(t)$ and determine $\boldsymbol{c}(t) = \int X(t)^{-1}\boldsymbol{s}(t)\mathrm{d}t$, for which we first compute:

$$X^{-1}(t)\,\boldsymbol{s}(t) = \begin{pmatrix}\mathrm{e}^{-t} & \mathrm{e}^{-t}(1 + \frac{2}{t} + \frac{2}{t^2})\\ 0 & \frac{1}{t^2}\end{pmatrix}\begin{pmatrix}-2\,\mathrm{e}^t\\ t^2\,\mathrm{e}^t\end{pmatrix} = \begin{pmatrix}t^2 + 2t\\ \mathrm{e}^t\end{pmatrix}.$$

Thus we obtain for $c(t)$:

$$c(t) = \int X^{-1}(t)s(t)\mathrm{d}t = \int \begin{pmatrix} t^2 + 2t \\ \mathrm{e}^t \end{pmatrix} \mathrm{d}t = \begin{pmatrix} \frac{1}{3}t^3 + t^2 \\ \mathrm{e}^t \end{pmatrix}.$$

A particulate solution is thus:

$$x_p(t) = X(t)c(t) = \begin{pmatrix} \mathrm{e}^t & -t^2 - 2t - 2 \\ 0 & t^2 \end{pmatrix} \begin{pmatrix} \frac{1}{3}t^3 + t^2 \\ \mathrm{e}^t \end{pmatrix} = \begin{pmatrix} \frac{1}{3}t^3\,\mathrm{e}^t - 2t\,\mathrm{e}^t - 2\,\mathrm{e}^t \\ t^2\,\mathrm{e}^t \end{pmatrix}.$$

(4) Finally, we can give the general solution of the inhomogeneous system:

$$x(t) = x_p + c_1 x_1 + c_2 x_2 = \begin{pmatrix} \frac{1}{3}t^3\,\mathrm{e}^t - 2t\,\mathrm{e}^t - 2\,\mathrm{e}^t \\ t^2\,\mathrm{e}^t \end{pmatrix} + c_1 \begin{pmatrix} \mathrm{e}^t \\ 0 \end{pmatrix} + c_2 \begin{pmatrix} -t^2 - 2t - 2 \\ t^2 \end{pmatrix}$$

with $c_1, c_2 \in \mathbb{R}$.

■

67.2 Stability

In the following we only consider *autonomous* ODE systems. Here we call a ODE system $\dot{x} = f(t, x(t))$ **autonomous**, if the right hand side only depends on x, i.e.

$$\dot{x} = f(x).$$

Such an autonomous system has in general *stationary solutions*, which are constant solutions $x(t) = a$. These describe a state which the system does not leave, provided there are no perturbations.

Example 67.2 We consider a pendulum consisting of a massless rigid rod and a mass point. The motion of the pendulum can be described by a 1st order ODE system with a given initial displacement and initial velocity. Obviously, there are two constant solutions: The mass point of the pendulum hangs vertically downward, or it is vertically upward. While the first constant solution is *stable* (at small deflections, the pendulum returns to its equilibrium position), the second solution is *unstable* (after a shock, the system will not return to this vertically upward rest position). ■

We formulate precisely these vague notions implied so far and consider for the time being only linear ODE systems. Such a linear ODE system $\dot{\boldsymbol{x}} = A(t)\,\boldsymbol{x} + \boldsymbol{s}(t)$ is autonomous exactly when $A(t) = A \in \mathbb{R}^{n\times n}$ and $\boldsymbol{s}(t) = \boldsymbol{s} \in \mathbb{R}^n$ are constant.

Equilibrium Points and Their Stability

We consider an autonomous linear ODE system $\dot{\boldsymbol{x}} = A\,\boldsymbol{x} + \boldsymbol{s}$. Each point $\boldsymbol{a} \in \mathbb{R}^n$ with

$$A\,\boldsymbol{a} + \boldsymbol{s} = \boldsymbol{0}$$

is called **equilibrium point** or **stationary** or **critical point** of the ODE system. For each equilibrium point $\boldsymbol{x}(t) = \boldsymbol{a}$ is a **stationary solution**, we have $\dot{\boldsymbol{x}} = \boldsymbol{0}$.

An equilibrium point $\boldsymbol{a}$ is called

- **stable**, if for every $\varepsilon > 0$ there exists a $\delta > 0$ such that for every solution $\boldsymbol{x}$ we have

$$|\boldsymbol{x}(t_0) - \boldsymbol{a}| < \delta \Rightarrow |\boldsymbol{x}(t) - \boldsymbol{a}| < \varepsilon \text{ for all } t > t_0\,,$$

- **attractive**, if there exists a $\delta > 0$ such that for each solution $\boldsymbol{x}$ we have

$$|\boldsymbol{x}(t_0) - \boldsymbol{a}| < \delta \Rightarrow \lim_{t\to\infty} \boldsymbol{x}(t) = \boldsymbol{a}\,,$$

- **asymptotically stable**, if it is stable and attractive,
- **unstable** if it is not stable.

The pictures in Fig. 67.1 represent what the symbolism means.

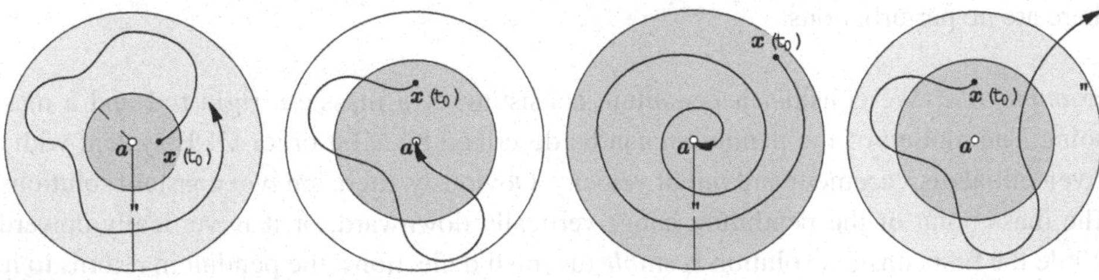

Fig. 67.1 A stable, attractive, asymptotically stable and unstable equilibrium point

The definitions of the different qualities of an equilibrium point are richly complicated. Of course, the question arises whether it is not possible in a simple way to tell an equilibrium point of the system $\dot{\boldsymbol{x}} = A\boldsymbol{x} + \boldsymbol{s}$ whether it is stable, asymptotically stable or unstable. Such a (at least theoretical) possibility exists. One can use the eigenvalues $\lambda_1, \ldots, \lambda_n \in \mathbb{C}$ of the matrix A about the stability behavior of the point $\boldsymbol{a}$:

The Stability Theorem for Linear Systems

If $\boldsymbol{a}$ is an equilibrium point of the linear autonomous ODE system $\dot{\boldsymbol{x}} = A\,\boldsymbol{x} + \boldsymbol{s}$ with $A \in \mathbb{R}^{n\times n}$ and $\boldsymbol{s} \in \mathbb{R}^n$ so with the eigenvalues $\lambda_1, \ldots, \lambda_n \in \mathbb{C}$ of A we have:

- $\mathrm{Re}(\lambda_i) < 0$ for all i $\Leftrightarrow$ $\boldsymbol{a}$ is asymptotically stable.
- $\mathrm{Re}(\lambda_i) > 0$ for one i $\Rightarrow$ $\boldsymbol{a}$ is unstable.
- $\mathrm{Re}(\lambda_i) \le 0$ for all i and if $\mathrm{Re}(\lambda_i) = 0$, then $\mathrm{alg}(\lambda_i) = \mathrm{geo}(\lambda_i)$ $\Leftrightarrow$ $\boldsymbol{a}$ is stable.

Example 67.3 The autonomous ODE system

$$\dot{\boldsymbol{x}} = \begin{pmatrix} -1 & 1 \\ \lambda & -1 \end{pmatrix} \boldsymbol{x}$$

has the zero point $\boldsymbol{a} = \boldsymbol{0}$ as equilibrium point.

- In the case $\lambda = 0$, $\boldsymbol{0}$ is asymptotically stable (the eigenvalues are -1 and -1).
- In the case $\lambda = 2$, $\boldsymbol{0}$ is unstable (the eigenvalues are 0.4142 and -2.4142).
- In the case $\lambda = 1$, $\boldsymbol{0}$ is stable (the eigenvalues are 0 and -2, where $\mathrm{alg}(0) = \mathrm{geo}(0)$).

■

We provide an overview of the behavior of solutions in the neighborhood of an equilibrium point $\boldsymbol{a}$, restricting ourselves for simplicity to the two-dimensional case, i.e. $n = 2$. Further we consider also only homogeneous systems, we set thus $\boldsymbol{s} = 0$ and thus always have $\boldsymbol{a} = \boldsymbol{0} = (0, 0)^\top$ as the equilibrium point. Finally we set $t_0 = 0$.

Since it is only the eigenvalues and their multiplicities that matter, we can get a good overview if we break down which possibilities we have as solutions of the system $\dot{\boldsymbol{x}} = A\boldsymbol{x}$ as a function of the eigenvalues and their multiplicities for a 2×2-matrix A. Using the methods from Chaps. 65 and 66 we obtain:

The General Solution of the System $\dot{x} = Ax$ in the Case $A \in \mathbb{R}^{2\times 2}$

The general solution x of the system $\dot{x} = A\,x$ with $A \in \mathbb{R}^{2\times 2}$ is:

- if A has two distinct real eigenvalues λ_1, λ_2 with corresponding eigenvectors v_1, v_2:

$$x(t) = c_1\,e^{\lambda_1 t}\,v_1 + c_2\,e^{\lambda_2 t}\,v_2\,.$$

- if A has a double eigenvalue of geometric multiplicity 2 with linearly independent eigenvectors v_1, v_2:

$$x(t) = c_1\,e^{\lambda t}\,v_1 + c_2\,e^{\lambda t}\,v_2$$

- if A has a double eigenvalue of geometric multiplicity 1 with eigenvector v_1 and principal vector v_2 (v_1, v_2 form Jordan basis):

$$x(t) = c_1\,e^{\lambda t}\,v_1 + c_2\,e^{\lambda t}\,(t\,v_1 + v_2)\,.$$

- if A has a non-real eigenvalue $\lambda = a + \mathrm{i}\,b \in \mathbb{C}$ with associated eigenvector v:

$$\begin{aligned} x(t) = {} & c_1\,e^{at}\,(\cos(b\,t)\,\mathrm{Re}(v) - \sin(b\,t)\,\mathrm{Im}(v)) \\ & + c_2\,e^{at}\,(\sin(b\,t)\,\mathrm{Re}(v) + \cos(b\,t)\,\mathrm{Im}(v)). \end{aligned}$$

Depending on the sign of the eigenvalues or the real part of a complex eigenvalue, there are 14 different solution curves of solutions near the equilibrium point $a = \mathbf{0}$. In the following we give a clear representation of these solution curves.

The respective image shows a **phase portrait**, i.e. a set of solution curves, around the equilibrium point.

- **Two different real eigenvalues:** $\lambda_1, \lambda_2 \in \mathbb{R}, \lambda_1 \neq \lambda_2$:
  - In the cases $\lambda_1 < \lambda_2 < 0$ (asymptotically stable) and $0 < \lambda_1 < \lambda_2$ (unstable) and $\lambda_1 < 0 < \lambda_2$ (unstable) we obtain as phase portraits for the solutions $x(t) = c_1\,e^{\lambda_1 t}\,v_1 + c_2\,e^{\lambda_2 t}\,v_2$ the portraits in Fig. 67.2.
  - In the cases $\lambda_1 < 0 = \lambda_2$ (stable) and $\lambda_1 = 0 < \lambda_2$ (unstable) we obtain the phase portraits for the solutions $x(t) = c_1\,e^{\lambda_1 t}\,v_1 + c_2\,e^{\lambda_2 t}\,v_2$ in Fig. 67.3.

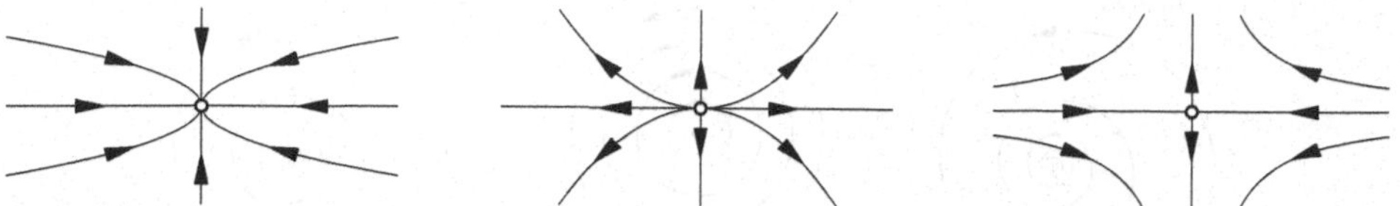

Fig. 67.2 The three portraits at two different eigenvalues $\neq 0$

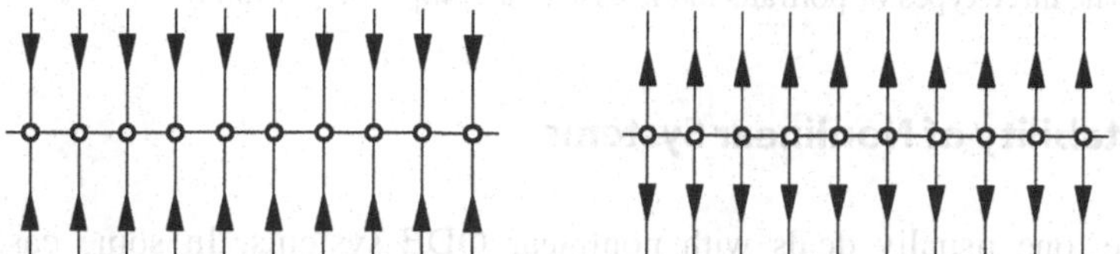

Fig. 67.3 The two portraits at two different eigenvalues, where one =0

Fig. 67.4 The three portraits at a double (real) eigenvalue

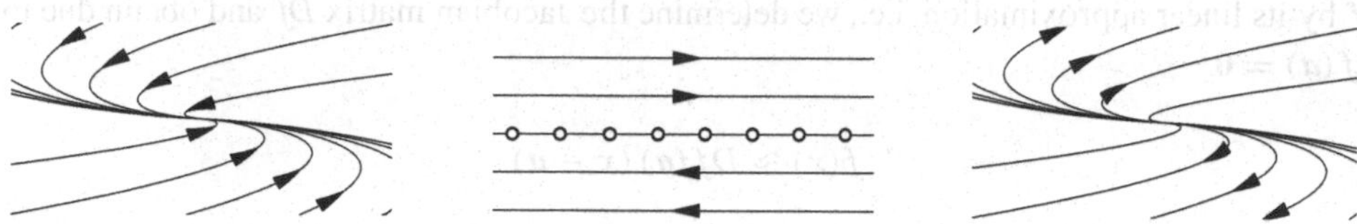

Fig. 67.5 The three portraits at double eigenvalue of geometric multiplicity 1

- **A double (real) eigenvalue:** $\lambda = \lambda_1 = \lambda_2 \in \mathbb{R}$:
  - If λ has geometric multiplicity 2: In the cases $\lambda < 0$ (asymptotically stable) and $\lambda = 0$ (stable) and $0 < \lambda$ (unstable), we obtain as phase portraits for the solutions $\boldsymbol{x}(t) = c_1\, \mathrm{e}^{\lambda_1 t}\, \boldsymbol{v}_1 + c_2\, \mathrm{e}^{\lambda_2 t}\, \boldsymbol{v}_2$ the potraits in Fig. 67.4.
  - In case λ has geometric multiplicity 1: In the cases $\lambda < 0$ (asymptotically stable) and $\lambda = 0$ (stable) and $0 < \lambda$ (unstable), we obtain the phase portraits for the solutions $\boldsymbol{x}(t) = c_1\, \mathrm{e}^{\lambda t}\, \boldsymbol{v}_1 + c_2\, \mathrm{e}^{\lambda t}(t\, \boldsymbol{v}_1 + \boldsymbol{v}_2)$ in Fig. 67.5.
- **Two nonreal eigenvalues:** $\lambda = a \pm \mathrm{i}\, b \in \mathbb{C} \setminus \mathbb{R}$:
  In the cases $a < 0$ (asymptotically stable) and $a = 0$ (stable) and $0 < a$ (unstable), we obtain as phase portraits for the solutions $\boldsymbol{x}(t) = c_1\, \mathrm{e}^{at}(\cos(b\,x)\,\mathrm{Re}(\boldsymbol{v}) - \sin(b\,t)\,\mathrm{Im}(\boldsymbol{v})) + c_2\, \mathrm{e}^{at}(\sin(b\,x)\,\mathrm{Re}(\boldsymbol{v}) + \cos(b\,t)\,\mathrm{Im}(\boldsymbol{v}))$ the portraits in Fig. 67.6.

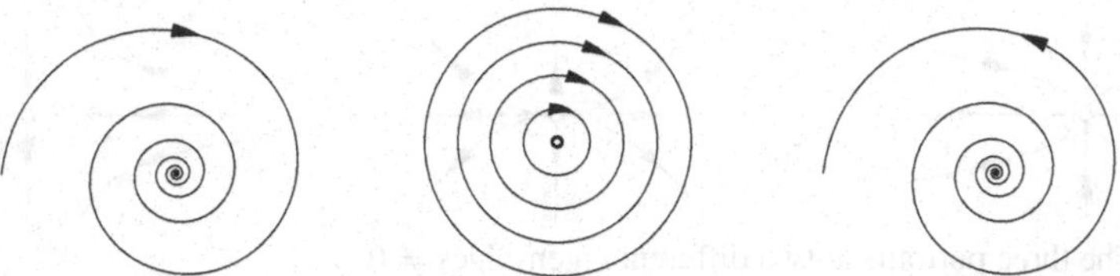

Fig. 67.6 The three types of portraits in the case of a complex eigenvalue

67.2.1 Stability of Nonlinear Systems

In practice, one usually deals with nonlinear ODE systems. In some cases, there is also an easy-to-handle criterion to obtain statements about the stability behavior in the neighborhood of equilibrium points. For this purpose we consider a nonlinear autonomous ODE system of the form

$$\dot{\boldsymbol{x}} = f(\boldsymbol{x}) .$$

The points $\boldsymbol{a}$ with $f(\boldsymbol{a}) = \boldsymbol{0}$ we again call equilibrium points. Now we replace the function f by its linear approximation, i.e., we determine the Jacobian matrix Df and obtain due to $f(\boldsymbol{a}) = \boldsymbol{0}$

$$f(\boldsymbol{x}) \approx Df(\boldsymbol{a})\,(\boldsymbol{x} - \boldsymbol{a}) .$$

Now we have:

The Stability Theorem for Nonlinear Systems
If $\boldsymbol{a}$ is an equilibrium point of the nonlinear autonomous ODE system $\dot{\boldsymbol{x}} = f(\boldsymbol{x})$ with $f : D \subseteq \mathbb{R}^n \to \mathbb{R}^n$, then with the eigenvalues $\lambda_1, \ldots, \lambda_n \in \mathbb{C}$ of $Df(\boldsymbol{a})$:

- $\mathrm{Re}(\lambda_i) < 0$ for all $i \Rightarrow \boldsymbol{a}$ is asymptotically stable.
- $\mathrm{Re}(\lambda_i) > 0$ for one $i \Rightarrow \boldsymbol{a}$ is unstable.

Example 67.4

- We consider the nonlinear autonomous ODE system

$$\begin{pmatrix} \dot{x}_1 \\ \dot{x}_2 \end{pmatrix} = \begin{pmatrix} x_1 - x_1x_2 - x_1^2 \\ x_1x_2 - 2x_2 - x_2^2 \end{pmatrix} = f(x_1, x_2) .$$

It is $(1, 0)^\top$ an equilibrium point. We form the Jacobian matrix of f and obtain

$$Df = \begin{pmatrix} 1 - x_2 - 2x_1 & -x_1 \\ x_2 & x_1 - 2 - 2x_2 \end{pmatrix} \text{ with } Df(1,0) = \begin{pmatrix} -1 & -1 \\ 0 & -1 \end{pmatrix}.$$

The eigenvalues of $Df(1,0)$ are both –1. Thus $(1,0)^\top$ is asymptotically stable.

- We now consider the nonlinear autonomous ODE system

$$\begin{pmatrix} \dot{x}_1 \\ \dot{x}_2 \end{pmatrix} = \begin{pmatrix} 2x_1 - x_1 x_2 \\ -x_2 + x_1 x_2 \end{pmatrix} = f(x_1, x_2).$$

It is $(1,2)^\top$ an equilibrium point. We form the Jacobian matrix of f and obtain

$$Df = \begin{pmatrix} 2 - x_2 & -x_1 \\ x_2 & -1 + x_1 \end{pmatrix} \text{ with } Df(1,2) = \begin{pmatrix} 0 & -1 \\ 2 & 0 \end{pmatrix}$$

The eigenvalues of $Df(1,2)$ are $\pm \mathrm{i}\sqrt{2}$ with the real part 0. Unfortunately our criterion does not give any information here.

■

If, as in the last example, we cannot use the stability theorem to obtain information about stability, we have to rely on more difficult methods for considering stability on a case-by-case basis.

67.3 Exercises

67.1 Determine the solutions of the following IVPs or ODE systems:

(a) $\dot{\boldsymbol{x}} = A\boldsymbol{x} + \boldsymbol{s}(t), \quad \boldsymbol{x}(0) = \begin{pmatrix} 0 \\ 0 \\ 0 \end{pmatrix}$ with $A = \begin{pmatrix} -4 & 1 & 0 \\ -3 & -1 & 1 \\ -2 & 1 & -2 \end{pmatrix}$ and $\boldsymbol{s}(t) = \cos(t) \begin{pmatrix} 10 \\ 10 \\ 10 \end{pmatrix}$.

(b) $\dot{\boldsymbol{x}} = A(t)\boldsymbol{x}, \quad \boldsymbol{x}(0) = \begin{pmatrix} 7 \\ 3 \\ -1 \\ 1 \end{pmatrix}$ with $A = \begin{pmatrix} 1 & -1 & 2 & -1 \\ 0 & 2 & 2 & -1 \\ 0 & 0 & 2 & 4\,\mathrm{e}^{-2t} \\ 0 & 0 & 0 & 2 \end{pmatrix}$.

Note: First find three solutions in order, where the last three, the last two, and then the last coordinate vanish.

67.2 Examine the equilibrium points of the ODE system $\dot{\boldsymbol{x}} = A\boldsymbol{x} + \boldsymbol{s}$, where

$$\text{(a) } A = \begin{pmatrix} 1 & 3 \\ -3 & -5 \end{pmatrix} \text{ and } s = \begin{pmatrix} 1 \\ 2 \end{pmatrix}, \quad \text{(b) } A = \begin{pmatrix} -1 & 2 & 1 \\ 2 & 1 & 1 \\ 1 & 0 & 3 \end{pmatrix} \text{ and } s = \begin{pmatrix} -2 \\ -4 \\ -4 \end{pmatrix}.$$

67.3 Determine the solutions of the following ODE systems:

$$\text{(a) } \dot{x} = \begin{pmatrix} \dot{x}_1 \\ \dot{x}_2 \end{pmatrix} = \begin{pmatrix} \frac{2t+1}{t^2+t} & -\frac{t}{t+1} \\ 0 & 0 \end{pmatrix} \begin{pmatrix} x_1 \\ x_2 \end{pmatrix} + \begin{pmatrix} t \\ 1 \end{pmatrix},$$

$$\text{(b) } \dot{x} = \begin{pmatrix} \dot{x}_1 \\ \dot{x}_2 \end{pmatrix} = \begin{pmatrix} -\frac{1}{t(t^2+1)} & \frac{1}{t^2(t^2+1)} \\ -\frac{t^2}{t^2+1} & \frac{2t^2+1}{t(t^2+1)} \end{pmatrix} \begin{pmatrix} x_1 \\ x_2 \end{pmatrix} + \begin{pmatrix} \frac{1}{t} \\ 1 \end{pmatrix}.$$

Note: There are $x_1 = (1, t)^\top$ and $x_2 = (-1/t, t^2)^\top$ solutions of the homogeneous system.

67.4 The mathematical pendulum is described by the following ODE for the deflection angle φ

$$\ddot{\varphi}(t) = -\tfrac{g}{l} \sin \varphi(t),$$

where is the acceleration due to gravity g and the length of the pendulum l.

(a) Convert the ODE into a 1st order ODE system and determine the stationary solutions.
(b) Characterize the stationary solutions in linear form.

Boundary Value Problems

68

In an initial value problem, the solution to a differential equation is sought which satisfies at time $t = t_0$ initial conditions. In practice, one often has to deal with *boundary value problems*: Here, a solution of a differential equation is sought which assumes given values at the boundary of a definition range.

68.1 Types of Boundary Value Problems

If an ODE and certain conditional equations are given, in which function and derivative values of the searched solutions at **two** points a and b, and one is only interested in solutions which satisfy the mentioned conditions, then one speaks of a **boundary value problem**. We abbreviate this to BVP.

For an ODE $\ddot{x} = f(t, x, \dot{x}) = 0$ of order 2, one of the following boundary conditions usually occurs:

- $x(a) = r, x(b) = s$,
- $\dot{x}(a) = r, \dot{x}(b) = s$,
- $a_1 x(a) + a_2 x'(a) = r, b_1 x(b) + b_2 \dot{x}(b) = s$,
- $x(a) = x(b), \dot{x}(a) = \dot{x}(b)$ (periodic boundary conditions).

A key difference with IVPs is that with BVPs one does not have an existence and uniqueness theorem; solutions need not exist no matter how *smooth* the function f is, and if a solution exists, it need not be unique at all. But of course BVPs occur frequently in practice, so one is interested in solutions even when it is not analytically possible. To solve such a problem one then chooses numerical methods that give approximate results; in the present chapter we give methods by which we obtain analytically closed solutions.

C. Karpfinger, *Calculus and Linear Algebra in Recipes*,
https://doi.org/10.1007/978-3-662-65458-3_68

68.2 First Solution Methods

There is a very obvious solution method: we do it like with the IVP—we determine the general solution of the ODE and determine the free parameters by inserting the boundary conditions. If this is possible, we have found a solution:

Recipe: Attempted Solution of an BVP

(1) Determine the general solution $x(t)$ of the ODE.
(2) Put the boundary conditions into $x(t)$ and obtain a system of equations.
(3) Try to find the free parameters of the general solution from the equations thus obtained.

In the third step several cases are possible: The system of equations has

- exactly one solution (that would be nice)
- no solution (then there is no solution of the BVP),
- many solutions (the solution of the BVP is then not unique).

Example 68.1

- Given is the BVP

$$\ddot{x} = 0 \text{ with } x(0) = 0\,,\ \dot{x}(1) = \dot{x}(1)\,.$$

(1) The general solution of the ODE is

$$x(t) = a\,t + b\,.$$

(2) Inserting the boundary conditions yields the system of equations

$$0 = b \text{ and } a = a + b\,.$$

We get an infinite number of solutions: For each $a \in \mathbb{R}$ is $y(x) = a\,x$ a solution of the given BVP.

- Given is the BVP

$$\ddot{x}(t) = 1 \text{ with } x(0) = 0\,,\ \dot{x}(1) = y(1)\,.$$

(1) The general solution of the ODE is

$$x(t) = \frac{1}{2}t^2 + a\,t + b\,.$$

(2) Inserting the boundary conditions yields the system of equations

$$0 = b \text{ and } 1 + a = \frac{1}{2} + a\,.$$

This system is not solvable, so no solution exists.

- Given is the BVP

$$\ddot{x}(t) = 1 \text{ with } x(0) = 0\,,\ \dot{x}(1) = 4\,x(1)\,.$$

(1) The general solution of the ODE is still

$$x(t) = \frac{1}{2}t^2 + a\,t + b\,.$$

(2) Inserting the boundary conditions yields the system of equations

$$0 = b \text{ and } 1 + a = 2 + 4\,a\,.$$

This system is with $b = 0$ and $a = -\frac{1}{3}$ uniquely solvable, the solution of the BVP is $x(t) = \frac{1}{2}t^2 - \frac{1}{3}t$.

■

68.3 Linear Boundary Value Problems

Given is a linear ODE of n-th order:

$$x^{(n)}(t) + a_{n-1}(t)\,x^{(n-1)}(t) + \cdots + a_0(t)\,x(t) = s(t)$$

with continuous functions $a_i,\ s : I \to \mathbb{R}$ and $a < b, a,\ b \in I$, and matrices $R,\ S \in \mathbb{R}^{n\times n}$ and the boundary conditions

$$R\begin{pmatrix} x(a) \\ \dot{x}(a) \\ \vdots \\ x^{(n-1)}(a) \end{pmatrix} + S\begin{pmatrix} x(b) \\ \dot{x}(b) \\ \vdots \\ x^{(n-1)}(b) \end{pmatrix} = \boldsymbol{r} \in \mathbb{R}^n\,.$$

This linear BVP is called

- **inhomogeneous**, if $s \neq 0$, $\boldsymbol{r} \neq 0$,
- **fully homogeneous**, if $s = 0$, $\boldsymbol{r} = 0$,
- **semi-homogeneous** if $s = 0$ or $\boldsymbol{r} = 0$,

Example 68.2 We consider the BVP

$$\ddot{x} + x = s(t) \text{ with } x(0) + 2\,\dot{x}(0) = a,\ 3\,x(0) - x(\pi/2) = b\,.$$

Because of

$$\begin{pmatrix} 1 & 2 \\ 3 & 0 \end{pmatrix} \begin{pmatrix} x(0) \\ \dot{x}(0) \end{pmatrix} + \begin{pmatrix} 0 & 0 \\ -1 & 0 \end{pmatrix} \begin{pmatrix} x(\pi/2) \\ \dot{x}(\pi/2) \end{pmatrix} = \begin{pmatrix} a \\ b \end{pmatrix}$$

we get here the matrices $R = \begin{pmatrix} 1 & 2 \\ 3 & 0 \end{pmatrix}$ and $S = \begin{pmatrix} 0 & 0 \\ -1 & 0 \end{pmatrix}$.

The BVP presented is in

- inhomogeneous, if $s \neq 0$ (i.e. $a \neq 0$ or $b \neq 0$),
- fully homogeneous if $s = 0$ (i.e. $a = 0 = b$),
- semi-homogeneous, if $s = 0$ or ($a = 0$ and $b = 0$).

■

We have the following criteria für semi-homogeneous BVPs with $s = 0$:

Solvability Criterion for Linear BVP
Given is the following linear BVP

$$x^{(n)}(t) + a_{n-1}(t)x^{(n-1)}(t) + \cdots + a_0(t)x(t) = s(t),$$

$$R \begin{pmatrix} x(a) \\ \dot{x}(a) \\ \vdots \\ x^{(n-1)}(a) \end{pmatrix} + S \begin{pmatrix} x(b) \\ \dot{x}(b) \\ \vdots \\ x^{(n-1)}(b) \end{pmatrix} = \boldsymbol{r}$$

(continued)

with continuous functions $a_i, s : I \to \mathbb{R}$ and $a < b$ from I and matrices $R, S \in \mathbb{R}^{n\times n}$ and $\boldsymbol{r} \in \mathbb{R}^n$.

If $x_1, \ldots, x_n$ is a fundamental system of the corresponding homogeneous ODE $x^{(n)}(t) + a_{n-1}(t)\,x^{(n-1)}(t) + \cdots + a_0(t)\,x(t) = 0$ and

$$\Phi(t) = \begin{pmatrix} x_1 & \ldots & x_n \\ \dot{x}_1 & \ldots & \dot{x}_n \\ \vdots & & \vdots \\ x_1^{(n-1)} & \ldots & x_n^{(n-1)} \end{pmatrix}$$

and $D = R\,\Phi(a) + S\,\Phi(b) \in \mathbb{R}^{n\times n}$, then we have:

1. In the case $\det D \neq 0$ the BVP has exactly one solution for each $\boldsymbol{r} \in \mathbb{R}^n$.
2. In the case $\det D = 0$ the BVP with $s = 0$ has a unique solution if and only if $\mathrm{rg}(D) = \mathrm{rg}(D\,|\,\boldsymbol{r})$.
3. In the case $\boldsymbol{r} = 0$ the BVP with $s = 0$ has a solution $x \neq 0$ if and only if $\det(D) = 0$.

We continue Example 68.2 above:

Example 68.3 We consider the semi-homogeneous BVP

$$\ddot{x} + x = 0 \text{ with } x(0) = 0\,,\ x(\pi/2) = 1\,.$$

Since $x_1(t) = \cos(t)$ and $x_2(t) = \sin(t)$ form a fundamental system of the ODE, the *fundamental matrix*

$$\Phi(t) = \begin{pmatrix} \cos(t) & \sin(t) \\ -\sin(t) & \cos(t) \end{pmatrix},$$

and thus we obtain for D the matrix

$$D = \begin{pmatrix} 1 & 2 \\ 3 & 0 \end{pmatrix}\begin{pmatrix} 1 & 0 \\ 0 & 1 \end{pmatrix} + \begin{pmatrix} 0 & 0 \\ -1 & 0 \end{pmatrix}\begin{pmatrix} 0 & 1 \\ -1 & 0 \end{pmatrix} = \begin{pmatrix} 1 & 2 \\ 3 & -1 \end{pmatrix}.$$

Because of $\det(D) \neq 0$ the BVP is uniquely solvable. ■

68.4 The Method with Green's Function

Green's function is a tool in solving inhomogeneous linear differential equations. It is said that a Green's function *propagates* inhomogeneity: one can set up the Green's function alone with the fully homogeneous BVP and obtain the solution of the semi-homogeneous or inhomogeneous BVP with its help. Thus, this method is particularly useful when a variation of the constants is too costly and a special perturbation approach is not possible.

We formulate this solution method for the case of a 2nd order ODE; a generalization of the method to higher order ODEs is straightforward, but hardly relevant for practical cases.

Solution with Green's Function

If x_1, x_2 is a fundamental system of the homogeneous ODE

$$\ddot{x}(t) + a_1(t)\,\dot{x}(t) + a_0(t)\,x(t) = 0\,,$$

if further $\det(D) \neq 0$ for $D = R\,\Phi(a) + S\,\Phi(b)$, where $\Phi = \begin{pmatrix} x_1 & x_2 \\ \dot{x}_1 & \dot{x}_2 \end{pmatrix}$, then the semi-homogeneous BVP

$$\ddot{x}(t) + a_1(t)\,\dot{x}(t) + a_0(t)\,x(t) = s(t)\,,\ R\begin{pmatrix} x(a) \\ \dot{x}(a) \end{pmatrix} + S\begin{pmatrix} x(b) \\ \dot{x}(b) \end{pmatrix} = 0$$

is uniquely solvable. The uniquely determined solution is

$$x(t) = \int_a^b g(t,\tau)\,s(\tau)\,\mathrm{d}\tau\,,$$

where the so-called **Green's function** $g : [a,b] \times [a,b] \to \mathbb{R}$ is given by the following properties:

(a) g solves for a fixed $\tau \neq t$ with respect to t solves the homogeneous ODE

$$\ddot{g}(t,\tau) + a_1(t)\,\dot{g}(t,\tau) + a_0(t)\,g(t,\tau) = 0\,.$$

(continued)

(b) g satisfies for a fixed τ , $a < \tau < b$ the homogeneous boundary conditions

$$R\begin{pmatrix} g(a,\tau) \\ g_t(a,\tau) \end{pmatrix} + S\begin{pmatrix} g(b,\tau) \\ g_t(b,\tau) \end{pmatrix} = 0 .$$

(c) g is continuous, and the partial derivative g_t is for $t = \tau$ discontinuous:

$$\frac{\partial}{\partial t} g(t^+, t) - \frac{\partial}{\partial t} g(t^-, t) = 1 .$$

Thus, to solve such a BVP, we proceed as follows: We determine the Green function from the fully homogeneous BVP and thus obtain a solution of the semi-homogeneous BVP with $s(t) \neq 0$:

Recipe: Determine the Green Function
Given the BVP

$$\ddot{x}(t) + a_1(t)\,\dot{x}(t) + a_0(t)\,x(t) = s(t) ,\; R\begin{pmatrix} x(a) \\ \dot{x}(a) \end{pmatrix} + S\begin{pmatrix} x(b) \\ \dot{x}(b) \end{pmatrix} = 0 .$$

We obtain the Green function $g(t, \tau)$ as follows

(1) Determine a fundamental system (x_1, x_2) of the homogeneous ODE and set for τ with $a < \tau < b$

$$g(t,\tau) = \begin{cases} (a_1(\tau) + b_1(\tau))\,x_1 + (a_2(\tau) + b_2(\tau))\,x_2, & \tau \le t \\ (a_1(\tau) - b_1(\tau))\,x_1 + (a_2(\tau) - b_2(\tau))\,x_2, & \tau \ge t \end{cases} .$$

(2) From the following system of linear equations, one determines b_1 and b_2:

$$b_1(t)x_1(t) + b_2(t)x_2(t) = 0 ,\; b_1(t)\dot{x}_1(t) + b_2(t)\dot{x}_2(t) = 1/2 .$$

(3) Now determine a_1 and a_2 from the system of equations obtained by substituting b_1 and b_2 in the approach for g in (1), after having inserted the boundary conditions.

Example 68.4 We consider the BVP

$$\ddot{x}(t) + x(t) = -t + 1\,,\; x(0) - x(\pi) = 0\,,\; \dot{x}(0) - \dot{x}(\pi) = 0\,.$$

(1) As is well known {cos, sin} is a fundamental system of the homogeneous ODE. We make the approach

$$g(t, \tau) = \begin{cases} (a_1(\tau) + b_1(\tau))\cos(t) + (a_2(\tau) + b_2(\tau))\sin(t), & \tau \leq t \\ (a_1(\tau) - b_1(\tau))\cos(t) + (a_2(\tau) - b_2(\tau))\sin(t), & \tau \geq t \end{cases}.$$

(2) The system of equations to be solved is

$$b_1(t)\cos(t) + b_2(t)\sin(t) = 0\,,\; -b_1(t)\sin(t) + b_2(t)\cos(t) = 1/2\,.$$

The uniquely determined solution is obviously $b_1(t) = -\frac{1}{2}\sin(t)$ and $b_2(t) = \frac{1}{2}\cos(t)$.

(3) Now we determine $a_1(\tau)$ and $a_2(\tau)$ from the system of equations

$$g(0, \tau) - g(\pi, \tau) = (a_1(\tau) - b_1(\tau)) + (a_1(\tau) + b_1(\tau)) = 0$$

$$g_t(0, \tau) - g_t(\pi, \tau) = (a_2(\tau) - b_2(\tau)) + (a_2(\tau) + b_2(\tau)) = 0\,.$$

Obviously $a_1 = 0 = a_2$ is the uniquely determined solution.

Thus, we obtain the Green function

$$g(t, \tau) = \begin{cases} \frac{1}{2}\sin(t - \tau), & \tau \leq t \\ -\frac{1}{2}\sin(t - \tau), & \tau \geq t \end{cases}.$$

Thus, the solution of the BVP is

$$x(t) = 1/2 \int_0^t \sin(t - \tau)(-\tau + 1)\mathrm{d}\tau - 1/2 \int_t^\pi \sin(t - \tau)(-\tau + 1)\mathrm{d}\tau$$

$$= 1 - t - \tfrac{\pi}{2}\cos(t)\,.$$

■

MATLAB With MATLAB one has the possibility to solve BVPe both analytically and numerically. For the analytical solution one uses the function dsolve. The solution for the above example is obtained as follows:

```
» syms x(t) » Dx = diff(x); » D2x = diff(x,2); » dsolve(D2x ==
-x-t+1, x(0) == x(pi),Dx(0) == Dx(pi)) ans = 1 - (pi*cos(t))/2 - t
```

68.5 Exercises

68.1 Find the solution set of the BVP $\ddot{x} + x = 0$, $x(0) = 1$, $x(b) = d$ for

(a) b=d=1,
(b) b= π, d= −1,
(c) b= π, d= −2.

68.2 Given is the inhomogeneous BVP

$$\ddot{x} + x = 1 + t + \cos t, \quad x(0) = 1, \quad x(b) = 1 + \pi\,.$$

For which $b \in \mathbb{R}$ is the BVP unsolvable, for which is it uniquely solvable? Determine a solution for $b = \pi/2$.

68.3 We consider the boundary value problem

$$\ddot{x} + Cx = g, \quad x(0) = x(1) = 0 \text{ with } C \in \mathbb{R}\,.$$

We *discretize* this equation: We consider the interpolation points $I_h = \{t_\nu \in [0, 1] \mid t_\nu = \nu h, h = 1/n, 0 \leq \nu \leq n\}$ and approximate the derivation $\ddot{x}$ by $\frac{1}{h^2}(x(t_{\nu+1}) - 2x(t_\nu) + x(t_{\nu-1}))$. What system of linear equations does this lead to?

Write a function in MATLAB that gives a plot of the obtained approximate solutions für $C = -1$ and $n = 2^3,\ 2^4,\ 2^5,\ 2^6$ and the function $g(t) = t^3$.

69 Basic Concepts of Numerics

Numerical mathematics, also called *numerics* for short, provides a numerical solution to a problem: Whether the value of a formula, the solution of an equation or a system of equations, the zero of a function, the solution of an optimization problem, the solution curve of an ordinary differential equation or possibly also the solution function of a partial differential equation is sought: In numerical mathematics, one develops algorithms that compute approximate solutions to these problems. The focus is on two things: The algorithms should deliver *exact* results and be *fast*.

Errors happen when calculating with the computer. A distinction is made:

- **Input errors** or **data error:** These are practically unavoidable errors that occur, for example, due to erroneous measured values.
- **Rounding error** or **procedural error:** These are errors whose influence can be avoided or reduced.

The *condition* provides a measure of the impact of input errors on the results obtained, while *stability* examines the impact of rounding errors or procedural errors on the results.

69.1 Condition

The results of a computer are subject to errors. The errors in the input data and the errors in the individual calculation steps lead to errors in the results. The determination of the result is called a *problem*. Each such *problem* can be assigned a number, called the *condition of* the *problem*. This *condition* is a measure of how strongly errors in the input data (if the calculation is correct) affect the results.

C. Karpfinger, *Calculus and Linear Algebra in Recipes*,
https://doi.org/10.1007/978-3-662-65458-3_69

We specify these terms: A **problem** is a mapping $f : X \to Y$, where X and Y are **normed spaces**, i.e., are vector spaces with norms (cf. Sect. 45.2.1: For $\boldsymbol{x},\ \delta\boldsymbol{x} \in X$ we write

$$\delta f = f(\boldsymbol{x} + \delta\boldsymbol{x}) - f(\boldsymbol{x}).$$

It stands $\delta\boldsymbol{x}$ for an *error* in the input data and δf for the corresponding error in the result. Now we ask the question, how large δf is compared to $\delta\boldsymbol{x}$. Since we have normed spaces X and Y, we have with the norm $\|\cdot\|$ on both X as well as on Y a possibility to use the elements $\delta\boldsymbol{x}$ and δf in terms of size:

Absolute and Relative Condition of a Problem

If $f : X \to Y$ is a problem, then for each $\boldsymbol{x} \in X$ the number

- $\kappa_{\text{abs}}(\boldsymbol{x}) = \lim\limits_{\delta\to 0} \sup\limits_{\|\delta\boldsymbol{x}\|\leq\delta} \dfrac{\|\delta f\|}{\|\delta\boldsymbol{x}\|}$ is called the **absolute condition** and
- $\kappa_{\text{rel}}(\boldsymbol{x}) = \lim\limits_{\delta\to 0} \sup\limits_{\|\delta\boldsymbol{x}\|\leq\delta} \dfrac{\|\delta f\|/\|f(\boldsymbol{x})\|}{\|\delta\boldsymbol{x}\|/\|\boldsymbol{x}\|}$ the **relative condition**

of f in the point $\boldsymbol{x} \in X$.

If $f : \mathbb{R} \to \mathbb{R}$ is differentiable, then for each $x \in \mathbb{R}$:

$$\kappa_{\text{abs}}(x) = |f'(x)| \text{ and } \kappa_{\text{rel}}(x) = \frac{|f'(x)|}{|f(x)|/|x|}.$$

The condition specifies the factor by which perturbations in the input variables affect the result; this number is independent of the algorithm by which the result is obtained. With the relative condition, the quantities $\delta\boldsymbol{x}$ and δf are seen in relation to the size of $\boldsymbol{x}$ and $f(\boldsymbol{x})$.

Example 69.1

- For the problem $f : \mathbb{R} \to \mathbb{R}$, $f(x) = 2x$ we get

$$\kappa_{\text{abs}}(x) = |f'(x)| = 2 \text{ and } \kappa_{\text{rel}}(x) = \frac{2}{|2x|/|x|} = 1.$$

Input errors are thus magnified by a factor of 2.

- For the problem $f : [0, \infty) \to [0, \infty)$, $f(x) = \sqrt{x}$ we get because of $f'(x) = \frac{1}{2\sqrt{x}}$

$$\kappa_{\text{abs}}(x) \to \infty \text{ for } x \to 0 \text{ and } \kappa_{\text{rel}}(x) = \frac{\frac{1}{2\sqrt{x}}}{\sqrt{x}/x} = \frac{1}{2}.$$

That is, the closer x is to zero, the more the input errors are amplified. Very close to 0, the calculation of $\sqrt{x}$ is extremely sensitive to input errors. When viewed relatively, the computation of the root is unproblematic.

- For the *subtraction* problem, i.e., compute to $x, y \in \mathbb{R}$, y fixed, the number $f(x, y) = x - y$ we get

$$\kappa_{\text{rel}}(x) = \frac{1}{|x - y|/|x|} = \frac{|x|}{|x - y|},$$

that is,the condition is large if $x \approx y$.

- We now consider the problem f of determining a zero of a quadratic polynomial: To do this, we decompose the polynomials $p_1 = x^2 - 2x + 1$ and $p_2 = x^2 - 2x + 0.9999$ in linear factors, we have

$$p_1 = x^2 - 2x + 1 = (x - 1)(x - 1) \text{ and } p_2 = x^2 - 2x + 0.9999 = (x - 0.99)(x - 1.01).$$

The constant coefficient of p_2 differs only by 0.0001 from the constant coefficient of p_1, while the zeros differ by 0.01, a difference by a factor of 100.
We look for the cause of this phenomenon: The polynomial $x^2 - 2x + q = 0$ has the zeros

$$x_{1,2} = 1 \pm \sqrt{1 - q}.$$

We can rephrase the problem to $f(q) = x_1 = 1 + \sqrt{1 - q}$. As an absolute condition of this problem we get:

$$\kappa_{\text{abs}}(q) = |f'(q)| = \frac{1}{2}(1 - q)^{-1/2}.$$

As a relative condition we obtain with $q = 1 - \delta q$:

$$\kappa_{\text{rel}}(q) = \frac{|f'(q)|}{|f(q)|/|q|} = \frac{\frac{1}{2\sqrt{\delta q}}}{(1 + \sqrt{\delta q})/(1 - \delta q)} = \frac{1}{2\sqrt{\delta q}(1 + \sqrt{\delta q})} \overset{\delta q \to 0}{\longrightarrow} \infty.$$

For $\delta q = 0.0001$ we get $\kappa_{\text{rel}}(q) \approx 50$.

■

The following terms *well conditioned* and *badly conditioned* are formulated somewhat woolly. What is *good* and *bad* is actually not a matter of taste, but of experience or empiricism.

Good and Bad Conditioned Problems
A problem $f : X \to Y$ is called in a point $\boldsymbol{x} \in X$

- **well conditioned** , if $\kappa_{\text{rel}}(\boldsymbol{x})$ is *small* (≤ 100), and
- **badly conditioned**, if $\kappa_{\text{rel}}(\boldsymbol{x})$ is *large* ($\geq 10^6$).

69.2 The Big O Notation

For two functions f, g defined on a subset of $\mathbb{R}$ means

$$f(x) = O(g(x)) \text{ for } x \to \infty \text{ or } x \to 0,$$

that there are constants $C > 0$ and $x_0 > 0$ such that

$$|f(x)| \leq C|g(x)| \text{ for all } x > x_0 \text{ or for all } |x| < x_0.$$

Thus, the function f will eventually be smaller in magnitude than g except for a constant factor C. The condition $x > x_0$ or $|x| < x_0$ says that we consider the functions *close to* ∞ or *close to 0*.

The O is meant to remind us of the word *Order*. However, it is more likely to be used in ways of speaking such as *f is O(g)*, in words *f is Big O of g*, is used.

Strictly speaking, $O(g)$ is a set of functions and the notation $f = O(g)$ a somewhat sloppy but useful abbreviation for $f \in O(g)$.

In practice, one often uses functions for comparison that are particularly *simple*. For example, one omits all coefficients.

Example 69.2

- Constant functions $X \to \mathbb{R}$, $x \mapsto c$ are in $O(1)$.
- For a polynomial function $f \in \mathbb{R}[x]$ with $n = \deg f$ we have $f(x) = O(x^n)$ for $x \to \infty$.
- For the function f with $f(x) = \sqrt{1+x^2}$ we have for $x \to \infty$

$$f(x) = O(x) \text{ and } f(x) = x + O\left(\frac{1}{x}\right).$$

Here the last equation is to be understood as follows:

$$f(x) - x = \sqrt{1+x^2} - x = o\left(\frac{1}{x}\right).$$

To prove it, examine the limits

$$\lim_{x\to\infty} \frac{f(x)}{x} \text{ and } \lim_{x\to\infty} x(f(x) - x).$$

■

69.3 Stability

Solving a problem with a computer consists of a sequence of arithmetic operations. Here, the arithmetic in the floating point numbers $\mathbb{G}_{b,t}$ (see Sect. 4.2) is not exact. Existing errors propagate from operation to operation, and new errors also arise.

In general, problems can be solved in different ways. And mostly different ways lead to different accumulations of errors: So it can happen that one way leads to usable results, while another way leads to completely useless results; and this, although both ways should lead to the same results if calculated exactly. The challenge is to formulate such an *algorithm* for solving a problem that yields usable results. Here we understand an **algorithm** a mapping $\tilde{f} : X \to Y$,

$$\tilde{f} = \tilde{f}_k \circ \tilde{f}_{k-1} \circ \cdots \circ \tilde{f}_1,$$

where $\tilde{f}_j$ only operations from $\{\oplus, \ominus, \odot, \oslash, \text{fl}\}$ (see floating point arithmetic, Sect. 4.2.2) contains. In fact, $\tilde{f}$ depends on the accuracy of the machine $\varepsilon_{b,t}$, i.e.

$$\tilde{f}(x) = \tilde{f}(x, \varepsilon_{b,t}) = \tilde{f}_{b,t}(x).$$

We use an algorithm $\tilde{f}$ to solve a problem *f*. In doing so, we do not solve the problem exactly, we make an error, we can now summarize this error mathematically as follows:

Absolute and Relative Error

If $f : X \to Y$ is a problem and $\tilde{f} : X \to Y$ an algorithm, then we call

- $\|\tilde{f}(x) - f(x)\|$ the **absolute error** of $\tilde{f}$ in $x \in X$ and
- $\frac{\|\tilde{f}(x) - f(x)\|}{\|f(x)\|}$ the **relative error** of $\tilde{f}$ in $x \in X$.

The goal is to find, for a given problem f, an algorithm $\tilde{f}$ so that the (absolute, relative) error is *small*, although this *small* still needs to be specified. One can demand that $\tilde{f}$ is *exact*. Thereby one calls an algorithm $\tilde{f}$ for a problem f **exactly** if for all $\boldsymbol{x} \in X$

$$\frac{\|\tilde{f}(\boldsymbol{x}) - f(\boldsymbol{x})\|}{\|f(\boldsymbol{x})\|} \leq O(\varepsilon_{b,t}),$$

which means that for all $\boldsymbol{x} \in X$ the errors generated in the course of the computation remain at most in the order of magnitude of the machine accuracy. However, this requirement has disadvantages: it is often too restrictive for practical problems. Somewhat weaker, on the other hand, is the *stability* or *backward stability of the* algorithm:

Stable and Backward Stable Algorithm
An algorithm $\tilde{f}$ for a problem f is called

- **stable** if for all $\boldsymbol{x} \in X$

$$\frac{\|\tilde{f}(\boldsymbol{x}) - f(\tilde{\boldsymbol{x}})\|}{\|f(\boldsymbol{x})\|} = O(\varepsilon_{b,t})$$

  for a $\tilde{\boldsymbol{x}}$ with $\frac{\|\tilde{\boldsymbol{x}}-\boldsymbol{x}\|}{\|\boldsymbol{x}\|} = O(\varepsilon_{b,t})$,
- **backward-stable** if for all $\boldsymbol{x} \in X$

$$\tilde{f}(\boldsymbol{x}) = f(\tilde{\boldsymbol{x}})$$

  for a $\tilde{\boldsymbol{x}} \in X$ with $\frac{\|\tilde{\boldsymbol{x}}-\boldsymbol{x}\|}{\|\boldsymbol{x}\|} = O(\varepsilon_{b,t})$; the number $\frac{\|\tilde{\boldsymbol{x}}-\boldsymbol{x}\|}{\|\boldsymbol{x}\|}$ is called (relative) **reverse error**.

Remark If $f : X \to Y$ is a problem with relative condition κ_{rel} and $\tilde{f} : X \to Y$ is a backward-stable algorithm for f, then

$$\frac{\|\tilde{f}(\boldsymbol{x}) - f(\boldsymbol{x})\|}{\|f(\boldsymbol{x})\|} = O(\kappa_{\text{rel}}(\boldsymbol{x})\, \varepsilon_{b,t}).$$

Thus, the relative error in computing a problem with a backward-stable algorithm is *small* if the relative condition is *small*, or even with a backward-stable algorithm, the relative error becomes *large* if the relative condition is *large*.

69.4 Exercises

69.1 Determine the absolute and relative condition of the problems $f(x) = x^3$, $g(x) = \frac{\sin x}{x}$.

69.2 Given an ODE $\dot{x} = v(x)$, $v : \mathbb{R} \to \mathbb{R}$, with initial value $x(0) = x_0$. The dependence of the solution $x(t)$ from the initial value is given by the notation $x(t) = x(t; x_0)$ is made clear. Calculate the *sensitivity of* the solution for a fixed t > 0 with respect to the initial value z, i.e. the absolute and relative condition of the problem $f : \mathbb{R} \to \mathbb{R}, x_0 \mapsto x(t, x_0)$ for a fixed $t > 0$ for

(a) $v(x) = \lambda x, \lambda \in \mathbb{R}$,
(b) $v(x) = x^2$.

69.3 Check whether the following statements are true or false:

(a) $\sin(x) = O(1)$ for $x \to \infty$.
(b) $\sin(x) = O(1)$ for $x \to 0$.
(c) $\text{fl}(\pi) - \pi = O(\varepsilon_{b,t})$ for $\varepsilon_{b,t} \to 0$.
(d) $x^3 + x^4 = O(x^4)$ for $x \to 0$.
(e) $A = O(V^{2/3})$ for $V \to \infty$, where A and V are the area and volume of a sphere, measured in square millimeters and cubic kilometers, respectively.

Fixed Point Iteration 70

Determining a solution x of an equation $F(x) = a$ is one of the most important and frequent problems in applied mathematics. In fact, however, it is often not even possible to state the solution of such an equation explicitly and exactly. Numerical mathematics provides iterative methods for the approximate solution of (linear and nonlinear) equations and systems of equations. These procedures are based on *fixed point iteration*, which is the content of the present chapter. However, we do not discuss procedures for solving equations or systems of equations now, but consider *fixed point iterations* as an object per se. In the next Chap. 71 we discuss in detail iterative solution methods for systems of linear equations.

70.1 The Fixed Point Equation

Solving the equation $F(x) = a$ to x is equivalent to determining an x with $f(x) = 0$, where $f(x) = F(x) - a$. And this **zero problem** in turn is equivalent to solving **fixed point equation**

$$\phi(x) = x\,,$$

where $\phi(x) = f(x) + x$. Indeed, with this ϕ we have

$$\phi(x) = x \;\Leftrightarrow\; f(x) = 0\,.$$

A point x with $\phi(x) = x$ is of course called a **fixed point** of ϕ.

C. Karpfinger, *Calculus and Linear Algebra in Recipes*,
https://doi.org/10.1007/978-3-662-65458-3_70

Remark Besides the function $\boldsymbol{\phi}(\boldsymbol{x}) = f(\boldsymbol{x}) + \boldsymbol{x}$ there may well be other functions $\boldsymbol{\psi}(\boldsymbol{x})$ whose fixed point is a zero of $f(\boldsymbol{x})$ e.g. for each $z \neq 0$ the function $\boldsymbol{\psi}(\boldsymbol{x}) = z\, f(\boldsymbol{x}) + \boldsymbol{x}$ or even more generally $\boldsymbol{\psi}(\boldsymbol{x}) = g(\boldsymbol{x})\, f(\boldsymbol{x}) + \boldsymbol{x}$ with a function g, which has no zero. Note also the following example.

Example 70.1 A zero of the function $f(x) = x^7 - x - 2$ is a fixed point of

$$\phi(x) = x^7 - 2 \text{ and } \psi(x) = (x + 2)^{1/7}.$$

■

Solving an equation or determining a zero of a function is thus reduced to determining a fixed point of a function $\boldsymbol{\phi}$. The essential advantage of this reformulation into a **fixed point problem**—*determine* $\boldsymbol{x}$ *with* $\boldsymbol{\phi}(\boldsymbol{x}) = \boldsymbol{x}$—lies in the fact that for the fixed point problem there is the easily formulated *fixed point iteration*, which in many cases, starting from a chosen initial value, gives a good approximation for a fixed point of the function $\boldsymbol{\phi}$. Moreover, *Banach's fixed point theorem* states under which conditions the fixed point iteration is guaranteed to converge to a fixed point; further, it gives an estimate of the error one makes when stopping the iteration to approximate a fixed point. The *fixed* point iteration is as follows:

Recipe: Fixed Point Iteration
Given a function $\boldsymbol{\phi}$. We are looking for a fixed point $\boldsymbol{x}$, i.e., an $\boldsymbol{x}$ with $\boldsymbol{\phi}(\boldsymbol{x}) = \boldsymbol{x}$.

(1) Choose a starting value $\boldsymbol{x}_0$ (in a neighborhood of $\boldsymbol{x}$).
(2) Determine

$$\boldsymbol{x}_{k+1} = \boldsymbol{\phi}(\boldsymbol{x}_k)\,,\ k = 0,\ 1,\ 2,\ \ldots$$

This **fixed point iteration** yields a sequence $(\boldsymbol{x}_k)_k$.

If a sequence $(\boldsymbol{x}_k)_k$ thus constructed iteratively by a function $\boldsymbol{\phi}$, the function $\boldsymbol{\phi}$ is also called **iteration method**.

The open question is whether the sequence $(\boldsymbol{x}_k)_k$ against a fixed point $\boldsymbol{x}$ converges. Before we address this question, let us consider an example:

Example 70.2 The Babylonian root extraction example shows that it is possible to find the zero of eq. $x^2 - a = 0$, $a > 0$, by the following fixed point iteration, for this we set

$$\phi(x) = \frac{1}{2}\left(x + \frac{a}{x}\right).$$

A x with $\phi(x) = x$ is a root of a. The fixed point iteration is:

$$x_0 = s\,,\ x_{k+1} = \phi(x_k)$$

with a starting value s. We set $a = 9$ and $s = 1$ and get the first sequence members

$$x_1 = 1.0000\,,\ x_2 = 5.0000\,,\ x_3 = 3.4000\,,\ x_4 = 3.0235\,,\ x_5 = 3.0001\,,$$
$$x_6 = 3.0000\,.$$

■

Remark Note that we have not yet said a word about the domain and values of the functions F, f and ϕ. It has been implicitly assumed so far that these are functions of subsets of $\mathbb{R}$ according to $\mathbb{R}$ is concerned. In fact, however, all this is possible in a much more general way: as a domain and values, we may use any normalized space X; a notion of distance will be necessary to speak of convergence. This is what the norm is for. In most cases X will be a subset of $\mathbb{R}^n$ with $n \in \mathbb{N}$, $X \subseteq \mathbb{R}^n$. In this sense we settle the case of systems of equations (linear or nonlinear) right away with; it is $(\boldsymbol{x}_k)_k$ then a sequence of vectors $\boldsymbol{x}_k \in \mathbb{R}^n$.

70.2 The Convergence of Iteration Methods

Let X be a normed space. If the iteration method $\phi : X \to X$ satisfies the conditions of Banach's fixed point theorem, then the sequence $(\boldsymbol{x}_k)_k$ is guaranteed to converge to a fixed point $\boldsymbol{x}$. But before we come to this very general fixed point theorem of Banach, we justify that in the case of convergence of $(\boldsymbol{x}_k)_k$ the limit is a fixed point, if ϕ is continuous:

On the Convergence of the Fixed Point Iteration
If $\phi : X \to X$, $X \subseteq \mathbb{R}^n$, is continuous and the sequence $(\boldsymbol{x}_k)_k$ which results from a fixed point iteration

$$\boldsymbol{x}_0 = \boldsymbol{s}\,,\ \boldsymbol{x}_{k+1} = \phi(\boldsymbol{x}_k)$$

converges to a $\boldsymbol{x} \in X$ then $\boldsymbol{x}$ is a fixed point of ϕ, i.e., $\phi(\boldsymbol{x}) = \boldsymbol{x}$.

This follows from:

$$x = \lim_{k\to\infty} x_{k+1} = \lim_{k\to\infty} \phi(x_k) = \phi(\lim_{k\to\infty} x_k) = \phi(x)\,.$$

We introduce suggestive terms for the concise formulation of the following results, where X is always a normed space:

- An iteration method $\phi : X \to X$ is called **globally convergent** against a fixed point x if $x_k \to x$ for all initial values $x_0 \in X$.
- An iteration method $\phi : X \to X$ is called **locally convergent** against a fixed point x if there is an environment $U \subseteq X$ of x with $x_k \to x$ for all initial values $x_0 \in U$.
- An iteration method ϕ is called a **contraction** if there is a $\theta \in [0, 1)$ such that

$$\|\phi(x) - \phi(y)\| \leq \theta\, \|x - y\| \text{ for all } x,\ y \in X\,.$$

The number θ is also called **contraction constant** or **Lipschitz constant.** If ϕ is a contraction, then there are $\phi(x)$ and $\phi(y)$ are closer together than x and y.

More generally, one speaks of a **lipschitz-continuous** function $\phi : X \to X$ if there exists a **Lipschitz constant** θ with

$$\|\phi(x) - \phi(y)\| \leq \theta\, \|x - y\| \text{ for all } x,\ y \in X\,.$$

This Lipschitz continuity is *something more* than (ordinary) continuity. A contraction is a lipschitz-continuous function with Lipschitz constant < 1. The crucial question is: how to know if a given function ϕ is a contraction? For this, the following result is useful:

Criterion for Contraction
If $\phi : X \to X$ is continuously differentiable on a closed, bounded and convex set $X \subseteq \mathbb{R}^n$, then ϕ is a contraction if there exists a number $\gamma < 1$ with

$$\|D\phi(x)\| < \gamma \text{ for all } x \in X\,.$$

Where $\|\cdot\|$ is a matrix norm induced by a vector norm of $\mathbb{R}^n$.

Indeed, under the above conditions, the Lipschitz constant is given by $\theta = \sup_{x\in X} \|D\phi(x)\|$.

One now rightly expects that, given a contracting iteration function $\phi : X \to X$ the convergence of the sequence $(x_k)_k$ generated by ϕ against a fixed point x. But at the same time, one probably expects that in general it will not be easy to find an iteration function $\phi : X \to X$ thas is *globally* a contraction ($\to$ *global convergence theorem*). We seek conditions that guarantee that ϕ ensures at least *locally* convergence. An obvious criterion is obtained with the absolute value of the derivative, i.e., with the norm of the Jacobian matrix at the fixed point $\boldsymbol{x}$ ($\to$ *local convergence theorem*). But this criterion depends on the norm used. We obtain a norm-independent version of the local convergence theorem using the *spectral radius* of the Jacobian matrix $D\phi$ ($\to$ *local norm-independent convergence theorem*). Here, the largest eigenvalue in absolute value of a matrix is called A the **spectral radius** $\rho(A)$,

$$\rho(A) = \max\{|\lambda| \mid \lambda \text{ is eigenvalue of } A\}.$$

The following global convergence Theorem is the already several times mentioned **Banach fixed point Theorem.**

Global and Local Convergence Theorems

- **Global convergence Theorem:** If $X \subseteq \mathbb{R}^n$ is closed and $\phi : X \to X$ a contraction with Lipschitz constant θ, then the fixed point iteration converges for each $\boldsymbol{x}_0 \in X$ against a **unique** fixed point $\boldsymbol{x} \in X$, and we have
  - $\|\boldsymbol{x}_k - \boldsymbol{x}\| \le \frac{\theta^k}{1-\theta} \|\boldsymbol{x}_1 - \boldsymbol{x}_0\|$—the **a priori error estimation,**
  - $\|\boldsymbol{x}_k - \boldsymbol{x}\| \le \frac{\theta}{1-\theta} \|\boldsymbol{x}_k - \boldsymbol{x}_{k-1}\|$—the **a posteriori error estimation.**
- **Local convergence Theorem:** If $X \subseteq \mathbb{R}^n$ is open, $\phi : X \to X$ continuously differentiable and $\boldsymbol{x} \in X$ a fixed point of ϕ with $\|D\phi(\boldsymbol{x})\| < 1$, then the fixed point iteration is locally convergent.
- **Local norm-independent convergence Theorem:** If $X \subseteq \mathbb{R}^n$ is open, $\phi : X \to X$ continuously differentiable and $\boldsymbol{x} \in X$ a fixed point of ϕ with $\rho(D\phi(\boldsymbol{x})) < 1$ for the Jacobian matrix $D\phi$ in $\boldsymbol{x}$, then the fixed point iteration is locally convergent to $\boldsymbol{x}$.

Note that the norms here are to be chosen uniformly: For example, in the a priori error estimate, the contraction constant θ is formed with respect to a norm $\|\cdot\|$: It must be this norm from the inequality.

The global convergence Theorem, i.e. Banach's fixed point Theorem, guarantees the existence and uniqueness of a solution to the fixed point equation. Further, it gives error estimates.

The a priori error estimation can be used to determine how long to iterate to achieve a desired accuracy ε: Choose k such that

$$\frac{\theta^k}{1-\theta}\,\|\boldsymbol{x}_1-\boldsymbol{x}_0\| \le \varepsilon \;\Leftrightarrow\; k \ge \ln\left(\frac{\varepsilon(1-\theta)}{\|\boldsymbol{x}_1-\boldsymbol{x}_0\|}\right)/\ln(\theta)\,.$$

The a posteriori error estimate does not take into account the errors of the first $k-1$ steps. One can use it to determine when to stop the procedure.

For the application of the global convergence Theorem, it is in general problematic to determine the contraction constant θ and to prove $\phi(X) \subseteq X$.

Example 70.3 We determine the solution $\sqrt{2}$ of the equation $x^2 = 2$ to seven decimal places using the iteration method

$$\phi(x) = -\frac{1}{3}(x^2-2)+x\,.$$

Because of

$$|\phi'(x)| = \left|\frac{-2x}{3}+1\right| \le \frac{1}{3} \quad \text{on the interval } [1,2]$$

the local convergence Theorem states due to $\sqrt{2} \approx 1.5$ that the iteration sequence $(\phi(x_k))_k$ for each starting value x_0 from $[1,2]$ converges to $\sqrt{2}$ for each initial value.

We choose $x_0 = 1.5$. The a priori error estimate states due to $x_1 = 1.41\overline{6}$:

$$|x_n-\sqrt{2}| \le \frac{1}{2\cdot 3^{n-1}}\,|x_1-x_0| < \frac{0.05}{3^{n-1}}\,.$$

Because of $\frac{0.05}{3^{13}} < 5\cdot 10^{-8}$ thus 14 steps are sufficient. We obtain:

$$x_2 = 1.414351852\,,\; x_3 = 1.414221465\,,\; x_4 = 1.414214014\,,$$

$$x_5 = 1.414213588\,,\; x_6 = 1.414213564\,.$$

The a posteriori error estimate shows for $n = 6$:

$$|x_6-\sqrt{2}| \le \frac{1}{2}\,|x_6-x_5| < 1.1\cdot 10^{-8}\,,$$

i.e.

$$1.41421355 < \sqrt{2} < 1.41421358\,.$$

■

70.3 Implementation

A fixed point iteration

$$x_{k+1} = \phi(x_k)\,,\ k = 0,\ 1,\ 2,\ \ldots$$

should be stopped in an implementation, if

- the fixed point x has been approximated sufficiently accurately or
- the iteration is not expected to converge.

The first problem is solved by specifying a **tolerance limit**: Specify a tol > 0 and STOP as soon as

$$\|x_k - x\| < \text{tol}\,.$$

Since x is unknown, one resorts to the a posteriori error estimation of the global convergence theorem: STOP as soon as

$$\frac{\theta}{1-\theta}\,\|x_k - x_{k-1}\| < \text{tol}\,.$$

The second problem is solved by checking the contraction: Contraction means

$$\|\phi(x_k) - \phi(x_{k-1})\| \le \theta\,\|x_k - x_{k-1}\|\,,\ \theta < 1\,.$$

Because of $\|\phi(x_k) - \phi(x_{k-1})\| = \|x_{k+1} - x_k\|$ we have: STOP, as soon as

$$\theta_k = \frac{\|x_{k+1} - x_k\|}{\|x_k - x_{k-1}\|} > 1\,.$$

70.4 Rate of Convergence

If $(x_k)_k$ is a sequence converging to the limit x, then in the applications, i.e. if for instance $(x_k)_k$ is the sequence of a fixed point iteration, one hopes that the convergence is *fast*, so that one may stop the iteration with only *small* error after only a *few* iterations. A measure for this rate of convergence of a sequence is given by the *order of convergence*:

Convergence Order
If $(\boldsymbol{x}_k)_k$ is a convergent sequence in $\mathbb{R}^n$ with limit $\boldsymbol{x} \in \mathbb{R}^n$ then the sequence $(\boldsymbol{x}_k)_k$ is said to has the **order of convergence** $p \in \mathbb{N}$ if a $C > 0$ exists with

$$\|\boldsymbol{x}_{k+1} - \boldsymbol{x}\| \leq C\,\|\boldsymbol{x}_k - \boldsymbol{x}\|^p \text{ for all } k \geq k_0$$

for a $k_0 \in \mathbb{N}$ where in the case $p = 1$ additionally $C < 1$ is required.
In the cases $p = 1$, $p = 2$ and $p = 3$ one also speaks of **linear, quadratic** and **cubic convergence.**

For a linear convergent fixed point iteration with $C = \theta \approx \frac{1}{2}$ about 52 iterations are necessary for 15 correct decimal places.
With quadratic convergence, the number of correct digits approximately doubles in each step: If $\|\boldsymbol{x}_k - \boldsymbol{x}_{k-1}\| \approx 10^{-k}$,

$$\|\boldsymbol{x}_{k+1} - \boldsymbol{x}_k\| \leq C\,\|\boldsymbol{x}_k - \boldsymbol{x}_{k-1}\|^2 \leq C\,10^{-2k}.$$

So for 15 correct digits you need only about 4 iterations. Cubic convergence, on the other hand, is hardly worthwhile: For 14 correct digits, you need about 3 iterations.

70.5 Exercises

70.1 Show that the system

$$6x = \cos x + 2y,$$
$$8y = xy^2 + \sin x$$

has a unique solution in $E = [0, 1] \times [0, 1]$. We want to determine the solution using the global convergence theorem to within 10^{-3} in the maximum norm $\|\cdot\|_\infty$ using the global convergence theorem. How many iteration steps are sufficient for this, if we start at the point $(0, 0)$?

70.2 To determine a zero of $f(x) = \mathrm{e}^x - \sin x$ we consider the fixed point equation $\phi(x) = x$ with

$$\phi_1(x) = \mathrm{e}^x - \sin x + x\,,$$
$$\phi_2(x) = \sin x - \mathrm{e}^x + x\,,$$

$$\phi_3(x) = \arcsin(e^x) \quad \text{for } x < 0,$$

$$\phi_4(x) = \ln(\sin x) \quad \text{for } x \in \left]-2\pi, -\pi\right[.$$

(a) In each case, determine the derivative of ϕ_i and sketch ϕ_i and ϕ_i'.
(b) Label the regions where the fixed point iteration converges with certainty.

70.3 Using trigonometric identities, we can show that the function

$$\phi : \mathbb{R}^2 \to \mathbb{R}^2, \quad \boldsymbol{x} \mapsto \frac{1}{2}\begin{pmatrix} \sin x_1 + \cos x_2 \\ \sin x_2 - \cos x_1 \end{pmatrix}$$

satisfies a global Lipschitz condition $\|\phi(\boldsymbol{x}) - \phi(\boldsymbol{y})\| \leq \theta \, \|\boldsymbol{x} - \boldsymbol{y}\|$ for $\boldsymbol{x},\ \boldsymbol{y} \in \mathbb{R}^2$, where $\theta = \frac{1}{2}\sqrt{2}$ and $\|\cdot\|$ denotes the Euclidean norm.

Specifically choose $\boldsymbol{x}_0 = (0, 0)$ and estimate, using only the first iteration, how many steps k of the fixed point iteration with ϕ are required to obtain an accuracy of 10^{-6}.

70.4 Discuss which conditions of the global convergence theorem are satisfied or violated for the following functions:

(a) $f_1 : \left]0, 1\right[\to \mathbb{R}, \quad x \mapsto x^2$,
(b) $f_2 : [0, 1] \to \mathbb{R}, \quad x \mapsto \frac{1}{2}(x + 1)$,
(c) $f_3 : [0, 1]^2 \to \mathbb{R}, \quad x \mapsto \frac{1}{2}\left(x_1^2 + x_2^2\right)$,
(d) $f_4 : [0, 1]^2 \to \mathbb{R}^2, \quad x \mapsto (x_2^2, 1 - x_1)$,
(e) $f_5 : \mathbb{R} \to \mathbb{R}, \quad x \mapsto \ln(1 + e^x)$.

Which functions have a unique fixed point in their domains?

Iterative Methods for Systems of Linear Equations

71

In many applications, such as equilibrium considerations in mechanical or electrical networks or the discretization of boundary value problems in ordinary and partial differential equations, one obtains very large systems of equations, sometimes with many millions of rows. The coefficient matrices of these systems of equations are typically *sparse*, i.e., most of the matrix entries are zero. To solve such systems, one uses iteration methods with an initial value $\boldsymbol{x}_0$ to get an approximate solution for the exact $\boldsymbol{x}$ that solves $A\,\boldsymbol{x} = \boldsymbol{b}$. Each step of the iteration $\boldsymbol{x}_0 \to \boldsymbol{x}_1 \to \boldsymbol{x}_2 \to \cdots \to \boldsymbol{x}_{k-1} \to \boldsymbol{x}_k$ should require little computational effort.

Since even *exact* solution methods are subject to rounding errors and input errors contribute further to inaccuracies in the *exact* solutions, the inaccuracies in the approximate solution $\boldsymbol{x}_k$ can be lived with.

71.1 Solving Systems of Equations by Fixed Point Iteration

We are faced with the task of solving a system of linear equations

$$A\,\boldsymbol{x} = \boldsymbol{b} \text{ with invertible } A \in \mathbb{R}^{n\times n} \text{ and } \boldsymbol{b} \in \mathbb{R}^n$$

where n is *large* and A is sparse.

We obtain an iterative solution procedure following the considerations in Sect. 70.1 by formulating the solution of the LGS $A\,\boldsymbol{x} = \boldsymbol{b}$ as a fixed point problem. To obtain a solution to the fixed point problem, we can then resort to fixed point iteration.

But first to the fixed point equation: We know (see the remark in Sect. 70.1 that one can convert an equation into a fixed point equation in different ways. We consider a general approach in advance, and then obtain various common iterative methods by special choices.

C. Karpfinger, *Calculus and Linear Algebra in Recipes*,
https://doi.org/10.1007/978-3-662-65458-3_71

We divide the matrix $A \in \mathbb{R}^{n\times n}$ into two matrices $M \in \mathbb{R}^{n\times n}$ and $N \in \mathbb{R}^{n\times n}$ where we require that M is invertible:

$$A = M - N\,, \quad \text{where } M^{-1} \text{ exists}\,.$$

Then we have

$$A\boldsymbol{x} = \boldsymbol{b} \Leftrightarrow (M - N)\boldsymbol{x} = \boldsymbol{b} \Leftrightarrow M\boldsymbol{x} = \boldsymbol{b} + N\boldsymbol{x} \Leftrightarrow \boldsymbol{x} = M^{-1}\boldsymbol{b} + M^{-1}N\boldsymbol{x}\,.$$

The LGS as a Fixed Point Problem
If A=$M-N$ for A, M, $N \in \mathbb{R}^{n\times n}$ with invertible M, then solving $A\boldsymbol{x} = \boldsymbol{b}$ is equivalent to the fixed point determination $\phi(\boldsymbol{x}) = \boldsymbol{x}$ for

$$\phi(\boldsymbol{x}) = M^{-1}\boldsymbol{b} + M^{-1}N\boldsymbol{x}\,.$$

Here the fixed point iteration

$$\boldsymbol{x}_0 = \boldsymbol{s}\,,\ \boldsymbol{x}_{k+1} = \phi(\boldsymbol{x}_k)$$

for each initial value $\boldsymbol{s} \in \mathbb{R}^n$ is convergent, if the spectral radius of the (constant) **iteration matrix** $M^{-1}N$ is smaller than 1,

$$\rho(M^{-1}N) < 1\,.$$

The statement on convergence is obtained directly from the local norm-independent convergence Theorem in Sect. 70.2 since the iteration matrix $M^{-1}N$ is just the Jacobian matrix $D\phi$.

By different choices of M and N we obtain different methods.

71.2 The Jacobian Method

For the **Jacobian method** one chooses the decomposition

$$A = D - (L + R)\,, \ \text{so} \ M = D \ \text{and} \ N = L + R\,.$$

with an invertible diagonal matrix D, a matrix L, which has only entries below the diagonal, and a matrix R, which has only entries above the diagonal:

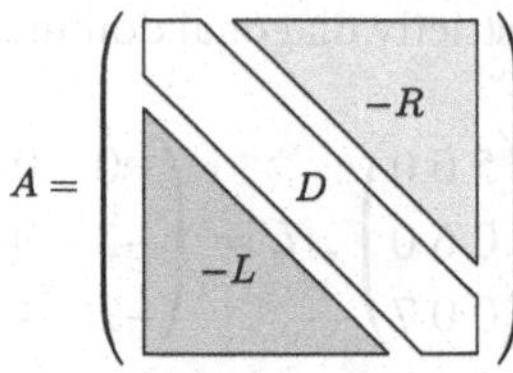

The function ϕ in the Jacobian method is as follows:

$$\phi(\boldsymbol{x}) = D^{-1}\boldsymbol{b} + D^{-1}(L + R)\,\boldsymbol{x}\,.$$

Thus, for the fixed point iteration we obtain

$$\boldsymbol{x}_0 = \boldsymbol{s}\,,\ \boldsymbol{x}_{k+1} = \phi(\boldsymbol{x}_k)$$

the explicit matrix-wise or component-wise formulation:

The Jacobian Method
Let $A = D - (L + R) \in \mathbb{R}^{n\times n}$ with invertible D, and L and R as indicated above. At **Jacobian method** one constructs from a starting vector $\boldsymbol{x}^{(0)} = (x_i^{(0)})_i \in \mathbb{R}^n$ the approximate solutions $\boldsymbol{x}^{(m+1)} = (x_i^{(m+1)})_i \in \mathbb{R}^n$ as follows:

$$\boldsymbol{x}^{(m+1)} = D^{-1}\boldsymbol{b} + D^{-1}(L + R)\,\boldsymbol{x}^{(m)}$$

or component-wise

$$x_i^{(m+1)} = a_{ii}^{-1}\left(b_i - \sum_{\substack{j=1\\ j\neq i}}^{n} a_{ij}x_j^{(m)}\right).$$

The Jacobian method converges for each starting value $\boldsymbol{x}^{(0)} \in \mathbb{R}^n$ if A is **strictly diagonally dominant**, i.e., if the absolute value of the diagonal entries are larger than the sums of the absolute values of the respective remaining row entries:

$$|a_{ii}| > \sum_{\substack{j=1\\ j\neq i}}^{n} |a_{ij}| \ \text{ for all } i\,.$$

We have posed the proof as an exercise problem.

Example 71.1 We consider the strictly diagonal dominant matrix

$$A = \begin{pmatrix} 5 & 1 & 2 \\ 2 & 6 & 3 \\ 2 & 4 & 7 \end{pmatrix}, \quad \text{also } D = \begin{pmatrix} 5 & 0 & 0 \\ 0 & 6 & 0 \\ 0 & 0 & 7 \end{pmatrix}, \; L = \begin{pmatrix} 0 & 0 & 0 \\ -2 & 0 & 0 \\ -2 & -4 & 0 \end{pmatrix}, \; R = \begin{pmatrix} 0 & -1 & -2 \\ 0 & 0 & -3 \\ 0 & 0 & 0 \end{pmatrix}.$$

For $\boldsymbol{b} = (3, -1, -5)^\top$ the exact solution is $\boldsymbol{x} = (1, 0, -1)^\top$ of the LGS $A\,\boldsymbol{x} = \boldsymbol{b}$. With the starting vector $\boldsymbol{x}^{(0)} = (3, 2, 1)^\top$ we obtain:

Iterated	$\boldsymbol{x}^{(0)}$	$\boldsymbol{x}^{(10)}$	$\boldsymbol{x}^{(20)}$	$\boldsymbol{x}^{(30)}$	$\boldsymbol{x}^{(40)}$	$\boldsymbol{x}^{(50)}$
1. entry	3.0000	0.8313	0.9867	0.9990	0.9999	1.0000
2. entry	2.0000	−0.2143	−0.0169	−0.0013	−0.0001	−0.0000
3. entry	1.0000	−1.2199	−1.0173	−1.0014	−1.0001	−1.0000
rel. error		0.4128	0.0390	0.0031	0.0002	0.0000

■

71.3 The Gauss-Seidel Method

The **Gauss-Seidel method** is a simple variation of the Jacobian method: For the component-wise determination of the approximation vector $x^{(m)} \in \mathbb{R}^n$ one uses the formula in the Jacobian procedure (see above):

$$x_i^{(m+1)} = a_{ii}^{-1} \left(b_i - \sum_{\substack{j=1 \\ j \neq i}}^{n} a_{ij} x_j^{(m)} \right).$$

Now it is only an obvious idea, in determining the i-th component $x_i^{(m+1)}$, $i \geq 2$, to use the previously determined components $x_1^{(m+1)}, \ldots, x_{i-1}^{(m+1)}$. This gives better values in calculating the next iterate, which improves convergence. This simple variation of the Jacobi method leads to the following procedure:

The Gauss-Seidel Method
With the Gauss-Seidel method, one obtains from a start vector $\boldsymbol{x}^{(0)} = (x_i^{(0)}) \in \mathbb{R}^n$ the components

$$x_i^{(m+1)} = a_{ii}^{-1}\left(b_i - \sum_{j=1}^{i-1} a_{ij}x_j^{(m+1)} - \sum_{j=i+1}^{n} a_{ij}x_j^{(m)}\right).$$

This approach corresponds to the following division of the matrix A:

$$A = \underbrace{(D-L)}_{=M} - \underbrace{R}_{=N}.$$

Here D has to be invertible. Explicitly, the iteration is:

$$\boldsymbol{x}^{(m+1)} = (D-L)^{-1}\boldsymbol{b} + (D-L)^{-1}R\,\boldsymbol{x}^{(m)}.$$

The Gauss-Seidel method converges

- for every strictly diagonally dominant matrix A,
- for any positive definite matrix A.

The fact that the Gauss-Seidel method converges for any strictly diagonally dominant matrix A follows from the convergence of the Jacobian method for these matrices: for the iteration matrix $D^{-1}(L+R)$ of the Jacobian method and $(D-L)^{-1}R$ of the Gauss-Seidel method we namely have

$$\|(D-L)^{-1}R\|_\infty \le \|D^{-1}(L+R)\|_\infty.$$

Example 71.2 We consider again the LGS from Example 71.1. We choose the same starting vector $\boldsymbol{x}^{(0)} = (3, 2, 1)^\top$ and give the iterates, which have approximately the same relative error as in the example for the Jacobian method; note that the number of iterations is much smaller:

Iterated	$\boldsymbol{x}^{(0)}$	$\boldsymbol{x}^{(4)}$	$\boldsymbol{x}^{(6)}$	$\boldsymbol{x}^{(8)}$	$\boldsymbol{x}^{(10)}$	$\boldsymbol{x}^{(12)}$
1. entry	3.0000	0.9740	0.9985	0.9985	0.9999	1.0000
2. entry	2.0000	−0.0967	−0.0087	−0.0008	−0.0001	−0.0000
3. entry	1.0000	−0.9374	−0.9946	−0.9995	−1.0000	−1.0000
rel. error		0.2000	0.0207	0.0018	0.0002	0.0000

■

71.4 Relaxation

By a variation of the Gauss-Seidel method we obtain the *SOR* method, also called the *relaxed Gauss-Seidel method*. Here, the Gauss-Seidel method introduces a relaxation parameter ω, which in many cases significantly improves the efficiency. We first formulate this relaxation in general and then consider it on the Gauss-Seidel method:

If ϕ is an iterative process, one forms a **convex combination** from $\boldsymbol{x}_{k+1} = \phi(\boldsymbol{x}_k)$ and $\boldsymbol{x}_k$:

$$\boldsymbol{x}_{k+1}^{\text{(new)}} = \omega\,\boldsymbol{x}_{k+1}^{\text{(old)}} + (1-\omega)\,\boldsymbol{x}_k\,.$$

The at first arbitrary number $\omega \in [0, 1]$ is called **relaxation parameter**. One thus obtains for the iteration ϕ a *family* $\{\phi_\omega \mid \omega \in [0, 1]\}$ of fixed point iterations: for each $\omega \in [0, 1]$ namely

$$\phi_\omega(\boldsymbol{x}) = \omega\phi(\boldsymbol{x}) + (1-\omega)\,\boldsymbol{x}\,.$$

Note that for $\omega = 1$ one obtains $\phi_\omega = \phi$. For the fixed point iteration with ϕ_ω instead of ϕ, one speaks of a **relaxed method**.

Since the spectral radius of the Jacobian matrix $D\phi_\omega(\boldsymbol{x})$ of ϕ is a measure of convergence, one now chooses ω such that

$$\rho(D\phi_\omega(\boldsymbol{x})) \text{ minimal}$$

is. We denote this ω by ω_{opt}. This optimal ω can be determined in the iterative methods we have discussed for solving an LGS; we have:

The Optimal ω

Let ϕ be an iterative procedure for solving an LGS $Ax = b$ with iteration matrix $M^{-1}N$ ($A = M - N$),

$$\phi(\boldsymbol{x}) = M^{-1}N\boldsymbol{x} + M^{-1}\boldsymbol{b}\,.$$

Let further

- $\rho(M^{-1}N) < 1$, and
- the eigenvalues $\lambda_1 \leq \cdots \leq \lambda_n$ of $M^{-1}N$ are all real.

(continued)

Then, for the relaxed method

$$\phi_\omega(x) = \omega(M^{-1}Nx + M^{-1}b) + (1-\omega)x = (\omega M^{-1}N + (1-\omega)E_n)x + \omega M^{-1}b$$

the optimal ω is given by

- $\omega_{\text{opt}} = \frac{2}{2-\lambda_1-\lambda_n}$, if ϕ is the Jacobian, and
- $\omega_{\text{opt}} = \frac{2}{1+\sqrt{1-\lambda_n^2}}$, if ϕ is the Gauss-Seidel method.

Note that it takes considerable effort to find the eigenvalues of $M^{-1}N$ to be calculated. In practice, this is not done. There one uses estimates for the eigenvalues. Sometimes one is much *coarser*; one distinguishes the two purposes *solver* and *pre conditioner* for relaxed iteration methods: If one uses such a procedure as a solver, one uses $\omega \approx 1.4$, but if one uses the procedure as a pre conditioner, one uses $\omega \approx 0.7$.

In the relaxed Gauss-Seidel method, we obtain the matrix decomposition

$$A = \underbrace{\left(\tfrac{1}{\omega}D - L\right)}_{=M} - \underbrace{\left(\left(\tfrac{1}{\omega} - 1\right) D + R\right)}_{=N}$$

and hence:

The SOR Method
The relaxed Gauss-Seidel method is also called the **SOR method** (*successive over-relaxation*). The procedure is explicitly:

$$x_i^{(m+1)} = \omega\, a_{ii}^{-1}\left(b_i - \sum_{j=1}^{i-1} a_{ij}x_j^{(m+1)} - \sum_{j=i+1}^{n} a_{ij}x_j^{(m)}\right) + (1-\omega)\, x_i^{(m)}\,.$$

In matrix-vector notation:

$$x^{(m+1)} = \left(E_n - \left(\tfrac{1}{\omega}D - L\right)^{-1} A\right) x^{(m)} + \left(\tfrac{1}{\omega}D - L\right)^{-1} b\,.$$

The SOR method converges for $0 < \omega < 2$ if A is positive definite.

Note that for $\omega = 1$ one gets back the Gauss-Seidel process. For $1 < \omega < 2$ one speaks of **over relaxation**.

Remark For all the methods discussed in this chapter, viz.

- Jacobian method,
- Gauss-Seidel method,
- SOR method

one iteration for a sparse matrix $A \in \mathbb{R}^{n\times n}$ corresponds approximately to a matrix-vector multiplication.

71.5 Exercises

71.1 Justify the convergence statement for the Jacobian method in Sect. 71.2.

71.2

(a) Use the Jacobi method to calculate the first three iterations $\boldsymbol{x}_1$, $\boldsymbol{x}_2$, $\boldsymbol{x}_3$ of the linear system of equations

$$\begin{aligned} 15x_1 + 2x_2 &= -1 \\ x_1 - 4x_2 &= -9\,, \end{aligned}$$

where $\boldsymbol{x}_0$ is the zero vector.

(b) Justify that the sequence $(\boldsymbol{x}_k)$ converges.

71.3 Repeat the previous task using the Gauss-Seidel method.

71.4 Using the Jacobi and Gauss-Seidel methods, determine the first two iterates $\boldsymbol{x}_1$, $\boldsymbol{x}_2$ of the following linear system of equations with $\boldsymbol{x}_0 = \boldsymbol{0}$

$$\begin{aligned} 3x_1 - \ \ x_2 + \ \ x_3 &= 1 \\ 3x_1 + 6x_2 + 2x_3 &= 0 \\ 3x_1 + 3x_2 + 7x_3 &= 4\,. \end{aligned}$$

71.5 Write a MATLAB PROGRAM that implements the Jacobi or Gauss-Seidel algorithm. Test this on matrices that are each strictly diagonally dominant.

Optimization

72

Optimization problems are manifold in nature. Whether the desire is for long battery life in a laptop or low fuel consumption in a car, there are always requirements placed on individual components: Minimize or maximize properties such as weight, size, performance, etc.

In fact, we can restrict ourselves to minimization *problems*, and we already know a criterion for a minimum of a function f in several variables: In a stationary place $\boldsymbol{x}^*$ is a minimum if the Hessian matrix $H_f(\boldsymbol{x}^*)$ is positive definite. However, we want to find a minimum without determining the zeros of the gradient, since this is too costly for realistic problems.

72.1 The Optimum

The abstract formulation of the **minimization problem** is as follows: Given is a **feasible region** $X \subseteq \mathbb{R}^n$ and a (continuous) **objective function** $f : X \to \mathbb{R}$. Determine the minimum of f on X, notation:

$$\min_{\boldsymbol{x} \in X} f(\boldsymbol{x}) .$$

Since the maximization of f just means the minimization of $-f$, it is therefore sufficient to treat the minimization problem.

As in Sect. 49.1 we call each point $\boldsymbol{x}^* \in X$ with $\nabla f(\boldsymbol{x}^*) = 0$ a **stationary point**. If $f \in C^1(X)$ in $\boldsymbol{x}^* \in X$ has a local minimum, then

$$\nabla f(\boldsymbol{x}^*) = 0 .$$

C. Karpfinger, *Calculus and Linear Algebra in Recipes*,
https://doi.org/10.1007/978-3-662-65458-3_72

72.2 The Gradient Method

How do you get from the mountain to the valley at night, in complete darkness? You slowly feel your way downhill: you check with the tip of your foot in which direction it goes downhill, and then you walk in this direction for a while before you stop again and choose a new direction of descent in the same way. The whole thing goes silly if you end up in a hollow: But at least you have then found a local minimum instead of the global minimum. The first idea is that you always wander in the direction of the steepest descent. You might think that's the fastest way to go downhill. This doesn't have to be so, it's easy to think about this now with the illustration we mentioned.

In the *gradient method* (it is also called the *steepest descent method*), one starts from an approximate value and progresses in the direction of the negative gradient, which is known to indicate the direction of steepest descent from that approximate value, until no numerical improvement is obtained.

This *gradient procedure* is a special case of the general *descent procedure*:

Recipe: The General Descent Procedure and the Gradient Procedure
Given are a C^1-function $f : \mathbb{R}^n \to \mathbb{R}$ and a starting vector $\boldsymbol{x}_0 \in \mathbb{R}^n$.

For $k = 1, 2, \ldots$ calculate iteratively $\boldsymbol{x}_1, \boldsymbol{x}_2, \ldots$ (as long as $\boldsymbol{x}_k$ is not approximatively stationary) by:

(1) Determine a **descent direction**, i.e. $\boldsymbol{v}_k \in \mathbb{R}^n$ with

$$\nabla f(\boldsymbol{x}_k)^\top \boldsymbol{v}_k < 0\,.$$

(2) Determine a step size, i.e. $h_k \in \mathbb{R}$ with

$$f(\boldsymbol{x}_k + h_k\, \boldsymbol{v}_k) < f(\boldsymbol{x}_k)\,.$$

(3) Set $\boldsymbol{x}_{k+1} = \boldsymbol{x}_k + h_k \boldsymbol{v}_k$.

Choosing $\boldsymbol{v}_k = -\nabla f(\boldsymbol{x}_k)$ one obtains the **Gradient method.**

In the case $\frac{\partial f}{\partial \boldsymbol{v}_k}(\boldsymbol{x}_k) = \nabla f(\boldsymbol{x}_k)^\top \boldsymbol{v}_k < 0$ we have $f(\boldsymbol{x}_k + t\, \boldsymbol{v}_k) < f(\boldsymbol{x}_k)$ for a t, i.e., in the direction $\boldsymbol{v}_k$ it goes downhill (note the box in Sect. 47.1). And the direction of the steepest descent is known to be $\boldsymbol{v}_k = -\nabla f(\boldsymbol{x}_k)$ (see Sect. 47.1).

Note that we still have a lot of leeway in the choice of the step size: The obvious choice is the following *exact* step size, where we move so long in the direction $\boldsymbol{v}_k$ as long as it goes downhill, but relevant for practice is the *Armijo step* size:

(a) **exact step size**: Consider the function

$$\varphi(h) = f(\boldsymbol{x}_k + h\boldsymbol{v}_k)$$

and determine $\min_{h>0} \varphi(h)$. This is difficult in practice.

(b) **Armijo step size**: For a $\gamma \in (0, 1)$ determine the largest number $h_k \in \{1, \frac{1}{2}, \frac{1}{4}, \frac{1}{8}, \ldots\}$ so that

$$f(\boldsymbol{x}_k + h_k\boldsymbol{v}_k) \leq f(\boldsymbol{x}_k) + h_k\gamma \, \nabla f(\boldsymbol{x}_k)^\top \boldsymbol{v}_k \,.$$

One can reason:

Convergence of the Gradient Method
The gradient method with the Armijo step either terminates in a stationary point or an infinite sequence $(\boldsymbol{x}_k)_{k\in\mathbb{N}}$ is generated for which we have:

$$f(\boldsymbol{x}_{k+1}) < f(\boldsymbol{x}_k) \text{ for all } k \,.$$

In the gradient method, one has only linear convergence.

72.3 Newton's Method

In the *Newton method*, as in the gradient method, the place $\boldsymbol{x}^*$ of a local minimum is tried to be determined approximately. Thereby one even achieves quadratic convergence, which, however, is only local. In a global version of Newton's method, on the other hand, quadratic convergence is lost; it is then only locally guaranteed.

Recipe: The Local Newton Method
Given is a C^2-function and a starting vector $\boldsymbol{x}_0 \in \mathbb{R}^n$.

For $k = 1, 2, \ldots$ calculate iteratively $\boldsymbol{x}_1, \boldsymbol{x}_2, \ldots$ (as long as $\boldsymbol{x}_k$ is not approximatively stationary) by:

(1) Determine $\boldsymbol{v}_k$ by solving

$$H_f(\boldsymbol{x}_k)\boldsymbol{v}_k = -\nabla f(\boldsymbol{x}_k) \,.$$

(continued)

(2) Set $\boldsymbol{x}_{k+1} = \boldsymbol{x}_k + \boldsymbol{v}_k$.

If $\boldsymbol{x}^*$ a local minimum of f, then there exists a $\delta > 0$, so that the above algorithm terminates for all $\boldsymbol{x}_0 \in B_\delta(\boldsymbol{x}^*)$; the convergence is quadratic.

In this local Newton method, using the ordinary Newton method (see recipe in Sect. 48.1) a zero of the gradient is approximated, the Hess matrix H_f is just the Jacobian matrix of the vector field ∇f, $D\nabla f = H_f$.

The Newton method converges only locally. By the following modification, we link the gradient method with the Newton method, we obtain a global version, which converges globally but only linearly. This linkage is done as follows: As with the Newton method, we compute a solution $\hat{\boldsymbol{v}}_k$ of the system of equations $H_f(\boldsymbol{x}_k)\hat{\boldsymbol{v}}_k = -\nabla f(\boldsymbol{x}_k)$ and now decide:

- If $\nabla f(\boldsymbol{x}_k)^\top \hat{\boldsymbol{v}}_k$ is much smaller than zero, then we do a Newton step.
- If $\nabla f(\boldsymbol{x}_k)^\top \hat{\boldsymbol{v}}_k$ is not much smaller than zero, then we do a gradient step with Armijo step size.

The decision of when a number is *much* smaller than zero is decided by practical experience. We formulate this globalized Newton procedure as a recipe with an indefinite $c > 0$:

Recipe: The Globalized Newton Method
Given is a C^2-function f and a starting vector $\boldsymbol{x}_0 \in \mathbb{R}^n$. Further let be given $c > 0$ and $\gamma \in (0, 1)$.

For $k = 1, 2, \ldots$ calculate iteratively $\boldsymbol{x}_1, \boldsymbol{x}_2, \ldots$ (as long as $\boldsymbol{x}_k$ is not approximately stationary):

(1) Determine $\hat{\boldsymbol{v}}_k$ by solving

$$H_f(\boldsymbol{x}_k)\hat{\boldsymbol{v}}_k = -\nabla f(\boldsymbol{x}_k)\,.$$

If

$$-\nabla f(\boldsymbol{x}_k)^\top \hat{\boldsymbol{v}}_k \geq c$$

set $\boldsymbol{v}_k = \hat{\boldsymbol{v}}_k$ otherwise $\boldsymbol{v}_k = -\nabla f(\boldsymbol{x}_k)$.

(2) $h_k =$ Armijo step size.

In this method, as in the gradient method with Armijo step size, global convergence is ensured, and the rate of convergence is globally linear and locally quadratic.

The major drawback is the evaluation of $H_f(\boldsymbol{x}_k)$ which for large problems ($n > 10^4$) is time-consuming. A remedy here is the

- *Inexact Newton method*: Solution of $H_f(\boldsymbol{x}_k)\boldsymbol{v}_k = -\nabla f(\boldsymbol{x}_k)$ is done (iteratively) only up to certain precision, or the
- *Quasi-Newton method*: Substitute $H_f(\boldsymbol{x}_k)$ by an (easier to calculate) matrix H_k.

We omit a more detailed exposition of these procedures.

72.4 Exercises

72.1 In the following, we consider the behavior of the gradient method with the minimization rule, i.e., the step size h_k is determined by

$$f(\boldsymbol{x}_k + h_k \boldsymbol{v}_k) = \min_{h \geq 0} f(\boldsymbol{x}_k + h\,\boldsymbol{v}_k)\,. \tag{M}$$

Let

$$f\colon \mathbb{R}^n \to \mathbb{R}, \quad \boldsymbol{x} \mapsto \boldsymbol{c}^\top \boldsymbol{x} + \frac{1}{2}\boldsymbol{x}^\top C\,\boldsymbol{x}$$

with $\boldsymbol{c} \in \mathbb{R}^n$ and $C \in \mathbb{R}^{n \times n}$ positive definite. Furthermore, let

$$\varphi(h) = f(\boldsymbol{x}_k + h\,\boldsymbol{v}_k).$$

(a) Calculate φ' and φ'' and conclude that $\varphi''(h) > 0$.
(b) By which equation is the solution of (M) uniquely determined? From this equation, determine the step size h_k.
(c) Let it be from now on

$$f\colon \mathbb{R}^2 \to \mathbb{R}\,,\ \boldsymbol{x} = (x_1, x_2) \mapsto x_1^2 + 3x_2^2$$

with $\boldsymbol{x}_0 = (3, 1)^\top$. Calculate the iterated $\boldsymbol{x}_k$ and the step sizes h_k.
(d) Show that $\boldsymbol{v}_k$ and $\boldsymbol{v}_{k+1}$ are perpendicular to each other.
(e) Based on (d), what can you say about the behavior of the gradient method?
(f) Determine the global minimum $\boldsymbol{x}^*$ of f.
(g) The rate of convergence γ of the method is defined by

$$\|\boldsymbol{x}_{k+1} - \boldsymbol{x}^*\| \leq \gamma\,\|\boldsymbol{x}_k - \boldsymbol{x}^*\|\,, \quad k \in \mathbb{N}_0.$$

Determine γ.

72.2 Consider $G(x) = 1/x - a$ specifically for $a = 2$ and the initial value $x_0 = 2$. Since $x_0 > 2/a$ diverges the *normal* Newton method for this initial value. To achieve convergence here as well, let us globalize Newton's method in a suitable way.

(a) For the determination of a zero of a continuously differentiable function $F : \mathbb{R}^n \to \mathbb{R}^n$ the Newton equation is

$$DF(\boldsymbol{x}_k)\boldsymbol{v}_k = -F(\boldsymbol{x}_k). \tag{72.1}$$

We now want to determine solutions of

$$\min_{\boldsymbol{x}\in\mathbb{R}^n} f(\boldsymbol{x}), \text{ where } f(\boldsymbol{x}) = \tfrac{1}{2}\|F(\boldsymbol{x})\|_2^2 .$$

Show that

$$\nabla f(\boldsymbol{x}_k)^\top \boldsymbol{v}_k < 0$$

(with $\boldsymbol{v}_k$ from (72.1) if $DF(\boldsymbol{x}_k)$ is regular).

(b) Formulate the Armijo condition for the descent direction $\boldsymbol{v}_k$ from (a) and calculate that it is equivalent to:
let $\gamma \in (0, 1)$, and choose the largest $h_k \in \{1, \frac{1}{2}, \frac{1}{4}, \ldots\}$ with

$$\|F(\boldsymbol{x}_k + h_k\boldsymbol{v}_k)\|_2^2 \le (1 - 2h_k\gamma)\|F(\boldsymbol{x}_k)\|_2^2 . \tag{72.2}$$

Use the Newton equation (72.1).

(c) Let us apply what we have said so far to the function G from above with $a = 2$. What is the initial direction, i.e., the initial Newton step v_0 here?

(d) Which step size $h_0 \in \{1, \frac{1}{2}, \frac{1}{4}, \ldots\}$ supplies the condition (72.2) for $\gamma \in]0, \frac{1}{2}[$?

(e) What do you notice about the iterated $x_1 = x_0 + h_0 v_0$?

72.3 Implement the gradient method and the Newton method for optimization problems in MATLAB. Compare the methods on the minimization problems $f_1(x, y) = \frac{1}{2}x^2 + \frac{9}{2}y^2 + 1$ and $f_2(x, y) = \frac{1}{2}x^2 + y^2 + 1$.

Numerics of Ordinary Differential Equations II 73

In Chap. 36 we have already discussed the essential methods for the numerical solution of ordinary differential equations; indeed, the methods given there are applicable unchanged to systems of differential equations as well. In the present chapter we talk about *convergence* and *consistency* of one-step methods and point out the importance of implicit methods for the solution of *stiff* differential equation systems.

73.1 Solution Methods for ODE Systems

We consider an IVP

$$\dot{\boldsymbol{x}} = \boldsymbol{f}(t, \boldsymbol{x}) \text{ with } \boldsymbol{x}(t_0) = \boldsymbol{x}_0$$

with a 1st order ODE system $\dot{\boldsymbol{x}} = \boldsymbol{f}(t, \boldsymbol{x})$. We assume that the IVP is uniquely solvable.

For the numerical solution of such a problem, the methods described in Sect. 36.3 i.e., the explicit and implicit Euler methods, the Runge-Kutta methods, or the explicit and implicit multistep methods from Sect. 36.3 can be used. All of these discussed methods work analogously for ODE systems. The programs from Exercise 36.1 can also be used unchanged for systems.

The Euler and Runge-Kutta Methods for ODE Systems

We obtain approximate solutions $\boldsymbol{x}_k$ with the following methods for the exact solution $\boldsymbol{x}(t_k)$ of the IVP

$$\dot{\boldsymbol{x}} = \boldsymbol{f}(t, \boldsymbol{x}) \text{ with } \boldsymbol{x}(t_0) = \boldsymbol{x}_0$$

(continued)

C. Karpfinger, *Calculus and Linear Algebra in Recipes*,
https://doi.org/10.1007/978-3-662-65458-3_73

at the points $t_k = t_0 + k\,h$ with $k = 0, 1, \ldots, n$ and $h = \frac{t-t_0}{n}$ for $n \in \mathbb{N}$.

- In the **explicit Euler method** the approximation points $\boldsymbol{x}_k$ for $\boldsymbol{x}(t_k)$ are recursively from $\boldsymbol{x}_0$ determined by

$$\boldsymbol{x}_{k+1} = \boldsymbol{x}_k + h\, f(t_k, \boldsymbol{x}_k)\,,\ k = 0, 1, 2 \ldots .$$

- At **implicit Euler method** one determines the $\boldsymbol{x}_k$ recursively according to

$$\boldsymbol{x}_{k+1} = \boldsymbol{x}_k + h\, f\,(t_{k+1}, \boldsymbol{x}_{k+1})\,,\ k = 0, 1, 2 \ldots .$$

- At **classical Runge-Kutta method** one determines the $\boldsymbol{x}_k$ recursively according to

$$\boldsymbol{x}_{k+1} = \boldsymbol{x}_k + \frac{h}{6}\,(\boldsymbol{k}_1 + 2\boldsymbol{k}_2 + 2\boldsymbol{k}_3 + \boldsymbol{k}_4)$$

with

$$\boldsymbol{k}_1 = f(t_k, \boldsymbol{x}_k)\,,\ \boldsymbol{k}_2 = f\left(t_k + \frac{h}{2}, \boldsymbol{x}_k + \frac{h}{2}\boldsymbol{k}_1\right),$$

$$\boldsymbol{k}_3 = f\left(t_k + \frac{h}{2}, \boldsymbol{x}_k + \frac{h}{2}\boldsymbol{k}_2\right),\ \boldsymbol{k}_4 = f\,(t_k + h, \boldsymbol{x}_k + h\,\boldsymbol{k}_3)\,.$$

We use the explicit Euler method in the *predator-prey model*:

Example 73.1 At **predator-prey model** we consider two time-dependent populations: The predators $r = r(t)$ and the prey $b = b(t)$ whose *coexistence* is described by the following ODE system:

$$\dot{b}(t) = a_1 b(t) - a_2 r(t) b(t)$$

$$\dot{r}(t) = -a_3 r(t) + a_4 b(t) r(t)\,.$$

We consider more concretely the system with the numbers $a_1 = 2$, $a_2 = a_3 = a_4 = 1$ with the initial condition $(b(0), r(0)) = (1, 1)$ i.e. $t_0 = 0$ and $\boldsymbol{x}_0 = (1, 1)^\top$.

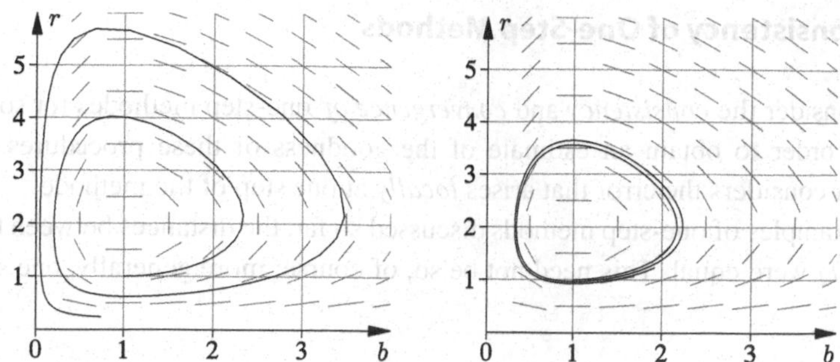

Fig. 73.1 Approximations with the different step sizes $h = 0.1$ and $h = 0.01$

In the explicit Euler method, we choose the step sizes $h = 0.1$ and $h = 0.01$ and select the method implemented in MATLAB from Exercise 36.1: First, we enter the right-hand side of the ODE as the function f,

```
f = @(x)  [2*x(1)-x(2)*x(1);  -x(2)+x(1)*x(2)];
```

and then call the program

```
h=0.1; N=100;
x=expl_euler(f,[1;1],h,N);
figure(1);plot(x(1,:),x(2,:));
grid on;
```

or

```
h=0.01; N=1000;
x=expl_euler(f,[1;1],h,N);
figure(2);plot(x(1,:),x(2,:));
grid on;
```

This gives us the plots in Fig. 73.1. ■

73.2 Consistency and Convergence of One-Step Methods

As before, we consider a uniquely solvable IVP

$$\dot{x} = f(t, x) \text{ with } x(t_0) = x_0 ,$$

where we assume the function $x = x(t)$ as vector-valued,

$$x : I \subseteq \mathbb{R} \to \mathbb{R}^n , \ x(t) = (x_1(t), \ldots, x_n(t))^\top .$$

73.2.1 Consistency of One-Step Methods

We now consider the *consistency* and *convergence of* one-step methodes for solving such an IVP in order to obtain an estimate of the goodness of these procedures. Here, the *consistency* considers the error that arises *locally* at one step of the methode.

In the examples of one-step methods discussed so far, the distances between times were $t_0, t_1, \ldots, t_d$ were equal. This need not be so, of course; more generally, one speaks of a **time grid**

$$\Delta = \{t_0, t_1, \ldots, t_d\} \subseteq \mathbb{R}$$

with the **step sizes** $h_j = t_{j+1} - t_j$ for $j = 0, \ldots, d-1$ and the **maximum step size**

$$h_\Delta = \max\{h_j \mid j = 0, \ldots, d-1\}.$$

In the approximate solution of an IVP, one determines a **lattice function**

$$\boldsymbol{x}_\Delta : \Delta \to \mathbb{R}^n$$

with $\boldsymbol{x}_\Delta(t_j) \approx x(t_j)$ for all $j = 0, \ldots, d-1$.

In a one-step method, successive

$$\boldsymbol{x}_\Delta(t_1), \boldsymbol{x}_\Delta(t_2), \ldots, \boldsymbol{x}_\Delta(t_d)$$

are calculated, whereby in the calculation of $\boldsymbol{x}_\Delta(t_{k+1})$ only $\boldsymbol{x}_\Delta(t_k)$ is used. This is suggestively abbreviated with the following notation:

$$\boldsymbol{x}_\Delta(t_0) = \boldsymbol{x}_0, \ \boldsymbol{x}_\Delta(t_{k+1}) = \Psi(\boldsymbol{x}_\Delta(t_k), t_k, h_k)$$

and also briefly referred to as the one-step method Ψ.

Consistency of a One-Step Method

A one-step method Ψ is called **consistent**, if

(i) $\Psi(\boldsymbol{x}_\Delta(t_k), t_k, 0) = \boldsymbol{x}_\Delta(t_k)$,
(ii) $\frac{\mathrm{d}}{\mathrm{d}h}\Psi(\boldsymbol{x}_\Delta(t_k), t, h)|_{h=0} = f(t, \boldsymbol{x})$.

By the **consistency error** of a one-step procedure is understood:

$$\varepsilon(\boldsymbol{x}, t, h) = \boldsymbol{x}(t+h) - \Psi(\boldsymbol{x}, t, h).$$

(continued)

A one-step procedure is said to have the **consistency order** $p \in \mathbb{N}$ if

$$\varepsilon(\boldsymbol{x}, t, h) = O(h^{p+1}) \quad (h \to 0).$$

That is, for every h_0 there is a constant $C = C(h_0)$ such that

$$|\varepsilon(\boldsymbol{x}, t, h)| \le Ch^{p+1} \text{ for } h \ge h_0 .$$

Thus, the consistency error gives the difference between the exact solution at $t+h$ and the approximate solution $\boldsymbol{x}_{t+h}$ of the one-step procedure Ψ, where here the exact value is assumed. The consistency thus describes a local behavior of the one-step procedure Ψ (note Fig. 73.2).

Thus, a high consistency order locally ensures that the error made by the one-step method Ψ quickly becomes vanishingly small when the step size h is reduced. So, in this view, it is desirable to have one-step methods with high consistency order at hand. Of course, we will immediately see what consistency order the methods we are considering have. Of course, this immediately raises the question of how on earth this quantity p can be calculated.

Example 73.2 The explicit Euler method

$$\Psi(x_0, t, h) = x_0 + h\, f(t, x_0)$$

has the consistency order $p = 1$; namely

$$x(t+h; t, x_0) = x_0 + \int_t^{t+h} f(s, x(s))\mathrm{d}s = x_0 + h\, f(t, x(t)) + O(h^2).$$

■

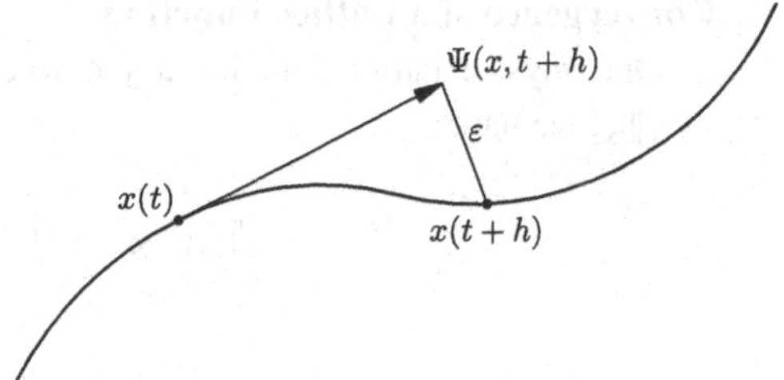

Fig. 73.2 The consistency error

Remark It can be shown that under suitable conditions the following three statements are equivalent:

(i) The one-step method Ψ is consistent.
(ii) $\frac{\varepsilon(x,t,h)}{h} \to 0$ for $h \to 0$.
(iii) The one-step method Ψ has the form

$$\Psi(x, t, h) = x + h\,\psi(x, t, h)$$

with a function ψ that is called **increment function** or **process function**.

73.2.2 Convergence of One-Step Method

Consistency provides a local assessment of the quality of a one-step method depending on the selected time grid. *Convergence*, on the other hand, provides a global estimate of the goodness of a one-step method.

We still assume the IVP $\dot{\boldsymbol{x}} = f(t, \boldsymbol{x})$, $\boldsymbol{x}(t_0) = \boldsymbol{x}_0$ and consider a time lattice $\Delta = \{t_0, t_1, \ldots, t_d\}$ with a lattice function $\boldsymbol{x}_\Delta$. The following terms describe the error obtained by the lattice function—once locally as a function in t, once globally as a numerical value: To a lattice function. $\boldsymbol{x}_\Delta : \Delta \to \mathbb{R}^n$ denotes

$$\varepsilon_\Delta : \Delta \to \mathbb{R}^n\,,\ \varepsilon_\Delta(t) = \boldsymbol{x}(t) - \boldsymbol{x}_\Delta(t)$$

the **lattice error**, and

$$\|\varepsilon_\Delta\|_\infty = \max_{t\in\Delta} \|\varepsilon_\Delta(t)\|_2$$

is called **discretization error**.

If the discretization error always becomes smaller when the step sizes are reduced, the function is said to *converge*, or more precisely:

Convergence of a Lattice Function

One says, a lattice function $\boldsymbol{x}_\Delta$ **converges** to $\boldsymbol{x}$, if for the discretization error $\|\varepsilon_\Delta\|_\infty$ we have:

$$\|\varepsilon_\Delta\|_\infty \to 0 \text{ for } h_\Delta \to 0\,.$$

(continued)

In this case we say, $\boldsymbol{x}_\Delta$ has the **order of convergence** p if

$$\|\varepsilon_\Delta\|_\infty = O(h_\Delta^p) \text{ for } h_\Delta \to 0\,.$$

Remarks *Roughly*, if Ψ is a one-step method with consistency order p, then $\boldsymbol{x}_\Delta$ with order p converges to $\boldsymbol{x}$. This is not quite true as it is, since another condition must be satisfied for this to be true. We do not specify this condition further, since the above *rough* rule holds for standard methods.

73.3 Stiff Differential Equations

The phenomenon of *stiffness* occurs only in the numerical solution of an IVP and is due to the ODE system: A ODE system can have a solution fraction that quickly becomes small and then unobservable with respect to another solution fraction. And yet, one may be forced to align the step size according to this rapidly vanishing solution fraction. This phenomenon is already observable in the one-dimensional. We consider the **Dahlquist's test equation**:

$$\dot{x} = -\lambda\, x\,,\ x(0) = 1\,,\ \lambda > 0\,.$$

As we know, the exact solution is $x(t) = \mathrm{e}^{-\lambda t}$. We solve this IVP (theoretically) using the explicit and implicit Euler methods and make an interesting observation:

- **Explicit Euler method:** We apply the explicit Euler method to Dahlquist's test equation, it reads:

$$\begin{aligned} x_{k+1} &= x_k + h(-\lambda x_k) = x_k(1 - h\lambda) = (x_{k-1} + h(-\lambda x_{k-1}))(1 - h\lambda) \\ &= (x_{k-1}(1 - h\lambda))(1 - h\lambda) \\ &= x_{k-1}(1 - h\lambda)^2 \\ &= x_0(1 - h\lambda)^{k+1}\,. \end{aligned}$$

Since $\mathrm{e}^{-\lambda t}$ is the exact solution, we expect $x_k \to 0$ for $k \to \infty$. For this is necessary:

$$|1 - h\lambda| < 1\,.$$

Because of $\lambda,\ h > 0$ this means

$$-(1 - h\lambda) < 1 \Leftrightarrow h\lambda < 2 \Leftrightarrow h < \frac{2}{\lambda}.$$

Thus, for a large λ, a small h is necessary. That is, we need many steps. The explicit Euler method is thus unsuitable for the approximate determination of the solution of this IVP, it becomes unstable for a large λ.

- **Implicit Euler method:** We now apply the implicit Euler method to Dahlquist's test equation, it reads:

$$x_{k+1} = x_k + h(-\lambda x_{k+1}).$$

Now follows:

$$\begin{aligned} x_{k+1} + \lambda h x_{k+1} = x_k &\Leftrightarrow x_{k+1}(1 + \lambda h) = x_k \\ &\Leftrightarrow x_{k+1} = \frac{1}{1 + \lambda h} x_k \\ &\Leftrightarrow x_{k+1} = \left(\frac{1}{1 + \lambda h}\right)^{k+1} x_0 . \end{aligned}$$

Now is for $x_k \to 0$ necessary

$$\left|\frac{1}{1 + \lambda h}\right| < 1 ;$$

but this is satisfied for all $\lambda > 0$ and $h > 0$. Thus we have no step size constraint.

This phenomenon is typical for so-called **stiff differential equations**: In the numerical solution of such a ODE system, one is forced to align the step size according to a rapidly vanishing solution fraction, which leads to a disproportionately large number of integration steps.

The stiffness of a ODE $\dot{\boldsymbol{x}} = f(t, \boldsymbol{x})$ is to be expected whenever the matrix $f_{\boldsymbol{x}}(t, \boldsymbol{x}(t))$, which is the $(n \times n)$-Jacobimatrix of f with the derivatives to $x_1, \ldots, x_n$, at the time t and at the position $\boldsymbol{x}(t)$ (i.e., on the solution curve) has eigenvalues λ with $\text{Re}(\lambda) << 0$, i.e, $\text{Re}(\lambda)$ is strongly negative. In principle:

Solution of Stiff Differential Equations

To solve stiff ODEs one has to rely on implicit methods.

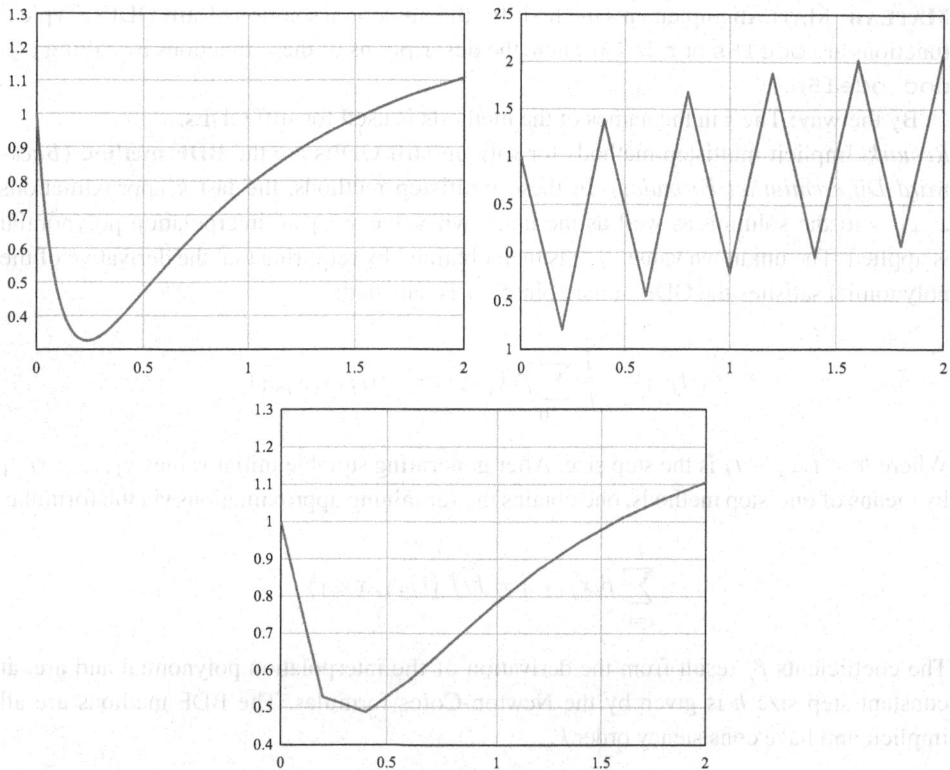

Fig. 73.3 The exact solution, the approximation with the explicit Euler, and the approximation with the implicit Euler

Indeed, it can be shown that the *stability region of* implicit methods is generally larger than that of explicit methods.

Example 73.3 We consider the initial value problem

$$\dot{x}(t) = -10\,(x(t) - \arctan(t)) + \tfrac{1}{1+t^2} \text{ with } x(0) = 1\,.$$

The exact solution of this IVP is

$$x = x(t) = \mathrm{e}^{-10t} + \arctan(t)\,.$$

Because of $f_x = -10$ we recognize it as stiff. For the numerical solution, we use the explicit and implicit Euler methods, each with the same parameters $h = 0.2$, $n = 10$. The result can be seen in Fig. 73.3. ■

MATLAB MATLAB implements methods for the numerical solution of stiff ODEs. Typical functions are `ode15s` or `ode23s`; note the descriptions of these functions by calling e.g. `doc ode15s`.

By the way: The *s* in the names of the methods is used for stiff ODEs.

Remark Implicit multistep methods for solving stiff ODEs are the **BDF method** (*Backward Differentiation Formulas*). In these multistep methods, the last k approximations x_{j+1-k} to the solution as well as the unknown value x_{j+1} an interpolation polynomial is applied. The unknown value x_{j+1} is then obtained by requiring that the derivative of the polynomial satisfies the ODE at the point t_{j+1} is satisfied:

$$\dot{x}(t_{j+1}) = \frac{1}{h}\sum_{i=0}^{k} \beta_i x_{j+1-i} = f(t_{j+1}, x_{j+1})\,.$$

Where $h = t_{j+1} - t_j$ is the step size. After generating suitable initial values $x_1, \ldots, x_{k-1}$ by means of one-step methods, one obtains the remaining approximations via the formula:

$$\sum_{i=0}^{k} \beta_i x_{j+1-i} = h\, f\left(t_{j+1}, x_{j+1}\right)\,.$$

The coefficients β_i result from the derivation of the interpolation polynomial and are, at constant step size h is given by the Newton-Cotes formulas. The BDF methods are all implicit and have consistency order k.

73.4 Boundary Value Problems

We consider a **boundary value problem** (BVP) with a 2nd order ODE, i.e.:

$$\ddot{x}(t) + b(t)\,\dot{x}(t) + c(t)\,x(t) = s(t) \;\text{ with }\; x(t_0) = x_{t_0}\,,\; x(t_1) = x_{t_1}\,.$$

For IVPs, we presented in Sect. 63.2 a far-reaching existence and uniqueness theorem. For BVPs there is no comparable theorem, more precisely: From Sect. 68.2 we already know that BVPs are often not solvable or have infinitely many solutions. If a BVP is not solvable with analytical solution methods (cf. also Chap. 68), one has to rely on numerical methods. We present three solution methods, where the first solution method is an analytical one.

73.4.1 Reduction of a BVP to an IVP

We start from the BVP

$$\ddot{x}(t) + b(t)\,\dot{x}(t) + c(t)\,x(t) = s(t) \;\text{ with }\; x(t_0) = x_{t_0}\,,\; x(t_1) = x_{t_1}\,.$$

For this BVP we now consider the two IVPs:

$$\ddot{u}(t) + b(t)\,\dot{u}(t) + c(t)\,u(t) = s(t) \text{ with } u(t_0) = x_{t_0}\,,\ \dot{u}(t_0) = 0\,,$$

$$\ddot{v}(t) + b(t)\,\dot{v}(t) + c(t)\,v(t) = 0 \text{ with } v(t_0) = 0\,,\ \dot{v}(t_0) = 1\,.$$

From these IVPs determine the functions $u = u(t)$ and $v = v(t)$. It is then

$$x(t) = u(t) + a\,v(t) \text{ with } a = \frac{x_{t_1} - u(t_1)}{v(t_1)}\,, \text{ if } v(t_1) \neq 0$$

a solution of the original boundary value problem, as can be easily checked.

Example 73.4 We consider the BVP

$$\ddot{x}(t) - \dot{x}(t) - 2\,x(t) = t \text{ with } x(0) = -1\,,\ x(5) = 1\,.$$

To do this, we now consider the two IVPs:

$$(1)\quad \ddot{u}(t) - \dot{u}(t) - 2\,u(t) = t \text{ with } u(0) = -1\,,\ \dot{u}(0) = 0\,,$$

$$(2)\quad \ddot{v}(t) - \dot{v}(t) - 2\,v(t) = 0 \text{ with } v(0) = 0\,,\ \dot{v}(0) = 1\,.$$

We start with (1): the characteristic polynomial is

$$p(\lambda) = \lambda^2 - \lambda - 2 = (\lambda - 2)\,(\lambda + 1)\,.$$

Thus the general solution of the homogeneous ODE is

$$u_h(t) = c_1\,\mathrm{e}^{2t} + c_2\,\mathrm{e}^{-t} \text{ with } c_1,\ c_2 \in \mathbb{R}\,.$$

By right-hand side type approach, we determine a particulate solution. Since there is no resonance, the approach is

$$u_p(t) = A\,t + B\,, \text{ so } -A - 2\,A\,t - 2\,B = t\,.$$

We obtain by coefficient comparison $A = -\frac{1}{2}$ and $B = \frac{1}{4}$. Thus a particulate solution is $u_p(t) = -\frac{1}{2}\,t + \frac{1}{4}$. We thus obtain the general solution

$$u_a(t) = u_p(t) + u_h(t) = -\frac{1}{2}\,t + \frac{1}{4} + c_1\,\mathrm{e}^{2t} + c_2\,\mathrm{e}^{-t} \text{ mit } c_1,\ c_2 \in \mathbb{R}\,.$$

Now we obtain from the initial conditions the equations

$$\frac{1}{4} + c_1 + c_2 = -1 \text{ and } -\frac{1}{2} + 2\,c_1 - c_2 = 0$$

and thus $c_1 = -\frac{1}{4}$ and $c_2 = -1$. The solution to the IVP (1) is thus

$$u(t) = -\frac{1}{2}\,t + \frac{1}{4} - \frac{1}{4}\,\mathrm{e}^{2t} - \mathrm{e}^{-t}\,.$$

We solve the IVP in (2): The general solution of the (homogeneous) ODE is

$$v_a(t) = v_h(t) = c_1\,\mathrm{e}^{2t} + c_2\,\mathrm{e}^{-t} \quad \text{with } c_1,\, c_2 \in \mathbb{R}\,.$$

Now we get from the initial conditions the equations

$$c_1 + c_2 = 0 \text{ and } 2\,c_1 - c_2 = 1$$

and thus $c_1 = \frac{1}{3}$ and $c_2 = -\frac{1}{3}$. The solution to the IVP (2) is thus

$$v(t) = \frac{1}{3}\,\mathrm{e}^{2t} - \frac{1}{3}\,\mathrm{e}^{-t}\,.$$

We determine

$$a = \frac{1 - u(5)}{v(5)} = -1.741371880668659...$$

and thus obtain the solution to the original BVP

$$x(t) = u(t) + a\,v(t) = -\frac{1}{2}\,t + \frac{1}{4} + \left(\frac{a}{3} - \frac{1}{4}\right)\mathrm{e}^{2t} - \left(1 + \frac{a}{3}\right)\mathrm{e}^{-t}\,.$$

■

Remark

In this method, a BVP is solved by solving two IVPs analytically. This is not easy or possible if the ODE is more complicated. Of course, the IVPs can also be solved numerically. In the following, we describe other numerical methods for solving boundary value problems.

73.4.2 Difference Method

In the following we assume a BVP with a 2nd order ODE. If a ODE of a different order is present, the procedure can be easily modified:

$$\ddot{x}(t) + b(t)\,\dot{x}(t) + c(t)\,x(t) = s(t) \text{ with } x(t_0) = x_{t_0}\,,\ x(t_1) = x_{t_1}\,.$$

We choose a (large) $N \in \mathbb{N}$ and obtain a (small) step size $h = \frac{t_1 - t_0}{N}$ and (closely spaced) $t_k = t_0 + k\,h$ for $k = 0, \ldots, N$. The function $x = x(t)$ we are looking for, but which is usually not analytically determinable, has at the boundaries of the equidistantly divided interval $[t_0, t_1]$ the function values known from the boundary conditions

$$x(t_0) = x_{t_0} \text{ and } x(t_1) = x_{t_1}\,.$$

With the **difference method** we determine approximate values $x_k \approx x(t_k)$ at the intermediate points t_k for $k = 1, \ldots, N-1$.

To do this, we substitute at the intermediate points the differential quotients $\dot{x} = \frac{\mathrm{d}x}{\mathrm{d}t}$ and $\ddot{x} = \frac{\mathrm{d}^2x}{\mathrm{d}t^2}$ by differential quotients with *small h* (note the box in Sect. 26.3), we have:

$$\dot{x}(t_k) \approx \frac{x(t_{k+1}) - x(t_k)}{h} \text{ and } \ddot{x}(t_k) \approx \frac{x(t_{k+1}) - 2\,x(t_k) + x(t_{k-1})}{h^2}\,.$$

Now we replace in above ODE the derivatives by these approximations and shorten expediently $x_{k-1} = x(t_{k-1})$, $x_k = x(t_k)$, $x_{k+1} = x(t_{k+1})$, $s_k = s(t_k)$,... Thus we obtain:

$$\frac{x_{k+1} - 2\,x_k + x_{k-1}}{h^2} + b_k \frac{x_{k+1} - x_k}{h} + c_k x_k = s_k\,.$$

We sort and obtain:

$$\frac{1}{h^2} x_{k-1} + \left(\frac{-2}{h^2} - \frac{b_k}{h} + c_k\right) x_k + \left(\frac{1}{h^2} + \frac{b_k}{h}\right) x_{k+1} = s_k\,.$$

For $k = 1$ we get as first summand $\frac{1}{h^2} x_0$ with the known from the boundary conditions x_0. And for $k = N-1$ we obtain analogously as the last summand $\left(\frac{1}{h^2} + \frac{b_{N-1}}{h}\right) x_N$ with the known x_N. We put these two known summands in the following system of equations to the right-hand side. With the abbreviations $l_k = \frac{1}{h^2}$, $d_k = \frac{-2}{h^2} - \frac{b_k}{h} + c_k$, $r_k = \frac{1}{h^2} + \frac{b_k}{h}$ this system of equations for the approximations $x_1, \ldots, x_{N-1}$ we are looking for is:

$$\begin{pmatrix} d_1 & r_1 & & & & \\ l_2 & d_2 & r_2 & & & \\ & l_3 & d_3 & r_3 & & \\ & & \ddots & \ddots & \ddots & \\ & & & l_{N-2} & d_{N-2} & r_{N-2} \\ & & & & l_{N-1} & d_{N-1} \end{pmatrix} \begin{pmatrix} x_1 \\ x_2 \\ x_3 \\ \vdots \\ x_{N-2} \\ x_{N-1} \end{pmatrix} = \begin{pmatrix} s_1 - \frac{1}{h^2} x_0 \\ s_2 \\ s_3 \\ \vdots \\ s_{N-2} \\ s_{N-1} - r_{N-1} x_N \end{pmatrix}.$$

Once we have determined the solutions $x_1, \ldots, x_{N-1}$ we plot the points (t_k, x_k) for $k = 0, \ldots, N$ and thus see approximately the course of the solution $x(t)$—note the following example.

Example 73.5 We consider the BVP

$$\ddot{x} - 2t\,\dot{x} - 2x = -4t \text{ with } x(0) = 1\,,\ x(1) = 1 + \mathrm{e}\,.$$

We choose $N = 5$ and obtain as k-th equation with above notation:

$$\frac{1}{h^2} x_{k-1} + \left(\frac{-2}{h^2} + \frac{2t_k}{h} - 2 \right) x_k + \left(\frac{1}{h^2} - \frac{2t_k}{h} \right) x_{k+1} = -4t_k\,.$$

The resulting LGS is thus:

$$\begin{pmatrix} -50 & 23 & 0 & 0 \\ 25 & -48 & 21 & 0 \\ 0 & 25 & -46 & 19 \\ 0 & 0 & 25 & -44 \end{pmatrix} \begin{pmatrix} x_1 \\ x_2 \\ x_3 \\ x_4 \end{pmatrix} = \begin{pmatrix} -\frac{4}{5} - 25 \cdot 1 \\ -\frac{8}{5} \\ -\frac{12}{5} \\ -\frac{16}{5} - \left(25 - \frac{40}{5}\right) \cdot (1 + \mathrm{e}) \end{pmatrix}.$$

As a solution we obtain (besides the boundary values x_0 and x_5)

$$x_0 = 0\,,\ x_1 = 1.2200\,,\ x_2 = 1.5305\,,\ x_3 = 1.9696\,,\ x_4 = 2.6284\,,\ x_5 = 3.7183\,.$$

In Fig. 73.4 we see this approximation for $N = 5$ next to the approximation for $N = 20$. The exact solution of the BVP is $x(t) = t + \mathrm{e}^{t^2}$. ■

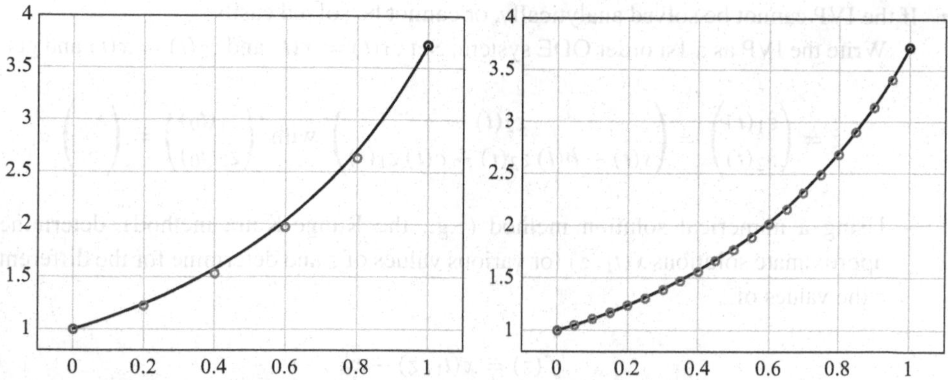

Fig. 73.4 The approximations for N= 5 and N= 20 next to the exact solution

73.4.3 Shooting Method

We start again from the BVP

$$\ddot{x}(t) + b(t)\,\dot{x}(t) + c(t)\,x(t) = s(t) \text{ with } x(t_0) = x_{t_0}\,,\ x(t_1) = x_{t_1}\,.$$

This BVP is now reformulated into an IVP as follows:

$$\ddot{x}(t) + b(t)\,\dot{x}(t) + c(t)\,x(t) = s(t) \text{ with } x(t_0) = x_0\,,\ \dot{x}(t_0) = z\,,$$

where the second, initially unknown, initial value z is to be determined in such a way that the solution x satisfies the (desired) boundary condition $x(t_1) = x_{t_1}$ satisfies.

To do this, the IVP is solved as a function of the initial value z. This can be done in two ways:

- If the IVP can be solved analytically: Let $x = x(t, z)$ be the solution of the IVP. Then consider the function

$$f(z) = x(t_1, z) - x_{t_1}$$

and determine a zero of f, i.e. a solution z^* of the generally nonlinear system of equations $f(z) = 0$ (e.g. using the bisection method or Newton's method). It is then

$$x = x(t) = x(t, z^*)$$

a solution of the original BVP, since it x on the one hand the IVP and on the other hand the second boundary condition $x(t_1) = x(t_1, z^*) = x_{t_1}$ satisfies.

- If the IVP cannot be solved analytically, or cannot be solved easily:
  - Write the IVP as a 1st order ODE system: Set $z_1(t) = x(t)$ and $z_2(t) = \dot{x}(t)$ and get

$$\dot{z} = \begin{pmatrix} \dot{z}_1(t) \\ \dot{z}_2(t) \end{pmatrix} = \begin{pmatrix} z_2(t) \\ s(t) - b(t)\, z_2(t) - c(t)\, z_1(t) \end{pmatrix} \text{ with } \begin{pmatrix} z_1(t_0) \\ z_2(t_0) \end{pmatrix} = \begin{pmatrix} x_{t_0} \\ z \end{pmatrix}.$$

  - Using a numerical solution method (e.g., the Runge-Kutta method), determine approximate solutions $x(t_1, z)$ for various values of z and determine for the different z the values of

$$f(z) = x(t_1, z) - x_{t_1}\,.$$

  - Choose two successive z_k, z_{k+1} with $f(z_k)\, f(z_{k+1}) < 0$ and determine by bisection method a z^* between z_k and z_{k+1} which approximately satisfies

$$f(z^*) = x(t_1, z^*) - x_{t_1} = 0\,.$$

  The approximate solution z^* is then accepted as an approximate solution for x.

In any case one speaks of a **(single) shooting method**: Consider a cannon which sends a bullet from (t_0, x_{t_0}) to (t_1, x_{t_1}). For this we test a *slope* z of the trajectory and see where the bullet lands. Maybe you shoot too far, maybe too short ... by successive corrections the bullet will land at the point (t_1, x_{t_1}).

Example 73.6

- We solve

$$\ddot{x}(t) - 2\,\dot{x}(t) + x(t) = 0 \text{ with } x(0) = 1\,,\ x(1) = 0\,.$$

  We convert the BVP into an IVP with initial value z:

$$\ddot{x}(t) - 2\,\dot{x}(t) + x(t) = 0 \text{ with } x(0) = 1\,,\ \dot{x}(0) = z\,.$$

  The solution of this IVP as a function of z is obviously

$$x = x(t, z) = \mathrm{e}^t + (z - 1)\, t\, \mathrm{e}^t\ .$$

We are looking for a zero z^* of the function

$$f(z) = x(t_1, z) - x_1 = x(1, z) - 0 = \mathrm{e} + (z - 1)\,\mathrm{e}\,.$$

This is very simple in the present case, apparently we have $z^* = 0$. Thus,

$$x = x(t) = x(t, 0) = \mathrm{e}^t - t\,\mathrm{e}^t$$

is a solution of the original BVP.

- We solve the BVP

$$\ddot{x}(t) + (1 + t^2)\,x(t) = -1 \ \text{ with } \ x(-1) = 0\,,\ x(1) = 0\,.$$

An analytical solution of the associated IVP is (not easily) possible, so we form the IVP per $z_1(t) = x(t)$ and $z_2(t) = \dot{x}(t)$ into an IVP with a 1st order ODE system:

$$\dot{z}(t) = \begin{pmatrix} \dot{z}_1(t) \\ \dot{z}_2(t) \end{pmatrix} = \begin{pmatrix} z_2(t) \\ -z_1(t) - t^2\, z_1(t) - 1 \end{pmatrix} \ \text{ with } \ \begin{pmatrix} z_1(-1) \\ z_2(-1) \end{pmatrix} = \begin{pmatrix} 0 \\ z \end{pmatrix}.$$

For $z = 0.0,\ 1.0,\ 2.0,\ 3.0$ we choose the Runge-Kutta method with the step size $h = 0.1$ and obtain the following approximate values for $x(1, z)$:

z	0	1	2	3
$x(1, z)$	−1.3193	−0.5596	0.2002	0.9600

Since obviously between $z = 1$ and $z = 2$ is a searched place z^* we proceed by bisection; we compute successively:

z	1.5	1.75	1.625	1.6875	1.71875	1.734375
$x(1, z)$	−0.1797	0.0103	−0.0847	−0.0372	−0.0135	−0.0016

We break off at $z^* = 1.734375$ with the bisection method and obtain an approximate solution—note the Fig. 73.5.

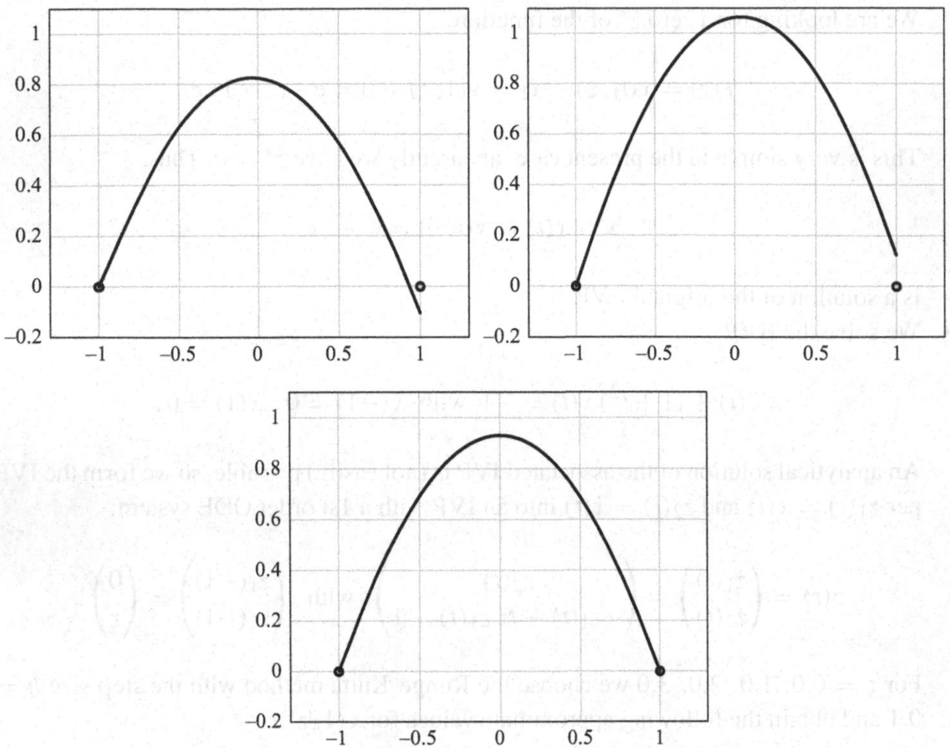

Fig. 73.5 At $z = 1.6$ we shoot too short, at $z = 1.9$ we shoot too far, at $z = 1.734375$ it fits

73.5 Exercises

73.1 *Martiniglase effect*: we consider the linear ODE system

$$\dot{\boldsymbol{x}} = \begin{pmatrix} 0 & 1 \\ -1 & 0 \end{pmatrix} \boldsymbol{x} .$$

(a) Show that for the solution $\boldsymbol{x}(t, x_0)$ of the IVP to the initial value $\boldsymbol{x}_0$ we have: $\|\boldsymbol{x}(t, x_0)\| = \|\boldsymbol{x}(t_0, x_0)\|$. What does this mean geometrically?

(b) Now consider a numerical approximation $\tilde{\boldsymbol{x}}(t)$ to the IVP, where once the approximation is calculated by the explicit Euler method and once by the implicit Euler method, in each case with constant step size h. Remains also here $\|\tilde{\boldsymbol{x}}(t, x_0)\|$ for all $t \geq t_0$ constant? Where does the name *Martiniglase effect* come from?

(c) What happens if you let the step size h approach 0? Does this avoid the effect just observed?

73.2 A cat chases a mouse in the x-y-plane and always runs with a constant velocity in absolute value $v_K = 2$ directly to the mouse. The mouse, for its part, wants to move directly with velocity (in absolute value) $v_M = 1$ into its hole at the point (0, 1). The mouse is at the time $t = 0$ at the point (0, 0) and the cat at the point (1, 0).

(a) Set up the ODEs describing the path of the cat and the path of the mouse, respectively.
(b) Using MATLAB, calculate when and where the cat approached the mouse to within 10^{-3} and when and where it approached the mouse to within 10^{-6}. Also use different step sizes for this.

73.3 We consider the damped harmonic oscillator

$$\ddot{x} + \mu \dot{x} + x = 0$$

with parameter $\mu > 0, \mu \neq 2$.

(a) Rewrite the ODE into a 1st order system.
(b) Depending on the parameter μ, determine the eigenvalues of the linear system.
(c) For what values of μ is the system considered *stiff*?
(d) Let it now be $\mu = 256 + \frac{1}{256}$. Then how small must the step size of the explicit Euler method be chosen so that the numerically approximated solution is bounded?
(e) We now apply the implicit Euler method to compute an approximated solution. Let the parameter μ be chosen as in (d). Now, for which step sizes is a bounded numerical solution guaranteed?

73.4 In the following we consider the temperature distribution $u(x)$, $x \in [0, 1]$, in a rod heated from the center and cooled at the edge. This leads to the following boundary value problem:

$$-u''(x) = f(x), \text{ where } f(x) = x(1-x), \text{ with boundary condition } u(0) = u(1) = 0.$$

(a) Discretize this problem by considering for any $N \in \mathbb{N}$
 1. in the interval [0, 1] choose the N+ 1 interpolation points $x_k := \frac{k}{N}, k = 0, \cdots, N$, substitute
 2. the derivative $u''(x)$ by $\frac{1}{h^2}(u(x+h) - 2u(x) + u(x-h))$ with $h = \frac{1}{N}$ and
 3. replace the functions $u(x)$ or $f(x)$ by the vectors $U := (u(x_k))_{0 \le k \le N}$ or $F := (f(x_k))_{0 \le k \le N}$ of their function values on the grid points.

 In this way you get an LGS of the form $A\tilde{U} = \tilde{F}$ where $\tilde{U} := (U_k)_{1 \le k \le N-1}$ and $\tilde{F} := (F_k)_{1 \le k \le N-1}$. What is A? Is A strictly diagonally dominant?
(b) Use the Jacobi and the Gauss-Seidel methods with $x_0 = 0$, `maxiter=1000` and `tol=1e-3` to calculate solutions of the discretization for $N = 10, 100, 1000$. Plot your result for U. Justify that your program converges in these cases.

(c) Solve the obtained LGS for $N = 10,\ 100,\ 1000$ also using the MATLAB operator \. Plot U and compare with (b). How do you explain the discrepancies?

73.5 Solve the boundary value problem

$$u''(x) = u(x)^2, \quad u(0) = 2,\ u(1) = 3,$$

using the shooting method.

(a) Convert the boundary value problem into an initial value problem by omitting the boundary condition $u(1) = 3$ and adding the initial condition $u'(0) = s$ instead. Determine numerically a solution $u(x; s)$ of the IVP for $s \in \{-1.4, -1.2, \cdots, 0.4, 0.6\}$ using the explicit Euler method with step size $h = 0.01$. Plot $u(x; s)$.
(b) Where do you suppose the correct s and why?
(c) Use the bisection method to determine s within two digits.

Fourier Series: Calculation of Fourier Coefficients

74

It is often possible to express a periodic function f as a *sum* or *series* of cosine and sine functions. This is based on the idea that a periodic signal, namely the function f, can be regarded as a superposition of many harmonic oscillations, namely cosine and sine functions. The determination of the individual harmonic oscillations corresponds to a decomposition of the periodic signal into its fundamental oscillations.

The mathematics behind this decomposition is the calculation of the Fourier coefficients for the fundamental oscillations. Behind this is a scalar product formation by means of an integral. We describe this calculation of the *Fourier coefficients* and thus represent periodic functions of all colors as superpositions of harmonic oscillations.

74.1 Periodic Functions

A function $f : \mathbb{R} \to \mathbb{C}$ is called **periodic** with the **period** $T > 0$, short T**-periodic**, if we have

$$f(x+T) = f(x) \text{ for all } x \in \mathbb{R}\,.$$

For a T-periodic function follows

$$f(x+kT) = f(x) \text{ for all } k \in \mathbb{Z} \text{ and } x \in \mathbb{R}\,.$$

In particular, a T-periodic function is also kT-periodic for all $k \in \mathbb{N}$. The following Fig. 74.1 shows a 2π-, 4π-, 6π-, ... periodic function:

C. Karpfinger, *Calculus and Linear Algebra in Recipes*,
https://doi.org/10.1007/978-3-662-65458-3_74

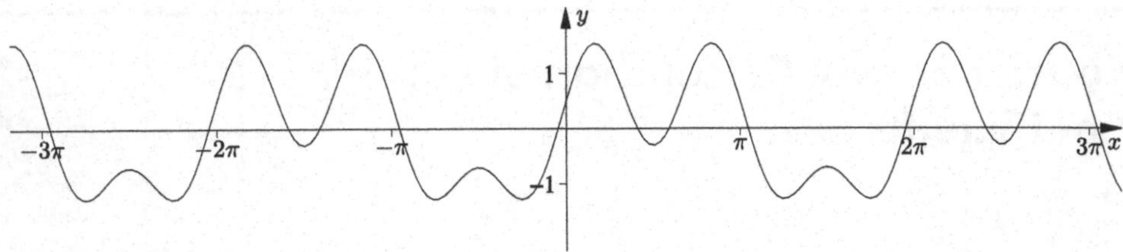

Fig. 74.1 The function is 2π-periodic, but also 4π-, 6π- ... periodic

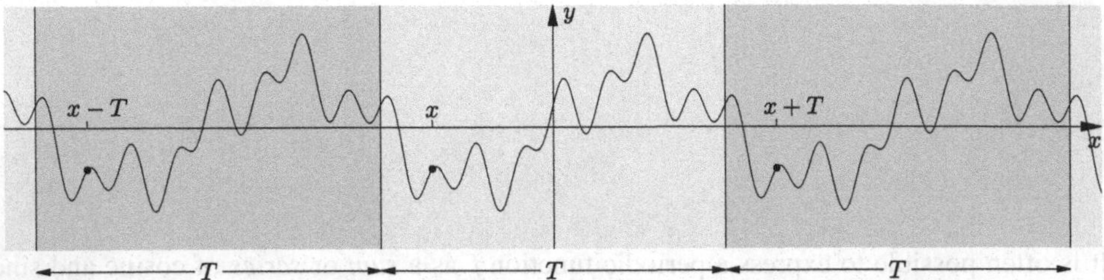

Fig. 74.2 Every interval of length T is, in the case of a T-periodic function a period interval

Example 74.1 The functions

$$x \mapsto \cos(x)\,,\ x \mapsto \sin(x)\,,\ x \mapsto \mathrm{e}^{\mathrm{i}\,x}$$

are 2π-periodic. For each $T > 0$ and each $k \in \mathbb{Z}$ the functions

$$x \mapsto \cos\left(k\frac{2\pi}{T}x\right)\,,\ x \mapsto \sin\left(k\frac{2\pi}{T}x\right)\,,\ x \mapsto \mathrm{e}^{\mathrm{i}\,k\frac{2\pi}{T}x}$$

are T-periodic. ■

We consider T-periodic functions $f : \mathbb{R} \to \mathbb{C}$, where of course $T > 0$. Any such function is uniquely determined by its values on any interval of length T, since the further values of f outside the interval under consideration are given by periodicity, see Fig. 74.2.

Thus we know a T-periodic function $f : \mathbb{R} \to \mathbb{C}$, if we only know what the function on, say, the period interval $I = [-T/2, T/2)$ or $I = [0, T)$ does. Therefore, we will such functions f often specify on only one such period interval, but we mean the function periodically extended on $\mathbb{R}$ (see Fig. 74.3).

If f is a T-periodic function, then for any $S > 0$, one can easily construct a function g with the help of f, which is S-periodic:

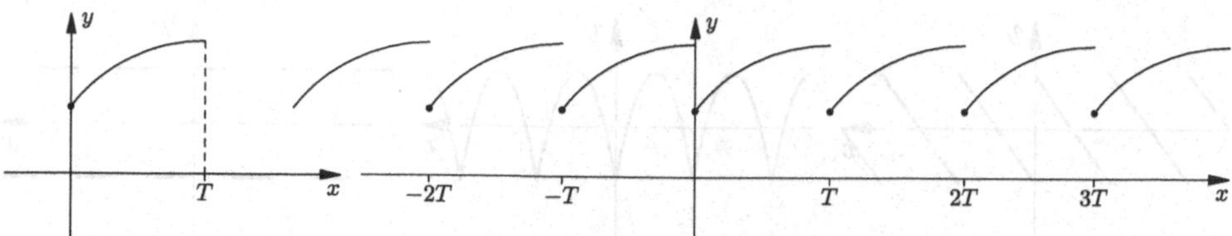

Fig. 74.3 If f is given on an interval I of length T, then f can be extended T-periodically

Conversion from T- to S-Periodicity

If $f : \mathbb{R} \to \mathbb{C}$ is a T-periodic function, then for every $S > 0$ the function $g : \mathbb{R} \to \mathbb{C}$ with $g(x) = f\left(\frac{T}{S}x\right)$ is a S-periodic function.

This can be shown quite simply, for it is true for all $x \in \mathbb{R}$:

$$g(x+S) = f\left(\frac{T}{S}(x+S)\right) = f\left(\frac{T}{S}x + T\right) = f\left(\frac{T}{S}x\right) = g(x).$$

Example 74.2 The cosine $\cos : x \mapsto \cos(x)$ is 2π-periodic. With $T = 2\pi$ and $S = 1$, we obtain that the function $g : x \mapsto \cos(2\pi x)$ then is 1-periodic. ■

Remarks

1. This possibility of converting T- to S-periodicity has a simple benefit: We will learn how to expand a T-periodic function $f : \mathbb{R} \to \mathbb{C}$, $x \mapsto f(x)$ in a *Fourier series* F. We will then also present a formula which allows us to calculate from the Fourier series F the Fourier series of $g : \mathbb{R} \to \mathbb{C}$, $x \mapsto f(\frac{T}{S}x)$. Thus, we do not need to again consider the Fourier series of g; we obtain this Fourier series of g from the Fourier series of f by applying a simple formula. In fact, we can immediately refer to e.g. 2π-periodic functions restriction. To remain more flexible, we do not make this restriction.
2. Also the possibility of freely choosing the period interval I of length T is useful: we will learn a formula that allows us to extract from the Fourier series of a T-periodic function f, which is set, for example, to $[0, T)$ given the Fourier series of the *shifted* function to, e.g. $[-T/2, T/2)$. So we can commit to an interval right away, we'll mostly use $[-T/2, T/2)$ occasionally also $[0, T)$.

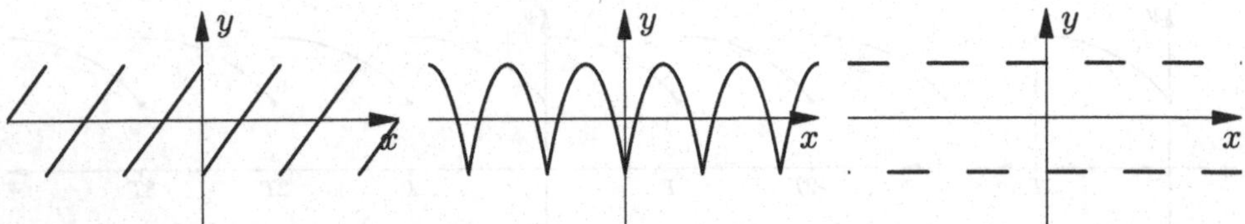

Fig. 74.4 Examples of admissible functions

74.2 The Admissible Functions

Not every function can be represented by a Fourier series. We choose functions that have properties that not only allow us to expand that function into a Fourier series, but also allow us to make strong statements about them that describe the relationship between f and its Fourier series F. From now on we only consider functions $f : \mathbb{R} \to \mathbb{C}$ with the properties

(i) f is T-periodic with period interval I, usually one chooses $I = [-T/2, T/2)$ or $I = [0, T)$.
(ii) The period interval I can be decomposed into finitely many subintervals on which f is continuous and monotone.
(iii) In the (finitely many) allowed discontinuities $a_1, \ldots, a_\ell$ there exist the left and right limits

$$f\left(a_k^+\right) = \lim_{x \to a_k^+} f(x) \quad \text{and} \quad f\left(a_k^-\right) = \lim_{x \to a_k^-} f(x)\,.$$

To avoid having to constantly repeat these preconditions on f, we write for the set of all functions $f : \mathbb{R} \to \mathbb{C}$ which have these properties (i) to (iii), in short $C(T)$. The notation $f \in C(T)$ thus abbreviates the fact that f is a function of $\mathbb{R}$ to $\mathbb{C}$ with the properties (i), (ii), (iii). Note that any such function is in particularly bounded. Figure 74.4 shows some graphs of such functions:

Remark The theory is also possible for more general functions, but we restrict ourselves to these special functions: On the one hand we have the functions of practical relevance, and on the other hand we can make the formalism simple.

Our aim is to approximate a function $f \in C(T)$ with a period $T > 0$ by linear combinations of T/k-periodic *basis functions* or to represent them by series. We distinguish between a complex and a real version:

	Basis functions	Series
Real version	$\sin\left(k\frac{2\pi}{T}x\right)$ and $\cos\left(k\frac{2\pi}{T}x\right)$	$\sum\limits_{k=0}^{\infty} a_k \cos\left(k\frac{2\pi}{T}x\right) + b_k \sin\left(k\frac{2\pi}{T}x\right)$
Complex version	$e^{i\,k\frac{2\pi}{T}x}$	$\sum\limits_{k=-\infty}^{\infty} c_k\, e^{i\,k\frac{2\pi}{T}x}$

We start with the real version.

74.3 Expanding in Fourier Series—Real Version

Any real function $f \in C(T)$ can be expanded in a *Fourier series* in the following sense:

Fourier Coefficients and Fourier Series: Real Version
Determine for the *T*-periodic function $f \in C(T)$, $f : [-T/2, T/2) \to \mathbb{R}$ the so-called **Fourier coefficients**

$$a_k = \frac{2}{T} \int\limits_{-T/2}^{T/2} f(x) \cos\left(k\frac{2\pi}{T}x\right) \mathrm{d}x\,,\ k \in \mathbb{N}_0\,,$$

$$b_k = \frac{2}{T} \int\limits_{-T/2}^{T/2} f(x) \sin\left(k\frac{2\pi}{T}x\right) \mathrm{d}x\,,\ k \in \mathbb{N}\,,$$

and set

$$F(x) = \frac{a_0}{2} + \sum_{k=1}^{\infty} a_k \cos\left(k\frac{2\pi}{T}x\right) + b_k \sin\left(k\frac{2\pi}{T}x\right);$$

one calls $F(x)$ the **Fourier series** to f; we write for this $f(x) \sim F(x)$.
If the period interval is $[0, T)$ then the above integrals are of course to be formed over this interval.

In the case $T = 2\pi$ the formulas for the Fourier coefficients a_k and b_k and for the Fourier series are

$$a_k = \frac{1}{\pi} \int\limits_{-\pi}^{\pi} f(x) \cos(kx)\, \mathrm{d}x\,,\ k \in \mathbb{N}_0\,,\ b_k = \frac{1}{\pi} \int\limits_{-\pi}^{\pi} f(x) \sin(kx)\, \mathrm{d}x\,,\ k \in \mathbb{N}\,,$$

and

$$F(x) = \frac{a_0}{2} + \sum_{k=1}^{\infty} a_k \cos(kx) + b_k \sin(kx)\,.$$

The relation between $f(x)$ and $F(x)$ is now as follows:

The Relation of f and the Corresponding Fourier Series F

Given is $f \in C(T)$, $f : [-T/2, T/2) \to \mathbb{R}$ where $\alpha_1, \ldots, \alpha_\ell$ denote the points of discontinuity, if any, of f on $I = [-T/2, T/2)$. Let F be the Fourier series of f, i.e. $f(x) \sim F(x)$. Then we have:

- If f is continuous in x, then $f(x) = F(x)$, i.e.

$$f(x) = \frac{a_0}{2} + \sum_{k=1}^{\infty} a_k \cos\left(k\frac{2\pi}{T}x\right) + b_k \sin\left(k\frac{2\pi}{T}x\right).$$

- If f is not continuous in x, then $x = \alpha_k$ for a $k = 1, \ldots, \ell$ and

$$F(\alpha_k) = \frac{f(\alpha_k^-) + f(\alpha_k^+)}{2}\,.$$

In the continuity places the value $F(x)$ of the Fourier series coincides with the function value $f(x)$. At the discontinuity points, i.e. at the points where jumps are present, the value $F(x)$ lies exactly in the middle of the values of the jumps of f.

Pay particular attention to the illustrations in the following examples:

Example 74.3

- We consider the function $f \in C(2\pi)$ with $f : [-\pi, \pi) \to \mathbb{R}$, $f(x) = x$. Its graph (see Fig. 74.5) is a **sawtooth**:

  For a_0 we obtain:

$$a_0 = \frac{1}{\pi}\int_{-\pi}^{\pi} x\,\mathrm{d}x = \frac{1}{2\pi}x^2\Big|_{-\pi}^{\pi} = 0\,.$$

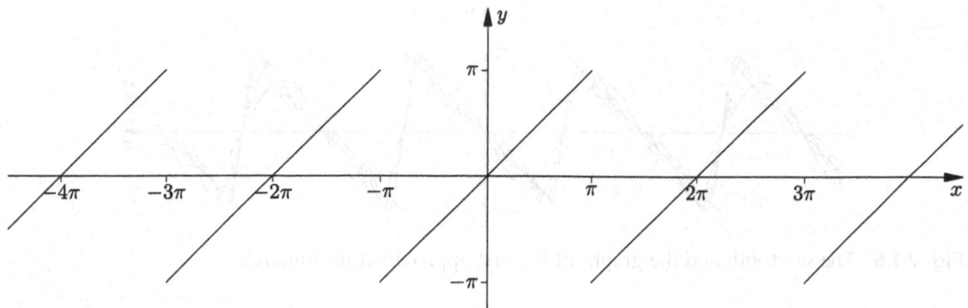

Fig. 74.5 The sawtooth is given by an admissible function

And for $k > 0$ we obtain by partial integration for the Fourier coefficients a_k:

$$a_k = \frac{1}{\pi}\int_{-\pi}^{\pi} x\cos(kx)\,\mathrm{d}x = \frac{1}{\pi}\left(\frac{x}{k}\sin(kx)\Big|_{-\pi}^{\pi} - \frac{1}{k}\int_{-\pi}^{\pi}\sin(kx)\,\mathrm{d}x\right)$$
$$= \frac{1}{\pi}\left(\frac{1}{k^2}\cos(kx)\Big|_{-\pi}^{\pi}\right) = \frac{1}{k^2\pi}\left((-1)^k - (-1)^k\right) = 0\,.$$

And for the Fourier coefficients b_k we have:

$$b_k = \frac{1}{\pi}\int_{-\pi}^{\pi} x\sin(kx)\,\mathrm{d}x = \frac{1}{\pi}\left(-\frac{x}{k}\cos(kx)\Big|_{-\pi}^{\pi} + \frac{1}{k}\int_{-\pi}^{\pi}\cos(kx)\,\mathrm{d}x\right)$$
$$= \frac{1}{\pi}\left(-\frac{\pi}{k}(-1)^k - \frac{\pi}{k}(-1)^k\right) = \frac{2}{k}(-1)^{k+1}\,.$$

The Fourier series of f thus has the form

$$F(x) = \sum_{k=1}^{\infty}\frac{2}{k}(-1)^{k+1}\sin(kx) = 2\left(\sin(x) - \frac{\sin(2x)}{2} + \frac{\sin(3x)}{3} - + \ldots\right).$$

In Fig. 74.6 are f and the first approximating functions are shown,

$$F_1(x) = 2\sin(x)\,,\ F_2(x) = 2\left(\sin(x) - \frac{\sin(2x)}{2}\right)\,,\ \ldots .$$

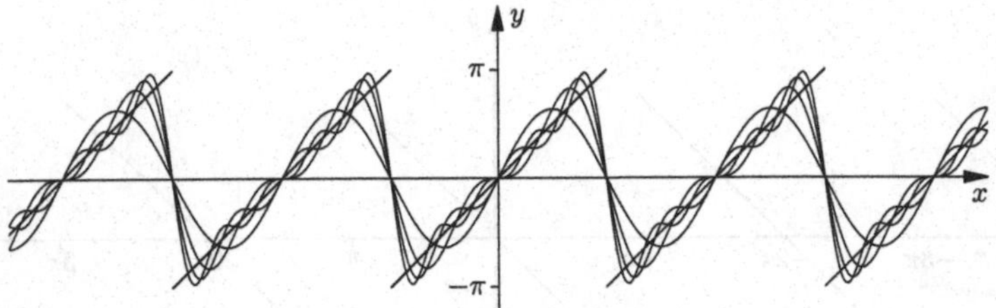

Fig. 74.6 The sawtooth and the graphs of the first approximating functions

- We now consider the **rectangular function** $f \in C(2\pi)$ with

$$f : [-\pi, \pi) \to \mathbb{R},\ f(x) = \begin{cases} 1, & 0 < x < \pi \\ 0, & x = 0 \\ -1, & -\pi \le x < 0 \end{cases}$$

Again we compute the Fourier coefficients: For all $k \in \mathbb{N}_0$ we have

$$a_k = \frac{1}{\pi} \int_{-\pi}^{\pi} f(x) \cos(kx)\, \mathrm{d}x = 0\,,$$

since $f(x)\cos(kx)$ is odd. For the coefficients b_k we obtain

$$b_k = \frac{1}{\pi} \int_{-\pi}^{\pi} f(x) \sin(kx)\, \mathrm{d}x = \frac{2}{\pi} \int_0^{\pi} f(x) \sin(kx)\, \mathrm{d}x = \frac{2}{\pi} \int_0^{\pi} \sin(kx)\, \mathrm{d}x$$

$$= \frac{2}{\pi} \frac{(-1)}{k} \cos(kx)\Big|_0^{\pi} = -\frac{2}{k\pi} \left((-1)^k - 1\right) = \begin{cases} 0\,, & \text{if } k \text{ even} \\ \frac{4}{k\pi}\,, & \text{if } k \text{ odd} \end{cases}.$$

The Fourier series of f is thus

$$F(x) = \sum_{k=0}^{\infty} \frac{4}{(2k+1)\pi} \sin\left((2k+1)x\right) = \frac{4}{\pi}\left(\sin(x) + \frac{\sin(3x)}{3} + \frac{\sin(5x)}{5} + \ldots\right).$$

Again we plot $f(x)$ and some approximating functions, see Figs. 74.7 and 74.8.

■

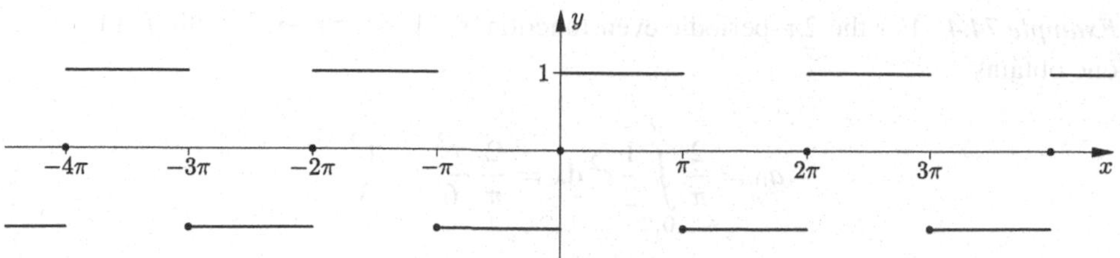

Fig. 74.7 The rectangular function is an admissible function

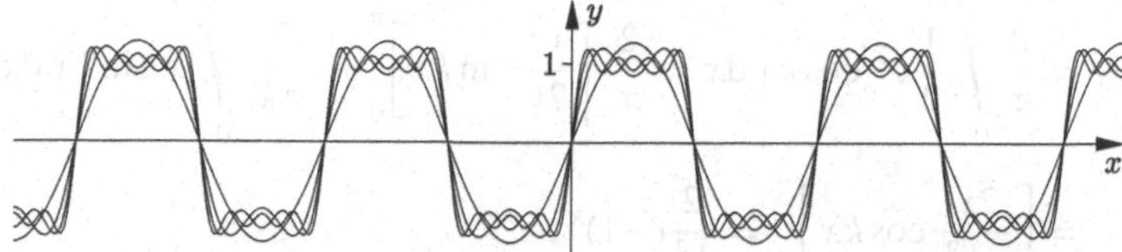

Fig. 74.8 The rectangular function and the graphs of the first approximating functions

We keep the following rules, which often greatly simplify the calculation work:

Computational Rules

- If $f \in C(T)$ odd, i.e. $f(-x) = -f(x)$ for all $x \in \mathbb{R}$, then we have:

$$a_k = 0 \quad \forall k \in \mathbb{N}_0 \text{ and } b_k = \frac{4}{T} \int_0^{T/2} f(x) \sin\left(k \frac{2\pi}{T} x\right) \mathrm{d}x \quad \forall k \in \mathbb{N}\,.$$

- If $f \in C(T)$ is even, i.e. $f(-x) = f(x)$ for all $x \in \mathbb{R}$, then we have:

$$b_k = 0 \quad \forall k \in \mathbb{N} \text{ and } a_k = \frac{4}{T} \int_0^{T/2} f(x) \cos\left(k \frac{2\pi}{T} x\right) \mathrm{d}x \quad \forall k \in \mathbb{N}_0\,.$$

74.4 Application: Calculation of Series Values

With the aid of the Fourier series expansion, values of series can be determined, such as the value of $\sum_{k=1}^{\infty} \frac{1}{k^2}$ (see the Remark in Sect. 20.2). And this works quite simply: We specify a function $f \in C(T)$ whose Fourier series F at a continuity point x has the value $F(x) = \sum_{k=1}^{\infty} \frac{1}{k^2}$. This series then has the value $f(x)$, which can be obtained by simply substituting x into f:

Example 74.4 For the 2π-periodic even function $f : [-\pi, \pi) \to \mathbb{R}$ with $f(x) = \frac{1}{2}x^2$ one obtains

$$a_0 = \frac{2}{\pi}\int_0^{\pi} \frac{1}{2}x^2 \,\mathrm{d}x = \frac{2}{\pi}\frac{\pi^3}{6} = \frac{\pi^2}{3}.$$

And further for $k \geq 1$:

$$a_k = \frac{2}{\pi}\int_0^{\pi} \frac{1}{2}x^2 \cos kx \,\mathrm{d}x = \frac{2}{\pi}\left[\frac{x^2}{2k}\sin kx\right]_0^{\pi} - \frac{2}{\pi k}\int_0^{\pi} x \sin kx \,\mathrm{d}x$$

$$= \left[\frac{2x}{\pi k^2}\cos kx\right]_0^{\pi} = \frac{2}{k^2}(-1)^k.$$

The Fourier series of f is thus

$$F(x) = \frac{\pi^2}{6} + 2\sum_{k=1}^{\infty} \frac{(-1)^k}{k^2} \cos kx\,.$$

For $x = 0$ and $x = \pi$ we obtain because of $f(0) = 0$ and $f(\pi) = \frac{\pi^2}{2}$ the equality

$$\sum_{k=1}^{\infty} \frac{(-1)^k}{k^2} = -\frac{\pi^2}{12} \quad \text{and} \quad \sum_{k=1}^{\infty} \frac{1}{k^2} = \frac{\pi^2}{6}.$$

■

74.5 Expanding in Fourier Series: Complex Version

There is another representation of Fourier series with *complex Fourier coefficients*. We start from a complex T-periodic function $f \in C(T)$, $f : [-T/2, T/2) \to \mathbb{C}$.

Fourier Coefficients and Fourier Series: Complex Version

Determine for the T-periodic function $f \in C(T)$, $f : [-T/2, T/2) \to \mathbb{C}$, the so-called **Fourier coefficients**

$$c_k = \frac{1}{T}\int_{-T/2}^{T/2} f(x)\, \mathrm{e}^{-\mathrm{i}k\frac{2\pi}{T}x} \,\mathrm{d}x$$

(continued)

and set

$$F(x) = \sum_{k=-\infty}^{\infty} c_k \, e^{\mathrm{i}\,k\frac{2\pi}{T}x} \,;$$

one calls $F(x)$ **Fourier series** to f; we write for this $f(x) \sim F(x)$.

If the period interval is $[0, T)$ then the above integrals are of course to be formed over this interval.

In the case $T = 2\pi$ the formula for the Fourier coefficients and for the Fourier series are

$$c_k = \frac{1}{2\pi} \int_{-\pi}^{\pi} f(x)\, e^{-\mathrm{i}kx}\, \mathrm{d}x \text{ and } F(x) = \sum_{k=-\infty}^{\infty} c_k \, e^{\mathrm{i}kx} \,.$$

Example 74.5 We determine the complex version of the Fourier series of the rectangular function f from example 74.3. To do this, we calculate the Fourier coefficients c_k:

$$\begin{aligned}
c_k &= \frac{1}{2\pi} \int_{-\pi}^{\pi} f(x)\, e^{-\mathrm{i}kx}\, \mathrm{d}x = \frac{1}{2\pi} \left(\int_{-\pi}^{0} -e^{-\mathrm{i}kx}\, \mathrm{d}x + \int_{0}^{\pi} e^{-\mathrm{i}kx}\, \mathrm{d}x \right) \\
&= \frac{1}{2\pi} \left[\frac{1}{\mathrm{i}k} e^{-\mathrm{i}kx} \right]_{-\pi}^{0} + \frac{1}{2\pi} \left[\frac{-1}{\mathrm{i}k} e^{-\mathrm{i}kx} \right]_{0}^{\pi} \\
&= \frac{1}{2\pi} \left(\frac{2}{\mathrm{i}k} - \frac{1}{\mathrm{i}k} \left(e^{\mathrm{i}k\pi} + e^{-\mathrm{i}k\pi} \right) \right) = \frac{-\mathrm{i}}{2\pi k} (2 - 2\cos(k\pi)) \\
&= \frac{-\mathrm{i}}{k\pi} \left(1 - (-1)^k \right) = \begin{cases} 0\,, & \text{if } k \text{ even} \\ \frac{-2\mathrm{i}}{k\pi}\,, & \text{if } k \text{ odd} \end{cases} .
\end{aligned}$$

The Fourier series is thus

$$\begin{aligned}
F(x) &= \sum_{k=-\infty}^{\infty} \frac{-2\mathrm{i}}{(2k+1)\pi} e^{\mathrm{i}(2k+1)x} \\
&= \cdots + \frac{2\mathrm{i}}{3\pi} e^{-3\mathrm{i}x} + \frac{2\mathrm{i}}{\pi} e^{-\mathrm{i}x} - \frac{2\mathrm{i}}{\pi} e^{\mathrm{i}x} - \frac{2\mathrm{i}}{3\pi} e^{3\mathrm{i}x} - \cdots .
\end{aligned}$$

■

The Fourier coefficients c_k and a_k, b_k are of course closely related: If one calculates the respective coefficients for a function $f : \mathbb{R} \to \mathbb{C}$ then for all $x \in \mathbb{R}$ and $n \in \mathbb{N}$:

$$\frac{a_0}{2} + \sum_{k=1}^{n} a_k \cos\left(k\frac{2\pi}{T}x\right) + b_k \sin\left(k\frac{2\pi}{T}x\right) = \sum_{k=-n}^{n} c_k \, \mathrm{e}^{\mathrm{i}\,k\frac{2\pi}{T}x} ,$$

the relationship between the coefficients is obtained with the following conversion formulas:

Conversion Formulas

We consider a T-periodic function $f : \mathbb{R} \to \mathbb{C}$, $f \in C(T)$, and determine for this function the Fourier coefficients c_k or a_k, b_k:

- Given are c_k for $k \in \mathbb{Z}$, i.e.

$$f(x) \sim F(x) = \sum_{k=-\infty}^{\infty} c_k \, \mathrm{e}^{\mathrm{i}\,k\frac{2\pi}{T}x} .$$

Then $F(x) = \frac{a_0}{2} + \sum_{k=1}^{\infty} a_k \cos\left(k\frac{2\pi}{T}x\right) + b_k \sin\left(k\frac{2\pi}{T}x\right)$ with

$$a_0 = 2\,c_0 , \; a_k = c_k + c_{-k} , \; b_k = \mathrm{i}\,(c_k - c_{-k}) \text{ fof } k \in \mathbb{N} .$$

- Given are a_0, a_k, b_k, $k \in \mathbb{N}$, i.e.

$$f(x) \sim F(x) = \frac{a_0}{2} + \sum_{k=1}^{\infty} a_k \cos\left(k\frac{2\pi}{T}x\right) + b_k \sin\left(k\frac{2\pi}{T}x\right) .$$

Then $F(x) = \sum_{k=-\infty}^{\infty} c_k \, \mathrm{e}^{\mathrm{i}\,k\frac{2\pi}{T}x}$ with

$$c_0 = \frac{a_0}{2} , \; c_k = \frac{1}{2}(a_k - \mathrm{i}\,b_k) , \; c_{-k} = \frac{1}{2}(a_k + \mathrm{i}\,b_k) \text{ for } k \in \mathbb{N} .$$

We have formulated the proof of these formulas as an exercise, it results quite simply with the help of Euler's formula

$$\mathrm{e}^{\mathrm{i}\,y} = \cos(y) + \mathrm{i}\sin(y) .$$

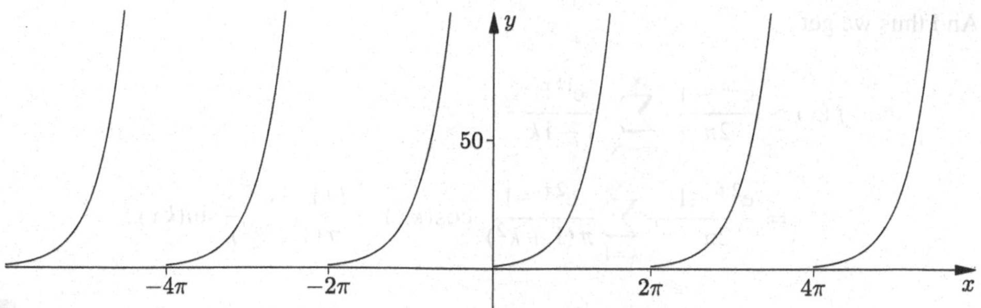

Fig. 74.9 The 2π-periodic continuation of f is an admissible function

If one wants to expands a function f with the Fourier coefficients a_k and b_k it is often easier to first determine the complex version c_k for $k \in \mathbb{Z}$ of the Fourier coefficients and then, using the conversion formulas given, to determine the coefficients a_k and b_k. Often the integrals over $f(x)\, e^{-ikx}$ are easier to determine than those over $f(x)\cos(kx)$ and $f(x)\sin(kx)$.

For example, we calculate the real Fourier coefficients a_k and b_k by first calculating the complex coefficients c_k and then using the conversion formulae to calculate the a_k and b_k:

Example 74.6 We consider the 2π-periodic function $f : [0, 2\pi) \to \mathbb{R}$ with $f(x) = e^x$ (Fig. 74.9).

To determine a_k and b_k, we would have to calculate the integrals

$$a_k = \frac{1}{\pi}\int_0^{2\pi} e^x \cos(kx)dx \text{ and } b_k = \frac{1}{\pi}\int_0^{2\pi} e^x \sin(kx)dx$$

the integrals. We simplify our work and determine

$$c_k = \frac{1}{2\pi}\int_0^{2\pi} e^x\, e^{-ikx}\, dx = \frac{1}{2\pi(1-ik)}\left[e^{(1-ik)x}\right]_0^{2\pi} = \frac{e^{2\pi}-1}{2\pi(1-ik)}\,.$$

With the conversion formulas we get

$$a_0 = 2\, c_0 = \frac{e^{2\pi}-1}{\pi}\,,$$

$$a_k = c_k + c_{-k} = \frac{e^{2\pi}-1}{2\pi(1-ik)} + \frac{e^{2\pi}-1}{2\pi(1+ik)} = \frac{e^{2\pi}-1}{\pi(1+k^2)}\,,$$

$$b_k = i(c_k - c_{-k}) = i\left(\frac{e^{2\pi}-1}{2\pi(1-ik)} - \frac{e^{2\pi}-1}{2\pi(1+ik)}\right) = \frac{k(1-e^{2\pi})}{\pi(1+k^2)}\,.$$

And thus we get

$$f(x) \sim \frac{e^{2\pi}-1}{2\pi} \sum_{k=-\infty}^{\infty} \frac{e^{ikx}}{1-ik}$$

$$= \frac{e^{2\pi}-1}{2\pi} + \sum_{k=1}^{\infty} \frac{e^{2\pi}-1}{\pi(1+k^2)} \cos(kx) + \frac{k(1-e^{2\pi})}{\pi(1+k^2)} \sin(kx) .$$

■

To distinguish between the two versions, i.e. the real and the complex version, the coefficients a_k and b_k of the real version are also called the Fourier coefficients of the **cos-sin representation** of the Fourier series and the coefficients c_k of the complex version also Fourier coefficients of the **exp representation** of the Fourier series.

74.6 Exercises

74.1 Determine the cos-sin-representations of the Fourier series of the following 2π-periodic functions:

(a) $f(x) = \left(\frac{x}{\pi}\right)^3 - \frac{x}{\pi}$ for $x \in [-\pi, \pi)$,
(b) $f(x) = (x-\pi)^2$ for $x \in [0, 2\pi)$,
(c) $f(x) = |\sin x|$ for $x \in [-\pi, \pi)$,
(d) $f(x) = \begin{cases} 0, & -\pi < x \le 0 \\ \sin x, & 0 \le x \le \pi \end{cases}$.

74.2 Given is the 2π-periodic function f with

$$f(x) = \pi - |x| \qquad \text{for} \quad -\pi \le x \le \pi .$$

(a) Calculate the coefficients of the associated cos-sin-representation $F(x)$.
(b) Using (a), determine the value of the infinite series

$$\frac{1}{1^2} + \frac{1}{3^2} + \frac{1}{5^2} + \frac{1}{7^2} + \ldots$$

74.3 Justify the calculation rules in the box in Sect. 74.3

74.4 Determine the Fourier coefficients a_k, b_k and c_k of the 2π-periodic function f with

$$f(x) = \begin{cases} 0\,, & x \in [-\pi, -\frac{\pi}{2}) \\ \cos(x)\,, & x \in [-\frac{\pi}{2}, \frac{\pi}{2}) \\ 0\,, & x \in [\frac{\pi}{2}, \pi) \end{cases} .$$

Fourier Series: Background, Theorems and Application

75

We again consider functions $f : \mathbb{R} \to \mathbb{C}$ with the properties (i) to (iii) (see Sect. 74.2) and write briefly for that as we have already done $f \in C(T)$. Every function $f \in C(T)$ can be expanded into a Fourier series, i.e.

$$f(x) \sim \frac{a_0}{2} + \sum_{k=1}^{\infty} a_k \cos(kx) + b_k \sin(kx) = \sum_{k=-\infty}^{\infty} c_k \, \mathrm{e}^{\mathrm{i}\,kx}$$

with the Fourier coefficients $a_0,\ a_k,\ b_k$ for $k \in \mathbb{N}$ or c_k for $k \in \mathbb{Z}$.

In this chapter we show what this expansion is based on. In addition, we give numerous useful rules and theorems which help further to determine the Fourier series of f *similar* functions, if one already knows the Fourier series of f. Furthermore, we discuss a typical application of Fourier series for the determination of particular solutions of differential equations.

75.1 The Orthonormal System $1/\sqrt{2}, \cos(kx), \sin(kx)$

In order not to lose the overview in the following presentation, we restrict ourselves to 2π-periodic, real functions, which we always specify on the period interval $[-\pi, \pi)$.

We now consider in the vector space $\mathbb{R}^{\mathbb{R}}$ the subvector space $V = \langle C(2\pi) \rangle$ which is generated of all 2π-periodic functions from $C(2\pi)$ and justify:

C. Karpfinger, *Calculus and Linear Algebra in Recipes*,
https://doi.org/10.1007/978-3-662-65458-3_75

The Orthonormal System B of $V = \langle C(2\pi)\rangle$

The set $B = \left(1/\sqrt{2},\ \cos(x),\ \cos(2x), \ldots, \sin(x),\ \sin(2x), \ldots\right)$ is an orthonormal system of V with respect to the scalar product

$$\langle f, g\rangle = \frac{1}{\pi}\int_{-\pi}^{\pi} f(x)\, g(x)\, \mathrm{d}x\,,$$

i.e., for all $f,\ g \in B$ we have:

$$\langle f, g\rangle = \begin{cases} 1\,, & \text{if } f = g \\ 0\,, & \text{otherwise} \end{cases}.$$

That the given product is a scalar product we have already shown in essence in Sect. 16.1. It remains to prove that each pair of different elements of B is orthogonal and each element of B has length 1 (note Sect. 16.3), we summarize these proofs clearly:

$$\langle 1/\sqrt{2}\,, 1/\sqrt{2}\rangle = \frac{1}{\pi}\int_{-\pi}^{\pi} \frac{1}{2}\,\mathrm{d}x = 1\,,$$

$$\langle 1/\sqrt{2}\,, \cos(kx)\rangle = \frac{1}{\sqrt{2}\pi}\int_{-\pi}^{\pi} \cos(kx)\,\mathrm{d}x = 0\,,$$

$$\langle 1/\sqrt{2}\,, \sin(kx)\rangle = \frac{1}{\sqrt{2}\pi}\int_{-\pi}^{\pi} \sin(kx)\,\mathrm{d}x = 0\,,$$

$$\langle \cos(kx)\,, \cos(\ell x)\rangle = \begin{cases} \frac{1}{2\pi}\left(\frac{\sin((k-\ell)x)}{k-\ell} + \frac{\sin((k+\ell)x)}{k+\ell}\Bigg|_{-\pi}^{\pi}\right) = 0 & k \neq \ell \\ \frac{1}{\pi}\left(\frac{1}{2}x + \frac{1}{4k}\sin(2kx)\Bigg|_{-\pi}^{\pi}\right) = 1 & k = \ell\,, \end{cases}$$

$$\langle \sin(kx)\,, \sin(\ell x)\rangle = \begin{cases} \frac{1}{2\pi}\left(\frac{\sin((k-\ell)x)}{k-\ell} - \frac{\sin((k+\ell)x)}{k+\ell}\Bigg|_{-\pi}^{\pi}\right) = 0 & k \neq \ell \\ \frac{1}{\pi}\left(\frac{1}{2}x - \frac{1}{4k}\sin(2kx)\Bigg|_{-\pi}^{\pi}\right) = 1 & k = \ell\,, \end{cases}$$

$$\langle \cos(kx)\,,\sin(\ell x)\rangle = \begin{cases} -\frac{1}{2\pi}\left(\frac{\cos((\ell-k)x)}{\ell-k} + \frac{\cos((\ell+k)x)}{k+\ell}\Big|_{-\pi}^{\pi}\right) = 0 & k \neq \ell \\ -\frac{1}{4k\pi}\left(\cos(2kx)\Big|_{-\pi}^{\pi}\right) = 0 & k = \ell\,. \end{cases}$$

If $f \in V$ is any 2π-periodic function, then (even if the appropriate conditions are not satisfied) we can proceed as in the second recipe in Sect. 16.4 to try to find the coefficients of a linear combination with respect to the orthonormal system B. If we do this, we obtain the coefficients a, a_k, b_k, $k = 1, 2, \ldots$:

$$a = \langle f\,, 1/\sqrt{2}\rangle, \quad a_k = \langle f\,, \cos(kx)\rangle \text{ and } b_k = \langle f\,, \sin(kx)\rangle\,.$$

These are now almost the Fourier coefficients a_0, a_k, b_k; we obtain these by unifying the formulas for a and a_k, we have:

$$a_k = \langle f\,, \cos(kx)\rangle\,,\ k \in \mathbb{N}_0\,,\ \text{ where } a = a_0/\sqrt{2}\,.$$

The Fourier series expansion is thus a kind of infinite linear combination of a function $f \in V$, where the coefficients of the expansion are obtained by projection onto the vectors $\cos(kx)$ and $\sin(kx)$ of the orthonormal system B.

Remark For the sake of simplicity, we have based all our considerations in this section on the cos-sin-representation of 2π-periodic functions. But the generalization to T-periodic functions as well as on the exp-representation is simple: In the cos-sin representation one considers the orthonormal system B with the functions $\cos\left(k\frac{2\pi}{T}x\right)$ and $\sin\left(k\frac{2\pi}{T}x\right)$ and with the exp-representation the orthonormal system $B = \{e^{\mathrm{i}k\frac{2\pi}{T}x} \mid k \in \mathbb{Z}\}$—that this set B is an orthonormal system, the following orthogonality relation states:

$$\frac{1}{T}\int_{-T/2}^{T/2} e^{\mathrm{i}k\frac{2\pi}{T}x} \cdot e^{-\mathrm{i}l\frac{2\pi}{T}x}\,\mathrm{d}x = \begin{cases} 0\,, & k \neq l \\ 1\,, & k = l \end{cases}\,.$$

75.2 Theorems and Rules

In Chap. 74 we have described in detail how to determine the Fourier series F to a function $f \in C(T)$. We also mentioned that the series value $F(x)$ at each continuity point x of f is exactly the function value of f at this point, i.e. $f(x) = F(x)$. The Fourier series is therefore a *representation* of the function f on the continuity intervals of f. In this respect, it is not a little surprising that we can derive from the Fourier series F of the function f a

Fourier series of a primitive function or derivative function of *f*, provided that the latter is again periodic or differentiable. Also of many other functions closely related to *f*, we can easily find the Fourier series with the help of the Fourier series *F* of *f*. We summarize the most important of these rules below:

Theorems and Rules for Fourier Coefficients or Fourier Series

Given are two *T*-periodic functions $f, g \in C(T)$ with the Fourier series $F \sim f$ and $G \sim g$, detailed

$$F(x) = \frac{a_0}{2} + \sum_{k=1}^{\infty} a_k \cos\left(k\frac{2\pi}{T}x\right) + b_k \sin\left(k\frac{2\pi}{T}x\right) = \sum_{k=-\infty}^{\infty} c_k\, \mathrm{e}^{\mathrm{i}k\frac{2\pi}{T}x} \text{ and}$$

$$G(x) = \frac{\tilde{a}_0}{2} + \sum_{k=1}^{\infty} \tilde{a}_k \cos\left(k\frac{2\pi}{T}x\right) + \tilde{b}_k \sin\left(k\frac{2\pi}{T}x\right) = \sum_{k=-\infty}^{\infty} \tilde{c}_k\, \mathrm{e}^{\mathrm{i}k\frac{2\pi}{T}x} .$$

We have the following theorems and rules:

- **Riemann's Lemma:** For the Fourier coefficients $a_k,\ b_k$ andc_k we have

$$a_k,\ b_k,\ c_k,\ c_{-k} \overset{k\to\infty}{\longrightarrow} 0 .$$

- **Linearity:** For all $\lambda,\ \mu \in \mathbb{C}$ we have: The Fourier series *H* of $\lambda f + \mu g$ is obtained as follows

$$\begin{aligned} H(x) &= \lambda F(x) + \mu G(x) \\ &= \frac{\lambda a_0 + \mu \tilde{a}_0}{2} + \sum_{k=1}^{\infty} (\lambda a_k + \mu \tilde{a}_k) \cos\left(k\frac{2\pi}{T}x\right) \\ &\quad + (\lambda b_k + \mu \tilde{b}_k) \sin\left(k\frac{2\pi}{T}x\right) \\ &= \sum_{k=-\infty}^{\infty} (\lambda c_k + \mu \tilde{c}_k)\, \mathrm{e}^{\mathrm{i}k\frac{2\pi}{T}x} . \end{aligned}$$

(continued)

- **Fourier series of the derivative:** If $f \in C(T)$ is continuous on $\mathbb{R}$ and piecewise continuous differentiable on $[-T/2, T/2)$, then we obtain by deriving the summands of $F(x)$ the Fourier series of $f'(x)$, i.e. $f'(x) \sim F'(x)$ where

$$F'(x) = \sum_{k=1}^{\infty} -k\frac{2\pi}{T}a_k \sin\left(k\frac{2\pi}{T}x\right) + k\frac{2\pi}{T}b_k \cos\left(k\frac{2\pi}{T}x\right)$$
$$= \sum_{k=-\infty}^{\infty} \mathrm{i}k\frac{2\pi}{T}c_k\, \mathrm{e}^{\mathrm{i}k\frac{2\pi}{T}x}\,.$$

- **Fourier series of the primitive function:** Is $a_0 = 0 = c_0$ so is any primitive function $\tilde{F}$ of f in $C(T)$, and we obtain by integration of $F(x)$ the Fourier series of the primitive function $\tilde{F}$ of $f(x)$

$$\tilde{F} \sim \frac{A_0}{2} + \sum_{k=1}^{\infty} \frac{T}{k2\pi}a_k \sin\left(k\frac{2\pi}{T}x\right) - \frac{T}{k2\pi}b_k \cos\left(k\frac{2\pi}{T}x\right)$$
$$= C_0 + \sum_{\substack{k=-\infty \\ k\neq 0}}^{\infty} \frac{T}{\mathrm{i}k2\pi}c_k\, \mathrm{e}^{\mathrm{i}k\frac{2\pi}{T}x}\,,$$

  where A_0 and C_0 is the zero Fourier coefficient of $\tilde{F}$.
- **Time scaling:** If f is a T-periodic function, then $h : x \mapsto f(cx)$ with $c \in \mathbb{R}_{>0}$ a T/c-periodic function; the function h has the Fourier series

$$h(x) = f(cx) \sim \sum_{k=-\infty}^{\infty} c_k\, \mathrm{e}^{\mathrm{i}kc\frac{2\pi}{T}x}\,.$$

- **Time reversal:** The Fourier series of the function $h : x \mapsto f(-x)$ is obtained from the Fourier series given by $f(x)$ by substituting c_k by c_{-k}, i.e.

$$h(x) = f(-x) \sim \sum_{k=-\infty}^{\infty} c_{-k}\, \mathrm{e}^{\mathrm{i}k\frac{2\pi}{T}x}\,.$$

- **Time shifting:** The Fourier series of $h : x \mapsto f(x+a)$ with $a \in \mathbb{R}$ is obtained from that of $f(x)$ by multiplication by $\mathrm{e}^{\mathrm{i}k\frac{2\pi}{T}a}$, i.e.

$$h(x) = f(x+a) \sim \sum_{k=-\infty}^{\infty} \mathrm{e}^{\mathrm{i}k\frac{2\pi}{T}a}\, c_k\, \mathrm{e}^{\mathrm{i}k\frac{2\pi}{T}x}\,.$$

(continued)

- **Frequency shifting:** The Fourier series of $h : x \mapsto \mathrm{e}^{\mathrm{i}\,n\frac{2\pi}{T}x}\, f(x)$ with $n \in \mathbb{Z}$ is obtained from that of $f(x)$ by shifting the coefficients c_k, i.e.

$$h(x) = \mathrm{e}^{\mathrm{i}\,n\frac{2\pi}{T}x}\, f(x) \sim \sum_{k=-\infty}^{\infty} c_{k-n}\, \mathrm{e}^{\mathrm{i}\,k\frac{2\pi}{T}x}\,.$$

- **Convolution:** The **convolution product** of f and g, which is the T-periodic function

$$(f * g)(x) = \frac{1}{T}\int_{-T/2}^{T/2} f(x-t)\, g(t)\mathrm{d}t\,,$$

has the Fourier coefficients $c_k\, \tilde{c}_k$, i.e.

$$(f * g)(x) \sim \sum_{k=-\infty}^{\infty} c_k\, \tilde{c}_k\, \mathrm{e}^{\mathrm{i}\,k\frac{2\pi}{T}x}\,.$$

Example 75.1

- The 2π-periodic function $f : [-\pi, \pi) \to \mathbb{R}$ with $f(x) = x$ has according to Example 74.3 the Fourier series expansion

$$f(x) \sim 2\sum_{k=1}^{\infty} \frac{(-1)^{k+1}}{k} \sin kx\,.$$

We now consider the function $g : [0, 2\pi) \to \mathbb{R}$ with $g(x) = \frac{1}{2}(\pi - x)$. The relationship between f and g is as follows:

$$g(x) = \frac{1}{2}\, f(\pi - x)\,.$$

From $f(x) \sim 2\sum_{k=1}^{\infty} \frac{(-1)^{k+1}}{k} \sin kx$ follows with the rules to *time shifting* and *linearity* because of $\mathrm{e}^{\mathrm{i}\,k\pi} = (-1)^k$:

$$g(x) \sim \frac{1}{2}\cdot 2\sum_{k=1}^{\infty} \frac{1}{k} \sin kx = \frac{\sin(x)}{1} + \frac{\sin(2x)}{2} + \frac{\sin(3x)}{3} + \cdots\,.$$

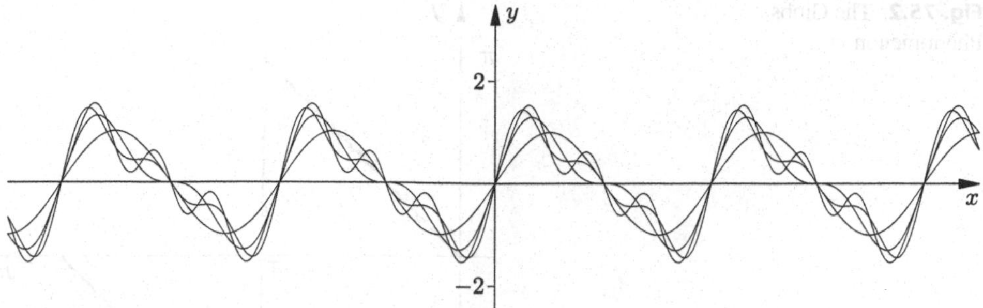

Fig. 75.1 A sawtooth and the graphs of the first approximating functions

Observe Fig. 75.1.

- The periodic continuation of the function $f : [-\pi, \pi) \to \mathbb{R}$ with $f(x) = \frac{1}{2}x^2$ is continuous. Its Fourier series is according to Example 74.4

$$F(x) = \frac{\pi^2}{6} + 2\sum_{k=1}^{\infty} \frac{(-1)^k}{k^2} \cos kx\,.$$

Because $f'(x) = x$ is therefore

$$F'(x) = 2\sum_{k=1}^{\infty} \frac{(-1)^{k+1}}{k} \sin kx$$

the Fourier series to the sawtooth (cf. Example 74.3).

■

At a discontinuity point of a real function $f \in C(T)$ the following phenomenon can be observed: The approximating functions $F_1(x)$, $F_2(x)$, ... show typical *overshoots* and whose *excursions* do not decrease even with the addition of further summands, note Fig. 75.2.

The occurrence of these overshoots or undershoots at jump points of about 18% of half the jump height is called the **Gibbs phenomenon**.

75.3 Application to Linear Differential Equations

We show how to use Fourier series to determine a *T*-periodic solution to a linear differential equation with a *T*-periodic perturbation function. We restrict ourselves to a 2nd order ODE, for higher orders we proceed analogously.

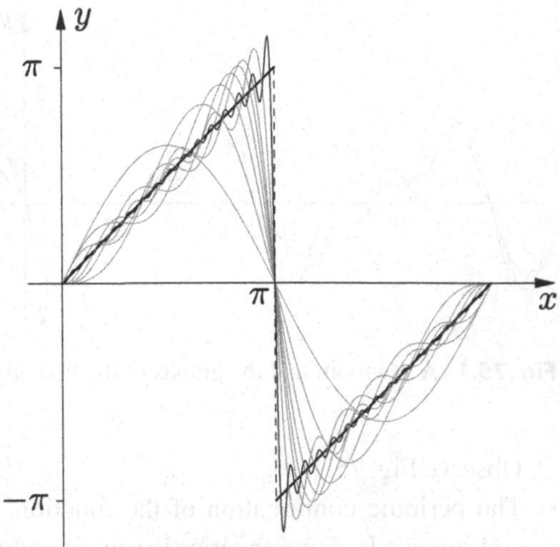

Fig. 75.2 The Gibbs Phenomenon

We denote (as in earlier chapters on ODEs) the searched solution function of a ODE in the variable t with x, $x = x(t)$.

Determination of a *T*-Periodic Solution of an ODE with *T*-Periodic Perturbation Function

To determine a *T*-periodic solution $x = x(t)$ of the linear ODE

$$a\,\ddot{x}(t) + b\,\dot{x}(t) + c\,x(t) = s(t)$$

with a continuous and *T*-periodic perturbation function $s = s(t)$ and real numbers a, b and c, proceed as follows:

(1) Develop s into a Fourier series $s(t) = \sum_{k=-\infty}^{\infty} d_k\, \mathrm{e}^{\mathrm{i}\,k\omega t}$ where $\omega = 2\pi/T$.
(2) Enter the approach $x(t) = \sum_{k=-\infty}^{\infty} c_k\, \mathrm{e}^{\mathrm{i}\,k\omega t}$ into the ODE and obtain with

$$\dot{x}(t) = \sum_{k=-\infty}^{\infty} \mathrm{i}\,k\,\omega\,c_k\,\mathrm{e}^{\mathrm{i}\,k\omega t} \quad \text{and} \quad \ddot{x}(t) = -\sum_{k=-\infty}^{\infty} k^2\,\omega^2\,c_k\,\mathrm{e}^{\mathrm{i}\,k\omega t}$$

the equation

$$\sum_{k=-\infty}^{\infty} (-a\,k^2\omega^2 c_k + \mathrm{i}\,b\,k\,\omega\,c_k + c\,c_k - d_k)\,\mathrm{e}^{\mathrm{i}\,k\omega t} = 0\,.$$

(continued)

(3) Equating coefficients yields

$$-a\,k^2\omega^2 c_k + \mathrm{i}\,b\,k\,\omega\,c_k + c\,c_k = d_k\,,\ \text{d. h. } c_k = \frac{d_k}{-a\,k^2\omega^2 + \mathrm{i}\,b\,k\,\omega + c}\,.$$

(4) Obtain the solution

$$x(t) = \sum_{k=-\infty}^{\infty} \frac{d_k}{c - a\,k^2\omega^2 + \mathrm{i}\,b\,k\,\omega}\,\mathrm{e}^{\mathrm{i}\,k\omega t}\,.$$

Note that to specify $x = x(t)$ in (4) besides the coefficients a, b and c of the ODE and $\omega = 2\pi/T$ only the Fourier coefficients d_k of the Fourier series of the perturbation function from (1) are necessary. So the steps (2) and (3) can be skipped. But by specifying steps (2) and (3), the procedure for finding the solution can also be easily transferred to ODEs of higher order.

Example 75.2 For the *RC* **lowpass** exists between the input voltage $U_e(t)$ and the output voltage $U(t)$ the differential equation

$$RC\,\dot{U}(t) + U(t) = U_e(t)\,.$$

As an example, we consider the input voltage (this is the disturbance function $s = s(t)$)

$$U_e(t) = U_0|\sin(\omega t)|\,.$$

Now we apply the above recipe to determine a *T*-periodic solution $U = U(t)$ with $T = 2\pi/\omega$.

(1) By Fourier expansion of the function $U_e(t)$ we obtain

$$U_e(t) = U_0|\sin(\omega t)| \sim -\frac{2U_0}{\pi}\sum_{k=-\infty}^{\infty}\frac{1}{4k^2-1}\,\mathrm{e}^{2\,\mathrm{i}\,k\omega t}\,.$$

(4) With $a = 0$, $b = RC$, $c = 1$ and $d_k = \frac{-2/\pi}{4k^2-1}U_0$ we obtain the *answer*

$$U(t) = -\frac{2U_0}{\pi}\sum_{k=-\infty}^{\infty}\frac{1}{(4k^2-1)(1+2\,\mathrm{i}\,RC\,k\,\omega)}\,\mathrm{e}^{2\,\mathrm{i}\,k\omega t}\,.$$

■

Remark By the above recipe, we determine a *T*-periodic solution function $x = x(t)$ of the ODE under consideration. Thus, we implicitly solve the following **boundary value problem**

$$a\,\ddot{x}(t) + b\,\dot{x}(t) + c\,x(t) = s(t)\,,\ x(0) = x(T)\,,\ \dot{x}(0) = \dot{x}(T)$$

with a continuous and *T*-periodic perturbation function $s = s(t)$ and real numbers a, b and c.

75.4 Exercises

75.1 By $f(x) = \sum_{k\in\mathbb{Z}\setminus\{0\}} \frac{e^{ikx}}{k^5}$ a 2π-periodic C^2-function $\mathbb{R} \to \mathbb{C}$ is defined. Determine for the two functions

(a) $g(x) = f(2x - 3)$,
(b) $h(x) = g''(x) + f(4x)$

the period T, the angular frequency $\omega = \frac{2\pi}{T}$ and the Fourier coefficients c_k.

75.2 For the convolution product $*$ confirm the formulas

$$\text{(a)}\quad \left(\sum_{k=1}^{\infty} b_k \sin kx\right) * \left(\sum_{k=1}^{\infty} \beta_k \sin kx\right) = -\frac{1}{2}\sum_{k=1}^{\infty} b_k\,\beta_k \cos kx\,.$$

$$\text{(b)}\quad \left(\frac{a_0}{2} + \sum_{k=1}^{\infty} a_k \cos kx\right) * \left(\frac{\alpha_0}{2} + \sum_{k=1}^{\infty} \alpha_k \cos kx\right) = \frac{a_0\alpha_0}{4} + \frac{1}{2}\sum_{k=1}^{\infty} a_k\,\alpha_k \cos kx\,.$$

75.3 Let be given a triple low-pass filter described by the differential equation

$$\left(\alpha\frac{d}{dt} + 1\right)^3 x(t) = s(t)$$

with $\alpha = RC > 0$ and 2π-periodic input voltage $s(t)$. Note here:

$$\left(\alpha\frac{d}{dt} + 1\right)^3 x(t) = \alpha^3\frac{d^3}{dt^3}x(t) + 3\alpha^2\frac{d^2}{dt^2}x(t) + 3\alpha\frac{d}{dt}x(t) + x(t)\,.$$

Calculate the Fourier series of the *answer* $x(t)$, if $s(t) = t$ for $t \in [0, 2\pi)$.

75.4 Consider the differential equation

$$\ddot{x}(t) + 2\dot{x}(t) + 2x(t) = s(t)$$

with 2π-periodic input $s(t) = \frac{\pi - t}{2}$ for $t \in [0, 2\pi)$. Determine the Fourier series of the 2π-periodic answer $x(t)$.

75.5 Let s be $s(x) = \frac{\pi - x}{2}$ for $x \in [0, 2\pi)$ a 2π-periodic sawtooth function.

(a) Show that the convolution $(s * s)(x)$ again gives a 2π-periodic function.
(b) Calculate the periodic convolution $(s * s)(x) = \frac{1}{2\pi} \int_0^{2\pi} s(x-t)s(t)\,\mathrm{d}t$ for $x \in \mathbb{R}$ directly.
(c) Determine the Fourier coefficients c_k of the function $s * s$ by direct calculation.

75.6 The Schrödinger equation in the form $\mathrm{i}\,\partial_t u(t, x) = -\partial_{xx} u(t, x)$ describes the wave function of a free particle. We assume the particle is moving in a ring, i.e., the function $u(t, x)$ is 2π-periodic in the second variable, i.e. $u(t, x + 2\pi) = u(t, x)$ for all $t \geq 0$. In the following, by Fourier series development, you will find a representation of the solution $u(t, x) : [0, \infty) \times \mathbb{R} \to \mathbb{C}$.

(a) Write the complex Fourier series of the function $u(t, x)$, where the Fourier coefficients $c_k(t)$ depend of t.
(b) Substitute the Fourier series into the equation and simplify until your equation is of the type $\sum_{k \in \mathbb{Z}} \ldots \mathrm{e}^{\mathrm{i}kx} = \sum_{k \in \mathbb{Z}} \ldots \mathrm{e}^{\mathrm{i}kx}$.
(c) By equating coefficients, obtain for each $c_k(t)$ an ODE. What is it?
(d) We assume that the function $u(0, x)$ is known and has the complex Fourier coefficients γ_k. Determine $c_k(t)$. What is the solution $u(t, x)$ of the original problem?
(e) What can you say about $|c_k(t)|$?

Fourier Transform I

76

In the Fourier series expansion, we have developed a piecewise continuous and monotone T-periodic function f into a Fourier series and thus decomposed the periodic signal f into a sum of harmonic oscillations with discrete amplitudes. The *Fourier transform* can be seen as a decomposition of a non-periodic signal f into harmonic oscillations with a continuous amplitude spectrum.

We cover the amazing applications of this transformation in the next chapter. In the present chapter we do the computational work: We transform functions and also consider the possibility of inverse transformation. But in view of the applications we take the considered functions f to be functions in time t, we will speak of *time* functions.

76.1 The Fourier Transform

We declare for a function f the *Fourier transform F*. Thereby we get each value $F(\omega)$ from F by an improper integral over an unbounded interval whose integrand has the function f as a component. Therefore there are restrictions on f since integrals over unbounded intervals are known not to exist for all integrands.

In order not to make the formalism expressing this unnecessarily complicated, we speak of a **fourier transformable** function, meaning a function f for which the integrals under consideration exist.

We assume a fourier transformable function $f : \mathbb{R} \to \mathbb{C}$ and calculate for $\omega \in \mathbb{R}$ the integral

$$\text{CHW} \int_{-\infty}^{\infty} f(t)\, \mathrm{e}^{-\mathrm{i}\,\omega t}\, \mathrm{d}t\,.$$

C. Karpfinger, *Calculus and Linear Algebra in Recipes*,
https://doi.org/10.1007/978-3-662-65458-3_76

Note the definition of the Cauchy principal value in Sect. 32.1. The value of this improper integral depends on ω. Since we can determine this value for any $\omega \in \mathbb{R}$ we thus obtain to the *time function* $f(t)$ is a *frequency function* F in ω:

The Fourier Transform
If $f : \mathbb{R} \to \mathbb{C}$ is a fourier transformable function, then the function is called $F : \mathbb{R} \to \mathbb{C}$ given by

$$F(\omega) = \text{CHW} \int_{-\infty}^{\infty} f(t)\, \mathrm{e}^{-\mathrm{i}\omega t}\, \mathrm{d}t ,$$

the **Fourier transform** of f. The notations

$$f(t) \circ\!\!-\!\!\bullet\ F(\omega) \quad \text{or} \quad \mathcal{F}(f(t)) = F(\omega)$$

are common. One calls $F(\omega)$ also **frequency function** or **spectral function** to the **time function** $f(t)$.

The assignment $f \to F$ which gives to a fourier transformable function f the Fourier transform F is called **Fourier transform.**

It is customary, and we will do the same from now on, to omit the term CHW for the Cauchy principal value. Within the Fourier transform, any improper integral over all of $\mathbb{R}$ is to be understood as a Cauchy principal value.

Example 76.1

- We determine the Fourier transform F of the square wave function

$$f : \mathbb{R} \to \mathbb{C},\ f(t) = \begin{cases} 1/2, & |t| \leq 1 \\ 0, & |t| > 1 \end{cases} .$$

We have

$$F(\omega) = \int_{-\infty}^{\infty} f(t)\, \mathrm{e}^{-\mathrm{i}\omega t}\, \mathrm{d}t = \frac{1}{2} \int_{-1}^{1} \mathrm{e}^{-\mathrm{i}\omega t}\, \mathrm{d}t .$$

Here we have to make a case distinction. In the case $\omega = 0$ we get

$$F(0) = \frac{1}{2} \int_{-1}^{1} 1 \mathrm{d}t = 1 ,$$

and in the case $\omega \neq 0$ we obtain

$$F(\omega) = \frac{1}{2} \int_{-1}^{1} \mathrm{e}^{-\mathrm{i}\omega t} \,\mathrm{d}t = \frac{1}{2} \left[\frac{-1}{\mathrm{i}\omega} \mathrm{e}^{-\mathrm{i}\omega t} \right]_{-1}^{1} = \frac{\sin(\omega)}{\omega} .$$

Let us explain the so-called **sinc function** sinc: $\mathbb{R} \to \mathbb{C}$ by

$$\operatorname{sinc}(\omega) = \begin{cases} 1 , & \omega = 0 \\ \frac{\sin(\omega)}{\omega} , & \omega \neq 0 \end{cases} ,$$

so it is shown that sinc is the Fourier transform of the rectangular function f, we have

$$f(t) \quad \circ\!\!-\!\!\bullet \quad \operatorname{sinc}(\omega) .$$

Fig. 76.1 shows the graphs of the functions under consideration.

- We consider for a $a \in \mathbb{R}_{>0}$ the function

$$f : \mathbb{R} \to \mathbb{C} , \ f(t) = \begin{cases} \mathrm{e}^{-at} , & \text{for } t \geq 0 \\ 0 , & \text{for } t < 0 \end{cases}$$

and determine their Fourier transforms $F(\omega)$, we have

$$F(\omega) = \int_{-\infty}^{\infty} f(t) \,\mathrm{e}^{-\mathrm{i}\omega t} \,\mathrm{d}t = \int_{0}^{\infty} \mathrm{e}^{-(a+\mathrm{i}\omega)t} \,\mathrm{d}t = \lim_{b \to \infty} \int_{0}^{b} \mathrm{e}^{-(a+\mathrm{i}\omega)t} \,\mathrm{d}t$$

$$= \lim_{b \to \infty} \frac{-1}{a + \mathrm{i}\omega} \left[\mathrm{e}^{-(a+\mathrm{i}\omega)t} \right]_{0}^{b} = \frac{1}{a + \mathrm{i}\omega} .$$

With this $F(\omega) = \frac{1}{a+\mathrm{i}\omega}$ is the Fourier transform of $f(t)$.

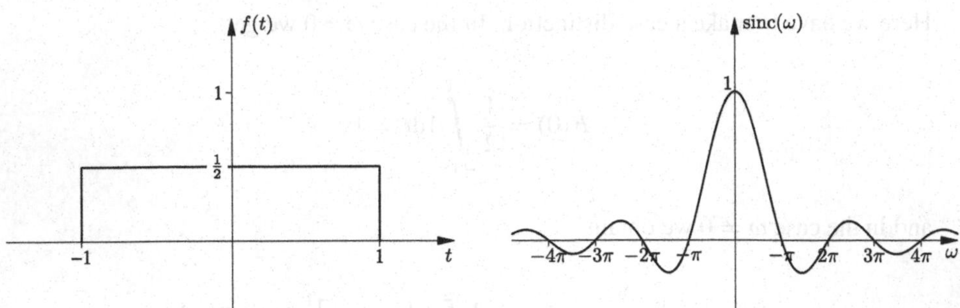

Fig. 76.1 The rectangular function f and its Fourier transform sinc

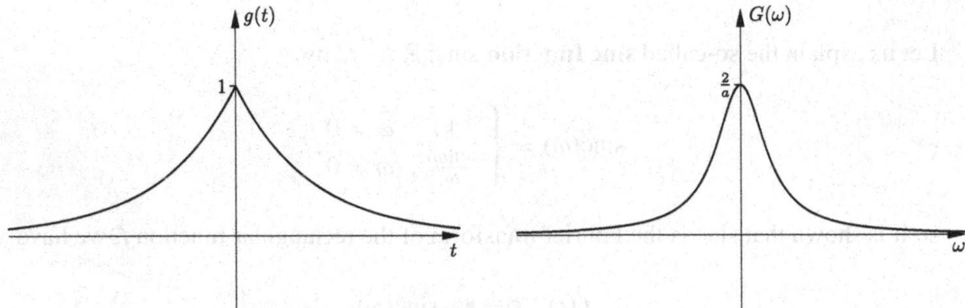

Fig. 76.2 The function f and its Fourier transform G

- Now we consider the figure $g : \mathbb{R} \to \mathbb{C}$ with $g(t) = \mathrm{e}^{-a|t|}$, where $a > 0$. We obtain as Fourier transform:

$$G(\omega) = \int_{-\infty}^{0} \mathrm{e}^{at}\, \mathrm{e}^{-\mathrm{i}\,\omega t}\, \mathrm{d}t + \int_{0}^{\infty} \mathrm{e}^{-at}\, \mathrm{e}^{-\mathrm{i}\,\omega t}\, \mathrm{d}t$$

$$= \frac{1}{a - \mathrm{i}\,\omega}\, \mathrm{e}^{(a-\mathrm{i}\,\omega)t}\Big|_{-\infty}^{0} + \frac{1}{-a - \mathrm{i}\,\omega}\, \mathrm{e}^{(-a-\mathrm{i}\,\omega)t}\Big|_{0}^{\infty}$$

$$= \frac{1}{a - \mathrm{i}\,\omega} + \frac{1}{a + \mathrm{i}\,\omega} = \frac{2\,a}{a^2 + \omega^2}\,.$$

Therefore $G(\omega) = \frac{2\,a}{a^2+\omega^2}$ is the Fourier transform of $g(t)$.

Figure 76.2 shows the function f and its Fourier transform G: ■

MATLAB Using the SYMBOLIC MATH TOOLBOX it is often possible with MATLAB to calculate the Fourier transforms $F(\omega)$ of $f(t)$. We show this exemplarily by the last example:

```
» syms t w; » f=exp(-abs(t)); » fourier(f,t,w) ans = 2/
(w^2 + 1)
```

In practice functions are important, which are themselves not Fourier transformable at all; and nevertheless one wants to determine their Fourier transforms. This is already the case with the so simple and important *Heaviside function* (see Example 76.2). A Fourier transform can be given with the aid of *Dirac's delta function*. This *Dirac delta function* is not in itself a *function* at all, it is a so-called *distribution*.

Example 76.2

- **Dirac's delta function.** We consider for a $t_0 \in \mathbb{R}$ and $\varepsilon \in \mathbb{R}_{>0}$ the *pulse function* $\delta_\varepsilon : \mathbb{R} \to \mathbb{R}$ given by

$$\delta_\varepsilon(t - t_0) = \begin{cases} 0\,, & t < t_0 \\ 1/\varepsilon\,, & t_0 \le t \le t_0 + \varepsilon\,. \\ 0\,, & t > t_0 + \varepsilon \end{cases}$$

This pulse function δ_ε (Fig. 76.3) is normalized in the following sense, we have

$$\int_{-\infty}^{\infty} \delta_\varepsilon(t - t_0)\mathrm{d}t = 1\,.$$

This is true for any real $\varepsilon > 0$. With $\varepsilon \to 0$ we get $\delta_\varepsilon \overset{\varepsilon \to 0}{\longrightarrow} \delta$; where the *function* δ has a *peak* at the point $t = t_0$, for $t \neq t_0$ it is equal to 0, we write this casually as

$$\delta(t - t_0) = \begin{cases} 0\,, & t \neq t_0 \\ \infty\,, & t = t_0 \end{cases}.$$

This δ is not a function in the classical sense, one speaks of a *distribution* and calls δ the **Dirac's delta function** . It has the following nice property: If g is any continuous function, then

$$\int_{-\infty}^{\infty} g(t)\delta(t - t_0)\,\mathrm{d}t = g(t_0)\,.$$

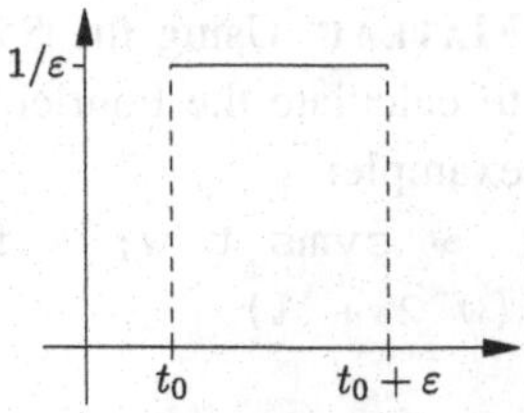

Fig. 76.3 The function δ_ε

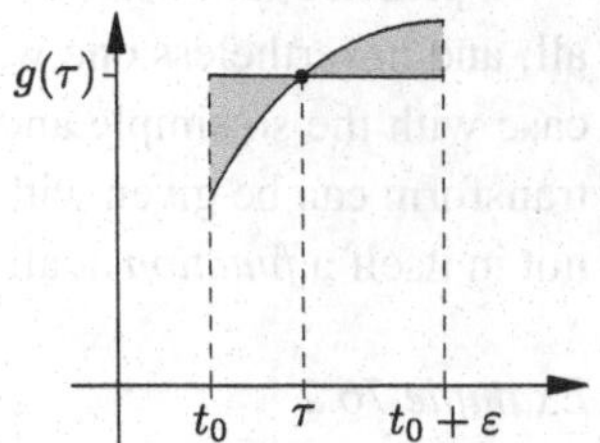

Fig. 76.4 The mean value theorem for δ_ε

This can be *justified as* follows:

$$\int_{-\infty}^{\infty} g(t)\,\delta(t-t_0)\,\mathrm{d}t = \lim_{\varepsilon\to 0}\int_{-\infty}^{\infty} g(t)\,\delta_\varepsilon(t-t_0)\,\mathrm{d}t = \lim_{\varepsilon\to 0}\int_{t_0}^{t_0+\varepsilon} g(t)\,\frac{1}{\varepsilon}\,\mathrm{d}t$$

$$= \lim_{\varepsilon\to 0}\left(g(\tau)\,\frac{1}{\varepsilon}\right)\,\varepsilon = g(t_0)\,,$$

Where we have used the mean value theorem of the integral calculus (see Fig. 76.4). Using this property, it is not difficult to determine the Fourier transform $\Delta_{t_0}(\omega)$ of the Dirac delta function $\delta(t-t_0)$.

We have

$$\Delta_{t_0}(\omega) = \int_{-\infty}^{\infty} \delta(t-t_0)\,\mathrm{e}^{-\mathrm{i}\omega t}\,\mathrm{d}t = \mathrm{e}^{-\mathrm{i}\omega t_0}\,.$$

For $t_0 = 0$ the Fourier transform is $\Delta_0(\omega)$ of $\delta(t)$ so that $\Delta_0(\omega) = 1$,

$$\delta(t-t_0) \;\circ\!\!-\!\!\bullet\; \mathrm{e}^{-\mathrm{i}\omega t_0}\,, \quad \text{in particular } \delta(t) \;\circ\!\!-\!\!\bullet\; 1\,.$$

- **Heaviside function.** This function is given by

$$u : \mathbb{R} \to \mathbb{C}\,,\; u(t) = \begin{cases} 1\,, & \text{for } t > 0 \\ 0\,, & \text{for } t < 0 \end{cases}.$$

The graph of this function can be seen in Fig. 76.5.

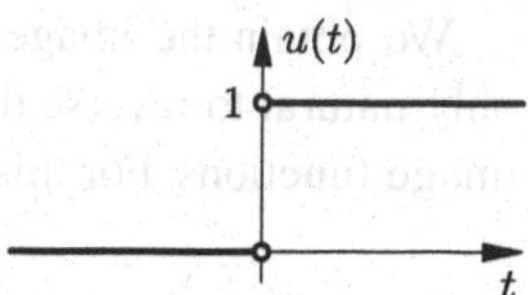

Fig. 76.5 The Heaviside function u

For the transformed one obtains in the case $\omega \neq 0$:

$$F(\omega)=\int_{-\infty}^{\infty} u(t)\,\mathrm{e}^{-\mathrm{i}\omega t}\,\mathrm{d}t=\int_{0}^{\infty} \mathrm{e}^{-\mathrm{i}\omega t}\,\mathrm{d}t=\lim_{b\to\infty}\int_{0}^{b} \mathrm{e}^{-\mathrm{i}\omega t}\,\mathrm{d}t=\lim_{b\to\infty}\frac{-1}{\mathrm{i}\,\omega}\left[\mathrm{e}^{-\mathrm{i}\omega t}\right]_0^b .$$

Here we have to stop the calculation, because this limit does not exist. But if we take the Heaviside function as a limit function of

$$f_a:\mathbb{R}\to\mathbb{C},\ f_a(t)=\begin{cases}\mathrm{e}^{-at}, & \text{for } t>0\\ 0, & \text{for } t<0\end{cases}$$

with $a>0$, i.e. $u(t)=\lim_{a\to 0} f_a(t)$, then we get according to Example 76.1, for $\omega \neq 0$ the function F with $F(\omega)=\frac{1}{\mathrm{i}\omega}$ as the Fourier transform of $u(t)$. Open is the problem, what $F(0)$ is. Using the inverse transform, one can reason that the Fourier transform of $u(t)$ for all ω is given by

$$F(\omega)=\frac{1}{\mathrm{i}\,\omega}+\pi\,\delta(\omega);$$

i.e.

$$u:\mathbb{R}\to\mathbb{C},\ u(t)=\begin{cases}1, & \text{for } t>0\\ 0, & \text{for } t<0\end{cases} \quad \circ\!\!-\!\!\bullet \quad \frac{1}{\mathrm{i}\,\omega}+\pi\,\delta(\omega).$$

The Dirac delta funktion and the Heaviside funktion aren't functions, but so-called **distributions**. These functions and their Fourier transforms play a fundamental role in the following. Note that these distributions are not actually Fourier transformable at all, although we have given Fourier transforms. ■

Every Fourier transformable time function $f = f(t)$ can be assigned by Fourier transformation to the Fourier transform $F = F(\omega)$. One calls the Fourier transformable time functions $f(t)$ also **original functions** and the set of all Fourier transformable time functions the **original domain**. The Fourier transforms $F(\omega)$ are also called **image functions** and the set of all Fourier transforms of original functions the **image domain**.

We obtain the image functions of the time functions by Fourier transform; now it is only natural to reverse this transformation: We try to recover the time functions from the image functions. For this purpose we consider the *inverse Fourier transform.*

76.2 The Inverse Fourier Transform

For each Fourier transformable function $f = f(t)$ we can determine the Fourier transform $F = F(\omega)$. Now we determine for a Fourier transform $F = F(\omega)$ by *inversion* of the Fourier transform a time function $\tilde{f} = \tilde{f}(t)$. The wish is of course to get f back: $\tilde{f} = f$, but unfortunately this does not have to be so, somewhat roughly we can say: $\tilde{f}$ and f differ only at the discontinuity points of f; the values of $\tilde{f}$ lie in the jumps from f at the midpoint of the step height; more precisely:

The Inverse Fourier Transform
If $F = F(\omega)$ is an image function, then the function $\tilde{f} : \mathbb{R} \to \mathbb{C}$ explained as follows

$$\tilde{f}(t) = \frac{1}{2\pi} \int_{-\infty}^{\infty} e^{i\omega t} F(\omega) d\omega$$

is an original function if this integral exists for every $t \in \mathbb{R}$. One calls $\tilde{f}$ the **inverse Fourier transform** of F.

The mapping $F \to \tilde{f}$ which assigns to an image function F a time function $\tilde{f}$ is called **inverse Fourier transform**.

If $f = f(t)$ is Fourier transformable and piecewise continuously differentiable with the Fourier transform $F = F(\omega)$ then we have for the inverse Fourier transform $\tilde{f}$ of F:

$$\tilde{f}(t) = \begin{cases} f(t), & \text{if } f \text{ in } t \text{ continuous} \\ \frac{f(t^+)+f(t^-)}{2}, & \text{if } f \text{ in } t \text{ discontinuous} \end{cases}.$$

In practice, one does not determine the original function with the inverse Fourier transform, since even simple image functions F lead to integrals that are difficult to determine, note the following examples (cf. also Example 76.1):

Example 76.3

- We try to determine the inverse Fourier transform $\tilde{f}$ of the sinc function

$$F : \mathbb{R} \to \mathbb{C}, \; F(\omega) = \operatorname{sinc}(\omega) = \begin{cases} 1, & \omega = 0 \\ \frac{\sin(\omega)}{\omega}, & \omega \neq 0 \end{cases}.$$

We have

$$\tilde{f}(t) = \frac{1}{2\pi} \int_{-\infty}^{\infty} e^{i\omega t} \operatorname{sinc}(\omega) d\omega = \frac{1}{2\pi} \int_{-\infty}^{\infty} e^{i\omega t} \frac{\sin(\omega)}{\omega} d\omega = \frac{1}{\pi} \int_{0}^{\infty} \cos(\omega t) \frac{\sin(\omega)}{\omega} d\omega .$$

- We try to find the inverse Fourier transform f of the function

$$F : \mathbb{R} \to \mathbb{C}, \; F(\omega) = \frac{1}{a + i\omega} .$$

We have

$$\tilde{f}(t) = \frac{1}{2\pi} \int_{-\infty}^{\infty} e^{i\omega t} \frac{1}{a + i\omega} d\omega .$$

With methods of Complex Analysis (see Chap. 84) these integrals can be determined, but the effort is not worthwhile. In practice, other methods are used to determine the inverse transforms of image functions. These methods result from the calculation rules for the Fourier transform, which we will discuss in the next chapter.

But according to the above theorem we know the inverse transforms $\tilde{f}$ (note Example 76.1), we see the graphs of the back-transformed in Fig. 76.6. ■

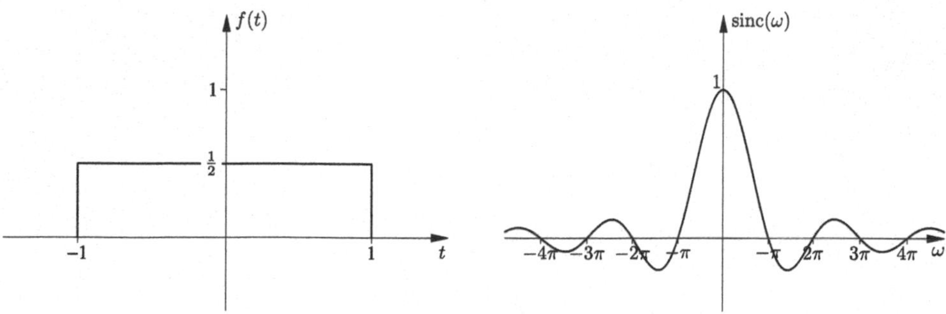

Fig. 76.6 The inverse transforms of $F(\omega) = \frac{1}{1+i\omega}$ and sinc

Remark Note that the inverse Fourier transform is not the inverse of the Fourier transform if one allows all original functions. However, if one restricts the Fourier transform to the so-called **Schwartz space**

$$S(\mathbb{R}) = \{f \in C^{\infty}(\mathbb{R}) \mid \int_{-\infty}^{\infty} x^p f^{(k)}(x)\mathrm{d}x \text{ exists for all } p, k \in \mathbb{N}_0\}$$

a (the elements of $S(\mathbb{R})$ are also called *fast decaying functions*), the Fourier transform $f \in S(\mathbb{R}) \to F \in S(\mathbb{R})$ is bijective; the inverse of this is the inverse Fourier transform.

76.3 Exercise

76.1 Determine the Fourier transform of the function

$$f(t) = \begin{cases} \frac{1}{2}(1 - |t|), & |t| \leq 1 \\ 0, & |t| > 1 \end{cases}$$

and confirm using the inverse transform $\int_{-\infty}^{\infty} \left(\frac{\sin x}{x}\right)^2 \mathrm{d}x = \pi$.

76.2 Show the correspondence

$$g(t) = \frac{1}{\sqrt{4\pi a}}\, \mathrm{e}^{-\frac{t^2}{4a}} \quad \circ\!\!-\!\!\bullet \quad G(\omega) = \mathrm{e}^{-a\omega^2} \quad \text{for each } a > 0\,.$$

Use the known integral $\int_{-\infty}^{\infty} \mathrm{e}^{-u^2}\, \mathrm{d}u = \sqrt{\pi}$.

Fourier Transform II

77

The Fourier transform offers the possibility to determine particulate solutions of linear differential equations. In this process, a differential equation is transformed into an equation. By solving this equation and back-transforming the solution, one obtains a desired solution of the original differential equation.

The essential tool is therefore the inverse Fourier transform. That the (direct) calculation of the inverse Fourier transform of an image function is not quite easy, we have noticed in the last chapter. Fortunately, the Fourier transform rules often spare us the direct computation of the inverse transform. We begin this chapter with an overview of the rules and theorems for the Fourier transform.

77.1 The Rules and Theorems for the Fourier Transform

If $f : \mathbb{R} \to \mathbb{C}$ is a Fourier transformable function, then the Fourier transform $F : \mathbb{R} \to \mathbb{C}$ of f is given by

$$F(\omega) = \int_{-\infty}^{\infty} f(t)\, \mathrm{e}^{-\mathrm{i}\omega t}\, \mathrm{d}t\,.$$

This relation between f and F we express by the notation:

$$f(t) \;\circ\!\!-\!\!\bullet\; F(\omega) \quad \text{or} \quad \mathcal{F}(f(t)) = F(\omega)\,.$$

We summarize the most important calculation rules and theorems for the Fourier transform.

C. Karpfinger, *Calculus and Linear Algebra in Recipes*,
https://doi.org/10.1007/978-3-662-65458-3_77

Rules and Theorems for the Fourier Transform

Let $F = F(\omega)$ and $G = G(\omega)$ are the image function of Fourier-transformable time functions $f = f(t)$ and $g = g(t)$, i. e.

$$f(t) \quad \circ\!\!-\!\!\bullet \quad F(\omega) \quad \text{and} \quad g(t) \quad \circ\!\!-\!\!\bullet \quad G(\omega)\,.$$

- **Continuity and improper convergence of F:** If the integral $\int_{-\infty}^{\infty} f(t)\,\mathrm{d}t$ exists then the image function F is continuous, and we have $F(\omega) \to 0$ for $\omega \to \pm\infty$; thus the values of F become small towards the *boundaries.*
- **Linearity:** For all $\lambda,\ \mu \in \mathbb{C}$ is $\lambda\, F + \mu\, G$ the Fourier transform of $\lambda f + \mu g$, i.e.

$$\lambda f(t) + \mu g(t) \quad \circ\!\!-\!\!\bullet \quad \lambda F(\omega) + \mu G(\omega)\,.$$

- **Conjugation:** The Fourier transform of $\overline{f(t)}$ is $\overline{F(-\omega)}$, i.e.

$$\overline{f(t)} \quad \circ\!\!-\!\!\bullet \quad \overline{F(-\omega)}\,.$$

- **Time scaling:** The Fourier transform of $f(c\,t)$ for $c \in \mathbb{R} \setminus \{0\}$ is $\frac{1}{|c|}\, F(\frac{\omega}{c})$, i.e.

$$f(c\,t) \quad \circ\!\!-\!\!\bullet \quad \frac{1}{|c|}\, F\left(\frac{\omega}{c}\right).$$

- **Time shifting:** The Fourier transform of $f(t-a)$ for $a \in \mathbb{R}$ is $\mathrm{e}^{-\mathrm{i}\omega a}\, F(\omega)$, i.e.

$$f(t-a) \quad \circ\!\!-\!\!\bullet \quad \mathrm{e}^{-\mathrm{i}\omega a}\, F(\omega)\,.$$

- **Frequency shifting:** The Fourier transform of $\mathrm{e}^{\mathrm{i}\,\tilde{\omega}t}\, f(t)$ for $\tilde{\omega} \in \mathbb{R}$ is $F(\omega - \tilde{\omega})$, i.e.

$$\mathrm{e}^{\mathrm{i}\,\tilde{\omega}t}\, f(t) \quad \circ\!\!-\!\!\bullet \quad F(\omega - \tilde{\omega})\,.$$

- **Differentiation in time:** If f is piecewise differentiable and f' is Fourier transformable, then

$$f'(t) \quad \circ\!\!-\!\!\bullet \quad \mathrm{i}\,\omega\, F(\omega)\,.$$

- **Differentiation in frequency:** If $t\, f(t)$ is Fourier transformable, then

$$t\, f(t) \quad \circ\!\!-\!\!\bullet \quad \mathrm{i}\, F'(\omega)\,.$$

(continued)

- **Convolution:** For the **convolution product of** f and g, which is the Fourier transformable function

$$(f * g)(t) = \int_{-\infty}^{\infty} f(t-\tau)\, g(\tau) \mathrm{d}\tau \,,$$

we have

$$(f * g)(t) \quad \circ\!\!-\!\!\bullet \quad F(\omega)\, G(\omega)\,.$$

Further we have $f * g = g * f$.

- **Inverse Theorem:** If f is continuous in t, then

$$F(t) \quad \circ\!\!-\!\!\bullet \quad 2\pi\, f(-\omega)\,.$$

- **Symmetry:** The original function f is even or odd if and only if the image function F is even or odd, respectively.

With the help of these rules of calculation we can infer from known correspondences $f(t) \circ\!\!-\!\!\bullet F(\omega)$ to further correspondences. A table of known correspondences thus yields the corresponding image functions for numerous original functions or vice versa.

How to use these rules to obtain the Fourier transform $F(\omega)$ or inverse Fourier transform $f(t)$ of original or image functions, we show in the following examples.

Example 77.1

- We determine the Fourier transform $R(\omega)$ for the momentum

$$r : \mathbb{R} \to \mathbb{C}\,,\ r(t) = \begin{cases} A\,, & |t-a| \le T \\ 0\,, & |t-a| > T \end{cases}$$

for $a \in \mathbb{R}$ and $A,\ T > 0$ using the rectangle function $f(t)$ from Example 76.1; we have $r(t) = f(t)$ in the case $a = 0, T = 1, A = 1/2$ and $f(t) \quad \circ\!\!-\!\!\bullet \quad \mathrm{sinc}(\omega)$. In general, we have between $f(t)$ and $r(t)$ the relation

$$r(t) = 2\,A\, f\left(\frac{t-a}{T}\right),$$

so that according to the rules *linearity*, *time shifting* and *time scaling* for the Fourier transform R of r we have

$$R(\omega) = 2\,A\,T\,\mathrm{e}^{-\mathrm{i}\omega a}\,\mathrm{sinc}(\omega T)\,.$$

- According to the second example in Example 76.1 we have the correspondence

$$f(t) = \begin{cases} \mathrm{e}^{-at}\,, & \text{for } t \geq 0 \\ 0\,, & \text{for } t < 0 \end{cases} \quad \circ\!\!-\!\!\bullet \quad F(\omega) = \frac{1}{a + \mathrm{i}\omega}\,.$$

For all $t \neq 0$ we have $g(t) = f(t) + f(-t)$ for $g(t) = \mathrm{e}^{-a|t|}$, where $a > 0$ (see the third example in Example 76.1). Thus we obtain according to the rule for *time scaling* again:

$$g(t) = \mathrm{e}^{-a|t|} \quad \circ\!\!-\!\!\bullet \quad G(\omega) = \frac{2\,a}{a^2 + \omega^2}\,.$$

- We determine the Fourier transform of $f(t) = \frac{1}{1+t^2}$. According to the third example in Example 76.1 we have

$$g(t) = \mathrm{e}^{-a|t|} \quad \circ\!\!-\!\!\bullet \quad G(\omega) = \frac{2\,a}{a^2 + \omega^2}\,.$$

With $a = 1$ we obtain because of $f(t) = \frac{1}{2}G(t)$ with the rule for *linearity* and the *Inverse Theorem*

$$f(t) = \frac{1}{2}\,G(t) \quad \circ\!\!-\!\!\bullet \quad \frac{1}{2}\,2\,\pi\,g(-\omega) = \pi\,\mathrm{e}^{-|\omega|}\,.$$

- With the last example, because of the *time scaling* rule, we simply obtain the Fourier transform of $f : \mathbb{R} \to \mathbb{C}$ with $f(t) = \frac{1}{a^2+t^2}$ for $a \in \mathbb{R} \setminus \{0\}$ and because of the rule for *differentiation in frequency* then also the Fourier transform of $g : \mathbb{R} \to \mathbb{C}$ with $g(t) = \frac{t}{a^2+t^2}$ for $a \in \mathbb{R} \setminus \{0\}$: We have namely

$$f(t) = \frac{1}{a^2 + t^2} = \frac{1}{a^2}\frac{1}{1 + \left(\frac{t}{a}\right)^2} \quad \circ\!\!-\!\!\bullet \quad F(\omega) = \frac{1}{a^2}\,|a|\,\pi\,\mathrm{e}^{-|a\omega|} = \frac{\pi}{|a|}\,\mathrm{e}^{-|a\omega|}\,.$$

And finally

$$g(t) = t f(t) \quad \circ\!\!-\!\!\bullet \quad \mathrm{i}\,F'(\omega) = \mathrm{i}\,\frac{\pi}{|a|}(-|a|\;\mathrm{sgn}(\omega))\,\mathrm{e}^{-|a\omega|} = -\,\mathrm{i}\,\pi\,\mathrm{sgn}(\omega))\,\mathrm{e}^{-|a\omega|}\,.$$

■

77.2 Application to Linear Differential Equations

Given the computational rules for the Fourier transform, we obtain an equation by transforming the left and right sides of a ODE. This equation can usually be solved. A back transformation of the solution of this equation then gives a solution of the original ODE. This principle is what makes Fourier transforms (but even more so the Laplace transform described in the next chapter) so important for applications.

Before discussing the concrete applications, we illustrate the principle with a simple example and, as usual, denote the solution function of a ODE by x, i.e. $x = x(t)$:

Example 77.2 We are looking for a function $x = x(t)$ which satisfies the ODE

$$\dot{x}(t) + x(t) = \begin{cases} \mathrm{e}^{-t}, & \text{for } t \geq 0 \\ 0, & \text{for } t < 0 \end{cases}$$

assuming that the functions x and $\dot{x}$ are Fourier transformable, and denote the Fourier transform of $x(t)$ with $X(\omega)$. Because of the rules *linearity* and the *differentiation in time*, we obtain, by Fourier transforming both sides of the ODE, the equation

$$\mathrm{i}\,\omega\, X(\omega) + X(\omega) = \frac{1}{1+\mathrm{i}\,\omega},$$

see Example 76.1. From this last equation we obtain the following representation of $X(\omega)$:

$$X(\omega) = \frac{1}{(1+\mathrm{i}\,\omega)^2}.$$

We determine an inverse Fourier transform $x(t)$ of $X(\omega)$ using our rules of calculation. Because of

$$\mathrm{i}\left(\frac{1}{1+\mathrm{i}\,\omega}\right)' = \mathrm{i}\,\frac{-\mathrm{i}}{(1+\mathrm{i}\,\omega)^2} = \frac{1}{(1+\mathrm{i}\,\omega)^2}$$

we obtain with the rule *differentiation in frequency* the inverse transform

$$x(t) = \begin{cases} t\,\mathrm{e}^{-t}, & \text{for } t \geq 0 \\ 0, & \text{for } t < 0 \end{cases}.$$

This is a solution of the differential equation under consideration—of which it is easy to convince ourselves. ■

The essential element in solving the ODE in this example was that the Fourier transform turns the ODE into an equation. Also linear ODEs of higher order become equations by Fourier transformation. We describe the procedure in detail in a recipe for a 2nd order linear ODE. For higher order ODEs one proceeds analogously.

Recipe: Solving an ODE with Fourier Transformation
To determine a solution $x(t)$ of the ODE

$$a\,\ddot{x}(t) + b\,\dot{x}(t) + c\,x(t) = s(t) \quad \text{with } a,\ b,\ c \in \mathbb{R}$$

and Fourier transformable functions x and s, proceed as follows:

(1) Denote the Fourier transform of $x(t)$ with $X(\omega)$ and those of $s(t)$ with $S(\omega)$, i.e.

$$x(t) \ \circ\!\!-\!\!\bullet\ X(\omega) \text{ and } s(t) \ \circ\!\!-\!\!\bullet\ S(\omega)\,.$$

(2) Fourier transformation of both sides of the ODE yields the equation

$$(\mathrm{i}\,\omega)^2 a\,X(\omega) + \mathrm{i}\,\omega\,b\,X(\omega) + c\,X(\omega) = S(\omega)\,.$$

(3) Resolve the equation to $X(\omega)$:

$$X(\omega) = \frac{1}{-\omega^2 a + \mathrm{i}\,\omega\,b + c}\,S(\omega)\,.$$

(4) Find the inverse Fourier transform $h(t)$ of

$$H(\omega) = \frac{1}{-\omega^2 a + \mathrm{i}\,\omega\,b + c}\,.$$

One calls $H(\omega)$ the **transfer function.**

(5) A solution $x(t)$ of the given differential equation is

$$x(t) = (h * s)(t) = \int_{-\infty}^{\infty} h(t-\tau)\,s(\tau)\mathrm{d}\tau\,.$$

Note that in the actual calculation of $x = x(t)$ using this recipe, steps (1)–(3) need not be performed. We have given them so that one can understand the procedure for higher order ODEs.

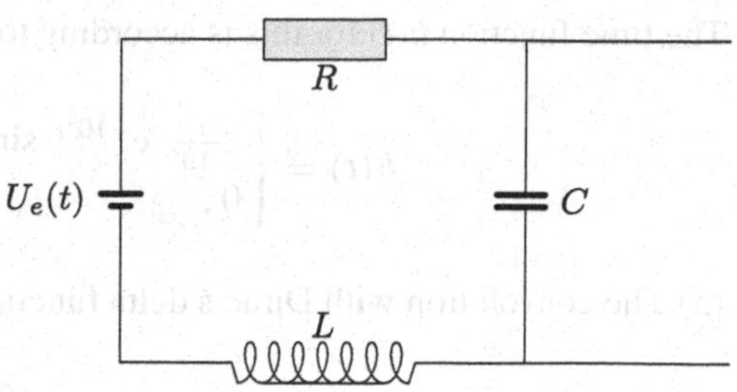

Fig. 77.1 The RCL circuit

Using the inverse Fourier transform, one could directly attempt to determine the inverse transform $x = x(t)$ in (3), i.e. the function x with

$$x(t) \quad \circ\!\!-\!\!\bullet \quad \frac{1}{-\omega^2 a + \mathrm{i}\,\omega\, b + c}\, S(\omega)\,.$$

This would require the determination of the following integral:

$$x(t) = \frac{1}{2\pi} \int\limits_{-\infty}^{\infty} \mathrm{e}^{\mathrm{i}\,\omega t}\, \frac{1}{-\omega^2 a + \mathrm{i}\,\omega\, b + c}\, S(\omega) \mathrm{d}\omega\,.$$

The detour via the rule to the convolution product is in general a clear simplification.

Remark In practice, it is generally sufficient to know the transfer function $H(\omega)$; this is given by the coefficients $a,\ b,\ c \in \mathbb{R}$ and can therefore be read directly from the ODE.

Example 77.3 (RCL Circuit) We consider an RCL circuit, note the figure (Fig. 77.1). Ohm's law gives for the resistance R, the inductance L and the capacitance C the differential equation

$$L\,C\,\ddot{U}(t) + R\,C\,\dot{U}(t) + U(t) = U_e(t)\,,$$

when the input voltage $U_e(t)$ is applied. With the values $L = 10^{-4}$[Henry], $R = 2\Omega$ and $C = 10^{-5}[F]$ and $U_e(t) = 10^{-5}\,\delta(t)$ we obtain after multiplication by 10^9 the differential equation (where we have $x = U$)

$$\ddot{x} + 2 \cdot 10^4\, \dot{x} + 10^9\, x = 10^4\, \delta\,.$$

(4) The transfer function is

$$\begin{aligned} H(\omega) &= \frac{1}{-\omega^2 + \mathrm{i}\,\omega\, 2 \cdot 10^4 + 10^9} \\ &= \frac{1}{(10^4 + \mathrm{i}\,\omega)^2 + 9 \cdot 10^8} = \frac{1}{3 \cdot 10^4}\, \frac{3 \cdot 10^4}{(10^4 + \mathrm{i}\,\omega)^2 + 9 \cdot 10^8}\,. \end{aligned}$$

The time function $h(t)$ for this is according to the following table in Sect. 77.2:

$$h(t) = \begin{cases} \frac{1}{3 \cdot 10^4}\, \mathrm{e}^{-10^4 t} \sin\left(3 \cdot 10^4 t\right) , & \text{for } t > 0 \\ 0 , & \text{for } t < 0 \end{cases} .$$

(5) The convolution with Dirac's delta function gives the solution x:

$$x(t) = 10^4 \, (\delta * h)(t) = 10^4 \int_{-\infty}^{\infty} \delta(t - \tau)\, h(\tau) \mathrm{d}\tau = 10^4 \, h(t) ,$$

note the property of the Dirac delta function mentioned in the in Example 76.2. Thus, the solution is:

$$x(t) = U(t) = 10^4 \, h(t) .$$

■

Solving differential equations is one of the essential applications of *transforms*. The Fourier transform is in principle somewhat less suitable for this application, since Fourier-transformable functions f must slope towards the *boundaries*, as the integral

$$\int_{-\infty}^{\infty} f(t)\, \mathrm{e}^{-\mathrm{i}\omega t} \, \mathrm{d}t$$

must exist. Already the simple functions $\mathrm{e}^{\lambda t}$ which are solutions of linear homogeneous differential equations with constant coefficients, do not satisfy this property. Initial value problems with differential equations that have such solutions can often be solved with the *Laplace transform* (see Chap. 79).

We summarize important functions and their Fourier transforms in the following table; we have proved many of these correspondences in the examples and the exercises.

$F(\omega)$	$f(t)$	$F(\omega)$	$f(t)$
$e^{-i\omega t_0}$	$\delta(t-t_0)$	$\frac{1}{i\omega}+\pi\delta(\omega)$	$\begin{cases}1, & t>0\\ 0, & t<0\end{cases}$
$\frac{1}{a+i\omega}$	$\begin{cases}e^{-at}, & t\geq 0\\ 0, & t<0\end{cases}$	$\operatorname{sinc}\omega$	$\begin{cases}\frac{1}{2}, & \|t\|\leq 1\\ 0, & \|t\|>1\end{cases}$
$2AT\,e^{-i\omega a}\operatorname{sinc}\omega T$	$\begin{cases}A, & \|t-a\|\leq T\\ 0, & \|t-a\|>T\end{cases}$	$\frac{2}{1+\omega^2}$	$e^{-\|t\|}$
$\frac{2(1-\omega^2)}{(1+\omega^2)^2}$	$\|t\|\,e^{-\|t\|}$	$\frac{4(1-3\omega^2)}{(1+\omega^2)^3}$	$t^2\,e^{-\|t\|}$
$\frac{a+i\omega}{(a+i\omega)^2+n^2}$	$\begin{cases}e^{-at}\cos nt, & t\geq 0\\ 0, & t<0\end{cases}$	$\frac{n}{(a+i\omega)^2+n^2}$	$\begin{cases}e^{-at}\sin nt, & t\geq 0\\ 0, & t<0\end{cases}$
$\frac{n!}{(1+i\omega)^{n+1}}$	$t^n\,e^{-t}\,\tilde{u}(t)$	$\pi\,e^{-\|\omega\|}$	$\frac{1}{1+t^2}$

Here is $\tilde{u}$ is the modified Heaviside function with $\tilde{u}(t)=u(t)$ for $t\neq 0$ and $\tilde{u}(0)=\frac{1}{2}$. Further we have $a>0$.

77.3 Exercises

77.1 Let $f(t)=e^{-|t|}$.

(a) Calculate the convolution $(f*f)(t)$. *Hint:* Case distinction $t\geq 0$ and $t<0$.
(b) Calculate the Fourier transform $\mathcal{F}(f(t))(\omega)$.
(c) With the aid of the convolution determine $\mathcal{F}(|t|\,e^{-|t|})(\omega)$.

77.2 Let be given a triple low-pass filter which is given by the differential equation

$$\left(\alpha\frac{d}{dt}+1\right)^3 x(t)=s(t)$$

with $\alpha=RC>0$ and Fourier transformable right hand side s (the *input*) is described. Here denotes

$$\left(\alpha\frac{d}{dt}+1\right)^3 x(t)=\alpha^3\frac{d^3}{dt^3}x(t)+3\alpha^2\frac{d^2}{dt^2}x(t)+3\alpha\frac{d}{dt}x(t)+x(t)\,.$$

Now let be given the correspondences $x(t)$ ○—● $X(\omega)$ and $s(t)$ ○—● $S(\omega)$.

(a) Formulate the differential equation given in the time domain in the image domain.
(b) Determine the transfer function H and the impulse response h.
(c) Calculate the *response* x for general s.
(d) Calculate x for the rectangular momentum $s(t) = \begin{cases} 1, & |t| < 1 \\ 1/2, & |t| = 1 \\ 0, & |t| > 1 \end{cases}$.

77.3 It denotes $F_n(\omega)$ the Fourier transform of $f_n(t) = \frac{1}{(1+t^2)^n}$ for $n = 1, 2, \ldots$

(a) Using *time scaling*, calculate the Fourier transform of $\frac{1}{(a^2+t^2)^n}$ for $a > 0$ by F_n for a > 0.
(b) Which function $g(t)$ has as its Fourier transform $G(\omega) = \frac{\mathrm{d}}{\mathrm{d}\omega}(\omega F_n(\omega))$?
(c) Confirm for F_n the recursion formula

$$F_{n+1}(\omega) = F_n(\omega) - \frac{1}{2n}\frac{\mathrm{d}}{\mathrm{d}\omega}(\omega F_n(\omega))$$

and calculate $F_2(\omega)$ from $F_1(\omega) = \pi\, \mathrm{e}^{-|\omega|}$.

77.4 Let $\widetilde{u}(t) = u(t)$ for $t \neq 0$ with $\widetilde{u}(0) = 1/2$ where u is the Heaviside function. One can show that then for all $n \in \mathbb{N}_0$ we have the relation

$$t^n\, \mathrm{e}^{-t}\, \widetilde{u}(t) \quad ○\!\!-\!\!\bullet \quad \frac{n!}{(1+\mathrm{i}\,\omega)^{n+1}}$$

between time and frequency domain. Determine a solution of each of the following LTI systems by means of Fourier transform:

(a) $\dot{x}(t) + x(t) = t^n\, \mathrm{e}^{-t}\, \tilde{u}(t)$,
(b) $\ddot{x}(t) - 2\,\dot{x}(t) + x(t) = s(t)$ with continuous and Fourier transformable $s : \mathbb{R} \to \mathbb{C}$.

77.5 For $a \neq 0$, what are the Fourier transforms of the following functions

$$\frac{1}{1+t^2},\ \frac{t}{a^2+t^2},\ \frac{t}{(a^2+t^2)^2},\ \frac{t^2}{(a^2+t^2)^2},\ \frac{1}{(a^2+t^2)^2}\,?$$

77.6 For $\lambda > 0$ and $a \in \mathbb{R}$ let $f(t) = \begin{cases} 0\,, & t < 0 \\ 1/2\,, & t = 0\,. \\ \exp((-\lambda + \mathrm{i}\,a)t) & t > 0 \end{cases}$

(a) Calculate the Fourier transform of $f(t)$.
(b) What are the Fourier transforms of the *damped oscillations*

$$x(t) = \mathrm{e}^{-\lambda t} \cos Nt \quad \text{and} \quad y(t) = \mathrm{e}^{-\lambda t} \sin Nt\,, \ N \in \mathbb{N}\,, \ t > 0?$$

Discrete Fourier Transform

78

In the discrete Fourier transform, the Fourier coefficients of a 2π-periodic function, which itself is not given, but whose values are known at discrete points, for example by *sampling a signal* are determined approximately. In this way, approximations are obtained for the amplitudes at certain frequencies of a signal. In applications, this discrete Fourier transform plays a role in the construction of digital filters.

The approximate values for the Fourier coefficients of a function, of which only the values at discrete points are known, are at the same time the coefficients of an (interpolating) trigonometric polynomial at these discrete points. We treat this trigonometric interpolation and also give the real version of it.

78.1 Approximate Determination of the Fourier Coefficients

In the following we assume that the considered 2π-periodic function f is expandable into a Fourier series, and, because of the simpler indexing, choose the periodic interval $[0, 2\pi)$.

The starting point is a 2π-periodic signal, which we take to be a function $f : \mathbb{R} \to \mathbb{C}$ which is not given concretely. We *test* this signal f at N equidistant points

$$x_0 = 0,\ x_1 = \frac{2\pi}{N},\ x_2 = 2\,\frac{2\pi}{N}, \ldots, x_{N-1} = (N-1)\,\frac{2\pi}{N}$$

and obtain the so-called *sample*, i.e. the N grid points (see Fig. 78.1):

$$(x_0, f(x_0)), \ldots, (x_{N-1}, f(x_{N-1}))\,.$$

C. Karpfinger, *Calculus and Linear Algebra in Recipes*,
https://doi.org/10.1007/978-3-662-65458-3_78

Although we do not know the function f we can determine the Fourier coefficients c_k for $k = 0, \dots, N-1$ *approximately*:

The Approximation of the Fourier Coefficients
Let f be a piecewise continuously differentiable 2π-periodic function with the Fourier coefficients c_k, $k \in \mathbb{Z}$. Let be given the equidistant interpolation points $x_\ell = \ell \frac{2\pi}{N}$ for $\ell = 0, \dots, N-1$ with the corresponding function values $f(x_\ell)$:

$$(x_0, f(x_0)), \dots, (x_{N-1}, f(x_{N-1}))\,.$$

We set $v_\ell = f(\ell \frac{2\pi}{N})$ and $\zeta = \mathrm{e}^{-2\pi\,\mathrm{i}/N}$; then we have:

$$c_k \approx \hat{c}_k = \frac{1}{N} \sum_{\ell=0}^{N-1} v_\ell\, \zeta^{k\ell} \quad \text{for } k = 0, \dots, N-1\,.$$

In this approximation, the integral for the Fourier coefficients c_k is calculated approximately. We have formulated this calculation as Exercise 78.4.
We fomulate the N equations for the coefficients $\hat{c}_0, \dots, \hat{c}_{N-1}$:

$$\begin{aligned}
\hat{c}_0 &= \frac{1}{N}\left(v_0\,\zeta^0 + v_1\,\zeta^0 + v_2\,\zeta^0 + \cdots + v_{N-1}\,\zeta^0\right)\\
\hat{c}_1 &= \frac{1}{N}\left(v_0\,\zeta^0 + v_1\,\zeta^1 + v_2\,\zeta^2 + \cdots + v_{N-1}\,\zeta^{N-1}\right)\\
\hat{c}_2 &= \frac{1}{N}\left(v_0\,\zeta^0 + v_1\,\zeta^2 + v_2\,\zeta^4 + \cdots + v_{N-1}\,\zeta^{2(N-1)}\right)\\
&\ \vdots\\
\hat{c}_{N-1} &= \frac{1}{N}\left(v_0\,\zeta^0 + v_1\,\zeta^{N-1} + v_2\,\zeta^{2(N-1)} + \cdots + v_{N-1}\,\zeta^{(N-1)(N-1)}\right),
\end{aligned}$$

we can write this as a matrix-vector product, namely as

$$\begin{pmatrix} \hat{c}_0 \\ \vdots \\ \vdots \\ \hat{c}_{N-1} \end{pmatrix} = \frac{1}{N} \begin{pmatrix} 1 & 1 & 1 & \cdots & 1 \\ 1 & \zeta & \zeta^2 & \cdots & \zeta^{N-1} \\ 1 & \zeta^2 & \zeta^4 & \cdots & \zeta^{2(N-1)} \\ \vdots & \vdots & \vdots & \vdots & \vdots \\ 1 & \zeta^{N-1} & \zeta^{2(N-1)} & \cdots & \zeta^{(N-1)(N-1)} \end{pmatrix} \begin{pmatrix} v_0 \\ \vdots \\ \vdots \\ v_{N-1} \end{pmatrix}$$

with $v_\ell = f(\ell \frac{2\pi}{N})$ for $\ell = 0, \dots, N-1$ and $\zeta = \mathrm{e}^{-2\pi\,\mathrm{i}/N}$.

To determine the approximate values $\hat{c}_k$ for c_k we only need to perform the matrix-vector-product mentioned here. We abbreviate:

$$\hat{c} = \begin{pmatrix} \hat{c}_0 \\ \vdots \\ \hat{c}_{N-1} \end{pmatrix}, \quad F_N = \begin{pmatrix} 1 & 1 & 1 & \cdots & 1 \\ 1 & \zeta & \zeta^2 & \cdots & \zeta^{N-1} \\ 1 & \zeta^2 & \zeta^4 & \cdots & \zeta^{2(N-1)} \\ \vdots & \vdots & \vdots & \vdots & \vdots \\ 1 & \zeta^{N-1} & \zeta^{2(N-1)} & \cdots & \zeta^{(N-1)(N-1)} \end{pmatrix}, \quad v = \begin{pmatrix} v_0 \\ \vdots \\ v_{N-1} \end{pmatrix}.$$

One calls the $N \times N$-matrix $F_N = (\zeta^{k\ell})_{k,\ell}$ the **N-th Fourier matrix**, because of $\zeta^{k\ell} = \zeta^{\ell k}$ the matrix F_N is symmetric, i.e. $F_N^{\top} = F_N$. The first Fourier matrices are $F_1 = (1)$,

$$F_2 = \begin{pmatrix} 1 & 1 \\ 1 & -1 \end{pmatrix}, \quad F_3 = \begin{pmatrix} 1 & 1 & 1 \\ 1 & -\frac{1}{2} - \mathrm{i}\frac{\sqrt{3}}{2} & -\frac{1}{2} + \mathrm{i}\frac{\sqrt{3}}{2} \\ 1 & -\frac{1}{2} + \mathrm{i}\frac{\sqrt{3}}{2} & -\frac{1}{2} - \mathrm{i}\frac{\sqrt{3}}{2} \end{pmatrix}, \quad F_4 = \begin{pmatrix} 1 & 1 & 1 & 1 \\ 1 & -\mathrm{i} & -1 & \mathrm{i} \\ 1 & -1 & 1 & -1 \\ 1 & \mathrm{i} & -1 & -\mathrm{i} \end{pmatrix}.$$

We obtain the procedure for **discrete Fourier transform**:

Recipe: Discrete Fourier Transform

Given are $N \in \mathbb{N}$, a data vector $v \in \mathbb{C}^N$ and the complex number $\zeta = \mathrm{e}^{-2\pi\,\mathrm{i}/N}$. To determine the **discrete Fourier coefficients** $\hat{c}_k$ proceed as follows:

(1) Set up the matrix $F_N = (\zeta^{k\ell})_{k,\ell} \in \mathbb{C}^{N \times N}$.
(2) Calculate $\hat{c} = \frac{1}{N} F_N\, v \in \mathbb{C}^N$.
(3) Obtain the discrete Fourier coefficients $\hat{c}_k$ from $\hat{c} = (\hat{c}_k)_k \in \mathbb{C}^N$ for $k = 0, \ldots, N-1$.

We call the mapping DFT: $\mathbb{C}^N \to \mathbb{C}^N$, $v \mapsto \hat{c}$ **discrete Fourier transform.**

It is a good idea to implement this discrete Fourier transform, see Exercise 78.4.

Example 78.1 Actually, we should not know the function f; but what good would an example be, if we could not estimate afterwards how good the described method is. Therefore we consider a function whose Fourier coefficients we already know. Given is the 2π-periodic continued f of the function $f : [0, 2\pi) \to \mathbb{C}$, $f(x) = x$. From Example 74.3 we obtain with the rules for shifting and the conversion formulas to complex Fourier coefficients:

$$c_0 = \pi \text{ and } c_k = \mathrm{i}/k \text{ for } k \in \mathbb{Z} \setminus \{0\}.$$

We sample the signal f at the four places $x_\ell = \ell 2\pi/4, \ell = 0, \dots, 3$ and obtain the sample

$$(0,0)\ ,\ (2\pi/4, 2\pi/4)\ ,\ (4\pi/4, 4\pi/4)\ ,\ (6\pi/4, 6\pi/4)\ .$$

With the vector $v = (0,\ 2\pi/4,\ 4\pi/4,\ 6\pi/4)^\top = (0,\ \pi/2,\ \pi,\ 3\pi/2)^\top$ and the Fourier matrix F_4 we obtain (with rounded values):

$$\hat{c} = \frac{1}{4}\,F_4\,v = \frac{1}{4}\begin{pmatrix} 1 & 1 & 1 & 1 \\ 1 & -\mathrm{i} & -1 & \mathrm{i} \\ 1 & -1 & 1 & -1 \\ 1 & \mathrm{i} & -1 & -\mathrm{i} \end{pmatrix}\begin{pmatrix} 0 \\ \pi/2 \\ \pi \\ 3\pi/2 \end{pmatrix} = \begin{pmatrix} 2.3576 \\ -0.7854 + 0.7854\,\mathrm{i} \\ -0.7854 \\ -0.7854 - 0.7854\,\mathrm{i} \end{pmatrix}.$$

Compare the values with the (rounded) values of the exact Fourier coefficients:

$$c_0 = 3.1416\,,\ c_1 = \mathrm{i}\,,\ c_2 = 0.5\,\mathrm{i}\,,\ c_3 = 0.3333\,\mathrm{i}\ .$$

If one chooses $N = 2^8$, one obtains the better approximation for the same function

$$\hat{c} = \begin{pmatrix} \hat{c}_0 \\ \hat{c}_1 \\ \hat{c}_2 \\ \hat{c}_3 \\ \vdots \end{pmatrix} = \frac{1}{2^8}\,F_{2^8}\,v = \begin{pmatrix} 3.1416 \\ -0.0123 + 1.0039\,\mathrm{i} \\ -0.0123 + 0.5019\,\mathrm{i} \\ -0.0123 + 0.3345\,\mathrm{i} \\ \vdots \end{pmatrix}.$$

■

To determine the amplitude c_k to a *high* frequency, i.e. for a large k, the number of equidistant interpolation points N of the equidistant interpolation points has to be chosen accordingly high. In other words: By specifying N, only the amplitudes up to a certain frequency are determined approximately.

Remark With the (naive) implementation of the discrete Fourier transform, as we also did in Exercise 78.4, one quickly encounters limitations when N is large. If $N = 2^p$, the computational effort for the calculation of c or v can be reduced considerably by clever *splitting of* c and v, which is done with the *fast Fourier transform* FFT.

78.2 The Inverse Discrete Fourier Transform

As you can easily calculate, we have the following:

$$F_N \overline{F}_N = N E_N ,$$

i.e., the matrix F_N is invertible, and the inverse is

$$F_N^{-1} = \frac{1}{N} \overline{F}_N .$$

Thus, from $\hat{c}$ the vector v can be recovered, because of $\hat{c} = \frac{1}{N} F_N v$ we get the **inverse discrete Fourier transform**:

Recipe: Inverse Discrete Fourier Transform
Given are $N \in \mathbb{N}$, the vector $\hat{c} \in \mathbb{C}^N$ and the complex number $\zeta = \mathrm{e}^{-2\pi \mathrm{i}/N}$.

(1) Set up the matrix $\overline{F}_N = (\overline{\zeta}^{k\ell})_{k,\ell} \in \mathbb{C}^{N \times N}$.
(2) Calculate $v = \overline{F}_N \hat{c} \in \mathbb{C}^N$.
(3) Obtain the data v_k from $v = (v_k)_k \in \mathbb{C}^N$ for $k = 0, \ldots, N-1$.

We call the mapping IDFT: $\mathbb{C}^N \to \mathbb{C}^N$, $\hat{c} \mapsto v$ **inverse discrete Fourier transform.**

78.3 Trigonometric Interpolation

With the determination of the coefficients $\hat{c}_k$ we have solved another problem which we have not yet addressed: Given, as in the initial situation, is a *sample* with N equidistant interpolation points $x_0, \ldots, x_{N-1}, x_\ell = \ell \frac{2\pi}{N}$ with $\ell = 0, \ldots, N-1$ and complex numbers $v_0, \ldots, v_{N-1}$ (see Fig. 78.1):

$$(x_0, v_0), \ldots, (x_{N-1}, v_{N-1}) .$$

We are looking for an interpolating **trigonometric polynomial** of degree $N - 1$:

$$p(x) = \sum_{k=0}^{N-1} d_k \, \mathrm{e}^{\mathrm{i}\, kx} ,$$

i.e. $p(x_\ell) = v_\ell$. One speaks of **trigonometric interpolation**. For this the coefficients d_k have to be determined. To do this we take the following approach: For each $x_\ell = \ell\frac{2\pi}{N}$, $\ell = 0, \ldots, N-1$ we have:

$$v_\ell = \sum_{k=0}^{N-1} d_k \, \mathrm{e}^{\mathrm{i}\,k x_\ell} = \sum_{k=0}^{N-1} d_k \, \mathrm{e}^{\mathrm{i}\,\frac{2\pi}{N} k\ell} = \sum_{k=0}^{N-1} d_k \overline{\zeta}^{k\ell} ,$$

where $\zeta = \mathrm{e}^{-2\pi\,\mathrm{i}/N}$. According to the inverse Fourier transform (see the recipe in Sect. 78.2) we can use $\hat{c}_k$ for d_k; so the problem of trigonometric interpolation is already solved. The procedure for **trigonometric interpolation** is as follows:

Recipe: Trigonometric Interpolation
For a *sample*

$$(x_0, v_0), \ldots, (x_{N-1}, v_{N-1})$$

with N equidistant grid points $x_i = \ell\frac{2\pi}{N}$, $\ell = 0, \ldots, N-1$ the interpolating trigonometric polynomial $p(x) = \sum_{k=0}^{N-1} \hat{c}_k \, \mathrm{e}^{\mathrm{i}\,kx}$ is found as follows:

(1) To N and the data vector $v = (v_k)_k \in \mathbb{C}^N$ determine the discrete Fourier coefficients $\hat{c}_k$, $k = 0, \ldots, N-1$ using the recipe for the discrete Fourier transform in Sect. 78.1.
(2) Obtain the interpolating trigonometric polynomial

$$p(x) = \sum_{k=0}^{N-1} \hat{c}_k \, \mathrm{e}^{\mathrm{i}\,kx} .$$

Example 78.2 We determine an interpolating trigonometric polynomial of degree 2 to the grid points

$$(0, 0) , \ (2\pi/3, 1) , \ (4\pi/3, 0) .$$

(1) We have N= 3, $x_\ell = \ell\frac{2\pi}{3}$ for $\ell = 0, 1, 2$ and $v_0 = 0$, $v_1 = 1$ and $v_2 = 0$. Using the Fourier matrix F_3 we obtain the coefficients $\hat{c}_0$, $\hat{c}_1$, $\hat{c}_2$ as components of:

$$\hat{c} = \frac{1}{3} F_3 \, v = \frac{1}{3} \begin{pmatrix} 1 & 1 & 1 \\ 1 & -\frac{1}{2} - \mathrm{i}\frac{\sqrt{3}}{2} & -\frac{1}{2} + \mathrm{i}\frac{\sqrt{3}}{2} \\ 1 & -\frac{1}{2} + \mathrm{i}\frac{\sqrt{3}}{2} & -\frac{1}{2} - \mathrm{i}\frac{\sqrt{3}}{2} \end{pmatrix} \begin{pmatrix} 0 \\ 1 \\ 0 \end{pmatrix} = \frac{1}{3} \begin{pmatrix} 1 \\ -\frac{1}{2} - \frac{1}{2}\sqrt{3}\,\mathrm{i} \\ -\frac{1}{2} + \frac{1}{2}\sqrt{3}\,\mathrm{i} \end{pmatrix} .$$

(2) The interpolating polynomial is thus (after conversion to sine-cosine form)

$$p(x) = \frac{1}{3}\left(\left(1 - \frac{1}{2}\cos x + \frac{1}{2}\sqrt{3}\sin x - \frac{1}{2}\cos 2x - \frac{1}{2}\sqrt{3}\sin 2x\right) + \right.$$

$$\left. +\mathrm{i}\left(-\frac{1}{2}\sqrt{3}\cos x - \frac{1}{2}\sin x + \frac{1}{2}\sqrt{3}\cos 2x - \frac{1}{2}\sin 2x\right)\right).$$

■

Because of the non-vanishing imaginary part, this representation is somewhat awkward: given real grid points, one also expects a real polynomial. To obtain such a polynomial, we recall the stories about converting the complex Fourier series into a real Fourier series: A trigonometric polynomial of the form

$$\sum_{k=-n}^{n} \hat{c}_k \, \mathrm{e}^{\mathrm{i}\,kx}$$

can be converted to a real trigonometric polynomial of the form

$$\frac{a_0}{2} + \sum_{k=1}^{n} a_k \cos(kx) + b_k \sin(kx)$$

using the conversion formulas. This second representation is called the **sine-cosine form** of the trigonometric polynomial. To be determined are the $2\,n+1$ coefficients $\hat{c}_{-n}, \ldots, \hat{c}_n$; the conversion formulas can then be used to obtain the real coefficients. The main advantage of this form is: If the grid points are all real, the sine-cosine form yields a real polynomial. Unfortunately, there is a small complication: We can only determine $N = 2n + 1$ coefficients if we also have $2\,n + 1$ interpolation points. So we would always have a problem if there are an even number of interpolation points. But this problem can also be solved by determining fewer coefficients. In the following overview we give the formulas for the determination of the complex and real coefficients in both cases separately.

(Real) Trigonometric Interpolation

Given are the N equidistant interpolation points in the interval $[0, 2\pi)$:

$$(x_0, v_0), \ldots, (x_{N-1}, v_{N-1}),$$

(continued)

where $x_\ell = \ell \frac{2\pi}{N}$ with $\ell = 0, \ldots, N-1$.

- If $N = 2n+1$ is odd, then

$$p(x) = \sum_{k=-n}^{n} \hat{c}_k \, e^{\mathrm{i}\,kx} = \frac{a_0}{2} + \sum_{k=1}^{n} a_k \cos(kx) + b_k \sin(kx)$$

  with

$$\hat{c}_k = \frac{1}{N} \sum_{\ell=0}^{N-1} v_l \, \zeta^{k\ell} \quad \text{for } k = -n, \ldots, 0, \ldots, n\,, \text{ where } \zeta = e^{-2\pi\,\mathrm{i}/N} \text{ and}$$

$$a_k = \frac{2}{N} \sum_{\ell=0}^{N-1} v_l \, \cos\left(2\pi k\ell/N\right) \text{ for } k = 0, \ldots, n \text{ and}$$

$$b_k = \frac{2}{N} \sum_{\ell=0}^{N-1} v_l \, \sin\left(2\pi k\ell/N\right) \text{ for } k = 1, \ldots, n$$

  is an interpolating trigonometric polynomial.
- If $N = 2n$ is even, then

$$p(x) = \sum_{k=-n}^{n-1} \hat{c}_k \, e^{\mathrm{i}\,kx} = \frac{a_0}{2} + \sum_{k=1}^{n-1} \left(a_k \cos(kx) + b_k \sin(kx)\right) + \frac{a_n}{2} \cos(nx)$$

  with

$$\hat{c}_k = \frac{1}{N} \sum_{\ell=0}^{N-1} v_l \, \zeta^{k\ell} \quad \text{for } k = -n, \ldots, 0, \ldots, n-1\,, \text{ where } \zeta = e^{-2\pi\,\mathrm{i}/N} \text{ and}$$

$$a_k = \frac{2}{N} \sum_{\ell=0}^{N-1} v_l \, \cos\left(2\pi k\ell/N\right) \text{ for } k = 0, \ldots, n \text{ and}$$

$$b_k = \frac{2}{N} \sum_{\ell=0}^{N-1} v_l \, \sin\left(2\pi k\ell/N\right) \text{ for } k = 1, \ldots, n-1$$

  is an interpolating trigonometric polynomial.

The sine-cosine representation is real if the values $v_0, \ldots, v_{N-1}$ are real.

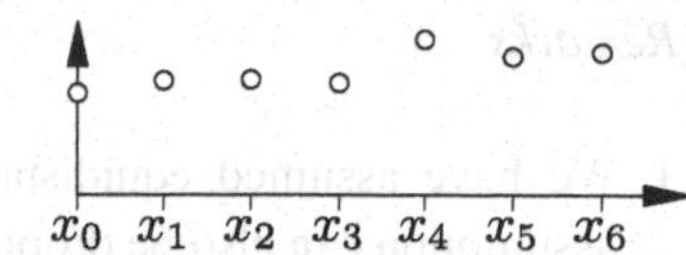

Fig. 78.1 A sample with $N = 7$

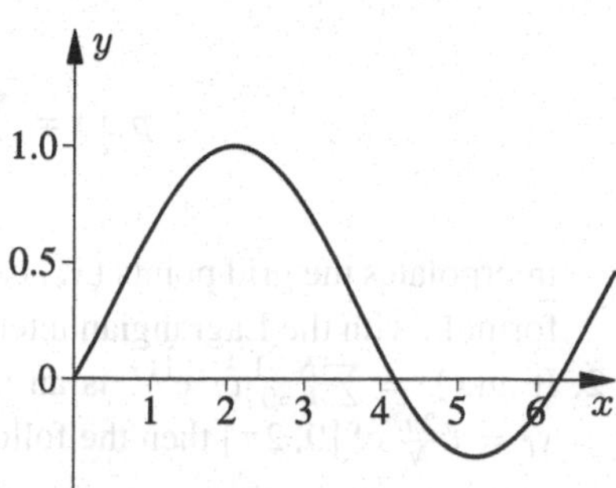

Fig. 78.2 The graph of the interpolating function

Remark Note that the coefficients a_k and b_k can particularly easy be determined in the case $v \in \mathbb{R}^N$ with the coefficients $\hat{c}_k$ of the discrete Fourier transform: Because of $\hat{c}_{-k} = \overline{\hat{c}_k}$ one obtains the sine-cosine form of the trigonometric polynomial by means of Euler's formula from the exp-form by the formulas:

$$a_0 = 2\,\hat{c}_0\,,\ a_k = 2\ \mathrm{Re}(\hat{c}_k)\,,\ b_k = -2\ \mathrm{Im}(\hat{c}_k)\ (\text{and } a_n = 2\,\hat{c}_n\,,\ \text{if } N = 2n \text{ even})\,.$$

Here only the coefficients $\hat{c}_0, \ldots, \hat{c}_n$ are needed which are obtained with the discrete Fourier transform (note the recipe in Sect. 78.1).

Example 78.3 We consider again Example 78.2: Given are the $N = 3$ grid points

$$(0,0),\ (2\pi/3, 1),\ (4\pi/3, 0)\,.$$

We immediately give the sine-cosine form, we get:

$$a_0 = \frac{2}{3}\,,\ a_1 = \frac{2}{3}\cos(2\pi/3) = \frac{-1}{3} \text{ and } b_1 = \frac{2}{3}\sin(2\pi/3) = \frac{1}{\sqrt{3}}$$

and hence the interpolating (real) trigonometric polynomial.

$$p(x) = \frac{1}{3} - \frac{1}{3}\cos(x) + \frac{1}{\sqrt{3}}\sin(x)\,,$$

Figure 78.2 shows the graph of this function. ■

Remarks

1. We have assumed equidistant grid points for the trigonometric interpolation. This assumption can also be dropped. One can show that the *polynomial*

$$p(x) = \sum_{k=1}^{2n+1} v_k \prod_{\ell=1,\ell\neq k}^{2n+1} \frac{\sin\frac{1}{2}(x - x_\ell)}{\sin\frac{1}{2}(x_k - x_\ell)}$$

interpolates the grid points (x_k, v_k) with $0 \leq x_0 < x_1 < \cdots < x_{2n} < 2\pi$. Compare this formula with the Lagrangian interpolation polynomial in Sect. 29.1.

2. If $p(x) = \sum_{k=0}^{N-1} \hat{c}_k\, \mathrm{e}^{\mathrm{i}\,kx}$ is an interpolating polynomial, $p(x_\ell) = v_\ell$, for the points $x_\ell = \ell\frac{2\pi}{N}$ of $[0, 2\pi]$ then the following polynomial

$$Q(x) = \sum_{k=0}^{N-1} \hat{c}_k\, \mathrm{e}^{\mathrm{i}\,k\frac{2\pi}{T}x}$$

solves the interpolation task $Q(x'_\ell) = v_\ell$ for the points $x'_\ell = \ell\,\frac{T}{N}$ of $[0, T]$.

78.4 Exercise

78.1 Prove the approximation of the Fourier coefficients in the box in Sect. 78.1.

78.2 Program the discrete Fourier transform in MATLAB.

78.3 The 2π-periodic function

$$f(x) = 3\sin(4x) + \tfrac{1}{2}\cos(7x) - 2\cos(3x)$$

gives a sample for the N points $x_k = k\frac{2\pi}{N}$, $k = 0, \ldots, N-1$.

(a) Determine the coefficients of the discrete Fourier transform for $N = 4$ and $N = 5$.
(b) Also determine the trigonometric interpolation polynomial for $N = 4$ and $N = 5$ in the sine-cosine form.
(c) Using MATLAB, determine the interpolation polynomial for 10 and 15 grid points.

78.4 The 2π-periodic square wave

$$f(x) = \begin{cases} 1, & 0 \leq x < \pi \\ 0, & \pi \leq x < 2\pi \end{cases}$$

is sampled at the 8 points $x_k = k\frac{2\pi}{8}$, $k = 0, \ldots, 7$. Determine the coefficients $c_0, \ldots, c_7$ of the discrete Fourier transform.

Also determine the trigonometric interpolation polynomial in the sine-cosine form.

78.5 The 2π-periodic function

$$f(x) = |\sin(x)|$$

is sampled at the 4 points $x_k = k\frac{2\pi}{4}$, $k = 0, \ldots, 3$. Determine the coefficients $c_0, \ldots, c_3$ of the discrete Fourier transform and compare these values with the exact Fourier coefficients.

78.6 Determine the trigonometric interpolation polynomial of degree 5 to the grid points

$$(0, 0), (2\pi/5, \sin(2\pi/5)), (4\pi/5, \sin(4\pi/5)), (6\pi/5, \sin(4\pi/5)),$$
$$(8\pi/5, \sin(2\pi/5)).$$

78.7 The 2π-periodic functions

(a) $f(x) = (x - \pi)^2$,
(b) $g(x) = ((x - \pi)/\pi)^3 - (x - \pi)/\pi$

are sampled at the 4 points $x_k = k\frac{2\pi}{4}$, $k = 0, \ldots, 3$. Determine in each case the coefficients $c_0, \ldots, c_3$ of the discrete Fourier transform.

Also determine the trigonometric interpolation polynomial in the sine-cosine form.

The Laplace Transformation

79

The procedure for the *Laplacian transformation* is analogous to that for the Fourier transformation: For a *Laplace transformable* function $f : [0, \infty) \to \mathbb{C}$ one declares a new function $F : D \to \mathbb{R}$ with

$$F(s) = \int_0^\infty f(t)\, \mathrm{e}^{-st}\, \mathrm{d}t\,.$$

For this *Laplace transformation* $f \to F$ numerous rules of calculation apply, which again make it possible to conclude from the knowledge of some correspondences $f(t) \circ\!\!-\!\!\bullet F(s)$ to infer many other correspondences. And again a linear differential equation transforms by Laplace transformation into an equation: By solving the equation and back transformation of the solution we get a solution of the differential equation considered at the beginning. This works analogously with systems of linear differential equations as well as with certain integral equations.

$$f(t) \circ\!\!-\!\!\bullet F(s)$$

79.1 The Laplacian Transformation

We proceed analogously to the Fourier transform: We declare for a function f the *Laplace transform* F, where we declare each value $F(s)$ of F by an improper integral. We again

C. Karpfinger, *Calculus and Linear Algebra in Recipes*,
https://doi.org/10.1007/978-3-662-65458-3_79

choose the easy way and call a function $f : [0, \infty) \to \mathbb{R}$ **Laplace transformable** if the integral

$$\int_0^\infty f(t)\, \mathrm{e}^{-st}\, \mathrm{d}t$$

for all elements s of a *admissibility domain* $D \subseteq \mathbb{R}$ exists.

Remark Since the function e^{-st} for $t \to \infty$ falls sharply, it suffices if f is of at most *exponential growth*. By this way of speaking, it is meant that f does not grow faster than the inverse exponential function falls, so that the integral under consideration for each $s \in D$ exists.

The Laplacian Transform
For a Laplace transformable function $f : [0, \infty) \to \mathbb{R}$ consider the function $F : D \to \mathbb{R}$ given by

$$F(s) = \int_0^\infty f(t)\, \mathrm{e}^{-st}\, \mathrm{d}t\,.$$

We call F the **Laplace transform** from f; here the domain $D \subseteq \mathbb{R}$ is maximally chosen. The notations

$$f(t) \quad \circ\!\!-\!\!\bullet \quad F(s) \text{ and } \mathcal{L}(f(t)) = F(s)$$

are common.

The mapping $f \to F$, a Laplace transformable function f is mapped to its Laplacian transform F is called the **Laplace transform**.

As by the Fourier transform, one speaks of the **original function** $f = f(t)$, the **image function** $F = F(s)$, the **original domain** (which is the set of all Laplace transformable functions) and of the **image domain** (which is the set of all Laplace-transformed functions).

Remark More generally, we can take s to be complex, i.e., it is $D \subseteq \mathbb{C}$ allowed. In this complex version, write $s = \rho + \mathrm{i}\,\omega$ with real part ρ and imaginary part ω; from this one can see that the Laplacian transform generalizes the Fourier transform in a certain way.

Example 79.1

- The function $f : [0, \infty) \to \mathbb{R}$, $f(t) = 1$ has because of

$$F(s) = \int_0^\infty \mathrm{e}^{-st}\,\mathrm{d}t = \lim_{b\to\infty} \int_0^b \mathrm{e}^{-st}\,\mathrm{d}t = \lim_{b\to\infty} \frac{-1}{s}\,\mathrm{e}^{-st}\Big|_0^b = \frac{1}{s}$$

for $s > 0$ the Laplace transform $F : (0, \infty) \to \mathbb{R}$, $F(s) = \frac{1}{s}$, i.e.

$$1 \circ\!\!-\!\!\bullet \frac{1}{s} \text{ or } \mathcal{L}(1) = \frac{1}{s}.$$

Note that this function f is essentially the heaviside function u.

- The function $f : [0, \infty) \to \mathbb{R}$, $f(t) = t^n$ has because of

$$F(s) = \int_0^\infty t^n\,\mathrm{e}^{-st}\,\mathrm{d}t = \lim_{b\to\infty} \int_0^b t^n\,\mathrm{e}^{-st}\,\mathrm{d}t$$

$$= \lim_{b\to\infty} -\frac{t^n}{s}\,\mathrm{e}^{-st}\Big|_0^b + \frac{n}{s}\int_0^b t^{n-1}\,\mathrm{e}^{-st}\,\mathrm{d}t = \cdots = \frac{n!}{s^{n+1}}$$

for $s > 0$ the Laplace transform $F : (0, \infty) \to \mathbb{R}$, $F(s) = \frac{n!}{s^{n+1}}$, i.e.

$$t^n \circ\!\!-\!\!\bullet \frac{n!}{s^{n+1}} \text{ or } \mathcal{L}(t^n) = \frac{n!}{s^{n+1}} \text{ for } n = 0, 1, 2, \ldots.$$

- The function $f : [0, \infty) \to \mathbb{R}$, $f(t) = \exp(at)$ has because of

$$F(s) = \int_0^\infty \mathrm{e}^{at}\,\mathrm{e}^{-st}\,\mathrm{d}t = \lim_{b\to\infty} \int_0^b \mathrm{e}^{(a-s)t}\,\mathrm{d}t = \lim_{b\to\infty} \frac{1}{a-s}\,\mathrm{e}^{(a-s)t}\Big|_0^b = \frac{1}{s-a}$$

for $s > a$ is the Laplace transform $F : (a, \infty) \to \mathbb{R}$, $F(s) = \frac{1}{s-a}$, i.e.

$$\mathrm{e}^{at} \circ\!\!-\!\!\bullet \frac{1}{s-a} \text{ or } \mathcal{L}(\mathrm{e}^{at}) = \frac{1}{s-a}.$$

- The function $f:[0,\infty)\to\mathbb{R}$, $f(t)=\sin(t)$ has because of

$$F(s)=\int_0^\infty \sin(t)\,\mathrm{e}^{-st}\,\mathrm{d}t=\lim_{b\to\infty}\frac{\mathrm{e}^{-st}}{s^2+1}(-s\,\sin(t)-\cos(t))\Big|_0^b=\frac{1}{s^2+1}$$

for $s>0$ the Laplace transform $F:(0,\infty)\to\mathbb{R}$, $F(s)=\frac{1}{s^2+1}$, i.e.

$$\sin(t)\ \circ\!\!-\!\!\bullet\ \frac{1}{s^2+1}\quad\text{or}\quad\mathcal{L}(\sin(t))=\frac{1}{s^2+1}\,.$$

- **Dirac's delta function.** As the Laplacian transform of the Dirac delta function

$$\delta(t-t_0)=\begin{cases}0\,, & t\neq t_0\\ \infty\,, & t=t_0\end{cases}$$

with $t_0\in\mathbb{R}$ we obtain because of

$$\Delta_{t_0}(s)=\int_0^\infty \delta(t-t_0)\,\mathrm{e}^{-st}\,\mathrm{d}t=\mathrm{e}^{-t_0 s}\,.$$

For $t_0=0$ the Laplacian transform of $\delta(t)$ is $\Delta_0(s)$ with $\Delta_0(s)=1$,

$$\delta(t-t_0)\ \circ\!\!-\!\!\bullet\ \mathrm{e}^{-t_0 s}\,,\ \text{ in particular }\ \delta(t)\ \circ\!\!-\!\!\bullet\ 1\,.$$

■

The domains D are always upwardly unbounded intervals of the form (a,∞); this is not a coincidence: If the integral $\int_0^\infty f(t)\,\mathrm{e}^{-st}\,\mathrm{d}t$ exists for a s, so also for all larger s, since e^{-st} with increasing s becomes smaller.

MATLAB Analogous to the Fourier transform, the Laplace transform is of course also possible with MATLAB, an example says more than any explanation:

```
» syms t s; » f = exp(-t)*cos(t); » laplace(f, s) ans =
(s + 1)/((s + 1)^2 + 1)
```

Remark As by the Fourier transform, there is also an inverse for the Laplacian transform, i.e. a **inverse Laplace transform**. In fact, however, this does not play nearly as large a role as one might first expect. We give the formula for the inverse transformation, although one can understand it at the earliest after working through the chapters on complex analysis:

$$f(t) = \frac{1}{2\pi\,\mathrm{i}} \int_{\gamma-\mathrm{i}\infty}^{-\gamma+\mathrm{i}\infty} F(s)\,\mathrm{e}^{st}\,\mathrm{d}s\,,$$

where here s is regarded as complex. In the applications one determines the inverse transform $f(t)$ of the image function $F(s)$ after a corresponding decomposition of $F(s)$ from a table like Table 79.1.

79.2 The Rules and Theorems for the Laplace Transformation

We summarize the most important calculation rules and theorems for the Laplace transformation.

Rules and Theorems for the Laplace Transformation
Let $F = F(s)$ with $s > a$ and $G = G(s)$ with $s > b$ are the image function of a Laplace transformable time function $f = f(t)$ and $g = g(t)$, i.e.

$$f(t) \circ\!\!-\!\!\bullet\ F(s) \text{ and } g(t) \circ\!\!-\!\!\bullet\ G(s)\,.$$

- **Continuity and improper convergence** of F: The function $F(s)$ is continuous and $F(s) \to 0$ for $s \to \infty$; thus, the values of F become small towards the *boundaries.*
- **Linearity:** For all $\lambda,\ \mu \in \mathbb{R}$ we have

$$\lambda\, f(t) + \mu\, g(t) \circ\!\!-\!\!\bullet\ \lambda\, F(s) + \mu\, G(s) \text{ with } s > \max\{a, b\}\,.$$

- **Time scaling:** For all $c \in (0, \infty)$ we have

$$f(c\,t) \circ\!\!-\!\!\bullet\ \frac{1}{c}\, F\left(\frac{s}{c}\right) \text{ with } s > a\,.$$

(continued)

Table 79.1 Laplace correspondences

$F(s)$	$f(t)$	$F(s)$	$f(t)$
1	$\delta(t)$	$\dfrac{s}{(s^2+a^2)^2}$	$\dfrac{t\sin at}{2a}$
e^{-as}	$\delta(t-a)$	$\dfrac{s^2}{(s^2+a^2)^2}$	$\dfrac{\sin at+at\cos at}{2a}$
$\dfrac{1}{s}$	1	$\dfrac{1}{(s^2+a^2)(s^2+b^2)}$, $a^2\neq b^2$	$\dfrac{b\sin at-a\sin bt}{ab(b^2-a^2)}$
$\dfrac{1}{s^2}$	t	$\dfrac{s}{(s^2+a^2)(s^2+b^2)}$, $a^2\neq b^2$	$\dfrac{\cos at-\cos bt}{b^2-a^2}$
$\dfrac{1}{s^n}$, $n\in\mathbb{N}$	$\dfrac{t^{n-1}}{(n-1)!}$	$\dfrac{1}{(s+a)^2+b^2}$	$\dfrac{1}{b}e^{-at}\sin bt$
$\dfrac{1}{s+a}$	e^{-at}	$\dfrac{s+a}{(s+a)^2+b^2}$	$e^{-at}\cos bt$
$\dfrac{1}{(s+a)^n}$, $n\in\mathbb{N}$	$\dfrac{t^{n-1}e^{-at}}{(n-1)!}$	$\dfrac{1}{s^4-a^4}$	$\dfrac{\sinh at-\sin at}{2a^3}$
$\dfrac{1}{(s+a)(s+b)}$, $a\neq b$	$\dfrac{e^{-at}-e^{-bt}}{b-a}$	$\dfrac{s}{s^4+a^4}$	$\dfrac{\sin at\sinh at}{2a^2}$
$\dfrac{1}{s^2+a^2}$	$\dfrac{1}{a}\sin at$	$\dfrac{1}{\sqrt{s}}$	$\dfrac{1}{\sqrt{\pi t}}$
$\dfrac{s}{s^2+a^2}$	$\cos at$	$\arctan\dfrac{a}{s}$	$\dfrac{\sin at}{t}$
$\dfrac{1}{s^2-a^2}$	$\dfrac{1}{a}\sinh at$	$\dfrac{1-e^{-ks}}{s}$	$u(t)-u(t-k)$
$\dfrac{s}{s^2-a^2}$	$\cosh at$	$\dfrac{1}{s(1-e^{-ks})}$	$\sum\limits_{n=0}^{\infty}u(t-nk)$
$\dfrac{1}{s(s^2+a^2)}$	$\dfrac{1-\cos at}{a^2}$	$\dfrac{1}{(s^2+1)(1-e^{-\pi s})}$	$\dfrac{1}{2}(\sin t+\lvert\sin t\rvert)$
$\dfrac{1}{s^2(s^2+a^2)}$	$\dfrac{at-\sin at}{a^3}$	$\dfrac{a\coth\frac{\pi s}{2a}}{s^2+a^2}$	$\lvert\sin at\rvert$
$\dfrac{1}{(s^2+a^2)^2}$	$\dfrac{\sin at-at\cos at}{2a^3}$	$\dfrac{1}{s}\tanh as$	$\operatorname{sgn}\left(\sin\left(\dfrac{\pi}{2a}t\right)\right)$

- **Derivation of the original function:** If f is differentiable and f' is also Laplace transformable, then

$$f'(t) \quad \circ\!\!-\!\!\bullet \quad s\,F(s) - f(0) \text{ for } s > a\,.$$

- **Integral of the original function:** For the primitive function $\int_0^t f(x)\mathrm{d}x$ of f we have

$$\int_0^t f(x)\mathrm{d}x \quad \circ\!\!-\!\!\bullet \quad \frac{1}{s}\,F(s) \text{ with } s > a\,.$$

- **Derivation of the image function:** If $F(s)$ is differentiable, then

$$t\,f(t) \quad \circ\!\!-\!\!\bullet \quad -\,F'(s) \text{ with } s > a\,.$$

- **Frequence shifting:** For $c \in \mathbb{R}$ we have

$$\mathrm{e}^{-ct}\,f(t) \quad \circ\!\!-\!\!\bullet \quad F(s+c) \text{ with } s > a - c\,.$$

- **Time shifting:** For $c \in \mathbb{R}_{>0}$ we have

$$f(t-c)\,u(t-c) \quad \circ\!\!-\!\!\bullet \quad \mathrm{e}^{-cs}\,F(s) \text{ with the Heaviside function } u\,.$$

- **Convolution:** For the **convolution product of** f and g, which is the Laplace transformable function

$$(f * g)(t) = \int_0^t f(t-\tau)\,g(\tau)\mathrm{d}\tau\,,$$

we have $f * g = g * f$ and

$$(f * g)(t) \quad \circ\!\!-\!\!\bullet \quad F(s)\,G(s) \text{ with } s > \max\{a, b\}\,.$$

Using these rules, we can now easily derive many more correspondences from the computed correspondences in Example 79.1

Example 79.2

- Since the cosine is the derivative of the sine, we get

$$f(t) = \sin(t) \quad ○\!\!-\!\!\bullet \quad F(s) = \frac{1}{s^2+1}$$

with the rule for *derivative of the original function*:

$$f'(t) = \cos(t) \quad ○\!\!-\!\!\bullet \quad s\,F(s) - f(0) = \frac{s}{s^2+1}\,.$$

- Because of $\cos(t) \quad ○\!\!-\!\!\bullet \quad \frac{s}{s^2+1}$ and $\sin(t) \quad ○\!\!-\!\!\bullet \quad \frac{1}{s^2+1}$ we obtain for each $\omega \in (0, \infty)$ with the rule to *time scaling*:

$$\cos(\omega t) \quad ○\!\!-\!\!\bullet \quad \frac{1}{\omega}\frac{s/\omega}{(s/\omega)^2+1} = \frac{s}{s^2+\omega^2}$$

and analogously

$$\sin(\omega t) \quad ○\!\!-\!\!\bullet \quad \frac{\omega}{s^2+\omega^2}\,.$$

- Using the rule *frequence shifting*, we obtain for $c \in \mathbb{R}$ and each $\omega \in (0, \infty)$ the correspondences

$$\mathrm{e}^{-ct}\cos(\omega t) \quad ○\!\!-\!\!\bullet \quad \frac{s+c}{(s+c)^2+\omega^2} \quad \text{and} \quad \mathrm{e}^{-ct}\sin(\omega t) \quad ○\!\!-\!\!\bullet \quad \frac{\omega}{(s+c)^2+\omega^2}\,.$$

- If we want to determine the inverse transform of $\frac{1}{(s+2)^2}$ there are two possible ways:
  - *Derivation of the image function*: For $F(s) = \frac{1}{s+2} \quad \bullet\!\!-\!\!○ \quad f(t) = \mathrm{e}^{-2t}$ we have $F'(s) = \frac{-1}{(s+2)^2}$ so that

$$t\,\mathrm{e}^{-2t} \quad ○\!\!-\!\!\bullet \quad \frac{1}{(s+2)^2}\,.$$

  - *Frequence shifting*: For $F(s) = \frac{1}{s^2} \quad \bullet\!\!-\!\!○ \quad f(t) = t$ we have

$$\mathrm{e}^{-2t}\,t \quad ○\!\!-\!\!\bullet \quad \frac{1}{(s+2)^2}\,.$$

  In any case $\frac{1}{(s+2)^2} \quad \bullet\!\!-\!\!○ \quad t\,\mathrm{e}^{-2t}$.

- First, we determine the back-transform of $F(s) = \frac{1-s}{s^2+2s+2}$. It is

$$\frac{1-s}{s^2+2s+2} = \frac{1-s}{(s+1)^2+1} = \frac{2}{(s+1)^2+1} - \frac{s+1}{(s+1)^2+1}.$$

With the Laplacian transforms of sin and cos we obtain with *frequence shifting*:

$$\mathrm{e}^{-t}\sin t \quad \circ\!\!-\!\!\bullet \quad \frac{1}{(s+1)^2+1} \text{ and } \mathrm{e}^{-t}\cos t \quad \circ\!\!-\!\!\bullet \quad \frac{s+1}{(s+1)^2+1}.$$

The back transform we are looking for is therefore $\mathrm{e}^{-t}(2\sin t - \cos t)$. ■

79.3 Applications

Similar to the Fourier transform, we obtain an equation by transforming a linear ODE, but an initial condition or initial conditions are also required. A back transformation of the solution to this equation then yields a solution to the original IVP. This works in a similar way for linear ODE systems as it does for certain integral equations. The typical applications of Laplace transform in engineering mathematics are:

- Solving IVPen with linear ODE systems,
- Solving IVP with linear ODE systems,
- Solving Volterra integral equations.

79.3.1 Solving IVPs with Linear ODEs

We discuss in detail the case of a 2nd order linear ODE with constant coefficients and give recipe-like the procedure for solving a corresponding IVP with such a ODE. The transfer of the principle to higher order ODEs is quite simple.

The starting point is an IVP with a linear second-order ODE with constant coefficients

$$a\,\ddot{x}(t) + b\,\dot{x}(t) + c\,x(t) = s(t) \text{ with } x(0) = d,\ \dot{x}(0) = e,$$

$a,\ b,\ c,\ d,\ e \in \mathbb{R}$ where the functions $x(t),\ \dot{x}(t),\ \ddot{x}(t)$ and $s(t)$ are Laplace transformable and we have the following correspondences:

$$x(t) \quad \circ\!\!-\!\!\bullet \quad X(s) \text{ and } s(t) \quad \circ\!\!-\!\!\bullet \quad S(s).$$

With double application of the rule *derivation of the original function* we obtain the further correspondences:

$$\dot{x}(t) \;\circ\!\!-\!\!\bullet\; sX(s) - x(0) \text{ and } \ddot{x}(t) \;\circ\!\!-\!\!\bullet\; s\,(sX(s) - x(0)) - \dot{x}(0)\,.$$

Substituting this into the above ODE, we obtain, because of the rule *linearity* first

$$a\,(s^2X(s) - s\,x(0) - \dot{x}(0)) + b\,(s\,X(s) - x(0)) + c\,X(s) = S(s)\,,$$

and after substituting the initial conditions we have:

$$a\,(s^2X(s) - s\,d - e) + b\,(s\,X(s) - d) + c\,X(s) = S(s)\,.$$

Finally, we solve for $X(s)$ to:

$$X(s) = \frac{S(s) + a\,(s\,d + e) + b\,d}{a\,s^2 + b\,s + c} = \frac{a\,(s\,d + e) + b\,d}{a\,s^2 + b\,s + c} + \frac{1}{a\,s^2 + b\,s + c}\,S(s)\,.$$

A back transformation, i.e., determining $x(t)$ with $x(t) \;\circ\!\!-\!\!\bullet\; X(s)$ gives a solution of the IVP. But if this expression for $X(s)$ is *complicated*, it will not be found in Table 79.1. It is then necessary to decompose $X(s)$ into simpler summands by a partial fraction decomposition; a subsequent inverse transformation of the summands then leads, because of the rule of *linearity*, in the sum to the inverse transform $x(t)$. We describe the procedure in a recipe-like way for a 2nd order ODE: On the one hand, the 2nd order is the most frequently needed case, on the other hand, the generalization to other orders is quite simple:

Recipe: Solve an IVP with a Linear ODE Using Laplace Transformation] To solve the IVP

$$a\,\ddot{x}(t) + b\,\dot{x}(t) + c\,x(t) = s(t) \text{ with } x(0) = d,\ \dot{x}(0) = e$$

by means of Laplace transformation, proceed as follows:

(1) Denote by $X(s)$ the Laplace transform of the function $x(t)$.
(2) Transform the ODE, find out the correspondence $s(t) \;\circ\!\!-\!\!\bullet\; S(s)$ and insert the initial conditions.
(3) Solve the equation in (2) for $X(s)$ and get $X(s) = \frac{S(s)+a\,(s\,d+e)+b\,d}{a\,s^2+b\,s+c}$.
(4) If necessary, perform a partial fraction decomposition of $X(s)$.

(continued)

(5) Determine the inverse transforms of the summands of $X(s)$ of the partial fraction decomposition.
(6) Obtain the solution $x(t)$.

Example 79.3

- We determine the solution $x(t)$ of the IVP

$$\dot{x}(t) + x(t) = \sin(t) \text{ with } x(0) = 1 .$$

(1) Let $X(s)$ denote the Laplacian transform of the function $x(t)$ we are looking for.
(2) Because of $s(t) = \sin(t) \quad \circ\!\!-\!\!\bullet \quad S(s) = \frac{1}{s^2+1}$ and $\dot{x}(t) \quad \circ\!\!-\!\!\bullet \quad sX(s) - x(0)$ we obtain the equation

$$sX(s) - 1 + X(s) = \frac{1}{s^2+1} .$$

(3) We have $X(s) = \frac{s^2+2}{(s^2+1)(s+1)}$.
(4) A partial fraction decomposition of this expression leads to:

$$X(s) = \frac{s^2+2}{(s+1)(s^2+1)} = \frac{3}{2}\frac{1}{s+1} + \frac{1}{2}\frac{1-s}{s^2+1} = \frac{3}{2}\frac{1}{s+1} + \frac{1}{2}\frac{1}{s^2+1} - \frac{1}{2}\frac{s}{s^2+1} .$$

(5) The inverse transforms of the summands are

$$\frac{1}{s+1} \quad \bullet\!\!-\!\!\circ \quad \mathrm{e}^{-t} \text{ and } \frac{1}{s^2+1} \quad \bullet\!\!-\!\!\circ \quad \sin(t) \text{ and } \frac{s}{s^2+1} \quad \bullet\!\!-\!\!\circ \quad \cos(t) .$$

(6) We obtain the solution $x(t) = \frac{3}{2}\,\mathrm{e}^{-t} + \frac{1}{2}\sin(t) - \frac{1}{2}\cos(t)$.

- We determine the solution $x(t)$ of the IVP

$$\ddot{x}(t) + \dot{x}(t) - 2\,x(t) = 2\,\mathrm{e}^{t} \text{ with } x(0) = 0,\ \dot{x}(0) = 0 .$$

(1) Let $X(s)$ be the Laplacian transform of the function $x(t)$ we are looking for.
(2) Transformation of the IVP yields : $s^2X(s) + s\,X(s) - 2\,X(s) = 2\frac{1}{s-1}$.
(3) We solve for $X(s)$: $X(s) = \frac{2}{(s-1)^2(s+2)}$.
(4) A partial fraction decomposition yields $X(s) = \frac{2}{(s-1)^2(s+2)} = \frac{2/9}{s+2} - \frac{2/9}{s-1} + \frac{6/9}{(s-1)^2}$.

(5) According to Table 79.1 and the above examples, the following applies

$$\frac{1}{s+2} \;\bullet\!\!-\!\!\circ\; \mathrm{e}^{-2t} \text{ and } \frac{1}{s-1} \;\bullet\!\!-\!\!\circ\; \mathrm{e}^{t} \text{ and } \frac{1}{(s-1)^2} \;\bullet\!\!-\!\!\circ\; t\,\mathrm{e}^{t}\,.$$

(6) This gives us the solution

$$x(t) = \frac{2}{9}(\mathrm{e}^{-2t} - \mathrm{e}^{t} + 3t\,\mathrm{e}^{t})\,.$$

■

79.3.2 Solving IVPs with Linear ODE Systems

We discuss in detail the case of a 1st order linear ODE system with two functions and constant coefficients and give recipe-like the procedure for solving a corresponding IVP with such an ODE system.

The starting point is an IVP with a linear ODE system of the form

$$\begin{aligned} \dot{x}(t) &= a\,x(t) + b\,y(t) + s_1(t) \\ \dot{y}(t) &= c\,x(t) + d\,y(t) + s_2(t) \end{aligned} \quad \text{with } x(0) = e \text{ and } y(0) = f$$

with real numbers $a, \ldots, f$. We are looking for the functions $x = x(t)$ and $y = y(t)$. In vector-matrix notation, the IVP is:

$$\begin{pmatrix} \dot{x}(t) \\ \dot{y}(t) \end{pmatrix} = \begin{pmatrix} a & b \\ c & d \end{pmatrix} \begin{pmatrix} x(t) \\ y(t) \end{pmatrix} + \begin{pmatrix} s_1(t) \\ s_2(t) \end{pmatrix} \quad \text{with} \quad \begin{pmatrix} x(0) \\ y(0) \end{pmatrix} = \begin{pmatrix} e \\ f \end{pmatrix}$$

or with the abbreviations

$$\boldsymbol{x}(t) = \begin{pmatrix} x(t) \\ y(t) \end{pmatrix} \text{ and } A = \begin{pmatrix} a & b \\ c & d \end{pmatrix} \text{ and } \boldsymbol{s}(t) = \begin{pmatrix} s_1(t) \\ s_2(t) \end{pmatrix} \text{ and } \boldsymbol{v} = \begin{pmatrix} x(0) \\ y(0) \end{pmatrix}$$

even more succinctly:

$$\dot{\boldsymbol{x}}(t) = A\,\boldsymbol{x}(t) + \boldsymbol{s}(t) \text{ with } \boldsymbol{x}(0) = \boldsymbol{v}\,.$$

In what follows, we assume that all involved functions $x(t)$, $y(t)$, $s_1(t)$, $s_2(t)$ are Laplace transformable and we have the following correspondences

$$x(t) \circ\!\!-\!\!\bullet\, X(s) \text{ and } y(t) \circ\!\!-\!\!\bullet\, Y(s) \text{ and } s_1(t) \circ\!\!-\!\!\bullet\, S_1(s) \text{ and } s_2(t) \circ\!\!-\!\!\bullet\, S_2(s)\,.$$

Because of $\dot{x}(t) \quad \circ\!\!-\!\!\bullet \quad s\,X(s) - x(0)$ and $\dot{y}(t) \quad \circ\!\!-\!\!\bullet \quad s\,Y(s) - y(0)$ the ODE system becomes the following system of equations by Laplace transformation

$$\begin{pmatrix} s\,X(s) - x(0) \\ s\,Y(s) - y(0) \end{pmatrix} = \begin{pmatrix} a & b \\ c & d \end{pmatrix} \begin{pmatrix} X(s) \\ Y(s) \end{pmatrix} + \begin{pmatrix} S_1(s) \\ S_2(s) \end{pmatrix}.$$

We transform this system of equations into a form we are used to and obtain the formula

$$\begin{pmatrix} s-a & -b \\ -c & s-d \end{pmatrix} \begin{pmatrix} X(s) \\ Y(s) \end{pmatrix} = \begin{pmatrix} S_1(s) \\ S_2(s) \end{pmatrix} + \boldsymbol{v}.$$

This system of equations is to be solved, i.e., there are $X(s)$ and $Y(s)$ from this system of equations. The inverse transforms $x(t)$ and $y(t)$ provide solutions we are looking for. This results in the following recipe:

Recipe: Solving an IVP with a Linear ODE System Using Laplace Transformation

To solve the IVP

$$\dot{x}(t) = a\,x(t) + b\,y(t) + s_1(t)$$

$$\dot{y}(t) = c\,x(t) + d\,y(t) + s_2(t)$$

$$\text{with } x(0) = e \quad \text{and} \quad y(0) = f$$

using Laplace transformation proceed as follows:

(1) Denote by $X(s)$ and $Y(s)$ the Laplace transforms of the searched functions $x(t)$ and $y(t)$ and determine the Laplace transforms $S_1(s)$ of $s_1(t)$ and $S_2(s)$ of $s_2(t)$.
(2) Transform the ODE system and set the initial conditions.
(3) Write the system of equations in (2) in the following form

$$\begin{pmatrix} s-a & -b \\ -c & s-d \end{pmatrix} \begin{pmatrix} X(s) \\ Y(s) \end{pmatrix} = \begin{pmatrix} S_1(s) + e \\ S_2(s) + f \end{pmatrix}$$

and determine the solution $(X(s), Y(s))^\top$.
(4) If necessary, perform a partial fraction decomposition of $X(s)$ and $Y(s)$.

(continued)

(5) Determine the inverse transforms of the summands of $X(s)$ and $Y(s)$ of the partial fraction decompositions.
(6) Get the solution $\boldsymbol{x}(t) = (x(t), y(t))^\top$.

Example 79.4 We consider the ODE system

$$\begin{aligned}\dot{x}(t) &= x(t) + y(t)\\ \dot{y}(t) &= -x(t) + y(t) + \mathrm{e}^t\end{aligned} \quad \text{with } x(0) = 0 \text{ and } y(0) = 0.$$

We start right away with (3):

(3) With the numbers $a = b = d = 1$, $c = -1$ and $e = f = 0$ and the Laplace correspondences

$$s_1(t) = 0 \;\circ\!\!-\!\!\bullet\; S_1(s) = 0 \text{ and } s_2(t) = \mathrm{e}^t \;\circ\!\!-\!\!\bullet\; S_2(s) = \frac{1}{s-1}$$

we obtain the system of equations

$$\begin{pmatrix} s-1 & -1 \\ 1 & s-1 \end{pmatrix}\begin{pmatrix} X(s) \\ Y(s) \end{pmatrix} = \begin{pmatrix} 0 \\ \frac{1}{s-1} \end{pmatrix}.$$

We solve the second equation to $X(s)$ and substitute the result into the first equation:

$$X(s) = \frac{1}{s-1} - (s-1)\,Y(s) \text{ and } Y(s) = \frac{1}{(s-1)^2+1}.$$

Thus we obtain the solution $(X(s), Y(s))^\top$ of the (nonlinear) system of equations:

$$X(s) = \frac{1}{s-1} - \frac{s-1}{(s-1)^2+1} \text{ and } Y(s) = \frac{1}{(s-1)^2+1}.$$

(4) not applicable.
(5) A look at our examples, or at Table 79.1, yields

$$x(t) = \mathrm{e}^t(1-\cos(t)) \;\circ\!\!-\!\!\bullet\; X(s) = \frac{1}{s-1} - \frac{s-1}{(s-1)^2+1} \text{ and}$$

$$y(t) = \mathrm{e}^t\sin(t) \;\circ\!\!-\!\!\bullet\; Y(s) = \frac{1}{(s-1)^2+1}.$$

(6) The solution is $\boldsymbol{x}(t) = (\mathrm{e}^t(1-\cos(t)), \mathrm{e}^t\sin(t))^\top$. ■

79.3.3 Solving Integral Equations

An equation in which a desired function $x = x(t)$ appears in the integrand of a given integral, is called an **integral equation**, e.g.

$$\int_0^t x(\tau)\mathrm{d}\tau = \sin(t)\,.$$

A solution $x(t)$ of this integral equation is easily guessed, namely $x(t) = \cos(t)$. The Laplace transformation provides a method to solve such *Volterra integral equations* systematically. Here, one calls an integral equation of the form

$$a\,x(t) + \int_0^t k(t-\tau)x(\tau)\mathrm{d}\tau = s(t) \ \text{ with } a \in \mathbb{R}$$

Volterra integral equation with the so-called **kernel** k. If the functions x, k and s can be Laplace transformed, then a solution is obtained by means of the Laplace transformation: We assrume the following correspondences:

$$x(t) \ \circ\!\!-\!\!\bullet \ X(s) \text{ and } k(t) \ \circ\!\!-\!\!\bullet \ K(s) \text{ and } s(t) \ \circ\!\!-\!\!\bullet \ S(s)\,,$$

then one obtains with the rule for convolution from this integral equation the equation

$$a\,X(s) + K(s)\,X(s) = S(s)\,, \ \text{ i.e. } X(s) = \frac{S(s)}{a + K(s)}\,.$$

The inverse transform $x(t)$ of $X(s) = \frac{S(s)}{a+K(s)}$ is a solution of the considered integral equation.

Recipe: Solving a Volterra Integral Equation with Laplacian Transformation
A solution of the Volterra integral equation

$$a\,x(t) + \int_0^t k(t-\tau)\,x(\tau)\mathrm{d}\tau = s(t) \ \text{ with } a \in \mathbb{R}$$

(continued)

with Laplace transformable functions x, k and s are obtained as follows:

(1) Denote by $X(s)$ the Laplace transform of $x(t)$.
(2) Determine the Laplace transforms $K(s)$ and $S(s)$ of $k(t)$ and $s(t)$.
(3) Get $X(s) = \frac{S(s)}{a+K(s)}$ and possibly perform a partial fraction decomposition.
(4) By back-transforming the summands of the partial fraction decomposition, obtain the solution $x(t)$ of the integral equation.

Example 79.5 We consider the Volterra integral equation

$$x(t) + \int_0^t (t - \tau)\, x(\tau) \mathrm{d}\tau = \sin(t)$$

with $a = 1$, the kernel is $k(t) = t$ and $s(t) = \sin(t)$.

(1) Let $x(s)$ be the Laplace transform of $x(t)$.
(2) $K(s) = 1/s^2$ and $S(s) = 1/(s^2 + 1)$ are the Laplace transforms of $k(t)$ and $s(t)$.
(3) We obtain $X(s) = \frac{s^2}{(s^2+1)^2}$.
(4) By back-transforming (see Table 79.1), we obtain the solution

$$x(t) = \frac{\sin(t) + t\cos(t)}{2} .$$

■

79.4 Exercises

79.1 Determine the Laplace transforms of

(a) $\int_0^t \sin(a\tau)\, \mathrm{d}\tau$,
(b) $\sin^2(t)$,
(c) $\mathrm{e}^{2t} - \mathrm{e}^{-2t}$,
(d) $\mathrm{e}^{2t} \cos(at)$,
(e) $\cos(at) - \cos(bt)$,

(f) $\int\limits_0^t (\cos(a\tau) - \cos(b\tau))\mathrm{d}\tau.$

79.2 Let $F(s)$ be the Laplacian transform of $f(t)$. Determine the inverse transforms $f(t)$ if $F(s)$ is given by:

(a) $\frac{s}{(s+a)(s+b)}$,
(b) $\frac{1}{(s+a)^3(s+b)}$,
(c) $\frac{1}{s^3(s^2+a^2)}$,
(d) $\frac{1}{(s+1)^3((s+1)^2+a^2)}$,
(e) $\frac{s^2+s+2}{(s-1)^2(s^2-2s+2)}$.

79.3 Using the Laplace transformation, calculate the solution of IVPe

(a) $\ddot{x} + 5\dot{x} + 6X = t\,\mathrm{e}^{-2t}, \quad x(0) = x_0, \quad \dot{x}(0) = x_1,$
(b) $\dot{\boldsymbol{x}} = \begin{pmatrix} -3 & -2 \\ 2 & 1 \end{pmatrix} \boldsymbol{x} + \mathrm{e}^{-t} \begin{pmatrix} 1 \\ 0 \end{pmatrix}, \quad \boldsymbol{x}(0) = 0.$

79.4 Using the Laplace transformation, solve the IVP

$$\ddot{x} + 2x = r(t)\ , \quad x(0) = 0\ , \quad \dot{x}(0) = 0\ , \quad \text{where } r(t) = \begin{cases} 1 & \text{if} \quad 0 < t < 1 \\ 0 & \text{if} \quad 1 \le t \end{cases}.$$

Note: Let $r(t)$ using the Heaviside function.

79.5 Using Laplace transform, determine the solutions of the integral equations:

(a) $x(t) + \int\limits_0^t x(t-\tau)\,\mathrm{e}^{\tau}\,\mathrm{d}\tau = \sin t \quad \text{for} \quad t \ge 0,$

(b) $x(t) + \int\limits_0^t \sin(t-\tau)x(\tau)\,\mathrm{d}\tau = 1 \quad \text{for} \quad t \ge 0.$

Holomorphic Functions

80

We consider complex-valued functions in a complex variable, that is, functions of the form $f : G \to \mathbb{C}$ with $G \subseteq \mathbb{C}$; so we do *Complex Analysis*. Complex Analysis has applications in engineering mathematics, for example in electrostatics or fluid mechanics. We develop this theory to the point where we can treat these engineering applications. In doing so, we begin with some examples of complex functions, and then quickly aim at the *differentiability of* complex functions, arriving at the essential notion of Complex Analysis, *holomorphism*. *Holomorphic* functions form the central object of interest.

80.1 Complex Functions

A complex function is a mapping from a set $G \subseteq \mathbb{C}$ to $\mathbb{C}$:

$$f : G \subseteq \mathbb{C} \to \mathbb{C}.$$

In principle, the set G is a completely arbitrary subset of $\mathbb{C}$. For us, however, especially important are *domains* G, therefore we consider complex functions always on *domains*.

80.1.1 Domains

In Complex Analysis, the domain (of definition) of a complex function is usually a **domain** G, i.e., G is an non-empty **open** and **connected** set, i.e.

- for each $z \in G$ there is a $\varepsilon > 0$ with $B_\varepsilon(a) \subseteq G$ (i.e. G is open),
- G cannot be divided into two disjoint non-empty open sets (i.e. G is connected).

C. Karpfinger, *Calculus and Linear Algebra in Recipes*,
https://doi.org/10.1007/978-3-662-65458-3_80

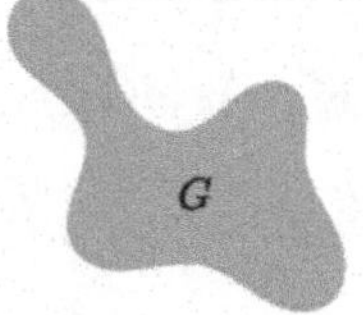

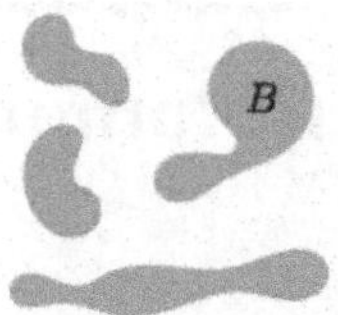

Fig. 80.1 A connected and a disconnected set

Note Fig. 80.1.

80.1.2 Examples of Complex Functions

We give numerous examples of **complex functions** f on domains G,

$$f : G \to \mathbb{C} \text{ with } G \subseteq \mathbb{C}.$$

Example 80.1

- **Polynomial functions:**

$$f : \mathbb{C} \to \mathbb{C},\ f(z) = a_n z^n + \cdots + a_1 z + a_0 \text{ with } a_k \in \mathbb{C}.$$

- **Rational functions:**

$$f : G \to \mathbb{C},\ f(z) = \frac{g(z)}{h(z)} \text{ with polynomials } g(z),\ h(z) \text{ and } G = \{z \in \mathbb{C} \mid h(z) \neq 0\}.$$

- **Power series functions:**

$$f : G \to \mathbb{C},\ f(z) = \sum_{k=0}^{\infty} a_k z^k \text{ with } a_k \in \mathbb{C} \text{ and } G = \left\{ z \in \mathbb{C} \mid \sum_{k=0}^{\infty} a_k z^k \text{ converges} \right\}.$$

The following examples are special power series functions. However, these functions play such a fundamental role that we state them separately.

- **Exponential function:**

$$\exp : \mathbb{C} \to \mathbb{C},\ e^z = \exp(z) = \sum_{k=0}^{\infty} \frac{z^k}{k!}.$$

- **Sine function:**

$$\sin : \mathbb{C} \to \mathbb{C}\,,\ \sin(z) = \sum_{k=0}^{\infty} \frac{(-1)^k}{(2k+1)!} z^{2k+1}\,.$$

- **Cosine function:**

$$\cos : \mathbb{C} \to \mathbb{C}\,,\ \cos(z) = \sum_{k=0}^{\infty} \frac{(-1)^k}{(2k)!} z^{2k}\,.$$

- **Tangent function:**

$$\tan : G \to \mathbb{C}\,,\ \tan(z) = \frac{\sin(z)}{\cos(z)} \ \text{ with } G = \{z \in \mathbb{C} \mid \cos(z) \neq 0\}\,.$$

- **Cotangent function:**

$$\cot : G \to \mathbb{C}\,,\ \cot(z) = \frac{\cos(z)}{\sin(z)} \ \text{ with } G = \{z \in \mathbb{C} \mid \sin(z) \neq 0\}\,.$$

- **Sine hyperbolic function:**

$$\sinh : \mathbb{C} \to \mathbb{C}\,,\ \sinh(z) = \sum_{k=0}^{\infty} \frac{1}{(2k+1)!} z^{2k+1}\,.$$

- **Cosine hyperbolic function:**

$$\cosh : \mathbb{C} \to \mathbb{C}\,,\ \cosh(z) = \sum_{k=0}^{\infty} \frac{1}{(2k)!} z^{2k}\,.$$

For the following two examples, note that any complex number $z \in \mathbb{C} \setminus \{0\}$ can be represented in polar form in exactly one way (see Sect. 24.3),

$$z = r\,\mathrm{e}^{\mathrm{i}\varphi} \ \text{ with } r = |z| \in \mathbb{R}_{>0} \text{ and } \varphi = \arg(z) \in (-\pi, \pi]\,.$$

- **Root function:**

$$\sqrt{\cdot} : \mathbb{C} \to \mathbb{C}\,,\ \sqrt{r\,\mathrm{e}^{\mathrm{i}\varphi}} = \sqrt{r}\,\mathrm{e}^{\mathrm{i}\varphi/2} \ \text{ and } \sqrt{0} = 0\,,$$

where $\sqrt{r}$ is the real (positive) root of the positive real number r.

Fig. 80.2 The roots z_0 and z_1 of z

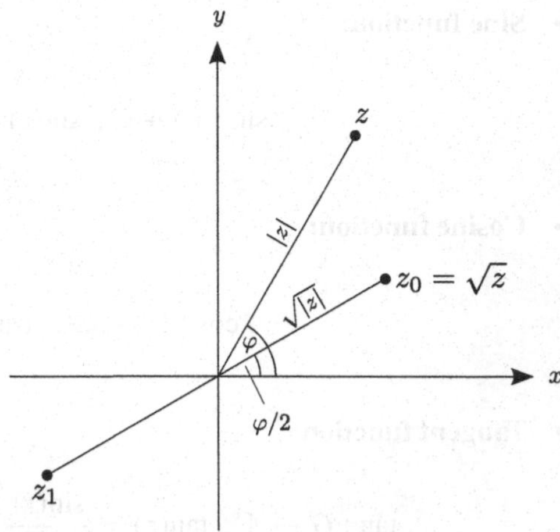

- **Logarithm function**

$$\mathrm{Log} : \mathbb{C} \setminus \{0\} \to \mathbb{C}\,,\ \mathrm{Log}(z) = \ln|z| + \mathrm{i}\arg(z)\,,$$

where $\ln|z|$ is the real logarithm of the positive real number $|z|$. ■

Note that according to the mnemonic box in Sect. 8.2 for every complex number $z \neq 0$ there are two different complex roots z_0 and z_1 (note the designations in Sect. 8.2); here $z_1 = -z_0$. We agree with the root function that $\sqrt{z} = z_0$, more precisely we speak of the **principal branch** of the root function. Note also Fig. 80.2.

Basically, one proceeds in the same way in the real: One declares $\sqrt{a}$ as the positive solution of the equation $x^2 = a$ for $a > 0$ and obtains as here now also in the complex a (unique) mapping $a \mapsto \sqrt{a}$.

Just as the root function is an inverse of the square function, so the logarithm function is said to be an inverse of the exponential function. But just as squaring is not a bijection, $z^2 = (-z)^2$ the complex exponential function is not a bijection, $\mathrm{e}^z = \mathrm{e}^{z+k2\pi\,\mathrm{i}}$, $k \in \mathbb{Z}$. In the case of the root function, we have committed to one of the two possible values of the root function by choosing the argument. In the case of the logarithm function, we also commit ourselves to one of the infinitely many possible values with the help of the argument: If $z \in \mathbb{C} \setminus \{0\}$ then the following infinitely many complex numbers

$$z_k = \ln|z| + \mathrm{i}(\arg(z) + k2\pi) \ \text{ with } \ k \in \mathbb{Z}$$

satisfy the equation $\mathrm{e}^{z_k} = z$. We have $k = 0$ selected (see Fig. 80.3), more precisely we also speak of the **principal branch** of the logarithm and when choosing another one $k \in \mathbb{Z}$ also from the ***k*-th branch of the logarithm**.

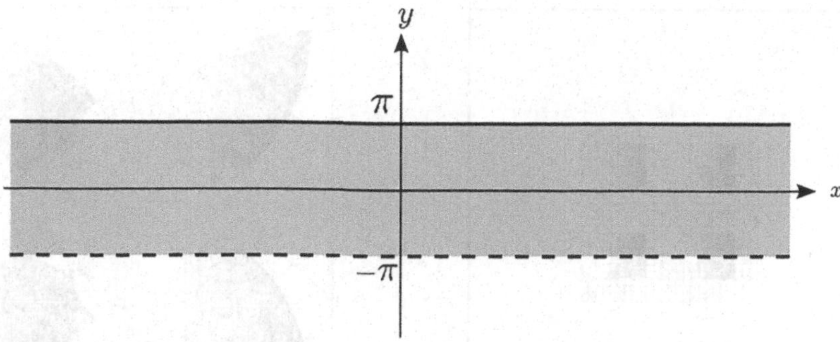

Fig. 80.3 For each $z \in \mathbb{C} \setminus \{0\}$ lies $z_k = \ln|z| + \mathrm{i}\arg(z)$ in this strip of width 2π

In the following box, we give formulas and rules that are familiar from the real and also apply to these complex versions of these functions:

Important Formulas

For all $z,\ w \in \mathbb{C}$:

- $\mathrm{e}^{\mathrm{i}\,z} = \cos(z) + \mathrm{i}\sin(z)$,
- $\mathrm{e}^{z}\,\mathrm{e}^{w} = \mathrm{e}^{z+w}$,
- $\cos(-z) = \cos(z)$, and $\sin(-z) = -\sin(z)$,
- $\sin(z) = \frac{\mathrm{e}^{\mathrm{i}\,z} - \mathrm{e}^{-\mathrm{i}\,z}}{2\mathrm{i}}$ and $\cos(z) = \frac{\mathrm{e}^{\mathrm{i}\,z} + \mathrm{e}^{-\mathrm{i}\,z}}{2}$,
- $\cos^2(z) + \sin^2(z) = 1$,
- $\cos(z) = \cos(x)\cosh(y) - \mathrm{i}\sin(x)\sinh(y)$ for $z = x + \mathrm{i}\,y,\ x,\ y \in \mathbb{R}$,
- $\sin(z) = \sin(x)\cosh(y) + \mathrm{i}\cos(x)\sinh(y)$ for $z = x + \mathrm{i}\,y,\ x,\ y \in \mathbb{R}$,
- $\cos(z + w) = \cos(z)\cos(w) - \sin(z)\sin(w)$,
- $\sin(z + w) = \sin(z)\cos(w) + \sin(w)\cos(z)$,
- $\sin(z) = 0 \Leftrightarrow z = k\pi$ with $k \in \mathbb{Z}$,
- $\cos(z) = 0 \Leftrightarrow z = (k + 1/2)\,\pi$ with $k \in \mathbb{Z}$,
- cos and sin are 2π-periodic, i.e.

$$\cos(z + 2\pi) = \cos(z) \text{ and } \sin(z + 2\pi) = \sin(z) \text{ for all } z \in \mathbb{C}\,.$$

- $\sin(z + \pi/2) = \cos(z)$,
- $\exp(\mathrm{Log}(z)) = z$ for all $z \in \mathbb{C} \setminus \{0\}$,
- $\mathrm{Log}(\exp(z)) = z$ for all $z \in \mathbb{C}$ with $\mathrm{Im}(z) \in (-\pi, \pi]$.

However, the complex functions also have properties that one is not used to from the real and therefore seem strange at first; we give the most important of these properties:

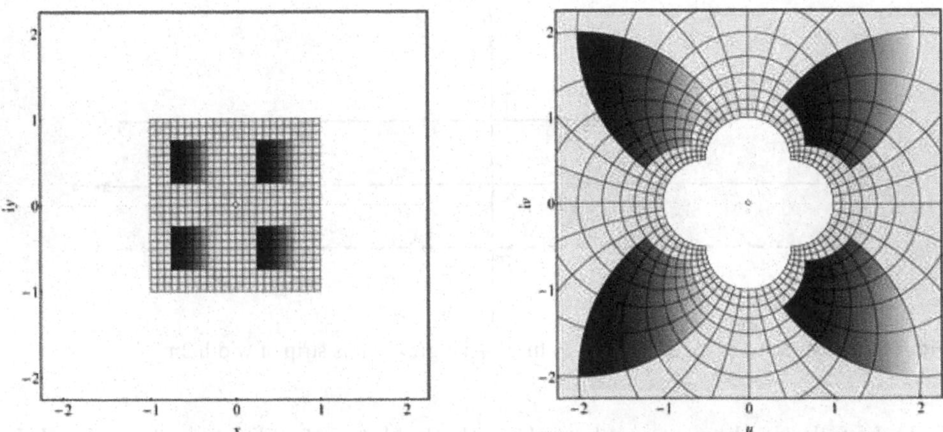

Fig. 80.4 Image and primal image of the inversion $f(z) = 1/z$

- The complex functions cos and sin are unrestricted, i.e., $\cos(z)$ and $\sin(z)$ become arbitrarily large and small, respectively, for example $\cos(\mathrm{i}\,y) = \cosh(y)$ for all $y \in \mathbb{R}$. Now notice the graph of cosh in Fig. 24.3.
- exp is 2π i-periodic and thus no longer bijective, namely $\mathrm{e}^{z+2\pi\,\mathrm{i}} = \mathrm{e}^z$ for all $z \in \mathbb{C}$.
- Euler's formula $\mathrm{e}^{\mathrm{i}\,z} = \cos(z) + \mathrm{i}\sin(z)$ does not provide a decomposition into real and imaginary parts, e.g. Euler's formula gives with $z = -\,\mathrm{i}\ln(2)$:

$$\mathrm{e}^{\mathrm{i}\,z} = 2 = \cos(-\,\mathrm{i}\ln(2)) + \mathrm{i}\sin(-\,\mathrm{i}\ln(2)) = \frac{5}{4} + \mathrm{i}\left(-\,\mathrm{i}\frac{3}{4}\right).$$

- $\mathrm{Log}(zw) \neq \mathrm{Log}(z) + \mathrm{Log}(w)$ for all $z, w \in \mathbb{C}$, e. g. we have with $z = -1$ and $w = \mathrm{i}$:

$$\mathrm{Log}(zw) = \mathrm{Log}(-\,\mathrm{i}) = \frac{-\,\mathrm{i}\,\pi}{2} \quad \text{and}$$

$$\mathrm{Log}(z) + \mathrm{Log}(w) = \mathrm{Log}(-1) + \mathrm{Log}(\mathrm{i}) = \mathrm{i}\,\pi + \frac{\mathrm{i}\,\pi}{2} = \frac{3\pi}{2}.$$

80.1.3 Visualization of Complex Functions

The graph $\Gamma_f = \{(z, f(z)) \,|\, z \in G\}$ of a complex function $f : G \subseteq \mathbb{C} \to \mathbb{C}$ is a subset of the $\mathbb{C}^2$, after identification of $\mathbb{C}$ with $\mathbb{R}^2$ thus a subset of the $\mathbb{R}^4$. So it is no longer possible to imagine the graph of a complex function. In order to get an idea of the function nevertheless, one likes to help oneself with the following *visualization of* a complex function $f : G \subseteq \mathbb{C} \to f(G) \subseteq \mathbb{C}$ You draw two planes $\mathbb{C} = \mathbb{R}^2$ next to each other and draw on the left G and right $f(G)$. For example, we obtain a visualization of the inversion $f(z) = 1/z$ (see Fig. 80.4). For a visualization of the sine function see Fig. 80.5.

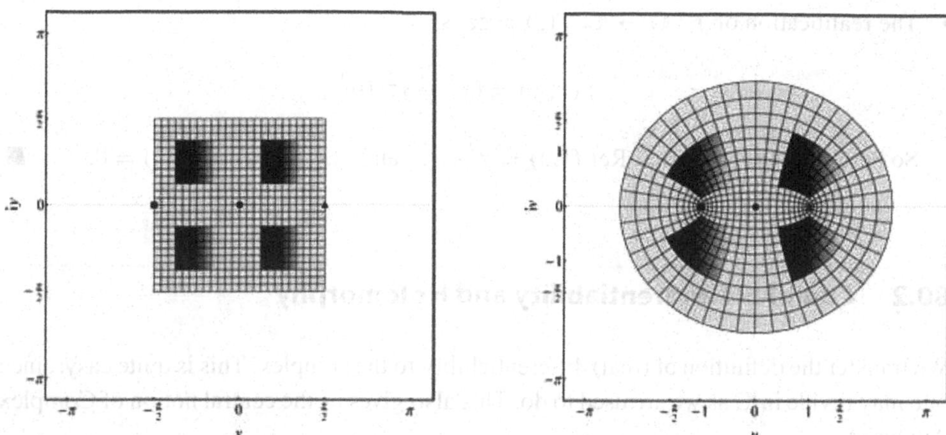

Fig. 80.5 Image and original image of the sine function

80.1.4 Realification of Complex Functions

By identifying

$$\mathbb{C} \longleftrightarrow \mathbb{R}^2 \text{ and } z = x + \mathrm{i}\, y \longleftrightarrow (x, y)$$

we can define a complex function $f : G \subseteq \mathbb{C} \to \mathbb{C}$ in the following sense as a *real* function $f : G \subseteq \mathbb{R}^2 \to \mathbb{R}^2$ function; we speak of a **realification**:

$$f : \begin{cases} G \subseteq \mathbb{C} \to & \mathbb{C} \\ \quad z & \mapsto f(z) \end{cases} \longleftrightarrow f : \begin{cases} G \subseteq \mathbb{R}^2 \to & \mathbb{R}^2 \\ (x, y) & \mapsto (u(x, y), v(x, y)) \end{cases}.$$

Here $f(z) = u(x, y) + \mathrm{i}\, v(x, y)$ with $u(x, y) = \mathrm{Re}(f(z))$ and $v(x, y) = \mathrm{Im}(f(z))$.

Example 80.2

- The realification of $f : \mathbb{C} \to \mathbb{C}$, $f(z) = z^2$ is

$$f(x, y) = (x^2 - y^2, 2xy)^\top .$$

  Here therefore we have $u(x, y) = \mathrm{Re}(f(z)) = x^2 - y^2$ and $v(x, y) = \mathrm{Im}(f(z)) = 2xy$.
- The realification of $f : \mathbb{C} \to \mathbb{C}$, $f(z) = z^2 - \bar{z}$ is

$$f(x, y) = (x^2 - y^2 - x, 2xy + y)^\top .$$

  So here we have $u(x, y) = \mathrm{Re}(f(z)) = x^2 - y^2 - x$ and $v(x, y) = \mathrm{Im}(f(z)) = 2xy + y$.

- The realification of $f : \mathbb{C} \to \mathbb{C}$, $f(z) = z\bar{z}$ is

$$f(x, y) = (x^2 + y^2, 0)^\top .$$

So here we have $u(x, y) = \mathrm{Re}(f(z)) = x^2 + y^2$ and $v(x, y) = \mathrm{Im}(f(z)) = 0$. ■

80.2 Complex Differentiability and Holomorphy

We transfer the definition of (real) differentiability to the complex. This is quite easy, since one may divide in $\mathbb{C}$ as we are used to do. This also gives us the central notion of Complex Analysis:

Complex Differentiability, Holomorphy and Entireness
Let there be $f : G \to \mathbb{C}$ a complex function on a domain G.

- We call $f : G \subseteq \mathbb{C} \to \mathbb{C}$ in $z_0 \in G$ **(complex) differentiable**, if the following limit exists:

$$f'(z_0) = \lim_{z \to z_0} \frac{f(z) - f(z_0)}{z - z_0} .$$

- One says, $f : G \subseteq \mathbb{C} \to \mathbb{C}$ is **on** $U \subseteq G$ **(complex) differentiable** if f in every $z_0 \in U$ (complex) is differentiable.
- One calls f **holomorphic** in $z_0 \in G$ if there exists an open neighborhood $U \subseteq G$ of z_0 on which f is (complex) differentiable.
- One calls f **holomorphic on** (an open set) $U \subseteq G$ if f is holomorphic in every $z_0 \in U$.
- We call f **entire** if f is on $G = \mathbb{C}$ holomorphic.

Note that holomorphy on an open set and complex differentiability on an open set are one and the same notions.

It is customary, and we will do so as well, to speak briefly of a *holomorphic function* f. By this we mean that $f : G \subseteq \mathbb{C} \to \mathbb{C}$ is on its domain G holomorphic. Instead of holomorphic one also says **analytic**.

If $f : G \to \mathbb{C}$ is holomorphic, then there exists the **derivative function** $f' : G \to \mathbb{C}$, $z \mapsto f'(z)$. For the calculation of the derivative function we have the known rules:

$$f'(z) = nz^{n-1} \text{ for } f(z) = z^n ,$$
$$f'(z) = g'(z) + h'(z) \text{ for } f(z) = g(z) + h(z) ,$$
$$f'(z) = g'(z)h(z) + g(z)h'(z) \text{ for } f(z) = g(z)h(z) ,$$
$$f'(z) = \frac{g'(z)h(z) - g(z)h'(z)}{(h(z))^2} \text{ for } f(z) = \frac{g(z)}{h(z)} ,$$
$$f'(z) = g'(h(z))h'(z) \text{ for } f(z) = g(h(z)) .$$

In particular, the sum, product, quotient and compositions of holomorphic functions are also holomorphic again.

Example 80.3

- Polynomial functions, exp, sin, cos, sinh and cosh are entire functions, since they are holomorphic on $\mathbb{C}$, namely they are complex differentiable in every $z_0 \in \mathbb{C}$, and we have:

$$(\mathrm{e}^z)' = \mathrm{e}^z , \ (\sin z)' = \cos z , \ (\cos z)' = -\sin z .$$

- Rational functions and tan and cot are holomorphic in their domain of definition. ■

To decide whether a function is holomorphic or not, one likes to resort to the following criterion:

Criterion for Holomorphy
We consider a complex function $f : G \to \mathbb{C}$ with the realification

$$f : G \to \mathbb{R}^2 , \ f(x, y) = u(x, y) + \mathrm{i}\, v(x, y) .$$

The function f is holomorphic on G if the functions

$$u : (x, y) \mapsto u(x, y) \text{ and } v : (x, y) \mapsto v(x, y)$$

(continued)

is continuously partial differentiable and the following **Cauchy-Riemann differential equations** at G are satisfied:

$$u_x(x, y) = v_y(x, y) \text{ and } u_y(x, y) = -v_x(x, y) .$$

Recipe: Proof of Holomorphy of a Complex Function
To prove whether a complex function $f : G \to \mathbb{C}$ is holomorphic, proceed as follows:

(1) Set $z = x + \mathrm{i}\, y$ and determine u and v with $f(x + \mathrm{i}\, y) = u(x, y) + \mathrm{i}\, v(x, y)$.
(2) Check whether u and v are continuously partial differentiable.

- if no: f is not holomorphic.
- if yes: continue in the next step.

(3) Check whether the Cauchy-Riemann differential equations $u_x = v_y$ and $u_y = -v_x$ are satisfied in a domain $U \subseteq G$.

- if no: f is nowhere holomorphic.
- if yes: f is holomorphic on U.

Even though we have long known that polynomial functions and the exponential function are holomorphic, we show this with the help of this criterion:

Example 80.4

- We examine the function $f : \mathbb{C} \to \mathbb{C}$, $f(z) = z$ on holomorphy: (1) Because of $f(x + \mathrm{i}\, y) = x + \mathrm{i}\, y$ we have here

  $$u(x, y) = x \text{ and } v(x, y) = y .$$

  (2) Since u and v are continuously partially differentiable and (3) the Cauchy-Riemann differential equations are satisfied,

  $$u_x = 1 = v_y \text{ and } u_y = 0 = -v_x ,$$

  f is holomorphic on $\mathbb{C}$ and thus an entire function.

- We study the function $\exp : \mathbb{C} \to \mathbb{C}$, $\exp(z) = \mathrm{e}^z$:
  (1) Because of $\exp(x + \mathrm{i}\,y) = \mathrm{e}^x \mathrm{e}^{\mathrm{i}\,y} = \mathrm{e}^x (\cos(y) + \mathrm{i}\sin(y)) = \mathrm{e}^x \cos(y) + \mathrm{i}\,\mathrm{e}^x \sin(y)$ we have

$$u(x, y) = \mathrm{e}^x \cos(y) \text{ and } v(x, y) = \mathrm{e}^x \sin(y)\,.$$

  (2) Since u and v are continuously partially differentiable and (3) the Cauchy-Riemann differential equations are satisfied,

$$u_x = \mathrm{e}^x \cos(y) = v_y \text{ and } u_y = -\,\mathrm{e}^x \sin(y) = -v_x\,,$$

  f is holomorphic on $\mathbb{C}$ and thus an entire function.
- We study the function $f : \mathbb{C} \to \mathbb{C}$, $f(z) = z\,\bar{z}$ on holomorphy: (1) Because of $f(x + \mathrm{i}\,y) = x^2 + y^2$ we have here

$$u(x, y) = x^2 + y^2 \text{ and } v(x, y) = 0\,.$$

  (2) The functions u and v are continuously partially differentiable, but (3) the Cauchy-Riemann differential equations are not fullfilled except at the point $(0, 0)$,

$$u_x = 2\,x \neq 0 = v_y \text{ and } u_y = 2\,y \neq 0 = -v_x \text{ for } (x, y) \neq (0, 0)\,,$$

  so that f is not holomorphic on any subset of $\mathbb{C}$. ∎

80.3 Exercises

80.1 Determine the sets of points in $\mathbb{C}$ each of which is defined by one of the following conditions:

(a) $|z + 1 - 2\,\mathrm{i}| = 3$, (d) $\mathrm{Re}\,(1/z) = 1$, (e) $|z - 3| + |z + 3| = 10$,
(b) $1 < |z + 2\,\mathrm{i}| < 2$, (c) $|z - 2| = |z + \mathrm{i}|$, (f) $\mathrm{Im}\,((z + 1)/(z - 1)) \leq 0$.

80.2 Determine the realifications of the following functions $f : \mathbb{C} \to \mathbb{C}$, $z \mapsto f(z)$ with:

(a) $f(z) = z^3$,
(b) $f(z) = \frac{1}{1-z}$,
(c) $f(z) = \mathrm{e}^{3z}$.

80.3 For which $z \in \mathbb{C}$ is $\sin z = 1000$?

80.4 Show that the function $f(z) = \frac{(2z-1)}{(2-z)}$ maps the unit circle of the complex plane onto itself.

80.5 Determine in which domains $G \subseteq \mathbb{C}$ the following functions are holomorphic:

$$\text{(a) } f(z) = z^3, \quad \text{(b) } f(z) = z \operatorname{Re} z, \quad \text{(c) } f(z) = |z|^2, \quad \text{(d) } f(z) = \bar{z}/|z|^2.$$

80.6 Calculate:

(a) $e^{2+i\pi/6}$,
(b) $\cosh(i\,t)$, $t \in \mathbb{R}$,
(c) $\cos(1 + 2i)$.

80.7 Determine and sketch the images of the areas

(a) $\{z \in \mathbb{C} \mid 0 < \operatorname{Re} z < 1,\ 0 < \operatorname{Im} z < 1\}$ at $w = e^z$,
(b) $\{z \in \mathbb{C} \mid 0 < \operatorname{Re} z < \pi/2,\ 0 < \operatorname{Im} z < 2\}$ at $w = \sin z$.

Complex Integration

81

The analogue of real integration in the complex is integration along curves. The complex integration of a function f along a curve γ is analogous to the real vectorial line integral:

$$\int_\gamma f \mathrm{d}s = \int_a^b f(\gamma(t))\,\dot{\gamma}(t)\mathrm{d}t\,.$$

To calculate the value of a complex line integral, as in the real, a primitive function of f may be useful, such a function exists, provided that the function f is holomorphic.

81.1 Complex Curves

We give *curves* in $\mathbb{C}$ with a real parameter $t \in [a, b] \subseteq \mathbb{R}$: We call for real numbers $a < b$ a continuous mapping

$$\gamma : [a, b] \subseteq \mathbb{R} \to \mathbb{C}\,,\ t \mapsto \gamma(t)$$

a **curve** in $\mathbb{C}$—one speaks of a **parameterization**. It is $\gamma(a)$ the **starting point** and $\gamma(b)$ the **endpoint** of the curve, and the curve is called **closed** if $\gamma(a) = \gamma(b)$. And it is called **double point free** if from $\gamma(t_1) = \gamma(t_2)$ with $t_1,\ t_2 \in (a, b)$ always $t_1 = t_2$ follows. The set of curvepoints $\gamma([a, b]) \subseteq \mathbb{C}$ is called the **trace** of the curve, but here one is often careless and also speaks of the *curve*.

It is said that a closed, double point free curve is **positively traversed**, if the enclosed region B is to the left of the trace. In Fig. 81.1 the closed curve is positively traversed.

We deal with the two simplest and most important curves in the following example.

C. Karpfinger, *Calculus and Linear Algebra in Recipes*,
https://doi.org/10.1007/978-3-662-65458-3_81

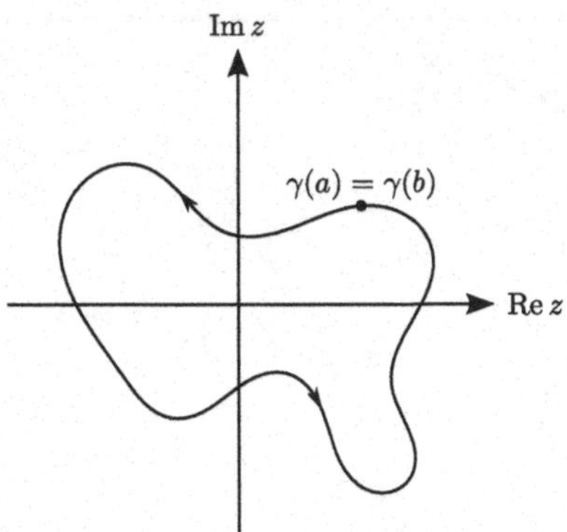

Fig. 81.1 The curve is positively traversed

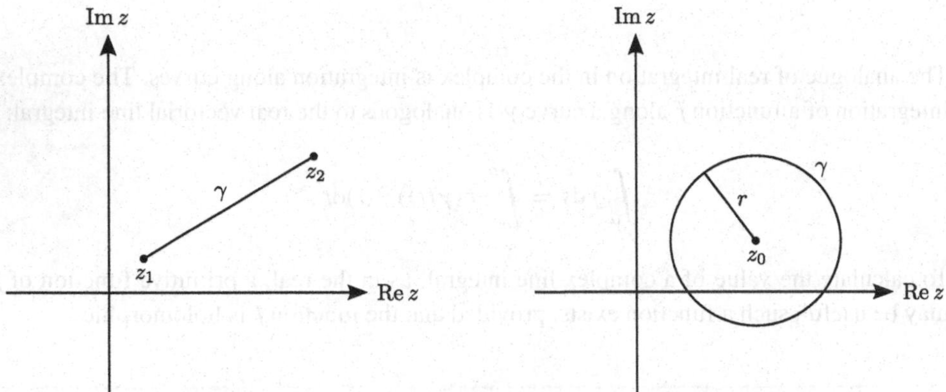

Fig. 81.2 The connecting line between two points and a circle around a point

Example 81.1 We give the two curves whose traces are, on the one hand, the connecting line between two points $z_1,\ z_2 \in \mathbb{C}$ on the other hand a circle of radius r around the center $z_0 \in \mathbb{C}$ (see Fig. 81.2).

The following curve γ_1 has as trace the connecting line between the points $z_1,\ z_2 \in \mathbb{C}$:

$$\gamma_1 : [0, 1] \to \mathbb{C}\,,\ \gamma_1(t) = z_1 + t\,(z_2 - z_1)\,.$$

The following curve γ_2 has as trace the circle of radius r around the center z_0:

$$\gamma_2 : [0, 2\pi] \to \mathbb{C}\,,\ \gamma_2(t) = z_0 + r\,\mathrm{e}^{\mathrm{i}\,t}\,.$$

The circle is traversed positively. If one chooses for γ_2 the interval $[0, \pi]$, an *upper semicircle* is obtained, and given the choice $[-\pi/2, \pi/2]$ you get a *right semicircle*. In this way, one can obtain arbitrary sections of circles. ■

Fig. 81.3 A composite curve

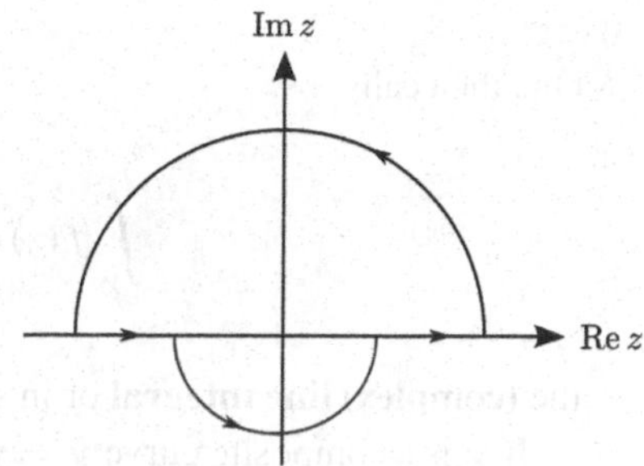

Let $\gamma_1, \ldots, \gamma_n$ be curves with the property, that the end point of one curve coincides with the curve starting point of the next curve. One then writes $\gamma = \gamma_1 + \cdots + \gamma_n$ and means exactly what one imagines by it, note Fig. 81.3.

When assembling curves, but also in other cases, one often wants to traverse a given curve in the opposite direction, i.e. not from the starting point $\gamma(a)$ to the end point $\gamma(b)$ but from $\gamma(b)$ to $\gamma(a)$. This can be done quite easily: If

$$\gamma : [a, b] \to \mathbb{C}, \; t \mapsto \gamma(t)$$

is a curve with starting point $\gamma(a)$ and end point $\gamma(b)$ then the curve

$$-\gamma : [a, b] \to \mathbb{C}, \; t \mapsto \gamma(a + b - t)$$

traverses the track in the opposite direction, starting at $\gamma(b)$ and ending at $\gamma(a)$.

In what follows, we will repeatedly consider, in addition to γ also $\dot{\gamma}$. To do this, γ must be differentiable. For the sake of simplicity we assume this, we even go one step further: All curves considered in the following shall be piecewise continuously differentiable.

81.2 Complex Line Integrals

We now introduce the complex line integral in analogy to the real vector-valued *line integral*. The role of the scalar product is taken over by the complex multiplication:

The Complex Line Integral

Given are

- a complex function $f : G \subseteq \mathbb{C} \to \mathbb{C}$ and
- a curve $\gamma : [a, b] \to \mathbb{C}$ whose trace lies in the domain G, $\gamma([a, b]) \subseteq G$.

(continued)

One then calls

$$\int_\gamma f(z)\,\mathrm{d}z = \int_a^b f(\gamma(t))\,\dot\gamma(t)\mathrm{d}t$$

the **(complex) line integral** or in short **integral of** f along γ.

If γ is a composite curve $\gamma = \gamma_1 + \cdots + \gamma_n$ then one sets

$$\int_\gamma f(z)\,\mathrm{d}z = \sum_{i=1}^n \int_{\gamma_i} f(z)\,\mathrm{d}z\,.$$

If γ is a closed curve, we also write $\oint_\gamma f(z)\,\mathrm{d}z$. And if $\gamma : [0, 2\pi] \to \mathbb{C}$, $\gamma(t) = z_0 + r\,\mathrm{e}^{\mathrm{i}t}$ is a circle around z_0 of radius r, then one also writes

$$\oint_{|z-z_0|=r} f(z)\,\mathrm{d}z = \int_\gamma f(z)\,\mathrm{d}z\,.$$

Note that in this notation the direction of traversal of the circle is agreed upon: We traverse the circle positively, that is, counter clock wise.

Recipe: Calculate a Complex Line Integral
To calculate the complex line integral $\int_\gamma f(z)\mathrm{d}z$ proceed as follows:

(1) Determine a parametrization of $\gamma = \gamma_1 + \cdots + \gamma_n$, i.e.

$$\gamma_i : [a_i, b_i] \to \mathbb{C}\,,\ t \mapsto \gamma_i(t) \ \text{ for } i = 1, \ldots, n\,.$$

(2) Set up the integrals:

$$\int_{\gamma_i} f(z)\,\mathrm{d}z = \int_{a_i}^{b_i} f(\gamma_i(t))\,\dot\gamma_i(t)\mathrm{d}t\,.$$

(continued)

(3) Calculate the integrals from (2) and sum up:

$$\int_\gamma f(z)\,\mathrm{d}z = \sum_{i=1}^{n} \int_{\gamma_i} f(z)\mathrm{d}z\,.$$

Example 81.2

- We determine for $r > 0$ and $m \in \mathbb{Z}$ the complex line integral

$$\oint_{|z-a|=r} (z-a)^m \mathrm{d}z\,.$$

(1) A parametrization of γ is $\gamma : [0, 2\pi] \to \mathbb{C}$, $\gamma(t) = a + r\,\mathrm{e}^{\mathrm{i}t}$.
(2) We obtain the integral

$$\int_\gamma f(z)\,\mathrm{d}z = \int_0^{2\pi} r^m\,\mathrm{e}^{\mathrm{i}mt}\,\mathrm{i}\,r\,\mathrm{e}^{\mathrm{i}t}\,\mathrm{d}t\,.$$

(3) We have

$$\int_\gamma f(z)\,\mathrm{d}z = \mathrm{i}\,r^{m+1} \int_0^{2\pi} \mathrm{e}^{\mathrm{i}(m+1)t}\,\mathrm{d}t = \begin{cases} \frac{\mathrm{i}\,r^{m+1}}{\mathrm{i}(m+1)}\,\mathrm{e}^{\mathrm{i}(m+1)t}\Big|_0^{2\pi}, & m \neq -1 \\ \mathrm{i}\,t\Big|_0^{2\pi}, & m = -1 \end{cases}$$

$$= \begin{cases} 0, & m \neq -1 \\ 2\pi\,\mathrm{i}, & m = -1 \end{cases}.$$

- We calculate the complex line integral

$$\int_\gamma f(z)\,\mathrm{d}z$$

for $f(z) = z$ and $g(z) = \bar{z}$ and the curve γ with the trace given in Fig. 81.4.
(1) As parameterization we obtain for $\gamma = \gamma_1 + \gamma_2$:

$$\gamma_1 : [0, 3\pi/2] \to \mathbb{C}\,,\ \gamma_1(t) = \mathrm{e}^{\mathrm{i}t}$$

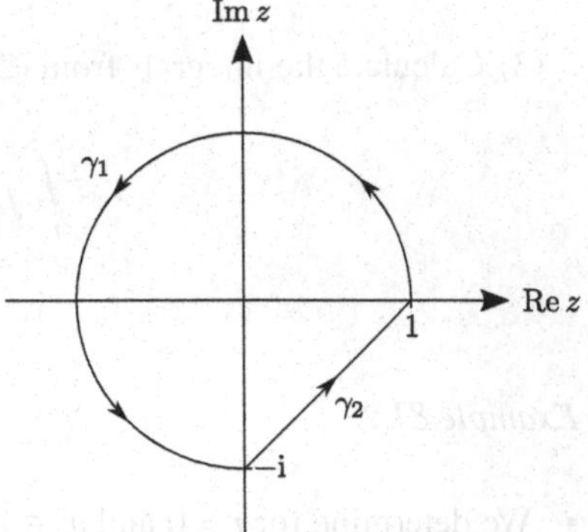

Fig. 81.4 Integral along $\gamma = \gamma_1 + \gamma_2$

and

$$\gamma_2 : [0, 1] \to \mathbb{C}\,,\ \gamma_2(t) = -\mathrm{i}+(1+\mathrm{i})t\,.$$

(2) For $f(z) = z$ and $g(z) = \bar{z}$ we obtain because of $\overline{\mathrm{e}^{\mathrm{i}t}} = \mathrm{e}^{-\mathrm{i}t}$:

$$\int\limits_{\gamma_1+\gamma_2} f(z)\mathrm{d}z = \int\limits_0^{3\pi/2} \mathrm{e}^{\mathrm{i}t}\,\mathrm{i}\,\mathrm{e}^{\mathrm{i}t}\,\mathrm{d}t + \int\limits_0^1 (-\mathrm{i}+(1+\mathrm{i})t)(1+\mathrm{i})\mathrm{d}t\,.$$

$$\int\limits_{\gamma_1+\gamma_2} g(z)\,\mathrm{d}z = \int\limits_0^{3\pi/2} \mathrm{e}^{-\mathrm{i}t}\,\mathrm{i}\,\mathrm{e}^{\mathrm{i}t}\,\mathrm{d}t + \int\limits_0^1 (\mathrm{i}+(1-\mathrm{i})t)(1+\mathrm{i})\mathrm{d}t\,.$$

(3) A quick calculation reveals:

$$\int\limits_{\gamma} f(z)\mathrm{d}z = \mathrm{i}\,\frac{1}{2\,\mathrm{i}}\,\mathrm{e}^{2\mathrm{i}t}\Big|_0^{3\pi/2} + (-\mathrm{i}+1)t\,\Big|_0^1 + \frac{1}{2}t^2(1+\mathrm{i})^2\Big|_0^1 = -1+(-\mathrm{i}+1)+\mathrm{i} = 0\,.$$

$$\int\limits_{\gamma} g(z)\mathrm{d}z = \mathrm{i}\,t\Big|_0^{3\pi/2} + (\mathrm{i}-1)t\Big|_0^1 + t^2\Big|_0^1 = \frac{3\pi\,\mathrm{i}}{2} + (\mathrm{i}-1)+1 = \left(\frac{3\pi}{2}+1\right)\mathrm{i}\,.$$

Note that the integral $\int_{|z|=1} g(z)\mathrm{d}z$ along the (full) unit circle has the value $2\pi\,\mathrm{i}$ so that $\int_\gamma g(z)\,\mathrm{d}z \neq \int_{|z|=1} g(z)\,\mathrm{d}z$, even though the two curves have the same starting and end points. ■

81.3 The Cauchy Integral Theorem and the Cauchy Integral Formula

The value of the complex line integral $\int_\gamma f(z)\mathrm{d}z$ generally depends on the integration path (see Example 81.2). However, there is a class of functions f for which it is essentially irrelevant which path is used to integrate them from the curve starting point $\gamma(a)$ to the curve end point $\gamma(b)$ integrated, the result is the same. These are the holomorphic functions. One only has to make sure that the area G in which the curve runs is simply connected (see Sect. 57.2). We become more precise and formulate the central theorems of the theory of holomorphic functions:

The Cauchy Integral Theorem, the Cauchy Integral Formula, and Corollaries

- **The Cauchy Integral Theorem:** If γ is a closed curve in a simply connected domain G and $f : G \to \mathbb{C}$ holomorphic, then

$$\oint_\gamma f(z)\,\mathrm{d}z = 0\,.$$

- **Cauchy Integral Formula:** If γ is a closed curve, free of double points and positively traversed, in a simply connected domain G and $f : G \to \mathbb{C}$ holomorphic, then for every z_0 in the interior of we have

$$f(z_0) = \frac{1}{2\pi\,\mathrm{i}} \oint_\gamma \frac{f(z)}{z - z_0}\,\mathrm{d}z\,.$$

- **Path independence of the line integral:** If $f : G \to \mathbb{C}$ holomorphic on the simply connected domain G, then for any curves γ_1 and γ_2 in G with the same starting and end points we have

$$\int_{\gamma_1} f(z)\mathrm{d}z = \int_{\gamma_2} f(z)\mathrm{d}z\,.$$

- **Existence of a primitive function:** If $f : G \to \mathbb{C}$ holomorphic on the simply connected domain G, then there exists for f there exists a primitive function F, and we have for every curve G with starting point $\gamma(a)$ and end point $\gamma(b)$:

$$\int_\gamma f(z)\mathrm{d}z = F(\gamma(b)) - F(\gamma(a))\,.$$

Fig. 81.5 γ_1 and γ_2 connect z_1 and z_2

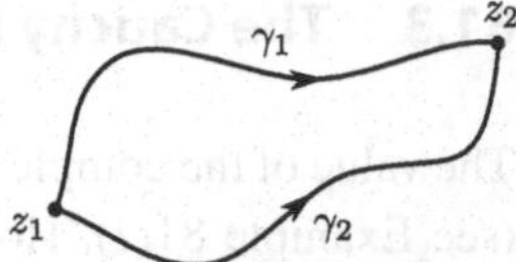

We dispense with the proofs of the Cauchy integral theorem and the integral formula.

The path independence follows so easily from the integral theorem that we do not want to dispense with this proof: We integrate f along the composite and closed curve $\gamma_1+(-\gamma_2)$ and obtain the value 0 according to the Cauchy Integral Theorem, see Fig. 81.5 Thus we have:

$$0 = \int_{\gamma_1} f(z)\mathrm{d}z + \int_{-\gamma_2} f(z)\mathrm{d}z = \int_{\gamma_1} f(z)\mathrm{d}z - \int_{\gamma_2} f(z)\mathrm{d}z\,.$$

From this follows the assertion $\int_{\gamma_1} f(z)\mathrm{d}z = \int_{\gamma_2} f(z)\mathrm{d}z$.

The proof of the existence of a primitive function is done as in the real.

The Cauchy Integral Formula takes a bit of getting used to, so a few comments on it: According to the Cauchy Integral Formula, the values of a holomorphic function f inside a circular disk are completely determined by the values on the edge of that circle. It is important to note that the function f in the interior of γ is holomorphic, the integrand $\frac{f(z)}{z-z_0}$ is not, this is not even explained inz_0, so it is not holomorphic there.

In the following examples we show typical applications of these rather theoretical theorems to make clear that applying these theorems makes the calculation of complex line integrals much easier in many cases.

Example 81.3

- We determine the complex line integral

$$\int_{|z|=1} \frac{\mathrm{e}^z}{z-2}\mathrm{d}z\,.$$

Since the function $g(z) = \frac{\mathrm{e}^z}{z-2}$ is holomorphic on a simple connected domain containing the unit circle $|z| = 1$ is holomorphic, it follows from the Cauchy Integral Theorem

$$\int_{|z|=1} \frac{\mathrm{e}^z}{z-2}\mathrm{d}z = 0\,.$$

- We determine the complex line integral

$$\int\limits_{|z|=3} \frac{e^z}{z-2} dz .$$

Since the function $f(z) = e^z$ is holomorphic on a simple connected domain containing the circle $|z| = 3$ is holomorphic, it follows from the Cauchy Integral Formula

$$\int\limits_{|z|=3} \frac{e^z}{z-2} dz = 2\pi \, i \, e^2 .$$

- We determine the complex line integral

$$\int\limits_{\gamma} z^5 dz \text{ along } \gamma : [0, 1] \to \mathbb{C}, \; \gamma(t) = -i + (1+i)t .$$

Since the function $f(z) = z^5$ is holomorphic on a simple connected domain containing the segment $\gamma([0, 1])$ is holomorphic, it follows from the existence of the primitive function $F(z) = z^6/6$

$$\int\limits_{\gamma} z^5 dz = F(\gamma(1)) - F(\gamma(0)) = F(1) - F(i) = 1/6 + 1/6 = 1/3 .$$

■

We list some important properties of holomorphic functions in the following box. These properties essentially follow from the Cauchy Integral Theorem:

Important Properties of Holomorphic Functions

- **Mean value property:** If $f : G \to \mathbb{C}$ is holomorphic and $\{z \in \mathbb{C} \mid |z - z_0| \leq r\} \subseteq G$ then the value $f(z_0)$ is the mean of the function values on the boundary of the circle.

$$f(z_0) = \frac{1}{2\pi} \int\limits_0^{2\pi} f(z_0 + r\, e^{it}) dt .$$

(continued)

- **Goursat's Theorem:** Any holomorphic function is differentiable (complex) any number of times.
- **Power series representation of holomorphic functions:** If $f : G \to \mathbb{C}$ holomorphic, $z_0 \in G$ and $B_r(z_0) \subseteq G$, then f in the interior of $B_r(z_0)$ can be developed into a Taylor series, i.e.

$$f(z) = \sum_{k=0}^{\infty} \frac{f^{(k)}(z_0)}{k!}(z - z_0)^k\,, \quad \text{where}$$

$$f^{(k)}(z_0) = \frac{k!}{2\pi\,\mathrm{i}} \oint\limits_{|z-z_0|=r} \frac{f(z)}{(z - z_0)^{k+1}}\mathrm{d}z\,.$$

- **The maximum principle:** If $f : G \to \mathbb{C}$ holomorphic and has $|f|$ in $z_0 \in G$ a maximum, then f constant on G.
- **Liouville's Theorem:** Any bounded integer function is constant.
- **The identity Theorem:** If $f, g : G \to \mathbb{C}$ holomorphic and $f(z_n) = g(z_n)$ for all sequence members of a nonconstant, in G convergent sequence $(z_n)_n$, then follows $f(z) = g(z)$ for all $z \in G$.

Note that with the formula of the power series representation of holomorphic functions, we again have a formula for calculating line integrals:

$$\oint\limits_{|z-z_0|=r} \frac{f(z)}{(z - z_0)^{k+1}}\mathrm{d}z = \frac{2\pi\,\mathrm{i}\, f^{(k)}(z_0)}{k!}\,.$$

With the help of these theorems we can derive other important statements:

- From Liouville's Theorem we can easily deduce the **Fundamental Theorem of Algebra**: *Every non-constant polynomial (over $\mathbb{C}$) has a zero in $\mathbb{C}$.* If $f = a_n z^n + \cdots + a_1 z + a_0$ is a non-constant polynomial, then it follows because of $|f(z)| \to \infty$ for $|z| \to \infty$ from the assumption, f would have no zero in $\mathbb{C}$ the entireness and boundedness of the function $g(z) = \frac{1}{f(z)}$. The function $g(z)$ and so $f(z)$ would then be constant according to Liouville's theorem—a contradiction.
- It follows from the Identity Theorem: For each zero z_0 of a holomorphic function on a domain G unequal to the zero function there is a circular disc $B_\varepsilon(z_0)$ in which no further zero lies (one says: the zeros of f are *isolated*). If this were not so, one could have a nonconstant, vs. z_0 convergent sequence (z_n) in G for which $f(z_n) = 0$ for all n. With the Identity Theorem follows $f = 0$, a contradiction.

In the following examples we determine the power series representations of some holomorphic functions. For the construction of the Taylor series one uses only in the rarest cases the formula in the box given above, one resorts rather to the well-known Taylor expansion of simple functions. Mostly the geometric series plays the decisive role.

Recipe: Determining Taylor Series of Holomorphic Functions
Mostly you get the Taylor series of a holomorphic function

$$f = \frac{p(z)}{q(z)} = \frac{\tilde{p}(z-z_0)}{\tilde{q}(z-z_0)}$$

(p and q or $\tilde{p}$ and $\tilde{q}$ are usually polynomials, but need not be) at the developing point z_0 in one of the following ways:

(a) Possibly f can be written in the form

$$f(z) = \frac{p_1(z-z_0)}{q_1(z-z_0)} + \cdots + \frac{p_r(z-z_0)}{q_r(z-z_0)}$$

where the sum can also consist of only one summand and each summand can be expressed by known power series.

(b) The approach $f(z) = \frac{\tilde{p}(z-z_0)}{\tilde{q}(z-z_0)} = \sum_{k=0}^{\infty} a_k(z-z_0)^k$ yields after multiplication by $\tilde{q}(z-z_0)$

$$\tilde{p}(z-z_0) = \tilde{q}(z-z_0)\sum_{k=0}^{\infty} a_k(z-z_0)^k .$$

From this one obtains in particular in the case of a polynomial $\tilde{p}(z-z_0)$ by equating coefficients the coefficients a_k and thus the Taylor series for $f(z)$.

Example 81.4

- The Taylor series of exp, sin and cos in $z_0 = 0$ are given by

$$\exp(z) = \sum_{k=0}^{\infty} \frac{z^k}{k!}, \quad \sin(z) = \sum_{k=0}^{\infty} \frac{(-1)^k}{(2k+1)!} z^{2k+1}, \quad \cos(z) = \sum_{k=0}^{\infty} \frac{(-1)^k}{(2k)!} z^{2k} .$$

- The power series representation of $f(z) = \frac{1}{1-z}$ in $z_0 = 0$ is on the circle $|z| < 1$ given by

$$\frac{1}{1-z} = \sum_{k=0}^{\infty} z^k .$$

- We determine the Taylor series to be $f(z) = \frac{1}{2+3z}$, $z \neq -2/3$, at the development point $z_0 = 2$. To do this, we use method (a) of the above recipe:

$$f(z) = \frac{1}{2+3z} = \frac{1}{2+3(z-2)+6} = \frac{1}{8+3(z-2)} = \frac{1}{8}\frac{1}{1-(-\frac{3}{8}(z-2))} .$$

We now substitute the expression $-\frac{3}{8}(z-2)$ into the formula for the geometric series and obtain:

$$f(z) = \frac{1}{8}\sum_{n=0}^{\infty}(-1)^n\left(\frac{3}{8}\right)^n (z-2)^n = \sum_{n=0}^{\infty}\frac{(-3)^n}{8^{n+1}}(z-2)^n .$$

- We determine the Taylor series of the holomorphic function $f(z) = \frac{z-1}{z^2+2}$, $z \neq \pm\sqrt{2}\,\mathrm{i}$, at the development point $z_0 = 0$. In doing so, we use method (b) of the above recipe by equating coefficients:

$$\begin{aligned} z-1 &= (z^2+2)\sum_{n=0}^{\infty} a_n z^n = \sum_{n=0}^{\infty} a_n z^{n+2} + \sum_{n=0}^{\infty} 2a_n z^n \\ &= \sum_{n=2}^{\infty} a_{n-2} z^n + \sum_{n=0}^{\infty} 2a_n z^n = 2a_0 + 2a_1 z + \sum_{n=2}^{\infty}(a_{n-2}+2a_n)\, z^n . \end{aligned}$$

Equating coefficients yields:

$$a_0 = -1/2\,,\ a_1 = 1/2\,,\ a_2 = 1/4\,,\ a_3 = -1/4\,,\ a_4 = -1/8\,,\ a_5 = 1/8\,,\ldots$$

■

81.4 Exercises

81.1 Compute $\int_\gamma \operatorname{Re} z \, \mathrm{d}z$ along the two sketched paths γ:

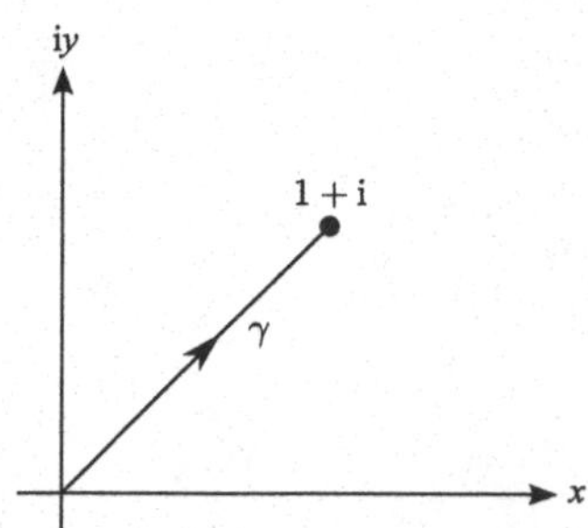

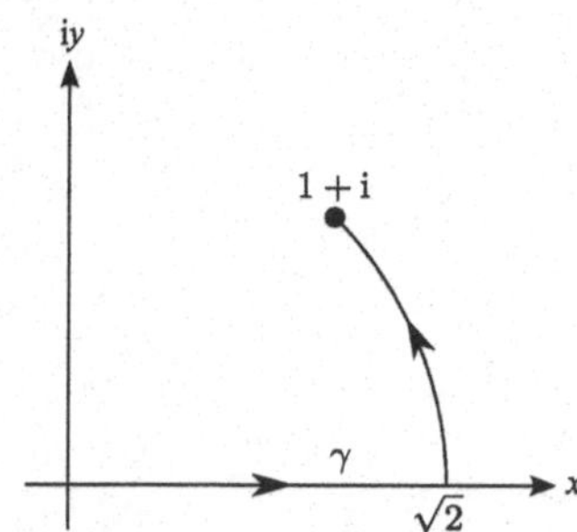

81.2 One computes $\oint_\gamma \frac{1}{z^2+1} \mathrm{d}z$ for the 4 circles

$$\gamma : \ |z| = \frac{1}{2}, \quad |z| = 2, \quad |z - \mathrm{i}| = 1, \quad |z + \mathrm{i}| = 1 .$$

81.3 Calculate

(a) $\oint_{|z-1|=1} \frac{z\,\mathrm{e}^z}{(z-a)^3} \mathrm{d}z, \ |a| < 1,$

(b) $\oint_\gamma \frac{5z^2 - 3z + 2}{(z-1)^3} \mathrm{d}z$, where γ is a double-point free closed curve containing the point $z = 1$ in the interior.

81.4 Show for $a, \ b \in \mathbb{R} \setminus \{0\}$:

$$\int_0^{2\pi} \frac{1}{a^2 \cos^2 t + b^2 \sin^2 t} \mathrm{d}t = \frac{2\pi}{ab} .$$

Note: Integrate $\frac{1}{z}$ along the ellipse $\frac{x^2}{a^2} + \frac{y^2}{b^2} = 1$ or along $|z| = 1$.

81.5 Determine the Taylor series around the developing point $z_0 = 0$ of the function

$$f(z) = \frac{1 + z^3}{2 - z}, \ z \in \mathbb{C} \setminus \{2\} .$$

Laurent Series

82

We generalize power series to *Laurent series* by allowing negative exponents as well,

$$\sum_{k=0}^{\infty} c_k (z - z_0)^k \longrightarrow \sum_{k=-\infty}^{\infty} c_k (z - z_0)^k .$$

We do not do this arbitrarily, there is a close connection with the functions described by it: Power series describe functions that are holomorphic on a circle around z_0, Laurent series describe functions that are holomorphic on a *circular ring* around z_0. With the Laurent *series development* we obtain a series representation of functions with singularities. The main application of this development is the *Residue Calculus*, which we introduce in the next chapter.

82.1 Singularities

A *singularity of* a function is, in a sense, a zero of a denominator. We will distinguish three kinds of singularities, namely singularities of *order* 0 or *finite order* or *infinite order*. We will be more precise:

C. Karpfinger, *Calculus and Linear Algebra in Recipes*,
https://doi.org/10.1007/978-3-662-65458-3_82

Removeable, Not Removable and Essential Singularities
If $f : G \setminus \{z_0\} \to \mathbb{C}$ is a holomorphic function not defined in $z_0 \in G$, then z_0 is called an **isolated singularity**. In this case, one calls

- z_0 a **removable singularity** if f is bounded on a punctured neighborhood U of z_0,
- z_0 a **pole** if $(z - z_0)^m f(z)$ has for a $m \geq 1$ a removeable singularity in z_0. The smallest such m is called the **order of the pole**,
- z_0 an **essential singularity** else.

The set $G \setminus \{z_0\}$ is also called a **punctured neighborhood** of z_0.

Example 82.1

- The function $f(z) = \frac{\sin(z)}{z}$ has in $z_0 = 0$ a removeable singularity, with the definition $f(0) = 1$ f is holomorphic on $\mathbb{C}$.
- The function $f(z) = \frac{z}{(z-2)^3}$ has in $z_0 = 2$ a pole of order 3.
- The function $f(z) = \sin(\frac{1}{z})$ has an essential singularity in $z_0 = 0$. Namely, if one chooses $z = \frac{-\mathrm{i}}{t}$, $t \in (0, \infty)$, then for all $m \in \mathbb{N}$:

$$|\sin\left(\frac{1}{z}\right)| = \frac{1}{2}\,|\mathrm{e}^{\mathrm{i}\frac{1}{z}} - \mathrm{e}^{-\mathrm{i}\frac{1}{z}}| = \frac{1}{2}\,|\mathrm{e}^{-t} - \mathrm{e}^{t}| \to \infty \text{ for } t \to \infty$$

and

$$|z^m \sin\left(\frac{1}{z}\right)| = \frac{1}{2}\,|t^{-m}(\mathrm{e}^{-t} - \mathrm{e}^{t})| \to \infty \text{ for } t \to \infty\,.$$

- The function $f(z) = \cot(\frac{\pi}{z}) : \mathbb{C}\setminus\{z \mid \frac{1}{z} \in \mathbb{Z}\}\cup\{0\} \to \mathbb{C}$ has the rational numbers of the form $\frac{1}{k}$, $k \in \mathbb{Z}\setminus\{0\}$, as isolated singularities. Note: The zero is not an isolated singularity of f, in fact there is no punctured neighborhood of zero in which f is holomorphic. ■

In the following we develop functions which are holomorphic in a punctured neighborhood of z_0 and have an isolated singularity in z_0 into a *Laurent series* around z_0. The nature of the singularity can then be easily seen from the *Laurent series* of this function.

82.2 Laurent Series

Holomorphy is a strong property: If f is holomorphic on G, then f is complex differentiable any number of times, and for every $z_0 \in G$ according to the theorem in Sect. 81.3 can be developed into a power series,

$$f(z) = \sum_{k=0}^{\infty} c_k (z - z_0)^k .$$

This development of holomorphic functions in power series around z_0 can be generalized to functions with isolated singularities z_0 by *Laurent series*. Here Laurent series are explained as follows:

Laurent Series
A **Laurent series** is a series of the form

$$\sum_{k=-\infty}^{\infty} c_k (z - z_0)^k .$$

One calls

- is the complex number z_0 is the **development point** of the Laurent series,
- the series $\sum_{k=-\infty}^{-1} c_k (z - z_0)^k$ the **principal part** of the Laurent series and
- the series $\sum_{k=0}^{\infty} c_k (z - z_0)^k$ the **regular part** of the Laurent series.

We want to introduce a meaningful notion of convergence for a Laurent series: To do this, we consider a Laurent series

$$\sum_{k=-\infty}^{\infty} c_k (z - z_0)^k$$

and note that

- the regular part $\sum_{k=0}^{\infty} c_k (z - z_0)^k$ is an ordinary power series (see Chap. 24). As is well known, every such power series has a radius of convergence. Let this be for the regular part of the Laurent series under consideration $R \in [0, \infty]$. The regular part converges for

$$|z - z_0| < R ,$$

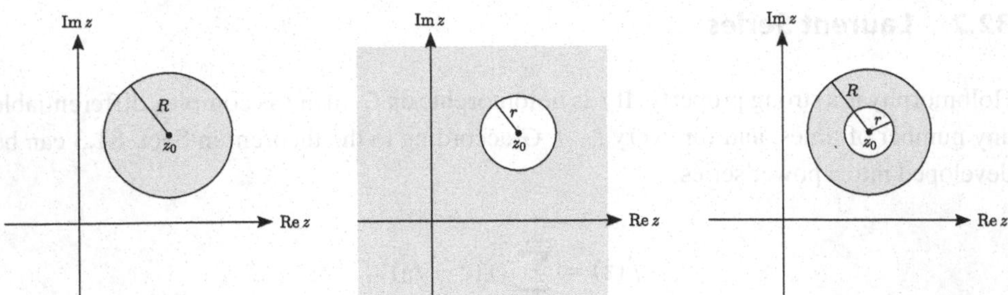

Fig. 82.1 The convergence region of a Laurent series is typically an annulus

- the principal part $\sum_{k=-\infty}^{-1} c_k(z-z_0)^k$ is a power series in $w = \frac{1}{z-z_0}$ that is

$$\sum_{k=-\infty}^{-1} c_k(z-z_0)^k = \sum_{k=1}^{\infty} d_k\, w^k \text{ with } d_k = c_{-k}\,.$$

Let the radius of convergence of this power series be $\frac{1}{r} \in [0, \infty]$ with $\frac{1}{0} = \infty$ and $\frac{1}{\infty} = 0$. The principal part converges for

$$|w| < \frac{1}{r} \Leftrightarrow |z - z_0| > r\,.$$

If now $0 \leq r < R \leq \infty$, then the principal and regular parts of the Laurent series converge on the **annulus** with center z_0,

$$K_{r,R}(z_0) = \{z \in \mathbb{C} \mid 0 \leq r < |z - z_0| < R \leq \infty\}\,.$$

Outside this annulus, either the principal or the regular part diverges. On the boundaries no general statement can be made.

Figure 82.1 shows the situation with the annulus.

Now we will call a Laurent series *convergent* if both the principal and the regular part converge in the sense just explained. Thus Laurent series explain functions on annuli, just as power series explain functions on circles. As with power series, the derivative or integral of a Laurent series is obtained by differentiating or integrating term-by-term. We also mention the designation of the coefficient c_{-1} of a Laurent series:

Laurent Series—Convergence, Derivative, Integral, Residual
It is said that the Laurent series

$$\sum_{k=-\infty}^{\infty} c_k (z - z_0)^k$$

converges if the principal and regular parts of the Laurent series converge.

If R is the radius of the circle of convergence of the regular and r is the radius of the convergence *circle* of the regular part, then the Laurent series converges on the annulus $K_{r,R}(z_0)$ with center z_0,

$$K_{r,R}(z_0) = \{z \in \mathbb{C} \,|\, 0 \le r < |z - z_0| < R \le \infty\}.$$

Further, we have for the function $f : K_{r,R}(z_0) \to \mathbb{C}$, $f(z) = \sum\limits_{k=-\infty}^{\infty} c_k(z - z_0)^k$:

- f is differentiable, the derivative function f' is obtained by differentiation term-by-term,

$$f'(z) = \sum_{k=-\infty}^{\infty} k\, c_k (z - z_0)^{k-1},$$

  and f' also converges on $K_{r,R}(z_0)$.
- f is in the case $c_{-1} = 0$ integrable, a primitive function F is obtained by integrating term-by-term,

$$F(z) = \sum_{\substack{k=-\infty \\ k \neq -1}}^{\infty} \frac{c_k}{k+1} (z - z_0)^{k+1},$$

  and F also converges to $K_{r,R}(z_0)$.
- The coefficient c_{-1} is called **residue** of the series $f(z) = \sum_{k=-\infty}^{\infty} c_k(z - z_0)^k$ at the point z_0; we write for it

$$c_{-1} = \mathrm{Res}_{z_0} f.$$

Example 82.2

- The Laurent series

$$f(z) = 2z^2 + \frac{1}{z} - \frac{3}{z^3}$$

has at $z_0 = 0$ the residue $\mathrm{Res}_0 f = c_{-1} = 1$.
- As is well known, we have with the geometric series:

$$\frac{-1}{z-1} = \frac{1}{1-z} = \sum_{k=0}^{\infty} z^k .$$

Thus the Laurent series $f(z) = \frac{-1}{z-1}$ has at the point $z_0 = 1$ the residue -1 and at the point $z_0 = 0$ the residue 0,

$$\mathrm{Res}_1 f = -1 \text{ and } \mathrm{Res}_0 f = 0 .$$

■

82.3 Laurent Series Development

We now generalize the Taylor development of a function holomorphic on a domain G to a *Laurent series development* of a function holomorphic on an annulus K:

Laurent Series Development
If the function $f : K \to \mathbb{C}$ is holomorphic on the annulus

$$K = K_{r,R}(z_0) = \{z \in \mathbb{C} \,|\, 0 \le r < |z - z_0| < R \le \infty\},$$

then f on $K_{r,R}(z_0)$ can be uniquely represented as a Laurent series,

$$f(z) = \sum_{k=-\infty}^{\infty} c_k (z - z_0)^k \text{ with } c_k \in \mathbb{C} .$$

(continued)

One says, f can be developed **into a Laurent series**, and speaks of the **Laurent series development** of f. The coefficients c_k, $k \in \mathbb{Z}$, are obtained by

$$c_k = \frac{1}{2\pi\,\mathrm{i}} \oint_{|z-z_0|=\rho} \frac{f(z)}{(z-z_0)^{k+1}}\mathrm{d}z\,,$$

where $\rho \in (r, R)$ can be chosen arbitrarily.

The coefficients c_k, $k \in \mathbb{Z}$, of the Laurent series expansion are usually not determined with the given formula. As in determining the Taylor development of a holomorphic function, one also traces the Laurent series development back to known series; for this purpose, the geometric series is particularly useful, whose Laurent series around $z_0 = 0$ is well known:

$$\frac{1}{1-z} = \sum_{k=0}^{\infty} z^k \text{ for } |z| < 1\,.$$

Note that here we have two Laurent series representations of the same function; these are developments about different points. On the left we have the Laurent series coefficients c_k with $c_{-1} = -1$ and $c_k = 0$ otherwise, on the right we have Laurent series coefficients $\tilde{c}_k = 1$ for $k \geq 0$ and $\tilde{c}_k = 0$ for $k < 0$.

Example 82.3

- We determine the Laurent series development of the function $f(z) = \frac{1}{z-a}$ which has a pole of order 1 in a in $z_0 = 0$. Since in the case $a = 0$ the function $f(z)$ is already given by its Laurent series $f(z) = 1/z$, we may assume $a \neq 0$.
  - Because of

$$\frac{1}{z-a} = -\frac{1}{a}\frac{1}{1-z/a}$$

we obtain the Laurent series development

$$f(z) = \frac{1}{z-a} = \sum_{k=0}^{\infty}\left(-\frac{1}{a^{k+1}}\right) z^k \text{ for } |\frac{z}{a}| < 1\,, \text{ d.h. } |z| < |a|$$

in the *annulus* $|z| < |a|$ (we have $R = |a|$ and $r = 0$).

– The function $f(z) = \frac{1}{z-a}$ is also holomorphic on the *annulus* $|z| > a$. Because of

$$\frac{1}{z-a} = \frac{1}{z}\frac{1}{1-a/z}$$

we obtain the Laurent series development

$$f(z) = \frac{1}{z-a} = \sum_{k=0}^{\infty} a^k \left(\frac{1}{z^{k+1}}\right) \quad \text{for } |\frac{a}{z}| < 1, \text{ i. e. } |z| > |a|$$

in the *annulus* $|z| > |a|$ (we have $R = \infty$ and $r = |a|$).

- We determine the Laurent series development of the function $f(z) = \frac{1}{(z-2)(z-3)}$ which has two poles of order 1 in 2 and 3 in $z_0 = 0$. Because of

$$\frac{1}{(z-2)(z-3)} = \frac{1}{z-3} - \frac{1}{z-2}$$

we obtain with the above example the Laurent series

$$\begin{aligned} f(z) &= \frac{1}{(z-2)(z-3)} = \sum_{k=0}^{\infty}\left(-\frac{1}{3^{k+1}}\right) z^k - \sum_{k=0}^{\infty}\left(-\frac{1}{2^{k+1}}\right) z^k \\ &= \sum_{k=0}^{\infty}\left(\frac{1}{2^{k+1}} - \frac{1}{3^{k+1}}\right) z^k \end{aligned}$$

in the *annulus* $|z| < |2|$ (we have $R = 2$ and $r = 0$).

- We determine the Laurent series development of the function $f(z) = \mathrm{e}^{1/z}$ in $z_0 = 0$ which has an essential singularity in z_0. Using the well-known Taylor development of the exponential function, we obtain

$$f(z) = \mathrm{e}^{1/z} = \sum_{k=0}^{\infty} \frac{(1/z)^k}{k!} = \sum_{k=-\infty}^{0} \frac{1}{(-k)!} z^k .$$

■

From the Laurent series development of a function f in a singularity z_0 of f, the type of the singularity can be read off, namely:

Classification of Singularities
If $\sum_{k=-\infty}^{\infty} c_k(z - z_0)^k$ is the Laurent series development of a function f in a singularity z_0 of f, then we have

- z_0 is removeable if $c_k = 0$ for all $k < 0$,
- z_0 is a pole m-*th* order, if $c_k = 0$ for all $k < -m$ and $c_{-m} \neq 0$,
- z_0 is an essential singularity if $c_k \neq 0$ for infinity many $k < 0$.

82.4 Exercises

82.1 Where do the series $\sum_{n=1}^{\infty} a_n z^{2n}$ and $\sum_{n=1}^{\infty} a_n z^{-n}$ converge for $a_n = n$ or $a_n = \frac{1}{(2n)!}$?

82.2 Give for $f(z) = \frac{1}{z^2 - \mathrm{i}z}$ all possible developments by powers of $z + \mathrm{i}$. Which representation converges for $z = 1/2$?

82.3 Calculate the Laurent series of

(a) $\cosh \frac{1}{z^2}$ in $z_0 = 0$,
(b) $\frac{1}{1-\cos z}$ for $0 < |z| < 2\pi$ in $z_0 = 0$ (the first summands other than 0 will do),
(c) $\frac{\mathrm{e}^z}{z-1}$ in $z_0 = 0$.

82.4 Determine in each case the Laurent series of $f(z)$ with the development point $z_0 = 0$ and give the areas of convergence:

(a) $f(z) = \frac{1}{z^2 - 3z + 2}$,
(b) $f(z) = \frac{\sin z}{z^3}$.

The Residual Calculus

83

The *residue* is a complex number associated with a function in a point z_0. The Residue Theorem provides a formula for calculating the sum of the residues of a complex function within a closed curve and thus, as an application, provides another method for calculating complex curve integrals. The essential tool here is the Laurent series expansion of the integrand.

Another application of the Residue Theorem is the determination of real integrals, where real methods often fail.

83.1 The Residue Theorem

We now introduce the residue of a function f at a point z_0. We do this with the help of the already defined notion of *the residue of a Laurent series of f in z_0* (see Sect. 82.2): Develop the holomorphic function $f : G \setminus \{z_0\} \to \mathbb{C}$ into a Laurent series around z_0, which is always possible according to the Theorem on Laurent series expansion in Sect. 82.3:

Residual of a Function in a Point z_0

If $f : G \setminus \{z_0\} \to \mathbb{C}$ is a holomorphic function with the Laurent series

$$f(z) = \sum_{k=-\infty}^{\infty} c_k (z - z_0)^k$$

(continued)

C. Karpfinger, *Calculus and Linear Algebra in Recipes*,
https://doi.org/10.1007/978-3-662-65458-3_83

in z_0, then the Laurent series coefficient c_{-1} is called the **residue** of f in z_0, one writes

$$\mathrm{Res}_{z_0} f = c_{-1}\,.$$

Determining the residue of a holomorphic function on a punctured neighborhood $G \setminus \{z_0\}$ is the central problem in the following considerations. We therefore give recipe-like the essential methods to determine this number:

Recipe: Determine the Residue of a Function f in z_0
The residue $\mathrm{Res}_{z_0} f$ of a holomorphic function $f : G \setminus \{z_0\} \to \mathbb{C}$ in z_0 is obtained in the following ways:

(a) Determine the Laurent series expansion

$$f(z) = \sum_{k=-\infty}^{\infty} c_k (z - z_0)^k\,.$$

Then $c_{-1} = \mathrm{Res}_{z_0} f$ is the residue we are looking for.

(b) Determine the line integral

$$c_{-1} = \frac{1}{2\pi\,\mathrm{i}} \oint_{\gamma} f(z)\mathrm{d}z$$

along any positively oriented and closed curve γ which is in the holomorphic domain $G \setminus \{z_0\}$ of f.

(c) If $f(z) = \frac{g(z)}{h(z)}$ and if $h(z)$ has in z_0 a zero of order 1, i.e. $h(z_0) = 0$ and $h'(z_0) \neq 0$, then

$$\mathrm{Res}_{z_0} f = \frac{g(z_0)}{h'(z_0)}\,.$$

(d) Is $f(z) = \frac{g(z)}{(z-z_0)^m}$ with $g(z_0) \neq 0$, then we have

$$\mathrm{Res}_{z_0} f = \frac{1}{(m-1)!}\, g^{(m-1)}(z_0)\,.$$

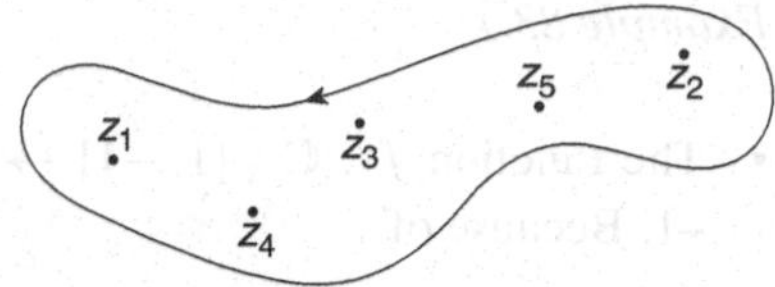

Fig. 83.1 The singularities lie in the interior of the positively oriented curve

We will try out the methods with examples. However, so that this immediately receives a sense and purpose, we first give the *Residue Theorem*. With this Theorem, complex line integrals are determined by determining residues. The idea is quite easy to follow: According to method (b) in the above recipe, we find the residue $c_{-1} = \text{Res}_{z_0} f$ of f in z_0 by determining the integral of the curve

$$\frac{1}{2\pi\,\mathrm{i}} \oint_\gamma f(z)\mathrm{d}z\,.$$

Now we simply turn the tables: We determine the residue $\text{Res}_{z_0} f$ using one of the other methods (a), (c), (d) and then, thanks to method (b), obtain the value of the line integral

$$\oint_\gamma f(z)\mathrm{d}z = 2\pi\,\mathrm{i}\,\text{Res}_{z_0} f\,.$$

This can be generalized to a function f, which on a domain G has finitely many singularities (see Fig. 83.1):

The Residue Theorem

If G is a domain and if $f : G\setminus\{z_1, \ldots, z_n\} \to \mathbb{C}$ for $z_1, \ldots, z_n \in G$ is holomorphic, then for any positively oriented, double point free and closed curve γ in G with $z_1, \ldots, z_n$ in its interior we have:

$$\oint_\gamma f(z)\mathrm{d}z = 2\pi\,\mathrm{i} \sum_{k=1}^{n} \text{Res}_{z_k} f\,.$$

In order to calculate the complex line integral $\oint_\gamma f(z)\mathrm{d}z$ of a holomorphic function on $G \setminus \{z_1, \ldots, z_n\}$, we have to determine *only* the residues $\text{Res}_{z_1} f, \ldots, \text{Res}_{z_n} f$.

Example 83.1

- The function $f : \mathbb{C} \setminus \{1, -1\} \to \mathbb{C}$, $f(z) = \frac{1}{z-1} - \frac{2}{z+1}$ has first-order poles in 1 and -1. Because of

$$\mathrm{Res}_1 f = 1 \text{ and } \mathrm{Res}_{-1} f = -2$$

we obtain for any positively oriented closed curve γ with 1 and -1 in its interior:

$$\oint_\gamma f(z)\mathrm{d}z = 2\pi\,\mathrm{i}\cdot(1 + (-2)) = -2\pi\,\mathrm{i}\,.$$

- The function $f : \mathbb{C} \setminus \{\mathrm{i}, -\mathrm{i}\} \to \mathbb{C}$, $f(z) = \frac{1}{(z-\mathrm{i})^2(z+\mathrm{i})^2}$ has 2nd order poles in i and $-$ i. Because of

$$f(z) = \frac{(z+\mathrm{i})^{-2}}{(z-\mathrm{i})^2}$$

we obtain with $g(z) = (z+\mathrm{i})^{-2}$, $g'(z) = -2(z+\mathrm{i})^{-3}$ and method (d) for $m = 2$:

$$\mathrm{Res}_{\mathrm{i}} f = \frac{1}{4\,\mathrm{i}} \text{ and analogously } \mathrm{Res}_{-\mathrm{i}} f = \frac{-1}{4\,\mathrm{i}}\,.$$

Thus, for any curve γ with i and $-$ i in its interior we have

$$\oint_\gamma f(z)\mathrm{d}z = 2\pi\,\mathrm{i}\left(\frac{1}{4\,\mathrm{i}} - \frac{1}{4\,\mathrm{i}}\right) = 0\,.$$

- The function $f : \mathbb{C} \setminus \{\mathrm{i}, -\mathrm{i}\} \to \mathbb{C}$, $f(z) = \frac{\mathrm{e}^z}{z^2+1}$ has first order poles in i and $-$ i. We obtain with $g(z) = \mathrm{e}^z$, $h(z) = z^2 + 1$ and $h'(z) = 2z$ and method (c)

$$\mathrm{Res}_{\mathrm{i}} f = \frac{\mathrm{e}^{\mathrm{i}}}{2\,\mathrm{i}} \text{ and analogously } \mathrm{Res}_{-\mathrm{i}} f = \frac{-\,\mathrm{e}^{-\mathrm{i}}}{2\,\mathrm{i}}\,.$$

Thus, for any curve γ that has encloses i and $-$ i positively oriented we have:

$$\oint_\gamma f(z)\mathrm{d}z = \frac{2\pi\,\mathrm{i}}{2\,\mathrm{i}}(\mathrm{e}^{\mathrm{i}} - \mathrm{e}^{-\mathrm{i}}) = \pi(\mathrm{e}^{\mathrm{i}} - \mathrm{e}^{-\mathrm{i}})\,.$$

■

If a curve goes around a singularity several times, not everything is lost. Here is only to count, how often the singularity is circled, correspondingly often you count the residue of f in this singularity to get the value of the complex curve integral.

The Residue Theorem provides another way to determine a complex line integral along a closed curve. We give an overview of the methods to determine a complex line integral:

Recipe: The Methods for Calculating a Complex Line Integral

To be determined is the value of the complex line integral $\int_\gamma g(z)\mathrm{d}z$.

To do this, note:

(1)

$$\int_\gamma g(z)\mathrm{d}z = 0\,,$$

if γ is closed and $g(z)$ is holomorphic in a simply connected domain where the curve γ lies.

(2)

$$\int_\gamma g(z)\mathrm{d}z = G(\gamma(b)) - G(\gamma(a))\,,$$

if $g(z)$ is holomorphic in a simply connected region in which the curve γ from the starting point $\gamma(a)$ to the endpoint $\gamma(b)$ lies. G is a primitive function of g.

(3)

$$\int_\gamma g(z)\mathrm{d}z = 2\pi\,\mathrm{i}\,f(z_0)\,,\ \text{ if } g(z) = \frac{f(z)}{z - z_0}\,,$$

where f is holomorphic in a simply connected domain containing the closed, double-point free and positively oriented curve γ enclosing the point z_0.

(4)

$$\oint_\gamma g(z)\mathrm{d}z = \frac{2\pi\,\mathrm{i}\,f^{(k)}(z_0)}{k!}\,,\ \text{ if } g(z) = \frac{f(z)}{(z - z_0)^{k+1}}\,,$$

where f is holomorphic in a simply connected domain G containing the closed, double-point free and positively oriented curve γ enclosing the point z_0.

(continued)

(5)

$$\oint_{\gamma} g(z)\mathrm{d}z = 2\pi\,\mathrm{i} \sum_{k=1}^{n} \mathrm{Res}_{z_k} g\,,$$

if g is holomorphic on $G \setminus \{z_1, \ldots, z_n\}$ and γ is a closed, double point free and positively oriented curve in G which has the points $z_1, \ldots, z_n$ in its interior.

83.2 Calculation of Real Integrals

The Residue Theorem can also be used to compute certain (even improper) real integrals. In particular, for the following two typical example classes, the Residue Theorem provides an elegant method for computing the value of a

- improper integral or Cauchy principal value,

$$\int_{-\infty}^{\infty} \frac{p(x)}{q(x)}\mathrm{d}x\,,$$

- and a trigonometric integral

$$\int_{0}^{2\pi} R(\cos t, \sin t)\mathrm{d}x$$

with a rational function R.

For the (exact) calculation of a real integral $I = \int_a^b f(x)\mathrm{d}x$ one proceeds as follows: One determines a primitive function $F(x)$ of $f(x)$ and then inserts the integration limits,

$$I = F(b) - F(a) \text{ or } I = \lim_{R\to\infty} F(R) - F(-R)\,.$$

In general, the difficult or impossible part is to determine the primitive function F. When calculating these integrals with the Residue Theorem, this determination of the primitive function is completely omitted: We compute the residues of the integrand f in the isolated

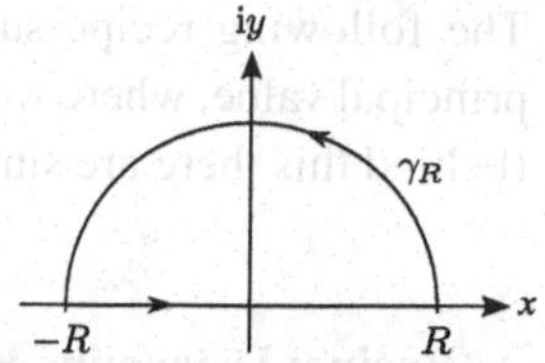

Fig. 83.2 γ_R encloses the singularities

singularities $z_1, \ldots, z_n$ within a closed, positively oriented curve γ, add them up and obtain the value I of the searched integral in the form

$$I = \int_a^b f(x)\mathrm{d}x = \int_\gamma f(z)\mathrm{d}z = 2\pi\,\mathrm{i} \sum_{i=1}^n \mathrm{Res}_{z_i} f\,.$$

We explain the idea for the computation of the Cauchy principal value: Here we consider a quotient $f(x) = \frac{p(x)}{q(x)}$ of two polynomials p and q with $\deg q \geq \deg p + 2$ and $q(x) \neq 0$ for all $x \in \mathbb{R}$.

We consider the integration path in Fig. 83.2 along the closed curve γ_R of $-R$ to R on the real axis and the arc, which includes all singularities $z_1, \ldots, z_n$ from f in the upper half-plane.

Now, for the complex curve integral with the Residue Theorem and because of the additivity of the line integral, we have

$$\int_{\gamma_R} \frac{p(z)}{q(z)}\mathrm{d}z = \int_{\gamma_1} \frac{p(z)}{q(z)}\mathrm{d}z + \int_{\gamma_2} \frac{p(z)}{q(z)}\mathrm{d}z = 2\pi\,\mathrm{i} \sum_{k=1}^n \mathrm{Res}_{z_k} f\,,$$

where γ_1 connects $-R$ and R on the real axis and γ_2 connects R and $-R$ along the semicircle. Now, because of the assumption $\deg q \geq \deg p + 2$ we have with a constant c:

$$\left|\int_{\gamma_2} \frac{p(z)}{q(z)}\mathrm{d}z\right| \leq \max_{z\in\gamma_2} \left|\frac{p(z)}{q(z)}\right| \pi\, R = \frac{c}{R^2}\pi\, R = \frac{c\pi}{R} \overset{R\to\infty}{\longrightarrow} 0\,.$$

Thus it follows

$$\int_{-\infty}^{\infty} \frac{p(x)}{q(x)}\mathrm{d}x = 2\pi\,\mathrm{i} \sum_{k=1}^n \mathrm{Res}_{z_k} f\,.$$

The following recipe summarizes how to perform this computation of such a Cauchy principal value, where we also indicate at once how to treat the second type of real integrals (behind this there are similar considerations as those just described):

Recipe: Determine Real Integrals with the Residue Calculus
The value of the integral

$$\int_{-\infty}^{\infty} \frac{p(x)}{q(x)} \mathrm{d}x \quad \text{or} \quad \int_{0}^{2\pi} R(\cos t, \sin t)\mathrm{d}t$$

is obtained as follows:

- $\int_{-\infty}^{\infty} \frac{p(x)}{q(x)} \mathrm{d}x$: If $\deg q \geq \deg p + 2$ and $q(x) \neq 0$ for all $x \in \mathbb{R}$ so:

  (1) Determine the singularities $z_1, \ldots, z_n$ of the complex function $f(z) = \frac{p(z)}{q(z)}$ in the upper half-plane, $\mathrm{Im}(z_i) > 0$.
  (2) Determine the residues of $f(z)$ in the singularities $z_1, \ldots, z_n$.
  (3) Get the value of the (real) integral

$$\int_{-\infty}^{\infty} \frac{p(x)}{q(x)} \mathrm{d}x = 2\pi \mathrm{i} \sum_{k=1}^{n} \mathrm{Res}_{z_k} f .$$

- $\int_{0}^{2\pi} R(\cos t, \sin t)\mathrm{d}t$: If the rational function R has no singularities on the unit circle $|z| = 1$, then:

  (1) Substitute

$$\frac{1}{2}\left(z + \frac{1}{z}\right) = \cos t\,, \quad \frac{1}{2\mathrm{i}}\left(z - \frac{1}{z}\right) = \sin t\,, \quad \frac{1}{\mathrm{i}z}\mathrm{d}z = \mathrm{d}t$$

  and obtain the complex rational function

$$f(z) = R\left(\frac{1}{2}\left(z + \frac{1}{z}\right), \frac{1}{2\mathrm{i}}\left(z - \frac{1}{z}\right)\right) \frac{1}{\mathrm{i}z}.$$

(continued)

(2) Determine the singularities $z_1, \ldots, z_n$ of the complex function $f(z) = \frac{p(z)}{q(z)}$ within the unit circle $|z| < 1$.
(3) Determine the residuals of $f(z)$ in the singularities $z_1, \ldots, z_n$.
(4) Get the value of the (real) integral

$$\int_0^{2\pi} R(\cos t, \sin t)\mathrm{d}t = 2\pi\,\mathrm{i}\sum_{k=1}^{n} \mathrm{Res}_{z_k} f .$$

Example 83.2

- We determine the value of the real integral

$$\int_{-\infty}^{\infty} \frac{1}{4+x^4}\mathrm{d}x .$$

(1) The complex function $f(z) = \frac{1}{4+z^4}$ has four 1st order poles, where the two poles $z_1 = 1+\mathrm{i}$ and $z_2 = -1+\mathrm{i}$ lie in the upper half plane.
(2) Using method (c) from the above recipe to determine the residues, we get

$$\mathrm{Res}_{z_1} f = \frac{1}{4\,(1+\mathrm{i})^3} = -\frac{1+\mathrm{i}}{16} \text{ and } \mathrm{Res}_{z_2} f = \frac{1}{4\,(-1+\mathrm{i})^3} = -\frac{-1+\mathrm{i}}{16} .$$

(3) This gives us the value of the (real) integral

$$\int_{-\infty}^{\infty} \frac{1}{4+x^4}\mathrm{d}x = 2\pi\,\mathrm{i}\,\frac{\mathrm{i}}{8} = \frac{\pi}{4} .$$

- We determine the value of the real integral

$$\int_0^{2\pi} \frac{1}{2+\sin t}\mathrm{d}t .$$

(1) Through the substitution $\sin t = \frac{1}{2\mathrm{i}}(z - 1/z)$ and $\mathrm{d}t = \frac{1}{\mathrm{i}z}$ we get the complex rational function

$$f(z) = \frac{1}{2 + \frac{1}{2\mathrm{i}}\left(z - \frac{1}{z}\right)} \frac{1}{\mathrm{i}\,z} = \frac{2}{z^2 + 4\,\mathrm{i}\,z - 1}.$$

(2) The function $f(z)$ has two 1st order poles, where only the pole $z_1 = (-2 + \sqrt{3})\,\mathrm{i}$ lies inside the unit circle.

(3) Using the method (c) from the above recipe to determine the residue, we get

$$\mathrm{Res}_{z_1} f = \frac{2}{2\,(-2 + \sqrt{3})\,\mathrm{i} + 4\,\mathrm{i}} = \frac{1}{\mathrm{i}\sqrt{3}}.$$

(4) This gives us the value of the (real) integral

$$\int_0^{2\pi} \frac{1}{2 + \sin t}\mathrm{d}t = 2\pi\,\mathrm{i}\,\frac{1}{\mathrm{i}\sqrt{3}} = \frac{2\pi}{\sqrt{3}}.$$

■

83.3 Exercises

83.1 Determine for the following functions $f(z)$, the location and nature of the isolated singularities, and the associated residues:

(a) $f(z) = \frac{z^2}{z^4 - 16}$,
(b) $f(z) = \frac{1 - \cos z}{z^n}$,
(c) $f(z) = \frac{1}{z} \cos \frac{1}{z}$,
(d) $f(z) = \frac{1}{\cos 1/z}$,
(e) $f(z) = \frac{z^4 + 18z^2 + 9}{4z(z^2 + 9)}$,
(f) $f(z) = \frac{z}{\sin z}$.

83.2 Using the Residue Theorem, calculate the integrals of

(a) $\displaystyle\oint_{\gamma_1} \frac{\mathrm{d}z}{\sin z}$ where γ_1 traverses the rectangle with the corners $\pm 4 \pm \mathrm{i}$ positively oriented,

(b) $\oint\limits_{\gamma_2} \frac{\mathrm{d}z}{\cosh z}$, where γ_2 is the sketched loop.

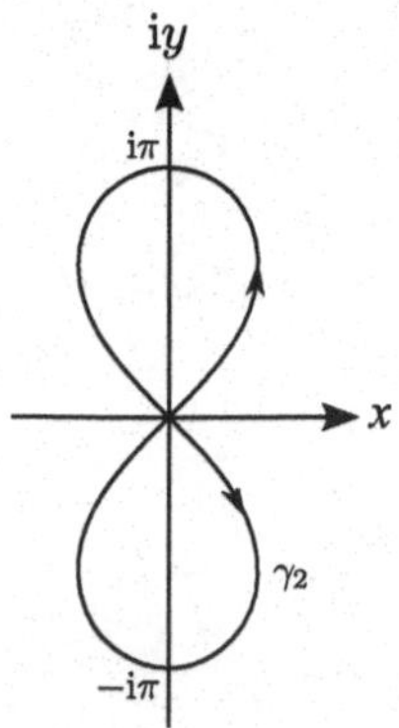

83.3 Calculate the following integrals:

(a) $\int\limits_{-\infty}^{\infty} \frac{\mathrm{d}x}{1+x^6}$,

(b) $\int\limits_{0}^{2\pi} \frac{\mathrm{d}t}{5+3\sin t}$,

(c) $\int\limits_{-\infty}^{\infty} \frac{x^2}{(x^2+4)^2}\mathrm{d}x$.

83.4 Determine for the function $f(z) = \frac{z\,\mathrm{e}^{\frac{1}{z}-1} - 1}{z-1}$

(a) location and nature of the singularities in $\mathbb{C}$,

(b) the value of $\oint\limits_{|z|=2} f(z)\mathrm{d}z$.

Conformal Mappings

84

The main applications of Complex Analysis in engineering concern *plane potential problems*, such as plane boundary value problems or problems in fluid dynamics. Typically, to solve such problems, one transforms the domain under consideration by means of a *conformal mapping* f to a domain in which the problem is *easier to* solve. A possible solution of the simple problem is then transformed back by f^{-1} to the original problem. This then provides a solution to the original problem. Before we move on to applications in the next chapter, we consider conformal mappings in their own right. A conformal mapping is a mapping that preserves angles and their orientation. Of particular interest are Möbius transformations, which allow the most important domains to be conformally mapped to each other in a straightforward manner.

84.1 Generalities of Conformal Mappings

Purely for the sake of the term, one will think of a *conformal* mapping as one that *preserves shapes*, such as angles between straight lines. In fact, this is how one usually introduces this term in mathematics as well: A mapping is called $f : G \subseteq \mathbb{C} \to \mathbb{C}$ *conformal*, if it is *angle- and orientation-faithful*. However, since it is difficult to introduce these two terms, and since it is often not so easy to see that a mapping is conformal in angle and orientation, we take a pragmatic approach:

C. Karpfinger, *Calculus and Linear Algebra in Recipes*,
https://doi.org/10.1007/978-3-662-65458-3_84

Conformal Mappings
A holomorphic function $f : G \subseteq \mathbb{C} \to \mathbb{C}$ is called **conformal**

- in $z_0 \in G$ if $f'(z_0) \neq 0$,
- on G, if f in every $z_0 \in G$ is conformal.

Remark If f is holomorphic and $f'(z_0) \neq 0$, then f can be replaced in a small neighborhood of z_0 by its linear approximation $f(z) = f(z_0) + f'(z_0)(z - z_0)$. This linear approximation is due to $f'(z_0) \neq 0$ *locally* around z_0, it represents a *rotational stretching*: An infinitesimally small figure is rotated and stretched and shifted; *angle* and *orientation* are preserved. A conformal mapping in our sense is thus *angle and orientation faithful*.

Example 84.1

- The **Möbius transformation**
$$f(z) = \frac{az+b}{cz+d} \text{ with } a, b, c, d \in \mathbb{C} \text{ and } ad - bc \neq 0$$
is in all points with $cz + d \neq 0$ conformal, namely we have $f'(z) = \frac{ad-bc}{(cz+d)^2}$.
- The **Joukowski mapping**
$$f(z) = \frac{1}{2}\left(z + \frac{1}{z}\right)$$
is in all points $z \neq \pm 1$ conformal, namely we have $f'(z) = \frac{1}{2}\left(1 - \frac{1}{z^2}\right)$. This mapping maps the circle around $z_0 = \frac{\mathrm{i}-1}{2}$ of radius $\sqrt{5/2}$ into the wing-shaped **Kutta-Joukowski profile**, which served as a model for the lift of airfoils, see Fig. 84.1.
- The square mapping $f(z) = z^2$ is for $z \neq 0$ conformal, since $f'(z) = 2z$, see Fig. 84.2. ■

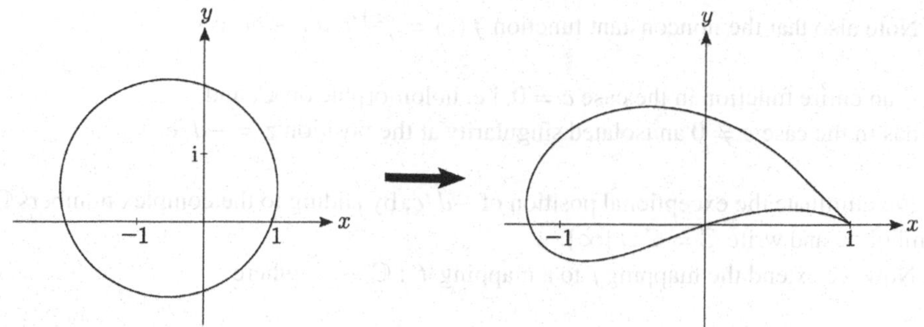

Fig. 84.1 The Kutta-Joukowski profile is the image of a circle under the Joukowski mapping

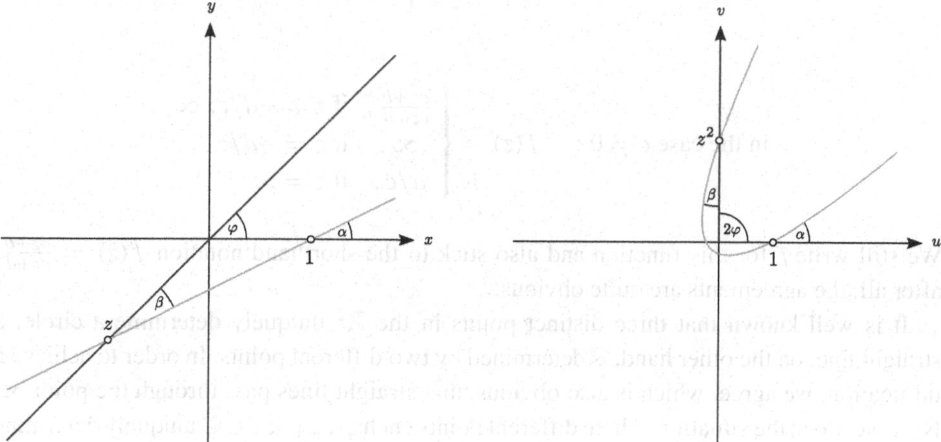

Fig. 84.2 Two straight lines and their images under the square figure

84.2 Möbius Transformations

Of the many conformal mappings that exist, we consider the Möbius transformations or **broken-linear transformations**, as they are also called, more detailed:

$$f(z) = \frac{az+b}{cz+d} \quad \text{with } a,\ b,\ c,\ d \in \mathbb{C} \text{ and } a\,d - b\,c \neq 0\,.$$

The condition $a\,d - b\,c \neq 0$ guarantees two things:

- The denominator $cz+d$ is not constant zero, since $(c, d) \neq (0, 0)$.
- The numerator $az+b$ is not a multiple of the denominator $cz+d$, since $a/c \neq b/d$. Therefore, f is not constant.

Note also that the nonconstant function $f(z) = \frac{az+b}{cz+d}$, $ad - bc \neq 0$,

- is an entire function in the case $c = 0$, i.e. holomorphic on $\mathbb{C}$, and
- has in the case $c \neq 0$ an isolated singularity at the position $z = -d/c$.

We eliminate the exceptional position of $-d/c$, by adding to the complex numbers $\mathbb{C}$ a symbol ∞ and write $\hat{\mathbb{C}} = \mathbb{C} \cup \{\infty\}$.

Now we extend the mapping f to a mapping $f : \hat{\mathbb{C}} \to \hat{\mathbb{C}}$ where

$$\text{in the case } c = 0: \quad f(z) = \begin{cases} \frac{az+b}{d}, & \text{if } z \neq \infty \\ \infty, & \text{if } z = \infty \end{cases},$$

$$\text{in the case } c \neq 0: \quad f(z) = \begin{cases} \frac{az+b}{cz+d}, & \text{if } z \neq -d/c, \infty \\ \infty, & \text{if } z = -d/c \\ a/c, & \text{if } z = \infty \end{cases}.$$

We still write f for this function and also stick to the shorthand notation $f(z) = \frac{az+b}{cz+d}$; after all, the agreements are quite obvious.

It is well known that three distinct points in the $\mathbb{R}^2$ uniquely determine a circle; a straight line, on the other hand, is determined by two different points. In order to achieve a unification, we agree, which is also obvious, that straight lines pass through the point ∞. Now we are in the situation: Three different points each $z_1, z_1, z_3 \in \hat{\mathbb{C}}$ uniquely determine a straight line or a circle. Now we conclude these agreements with another agreement in order to be able to keep the formulations short and concise. From now on we speak of **generalized circles** and thus mean a straight line or a circle.

Möbius Transformations

A mapping of the form $f : \hat{\mathbb{C}} \to \hat{\mathbb{C}}$, $f(z) = \frac{az+b}{cz+d}$ with $ad - bc \neq 0$ is called **Möbius transformation**. For any such mapping, we have:

- f is bijective, the inverse function is

$$f^{-1}(z) = \frac{dz - b}{-cz + a}.$$

- If, in addition to $f(z)$ also $g(z) = \frac{a'z+b'}{c'z+d'}$ is a Möbius transformation, so also $f(g(z))$ and $g(f(z))$.

(continued)

- f maps generalized circles to generalized circles.
- To each of three different points z_1, z_2, z_3 and w_1, w_2, w_3 from $\hat{\mathbb{C}}$ there is exactly one Möbius transformation f with $f(z_i) = w_i$ for all $i = 1, 2, 3$.

Pay close attention to the formulation for *circle fidelity*: A Möbius transformation does not necessarily transform straight lines into straight lines and circles into circles, but rather straight lines are mapped onto straight lines or circles and likewise circles onto straight lines or circles.

According to the last point in the theorem above, there are to every two triples

$$(z_1, z_2, z_3) \text{ and } (w_1, w_2, w_3)$$

of different points from $\hat{\mathbb{C}}$ there is exactly one Möbius transformation f with $f(z_i) = w_i$. The justification of uniqueness is simple: if besides f also g is a Möbius transformation with the property $g(z_i) = w_i$, then

$$g^{-1}(f(z_i)) = g^{-1}(w_i) = z_i ,$$

so that the Möbius transformation $h(z) = g^{-1}(f(z))$ has at least the three different fixed points z_1, z_2, z_3. But the equation

$$h(z) = \frac{az+b}{cz+d} = z$$

is a quadratic equation with at most two solutions: A Möbius transformation has at most two fixed points. So it must be $g = f$.

We give a recipe that can be used to easily determine the uniquely determined Möbius transformation to two triples of different elements from $\hat{\mathbb{C}}$:

Recipe: Determine a Möbius Transformation with the 6-Point Formula
If z_1, z_2, z_3 and w_1, w_2, w_3 are three different elements of $\hat{\mathbb{C}}$ then one obtains the uniquely determined Möbius transformation $f(z) = \frac{az+b}{cz+d}$ with $f(z_i) = w_i$ as follows with the **6-point formula**:

(1) Set

$$\frac{(w-w_1)(w_2-w_3)}{(w-w_3)(w_2-w_1)} = \frac{(z-z_1)(z_2-z_3)}{(z-z_3)(z_2-z_1)},$$

(continued)

where in the case $z_i = \infty$ or $w_j = \infty$

$$\frac{(u - z_i)}{(v - z_i)} = 1 \text{ and } \frac{(u - w_j)}{(v - w_j)} = 1$$

for the corresponding quotient in the above 6-point formula.

(2) Solve the equation from (1) to get $w = f(z)$.

Since Möbius transformations are conformal mappings, roughly speaking, *shapes are preserved*. We will use this in the following way: Let us determine, say, to three points z_1, z_2, z_3 which lie in that order on a straight line, and three points w_1, w_2, w_3 which lie in this order on a circle, the uniquely determined Möbius transfromation $w = f(z)$ then the area to the left of the direction of the straight line is mapped to the area to the left of the direction of the circle, see Fig. 84.3.

We use this in the following examples.

Example 84.2

- We determine a Möbius transformation $w = f(z)$ which maps the upper half-plane $\text{Im}(z) > 0$ onto the interior of the unit circle $|z| < 1$ see Fig. 84.4.

  We choose points z_1, z_2, z_3 and w_1, w_2, w_3, which are boundary points of the upper half-plane and the unit circle, respectively, arranging them so that the regions under consideration are on the left in the direction of passage:

z_i	0	1	∞
w_i	1	i	-1

.

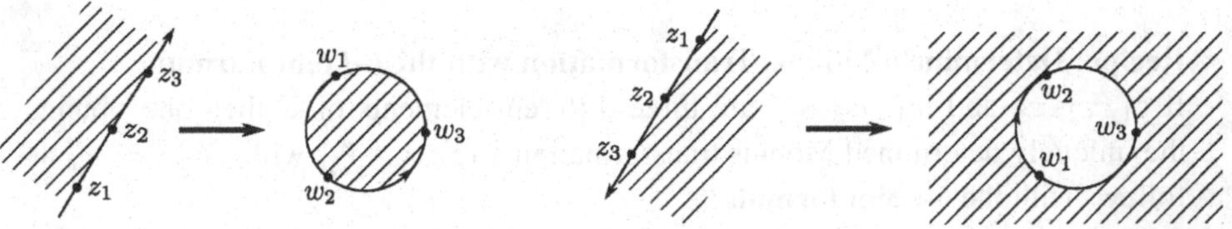

Fig. 84.3 The area to the left of the throughput direction remains to the left of the throughput direction

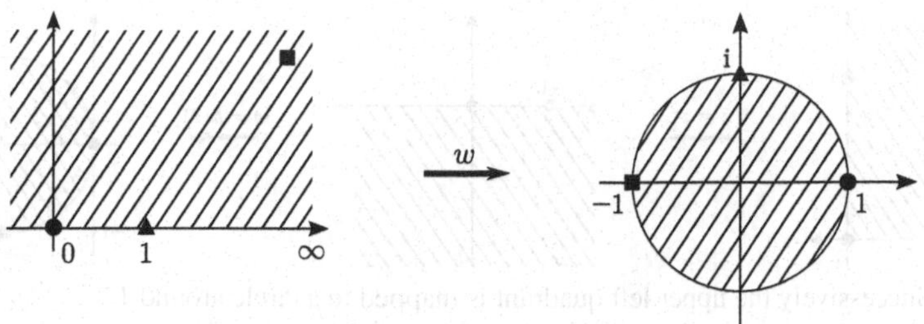

Fig. 84.4 We are looking for a mapping that maps the upper half-plane to the unit circle

We apply to these points the above prescription to the 6-point formula and obtain:

(1) The 6-point formula is:

$$\frac{(w-1)(\mathrm{i}+1)}{(w+1)(\mathrm{i}-1)} = \frac{z-0}{(1-0)},$$

where, because of $z_3 = \infty$ we set the quotient $\frac{(z_2-z_3)}{(z-z_3)}$ equal to 1.

(2) Solving the equation in (1) for $w = f(z)$ yields

$$w = f(z) = \frac{\mathrm{i}\,z+1}{1-\mathrm{i}\,z}.$$

- We determine a function $f\colon \hat{\mathbb{C}} \to \hat{\mathbb{C}}, z \mapsto w = f(z)$, which covers the area

$$G = \left\{z \in \hat{\mathbb{C}} \mid \operatorname{Re} z < 0,\ \operatorname{Im} z > 0\right\}$$

bijective and conforming to the area

$$H = \left\{w \in \hat{\mathbb{C}} \mid |w-1| < 1\right\}$$

and thereby the points $z_1 = 0$, $z_2 = \mathrm{i}$, $z_3 = \infty$ into the points $w_1 = 0$, $w_2 = 1 - \mathrm{i}$, $w_3 = 2$: Since a Möbius transformation maps generalized circles to generalized circles, the function f we are looking for will not be a Möbius transformation: The boundary of the second quadrant G is not a generalized Möbius circle. The trick now is to use an upstream function to *bend up* the angle $\pi/2$ to an angle π to obtain a straight line, i.e., a generalized circle. We then map this straight line to the desired circle using a Möbius transformation:

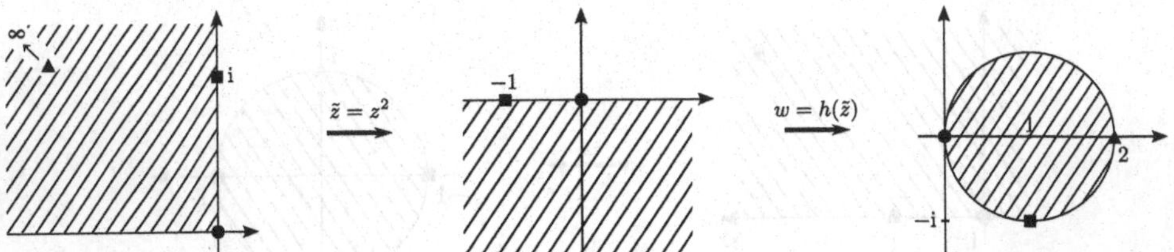

Fig. 84.5 Successively the upper left quadrant is mapped to a circle around 1

- **Step 1:** The function $\tilde{z} = g(z) = z^2$ transforms the second quadrant G on the lower half plane $\text{Im}(z) < 0$. Thereby the points $z_1 = 0, z_2 = \text{i}, z_3 = \infty$ pass over to the points $\tilde{z}_1 = 0, \tilde{z}_2 = -1, \tilde{z}_3 = \infty$.
- **Step 2:** We now use the 6-point formula to determine the Möbius transformation $h(z) = \frac{a\tilde{z}+b}{c\tilde{z}+d}$ to

$$\begin{array}{c|ccc} \tilde{z}_i & 0 & -1 & \infty \\ \hline w_i & 0 & 1-\text{i} & 2 \end{array}.$$

(1) The 6-point formula is:

$$\frac{(w-0)(1-\text{i}-2)}{(w-2)(1-\text{i}-0)} = \frac{\tilde{z}-0}{(-1-0)},$$

where, because of $z_3 = \infty$ we have set the quotient $\frac{(\tilde{z}_2-\tilde{z}_3)}{(\tilde{z}-\tilde{z}_3)}$ equal to 1.

(2) Solving the equation in (1) to $w = h(\tilde{z})$ yields

$$w = h(\tilde{z}) = \frac{2\tilde{z}}{\tilde{z}-\text{i}}.$$

- **Step 3:** By putting them together we get the function f we are looking for, which maps G to H:

$$w = f(z) = h(g(z)) = h(z^2) = \frac{2z^2}{z^2-\text{i}}.$$

Note also the following Fig. 84.5. ■

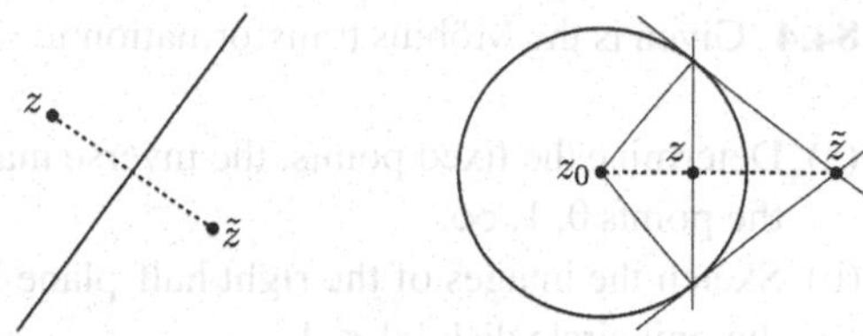

Fig. 84.6 The points z and $\tilde{z}$ are symmetric about a straight line and a circle, respectively

Remarks

1. The **Riemann mapping theorem** says that every simply connected region $G \neq \mathbb{C}$ can be conformally mapped onto the unit circle. Unfortunately, this theorem does not specify how to choose the mapping that accomplishes this. In order to map a complicated area G onto, say, the unit circle conformally, numerical methods are used.
2. Möbius transformations also preserve symmetries to generalized circles: We say two points z and $\tilde{z}$ are **symmetric to a straight line** or **to a circle**, if $\tilde{z}$, as shown in Fig. 84.6, by **mirroring** at a straight line or a circle arises of z.

Now if z and $\tilde{z}$ are symmetric to a generalized circle, so do the images $f(z)$ and $f(\tilde{z})$ to the generalized circle obtained with a Möbius transformation f.

With these symmetry considerations it is possible not only to map boundary points to boundary points, but also to map interior points to prescribed points.

84.3 Exercises

84.1 Determine a Möbius transformation $f(z) = \frac{az+b}{cd+d}$ with the property $f(0) = 1$, $f(1) = -\mathrm{i}$, $f(\mathrm{i}) = \infty$.

84.2 Which conformal mapping $w = f(z)$ maps the interior of the right half of the unit circle to the interior of the unit circle with $f(\mathrm{i}) = \mathrm{i}$, $f(1) = 1$ and $f(-\mathrm{i}) = -\mathrm{i}$? Hint: Note the angles at the boundary points i and $-$i.

84.3 Let $f(z) = \frac{\mathrm{i}(z-1)}{z+\mathrm{i}}$ and $w = h(z)$ is the Möbius transformation for which $h(0) = \mathrm{i}$, $h(\mathrm{i}) = \infty$ and $h(\infty) = 1$.

(a) Determine $h(z)$.
(b) Determine the representation and the fixed points of $g(z) = h(f(z))$.
(c) Sketch the images of the 4 quadrants under $w = f(z)$.
(d) Which straight lines are mapped back to straight lines by f?
(e) What is the original image of the semicircular disk $\{w \in \mathbb{C} \mid |w| \leq 1,\ \mathrm{Re}\, w \geq 0\}$ under the mapping $w = f(z)$?

84.4 Given is the Möbius transformation $w = \frac{z}{z-\mathrm{i}}$.

(a) Determine the fixed points, the inverse mapping and the images and inverse images of the points $0, 1, \infty$.
(b) Sketch the images of the right half-plane $\operatorname{Re} z \geq 0$ the upper half-plane $\operatorname{Im} z \geq 0$ and the unit circle disk $|z| \leq 1$.
(c) Which curves of the z-plane are mapped to straight lines in the w-plane and which of them are mapped to straight lines through $w = 0$?

85 Harmonic Functions and the Dirichlet Boundary Value Problem

In the *Dirichlet boundary value problem*, a function $u = u(x, y)$ is sought, which on a domain D solves the Laplace equation $\Delta u = u_{xx}+u_{yy} = 0$ and assumes given (boundary) values on the boundary of D. The solutions of the Laplace equation $\Delta u = 0$ are the *harmonic functions*. These are the real and imaginary parts of holomorphic functions.

We first show the connections between holomorphic and harmonic functions and then give a concrete solution of Dirichlet's boundary value problem on the circle. This also solves other boundary value problems: namely, if one knows the solution on the unit circle, one can determine this solution for other domains by means of conformal mappings.

85.1 Harmonic Functions

We consider a holomorphic function $f : G \to \mathbb{C}$ on a simply connected domain G. According to the criterion for holomorphy in Sect. we have for the real and imaginary parts $u = u(x, y)$ and $v = v(x, y)$ of f the Cauchy-Riemann differential equations:

$$u_x = v_y \text{ and } u_y = -v_x \,.$$

We differentiate the two equations by x and y and obtain with Schwarz's Theorem (see in Sect.):

$$u_{xx} + u_{yy} = v_{yx} - v_{xy} = 0 \text{ and } v_{xx} + v_{yy} = -u_{yx} + u_{xy} = 0\,,$$

i.e.

$$\Delta u = 0 \text{ and } \Delta v = 0\,.$$

C. Karpfinger, *Calculus and Linear Algebra in Recipes*,
https://doi.org/10.1007/978-3-662-65458-3_85

Real and imaginary parts of holomorphic functions are thus solutions of the following partial differential equation, which, because of the occurrence of the Laplace operator, is also called **Laplace's equation** or also **potential equation**:

$$\Delta u = u_{xx} + u_{yy} = 0\,.$$

Harmonic Functions

Every function $u : G \subseteq \mathbb{R}^2 \to \mathbb{R}$, G simply connected, with

$$\Delta u = u_{xx} + u_{yy} = 0$$

is called **harmonic function**. We have:

- Real and imaginary part of a holomorphic function $f : G \to \mathbb{C}$ are harmonic functions.
- For each harmonic function $u : G \to \mathbb{R}$, $(x, y) \mapsto u(x, y)$ there is a harmonic function $v : G \to \mathbb{R}$, $(x, y) \mapsto v(x, y)$ which is uniquely determined up to an additive constant so that

$$f : G \to \mathbb{C},\ f(z) = u(x, y) + \mathrm{i}\, v(x, y)\,,\ z = x + \mathrm{i}\, y\,,$$

  is holomorphic. One calls v **harmonically conjugate function** of u.
- If $f : G \to \mathbb{C},\ f(z) = u(x, y) + \mathrm{i}\, v(x, y)$ is holomorphic, then the sets of curves $u(x, y) = c$ and $v(x, y) = d$ are perpendicular to each other.

Example 85.1

- The function $f : G \subseteq \mathbb{C} \to \mathbb{C}$ with $f(z) = 1/z$ is holomorphic on any simply connected domain G of $\mathbb{C}$ which does not contain 0. Because of

$$\frac{1}{z} = \frac{x - \mathrm{i}\, y}{x^2 + y^2} = \frac{x}{x^2 + y^2} + \mathrm{i}\,\frac{-y}{x^2 + y^2}$$

  the functions

$$\mathrm{Re}(f) = u(x, y) = \frac{x}{x^2 + y^2} \text{ and } \mathrm{Im}(f) = v(x, y) = \frac{-y}{x^2 + y^2}$$

  are harmonic functions. The function v is harmonically conjugate to u.

- Other examples of harmonic functions are

$$u_1 = x^3 - 3\,x\,y^2\,,\ v_1 = 3x^2y - y^3\,,\ u_2 = \cos(x)\,\cosh(y)\,,\ v_2 = -\sin(x)\,\sinh(y)\,.$$

The functions v_1 and u_1 are harmonically conjugated, and also v_2 and u_2, because:

$$u_1 = \mathrm{Re}(z^3)\,,\ v_1 = \mathrm{Im}(z^3)\ \text{ and }\ u_2 = \mathrm{Re}(\cos(z))\,,\ v_2 = \mathrm{Im}(\cos(z))\,.$$ ■

The question arises whether, and if so, how, for a given harmonic function u a harmonically conjugate function v can be determined. We describe the procedure in a recipe:

Recipe: Determine the Harmonically Conjugate Function
Given is a harmonic function $u : G \to \mathbb{R}$, $(x, y) \mapsto u(x, y)$ on a simply connected domain G. A to u harmonically conjugate function $v : G \to \mathbb{R}$ (it is then $f : G \to \mathbb{C}$, $f(z) = u(x, y) + \mathrm{i}\,v(x, y)$ holomorphic) is found by integrating the Cauchy-Riemann differential equations $v_y = u_x$ and $v_x = -u_y$:

(1) Determine $v(x, y) = \int u_x \mathrm{d}y$ with the integration *constant* $h(x)$.
(2) Derive v partial to x and obtain from $v_x = -u_y$ a representation for $h'(x)$.
(3) Obtain by integration of h' the function v from (1) up to a constant.

Example 85.2 The function $u : \mathbb{R}^2 \to \mathbb{R}$, $u(x, y) = x^2 - y^2 + \mathrm{e}^x \sin(y)$ is harmonic on $\mathbb{R}^2$ because

$$u_{xx} = 2 + \mathrm{e}^x \sin y\,, \qquad u_{yy} = -2 - \mathrm{e}^x \sin y\,.$$

Thus

$$\Delta u = u_{xx} + u_{yy} = 0\,.$$

(1) Integration of u_x to y yields:

$$v(x, y) = \int u_x \mathrm{d}y = \int 2x + \mathrm{e}^x \sin y\,\mathrm{d}y = 2xy - \mathrm{e}^x \cos y + h(x)\,.$$

(2) Differentiation of v to x and equating $v_x = -u_y$ delivers:

$$v_x = 2y - \mathrm{e}^x \cos y + h'(x) = -u_y = 2y - \mathrm{e}^x \cos y \;\Rightarrow\; h'(x) = 0\,.$$

(3) Integration of $h'(x)$ yields:

$$h(x) = c \Rightarrow v(x, y) = 2xy - e^x \cos y + c\,.$$

The associated holomorphic function is thus

$$\begin{aligned} f(z) &= x^2 - y^2 + e^x \sin y + i\left(2xy - e^x \cos y + c\right) \\ &= z^2 - i(e^x(\cos y + i \sin y) - c) = z^2 - i(e^z - c)\,. \end{aligned}$$ ■

85.2 The Dirichlet Boundary Value Problem

In applications, especially in electrostatics and heat transport phenomena, one often looks for a function $u = u(x, y)$ which is inside a region D harmonic and on the boundary ∂D of the domain D takes on prescribed (boundary) values. See Fig. 85.1.

We are looking for a solution u of the partial differential equation with boundary condition:

$$\Delta u(x, y) = 0 \text{ for all } (x, y) \in D$$

and

$$u(x, y) = g(x, y) \text{ for all } (x, y) \in \partial D\,,$$

where the continuous boundary function g is given. This is called the **Dirichlet boundary value problem**. By means of Complex Analysis a very general and straightforward solution of this boundary value problem succeeds.

We begin by describing the solution method for this problem on a circle G. We then trace the general case back to this case of a circle by means of conformal mappings.

To solve Dirichlet's boundary value problem for a circle $D = \{(x, y) \mid x^2 + y^2 < R^2\}$ of radius R, it is of course advantageous to solve the problem in polar coordinates (r, φ): The circle D and its boundary ∂D in polar coordinates is

$$D = \{(r, \varphi) \mid 0 \leq r < R\,,\ 0 \leq \varphi < 2\pi\} \text{ and } \partial D = \{(R, \varphi) \mid 0 \leq \varphi < 2\pi\}\,,$$

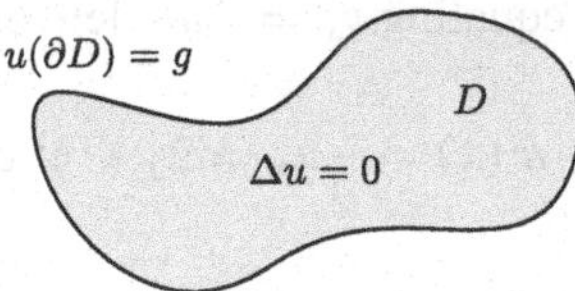

Fig. 85.1 The Dirichlet boundary value problem

the boundary function $g = g(\varphi)$ is thus a function in φ, and the Laplace operator Δ is according to Sect.

$$\Delta u(r, \varphi) = u_{rr} + \frac{1}{r} u_r + \frac{1}{r^2} u_{\varphi\varphi} .$$

Dirichlet's Boundary Value Problem on the Unit Circle and its Solution
If $g = g(\varphi)$ is a continuous function on the boundary ∂D of a circle $D = \{(r, \varphi) \,|\, 0 \leq r < R,\ 0 \leq \varphi < 2\pi\} \subseteq \mathbb{R}^2$ of radius R, then there is exactly one harmonic function $u = u(r, \varphi)$ which solves the **Dirichlet's boundary value problem**

$$\Delta u(r, \varphi) = 0 \text{ on } D \text{ and } u(R, \varphi) = g(\varphi) \text{ on } \partial D .$$

The solution function $u = u(r, \varphi)$ is given by **Poisson's integral formula:**

$$u(r, \varphi) = \frac{R^2 - r^2}{2\pi} \int_0^{2\pi} \frac{g(t)}{R^2 + r^2 - 2rR\cos(\varphi - t)} \mathrm{d}t .$$

Thus we have a solution formula with which, at least for theoretical purposes, Dirichlet's boundary value problem is solved. But for practical purposes it does not give much: Although the formula shows wonderfully that at the zero point, i.e. for $r = 0$, in a sense the mean value

$$u(0, \varphi) = \frac{1}{2\pi} \int_0^{2\pi} g(t)\mathrm{d}t$$

of the boundary function g is assumed. But if one wants to know what the function value of u at a concrete point (r_0, φ_0) is then a line integral, which may not be easy to solve, is to be evaluated.

Using the following recipe, we obtain the solution in many cases concretely, or at least an approximation. Here we use without justification that the function $u = u(r, \varphi)$ which is obtained by Poisson's integral formula, is harmonic on G and can be developed into an infinite series

$$u(r, \varphi) = \frac{a_0}{2} + \sum_{k=1}^{\infty} \left(\frac{r}{R}\right)^k (a_k \cos(k\varphi) + b_k \sin(k\varphi)) .$$

Here the coefficients a_k and b_k are just the Fourier coefficients of the 2π-periodic continuous function $g = g(\varphi)$:

Recipe: Solving a Dirichlet Boundary Value Problem for a Circle
Given is the Dirichlet boundary value problem for a circle $D = \{(r, \varphi) \,|\, 0 \le r < R\,, 0 \le \varphi < 2\pi\}$ of radius R

$$\Delta u(r, \varphi) = 0 \text{ on } D \text{ and } u(R, \varphi) = g(\varphi) \text{ for } 0 \le \varphi < 2\pi\,,$$

where $g = g(\varphi)$ is continuous.

(1) Determine the Fourier series expansion $G = G(\varphi)$ of $g = g(\varphi)$ (see Sect. 74.3):

$$G(\varphi) = \frac{a_0}{2} + \sum_{k=1}^{\infty} a_k \cos(k\,\varphi) + b_k \sin(k\,\varphi)\,.$$

(2) Obtain the solution:

$$u(r, \varphi) = \frac{a_0}{2} + \sum_{k=1}^{\infty} \left(\frac{r}{R}\right)^k (a_k \cos(k\,\varphi) + b_k \sin(k\,\varphi))\,.$$

Example 85.3 We consider a cylinder whose intersection with the *x*-*y*-plane is the unit circle $\mathbb{E}$ around the zero point (see Fig. 85.2). Let the surface temperature of the cylinder be independent of time. Thus we have a boundary temperature $g = g(\varphi)$ given on the boundary of the unit circle. The temperature $u = u(r, \varphi)$ satisfies the heat conduction equation $u_t = c^2 \Delta u$, which is just Laplace's equation because of its independence of temperature with respect to time (cf. Remark in Sect. 89.1). So we have Dirichlet's boundary value problem for the unit circle $\mathbb{E}$ with boundary $\partial\mathbb{E}$ to solve:

$$\Delta u(r, \varphi) = 0 \text{ on } \mathbb{E} \text{ and } u(1, \varphi) = g(\varphi) \text{ on } \partial\mathbb{E}\,.$$

- We consider the boundary function $g(\varphi) = \sin^3(\varphi)$.

  (1) The Fourier series expansion of $g(\varphi) = \sin^3(\varphi)$ is

$$G(\varphi) = \frac{3}{4}\sin(\varphi) - \frac{1}{4}\sin(3\varphi)\,.$$

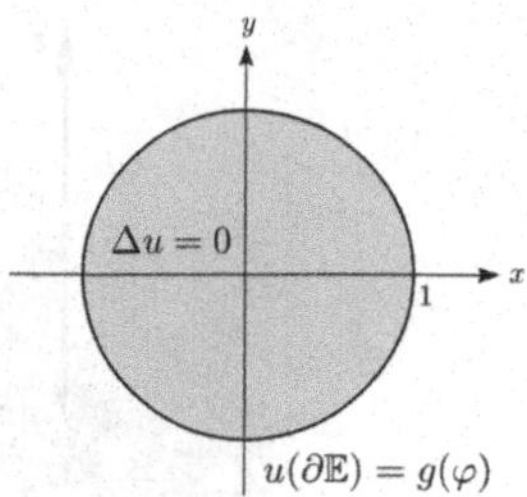

Fig. 85.2 Cross section through cylinder

(2) Obtain the solution:

$$u(r, \varphi) = \frac{3}{4} r \sin(\varphi) - \frac{1}{4} r^3 \sin(3\varphi) .$$

- We consider the boundary function

$$g : [0, 2\pi) \to \mathbb{R}, \; g(x) = \begin{cases} 1, & 0 < x < \pi \\ 0, & x = 0 \\ -1, & \pi \leq x < 2\pi \end{cases} .$$

This boundary function, although not continuous, can still be imagined as an approximation of a real problem. Although the existence and uniqueness of a solution is not guaranteed, we apply our prescription in the hope of finding a reasonable solution.

(1) The Fourier series expansion of $g(\varphi)$ according to the example in Sect. 74.3 is:

$$G(\varphi) = \sum_{k=0}^{\infty} \frac{4}{(2k+1)\pi} \sin\left((2k+1)\varphi\right)$$
$$= \frac{4}{\pi} \left(\sin(\varphi) + \frac{\sin(3\varphi)}{3} + \frac{\sin(5\varphi)}{5} + \ldots \right) .$$

(2) Obtain the solution:

$$u(r, \varphi) = \frac{4}{\pi} \left(r \sin(\varphi) + \frac{r^3 \sin(3\varphi)}{3} + \frac{r^5 \sin(5\varphi)}{5} + \ldots \right) .$$

The graphs of the two solutions are plotted in Fig. 85.3, where we have the (infinite) series for the second case of the discontinuous boundary function following the summand $\frac{r^{15} \sin(15x)}{15}$ cancelled. ■

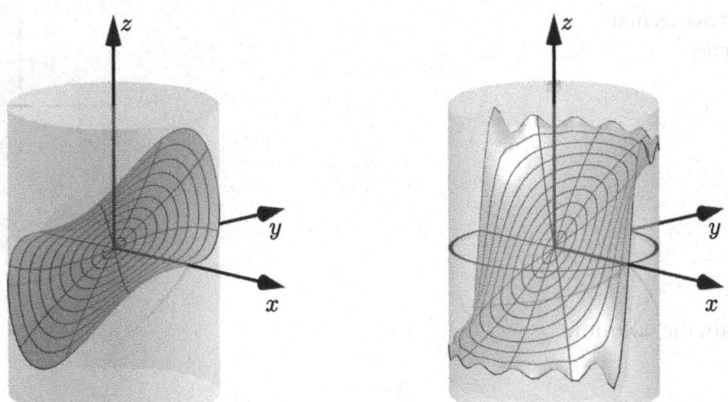

Fig. 85.3 Graphs of the solutions

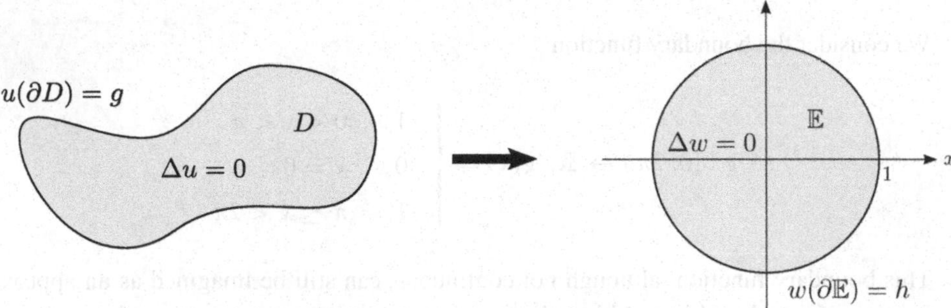

Fig. 85.4 Transformation of Dirichlet's BVP to the unit circle

The solution of Dirichlet's boundary value problem for the circle D can be used to solve many other Dirichlet boundary value problems for domains of different types $\tilde{D}$. If such a domain $\tilde{D}$ can be conformally and bijectively mapped on a circle D of radius R, then one can transform the entire Dirichlet boundary value problem to this circle, solve it there, and back-transform the solution, where the back-transformed solution is then the sought solution to the original problem, see Fig. 85.4.

The procedure for the solution is described in the following recipe:

Recipe: Solving Dirichlet's Boundary Value Problem
A solution $u = u(x, y)$ of the Dirichlet boundary value problem

$$\Delta u(x, y) = 0 \text{ for all } (x, y) \in D \text{ and } u(x, y) = g(x, y) \text{ for all } (x, y) \in \partial D\,,$$

(continued)

where the boundary function g on the double-point free and generally closed boundary ∂D is continuous and bounded, is obtained as follows:

(1) Determine a bijective and conformal mapping $f : D \to \mathbb{E}$, which maps ∂D on $\partial \mathbb{E}$. Give the inverse mapping f^{-1}.
(2) Solve the (transformed) Dirichlet's boundary value problem on the unit circle

$$\Delta w(r, \varphi) = 0 \text{ on } \mathbb{E} \text{ and } w(1, \varphi) = h(\varphi) = g(f^{-1}(\mathrm{e}^{\mathrm{i}\varphi}))$$

$$\text{for all } 0 \le \varphi < 2\pi$$

with the recipe in Sect. 85.2 and obtain $w = w(r, \varphi)$.
(3) Obtain by back transformation the solution $u = u(x, y)$ of the original Dirichlet boundary value problem to D:

$$u(x, y) = w(f(x, y)) \text{ for all } (x, y) \in D \cup \partial D\,.$$

Example 85.4 We solve Dirichlet's boundary value problem in the upper half-plane $D = \{(x, y) \in \mathbb{R}^2 \mid y > 0\}$ with the x-axis as boundary ∂D and the continuous function $g(x) = \sin^2(\arctan(x))$.

(1) It is $f(z) = \frac{z-\mathrm{i}}{-z-\mathrm{i}}$ a bijective and conformal mapping that maps the upper half-plane to the unit circle. Thereby ∂D is mapped on $\partial \mathbb{E}$. The inverse mapping is $f^{-1}(z) = -\mathrm{i}\,\frac{z-1}{z+1}$.
(2) We solve the (transformed) Dirichlet's boundary value problem on the unit circle:

$$\Delta w(r, \varphi) = 0 \text{ on } \mathbb{E} \text{ and } w(1, \varphi) = h(\varphi) = g(f^{-1}(\mathrm{e}^{\mathrm{i}\varphi})) = \frac{1}{2} - \frac{\cos(\varphi)}{2}$$

for all $0 \le \varphi < 2\pi$ where we have used for the computation of $h(\varphi)$ we used some well-known trigonometric identities, in particular

$$\sin^2\left(\arctan\left(\frac{\sin(\varphi)}{1 + \cos(\varphi)}\right)\right) = \sin^2(\varphi/2) = \frac{1}{2} - \frac{\cos(\varphi)}{2}\,.$$

Now we apply the prescription of Sect. 85.2:
(1) The function $h = h(\varphi)$ is already given as a Fourier series:

$$H(\varphi) = \frac{1}{2} - \frac{\cos(\varphi)}{2}\,.$$

(2) Obtain the solution on the unit circle

$$w(r, \varphi) = \frac{1}{2} - \frac{r\cos(\varphi)}{2}.$$

(3) For the inverse transformation, we first compute $f(z)$ for $z = x + \mathrm{i}\,y$:

$$f(z) = \frac{-x^2 - (y^2 - 1)}{x^2 + (y+1)^2} - \mathrm{i}\,\frac{2x}{x^2 + (y+1)^2}.$$

From this we obtain

$$r = r(x, y) = \left(\left(\frac{-x^2 - (y^2 - 1)}{x^2 + (y+1)^2}\right)^2 + \left(\frac{2x}{x^2 + (y+1)^2}\right)^2\right)^{1/2} \quad \text{and}$$

$$\varphi = \varphi(x, y) = \arctan\left(\frac{\frac{-2x}{x^2+(y+1)^2}}{\frac{-x^2-(y^2-1)}{x^2+(y+1)^2}}\right),$$

from which, using a computer algebra system, we obtain the following solution:

$$u(x, y) = \frac{1}{2} - \frac{r(x, y)\cos(\varphi(x, y))}{2} = \frac{x^2 + y + y^2}{x^2 + (1+y)^2}. \qquad \blacksquare$$

As this simple example already shows, the calculations or expressions quickly become confusing because of the rational transformation. Fortunately, it is also easier: In fact, one can also proceed the other way round: One can show that Dirichlet's boundary value problem in the upper half plane can be solved with a corresponding *Poisson integral formula* for this upper half-plane and then give a solution algorithm for this problem. Then, analogously, one could solve the general Dirichlet boundary value problem for a domain D with a conformal and bijective mapping back to the problem in the upper half-plane. We started with the Dirichlet boundary value problem on the circle. This was purely arbitrary. By the way, for the Dirichlet boundary value problem

$$\Delta u(x, y) = 0 \text{ for all } (x, y) \in H \text{ and } u(x, 0) = g(x) \text{ for all } x \in \mathbb{R}$$

in the upper half plane $H = \{(x, y) \in \mathbb{R}^2 \mid y > 0\}$ one obtains the solution $u = u(x, y)$ for continuous and bounded g by the **Poisson's integral formula**

$$u(x, y) = \frac{y}{\pi}\int_{-\infty}^{\infty} \frac{g(t)}{(x-t)^2 + y^2}\,\mathrm{d}t \text{ for } y > 0.$$

Remark Besides the Dirichlet boundary value problem, in practice also the **Neumann boundary value problem** plays a role: given is a continuous function g on the boundary ∂D of a domain D. We are looking for a function u, which solves on D the Laplace equation $\Delta u = 0$ and which normal derivative on the boundary matches with g:

$$\Delta u = 0 \text{ on } D \text{ and } \frac{\partial u}{\partial \boldsymbol{n}} = g \text{ on } \partial D .$$

Here $\boldsymbol{n}$ is the normal unit vector on ∂D, pointing outwards. We do not solve this problem any more, but we do not want to miss to remark that this Neumann boundary value problem can be traced back to a Dirichlet boundary value problem.

85.3 Exercises

85.1 For which $a \in \mathbb{R}$ is the function $u : \mathbb{R}^2 \to \mathbb{R}$ with $u(x, y) = x^3 + a\,x\,y^2$ harmonic? In this case, determine one to u harmonically conjugate function.

85.2 In which domains $G \subseteq \mathbb{R}^2$ are the following functions harmonic?

(a) $u(x, y) = x^3 - 3xy^2 + 2x + 5$,
(b) $u(x, y) = x + \frac{x}{x^2+y^2}$.

Calculate in each case the harmonically conjugate function $v(x, y)$ and the corresponding holomorphic function $f(z)$.

85.3 Which holomorphic functions $f(z) = u(x, y) + \mathrm{i}\, v(x, y)$ have the imaginary part $v(x, y) = x^2 - y^2 + \mathrm{e}^x \sin y$?

85.4 Given is the boundary value problem

$$\delta u = u_{xx} + u_{yy} = 0 \text{ for } x^2 + y^2 < 1$$

and

$$u(\cos\varphi, \sin\varphi) = \begin{cases} 1 & \text{for } 0 < \varphi < \frac{\pi}{2} \\ 2 & \text{for } \frac{\pi}{2} < \varphi < \pi \\ 0 & \text{for } \pi < \varphi < 2\pi \end{cases} .$$

Using a conformal transformation and Poisson's integral formula for the upper half-plane, determine the solution.

Partial Differential Equations of First Order

86

In the case of an ordinary differential equation (ODE), a function $x = x(t)$ in one variable t is sought, which solves an equation in which the function x and derivatives of x to t appear. In a partial differential equation (PDE), a function u in several variables is sought, usually $u = u(x, t)$ or $u = u(x, y)$ or $u = u(x, y, z, t)$, where u satisfies an equation which, in addition to u also partial derivatives of u with respect to the different variables apear.

Maxwell's equations, the Navier-Stokes equation, Schrödinger's equation, ... these are partial differential equations that underlie entire sciences. Besides these, there are many other partial differential equations that arise in all sorts of problems in engineering and science.

It is certainly no exaggeration to claim that the broad topic of *partial differential equations* is one of the most important and fundamental areas of applied mathematics and is included in almost all basic subjects of engineering and science courses.

As important and fundamental as the field is, the theory and solution methods of partial differential equations seem impenetrable or endless. Within the scope of this book, we can only briefly touch on the subject and, in a sense, only form a springboard to release the reader into a sea of partial differential equations and possible solution methods.

Usually, one starts the topic of *partial differential equations* with a classification of types or derivations. We deviate from this tradition. We want to start the topic with a positive signal and in this first chapter we provide solution methods for the simplest types of partial differential equations. In practice one has to deal with more complicated equations. But one learns some essential aspects already with these simple types.

C. Karpfinger, *Calculus and Linear Algebra in Recipes*,
https://doi.org/10.1007/978-3-662-65458-3_86

86.1 Linear PDEs of First Order with Constant Coefficients

In the case of a partial differential equation, a function $u = u(x_1, \ldots, x_n)$ is sought which satisfies an equation in which $u, x_1, \ldots, x_n$ and partial derivatives of u according to the variables $x_1, \ldots, x_n$ appear. We make life easier for ourselves and usually only consider the case of two variables, i.e. $u = u(x, y)$. In the applications are often the time t and the location x the variables, we then have it with $u = u(x, t)$ to do.

The simplest type of a PDE is the 1st order linear PDE with constant coefficients, it has the form.

$$a\,u_x + b\,u_y = f(x, y) \text{ with } a,\ b \in \mathbb{R} \text{ and a function } f = f(x, y)\,.$$

We are looking for a function $u = u(x, y)$ that satisfies this equation. In the case $a = 0$ or $b = 0$, solutions are obtained by integration, e.g. for $b = 0$:

$$u(x, y) = \frac{1}{a} \int f(x, y)\,\mathrm{d}x + g(y)$$

with any function $g(y)$ is a solution. A little more interesting is the case $a \neq 0 \neq b$:

Recipe: Solve a Linear PDE of 1st Order with Constant Coefficients
The solution $u = u(x, y)$ of the PDE $a\,u_x + b\,u_y = f(x, y)$ with $a \neq 0 \neq b$ is obtained as follows:

(1) Perform variable substitution:

$$r = r(x, y) = b\,x + a\,y \text{ and } s = s(x, y) = b\,x - a\,y\,.$$

(2) Set

$$U(r, s) = u\left(\frac{r+s}{2b}, \frac{r-s}{2a}\right) = u(x, y)$$

and

$$F(r, s) = f\left(\frac{r+s}{2b}, \frac{r-s}{2a}\right) = f(x, y).$$

(continued)

(3) Substituting U and F into the PDE yields the ODE

$$U_r = \frac{1}{2ab} F(r, s) .$$

(4) Get the solution $U = U(r, s)$ of the ODE in (3):

$$U(r, s) = \frac{1}{2ab} \int F(r, s) \, \mathrm{d}r + G(s)$$

with a differentiable function $G = G(s)$.

(5) A back substitution yields $u(x, y)$.

(6) By a possible boundary (initial) condition like $u(x, 0) = g(x)$ the function G is specified in (4).

By the given transformation the PDE becomes an ODE. Behind this is essentially the chain rule, see Exercise 86.5.

We have not explained what the terms *1st order* and *linear* mean in the context of the PDE under consideration; but that goes without saying: *1st order* means that at most 1st order partial derivatives appear, and *linear* means that the function u and the partial derivatives of u appear only in 1st power and not in nonlinear functions like sin or exp. Note also: Instead of an arbitrary *integration constants* c, as one usually gets when solving an ODE, one gets here a function G: Just as c is determined by the initial condition of an ODE, here the function G is determined by an boundary (initial) condition.

Example 86.1 Let $u(x, t)$ is the traffic density at the location x at time t along a road. All cars travel at the constant speed $v > 0$. There is an inflow or outflow of cars through side roads, which is given by $f(x, t)$. The function $u(x, t)$ satisfies the partial differential equation

$$u_t + v \, u_x = f(x, t) .$$

We first consider the simple case $f(x, t) = 0$ i.e., there are neither cars flowing in nor out—at any time and at any place.

(1) We perform variable substitution:

$$r = r(x, t) = x + v\,t \quad \text{and} \quad s = s(x, t) = x - v\,t .$$

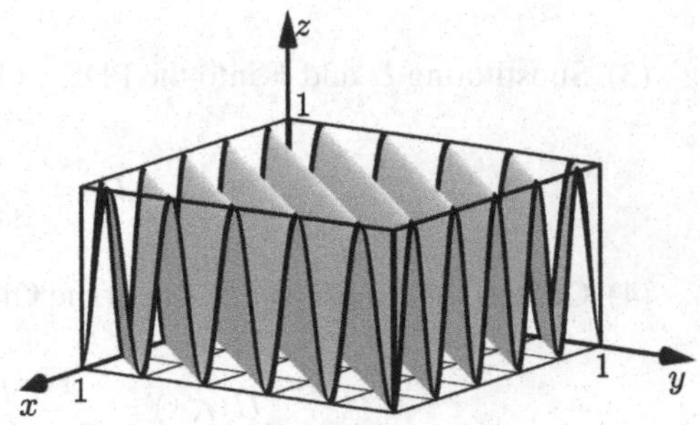

Fig. 86.1 The graph of the solution

(2) Set

$$U(r,s) = u\left(\frac{r+s}{2}, \frac{r-s}{2v}\right) = u(x,t) \text{ and } F(r,s) = f\left(\frac{r+s}{2}, \frac{r-s}{2v}\right) = f(x,t).$$

(3) Substituting U and F into the PDE yields the ODE

$$U_r = \frac{1}{2v} F(r,s) = 0\,.$$

(4) The solution of this last ODE is

$$U(r,s) = G(s)\,.$$

(5) Back substitution yields the solution $u(x,t) = G(x - vt)$.
(6) If an *initial density* $u(x,0) = \sin^2(5\pi x)$ is given, the function G is determined by:

$$u(x,0) = \sin^2(5\pi x) = G(x) \Rightarrow G(x) = \sin^2(5\pi x)\,.$$

Now that we know what G is, we can now concretely find the solution $u = u(x,t)$ we are looking for:

$$u(x,t) = \sin^2(5\pi(x - vt))\,.$$

In Fig. 86.1 we see from the graph of this function with $v = 1$, how this initial density $u(x,0)$ unchanged over time along the *road* x shifts; clearly, there is neither an outflow nor an inflow of cars, moreover, all cars have the same speed $v = 1$.

Now we consider the case $f(x,t) = \frac{1}{1+x^2}$ and $u(x,0) = 0$: So there is a positive inflow, which with increasing x becomes smaller and smaller; and at the beginning of the

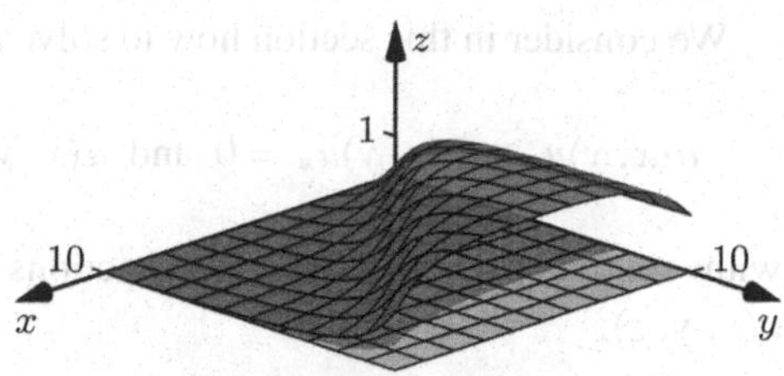

Fig. 86.2 The graph of the solution

observation there is no car on the road. We use the above recipe to determine the solution, first noting that steps (1) and (2) are valid without change.
(3) Substituting U and F into the PDE yields the ODE

$$U_r = \frac{1}{2v} F(r,s) = \frac{1}{2v} \frac{1}{1+(\frac{r+s}{2})^2}\,.$$

(4) We obtain the solution of the ODE in (3) by applying the substitution rule for the integration according to r, for which we substitute $u = \frac{r+s}{2}$, $du = \frac{1}{2}dr$:

$$U(r,s) = \frac{1}{2v} \int \frac{1}{1+(\frac{r+s}{2})^2} dr = \frac{1}{v} \int \frac{1}{1+u^2} du = \frac{1}{v} \arctan\left(\frac{r+s}{2}\right) + G(s)\,.$$

(5) A back substitution yields $u(x,t) = \frac{1}{v}\arctan(x) + G(x-vt)$.
(6) Substituting the initial condition $0 = u(x,0) = \frac{1}{v}\arctan(x) + G(x)$ yields $G(x) = -\frac{1}{v}\arctan(x)$, so that

$$u(x,t) = \frac{1}{v}\left(\arctan(x) - \arctan(x-vt)\right)$$

is a solution.

In Fig. 86.2 we see the graph of this function with $v = 1$, how this initial density $u(x,0) = 0$ in the course of time along the *road* x increases; the steady influx of cars over time causes the road to become crowded. Don't forget to make some plots of this graph with MATLAB. ■

86.2 Linear PDEs of First Order

Every linear PDE of 1st order can be reduced to a system of ODEs. By solving this system of ODEs we obtain solutions of the linear PDE.

We consider in this section how to solve a PDE of the form

$$a(x, y)u_x + b(x, y)u_y = 0 \text{ and } a(x, y, z)u_x + b(x, y, z)u_y + c(x, y, z)u_z = 0$$

with continuously differentiable functions $a = a(x, y, z)$, $b = b(x, y, z)$ and $c = c(x, y, z)$:

Recipe: Solving a Linear Homogeneous PDE of 1st Order
To determine a solution u of the PDE

$$\text{(i) } a(x, y)u_x + b(x, y)u_y = 0 \text{ or}$$

$$\text{(ii) } a(x, y, z)u_x + b(x, y, z)u_y + c(x, y, z)u_z = 0$$

proceed as follows:

(1)

- Set in case (i): $\frac{dy}{dx} = \frac{b(x,y)}{a(x,y)}$—this is an ODE.
- Set in case (ii): $\frac{dy}{dx} = \frac{b(x,y,z)}{a(x,y,z)}$ and $\frac{dz}{dx} = \frac{c(x,y,z)}{a(x,y,z)}$—that is a system of ODEs.

(2)

- In case (i), solve the ODE from (1) and get $y = y(x) = F(x, c)$.
- In case (ii), solve the system of ODEs from (1) and get $y = y(x) = F(c_1, x)$ and $z = z(x) = G(c_2, x)$.

(3)

- In case (i), solve the equation $y(x) = F(x, c)$ to $c = c(x, y)$ (if possible).
- In case (ii), solve the system $y(x) = F(c_1, x)$ and $z(x) = G(c_2, x)$ to $c_1 = c_1(x, y, z)$ and $c_2 = c_2(x, y, z)$ (if possible).

(4)

- Then in case (i) $u(x, y) = f(c(x, y))$ for every continuously differentiable function f is a solution of the PDE.
- In case (ii) then $u(x, y, z) = f(c_1(x, y, z), c_2(x, y, z))$ for every continuously differentiable function f is a solution of the PDE.

(continued)

(5) The function f is then determined by an initial condition, if any, given.

In step (1), one can also use the quotients

$$\frac{dx}{dy} \text{ in case (i) and } \frac{dx}{dy}, \frac{dz}{dy} \text{ or } \frac{dx}{dz}, \frac{dy}{dz} \text{ in case (ii).}$$

Possibly by another choice the system of ODEs in (2) becomes simpler.

Behind this solution procedure is what is called the *method with characteristics*. We refrain from describing this method in more detail here and simply use the above recipe for solving such PDEs:

Example 86.2

- We search for solutions of the PDE

$$x\, u_x + y\, u_y = 0\,.$$

We obtain step by step:

$$(1)\ \frac{dy}{dx} = \frac{y}{x} \quad (2)\ y = c\,x \quad (3)\ c = \frac{y}{x}\,.$$

(4) For every differentiable function f is

$$u = u(x, y) = f\left(\frac{y}{x}\right)$$

a solution of the PDE.

(5) The initial condition $u(x, 1) = \sin(x)$ provides

$$\sin(x) = u(x, 1) = f\left(\frac{1}{x}\right), \quad \text{so that} \quad f(x) = \sin\left(\frac{1}{x}\right).$$

Thus $u = u(x, y) = \sin(x/y)$ a solution of the corresponding initial value problem.

- We search for solutions of the PDE

$$\frac{1}{x}\, u_x + y^3\, u_y = 0\,.$$

We obtain step by step:

$$(1)\ \frac{\mathrm{d}y}{\mathrm{d}x} = xy^3 \quad (2)\ y = \frac{1}{\sqrt{-x^2 - 2c}} \quad (3)\ c = -\frac{1}{2}\left(x^2 + \frac{1}{y^2}\right).$$

(4) For every differentiable function f is

$$u = u(x, y) = f\left(-\frac{1}{2}\left(x^2 + \frac{1}{y^2}\right)\right)$$

a solution of the PDE. ■

We thus have a systematic to solve arbitrary linear homogeneous PDEs of 1st order (provided we can solve the resulting systems of ODEs). We will be able to take advantage of this: In the next section, we consider *quasilinear* 1st-order PDEs. Every such PDE can be traced back to a linear homogeneous 1st-order PDE. And this one we can now solve. But this also gives us a scheme to solve arbitrary quasilinear PDEs.

86.3 The First Order Quasi Linear PDE

A PDE of the form

$$a(x, y, u(x, y))\, u_x + b(x, y, u(x, y))\, u_y = c(x, y, u(x, y))$$

with differentiable functions $a = a(x, y, u)$, $b = b(x, y, u)$, $c = c(x, y, u)$ is called **quasi linear differential equation** (first order). The difference to the linear PDE is that the searched function $u = u(x, y)$ may appear arbitrarily complicated in the coefficient functions a and b and in the inhomogeneity c. To solve a quasi linear PDE, proceed as follows:

Recipe: Solution of a First Order Quasi Linear PDE
Given is the quasilinear PDE

$$a(x, y, u)\, u_x + b(x, y, u)\, u_y = c(x, y, u).$$

(1) Consider the linear PDE in three variables x, y, u:

$$a(x, y, u)\, F_x + b(x, y, u)\, F_y + c(x, y, u)\, F_u = 0.$$

(continued)

(2) Solve the linear PDE from (1) using the recipe of Sect. 86.2 (case (ii)) and obtain $F = F(x, y, u)$.
(3) By $F(x, y, u) = 0$ implicitly a solution $u = u(x, y)$ is given.

Example 86.3 We consider the quasi linear PDE

$$y\, u_x - x\, u_y = x\, u^2 .$$

(1) We proceed to the linear PDE

$$y\, F_x - x\, F_y + x\, u^2\, F_u = 0 .$$

(2) We follow the prescription of Sect. 86.2, where we have $\frac{dx}{dy}$, $\frac{du}{dy}$:

$$\frac{dx}{dy} = -\frac{y}{x} \text{ and } \frac{du}{dy} = -u^2 .$$

From this we obtain by solving these ODEs:

$$c_1 = \frac{1}{2}(x^2 + y^2) \text{ and } c_2 = \frac{1}{u} - y$$

and herewith for any differentiable function f is the solution

$$F(x, y, u) = f\left(\frac{1}{2}(x^2 + y^2),\ \frac{1}{u} - y\right) .$$

(3) By

$$f\left(\frac{1}{2}(x^2 + y^2),\ \frac{1}{u} - y\right) = 0$$

implicitly a solution $u = u(x, y)$ given. ■

86.4 The Characteristics Method

With the **characteristics method** it is often possible to find the solution $u = u(x, y)$ of a *boundary* or *initial value problem* with a PDE explicitly. We show how to apply the procedure to a problem of the following form, a generalization of the procedure to similar problems is then easily possible. We consider the boundary value problem

$$(*) \qquad a(x, y)\, u_x + b(x, y)\, u_y = c(x, y) \text{ on } B \subseteq \mathbb{R}^2 \text{ with } u|\partial B = g(x)\,.$$

The decisive role for the determination of the searched function $u = u(x, y)$ play (still to be determined) curves

$$\gamma : [0, l] \to B \subseteq \mathbb{R}^2\,,\ \gamma(s) = \begin{pmatrix} x(s) \\ y(s) \end{pmatrix}.$$

These curves in B are called **characteristics**. We make the approach

$$z(s) = u(x(s), y(s))\,.$$

Note: $z = u(x(s), y(s))$ is the solution on the characteristic $(x(s), y(s))^\top$ (in B), but not yet the solution u on B we are looking for. But sometimes it is possible to deduce from this solution z on the characteristic to the solution u on the whole B.

We now differentiate the approach $z(s) = u(x(s), y(s))$ to s and obtain:

$$\dot{z}(s) = u_x\, \dot{x}(s) + u_y\, \dot{y}(s)\,.$$

Note the similarity of this equation to the original PDE in (∗). This comparison suggests, $\dot{x} = a$ and $\dot{y} = b$ which yields $\dot{z} = c$. More precisely, we obtain the following system of differential equations, which is also called the **system of characteristic differential equations**:

$$\begin{aligned} \dot{x}(s) &= a(x, y) \ \text{ with } x(0) = c_1\,, \\ \dot{y}(s) &= b(x, y) \ \text{ with } y(0) = c_2\,, \\ \dot{z}(s) &= c(x, y) \ \text{ with } z(0) = u(x(0), y(0)) = u(c_1, c_2)\,. \end{aligned}$$

Here the initial point (c_1, c_2) of the characteristic $(x(s), y(s))^\top$ is a (general) point of the boundary ∂B of B.

If one can determine $z(s) = u(x(s), y(s))$ from this system, then this solution depends on s and of the initial values $(c_1, c_2) \in \partial B$. If one manages to determine both the curve parameter s as well as the initial conditions c_1, c_2 by x and y, we obtain an explicit representation of the function we are looking for $u = u(x, y)$. In the following examples we show typical procedures:

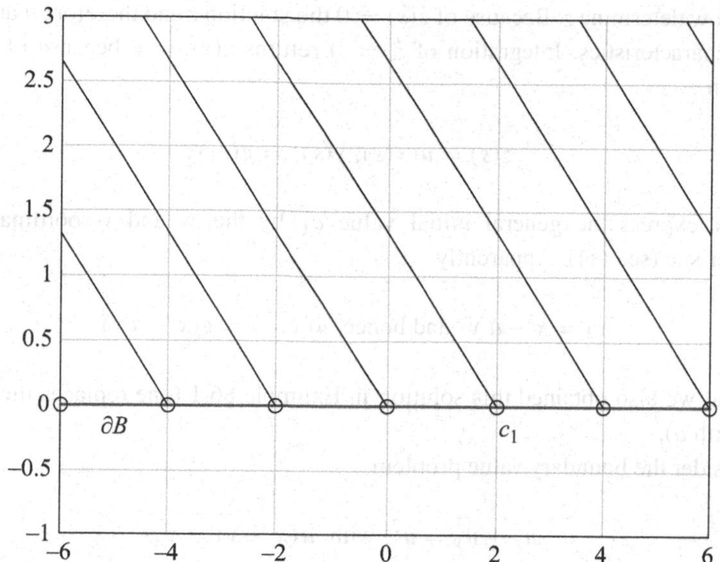

Fig. 86.3 The characteristics are *half-straights* with slope a, which start at the edge of B at the points $(c_1, 0)$

Example 86.4

- We consider the boundary value problem

$$a\,u_x + u_y = 0 \text{ with } u(x, 0) = g(x)$$

with a constant $a \in \mathbb{R}$. The system of characteristic differential equations is

$$\begin{aligned}
\dot{x}(s) &= a \text{ with } x(0) = c_1\,,\\
\dot{y}(s) &= 1 \text{ with } y(0) = c_2 = 0\,,\\
\dot{z}(s) &= 0 \text{ with } z(0) = u(x(0), y(0)) = u(c_1, 0) = g(c_1)\,.
\end{aligned}$$

The characteristics have due to $x(s) = a\,s + c_1$ and $y(s) = s$ the form

$$(*) \qquad \gamma(s) = \begin{pmatrix} x(s) \\ y(s) \end{pmatrix} = \begin{pmatrix} a\,s + c_1 \\ s \end{pmatrix},$$

note Fig. 86.3.

We now determine z. Because of $\dot{z}(s) = 0$ the solution z and therefore u are constant on the characteristics. Integration of $\dot{z} = 0$ returns $z(s) = c$ because of the initial condition

$$z(s) = u(x(s), y(s)) = g(c_1) .$$

We now express the general initial value c_1 by the x- and y-coordinates of the characteristic (see $(*)$): Apparently

$$c_1 = x - a\,y \quad \text{and hence} \quad u(x, y) = g(x - a\,y) .$$

Note that we also obtained this solution in Example 86.1 (one replaces there t with y and v with a).

- We consider the boundary value problem

$$u_x + u_y = u^2 \quad \text{with} \quad u(x, -x) = x .$$

The system of characteristic differential equations is

$$\begin{aligned} \dot{x}(s) &= 1 \quad \text{with } x(0) = c_1 , \\ \dot{y}(s) &= 1 \quad \text{whith } y(0) = -c_1 , \\ \dot{z}(s) &= z^2 \quad \text{with } z(0) = u(x(0), y(0)) = u(c_1, -c_1) = c_1 . \end{aligned}$$

The characteristics have due to $x(s) = s + c_1$ and $y(s) = s - c_1$ the form

$$(*) \qquad \gamma(s) = \begin{pmatrix} x(s) \\ y(s) \end{pmatrix} = \begin{pmatrix} s + c_1 \\ s - c_1 \end{pmatrix} ,$$

note Fig. 86.4.

We now determine z. The ODE $\dot{z}(s) = z^2$ is separable:

$$\frac{\mathrm{d}z}{z^2} = \mathrm{d}s \Rightarrow -\frac{1}{z} = s + c \Rightarrow z = \frac{-1}{s + c} .$$

If we now add the initial condition $z(0) = c_1$ one obtains $c = -\frac{1}{c_1}$. Thus the solution we are looking for is

$$z(s) = u(x(s), y(s)) = \frac{-1}{s - \frac{1}{c_1}} .$$

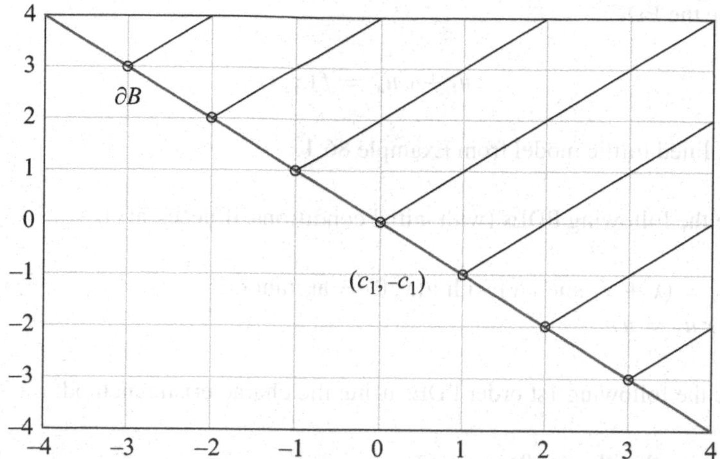

Fig. 86.4 The characteristics are *half-straights* with slope 1, which start at the edge of B at the points $(c_1, -c_1)$

We now express the general initial value c_1 and the curve parameter s by the x- and y-coordinates of the characteristic (see (∗)): Obviously

$$c_1 = \frac{x-y}{2} \text{ and } s = \frac{x+y}{2}.$$

Replacing c_1 and s in the above solution gives the general solution

$$u(x, y) = \frac{-1}{\frac{x+y}{2} - \frac{2}{x-y}}.$$

■

86.5 Exercises

86.1 Justify why the recipe for solving linear 1st order PDEs with constant coefficients in Sect. 86.1 works.

86.2 Solve the following PDEs:

(a) $2u_x + 3u_t = e^{x+t}$,
(b) $yu_x - xu_y = 0$,
(c) $(x + u)u_x + (y + u)u_y = -u$,
(d) $xu_x + yu_y = xu$.

86.3 Derive the PDE

$$u_t + v\,u_x = f(x,t)$$

for the simplified traffic model from Example 86.1.

86.4 Solve the following PDEs (with initial conditions, if necessary):

(a) $u_x + u_y = (x+y)\sin(x\,y)$ with $u(x,0) = \arctan(x)$.
(b) $y\,u_x + x\,u_y = u$.

86.5 Solve the following 1st order PDEs using the characteristic method:

(a) $u_x + 2u_y = 0$ with $u(x,0) = u_0(x)$.
(b) $xu_x + yu_y + u_z = u$ with $u(x,y,0) = xy$.

86.6 We consider for $u : [0,\infty) \times \mathbb{R} \to \mathbb{R}$ the initial value problem of the Burgers equation,

$$u_t + uu_x = 0 \quad \text{with} \quad u(0,x) = u_0(x) = \begin{cases} 1, & x \leq -1, \\ -x, & -1 < x \leq 0, \\ 0, & 0 < x. \end{cases}$$

(a) Apply the method of characteristics to transform the PDE into a system of ODEs.
(b) Solve the system of ODEs obtained in (a).
(c) Sketch the characteristic curves in the x-t-plane and write down the solution $u(t,x)$ obtained in (b) as explicitly as possible.
(d) Determine a point in time t_* at which two different characteristic curves intersect.
(e) Use (d) to justify that the found solution $u(t,x)$ is not valid for all $t > 0$ can be continuous.

Partial Differential Equations of Second Order: General

87

The 2nd order partial differential equations are those which are essential for the applications. We can only give a tiny glimpse of the extensive theory of these differential equations here. Analytic solution schemes, such as we considered in the last chapter on 1st order equations, also exist for 2nd order equations. Only these are considerably more complicated; we dispense with these representations and give systematic solution procedures only for special differential equations in the further chapters.

In order to get a certain overview of the rather complicated situation of 2nd order differential equations, we introduce in the present chapter some terms concerning the classification as well as solution methods of such differential equations.

87.1 First Terms

We have also distinguished different types of ODEs, such as *linear* and *nonlinear*. In the case of PDEs, even further distinctions are possible.

87.1.1 Linear-Nonlinear, Stationary-Nonstationary

We distinguish linear and nonlinear PDEs. A PDE is called **linear**, if the searched function u and its partial derivatives only in 1st power and not in sin-, exp- ... functions appear.

C. Karpfinger, *Calculus and Linear Algebra in Recipes*,
https://doi.org/10.1007/978-3-662-65458-3_87

Example 87.1 Linear PDEs are:

- **Laplace equation** $-\Delta u = 0$: This describes a potential, such as a gravitational potential or an electric field strength. The Laplace equation is **stationary** i.e., a time-independent process is described.
- **Heat conduction equation** $u_t - c^2 \Delta u = 0$ This describes a heat conduction or diffusion process. The heat conduction equation is **nonstationary** i.e., a time-dependent process is described.
- **Wave equation** $u_{tt} - c^2 \Delta u = 0$ The wave equation describes oscillation processes, such as the propagation of sound waves or electromagnetic waves. The wave equation is nonstationary, i.e. time-dependent.
- **Schrödinger equation** $\mathrm{i}\, u_t + \Delta u = 0$ This equation describes the motion of elementary particles. The Schrödinger equation is nonstationary, i.e. time-dependent.
- **Maxwell's equations**

$$\nabla \cdot E = \frac{\rho}{\varepsilon_0}, \quad \nabla \cdot B = 0, \quad \nabla \times E = -\frac{\partial B}{\partial t}, \quad \nabla \times B = \mu_0 j + \mu_0 \varepsilon_0 \frac{\partial E}{\partial t}.$$

  This system of linear PDEs describes the connection of electric and magnetic fields with electric charges and electric current.

Nonlinear PDEs are:

- **Burgers equation** $u_t + u\, u_x = 0$: This equation occurs with conservation laws. The Burgers equation is nonstationary.
- **Navier-Stokes equations**

$$\rho(\partial_t u + (u \cdot \nabla)u) = -\nabla p + \mu \delta u\,, \quad \nabla \cdot u = 0\,.$$

  This system of nonlinear PDEs describes a transient flow of an incompressible viscous fluid. The quantities sought are the velocity field u: $D \subseteq \mathbb{R}^3 \to \mathbb{R}^3$, the density ρ and the pressure p. ■

87.1.2 Boundary Value and Initial Boundary Value Conditions

We recall the notion of *initial value problem*: that was an (ordinary) ODE with an initial condition: This initial condition was used to select a (usually uniquely) determined solution from the variety of solutions to the ODE, which then satisfied the ODE and the initial condition. By the initial condition usually a constant c was nailed down.

For PDEs, the situation is similar: instead of an initial condition, one usually has *boundary value conditions* or *initial boundary value conditions*, more precisely:

- **Boundary value condition:** For a stationary PDE is that solution $u : B \subseteq \mathbb{R}^n \to \mathbb{R}$ sought, which on the boundary ∂B assumes given values.
- **Initial boundary value condition:** For a nonstationary PDE is that solution $u : B \subseteq \mathbb{R}^{n+1} \to \mathbb{R}$ sought, which on the boundary ∂B assumes given values and fulfills at the starting time $t = 0$ given conditions.

Among the usually many solutions of a PDE, one searches for that solution which satisfies given boundary value conditions or initial boundary value conditions. In the optimal case, this solution is then unique. Typically, however, one does not nail down constants c, but functions g, since in a PDE one generally has free-choice functions in the solution manifold (note the examples in Chap. 86).

87.1.3 Well-Posed and Ill-Posed Problems

A boundary value or initial boundary value problem is called **well posed**, if

- a solution u exists,
- the solution u is unique,
- the solution u depends continuously on the data.

If one of these three requirements is not fulfilled, one speaks of a **ill-posed** problem.

Continuous dependence on the data expresses *stability*: Error-prone input data, whose errors are to be sought in inaccuracies of measurements, also result in only relatively small deviations in the solutions. If this continuous dependence is not given, small inaccuracies in the PDE or in the boundary or initial values can lead to the fact that a possibly determined *solution* has almost nothing to do with the problem.

87.2 The Type Classification

Before we get down to solving 2nd order PDEs, we distinguish the types of 2nd order PDEs. In doing so, we make two major simplifications:

- The function u depends only on two variables x and y, so we have $u = u(x, y)$.
- We consider only linear PDEs.

The general linear PDE of 2nd order in two variables has the following appearance:

$$a(x, y)\, u_{xx} + 2b(x, y)\, u_{xy} + c(x, y)\, u_{yy} + d(x, y)\, u_x + e(x, y)\, u_y + f(x, y)\, u = g(x, y)\,.$$

Note that the solution u is supposed to be twice continuously partial differentiable, therefore, by Schwarz's Theorem $u_{xy} = u_{yx}$; this also explains the 2 before u_{xy}.

We call

$$a(x, y)\, u_{xx} + 2b(x, y)\, u_{xy} + c(x, y)\, u_{yy}$$

the **principal part** of the PDE, so the principal part is that part which includes the second partial derivatives. Only the principal part is crucial for the type classification of a PDE:

Type Classification of Second Order Linear PDEs

We consider the linear PDE

$$a(x, y)\, u_{xx} + 2b(x, y)\, u_{xy} + c(x, y)\, u_{yy} + d(x, y)\, u_x + e(x, y)\, u_y + f(x, y)\, u = g(x, y)\,.$$

One calls this PDE

- **elliptic** on $D \subseteq \mathbb{R}^2$ if $a(x, y)\, c(x, y) - b(x, y)^2 > 0$ for all $(x, y) \in D$,
- **parabolic** on $D \subseteq \mathbb{R}^2$ if $a(x, y)\, c(x, y) - b(x, y)^2 = 0$ for all $(x, y) \in D$,
- **hyperbolic** on $D \subseteq \mathbb{R}^2$ if $a(x, y)\, c(x, y) - b(x, y)^2 < 0$ for all $(x, y) \in D$,
- **of mixed type** on $D \subseteq \mathbb{R}^2$ if it has different behavior for different $(x, y) \in D$ has different behavior.

Remark These notations originate from the theory of quadrics: We have

$$\det \begin{pmatrix} a(x, y) & b(x, y) \\ b(x, y) & c(x, y) \end{pmatrix} = a(x, y)\, c(x, y) - b(x, y)^2\,,$$

which is why the quadric defined by the given symmetric matrix A is represented, in the case $\det(A) > 0$ is an ellipse, in the case $\det(A) = 0$ a parabola and in the case $\det(A) < 0$ a hyperbola.

Example 87.2

- The Laplace equation $-\Delta u = -u_{xx} - u_{yy} = 0$ is elliptic on $\mathbb{R}^2$.
- The heat conduction equation $u_t = c^2 u_{xx}$ is parabolic on $\mathbb{R}^2$.

- The wave equation $u_{tt} - c^2 u_{xx} = 0$ is hyperbolic on $\mathbb{R}^2$.
- The **Tricomial equation** $-y u_{xx} + u_{yy} = 0$ is on $D = \mathbb{R}^2$ of mixed type, it is elliptic for $y < 0$, parabolic for $y = 0$, and hyperbolic for $y > 0$. ■

We conclude this section with a brief overview (IV = initial value; BV = boundary value):

Type PDE	well posed	Representative	Application
Elliptic	BV	$-\Delta u = f(x, y)$	Mechanics, electr. engin., statics
Hyperbolic	IV-BV	$u_{tt} = \Delta u$	Waves, propagation
Parabolic	IV-BV	$u_t = \Delta u$	Heat conduction, diffusion
Mixed types	–	–	Semiconductors, fracture mechanics

87.3 Solution Methods

In practice, boundary value or initial value problems with PDEs are solved numerically. However, there are also a number of exact solution methods. Mostly these methods are very deep, a presentation of these methods would go beyond the scope of this book. An exact solution method, which occasionally also leads to interesting solutions and can be presented simply, is the *separation approach.*

87.3.1 The Separation Method

With the **separation method** one determines solutions u of the PDE which are of the form

$$u(x, y) = f(x)\,g(y)\,.$$

Entering this approach into the PDE, one then obtains two ODEs; one for f and one for g. One solves these; it is then $u(x, y) = f(x)\,g(y)$ a solution of the PDE. Thus one obtains in this way only solutions which are in this special form, namely products of functions in x and functions in y. For example, one obtains the simple solution $u(x, y) = x + y$ of the Laplace equation $u_{xx} + u_{yy} = 0$ not in this way. The solution of a PDE by the separation method can be easily formulated:

Recipe: Solving a PDE with the Separation Method

(continued)

To find solutions of a PDE by the separation method, proceed as follows:

(1) Put $u(x, y) = f(x)\, g(y)$ into the PDE and get two ODEs for f and g.
(2) Solve the two ODEs and get $f = f(x)$ and $g = g(y)$.
(3) It is $u = u(x, y) = f(x)\, g(y)$ a solution of the PDE.

Example 87.3 We determine solutions of the Laplace equation $-\Delta u(x, y) = 0$:

(1) We set $u(x, y) = f(x)\, g(y)$ into the PDE and obtain

$$-f'(x)g(y) - f(x)g'(y) = 0 \text{ i. e. } \frac{f''(x)}{f(x)} = -\frac{g''(y)}{g(y)}\,.$$

Now notice: We fix x and thus have a constant k on the left. But this means that for every y the right side has the value k. We can now do this analogously with a fixed y and variable x. We get:

$$\frac{f'(x)}{f(x)} = k \text{ and } -\frac{g'(y)}{g(y)} = k \text{ for } k \in \mathbb{R}\,.$$

Thus we have separated the functions f and g or the variables x and y.

(2) We solve the two ODEs $f' = k\, f$ and $-g' = k\, g$ with $k \in \mathbb{R}$ from (1) and obtain

$$f = f(x) = \begin{cases} c_1\, e^{\sqrt{k}x} + c_2\, e^{-\sqrt{k}x}\,, & \text{if } k > 0 \\ c_1 + c_2\, x\,, & \text{if } k = 0 \\ c_1 \cos(\sqrt{-k}\, x) + c_2 \sin(\sqrt{-k}\, x)\,, & \text{if } k < 0 \end{cases}$$

with $c_1, c_2 \in \mathbb{R}$ and

$$g = g(y) = \begin{cases} d_1 \cos(\sqrt{k}\, y) + d_2 \sin(\sqrt{k}\, y)\,, & \text{if } k > 0 \\ d_1 + d_2\, y\,, & \text{if } k = 0 \\ d_1\, e^{\sqrt{-k}y} + d_2\, e^{-\sqrt{-k}y}\,, & \text{if } k < 0 \end{cases}$$

with $d_1, d_2 \in \mathbb{R}$.

(3) It is $u = u(x, y) = f(x)\, g(y)$ a solution of the PDE. ■

In the following chapters, we consider Laplace's equation, the heat conduction equation and the wave equation. We will use the separation method to find solutions of these equations in the appropriate chapters.

Once one has a set of solutions of a PDE, it is then a matter of selecting from this multitude of solutions those that satisfy given boundary value or boundary and initial conditions. Essential for the solution of such a boundary value or boundary initial value problem is the **superposition principle** which states that any *superposition* of solutions of a linear homogeneous PDE is again a solution of the PDE.

87.3.2 Numerical Solution Methods

For the boundary or boundary initial value problems from practice, one usually relies on numerical solution methods: There does not exist an exact solution method for every type of a PDE. However, numerical methods are not universally suitable either, since they are often unstable, i.e., small errors in the initial data lead to strong fluctuations in the solutions; and the data of the problems from practice are by nature error-prone.

The most commonly used numerical methods are

- **finite difference method (FDM)**: This is an easy-to-understand method in which the derivatives, i.e., the differential quotients that appear in the PDE, are approximated by difference quotients.
- **finite element method (FEM)**: This method is probably the most popular method in applications. It is superior to the finite difference method for complicated geometric domains and is based on the functional analysis representation of the solution in special function spaces.
- **finite volume method (FVM)**: In this method, the PDE is written as an integral equation using, for example, Gauss' divergence theorem. This integral equation is then discretized on small standard volume elements.

These methods have one thing in common: they are all based on a discretization of the considered partial differential equation. In fact, however, each method is a science in itself. To give an overview of these methods is not possible within one book, we offer a little insight into FDM in the following chapters.

87.4 Exercises

87.1 Determine the types of the PDEs and sketch in the $\mathbb{R}^2$ if necessary the regions of different types:

(a) $2u_{xx} + 4u_{xy} + 2u_{yy} + 2u_x + 4u_y = 2u$,
(b) $x^3 u_{xx} + 2u_{xy} + y^3 u_{yy} + u_x - yu_y = \mathrm{e}^x$,
(c) $yu_{xx} + 2xu_{xy} + yu_{yy} = y^2 + \ln(1 + x^2)$.

87.2 Using the separation method, find solutions of the partial differential equations

(a) $x^2\, u_x + \frac{1}{y}\, u_y = u$,
(b) $x^2\, u_{xy} + 3y^2\, u = 0$.

87.3 The telegraph equation

$$u_{tt} - u_{xx} + 2\, u_t + u = 0$$

describes (qualitatively) the time course of a signal voltage u at location $x > 0$ in a long transmission cable. We are looking for the signal voltage $u(x, t)$ if at the boundary $x = 0$ of the transmission cable a periodic signal of the form $u(0, t) = 3\sin(2t)$ is fed in.

(a) Try to find a reasonable solution with the separation method $u(x, t) = f(x)\, g(t)$ to find a reasonable solution. (Note: You can determine g here already from the boundary condition).
(b) Calculate that the approach $u(x, t) = a\ \mathrm{e}^{-bx}\ \sin(2t - cx)$ with constants $a, b, c \in \mathbb{R}$, $b > 0$, leads to the goal.

87.4 Determine all rotationally symmetric (with respect to the origin) harmonic functions $u(x, y)$. A function $u = u(x, y)$ is called harmonic, if $u_{xx} + u_{yy} = 0$.
Hint. Consider the Laplace operator in polar coordinates.

87.5 Solve the following PDE using the given approach:

(a) $u_x + 2u_y = 0$ with $u(x, 0) = u_0(x)$ (separation method),
(b) $y^2(u_x)^2 + x^2(u_y)^2 = (xyu)^2$ (separation method),
(c) $yu_x + xu_y = 0$ (approach $u(x, y) = f(x) + g(y)$),
(d) $u_t + 2uu_x = u_{xx}$ (approach $u(t, x) = v(x - 2t)$ with $\lim\limits_{\xi \to -\infty} v(\xi) = 2$).

The Laplace or Poisson Equation

88

We consider some aspects of what is undoubtedly one of the most important partial differential equations, Laplace's or *Poisson's equation*. These stationary differential equations are elliptic. They typically describe a (stationary) temperature distribution or an electrostatic charge distribution in a body and thus more generally a state of equilibrium.

88.1 Boundary Value Problems for the Poisson Equation

The (stationary) Laplace equation $-\Delta u = 0$ is well known, the **Poisson's equation** is the inhomogeneous variant of it:

$$-\Delta u = f \text{ with a function } f .$$

Usually one considers two- or three-dimensional problems, i.e. $u = u(x, y)$ or $u = u(x, y, z)$. One considers boundary value problems to elliptic the PDE.

For the Poisson equation one distinguishes the following types of boundary value problems, in each case a function $u : D \cup \partial D \subseteq \mathbb{R}^n \to \mathbb{R}$ with $n = 2$ or $n = 3$ is sought:

Boundary Value Problems for the Poisson Equation

- **Dirichlet's boundary value problem:**

$$-\Delta u(\boldsymbol{x}) = f(\boldsymbol{x}) \text{ for all } \boldsymbol{x} \in D \text{ and } u(\boldsymbol{x}) = u_0(\boldsymbol{x}) \text{ for all } \boldsymbol{x} \in \partial D .$$

(continued)

C. Karpfinger, *Calculus and Linear Algebra in Recipes*,
https://doi.org/10.1007/978-3-662-65458-3_88

- **Neumann's boundary value problem:**

$$-\Delta u(\boldsymbol{x}) = f(\boldsymbol{x}) \text{ for all } \boldsymbol{x} \in D \text{ and } \frac{\partial u}{\partial \boldsymbol{n}}(\boldsymbol{x}) = u_0(\boldsymbol{x}) \text{ for all } \boldsymbol{x} \in \partial D\,.$$

- **The mixed boundary value problem:**

$$-\Delta u(\boldsymbol{x}) = f(\boldsymbol{x}) \text{ for all } \boldsymbol{x} \in D \text{ and}$$

$$\frac{\partial u}{\partial \boldsymbol{n}}(\boldsymbol{x}) + k(\boldsymbol{x})\,u(\boldsymbol{x}) = u_0(\boldsymbol{x}) \text{ for all } \boldsymbol{x} \in \partial D\,.$$

Here $\boldsymbol{n}$ is in each case a normal unit vector pointing out of D, and $k = k(\boldsymbol{x})$ is a continuous function. Furthermore one distinguishes:

- **The inner boundary value problem**, if D is a bounded domain, and
- **the outer boundary value problem**, if D is the complement of a bounded domain (in this case further boundary conditions must be imposed on the solutions).

88.2 Solutions of the Laplace Equation

Using the separation method from the recipe in Sect. 87.3.1 we obtain solutions of Laplace's equation. In doing so, we have several options. In the two-dimensional case we can formulate the Laplace equation in Cartesian or polar coordinates and separate them according to the respective variables, in the three-dimensional case we can use not only cartesian but also cylindrical or spherical coordinates. As an example we choose polar coordinates in the $\mathbb{R}^2$:

The Laplace equation in polar coordinates (r, φ) is

$$u_{rr} + \frac{1}{r}\,u_r + \frac{1}{r^2}\,u_{\varphi\varphi} = 0\,.$$

We now use the separation method from the recipe in Sect. 87.3.1 and find solutions u of Laplace's equation of the form

$$u = u(r, \varphi) = f(r)\,g(\varphi)\,,$$

where we specialize in g functions which are 2π-periodic; we want to solve Dirichlet's boundary value problem for a circle in the next section. Note the recipe in Sect. 87.3.1.

(1) Substituting $u(r, \varphi) = f(r)g(\varphi)$ in the PDE yields two ODEs for f and g:

$$0 = u_{rr} + \frac{1}{r}u_r + \frac{1}{r^2}u_{\varphi\varphi} = \left(f''(r) + \frac{1}{r}f'(r)\right)g(\varphi) + \frac{1}{r^2}f(r)g''(\varphi)$$

and thus leads because of $r > 0$ to

$$\frac{r^2 f'(r) + rf'(r)}{f(r)} = -\frac{g'(\varphi)}{g(\varphi)},$$

and finally yields the two ODEs:

$$r^2 f'(r) + rf'(r) - k\, f(r) = 0 \text{ and } g'(\varphi) = -k\, g(\varphi) \text{ for } k \in \mathbb{R}.$$

(2) We solve the two ODEs from (1):

- First, we take care of the function g:
  - In the case $k < 0$, the ODE has no 2π-periodic solutions g.
  - In the case $k = 0$, only the constant solution $g(\varphi) = c$ is periodic.
  - In the case $k > 0$ the ODE for g has the fundamental solutions $\sin(\sqrt{k}\,\varphi)$ and $\cos(\sqrt{k}\,\varphi)$, which are 2π-periodic if $k = n^2$ for a $n \in \mathbb{N}$, because the 2π-periodicity says

  $$\sin(\sqrt{k}\,(\varphi + 2\pi))) = \sin(\sqrt{k}\,\varphi) \text{ for all } \varphi \in \mathbb{R},$$

  analogously for the cosine.
  The 2π-periodic solutions g are thus

  $$g = g(\varphi) = a_n \cos(n\varphi) + b_n \sin(n\varphi) \text{ with } a_n,\, b_n \in \mathbb{R} \text{ for all } n \in \mathbb{N}_0.$$

- Now we care about f: Since $k = n^2$ for a $n \in \mathbb{N}_0$, we obtain for f the Eulerian ODE

  $$r^2 f'' + rf' - n^2 f = 0.$$

  The general solution of this ODE is according to Example 35.3

  $$f(r) = \begin{cases} c_1 r^n + c_2 r^{-n}, & \text{if } n \neq 0 \\ c_1 + c_2 \ln(r), & \text{if } n = 0 \end{cases}.$$

(3) Thus we have now found the following solutions:

$$u = u(r,\varphi) = \begin{cases} u_n(r,\varphi) = (a_n\cos(n\varphi) + b_n\sin(n\varphi))r^n & \text{for } n \in \mathbb{Z}\setminus\{0\} \\ a + b\,\ln(r) & \text{for } n = 0 \end{cases}$$

for real a_n or b_n and a and b.

We summarize:

Solutions of the Laplace Equation
For any real a_n and b_n or a and b is

$$u(r,\varphi) = \begin{cases} u_n(r,\varphi) = (a_n\cos(n\varphi) + b_n\sin(n\varphi))r^n & \text{for } n \in \mathbb{Z}\setminus\{0\} \\ a + b\,\ln(r) & \text{for } n = 0 \end{cases}$$

a solution of the Laplace equation.

88.3 The Dirichlet Boundary Value Problem for a Circle

We now consider more concretely the inner Dirichlet boundary value problem for a circle of radius R around the point 0, i.e.

$$D = \{(x,y) \mid x^2 + y^2 < R^2\} \text{ and } \partial D = \{(x,y) \mid x^2 + y^2 = R^2\}.$$

If we search for solutions of a boundary value problem on a circle D around the zero point, then all singular terms ($n < 0$) must vanish. Then, by superposition of the remaining solutions, one obtains

$$u(r,\varphi) = \frac{a_0}{2} + \sum_{n=1}^{\infty}(a_n\cos(n\varphi) + b_n\sin(n\varphi))r^n\,,$$

where we have written $\frac{a_0}{2}$ in the case $n = 0$ instead of a (the solution in this case is $a + b\,\ln(r)$ with $b = 0$). Substituting a boundary condition of the form $u(r,\varphi) = u_0(\varphi)$ for $r = R$, where R is the radius of the circle D, provides a sine-cosine representation of the boundary condition:

$$u_0(\varphi) = \frac{a_0}{2} + \sum_{n=1}^{\infty}(a_nR^n)\cos(n\varphi) + (b_nR^n)\sin(n\varphi)\,.$$

Since the sine-cosine representation is unique, we thus obtain the (real) Fourier coefficients of the 2π-periodic function $u_0(\varphi)$ we obtain the uniquely determined solution of Dirichlet's boundary value problem for a circle:

Recipe: Solving a Dirichlet Boundary Value Problem for a Circle
The solution $u = u(r, \varphi)$ in polar coordinates of Dirichlet's boundary value problem

$$-\Delta u(x, y) = 0 \text{ for } x^2 + y^2 < R^2 \text{ and } u(x, y) = u_0(x, y) \text{ for } x^2 + y^2 = R^2$$

is obtained as follows:

(1) Determine the coefficients a_n and b_n of the sine-cosine representation of the 2π-periodic function $u_0(\varphi) : [0, 2\pi) \to \mathbb{R}$.
(2) Obtain the solution $u = u(r, \varphi)$ as a series representation in polar coordinates:

$$u(r, \varphi) = \frac{a_0}{2} + \sum_{k=1}^{\infty} (a_k \cos(k\varphi) + b_k \sin(k\varphi)) \left(\frac{r}{R}\right)^k .$$

Note that we have already treated this solution with the harmonic functions together with an example in Sect. 85.2.

Also, the *associated* outer space problem

$$-\Delta u(x, y) = 0 \text{ for } x^2 + y^2 > R^2 \text{ and } u(x, y) = u_0(x, y) \text{ for } x^2 + y^2 = R^2$$
$$\text{and } u(x, y) \text{ is bounded for } x^2 + y^2 \to \infty$$

can be solved by this method, we just need to exchange the roles of r and R in (2), thus one obtains the series representation of the solution of the outer space problem with

$$u(r, \varphi) = \frac{a_0}{2} + \sum_{k=1}^{\infty} (a_k \cos(k\varphi) + b_k \sin(k\varphi)) \left(\frac{R}{r}\right)^k .$$

88.4 Numerical Solution

As already mentioned several times, in general it is not possible to give an exact solution of a boundary value problem. One has to rely on numerical solution methods in this case. An obvious method for the approximate solution of a Dirichlet boundary value problem

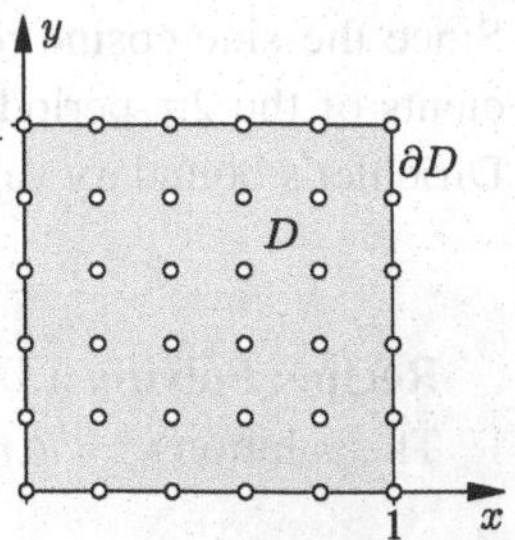

Fig. 88.1 A grid with $n = 4$

in the $\mathbb{R}^2$ is the one described in the following **difference method**. For this purpose, we consider the Dirichlet boundary value problem (for simplicity) on the square $D = [0, 1]^2$ in the first quadrant:

$$-\Delta u(x, y) = f(x, y) \text{ for } (x, y) \in D = (0, 1)^2 \text{ and } u(x, y) = u_0(x, y) \text{ for } (x, y) \in \partial D.$$

As in the numerical solution of an ODE we **discretize** we cover the domain D by a grid (see Fig. 88.1): We choose the *step size* $h = \frac{1}{n+1}$ with a $n \in \mathbb{N}$ in x- and y-direction and obtain grid points

$$(x_i, y_j) \text{ with } x_i = ih \text{ and } y_j = jh \text{ with } i, j = 0, \dots, n+1 .$$

Now we approximate the partial derivatives u_{xx} and u_{yy} of the function $u = u(x, y)$ in the inner grid points (x_i, y_j) by corresponding difference quotients, i.e., the second partial derivatives $u_{xx}(x_i, y_i)$ and $u_{yy}(x_i, y_i)$ are approximated by the quotients

$$u_{xx}(x_i, y_j) \approx \frac{u_{i+1,j} - 2u_{i,j} + u_{i-1,j}}{h^2} \text{ and } u_{yy}(x_i, y_j) \approx \frac{u_{i,j+1} - 2u_{i,j} + u_{i,j-1}}{h^2} ,$$

here are

$$u_{i+1,j} = u(x_{i+1}, y_j),\ u_{i,j} = u(x_i, y_j),\ u_{i-1,j} = u(x_{i-1}, y_j),\ \dots$$

the values sought—note the box with the formulas for numerical differentiation in Sect. 26.3. The *discretized* Poisson equation, with these abbreviations, is in a (x_i, y_j)

$$f(x_i, y_j) = -\Delta u(x_i, y_j) \approx -\frac{1}{h^2} \left(u_{i+1,j} + u_{i-1,j} + u_{i,j+1} + u_{i,j-1} - 4u_{i,j}\right) ,$$

where the numbers $f(x_i, y_j)$ as well as the boundary values $u_{0,j}$, $u_{i,0}$, $u_{n+1,j}$ and $u_{i,n+1}$ are known. We now replace $\approx$ by $=$ and thus obtain n^2 equations, $i, j = 1, \dots, n$ whose solutions u_{ij} are approximate solutions for the searched values $u(x_i, y_j)$.

Fig. 88.2 The 5-point star

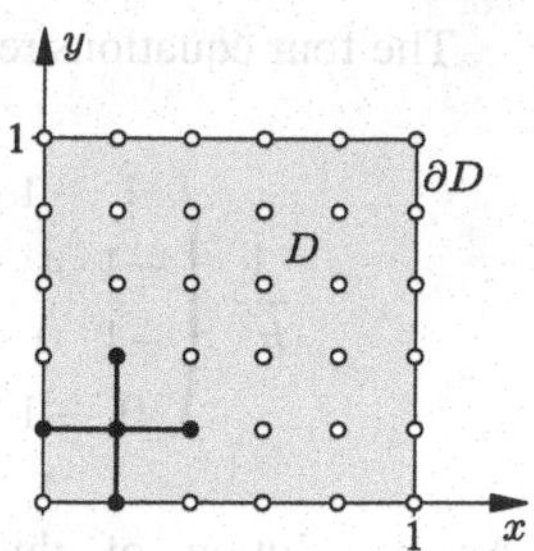

Because of the form of the equations

$$f(x_i, y_j) = \frac{1}{h^2}\left(4u_{i,j} - u_{i+1,j} - u_{i-1,j} - u_{i,j+1} - u_{i,j-1}\right)$$

one speaks of the **5-point star** (Fig. 88.2).

From the boundary condition $u(x, y) = u_0(x, y)$ for all $(x, y) \in \partial D$ we know which values the function u on the boundary assumes. We now, in a sense, set the 5-point star from lattice point to lattice point, grazing the boundary where we know the values of the *function u*.

We demonstrate this with a simple example:

Example 88.1 In the case $n = 2$, the four equations for (x_i, y_j) with $i, j = 1, 2$ with $h = \frac{1}{3}$ are as follows:

$$f(x_1, y_1) = \frac{1}{h^2}\left(4\,u_{1,1} - u_{2,1} - u_{0,1} - u_{1,2} - u_{1,0}\right),$$

$$f(x_1, y_2) = \frac{1}{h^2}\left(4\,u_{1,2} - u_{2,2} - u_{0,2} - u_{1,3} - u_{1,1}\right),$$

$$f(x_2, y_1) = \frac{1}{h^2}\left(4\,u_{2,1} - u_{3,1} - u_{1,1} - u_{2,2} - u_{2,0}\right),$$

$$f(x_2, y_2) = \frac{1}{h^2}\left(4\,u_{2,2} - u_{3,2} - u_{1,2} - u_{2,3} - u_{2,1}\right).$$

Known here are the values $f(x_i, y_j)$ and the boundary values $u_{0,j}$, $u_{3,j}$, $u_{i,0}$, $u_{i,3}$.

The four equations result in the following linear system of equations:

$$\frac{1}{h^2}\begin{pmatrix} 4 & -1 & -1 & 0 \\ -1 & 4 & 0 & -1 \\ -1 & 0 & 4 & -1 \\ 0 & -1 & -1 & 4 \end{pmatrix}\begin{pmatrix} u_{1,1} \\ u_{1,2} \\ u_{2,1} \\ u_{2,2} \end{pmatrix} = \begin{pmatrix} f(x_1, y_1) + \frac{1}{h^2}(u_{0,1} + u_{1,0}) \\ f(x_1, y_2) + \frac{1}{h^2}(u_{0,2} + u_{1,3}) \\ f(x_2, y_1) + \frac{1}{h^2}(u_{3,1} + u_{2,0}) \\ f(x_2, y_2) + \frac{1}{h^2}(u_{3,2} + u_{2,3}) \end{pmatrix}.$$

As a solution of this LGS we obtain the approximations we are looking for $u_{1,1}, u_{1,2}, u_{2,1}, u_{2,2}$ for the values of u in the inner grid points. ■

In general we obtain n^2 equations, which can be derived by appropriately numbering the inner grid points, e.g. $u_1 = u_{1,1},\ u_2 = u_{1,2}, \ldots, u_{n^2} = u_{n,n}$ can be formulated in an uniquely solvable LGS. This system of equations reads in the case of the **zero boundary condition** $u(x, y) = 0$ for all $(x, y) \in \partial D$, i.e.

$$u_{0,j} = u_{n+1,j} = u_{i,0} = u_{i,n+1} = 0$$

as follows:

$$A_h\, u_h = f_h \text{ with } u_h = \begin{pmatrix} u_1 \\ \vdots \\ u_{n^2} \end{pmatrix} \text{ and } f_h = \begin{pmatrix} f(x_1, y_1) \\ \vdots \\ f(x_n, y_n) \end{pmatrix}$$

and

$$A_h = \frac{1}{h^2}\left.\begin{pmatrix} B_n & -E_n & & \\ -E_n & B_n & -E_n & \\ & -E_n & B_n & \ddots \\ & & \ddots & \end{pmatrix}\right\} n\,,\quad B_n = \left.\begin{pmatrix} 4 & -1 & & \\ -1 & 4 & -1 & \\ & -1 & 4 & \ddots \\ & & \ddots & \end{pmatrix}\right\} n\,.$$

Note that the system of equations can be stated so concretely only because the domain we are considering is $D = [0, 1]^2$ is so simple. The appearance of the matrix A_h depends strongly on the considered area D and becomes *unpleasant* in realistic cases. In these cases the matrix and the right side are generated programmatically.

Indeed, if the region under consideration is not a rectangle, usually one corner of the 5-point star no longer falls on the boundary. Note that our unit square is an exceedingly lucky special case, here such a thing does not happen. In other cases, various boundary approximation methods are common.

The linear system of equations to be solved is very large and sparse. To solve it, one uses the iterative procedures that we described in Chap. 71 which are tailor-made for such

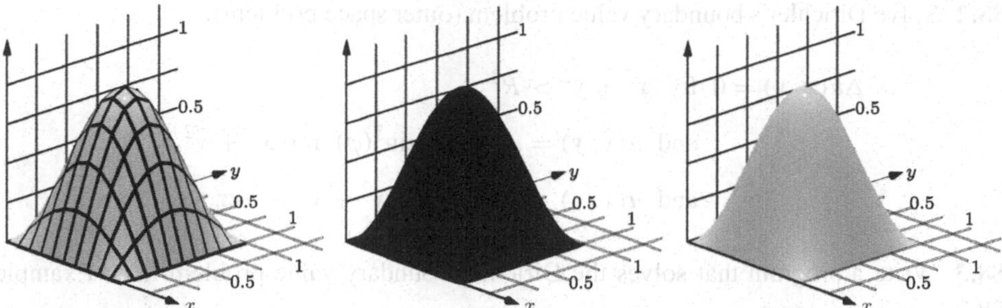

Fig. 88.3 Approximate solutions with $n = 10$, $n = 100$ and the exact solution

systems of equations. And yet there is a problem: The larger the number of unknowns u_{ij} the worse conditioned is the matrix A_h. Therefore, in practice, one needs other methods after all. One uses other discretizations, e.g. *multigrid methods*.

Example 88.2 We consider Dirichlet's boundary value problem.

$$-\Delta u(x, y) = -\sin(\pi x)\sin(\pi y) \text{ on } D = (0, 1)^2 \text{ with } u(x, y) = 0 \text{ on } \partial D.$$

To compute approximate solutions, we use MATLAB (see Exercise 88.5). The following Fig. 88.3 shows approximate solutions for $n = 10$ and $n = 100$ as well as the exact solution $u(x, y) = \sin(\pi x)\sin(\pi y)$. ■

Remark The Laplace equation is also called **potential equation**. This naming is obvious. For if $\boldsymbol{v}$ is a vortex- and source-free vector field, i.e. $\operatorname{rot} \boldsymbol{v} = \boldsymbol{0}$ and $\operatorname{div} \boldsymbol{v} = 0$, i.e. the velocity field of a stationary flow of incompressible fluids, then there exists a potential $-U$ of $\boldsymbol{v}$ due to $\operatorname{rot} \boldsymbol{v} = \boldsymbol{0}$, i.e. $-\nabla U = \boldsymbol{v}$. Because of $\operatorname{div} \boldsymbol{v} = 0$ one obtains with $\operatorname{div} \nabla = \Delta$:

$$-\Delta U = 0.$$

88.5 Exercises

88.1 Solve Dirichlet's boundary value problem (interior problem):

$$-\Delta u(x, y) = 0 \text{ for } x^2 + y^2 < 4 \text{ and } u(x, y) = u_0(\varphi) = \sin^3(\varphi) \text{ for } x^2 + y^2 = 4.$$

88.2 Solve Dirichlet's boundary value problem (outer space problem):

$$-\Delta u(x, y) = 0 \text{ for } x^2 + y^2 > R^2$$
$$\text{and } u(x, y) = u_0(\varphi) = \sin^3(\varphi) \text{ for } x^2 + y^2 = R^2$$
$$\text{and } u(x, y) \text{ bounded for } x^2 + y^2 \to \infty\,.$$

88.3 Write a program that solves the Dirichlet boundary value problem from Example 88.2.

88.4 We consider the following Poisson equation with zero boundary condition on the unit square.

$$\begin{aligned} -\left(u_{xx} + u_{yy}\right) &= 5x && \text{for } x \in (0, 1),\ y \in (0, 1), \\ u(x, 0) = u(x, 1) &= 0 && \text{for } x \in [0, 1], \\ u(0, y) = u(1, y) &= 0 && \text{for } y \in [0, 1]. \end{aligned}$$

We are looking for approximations $u_{i,j} \approx u(x_i, y_j)$ at the 16 points

$$\left\{(x_i, y_j) \,|\, x_i = \tfrac{i}{5},\, y_j = \tfrac{j}{5},\ i, j = 1, \ldots, 4\right\}$$

using the 5-point star.

Set up a linear system of equations for the unknowns $u_{i,j}$.

88.5 We consider the Laplace equation $-\Delta u = 0$ on the set $D = [0, 1] \times [0, 1]$.

(a) Find a solution to the equation that has the boundary value $u(x, 0) = \sin(\pi x)$ for $x \in [0, 1]$ and $u(x, 1) = u(0, y) = u(1, y) = 0$ for $x, y \in [0, 1]$.
(b) Find a solution to the equation that takes the boundary value $u(x, 0) = \sin(2\pi x)$ for $x \in [0, 1]$ and $u(x, 1) = u(0, y) = u(1, y) = 0$ for $x, y \in [0, 1]$.
(c) Verify that the equation satisfies the superposition principle: If u_1 and u_2 solutions of the equation and $c_1, c_2 \in \mathbb{R}$ then also $c_1 u_1 + c_2 u_2$ are solution of the equation.
(d) Use (a)-(c) to give a solution that fulfills the boundary values $u(x, 0) = \sin(\pi x)(1 + 2\cos(\pi x))$ for $x \in [0, 1]$ and $u(x, 1) = u(0, y) = u(1, y) = 0$ for $x, y \in [0, 1]$.

The Heat Conduction Equation

89

The heat conduction equation is a typical representative of a parabolic partial differential equation. The heat conduction equation is nonstationary and describes the relationship between the temporal and the spatial change of the temperature at a location in a heat-conducting body.

The heat conduction equation can also be interpreted as a diffusion equation, where *heat* is to be interpreted as a *concentration*. The solution u then describes the concentration distribution of a diffusing substance instead of the heat distribution in a heat-conducting body.

89.1 Initial Boundary Value Problems for the Heat Conduction Equation

Since the **heat conduction equation**

$$u_t = c^2 \Delta u$$

with a constant $c > 0$, the **thermal diffusivity coefficient**, is nonstationary, it typically occurs in the context of an initial boundary value problem:

The Initial Boundary Value Problem for the Heat Conduction Equation
A solution $u = u(\boldsymbol{x}, t)$ is sought which satisfies the heat conduction equation

$$u_t(\boldsymbol{x}, t) = c^2 \Delta u(\boldsymbol{x}, t) \quad \text{for } \boldsymbol{x} \in D \text{ and } t \in I$$

(continued)

C. Karpfinger, *Calculus and Linear Algebra in Recipes*,
https://doi.org/10.1007/978-3-662-65458-3_89

and the boundary and initial conditions:

$$u(\boldsymbol{x}, t) = f(\boldsymbol{x}) \text{ for } \boldsymbol{x} \in \partial D, \ t \in I \text{ and } u(\boldsymbol{x}, 0) = g(\boldsymbol{x}) \text{ for } \boldsymbol{x} \in D.$$

Note that our formulation is quite general, in general $\boldsymbol{x}$ is an element of $\mathbb{R}$ or $\mathbb{R}^2$ or $\mathbb{R}^3$:

- $\boldsymbol{x} \in \mathbb{R}$: Then D is typically an interval $D = [a, b]$, ∂D is then the boundary of the interval, i.e. the points a and b. Interpretation: What needs to be determined is the temperature distribution $u = u(x, t)$ in a bar of length $b - a$, where the temperature at the beginning and end of the bar is fixed and $g(x)$ is an initial distribution of the temperature in the bar.
- $\boldsymbol{x} \in \mathbb{R}^2$: Then D typically a circle $D = \{(x, y) \mid x^2 + y^2 \leq R^2\}$, ∂D is then the boundary of the circle, i.e. the points (x, y) with $x^2 + y^2 = R^2$. Interpretation: The temperature distribution $u = u(\boldsymbol{x}, t)$ in a plate, where the temperature is fixed at the boundary of the plate and is $g(\boldsymbol{x})$ is an initial distribution of the temperature in the plate.
- $\boldsymbol{x} \in \mathbb{R}^3$: Then D typically a sphere or a cylinder, ∂D is then the spherical or cylindrical surface. Interpretation: To be determined is the temperature distribution $u = u(\boldsymbol{x}, t)$ in the sphere or in the cylinder, where the temperature is fixed at the surface and $g(\boldsymbol{x})$ is an initial distribution of the temperature in the sphere or cylinder.

Remark The stationary variant of the heat conduction equation is Laplace's equation: it is then $u_t = 0$.

89.2 Solutions of the Equation

Using the separation methode from the recipe in Sect. 87.3.1. we obtain solutions of the heat conduction equation. In principle we have in the case $\boldsymbol{x} \in \mathbb{R}^n$ with $n = 2$ or $n = 3$, as in the case of the Laplace equation, the choice between different coordinate systems. However, for simplicity, we restrict ourselves to the one-dimensional case. We thus consider the heat conduction equation

$$u_t - c^2 u_{xx} = 0.$$

We use now the separation method from the recipe in Sect. 87.3.1 and find solutions u to the above heat conduction equation of the form

$$u = u(x, t) = f(x)\, g(t).$$

Note the recipe in Sect. 87.3.1:

(1) Substitution of $u(x,t) = f(x)g(t)$ into the PDE yields:

$$0 = u_t - c^2 u_{xx} = f(x)\,g'(t) - c^2 f''(x)\,g(t)\,,\ \text{i.e.}\ \frac{g'(t)}{g(t)} = c^2\frac{f''(x)}{f(x)}\,.$$

From this we obtain the two ODEs:

$$\frac{g'(t)}{g(t)} = k \ \text{ and } \ c^2\frac{f'(x)}{f(x)} = k \ \text{ i.e., } \ g' = k\,g \ \text{ and } \ f' = \frac{k}{c^2}\,f\,,$$

where $k \in \mathbb{R}$.

(2) We solve the two ODEs from (1):

- First, we take care of the function g: for every $k \in \mathbb{R}$ is

$$g = g(t) = c\ \mathrm{e}^{kt} \ \text{ with } \ c \in \mathbb{R}$$

a solution.

- Now we take care of f: the solutions f are

$$f = f(x) = \begin{cases} c_1\,\mathrm{e}^{\sqrt{k}x/c} + c_2\,\mathrm{e}^{-\sqrt{k}x/c}\,, & \text{if } k > 0 \\ c_1 + c_2\,x\,, & \text{if } k = 0 \\ c_1\,\cos(\sqrt{|k|}x/c) + c_2\,\sin(\sqrt{|k|}x/c)\,, & \text{if } k < 0 \end{cases}$$

with $c_1, c_2 \in \mathbb{R}$.

(3) We have now found the following solutions:

$$u = u(x,t) = \begin{cases} a\,\mathrm{e}^{\sqrt{k}x/c+kt} + b\,\mathrm{e}^{-\sqrt{k}x/c+kt}\,, & \text{if } k > 0 \\ a + b\,x\,, & \text{if } k = 0 \\ \mathrm{e}^{kt}\left(a\,\cos(\sqrt{|k|}x/c) + b\,\sin(\sqrt{|k|}x/c)\right)\,, & \text{if } k < 0 \end{cases}$$

for real a and b.

Each of these given functions u is a solution of the heat conduction equation. But some of these solutions are not interesting: In the case $k > 0$ we would have an exponential increase of the temperature in time t, in the case $k = 0$ we would have only stationary solutions. What is interesting for us is the case $k < 0$. We simplify the notation in this case by writing $\tilde{k} = \sqrt{|k|}/c$. Then we have $k = -c^2\tilde{k}^2$. The physically meaningful solutions are thus, where we again write k instead of $\tilde{k}$:

Solutions of the Heat Conduction Equation
For any real a and b and for any $k \in \mathbb{R}$ is

$$u = u(x, t) = \mathrm{e}^{-c^2k^2t}\ (a\ \cos(k\,x) +\ b\ \sin(k\,x))$$

is a solution of the heat conduction equation.

89.3 Zero Boundary Condition: Solution with Fourier Series

We now consider somewhat more concretely the initial boundary value problem with the heat conduction equation for a thermally conductive bar of length l, with ends *on ice*: We are looking for the function $u = u(x, t)$ with

$$u_t - c^2 u_{xx} = 0 \ \text{ with } \ u(x, 0) = g(x) \ \text{ and } \ u(0, t) = u(l, t) = 0\,,$$

is present $g(x)$ is the initial temperature distribution. We explain the solution in 3 steps:

Step 1. *Determine solutions of the heat conduction equation:* we consider as large a set of solutions of the heat conduction equation as possible. We have already compiled these in the box above:

$$u = u(x, t) = \mathrm{e}^{-c^2k^2t}\ (a\ \cos(k\,x) +\ b\ \sin(k\,x))\ , \quad a,\ b,\ k \in \mathbb{R}\,.$$

Step 2. *The boundary conditions impose constants, we obtain a general solution:* among the solutions from step 1, we determine the one that also satisfies the boundary conditions $u(0, t) = u(l, t) = 0$:

- From $u(0, t) = 0$ for all t follows $a = 0$.
- From $a = 0$ and $u(l, t) = 0$ for all t follows $b \sin(k\,l) = 0$ for all k. This means $b = 0$ or $\sin(k\,l) = 0$ for all k. The case $b = 0$ would lead to the trivial solution $u = 0$. We are not interested in this one. The case $\sin(k\,l) = 0$ for all k, on the other hand, leads to nontrivial solutions:

$$\sin(k\,l) = 0 \Leftrightarrow k\,l = n\,\pi \ \text{ for } n \in \mathbb{N} \Leftrightarrow k = \frac{n\,\pi}{l} \ \text{ for } n \in \mathbb{N}\,.$$

Thus for each $n \in \mathbb{N}$ the function

$$u_n(x,t) = b_n \, \mathrm{e}^{-c^2 (\frac{n\pi}{l})^2 t} \sin\left(\frac{n\,\pi}{l} x\right)$$

is a solution to the heat conduction equation that also satisfies the boundary condition.

By superposing these solutions, we obtain a general solution that satisfies the heat conduction equation and the boundary condition:

$$u(x,t) = \sum_{n=1}^{\infty} u_n(x,t) = \sum_{n=1}^{\infty} b_n \, \mathrm{e}^{-c^2 (\frac{n\pi}{l})^2 t} \sin\left(\frac{n\,\pi}{l} x\right).$$

Step 3. *Nailing down the coefficients b_n of the general solution by the initial conditions:* We now use the general solution from step 2 to determine a solution that also satisfies the initial condition $u(x,0) = g(x)$. For this purpose we insert the initial condition: The initial distribution $u(x,0) = g(x)$ provides:

$$g(x) = u(x,0) = \sum_{n=1}^{\infty} b_n \sin\left(\frac{n\,\pi}{l} x\right) = \sum_{n=1}^{\infty} b_n \sin\left(n \frac{2\pi}{T} x\right) \quad \text{with } T = 2l\,.$$

These representations of g we know from the chapter on Fourier series expansion (see Sect. 74.3): The coefficients b_n are the Fourier coefficients of the function g, if g is an odd function on the interval $[-l, l)$ of length $T = 2l$.

But these last conditions are now easily fulfilled: Since g is only defined on the interval $[0, l]$, we extend g to an odd T-periodic function, by first extending the function oddly to the interval $[-l, l)$: $g(-x) = -g(x)$ for $x \in [0, l]$. And then we extend this function again T-periodic on $\mathbb{R}$. Now we are faced with a calculation of the b_n as *Fourier coefficients* nothing more stands in the way. With the b_n we get the solution $u = u(x,t)$ as follows (the mentioned extension of the functions g is not to be carried out explicitly at all):

Recipe: Solve a Zero Bound Problem for a Bar

The solution $u = u(x,t)$ of the zero bound problem

$$u_t = c^2 u_{xx} \text{ for } x \in (0, l),\ t \geq 0 \text{ and } u(x,0) = g(x) \text{ and } u(0,t) = 0 = u(l,t)$$

(continued)

is obtained as follows:

(1) Determine the coefficients b_n by:

$$b_n = \frac{2}{l} \int_0^l g(x) \sin\left(n \frac{\pi}{l} x\right) \mathrm{d}x \quad \text{for } n = 1, 2, 3 \ldots .$$

(2) Obtain the solution $u = u(x, t)$ as a series representation:

$$u(x, t) = \sum_{n=1}^{\infty} b_n \, \mathrm{e}^{-c^2 (\frac{n\pi}{l})^2 t} \sin\left(n \frac{\pi}{l} x\right) .$$

Note that the initial temperature distribution *dissipates* over time, for $t \to \infty$ the constant final temperature 0 sets in.

Example 89.1 The function sought is $u = u(x, t)$ for a bar of length $l = \pi$, where

$$u_t - u_{xx} = 0 \text{ with } u(x, 0) = x\,(x^2 - \pi^2) \text{ and } u(0, t) = u(\pi, t) = 0 .$$

(1) The Fourier series expansion of the function $g = g(x)$ is

$$g(x) \sim \sum_{n=1}^{\infty} \frac{12(-1)^n}{n^3} \sin(nx) .$$

(2) Therefore

$$u(x, t) = \sum_{n=1}^{\infty} \frac{12\,(-1)^n}{n^3} \, \mathrm{e}^{-n^2 t} \sin(nx)$$

A solution to the initial boundary value problem.

Note that as time passes, the temperature is uniformly distributed across the bar. ■

89.4 Numerical Solution

One can also solve initial boundary value problems with the heat conduction equation numerically using a difference method. We consider the (simple) zero boundary value problem

$$u_t = u_{xx} \text{ for } x \in (0,1),\ t \geq 0 \text{ and } u(x,0) = g(x) \text{ and } u(0,t) = 0 = u(1,t)\,.$$

As in the numerical solution of Dirichlet's boundary value problem, we discretize the domain $D = [0,1] \times [0,T]$ with a $T > 0$ by covering it with a grid: We choose the (small) step size $h = \frac{1}{n+1}$ with $n \in \mathbb{N}$ in x-direction and a (small) step size $k = \frac{T}{m}$ with $m \in \mathbb{N}$ in t-direction and obtain the lattice points (see Fig. 89.1):

$$(x_i, t_j) \text{ with } x_i = i\,h \text{ and } t_j = j\,k \text{ with } i = 0, \ldots, n+1,\ j = 0, \ldots, m\,.$$

Now we approximate in the inner grid points (x_i, t_j) the partial derivatives u_{xx} and u_t of the function $u = u(x,t)$ by corresponding difference quotients:

$$u_{xx}(x_i, t_j) \approx \frac{u_{i+1,j} - 2u_{i,j} + u_{i-1,j}}{h^2} \text{ and } u_t(x_i, t_j) \approx \frac{u_{i,j+1} - u_{i,j}}{k}\,.$$

Here are

$$u_{i+1,j} = u(x_{i+1}, t_j),\ u_{i,j} = u(x_i, t_j),\ u_{i-1,j} = u(x_{i-1}, t_j),\ \ldots$$

the values sought—note the box with the formulas for numerical differentiation in Sect. 26.3. The *discretized* heat conduction equation, with these abbreviations, reads in a (x_i, t_j)

$$0 = u_t(x_i, t_j) - u_{xx}(x_i, t_j) \approx \frac{1}{k}\left(u_{i,j+1} - u_{i,j}\right) - \frac{1}{h^2}\left(u_{i+1,j} - 2u_{i,j} + u_{i-1,j}\right),$$

which, after replacing $\approx$ by $=$ with the abbreviation $r = k/h^2$ can be written more simply as

$$(*)\quad u_{i,j+1} = r\,u_{i-1,j} + (1-2r)\,u_{i,j} + r\,u_{i+1,j} \text{ for } i = 1, \ldots, n \text{ and } j \geq 0\,.$$

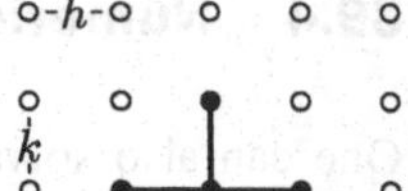

Fig. 89.1 Difference star

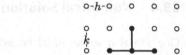

The initial and boundary conditions still yield the following equations:

- **The initial conditions**:

$$u_{i,0} = g(x_i) \text{ for all } i = 1, \ldots, n\,.$$

- **The boundary conditions**:

$$u_{0,j} = 0 \text{ and } u_{n+1,j} = 0 \text{ for all } j \geq 0\,.$$

The equation (∗) thus provides a method of determining with the help of the known values $u_{i,0}$ of the initial distribution (zero time series) the approximate values $u_{i,1}$ (first time series). This is continued from time series to time series.

The procedure can again be represented with a *difference star*, see Fig. 89.1.

Example 89.2 In the case $n = 2$, $m = 3$ and $T = 1$ we get $h = \frac{1}{3}$, $k = \frac{1}{3}$ and $r = 3$ and continue with the initial distribution g:

$$\text{as approximation for } u(x_1, t_1):\ u_{11} = r\,u_{00} + (1 - 2r)\,u_{10} + r\,u_{20} = -5\,g(x_1) + 3\,g(x_2),$$

$$\text{as approximation for } u(x_2, t_1):\ u_{21} = r\,u_{10} + (1 - 2r)\,u_{20} + r\,u_{30} = 3\,g(x_1) - 5\,g(x_2)\,.$$

With these values in the first time series we now obtain approximations in the second time series:

$$\text{as approximation for } u(x_1, t_2):\ u_{12} = r\,u_{01} + (1 - 2r)\,u_{11} + r\,u_{21} = -5\,u_{11} + 3\,u_{21}\,,$$

$$\text{as approximation for } u(x_2, t_2):\ u_{22} = r\,u_{11} + (1 - 2r)\,u_{21} + r\,u_{31} = 3\,u_{11} - 5\,u_{21}\,.$$

Using these values in the second time series, we obtain analogous approximations in the third and then final time series. ■

In fact, such an *explicit* procedure is often not suitable for numerical purposes. Better, since *more stable*, are usually *implicit* methods. Such a method is obtained by using other difference quotients as approximations to the derivative. If, for example, one retains

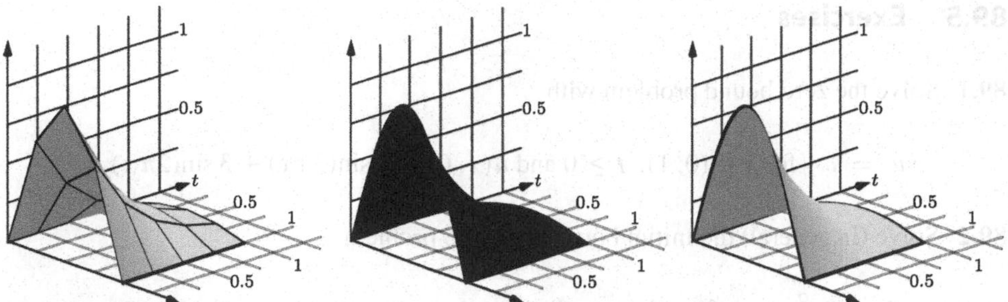

Fig. 89.2 Approximate solutions with $m = n = 5$, $m = n = 50$ and the exact solution

the central second difference quotient for the spatial derivative and uses the *backward difference quotient* for the time derivative, thus

$$u_{xx}(x_i, t_j) \approx \frac{u_{i+1,j} - 2u_{i,j} + u_{i-1,j}}{h^2} \text{ and } u_t(x_i, t_j) \approx \frac{u_{i,j} - u_{i,j-1}}{k},$$

then one obtains an implicit method with a system of equations in which there are at most four unknowns per line. Therefore, the coefficient matrix of this system is sparse. To solve this large system of equations, *sparse methods* for *sparse* matrices or the iterative methods that we have given in Chap. 71 are used. We have also used such a method in the following example.

Example 89.3 We consider the zero bound problem

$$u_t = u_{xx} \text{ for } x \in (0, 1),\ t \geq 0 \text{ and } u(x, 0) = \sin(\pi x) \text{ and } u(0, t) = 0 = u(1, t).$$

For the computation of approximate solutions we use MATLAB (see Exercise 89.5). The following Fig. 89.2 shows approximate solutions for $m = n = 5$ and $m = n = 50$ as well as the exact solution $u(x, t) = \exp(-\pi^2 t) \sin(\pi x)$. ■

Our procedure, i.e. discretization in x- and t-direction, is also called **global discretization**. However, this method is only useful for the one-dimensional case, as in our example. In the plane or in space, this discretization results in systems of equations that are too large. Here, other discretizations are appropriate, e.g., discretize only in space; one then obtains an ODE-system in time.

89.5 Exercises

89.1 Solve the zero bound problem with

$$u_t = u_{xx} \text{ for } x \in (0, 1),\ t \geq 0 \text{ and } u(x, 0) = 2\sin(3\pi x) + 3\sin(2\pi x)\,.$$

89.2 Solve (in general) the initial boundary value problem

$$u_t - c^2 u_{xx} = 0 \text{ with } u(x, 0) = g(x) \text{ and } u_x(0, t) = u_x(l, t) = 0$$

for a bar of length l, whereby at the boundary no *heat transport* takes place, $u_x = 0$.

89.3 Write a program that solves the zero bound problem from Example 89.3.

89.4 Find a solution to the initial boundary value problem

$$u_{xx}(x, t) - 4\,u_t(x, t) - 3\,u(x, t) = 0 \quad \text{for } x \in [0, \pi],\ t \in [0, \infty), \qquad (1)$$

$$u(x, 0) = x\left(x^2 - \pi^2\right) \quad \text{for } x \in [0, \pi] \quad \text{(initial values)}, \qquad (2)$$

$$u(0, t) = u(\pi, t) = 0 \quad \text{for } t \in [0, \infty) \quad \text{(boundary values)}. \qquad (3)$$

To do this, proceed as follows:

(a) Find as many real solutions to (1) as possible using the separation approach.
(b) Among them, identify those solutions $u_n(x, t)$ that satisfy the boundary condition (3).
(c) Develop the initial condition $g(x) := x\left(x^2 - \pi^2\right)$ at $[-\pi, \pi]$ into a Fourier series.
(d) Make the superposition approach $u(x, t) = \sum u_n(x, t)$ with the u_n from (b) and thus find a solution to the initial boundary value problem (1)–(3).

89.5 We consider the initial boundary value problem of the heat conduction equation $u_t = u_{xx}$ at $\mathbb{R}$ with the initial condition

$$u(0, x) = \begin{cases} -1, & \text{for } x < 0, \\ 0, & \text{for } x = 0, \\ 1, & \text{for } x > 0. \end{cases}$$

Find a solution of the form $u(t, x) = \varphi\left(\frac{x}{\sqrt{t}}\right)$. Sketch in the x-t-plane curves along which u is constant.

The Wave Equation

90

The wave equation is a classical example of a hyperbolic partial differential equation. It is nonstationary and describes wave phenomena or oscillations.

90.1 Initial Boundary Value Problems for the Wave Equation

Since the **wave equation**

$$u_{tt} = c^2 \Delta u$$

with a constant $c > 0$, the **wave velocity**, is nonstationary, it typically occurs in the context of an initial boundary value problem. We formulate a concrete such problem for the one-dimensional wave equation: consider a string that is fixed at the two ends $x = 0$ and $x = l$. At the time $t = 0$ this string is deflected from an initial oscillation $g(x)$ and with the initial velocity $v(x)$ to oscillate:

The Vibrating String—An Initial Boundary Value Problem for the One-Dimensional Wave Equation
A solution $u = u(x, t)$ is sought which satisfies the wave equation

$$u_{tt}(x, t) = c^2 \Delta u(x, t) \quad \text{for } x \in D \text{ and } t \geq 0$$

(continued)

C. Karpfinger, *Calculus and Linear Algebra in Recipes*,
https://doi.org/10.1007/978-3-662-65458-3_90

and the boundary and initial conditions:

- Initial conditions:
  - $u(x, 0) = g(x)$ for $0 \leq x \leq l$,
  - $u_t(x, 0) = v(x)$ for $0 \leq x \leq l$.
- Boundary conditions:
  - $u(0, t) = 0$ for $t \geq 0$,
  - $u(l, t) = 0$ for $t \geq 0$.

90.2 Solutions of the Equation

Using the separation method from the recipe in Sect. 87.3.1 we obtain solutions of the wave equation, in principle even in each dimension. However, for simplicity, we stick to the one-dimensional case. We consider the wave equation

$$u_{tt} - c^2 u_{xx} = 0 .$$

We now use the separation method from the recipe in Sect. 87.3.1 and find solutions u of the form

$$u = u(x, t) = f(x)\, g(t) .$$

Note the recipe in Sect. 87.3.1.

(1) Substituting $u(x, t) = f(x)g(t)$ into the PDE yields:

$$0 = u_{tt} - c^2 u_{xx} = f(x)\, g''(t) - c^2 f''(x)\, g(t) , \text{ i.e. } \frac{1}{c^2} \frac{g''(t)}{g(t)} = \frac{f''(x)}{f(x)} .$$

From this we obtain the two ODEs:

$$\frac{g'(t)}{g(t)} = c^2 k \text{ and } \frac{f'(x)}{f(x)} = k \text{ i.e., } g' = c^2 k\, g \text{ and } f' = k\, f$$

with $k \in \mathbb{R}$.

(2) We solve the two ODEs from (1).

- The solutions for g are:

$$g = g(t) = \begin{cases} c_1 \, \mathrm{e}^{c\sqrt{k}t} + c_2 \, \mathrm{e}^{-c\sqrt{k}t}, & \text{if } k > 0 \\ c_1 + c_2 \, t, & \text{if } k = 0 \\ c_1 \, \cos(c\sqrt{|k|}t) + c_2 \, \sin(c\sqrt{|k|}t), & \text{if } k < 0 \end{cases}$$

with $c_1, c_2 \in \mathbb{R}$.

- The solutions for f are:

$$f = f(x) = \begin{cases} c_1 \, \mathrm{e}^{\sqrt{k}x} + c_2 \, \mathrm{e}^{-\sqrt{k}x}, & \text{if } k > 0 \\ c_1 + c_2 \, x, & \text{if } k = 0 \\ c_1 \, \cos(\sqrt{|k|}\, x) + c_2 \, \sin(\sqrt{|k|}\, x), & \text{if } k < 0 \end{cases}$$

with $c_1, c_2 \in \mathbb{R}$.

(3) We have now found the following solutions:

$$u = u(x, t) = f(x) \, g(t)$$

with f and g from (2).

Each of these given functions u is a solution of the wave equation. However, for the initial and boundary conditions usually considered, some of these solutions are not useful: the case $k \geq 0$ leads, for many relevant problems, to the uninteresting trivial solution $u = 0$.

Usually, the case $k < 0$ leads to the time-dependent solutions we are interested in. These solutions are given by k instead of $\sqrt{|k|}$:

Solutions of the Wave Equation

For any real a_1, a_2 and b_1, b_2 and for any $k \in \mathbb{R}$

$$u = u(x, t) = (a_1 \cos(k \, x) + b_1 \sin(k \, x)) \, (a_2 \cos(c \, k \, t) + b_2 \sin(c \, k \, t))$$

is a solution of the wave equation.

90.3 The Vibrating String: Solution with Fourier Series

We now consider again concretely the initial boundary value problem of a vibrating string of length l, with the ends fixed: We are looking for a function $u = u(x, t)$ with

$$u_{tt} - c^2 u_{xx} = 0 \text{ where } u(x, 0) = g(x),\ u_t(x, 0) = v(x),\ u(0, t) = u(l, t) = 0\,.$$

We explain the solution in 3 steps:

Step 1 *Determining solutions of the wave equation:* We consider as large a set of solutions of the wave equation as possible. We have already compiled these in the box above:

$$u = u(x, t) = (a_1 \cos(kx) + b_1 \sin(kx))\ (a_2 \cos(ckt) + b_2 \sin(ckt))\,,\ a_i, b_i, k \in \mathbb{R}\,.$$

Step 2 *The boundary conditions impose constants, we obtain a general solution:* Among the solutions from step 1, we determine the one that also satisfies the boundary conditions $u(0, t) = u(l, t) = 0$:

- From $u(0, t) = 0$ for all t follows $a_1 = 0$.
- From $a_1 = 0$ and $u(l, t) = 0$ for all t follows $b_1 \sin(k\,l) = 0$ for all k. This means $b_1 = 0$ or $\sin(k\,l) = 0$ for all k. The case $b_1 = 0$ would lead to the trivial solution $u = 0$. We are not interested in this one. The case $\sin(k\,l) = 0$ for all k, on the other hand, leads to nontrivial solutions:

$$\sin(k\,l) = 0 \Leftrightarrow k\,l = n\pi \text{ for } n \in \mathbb{N} \Leftrightarrow k = \frac{n\pi}{l} \text{ for } n \in \mathbb{N}\,.$$

Thus for each $n \in \mathbb{N}$ the function

$$u_n(x, t) = b_n \sin\left(\frac{n\pi}{l}x\right)\left(a_2 \cos\left(c\frac{n\pi}{l}t\right) + b_2 \sin\left(c\frac{n\pi}{l}t\right)\right)$$

is a solution to the wave equation that also satisfies the boundary condition.

By superposition of these solutions, we obtain a general solution satisfying the wave equation and the boundary condition:

$$u(x, t) = \sum_{n=1}^{\infty} u_n(x, t) = \sum_{n=1}^{\infty} \sin\left(\frac{n\pi}{l}x\right)\left(a_n \cos\left(c\frac{n\pi}{l}t\right) + b_n \sin\left(c\frac{n\pi}{l}t\right)\right),$$

Here the b_n above has not been forgotten, but has been substituted into the new coefficients a_n and b_n.

Step 3 *Nailing down the coefficients a_n and b_n of the general solution by the initial conditions:* We now use the general solution from step 2 to determine a solution that also satisfies the initial conditions $u(x,0) = g(x)$ and $u_t(x,0) = v(x)$. To do this, we set the initial conditions:

- **The initial oscillation** $u(x,0) = g(x)$ yields:

$$g(x) = u(x,0) = \sum_{n=1}^{\infty} a_n \sin\left(\frac{n\pi}{l}x\right) = \sum_{n=1}^{\infty} a_n \sin\left(n\frac{2\pi}{T}x\right) \quad \text{mit } T = 2l\,.$$

- **The initial velocity** $u_t(x,0) = v(x)$ yields after differentiation of the series:

$$v(x) = u_t(x,0) = \sum_{n=1}^{\infty} \left(b_n\, c\,\frac{n\pi}{l}\right) \sin\left(n\frac{2\pi}{T}x\right) \quad \text{with } T = 2l\,.$$

These representations of g and v should be familiar (see Sect. 74.3): The coefficients a_n and b_n are essentially just the Fourier coefficients of the function g and v. It must be taken into account that both g as well as v must be odd, since the representations are pure sinusoidal representations. Furthermore, both g as well as v must have the period $T = 2l$; but both functions are *only* defined on the interval $[0, l]$. The problem is easily solved: We extend g and v to odd T-periodic functions, by first extending the functions oddly to the interval $[-l, l)$: $g(-x) = -g(x)$ and $v(-x) = -v(x)$ for $x \in [0, l]$. Finally, we extend these functions again T-periodic on $\mathbb{R}$. Now we can calculate the coefficients a_n and b_n. Only with the b_n we have to be a little bit careful, because there is a prefactor to consider. With the a_n and b_n we get the solution $u = u(x,t)$ as follows (the mentioned extension of the functions g and v are not to be performed explicitly at all here):

Recipe: Solve the Initial Boundary Value Problem for the Vibrating String

The solution $u = u(x,t)$ of the initial boundary value problem

$$u_{tt} - c^2 u_{xx} = 0 \text{ where } u(x,0) = g(x),\ u_t(x,0) = v(x),\ u(0,t) = u(l,t) = 0$$

is obtained as follows:

(1) Determine the coefficients a_n and b_n by:

$$a_n = \frac{2}{l}\int_0^l g(x)\sin\left(n\frac{\pi}{l}x\right)\mathrm{d}x \text{ for } n = 1, 2, 3 \ldots$$

(continued)

$$b_n = \frac{2}{n\pi c}\int_0^l v(x)\sin\left(n\frac{\pi}{l}x\right)\mathrm{d}x \text{ for } n = 1, 2, 3 \ldots$$

(2) Obtain the solution $u = u(x, t)$ as a series representation:

$$u(x,t) = \sum_{n=1}^{\infty} \sin\left(\frac{n\pi}{l}x\right)\left(a_n\cos\left(c\frac{n\pi}{l}t\right) + b_n\sin\left(c\frac{n\pi}{l}t\right)\right).$$

Example 90.1 What we are looking for is the function $u = u(t, x)$ for a string of length $l = 2\pi$, where

$$u_{tt} - u_{xx} = 0 \text{ where } u(x,0) = \pi - |x-\pi|,\ u_t(x,0) = 0,\ u(0,t) = u(l,t) = 0.$$

(1) Since $u_t(x, 0) = v(x)$ is the zero function, all coefficients b_n are equal to zero. And the Fourier series expansion of the function $g = g(x)$ is

$$g(x) \sim \sum_{n=1}^{\infty} \frac{(-1)^{n-1}8}{(2n-1)^2\pi}\sin\left(\frac{2n-1}{2}x\right).$$

(2) Therefore

$$u(x,t) = \sum_{n=1}^{\infty} \frac{(-1)^{n-1}8}{(2n-1)^2\pi}\sin\left(\frac{2n-1}{2}x\right)\cos\left(\frac{2n-1}{2}t\right)$$

is a solution to the initial boundary value problem. ■

90.4 Numerical Solution

We discuss a difference method for numerically solving an initial boundary value problem with the wave equation. For this purpose, we consider the problem of the vibrating string on the interval $D = [0, 1]$ with the wave velocity $c = 1$:

$$u_{tt}(x,t) = u_{xx}(x,t) \text{ for } x \in (0,1) \text{ and } t \geq 0$$

with the boundary and initial conditions:

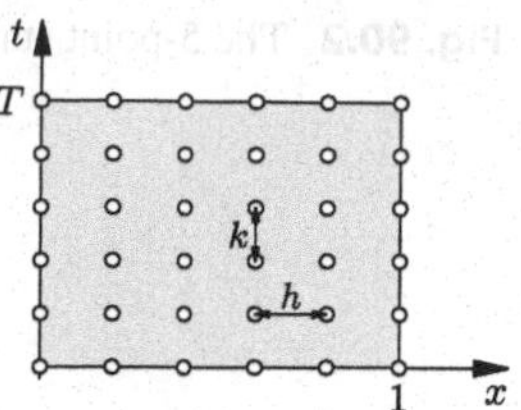

Fig. 90.1 Grid on $[0, 1] \times [0, T]$

- $u(x, 0) = g(x)$ for $0 \leq x \leq 1$,
- $u_t(x, 0) = v(x)$ for $0 \leq x \leq 1$.
- $u(0, t) = 0$ for $t \geq 0$,
- $u(1, t) = 0$ for $t \geq 0$.

To do this, we discretize the domain $D = [0, 1] \times [0, T]$ with a $T > 0$ by covering it with a grid: We choose in x-direction the step size h and in t-direction the step size k and approximate the second partial derivative with the corresponding difference quotient:

- Choose a (large) $n \in \mathbb{N}$ and set $h = \frac{1}{n+1}$.
- Choose a (large) $m \in \mathbb{N}$ and set $k = \frac{T}{m+1}$.

Note Fig. 90.1. As already indicated in the figure m should be greater than n, i.e. k should be less than h.

This gives us the grid points:

$$(x_i, t_j) \text{ with } x_i = i\,h$$

and

$$t_j = j\,k \text{ with } i = 0, \ldots, n+1, \; j = 0, \ldots, m+1\,.$$

Now we approximate in the inner lattice points (x_i, t_j), $x_i = i\,h$, $t_j = j\,k$, the partial derivatives u_{xx} and u_{tt} of the function $u = u(x, t)$ by corresponding difference quotients:

$$u_{xx}(x_i, t_j) \approx \frac{u_{i+1,j} - 2u_{i,j} + u_{i-1,j}}{h^2} \text{ and } u_{tt}(x_i, t_j) \approx \frac{u_{i,j+1} - 2u_{i,j} + u_{i,j-1}}{k^2},$$

where

$$u_{i+1,j} = u(x_{i+1}, t_j),\; u_{i,j} = u(x_i, t_j),\; u_{i-1,j} = u(x_{i-1}, t_j),\; \ldots$$

the values sought—note the box with the formulas for numerical differentiation in Sect. 26.3. We memorize: The first index i at $u_{i,j}$ refers to the x-coordinate, the second index j on the other hand refers to the t-coordinate.

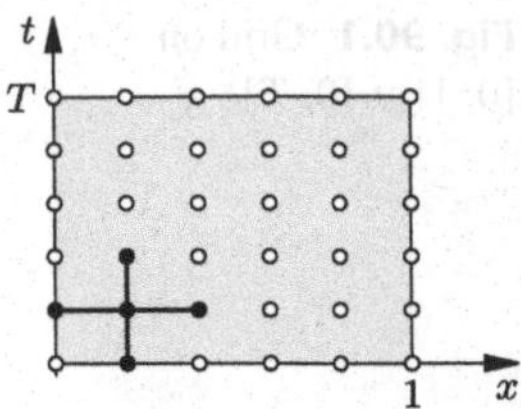

Fig. 90.2 The 5-point star

Using these abbreviations, the *discretized* wave equation for (x_i, t_j) reads as follows

$$\frac{u_{i+1,j} - 2u_{i,j} + u_{i-1,j}}{h^2} = \frac{u_{i,j+1} - 2u_{i,j} + u_{i,j-1}}{k^2}.$$

As with the Poisson equation, you get a **5-point star** (Fig. 90.2).

Using this 5-point star, we can *carry* the values at the boundaries into the interior, i.e. to the inner grid points (x_i, t_j). Note, however, that in contrast to the Poisson equation we have no *upper* boundary, but we have an initial velocity. The boundary conditions and the initial oscillation yield the values for $i = 0$ (t-axis), $j = 0$ (x-axis) and $i = n+1$ (parallel to the t-axis).

The number j numbers through the time series: We speak in the case $j = 0$ (i.e. for the values $u_{i,0}$, $i = 0, \ldots, n+1$) from the zeroth time series, here the function values of u by the initial oscillation $g(x)$. The values in the first time series (i.e. the values $u_{i,1}$, $i = 0, \ldots, n+1$) we will now obtain approximately by Taylor expansion with the initial velocity. Once we have these values, we obtain (by means of the 5-point star; at least one can observe it well from this, see Fig. 90.2) the approximate values in the second time series explicitly from the values of the zeroth and first time series. This is then continued time series for time series:

Obtain the approximations $u_{i,1}$:
The 2nd order Taylor expansion in the time direction is:

$$u(x,k) = u(x,0) + k\,u_t(x,0) + \frac{k^2}{2}\,u_{tt}(x,0) + \cdots \approx g(x) + k\,v(x) + \frac{k^2}{2}\,g''(x),$$

where we have used $u_{tt}(x, 0) = u_{xx}(x, 0) = g''(x)$. We discretize this approximation and obtain the following approximation for every $i = 1, \ldots, n$:

$$u_{i,1} = u_{i,0} + k\,v(x_i) + \frac{k^2}{2h^2}(u_{i-1,0} - 2u_{i,0} + u_{i+1,0}).$$

This now gives approximations in the zeroth and first time series:

$$u^{(0)} = (u_{1,0}, \ldots, u_{n,0}) = (g(x_1), \ldots, g(x_n)) \text{ and } u^{(1)} = (u_{1,1}, \ldots, u_{n,1}).$$

Obtain the approximations in the further time series $u_{i,j}$ with $j \geq 2$:
We solve the discretized wave equation (see above) for $u_{i,j+1}$ and get for $i = 1, \dots, n$:

$$u_{i,j+1} = \frac{k^2}{h^2}(u_{i-1,j} - 2\,u_{i,j} + u_{i+1,j}) - u_{i,j-1} + 2\,u_{i,j}\,.$$

We obtain the approximations in the second, third and further time series by the following explicit formula.

$$\begin{pmatrix} u_{1,j+1} \\ \vdots \\ u_{n,j+1} \end{pmatrix} = \left[\begin{pmatrix} 2 & 0 & \cdots & 0 \\ 0 & \ddots & & \vdots \\ \vdots & & \ddots & 0 \\ 0 & \cdots & 0 & 2 \end{pmatrix} - \frac{k^2}{h^2} \begin{pmatrix} 2 & -1 & \cdots & 0 \\ -1 & \ddots & \ddots & \vdots \\ \vdots & \ddots & \ddots & -1 \\ 0 & \cdots & -1 & 2 \end{pmatrix} \right] \begin{pmatrix} u_{1,j} \\ \vdots \\ u_{n,j} \end{pmatrix} - \begin{pmatrix} u_{1,j-1} \\ \vdots \\ u_{n,j-1} \end{pmatrix},$$

where $j = 2, 3, \dots, m$. Using

$$A_h = \frac{1}{h^2} \begin{pmatrix} 2 & -1 & \cdots & 0 \\ -1 & \ddots & \ddots & \vdots \\ \vdots & \ddots & \ddots & -1 \\ 0 & \cdots & -1 & 2 \end{pmatrix}$$

the explicit formula can be succinctly formulated as

$$u^{(j+1)} = (2\,E_n - k^2\,A_h)\,u^{(j)} - u^{(j-1)}\,.$$

Remark Obviously, the amplitude of a vibrating string (general wave) does not increase. Therefore, when solving a corresponding initial boundary value problem numerically, one also requires that the numerical method does not increase the amplitude when progressing in the positive time direction. In explicit methods, this request imposes restrictions on the two discretization parameters k and h (k is the step width in t-direction and h is the step size in x-direction). For our chosen discretization

$$\frac{k}{h} \leq 1$$

must be fulfilled. If k and h are chosen such that this inequality is not satisfied, the method is unstable; if k and h, on the other hand, are chosen appropriately, the method need not be stable at all. It is in general very difficult to formulate a necessary and sufficient criterion in general wave equations.

In practice, one typically uses finite elements in place and obtains an ODE-system in time.

For finite elements there is the package COMSOL Multiphysics, formerly FEMLAB. In MATLAB there is the simple variant `pdetool`. A description can be found under `doc pdetool`.

90.5 Exercises

90.1 Determine a solution to the following initial boundary value problem for an oscillating string of length $l = \pi$, where

$$u_{tt} - u_{xx} = 0$$

with

$$u(x,0) = \tfrac{\pi}{2} - |x - \tfrac{\pi}{2}|, \; u_t(x,0) = 0, \; u(0,t) = u(l,t) = 0 .$$

90.2 Determine a solution to the following initial boundary value problem for an oscillating string of length $l = 2$, where

$$u_{tt} - u_{xx} = 0$$

with

$$u(x,0) = \sin\left(\tfrac{\pi}{2}x\right) + \sin^3\left(\tfrac{\pi}{2}x\right), \; u_t(x,0) = 0, \; u(0,t) = u(l,t) = 0 .$$

90.3 Find the solution $u(x,t)$ for the following initial boundary value problem for a vibrating string of length $l = 3$ with the boundary conditions $u(0,t) = u(3,t) = 0$:

$$u_{tt} - 9\,u_{xx} = 0, \; u(x,0) = \sin\left(\tfrac{2\pi}{3}x\right)\cos\left(\tfrac{2\pi}{3}x\right) + \sin(\pi x), \; u_t(x,0) = \sin\left(\tfrac{4\pi}{3}x\right) .$$

91 Solving PDEs with Fourier and Laplace Transforms

In Chaps. 77 and 79 we solved linear (ordinary) differential equations and initial value problems with linear (ordinary) differential equations, respectively, using Fourier and Laplace transforms. Here we made use of the fact that the transformation turns a differential equation into an algebraic equation. This equation is then usually easy to solve, and the inverse transform is then a solution of the original differential equation. This principle can also be successfully applied to partial differential equations: In this case, a partial differential equation becomes an ordinary differential equation by means of an integral transformation. This equation can then be solved using conventional methods, and a reverse transformation yields a solution of the original partial differential equation. In general, one obtains solution formulas for the considered partial differential equation with possibly given initial conditions.

91.1 An Introductory Example

We briefly recall the Fourier transform: to a function $f : \mathbb{R} \to \mathbb{C}$ in time t, i.e. $f = f(t)$ consider the Fourier transform $F : \mathbb{R} \to \mathbb{C}$ from f, given by

$$F(\omega) = \int_{-\infty}^{\infty} f(t)\, \mathrm{e}^{-\mathrm{i}\omega t}\, \mathrm{d}t\,, \text{ and writes } f(t) \quad \circ\!\!-\!\!\!-\!\!\bullet \quad F(\omega) \text{ or } \mathcal{F}(f(t)) = F(\omega)\,.$$

In the following we will typically Fourier transform a searched solution $u = u(x, t)$ of a PDE and, for example, an initial oscillation $g(x)$. Here we will transform according to the variable x and not as before according to the time t. For the transformed variable we do not choose ω, this symbol is reserved for the transform of t, we choose the symbol k (besides k, the symbol ξ is also common). Moreover, as before, we write the transforms with capital letters. Thus, we have exemplary correspondences of the type.

C. Karpfinger, *Calculus and Linear Algebra in Recipes*,
https://doi.org/10.1007/978-3-662-65458-3_91

$$u(x,t) \quad \circ\!\!-\!\!\bullet \quad U(k,t), \quad g(x) \quad \circ\!\!-\!\!\bullet \quad G(k), \ldots$$

By Fourier transforming a PDE we obtain an ODE. We solve this and then form the inverse transform, which is a solution of the PDE. We show this principle for solving a PDE by Fourier transform on an example and will then quickly take up this idea to give formal solutions of many a well-known PDE:

Example 91.1 We consider the PDE (a wave equation) with initial conditions:

$$u_{tt} - c^2 u_{xx} = 0 \text{ with } u(x,0) = g(x) \text{ and } u_t(x,0) = v(x). \qquad \text{(pIVP)}$$

Here $g(x)$ is the initial oscillation and $v(x)$ the initial velocity.

In what follows, we assume that the solution we are looking for is $u = u(x,t)$ as well as the functions $g = g(x)$ and $v = v(x)$ with respect to the variable x are Fourier transformable, i.e., there exist the Fourier transforms $U = U(k,t)$, $G = G(k)$ and $V = V(k)$. Thus we have the correspondences:

$$u(x,t) \quad \circ\!\!-\!\!\bullet \quad U(k,t), \quad g(x) \quad \circ\!\!-\!\!\bullet \quad G(k), \quad v(x) \quad \circ\!\!-\!\!\bullet \quad V(k).$$

Known are the rules *linearity* and *differentiation in time* (see Sect. 77.1), which say (with k instead of ω and x instead of t):

$$u_{xx} \quad \circ\!\!-\!\!\bullet \quad -k^2 U \text{ and thus } u_{tt} - c^2 u_{xx} \quad \circ\!\!-\!\!\bullet \quad U_{tt} + c^2 k^2 U.$$

Here we have exploited the fact that the Fourier transform to x is *blind* for differentiating to t (we explain this in more detail in a remark below).

Thus we obtain from the above wave equation together with the initial conditions for each k the following ODE together with the initial conditions

$$U_{tt} + c^2 k^2 U = 0 \text{ with } U(k,0) = G(k) \text{ and } U_t(k,0) = V(k). \qquad \text{(oIVP)}$$

Compare this with the usual notation (in earlier chapters on ODEs) for IVP with second order ODEs:

$$\ddot{x} + c^2 k^2 x = 0 \text{ with } x(0) = a \text{ and } \dot{x}(0) = b.$$

We solve (oIVP) using the conventional methods (see recipe from Sect. 34.1): The characteristic equation is $p(\lambda) = \lambda^2 + c^2 k^2 = 0$, thus

$$p(\lambda) = (\lambda - \mathrm{i}\,c\,k)\,(\lambda + \mathrm{i}\,c\,k) = 0.$$

Thus we have the general solution of the ODE:

$$U(k,t) = c_1 \cos(c\,k\,t) + c_2 \sin(c\,k\,t)\,.$$

Substituting the initial conditions $U(k,0) = G(k)$ and $U_t(k,0) = V(k)$ yields the solution $U(k,t)$ of (oIVP):

$$U(k,t) = G(k)\,\cos(c\,k\,t) + V(k)\,\frac{\sin(c\,k\,t)}{c\,k}\,.$$

The back transformation, i.e., finding $u(x,t)$ with $u(x,t) \;\circ\!\!-\!\!\bullet\; U(k,t)$ (see Exercise 91.3), yields the solution $u = u(x,t)$ of (pIVP):

$$u(x,t) = \frac{1}{2}\,\left(g(x+c\,t) + g(x-c\,t)\right) + \frac{1}{2\,c}\int_{x-ct}^{x+ct} v(\xi)\mathrm{d}\xi\,.$$

This formula for the solution of the wave equation with initial conditions $u(x,0) = g(x)$ and $u_t(x,0) = v(x)$ is also called the **D'Alembert's solution formula.** ■

Remark We used the following rule in the example: If $U(k,t) = \mathcal{F}(u(x,t))$ is the Fourier transform of $u(x,t)$, then $U_{tt}(k,t) = \mathcal{F}(u_{tt}(x,t))$ is the Fourier transform of $u_{tt}(x,t)$. Behind this rule is the fact that partial differentiation according to t can be exchanged with the integration to x, more precisely:

$$\begin{aligned}\mathcal{F}(u_{tt}(x,t)) &= \int_{-\infty}^{\infty} u_{tt}(x,t)\,\mathrm{e}^{-ikx}\,\mathrm{d}x = \frac{\partial^2}{\partial t^2}\int_{-\infty}^{\infty} u(x,t)\,\mathrm{e}^{-ikx}\,\mathrm{d}x \\ &= \frac{\partial^2}{\partial t^2} U(k,t) = U_{tt}(k,t)\,.\end{aligned}$$

91.2 The General Procedure

The introductory example is already sufficiently general to see the Fourier transform as a useful tool to give general solutions or solution formulas for PDEs. The Laplacian transform can do this as well. We briefly recall this integral transform:

For a function $f : [0,\infty) \to \mathbb{C}$ in time t, i.e. $f = f(t)$ consider the Laplacian transform $F : (a,\infty) \to \mathbb{C}$ of f given by

$$F(s) = \int_0^{\infty} f(t)\,\mathrm{e}^{-st}\,\mathrm{d}t\,, \quad \text{one writes } f(t) \;\circ\!\!-\!\!\bullet\; F(s) \text{ or } \mathcal{L}(f(t)) = F(s)\,.$$

Again we will transform a sought solution $u = u(x, t)$ of a PDE. In the Laplace transform, we will look for the variable t and denote the transformed variable as before by s. Also, as before, we write the transforms with capital letters. So we typically have a correspondence

$$u(x, t) \quad \circ\!\!-\!\!\bullet \quad U(x, s) .$$

Somewhat strikingly, the choice of whether to perform a Fourier or a Laplace transform can be described as follows: If one wants to solve a PDE with (possibly several) initial and boundary conditions and the solution is $u = u(x, t)$

- is unbounded up and down in the locus, $x \in (-\infty, \infty)$, then choose the Fourier transform, $x \to k$,
- unbounded upwards in time, $t \in [0, \infty)$ then choose the Laplacian transform, $t \to s$.

Thus, in the Fourier transform one typically transforms the variable x, and in the Laplacian transform one transforms the time t. Typical applications for the Fourier transform are, for example, the heat conduction equation for an infinitely long bar, the Laplace equation for the upper half plane with the x-axis as boundary etc. The Laplace transform, on the other hand, is typically applied to locally restricted problems: a finite vibrating string, a finite bar, etc.

Just as the Fourier transform to x is *blind* to the partial derivative to t, the Laplacian transform to t is *blind* to the partial derivative with respect to x. This is the basis for the fact that by transforming a PDE, one *eliminates* a partial derivative and is left with an ODE. The general procedure for solving a PDE by integral transformation can be formulated recipe-like as follows:

Recipe: Solve a Linear PDE Using Fourier or Laplace Transform

For the solution of a linear PDE, e.g.

$$u_t - u_{xx} = s(x, t) \text{ with initial and boundary conditions if necessary}$$

by means of a Fourier or Laplace transformation, proceed as follows:

(1) Designate with $U = U(k, t)$ or $U = U(x, s)$ the Fourier- or Laplacetransform of the searched function u.

(2) Transform the PDE (including the boundary and initial conditions, if necessary) and obtain an ODE or IVP for U, e.g.

$$U_t(k, t) + k^2 U(k, t) = S(k, t) \text{ or } s\,U(x, s) - u(x, 0) - U_{xx}(x, s) = S(x, s)$$

(continued)

with initial and boundary conditions if necessary.

(3) Solve the ODE or IVP from (2) and obtain U. Plausibility considerations often limit the variety of solutions.

(4) By back transformation, i.e. determination of $u = u(x,t)$ with $u(x,t)$ ○—● $U(k,t)$ or $u(x,t)$ ○—● $U(x,s)$ you get a solution formula or solution u.

If the initial conditions or inhomogeneities are given concretely, U can sometimes be transformed back directly. Often, however, the rule *convolution* helps, in which case the solution is then usually given in an integral representation.

We test the recipe on two detailed examples:

Example 91.2

- We determine the solution of the Laplace equation on the half-plane $\mathbb{H} = \{(x, y)^\top \mid y \geq 0\}$ i.e. the solution of the **Dirichlet's boundary value problem** (cf. Chaps. 85 and 88):

$$-u_{xx} - u_{yy} = 0 \text{ with } x \in \mathbb{R},\ y \geq 0 \text{ and } u(x,0) = u_0(x)\,.$$

(1) With $U = U(k, y)$ is the Fourier transform of the searched solution $u = u(x, y)$ denoted.

(2) Fourier transform of the PDE including the boundary condition yields the IVP

$$k^2 U - U_{yy} = 0 \text{ with } U(k,0) = U_0(k)\,.$$

(3) The ODE has the solution $U(k, y) = c_1\,\mathrm{e}^{|k|y} + c_2\,\mathrm{e}^{-|k|y}$ with $c_1,\ c_2 \in \mathbb{R}$. We only want to consider solutions that remain bounded for $y \to \infty$, therefore we have $c_1 = 0$. With the initial condition we get the solution

$$U(k, y) = U_0(k)\,\mathrm{e}^{-|k|y}\,.$$

(4) Given the table in Sect. 77.2 and the rule *times scaling* we have:

$$\mathrm{e}^{-|k|y} \quad \bullet\!\!-\!\!\circ \quad \frac{1}{2\pi}\frac{2y}{y^2 + x^2}\,.$$

With the convolution product we finally have

$$U(k, y) = U_0(k)\,\mathrm{e}^{-|k|y} \quad \bullet\!\!-\!\!\circ \quad u(x, y) = u_0(x) * \frac{1}{2\pi}\frac{2y}{y^2 + x^2}\,,$$

so

$$u(x, y) = \frac{y}{\pi} \int_{-\infty}^{\infty} \frac{u_0(\tau)}{(x-\tau)^2 + y^2} \mathrm{d}\tau .$$

This representation is **Poisson's integral formula** for the upper half plane from Sect. 85.2.

- We consider the PDE (an inhomogeneous wave equation):

$$u_{tt}(x, t) - c^2 u_{xx}(x, t) = \sin(\pi\, x) \ \text{ for } x \in (0, 1)\,,\ t \in (0, \infty)\,.$$

Further, the following initial and boundary conditions are given (where, in every case $x \in (0, 1)\,,\ t \in (0, \infty)$):

$$u(x, 0) = 0\,,\ u_t(x, 0) = 0\,,\ u(0, t) = 0\,,\ u(1, t) = 0\,.$$

(1) Let $U = U(x, s)$ be the Laplacian transform of the solution $u = u(x, t)$ we are looking for. We thus have the correspondences:

$$u(x, t) \ \circ\!\!-\!\!\bullet\ U(x, s)\,,\ u(0, t) \ \circ\!\!-\!\!\bullet\ U(0, s) \text{ and } u(1, t) \ \circ\!\!-\!\!\bullet\ U(1, s)\,.$$

(2) Known are the rules *linearity* and *derivation of the original function* (see Sect. 79.2) and the correspondence $1 \ \circ\!\!-\!\!\bullet\ \frac{1}{s}$ from which, because of $u_{tt} \ \circ\!\!-\!\!\bullet\ s^2U - s\,u(x, 0) - u_t(x, 0)$ and $u(x, 0) = 0 = u_t(x, 0)$ follows:

$$u_{tt} - c^2 u_{xx} \ \circ\!\!-\!\!\bullet\ s^2U - c^2U_{xx} \text{ and } \sin(\pi\, x) \ \circ\!\!-\!\!\bullet\ \frac{\sin(\pi x)}{s}\,.$$

Thus we obtain from the above wave equation the following ODE.

$$s^2U - c^2U_{xx} = \frac{\sin(\pi x)}{s} \ \text{ with } U(0, s) = 0 = U(1, s)\,.$$

(3) The solution of this linear ODE is obtained with the recipe from Sect. 34.2.2. The general solution U_h of the corresponding homogeneous ODE $\frac{s^2}{c^2}U - U_{xx} = 0$ is

$$U_h = c_1\, \mathrm{e}^{\frac{s}{c}x} + c_2\, \mathrm{e}^{-\frac{s}{c}x} \ \text{ mit } c_1,\ c_2 \in \mathbb{R}\,.$$

We make the right-hand side type approach $U_p = a\cos(\pi x)+b\sin(\pi x)$ and enter this in the inhomogeneous ODE. In doing so, we obtain $a = 0$ and $b = \frac{1}{s(s^2+c^2\pi^2)}$, thus

$$U_p = \frac{\sin(\pi x)}{s(s^2+c^2\pi^2)} .$$

The solution $U = U(x, s)$ of the ODE is

$$U(x, s) = \frac{\sin(\pi x)}{s(s^2+c^2\pi^2)} + c_1\,\mathrm{e}^{\frac{s}{c}x} + c_2\,\mathrm{e}^{-\frac{s}{c}x} .$$

We now insert the boundary conditions $U(0, s) = 0 = U(1, s)$ and obtain $c_1 = 0 = c_2$ and with a partial fraction decomposition finally

$$U(x, s) = \frac{\sin(\pi x)}{s(s^2+c^2\pi^2)} = \frac{1}{c^2\pi^2}\left(\frac{1}{s} - \frac{s}{s^2+c^2\pi^2}\right)\sin(\pi x) .$$

(4) A back transformation yields the solution to the original initial boundary value problem

$$u(x, t) = \frac{1}{c^2\pi^2}(1-\cos(c\pi t))\sin(\pi x) . \qquad \blacksquare$$

91.3 Exercises

91.1 Prove the Fourier correspondence used in Example 91.1:

$$G(k)\cos(c\,k\,t) + V(k)\frac{\sin(c\,k\,t)}{ck} \quad \bullet\!\!-\!\!\circ \quad \frac{1}{2}\big(g(x+c\,t)+g(x-c\,t)\big) + \frac{1}{2c}\int_{x-ct}^{x+ct} v(\xi)\mathrm{d}\xi .$$

Hint: Set $\cos(c\,k\,t) = \frac{1}{2}(\mathrm{e}^{\mathrm{i}ckt} + \mathrm{e}^{-\mathrm{i}ckt})$ and use the rules *time shifting* and *convolution*.

91.2 For $c > 0$, solve the initial value problem of the heat conduction equation

$$u_t = c^2\,u_{xx} \qquad \text{for } x \in \mathbb{R},\ t \in (0, \infty),$$
$$u(x, 0) = g(x), \qquad \text{for } x \in \mathbb{R},$$

using Fourier transform with respect to x. Here g describes a (given) initial state. *Note:* See Exercise 76.2.

91.3 Solve the following initial value problem using Laplace transform:

$$u_t = 4\,u_{xx} \text{ for } x \in (0, 3),\ t \geq 0 \text{ and } u(x, 0) = 12\sin(2\pi x) - 3\sin(4\pi x)\,.$$

Note: The inhomogeneous linear ODE for the Laplacian transform of the solution can be solved by the general approach

$$U(x, s) = a(s)\sin(2\pi x) + b(s)\sin(4\pi x)\,.$$

91.4 Solve the following initial boundary value problem using Laplacian transformation:

$$x\,u_t + u_x = x \text{ with } u(x, 0) = 0\,,\ x > 0\,,\ u(0, t) = 0\,, t \geq 0\,.$$

91.5 The Schrödinger equation in the form $\mathrm{i}\,\partial_t u(t, x) = -\partial_{xx}u(t, x)$ describes the wave function of a free particle. We assume the particle is moving on the real axis. In the following, using the Fourier transform, you will find a representation of the solution $u(t, x) : [0, \infty) \times \mathbb{R} \to \mathbb{C}$.

(a) Now, construct the function $u(t, x)$ using its Fourier transform $U(t, \xi)$ (with respect to x).
(b) Use the formula for $u(t, x)$ to find the derivatives $\partial_t u(t, x)$ and $\partial_{xx}u(t, x)$ using $U(t, \xi)$.
(c) Use the expressions from (b) to derive from the Schrödinger equation an ODE for $U(t, \xi)$.
(d) Assume that the function $U(0, \xi)$ is known. Solve the ODE from (c) and give an integral representation for $u(t, x)$.
(e) What can you say about $|U(t, \xi)|$?

Recipe: Collapsing the Book
To collapse the book, do the following:

(1) Have you read all the pages?

- If yes: Continue with the next step.
- If no: Please read the omitted pages.

(continued)

(2) Have you understood everything?

- If yes: Continue with the next step.
- If no: Continue with the next step.

(3) DONE. Please close the book.

Index

Symbols
LR-decomposition, 106
QR-method, 454
π, 49
ε-ball, 519
n-th root, 22
C^1-curve, 616
C^2-curve, 616
k-times continuous partial differentiable, 533
k-times continuously partial differentiable, 535
Im, 61
Re, 61
cos-sin-representation, 826
$\mathbb{K}$-vector space, 123
exp-representation, 826
$\mathbb{R}$, 19
$\mathbb{N}$, 11
$\mathbb{Q}$, 15
$\mathbb{Z}$, 15
1-norm, 508
5-point star, 991, 1012
6-point formula, 945

A
Absolute condition, 762
Absolute convergence
  of a series, 218
Absolute error, 765
Absolutely convergent series, 218
Absolute value
  of a complex number, 62
  of a real number, 21
Accuracy, 31
Adams-Bashforth procedure, 403
Adams-Moulton procedure, 403
Addition
  of matrices, 90
  theorems, 52, 54
Admissible point, 566
Affine coordinates, 606
Affine-linear mapping, 410
Algebraic multiplicity, 434
Algorithm, 765
  backward-stable, 766
  more exactly, 766
  stable, 766
Alternating harmonic series, 219, 221
Analytic function, 900
AND junctor, 2
Angle, 156
Angle doubling, 52, 54
Annulus, 922
Antiderivative, 326
A posteriori error estimation, 773
Approximation
  linear, 272
A-priori error estimation, 773
Arc cosine, 56
Arc cotangent, 56
Arc length, 623
Arc length function, 623
Arc sine, 56
Arc tangent, 56
Area integral
  scalar , 679
  vectorial , 679
Argument
  of a complex number, 67
Armijo step size, 789

C. Karpfinger, *Calculus and Linear Algebra in Recipes*,
https://doi.org/10.1007/978-3-662-65458-3

Associative law, 20
Asymptote, 259
Asymptotically stable equilibrium point, 744
Attractive equilibrium point, 744
Autonomous differential equation system, 743
Average of sets, 7

B

Babylonian method, 214
Backward-stable, 766
Backward substitution, 76, 102
b-adic number representation, 29
Badly posed problem, 979
Ball, 519
Banach's fixed point Theorem, 773
Base, 31
  ordered, 412
Base transformation, 420
Base transformation formula, 421
Basis, 141
BDF methods, 802
Bernoulli's differential equation, 386
Best fit straight line, 182
Bijective, 231
Bilinear, 153
Binary representation, 29
Binomial coefficient, 14
Binomial formula, 15
Binormal unit vector, 630
Bisection method, 264
Block triangle shape, 114
Bolzano's Theorem, 263
Boundary, 520
Boundary conditions
  complete, 316
  natural, 315
  not-a-knot-, 316
Boundary point, 520
Boundary value problem, 751, 802, 838
  Dirichlet's, 954, 955, 985, 1019
  fully homogeneous, 754
  inhomogeneous, 754
  mixed, 986
  Neumann's, 961, 986
  semi-homogeneous, 754
Bounded, 202
  from above, 25, 202, 237
  from below, 25, 202, 237
  function, 237
  interval, 20
  set, 520
Burgers equation, 978
Butcher scheme, 400

C

Canonical basis, 142
Canonical scalar product, 154
Cardinality, 7
Cardinal sinus, 257
Cardioid, 634
Cartesian product, 7
Cauchy criterion, 206
Cauchy integral formula, 911
Cauchy integral theorem, 911
Cauchy principal value, 348
Cauchy product, 224
Cauchy-Riemann differential equations, 902
Cauchy-Schwarz inequality, 156, 164
Centre of gravity, 642, 671
  geometric, 671
Chain rule, 273, 537
Change of sign, 284
Characteristic, 972
  differential equations system, 972
  equation, 368
  method, 971
  polynomial, 368, 433
Charge, 642
Charge density, 642
Circle
  generalized, 944
  number, 49
Circulation, 639
Classification
  of partial differential equations, 979
Closed, 520
  curve, 616, 905
  interval, 21
Closure, 520
Codomain, 227
Coefficient, 87, 241
  highest, 38
  of a polynomial, 37
Coefficient matrix, 78
  extended, 78
Column, 88

index, 87
pivot search, 108
rank, 146
space, 146
vector, 88
Commutative law, 20
Compact, 520
Comparison test, 220
Compatible, 509
Complement
of a set, 7
orthogonal, 172
Complete boundary conditions, 316
Complex differentiable, 900
Complex function, 894
Complex line integral, 908
Complex numbers, 59, 60
Complex power series, 246
Complex sequence, 201
Component, 87
Composition, 229
Concave function, 290
Condition
absolute, 762
relative, 762
Conditionally convergent, 219
Conditioned
bad, 764
good, 764
Cone, 462
Conformal mapping, 942
Conjugate
complex number, 62
harmonic, 952
Connected, 893
Conservative field, 646
Consistency error, 796
Consistency order, 797
Consistent, 796
Continuous in *a*, 261, 523
Continuously differentiable, 275
Continuously extended, 262
Continuously partial differentiable, 533
Continuous on *D*, 261, 523
Contour line, 531
Contraction, 772
Contraction constant, 772
Convergence, 204
conditional, 219
cubic, 776
linear, 776
of a Laurent series, 923
order, 776, 799
quadratic, 297, 543, 776
of a series, 218
Convergent majorant, 220
Convergent sequence, 204
Convergent series, 218
Convex combination, 784
Convex function, 289
Convex set, 520
Convolution, 834
Convolution product, 834
Coordinates
affine, 606
cylinder-, 605
elliptical, 606
parabolic, 606
polar-, 604
spherical-, 605
transformation, 603
vector, 412
Cos, 51
Cosine
complex, 895
function, 51, 895
hyperbolicus, 895
Cotangent, 53
complex, 895
function, 895
Cramer's rule, 119
Critical point, 284, 554, 744
Cubic convergence, 776
Cubic spline function, 315
Curvature
of a plane curve, 631
of a space curve, 631
Curve, 516, 616, 905
closed, 616, 905
continuously differentiable, 616
double point free, 905
piecewise continuously differentiable, 616
planar, 516
plane, 616
regular, 620
space-, 516, 616
Cylinder coordinates, 605

D

Dahlquist's test equation, 799
Data error, 761
Decimal representation, 16, 29
Defect of a linear mapping, 410
Degree
  of a polynomial, 38
  Theorem, 39
Deletion matrix, 111
Delta function, 845
  Dirac's, 878
Derivative, 270
  function, 270, 901
  partial, 528
Descent direction, 788
Determinant, 111
Determinant Multiplication Theorem, 114
Development according to *i-th* row, 113
Development according to *j-th* column, 113
Development point, 241, 300, 921
Diagonal form, 431
Diagonalizable, 431
Diagonalizing, 431
Diagonalizing matrix, 431
Diagonally dominant
  strict, 781
  strictly, 446
Diagonal matrix, 88
Difference
  method, 805, 990, 1010
  of sets, 7
Differentiable, 270
  complex, 900
  in a point, 270
  partial, 528, 535
  total, 580
Differential, 582
  total, 580, 582
Differential equation
  Bernoulli's, 386
  Eulerian, 383
  exact, 713
  homogeneous, 367, 381
  inhomogeneous, 367
  linear, 356
  logistic, 386
  ordinary, 356
  partial, 535
  Riccati's, 387
  separable, 357
  stiff, 800
Differential equation system, 708
Differential operator, 583
Differentiation, 269
  numerical, 277
Dimension, 143
Dimension formula
  for linear mappings, 410
Dirac's delta function, 845, 878
Directional derivative, 528
Directional field, 705
Dirichlet's boundary value problem, 955, 985, 1019
Discrete Fourier coefficients, 865
Discrete Fourier transform, 865
Discretization, 396, 990
Discretization error, 798
Disjoint, 8
Displacement, 52, 54
Distance, 156
  minimum, 173
Distribution, 847
Distributive law, 20
Divergence, 204, 584
Divergence Theorem, 694
Divergent minorant, 220
Divergent sequence, 204
Division with remainder, 39
Divisor, 39
Domain, 227, 645, 893
Domain of convergence, 242
Double integral, 660
Double point, 620
Double point free curve, 905
Dual representation, 29

E

Eigenspace, 430
  generalized, 484
Eigenvalue
  criterion, 505
  of a matrix, 429
Eigenvector
  of a matrix, 429
Element, 7
Elementary function, 231
Elementary matrices, 94

Elementary row transformations, 77
Elimination method
of Gauss, 75
Elliptic coordinates, 606
Elliptic differential equation, 980
Empty set, 7
End point, 616, 905
Entire function, 900
Entry, 87
Equality
of polynomials, 39
of sets, 7
Equalization problem
linear, 177
Equating coefficients, 40, 247
Equilibrium point, 744
Equivalence, 2
Error
absolute, 765
linear, 582
maximum linear, 582
relative, 765
Essential singularity, 920
Euclidean length, 157
Euclidean norm, 157, 508
Euclidean vector space, 154
Euler method
explicit, 396, 794
implicit, 396, 794
Euler's differential equation, 383
Euler's formula, 73, 248
Even function, 128, 279
Exact algorithm, 766
Exact differential equation, 713
Exact step size, 789
Explicit differential equation, 705
Explicit Euler method, 396, 794
Explicit midpoint rule, 403
Explicit sequence, 202
Exponent, 31
Exponential function, 894
for matrices, 721
real, 248
Exponential series, 248
Extend continuously, 262
Extended coefficient matrix, 78
Extreme value problem
with constraints, 566

F
Factorial, 13
Feasible region, 787
Field, 60
Field line, 706
Finite differences, 983
Finite elements, 983
Finite volumes, 983
Fixed point, 769
equation, 212, 769
problem, 770
Theorem, 263
Fix point iteration, 770
Floating point arithmetic, 34
Floating point number, 31
Flops, 35
Flux, 679
Forward substitution, 102
Fourier coefficients
discrete, 865
Fourier series, 817, 823
Fourier transform, 842
discrete, 865
inverse, 848
inverse discrete, 867
Fourier transformable, 841
Fractional-linear transformations, 943
Frenet–Serret frame, 630
Frequency function, 842
Frequency shifting, 834
Frobenius norm, 508
full QR-decomposition, 190
Function, 227
analytic, 900
bounded, 237
complex, 894
elementary, 231
entire, 900
even, 128, 279
harmonic, 952
holomorphic, 900
implicit, 228, 593
monotone, 237
odd, 128, 279
in several variables, 515
Functional determinant, 603
Functional equation, 248
Functional matrix, 536

Fundamental system, 368, 740
Fundamental Theorem of Algebra, 63, 914

G
Gaussian elimination method, 75, 78
Gaussian number plane, 67
Gaussian summation formula, 15
Gauss-Seidel method, 782
General harmonic series, 222, 225
Generalized circle, 944
Generalized eigenspace, 484
Generate, 134
Generating system, 134
Geometric centre of gravity, 642, 671
Geometric multiplicity, 430
Geometric series, 218, 244
Geometric summation formula, 15
Gerschgorin circles, 445
Gibbs phenomenon, 835
Globalized Newton method, 790
Globally convergent iteration method, 772
Global maximum, 282, 553
Global minimum, 282, 553
Goursat
  Theorem of, 914
Gradient, 529, 583
  field, 646
  method, 788
Gram-Schmidt orthonormalization method, 165
Graph
  of a mapping, 227
Grassmann identity, 169
Green's function, 756
Green's integral formulas, 701

H
Half-open interval, 21
Harmonically conjugate, 952
Harmonic function, 952
Harmonic series, 218
Hauptspace, 484
Heat conduction equation, 978, 995
Height, 618
Helical, 618
Helical surface, 677
Hessian matrix, 534
Higher partial derivative, 533
Highest coefficient, 38
Holomorphic function, 900
Homogeneous differential equation, 367, 381
Homogeneous linear system of equations, 82
Homomorphism, 407
Horizontal asymptote, 259
Horner scheme, 43
Householder reflection, 191
Householder transformation, 191
Hull
  linear, 134
Hyperbolic cosine, 250
Hyperbolic cotangent, 250
Hyperbolic differential equation, 980
Hyperbolic functions, 251
Hyperbolic sine, 250
Hyperbolic tangent, 250

I
Identical mapping, 232
Identity, 232
Identity Theorem, 247, 914
Ill-conditioned, 764
Image domain, 847, 876
Image function, 847, 876
Imaginary part, 61
Imaginary unit, 61
Implication, 2
Implicit Euler method, 396, 794
Implicit function, 228, 593
Improper integral, 347
Increment function, 798
Indefinite, 504
Indeterminate, 241
Induced matrix norm, 509
Induction, 11
Infimum, 26
Inhomogeneous differential equation, 367
Initial condition, 360
Initial value problem (IVP), 360
Injective, 231
Input error, 761
In-situ storage, 104
Integers, 15
Integrability condition, 648
Integral, 327
  double, 660
  equation, 889

improper, 347
triple, 660
Integral formula
Cauchy, 911
Green's, 701
Poisson's, 955, 960, 1020
Integral Theorem
Cauchy, 911
Integrand, 327
Integrate, 327
Integration
logarithmic, 328
numerical, 341
by parts, 328
of power series, 328
of rational functions, 335
of rational functions in sin and cos, 338
by substitution, 328
Interior, 519
Interior point, 519
Intermediate value Theorem, 263
Interpolation, 311
points, 395
trigonometric, 868
Interpolation formula
of Lagrange, 312
Interval, 20
Interval halving method, 265
Inverse
of a matrix, 94
Inverse discrete Fourier transform, 867
Inverse Fourier transform, 848
Inverse function, 56
Inverse hyperbolic cosine, 251
Inverse hyperbolic sine, 251
Inverse Laplace transform, 878
Inverse mapping, 234
Inverse vector, 124
Invertibility criterion, 117
Invertible, 234
Invertible matrix, 94
Irrational numbers, 19
Isoclines, 706
Isolated singularity, 920
Iterated integral, 656
Iteration matrix, 780
Iteration method, 770
globally convergent, 772
locally convergent, 772

J
Jacobian matrix, 536
Jacobian method, 780, 781
Jacobi determinant, 603
Jacobi identity, 169
Jacobi rotation, 451
Jordan base, 483
Jordan box, 481
Jordan matrix, 482
Jordan normal form, 483
Joukowski mapping, 942
Junctors, 1

K
Kernel
of a linear mapping, 410
of a matrix, 148
of an integral equation, 889
Kronecker delta, 158
K-times continuously differentiable, 275
Kutta-Joukowski profile, 942

L
Lagrange function, 570, 574
Lagrange multipliers, 566, 570, 574
Lagrange's interpolation formula, 312
Lagrangian identity, 169
Laplace equation, 535
Laplace operator, 584
Laplace's equation, 952, 978
Laplace transform, 876
Laplace transformable, 876
Laplace transformation
inverse, 878
Lattice error, 798
Lattice function, 398, 796
Laurent series, 921
convergence, 923
series development, 925
Leibniz criterion, 220
Leibniz sector formula, 633
Lemma of Riemann, 832
Lemniscate, 591
Length
of a complex number, 62
of a curve, 623
Euclidean, 157

of a vector, 156
Length-preserving, 161
Levels, 531
Limit, 204
Linear approximation, 272
Linear combination, 131
Linear convergence, 776
Linear differential equation, 356
Linear equalization problem, 177
Linear error, 582
Linear factors
splits into, 42
Linear hull, 134
Linear image
rank, 410
Linear map
defect, 410
Linear mapping, 407
Linear part, 457
Linear system of equations, 75
Linearly dependent, 135
Linearly independent, 135
Line element, 705
Line integral
complex, 908
scalar, 637
vectorial, 638
Liouville
Theorem of, 914
Lipschitz constant, 772
Lipschitz-continuous, 772
Locally convergent iteration method, 772
Local maximum, 282, 554
Local minimum, 282, 554
Local Newton method, 789
Logarithm
function, 250, 896
natural, 249
Logarithmic integration, 328
Logarithmic spiral, 624
Logistic differential equation , 386
Loss of significance, 35
Lower sum, 322
Lower triangular matrix, 89
Low-pass filter, 837

M

Möbius transformation, 942, 944
Machine epsilon, 33
Machine numbers, 31
Mantissa, 31
Mapping, 227
affine-linear, 410
bilinear, 153
conformal, 942
identical, 232
linear, 407
positive definite, 154
rule, 227
symmetric, 153
Mass, 642, 671
Mass density, 642
Matrix, 87
diagonalizing, 431
exponential function, 721
indefinite, 504
invertible, 94
negative definite, 504
negative semi definite, 504
positive definite, 504
positive semi definite, 504
similar, 422
skew-symmetric, 91
symmetric, 91
unitary, 470
Matrix norm, 508, 509
induced, 509
natural, 509
Maximum, 25
global, 282, 553
linear error, 582
local, 282, 554
norm, 508
point, 263
principle, 914
solution, 707
step size, 398, 796
Maxwell's equations, 701, 978
Mean value property, 913
Method
of finite differences, 983
of finite elements, 983
of the finite volumes, 983
of Lagrange multipliers, 566, 570, 574
of least squares, 181
Midnight formula, 63
Midpoint quadrics, 462

Midpoint rule, 396
explicit, 403
Minimization problem, 173, 787
Minimum, 25
distance, 173
global, 282, 553
local, 282, 554
point, 263
Mises iteration, 447
Mixed boundary value problem, 986
Mixed type, 980
Moivre's formula, 70
Monkey saddle, 555
Monotone function, 237
Monotonically decreasing, 202
Monotonically increasing, 202
Monotonicity, 237
Monotonicity criterion, 206
Monotonicity test
natural, 542
Multiple continuously differentiable, 275
Multiple continuous partial differentiable, 533, 535
Multiplication
of matrices, 90
by scalars, 90
with scalars, 124
Multiplicity, 42
algebraic, 434
geometric, 430
Multistep method, 396, 402

N

Nabla operator, 529
Natural boundary conditions, 315
Natural matrix norm, 509
Natural monotonicity test, 542
Natural numbers, 11
Natural parameterization, 628
Navier-Stokes equations, 978
Negation, 2
Negative definite, 504
Negative semi definite, 504
Negative vector, 124
Neumann boundary value problem, 961, 986
Newton-Cotes formulas, 342
Newton method, 295
globalized, 790
local, 789
multidimensional, 542
one-dimensional, 297
Node, 315
Nonstationary, 978
Norm, 507
Euclidean, 157, 508
maximum norm, 508
submultiplicative, 509
of a vector, 156
Normal domain, 658
Normal equation, 177
Normal form
of a quadric, 462
Normalized, 31
Normalizing, 158
Normal vector, 630
Normed space, 762
Normed vector space, 507
Not-a-knot boundary conditions, 316
n-tuple, 75
Null vector, 124
Number line, 19
Number of stages, 400
Number of turns, 618
Numbers
complex, 59
Numerical differentiation, 277
Numerical integration, 341

O

Objective function, 787
Oblique asymptote, 260
Odd function, 128, 279
ONB, 158
Open, 520
Open interval, 21
Open set, 893
Opposite vektor, 124
Optimal solution, 180
OR-junctor, 2
Order
of a differential equation, 356
of the pole, 920
Ordered basis, 412
Ordinary differential equation, 356
Original domain, 847, 876
Original function, 847, 876

Orthogonal basis, 158
Orthogonal complement, 172
Orthogonal decomposition of a vector, 159
Orthogonal matrix, 161
Orthogonal projection, 173
Orthogonal system, 158
Orthogonal vectors, 156
Orthonormal basis, 158
Orthonormalization method of Gram and Schmidt, 165
Orthonormal system, 158
Osculating plane, 631
Over-determined linear system of equations, 179
Over relaxation, 785

P
Parabolic coordinates, 606
Parabolic differential equation, 980
Paraboloid, 462
Parallelepiped, 168
Parameterization, 676
  by the arc length, 628
  natural, 628
Partial derivative, 528
Partial differential equation, 535
Partial fraction decomposition, 45
Partially differentiable, 528, 535
Partial sum, 217
Particulate solution, 362
Path independent, 647
Period, 813
Periodic, 16
Periodic function, 813
Periodicity, 51, 53
Perpendicular vectors, 156
Perturbation function, 361, 367, 721
Phase portrait, 746
Piecewise, 616
Pivot element, 105
Place
  of a matrix, 87
Plane curve, 516, 616
Point
  boundary-, 520
  interior, 519
  permissible, 566
  singular, 620
Poisson's equation, 985
Poisson's integral formula, 955, 960, 1020
Polar coordinates, 67, 604
Polar representation, 68
Pole, 920
Polynomial, 125
  characteristic, 433
  division, 40
  equality, 39
  function, 894
  interpolation, 311
  quadratic, 457
  real, 37
  trigonometric, 867
Positive definite, 154, 504
Positively oriented, 697
Positively parameterized, 697
Positively traversed, 905
Positive semi definite, 504
Potential, 646
  equation, 952, 993
  field, 646
Power series
  approach, 389
  complex, 246
  function, 894
  real, 241
Predator-prey model, 794
Primitive function, 326
  of a gradient field, 646
Principal argument, 67
Principal axis transformation, 461
Principal branch, 896
Principal minor, 504
Principal minors criterion, 505
Principal normal unit vector, 630
Principal part, 921, 980
Problem, 762
Process error, 761
Process function, 798
Product
  cartesian, 7
  of matrices, 90
  rule, 272
Projection
  orthogonal, 173
Proper subset, 7
Punctured neighborhood, 920
Purely imaginary, 61

Q

QR-decomposition
  full, 190
  reduced, 190
Quadratic convergence, 297, 543, 776
Quadratic part, 457
Quadratic polynomial, 457
Quadric, 457
Quotient rule, 272

R

Radially symmetric, 538
Radians, 49
Radioactive decay, 356
Radius of convergence, 243
Range
  of a linear mapping, 410
  of a mapping, 227
Rank of a linear mapping, 410
Rank of a matrix, 80
Ratio test, 221
Rational function, 894
Rational numbers, 15
Realification, 899
Real numbers, 19
Real part, 61
Real power series, 241
Real sequence, 201
Rectangular function, 820
Recursion rule, 212
Recursive sequence, 202
reduced QR-decomposition, 190
Reduced row step form, 78
Reflection matrix, 162
Regular curve, 620
Regular part, 921
Regular surface, 676
Relative condition, 762
Relative error, 765
Relaxation, 784
Relaxation parameter, 784
Relaxed method, 784
Remainder, 300
Removeable singularity, 920
Reparameterization, 627, 628
Representation, 131
Residual, 83
Residue, 923, 930
Residue Theorem, 931
Residuum, 180
Reverse error, 766
rg, 80
Riccati's differential equation, 387
Riemann
  integrable, 323
  integral, 323
  Lemma of, 832
  mapping theorem, 949
Right handed coordinate system, 169
Right hand rule, 169
Right-hand side type approach, 375
Root, 22
  function, 895
  test, 221
  of unity, 73
Rotation, 584
Rounding, 33
Rounding error, 761
Row, 88
  index, 87
  rank, 146
  space, 146
  vector, 88
Row step form, 76
  reduced, 78
Row transformations
  elementary, 77
Rule of Sarrus, 112
Runge-Kutta method, 400
  s-stage, 400
Runge-Kutta procedure, 400, 794
  classical, 400, 794

S

Saddle point, 284, 554
Sarrus
  rule of, 112
Scalar field, 516
Scalar line integral, 637
Scalar multiple, 90
Scalar multiplication, 124
Scalar product, 154
  canonical , 154
Scalar surface integral, 679
Schmidt orthonormalization method, 165
Schrödinger equation, 978

Schur decomposition, 470
Schwartz space, 850
Sector formula
  Leibniz, 633
Separable differential equation, 357
Separation method, 981
Sequence, 201
  bounded, 202
  bounded from above, 202
  bounded from below, 202
  complex, 201
  elements, 201
  explicit, 202
  monotonically decreasing, 202
  monotonically increasing, 202
  real, 201
  recursive, 202
  of vectors, 522
Series, 217
  absolute convergent, 218
  convergent, 218
  elements, 217
  geometric, 244
Set, 6
  bounded, 520
  closed, 520
  compact, 520
  convex, 520
  difference, 7
  empty, 7
  open, 520, 893
Shooting method, 808
Sign, 31
Significant digits, 31
Similar matrices, 422
Simply connected, 647
Simpson's rule, 342
Sin, 51
Sinc function, 843
Sine
  complex, 895
  function, 51, 895
Single shooting method, 808
Single-step method, 396
Singularity
  essential, 920
  isolated, 920
  removeable, 920
Singular point, 620
Singular value, 474
Singular value decomposition, 474
Sinus hyperbolicus, 895
Skew-symmetric matrix, 91
Solid of revolution, 343
Solution
  maximum, 707
  optimal, 180
  particulate, 362
  trivial, 82
SOR method, 785
Space
  curve, 516, 616
  normed, 762
Span, 134
Spectral function, 842
Spectral norm, 510
Spectral radius, 447, 773
Spectrum, 513
Spherical coordinates, 605
Spline function
  cubic, 315
Spline interpolation, 311
Splits into linear factors, 42
Splitting off zeros, 39
Square matrix, 88
Squeeze Theorem, 210
Stability, 744
Stability theorem, 708
Stable algorithm, 766
Stable equilibrium point, 744
Standard
  basis, 142
  compatible, 509
  scalar product, 154
  unit vectors, 88
Starting point, 616, 905
Stationary, 978
  point, 284, 554, 744, 787
  solution, 744
Steps, 400
Step size, 395, 398, 796
  exact, 789
  maximum, 398, 796
Stiff differential equation, 800
Strict extremum, 282, 554
Strictly diagonally dominant, 446, 781
Strictly monotone, 202
Strictly monotonic, 237

Submultiplicative, 509
Subset, 7
  proper, 7
Subspace, 126
  trivial, 128
Substitution
  backward, 76
  method, 566
  for multiple variables, 667
Sum
  formula, 15
  of matrices, 90
Superposition principle, 379, 983
Supremum, 26
Surface, 516, 676
Surjective, 231
Symmetric bilinear, 153
Symmetric matrix, 91
Symmetry
  to a circle, 949
  to a straight line, 949
System of characteristic differential equations, 972
System of differential equations
  autonomous, 743
System of equations
  linear, 75
System of linear equations
  over-determined, 179

T

Tangent, 53
  function, 895
  unit vector, 630
Tangent vector
  of a curve, 619
  to a surface, 676
Taylor expansion, 299
Taylor formula, 550
Taylor polynomial, 300
Taylor series, 300
Tend to infinity, 207
Theorem
  about implicit functions, 594, 597
  divergence Theorem of Gauss, 694
  of Fubini, 656
  Fundamental Theorem of Algebra, 63, 914
  of Gauss (plane), 689, 691
  of Gerschgorin, 445
  of Goursat, 914
  of Green (even), 688
  of Green (plane), 686
  intermediate value, 263
  of Liouville, 914
  of maximum and minimum, 263, 525
  of Pythagoras, 52
  of Stokes, 698, 700
  of Taylor, 550
  for the Schur decomposition, 470
  of the singular value decomposition, 474
Thermal diffusion coefficient, 995
Time function, 842
Time grid, 398, 796
Time reversal, 833
Time scaling, 833
Time shifting, 833
Tolerance limit, 775
Torsion, 631
Torus, 678
Total differential, 580, 582
Totally differentiable, 580
Trace, 437, 905
Track
  of a curve, 616
Transfer function, 856
Transformation
  coordinate-, 603
  fractional-linear, 943
  matrix, 414, 603
Translation, 461
Transposed, 90
Trapezoidal rule, 342
Triagonalizable, 469
Triangle inequality, 21, 62
Triangular matrix, 89
Tricomial equation, 981
Tridiagonal matrix, 452
Trigonometric function , 49
Trigonometric interpolation, 868
Trigonometric polynomial, 867
Triple integral, 660
Triple product, 168
Trivial solution, 82
Trivial subspace, 128
Tuples, 7, 75
Type classification
  of partial differential equations, 979

U
Unbounded interval, 21
Union
  of sets, 7
Unitary matrix, 470
Unit matrix, 88
Unity
  roots of, 73
Unstable equilibrium point, 744
Upper sum, 322
Upper triangular matrix, 89

V
Value
  of a series, 218
Variation of parameters, 373, 741
Vector, 124
Vector field, 516
  vortex-free, 587
Vectorial line integral, 638
Vectorial surface integral, 679
Vector iteration, 447
Vector product, 168
Vector space, 123
  axioms, 97
  Euclidean, 154
  normed, 507
Vector-valued function, 515
Velocity, 620
Velocity vector, 619
Vertical asymptote, 260
Volterra integral equation, 889
Von Mises iteration, 447
Vortex density, 587
Vortex flux, 698
Vortex-free, 587

W
Wave equation, 978, 1005
Wave velocity, 1005
Weights, 400
Well-conditioned, 764
Well posed problem, 979
Wronski determinant, 371, 741

Z
Zero
  of the gradient, 554
  of a vector field, 541
Zero boundary condition, 992
Zero matrix, 88
Zero point problem, 769
Zero polynomial, 37
Zero sequence, 204
Zero sequence criterion, 220

Printed in the United States
by Baker & Taylor Publisher Services